The series "Advances in Intelligent Systems and Computing" contains publications on theory, applications, and design methods of Intelligent Systems and Intelligent Computing. Virtually all disciplines such as engineering, natural sciences, computer and information science, ICT, economics, business, e-commerce, environment, healthcare, life science are covered. The list of topics spans all the areas of modern intelligent systems and computing such as: computational intelligence, soft computing including neural networks, fuzzy systems, evolutionary computing and the fusion of these paradigms, social intelligence, ambient intelligence, computational neuroscience, artificial life, virtual worlds and society, cognitive science and systems, Perception and Vision, DNA and immune based systems, self-organizing and adaptive systems, e-Learning and teaching, human-centered and human-centric computing, recommender systems, intelligent control, robotics and mechatronics including human-machine teaming, knowledge-based paradigms, learning paradigms, machine ethics, intelligent data analysis, knowledge management, intelligent agents, intelligent decision making and support, intelligent network security, trust management, interactive entertainment, Web intelligence and multimedia.

The publications within "Advances in Intelligent Systems and Computing" are primarily proceedings of important conferences, symposia and congresses. They cover significant recent developments in the field, both of a foundational and applicable character. An important characteristic feature of the series is the short publication time and world-wide distribution. This permits a rapid and broad dissemination of research results.

** Indexing: The books of this series are submitted to ISI Proceedings, EI-Compendex, DBLP, SCOPUS, Google Scholar and Springerlink **

More information about this series at http://www.springer.com/series/11156

Advances in Intelligent Systems and Compu

Volume 941

Series Editor

Janusz Kacprzyk, Systems Research Institute, Polish Academy of Sciences,
Warsaw, Poland

Advisory Editors

Nikhil R. Pal, Indian Statistical Institute, Kolkata, India
Rafael Bello Perez, Faculty of Mathematics, Physics and Computing,
Universidad Central de Las Villas, Santa Clara, Cuba
Emilio S. Corchado, University of Salamanca, Salamanca, Spain
Hani Hagras, Electronic Engineering, University of Essex, Colchester, UK
László T. Kóczy, Department of Automation, Széchenyi István University,
Gyor, Hungary
Vladik Kreinovich, Department of Computer Science, University of Texas
at El Paso, El Paso, TX, USA
Chin-Teng Lin, Department of Electrical Engineering, National Chiao
Tung University, Hsinchu, Taiwan
Jie Lu, Faculty of Engineering and Information Technology,
University of Technology Sydney, Sydney, NSW, Australia
Patricia Melin, Graduate Program of Computer Science, Tijuana Institute
of Technology, Tijuana, Mexico
Nadia Nedjah, Department of Electronics Engineering, University of Rio de Janeiro,
Rio de Janeiro, Brazil
Ngoc Thanh Nguyen, Faculty of Computer Science and Management,
Wrocław University of Technology, Wrocław, Poland
Jun Wang, Department of Mechanical and Automation Engineering,
The Chinese University of Hong Kong, Shatin, Hong Kong

Ajith Abraham · Aswani Kumar Cherukuri · Patricia Melin · Niketa Gandhi
Editors

Intelligent Systems Design and Applications

18th International Conference on Intelligent Systems Design and Applications (ISDA 2018) held in Vellore, India, December 6–8, 2018, Volume 2

 Springer

Editors
Ajith Abraham
Machine Intelligence Research Labs
Auburn, WA, USA

Patricia Melin
Tijuana Institute of Technology
Tijuana, Mexico

Aswani Kumar Cherukuri
School of Information Technology
and Engineering
Vellore Institute of Technology
Vellore, Tamil Nadu, India

Niketa Gandhi
Machine Intelligence Research Labs
Auburn, WA, USA

ISSN 2194-5357   ISSN 2194-5365   (electronic)
Advances in Intelligent Systems and Computing
ISBN 978-3-030-16659-5   ISBN 978-3-030-16660-1   (eBook)
https://doi.org/10.1007/978-3-030-16660-1

Library of Congress Control Number: 2019936140

This Springer imprint is published by the registered company Springer Nature Switzerland AG
The registered company address is: Gewerbestrasse 11, 6330 Cham, Switzerland

Preface

Welcome to the Proceedings of the Joint Conferences on 18th International Conference on Intelligent Systems Design and Applications (ISDA) and 10th World Congress on Nature and Biologically Inspired Computing (NaBIC), which is held in VIT University, India, during December 6–8, 2018. ISDA - NaBIC 2018 is jointly organized by the VIT University, India, and Machine Intelligence Research Labs (MIR Labs), USA. ISDA - NaBIC 2018 brings together researchers, engineers, developers, and practitioners from academia and industry working in all interdisciplinary areas of intelligent systems, nature-inspired computing, big data analytics, real-world applications and to exchange and cross-fertilize their ideas. The themes of the contributions and scientific sessions range from theories to applications, reflecting a wide spectrum of the coverage of intelligent systems and computational intelligence areas. ISDA 2018 received submissions from 30 countries, and each paper was reviewed by at least five reviewers in a standard peer-review process. Based on the recommendation by five independent referees, finally 189 papers were accepted for ISDA 2018 (acceptance rate of 48%). NaBIC 2018 received submissions from 11 countries, and each paper was reviewed by at least five reviewers in a standard peer-review process. Based on the recommendation by five independent referees, finally about 23 papers were accepted for NaBIC 2018 (acceptance rate of 37%). Conference proceedings are published by Springer Verlag, Advances in Intelligent Systems and Computing Series.

Many people have collaborated and worked hard to produce the successful ISDA - NaBIC 2018 conference. First, we would like to thank all the authors for submitting their papers to the conference, for their presentations and discussions during the conference. Our thanks go to program committee members and reviewers, who carried out the most difficult work by carefully evaluating the submitted papers. Our special thanks to Raija Halonen, University of Oulu, Finland, Junzo Watada, Universiti Teknologi Petronas, Malaysia, and Nelishia Pillay, University of Pretoria, South Africa, for the exciting plenary talks. We express our

sincere thanks to the session chairs and organizing committee chairs for helping us
to formulate a rich technical program.

Enjoy reading the articles!

Ajith Abraham
Aswani Kumar Cherukuri
General Chairs

Patricia Melin
Emilio Corchado
Florin Popentiu Vladicescu
Ana Maria Madureira
Program Chairs

Organization

Chief Patron

G. Viswanathan (Chancellor)　　　　Vellore Institute of Technology

Patrons

Sanakar Viswanathan (Vice President)　Vellore Institute of Technology, Vellore
Sekar Viswanathan (Vice President)　　Vellore Institute of Technology, Vellore
G. V. Selvam (Vice President)　　　　Vellore Institute of Technology, Vellore

Advisors

Anand A. Samuel (Vice Chancellor)　　Vellore Institute of Technology, Vellore
S. Narayanan (Pro-vice Chancellor)　　Vellore Institute of Technology, Vellore

General Chairs

Ajith Abraham　　　　　　　　　　Machine Intelligence Research Labs
　　　　　　　　　　　　　　　　　　(MIR Labs), USA
Aswani Kumar Cherukuri　　　　　　Vellore Institute of Technology, India

Program Chairs

Patricia Melin　　　　　　　　　　Tijuana Institute of Technology, Mexico
Emilio Corchado　　　　　　　　　　University of Salamanca, Spain
Florin Popenţiu Vlădicescu　　　　　University Politehnica of Bucharest,
　　　　　　　　　　　　　　　　　　Romania
Ana Maria Madureira　　　　　　　　Instituto Superior de Engenharia do
　　　　　　　　　　　　　　　　　　Porto, Portugal

International Advisory Board

Albert Zomaya	University of Sydney, Australia
Bruno Apolloni	University of Milano, Italy
Imre J. Rudas	Óbuda University, Hungary
Janusz Kacprzyk	Polish Academy of Sciences, Poland
Marina Gavrilova	University of Calgary, Canada
Patrick Siarry	Université Paris-Est Crétcil, France
Ronald Yager	Iona College, USA
Sebastian Ventura	University of Cordoba, Spain
Vincenzo Piuri	Universita' degli Studi di Milano, Italy
Francisco Herrera	University of Granada, Spain
Sankar Kumar Pal	ISI, Kolkata, India

Publication Chair

Niketa Gandhi	Machine Intelligence Research Labs (MIR Labs), USA

Web Master

Kun Ma	Jinan University, China

Publicity Committee

Mayur Rahul	C.S.J.M. University, Kanpur, India
Sanju Tiwari	National Institute of Technology, Kurukshetra, Haryana, India

Local Organizing Committee

Advisory Committee

R. Saravanan	SCOPE
Jasmine Norman	Department of IT, SITE
Sree Dharinya	Department of SSE, SITE
T. Ramkumar	Department of CACM, SITE
Agilandeswari	Department of DC, SITE
Lakshmi Priya	Department of MM, SITE
Ajit Kumar Santra	SITE
Bimal Kumar Ray	SITE
K. Ganesan	SITE
Hari Ram Vishwakarma	SITE

Daphne Lopez	SITE
Shantharajah	SITE
Subha	SITE

Organizing Chair

E. Sathiyamoorthy

Finance Committee

S. Prasanna (In-charge)
J. Karthikeyan

Event Management Committee

R. Srinivasa Perumal (In-charge)
R. Sujatha
S. L. Arthy
S. Siva Ramakrishnan
S. Jayakumar
G. Kavitha
M. Priya

Hospitality and Guest Care Committee

G. Jagadeesh (In-charge)
P. Thanapal
J. Prabu
Vijaya Anand
J. Gitanjali

Printing Committee

N. Deepa (In-charge)
U. Rahamathunnisa
P. Jayalakshmi
L. B. Krithika

Registration Committee

N. Mythili (In-charge)
Divya Udayan
K. Brindha
S. Sudha
M. Deepa
K. Shanthi

Technical Program Committee

Abdelkrim Haqiq	GREENTIC, FST, Hassan First University, Settat, Morocco
Alberto Cano	University of Córdoba, Spain
Ali Yakoob	University of Babylon, Iraq
Amit Kumar Shukla	South Asian University, Delhi, India
Amparo Fuster-Sabater	Institute of Physical and Information Technologies (CSIC), Spain
Antonio J. Tallón-Ballesteros	University of Seville, Spain
Angel Garcia-Baños	Universidad del Valle/Cali, Colombia
Anjali Chandavale	Dr. Vishwanath Karad MIT World Peace University, India
Aswani Cherukuri	Vellore Institute of Technology, India
Atta Rahman	University of Dammam, Dammam, Saudi Arabia
Azah Kamilah-Muda	UTeM, Malaysia
Aurora Ramírez	University of Córdoba, Spain
Cesar Hervas	Martínez, University of Córdoba, Spain
Daniela Zaharie	West University of Timisoara, Romania
Denis Felipe	Federal University of Rio Grande do Norte, Brazil
Durai Raj Vincent	Vellore Institute of Technology, India
Elizabeth Goldbarg	Universidade Federal do Rio Grande do Norte, Brazil
Francisco Chicano	Universidad de Málaga, Spain
Frantisek Zboril	Brno University of Technology, Czechia
Giovanna Castellano	Università degli Studi di Bari Aldo Moro, Italy
Givanaldo Rocha de Souza	Federal University of Rio Grande do Norte, Brazil
Gregorio Sainz-Palmero	Universidad de Valladolid, Spain
Igor Medeiros	Federal University of Rio Grande do Norte, Brazil
Isaac Chairez	Instituto Politécnico Nacional, Mexico
Isabel S. Jesus	Instituto Superior de Engenharia do Porto, Portugal
Ishwarya Srinivasan	Vellore Institute of Technology, India
Jagadeesh Kakarla	IIITDM Kancheepuram, Chennai, India
Jaroslav Rozman	Brno University of Technology, Czech Republic
Javier Ferrer	University of Málaga, Spain
Jerry Chun-Wei Lin	Western Norway University of Applied Sciences (HVL), Bergen, Norway

Jesus Alcala-Fdez	University of Granada, Spain
José Everardo Bessa Maia	State University of Ceará, Brazil
José Raúl Romero	University of Córdoba, Spain
Jose Tenreiro Machado	ISEP, Portugal
Kathiravan Srinivasan	Vellore Institute of Technology, India
Kaushik Das Sharma	University of Calcutta, India
Lin Wang	Jinan University, China
Mario Giovanni C. A. Cimino	University of Pisa, Italy
Martin Hruby	Brno University of Technology, Czech Republic
Matheus Menezes	Universidade Federal Rural do Semi-Árido, Brazil
Mohammad Shojafar	Sapienza University of Rome, Italy
Nadesh Rk	Vellore Institute of Technology, India
Niketa Gandhi	Machine Intelligence Research Labs (MIR Labs), USA
Oscar Castillo	Tijuana Institute of Technology, Tijuana
Ozgur Koray Sahingoz	Istanbul Kultur University, Turkey
P. E. S. N. Krishna Prasad	S V College of Engineering, Tirupati, India
Paolo Buono	Università degli Studi di Bari Aldo Moro, Italy
Patrick Siarry	Université Paris-Est Créteil, France
Prabukumar Manoharan	Vellore Institute of Technology, India
Pranab Muhuri	South Asian University, Delhi, India
Radu-Emil Precup	Politehnica University of Timisoara, Romania
Raghavendra Kumar Chunduri	Vellore Institute of Technology, India
Ricardo Tanscheit	PUC-Rio, Brazil
Saravanakumar Kandasamy	Vellore Institute of Technology, India
Simone Ludwig	North Dakota State University, USA
Sustek Martin	Brno University of Technology, Czechia
Thatiana C. N. Souza	Federal University Rural Semi-Arid, Brazil
Thomas Hanne	University of Applied Sciences Northwestern Switzerland, Switzerland
Varun Ojha	Swiss Federal Institute of Technology, Switzerland

Contents

Differential Evolution Trained Fuzzy Cognitive Map: An Application to Modeling Efficiency in Banking

Gutha Jaya Krishna[1,3], Meesala Smruthi[2], Vadlamani Ravi[3(✉)], and Bhamidipati Shandilya[4]

[1] School of Computer and Information Sciences, University of Hyderabad, Hyderabad 500046, India
krishna.gutha@gmail.com
[2] Department of Electrical and Electronics Engineering, Birla Institute of Technology and Science Pilani, Hyderabad 500078, India
smruthimvk@gmail.com
[3] Center of Excellence in Analytics, Institute for Development and Research in Banking Technology, Hyderabad 500057, India
vravi@idrbt.ac.in
[4] Institute for Development and Research in Banking Technology, Hyderabad 500057, India
bshandilya@idrbt.ac.in

Abstract. In this work, we developed a Differential Evolution (DE) trained Fuzzy Cognitive Map (FCM) for predicting the bank efficiency. We developed two modes of training namely (i) sequential and (ii) batch modes. We compared the DE trained FCM models with the conventional Hebbian training in both modes. We employed Mean Absolute Percentage Error (MAPE) as an error measure while predicting the efficiency from Return on Assets (ROA), Return on Equity (ROE), Profit Margin (PM), Utilization of Assets (UA), and Expenses Ratio (ER). We employed 5x2-fold cross-validation framework. In the first case i.e. sequential mode of training, the DE trained FCM statistically outperformed the Hebbian trained FCM and in the second case i.e. batch mode of training, DE trained FCM is statistically the same as the Hebbian trained FCM. To break the tie in the batch mode, the training time is compared where DE trained FCM turned to be 19% faster than the Hebbian trained FCM. The proposed model can be applied to solving similar banking and insurance problems.

Keywords: Bank efficiency prediction · Financial ratios · Differential Evolution · Fuzzy Cognitive Map · Hebbian learning

1 Introduction

Kosko in the seminal work [12] introduced Fuzzy Cognitive Maps (FCM) in the year 1986. An FCM is a map of the cognitive process wherein the interconnection of the components i.e. concepts are utilized to evaluate the degree of the effect of the components in the psychological scene or area of the imagination. FCM is mainly

© Springer Nature Switzerland AG 2020
A. Abraham et al. (Eds.): ISDA 2018, AISC 941, pp. 1–11, 2020.
https://doi.org/10.1007/978-3-030-16660-1_1

employed for causal knowledge representation and reasoning process. FCMs come under the family of neuro-fuzzy systems that are employed for decision making and are also utilized for modelling and simulation of complex systems. FCM is also employed for the prediction, forecasting and classification tasks to acquire the causal knowledge.

Various training methods are employed for the dynamic adaptation of the FCM model and to tune the weights. Most of the training methodologies come from the field of study of Artificial Neural Networks (ANN). In [4], the Differential Hebbian Learning (DHL) was employed to train the FCM. A wide range of modifications is proposed to this Hebbian algorithm which is, by nature, deterministic. Another paradigm of training algorithm includes the Swarm Intelligence and Evolutionary Computation (EC)-based approaches which are, by nature, stochastic. The above-mentioned training algorithms are employed to decrease the human intervention in the FCM, and to make the FCMs adaptive and non-black box, unlike ANNs.

Our focus is on the application of Evolutionary algorithms (i.e. Differential Evolution) which are, population-based stochastic search strategies, where a population of variables gets updated iteratively utilizing heuristics until the point where optimum or near optimum is achieved. These methods contrast in how the variables are encoded and the heuristics they utilize when refreshing the variables. Genetic Algorithms [6], Differential Evolution [21, 23, 24], Particle Swarm Optimization [9], Ant Colony Optimization [5], Firefly Algorithm [22], are a portion of the meta-heuristics from this family.

Banking, Financial Services and Insurance (BFSI) sector is the backbone of the economy of any country, and it does have many operational issues as well as financial issues. BFSI sector is moving away from conventional ways towards more automated and robust methods. EC based techniques play a vital role in solving the above-mentioned operational issues because they yield global or near-global optimal results [11]. For instance, how EC techniques are useful in solving various customer relationship management problems is captured very well in a survey paper by Krishna and Ravi [10]. Finding the causal knowledge from the data using the cognitive map for the bank efficiency prediction is an important operational issue.

In this section, a brief introduction of the FCMs, Evolutionary Computation (EC) and the role of ECs in solving the operational issues are provided. Motivation and contributions of the current work are presented in Sects. 2 and 3. Related literature of the present work is reported in Sect. 4. The proposed methodology is described in Sect. 5. Description of the bank efficiency dataset is given in Sect. 6. Experimental design of the proposed method is illustrated in Sect. 7. Results and discussion are presented in Sect. 8. Finally, we conclude with some future directions in Sect. 9.

2 Motivation

The motivation for the work is as follow:

– To the best of our knowledge, bank efficiency is never predicted as a function of the important financial ratios using FCM.
– To find the causal knowledge and the cognitive process behind the bank efficiency and its relationship with Return on Assets (ROA), Return on Equity (ROE), Profit Margin (PM), Utilization of Assets (UA), and Expenses Ratio (ER).

- To investigate the sequential and batch mode of training in FCMs, which may be useful to future researchers.
- To automate the decision-making process in the banking domain with multiple solutions using Differential Evolution (DE) trained Fuzzy Cognitive Map (FCM) (i.e. DE-FCM).

3 Contribution

The contributions of the work are as follows:

- Developed DE trained and Hebbian trained FCM models separately for the bank efficiency prediction.
- Developed FCM training algorithms in two modes i.e. sequential and batch modes.
- Modeled Efficiency of a bank with respect to five financial ratios using DE and Hebbian trained FCM.

4 Related Work

In [20], a seminal work on utilizing ECs for training FCMs i.e. a genetic learning trained FCM is developed. In [19], a data-driven non-linear Hebbian learning (DD-NHL) was proposed for training multiple samples of data. Multiple applications like precision agriculture, geo-spatial dengue outbreak risk prediction of tropical regions, predicting yield in cotton crop production, Radiation Therapy Systems where FCM based prediction is employed is studied in [7, 8, 13, 16]. Two surveys [13, 14] were performed by the author Papageorgiou on FCMs. One of the surveys describes [14] the types of training i.e. Hebbian trained, EC trained and a hybrid of both as well as different applications of FCMs. Another survey describes [15] the different types of cognitive maps and possible applications areas of FCMs. In [1, 2], stability, parameter convergence, existence and uniqueness of solutions is studied and analyzed. In [3], the behaviour of various activation functions i.e. sigmoid, hyperbolic, step function, threshold linear function is studied.

In [17], Data Envelopment Analysis and Fuzzy Multi-Attribute Decision Making hybrid was developed for ranking banks based on efficiency. In [18], a review of sequential learning is performed where the terminology of "catastrophic forgetting" for sequential learning is described.

5 Proposed Methodology

The continuous FCM is utilized, with modified Kosko activation function and sigmoid threshold function as described in Sect. 5.1. The objective is to minimize the MAPE and the convergence criterion is either reaching a pre-specified number of iterations or MAPE value falling below 0.1%. The procedure for training the DE-FCM, where both sequential and batch modes of training are presented in Sect. 5.3. The Hebbian learning with Oja's weight update rule [19], where both sequential and batch modes of training is performed, is described in Sect. 5.4.

5.1   Activation Function

Concepts of the model are denoted by C_i where i = 1...N where N is the total number of concepts. Each concept has an activation value $A_i \in [0, 1]$, i = 1...N and signed fuzzy weights W_{ij}, which takes the values in the range [−1, 1], of the edge between C_i and C_j where j = 1...N.

The value of each concept A_i at any occurrence 'k' is calculated by the sum of the previous value A_i in a precedent occurrence 'k − 1' with the product of the value of A_i of the concept node C_i in the precedent occurrence 'k − 1' and the value of concept effect link weight W_{ij}. The mathematical formula is given below.

$$A_i^k = Sigmoid(A_i^{k-1} + \sum_{\substack{j \neq i \\ J = 1}}^{N} A_j^{k-1} * W_{ji}^{k-1})$$

$$Sigmoid(x) = \frac{1}{1 + e^{-\lambda x}}, \text{ where } \lambda = 1.$$

5.2   MAPE Measure

The Mean Absolute Percentage Error (MAPE), is a measure of prediction accuracy in statistics. It usually expresses accuracy as a percentage, and is defined by the formula:

$$MAPE = \frac{100}{n} \sum_{t=1}^{n} \left| \frac{Actual_t - Predicted_t}{Actual_t} \right|$$

5.3   RMSE Measure

The Root Mean Squared Error (RMSE), is a measure of prediction in statistics. It is defined by the formula:

$$RMSE = \sqrt{\frac{1}{n} \sum_{t=1}^{n} (Actual_t - Predicted_t)^2}$$

5.4   DE-FCM

DE trained FCM in 2 modes: (i) sequential & (ii) batch mode as described in Algorithm 1.

```
 1: procedure SEQ-DE-FCM
 2:   no_itr ← fixed_value
 3:   ncv_train ← nrow(concept_vec_train)
 4:     while i ≤ ncv_train do
 5:         cv_train ← concept_vec_train[i]
 6:         itr ← 1
 7:         while itr ≤ no_itr do
 8:             Step-1: Generate weights
 9:             of FCM using DE heuristic
10:             Step-2: Apply Modified
11:             Kosko with Sigmoid on
12:             the cv_train for the
13:             weights generated by DE
14:             Step-3: Compute
15:             MAPE_new on actual and
16:             predicted productivity
17:             concept values of cv_train
18:             Step-4: If MAPE of new
19:             weight vector less than the
20:             MAPE value of current
21:             weight vector in the DE
22:             population then replace
23:             the current weight vector
24:             and MAPE value in DE
25:             with the new weight vector
26:             and new MAPE value
27:         end while
28:     end while
29:   ncv_test ← nrow(concept_vec_test)
30:   no_pop ← nrow(DE_Population)
31:   while pop ≤ no_pop do
32:       while j ≤ ncv_test do
33:           cv_test ← concept_vec_test[j]
34:           Apply Modified
35:           Kosko with Sigmoid on
36:           the cv_test for the
37:           weight vector in the DE
38:           population i.e. DE[pop]
39:       end while
40:       For all the samples in
41:       in concept_vec_test compute
42:       MAPE on actual and
43:       predicted productivity
44:       concept values
45:   end while
46:   Pick the weight vector with
47:   less MAPE value on cv_test
48:   from the DE_Population
49: end procedure
```

(a) Sequential mode DE-FCM algorithm

```
 1: procedure BATCH-DE-FCM
 2:   no_itr ← fixed_value
 3:   ncv_train ← nrow(concept_vec_train)
 4:     itr ← 1
 5:     while itr ≤ no_itr do
 6:         Step-1: Generate weights
 7:         of FCM using DE heuristic
 8:         while i ≤ ncv_train do
 9:             cv_train ← concept_vec_train[i]
10:             Step-2: Apply Modified
11:             Kosko with Sigmoid on
12:             the cv_train for the
13:             weights generated by DE
14:         end while
15:         Step-3: For all the samples in
16:         in concept_vec_train compute
17:         MAPE_new on actual and
18:         predicted productivity
19:         concept values
20:         Step-4: If MAPE of new
21:         weight vector less than the
22:         MAPE value of current
23:         weight vector in the DE
24:         population then replace
25:         the current weight vector
26:         and MAPE value in DE
27:         with the new weight vector
28:         and new MAPE value
29:     end while
30:   ncv_test ← nrow(concept_vec_test)
31:   no_pop ← nrow(DE_Population)
32:   while pop ≤ no_pop do
33:       while j ≤ ncv_test do
34:           cv_test ← concept_vec_test[j]
35:           Apply Modified
36:           Kosko with Sigmoid on
37:           the cv_test for the
38:           weight vector in the DE
39:           population i.e. DE[pop]
40:       end while
41:       For all the samples in
42:       in concept_vec_test compute
43:       MAPE on actual and
44:       predicted productivity
45:       concept values
46:   end while
47:   Pick the weight vector with
48:   less MAPE value on cv_test
49:   from the DE_Population
50: end procedure
```

(b) Batch mode DE-FCM algorithm

Alg. 1: Two modes of Training

5.5   Hebbian Trained FCM

Oja's learning rule [19] is a modification of the standard Hebb's Rule that, through multiplicative normalization, which solves all stability problems.

$$w_{ij}^{n+1} = w_{ij}^n + \eta * A_j^{new} * \left(A_j^{old} - w_{ij}^n * A_j^{new} \right)$$

The above mentioned Oja's weight update rule is employed for training the FCM in both sequential and batch modes. This weight update rule is a nonlinear weight update rule, employed to capture the non-linearity in data-driven approaches.

5.6   Sequential and Batch Modes of Training

In the sequential mode of training, the FCM is trained with a single concept vector one after another. One main disadvantage of the sequential mode of training is the lack of memory of the previous trained values of the model. It has an advantage of fast training. But, repeated training with individual concept vectors also makes it slow. In the batch mode of training, the FCM is trained with all the concept vectors in one go. This mode of training is good for small datasets. But, takes a lot of training time when training data is large. Both modes are described in Algorithm 1-a and 1-b.

6   Description of the Dataset

The data of the five financial ratios along with the corresponding efficiency levels as computed by the DEA-FMADM model are taken from [17]. There are 5 input and one output concept in the dataset of 27 different banks. The five ratios viz., Return on Assets (ROA), Return on Equity (ROE), Profit Margin (PM), Utilization of Assets (UA), and Expenses Ratio (ER) are presented in Table 1. The output concept is efficiency. Out of the 15 possible concept linkages among the 6 concepts, only 13 were considered logically plausible by the domain expert.

Table 1. Concept description

S. no.	Input ratio name	Definition
1	Return on Assets (ROA)	Assets/Net profit
2	Return on Equity (ROE)	Net profit/Owned funds
3	Profit Margin (PM)	Net profit/Interest income + Non-interest income
4	Utilization of Assets (UA)	Interest income + Non-interest income/Assets
5	Expenses Ratio (ER)	Operating Expenses/Spread
Output: Efficiency		

7 Experimental Setup

Training and testing are performed on an Intel (R) Core (TM) i7-6700 processor with 32 GB RAM in fastR. The data is divided into 60:40 percent training and testing respectively randomly for 5 times i.e. the data is divided into 16 concept vectors (samples) of training and 11 concept vectors (samples) of testing. On each of the five training and test sets, two models i.e. DE-FCM and Hebbian trained FCM are built in each of the two modes of training i.e. sequential and batch modes of training. In DE-FCM the best weight matrix out of the population of 25 on the test data is picked. The above experimentation for DE-FCM in both modes of training is repeated for 30 runs, as DE is stochastic by nature and the best MAPE values out of 30 runs are considered. Hebbian-FCM is deterministic by nature and so the above procedure considered for DE-FCM is not followed for Hebbian trained FCM.

8 Results and Discussion

The 5x2 fold cross validation (5x2 FCV) is performed on the dataset and the results are reported in Tables 2 and 3. In the sequential and batch modes, we assessed the statistical significance of the performance of Hebbian-FCM versus DE-FCM. Thus, we performed t-test on the average best (out of 30 runs of the DE-FCM) MAPE value over 5 folds versus average MAPE of the Hebbian-FCM value at 1% level of significance.

Table 2. Sequential trained FCM

5x2 FCV	Hebbian-FCM (MAPE)	DE-FCM (MAPE)	Hebbian-FCM (RMSE)	DE-FCM (RMSE)
1	30.9878	5.2727	0.23562	0.04811
2	29.2433	3.9027	0.22489	0.03765
3	16.1033	14.8488	0.14209	0.12802
4	20.51599	12.8809	0.15409	0.11803
5	33.12868	4.1504	0.24733	0.04473
Mean	25.995814	8.2111	0.200804	0.075308
Variance	53.558542	27.387970	0.002396	0.001924
t-Value	4.420107317		4.269079683	
p-value	0.002226		0.002727	

It turned out that the DE-FCM is statistically significant compared to Hebbian trained FCM with a p-value of 0.002. But, in a batch mode of training, DE trained FCM turned out to be statistically the same as Hebbian trained FCM with a p-value of 0.663. Similar results were noticed in the case of RMSE too.

It is worth noticing that the proposed method employs MAPE as the objective function, while RMSE is just computed on the convergence of the algorithm. Thus, MAPE and not the RMSE drives the entire learning process.

Table 3. Batch trained FCM

5x2 FCV	Hebbian-FCM (MAPE)	DE-FCM (MAPE)	Hebbian-FCM (RMSE)	DE-FCM (RMSE)
1	6.446347	5.525913	0.05554	0.05229
2	5.18635	5.231981	0.04297	0.05401
3	15.3066	15.213922	0.13614	0.12734
4	18.95404	15.201556	0.14485	0.12717
5	7.921394	4.337817	0.06182	0.04286
Mean	10.7629562	9.1022378	0.088264	0.080734
Variance	36.386283	31.255764	0.002328	0.001821
t-Value	0.4515155		0.2613553	
p-value	0.66361		0.80042	

To break the tie in batch mode, the training time is compared where DE trained FCM is 19% (i.e. 0.2x) faster than Hebbian trained FCM. Though batch mode DE-FCM may not seem faster (0.2x) in the current setup of multi-threaded and high-end processor setup. But, DE-FCM will show its significance in the low end and single threaded configurations.

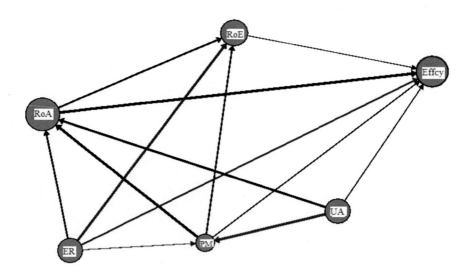

Fig. 1. FCM of the banks' efficiency data

In order to obtain the cognitive knowledge from the given data, both Hebbian trained and DE trained FCM were built on the whole dataset of five financial ratios and the corresponding efficiency level of 27 banks in both sequential and batch modes of training for visual and cognitive understanding. This is because the given data set is too small in size, thereby building FCM under 5x2 FCV is fraught with inaccuracies. However, when FCM is used for prediction purpose, 5x2 FCV is considered in order to have an authentic method of testing.

In order to generate the FCM graph (see Fig. 1), owing to the small size of the dataset, we trained the FCM with Hebbian and DE in both sequential and batch modes on the entire data without resorting to 5x2 cross validation. In Fig. 1, a positive correlation is represented with a positive weight (i.e. black edge) and negative correlation is represented with a negative weight (i.e. red edge), though weights are not mentioned in the Fig. 1. It turned out that graph of the DE trained FCM with sequential training (depicted in Fig. 1) conformed to the domain expert's knowledge with the least number of violations (i.e. two) of both the positive and negative correlations between the nodes of the FCM. According to domain expert only ER should have a negative correlation with other concepts including Efficiency, especially when expenses do not lead to profits. But, the weights obtained by the DE-FCM in sequential mode of training has three negative correlations for the ER. Except for ER all the remaining nine edges should have positive correlations, but DE-FCM in sequential mode obtained eight positive correlations. It is noteworthy that these the violations are not very serious because we considered only 5 input concepts and it would have matched with the domain expert's knowledge, has we considered an exhaustive list of ratios. Another interesting aspect of Fig. 1 is that the thickness of the edges between two nodes in the FCM is proportional to the level of correlation between of the two concepts in question and the black lines indicate a positive correlation, while red lines indicate negative correlation. It should be noted that the present work can be used for both descriptive and predictive analytics purposes. Firstly, the FCM can be used as an analytical tool to predict the efficiency of a bank based on five financial ratios taken from 27 banks' data corresponding to the year 2004. Secondly, the FCM obtained by DE in the sequential mode of training and presented in Fig. 1 can also be used for causal visualization of the influence of five financial ratios on the efficiency of a bank.

9   Conclusion and Future Directions

In the current work, we proposed DE trained and Hebbian trained FCM training algorithms separately for predicting bank efficiency. In both cases, FCM was trained in two modes i.e. sequential and batch modes. Minimizing the Mean Absolute Percentage Error (MAPE) error measure (i.e. on the efficiency concept values) is chosen as the convergence criteria. To automate and to help the decision-maker with multiple solutions of the population of EC-based techniques, we employed Differential Evolution (DE) for training the Fuzzy Cognitive Map (FCM).

One limitation of the study from the domain perspective, is that a few number of financial ratios are considered to model efficiency. In future, we plan to extend this study by including more financial ratios that directly or indirectly influence efficiency. It will hopefully, present better picture of the ground realities.

In the future, better convergence criteria may be developed and studied with the success of this approach. Better training algorithms may be proposed by hybridizing both the training algorithms implemented in the current work. Even better modes of training may be suggested in future to handle larger datasets. Better data-driven approaches for FCMs can be developed by formulating the problem in multi-objective optimization framework. Further, 'if-then' rules can be generated while building the FCM which would act as an expert system.

References

1. Boutalis, Y., Kottas, T., Christodoulou, M.: Adaptive estimation of fuzzy cognitive maps with proven stability and parameter convergence. IEEE Trans. Fuzzy Syst. **17**(4), 874–889 (2009)
2. Boutalis, Y., Kottas, T., Christodoulou, M.: On the existence and uniqueness of solutions for the concept values in fuzzy cognitive maps. In: 47th IEEE Conference on Decision and Control, Cancun, Mexico, pp. 98–104. IEEE (2008)
3. Bueno, S., Salmeron, J.L.: Benchmarking main activation functions in fuzzy cognitive maps. Expert Syst. Appl. **36**(3), 5221–5229 (2009)
4. Dickerson, J.A., Kosko, B.: Virtual worlds as fuzzy cognitive maps. Presence: Teleoperators Virtual Environ. **3**(2), 173–189 (1994)
5. Dorigo, M., Maniezzo, V., Colorni, A.: Ant system: optimization by a colony of cooperating agents. IEEE Trans. Syst. Man Cybern. Part B (Cybern.) **26**(1), 29–41 (1996)
6. Goldberg, D.E.: Genetic Algorithms in Search, Optimization, and Machine Learning. Addison-Wesley Longman Publishing Co., Boston (1989)
7. Jayashree, L.S., Palakkal, N., Papageorgiou, E.I., Papageorgiou, K.: Application of fuzzy cognitive maps in precision agriculture: a case study on coconut yield management of southern India's Malabar region. Neural Comput. Appl. **26**(8), 1963–1978 (2015)
8. Jayashree, L., Lakshmi Devi, R., Papandrianos, N., Papageorgiou, E.I.: Application of fuzzy cognitive map for geospatial dengue outbreak risk prediction of tropical regions of Southern India. Intell. Decis. Technol. **12**(2), 231–250 (2018)
9. Kennedy, J., Eberhart, R.: Particle swarm optimization. In: International Conference on Neural Networks (ICNN 1995), Piscataway, NJ, pp. 1942–1948. IEEE (1995)
10. Krishna, G.J., Ravi, V.: Evolutionary computing applied to customer relationship management: a survey. Eng. Appl. Artif. Intell. **56**, 30–59 (2016)
11. Krishna, G.J., Ravi, V.: Evolutionary computing applied to solve some operational issues in banks. In: Datta, S., Davim, J. (eds.) Optimization in Industry. Management and Industrial Engineering, pp. 31–53. Springer, Cham (2019)
12. Kosko, B.: Fuzzy cognitive maps. Int. J. Man Mach. Stud. **24**(1), 65–75 (1986)
13. Papageorgiou, E., Markinos, A., Gemtos, T.: Fuzzy cognitive map based approach for predicting yield in cotton crop production as a basis for decision support system in precision agriculture application. Appl. Soft Comput. **11**(4), 3643–3657 (2011)
14. Papageorgiou, E.I.: Learning algorithms for fuzzy cognitive maps: a review study. IEEE Trans. Syst. Man Cybern. Part C (Appl. Rev.) **42**(2), 150–163 (2012)
15. Papageorgiou, E.I., Salmeron, J.L.: A review of fuzzy cognitive maps research during the last decade. IEEE Trans. Fuzzy Syst. **21**(1), 66–79 (2013)
16. Parsopoulos, K.E., Papageorgiou, E.I., Groumpos, P.P., Vrahatis, M.N.: Evolutionary computation techniques for optimizing fuzzy cognitive maps in radiation therapy systems. In: Deb, K. (ed.) Genetic and Evolutionary Computation Conference, pp. 402–413. Springer, Heidelberg (2004)

17. Pramodh, C., Ravi, V., Nagabhushanam, T.: Indian banks' productivity ranking via data envelopment analysis and fuzzy multi-attribute decision-making hybrid. Int. J. Inf. Decis. Sci. 1(1), 44 (2008)
18. Robins, A.: Sequential learning in neural networks: a review and a discussion of pseudorehearsal based methods. Intell. Data Anal. 8(3), 301–322 (1997)
19. Stach, W., Kurgan, L., Pedrycz, W.: Data-driven nonlinear Hebbian learning method for fuzzy cognitive maps. In: IEEE International Conference on Fuzzy Systems (IEEE World Congress on Computational Intelligence), Hong Kong, China, pp. 1975–1981. IEEE, June 2008
20. Stach, W., Kurgan, L., Pedrycz, W., Reformat, M.: Genetic learning of fuzzy cognitive maps. Fuzzy Sets Syst. 153(3), 371–401 (2005)
21. Storn, R., Price, K.: Differential evolution-a simple and efficient heuristic for global optimization over continuous spaces. J. Glob. Optim. 11(4), 341–359 (1997)
22. Yang, X.S.: Nature-Inspired Metaheuristic Algorithms. Luniver Press (2010)
23. Dasgupta, S., Das, S., Biswas, A., Abraham, A.: On stability and convergence of the population-dynamics in differential evolution. AI Commun. 22(1), 1–20 (2009)
24. Ali, M., Pant, M., Abraham, A.: Simplex differential evolution. Acta Polytechnica Hungarica 6(5), 95–115 (2009)

Novel Authentication System for Personal and Domestic Network Systems Using Image Feature Comparison and Digital Signatures

Hrishikesh Narayanankutty[1]([✉]) and Chungath Srinivasan[2]([✉])

[1] Amrita Center for Cyber Security Systems and Networks,
Amrita School of Engineering, Amritapuri Campus,
Amrita Vishwa Vidyapeetham, Coimbatore, India
hrishikesh9409@gmail.com
[2] TIFAC-CORE in Cyber Security, Amrita School of Engineering,
Amrita Vishwa Vidyapeetham, Coimbatore, India
c_srinivasan@cb.amrita.edu

Abstract. With the advent of the digital age, there has been an increased need for the robustness and the versatility of authentication systems. Many techniques have been developed since the last few decades in digital security to secure and encrypt data. As much as these techniques are versatile and robust, authentication systems remain the weakest link in any given cyber-physical system. Human intervention does not necessarily make a system as robust as it ought to be, more so often, the chances are it might result in adverse and also opposite of the desired effect. In order for a system to be robust and secure, all forms of precaution must be in place for it to function securely. Authentication systems, since its inception are designed largely in order to minimize human errors and to securely authenticate and verify an individual's identity in question. Traditional authentication systems are either not efficient and or can be easily bypassed. This could be largely rectified or reduced using modern authentication systems like bio-metrics, iris detection systems etc. However, these systems are not easy to setup and configure, being highly sophisticated and require trained personnel to maintain. Through this paper, we propose a novel authentication system which is relatively inexpensive, easily made mobile and secure to setup. The system targets domestic networks and aims to secure personal and small scale systems. This novel authentication scheme proposed operates based on image feature comparison and digital signatures. The image comparison is used for the authentication process itself while a digital signature is used to provide non-repudiation. Through this novel approach, we aim to demonstrate and highlight the efficiency of the proposed system to the other existing authentication schemes used today. This approach not only provides an easy way to authenticate, it provides a way to determine if the person has indeed the authorization to access the system.

https://github.com/hrishikesh9409/ImageAuthentication/implementation.

© Springer Nature Switzerland AG 2020
A. Abraham et al. (Eds.): ISDA 2018, AISC 941, pp. 12–23, 2020.
https://doi.org/10.1007/978-3-030-16660-1_2

1   Introduction

Since time immemorial, mankind has always been profoundly interested in the transference and the storage of data in secret, without detection and with utmost reliability. The race to safeguard data has led to many sophisticated modern encryption schemes such as AES, RSA, Blowfish etc.; complex mathematical models and methods also have been formulated for the transference of keys to ensure total privacy. However, for any security system, the "weakest link" is usually considered to be users themselves and subsequently user interactions and actions. In most cases, authentication systems are considered to be the "security bottleneck" of any secure system. Unlike encryption, encoding etc. authentication systems are relatively insecure as human users are involved. Instead of selecting an arbitrary string of alphanumeric and other special characters, users often choose passwords that are short, pattern oriented and easy to remember numbers and words. Traditional password-based authentication systems are also subject to various attacks like dictionary attacks, and brute force attacks, not to mention phishing techniques through key logging and social engineering attacks by shoulder surfing. Similarly, many bio-metric systems can be easily bypassed using fraudulent means, smudged fingerprints etc., thus rendering it vulnerable. In any case, more secure authentication systems like iris detection etc. are harder to implement as they require the specific hardware, are more expensive and also are fixed to one particular area, a passage door etc. Hence a more secure, easy-to-setup and reliable system is necessary for authentication of users.

The following pipeline for the proposed system is employed in order to improve the reliability and the security of authentication systems. The proposed novel system works on image feature detection and comparison in combination with digital signatures. The image feature detection and comparison provides a robust method in order to authenticate users and digital signatures provide means for non-repudiation. Users capture images that serve as password which is then sent to the server for authentication. The user also sends his/her digital signature attached to this password, both of which are then encrypted and sent to the server side for verification and validation. The AES encryption scheme as part of the OpenSSL library has been used for the encryption of data and RSA Digital signature scheme has been used for digitally signing the data. Through this paper, a novel authentication system is proposed which can overcome the traditional attacks on authentication systems as aforementioned. First, we explain a general implementation of the proposed novel system. Then the various stages of the pipeline used in the implementation are explained in greater detail and depth. Third, a feasibility and risk study is presented for the real-time implementation of the system.

The paper is structured as follows: Sect. 2 presents the literature survey and other work related to this field. Section 3 explains the detailed structure of the various stages of implementation. Section 4 deals with the feasibility and risk analysis of the system. Section 5 draws some conclusions based on the comparisons of other existing systems, feasibility and risk analysis.

Lastly, it may be noted that this system has made use of the Speeded Up Robust Features (SURF) Algorithm for image feature detection and comparison. This algorithm had been chosen for its high accuracy, speed and efficiency in detecting similar images regardless of their orientation.

2   Related Work

2.1   Background

Password-based techniques to protect data are the most popular and by far the most susceptible to a volley of attacks. Though modern encryption schemes are secure themselves, if the person authenticating the process is compromised, so is the entire system of exchange and information. Modern state of the art authentication systems can be subdivided into the following areas [12]:

1. Token-Based Authentication
2. Bio-metric-Based Authentication
3. Knowledge-Based Authentication

Token-based authentication technique includes keycards, smart cards, ATM cards etc. in conjunction with a time-bound token, which is used to authenticate the user. In other words, it provides one level of indirection for authentication - instead of having to authenticate with username and password for each protected resource, the user authenticates that way once (within a session of limited duration), obtains a time-limited token in return, and uses that token for further authentication during the session.

Bio-metric-based authentication systems use exploits the unique biological traits and deals with statistical data. There are several techniques that further fall into the category of bio-metric-based authentication scheme. The two major ones are [8]:

1. Physiological based techniques - this includes facial features and analysis, retinal patterns, fingerprints, hand geometry, DNA etc.
2. Behavior based techniques - this includes the common behavioral characteristics such as keystroke, voice, smell, signatures etc.

The above-mentioned techniques work based on identification or authentication. Identification mode is when the systems identify the user after finding a match in the database for enrolled for a match. The authentication mode verifies the user by matching with an earlier enrolled pattern.

Knowledge-based techniques combine the usage of both text-based and graphical imagery to achieve successful authentication. This technique also can be further classified into the following [12]:

1. Recognition based – where the user is authorized upon recognition of the right image from a set of images.
2. Recall based – where the user is required to create or reproduce something that the user created or selected during the registration stage.

2.2   Disadvantages of Existing Systems

The token-based authentication mechanism can have either a software or a hardware-based token mechanism. In the case of a hardware token, the token assigned to the user has the pseudo-random generated number keyed in by the user. This incurs additional costs for hardware and not to mention the token has to be carried with the user at all times. In case of software-based tokens, recurring expense of user's subscription for SMS is a key factor. Low coverage also issues in these cases. Whilst digital certificates can be issued by a third party for certification, it further involves additional costs, such as the certifying authority's subscription cost for issuing the digital certificates. Although few bio-metric security systems today are fast and accurate, most of today's systems are suitable for verification only as false acceptance rates are too high [7]. People who are paraplegic cannot use the fingerprint of hand geometry-based systems. Further more bio-metric data might also be considered as a violation of privacy. This data might contain sensitive and personal information. Hence many users consider such authentication scheme as intrusive or even invasive [7]. One final disadvantage that is prone to bio-metric schemes is that the bio-metric input device must be trusted. Its authenticity should be verified and the user's liveness should be checked [7]. It should also be noted further bio-metrics of people can change with age, or suffer physical injuries or disease. Integrating such system into corporate infrastructure can also be challenging [4].

A bio-metric system consists of mainly the following modules – sensors, feature extractors, a template database, matching and decision-making module. It is observed and recorded that an attack can be performed in any of the above mentioned five modules. Further, the channels connecting the modules can also be attacked [1]. This causes severe high risk in not only the setup phase, but also during the authentication phase.

Knowledge-based techniques also is prone to disadvantages. It is, like any password authentication system, prone to shoulder surfing. Graphical passwords are also often predictable, enabling attackers to simply guess the image based password. Further, the Passface technique elaborates on how users tend to choose weak and predictable graphical passwords [3]. Nali et al. have also established similar results on predictability among graphical passwords created with the DAS technique. The DAS scheme produces a larger input of password space than traditional password schemes by decoupling positions of input from the temporal order. DAS disallows for the repetition of passwords that are considered difficult to repeat [9].

However, the DAS scheme is also subjected to limitations like shoulder surfing, risk in public environments, brute force and also for the system to completely fail in the event of an attacker obtaining a copy of the shared secret [5].

3   Pipeline and Working

3.1   Motivation

One of the other primary disadvantages of existing systems is that they do not provide non-repudiation, i.e. the system cannot accurately and effectively know for sure that the user is indeed authorized to log in. This causes many problems, making existing systems to be vulnerable to a volley of attacks - the most common types of attacks being masquerading attacks, replay, and man in the middle attacks. Hence through this paper, we evolve a novel pipeline for authentication that allows for a system to not only validate and authenticate a user, but also to provide means with which the system can effectively provide non-repudiation, i.e., the system recognizes the person and can for certainty say the user is who he claims to be.

3.2   Pipeline

The way the system is designed is twofold – one part of the pipeline handles the actual verification of the password whilst the other part of the pipeline handles non-repudiation, i.e., validating the actual identity of the user. The way non-repudiation is handled is by verification of digital signatures. The password here is taken to be an image of the user's choosing. An average hash of the image is also calculated at the client side. The average hash algorithm is a very simple algorithmic version of a perceptual hash. It's a good algorithm to use to find images that are very similar and haven't had localized corrections performed (like a border or a watermark). Hence the more similar the images are, the difference between the two generated hashes are minimal and almost identical. The main advantages of this hashing algorithm are its speed and simplicity.

The average hash value comparison involves in checking how similar the two images are. One another comparison between the similarities of the images uses the feature detection algorithm called SURF. This provides further, a robust way of determining the similarity of the two images. The novelty and the effectiveness of this system are through the means of feature comparison of the password and the verification of the digital certificate at the server side as provided by the user whilst logging in. The novel pipeline for the authentication system is shown with the help of the given figure. The stages are designed for both server and client side. More details on server and client side are discussed below.

Client Side. The various stages proposed for the client side of the novel authentication system consists of the following:

- The user takes a picture that he that he had set during the registration of credentials – the picture password itself can be anything, the surrounding of his workplace, a particular object of his choosing etc.
- The user also provides with a time-stamp. This time-stamp acts as a nonce value, which is verified at the server side. This serves to act as a server-side challenge.

The user then generates a private key for the generation of a digital signature, which is then signed against the password together with the time-stamp.

- An average hash of the image is calculated which is used to further strengthen verification of the password.
- In order to reduce the footprint and the size of the digital signature that is to be generated, a SHA 3 hash of the picture and the time-stamp is first calculated.
- The digital signature is then generated based on the hash value obtained.
- The SHA 3 hash, the signature, the time-stamp and the average hash value together are then compressed.
- This compressed file is then encrypted with a symmetric encryption scheme and sent to the server side.

The digital signature used in the development of this prototype is the Advanced Encryption Standard (AES) scheme, as it is highly secure and can be sent over any public channel. The digital signature used here for non-repudiation exploits the RSA public signature scheme. The combination of AES scheme and RSA digital signature provides the much-needed security against attackers who seek to impersonate, replay or perform man in the middle attacks (MITM). Hence in total, 4 parameters are sent to the server side for verification and validation – the image password, the time-stamp, the average hash and the digital signature which was signed by the user.

The SHA 3 hash of the image and the time-stamp is initially calculated. This hash value is fed into the digital signature algorithm to generate a certificate. The OpenSSL library is used for generating the certificate. It is to be noted that the initial transference of the key value pair between the client and the server is not specified in this paper, as it is beyond the scope of this paper and it is assumed that the key value pair is already communicated between the server and the client and is ready for further transference of encrypted data from the client side and the decryption from the server side (Fig. 1).

The sample signing algorithm that is used during the creation of the prototype at the client side is given below:

Client Side Signature Algorithm
function sign(hash, private_key, hashAlg = "SHA-512"):
signer = PKCS1_v1_5.new(private_key) // formulating signer according to the RSASSA-PKCS1-v1_5 protocol
digest = SHA512.new() // step to formulate digest
digest.update (message)
return signer.sign()

Sample Usage
Signature = b64encode(rsa.sign(hash, private_key, "SHA-512"))

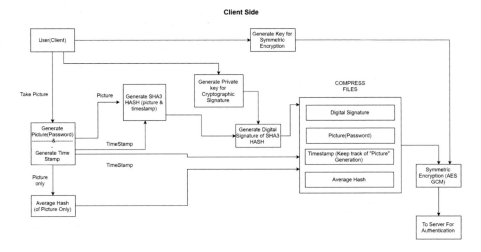

Fig. 1. Client side

Server Side

The stages as part of the server side is as follows:

- Decrypt the encrypted file that was sent by the client.
- Decompress the file and verify the signature – this validates the claim of the identity of the user.
- Verifies the SHA 3 hash of the client side image.
- The server also in addition computes the average hash value using the base image that was pre-setup during the registration phase.
- After the validation of the signature and the computation of the average hash, 3 parameters are then compared to offer further acknowledgment for authentication – time stamp is firstly verified, this greatly reduces any chances for man in the middle attacks and other forms of replay attacks as any delay beyond a threshold value for time causes the authentication to fail.
- The Hamming distance is then calculated for the two average hash values. Hamming distance provides the minimum number of substitutions required to change one string to the other, or the minimum number of errors that could have effectively transformed one string to another. A certain threshold value is assumed and beyond this value, the authentication fails.
- The final comparison is then performed using the SURF algorithm, used for fast feature detection; checking for similarities of the two images itself, to compare and to contrast how similar or dissimilar the two images are.

The hash value is computed using the time-stamp and the client side image. This prevents the possibility of collisions and further renders hashes unique. With this modification, it is almost impossible to construct a hash with collisions.

To detect false positives and negatives, two key steps and two levels of verification is performed. The first check is the comparison between the two average hash values. Although these detect the similarity of two images using the Hamming distance calculation, it can generate many false positives.

The detailed signature verification algorithm is given below

Server Side Signature Verification Algorithm
function verify(message, signature, pub_key):
signer = PKCS1_v1_5.new(pub_key) //according to the protocol standard
digest = SHA512.new()
digest.update(message)
return signer.verify(digest, signature)

Sample Usage
verify = rsa.verify(msg1, b64decode(signature), public)
The following are detailed steps involved in the generation of the average hash
is as follows:

1. Reduce size of the image
   - Image size is reduced to remove high frequencies and details. This step
     ignores the original size and ratio of the image and will always have a
     standard fixed dimension.
2. Reduce the color
   - The color of the reduced image is reduced by converting it into a grayscale
     image.
3. Calculation of the average color
   - The average color value from the color values from the previous step is
     calculated.
4. Calculation of the hash value
   - The final fingerprint is calculated based on whether the pixel intensity
     is brighter or darker than the average grayscale value from the previous
     step.

As it can be seen from the above steps, the average hash value generation
appears to be too simple and inaccurate. A large probability of false positives
is possible by carefully constructing images that have similar color histogram
frequencies. This is, however, is a crucial step in negating all the dissimilar
images without costing too much in terms of computation power. The next level
of security comes from the actual image feature comparison, implemented by
means of the SURF algorithm (Fig. 2).

The Speeded Up Robust Feature (SURF) by Herbert Bay et al. is a feature detection algorithm that is more optimized and is a variant of the Scale
Invariant Feature Transform algorithm, originally developed by David Lowe [6].
Both algorithms perform feature detection. SURF, however, performs better
and faster compared to the SIFT algorithm. From a tabulated study Guerrero
and Maridalia [11], the 3 most popular image feature comparison algorithms
were considered and the most optimal choice that does not compromise much of
either feature detection points or performance came to be the SURF algorithm.

Feature detection algorithms like SIFT work based on feature descriptors.
The neighborhood of every "interest" point in an image is represented by a
feature vector. This is called a descriptor. The SIFT descriptor remains a very
appealing descriptor for practical uses. However, for higher dimensions, it serves

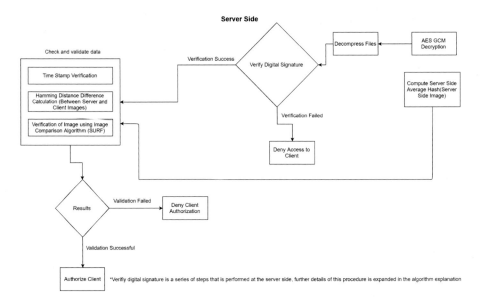

Fig. 2. Server side

as a drawback for SIFT, during the matching step [2]. The SURF algorithm uses a faster and novel descriptor called the Fast Hessiandescriptor that drastically reduces the time taken for feature computation and matching, also increasing robustness [2]. However, although this reduces the time taken for feature computation, it is still vastly slower compared to the average hashing technique. Hence SURF algorithm is used to detect false positives and to act as a secondary checking mechanism. This approach drastically increases not only the robustness and security of the system, it also serves to eliminate any false positives that may slip through whilst avoiding unnecessary computation.

With all of these checks and verification, the system provides a unique and a robust method to not only authenticate a user but also to effectively discourage any possible threats against this system.

A detailed summary description of the aforementioned protocol of the system is given below.

Protocol

Client U wants to authenticate to the server S. The structure of the communication sessions for the protocol is as follows:

1. $U \rightarrow S : U$, Cert U
2. $S \rightarrow U : S$, Cert S
3. $U \rightarrow S : \{Sig_U(H(P,\ T_U)),\ P,\ T_U,\ A_H(P,T_U)\}_{K_{US}}, E_S(K_{US})$
4. $S \rightarrow U :$ verified/access granted or verification failed/access denied

Server side verfication algorithm

1. Decrypt $E_S(K_{US})$ using private key of the server to get K_{US}

2. Decrypt the ciphertext $\{Sig_U(H(P,\ T_U)),\ P,\ T_U,\ A_H(P,T_U)\}_{K_{US}}$
3. $r = T_S - T_U$
   if $(r \geq \gamma)$
   client access denied
   else go to step 4
4. $\alpha = A_H(P'),\ \beta = A_H(P)$
5. $d = d_H(\alpha, \beta)$
   if $(d \geq t)$
   client access denied
   else go to step 6
6. $f = \text{SURF}(P, P')$
   if $(f \geq t')$
   client access denied
   else
   Client access granted

(where t, t' and γ are thresold parameters)

Protocol notations

- U: Client
- S: Authenticating Server
- Cert U: Digital Certificate of U
- Cert S: Digital Certificate of S
- T_U: Time-stamp generated by U
- P: Client side picture password; P': Server side picture password
- $H(\cdot)$: Cryptographic hash function
- $A_H(\cdot)$: Average hash function
- K_{US}: Symmetric key shared between entities U and S
- $E_S(\cdot)$: Public key encryption of a message with public key of the server S
- $Sig_U(\cdot)$: Digital signature by the entity U
- d_H: Hamming distance

4 Risk Analysis and Feasibility

Although the proposed novel system has many advantages over the traditional and other modern authentication systems, it would be amiss to deny that this system too suffers from some risks of its own. The proposed system is still in the development and the testing phase. The scenarios where this novel technique can be applied to is within the domains of domestic and personal usage, and not for an industrial or enterprise scenario. After some preliminary research and examinations, some analysis will be discussed in this section, where the SURF algorithm is presented, also the reason for its usage, over the other feature detection algorithms.

At the core of the authentication scheme, the password (the image the client sends over to the server side), feature points are first detected and the closest to those matching points are then mapped to the base image found at the server side. If the features cannot be matched precisely, or the number of detected points fall short of the threshold value, the authentication fails. SURF algorithm works best on texture-based images [11].

Some of the important criteria required for the proper and successful authentication of the user using this novel system is given as follows:

1. Since the system heavily relies on the actual comparison of two images, the image sent by the client to the server should be a clear image.
   - Lighting whilst taking the photo should be taken into consideration.
   - When taken in low lighting, the lower contrasted regions do not bear SURF descriptors and are hence not detected as features [10].
2. The exact nature of the picture itself must be taken into consideration. An image with more features, more details that are part of the picture would be more ideal, as compared to an image that contains lesser feature points.
3. Although this scheme is more or less likely for attackers to perform shoulder surfing, the user must be aware of keeping this picture password safe and private, as a potential attacker could try constructing his own image, from guessing what picture the user chooses.
4. There is one attack, which is possible by brute force. Attackers can make a video of the entire environment the user takes a picture in, a 4 pi steradian video, which he can then decompose into many frames of high quality and look for possible matches. This could be a potential threat to the proposed system.
5. Lastly, it must be mentioned that the base image present in the system must be updated periodically, in order have faster and better feature matching, as the environment and the user himself is subject to many changes, thus keeping the server image as up to date with the environment as possible.
   A heavy emphasis and assumption is placed on the user himself to be discrete. A password is only as good as it remains secret. In the same way the image that the user uses as the password must be kept safe. If the user wants to take a picture to send to the server for the specific purpose of authentication, it is assumed that the user is discrete in taking the picture. Further more it is helpful to also include that the picture that is used as the password can be inherently present a phone or a mobile device and it can be simply encrypted and transferred to the server via the above mentioned pipeline. There is no need to explicitly take a picture to send to the server. The image however has to then be time-stamped by the use before sending to the server. The advantage to this small modification is that the authentication can be done remotely and it is unnecessary for the user to be physically present near the environment of his choosing.

5   Conclusion

We propose a reliable and secure authentication system whose operations are based on image feature comparison and digital signatures. This system uses the

Speeded Up Robust Features (SURF) Algorithm for image feature detection and comparison due to its high speed and accuracy. In terms of cost, efficiency, and security the proposed system is found to be better than most authentication schemes based on the different and various parameters. A feasibility and risk study for the real-time implementation of the system and various stages of the pipeline used in the implementation are explained in greater detail and depth. The system is designed to provide a unique and a robust method to not only authenticate a user but also to effectively discourage any possible threats against this system.

On a final note, it should be noted that the source to this research work is made available on public domain, github.

References

1. Ashok, A., Poornachandran, P., Achuthan, K.: Secure authentication in multimodal biometric systems using cryptographic hash functions. In: International Conference on Security in Computer Networks and Distributed Systems, pp. 168–177. Springer, Heidelberg (2012)
2. Bay, H., Tuytelaars, T., Van Gool, L.: Surf: speeded up robust features. In: European Conference on Computer Vision, pp. 404–417. Springer, Heidelberg (2006)
3. Davis, D., Monrose, F., Reiter, M.K.: On user choice in graphical password schemes. In: USENIX Security Symposium, vol. 13, p. 11 (2004)
4. Kaschte, B.: Biometric authentication systems today and in the future. University of Auckland (2005)
5. Kumar, E.A., Bilandi, E.N.: A graphical password based authentication based system for mobile devices. Int. J. Comput. Sci. Mob. Comput. $3(4)$, 744–754 (2014)
6. Lowe, D.G.: Distinctive image features from scale-invariant keypoints. Int. J. Comput. Vis. $60(2)$, 91–110 (2004)
7. Matyáš, V., Říha, Z.: Biometric authentication–security and usability. In: Advanced Communications and Multimedia Security, pp. 227–239. Springer, Heidelberg (2002)
8. Mudholkar, S.S., Shende, P.M., Sarode, M.V.: Biometrics authentication technique for intrusion detection systems using fingerprint recognition. Int. J. Comput. Sci. Eng. Inf. Technol. (IJCSEIT) $2(1)$, 57–65 (2012)
9. Nali, D., Thorpe, J.: Analyzing user choice in graphical passwords. School of Computer Science, Carleton University, Technical report, TR-04-01 (2004)
10. Oyallon, E., Rabin, J.: An analysis of the surf method. Image Process. On Line 5, 176–218 (2015)
11. Pena, M.G.: A comparative study of three image matching algorithms: SIFT, SURF, and FAST. Utah State University (2011)
12. Suo, X., Zhu, Y., Owen, G.S.: Graphical passwords: a survey. In: Computer Security Applications Conference 21st Annual, pp. 10–pp. IEEE (2005)

Detecting Helmet of Bike Riders in Outdoor Video Sequences for Road Traffic Accidental Avoidance

N. Kumar$^{(\boxtimes)}$ and N. Sukavanam

Department of Mathematics, I.I.T. Roorkee, Roorkee 247667, India
{atrindma,nsukvfma}@iitr.ac.in

Abstract. In metro cities of all over the world, the growing number personal vehicles and fast life style of the people frequently meet very serious accidents. Due to deeply regretted reports from the loss of manpower and economy, accidental avoidance becomes a hot challenging research topic. In this paper, we consider specifically the accidents that happen due to bike rider's involvement. Focusing on detecting helmet test, we proposed a computer vision based model that exploits HOOG descriptor with RBF kernel based SVM classification. Our experiments have two tier classifications, first is between bike riders and non-bike rider's detection and second is to determine whether the bike riders in the first phases wearing a helmet or not. The initial phase uses video surveillance for detecting the bike riders by using background modeling and bounding based object segmentation. The performance comparison of our model on three widely used kernels ensures the validation of the satisfactory results. We achieved helmet detection accuracy with radial basis kernel 96.67%. Our model can detect any type of helmets in the outdoor video sequences and help security and safety aspects of bike riders.

Keywords: Collision Avoidance System (CAS) ·
Driving assistance system (DAS) · Gaussian Mixture Model (GMM) ·
Histogram of oriented gradients (HOOG) · Support vector machine (SVM) ·
Traffic monitor system

1 Introduction

The manpower loss is one of the most fatal and irrecoverable loses to any nation. Recently observed from World Health Organization (WHO) yearly road accident report that 1.2 million people die in fatal road accidents and 20 to 50 million comprised to survive their lives due to irrecoverable long injury. With the continuation of such a terrible trends, the statisticians noticed that the road accidents will be increased by 65% and counted as fifth major reason of unnatural deaths by 2030. In terms of economy, this fact presents an amazing loss of 518 billion dollars which is big enough amount for affecting the progress of any country. With major focus to save the nations from such a terrible fatality and high economy loss due to road accidents, United Nation (UN) government set up a program "Decade of Action for Road Safety" from 2011 to 2020 [11, 13]. The data statistics reported by Baker et al. [10] using the data collected

© Springer Nature Switzerland AG 2020
A. Abraham et al. (Eds.): ISDA 2018, AISC 941, pp. 24–33, 2020.
https://doi.org/10.1007/978-3-030-16660-1_3

from Reserve Bank of Australia (RBA) present that Chinese automobiles industries have remarkably more growth than US, Germany and Japan but US is at higher level for having number of automobiles per person. Such countries either should have more numbers of accidents due to motorcyclists or they are lass with embedded system technology in transport services.

The computer vision community has been established to ensure a milestone for solving many such real-time challenging problems in which the objective is to characterize the things that are fully obscured during whole the process. This happens almost in all the video analytics problems due to having high occlusion, camera calibration and viewpoints conditions. Detecting helmet is one of the interestingly hot research problems which outcomes to maintain the safety and security protocols in road traffic [1, 2, 5, 27]. As the context reflects, a helmet is very common term which simply can refer a shield during any process to accomplish that process successfully. In this paper we mean the helmet for motorbike rider and put an effort to detect it, in video sequences of outdoor environment. The most sensitive reasons to commit an accident by motor bike riders can be categorized as an active and passive class reasons. The causes of the list of active class are person or situation specific while the accidents in passive class happened due to the environmental conditions like roads problem, abnormal whether etc. [9]. Some of the regions may be most prone to happen the accidents due to high speed traffic in that area of its geographical architecture e.g. hill area. Apart from the Road Traffic Office (RTO) rules, there is possibility of situational understanding (someone may be in health emergency) between the bike riders and motor drivers to make the path clear. So, in case of active reasons of road accidents, the visual and cognitive observations are very specific for the old age as well as immature bike riders. This is conceptualized in Table 1.

Table 1. Categorization the reasons of road accidents

Passive reasons	Active reasons
• Environmental conditions fog etc.	• Driving with drinking
• Narrow road or irregular roads	• Talking at during driving
• Lack of strict law and order	• Office hours in congested city
• Defective vehicles	• Some function or rally
• Lack of multiple sided flyovers	• Inexperienced risk taking nature
• Red lights and fixed speed regions	• Age limit, health condition

Another reasonable side of active road accidents is noticed as the inadequacy of strict law and order regarding use of drugs and alcohol on driving, mental disorder and fatigue detection. In the age of multidisciplinary research the completely robust and cost effective vehicle or helmet detection is possible with developing the Driver Assistant System (DAS), Collision Avoidance System (CAS) [10]. Both these systems are automated with large scale deep intelligence experiments and developing such a highly embedded vehicle detection system requires the experts from various discipline. In this paper, we only focus on helmet detection experiments for developing an accident free bike-riding system. The processing experiments for the stated problem can have multimodal phases. Formally domain specifics research considers the issues from three things: (1) Environmental conditions; (2) Geographic situation issues; (3) Human health and age prospective; (4) Architecture and quality of vehicles; (5) Lack of enthusiasm towards accident avoidances. All these issues connects itself a big research domain. For example geographical issues are concerned with civil engineering problem of road manufacturing in sensitive areas like hilly regions and fly over bridges the canals. Environmental condition connects the research of atmospheric science for detecting natural disasters like flood, orphan etc. In case of humankind, old ages with eye sight problem and unexperienced young age are the key reason to meet the accidents. Size and architecture of vehicles is also one of the important reasons prone to happen the accidents. Developing an accident avoidant vehicle is the research problem in mechanical engineering. Finally, the drivers with lower the accidents should be encouraged by making the records in police and RTO offices.

Computational intelligence algorithms like machine learning algorithms to classify the objects are also used [31]. The algorithm used here detects any changes caused by moving objects. The issue is that movement in objects, for example, trees and bushes exist, that implies the background isn't static constantly. So our system has to differentiate which movements are due to really moving objects we are occupied with and which movements are due to background objects. All the static regions and the areas of dynamic background calculated using probabilistic model, form the background mask. All the rest is defined as foreground. Playing out this sort of separating makes it conceivable to have real moving objects and the background independently in various formats. Lighting changes can be solved by applying techniques such as contrast stretching and singular value equalization. Noise caused by weather changes (rain or wind) can be removed using blurring followed by morphological operations. Removal of shadow can be achieved after modelling the background mask followed by morphological operations using thresholding.

1.1 Motivation and Challenges

The message presented by the research on helmet detection cannot be consider just earning a credit but it must give an immense consideration towards the safety and wellness to the manpower of every nation. Government should have strict protocols for safe driving and encourage the public to come forward to maintain the safety and security of nation. Although the literature in the research of helmet detection is not rich enough but the problem is quite interesting as well as tough to process higher occluded video surveillance data captured from unconditional outdoor traffic. The sounding challenge in developing an accident accordance system requires to develop a reliable and robust real-time vehicle detection system.

2 Related Work

Road traffic monitoring is a hot research problem connected to rural or big metropolitan city Los Angeles, New York, Moscow, London and in India, Delhi, Mumbai, Kolkata and Jaipur. The research of helmet detection can be considered a very particular case of the research in object detection. Although this a computer vision problem in video analytics, the reviews reports noticed that the literature on helmet detection is not found very rich.

Chiverton et al. [1] presented an automatic tracking and classification of moving objects and extracted the motorcyclists by background segmentation. Wearing helmet during riding a motor bike is supposed a necessary safety cover in any serious road accidental event [14]. But due to several traditional but ignorable issues, unfortunately wearing a helmet is hard to enforce in the locality where the generation is not ready to adapt [4, 15, 16]. Silva et al. [5] had experimented with HOG and LBP to detect motorcyclist without helmet. And left a research gaps of multiple bike-riders detection. Chiu et al. [17] and Liu et al. [18] proposed helmet detection based on extracting circular and curve features from the data samples of bike riders. They used vertical histogram for detecting motor bike as a foreground image and computed projection of silhouette of the object in motion. The research work of helmet detection in video sequences connects three more classes, face detection in [19] AdaBoost algorithm is used, motor bike detection in [20] real time tracking algorithm is proposed and human detection in [6] is based on localized HOG features and SVM classifier. To develop an intelligently robust and cost effective accident avoidance system major amount of work is reported on global traffic monitoring system rather than very being specific like helmet detection. An automated video based traffic monitoring system enhances the state of the art with the system which can count the number of vehicles in live traffic, detect the high traffic region and measure the temperature of the vehicle [21–24]. Furthermore, video surveillance based traffic monitoring is focused to develop an automated motor bike monitoring which is capable to consider a motor bike like sensor body that can communicate for the location and architecture to the other motor bikes [7, 25, 26]. Developing such intelligent systems requires the majority of efforts from the experts of architectural network of CAS. A recent work presented in [12] gives every minor details of active reasons to happen the road accidents and raised research gaps for law enforcement to determine real test of driver distraction, alcoholic and drug addicted conditions, eyesight and health issues in old age using the data collected from polices official and public driving data. The research of helmet detection connects medical science also specifically by focusing on epidemiological study to provide health statistics for prefect driving conditions [28]. Motivated by the popularity of spatiotemporal profile of moving objects, Jazayeri et al. [29] presented a HMM based model for tracking a car in public video. The state-of-the-art on roads accidents is a little bit rich due to related hot research problems like real-time traffic monitoring in metro city and challenging transportation in hilly regions etc. But the literature on motorcyclist detection with securities privileged like helmet, maintenance of motor bike and rider's health issues etc. can have more importance in safety and security of individual life.

3 Proposed Methodology

Helmet detection for bike rider in outdoor video sequences can exploit several computer vision based methodologies. To process raw video sequences we can have preprocessing and feature classification blocks as presented in Fig. 1. In the preprocessing block, there are four sub-blocks as listed by English alphabets. Each of the block is referred to process very specific task. Preprocessing block refers to backgrounding modeling to remove the undesirable data and recover the foreground objects. Due to high occlusion and several environmental conditions of real life traffic video sequences, noise and shadow removal are also performed along with morphological filtering. Feature classification block is focused to refined the specified (human, bike and helmet) objects. In the last phase of our model, multiclass kernel based SVM classification is performed to determine human with or without bike and helmet utilizing HOG features of the upper body parts (25%) of detected human in traffic videos.

3.1 Background Modeling

First step in preprocessing block is to convert RGB samples into grey scale samples which are used for background modeling utilizing the Gaussian Mixture Model (GMM) [3, 8]. Since the real life data is full of noise, due to many natural facts like shrubs of blowing wind, raining, snow fall and foggy vision. All such natural obstacles are considered as background part of the outdoor vision.

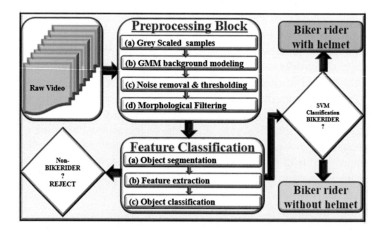

Fig. 1. Proposed model for bike rider detection with helmet

We used Gaussian filter with 5×5 window size and variance 0.8 to remove such undesirable components from our data. Effective details of background modeling and Gaussian filters on experimental data are presented in Fig. 2. This figure represents several others subsequent experiments like threshold by Otsu method and shadow removal etc. by morphological operations.

Fig. 2. First row: RGB sample data (left), Grey scale data (middle), Binary mask by GMM (right); Second row: shadow removal (right), Otsu thresholding followed by Gaussian filter, closing morphology on thresholded foreground mask (left)

3.2 Object Detection and Its Segmentation

Out of all the foreground objects in a video sequence, human body is considered an object with highly rich and sensitive (easier to detect) feature set due the face, skin color etc. We used Histogram of Oriented Gradient (HOOG) model to detect the human body in the given video sequence [6, 20k]. HOOG descriptor has been proved better state of art than HOG in feature detection community. The statistics of bounding box for human detection is 641 in numbers, 9 bins. Therefore, the feature size is 3780. Figure 3 represents the visualization. The feature descriptor is visualized by plotting 9×1 normalized histogram in 8×8 cells. The dominant direction of HOOG histogram is shown in Fig. 4.

Fig. 3. Bounding box of the foreground objects and segmented frame of objects (a) without helmet, (b) with helmet and (c) Non-bike rider

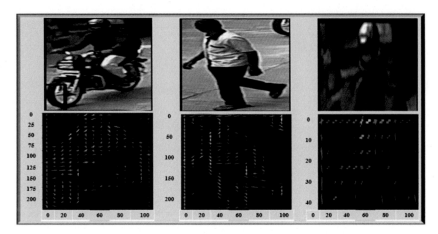

Fig. 4. Histogram of gradients for human with bike (left) and without bike (middle) and 25% of upper body part for helmet (right)

So, for detecting bike rider with helmet we focus at bike-rider's head and utilized the concept of helmet fitting exactly the area of upper territories of human head. Therefore, we take upper fourth part (25%) of object obtained in background modeling step given in Fig. 4 right.

3.3   Classification

Support vector machine (SVM) [30] is most famous and sounding optimization techniques in machine learning community. On the basis of training samples, the margin of hyperplane categorized SVM into hard margin classification and soft margin classification. Generally, outliers and high level of noise modify the data samples such a way that margin disappeared and SVM fails to separate the data linearly. Soft margin classifiers uses higher order non-linear polynomial to deal with unconstraint data sample. Almost non-linear function requires high computational cost. Some function which reduce the data linearly separable called kernels. The kernels of SVM take low dimensional input, produce high dimensional output. The most use kernel associated SVM are polynomial and Radial Basis Function (RBF) represented by Eqs. (3.1) and (3.2) respectively.

$$k(x_i, y_i) = ((x_i, y_i) + 1)^p \tag{3.1}$$

$$k(x_i, y_i) = exp\left(-\gamma \|x_i - y_j\|^2\right) \tag{3.2}$$

4   Experimental Analysis and Result Discussion

We follow two phases' strategies to get the results from SVM classifier. In first phase we test the comparison between bike riders and non-bike riders. This helps us to collect the data of all the bike riders in a video sequence. In the second phase from all the bike riders we get out final objective of bike riders with helmets. The experiments are performed on i-7 Intel processor with 4 GB RAM under Window 8 environment. The software used are OpenCV 2.7, scikit learn in Python 2.7. Experimental results for testing bike rider and non-bike rider are shown in Tables 2 and 3 is used to represent desired results of the said problem to test bike rider detection accuracy against non-bike rider.

Table 2. Phase-I classification for bike-rider detection

SVM (Kernel)	Accuracy	Training time(sec)
Linear	96.67	138.98
MLP	94.45	299.16
RBF	93.37	250.56
Average	94.83	229.57

Table 3. Phase-II: motorcyclist class of helmet and without helmet

SVM (Kernel)	Accuracy	Training time(sec)
Linear	96.83	136.69
MLP	95.45	303.16
RBF	93.99	280.45
Average	95.42	240.1

5   Conclusion and Future Issues

In this research paper, the experiments were performed to develop safety and security parameters to avoid outdoor bike riding accidents with main focus on helmet detection. The data samples of video sequence are collected from outdoor road traffic. Our model meets the desired standards without any assistance from video surveillances efficient. The experimental work incorporates the technology of classification and segmentation during motion of the objects. We used Gaussian filter and BMM for background modeling followed by several preprocessing steps like morphological filtering operations Otsu thresholding. The foreground object (human and bikes etc.) are processed using HOOG features. Finally SVM with several kernels is utilized for motorcyclist detection without helmet. The proposed model gives satisfactory results with 96.67% accuracy.

5.1   Future Scopes

The model is required to improve for handling high occlusion in complex video sequence of real-time traffic. Near future prospective of the research in accidents avoidance is supposed to have large scale experiments of real time data analytics. The key resources of training/testing samples data will be RTO office, Police station reports regarding road accidents and manufacturing companies (Hero, Honda, Maruti Suzuki etc.). Furthermore, motivated from the innovation of intelligently deep embedded system, the research can be improved up to an outstanding benchmark by developing sensor (SONY ECM-77B, Delphi radar and SV-625B optical camera) based intelligent vehicle detection system. More precisely, the research work on helmet detection for individual safety connects vision problems in old ages, drug addiction check, road manufacturing for civil engineering and developing an intelligent accident regressive mechanical device.

References

1. Chiverton, J.: Helmet presence classification with motorcycle detection and tracking. Intell. Transport Syst. (IET) **6**(3), 259–269 (2012)
2. Silva, R., Aires, K.R.T., Santos, T., Lima, K.A.B.: Automatic detection of motorcyclists without helmet. In: XXXIX Latin American Computing Conference (CLEI), pp. 1–7, October 2013. Author, F.: Article title. Journal **2**(5), 99–110 (2016)
3. Zivkovic, Z.: Improved adaptive Gaussian mixture model for background subtraction. In: Proceedings of the 17th International Conference on Pattern Recognition, ICPR 2004, vol. 2, pp. 28–31. IEEE, August 2004
4. Kulkarni, P., Sangam, V.G.: Smart helmet for hazardous event detection and evaluation in mining industries using wireless communication. J. Commun. Eng. Innovations **3**(1), 11–16 (2017)
5. E Silva, R.R.V., Aires, K.R.T., Veras, R.D.M.S.: Helmet detection on motorcyclists using image descriptors and classifiers. In: 2014 27th SIBGRAPI Conference on Graphics, Patterns and Images (SIBGRAPI), pp. 141–148. IEEE, August 2014
6. Dalal, N., Triggs, B.: Histograms of oriented gradients for human detection. In: International Conference on Computer Vision Pattern Recognition, pp. 886–893. IEEE (2005)
7. Dahiya, K., Singh, D., Mohan, C.K.: Automatic detection of bike-riders without helmet using surveillance videos in real-time. In: 2016 International Joint Conference on Neural Networks (IJCNN), pp. 3046–3051. IEEE, July 2016
8. Stauffer, C., Grimson, W.: Adaptive background mixture models for real-time tracking. In: Proceedings of the IEEE Conference on Computer Vision and Pattern Recognition (CVPR), vol. 2, pp. 246–252 (1999)
9. Ranney, T.A., Mazzae, E., Garrott, R., Goodman, M.J.: Driver Distraction Research: Past, Present, and Future. Transportation Research Center Inc., East Liberty (2000)
10. Baker, M., Hyvonen, M.: The emerging of the Chinese automobile sector. RBA Bulletin, March, pp. 23–29 (2011)
11. Mukhtar, A., Xia, L., Tang, T.B.: Vehicle detection techniques for collision avoidance systems: a review. IEEE Trans. Intell. Transp. Syst. **16**(5), 2318–2338 (2015)
12. Rolison, J.J., Regev, S., Moutari, S., Feeney, A.: What are the factors that contribute to road accidents? An assessment of law enforcement views, ordinary drivers' opinions, and road accident records. Accid. Anal. Prev. **115**, 11–24 (2018)
13. World Health Organization: Global Status Report on Road Safety 2015 (2015). http://www.who.int/violence_injury_prevention/road_safety_status/2015/en/
14. Bayly, M., Regan, M., Hosking, S.: 'Intelligent transport systems and motorcycle safety' (Monash University, Accident Research Centre), p. 260 (2006)

15. Pitaktong, U., Manopaiboon, C., Kilmarx, P.H., et al.: Motorcycle helmet use and related risk behaviors among adolescents and young adults in Northern Thailand. SE Asian J. Trop. Med. Public Health **35**(1), 232–241 (2004)
16. Bianco, A., Trani, F., Santoro, G., Angelillo, I.F.: 'Adolescents' attitudes and behaviour towards motorcycle helmet use in Italy. Eur. J. Pediatr. **164**(4), 207–211 (2005)
17. Chiu, C.C., Ku, M.Y., Chen, H.T.: Motorcycle detection and tracking system with occlusion segmentation. In: IEEE CS Eighth International Workshop on WIAMIS 2007, p. 32 (16g) (2007)
18. Liu, C.C., Liao, J.S., Chen, W.Y., Chen, J.H.: The full motorcycle helmet detection scheme using canny detection. In: IPPR 18th Conference CVGIP, pp. 1104–1110 (2005)
19. Viola, P., Jones, M.: Robust real-time face detection. Int. J. Comput. Vis. **57**(2), 137–154 (2004)
20. Stojmenovic, M.: Algorithms for real-time object detection in images. In: Nayak, A., Stojmenevic, I. (eds.) Handbook of Applied Algorithms, pp. 317–346. Wiley (2008)
21. Minge, E.: Evaluation of non-intrusive technologies for traffic detection. Minnesota Department of Transportation, Office of Policy Analysis, Research and Innovation, SRF Consulting Group, US Department of Transportation, Federal Highway Administration, pp. 2010–2036 (2010)
22. Semertzidis, T., Dimitropoulos, K., Koutsia, A., Grammalidis, N.: Video sensor network for real-time traffic monitoring and surveillance. IET Intell. Transp. Syst. **4**(2), 103–112 (2010)
23. Buch, N., Velastin, S.A., Orwell, J.: A review of computer vision techniques for the analysis of urban traffic. IEEE Trans. Intell. Transp. Syst. **12**(3), 920–939 (2011)
24. Song, K.T., Tai, J.C.: Image-based traffic monitoring with shadow suppression. Proc. IEEE **92**(2), 413–426 (2007)
25. Morris, B.T., Trivedi, M.M.: Learning, modeling, and classification of vehicle track patterns from live video. IEEE Trans. Intell. Transp. Syst. **9**(3), 425–437 (2008)
26. Kanhere, N.K., Birchfield, S.T., Sarasua, W.A., Khoeini, S.: Traffic monitoring of motorcycles during special events using video detection. Transp. Res. Record: J. Transp. Res. Board **2160**(10–3933), 69–76 (2010)
27. Rubaiyat, A.H., Toma, T.T., Kalantari-Khandani, M., Rahman, S.A., Chen, L., Ye, Y., Pan, C.S.: Automatic detection of helmet uses for construction safety. In: 2016 IEEE/WIC/ACM International Conference on Web Intelligence Workshops (WIW), pp. 135–142. IEEE, October 2016
28. Jha, N., Srinivasa, D.K., Roy, G., Jagdish, S., Minocha, R.K.: Epidemiological study of road traffic accident cases: a study from South India. Indian J. Commun. Med. **29**(1), 20–24 (2004)
29. Jazayeri, A., Cai, H., Zheng, J.Y., Tuceryan, M.: Vehicle detection and tracking in car video based on motion model. IEEE Trans. Intell. Transp. Syst. **12**(2), 583–595 (2011)
30. Hsu, C.W., Lin, C.J.: A comparison of methods for multiclass support vector machines. IEEE Trans. Neural Netw. **13**(2), 415–425 (2002)
31. Tesema, T., Abraham, A., Grosan, C.: Rule mining and classification of road accidents using adaptive regression trees. Int. J. Simul. Syst. Sci. Technol. **6**(10–11), 80–94 (2005)

Strategies and Challenges in Big Data: A Short Review

D. K. Santhosh Kumar[(✉)] and Demian Antony D'Mello

Department of Computer Science and Engineering,
Canara Engineering College Mangalore, VTU, Belagavi, India
santhosh.dk.kumar@gmail.com, demain.antony@gmail.com

Abstract. The Big Data is the new trending technology in the field of research in recent years and is not only big in size, but also generated at brisk rate and variety, which endeavors the research upsurge in multidisciplinary fields like Government, Healthcare and business performance applications. Due to the key features (Volume, Velocity, and Variety) of Big Data it's difficult to store and analyse with conventional tools and techniques. It acquaints unique challenges in scalability, storage, computational complexity, analytical, statistical correlation and security issues. Hence we describe the salient features of big data and how these affects the storage technologies and analytical techniques. We then present the taxonomy of Big Data sub-domains and discuss the different datasets based on data characteristics, privacy concern, and domain and application knowledge. Furthermore, we also explore research issues and challenges in big data storage technologies, privacy of data and data analytics.

Keywords: Big Data · Analytics · Data science · Big Data domains ·
Data security · Machine learning · Hadoop

1 Introduction

The Big Data is the most discussed phrase of today's era and it is believed that due to the influence of Big Data there will be a big change in Science, Industry, Society, government [1, 2]. There is a thought that the information economy future is powered by new petroleum called Big Data [1]. The fundamental reason for the Big Data is Digitization, Social networking, Automation, Health care, IoT, etc. The generated data is either in the form of structured, unstructured or semi-structured or in the form of multi-media content i.e. images or audio or videos of multiplicity platforms [2–4]. The 2.5 Exabytes of data is being generated per day, out of which 90% is unstructured and by 2020 will exceed 40 Zettabytes. The Big Data is presented as the top 10 technology trend and top 10 critical tech trends in the next five years [3]. Irrespective of how Big Data is generated, from where it is generated, the challenging part is how it can be stored, how it can be shared, how to provide security to that data and most importantly, how to analyse the data so that it will be converted to a valuable one.

© Springer Nature Switzerland AG 2020
A. Abraham et al. (Eds.): ISDA 2018, AISC 941, pp. 34–47, 2020.
https://doi.org/10.1007/978-3-030-16660-1_4

1.1   Definition and Characteristics of Big Data

The universally accepted definition of Big Data is not known till date [5]. The Wikipedia defines it as "*an all-encompassing term for any collection of data sets so large and complex that it becomes difficult to process using traditional data processing applications*" [1]. It is evident that Big Data definition is nascent and has uncertain origins.

In the mid of 1990 the term Big Data was originated at Silicon Graphics Inc. in lunch-table conversations, but the Fig. 1 illustrates Big Data is widespread after 2011 onwards. Figure 2 illustrates the online survey of 154 C-suite global executives conducted by Harris on behalf of SAP, in which many definitions raised, leads to the many confusion. The promotional initiatives by leading technological companies like IBM and others who are niche analytics market building investors. Different sources participate in the generation of huge volume data of different types in terms of Petabytes and Exabytes. The Fig. 1 shows the evolution of the term Big Data [6, 7].

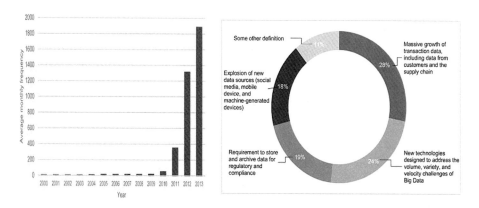

Fig. 1.   Evolution of Big Data term and various definitions of Big Data [6].

1.2   Classification of Big Data

We categorize data into (i) Data Source (ii) Data formats (iii) Data Store (iv) Data Staging and (v) Data Processing based on the characteristics and complexities of Big Data. Each of these categories has its trait as discussed in the Table 1. This Categorization plays a salient role in deciding the domain, approach, methodology and tools of analytics.

Table 1. Categories and classification of Big Data.

Category	Classification	Description
Data sources [8, 9]	Social media data [10, 11]	Social medias like Twitter, Facebook, blogs and others are prominent source of data generation. They provides platform for discussion on social affairs and business trends
	Machine-generated data [10]	This data is generated without human intervention by servers medical devices, airplanes etc., in terms of logs
	Sensor data [10]	Scathing vicinity can be better conducted by sensors than awaiting human intervention, where rate of data generation is high
	Transaction data [10]	Most generated by financial, Retail industries events which involves time dimension data description
	IoT data [10]	The automation of digital world and communication between devices over internet generates huge volume of data at high rate
Content format [8, 9]	Structured [12]	Structured data is usually stored and maintained by conventional RDBMS and has respective programming language to manage it
	Semi- structured [12]	This data is not from conventional RDBMS. This might be a structured but not well organized in tables, database models etc., and no respective language to manage it
	Unstructured [12]	Unstructured data will not follow any standard format like videos, audios, text messages, social media data, location information etc. and 90% of world data is in this unstructured format
Data store [8, 9]	Document-oriented [8, 13]	This data store stores the data in complex standard format such as XML, JSON and binary formats (pdf, doc etc.). In document-oriented data store the document is stored and retrieved on the basis of contents (CouchDB, MongoDB, and SimpleDB)
	Column-oriented [8, 13]	In this database instead of rows data is stored in columns contiguously with their respective attributes and this column orientation such as BigTable is completely different from traditional DBMS
	Graph database [8, 14]	A Graph database like Neo4j stores and represents data in graph with vertex, edges and its properties of related graph
Data staging [8, 9]	Cleaning [8]	In cleaning incomplete and unreasonable data is identified
	Transform [8]	Here the data will be transformed to a suitable form for analysis
	Normalization [8]	Normalization is a technique reducing the redundancy of database by structuring it

(continued)

Table 1. (*continued*)

Category	Classification	Description
Data processing [8, 9]	Batch [13, 15]	Most of the organizations from past few years have adopted one of the best batch processing system, map reduce for their long running batch jobs. This system is best suited for scaling of applications across huge clusters of nodes
	Real time [10, 13, 16]	The Simple Scalable Streaming System (S4) is the powerful and most famous real-time process-based Big Data tool. S4 is distributed in nature and allows to process continuous unbounded stream of data

2 Research Challenges in Big Data

The Big Data domain leads to many open challenges that can be mainly categorized to three areas, (1) Data accessing and mining platform [13, 17–19], (2) Semantics and Application domain knowledge [19–21] and (3) Big Data Analytics [13, 19]. The Fig. 2 represents the taxonomy of Big Data domains.

2.1 Data Accessing and Mining Platform

Big Data deals with huge magnitude of data which is generated at high rate by variety of source and platforms; its big challenge is to where to store the data and how to access it. The storage and accessing is completely different from traditional data. The data storage organizes the collected data in a convenient format for value extraction and analysis. To achieve this it should accommodate data persistently along with the reliability, also it should provide a scalable interface to access, query and analyze. It's a challenge to Storage Infrastructure and software tools.

2.1.1 Storage Infrastructure

The collected data is physically stored on storage devices like RAM, Magnetic Disks, Disk Arrays and Storage class Memory etc. These devices have their own performance metrics which leads to building high performance and scalable storage systems. New architectures like Solid-State Drives (SSD), Serial Attached SCSI (SAS) have been introduced with virtualization techniques to boost the performance.

Next these storage infrastructures are networked with Direct Attach Storage (DAS), Network Attached Storage (NAS) and Storage Area Network (SAN) to support huge storage space, data backup and recovery and also to provide cloud infrastructure. Storage infrastructure is among the areas which provide various research challenges [13, 19].

2.1.2 Data Management Tools

Data Management tools play a major role in organizing data over Storage Infrastructure so that data can be accessed in convenient manner for processing. Even before the

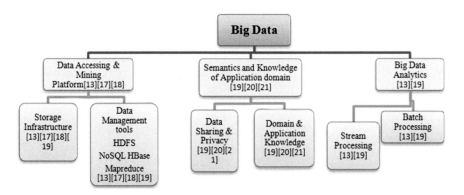

Fig. 2. An overview of Big Data domain taxonomy.

origin of BD such tools were actively researched. This research can be classified into three layers. (1) File System, (2) Database Technology and (3) Programming Models.

The basis of data storage is given by file system, which attracts the attention of academic and industry. The Google File System (GFS) is the first file system for Big Data. GFS is a distributed and scalable file system which runs over commodity servers. Main drawback of GFS is it provides poor performance for small size file and suffers with single point failure which has been overcome in Colossus. The open source derivatives of GFS are Hadoop Distributed File System (HDFC) and Kosmosfs. Cosmos was created by Microsoft to support its advertisement and search business. Haystack was invented by Facebook to store large volume of small-file photos. Taboos implemented Tao File System (TFS) and FastDFS [13, 19, 22].

For more than four decades the Database Technologies have gone thorough development. Due to the characteristics of Big Data traditional relational databases are not suitable. The NOSQL database standardizes the Big Data problem. The NoSQL databases are organized by Document databases, Column-oriented database and Key-value store models [13, 19, 23]. The characteristics of Big Data demands parallel distributed programming models to provide analytical results at high rate. The famous programming models like OpenMP, MPI, Map-reduce, Dryad, Pregel, GraphLab, Storm, S4 are widely used in Big Data analytics [13, 19, 22].

2.2   Semantic and Knowledge of Application Domain

The Semantic application and domain Knowledge refers to many aspects like policies, domain information, regulations and user knowledge. The most significant issues are (1) Data Privacy while sharing and (2) Knowledge of Application Domain. These two issues need be handled carefully, as they decide the dos and don'ts on Big Data.

2.2.1   Data Privacy While Sharing
The ultimate goal of digital world is to share the data and it has got such growth due to data sharing. The motivation for sharing data over multiple systems is clear, but true concern is emphasized on the sensitive data, involving banking transaction and medical

records processed by Big Data application. Two common approaches to protect privacy are (1) Restrict data access: The security certificates can be added to data or access control can be imposed on data so that sensitive data is accessed only by limited group of users. In this approach challenges are to address the designing of security certificates and access control framework, so that sensitive information can't be accessed by unauthorized users. Privacy concern must be provided, but care must be taken to protect the data integrity as well. (2) Data Field Anonymization: It is a process where individual record in sensitive data cannot be pinpointed. The main objective of anonymization is to inject randomness to data to ensure variety of privacy goals. Anonymized data is exempted from restrictive access controls and can be shared freely across different users [19–21].

2.2.2 Knowledge of Application Domain

The vital requirement for the Big Data analytics is knowledge of application domain, which plays a vital role to decide and design Big Data mining algorithms and framework and also facilitates veracious features for modeling the underlying data. This knowledge eases up the designing of Big Data analytical techniques which helps in determining achievable business objectives. Without a prior knowledge of domain and application it's a big challenge to extract the effective patterns from the underlying Big Data [19, 24].

2.3 Big Data Analytics

"The process of identifying the hidden patterns and unknown correlations by using analytical algorithms running on powerful machines is termed as Big Data analytics". It can be classified based on their processing requirements: (1) Incremental/Stream Processing, (2) Batch Processing.

2.3.1 Incremental/Stream Processing

The Incremental/Stream processing starts with an assumption that the data freshness decides the potential value of data. Due to this main impact of freshness over the value of data streaming, processing models should analyze the data at earliest otherwise data will be outdated. In this huge amount of data is streamed with high rate, limited memory is capable of storing only a chunk of stream. Technically it's a significant challenge to analyze partial data online as soon as possible without waiting for complete data. The Storm and S4 are from Apache which works on distributed streaming model can be scale to huge number of messages in a node per second. These models support any programming language. On the other side Storm adopts pull model, receiver pulls the data when they process it. The Hadoop 2.0 integrated with Apache Samza guarantees no data loss by using Apache Kafka [13, 19].

Microsoft Trill works on temporal streaming data. Microsoft also designed Stat using Microsoft Stream Insight for progressive computations of temporal streaming engine and designed a new solution Naiad to handle iterative stream scenario with changing input. The Infosphere Streams is from IBM analytical solution, which analyzes stream data at faster rate by applying real-time machine learning techniques. The Streams is a dynamic tool which adds input stream, output stream and operations dynamically without restoring. The Mill wheel from Google supports the user to create graph streams and provides code for each graph node which efficiently addresses failure recovery, continuous data rate and data persistence [13, 19].

2.3.2 Batch Processing

In batch processing store and analyse technique is used, where it stores the data first then analysis will be done. In batch processing a chain of jobs are executed without human intervention. Batch processing is the most popular scenario which processes huge data in a single run. The popular open source tool Map-Reduce supports batch processing which is a part of Hadoop framework. Hadoop is scalable, fault tolerant, flexible and easy-to-code. Map-Reduce is distributed parallel programming model which works on Single Program Multiple Data concept [13, 19].

3   Big Data Analytic Methods

The traditional data analysis techniques are not meant for high volume, complex and high rate data, it's almost impossible for traditional algorithms to analyse the Big Data. In Fig. 3 the algorithms for Big Data analysis is categorized as (a) Mining Algorithms for Specific Problems [25] and (b) Big Data Machine Learning algorithms [25].

3.1   Mining Algorithms for Specific Problems

As the name mining inherits the basics of data mining, in this category the algorithms used some basic features of data mining algorithms or they are up-gradation of mining algorithms. These algorithms can be divided into four categories as like mining algorithms. In Fig. 3 we have presented the same and in Table 2 we have reviewed some of the analytical algorithms.

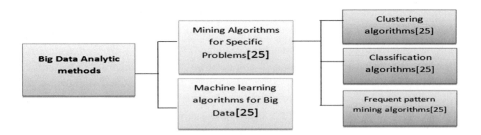

Fig. 3. An overview of Big Data analytic methods.

Clustering Algorithms: The conventional clustering algorithms are limited due to the large volume of data and data with high-dimension has been attracted large numbered researcher of various disciplines from past decades, and in recent years many proposal and solutions are proposed and presented. Due to Big Data characteristics and diversity still news issues are rising for data clustering. Reducing the data complexity is one of the major issues for clustering. The clustering can be divided into two categories. (1) Single-machine clustering: Concentrates on the sampling of data and provides solutions for dimension reduction. (2) Several-machine clustering: supports parallel processing. The traditional reduction solutions also suits BD era, because data analysis process time and space complexity can be reduced by applying sampling and dimension reduction methods [25, 41].

Classification Algorithms: Modifying conventional classification algorithms to support parallel computing environment similar to clustering algorithms several research studies has been conducted. During this research the input data is taken into account because this data is gathered by variety of distributed sources and heterogeneous set of learners are used to process it. The classification learner works in two ways to process the input data in distributed data classification system. For a given input data one will execute classification function itself and the other one will route to another learner to label it. To support the same the learner should be designed as cooperative learners, along with information exchange it improves the accuracy in providing the solution for big Data classification problem [25].

Frequent Pattern Mining Algorithms: To deal with large-scale dataset many researchers concentrated on mining of frequent pattern which included sequential pattern and association rule mining. In the early days frequent pattern mining is used to analyse super market transactional data where the number of transactions were in terms of thousands but the real challenge is in handling real-time huge quantity data in various format using frequent pattern mining algorithm. Along with conventional frequent pattern algorithms cloud and parallel computing also attracted the researchers to improve its performance by adopting the map-reduce models [25].

3.2 Machine Learning Algorithms for Big Data

The machine learning algorithm can be used for Big Data analytics. The machine learning algorithms typically employed as search algorithms and most of the algorithms are adopted to get an approximation solution for optimization problem. This highlights machine learning algorithms are used for data analytics only if problems are formulated to optimization. Consider, if Genetic algorithm is used to solve clustering and also frequent pattern mining problem, Then it shows by enhancing the features of machine learning algorithms the same algorithm can be used for different mining solutions. The machine learning algorithms not only addresses Big Data analytics problem but it also enhances the performance of data analytics with respect to feature reduction for input operators [12, 25, 26].

The machine learning algorithms for Big Data analytics can be categorized as: (1) Attempts to run learning algorithms on parallel machines like PIMRU, Mahout and Radoop. (2) Modifies the learning algorithms so that they are suitable for parallel computations like ant-based algorithms for grid computing and neural network algorithms for GPU. Both of these categories try to perform Big Data analytics using machine learning algorithms on parallel machines [12, 26].

Table 2. List of mining algorithms for specific problems.

Algorithm	Mining technique	Description	Advantage	Disadvantage
BIRCH [27]	Clustering	Incremental and dynamic clustering of incoming dataset	Scalable, minimizes I/O overhead	Limitation of CF-tree size due to memory
DBSCAN [28]	Clustering	The high density regions are located, are separated by low density region	Robust to outliers, less parameters	Not complete deterministic, purely dependent on distance metric. Not suitable for high dimensional dataset
RKM [29]	Clustering	Manages big clustering problem by integrating programming inside RDBMS	No limitation over memory and performance affected by dimensionality is minimal	Does not suit for categorical dataset. Not memory intensive
SLIQ [30]	Classification	Addresses categorical and numeric attributes. Tree-growth is handled pre-sorting	Better accuracy, small execution times. suits for large number attribute datasets	Produces small decision trees
TLAESA [31]	Classification	To work in sub linear time distance metric properties are used	The distance computation takes less time	Takes more time to compute for smaller distance
FastNN [32]	Classification	Set of labeled samples are used to find unknown sample the nearest neighbor just by applying moderate effort on preprocessing	Optimum number of cells are generated in few feature space	Demands some storage of the coordinates of the cell centers
FP-tree [33]	Association Rule	On crucial and compressed information pattern growth method applied to extract patterns	The cost of database scans in the subsequent mining processes is saved by highly compact FP-tree. costly candidate generation avoided by applying pattern growth method	Need to address high scalable FP-tree for SQL-based, constraint-based mining

(continued)

Table 2. (*continued*)

Algorithm	Mining technique	Description	Advantage	Disadvantage
FAST [34]	Association Rule	Applies two step for data reduction 1. The support of each item is used to estimate a large simple random sample 2. Final sample is generated either by trimming outliers or by choosing representative transaction	90–95% accuracy is achieved. Speedup by roughly a factor of 10 over algorithms	Combining sampling model with more association rule to be improved
SPADE [35]	Sequential patterns	To decompose the problem combinatorial properties are used. Solves sub-problems independently in main memory by applying lattice search technique	Performance is increased by order magnitude, factor of two Scaled up by number of parameters	Many frequent sequences are trivial and useless
CloSpan [36]	Sequential patterns	Lexicographic ordering system formulated CloSpan to mine frequent closed sequences efficiently	CloSpan adopts a novel pruning technology to provide new insight for scalable mining of long patterns	Not utilizes the full search space pruning property Not interactive
PrefixSpan [37]	Sequential patterns	Only the prefix subsequences are examined and corresponding postfix subsequences are projected. Two types of databases will improve efficiency	The disk-based processing better due to bit-level projection and main memory fit issue is handled by pseudo-projection	Does not support DNA databases, time constraint datasets and sequential patterns

(*continued*)

Table 2. (*continued*)

Algorithm	Mining technique	Description	Advantage	Disadvantage
ISE [38]	Sequential patterns	The iterative approach is used to discovers frequent sequences, large databases are handled without maintain negative border	More efficient for re-run over updated data. The increment from the original database is quickly in extracted	In updated database the new frequent sequences need to recognize
DPSP [39]	Sequential patterns	Works over Hadoop to address scalability issue for the progressive sequential patterns	The number of computing nodes are easily increased in the cluster	Single point failure

4   Research Challenges and Issues in Big Data

Many researchers are focused on addressing the issues in Big Data storage, access, security and analysis. But there are several exiting issues and challenges where researchers can look into. The challenges are as follows.

- **Heterogeneity and Incompleteness**

Large amount of data is being generated from sources. Data is in the form of text, data logs, videos, images, audios, structured, semi structured, unstructured data from sources like sensors, airplanes, social networks, retail industry, mobiles etc.,. Also uncertainty is created by incomplete data during analysis which should be managed correctly.

- **Scale and complexity**

Managing rapidly increasing huge volume of data is a challenging issue. The conventional mechanisms are not suitable for managing analyzing and retrieving of Big Data, which is an open challenge due to its complexity and scalability.

- **Timeliness**

The time required to analyze the data will increase due to rapidly increasing its generation. Hence there are some situations where misuse of data need to be addressed.

- **Security and Privacy**

Huge amount of data is being generated, processed and analyzed. During this process, the users and organizations are worried about data privacy and security related issued. Hence these issues need to be addressed to avoid malicious activates performed by internal and external adversaries.

- **Failure handling and Fault tolerance**

Fault tolerance is most important issue that needs to be addressed in big data. When a process started by involving many network nodes in the entire computation process, becomes cumbersome. The great concern is in deciding threshold for each level, retaining check points and restarting in case of failure.

- **Legal/regulatory and Governance**

To preserve the sensitive data of the users, a specific law and regulations must be established. Each county has its own laws and regulations, are these laws and regulations offers adequate protection over sensitive data? Or these laws and regulations provide complete liability over sensitive data to perform desired analysis? The data governance represents authority over data-related policies. These polices indicates what to be stored, how to access data and how fast anyone can access it.

- **Data analysis**

Deciding an appropriate technique to analyse a huge data is crucial phase of data analytics. To extract desired patterns from huge data there must be analytical algorithms to generate results. However the desired patterns purely depend on the timely requirements so the existing algorithms are not suitable for Big Data analysis. The data analytics is not only reserved to particular platform/domain, other trendy domains highly demands Big Data analytics should be performed on their platforms, such as IoT, Cloud Computing and Bio-inspired Computing [40]. Therefore there is a huge scope for novel data analytical techniques and models.

- **Knowledge Discovery and Computational Complexities**

The prime issue of big data is Knowledge discovery and representation of data which includes types of sub domains: preserving, archiving, and retrieving of information, authentication and data management. There are many existing tools and techniques to address these issues. But most of these techniques are specific to some problems; some of these are not suitable for handle huge amount of datasets. Due to this computational cost will be high while analyzing the large dataset.

5 Conclusions

Presently the volume of data being generated is high and continues to increase timely. There is an expansion in data variety and the velocity of data being generated and is high due to automation, social media, smart phones, sensor connected devices and internet. This endeavors the research upsurge in multidisciplinary fields as well as Government, Healthcare and business performance applications. The characteristic of Big Data makes it difficult to store and analyse with conventional tools and techniques. It acquaints unique challenges in scalability, storage, computational, analytical, statistical correlation and security issues. Hence we explore the salient features of Big Data and present and discuss the classification of datasets based on their behavior and respective available tools to address it.

We also addressed the research challenges in the field of big data domain. We discussed some of the available platforms to address the Big Data managing tools, analytical frameworks and also presented the taxonomy of Big Data domains and discussed the privacy concern, domain knowledge. We also presented some of available data analytical algorithms and reviewed, then identified some of the challenges in Big Data storage, Processing and Privacy concern and also highlighted the demand for novel Big Data analytical techniques for trendy research domains. The big data is the current trend and next era is ruled by big data and there is huge scope for open research challenges. We predict that, the research on big data storage, privacy of data and analytical techniques will continue to grow in forthcoming years.

References

1. Jin, X., Wah, B.W., Cheng, X., Wang, Y.: Significance and challenges of big data research. Big Data Res. **2**(2), 59–64 (2015)
2. Mining, D.: Big-data analytics : a critical review and some future directions. Uroš Jovanovi č, Aleš Štimec Daniel Vladuši č Gregor Papa * Jurij Šilc. **10**, 337–355 (2015)
3. Sivarajah, U., Kamal, M.M., Irani, Z., Weerakkody, V.: Critical analysis of big data challenges and analytical methods. J. Bus. Res. **70**, 263–286 (2017)
4. Du, D., Li, A., Zhang, L.: Survey on the applications of big data in Chinese real estate enterprise. Procedia Comput. Sci. **30**, 24–33 (2014)
5. De Mauroandrea, A., Greco, M., Grimaldim, M., Table, V.: What is big data? A consensual definition and a review of key research topics, p. 97 (2015)
6. Gandomi, A., Haider, M.: Beyond the hype: big data concepts, methods, and analytics. Int. J. Inf. Manag. **35**, 137–144 (2015)
7. Özköse, H., Uõ, P.L.Q., Gencer, C.: Yesterday, today and tomorrow of big data. Procedia-Soc. Behav. Sci. **195**, 1042–1050 (2015)
8. Abaker, I., Hashem, T., Yaqoob, I., Badrul, N., Mokhtar, S., Gani, A., Ullah, S.: The rise of "big data" on cloud computing: review and open research issues. Inf. Syst. **47**, 98–115 (2015)
9. Marr, B.: Big Data: 33 Brilliant and Free Data Sources for 2016. https://www.forbes.com/sites/bernardmarr/2016/02/12/big-data-35-brilliant-and-free-data-sources-for-2016/#4166dcf7b54d
10. Cao, L.: Data science: a comprehensive overview. ACM Comput. Surv. (CSUR) **50**(3), 43 (2017)
11. Tan, W., Blake, M.B., Saleh, I., Dustdar, S.: Social-network-sourced big data analytics. IEEE Internet Comput. **17**, 62–69 (2013)
12. Williams, G.J., Office, A.T.: Big data opportunities and challenges: discussions from data analytics persoectives. Comput. Intell. Mag. IEEE. **9**, 62–74 (2014)
13. Hu, H., Wen, Y., Chua, T.-S., Li, X.: Toward scalable systems for big data analytics: a technology tutorial. IEEE Access. **2**, 652–687 (2014)
14. Kaliyar, R.: Graph databases: a survey, pp. 785–790 (2015)
15. Assunção, M.D., Calheiros, R.N., Bianchi, S., Netto, M.A.S., Buyya, R.: Big data computing and clouds: trends and future directions. J. Parallel Distrib. Comput. **79–80**, 3–15 (2015)
16. Chen, M., Mao, S., Liu, Y.: Big data: a survey. Mob. Netw. Appl. **19**(2), 171–209 (2014)
17. Singh, D., Reddy, C.K.: A survey on platforms for big data analytics. J. Big Data **2**, 8 (2014)

18. Khalifa, S., Elshater, Y., Sundaravarathan, K., Bhat, A.: The six pillars for building big data analytics ecosystems. ACM Comput. Surv. **49**, 1–36 (2016)

19. Wu, X., Zhu, X., Wu, G.-Q., Ding, W.: Data mining with big data. IEEE Trans. Knowl. Data Eng. **26**, 97–107 (2014)

20. Colombo, P., Ferrari, E.: Privacy aware access control for big data: a research roadmap. Big Data Res. **2**, 145–154 (2015)

21. Rumbold, J.M.M., Pierscionek, B.K.: What are data? A categorization of the data sensitivity spectrum. Big Data Res. **12**, 49–59 (2017)

22. Zhang, Y., Ren, J., Liu, J., Xu, C., Guo, H., Liu, Y.: A survey on emerging computing paradigms for big data. Chin. J. Electron. **26**(1), 1–12 (2017)

23. Khan, N., Yaqoob, I., Abaker, I., Hashem, T., Inayat, Z., Kamaleldin, W., Ali, M., Alam, M., Shiraz, M., Gani, A.: Big Data: Survey, Technologies, Opportunities, and Challenges (2014)

24. Samuel, S.J., Rvp, K., Sashidhar, K., Bharathi, C.R.: A survey on big data and its research challenges. ARPN J. Eng. Appl. Sci. **10**, 3343–3347 (2015)

25. Tsai, C.W., Lai, C.F., Chao, H.C., Vasilakos, A.V.: Big data analytics: a survey. J. Big Data **2**, 1–32 (2015)

26. L'Heureux, A., Grolinger, K., Elyamany, H.F., Capretz, M.A.M.: Machine learning with big data: challenges and approaches. IEEE Access. **5**, 7776–7797 (2017)

27. Zhang, T., Ramakrishnan, R., Livny, M.: BIRCH: a new data clustering algorithm and its applications. Data Mining Knowl. Discov. **1**(2), 141–182 (1997)

28. Kisilevich, S., Mansmann, F., Keim, D.: P-DBSCAN: a density based clustering algorithm for exploration and analysis of attractive areas using collections of geo-tagged photos, pp. 1–4 (2010)

29. Ordonez, C., Omiecinski, E.: Efficient disk-based k-means clustering for relational databases. IEEE Trans. Knowl. Data Eng. **16**, 909–921 (2004)

30. Mehta, M., Agrawal, R., Rissanen, J.: SLIQ: A Fast Scalable Classifier for Data Mining (1996)

31. Mico, L., Oncina, J., Mic, L., Oncina, J.: Dynamic Insertions in TLAESA fast NN search algorithm (2016)

32. Djouadi, A., Bouktache, E.: A fast algorithm for the nearest-neighbor classifier. IEEE Trans. Pattern Anal. Mach. Intell. **19**, 277–281 (1997)

33. Han, J., Pei, J., Yin, Y.: Frequent Pattern Tree: Design and Construction, pp. 1–12 (2000)

34. Chen, B., Way, H., Francisco, S.S., Haas, P., Jose, S., Scheuermann, P.: A New Two-Phase Sampling Based Algorithm for Discovering Association Rules (2002)

35. Zaki, M.J.: SPADE: an efficient algorithm for mining frequent sequences. Mach. Learn. **42**(1–2), 31–60 (2001)

36. Zaki, M.J., Hsiao, C.: Efficient algorithms for mining closed itemsets and their lattice structure. IEEE Trans. Knowl. Data Eng. **17**, 462–478 (2005)

37. Pei, J., Han, J., Mortazavi-Asl, B., Pinto, H., Chen, Q., Dayal, U., Hsu, M.C.: PrefixSpan: Mining Sequential Patterns Efficiently by Prefix-Projected Pattern Growth (2001)

38. Masseglia, F., Poncelet, P., Teisseire, M.: Incremental mining of sequential patterns in large databases. Data Knowl. Eng. **46**(1), 97–121 (2003)

39. Huang, J., Lin, S., Chen, M.: DPSP: Distributed Progressive Sequential Pattern Mining on the Cloud, pp. 27–34 (2010)

40. Acharjya, D.P.: A survey on big data analytics: challenges, open research issues and tools. Int. J. Adv. Comput. Sci. Appl. **7**, 511–518 (2016)

41. Izakian, H., Abraham, A., Snášel, V.: Fuzzy clustering using hybrid fuzzy c-means and fuzzy particle swarm optimization. In: World Congress on Nature and Biologically Inspired Computing (NaBIC 2009), India, pp. 1690–1694. IEEE Press (2009). ISBN 978-1-4244-5612-3

Autonomous Water Surveillance Rover

Abhishek Rai, Chirag Shah$^{(\boxtimes)}$, and Nirav Shah

K. J. Somaiya College of Engineering, Mumbai 400077, India
{a.rai,chirag.ms,nirav.ms}@somaiya.edu

Abstract. A significant increase in Maritime Autonomous Systems can be observed across different sectors for various applications. Lake monitoring systems have been studied prior to the implementation of this project. This paper presents implementation of an integrated system focusing on a cost-effective, energy-efficient Autonomous Water Surveillance Rover (AWSR) with multiple capabilities. AWSR can be manually controlled through a web interface. Data of lake water parameters like temperature, pH, and turbidity are being logged and are available on the web interface. SONAR sensor is used to demonstrate depth measurement and fish finding applications. Two cameras are mounted for video surveillance of the water body on the surface and below. Simplistic and adequate design provides additional payload for emergency relief after-floods. The goal of the project is to implement autonomy and it's integration with mentioned applications.

Keywords: Autonomous surface vehicle · Lake monitoring · Lake surveillance · Ground Control Station · Web interface

1 Introduction

At present in India, lakes and wetlands, particularly natural and old man-made, are among the most threatened habitats today mainly due to drainage, land reclamation, pollution, and over exploitation of wetland species [1]. The groundwater table in intensely urbanized areas such as Whitefield, Bengaluru has declined to 400 to 500 m [2]. Today these water bodies are at a high risk due to sewage and garbage disposal. Pollution in large water bodies has resulted in a lot of disasters affecting aquatic life and the complete biocoenosis. Thus, there is a need to monitor the health of such large water bodies in order to maintain balance in the ecosystem.

Coastal surveillance presents the problem which is faced by the port authorities to maintain the channel depths to prevent grounding of ships. In most ports, channel depth is identified separately by a coast guard. A surface water vehicle which uses SONAR, models the underwater bed of the water body so that we can keep a track and maintain a clear pathway for ships which may prove to be a cost-effective and more reliable alternative.

Flooding of cities and villages happens due to various reasons. Open manholes pose as a serious threat post floods when people try to mobilize to safer grounds.

© Springer Nature Switzerland AG 2020
A. Abraham et al. (Eds.): ISDA 2018, AISC 941, pp. 48–57, 2020.
https://doi.org/10.1007/978-3-030-16660-1_5

In the aftermath of floods, an autonomous rover will be capable of detecting such manholes via SONAR and pave a road where it is safe to walk. Also, after flash-floods or when rivers abruptly changes course, roads get cut-off and accessibility to certain areas are hindered. During such situations emergency supplies like medical aid/food can be delivered to stranded people using the rover operated by rescue officials.

(a)   (b)

Fig. 1. (a) Autonomous water rover (b) AWSR deployed in lake

2 Prior Art

The problem of pollution in water bodies of metro cities like Mumbai has been studied previously by sensing various water parameters. In [3] the authors have highlighted the variation in the water parameters through different seasons by collecting samples of water from Powai lake, Mumbai. The samples were collected manually from six different sampling stations and were analyzed every month for a period of six months. The data collected showed variation in dissolved oxygen, nitrogen and chemical oxygen demand to a greater extent than other parameters over a period of six months.

In [4], the authors have demonstrated a working autonomous unmanned vehicle wirelessly controlled through a telemetry module from base station with Zigbee for communication. The model employs a conventional propeller-rudder assembly for propulsion and maneuvering. Various other implementations have targeted shape, support for navigation of underwater vehicles and autonomy of surface crafts [5–8].

The present research shows a potential scope for active monitoring and data acquisition. To increase the number of samples and check the variation in the water parameters with space, having a mobile system to collect data is realized to be more feasible. This also enables to locate the source of pollution. The water in these bodies is generally still and can be very shallow at some places. Hence a surface vehicle with shallow draught can prove to be very effective.

3   Mechanical System

AWSR is a prototype surface water vehicle that tracks and analyses the typical problems encountered by still water bodies. AWSR as shown in Fig. 1(a), is configured to carry out autonomous and remote controlled surveillance while gathering data and delivering payloads. The AWSR is configured to work under certain specific constraints to address lake water monitoring and flood relief:

- Stability: The rover is required to traverse still water bodies and provide pitch stability upto 30°
- Sensors: The rover is required to accommodate sensors to sense the water parameters
- Surveillance: The rover should be able to capture underwater and above water images
- Navigation: It is required to have the ability to travel autonomously
- Payload: The rover should be able to accommodate some payload for emergency reliefs
- Low budget: The rover is aimed to be economical

3.1   Mechanical Design

The design of the rover acquires it structure from rafts. It is a flat support structure with no hull. It aims to maximize the payload capacity while keeping the base cost to a minimum. The support structure is made of polyvinyl chloride pipes while all the electronics is contained in a waterproof box on top of the structure. The polyvinyl chloride material of the pipes are impervious to thermal changes and biological attacks which is a necessary feature for a hull in waste waters. The base consists of three pipes connected in the horizontal plane. The pipes are enclosed in a structural housing. The structural elements are primarily of aluminium construction. On the top of the housing, an enclosure box contains all the electronics, computing, battery and some sensors. The rover is provided with two bilge pumps in the rear for forward motion and one pump in the front for braking. The modeled structure provides high stability in pitch as well as roll. The high stability in still water is provided to address the problem of birds flocking on the top of the rover when it's stationary which significantly shifts the center of gravity of the rover, the problem was realized in the initial stages of testing.

3.2   Propulsion System

Conventionally surface rovers employ propeller and rudder systems. The rover has a reaction drive for propulsion and utilizes differential drive for steering. The bilge pumps generates the propulsive thrust when the water is forced in the backward direction. The water enters the pumps mounted on the support structure of the rover through the intake unit at the bottom. Here the impeller increases the pressure of the water inside. This pressure is then converted into high velocity at the nozzle. In other words, the discharge of the high velocity stream generates a reaction force in the opposite direction, which is transferred through the body of the pump to the raft, propelling it forward. The pumps provide with much quieter motion. Clogging is prevented due to the presence of the filter, this enables the rover to maneuver in eutrophicated water bodies.

3.3 Buoyancy

Buoyancy is the upward force exerted by a fluid that opposes the weight of an immersed object. Since the AWSR also addresses the application of post flood reliefs, the rover was designed to be significantly positive buoyant. The maximum buoyant force for the rover is about 30.6 kgf. While the rover along with the solar panel has a mass of 15.4 kg, payload of about 10 kg can be safely added on the rover.

4   System Configuration of AWSR

In the previous section mechanical structure of the AWSR was described. The other systems required for the working of the AWSR are shown in Fig. 2 and explained below.

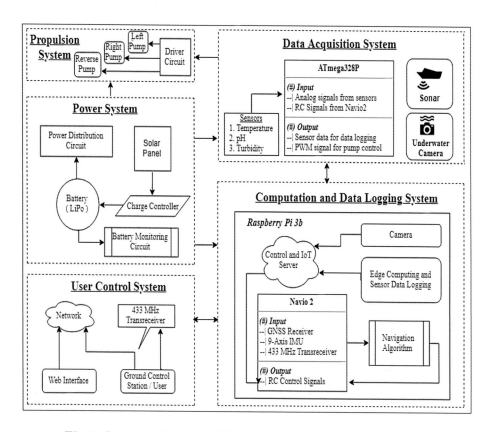

Fig. 2. System architecture of Autonomous Water Surveillance Rover

4.1 Power System

A 12 V DC 8000 mAh, 3 S 30 C/60 C Lithium Polymer battery is used as the powerhouse of the AWSR. The battery is connected to the battery monitoring circuit. The forward thrusting bilge pumps consume 3 A each at full capacity. A LM2596 Buck converter provides a stable 5 V, 3 A (max) output for the sensors and all micro-controllers, forming the power distribution circuit. The Raspberry Pi 3b consumes 2 A at maximum load of our system. The total current required by the whole system is approximately 10 A (measurements made practically). This system is self-sufficient for a single lap of a large lake, covering 1.8 Km in 1 h. A solar panel is mounted on the AWSR to increase it's mission endurance.

Power Calculations for a Solar Panel -

Power available in Watt Hours $(P) = V * A = 12 V * 10 Ah = 120 WH$

Considering 3 one hour trials/day,

Per day power consumption $= 360 WH$

Battery Current Rating $= 360 WH/12 V = 30 AH$

Battery of 12 V, 30 AH is needed for running the AWSR, which can be very costly.

Average duration of sunlight $= 4 h$ (depending on the period of the year)

Panel Requirement $=$ Watts/Sunlight hours $= 360/4 = 90 W$ panel

Thus, a Solar Panel of 12 V, 100 W would be available and suitable for higher efficiency, and can be easily mounted on the AWSR without requiring any changes in the structural design. A 12 V, 7.5 A charge controller is connected to the solar panel and to the LiPo battery. The charge controller and battery are connected in parallel so that it keeps charging during the day time. As soon as the voltage of the battery goes above pre-decided voltage on the charge controller, it will inhibit the solar panel from charging it any further. It helps to maintain a constant high battery voltage required for multiple missions.

4.2 User Control System

The AWSR supports both manual and autonomous modes for navigation. The AWSR can be operated using Mission Planner, an open-source Ground Control Station (GCS) or through a custom website which was developed to simplify the manual control of the AWSR.

Autonomous Mode. In autonomous mode, the user plans a mission by placing waypoints on a map in a chronological order using the GCS as shown in Fig. 3. A wireless local area network (LAN) communicating over TCP or UDP protocols or a radio telemetry transceiver can be used for connecting the AWSR with the GCS. The 'mRo SiK Telemetry Radio V2' which operates on 433 Mhz was used as it can provide a range of 300 m, which can be further increased to several

Fig. 3. Ground Control Station for Autonomous Water Surveillance Rover

kilometers by changing the antenna on the ground. Figure 3 shows the mission planned in the GCS for testing at Teen Talao which is a holding pond in Mumbai, India.

User Mode. In manual mode the user can operate the AWSR over a Wi-Fi LAN or over the internet (provided the region of deployment lies in a cellular network). Figure 4 shows the interface used for controlling the AWSR. The interface consists of the location of the AWSR in real time, first person view of the AWSR and interface for controlling the motion of the AWSR.

4.3   Computation and Data Logging System

Raspberry Pi 3 (RPi), a single-board computer is used for path planning, communication, data logging, motion control, camera interfacing, interfacing of localization sensors, generating radio control signals and hosting the server for the website.

Localization Sensors. Navio2, is a sensor hat for the RPi having a GNSS receiver which tracks multiple satellites, is used for global positioning. It also has a 9-axis IMU with an accelerometer, gyroscope and magnetometer for orientation and motion sensing.

Communication Technology. A UART radio transceiver, configured for a baud rate of 57600 is connected to the RPi through the Navio 2 sensor hat for communication between the GCS and AWSR in autonomous mode. The onboard Wi-Fi module on RPi, allows the AWSR to connect to a network created using a cellular network dongle and host a web server enabling the user to control it via internet. The water quality sensor data obtained from the Data Acquisition System is sent through the Universal Serial Bus (USB) to the RPi.

Data Logging. Water quality parameters like temperature, pH and turbidity are logged locally in a Sqlite3 database on the AWSR along with the time stamp. The latitude and longitude of the AWSR are also logged to provide real time location tracking.

Motion Control. Ardupilot, an open source firmware for autonomous rovers, is configured to work on the AWSR. The library at its core uses extended kalman filter for position estimation and a PID controller for smooth maneuvering. After computation, three PWM output channels are used to generate 1 ms to 2 ms pulses for controlling the thrust of the three pumps of the AWSR.

Fig. 4. Web interface for Autonomous Water Surveillance Rover

Camera. A Sony IMX219 8-megapixel sensor is interfaced to get the first person view of the AWSR which can be used for surveillance and for visual feedback in manual mode as seen in Fig. 4.

Web Server. The RPi hosts a web server developed using flask, which is a micro web framework written in Python. The front end of the web interface is developed using HTML, CSS and JavaScript. The current sensor readings, surveillance video, global position of AWSR and historic sensor data, are also accessible through the web interface hosted on the server.

4.4 Data Acquisition System

The AWSR has sensors to measure temperature, pH and turbidity. ATmega328P, an 8-bit microcontroller is used to interface these sensors and transmit them to the RPi through a USB to TTL converter, for further computation. The microcontroller is also used to convert the RC control signals to pulse width modulated signals which are required for varying the thrust of each pump. In order to measure the depth and detect fish, a sonar module with a dedicated receiver, as shown in Fig. 5(b) is used. The sonar can measures depths upto 120 ft and operates at a frequency of 125 KHz. An underwater camera is attached to the AWSR to obtain underwater images in shallow water areas as shown in Fig. 6. These images are of utmost importance while carrying out geological surveys of coastal areas. Refer Fig. 6(b).

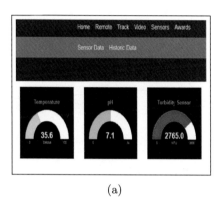
(a)

(b)

Fig. 5. (a) Sensor data visualization via web interface (b) Receiver showing depth and contours

4.5   Propulsion System

IRFZ544 Mosfets are used to drive the 3 bilge pumps. The ULN-2803 IC has Darlington pair transistors which are being used to provide the gate of the n-channel mosfets with the corresponding voltages. As ULN-2803 complements the output, thus a low i/p is given for a high output and vice versa, to the gate of the mosfet. The gate terminal is grounded with a high value resistor (10k ohms) since the Mosfet is a voltage controlled device. The resistor strip A103 helps to keep the output of the ULN-2803 at a stable output. Thus, by giving complementary and pulse width modulated signals to the ULN-2803, we can control the pumps.

5   Testing and Results

We tested the Autonomous Water Surveillance Rover at two locations, first at Mini Seashore, Vashi, Navi Mumbai, 1904′57.7″N 7259′38.3″E on 6th April, 2018 and then at Teen Talao, Chembur, Mumbai, 1903′03.2″N 7253′33.7″E on 17th April, 2018. Figure 6(a) summarizes the results obtained after surveying these

Location	Temperature (°C)	pH	Turbidity (NTU)
Mini Seashore, Vashi	31.097	7.129	2873.145
Teen Talao, Chembur	33.493	8.205	2728.041

(a)

(b)

Fig. 6. (a) Water parameters (b) Underwater image

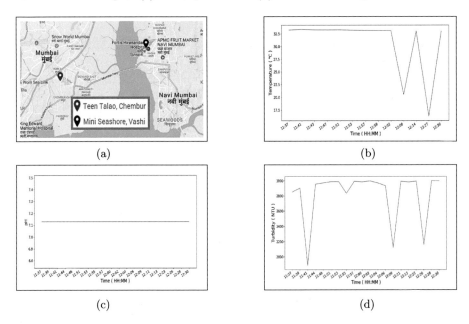

(a)

(b)

(c)

(d)

Fig. 7. (a) Testing locations (b) Temperature (°C) (c) pH (d) Turbidity (NTU)

water bodies. Figure 7(b), (c), and (d) depict the time series plot of temperature, pH and turbidity respectively, measured at Mini Seashore, Vashi through the AWSR. Figure 7(a) pinpoints the two testing sites on a map.

6   Conclusion

This paper has discussed Autonomous Water Surveillance Rover as a working prototype and has shown the possibility of successfully implementing such a system in inland water bodies like lakes and ponds. The rover has been designed to carry out missions in lake and estuarine environments. Lake water monitoring and sensor data acquisition are the most important objective for the system.

The mechanical design was validated and onboard processing and navigation was integrated in the system. This feature is implemented using both remote control and autonomous modes while using GPS and IMU data for navigation. Sensors to measure pH, temperature, turbidity, conductivity and onboard cameras were integrated and tested in two locations in Mumbai. Payload delivery was tested in a semi-controlled environment in a swimming pool. We successfully tested data acquisition for bathymetry and environment monitoring. The project is an ongoing research activity for development of low cost solution for lake monitoring in Indian cities and villages. The different abilities of the AWSR can be enhanced and upgraded to suit the challenges in different ecosystems and for various applications. Marine robotics is a fairly new and upcoming field in India with large implications to understanding societal problems. Emerging marine robotics developments can empower engineers and scientists with advanced tools to explore and exploit oceans, in a sustainable fashion. The smaller water ecosystems are equally important and therefore must be conserved responsibly, the availability of such low-cost systems can prove to be a very powerful tool in a developing country like India.

References

1. Samant, J.: Wetland Conservation in Maharashtra: Need, Threats, and Potential. Development Research Awareness and Action Institute (DEVRAAI) (2012)
2. Ramachandra, T.V., Aithal, B.H.: Decaying lakes of Bengaluru and today's irrational decision makers (2016). http://ces.iisc.ernet.in/energy
3. Koliyar, J.G., Rokade, N.S.: Water quality in Powai Lake: Mumbai, Maharashtra. In: Proceedings of Taal2007: The 12th World Lake Conference, pp. 1655–1659 (2007)
4. Ahmad, N., Bajwa, S.I.S., Ahmad, S., Khan, S.L., Malik, T.A., Tahir, Y.: Autonomous unmanned surface vehicle. Ghulam Ishaq Khan Institute of Engineering Sciences and Technology, Topi, Pakistan (2011)
5. Ferreira, H., Martins, A., Dias, A., Almeida, C., Almeida, J.M., Silva, E.P.: ROAZ autonomous surface vehicle design and implementation. Encontro Científico - Robótica 2006, Pavilhão Multi-usos, Guimarães, Portugal, 28 Abril 2006
6. Manley, J.: Development of an autonomous surface craft "ACES". In: Proceedings of MTS/IEEE Oceans 1997 Conference (1997)
7. Leonessa, A., Mandello, J., Morel, Y., Vidal, M.: Design of a small, multi-purpose, autonomous surface vessel. In: Proceedings of MTS/IEEE Oceans 2003 Conference, San Diego, USA (2003)
8. Cruz, N., Matos, A., Martins, A., Silva, J., Santos, D., Boutov, D., Ferreira, D., Pereira, F.L.: Estuarine environment studies with Isurus, a REMUS class AUV. In: Proceedings of MTS/IEEE Oceans 1999 Conference, Seattle, WA, USA (1999)

Bidirectional LSTM Joint Model for Intent Classification and Named Entity Recognition in Natural Language Understanding

Akson Sam Varghese[1](✉), Saleha Sarang[1], Vipul Yadav[1],
Bharat Karotra[1], and Niketa Gandhi[2]

[1] Technology and Research Group, Depasser Infotech, Mumbai, India
aksonsam@gmail.com, saleha@depasserinfotech.in,
{vipulyadav,bharatkarotra}@depasserinfotech.com
[2] Machine Intelligence Research Labs (MIR Labs),
Scientific Network for Innovation and Research Excellence,
Auburn, WA 98071, USA
niketa@gmail.com

Abstract. The aim of this paper is to present a Simple LSTM - Bidirectional LSTM in a joint model framework, for Intent Classification and Named Entity Recognition (NER) tasks. Both the models are approached as a classification task. This paper discuss the comparison of single models and joint models in the respective tasks, a data augmentation algorithm and how the joint model framework helped in learning a poor performing NER model in by adding learned weights from well performing Intent Classification model in their respective tasks. The experiment in the paper shows that there is approximately 44% improvement in performance of NER model when in joint model compared to when tested as independent model.

Keywords: LSTM · Joint model · Bidirectional LSTM · Intent Classification · Named Entity Recognition · Natural Language Understanding

1 Introduction

Higher level organisms in the animal kingdom are able to react to the situation taking into consideration all the inputs available from their surroundings. Humans possess the five sensory inputs on which they react or take a decision upon. For language, speech, style of the speech, pitch of the speech, body language all are taken into consideration while having a conversation. This research is similar in this sense of joining multiple inputs and producing output or being able to carry out a decision. It is quite evident from many good works on Recurrent Neural Networks (RNN) that they outperform in the task which involves the sequential data [26, 27]. RNN have claimed to achieve the state of the arts results in some cases, better performances than humans could have, especially RNN - Long Short Term Memory (LSTM) and RNN - Bidirectional LSTM, Attention based LSTM encoder - decoder networks in the domains of Speech Recognition, Sequence Labeling, Text Classification, Image Caption Generation and many more. It has been extensively used in tasks like Spoken Language Understanding

© Springer Nature Switzerland AG 2020
A. Abraham et al. (Eds.): ISDA 2018, AISC 941, pp. 58–68, 2020.
https://doi.org/10.1007/978-3-030-16660-1_6

(SLU), Natural Language Understanding (NLU) etc. When humans read a particular sentence may be on a flyer or some magazine, are able to understand the context/intent of the sentence as well as pick up the keywords in the sentence. This event of understanding the nature of the sentence and picking up the entities seems to be occurring simultaneously. This observation gave the thought that if it would be possible to combine multiple tasks in a single model, how better it can perform in comparison to independent classifiers and does understanding the context/intent of the sentence has any influence on picking up the entities. Although the concept of joint models is not new, the model proposed here have a different architecture [1–11], to accomplish the task of Intent Classification and NER two separate models are used generally. The paper would present the results of using the LSTM - Bidirectional LSTM independently and in a joint approach. There are other papers [1–5] which carry out the task of NER as sequence labeling, prediction task as a part of research in SLU and other domains while in this paper Intent Classification and NER are considered as a classification task on text data.

2 Related Work

There has been extensive line of research for joint models such as Joint-Models for Slot Filling and Intent Detection [6] which demonstrates the use of Bidirectional GRUs to learn the sequence representations shared by the two important tasks in SLU, intent determination and slot filling out performing over state of the art separate models for the respective task. Multi-Task Text Classification models in RNN [7] proposed three types of RNN architectures along with a learning framework that maps arbitrary text into semantic vector representations with both specific and shared layers whose main focus lies on the text classification task. Sequential Labelling and Classification with Attention Neural Networks [8] proposed a joint model approach for topic classification and slot filling, two tasks most important in SLU, introduced a mechanism of sparse attention to weigh words based on the semantic correlation to sentence level classification in a LSTM based model after realizing the shortcomings from other models such as Conditional Random Field (CRF) [9] and Convolutional Neural Network (CNN) [10] to capture the sequence at a time step. Joint Syntactic and Semantic Analysis with Multi-Task Deep Learning framework for Speech Language Understanding [11] have proposed the use of Bidirectional RNN models with LSTM and adding explicit weights of each task to minimize the loss at every iteration to solve multiple tasks in SLU like Parts of Speech (POS) tagging, syntactic dependency parsing, named entity tagging and FrameNet parsing. OneNet [23] addressing the issues of pipeline approach and adopting a principled architecture for multitask learning on domain classification, intent classification and slot filling. Joint Online Spoken Language Understanding and Language Modeling with RNN [24] describes a joint model that performs intent detection, slot filling and language modeling where the intent prediction is used as contextual features. Joint semantic Utterance classification and slot filling with Recursive Neural Networks (RecNN) provides a mechanism for incorporating discrete syntactic structure and continuous space by extending a continuous space model framework for domain detection, intent determination and slot filling. Most of these works are related to the domain of SLU and NLU. Most of these works consider slot filling in conjunction with syntactic type labels. Although, NER falls in same family of sequence tagging problem, here NER model is considered as a classification task. Our the contributions through the experiments presented are:

1. Data Transformation Technique: an algorithm to augment the data
2. Simple LSTM - Bidirectional LSTM - joint model architecture

The idea is to share the learned weights of Intent Classification model to the NER model and compare with the results produced jointly and independently. Since, the task of Intent Classification and NER are both crucial in systems where user text inputs are more closely investigated in real time, deploying a single model capable of carrying out both the tasks that can produce joint predictions can be beneficial.

3 Dataset

Dataset used here is Airline Travel Information System (ATIS) text corpus public dataset [12]. The dataset consists of sentences represented in IOB representation format. An example sentence "I want to fly from baltimore to dallas round trip" states that the user or input provider has the intention to book flight, thus the intent of "atis_flight" is assigned to the sentence. For each corresponding word there is an IOB representation where keywords required for the task completion that is to book flight such as "baltimore" the starting location is labelled as "B-fromloc.city_name", "dallas" the destination point is "B-toloc.city_name" and whether the ticket intended to book is one way trip or round trip is labelled as "B-round_trip" and "I-round_trip" for the keyword "round trip". "O" is used for non-important words like I, want, to, fly as shown in Table 1. The data is actually in sequential format. Since the approach of the NER model is as of a classification problem before using the actual data, the data is columned such that the every word is taken in as input as vector representations of the characters of the word and the corresponding IOB representation becomes the label for the input for the NER task.

Table 1. An example from the ATIS text corpus dataset

Text	BOS		I	want	to	fly	from
IOB	O		O	O	O	O	O
Text	baltimore		to	dallas	round	trip	EOS
IOB	B-fromloc.city_name		O	B-toloc.city_name	B-round_trip	I-round_trip	O
Label	atis_flight						

The input can be viewed as vector representation of words in the sentence with the corresponding manually labelled label for the Intent Classification task and a vector representation of a character sequence of a word [13] with the corresponding IOB represented label.

Algorithm 1. Simple Repetitive Random Sampling with Replacement Data Augmentation

---

1. SRRSRDA(list, count)
2. *length_list ← length(list) - 1*
3. while *length_list ≠ count*
4.   if *length_list > count*
5.   *list ← list upto count*
6.   if *length_list < count*
7.   for *i ← int(count - length_list)*
8.   *list += random (list)*
9.   *length_list = length(list)*
10. end SRRSRDA

There are 127 unique labels for NER and 26 unique labels for Intent Classification. The data is unbalanced, as the number of examples for each class in the sequence and the intent corresponding to each sequence are not equal. A simple repetitive random sampling with replacement data augmentation technique is introduced here to equalize the data corresponding to the count of maximum number of example present in the dataset. Equalization means, every class has equal number of examples. This algorithm selects a random sequence from the dataset and populates the class with minority number of samples as shown in Algorithm 1. Input is a list of sequences and the count to be equalized to. Output a list of sequences of the given count. ATIS text corpus contained large samples of 'O' labelled words and contained large samples of 'atis_-flight' labelled sequences. The data is represented in the dataset as word vectors in embedded sparse matrix. The approach for vector representation is different for NER and Intent Classification. The representation for Intent Classification input data will be word level and for NER input data will be represented on character level, i.e. the input is considered as a sequence of word representations for Intent Classification task and the input is considered as a sequence of character representations for NER task, both in the same sparse embedded space [14]. The length of the sequences are normalized to equal lengths by pre-padding and pre-truncating the excess data.

4 Model

Model consists of 14 layers as shown in Fig. 1. The first layer is the input layer which takes in a 3D vector representation [15] of input data for intent classification. Data is feed into the network in the form of sequences.

$$S^l = \left\{ S_1^l, S_2^l, S_3^l \ldots S_k^l \right\} \tag{1}$$

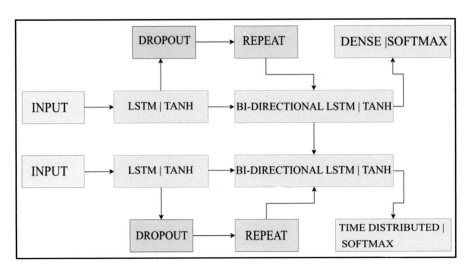

Fig. 1. Minimalistic model representation

$$S_k^l = [e(w_1), e(w_2), e(w_3), \ldots e(w_l)] \tag{2}$$

The Eq. 1 represents the set of sequences from intent dataset, k is the number of total sequences in the set S^l. The components of Eq. 2 are for every sequence S_k^l is a sequence of $e(w_l)$, e is the embedded vector representation of the word such that (w_l) ε R^m, R is the embedded vector space of dimension m and l are the number of words in the sequences. Second layer is the LSTM [16] layer with the hyperbolic tangent function [17] as its activation function.

$$LSTM(h_t) \tag{3}$$

$$h_t = tanh\left(W^{hh}h_{t-1} + W^{hx}x_t\right)W^{hh} \tag{4}$$

The Eq. 3 represents the *LSTM* layer with input h_t at time t. The components of Eq. 4 are W^{hh} is the previous to previous hidden state to previous hidden state at time step $t - 1$. W^{hx} is the weight of the previous hidden state to the current input x at time step t. Third layer is the dropout layer with a given dropout ratio. Fourth layer repeats the input data for a given number of times. Fifth layer is a Bidirectional LSTM [18] layer with hyperbolic tangent function as the activation function.

$$biLSTM\left(h_t^{\rightarrow}, h_t^{\leftarrow}\right)_1 \tag{5}$$

$$h_t^{\rightarrow} = tanh\left(W^{\rightarrow}x_t + V^{\rightarrow}h_{t-1}^{\rightarrow} + b^{\rightarrow}\right) \tag{6}$$

$$h_t^{\rightarrow} = tanh\left(W^{\leftarrow}x_t + V^{\leftarrow}h_{t+1}^{\leftarrow} + b^{\leftarrow}\right) \tag{7}$$

The Eq. 5 represents the Bidirectional LSTM layer with input sequences $h_t^{\rightarrow}$ in forward direction and $h_t^{\leftarrow}$ in reverse direction. Sixth layer flattens the output from the Bidirectional LSTM layer.

$$x = \prod h_i \tag{8}$$

The components of Eq. 8 are h_i output hidden state from Eq. 5, x is the input to each neuron in the fully connected dense layer. Seventh layer is the output layer with dense fully connected layer with density corresponding to the classes and softmax activation that squashes the output from the fully connected layer.

$$\hat{y} = argmax_y\, \hat{p}(y|S_k^l) \tag{9}$$

$$\hat{p}(y|S) = softmax(W^s x + b^s) \tag{10}$$

In Eq. 9 $\hat{y}$ is the argmax of the predicted probabilities at each node given the real value y and input sequence S_k^l. The components of Eq. 10 can be described as the sum of W^s weight of the sequence and the b^s bias of the sequence with *softmax* as the activation function. The Eighth layer is the input layer which takes in a 3D vector representation of input data for NER. Data is feed as,

$$S^N = \left\{S_1^N, S_2^N, S_3^N \ldots S_k^N\right\} \tag{11}$$

$$S_k^N = [e(c_1), e(c_2), e(c_3), \ldots e(c_l)] \tag{12}$$

In Eq. 11, S^N is the set of sequences of NER and k is the number of total sequences in S^N. In Eq. 12 every sequence S_k^N is a sequence of $e(c_l)$, e is the embedded vector representation of the characters of the words of the sequence, such that $(c_l)\ \varepsilon\ R^m$, R is the embedded vector space of dimension m. l is the number of characters in the word. The number of inputs for the NER input data and for Intent classification should be equal. Ninth layer is a LSTM layer which is stateless and more dense than that of the second layer, followed by tenth layer as dropout [19] layer.

$$LSTM(h_t) \tag{13}$$

$$h_t = tanh\left(W^{hh}h_{t-1} + W^{hx}x_t\right)W^{hh} \tag{14}$$

The Eq. 13 represents the LSTM layer with input as h at time step t. Equation 14 describes $W^{hh}h_{t-1}$ is the weight of previous to previous hidden state to previous hidden state at time step $t-1$. W^{hx} is the weight of previous hidden state to current input x at time step t. Eleventh layer repeats the input vector given number of times. The output from the Eq. 14 is fed into twelfth layer which is Bidirectional LSTM Eq. 15 with hyperbolic tangent as the activation function.

$$biLSTM\left(h_t^{\rightarrow}, h_t^{\leftarrow}\right)_2 \tag{15}$$

$$h_t^{\rightarrow} = tanh\left(W^{\rightarrow}x_t + V^{\rightarrow}h_{t-1}^{\rightarrow} + b^{\rightarrow}\right) \tag{16}$$

$$h_t^{\rightarrow} = tanh\left(W^{\leftarrow}x_t + V^{\leftarrow}h_{t+1}^{\leftarrow} + b^{\leftarrow}\right) \tag{17}$$

The thirteenth layer is the concatenation of the fifth Bidirectional layer and the twelfth Bidirectional LSTM.

$$q = biLSTM\left(h_t^{\rightarrow}, h_t^{\leftarrow}\right)_1 + biLSTM\left(h_t^{\rightarrow}, h_t^{\leftarrow}\right)_2 \tag{18}$$

The Eq. 18 represents the addition of Eq. 5 and the Eq. 15. The fourteenth layer is the output layer time distributed wrapping up a dense fully connected layer with density equal to that of the number of classes for named entity recognition and softmax function is used as an activation function that squashes the output from the fully connected layer.

$$\hat{y} = argmax_y \, \hat{p}(y|S_k^N) \tag{19}$$

$$\hat{p}(y|S_k^N) = softmax(W^s q + b^s) \tag{20}$$

The Eq. 19 represents the $\hat{y}$ is the argmax of the predicted probabilities at each node given the real value y and input sequence S_k^N. The Eq. 20 represents the softmax function which takes the sum of W^s is the weight of the sequence, b^s is the bias of the sequence as input and predicts the probabilities. Independent models are built of the same architecture as that of the joint model, except that they are disconnected to each other. The density, dropout ratios and activation functions used in the Joint Model is same as that in the Independent Models.

5 Training Method

During the training process, k - fold stratified cross validation [20] method is used on the network. In stratification process, data is rearranged such that each fold becomes a good representative of the entire dataset. Stratification is generally a better scheme when it comes to bias and variance compared to regular cross validation method. In classification tasks, overrepresented classes get too much weight as it's the nature of the classification algorithm to represent overrepresented classes with larger weight compared to the underrepresented classes. Even though, the data is equalized such that there are an equal number of classes, still when dataset is split into training set, testing set and validation set, split must contain equal samples of all the classes in the respective sets. Dataset is split into K folds. The stratified folds helps in selection of training and testing instances from the dataset which is from the 80% of the dataset and the rest 20% is used as validation sets. Same training process is carried out on both Joint Model and

Independent Model. RMSprop [21] is used as the optimizer and Categorical Cross Entropy [22] as the loss function. Optimizer is defined as,

$$\theta_{t+1} = \theta_t - \frac{\gamma}{\sqrt{E[g^2]_t + \varepsilon}} g_t \tag{21}$$

The components of Eq. 21 are γ the learning rate, $E[g^2]$ is the average of the past squared gradients g_t and ε as the constant. Loss function/cost function is defined as,

$$-\frac{1}{N} \sum_{i=1}^{N} \sum_{c=1}^{C} l_{y_i \, \varepsilon \, C_c} \log p_{model}[y_i \, \varepsilon \, C_c] \tag{22}$$

The components of Eq. 22 are N is the number of observations, C number of classes, $l_{y_i \, \varepsilon \, C_c}$ is the indicator function of the ith observation belonging to the cth class. $\log p_{model}[y_i \, \varepsilon \, C_c]$ is the probability prediction by the model. Since this a multiclass classification and the Categorical Cross Entropy and RMSprop seems to be a good general choice for the task.

6   Results

In total six experiments were carried out with the hyper-parameters within a random range. The hyper parameters were selected manually with randomness from model architecture designing experience. Three experiments with the specific set of chosen hyper-parameters for the joint model and similar three experiments for Independent models. During the experimentation dropout ratio, number of epochs, number folds used for training via stratified cross validation, validation ratio, random state to which the random seed is initialized. For reviewing the results validation loss and validation accuracy are taken into consideration.

6.1   Joint Model Results

Table 2. Results for joint model experiments

Model type	Exp.	Model name	Epoch	KFold	Batch size	Validation loss	Validation accuracy
Joint model	E1	Intent	20	5	5000	0.0062	0.9989
		NER				0.0135	0.9944
	E2	Intent	10	3	10000	0.0453	0.9866
		NER				0.1395	0.9690
	E3	Intent	5	7	15000	0.0354	0.9901
		NER				0.1444	0.9690

For all experiments in the Joint Model dropout ratio, depth, random state and validation ratio were kept constant with 30%, 32, 16, 20% as respective values.

For the first experiments the configuration settings were 20 epochs batches with batch size of 5000 feed to the network, 5 folds for cross validation. It resulted in validation loss as 0.0062 and 0.0135, validation accuracy as 0.9989 and 0.9944 for Intent and NER respectively. For the second experiment the configuration settings were number of epochs as 10, batches with batch size of 10000 feed to the network, 3 folds for cross validation. It resulted in validation loss as 0.0453 and 0.1395, validation accuracy as 0.9866 and 0.9690 for Intent and NER respectively. For the third experiment the settings were number of epochs as 5, batches with batch size of 15000 feed to the network, 7 folds for cross validation. It resulted in validation loss as 0.0354 and 0.1444, validation accuracy as 0.9901 and 0.9690 for Intent and NER as shown in Table 2.

6.2   Independent Model Results

Table 3. Results for independent model experiments

Model type	Exp.	Model name	Epoch	KFold	Batch size	Validation loss	Validation accuracy
Independent model	E1	Intent	20	5	5000	0.0020	0.9997
		NER				0.7689	0.6751
	E2	Intent	10	3	10000	0.0253	0.9931
		NER				0.9032	0.6417
	E3	Intent	5	7	15000	0.0283	0.9911
		NER				0.7761	0.6670

For all experiments in the Independent Model dropout ratio, depth, random state and validation ratio were kept constant with 30%, 32, 16, 20% as respective values.

For the first experiment the configuration settings were 20 epochs batch size of 5000 feed to the network, 5 folds for cross validation with validation loss as 0.0020 and 0.7689, validation accuracy as 0.9997 and 0.6751 for Intent and NER respectively. For the second experiment the settings were number of epochs as 10, batches with batch size of 10000 feed to the network, 3 folds for cross validation. It resulted in validation loss as 0.0253 and 0.9032, validation accuracy as 0.9931 and 0.6417 for Intent and NER respectively. For the third experiment the settings were number of epochs as 5, batches with batch size of 15000 feed to the network, 7 folds for cross validation. It resulted in validation loss as 0.0283 and 0.7761 for Intent and NER, validation accuracy as 0.9911 and 0.6670 respectively as shown in Table 3.

7 Conclusion and Future Work

Independently LSTM-Bidirectional LSTM performs Intent classification task quite well. Whereas for the Independent Bidirectional LSTM model, the performance is low with the given set of parameters in comparison to the Joint Bidirectional LSTM Model. The experiment 1 with the given set of parameters performs the best among the three for the total experiments conducted on Joint Model and in Independent Model the experiment 1 with its given set of parameters performs good. This shows that when the learning weights of Bidirectional LSTM model are added to training weights of the NER Bidirectional LSTM model, the model is capable of learning much better. We can conclude that when weights of Intent Classification are added to the weights of the NER Bidirectional LSTM, it helps in generalizing the model and thus results in better performance. It is also to be noted that the models can be tuned to produce much better result with grid wide search of hyper-parameters. In future, the aim would be to implement this type of network on different datasets and obtain a comparison table of all the results. It would be even possible to join more tasks like Sentiment Analysis, Semantic Analysis, Emotion Analysis in a joint model framework and run experiments on them. Like humans can intake multiple sensory inputs even machine learning models could be able to intake multiple inputs as a single input in a single joint model and output the necessary decisions on the inputs, finally becoming a machine learning system similar to human sensory system.

References

1. Young, T., Hazarika, D., Poria, S., Cambria, E.: Recent trends in deep learning based natural language processing. IEEE Comput. Intell. Mag. **13**(3), 55–75 (2018)
2. Mesnil, G., He, X., Deng, L., Bengio, Y.: Investigation of recurrent-neural-network architectures and learning methods for spoken language understanding. In: INTERSPEECH, pp. 3771–3775, August 2013
3. Tang, D., Qin, B., Liu, T.: Document modeling with gated recurrent neural network for sentiment classification. In: Proceedings of the 2015 Conference on Empirical Methods in Natural Language Processing, pp. 1422–1432 (2015)
4. Graves, A., Mohamed, A.R., Hinton, G.: Speech recognition with deep recurrent neural networks. In: 2013 IEEE International Conference on Acoustics, Speech and Signal Processing (ICASSP), pp. 6645–6649. IEEE, May 2013
5. Mikolov, T., Kombrink, S., Burget, L., Černocký, J., Khudanpur, S.: Extensions of recurrent neural network language model. In: 2011 IEEE International Conference on Acoustics, Speech and Signal Processing (ICASSP), pp. 5528–5531. IEEE, May 2011
6. Zhang, X., Wang, H.: A joint model of intent determination and slot filling for spoken language understanding. In: IJCAI, pp. 2993–2999, July 2016
7. Liu, P., Qiu, X., Huang, X.: Recurrent neural network for text classification with multi-task learning. arXiv preprint arXiv:1605.05101 (2016)
8. Ma, M., Zhao, K., Huang, L., Xiang, B., Zhou, B.: Jointly trained sequential labeling and classification by sparse attention neural networks. arXiv preprint arXiv:1709.10191 (2017)
9. Jeong, M., Lee, G.G.: Triangular-chain conditional random fields. IEEE Trans. Audio Speech Lang. Process. **16**(7), 1287–1302 (2008)

10. Xu, P., Sarikaya, R.: Convolutional neural network based triangular CRF for joint intent detection and slot filling. In: 2013 IEEE Workshop on Automatic Speech Recognition and Understanding (ASRU), pp. 78–83. IEEE, December 2013

11. Tafforeau, J., Bechet, F., Artières, T., Favre, B.: Joint syntactic and semantic analysis with a multitask deep learning framework for spoken language understanding. In: INTERSPEECH, pp. 3260–3264 (2016)

12. ATIS Dataset Source, yvchen. https://github.com/yvchen/JointSLU/tree/master/data. Accessed Oct 2018

13. Klein, D., Smarr, J., Nguyen, H., Manning, C.D.: Named entity recognition with character-level models. In: Proceedings of the Seventh Conference on Natural Language Learning at HLT-NAACL 2003, vol. 4, pp. 180–183. Association for Computational Linguistics, May 2003

14. Collobert, R., Weston, J.: A unified architecture for natural language processing: deep neural networks with multitask learning. In: Proceedings of the 25th International Conference on Machine Learning, pp. 160–167. ACM, July 2008

15. Mikolov, T., Sutskever, I., Chen, K., Corrado, G.S., Dean, J.: Distributed representations of words and phrases and their compositionality. In: Advances in Neural Information Processing Systems, pp. 3111–3119 (2013)

16. Greff, K., Srivastava, R.K., Koutník, J., Steunebrink, B.R., Schmidhuber, J.: LSTM: a search space odyssey. IEEE Trans. Neural Netw. Learn. Syst. **28**(10), 2222–2232 (2017)

17. Salehinejad, H., Baarbe, J., Sankar, S., Barfett, J., Colak, E., Valaee, S.: Recent Advances in Recurrent Neural Networks. arXiv preprint arXiv:1801.01078 (2017)

18. Cho, K., Van Merriënboer, B., Gulcehre, C., Bahdanau, D., Bougares, F., Schwenk, H., Bengio, Y.: Learning phrase representations using RNN encoder-decoder for statistical machine translation. arXiv preprint arXiv:1406.1078 (2014)

19. Srivastava, N., Hinton, G., Krizhevsky, A., Sutskever, I., Salakhutdinov, R.: Dropout: a simple way to prevent neural networks from overfitting. J. Mach. Learn. Res. **15**(1), 1929–1958 (2014)

20. Kohavi, R.: A study of cross-validation and bootstrap for accuracy estimation and model selection. In: IJCAI, vol. 14, no. 2, pp. 1137–1145, August 1995

21. Ruder, S.: An overview of gradient descent optimization algorithms. arXiv preprint arXiv: 1609.04747 (2016)

22. Janocha, K., Czarnecki, W.M.: On loss functions for deep neural networks in classification. arXiv preprint arXiv:1702.05659 (2017)

23. Kim, Y., Lee, S., Stratos, K.: ONENET, joint domain, intent, slot prediction for spoken language understanding. In: 2017 IEEE Automatic Speech Recognition and Understanding Workshop (ASRU) (2017)

24. Liu, B., Lane, I.: Joint online spoken language understanding and language modeling with recurrent neural networks. In: Proceedings of the 17th Annual Meeting of the Special Interest Group on Discourse and Dialogue, pp. 22–30. Association for Computational Linguistics, September 2016

25. Guo, D., Tur, G., Yih, W., Zweig, G.: Joint semantic utterance classification and slot filling with recursive neural networks. In: Spoken Language Technology Workshop (SLT), pp. 554–559. IEEE (2014)

26. Yao, K., Zweig, G., Hwang, M., Shi, Y., Yu, D.: Recurrent neural networks for language understanding. In: INTERSPEECH (2013)

27. Yao, K., Peng, B., Zweig, G., Yu, D., Li, X., Gao F.: Recurrent conditional random fields for language understanding. In: ICASSP (2014)

Analysis on Improving the Performance of Machine Learning Models Using Feature Selection Technique

N. Maajid Khan, Nalina Madhav C, Anjali Negi,
and I. Sumaiya Thaseen[✉]

School of Information Technology and Engineering,
Vellore Institute of Technology, Vellore, India
sumaiyathaseen@gmail.com

Abstract. Many organizations deploying computer networks are susceptible to different kinds of attacks in the current era. These attacks compromise the confidentiality, integrity and availability of network systems. It is a big challenge to build a reliable network as several new attacks are being introduced by the attackers. The aim of this paper is to improve the performance of the various machine learning algorithms such as KNN, Decision Tree, Random Forest, Bagging Meta Estimator and XGBoost by utilizing feature importance technique. These classifiers are chosen as they perform superior to other base and ensemble machine learning techniques after feature selection. Feature Importance technique is utilized to obtain the highest ranked features. Reduced attributes improve the accuracy as well as decrease the computation time and prediction time. The experimental results on UNSW-NB dataset show that there is a drastic decrease in the computation time with reduced attributes compared to evaluating the model using the dataset with the entire set of attributes.

Keywords: Accuracy · Attributes · Feature selection · Machine learning · Computation time

1 Introduction

With the epic development of computers and network, most of the systems experience security vulnerabilities. Also due to increase in the cyber-crime rate in current ecommerce world, security is the most important factor in a high performance networking system. An internet-attack is a planned way to exploit computer systems that challenges the confidentiality, integrity and availability of the computer network and is an international concern. The attackers use malicious ways to enter into the system and use the data for detrimental purposes leading to system infiltration, information modification etc. These attacks can be monitored using an intrusion detection system (IDS). IDS are used to monitor any type of suspicious activity that might happen in the network traffic. Suspicious activities include attacks like denial of service (DOS), fuzzers, backdoor, generic etc. The analysis for anomaly detection can be done using machine learning techniques on different train and test datasets. As researchers start

© Springer Nature Switzerland AG 2020
A. Abraham et al. (Eds.): ISDA 2018, AISC 941, pp. 69–77, 2020.
https://doi.org/10.1007/978-3-030-16660-1_7

finding out different ways to increase accuracy and improve the model, there is still enough scope of increasing the accuracy and efficiency of the model.

The methods developed earlier were not potential enough to detect the attacks. Systems used to execute days to track and find the attack that has occurred. Nowadays machine learning algorithms can detect these suspicious activities and categorize the attack that causes the most damage to the system. The system learns by itself based on the past experiences and can process and classify the attack category for the unseen data.

2 Related Work

Many intrusion detection systems (IDS) have been developed using machine learning techniques. An IDS [4] was proposed by using Association Mining Rule (AMR) for data pre-processing. They used Naive Bayes and Logistic Regression algorithms for classifying the normal and attack data. The authors [4] suggested that removal of redundant data minimizes the processing time and increases accuracy. The characteristics of features of UNSW-NB15 and KDD99 datasets are analyzed by the authors [5]. AMR algorithm is used to select the best features on which classifiers are utilized to evaluate accuracy. The results showed that KDD99 attributes were not too efficient compared to UNSW-NB15 attributes but the accuracy produced by KDD99 dataset was better. Machine learning is utilized for detecting network anomalies. Belouch [8] evaluated the accuracy, model building time and prediction time of three classification algorithms using Apache Spark and addresses the issue related to low accuracy and prediction time on UNSW-NB15 dataset when all 42 attributes were utilized. Experiment results showed that the accuracy of Random Forest Classifier was high with 97.49% and fastest detection time which was 0.08 s for binary classification. According to [9], the results showed that Random Forest ensemble classifier outperformed other base classifiers. They proposed a model fusing Random Forest classifier and was efficient with high detection rate and low false rate on NSL-KDD dataset. Random Forest classifier eliminates over fitting problem.

Muniyal [3] states that feature selection has to be implemented as a two-step process. Random Forest classifier is used and features with high importance score is selected and sorted based on the importance score which is then used for the classification. Due to the redundant data in the NSL-KDD dataset, the data is pre-processed to avoid biased accuracy towards the repeated data. The results show that Random forest classifier produced better accuracy with reduced attributes with increase in performance rate and reduction in False Positive Rate (FPR).

In this literature work [6], several feature selection methods have been studied and analysed. They [6] used filter, wrapper and hybrid approaches for choosing the best feature subset. Different feature selection techniques were applied to KDD99 dataset and determined the evaluation criteria for each method. Using these methods, they eliminated the repetitive attributes or features to increase the accuracy of the intrusion detection system.

Feature fusion and the decision fusion techniques were used by [7] in NIDSs. They proposed evaluation criteria for evaluating different data fusion techniques. They suggest that hybrid fusion techniques seem efficient compared to SVM and other methods. They [7] also found that ensemble algorithms like Adaboost can combine multiple decisions than any other method. The experiment was carried out on KDD99 dataset. They [7] concluded that for developing an excellent network intrusion detection system, data fusion techniques prove to be better as it help in improving the performance of the detection system.

Three important factors for developing an IDS system are pre-processing, feature selection and algorithm design. This paper [2] proposes a modified Naïve Bayes algorithm that improves accuracy and reduces the prediction time. In the proposed system, the algorithm approximates the attribute interactivity using conditional probabilities. They compared the proposed algorithm with Naïve Bayes, J48 and REPTree using several measurement parameters. Results show that good results in detecting intrusion.

In this paper [1], the authors have focused on false positive and false negative performance metrics to increase the detection rate of the IDS. Various machine learning classifiers such as J48, Random Forest, Random Tree, Decision Table, MLP, Naive Bayes, and Bayes Network were evaluated based on several tests on KDD intrusion detection dataset. Random forest classifier was successful in achieving the highest average rate.

3   Preliminary Examination of Data

3.1   Data Description

The UNSW-NB 15 ("UNSW-NB15 data set," 2015) data set was created for producing both normal and attack behaviours generated using tcpdump tool to capture raw traffic. This dataset contains 9 attack types. 42 attributes are considered for our research work. The dataset is divided into testing set and training set. The class distribution is shown in Table 1.

Table 1. Dataset class distribution

Category	Training set	Testing set
Normal	37,000	56,000
Analysis	677	2000
DoS	4089	12,264
Backdoor	583	1,746
Exploits	11,132	33,393
Fuzzers	6,062	18,184
Generic	18,871	40,000
Shell code	378	1,133
Worms	44	130
Reconnaissance	3,496	10,491
Total records	82,332	175,341

The UNSW-NB 15 dataset includes complex patterns compared to NSL KDD99 dataset and contains 9 different attack types unlike NSL KDD99 dataset with only 5 different attack types. Hence for the analysis, UNSW-NB 15 dataset is used to evaluate various classification methods.

3.2   Data Preprocessing

Feature Scaling. Standardization of a dataset is very important for many machine learning algorithms which uses Euclidean distance. If they are not standardized, then there is a possibility that attributes which have values in larger range may be given higher importance. Since each feature may not be represented in same unit of measurement, features with varying magnitude will have an adverse effect on the machine learning algorithms. This can be avoided by transforming the data to same range through standardization. Standard Scalar is a standardization technique utilizing centring and scaling. Standard deviation is calculated for each feature by computing the statistics on the samples in the training set. This will distribute in such a way that mean will be 0 and standard deviation will be 1.

Every feature in the dataset is scaled as

$$\text{Mean, } \bar{X} = \frac{1}{N} \sum_{i=1}^{N} x_i \tag{1.1}$$

$$\text{Standard Deviation, } \sigma_X = \sqrt{\frac{1}{N} \sum_{i=1}^{N} (X - \bar{X})^2} \tag{1.2}$$

$$\text{Standardization, } X' = \frac{(X - \bar{X})}{\sigma_X} \tag{1.3}$$

Where, X is the attribute, N is the total number of instances, X' is the new rescaled, $\bar{X}$ is the mean and σ_X is the standard deviation.

4   Proposed Methodology

4.1   System Architecture

In this section, we describe the architecture of adapting the important features from the dataset which would improve the accuracy of the model and also drastically improve the speediness of training and computation time of all the models. Figure 1 shows the proposed architecture to improve the performance of the model and minimizing the computation time. The step-by step approach of the proposed model is discussed below:

- Choose an input data set, UNSW-NB15.
- Load the dataset as Training and Testing datasets separately.

- Apply Feature scaling technique. Determine the mean (1.1) and standard deviation (1.2) for the feature and the features are further scaled using the formula (1.3).
- Evaluate the models-KNN, Decision tree, Random forest, bagging Meta estimator and XG Boost, determine the accuracy, training and prediction time of these models.
- Apply Feature selection by using Feature Importance model applied on Random Forest classifier.
- Consider only the reduced attributes which are important for the model obtained from the feature selection technique.
- Execute all the models again with reduced features and evaluate the performance in terms of accuracy, training and prediction time.
- Compare the results.

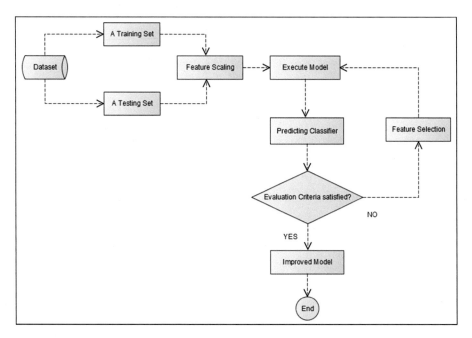

Fig. 1. The proposed architecture for improving computation speed of the models

4.2   Feature Selection

Feature Importance. Feature importance is implemented in our model by utilizing model.feature_importances in Sklearn Random Forest package to analyze about the essential features of the data. The features that are all the more closely related with dependent variable and contribute more for variation of the dependent variable are identified as important features. A large number of features are added to the random forest model and the classifier evaluates and returns back those features which contribute more for improving the performance of the model.

Steps in identifying the important features:

1. Train the random forest model assuming that right hyper-parameters have been considered.
2. Discover prediction score of models and highlight as a benchmark score.
3. Assess prediction scores 'p' more times where p is number of features and shuffle randomly across the column of jth feature.
4. Evaluate all the 'p' scores by a comparison with the already marked benchmark score. If the result of jth column after shuffling it randomly is troubling the score then the feature is required for the model.
5. Eliminate the features that do not trouble the benchmark score and retrain the model with newly reduced subset of features.

4.3   Evaluation Criteria

Accuracy: It is the most widely used metric for classification models which is determined as the fraction of samples predicted correctly.

$$Accuracy = \frac{(TN + TP)}{(TN + TP + FN + FP)} \tag{1.4}$$

The accuracy score is obtained by considering the input as actual labels and predicted labels.

Precision: It is a measure of result relevancy. It makes sure that the negative sample is never identified as positive.

$$precision = \frac{TP}{TP + FP} \tag{1.5}$$

Recall: It is the fraction of relevant instances that have been retrieved over the total amount of relevant instances.

$$recall = \frac{TP}{TP + FN} \tag{1.6}$$

F-measure: It is a measure that combines both precision and recall measures. It is measured as weighted average of precision and recall.

$$F1score = 2 * \left(\frac{precision * recall}{precision + recall}\right) \tag{1.7}$$

Where, TN = True Negative, TP = True Positive, FN = False Negative and FP = False Positive.

Computation Time: There are a lot of high-performance computing (HPC) platforms which support heterogeneous hardware resources (CPUs, GPUs, storage, etc.) in the current era. The prediction of application execution times over these devices is a great challenge and is crucial for efficient job scheduling.

5 Results

The proposed model is developed using Anaconda Navigator tool which has a Spyder interface to compute all the models. The different models chosen for our analysis are KNN, Decision tree, Random forest, bagging Meta estimator and XG Boost. These models obtained higher accuracy compared to other existing models in the literature and also in our analysis. Hence, these classifiers are retained for further improving the computation and prediction time using the proposed approach. Table 2 shows the overall accuracy for KNN, Decision tree, Random forest, bagging Meta estimator and XG Boost after feature scaling. Thus it is inferred from Table 2 that Random forest gave the highest accuracy with 74.875% followed by bagging Meta estimator at 74.641%, Decision tree at 74.227%, Boost at 71.437% and finally KNN at 71.103%.

Table 2. Results of the models before feature selection

Classifier	Accuracy	Training time(s)	Predict time(s)	Precision	Recall	F-score
Random forest	74.875	3.656	0.532	0.74	0.75	0.72
XG Boost	71.437	109.946	6.331	0.69	0.71	0.67
Bagging meta estimator	74.641	9.817	1.144	0.73	0.75	0.71
Decision tree	74.227	2.753	0.06	0.74	0.74	0.72
KNN	71.103	25.189	118.457	0.70	0.71	0.68

Table 2 shows the computation time of the models such as training and prediction times. It is inferred that the training time of XG Boost model is very high (109.946 s) while decision tree model showed more efficient in computation time (2.753). This is the reason for not utilizing few ensemble models for classification. Hence feature selection technique is utilized to obtain the important features useful for predicting the models and enhance their performance.

In (Fig. 2), X-axis represents the attribute names and Y-axis represents the level of importance of attributes. The 41 attributes were reduced to 11 attributes based on their importance by using feature importance on Random Forest classifier.

The reduced features of **UNSW-NB15 dataset** are Service, State, Sbytes, Dbytes, Rate, Sttl, Ackdat, ct_dst_ltm, ct_src_dport_ltm, ct_dst_sport_ltm, and ct_src_ltm.

These reduced features would be used for computing the accuracy and performance of the models.

Table 3 shows the accuracy results of all the models with 11 important attributes obtained by feature importance. Random forest results in the highest accuracy with

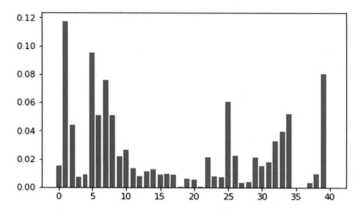

Fig. 2. Feature importance score of 41 attributes

Table 3. Results of the models with reduced attributes

Classifier	Accuracy	Training time(s)	Predict time(s)	Precision	Recall	F-score
Random forest	75.656	0.888	0.447	0.75	0.76	0.73
XG Boost	71.747	34.238	5.876	0.69	0.72	0.67
Bagging meta estimator	75.085	1.901	0.594	0.74	0.75	0.72
Decision tree	74.497	0.35	0.037	0.75	0.74	0.72
KNN	74.395	0.929	8.657	0.74	0.74	0.72

76.656% followed by bagging meta estimator at 75.085%, Decision tree at 74.497%, KNN at 74.395% and XGBoost at 71.747%. It can be observed that there is an improvement in the model accuracy. Table 3 shows the computation time of the models in terms of training and prediction times evaluated on dataset with reduced attributes.

From Table 3, it can be observed that the computation time reduced thereby improving the performance of the model. Thus an analysis of the classifiers that perform better with and without feature importance is studied and results are shown.

6 Conclusion

Accuracy and computation time are critical parameters to assess the performance of the machine learning models. The highest ranked important features are selected utilizing feature importance technique of Random forest classifier. The features with the highest importance score were selected to evaluate other existing machine learning models. The reduction of 41 features to 11 important features improves the XG Boost performance. Performance of Decision tree model also proved to be efficient in comparison to other four models that are used for our analysis. The accuracy of the models increased with reduced attributes and the performance of the model increased by a very big factor. Thus feature importance has been a key factor in decreasing the computation time of the classifier which has not been studied extensively in literature.

7   Future Work

Multiple machine learning models will be integrated to build hybrid architecture of the model. This can further improve the accuracy of the model as well as to speed up the computation time of the models.

References

1. Almseidin, M., Alzubi, M., Kovacs, S., Alkasassbeh, M.: Evaluation of machine learning algorithm for intrusion detection. Department of Information Technology, University of Miskolc, Hungary (2018)
2. Kumar, K., Batth, J.S.: Network intrusion detection with feature selection techniques using machine-learning algorithms. Int. J. Comput. Appl. **150**(12), 1–13 (2016)
3. Belavagi, M.C., Muniyal, B.: Performance evaluation of supervised machine learning algorithms for intrusion detection. Procedia Comput. Sci. **89**, 117–123 (2016)
4. Mogal, D.G., Ghungrad, S.R., Bapusaheb, B.B.: NIDS using machine learning classifiers on UNSW-NB15 and KDDCUP99 datasets. In: IJARCCE (2017)
5. Moustafa, N., Slay, J.: The significant features of the UNSW-NB15 and the KDD99 data sets for network intrusion detection systems. In: 4th International Workshop on Building Analysis Datasets and Gathering Experience Returns for Security (BADGERS), Kyoto, Japan (2015)
6. Kumar, D., Singh, H.: A study on performance analysis of various feature selection techniques in intrusion detection systems, vol. 3, no. 6, pp. 50–54 (2015)
7. Li, G., Yan, Z., Fu, Y., Chen, H.: Data fusion for network intrusion detection: a review. Secur. Commun. Netw. **2018**, 16 (2018)
8. Belouch, M., El Hadaj, S., Idhammad, M.: Performance evaluation of intrusion detection based on machine learning using Apache Spark. Procedia Comput. Sci. **127**, 1–6 (2018)
9. Farnaaz, N., Jabbar, M.A.: Random forest modeling for network intrusion detection system. Procedia Comput. Sci. **89**, 213–217 (2016)

Runtime UML MARTE Extensions
for the Design of Adaptive RTE Systems

Nissaf Fredj[1,2(✉)], Yessine Hadj Kacem[2], and Mohamed Abid[2]

[1] ISITCOM, University of Sousse, Sousse, Tunisia
nissaffredj@yahoo.com
[2] CES Laboratory, ENIS, University of Sfax, Sfax, Tunisia

Abstract. The Adaptive Real-Time Embedded Systems (A-RTES) should change their behavior as a response to the execution context and system constraints. They must be always consistent and available to preserve their usefulness and feasibility. For this reason, their adaptation should be applied at runtime whenever is required. High-level design-based approaches have merged to deal with the complexity of such systems, especially the use of Modeling and Analysis of Real-Time Embedded (MARTE) profile. However, MARTE standard doesn't offer specific concepts for designing the evolution of adaptive real-time resources at runtime. It only reasons the system behavior at a specific instant without taking into account historical data. In this work, we define a new package that extends UML MARTE modeling language to design the A-RTES structure and behavior at runtime and support the reasoning about historical information as well as enable the evaluation of system real-time constraints. Our contribution is evaluated through a Flood Prediction System (FPS) case study.

Keywords: A-RTES · Runtime adaptation · MARTE · Temporal reasoning · Real-time constraints

1 Introduction

The RTES are more complex, challenging to develop and require a multitude of constraints such as resource limitations and execution time. These systems may reconfigure their behavior in response to the execution context, user requirements and system constraints. Therefore, their reconfiguration should be performed at runtime whenever it is needed to evolve the system state between a current configurations and a new produced one in order to optimize the system resource allocations. Recently, high-level design approaches have merged based on UML MARTE profile [1] to provide simplified designs and build efficient adaptive systems with lower complexity. Some of these works have taken advantage of the runtime models [2] to design the running adaptive systems. However, existing runtime model-based approaches for A-RTES are limited to the modeling of coarse-grain reconfiguration which modifies the whole system. They don't explore the design of the fine-grain adaptation which adapts partially the application layer of the running system and constitutes an important ability of embedded systems to deal with applications complexity by scaling their output quality according to system resources limitations. Additionally, MARTE standard doesn't offer

© Springer Nature Switzerland AG 2020
A. Abraham et al. (Eds.): ISDA 2018, AISC 941, pp. 78–87, 2020.
https://doi.org/10.1007/978-3-030-16660-1_8

semantics and sufficient concepts for the design of running system and reasoning about its behavior. It only supports the reasoning about the behavior states at a current time and doesn't bring runtime concepts for designing the evolution of system states and reasoning about historical data which are necessary to take more stable future adaptation decisions. As a response to these weaknesses, we define a new package which extends MARTE profile to support the specification of adaptive resources structure and behavior at runtime and reason about its behavior by taking into account historical information. Our extensions are devoted especially for the fine-grain adaptation modeling. Since, the adaptation behavior aims at meeting resource and system constrains which present an inherent challenge in the adaptive RTES design, our modeling focus also on evaluating the system non functional properties under different instant of reconfiguration.

Our objective is to propose a runtime modeling-based approach that provides explicit and complete semantics for adaptive real-time resources and permits the automatic generation of the A-RTES from a high-level specification. This design will be integrated in a model driven-based process for the automatic generation of adaptive RTES. Hence, our idea constitutes a promote solution based on standardized UML MARTE modeling language to simplify the designer's tasks, improve productivity and promote reusability. Our proposal is evaluated through a Flood Prediction System (FPS) [3] case study which adapts itself to the execution context, depending on runtime events. The reminder of this work is divided into four parts; the second section summarizes the research studies about the modeling-based approaches of the adaptive RTES. In the third section, we present our proposal which is based on a new extension in UML MARTE language. Our contribution is validated by a flood prediction system case study in forth section. Finally, we conclude the work and present our future works.

2   Related Works

2.1   The High-Level Design of the Adaptive RTE Systems

The high-level modeling-based methods for RTES have received considerable attention recently to decrease the design complexity of embedded systems. Most of existing works were interested in the specification of dynamic reconfigurable systems using UML MARTE profile. A model-based design flow for dynamic reconfiguration has been presented in [4]. The reconfiguration behavior is modeled using specific semantics of automata-based technique. The latter is composed of a set of modes connected to each other with transitions. It describes the different states of reconfigurable system and performs a switch functionality to choose the next executing mode. The switch is related to some conditions that should be met to trigger events change.

A model-based process for the dynamic reconfiguration in distributed real-time systems is also proposed in [5]. The solution is based on a non-predefined set of configurations specified by new extensions in MARTE and AADL (Architecture Analysis and Design Language) meta-models. The dynamically reconfigurable RTE system is modeled by a set of modes. Each mode is associated to a configuration described by a set of components and connections between them. Besides, the behavior of semi-distributed control is designed in [6] by a state machine annotated by MARTE stereotypes. The authors proposed new concepts to perform the control and the adaptation features as well as the coordination between the controllers. On the other hand, the contribution of [7] has given an abstraction of self-adaptive systems based on design patterns for each of the adaptation loop modules annotated by MARTE profile stereotypes. These patterns take into account the monitoring, analyze and decision making as well as the evaluation of real-time system constraints during adaptation.

As a conclusion, the previously presented works [4–7] are limited to design the adaptation behavior statically at design time. They suffer from shortcomings linked to the running system modeling. These research studies don't give attention to follow the dynamic evolution of the reconfigurable system behavior and don't provide a reasoning feature that guides the reconfiguration and defines the future state to which the system should transit. As a response to these failing, certain approaches have taken advantage of runtime models to design the executing adaptive systems.

2.2   Runtime Models for the Design of Adaptive RTE Systems

In the context of DiVA project [8], Morin presented different UML runtime meta-models to design the dynamic adaptive software systems. The variability meta-model describes the features of the system. The context meta-model specifies the execution context that should be observed. The reasoning meta-model defines the features to be chosen from the variability meta-model based on the current context. The architecture meta-model defines the system configuration at runtime regarding components. Second, the work of [9] deals with the system configurations based on models at runtime. The objective is to generate the adaptation logic based on a comparison between the current system configuration and a newly produced one. They proposed an adaptation model processed at runtime to manage the system and containing invariant constraints to validate the adaptation before the execution. The authors of [10] proposed a runtime model-based approach for the development of adaptive systems. This approach helps both to specify a dynamic adaptation using AADL and check the system integrity at runtime. In this work, only the coarse grain global reconfiguration of the system is modeled. Besides, the design doesn't deal with the reasoning about execution changes and the evaluation of the real-time constraints of the running system. In [11], the authors presented the design and the implementation of a runtime pattern for the fine-grain adaptation performed partially in the system. The proposed solution copes with the unanticipated adaptation of programming languages at runtime. In order to introduce the adaptation feature in the infrastructure of a programming language, the pattern introduces an adaptation class associated to the class instance and which should be applied to a structural feature such as an operation call or an attribute of a given object. At runtime, in order to enable a dynamic adaptation, the pattern introduces an instance of the class adaptation.

To summarize, the proposed runtime models in [8–11] adapt the system at runtime however; these models only reflect the system state at a specific time (current time). In fact, it is necessary to reason about historical information to compare past versions of system states and take stable future reconfiguration decisions. Besides, the previously presented works don't give attention to the evaluation of system constraints and resources allocations which constitutes an intrusive challenge in the design of real-time embedded systems.

In the present work, we contribute by proposing a new MARTE package for the specification of adaptive RTES behavior and structure at runtime; especially we are interested in the fine grain application adaptation of these systems. In this package, we define new adaptation and runtime extensions to support the reasoning about the running system dynamically under different instants also to evaluate the real-time constrains. The proposed idea simplifies the designer's tasks and promotes the reusability of these systems. Unlike the existing runtime model-based approaches for adaptive RTE system, this work provides an innovative idea to bring the UML MARTE modeling language with new runtime semantics and reasoning features.

3   The Proposed Extensions

As a response to the current runtime model-based approaches weakness, we propose an extension in MARTE profile to design the adaptive real-time systems at runtime; especially we address the fine grain adaptation [12]. The latter is applied locally on the RTE system to modify the behavior of their application layer. The adaptation consists in algorithmic or application parameters modification where each parameter modification requires different amount of non-functional properties and temporal resources. Therefore, the proposed extensions help with the modeling of these systems behaviors by taking into account the real-time constraints evaluation under different instants of configuration.

For our extension, we need four sub-profiles of MARTE. First, since, we are interested in the fine-grain adaptation of RTE systems composed of adaptive software tasks; we use concepts from the Software Resource Modeling (SRM) package of MARTE. The SRM is a specialization of the Generic Resource Modeling (GRM) package of MARTE to describe the structure of the system which is composed of software multi-tasking. Second, in order to follow the system behavior changes at runtime, we need behavioral modeling and runtime semantics concepts provided respectively by the CommonBehavior and the RuntimeContext sub-packages. The latter are a specialization from the Causality package of MARTE which provides concepts related to the system behavior described by modes and configurations. Finally, we need to use the timing concepts offered by the Timed package of MARTE to define the adaptive resource execution period.

Thanks to the RuntimeContext sub-package (see Fig. 1) of MARTE, the behavior of the running system is represented by the BehaviorExecution stereotype. The latter is specialized into two types of runtime instances which are the action and the composite behavior presented respectively by the ActionExecution and the CompBehaviorExecution stereotypes. However, the RuntimeContext sub-package doesn't provide specific concepts for designing the evolution of the system at runtime. It only represents the system state at a specific time, while at runtime a resource may have several versions (instances) under different instants. Each version is created after changes (configurations). In order to overcome these challenges, we define a new meta-model (package) named RuntimeReconfiguration which extends the Software Resource Modeling (SRM) package of MARTE.

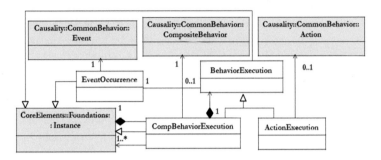

Fig. 1. The capabilities of UML MARTE profile to represent the runtime context [1]

3.1 The RuntimeReconfiguration Package

Our proposed package (see Fig. 2) enables the design of the structure and behavior of software adaptive resources (application layer) in the A-RTES. We have distinguished the standard stereotypes of MARTE from our own ones by the green classes.

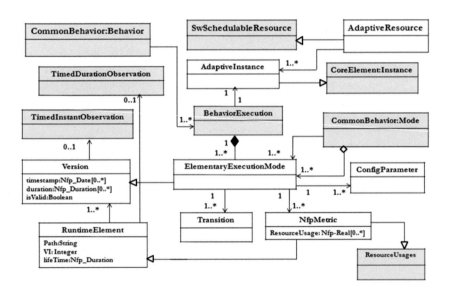

Fig. 2. The proposed RuntimeReconfiguration package

Structural View. The meta-model contains a proposed AdaptiveResource stereotype to specify the reconfigurable adaptive resource. The latter should be linked to a software scheduler that determines the order in which those should be executed. Thus, it inherits from SwSchedulableResource stereotype of MARTE which specifies CPU competing and schedulable resources. At runtime, an adaptive resource may be divided into several instances defined by the proposed AdaptiveInstance stereotype that inherits from the MARTE:CoreElement:Instance stereotype. At each instant of the execution, an adaptive instance is characterized by an execution mode which represents its current operational mode. Each execution mode requires a resource usage necessary for the evaluation of the system reconfiguration behavior.

Behavioral View. The adaptive resource behavior at runtime is specified by the BehaviorExecution stereotype which specifies runtime instances of the behavior. A whole system mode is represented in MARTE by the CommonBehavior: Mode. It may be composed of a set of elementary adaptation modes. In order to enable MARTE to support the fine grain adaptation which adapts partially the running system based on a set of elementary execution modes, we have proposed the ElementaryExecutionMode stereotype. Besides, we have defined two associations between the ElementaryExecutionMode and the CommonBehavior:Mode of MARTE. The first one means that an elementary mode may be part of one or several system global modes. The second association means that each mode may be instantiated to several execution modes. The execution behavior is composed of one or more elementary execution modes. The latter is a combination of one or more algorithmic parameters (ConfigParameter stereotype) values and it has a corresponding non functional property metric (NfpMetric). The NfpMetric stereotype represents the evaluation of reconfiguration output quality related to the execution mode non functional properties usage. Therefore, the NfpMetric inherits from the ResourceUsage stereotype of MARTE.

At runtime, each resource may pass by different configuration (versions). In order to keep track of the history changes; we have proposed the RuntimeElement stereotype. It is specified by a path and a version identifier and has a lifetime attribute which specifies the period of time when it is activated. It has one or more version under different instants. Hence, a Version stereotype has a timestamp and duration attribute which must have a multiplicity of [0..*]. The RuntimeElement and the Version stereotypes enable to reason about the system state and take more stable reconfiguration decisions by taking into account historical information. Thus, the ElementaryExecutionMode extends Version stereotype, the inherited attributes (duration, isValid, timestamp) make it easier to reason about the system execution modes. The RuntimeElement and Version stereotypes are associated with the TimedDurationObservation stereotype of MARTE:TimedElement package to define its lifetime. Likewise, the Version stereotype is associated with the MARTE:TimedInstantObservation to specify the instant of each version.

The consumed resources of the reconfigurable system are defined by a set of values in a precise period of time. A formal estimation of the system resource usage is specified by the ResourceUsage attribute. It can change its value and has several versions, thus, it must has a multiplicity of [0..*]. To obtain a good system performance during reconfiguration, the resource usage value at an instant t_i must be lower than the resource usage value at an instant $t_i - 1$ (see Eq. 1).

$$ResourceUsage(ti) - ResourceUsage(ti-1) < 0 \qquad (1)$$

The consumed resources of a reconfigurable system in a period of time T is an average value which is computed by the sum of the ResourceUsage values of all the versions created in T divided by N (number of versions).

$$ResourceUsage_{average}(T) = \sum_{i=1}^{N} ResourceUsage/N \qquad (2)$$

To summarize, thanks to the proposed extensions, the modeling solution of the executing A-RTES can be specified. This solution is based mainly on a proposed runtime state machine for the design of the system behavior and which will be illustrated in what follows.

3.2 Runtime State Machine

The state machine model illustrated by Fig. 3 is based on one or more states connected to each other by transitions. The class ChangeableElement stereotyped by AdaptiveInstance, represents an adaptive element from the system that updates its behavior. An execution behavior is composed of a set of state instances. A runtime state has three methods activate (), deactivate (), updateState(). The Condition class has an attribute isFulfilled and a method setFulfilled () to test whether a condition is met at runtime.

The history of system states need to be kept for taking more stable reconfiguration decisions. The designer should reason about the future state by comparing the historical versions values. To specify the reasoning in a formal and semantic way, we have associated to the state machine model, the OCL (Object Constraint Language) constraint of Fig. 4. Indeed, the constraint shows that a condition must be met (Event. Condition.isFulfilled = true) before calling the checkState() method of the ChangeableElement class. After calling the checkState() method, we have collected the valid state versions (State->exists(StVersions:State|(stVersions.isValid = true))). If the compareStValue() method of the class State returns true, then a future state is activated (st.activate()).

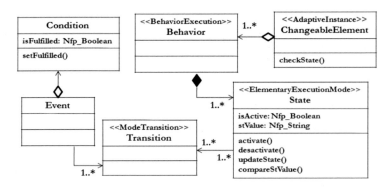

Fig. 3. Runtime state machine

The proposed RuntimeReconfiguration package has been integrated into MARTE component of Papyrus modeling environment, in which, we have specified the runtime state machine annotated by the proposed extensions and associated to the reasoning OCL constraint (see Fig. 4). The system modeling based mainly on the runtime state machine, will go through a model to text transformation for the automatic generation of the running adaptive RTES from a high-level specification as well as the evaluation of the real-time constraints.

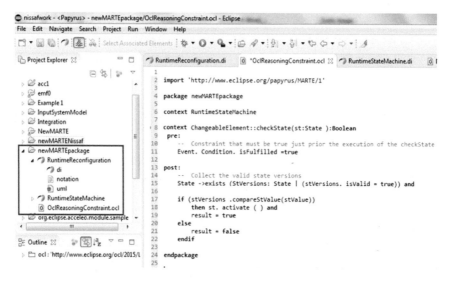

Fig. 4. The OCL reasoning constraint about future reconfiguration state

4   Case Study: An Adaptive Flood Prediction System

To evaluate our proposal, we present the specification of the Flood Prediction System (FPS) case study. The FPS updates its behavior at runtime depending on runtime events such as memory is low and energy is high. It is composed of a set of nodes which contains sensors to evaluate water speed and depth, a memory, battery and communication equipments. Each node computes the flood prediction level according to the values provided by sensors and communicates the result to a central node. Since, our proposal is devoted for the application fine grain application adaptation; we are interested in the application layer of the FPS system presented by the speed and level controller tasks which evaluate respectively the speed and the water level. We have supposed that only the speed controller is an adaptive task which changes its behavior as a response to user requirements and system constraints.

The FPS system model is displayed in Fig. 5 composed of SpeedController and LevelController tasks. Each task is allocated in one processor during its execution. The adaptive SpeedController task is divided into adaptive instances at runtime annotated by the AdaptiveInstance stereotype of the proposed RuntimeReconfiguration. Its behavior is specified by the state machine of Fig. 6.

The adaptive speed controller task has three elementary execution modes; High-Speed, LowSpeed and MediumSpeed. These modes indicate three levels from the highest to the lowest water speed. As shown by Fig. 6, when the condition 'speedlevel > highValue' is met at runtime (isFulfilled = true), a gotoHigh transition is performed and the HighSpeed execution mode is activated in order to be executed.

The execution of HighSpeed mode requires a set of system resources such as the energy usage and execution time. As a whole, in a period T, a set of modes can be executed. Thus, the execution time and the consumed energy average values of the adaptive speed controller task in T are quantified by Eq. 2 (see Sect. 3.1).

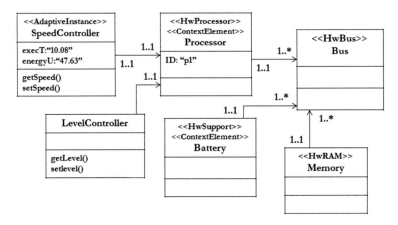

Fig. 5. FPS system modeling

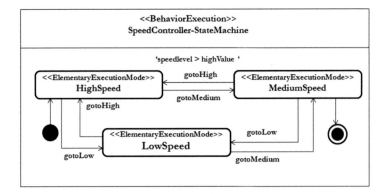

Fig. 6. Speed controller state machine

5 Conclusion and Future Work

In this work, we have brought the UML MARTE standard with runtime and adaptation concepts to design the fine-grain adaptation of A-RTES at runtime. These extensions offer a basis for reasoning about the system configurations and the estimation of real-time constrains. A formal validation of our contribution is performed by a flood prediction system case study which provides the specification of the structure and behavior of the adaptive FPS system. As future works, we plan to build a complete modeling of the adaptation engine composed by the monitoring, decision making and reconfiguration features which are crucial for the control of the running adaptive RTE system. Then, we aim to perform a Model-Driven Engineering [13]-based process for an automatic generation of A-RTES from a high-level specification.

References

1. OMG Object Management Group. A UML Profile for MARTE: Modeling and Analysis of Real-Time Embedded systems, ptc/2011-06-02. Object Management Group, June 2011
2. Blair, G., Bencomo, N., France, R.B.: Models@ run.time. Computer **42**(10), 22–27 (2009)
3. Gandla, S., Al-Assadi, W.K., Sedigh, S., Rao, R.A.R.: Design and FPGA prototyping of a flood prediction system. In: 2008 IEEE Region 5 Conference (2008)
4. Quadri, I.R., Meftali, S., Dekeyser, J.-L.: A model based design flow for dynamic reconfigurable FPGAs. Int. J. Reconfigurable Comput. (2009)
5. Krichen, F., Hamid, B., Zalila, B., Jmaiel, M.: Towards a model based approach for reconfigurable DRE systems. In: ECSA, pp. 295–302 (2011)
6. Trabelsi, C., Meftali, S., Dekeyser, J.L.: Semi-distributed control for FPG based reconfigurable systems (2012)
7. Said, M.B., Kacem, Y.H., Kerboeuf, M., Amor, N.B., Abid, M.: Design patterns for self-adaptive RTE systems specification. Int. J. Reconfigurable Comput. (2014)
8. Morin, B.: Leveraging models from design time to runtime to support dynamic variability. University of Rennes, Ph.D. thesis (2010)
9. Fleurey, F., Dehlen, V., Bencomo, N., Morin, B., Jézéquel, J.-M.: Modeling and validating dynamic adaptation. Computing Department Lancaster University Lancaster UK 3 IRISA INRIA Rennes (2010)
10. Loukil, S., Kallel, S., Jmaiel, M.: An approach based on runtime models for developing dynamically adaptive systems. In: IEEE International Conference on the Engineering of Computer Based Systems (2016)
11. Costiou, S., Kerboeuf, M., Cavarle, G., Plantec, A.: Lub: a pattern for fine grained behavior adaptation at runtime. Sci. Comput. Program. **161**, 149–171 (2017)
12. Vardhan, V., Sachs, D.G., Yuan, W., Harris, A.F., Adve, S.V., Jones, D.L., Kravets, R.H., Nahrstedt, K.: Integrating finegrained application adaptation with global adaptation for saving energy. University of Illinois at Urbana-Champaign (2009)
13. Schmidt, D.C.: Model-driven engineering. IEEE Comput. **39**(2), 25 (2006)

A Normalized Rank Based A* Algorithm for Region Based Path Planning on an Image

V. Sangeetha[✉], R. Sivagami, and K. S. Ravichandran

School of Computing, SASTRA Deemed University, Thirumalaisamudram,
Thanjavur 613401, Tamilnadu, India
sangeetha@sastra.ac.in

Abstract. With the development of many autonomous systems, the need for efficient and robust path planners are increasing every day. Inspired by the intelligence of the heuristic, a normalized rank-based A* algorithm has been proposed in this paper to find the optimal path between a start and destination point on a classified image. The input image is classified and a normalized rank value based on the priority of traversal on each class is associated with each point on the image. Using the modified A* algorithm, the final optimal path is obtained. The obtained results are compared with the traditional method and results are found to be far better than existing method.

Keywords: Path planning · Normalized rank · A* algorithm ·
Classified image · Autonomous systems

1 Introduction

Advancements in the applications of autonomous systems has led to the need for robust path planners. For example, an area of application is the path planning of cruise missiles. The cruise missile path needs to be planned cleverly to take into account fuel, regional traversability, sneaking or low detectability and the best way of target approach during terminal homing. Scientists has sought inspiration from every day intelligence to develop many planners for efficient navigation. Path planning [1] is an optimization process where an optimal path is plotted between a starting and destination with least cost and no-collision with obstacles. Cost is measured in a path by not only the time taken but also the resources used. There are certain applications [2] which require not only the shortest path but also the one that utilizes the least resources. In this case, the configuration space in which planning is carried out must be prioritized. A prioritized configuration space will have points associated with a priority based on its ease of traversability. Path planning can be either local or global. The former does not require any environment knowledge while the later requires a prior knowledge of the environment. Many methods are available in the literature for path planning which can be classified as Classical methods [3, 4] and Heuristic methods [5, 6]. Classical methods find the optimal solutions using accurate algorithms but consume large time in a large environment. Heuristic methods find near optimal solutions using approximation procedures and are more effective even in large environment. Over the years, many efforts are being put in searching in huge search spaces. The drawbacks of the

© Springer Nature Switzerland AG 2020
A. Abraham et al. (Eds.): ISDA 2018, AISC 941, pp. 88–97, 2020.
https://doi.org/10.1007/978-3-030-16660-1_9

traditional fixed order methods like breadth first and depth first search are overcome by best-first search. A prominent guided best first search is the A* algorithm. A* algorithm operates using two important list- OPEN and CLOSED list. The cost function calculates the cost value of each node. The guided best first search expands the OPEN minimum cost node. Upon expanding, the child nodes are generated and the next possible node is inserted into OPEN list based on the order of cost. There are many variations of A* search to overcome the memory bound limitation leading to exponential increase in time. Incremental A*, Field A*, Iterative A*, Thetha A* are some of the versions of A*. A* is used in many applications like computational biology, robotics, hardware verification, terrain matching. There are many works in the literature regarding Path planning. Zeng et al. [7] has discussed A* algorithm in terms of the direction of traversal from the current grid. The 8-neighbourhood principle is used. Sadrpour et al. [8], proposed a nature inspired based approach to predict the optimal path and also the mission energy. Real time measurements of Unmanned Ground Vehicles were used for the simulation. In [9], A* algorithm is been devised for the uniformly gridded space. It is showed that it performs A* over the uniformly gridded space. Opoku et al. [10] discussed robot path planner that differs from existing stereo based path planners. It uses many cost functions that is based on occupancy grid representations. In [11], Agnus proposed an Ant Colony based approach considering parameters like pheromone quantity, visibility and cost. The method was implemented on mounds, valleys, hills and tested for efficiency. Also in [12], path planning is performed in an overhead imagery. The image is initially classified using Gaussian Classifier. Then upon the image, the path is planned using Probability Density Function. The image is classified as low cost traversal and high cost traversal. Then the path is planned accordingly. There are many other works in the existing literature regarding path planning according to the type of terrain. Such planners will help in making out a path that consumes less energy and cost. In this proposed work, Normalized Rank based A* has been proposed. The purpose of the proposed approach is to find an optimal path between start and destination points on a prioritized configuration space. The configuration space has been taken as a classified image. The image is classified into different classes based on the terrain information on it. Each class is assigned a rank based on its ease of traversability. Upon normalization, a bias value is obtained and it is multiplied with the cost function of A* algorithm. The main aim of the proposed Normalized Rank based A* algorithm is to reduce the total cost of the path thereby leading to optimal resource usage.

2   Problem Formulation

Let R be the 2D image spread over the Cartesian coordinates X and Y. $P^{optimal}$ be the final path in the 2D space R. Let S be the starting point and D be the destination point. Then W be the sub grid confined by the S and D and {W} be the size of sub grid. The path $P^{optimal}$ is made of many connected segments $P_1....P_n$ where n depends upon the total number of grids. The region R is classified into N regions and each N is assigned a weight W. Upon the region R, $N \times N$ number of grids are made and each grid has a unit of 1-pixel i.e., 1 pixel of the image represents one grid.

Let $S_{X \times Y}$ be the sub window in R with (s, t) being the current point. Then the neighbors of (s, t) in $S_{X \times Y}$ are (s, t − 1) * b, (s, t + 1) * b, (s + 1, t) * b, (s − 1, t) * b, (s − 1, t − 1) * b, (s − 1, t + 1) * b, (s + 1, t − 1) * b, (s + 1, t + 1) * b. The sub path P_i will be the minimum cost path from the current point location to the minimum weighted neighbor. The final path $P^{optimal}$ will be the projection of sub path P_1 towards P_2 and so on till P_n with minimum cost. P_n are subpaths of $P^{optimal}$ and b is the bias value obtained through normalization. Figure 1 shows a well-structured configuration space.

The problem can be thus formulated as shown in Eq. (1).

Since the path is represented as sorted grid coordinates, a list U, is used to represent the final $P^{optimal}$. The first element of the list is the start and the last element is the destination.

The path planning problem is to find the optimal path from S to D which is collision free. The problem can be formulated as a multi-objective optimization problem as,

$$\textbf{\textit{Minimise}} f(x) = \sum_{i=1}^{N} \sum_{j=1}^{N} x_{ij} d_{ij} \tag{1}$$

Subject to

$$\sum_{i=1, i \neq j}^{N} x_{ij} = 1 \qquad j = 1.....N, \forall j \in \varLambda \backslash \{S\} \tag{2}$$

$$\sum_{j=1, j \neq i}^{N} x_{ij} = 1 \qquad i = 1.....N, \forall i \in \varLambda \backslash \{D\} \tag{3}$$

$$d_{ij} = \sqrt{(x_i - x_{i+1})^2 + (y_i - y_{i+1})^2} \tag{4}$$

$$x_{ij} = \begin{cases} 1 & for\ path\ between\ i\ and\ j \\ 0 & otherwise \end{cases} \tag{5}$$

Equations (1) represent the objective function to minimize the total distance. Equations (2) and (3) indicates that there can be only one arrival (except start S) and one departure (except Destination D) at a time. Equation (4) represent the distance function to calculate distance. Equation (5) indicates that either the node is included or discarded from the list.

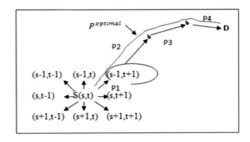

Fig. 1. Well-structured configuration space

The above formulated problem is solved through Normalized Rank based A* algorithm.

3   Materials and Methods

3.1   A* Algorithm

An easiest way to find the path between two points is to take a straight-line path. But this does not work when there are obstacles around. One of the well-known algorithm in such cases for path planning is the A* algorithm [13]. A* is a heuristic search method that uses the best first search to traverse from the start and final position. It uses the shortest path principle of Dijikstra's algorithm (looking for points near the starting point) and the guess - work principle of Greedy Best First Search (looking for points near the destination point). The algorithm expands at each minimum cost node and builds a tree of nodes (points) as it traverses each point. The leaves of the tree are put up in the priority queue. Starting from start state S, it expands itself towards goal state D using two components:

1. Actual edge cost from S to current node O
2. Estimate cost from O to D

These two components combine to form the cost function as given in Eq. (6)

$$f(o) = g(o) + h(o) \tag{6}$$

Where

$g(o)$ = movement cost from the starting point to the current point (Actual cost) (G)
$h(o)$ = movement cost from the current point to the destination point (Estimate Cost) (H)

The function h is the heuristic to calculate estimated distance. The more the estimation is nearer to the actual distance, the more will be the accuracy.

A* uses two main data structures [14] for processing namely the OPEN list and the CLOSED list. OPEN list maintains the list of points that are being considered for the shortest path(visited but not expanded). CLOSED list maintains the list of nodes that need not be visited again. Initially, A* algorithm starts with S. The S is now put into the CLOSED list. It calculates the actual distance from the S to the O. This calculation will compute the G value of the path score. Then the estimated cost from O to D is calculated and made as the H value of path score. The point that has the least path score is taken for the shortest path. Once the least score point is found, it is added to the CLOSED list. When more than one neighbors of O have the same value for G, the actual purpose of H comes into use. The node with least H value is considered. The algorithm terminates when the next to be expanded is D and H becomes 0. The final path P with minimum f is the optimal path.

3.1.1 Admissibility, Monotonicity, Informedness

Generally, an algorithm must not only determine an optimal solution, but find the solution efficiently. The efficiency of the algorithm is crucial when dealing with time and space complexity. Efficiency of A* depends on heuristic used. The heuristic should be admissible, monotone and informed. Informedness of a heuristic is that if the heuristic is able to prove itself for its better performance. For all search algorithms, a heuristic is admissible if it is able to find the minimum cost path whenever one such exists [13]. For A*, if $h(o) \leq h * (o)$, then h is admissible.

For all admissible heuristics, if the cost of f does not decrease on its path from the root, then h is monotone [13] as given in Eq. (7).

$$h(o_i) - h(o_j) \leq \text{cost}(o_i, o_j) \tag{7}$$

where O_j is the decedent of O_i, $cost(o_i, o_j)$ is the cost of moving from O_i to O_j.

3.1.2 Distance Metric

The correct working of A* algorithm [13] depends on the choice of distance metric. In this paper, Diagonal distance, a combination of Manhattan and Euclidean distance is used. The main advantage of this hybrid distance metric is that, it uses the 4-diagonal search of Euclidean and 4-linear search of Manhattan metric.

3.2 Normalized Rank-Based A* Algorithm

For certain environments, the planned path must not only be the shortest one but also must use the least resources like time, energy, cost, etc., [15]. In such cases, the points considered for the optimal path should be ranked and chosen accordingly for the path. Each point in the input image belongs to a class of terrain [16–18]. The classes are ranked based on the ease of traversability on it from 1.... n, where n = total number of classes. Each class is assigned a rank based on human perception. Normalized rank is calculated as given by Eq. (8),

$$NR_i = \frac{R_i}{\sum_{i=1}^{n} R_i} \tag{8}$$

Where, R_i is the rank assigned to class i, NR_i is the normalized rank of the class i.

Each point in the class i, is now assigned with this NR value as a bias. The bias value is then multiplied with the distance computation part of A* algorithm. Lower the bias value, higher is the priority of the point. So, while deciding the next point to visit, the point with lower bias value will be given first priority.

The distance cost function of Normalized Rank based A* algorithm is given by Eq. (9),

$$D = \left(\sqrt{\sum_{i=1}^{n} (x_i - y_i)^2} + \sum_{i=1}^{n} |x_i - y_i| \right) * NR_i \tag{9}$$

A lower NR value will decrease the distance from X to Y, thus making Y as the next point to be chosen.

The pseudo-code for Normalized rank-based A* algorithm is as given below:
Pseudo code for Normalized Rank-based A*

```
Input: S, D, OPEN, CLOSED
Output: P
```

```
Begin
1  Add S to OPEN; f(s)=h(s)
2  while OPEN not empty
        2.1 Take from OPEN, a Xcurrent with least F
        2.2 If Xcurrent == D
                        Break
        2.3 Generate adjacent nodes for Xcurrent
        2.4 For each adjacent node X of Xcurrent
            2.4.1 If X ∈ OPEN
                        Calculate g(X), g(Xcurrent)
                        If g(X)<g(Xcurrent)
                        |  Update X , Add Xcurrent to CLOSED
                        end
            2.4.2 end
            2.4.3 If X ∈ CLOSED
                        Calculate g(X), g(Xcurrent)
                        If g(X)< g(Xcurrent)
                        |  Move X from CLOSED to OPEN
                        end
            2.4.4 end
            2.4.5 If X ∉ OPEN
                        |  Add X to OPEN
                        |  Calculate F, Call (Evaluation_code)
            2.4.6 end
            2.4.7 Add Xcurrent to CLOSED
        2.5 end
3 end
4 If (Xcurrent = = D),
          Exit // OPEN list empty
5 Plot P
End
```

The evaluation function calculates the cost. The final path score, F is returned by evaluation function.

Pseudo code for Evaluation function:

```
Input: Xprevious, Xcurrent, X, D
Output: F (Path score)

Begin
    1.   For Xcurrent
                1.1 Calculate g(Xcurrent)
                1.2 Calculate h(Xcurrent)
                1.3 If (X == blocked_point)
                    Add X to CLOSED
                        h (Xcurrent) = h (Xcurrent)*obstacle_threshold
                            else
                            calculate h(Xcurrent)
                    end
                1.4 F= G+H
                1.5 end
    2.   end
end
```

4   Results and Discussion

The above said experiments are performed in the MATLAB R2017b [19]. The results are evaluated through a comparison between traditional A* algorithm, Ant colony optimization and the proposed Normalized rank-based A* algorithm. Experiments have been conducted on satellite images. The input images are classified using K-means classifier into different classes based on the information available on it. Results obtained using one such image is discussed in this section. The image is classified into three classes as forests, damp roads, shore. First priority is given to shore, second priority is given to damp roads, last priority is given to forests. Priority is assigned based on human perception and the normalized rank of each class is obtained. All pixels belonging to the class will carry the normalized rank value. The start and destination points are chosen. The result analysis between Traditional A*, ACO and proposed approach is numerically analyzed in Table 1. The pictographic result of Normalized rank-based A* algorithm and traditional A* algorithm is as shown in Fig. 2(a) & (b).

In Fig. 2(a) & (b), the blue squares are obstacles defined while simulation. The path planned using both the method may look the same but the cost of both the path will vary and is shown in Table 1. Figure 3(a) & (b) shows results with a different start and destination points.

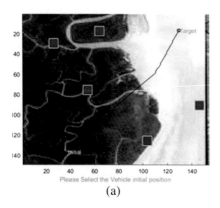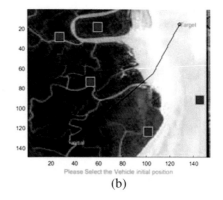

(a)   (b)

Fig. 2. (a) Path planned using normalized rank based A* algorithm (b) Path planned using traditional A* algorithm

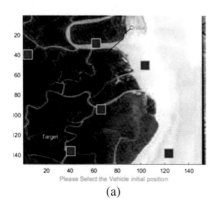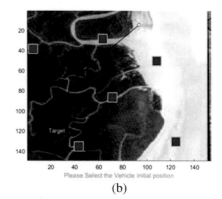

(a)   (b)

Fig. 3. (a) Path planned using normalized rank based A* algorithm (b) Path planned using traditional A* algorithm

The comparison of results between the traditional method and proposed method is given in Table 1.

Table 1. Comparison between traditional A* algorithm, ACO and proposed normalized rank based A* algorithm

	Fig. no.	Traditional A* algorithm	Traditional ACO	Proposed normalized rank based A* algorithm
Time taken (secs)	2	108.78	107.76	60.98
	3	61.64	80.96	81.96
Length of the path (points)	2	156.69	157.03	156.69
	3	136.06	138.50	138.38
Cost of the path (G + H)	2	276.10	275.92	197.63
	3	241.48	215.69	174.64

From Table 1 it can be inferred that the time taken and length of the path found through Normalized rank-based A* algorithm is greater than traditional A* algorithm in some cases. This because the cost function of traditional A* algorithm is multiplied with the Normalized rank value. The main objective of the proposed algorithm is to achieve reduced cost when compared with the traditional methods. From Table 1, it can be clearly seen that the cost of path obtained through the proposed method has reduced by 28% when compared with the traditional A* method and has reduced by 23% when compared with traditional ACO Method. This is because of the priorities assigned to each pixel. The NR value assigned to each pixel decides the cost of each pixel. By choosing correct pixels, the proposed method reduces the cost of the path.

5 Conclusion

In this paper, Normalized Rank based A* algorithm has been proposed to find optimal path between two points with reduced cost. The proposed approach has been compared with traditional A* algorithm, ACO method and it has been found that the overall cost of the path has been reduced by an average of 28% and 23% than the traditional methods. The algorithms have been implemented in MATLAB 2017b. The proposed algorithm may be improved in the future with respect to time by parallel processing to find the optimal path. The search mechanism can be parallelized.

Acknowledgement. The authors would like to thank DRDO-ERIPR for their funding under research grant no: ERIP/ER/1203080/M/01/1569. The first author and second author would like to thank CSIR for their funding under grant no: 09/1095(0026)18-EMR-I, 09/1095(0033)18-EMR-I.

References

1. Lavalle, S.M.: Planning Algorithms. Cambridge University Press, Cambridge (2006)
2. Canny, J.F.: The Complexity of Robot Motion Planning. MIT Press, Cambridge (1998)
3. Foskey, M., Garber, M., Lin, M.C., Manocha, D.: A Voronoi-based hybrid motion planner. In: Proceedings of IEEE/RSJ International Conference on Intelligent Robots and Systems, vol. 1, pp. 55–60 (2001)
4. Lopez, A.S., Zapata, R., Lama, M.O.: Sampling-based motion planning: a survey. Comput. Sist. **12**(1), 5–24 (2008)
5. Tozour, P.: Building a near-optimal navigation mesh. In: AI Game Programming Wisdom, pp. 171–185. Charles River Media, America (2002)
6. Rabin, S.: A* speed optimizations. In: Game Programming GEMS, pp. 264–271. Charles River Media, America (2000)
7. Zeng, C., Zhang, Q., Wei, X.: GA-based global path planning for mobile robot employing A* algorithm. J. Comput. **72**, 470–474 (2012)
8. Sadrpour, A., Jin, J., Ulsoy, A.G.: Mission energy prediction for unmanned ground vehicles using real-time measurements and prior knowledge. J. Field Robot. **30**(3), 399–414 (2013)
9. Hoang, V.-D., Hernández, D.C., Hariyono, J., Jo, K.-H.: Global path planning for unmanned ground vehicle based on road map images. In: 2014 7th International Conference on Human System Interactions (HSI) (2014)

10. Opoku, D., Homaifar, A., Tunstel, E.: The A-r-Star (Ar*) pathfinder. Int. J. Comput. Appl. **67**(8), 0975–8887

11. Otte, M.W., Richardson, S.G., Mulligan, J., Grudic, G.: Local path planning in image space for autonomous robot navigation in unstructured environments. In: Proceedings of the 2007 IEEE/RSJ International Conference on Intelligent Robots and Systems San Diego, CA, USA (2007)

12. Angus, D.: Solving a unique shortest path problem using ant colony optimization. Commun. T. Baeck (January), 1–26 (2005). https://pdfs.semanticscholar.org/665d/535400bd21077c07 56d3a4e151a7d64ead07.pdf

13. Ebendt, R., Drechsler, R.: Weighted A* search-unifying view and application. Artif. Intell. **173**(14), 1310–1342 (2009)

14. Xue, Q., Chien, Y.-P.: Determining the path search graph and finding a collision-free path by the modified A* algorithm for a 5-link closed chain. Appl. Artif. Intell. **92**, 235–255 (1995)

15. Plagemann, C., Mischke, S., Prentice, S., Kersting, K., Roy, N., Burgard, W.: Learning predictive terrain models for legged robot locomotion. In: Proceedings of the IEEE/RSJ International Conference on Intelligent Robots and Systems (IROS), Nice, France (2008)

16. Shair, S., Chandler, J., Gonzalez-Villela, V., Parkin, R., Jackson, M.: The use of aerial images and GPS for mobile robot waypoint navigation. IEEE/ASME Trans. Mechatron. **13**(6), 692–699 (2008)

17. Brenner, C., Elias, B.: Extracting landmarks for car navigation systems using existing GIS databases and laser scanning. Int. Arch. Photogram. Remote Sens. Spatial Inf. Sci. **34**(3/W8), 131–138 (2003)

18. Sun, W., Messinger, D.W.: An automated approach for constructing road network graph from multispectral images. In: Proceeding of SPIE 8390, Algorithms and Technologies for Multispectral, Hyperspectral, and Ultraspectral Imagery, p. 83901W-1 (2012)

19. MATLAB version 9.3.0. Natick, Massachusetts: The MathWorks Inc., r2017b

Quantum Inspired High Dimensional Conceptual Space as KID Model for Elderly Assistance

M. S. Ishwarya and Ch. Aswani Kumar[✉]

School of Information Technology and Engineering,
Vellore Institute of Technology, Vellore, India
cherukuri@acm.org

Abstract. In this paper, we propose a cognitive system that acquires knowledge on elderly daily activities to ensure their wellness in a smart home using a Knowledge-Information-Data (KID) model. The novel cognitive framework called high dimensional conceptual space is proposed and used as KID model. This KID model is built using geometrical framework of conceptual spaces and formal concept analysis (FCA) to overcome imprecise concept notation of conceptual space with the help of topology based FCA. By doing so, conceptual space can be represented using Hilbert space. This high dimensional conceptual space is quantum inspired in terms of its concept representation. The knowledge learnt by the KID model recognizes the daily activities of the elderly. Consequently, the model identifies the scenario on which the wellness of the elderly has to be ensured.

Keywords: Cognition · Concepts · Conceptual spaces ·
Formal concept analysis · Quantum theory

1 Introduction

Health decline is an increasing problem in elderly people leading to the need for special attention. With the current advancements in robotics and wearable technology, feasible devices are developed for elderly care [1]. However, most of proposed solutions are expensive, awkward and troublesome and thereby creating a cognitive health decline [2]. One possible solution of monitoring elderly wellness without disturbing daily activities is Active Assisted Living using IOT [3]. In this regard, data from smart home is collected for analysis and action at critical time. Using this acquired data, activities of the elderly are analyzed via activity recognition techniques [4]. Inspired by human cognition process, literature reveals that human activities can be represented as concepts in cognitive system [3]. The models that transform information acquired from the real world data to knowledge are called as Knowledge-Information-Data (KID) model [5]. Any model that could perceives data, extract information, builds knowledge just like humans learning process can be regarded as KID models [6].

Various cognitive frameworks have been proposed to represent knowledge and treat uncertainty (inadequate data) [7]. Among those, our particular interest is on

© Springer Nature Switzerland AG 2020
A. Abraham et al. (Eds.): ISDA 2018, AISC 941, pp. 98–107, 2020.
https://doi.org/10.1007/978-3-030-16660-1_10

conceptual spaces and FCA as they both can model human activity recognition [3, 8–10]. However, notion of concept is imprecise in the conceptual spaces framework requiring topological definition while topology oriented FCA is limited to classical probability theory.

In this paper, we propose a cognitive system to assist elderly by adapting a novel KID model called high dimensional conceptual space that has been built using both geometrical framework of conceptual space and FCA. By this treatment, the modified conceptual space framework called high dimensional conceptual space is represented in Hilbert space and shows quantum inspiration [11–13]. The proposed cognitive system alerts the kin of the elderly to ensure their wellness when the adapted KID attains uncertainty that is not resolved with time. The rest of the paper is organized as follows. Section 2 of the paper discusses a short literature review and in Sect. 3 we have provided our detailed derivation of the proposed high dimensional conceptual space with its quantum inspiration. Section 4 discusses the proposed cognitive system to assist elderly and the successive section elaborates on experimental analysis.

2 Literature Review

Cognitive frameworks can well be regarded as KID model since they provide a pragmatic transformation of data to knowledge in the form of concepts [5]. Concepts are learnt based on the cognitive contextual data. Generally, a context possess information relating to 'what', 'where', 'when', 'with' and 'who' against 'why' [14]. Uncertainty in a cognitive system can be defined as absence or inadequate information that is required to perform reasoning. Conceptual spaces, FCA and relation algebra are cognitive frameworks which can used for modeling human activity recognition [10, 15, 16]. Wang et al. suggests human activity can be represented as relational concepts [16]. According to Gärdenfors, actions can be represented using geometrical framework of conceptual space by adding a force pattern [10]. However, he suggests that conceptual space framework has imprecise notation of concepts and requires a topology based definition. FCA is a mathematics based framework that acquires knowledge in the form of concepts from a context consisting objects described by their attribute [17–20]. FCA can be modeled using associative memories [21]. Literature reveals that associative memories are treated as a KID model and can recognize human activity [3]. Therefore, FCA can well be used for modeling activity recognition considering the ability of associative memories for modeling activity recognition. On the other hand, machine learning algorithms such as artificial neural network (ANN), support vector machines (SVM) and hidden markov model (HMM) can also be used for recognizing daily activities of elderly [22]. Cognitive vision system understands the visual information in the environment by integrating knowledge acquired from environment [23]. One potential application of cognitive vision system is human activity recognition. Adding to these, mining algorithms such as ProM tool and cognitive robots are also used for recognizing human activities [24]. In order to detect the health decline in elderly, it is necessary to detect the anomalies or abnormal behavior in their daily activities [25]. Uncertainty is an anomaly in a cognitive system. Based on these observations, in this paper, we propose a cognitive system by adapting a novel KID model for assisting elderly by recognizing their daily activities. Following this brief review on literature, we propose a novel KID model called high dimensional conceptual space in the next section of the paper.

3   High Dimensional Conceptual Space

Let R be the smart home scenario in which we wish to assist elderly and E be the set of all activities of their daily living. According to geometrical framework of conceptual spaces, an event $e_i \subset E$ in R is represented by set of quality dimensions D such that $D = \sum_{j=1}^{n} d_j$ [8]. Each $d_j \subset D$ portrays a group of attributes of the corresponding quality dimension required to accomplish event e_i [10]. A set of separable and inter-related quality dimensions forms a domain. The set of quality dimensions under different domains form a region in a conceptual space and this region is called a concept in geometrical framework of conceptual spaces. According to Gärdenfors, this notion of concept is imprecise with just quality dimensions and without topological definition [8]. To overcome this, we see through the notions of concepts in other cognitive frameworks for topology oriented definition [7]. Our particular interest is on FCA since it is topology oriented and it has connections with human cognitive processes [26]. According to FCA, a concept is pair consisting of objects (extent) and its associated set of attributes (intent). Three-way FCA is special kind of FCA, in which concept is a triple represented by set of objects, contributing set of attributes and non-contributing set of attributes [27]. Let H be the proposed high dimensional conceptual space in which we represent a concept's extent (event x) in terms of its intent consisting of contributing as well non-contributing attributes under different quality dimensions as shown Eq. 1.

$$x = \sum_{i=1}^{n} \sum_{j=1}^{m} d_i a_j I_j \qquad (1)$$

We regard that an attribute a_j as a contributing attribute based on the binary relation I_j, i.e. if I_j is 1 (0), then the corresponding attribute a_j is contributing (non-contributing). The proposed high dimensional conceptual adapts the redefined concept denotation along with the geometrical framework of conceptual spaces.

In order to obtain the quantum inspiration behind the proposed high dimensional conceptual space, let us compare it with a quantum system. Let us consider a N qubit quantum system S formed by the set of basis vectors namely $|\phi_1>, |\phi_2>, \ldots |\phi_n>$ whose corresponding probability amplitudes are $\alpha_1 \alpha_2 \ldots \alpha_n$. A quantum state $|\psi>$ in S is represented by linear superposition of its basis vectors as shown in Eq. 2.

$$|\psi> = \sum_{k=1}^{N} \alpha_k |\phi_k> \qquad (2)$$

In the above equation, α_k represents the probability amplitude of the projection $|\phi_k>$. Let us compare term by term of Eqs. 1 and 2 for quantum inspiration behind H. The term $a_j I_j$ is the projection corresponding to different attributes under a quality dimension d_i and it is column vector. This projection vector $a_j I_j$ in Eq. 1 is analogous to the basis vector $|\phi_k >$ in Eq. 2. Similarly, the probability amplitude (angle) α_i of projection determines its importance to obtain the quantum state $|\psi >$. The probability amplitude term α_i is analogous to the quality dimension d_i which portrays the qualities of the concept (event). Inspired by this analogy, we propose the quantum inspiration behind the high dimensional conceptual space H as follows: (1) A concept in a high dimensional conceptual space is equivalent to a state in a quantum system. (2) An attribute in a high dimensional conceptual space is equivalent to the basis state in quantum system. (3) Relation between the attributes of a quality dimension with an event is equivalent to projection of a basis vector. In the following, we propose a formal method to assist elderly by recognition their daily activities.

4 Proposed Work

The proposed model adapts the high dimensional conceptual space as the KID model to assist elderly in a smart home. It is assumed that the sensors are equipped wherever necessary and useful in a smart home and these sensor data are centrally collected appropriately from the smart home. A set of time-driven sensor data that describes all possible actions or actions of interest is pre-processed and its corresponding context is created. We regard the aforementioned set of time-driven sensor data as training sensor data. This training sensor data is fed to KID model. Each activity of the elderly is learnt in the form of concepts. We regard the high dimensional conceptual space as KID model since it processes information obtained from the incoming data to acquire knowledge. Each action is represented by its sensor (quality dimensions), sub-type sensor data (attributes), and its relation with the action. Consequently, a high dimensional conceptual space (memory) is created consisting of concepts as geometrical structures formed by vector of associated quality dimensions and its attributes. Upon learning the list the concepts based on the training sensor data, the high dimensional conceptual space can now assist elderly based on real time sensor data acquired from the smart home. As represented Fig. 1, the assistance model is cyclic evolutionary model. This nature of model as well as modeling element (action) introduces a time space upon initial learning.

Let A be the set of time driven activities that are learnt by KID model from the training sensor data such that $A = a_1, a_2, \ldots a_n$ where n is the number of activities learnt. Let E be the time taken by each activity such that $E = e_1, e_2, \ldots e_n$. Let T be reference time and it is set to maximum time taken by longest activity in training sensor data. With this initial setup, a real time sensor data is given as input to the conceptual space, i.e. pieces of description cue corresponding to a single action are received at different time granules ΔT, such that $\Delta T = 1, 2, \ldots T$. Upon receiving each piece of the description cue, the proposed model ensures the certainty of the KID model. We regard the model is certain if the model possess a matching geometric structure (associated vector of quality dimensions and attributes) similar to description cue in its conceptual space ΔC at time ΔT else the model is uncertain [28]. The conceptual space (memory) ΔC coupled with time space ΔT produces a Heisenberg's like uncertainty. For a given pieces of description cue, the proposed model can take three different ways to assist elderly.

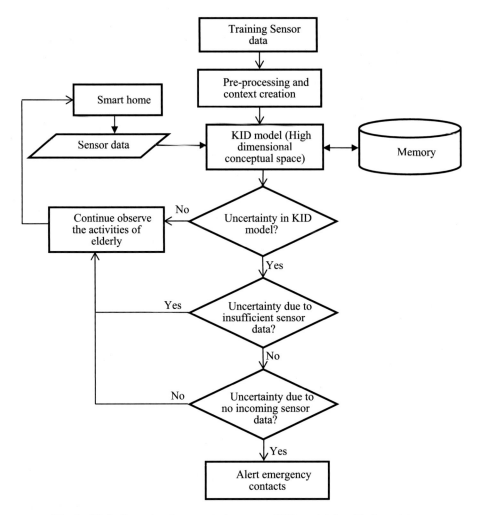

Fig. 1. High dimensional conceptual space as KID model for elderly assistance

1. No uncertainty: If $\Delta T < T$ and ΔC have a geometrical projection equivalent to the given description cue, then the model is certain at time ΔT. The recognized elderly activity is the activity corresponding to the aforementioned geometrical projection. This condition indicates that the elderly is currently pursuing an activity in wellness which is still need to be completed. In this case, the proposed model adjusts the reference time ΔT in accordance to the most similar description in ΔC until the time ΔT [29]. Based on the aforementioned literature, the proposed model regards the smallest superset corresponding to the given cue until time ΔT as the most similar description in ΔC. If there are no smallest supersets, the proposed model regards the largest subset corresponding to the given cue until time ΔT is regarded as the most similar description in ΔC. In this case, the proposed model continues to observe the sensor data until its completion or until receiving new sensor data.

2. Uncertainty due to insufficient sensor data: If $\Delta T < T$ and ΔC do not have geometrical projection equivalent to the given description cue, then the model is uncertain at time ΔT. The condition $\Delta T < T$ implies that all the pieces of description cue with regard to the action are not received. In this case, the proposed model adjusts its reference time in accordance to most similar pattern. The most similar pattern is chosen in the same way as in case 1. By adjusting the time reference, the proposed model is informed about the expected time of completion of the current activity considering received pieces of description. In this case, the received pieces of information are not sufficient to recognize the elderly and the proposed model continues to observe the next description cue with regard to current activity.

Table 1. Training sensor data of activities of daily living dataset

	PIR			Magnetic				Flush	Pressure		Electric		
	Shower	Basin	Cooktop	Main door	Fridge	Cabinet	Cupboard	Toilet	Seat	Bed	Microwave	Toaster	TV
Leaving	0	0	0	1	0	0	0	0	0	0	0	0	0
Toileting	1	0	0	0	0	0	0	1	0	0	0	0	0
Showering	1	0	0	0	0	0	0	1	0	0	0	0	0
Sleeping	0	0	0	0	0	0	0	0	0	1	0	0	0
Breakfast	0	0	0	0	1	0	0	0	1	0	0	1	0
Lunch	0	0	1	0	1	0	0	0	0	0	1	0	0
Dinner	0	0	1	1	1	0	0	0	0	0	1	0	0
Snacks	0	0	0	0	1	0	0	0	1	0	1	0	0
Spare_time	0	0	0	1	0	0	0	0	1	0	0	0	1
Grooming	0	0	0	0	0	0	1	0	0	1	0	0	0

3. Uncertainty due no incoming sensor data: If $\Delta T = T$, ΔC do not have geometrical projection equivalent to the given description cue, then the model is uncertain at time ΔT and no piece of description cues received till estimated time of completion adjusted time reference indicates that wellness of the elderly has to ensured. Under this case, the proposed model alerts the kin to ensure the wellness of the elderly. Based on these three conditions, the proposed model assists elderly by identifying the situations based on its knowledge on which the kin of the elderly has to be alerted.

5 Experimental Analysis

In this section of the paper, we have elaborated on the experimental analysis conducted on the proposed work using a sample data on human activity recognition. The dataset we have considered for this purpose is Activities of Daily Living dataset – A [22]. In this paper, we assume that the elderly we wish to assist lives in a smart home as in the dataset and the sensor data is centrally acquired by the proposed model via wireless link. The training sensor data lists the descriptions of 10 activities of the dataset in terms of 5 quality dimensions, 13 attributes and its relation as shown in Table 1. For this purpose, we have regarded the categories of sensors as quality dimensions, individual sensors under each category as attributes and their relation is represented with binary value [8]. The maximum of maximum time required for completion by the sample recordings corresponding to each activity in the dataset is the reference completion time for each activity as shown in Table 2.

The high dimensional conceptual space as a KID model learns the time driven activities in the terms of quality dimensions (sensors), attributes (types), and its relation (binary value) as geometrical structures. The proposed model stores the concepts obtained from KID model as geometrical structures in its memory as shown in Fig. 2. It can be observed from Fig. 2 that each concept is a geometrical structure consisting of projection formed by attributes under each quality dimension.

Table 2. Reference time of the proposed model

Sleeping	Leaving	Toileting	Showering	Breakfast
10:13:45 (T)	4:30:25	0:13:27	0:15:46	0:12:44
Lunch	Dinner	Snacks	Spare time	Grooming
0:35:56	Not available	0:04:51	6:03:10	0:13:41

Upon learning the activities as concepts, the proposed model can now assists the elderly by continuously monitoring their activities in reference to its memory as mentioned in the previous section. Subsequently, experiments are conducted to check the activity recognizing ability of the proposed model. Please note that 24-hours format of time is followed in this section. As mentioned in the previous section, a total of three different cases are possible during activity recognizing. In the following, we have illustrated an example for each of the case.

Case 1 - No uncertainty: Input from pressure sensor located in the bed is received at time 02:27 is received by the proposed model. The model detects the activity as sleeping. However, the reference time for sleeping activity is 10 h 13 min and 45 s. This implies that the elderly is just settled to sleep. The proposed model continues to observe the sensor input till the reference time.

Case 2 - Uncertainty due to insufficient sensor data: Input from magnetic sensor located in the fridge is received at time 10:34 is received by the proposed model. With this given information, the model is uncertain on the activity of the elderly. However, maximum reference time of activities involving fridge is 35 min and 56 s. The proposed model adjusts its reference time to 35 min and 56 s and continues to observe next sensor input. Similarly, the model receives successive inputs from magnetic sensor of the cupboard. With this current information, the model is still uncertain on the activity of the elderly. The proposed model continues observe the sensor data since it already has the maximum time reference of activities involving cupboard and fridge. Finally, an input from electric sensor of the toaster is received. With this information, the model is able to recognize the activity of the elderly is in wellness making his/her breakfast.

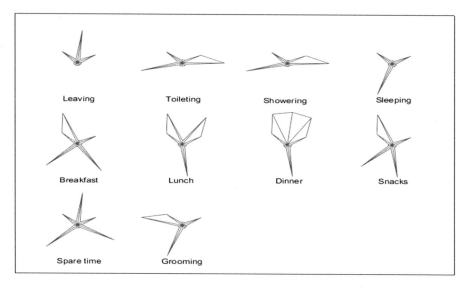

Fig. 2. Learning of activities by KID model as concepts

Case 3 - Uncertainty due to no sensor data: There are no explicit examples for this case in the dataset. Let us provide an imaginary example for this case. Input from the magnetic sensor located in the main door is received at 13:09 is received by the proposed model. With this information, the model is certain that the elderly is leaving the house. The maximum reference time is adjusted from 10 h 13 min and 45 s to 4 h 30 min and 25 s. This is the maximum time the elderly as spent out as per the training data. If the elderly is not home by this time reference time, the kin is notified to check on the wellness of the elderly.

6   Discussion

In this paper, we have used the proposed cognitive system for elderly assistance by monitoring their daily activities. However, this proposed cognitive system can be slightly modified and can be used for other applications where anomaly (uncertainty) detection can be of much importance such as surveillance, system health monitoring, fault detection and so on. As mentioned above, the concept representation of high dimensional conceptual space is analogous to state representation of quantum theories. Similarly, projections formed by attributes of quality dimensions are analogous to basis vectors of quantum system. This derivation leaves us on an insight that slight refinement on geometrical framework of conceptual space lands us in relevance between conceptual space based cognition and quantum processes. On a straight forward note, geometry is the key for cognition in human brain, conceptual space and quantum theories. In this paper, we have shared our insights on quantum static information representation. However, dynamic quantum aspect on high dimensional conceptual space is one potential future work we wish to endeavor.

7 Conclusions

In this paper, we have proposed and modeled a cognitive system that assist elderly for their active living. The novelty of the cognitive system is adapting the proposed high dimensional conceptual space as KID model to acquire knowledge from the smart home sensor data and to recognize the activities of the elderly. Also, this cognitive system is capable of identifying and resolving constrained uncertainties. We have demonstrated via experimental analysis that adapting the proposed cognitive system can identify the uncertain situations on which wellness of the elderly has to be ensured.

Acknowledgement. This research has received financial support from Department of Science and Technology, Government of India under the scheme Cognitive Science Research Initiative with grant number: SR/CSRI/118/2014.

References

1. Wang, Z., Yang, Z., Dong, T.: A review of wearable technologies for elderly care that can accurately track indoor position, recognize physical activities and monitor vital signs in real time. Sensors **17**(2), 341 (2017)
2. Wiederhold, M.D., Salva, A.M., Sotomayor, T., Coiro, C., Wiederhold, B.K.: Next Generation Stress Inoculation Training for Life Saving Skills Using Prosthetics, vol. 7, no. 1 (2009)
3. Huang, R., Mungai, P.K., Ma, J., Wang, K.I.-K.: Associative memory and recall model with KID model for human activity recognition. Futur. Gener. Comput. Syst. **92**, 312–323 (2018)
4. Ranasinghe, S., Al MacHot, F., Mayr, H.C.: A review on applications of activity recognition systems with regard to performance and evaluation. Int. J. Distrib. Sens. Netw. **12**(8), 1550147716665520 (2016)
5. Sato, A., Huang, R.: A generic formulated KID model for pragmatic processing of data, information, and knowledge. In: Proceedings - 2015 IEEE 12th International Conference on Ubiquitous Intelligence and Computing. 2015 IEEE 12th International Conference on Autonomic and Trusted Computing. 2015 IEEE 15th International Conference on Scalable Computing and Communications, vol. 20, pp. 609–616 (2016)
6. Sato, A., Huang, R.: From data to knowledge: a cognitive approach to retail business intelligence. In: 2015 IEEE International Conference on Data Science and Data Intensive Systems, pp. 210–217 (2015)
7. Goguen, J.: What is a concept? In: Conceptual Structures: Common Semantics for Sharing Knowledge. Lecture Notes in Computer Science, vol. 3596, no. April, pp. 52–77 (2005)
8. Gärdenfors, P.: Conceptual Spaces: The Geometry of Thought, vol. 106, no. 3 (2000)
9. Belohlavek, R.: Introduction to formal concept analysis (2008)
10. Gärdenfors, P.: Representing actions and functional properties in conceptual spaces. In: Body, Language and Mind, no. Gärdenfors 2000, pp. 167–195 (2007)
11. Yearsley, J.M., Busemeyer, J.R.: Quantum cognition and decision theories: a tutorial. J. Math. Psychol. **74**, 99–116 (2015)
12. Yearsley, J.M.: Advanced tools and concepts for quantum cognition: a tutorial. J. Math. Psychol. **78**, 24–39 (2017)
13. Bruza, P.D., Busemeyer, J.R.: Quantum Models of Cognition and Decision (2012)

14. Lunardi, G.M., Al Machot, F., Shekhovtsov, V.A., Maran, V., Machado, G.M., Machado, A., Mayr, H.C., de Oliveira, J.P.M.: IoT-based human action prediction and support. Internet Things **3-4**, 52–68 (2018)
15. Belohlavek, R.: Introduction to formal concept analysis. Palacky University, Department of Computer Science, Olomouc, p. 47 (2008)
16. Wang, Y.: On relation algebra: a denotational mathematical structure of relation theory for knowledge representation and cognitive computing. J. Adv. Math. Appl. **6**(1), 43–66 (2017)
17. Yang, K., Kim, E., Hwang, S., Choi, S.: Fuzzy concept mining based on formal concept analysis. Int. J. Comput. **2**(3), 279–290 (2008)
18. Ravi, K., Ravi, V., Prasad, P.S.R.K.: Fuzzy formal concept analysis based opinion mining for CRM in financial services. Appl. Soft Comput. **60**, 786–807 (2017)
19. Aswani Kumar, C., Srinivas, S.: Concept lattice reduction using fuzzy K-Means clustering. Expert Syst. Appl. **37**(3), 2696–2704 (2010)
20. Kumar, C.A.: Fuzzy clustering-based formal concept analysis for association rules mining. Appl. Artif. Intell. **26**(3), 274–301 (2012)
21. Belohlavek, R.: Representation of concept lattices by bidirectional associative memories, pp. 1–10 (1999)
22. Morales, F., de Toledo, P., Sanchis, A.: Activity recognition using hybrid generative/discriminative models on home environments using binary sensors. Sensors **13**(5), 5460–5477 (2013)
23. de Souza Alves, T., de Oliveira, C.S., Sanin, C., Szczerbicki, E.: From knowledge based vision systems to cognitive vision systems: a review. Procedia Comput. Sci. **126**, 1855–1864 (2018)
24. Lemaignan, S., Warnier, M., Sisbot, E.A., Clodic, A., Alami, R.: Artificial cognition for social human–robot interaction: an implementation. Artif. Intell. **247**, 45–69 (2017)
25. Meng, L., Miao, C., Leung, C.: Towards online and personalized daily activity recognition, habit modeling, and anomaly detection for the solitary elderly through unobtrusive sensing. Multimed. Tools Appl. **76**(8), 10779–10799 (2017)
26. Aswani Kumar, C., Ishwarya, M.S., Loo, C.K.: Formal concept analysis approach to cognitive functionalities of bidirectional associative memory. Biol. Inspired Cogn. Archit. **12**, 20–33 (2015)
27. Shivhare, R., Cherukuri, A.K.: Establishment of cognitive relations based on cognitive informatics. Cogn. Comput. **9**(5), 721–729 (2017)
28. Arecchi, F.T.: A quantum uncertainty entails entangled linguistic sequences, arXiv preprint arXiv:1807.03174, pp. 1–14 (2018)
29. Muthukrishnan, A.K.: Information Retrieval using Concept Lattices, Dissertation University of Cincinnati (2006)

Hybrid Evolutionary Algorithm for Optimizing Reliability of Complex Systems

Gutha Jaya Krishna[1,2] and Vadlamani Ravi[1(✉)]

[1] Center of Excellence in Analytics, Institute for Development
and Research in Banking Technology, Hyderabad 500057, India
krishna.gutha@gmail.com, rav_padma@yahoo.com
[2] School of Computer and Information Sciences,
University of Hyderabad, Hyderabad 500046, India

Abstract. In this paper, we propose a hybrid optimization algorithm of Harmony Search and Differential applied to three reliability complex system with static, extinctive constraint treatment. The proposed hybrid is contrasted with Harmony Search, Improved Modified Harmony Search, Differential Evolution, Modified Differential Evolution and other algorithms previous employed for Reliability of Complex Systems in the literature. We experimentally found that the proposed hybrid i.e. Improved Modified Harmony Search + Modified Differential Evolution needs less function evaluations as to the contrasted algorithms.

Keywords: Constraint handling · Improved Modified Harmony Search ·
Meta-heuristic · Modified Differential Evolution ·
Reliability of Complex Systems

1 Introduction

Optimization consists of minimizing or maximizing a real output objective function for real input decision variables within the specified bounds and may or may not include constraints. Optimization without constraints is termed unconstrained optimization and optimization with constraints is termed constrained optimization. Optimization has many sub areas which may include multiple objectives, i.e., multi-objective optimization, or which may yield multiple good solutions for the same objective function, i.e., multimodal optimization. Optimization has a wide range of applications in mechanics, economics, finance, electrical engineering, operational research, control engineering, geophysics, molecular modeling, etc. Optimization methods are classified into three sub-categories, namely (1) Classical optimization, (2) Heuristic-based optimization and (3) Metaheuristic-based optimization. Classical optimization techniques applied only to convex, continuous and differential search problems, but heuristic and metaheuristic-based optimization can also be applied to non-convex, discontinuous and non-differential search problems. Heuristic-based optimization has some inherent assumption which is specific to problem, but metaheuristic-based optimization has no problem-specific assumption [10, 14].

Metaheuristic-based optimization techniques are classified into Evolutionary Computing (EC), Swarm Intelligence-based optimization (SI), Stochastic-based

© Springer Nature Switzerland AG 2020
A. Abraham et al. (Eds.): ISDA 2018, AISC 941, pp. 108–119, 2020.
https://doi.org/10.1007/978-3-030-16660-1_11

optimization, Physics-based Optimization, Artificial Immune System (AIS)-based optimization, etc. Metaheuristic algorithms are further classified into the population, or point-based, optimization algorithms. Population based techniques include the Genetic Algorithm [13], Particle Swarm Optimization (PSO) [17], Differential Evolution (DE) [33], Harmony Search (HS) [9], Ant Colony Optimization (ACO) [5, 7], etc., which are all single objective optimization algorithms. Point-based optimization algorithms include Threshold Accepting (TA) [8], Tabu Search (TS) [11, 12], Simulated Annealing (SA) [19], etc. Bearing in mind the theory of no-free-lunch theorem [35], numerous novel as well as hybrid optimization procedures originated in the area of metaheuristic-based optimization.

Reliability of Complex Systems (RCS) holds a huge implication for reliability engineering. The reliability goal is to achieve a continuous system reliability and cost value from the available reliability parts. The RCS determines the optimal component reliabilities in each system, subject to certain cost or resource conditions, for achieving the overall system reliability [25, 28, 34]. The three RCS problems considered in this current work are with variables of continuous nature.

The article is organized in the following manner. Survey of the related literature is presented in Sect. 2. Details of the proposed optimization algorithm are specified in Sect. 3. In Sect. 4, the constrained handling approach employed in the current study is stated. In Sect. 5, description of the RCS problems is presented. Results are discussed in Sect. 6. Finally, we conclude in Sect. 7.

2   Literature Survey

Reliability of Complex Systems (RCS) holds a huge implication for reliability engineering RCS are devised as difficult non-convex problems [34]. Numerous optimization procedures are employed for addressing the RCS. Improved non-equilibrium simulated annealing (INESA) [28] was devised by Ravi et al. to address on some of the RCS problems. In [27], Ravi et al. also employed TA [8] to solve the fuzzy global optimization problems. Shelokar et al. in [31] made a major contribution by finding better optima of the problems in [28] with ACO algorithm. Later in [29], RCS was solved with even better results than in [31] by Modified Great Deluge Algorithm (MGDA). Later, Jaya Krishna and Ravi [20], solved the reliability redundancy allocation problems employing Improved Modified Harmony Search (IMHS) with better function evaluations.

There is also another class of meta-heuristics that combine the power of more than one meta-heuristic. These are called Memetic Algorithms (MA). MAs have evolved over three generations. First generation MAs are the hybrid optimization algorithms. These use the power of one class of metaheuristic to carry out the exploration and the power of another class of metaheuristic to do a local search. In a survey, presented by Chen et al. in [3] multi-facets of memetic algorithms are discussed. In [26], the past, present, and future of memetic algorithms is discussed.

Several hybrid optimization algorithms are developed employing either HS or DE like DE + TS [32], DE + TA [2], HS and PSO (NHPSO) [22], HS, PS and ACO

(HPSACO) [16], HS and MGDA (MHS + MGDA) [29], HS and TA (MHSTA) [24]. IMHS + MDE hybrid is proposed for key generation by Jaya Krishna et al. [15].

In Bhat et al. [1], improved DE is proposed in combination with the reflection property of the simplex method for the efficient parameter estimation of biofilter modeling. In Chauhan and Ravi [2], a hybrid, based on DE and TA named DETA, is developed for the unconstrained problems and is compared with DE. The results regarding function evaluations show that DETA outperforms DE. Choudhuri and Ravi [4], proposed a hybrid combining Modified Harmony Search (MHS) and the Modified Great Deluge Algorithm (MGDA) for the unconstrained problems and compared them with MHS. The results prove that MHS + MGDA perform better than MHS. In Maheshkumar and Ravi [24], a modified harmony search and threshold accepting hybrid was proposed and compared with HS and Modified HS on the unconstrained problems. In Maheshkumar et al. [23], a hybrid combining both PSO and TA, called PSOTA was developed for unconstrained optimization problems, which gave better optimization results than PSO. Finally, a hybrid optimization algorithm combining ACO, and the classical optimization algorithm called Nelder-Mead simplex was developed to train a neural network for bankruptcy prediction [30].

3 Improved Modified Harmony Search + Modified Differential Evolution

3.1 Improved Modified Harmony Search

Harmony Search (HS) is a process mimicking metaheuristic proposed by Geem et al. [9]. It mimics the musician's improvisation process of playing an old piece of music or generating new music from the piece of old music or generating a new music from scratch. Four parameters used in Harmony Search are Pitch Adjusting Rate (PAR), Harmony Memory Considering Rate (HMCR), Bandwidth (BW) and Harmony Memory Size (HMS).

Harmony update is an important part of HS, which generates new solution vectors. Below are the steps that describe Harmony Update:

1. Pick a solution from harmony memory if the random number generated between 0 and 1, is less than the HMCR.
   (i) Search in the local neighbourhood of the picked solution vector if the random number, generated between 0 and 1, is less than PAR.
   (ii) Else if random number is greater than the PAR, use the same picked solution vector.
2. Else if the random number is greater than the HMCR, generate random values in between the bounds of the input decision variable.

The general steps for Harmony Search are described below:

1. Firstly, fix the size of harmony memory (HM).
2. Initialize the HM number of solution vectors within bounds randomly and compute corresponding fitness values for the solution vectors.

3. HMCR is linearly increased between 0.7 and 0.99, and the PAR fixed at a constant is between 0 and 1.0.
4. The harmony update is performed i.e. using old solution vectors or generating a new solution vector from the old solution vector or generating a totally new solution vector utilizing the concerned HMCR and PAR values.
5. If the fitness value of the harmony updated solution is better than the worst solution in the harmony memory, then replace the worst solution vector and fitness value with the new solution vector and corresponding new fitness value from the harmony update.
6. Repeat steps 3, 4, and 5 until the maximum number of iterations.

3.2   Modified Differential Evolution

Differential Evolution (DE) is a population-based stochastic evolutionary metaheuristic [33]. DE is developed for real input parameter-based optimization problems. It includes four important sections, which are mainly initialization, mutation, recombination and selection. New solutions are generated from current set of solutions using 'mutation' and 'crossover'. Then 'selection' is performed to select the fittest. DE has two parameters which are CR and F. The following steps briefly describe the working of Modified DE (MDE).

Step-1: Initialization

    1. Initialize NP, the number solution vectors of dimensionality size, Dim.
    2. Check whether the solution vectors are within bounds.
    3. Compute the fitness values of the NP number of solution vectors.

Step-2: Mutation

    4. Let x_1, x_2, x_3 and $x^{current}$ be the three unique solution vectors chosen randomly from the population NP.
    5. Perform the mutation using the following equation:

$$u'_i = x_{1,i} + F * \left(x_{2,i} - x_{3,i} \right)$$

The above equations are of a variant of DE i.e. DE/rand/1/bin. Differential weight 'F' is between 0 and 2.

Step-3: Recombination

    6. Using the mutation vector and crossover rate parameter CR recombination is performed as follows:

$$v_i = \begin{cases} u_i & if\,(rand(0,1) \le CR | i = I_{rand}) \\ x_i^{current} & if\,(rand(0,1) > CR\,and\,i \ne I_{rand}) \end{cases}$$

CR is between 0 and 1. I_{rand} is a random dimension between 1 to Dim and I_{rand} is used so that v and $x^{current}$ are different.

Step-4: Selection

    7. If Fitness(v) $\leq$ Fitness(x^{worst}) replace current vector with solution vector v in the population.

    8. Else do not replace the current vector.

    9. Repeat steps 4–8, until a maximum number of iterations.

3.3   Improved Modified Harmony Search + Modified Differential Evolution

Das et al. in [6], show that the explorative power of HS is initially is very good, but that it falls, as the number of iterations increase, and it gets stuck at the local optimal basin, without progressing further. To overcome this, HS is to be perturbated with a good and powerful meta-heuristic. In [18], various performance measures such as dimensionality, the number of local optima, interval span of side constraints, the ratio of local optima, valley structure coefficient, and peak density ratio were compared to a set of good meta-heuristics. These compared meta-heuristics are as follows: Random Search (RS), Simulated Annealing (SA), Particle Swarm Optimization (PSO), Water Cycle Algorithm (WCA), Genetic Algorithms (GAs), Differential Evolution (DE), Harmony Search (HS), and Cuckoo Search (CS). The compared DE gave better results than all those with a good coverage, and HS gave comparatively better results than some with second-best coverage when a radar plot was created with various performance measures. This gave us the motivation to combine HS with DE to obtain a better optimization algorithm as DE gave better results for all performance measures.

We combined IMHS, which was developed for complex reliability systems in [20], with differential evolution. We also developed a modified DE with an alternate selection strategy in [21]. This MDE does the exploitation well as the selection of DE is replaced with that of HS where a better solution replaces the worst solution instead of the current solution vector. Selection operation usually suggests the exploitation power of any evolutionary algorithm as suggested in Das et al. in [6]. So, we replaced the DE selection strategy with that of HS, as it replaces the worst solution vector, which is better than DE regarding exploitation.

The proposed IMHS + MDE hybrid optimization algorithm runs in two critical stages cyclically. In the first stage, IMHS is run for a fixed number of generations. The resulting population from IMHS is then passed to MDE in the second stage, where MDE is run for a fixed number of generations. The population from MDE is then passed on to IMHS. Both these stages are repeated (cycled) till the maximum number of iterations is reached or optimum is reached (see Fig. 1).

4   Constraint Handling

For the optimization problems with constraints, the search space is divided into a feasible region if constraints are satisfied and an infeasible region if constraints are not satisfied. To handle constraints the following approaches are employed: 1. Repair operators, 2. Penalty functions, 3. Hybrid methods, 4. Special operators, etc.

Penalties are commonly employed techniques for constrained problems in evolutionary computing techniques. The concept of employing penalties is to regulate constraints i.e. we impose penalties for the constraints that are violating the optimization problem. The common approaches utilized for employing penalties are the following: 1. Static Penalty 2. Adaptive Penalty 3. Dynamic Penalty 4. Death Penalty.

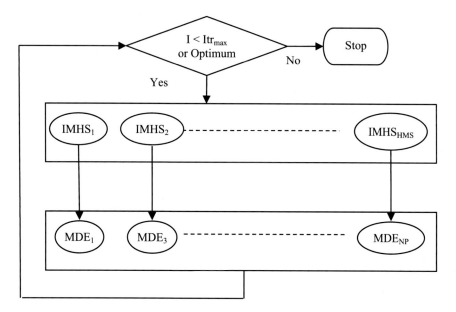

Fig. 1. The process of coupling between IMHS memory and MDE population vectors

4.1 Static Extinctive Penalty

The penalty employed is independent of the present generation number and remain constant for entire process. Generally, the penalty factor is considered as problem dependent. Static penalties are employed in three different methods (i) extinctive penalties, (ii) binary penalties, and (iii) distance-based penalties. We generally employ high extensive penalties for the infeasible solutions [14].

To address the issue of constraint handling, we utilized in this problem static extinctive penalties. The proposal to deal with constraints is defined as follows, if the solution is infeasible, a penalty with large enough static value is employed. For maximizing of the objective functions, a lower penalty value is employed. Similarly, for minimizing the objective function, higher penalty values are employed. In this problem, a static constant, either too small or too large is utilized as a penalty value for maximization and minimization problems respectively. This process results in removal of the infeasible solutions in further evaluations [14].

5 Problem Definitions

Problem-1: Life support system in a space capsule [34] is shown in Fig. 2.

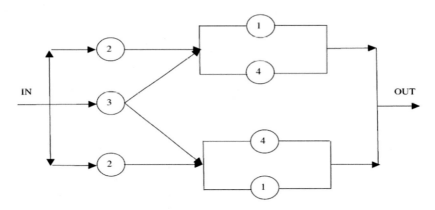

Fig. 2. Complex system of life support system in a space capsule

The component and system reliabilities are, R_i, i = 1, 2, ..., 4 and R_s, respectively.

$$R_s = 1 - R_3\left[(1 - R_1)(1 - R_4)\right]^2 - (1 - R_3)\left[1 - R_2\{1 - (1 - R_1)(1 - R_4)\}\right]^2$$

Objective function is C_s i.e. system cost which is to be minimized. The constraints are the component and system reliabilities which are R_i i.e. between $0.5 \leq R_i \leq 1.0$, $i = 1, 2, 3, 4$ and R_s i.e. between $R_{s,min} \leq Rs \leq 1.0$ respectively. There are two variants of system cost for problem-1, which are given below:

Case (i):

$$C_s = 2M_1R_1^{\delta_1} + 2M_2R_2^{\delta_2} + M_3R_3^{\delta_3} + 2M_4R_4^{\delta_4}$$

where, $M_1 = 100$, $M_2 = 100$, $M_3 = 200$, $M_4 = 150$, $R_{s,min} = 0.9$ and $\delta_i = 0.6$ for $i = 1, 2, 3, 4$.
Case (ii):

$$C_s = \sum_{i=1}^{4} M_i\left[\tan\left(\frac{\pi}{2}R_i\right)\right]^{\delta_i}$$

where, $M_1 = 25$, $M_2 = 25$, $M_3 = 50$, $M_4 = 37.5$, $R_{s,min} = 0.99$ and $\delta_i = 1.0$ for $i = 1, 2, 3, 4$.

Problem-2: Complex Bridge System [25] is shown in Fig. 3.

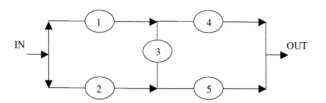

Fig. 3. Reliability complex bridge system

The component and system reliabilities are, R_i, i = 1, 2, ..., 5 and R_s, respectively.

$$R_s = R_1R_4 + R_2R_5 + R_2R_3R_4 + R_1R_3R_5 + 2R_1R_2R_3R_4R_5 - R_1R_2R_4R_5$$
$$-R_1R_2R_3R_4 - R_1R_3R_4R_5 - R_2R_3R_4R_5 - R_1R_2R_3R_5$$

Minimize: $C_s = \sum_{i=1}^{5} l_i \left[\exp\left(\frac{m_i}{1-R_i}\right) \right]$

Objective function is C_s i.e. system cost which is to be minimized. The constraints are the component and system reliabilities which are R_i i.e. between $0 \leq R_i \leq 1.0$, i = 1, 2, 3, 4, 5 and R_s i.e. between $0.99 \leq Rs \leq 1.0$ respectively. Here l_i = 1 and m_i = 0.0003 for i = 1, 2, 3, 4, 5.

6 Results and Discussion

The results of the problem-1 (i.e. case (i) and case (ii)) and problem-2 are summarized in Tables 1, 2 and 3. IMHS + MDE achieved optimums in problem-1: case (i) and problem-2 with a better number of function evaluations (FE). IMHS + MDE achieved 2.6%, 27.6%, 99.4%, 99.4%, 92.3%, 75.1% improvement for the problem-1: case (i) in terms of FE, compared to MDE, DE, IMHS, HS, MGDA, ACO respectively. IMHS + MDE achieved 24.4%, 22.9%, 99.7%, 99.7%, 98.8%, 96.0% improvement for the problem-1: case (ii) in terms of FE, compared to MDE, DE, IMHS, HS, MGDA, ACO respectively. IMHS + MDE achieved 57.0%, 89.5%, 97.5%, 91.1%, 93.7%, 96.0% improvement for the problem-2 in terms of FE, compared to MDE, DE, IMHS, HS, MGDA, ACO respectively.

The statistics in terms of median function evaluations (FE), success rate and standard deviation of the 25 runs results of the converged optima are presented in Table 4. Even though comparing the standard deviation and success rate of IMHS + MDE with DE, MDE, IMHS, HS doesn't convey much except in one case of MDE and in two cases of IMHS, HS. Therefore, FEs is the apt metric in this case and IMHS + MDE performed exceedingly better than all the compared.

Table 1. Summary of the results: problem-1 case (i)

Algorithm	R_1^b	R_2^b	R_3^b	R_4^b	R_s^b	C_s^b	FE^a
IMHS + MDE	0.5	0.838920	0.5	0.5	0.9	641.823562	5,003
MDE	0.5	0.838920	0.5	0.5	0.9	641.823562	5,133
DE	0.5	0.838920	0.5	0.5	0.9	641.823562	6,911
IMHS [20]	0.506316	0.832600	0.5	0.5	0.9	642.005607	1000000
HS [9]	0.505804	0.832803	0.5	0.500311	0.9	642.025896	1000000
MGDA [29]	0.50001	0.838919	0.5	0.5	0.9	641.823608	65,603
ACO [31]	0.5	0.838920	0.5	0.5	0.9	641.823562	20,100
INESA [28]	0.50006	0.83887	0.50001	0.50002	0.90001	641.8332	NA
SA [28]	0.50095	0.83775	0.50025	0.50015	0.90001	641.903	NA
[34]	0.50001	0.84062	0.5	0.5	0.90005	642.04	NA

^{a}Function evaluations (FE), ^{b} Median value of 25 runs

Table 2. Summary of the results: problem-1 case (ii)

Algorithm	R_1^b	R_2^b	R_3^b	R_4^b	R_s^b	C_s^b	FE^a
IMHS + MDE	0.825570	0.890103	0.628105	0.728733	0.99	390.570889	2,121
MDE	0.825436	0.890103	0.627929	0.729073	0.99	390.570881	2,808
DE	0.825639	0.890207	0.627411	0.728745	0.99	390.570867	2,752
IMHS [20]	0.825437	0.890179	0.627901	0.728797	0.99	390.570880	887,718
HS [9]	0.852543	0.890114	0.627904	0.729057	0.99	390.570876	720,236
MGDA [29]	0.825808	0.890148	0.627478	0.728662	0.99	390.570190	188,777
ACO [31]	0.825895	0.890089	0.627426	0.728794	0.99	390.570893	54,140
INESA [28]	0.82516	0.89013	0.62825	0.72917	0.99	390.572	NA
SA [28]	0.82529	0.89169	0.62161	0.72791	0.990003	390.6327	NA
[34]	0.825895	0.890089	0.627426	0.728794	0.99041	397.88	NA

^{a}Function evaluations (FE), ^{b} Median value of 25 runs

Table 3. Summary of the results: problem-2

Algorithm	R_1^b	R_2^b	R_3^b	R_4^b	R_5^b	R_s^b	C_s^b	FE^a
IMHS + MDE	0.934708	0.935126	0.792849	0.934505	0.935168	0.99	5.019918	3,192
MDE	0.934283	0.935384	0.791964	0.935249	0.934672	0.99	5.019918	7,432
DE	0.934948	0.935064	0.791613	0.935035	0.934582	0.99	5.019918	30,593
IMHS [20]	0.934791	0.934732	0.796893	0.934792	0.934792	0.99	5.019918	132,104
HS [9]	0.935206	0.934877	0.788511	0.935146	0.934699	0.99	5.019918	36,196
MGDA [29]	0.935400	0.935403	0.788027	0.93506	0.934111	0.99	5.019919	50,942
ACO [31]	0.933869	0.935073	0.798365	0.935804	0.934223	0.990001	5.019923	80,160
INESA [28]	0.93747	0.93291	0.78485	0.93641	0.93342	0.99	5.01993	NA
SA [28]	0.93566	0.93674	0.79299	0.93873	0.92816	0.99001	5.01997	NA
[25]	0.93924	0.93454	0.77154	0.93938	0.92844	0.99004	5.02001	NA

^{a}Function evaluations (FE), ^{b} Median value of 25 runs

Table 4. Statistics of the employed optimizations algorithms

Method	RCS problem	Median NFEs[a]	Success Rate[c]	SD[b]	Max. NFEs
HS [9]	Problem-1: case (i)	1,000,000	0.12	0.2660	1000000
	Problem-1: case (ii)	720,236	0.84	7.01e−05	1000000
	Problem-2	36,196	1.0	1.38e−07	1000000
IMHS [20]	Problem-1: case (i)	1,000,000	0.04	2.2250	1000000
	Problem-1: case (ii)	887,718	0.76	8.05e−05	1000000
	Problem-2	132,104	1.0	1.13e−07	1000000
DE	Problem-1: case (i)	6,911	1.0	7.93e−08	1000000
	Problem-1: case (ii)	2,752	1.0	3.08e−05	1000000
	Problem-2	30,593	1.0	1.29e−07	1000000
MDE	Problem-1: case (i)	5,133	1.0	8.18e−08	1000000
	Problem-1: case (ii)	2,808	1.0	3.35e−05	1000000
	Problem-2	7,432	0.52	0.3577	1000000
IMHS + MDE	Problem-1: case (i)	5,003	1.0	8.70e−08	1000000
	Problem-1: case (ii)	2,121	1.0	2.32e−05	1000000
	Problem-2	3,192	1.0	1.42e−07	1000000

[a]Median value of the function evaluations of 25 runs
[b]Standard deviation of the converged optima for 25 runs
[c]Percentage of the number times optimum is converged for 25 runs

7 Conclusion

In this study, three continuous RCS problems with four and five component reliabilities for the RCSs' are considered. Optimization with IMHS + MDE which was tested on these RCS problems achieved improved optima along with better function evaluations. We conclude that the proposed hybrid meta-heuristic proved its effectiveness on RCS problems. The experimental results consistently show that our proposed hybrid meta-heuristic for RCS, performs statistically on par with some algorithms in some problem cases, while it turned out to be the best in some problem cases. In future, better hybrids and adaptive versions of optimization algorithms can be employed to address the three considered RCS problems. Apart from the three employed RCS optimization problems a different set of RCS problems can be utilized by the IMHS + MDE.

References

1. Bhat, T.R., Venkataramani, D., Ravi, V., Murty, C.V.S.: An improved differential evolution method for efficient parameter estimation in biofilter modeling. Biochem. Eng. J. **28**(2), 167–176 (2006)
2. Chauhan, N., Ravi, V.: Differential evolution and threshold accepting hybrid algorithm for unconstrained optimisation. Int. J. Bio-Inspired Comput. **2**(3/4), 169 (2010)
3. Chen, X., Ong, Y.S., Lim, M.H., Tan, K.C.: A multi-facet survey on memetic computation. IEEE Trans. Evol. Comput. **15**(5), 591–607 (2011)

4. Choudhuri, R., Ravi, V.: A hybrid harmony search and modified great deluge algorithm for unconstrained optimisation. Int. J. Comput. Intell. Res. **6**(4), 755–761 (2010)
5. Colorni, A., Dorigo, M., Maniezzo, V.: Distributed optimization by ant colonies. In: Proceedings of the European Conference on Artificial Life, pp. 134–142 (1991)
6. Das, S., Mukhopadhyay, A., Roy, A., Abraham, A., Panigrahi, B.K.: Exploratory power of the harmony search algorithm: analysis and improvements for global numerical optimization. IEEE Trans. Syst. Man Cybern. Part B **41**(1), 89–106 (2011)
7. Dorigo, M., Maniezzo, V., Colorni, A.: Ant system: optimization by a colony of cooperating agents. IEEE Trans. Syst. Man Cybern. Part B **26**(1), 29–41 (1996)
8. Dueck, G., Scheurer, T.: Threshold accepting: a general purpose optimization algorithm. J. Comput. Phys. **90**, 161–175 (1990)
9. Geem, Z.W., Kim, J.H., Loganathan, G.V.: A new heuristic optimization algorithm: harmony search. Simulation **76**(2), 60–68 (2001)
10. Gendreau, M., Potvin, J.Y.: Handbook of Metaheuristics. Springer, Heidelberg (2010)
11. Glover, F.: Tabu search - part II. ORSA J. Comput. **2**(1), 4–32 (1989)
12. Glover, F.: Tabu search - part I. ORSA J. Comput. **1**(3), 190–206 (1989)
13. Goldberg, D.E.: Genetic Algorithms in Search, Optimization, and Machine Learning. Addison-Wesley Longman Publishing Co., Boston (1989)
14. Horst, R., Pardalos, P.M.: Handbook of Global Optimization. Kluwer Academic Publishers (1995)
15. Jaya Krishna, G., Vadlamani, R., Nagesh, B.S.: Key generation for plain text in stream cipher via bi-objective evolutionary computing. Appl. Soft Comput. **70**, 17 (2018)
16. Kaveh, A., Talatahari, S.: Particle swarm optimizer, ant colony strategy and harmony search scheme hybridized for optimization of truss structures. Comput. Struct. **87**(5–6), 267–283 (2009)
17. Kennedy, J., Eberhart, R.: Particle swarm optimization. In: International Conference on Neural Networks (ICNN 1995), Piscataway, NJ, pp. 1942–1948. IEEE (1995)
18. Kim, J.H., Lee, H.M., Jung, D., Sadollah, A.: Performance measures of metaheuristic algorithms (2016)
19. Kirkpatrick, S., Jtr, C.G., Vecchi, M.: Optimization by simulated annealing (1994)
20. Jaya Krishna, G., Ravi, V.: Modified harmony search applied to reliability optimization of complex systems. In: Kim, J., Geem, Z. (eds.) Advances in Intelligent Systems and Computing, pp. 169–180. Springer, Berlin, Heidelberg (2015)
21. Jaya Krishna, G., Ravi, V.: Outlier detection using evolutionary computing. In: Proceedings of the International Conference on Informatics and Analytics – ICIA 2016, pp. 1–6. ACM Press, New York (2016)
22. Li, H., Li, L.: A novel hybrid particle swarm optimization algorithm combined with harmony search for high dimensional optimization problems. In: The 2007 International Conference on Intelligent Pervasive Computing (IPC 2007), Jeju City, South Korea, pp. 94–97. IEEE (2007)
23. Maheshkumar, Y., Ravi, V., Abraham, A.: A particle swarm optimization-threshold accepting hybrid algorithm for unconstrained optimization. Neural Netw. World **23**(3), 191–221 (2013)
24. Maheshkumar, Y., Ravi, V.: A modified harmony search threshold accepting hybrid optimization algorithm. In: Sombattheera, C., et al. (eds.) Multi-disciplinary Trends in Artificial Intelligence (MIWAI), pp. 298–308. Springer, Hyderabad (2011)
25. Mohan, C., Shanker, K.: Reliability optimization of complex systems using random search technique. Microelectron. Reliab. **28**(4), 513–518 (1988)
26. Ong, Y.S., Lim, M., Chen, X.: Memetic computation—past, present and future research frontier. IEEE Comput. Intell. Mag. **5**(2), 24–31 (2010)

27. Ravi, V., Reddy, P.J., Zimmermann, H.J.: Fuzzy global optimization of complex system reliability. IEEE Trans. Fuzzy Syst. **8**(3), 241–248 (2000)
28. Ravi, V., Murty, B.S.N., Reddy, J.: Nonequilibrium simulated-annealing algorithm applied to reliability optimization of complex systems. IEEE Trans. Reliab. **46**(2), 233–239 (1997)
29. Ravi, V.: Optimization of complex system reliability by a modified great deluge algorithm. Asia-Pacific J. Oper. Res. **21**(04), 487–497 (2004)
30. Sharma, N., Arun, N., Ravi, V.: An ant colony optimisation and Nelder-Mead simplex hybrid algorithm for training neural networks: an application to bankruptcy prediction in banks. Int. J. Inf. Decis. Sci. **5**(2), 188 (2013)
31. Shelokar, P.S., Jayaraman, V.K., Kulkarni, B.D.: Ant algorithm for single and multiobjective reliability optimization problems. Qual. Reliab. Eng. Int. **18**(6), 497–514 (2002)
32. Srinivas, M., Rangaiah, G.P.: Differential evolution with tabu list for global optimization and its application to phase equilibrium and parameter estimation problems. Ind. Eng. Chem. Res. **46**(10), 3410–3421 (2007)
33. Storn, R., Price, K.: Differential evolution-a simple and efficient heuristic for global optimization over continuous spaces. J. Glob. Optim. **11**(4), 341–359 (1997)
34. Tillman, F.A., Hwang, C.L., Kuo, W.: Optimization of Systems Reliability. Marcel Dekker, New York (1980)
35. Wolpert, D.H., Macready, W.G.: No free lunch theorems for optimization. IEEE Trans. Evol. Comput. **1**(1), 67–82 (1997)

Identification of Phishing Attack in Websites Using Random Forest-SVM Hybrid Model

Amritanshu Pandey, Noor Gill, Kashyap Sai Prasad Nadendla,
and I. Sumaiya Thaseen$^{(\boxtimes)}$

School of Information Technology and Engineering,
Vellore Institute of Technology, Vellore, India
sumaiyathaseen@gmail.com

Abstract. Phishing attacks have increased in the last few years with the rapid growth of economy and technology. Attackers with less technical knowledge can also perform phishing with sources that are available in public. Financial losses are experienced by businesses and customers thus decreasing confidence in e-commerce. Hence there is a necessity to implement countermeasures to overcome phishing attacks in the website. In this paper, a hybrid model is proposed integrating Random Forest and Support Vector Machine (SVM) techniques. Machine learning models are efficient in prediction and analyze large volumes of data. Experimental results on the phishing datasets how that an accuracy of 94% is obtained by the hybrid model in comparison to the base classifier SVM accuracy of 90% and Random Forest accuracy of 92. 96%. Thus, the model is superior in classifying the phishing attacks.

Keywords: Accuracy · Feature Selection · Phishing · Random Forest · Support Vector Machine

1 Introduction

The process of cloning a website with unethical intentions to utilize the website result by the malicious, unethical entity is known as phishing. This form of an attack is different from the conventional methods due to the user being attacked and not the system. Websites that support phishing often target unsuspecting victims by imitating or by creating websites that are similar to the original versions of the website. Thus a user who assumes that such a website is a secure one will easily reveal important information such as their bank details, passwords of various accounts and other information which in normal circumstances should be kept private and confidential. Many organizations such as banks and other financial companies are maximizing their effort to minimize such incidents by creating awareness among their users about phishing and how to avoid it. However, there are many attackers still out with users as their prey. The number of websites that extract sensitive information is increasing day by day. This is also a highly intricate issue and can take a lot of time to grasp and rectify the problems that are caused by these websites [11]. Thus, in this paper, a novel hybrid technique is proposed to identify the phishing attacks on the website.

© Springer Nature Switzerland AG 2020
A. Abraham et al. (Eds.): ISDA 2018, AISC 941, pp. 120–128, 2020.
https://doi.org/10.1007/978-3-030-16660-1_12

2 Related Work

Phishing attacks have been analyzed in the literature by machine learning techniques. Some of the techniques are discussed in brief. Moghimi et al. [10] presented a new rule-based method to detect phishing attacks in internet banking. The proposed system was based on two feature sets. The feature sets contained 4 features to evaluate page resources identity and another 4 features to identify the access protocol for that page resource element. The SVM classifier was used to perform string matching and identify which phishing pages are legitimate and which page is the authentic. Abu-Nimeh et al. [2] has compared multiple machine learning algorithms to determine which algorithms are efficient in detecting phishing emails. The experiments were conducted on a dataset of 2889 emails. They used 43 features to train and test the classifiers. They used Logistic Regression, Classification and Regression Trees, Bayesian Additive Regression Trees, Support Vector Machines, Random Forests, and Neural Networks for predicting phishing emails. Abdelhamid et al. [1] implemented prevention of phishing in websites using associative classification called Multi-Label Classifier based Associative Classification (MCAC). The authors identified features that differentiate phishing websites from legitimate ones and analyzed intelligent approaches used to handle the phishing problem. Aburrous et al. [3] studied 3 kinds of phishing experiment case studies. The authors created awareness about phishing among users and developed phishing prevention techniques. The paper concluded that normal standard phishing factors are not always a good measure to detect phishing websites and therefore need alternate intelligent phishing detection approaches. Alkhozae and Batarfi [4] proposed a phishing detection approach based on the webpage source code. If there is a phishing character, the initial secure weight is decreased. The security percentage is calculated based on the final weight thus a high percentage indicates secure website and others indicates the website is most likely to be a phishing website. Barraclough et al. [5] performed phishing detection and protection against phishing attacks for online transactions. A Neuro-fuzzy scheme with 5 inputs was adopted to detect phishing websites. A total of 288 features with 5 inputs was analyzed. The main objective was to detect phishing websites in real time. Barakat and Bradley [6] conducted a review of the area of rule extraction by SVM. The aim was to provide an understanding and to summarize the main features of individual algorithms. Thus, the analysis is a comparative evaluation of the algorithm features and their relative performance. Rami et al. [7] developed a rule-based classification model for predicting and detecting phishing websites. Huang et al. [8] proposed a method to detect phishing URL based on Support Vector Machine Classifier. The feature vector for this model was constructed with 23 features to model the SVM in which 4 features are the structured feature of the phishing URL, 9 features are lexical feature and 10 features are mostly target phished brand name of website. Lakshmi and Vijaya [9] used machine-learning technique for modelling the prediction and used supervised learning algorithms namely Multi-Layer Perceptron. They also used Naive Bayes and Decision Trees for exploring the results. From the results of this paper, it was concluded the decision tree predicts the phishing websites more accurately compared to other learning algorithms. Thus, from the literature it is evident that phishing is a serious concern and machine learning approaches have been utilized for detecting and preventing phishing in websites. However, hybrid models have not been analyzed in the literature for improving the accuracy of phishing detection. This paper proposes a novel hybrid approach integrating Random Forest and SVM for detecting and analyzing phishing in websites.

3 Background

3.1 Random Forest (RF)

This is a supervised learning algorithm that creates a forest which is built randomly. The idea is that the learning model performs bagging to predict the result better. Random Forest builds many decision trees to produce a superior result in comparison to few decision trees. To understand the Random Forest better, we must first understand decision trees. Decision Trees use an algorithm that is preset and then they follow certain other methods for training. The basic point of the decision tree is to find the value that is present in the training in a way that produces child subsets from a training set. When we wish to 'train' a tree, we use the CART (Classification and Regression Tree). This algorithm features finding the optimum way to split a particular 'parent' node into two different parts. These trees are collectively called as the 'ensemble learning algorithm', which helps to figure out which set of the data are weak and which are strong and then based on this information, we take an average of the individual predictions' outcome to reduce the extra data in the training set. The RF is built by putting together decision trees. The RF now has multiple trees that are ensembled together and uses the bagging (or) Bootstrap Aggregating method to train its data. This method basically randomly selects subsets of the training data and makes smaller sets that are used for training repeatedly.

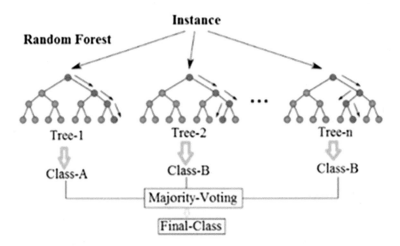

Fig. 1. Simplified random forest model

3.2   Support Vector Machine

Support Vector Machine (SVM) is a supervised machine learning algorithm which can be used for both classification and regression challenges. However, it is mostly used in classification problems. In this algorithm, we plot each data item as a point in n-dimensional space (where 'n' is the number of features) with the value of each feature being the value of a coordinate. Then, a classification is performed by determining the hyperplane that differentiates the two classes very well. Support vectors are data points that if removed from a plane of data, alter the position of the dividing plane. They are crucial to the data set (Fig. 2).

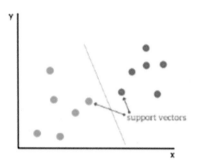

Fig. 2. Support vector approach

A hyperplane is a linear division that classifies a set of data. The more distance exist between vectors and this line, the more confidence we can gain about the classification of our data. When the data seems difficult to differentiate, we need to change our perception to a 3D level.

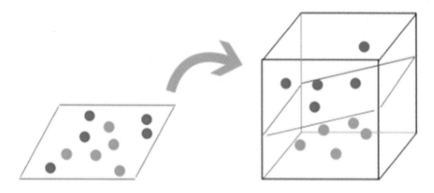

Fig. 3. Visualization of 2D data in 3D design.

In this manner, we utilize the SVM method for efficient training of data sets. As this method uses a subset of training points, it is more accurate and has a good result on small datasets as well.

4 Proposed Model

A novel hybrid model is developed by integrating Random Forest and SVM for detecting the phishing attacks in the website. The proposed model is shown in Fig. 1. The RF model will be used to train the dataset by dividing into multiple units wherein each subunit is bagged to obtain the final decision unit. Every subunit of RF is reclassified by the SVM Classifier to increase the accuracy of the model. A bagging is performed to merge the results of SVM and RF. The motivation to choose these algorithms for building a hybridized model is in our analysis, the individual classifiers namely the SVM classifier and the Random Forest model resulted in a higher accuracy, F-measure, and precision compared to other supervised classifiers. First, we import the dataset and store it in a table format. Before the data is split into training and testing datasets; we perform a pre-processing operation to identify which features are suitable for generating the model. We utilize the Correlation Matrix Feature Selection method, to identify features having a positive correlation with the class value. We then randomly select records from the table to be split into training and testing data. We utilize the training labels for training data i.e., the selected features which were selected in the previous pre-processing procedure. The training labels and the training data are used to train the RF classifier. In the RF classifier, the bagging process breaks the training data into multiple units which are basically multiple decision trees. These 'units', are further broken down into multiple subunits, where we use the SVM classifier on each of the 'sub-units'. The sub-units are then used to train the SVM classifier. In the SVM Classifier, we have 3 stages, the Feature Extraction, Feature Selection and finally, the Classification. Once the SVM classifies each of the subunits, a voting technique is used to perform a decision. The decision is based on majority rule. The class value which has the highest vote will become the output of the 'Decision Fusion' from the SVM Classifier. This is the output of one single unit of the Random Forest Classifier. The 'Final Decision Output' is also based on the Majority Rule. The class value which has the highest number of votes among the units will be the output of the final decision unit. Finally, from the 'Output Unit', we will get the trained RF-SVM Hybrid model as output. The trained RF-SVM hybrid model is utilized to predict the accuracy, precision and F1 values of the testing data (Fig. 4).

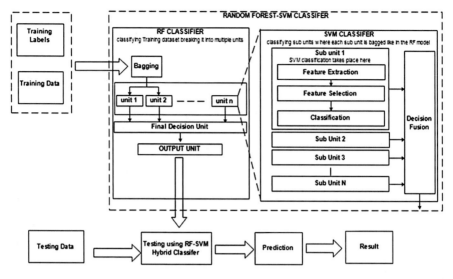

Fig. 4. The proposed architecture

5 Experimental Results

5.1 Dataset

Experiments are conducted on the website phishing Dataset, which is available in the Online Repository of the University of California, Irvine's Machine Learning Repository. The dataset has 10 Attributes and 1353 instances recorded. To generate this dataset, a collection of phishing websites was taken from Phish tank Data Archive. It has 548 legitimate websites, along with 702 phishing URLs and 103 Suspicious URLs. The dataset has 3 values in it (-1, 0 and 1). Malicious or phishing URL is represented by -1, suspicious URL is represented by 0 and legitimate URL is represented as 1.

The attributes used for our analysis are URL Anchor, Request URL, SFH, URL Length, Popup Window, SSL Final State, Web Traffic, Age of Domain, Having IP Address and Result (Class label). The experiments utilized 60% of the website phishing dataset as our training dataset and the rest 40% as our testing dataset. Table 1 shows the features selected by SVM classifier after which a prediction is done on the eight attributes of the dataset. The two features namely "age of domain" and "Having IP Address" are removed because these 2 features have the least impact on the actual dataset and therefore their influence on the result is not affected.

Table 1. Attributes selected by correlation matrix feature selection method

S. No.	Feature selection
1	SFH
2	popUpWindow
3	SSLfinal_State
4	Request_URL
5	URL_of_Anchor
6	wcb_traffic
7	URL_Length
8	Result

5.2   Comparative Study

It is inferred from Table 2 that the individual accuracy of SVM ranges from 89% to 92% for the individual class labels and for Random Forest ranges from 90% to 98%. The average accuracy for SVM is 91.18% and for Random Forest is 92.99%. The individual f-measure of SVM ranges from 25 to 91% and for Random Forest ranges from 85 to 90.90%. The average f-measure for SVM is 68.44% and Random Forest is 88.17%. The individual precision of SVM ranges from 70.00 to 88.50% and for Random Forest ranges from 82.86 to 91.39%. The average f-measure for SVM is 70.00% and Random Forest is 88.50%. The average precision for SVM is 81.3% and for Random Forest is 87.46%. The hybrid model results in an average accuracy of 94.74%, average f-measure of 91.75% and average precision of 90.08%. It is evident that the hybrid model is superior to the accuracy of SVM and Random Forest due to the integration of classifiers by bagging technique (Table 3).

Though the model execution time is slightly higher by 0.455 s in comparison to 0.1 of SVM and 0.38 of Random Forest, the time is closely related to the RF model. Figure 3 shows the graphical illustration of the performance metrics such as accuracy, precision, f-measure, and model execution time which are compared with the classifiers namely SVM, Random Forest, and proposed model (Fig. 5).

Table 2. Comparison of performance metrics of SVM and random forest model.

Measurement metrics	Class values	Support vector machine	Random forest
Accuracy	−1	91.43%	90.43%
	0	92.36%	98.12%
	1	89.76%	90.43%
F1	−1	91.70%	90.90%
	0	25.45%	85.29%
	1	88.17%	88.13%
Precision	−1	88.50%	91.39%
	0	70.00%	82.86%
	1	85.42%	88.13%

(continued)

Table 2. (*continued*)

Measurement metrics	Class values	Support vector machine	Random forest
Correlation coefficient (Mathews)	−1	0.8167	0.8292
	0	0.3810	0.7444
	1	0.7919	0.8471
True positive rate (sensitvity)	−1	88.70%	90.72%
	0	30.00%	67.50%
	1	90.56%	91.85%
True negative rate (specificity)	−1	90.84%	90.47%
	0	97.09%	98.96%
	1	86.45%	91.97%
Kappa	-	73.15692	81.20%
Execution time	-	0.165571	0.3650241

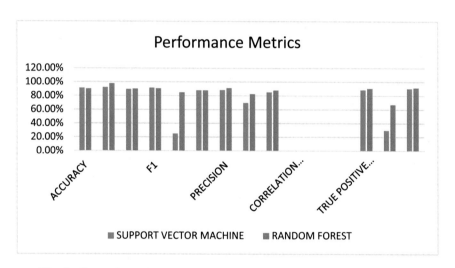

Fig. 5. Comparison of performance metrics of SVM and random forest model.

Table 3. Comparison of error metrics on SVM and random forest classifiers.

Error metrics	Support vector machine	Random forest
Classification error	14.97%	10.56%
Mean absolute error	0.22%	0.18%
Mean square error	6.14%	5.47%

6 Conclusion

Phishing is an important concern for many internet users as sensitive data is transferred over the web. In this paper, a hybrid model is developed using Random Forest and SVM to identify phishing attacks. An analysis on base classifier SVM and ensemble

Random Forest is also done to compare the hybrid model with the existing classifier results. In comparison the classifier SVM and Random Forest, the hybrid model results in better accuracy, precision, and f-measure. Thus, the model can be implemented on a network to identify phishing websites with high success rate.

References

1. Abdelhamid, N., Ayesh, A., Thabtah, F.: Phishing detection based associative classification data mining. Expert Syst. Appl. **41**(13), 5948–5959 (2014). https://doi.org/10.1016/j.eswa. 2014.03.019
2. Abu-Nimeh, S., Nappa, D., Wang, X., Nair, S.: A comparison of machine learning techniques for phishing detection. In: Proceedings of the Anti-Phishing Working Group eCrime Researchers Summit, pp. 60–69. ACM, New York (2007). https://doi.org/10.1145/ 1299015.1299021
3. Aburrous, M., Hossain, M.A., Dahal, K.: Experimental case studies for investigating e-banking phishing techniques and attack strategies. Cogn. Comput. **2**(3), 242–253 (2010). https://doi.org/10.1007/s12559-010-9042-7
4. Alkhozae, M.G., Batarfi, O.A.: Phishing websites detection based on phishing characteristics in the webpage source code. Int. J. Inf. Commun. Technol. Res. **1**(9), 238–291 (2011)
5. Barraclough, P., Hossain, M., Tahir, M., Sexton, G., Aslam, N.: Intelligent phishing detection and protection scheme for online transactions. Expert Syst. Appl. **40**(11), 4697–4706 (2013). https://doi.org/10.1016/j.eswa.2013.02.009
6. Barakat, N., Bradley, A.: Rule extraction from support vector machines: a review. Neurocomputing **74**(1–3), 178–190 (2010). https://doi.org/10.1016/j.neucom.2010.02.016
7. Rami, M., Thabtah, F.A., McCluskey, T.: Intelligent rule-based phishing websites classification. Inf. Secur. IET **8**(3), 153–160 (2014). https://doi.org/10.1049/iet-ifs.2013. 0202
8. Huang, H., Qian, L., Wang, Y.: An SVM-based technique to detect phishing URLs. Inf. Technol. J. **11**(7), 921–925 (2012). https://doi.org/10.3923/itj.2012.921.925
9. Lakshmi, V.S., Vijaya, M.S.: Efficient prediction of phishing websites using supervised learning algorithms. Procedia Eng. **30**, 798–805 (2012). https://doi.org/10.1016/j.proeng. 2012.01.930
10. Moghimi, M., Varjani, A.Y.: New rule-based phishing detection method. Expert Syst. Appl. **53**, 231–242 (2016)
11. Yue, X., Abraham, A., Chi, Z.X., Hao, Y.Y., Mo, H.W.: Artificial immune system inspired behavior based anti-spam filter. Soft Comput.: Soft Comput. - Fusion Found. Methodol. Appl. **11**(8), 729–740 (2007)

Conflict Detection and Resolution with Local Search Algorithms for 4D-Navigation in ATM

Vitor Filincowsky Ribeiro, Henrique Torres de Almeida Rodrigues,
Vitor Bona de Faria, Weigang Li$^{(\boxtimes)}$, and Reinaldo Crispiniano Garcia

Universidade de Brasília (UnB), Brasilia, DF, Brazil
weigang@unb.br

Abstract. Implementation of Trajectory Based Operations (TBO) has been updating the structure of the advanced Air Traffic Management (ATM). Although several methodologies for conflict detection and resolution (CDR) have been developed to the aviation community, the legacy problem is to find an efficient scheme to present the trajectories in this complex network with massive data and further to detect and resolve the conflicts. In this research we develop a CDR framework based on the management of predicted 4D-trajectories using a Not Only SQL (NoSQL) database and local search algorithms for conflict resolution. This paper describes the architecture and algorithms of the proposed solution in 4-Dimensional Trajectory (4DT). With the application of Trajectory Prediction (TP) simulator using the Brazilian flight plan database, the results from case study show the effectiveness of the proposed methods for this sophisticated problem in ATM.

Keywords: 4D trajectory · Air Traffic Management ·
Local search algorithms · Trajectory-based operations

1 Introduction

The amount of air transport users has been consistently growing through the last decades. According to International Air Transport Association (IATA) this demand is expected to double in the next 20 years [1]. Current systems and procedures for Air Traffic Management (ATM) need to evolve according to the growing air space usage demand, aiming to alleviate the workload of Air Traffic Control Officers (ATCOs), and at the same time enhancing the air traffic capacity and safety with adequate supporting structures [2].

Efforts like NextGen in United States and SESAR in Europe arise in order to follow the pace of the demands. The goal is to update the advanced ATM structure with the development of new supporting systems and investment in more modern and precise equipment [3,4]. These proposed changes enable to consider the 4-Dimensional Trajectory (4DT) in the implementation of Trajectory Based Operations (TBO).

© Springer Nature Switzerland AG 2020
A. Abraham et al. (Eds.): ISDA 2018, AISC 941, pp. 129–139, 2020.
https://doi.org/10.1007/978-3-030-16660-1_13

TBO systems are developed in order to support ATCOs in the management of every performed flight, favoring the development of efficient flight paths and the forecast of conflicts. The Trajectory Prediction (TP) technologies are paramount to this new paradigm, since the control is based on aircraft trajectories instead of Air Traffic Control (ATC) clearance. Therefore, aircraft trajectories must be completely modeled and interpreted by supporting systems for decision making available for ATCOs [5]. Such systems must enable a consolidated awareness of the air traffic scenario for all users through the provision of a common view of aircraft trajectories [6].

State-of-the-art procedures for ATM still have the ATCOs as centerpiece, which performs the control tasks with little to none decision support, therefore depending on their own expertise. The growth in the number of conflicts makes the ATM activity even more complex not only in terms of workload, but also raises the operation costs related to delays and fuel efficiency. This condition justifies the efforts for implementation of new automated supporting systems to ATCOs that aims at a more efficient air traffic scenario.

The Brazilian Airspace Control Department (DECEA) is currently working on the implementation plan for the National ATM. The main objective is to establish the strategic evolution of the performance-based National ATM System in order to comply with the national needs and ensure harmonic compliance with the international requirements [7]. The main contribution of this work is the development of a supporting tool for ATCO operations for maintaining a safe, efficient air traffic transport system in Brazil. Furthermore, we add to the National ATM system the support to 4D Navigation concepts, which are not yet fully implemented in the country.

Efficient and safe management of 4D trajectories is only possible if the involved agents (humans and systems) are provided adequate support to decision making. The System Wide Information Management (*SWIM*) program is an effort from NextGen to provide access to complete, up-to-date and reliable information shared through the network, so that every stakeholder has a consolidated view about the trajectories of aircraft and about the several operational scenarios inherent to the air transportation environment [8]. Our framework is developed in order to incorporate to SWIM architecture the capability of CDR for ATM.

Approaches that diminish the data volume transferred or that use efficient storage methods are part of the efforts from the global aviation community in order to adequate the operations to the SWIM architecture paradigm. Data compression and data mining techniques are proven to be effective on this workload [9,10]. Our technique advances the field by taking advantage of NoSQL concepts to store and retrieve data without the need of pre or post processing data for identifying conflicts.

Agent-based models were developed in order to represent the pilot's behavior and ATCO's directives to aircraft in [11]. Conflict detection is performed by pairwise comparisons between all aircraft's trajectories within a local section of the airspace and conflict resolution is done by flight path optimizations through re-routing and flight level updates. Multiobjective simulated annealing [12] was also used to search the optimal aircraft attitude during flight. Such methodologies are specifically designed to tactical CDR, with eventual adjustments to trajectories. This leaves space for enhancement on strategic deconfliction, which our model is able to address.

Methodologies for strategic CDR were already presented to the aviation community, ranging from extensive search to Integer Programming and Evolutionary algorithms [13,14]. The size of the search space may be a matter of concern in such approaches, hence solution attempts are limited to parameterized look-ahead periods in time. Furthermore, pairwise comparisons between trajectories tend to be computationally expensive, which demands the appropriate definition of time windows to constrain the processing scope.

Other techniques for collaborative conflict resolution may regard user-preferred trajectories. Narrowing down the search space is paramount due to computational resources limitations, therefore storing reference trajectories and defining sampling neighborhood for solution sets are also a matter of concern for conflict resolution [15,16]. The hereby proposed methodology does not compare reference trajectories, and narrows even more the search space by checking time-sequenced waypoints without the need to pre-shape the airspace.

2   Notions on Simulated Annealing

Optimization problems demand that the best alternative within a finite set of optimal solutions is selected. However, in some search spaces, the number of steps needed to ensure optimality grows exponentially in relation to the size of the problem [17]. Annealing is the process applied to temper solid materials like metals and glass, leading the material to a very low energy state by slowly cooling down from the material's fusion temperature [18]. The simulated annealing is a heuristic developed as a combinatorial optimization where high-quality solutions are obtained from the simulated reproduction of the behavior observed in materials when they are subject to a controlled cooling process.

In this model, the material is regarded as a system composed of organized atoms, where every possible atomic configuration has an associated energy amount that can be calculated. In each step of the processing, an atom is moved and the energy variation E of the system is calculated. If $E < 0$, then the new configuration is immediately accepted; else, this configuration is accepted with a probability $P(E)$, calculated according to Eq. 1, where k is called the Boltzmann constant and T is the current temperature:

$$P(\delta E) = exp\left(\frac{-\delta E}{kT}\right). \tag{1}$$

The algorithm simulates the annealing process by repeatedly reducing the value of T until the stabilization temperature. Small temperature variations imply greater acceptance of the current solution, thereafter the acceptance probability tends to 0 as T approaches the minimum (0) temperature [17].

The instantaneous energetic state of the subject material is regarded as a cost function in this algorithm, given the current configuration. Hence, every possible configuration is equal to a solution whose cost is measurable. Therefore, bringing the material to the lowest energetic state means minimizing the cost of the problem solution.

3   Proposed Model

We hereby present an innovative model to detect and resolve conflicts among aircraft in the strategic planning phase of the flight. The available flight intention documents are used to predict the trajectories, then the calculated trajectories are subject to evaluation of potential conflicts. If conflicts are detected, a conflict resolution procedure is triggered and the parameters described in the flight plans are updated in order to change the primary flight profile.

3.1   Trajectory Prediction

The main document used in order to perform the flight intent predictions for the aircraft is the Flight Plan. This document aggregates all the information inherent to the desired flight performance and enables the allocation of a ATC slot to the aircraft. Then, the ATC can plan the air traffic resources needed to the described operation.

Flight intent information can be defined as the flight path and the constraints that dictate the future performance of the aircraft, including air traffic control actions and airline preferences [19]. The atomic unity of a projected trajectory is a two-dimensional segment composed of an initial and a final point. A set $W = (w_1, w_2, ...w_n)$ composed of several 2D points describes a full trajectory for the aircraft, provided that w_1 is the starting point and w_n is the final point, w_i is connected to w_j if, and only if, $i = j - 1$ and there is always a connected path between w_i and w_j, $i \neq j$. Nevertheless, w_i must be remodeled to accommodate the altitude and a chronological constraint for an unambiguous description of the intended flight path, enabling the interoperability among the involved stakeholders, including CDR systems.

The core of the trajectory computation process is a set of movement equations based on the initial state and in the primary description of the aircraft flight intent. Specific performance attributes are also considered [5]. The Aircraft Intent Description Language (AIDL) is used to mathematically model the aircraft trajectories. AIDL regards aircraft as point-mass objects located on a three-dimensional space. The language alphabet in AIDL is built according to mathematical rules that define the possible combinations among the instructions, which are grouped according to their effects on the aircraft's behavioral profile [6].

The dynamics of the aircraft is described in terms of attitude parameters such as thrust, drag and lift, plus wind and mass variation due to fuel consumption. This model is composed of six parallel instruction sets (two Longitudinal, one

Lateral and three configuration threads) executed simultaneously to describe the aircraft attitude through space and time. This framework is the kernel to the trajectory computation in the proposed tool. Detailed description of the motion equations and their results can be found in the work of López-Leonés et al. [5].

3.2 Conflict Detection

A conflict in the airspace is the situation in which two or more en-route aircraft violate the minimum separation among each other. This separation must be held both in vertical and horizontal axis, provided that the minimum vertical distance is not less than 1000 ft for flights below the altitude at 29000 ft and 2000 ft above this altitude, and the minimum horizontal distance is not less than five nautical miles. This longitudinal distance corresponds to approximately 15 min of separation between two aircraft at the same route. These constraints create a safe volume around every aircraft, as depicted in Fig. 1.

Fig. 1. Safe separation volume around an aircraft.

In a 4D-scenario, the time restriction must be added. The conflict needs then to be redefined as the violation of minimum separation among aircraft in a given space fragment, within the same time window [20]. Then, it is possible to identify the logical relation that defines the existence of a conflict c between aircraft A_i and A_j at an instant t, according to Eq. 2:

$$c^{A_i, A_j}(t) \iff \left(d_h^{A_i, A_j}(t) < S_h\right) \wedge \left(d_v^{A_i, A_j}(t) < S_v\right). \qquad (2)$$

Here, $d_h^{A_i, A_j}(t)$ is the longitudinal distance between A_i and A_j at instant t, $d_v^{A_i, A_j}(t)$ is the vertical distance between A_i and A_j at instant t, and S_h and S_v are respectively the minimum longitudinal and vertical required separations.

A core feature in the proposed conflict detection model is the implementation of a non-relational database, Not Only SQL (*NoSQL*), to store and recover the predicted trajectories. NoSQL databases are developed to manage large amount of data distributed in several nodes on a network [21].

Traditional approaches try to calculate a function to trajectories by means of interpolation over the sampled points. This should involve the trajectory representation on a 4D-space for later integration. Nevertheless, the conflict detection procedure proposed in this implementation takes advantage of this column-based architecture by neglecting the specifics of the performed trajectories, due to the fact that only the sampled waypoints are stored in the database in an orderly fashion. Therefore, the whole mathematical effort is reserved to the Trajectory Predictor (TP) developed by Besada et al. in [19]. The waypoints table is queried in a way in which the returned result is a collection of consecutive points in a given time window. If separation constraints between two consecutive points is violated, then the referred trajectories are in conflict.

3.3   Conflict Resolution

If conflicts are detected, the conflict resolution algorithm is activated. The ATC must search a solution in which the conflicts are completely solved, but at the smallest cost as possible. Obviously, changes in the intended flight paths or insertion of delays will invariably increase the cost for some aircraft, therefore increasing the total cost to the aircraft scenario. However, the cost distribution must be fair among the affected aircraft.

The resolution algorithm consists of an adaptation of the Simulated Annealing (SA) algorithm. When the SA optimizer is started, the conflicts are coded into simulation cells, which hold information about the flight parameters. The total energy is conceived as the total cost for the conflict set. The algorithm iteratively adjusts the flight parameters of conflicting flights in order to reduce the system energy to the lowest, stable state, which means to resolve the scheduling scenario to the less expensive configuration in terms of aggregated cost. The basic algorithm can be viewed in Algorithm 1.

Algorithm 1. Simulated Annealing algorithm for aircraft conflict resolution

---

Data: conflict set
Result: solution with updated flight parameters
1 $currSolution \leftarrow initialize(conflicts)$;
2 $currEnergy \leftarrow calcEnergy(currSolution)$;
3 **for** $T = ini(T); \neg stopCondition(T); upd(T)$ **do**
4 $\quad solution \leftarrow selectNeighbor(currSolution)$;
5 $\quad energy \leftarrow calcEnergy(solution)$;
6 $\quad delta \leftarrow energy - currEnergy$;
7 $\quad$ **if** $delta < 0 \vee rand() \leq probability(delta, T)$ **then**
8 $\quad\quad currSolution \leftarrow solution$;
9 $\quad\quad currEnergy \leftarrow energy$;
10 $\quad$ **end**
11 **end**

The neighboring solution is selected by randomly picking one of the conflicting aircraft and adjusting the flight parameters so that the given trajectory is changed. If the selected solution has a lower energy state than the instantaneous best solution, then the current solution is updated. Else, the current solution is updated with probability given in Eq. 1. This is done to avoid the system to be stuck in sub-optimal results. Eventually, this probability is low enough to guide the solution to an immutable state, given the transition constraints.

In our work, the initial temperature for the annealing procedure is configured so that the system starts with an acceptance probability of 95% for bad solutions. Once the annealing result is ready, the yielded solution is applied to the aircraft set and the altered trajectories are recalculated and a new evaluation is performed on the sampled waypoints. Eventually, new conflicts might arise, and the conflict resolution task is performed again, with the updated conflict set as input.

We developed an alternate approach to SA as well [22], provided that local search algorithms are not deterministic. The second resolution algorithm implements a more direct approach. Instead of randomly selecting the aircraft to be updated, the conflicting aircraft are selected according to their Direct Operating Cost (DOC). The DOC describes the aggregated cost of a flight regarding crew, fuel, maintenance, aircraft insurances etc. Roskam algorithm is used to calculate the DOC in our experiment [23]. By regarding DOC in the conflict resolution procedure, we expect to optimize the costs associated to aircraft delays.

The operational costs are calculated and serve as input to the decision making process by a Binary Integer Programming (BIP) machine. Given the set A formed by aircraft $A_1, ..., A_n$ involved in conflicts, the BIP model is shown in Eq. 3, where DOC_i is the DOC inherent to aircraft i and $x_1, ..., x_n$ are mutually excluding variables, provided that $x_j = 1 \rightarrow x_k = 0, \forall k \neq j$.

$$minimize, Z = \sum_{i=1}^{n} DOC_i.x_i \Rightarrow \begin{cases} x_i \in \{0, 1\}, \\ x_1 + ... + x_n = 1 \end{cases} . \tag{3}$$

The BIP model works on the ordered set of conflicting flights and for each minimum subset of two consecutive elements it selects the aircraft with the least DOC for delaying the departure time in one minute. Therefore, the outcome is always the same for a given set of conflicts, which makes this approach deterministic. When the whole conflict set is traveled, the aircraft set is once more evaluated for conflicts, and the resolution procedure is triggered again if needed. The flowchart for CDR is shown in Fig. 2.

4   Simulation and Results

Nine airports in Brazil compose the simulation scenario, namely: (a) SBBR (Brasília-DF), (b) SBGR (Guarulhos-SP), (c) SBSP (São Paulo-SP), (d) SBKP (Campinas-SP), (e) SBGL (Rio de Janeiro-RJ), (f) SBRJ (Rio de Janeiro-RJ), (g) SBCF (Confins-MG), (h) SBSV (Salvador-BA) and (i) SBPA (Porto Alegre-RS). Brazil is the second country in the world in number of airports, and the selected airports serve the majority of the Brazilian population. In fact, in 2017 they responded for 56% of the national air traffic [24], hence they are responsible for the heaviest workload in the national Air Traffic Flow Management.

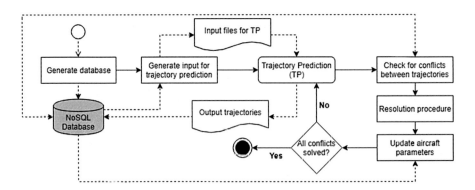

Fig. 2. Sequence diagram for conflict detection and resolution

The repeated flight plans (RPL) database was collected from the freely accessible site of the Brazilian Air Navigation Management Center (CGNA) and comprehends the valid flights to June 2017. Among them, 310 subject flights were filtered for simulation, provided that they have as origin and destination airports among the set of selected airports and are performed by the same manufacturer.

The developed application does not embed TP itself, but invokes an external HTTP service for such task. The application generates a set of XML files as input to TP containing information about the initial conditions of the aircraft at the beginning of the flight execution and the flight intent of the aircraft compiled from the RPL. Further input are the weather model, the characteristics of the airspace where the flight must be executed and the specific performance models for each aircraft type.

Main parameters for prediction procedure are the intended route, aircraft speed, flight level and departure time. TP calculates the evolution of aircraft so they comply with the requirements described in the flight intent. As result, TP outputs the 4D trajectory predicted for every aircraft, including climb and descent profiles. The underlying usage of AIDL within TP tool produces the aircraft intent, ensuring that the trajectories are unambiguously, completely described in low-level detail. Therefore, any client assessing the output intent would have the same view of aircraft trajectory.

A trajectory is presented as a sequence of points that form the flight path of aircraft. Each sampled point embeds information about latitude, longitude, altitude and predicted time, fulfilling the 4D requirements, but also information about instantaneous speeds, acceleration, mass and other parameters not accounted in our solution. The sampled trajectory is detailed in a XML file and simplified in a KML file, in which the waypoint coordinates are featured and can be graphically viewed in display tools such Google Maps. Sampling time interval between consecutive waypoints is about 15 s.

Table 1 summarizes the results after the application of the proposed algorithms. All the conflicts were properly solved by successive adjustments in the departure times after 2068 iterations in SA, provided that the algorithm was automatically reinitialized for three times due to the fact that there were remaining conflicts when the temperature was in the minimum value. On the other hand, the BIP approach was able to solve all conflicts in 5 iterations after traveling the whole conflict queue, consuming 36 s in execution.

Table 1. Results after execution of resolution algorithms for 310 flights

Method	Sampled points	Conflicts found	Iterations	Remaining conflicts	Updated FPL	Avg delay (min)
SA	43615	21	2068	0	47	4,8
BIP	43615	21	5	0	21	1,9

Ground holding is the default action to be taken in strategic conflict resolution. The main advantage in this kind of solution is that the CDR procedures may be performed so that the aircraft may still accomplish their desired flight performance.

5   Conclusion

In this proposed model, the neighborhood of solutions is designed to adjust some aircraft which is already involved in a conflict. Hence, the evaluation of the energetic state of solutions embeds the attribution of an uniform probability for the aircraft which fulfill this requirement. Therefore, every solution is an effective attempt to resolve a conflict between two aircraft.

The BIP model presented better performance than the SA model. The annealing process has the inherent property to imprint worse solutions under a given probability to avoid sub-optimal solutions [12], whilst the binary approach is driven to the minimum perceived cost at each step. Hence, fewer updates are necessary and the yielded scenario presents smaller costs. Furthermore, it was observed that SA does not comply with the fairness property, which means that the cost distribution is uneven among the updated aircraft.

5.1   Discussion and Future Work

The ATC conception about the operating costs for airlines is directly related to the estimated cost to the flights. This research is focused in the decision making process of ATC, trying to eliminate all the conflicts by adjusting the departure time of aircraft.

The main contribution in this research is the incorporation of a trajectory predictor and a database specifically designed for big data. The developed application consumes information from several sources conveniently aggregated in a NoSQL database, and invokes an external service for trajectory predictions, in a successful attempt to adequate the ATC operations to the SWIM architecture paradigm. Finally, this application is not supposed to replace the ATCO whatsoever, but serves as a decision-making supporting tool for the human operator.

As a future work we intend to extend this prototype solution to manage the aircraft motion profile by dynamically updating the flight parameters. Furthermore, we intend to investigate the usage of several NoSQL implementations and their impact on the storage and data evaluation performance. Also, the notion of conflict herein presented does not include capacity constraints, therefore this is a pending matter of concern for our future implementations.

Acknowledgements. This work has been partially supported by the Brazilian National Council for Scientific and Technological Development (CNPq) under the grant number 311441/2017-3 and also by Boeing Research & Technology/Brazil.

References

1. IATA: 2036 Forecast reveals air passengers will nearly double to 7.8 billion. 2017, press Release No.: 55, 24 October 2017
2. Radiöic, T.: The effect of trajectory-based operations on air traffic complexity. Ph.D. dissertation, University of Zagreb, Faculty of Transport and Traffic Sciences (2014)
3. JPDO: Concept of Operations for the Next Generation Air Transportation System. Joint Planning and Development Office, no. Version 3.2 (2011)
4. SESAR: Mission Trajectory Detailed Concept. Single European Sky ATM Research, Technical report, Document identifier DDS/CM/SPM/SESAR/12-042 (2012)
5. López-Leonés, J., Vilaplana, M., Gallo, E., Navarro, F., Querejeta, C.: The aircraft intent description language: a key enabler for air-ground synchronization in trajectory-based operations. In: IEEE/AIAA 26th Digital Avionics Systems Conference, Dallas (2007)
6. Frontera, G., Besada, J., Bernardos, A., Casado, E., López-Leonés, J.: Formal intent-based trajectory description languages. IEEE Trans. Intell. Transp. Syst. **15**(4), 1550–1566 (2014)
7. DECEA: Implementação Operacional do Conceito de Navegação Baseada em Performance (PBN) no Espaço Aéreo Brasileiro. Airspace Control Department, Aeronautical Information Report 24/13 (2013)
8. Dieudonne, J., Crane, H., Jones, S., Smith, C., Remillard, S., Snead, G.: NEO (NextGen 4D TM) Provided by SWIM's Surveillance SOA (SDN ASP for RNP 4D Ops). In: Integrated Communications, Navigation e Surveillance Conference (ICNS), Herndon (2007)
9. Wandelt, S., Sun, X.: Efficient compression of 4D-trajectory data in air traffic management. IEEE Trans. Intell. Transp. Syst. **16**, 844–853 (2015)
10. Song, Y., Cheng, P., Mu, C.: An improved trajectory prediction algorithm based on trajectory data mining for air traffic management. In: IEEE International Conference on Information and Automation, Shenyang, pp. 981–986 (2012)
11. Bongiorno, C., Micciché, S., Mantegna, R.N.: An empirically grounded agent based model for modeling directs, conflict detection and resolution operations in air traffic management. PLoS ONE **12**(4), e0175036 (2017)

12. Mateos, A., Jiménez-Martín, A.: Multiobjective simulated annealing for collision avoidance in ATM accounting for three admissible maneuvers. Math. Prob. Eng. **2016**, Article ID 8738014 (2016)
13. Durand, N., Alliot, C., Barnier, N.: A ground holding model for aircraft deconfliction. In: 29th Digital Avionics Systems Conference, Salt Lake City (2010)
14. Bertsimas, D., Lulli, G., Odoni, A.: An integer optimization approach to large-scale air traffic flow management. Oper. Res. **59**(1), 211–227 (2011)
15. Ruiz, S., Piera, M., Nosedal, J., Ranieri, A.: Strategic de-confliction in the presence of a large number of 4D trajectories using a causal modeling approach. Transp. Res. Part C: Emerg. Technol. **39**, 39:129–39:147 (2014)
16. Berling, J., Lau, A., Gollnick, V.: European air traffic flow management with strategic deconfliction. In: Operations Research Proceedings, pp. 279–286 (2017)
17. Dowsland, K.A., Thompson, J.M.: Simulated annealing, pp. 1623–1655. Springer, Heidelberg (2012)
18. Russell, S., Norvig, P.: Artificial Intelligence: A Modern Approach, 3rd edn. Prentice Hall, Upper Saddle River (2010)
19. Besada, J., Frontera, G., Crespo, J., Casado, E., López-Leonés, J.: Automated aircraft trajectory prediction based on formal intent-related language processing. IEEE Trans. Intell. Transp. Syst. **14**(3), 1067–1082 (2013)
20. Ribeiro, V.F., Pamplona, D.A., Fregnani, J.A.T.G., de Oliveira, I.R., Weigang, L.: Modeling the swarm optimization to build effective continuous descent arrival sequences. In: 2016 IEEE 19th International Conference on Intelligent Transportation Systems (ITSC), Rio de Janeiro, pp. 760–765, November 2016
21. Ferreira, G.R., Felipe Junior, C., de Oliveira, D.: Uso de SGDBs NoSQL na Gerência da Proveniência Distribuída em Workflows Científicos. In: 29th Brazilian Symposium on Databases (SBBD). Brazilian Computing Society, Curitiba (2014)
22. Rodrigues, H.T.A., de Faria, V.B.: Detecção e Resolução de Conflitos para Gerenciamento de Tráfego Aéreo em Trajetórias 4D. Bachelor's thesis, TransLab, University of Brasilia (2018)
23. Roskam, J.: Airplane Design Part VIII: Airplane Cost Estimation: Design, Development, Manufacturing and Operating, 1st edn. Roskam Aviation and Engineering Corporation (1990)
24. DECEA: Anuário Estatístico de Tráfego Aéreo. Headquarters of Air Navigation Management (CGNA), Technical report (2017)

Using Severe Convective Weather Information for Flight Planning

Iuri Souza Ramos Barbosa, Igor Silva Bonomo, Leonardo L. Cruciol,
Lucas Borges Monteiro, Vinicius R. P. Borges, and Weigang Li[(✉)]

TransLab - Department of Computer Science,
University of Brasilia, Brasilia, DF, Brazil
weigang@unb.br

Abstract. Aircraft fly in an environment that is subject to constant
weather changes, which considerably influences the decision-making process in Air Traffic Management (ATM). The stakeholders in ATM track
weather conditions to appropriately respond against new environmental
settings. The proposed work builds an intelligent system to constantly
monitor the impact of severe weather on airways, which are corridors with
specific width and height connecting two locations in the airspace. The
proposed approach integrates weather information on convection cells
obtained from ground-based weather radars, and flight tracking information detailing flight positions in real time. To delimit the boundaries
of airways, the set of flight positions is transformed to a more convenient
one using linear interpolation. Then, a cluster analysis via Density-Based
Spatial Clustering of Applications with Noise (DBSCAN) is performed
into this new set. The algorithm compares the positions of the clusters
found with the positions of convection cells to identify possible intersections between them. A case study was set up consisting of two phases:
(1) flight positions were tracked, and the boundaries of the airways were
identified; and (2) convection cell locations were monitored, and compared against the airways in order to identify possible intersections. The
solution showed the clusters representing the boundaries of the underlying airways, and some intersections were found during the case study.

Keywords: Aeronautical Meteorology · Air Traffic Management ·
Intelligent systems design · Unsupervised learning

1 Introduction

Air Traffic Management (ATM) is a complex activity that depends on a variety
of factors in order to work properly. A key factor to support the ATM decision-
making process is information. More specifically, meteorological information has
a substantial impact on ATM operations. The stakeholders in ATM (air traffic
controllers, aircraft pilots, flight dispatchers, etc.) track weather conditions to
appropriately respond and adapt against new environmental settings, such as
severe convective weather, one of the leading causes of airplane crashes around

© Springer Nature Switzerland AG 2020
A. Abraham et al. (Eds.): ISDA 2018, AISC 941, pp. 140–149, 2020.
https://doi.org/10.1007/978-3-030-16660-1_14

the world. Thus, making optimal use of ATM resources, such as air space, airports, aircraft, infrastructure, human resources, etc.

The proposed work builds an intelligent system to constantly monitor the impact of severe convective weather on airways, which are corridors with specific width and height connecting two locations in the airspace. The solution integrates weather information on convection cells obtained from ground-based weather radars, and flight tracking information detailing flight positions in real time. First, flight positions are transformed into a more convenient set using linear interpolation, and then, Density-Based Spatial Clustering of Applications with Noise (DBSCAN) is used as cluster model to delimit the boundaries of airways. Finally, the positions of the clusters found are compared with the positions of convection cells in order to identify intersections between them.

Implementing this solution as a service, that provides information on the impact of severe convective weather on airways, agrees with the concept of System Wide Information Management (SWIM) defined by the International Civil Aviation Organization (ICAO). SWIM consists of a set of standards, infrastructure, and governance enabling the management and exchange of ATM information between ATM stakeholders through inter-operable services [1]. Another initiative in accordance with the proposed solution is the Data-driven AiRcraft Trajectory prediction research (DART) project [2], which uses data-driven techniques in order to increase trajectory prediction accuracy and efficiency.

Many works with respect to aircraft trajectory prediction are based on exploiting surveillance data, such as flight tracking data, weather data, etc., combined with machine learning techniques. In [3] weather observations (such as temperature, wind speed, and wind direction) and trajectory information are used together with Hidden Markov Model (HMM) as stochastic model to predict aircraft trajectory. In [4] a system is built to predict the Estimated Time of Arrival (ETA) for commercial flights using multiple data sources: historical trajectory, meteorology, airport and air traffic data, in conjunction with regression models and Recurrent Neural Network (RNN). Likewise, the proposed solution also uses weather information in combination with flight tracking data to support the ATM decision-making process. However, the proposed solution uses severe convective weather information while the related works use the combination of different meteorological parameters [3,4]. Although, the proposed work does not directly predict aircraft trajectory, it supports such task by providing early detection of intersections on airways. While the related works [3,4] focus on predicting aircraft trajectory and estimated time of arrival, respectively, the proposed solution aims to provide information on the impact of convection cells on airways, acting as a new source of information.

2   Air Traffic Management

The flight phases of commercial flights can be defined as follows: flight planning, taxing, takeoff, climb, cruise, descent, final approach and landing [5]. In the flight planning phase, a flight plan is issued with the local Civil Aviation Authority, that includes general information about the flight: departure/destination airports; airways that will be used in the flight and their altitudes; meteorological forecasts of the airports and airways; etc. After the aircraft takes off, it moves towards the airways defined in its flight plan. When the aircraft starts to fly in its airways, it reaches the cruise phase, which is the longest phase of a flight, where the aircraft keeps constant speed and altitude. Due to the existence of predetermined airways in the airspace, tracking flight positions of flights flying through the same airways creates sets of similar positions. The proposed solution is designed to be used during flight planning, but it can also be used to support aircraft in flight since weather conditions can change at any time. Air traffic controllers and flight dispatchers are the target audience of this work.

3   Aeronautical Meteorology

Aeronautical Meteorology is a key factor for the safety of air operations. Aircraft pilots monitor weather conditions with the help of the on-board weather radar, which searches for severe convective weather (convection cells) in the airspace. For safety reasons, when facing severe convective weather, pilots usually decide on changing the heading of the aircraft instead of maintaining its course. However, aircraft weather radars do not provide a far-sighted view of weather conditions, instead, they provide a myopic view. For this reason, the proposed work employs weather information obtained from ground-based radar stations in the Brazilian airspace via STSC [6].

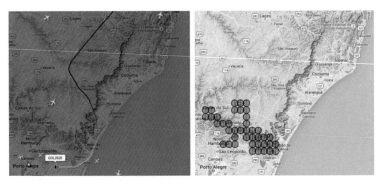

(a) Trajectory of flight GOL2026 in 26 August 2015 at 18:08 GMT-3.

(b) Convection cells via STSC in 26 August 2015 at 18:08 GMT-3.

Fig. 1. Impact of severe convective weather on ATM operation.

Figure 1a shows the trajectory of an aircraft that flew in the Brazilian airspace, where its trajectory is colored purple. Figure 1b shows the weather conditions at the same time the flight took place, where the red circles represent the convection cells located by ground-based weather radars. One can see that having the second source of information available would result in an early detection of convection cells followed by an anticipated change in the flight course. Thus, resulting in shorter flight time, a reduction of fuel consumption, increase of safety, etc. For this reason, the proposed approach aims to integrate severe convective weather information from ground-based radars into the ATM decision-making process.

4   Tracking Intersections Between Airways and Convection Cells with DBSCAN

To measure the impact of severe convective weather on ATM operation, the devised algorithm: (1) collects information of flight positions and convection cell locations; (2) normalizes flight positions using linear interpolation; and (3) searches for intersections between airways and convection cells using DBSCAN. The system takes in flight tracking information, real-time weather information, and the departure/destination airports, and outputs the intersections (represented by Impact Factors) between the current convection cells in the airspace and the airways connecting the given airports. The system overview is shown in Fig. 2, and its parts are discussed in the next sections.

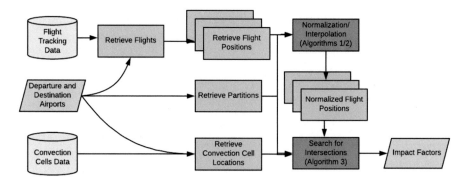

Fig. 2. System overview. Input/output information is colored green, data storage is colored yellow, retrieved/temporary data is colored blue, and the proposed algorithms are colored red.

4.1   Data Sources

Flights can be tracked via Automatic Dependent Surveillance - Broadcast (ADS-B). This technology can be used as a low cost replacement for conventional radars, providing flight positions information in real time. ADS-B is also used in some platforms consisting of crowd-sourced distributed networks of ADS-B receivers [7]. These platforms broadcast information such as: (a) flight status - flight number, departure/destination airports, etc.; (b) aircraft details - aircraft identifier, aircraft type, etc.; (c) flight details - current flight position of the aircraft (latitude, longitude, altitude), speed, heading, etc. .

STSC (Severe Convective Weather System, literal translation) tracks severe convective weather detected via meteorological radars stations in the Brazilian airspace. STSC detects and monitors convection cells in their most active stage: cumulonimbus clouds, which significantly impact the airspace and its operation. STSC provides information on convection cell such as: geographical position (latitude, longitude), time of occurrence, radius of the convection cell, etc. [6].

4.2   Normalizing Flight Positions with Linear Interpolation

Let F be a flight of an aircraft flying from the departure airport A to the destination airport B, and $(F.position_0, F.position_1, F.position_2, ...)$ be the set of flight positions of F retrieved during the flight tracking process, where $F.position_i$ is the triplet $(F.position_i.latitude, F.position_i.longitude, F.position_i.altitude)$.

In the normalization process (see Fig. 2), a new set of normalized flight positions is estimated from the original set of flight positions via linear interpolation [8]. The process takes in the original set of flight positions, and the set of values $PART(A, B) = PARTITIONS(A, B)$, representing the set of latitudes of interest. The set $PART$ is represented by values evenly spaced by some distance $PART\_DISTANCE$, values that start from the airport A latitude and that end with the airport B latitude as shown in Eq. 1:

$$PART(A, B) = (P_0 = A.latitude, P_1, ..., P_i, P_{i+1}, ..., P_n = B.latitude) \quad (1)$$

where $P_{i+1} - P_i = PART\_DISTANCE$ is constant for every index i. The interpolation/normalization processes are shown in the Algorithms 1 and 2.

Although flight positions are represented in the geographic coordinate system, the interpolation (Algorithm 2) is a good approximation method for estimating new flight positions. It also provides good approximation due to the fact

---

Algorithm 1. Normalize flight *positions* given *partitions* set, representing $(F.position_0, F.position_1, F.position_2, ...)$ and $PART(A, B)$, respectively.

```
 1: procedure NORMALIZE(positions, partitions)
 2:     normalized_positions ← List()                    ▷ Normalized flight positions
 3:     i ← 0                                            ▷ Index of flight positions
 4:     j ← 0                                            ▷ Index of partitions
 5:     while i + 1 < positions.length & j < partitions.length do
 6:         if positions_i.latitude ≤ partitions_j ≤ positions_{i+1}.latitude then
 7:             new_position ← INTERPOLATE(positions_i, positions_{i+1}, partitions_j)
 8:             INSERTINTO(new_position, normalized_positions)
 9:             j ← j + 1
10:         else if positions_{i+1} < partitions_j then
11:             i ← i + 1
12:         else if partitions_j < positions_i then
13:             j ← j + 1
14:     return normalized_positions
```

Algorithm 2. Perform linear interpolation on successive flight positions p_1 and p_2 to generate a new normalized flight position p such that $p.latitude = lat$.

1: **procedure** INTERPOLATE(p_1, p_2, lat)
2:   ▷ Longitude, latitude and altitude attributes are simplified to lon., lat. and alt.
3:   $p.lat. \leftarrow lat$
4:   $p.lon. \leftarrow \dfrac{(p_2.lon. - p_1.lon.)}{(p_2.lat. - p_1.lat.)} \times (lat - p_1.lat.) + p_1.lon.$
5:   $p.alt. \leftarrow \dfrac{(p_2.alt. - p_1.alt.)}{(p_2.lat. - p_1.lat.)} \times (lat - p_1.lat.) + p_1.alt.$
6:   **return** p

that positions are tracked each 10 s [7], and therefore they are close to each other. The normalization process (Algorithm 1) provides normalized flight positions with latitudes sampled from the partition set $PART$. Moreover, applying the normalization to every flight will make their normalized flight positions share the same values of latitudes. It is also worth mentioning that the Algorithm 1 assumes, for the sake of clarity, that *partitions* and *positions* sequences are increasing. A robust implementation should take care of that, and adapt Algorithms 1 and 2 accordingly.

Finally, applying normalization/interpolation aims to replace a set of flight positions for a more convenient one. This new set along with other sets of flight positions (generated from other flights) are used to delimit the boundaries of the airways on which these flights navigated.

4.3   Delimiting the Boundaries of Airways

Sets of flight positions from different flights might seem unrelated to one another at first, even if the airplanes flew through the exact same airways. To compare and analyze the similarity of these sets, they have to go through a normalization process so that they share a set of similar latitude values as shown in Fig. 2. Applying the normalization process to all flights ($F_0, F_1, F_2, ...$) sharing the same departure/destination airports A and B, the normalized flight positions of these flights will have their latitudes distributed throughout the partitions set. Hence, each partition P_k in $PART(A, B)$ can be associated with a set of normalized flight positions ($P_k.flight\_position_0, P_k.flight\_position_1, ...$), such that $P_k.flight\_position_i.latitude = P_k.flight\_position_j.latitude = P_k.value$ for every i and j.

Given that aircraft follow predefined airways in the airspace, each partition P_k in $PART$ will have clusters of flight positions, and other flight positions isolated from the rest (outliers). That happens because airplanes flying on the same airways (in cruise phase) generate similar trajectories, and therefore generate similar flight positions for a given latitude $P_k.value$. The clusters found will delimit the boundaries of the underlying airways crossing this specific partition. To discover the clusters in each partition, the Algorithm 3 runs a data clustering algorithm for each partition P_j as a preprocessing step. It aims to divide the normalized flight positions into clusters $(P_j.cluster_0, P_j.cluster_1, ...)$, where each cluster i consists of a set of positions $(P_j.cluster_i.flight\_position_0, P_j.cluster_i.flight\_position_1, ...)$.

The proposed solution uses Density-Based Spatial Clustering of Applications with Noise (DBSCAN) as the data clustering algorithm [10]. For every partition, DBSCAN takes in the normalized flight positions related to the underlying partition (samples), and separates them in areas of high density and areas of low density (noise). Two parameters are necessary to run the algorithm: $min\_samples$ and eps. A cluster consists of a set of core samples, and a set of non-core samples. A sample is a core sample if there exist $min\_samples$ other neighbor samples within a distance eps from the underlying sample. A sample that is not core sample, but is close to a core sample within a distance eps, is called a non-core sample. Samples that are neither core samples nor non-core samples are called outliers.

4.4 Search for Intersections Between Airways and Convection Cells

The intersections between convection cells and airways are identified by checking the intersections of the convection cells against each partition P_i of $PART$, more specifically, against each normalized flight position of the underlying partition. Let $(CC_0, CC_1, ...)$ be the convection cells, where CC_i is represented by the triplet $(CC_i.latitude, CC_i.longitude, CC_i.radius)$. Let P_j be a partition from $PART$ and $(P_j.flight\_position_0, P_j.flight\_position_1, ...)$ the normalized flight positions related to this partition. The distance between the convection cell CC_i and a normalized flight position $P_j.flight\_position_k$ is defined by the haversine formula [9]. The convection cell CC_i and the normalized flight position $P_j.flight\_position_k$ intersect if the distance between them is less than or equal to $CC_i.radius$. Then, the definition of the intersection of the convection cell CC_i and the flight position is extended to: (1) the intersection of CC_i and the cluster related to the flight position, and (2) the intersection of CC_i and the airway related to the cluster.

As a preprocessing step, DBSCAN is run for every partition P_i of $PART$. DBSCAN takes in the flight positions of a partition, and returns their clusters, which are then used in the Algorithm 3. The algorithm calculates the impact factor of a convection cell, which consists of the maximum percentage of airways that are blocked by the underlying cell. The impact factor measure takes into account the frequency at which the airways are used: (a) a high value if the airway being blocked was navigated by many flights (larger clusters), or (b) a short value if the airway was not commonly navigated (smaller clusters). The Algorithm 3 emphasizes the importance of airways as opposed to isolated flight locations by excluding flight positions that are not part of clusters (outliers), and therefore not within the boundaries of airways.

Algorithm 3. Identify the impact factor of convection *cells* on the airways

---

1: **procedure** MEASUREIMPACTFACTOR(*cells, partitions*)
2:     **for each** *partition* : *partitions* **do** DBSCAN(*partition*)
3:     *impact_factors* ← *List*()
4:     **for each** *cell* : *cells* **do**
5:         *max_impact_factor* ← 0
6:         **for each** *partition* : *partitions* **do**
7:             *all_flight_positions* ← *List*()   ▷ All flight positions in *partition*
8:             *intersected_flight_positions* ← *List*()
9:             **for each** *cluster* : *partition.clusters* **do**
10:                 INSERTINTO(*cluster.flight_positions, all_flight_positions*)
11:                 **for each** flight_position: cluster.flight_positions **do**
12:                     **if** HAVERSINE(*cell, flight_position*) ≤ *cell.radius* **then**
13:                         INSERTINTO(*cluster.flight_positions, intersected_flight_positions*)
14:                     **break**
15:             **if** *all_flight_positions.length* ≠ 0 **then**
16:                 $curr\_impact\_factor \leftarrow \dfrac{intersected\_flight\_positions.length}{all\_flight\_positions.length}$
17:             *max_impact_factor* ← MAX(*max_impact_factor, curr_impact_factor*)
18:         **if** *max_impact_factor* > 0 **then**
19:             *tuple* ← (*cell, max_impact_factor*)
20:             INSERTINTO(*tuple, impact_factors*)
21:     **return** *impact_factors*

---

5 Case Study

For the case study, an application was implemented to collect data from commercial flights and convection cells that had taken place in the Brazilian airspace. OpenSky Network [7] was the source of information used to track flights and their positions, while STSC [6] was used to track convection cells information.

The proposed scenario consists of analyzing the impact of severe convective weather on the airways connecting the Brasília International Airport (BSB) to the Guarulhos International Airport (GRU) (in both directions). The case study was set up in two phases described as follows: (1) Collection Phase (Algorithms 1 and 2) - Tracked positions of flights flying from/to the underlying airports from 25-07-2018 to 01-08-2018 (13,404 flight positions collected); (2) Tracking Phase (Algorithm 3) - Monitored actual severe convective weather in the Brazilian airspace from 19-09-2018 to 20-09-2018 (4 convection cells analyzed were blocking airways). Some specific settings needed to be set beforehand: (a) Distance between partitions $PART\_DISTANCE$ was set to 0.03°, small enough to find intersections of convection cells and airways; (b) DBSCAN *min_samples* was set to 10; (c) DBSCAN distance *eps* was set to 250 m. The collection phase (Algorithms 1 and 2) can be seen as a preprocessing step of the tracking phase, as shown in Fig. 2. The tracking phase is responsible for generating new information to ATM stakeholders, and its findings are discussed below.

Running DBSCAN for each partition, the algorithm discovered the clusters delimiting the airways connecting GRU to BSB airports, and connecting BSB to GRU airports, as shown in Figs. 3a and b, respectively. In the clustering analysis, clusters delimiting the airways were not found in partitions close to the departure or destination airports. However, as the algorithm continues, clusters started to be found in partitions that were relatively distant from these airports. Figures 3a and b describe the latter scenario. One possible explanation for that fact is that when airplanes are close to their departure airport, they are in their takeoff or climb phases, and when close to their destination, they are in their descent or final approach phases, therefore not following specific airways. When aircraft are in their cruise phase, they start to fly in predefined airways connecting the departure and destination airports, clusters therefore happen to be discovered by the algorithm. The algorithm was able to find three clusters for the partition from GRU to BSB shown in Fig. 3a, around altitudes 10, 900 m, 11, 500 m and 12, 100 m, On the other hand, the algorithm found only one cluster in the partition from BSB to GRU shown in Fig. 3b, around altitude 11, 900 m. The difference between the results found might be related to the fact there were more flight positions registered from GRU to BSB than flight positions registered from BSB to GRU. Figure 3b suggests that other clusters may be found with more trained data, i.e., the purple group (outliers) around altitude 11, 200 m.

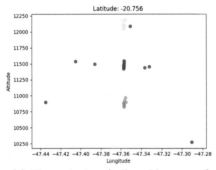 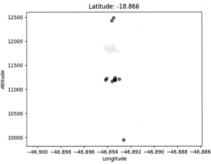

(a) Three clusters (yellow, blue, green) found in a partition from GRU to BSB. (b) One cluster (yellow) found in a partition from BSB to GRU.

Fig. 3. Clusters of flight positions found after running DBSCAN.

Algorithm 3 identified four convection cells blocking the airways connecting BSB to GRU. Their coordinates were: $(-19.45°, -46.75°)$, $(-20.75°, -46.55°)$, $(-21.05°, -46.55°)$, and $(-21.15°, -46.55°)$, each one having 5600 m of radius. The algorithm didn't identify any convection cells blocking the underlying airways in the opposite direction (from GRU to BSB). In conclusion, the solution produced the following results: (1) All convection cells output an impact factor of 1.0, which means that the underlying convection cell blocked all the airways found by the algorithm; (2) The convection cells impacted 11 flights in total; and (3) Checking the flight trajectories of the 11 flights indicated that the airplanes passed through the underlying convection cells. Finally, analyzing the results obtained: (1) Due to the small number of airways found by the algorithm connecting BSB to GRU (as shown in Fig. 3b), a simple intersection derives a high impact factor value; (2) The results suggest that the convection cells were not in enough number (such as in Fig. 1b) to support a change of course flight.

6 Conclusions

This work designs an intelligent system to constantly monitor the impact of severe convective weather on airways in the Brazilian airspace.

The normalization process was particularly useful at estimating flight positions on specific locations, and supporting therefore cluster analysis. The results also suggest that aircraft fly in their airways during the cruise phase, supported by the evidence of flight positions forming clusters. The clusters identified represent the boundaries of the airways connecting the departure/destination airports. Another important finding is that, depending on the locations of the convection cells, aircraft might not change their courses to overcome the cells. For further work: new case studies; comparison of different clustering algorithms; comparison between airways boundaries found and actual aeronautical charts, etc.

References

1. ICAO: Manual on system wide information management (swim) concept, (doc 10039). International Civil Aviation Organization (ICAO) (2015)
2. Data-Driven Aircraft Trajectory Prediction Research. http://dart-research.eu/the-project/
3. Ayhan, S., Samet, H.: Aircraft trajectory prediction made easy with predictive analytics. In: Proceedings of the 22nd ACM SIGKDD International Conference on Knowledge Discovery and Data Mining - KDD 2016 (2016)
4. Ayhan, S., Costas, P., Samet, H.: Predicting estimated time of arrival for commercial flights. In: Proceedings of the 24th ACM SIGKDD International Conference on Knowledge Discovery & Data Mining - KDD 2018 (2018)
5. Phases of a flight. https://www.fp7-restarts.eu/index.php/home/root/state-of-the-art/objectives/2012-02-15-11-58-37/71-book-video/parti-principles-of-flight/126-4-phases-of-a-flight.html
6. STSC. https://www.redemet.aer.mil.br/stsc/public/produto
7. OpenSky Network. https://opensky-network.org/
8. Phillips, G.: Interpolation and Approximation by Polynomials. Springer, New York (2005)
9. Veness, C.: Calculate distance and bearing between two Latitude/Longitude points using haversine formula in JavaScript. https://www.movable-type.co.uk/scripts/latlong.html
10. Ester, M., Kriegel, H., Sander, J., Xu, X.: A density-based algorithm for discovering clusters in large spatial databases with noise. In: Proceedings of the Second International Conference on Knowledge Discovery and Data Mining, pp. 226–231. AAAI Press (1996)

Fault Tolerant Control Using Interval Type-2 Takagi-Sugeno Fuzzy Controller for Nonlinear System

Himanshukumar Patel[(✉)] and Vipul Shah

Department of Instrumentation and Control Engineering,
Dharmsinh Desai University, Nadiad 387001, Gujarat, India
{himanshupatel.ic,vashah.ic}@ddu.ac.in

Abstract. A novel Fault Tolerant Control (FTC) strategy is proposed based on interval type-2 Takagi-Sugeno controller and conventional PID controller, in this paper it has been presented for a nonlinear system. The main advantage of this strategy is, it can handle the faults, uncertainties in describing the nonlinear systems and give rigorous fuzzy rules. This study assesses the design of the robust controller which can tolerate the system component and actuator faults into the system and maintain the performance of the system at an acceptable level. To validate the proposed FTC strategy three-tank interacting level control system is considered. Finally, the simulation of the three-tank interacting level process model demonstrates the superiority of the proposed strategy using two integral error performance index Integral Absolute Error (IAE) and Integral Square Error (ISE).

Keywords: Actuator fault · Interval Type-2 T-S Fuzzy logic control · Nonlinear system · PID controller · System component fault

1 Introduction

More and more attention has been paid to industrial systems as the industrial system have become more sophisticated and complex. This complexity leads to an increased demand for reliability, profitability and safety of the system as well as for human beings. In order to guarantee the above mentioned proprieties, it is essential to develop modern methods of supervision such as Fault Detection and Diagnosis (FDD) and Fault Tolerant Control (FTC) (Yang et al. 2014). Since two decades, the importance of FTC systems becomes increasingly apparent, and considerable amount of research has already been done in this area (Patton 1997; Korbicz *et al.* 2004; Guerra *et al.* 2006; Zhang and Jiang 2008; Noura *et al.* 2009; Puig 2010; Hao *et al.* 2011; Hao *et al.* 2015; Hupo and Yan 2017). A fault-tolerant control system is a control system that possesses the ability to accommodate system failures automatically, and to maintain overall system stability and acceptable performance (Patel and Shah 2018b). Two approaches are used to tolerate the faults into systems: the hardware redundancy method and the analytical redundancy method. In hardware redundancy method adding extra cost for mounting multiple sensors/actuators, and control algorithm is very complicated due to

© Springer Nature Switzerland AG 2020
A. Abraham et al. (Eds.): ISDA 2018, AISC 941, pp. 150–164, 2020.
https://doi.org/10.1007/978-3-030-16660-1_15

complexity of equipments and the management process (Boskovic and Mehra 2002). The advantage of the analytical redundancy method is that it makes possible to identify the occurrence of the fault and to deal with the unsafe situation using the information of the extent and class of the faults. To design point of view FTC scheme has classified in two types: Active Fault Tolerant Control (AFTC) and Passive Fault Tolerant Control (PFTC).

The AFTC method is based on the fact that the controller can reset its parameters or even change its structure, as well as implement rapid dynamic compensation control output to keep system stable after the fault occurs. To overcome this problem, the AFTC method must have the ability to actively acquire the information of the faults or system state changes. Currently, Fault Diagnosis and Isolation (FDI) approach is an important strategy for the design of AFTC. The AFTC is overly dependent on the FDI performance, and it may lead to failure of fault-tolerant control system, due to the mistake detection, omission, long delay or large diagnosis error caused by fault diagnosis mechanism.

The PFTC method is based on robust control ideas. It attributes the fault to the model parameter perturbation issue, and compensates the fault by the kind of fixed gain controller with strong robustness to ensure the system is insensitive to certain faults. It can be found that the PFTC method require designers to predict as many faults as possible before the controller design. According to the changing characteristics of the dynamic system model parameters in the fault states, a robust controller with fixed structure and gain are worked out, which it does not need to be adjusted on-line for the particular predicted fault modes. Its advantage is that the control law is fixed and easy to implement, and it can guarantee the system stability and achieve the preset performance whether or not a fault. However, the controller must calculate the parameters for the most serious fault condition that the system can take. It often lead to the control law too conservative, and with the increase of the system fault complexity, the controller design process will become more difficult.

According to a variety and severity of different faults that may affect the system, different levels of performance have to be considered in different fault scenarios. From safety region to danger region, degraded performances are often acceptable. In addition, to ensure that the closed-loop system be able to track a reference model/trajectory even in the event of faults. In the case of performance degradation, actuator faults (loss of effectiveness) and sensor faults avoidance being required, design controller needs to adjust the output that maintain the acceptable control performance in the event of faults.

In recent years, because of its combination of continuous and discrete dynamics researchers are attracted to develop progressively efficient control strategies for multi-tank level system. To the best of our knowledge, few works have published on the topic of fuzzy logic based fault tolerant system for interacting and non-interacting level control process. However, there are still many open issues which have to be addressed.

Although traditional fuzzy theory can resolve a number of problems, its robustness can be improved. Some researchers have proposed interval type-2 T-S fuzzy theory to improve the robustness of the fuzzy control systems. Type-2 fuzzy sets and their related definitions have been provided in related literature (Castillo and Melin 2008; Wang 2011; Sudha and Santhi 2011; Lam and Seneviratne 2008). In (Patel and Shah 2018a; Patel and Shah 2018c; and Patel and Shah 2018d) authors are investigate the

Passive FTC responses for interacting and non-interacting level control system for actuator, sensor and system component faults using fuzzy logic control. In (Patel and Shah 2018e) authors proposed and simulate the FTC for highly nonlinear level control system (i.e. two-tank non-interacting conical tank system) subject to actuator and system component faults, and found significant results. But less research and development on FTC using Interval Type-2 fuzzy Control system has been done before. Due to the advantages of Interval Type-2 fuzzy system and T-S fuzzy dynamic model, a FTC based on interval Type-2 T-S fuzzy control is proposed in this paper. It can show a great potential in handling various modeling as well as control applications.

This paper proposes a FTC strategy to tolerate the actuator and system component faults for interacting level control system based on Interval Type-2 T-S Fuzzy logic control (IT2FLC). In the present work, the capabilities of PFTC to handle two faults and reference trajectory management are exploited to design a fault tolerant controller.

The main contribution of this paper is to introduce a novel FTC strategy based on IT2FLC for interacting level control process subject to actuator and system component faults. The simulation results are shown. The paper is organized as follows. Design of FTC design using IT2FLC and PID controller are considered in Sect. 2. Problem description and mathematical model is detailed in Sect. 3. Section 4 is dedicated to the simulation results with different types of fault and magnitude, followed by conclusions in Sect. 5.

2   FTC Design Using Interval Type-2 T-S Fuzzy Logic Controller

2.1   Background of the Type-2 Fuzzy Logic Control

The concept of type-II fuzzy sets was also proposed by (Zadeh 1975) to overcome this limitation (Zadeh 1975). However, (Karnik and Mendel 1998) has provided more insight by developing and presenting the first type-II fuzzy logic system (T2FLS) in (Karnik and Mendel 1998). T2FLCs are able to outperform their colleague T1FLC, since their fuzzy sets are characterized by membership functions (MFs) which themselves are fuzzy compared to fuzzy sets of T1FLS, containing MFs having crisp values. Type-II fuzzy sets consists of Foot-print of Uncertainty (FoU), it is a bounded region of the uncertainty in the primary membership grades of a type-II MF. It is the union of all primary membership grades. Which makes it generally more robust and well capable to remove high oscillations. The interval type-II fuzzy logic system (IT2FLS), currently the most extensively used for their minimal computational cost is a special case of type-II FLS (Liang and Mendel 2000). Type-1 and type-2 fuzzy logic are mainly similar. The only essential difference between them which is the membership functions shape, besides the output process. Indeed, an interval type-2 fuzzy controller is consisting of: a fuzzifier, an inference engine, a rules base, a type reduction and a defuzzifier (Mendel 2011; Castillo and Melin 2012; Juan et al. 2009; Martínez et al. 2009; Wu and Tan 2006), the block diagram of the type-2 FLC is presented in Fig. 1.

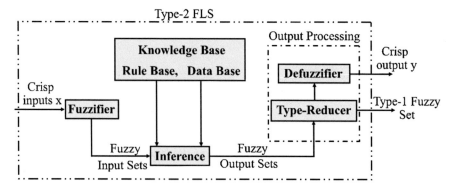

Fig. 1. Type-2 FLS structure block diagram.

2.2   Type-2 Fuzzy Logic Control

(A) Fuzzifier

The fuzzifier maps the crisp input vector $(e_1, e_2 \ldots, e_n)^T$ to a type-2 fuzzy system $\tilde{A}_x$, which very similar to the procedure performed in a type-1 fuzzy logic system.
Rules
The general form of the ith rule of the type-2 fuzzy logic system can be written as:

$$\text{If } e_1 \text{ is } \tilde{F}_1^i \text{ and } e_2 \text{ is } \tilde{F}_2^i \text{ and } \ldots e_n \text{ is } \tilde{F}_n^i, \text{Than } y^i = \tilde{G}^i \ i = 1, \ldots, M \tag{1}$$

Where:

$\tilde{F}_j^i$ represents the type-2 fuzzy system of the input state j of the ith rule, $x_1, x_2, \ldots, x_n$ are the inputs, $\tilde{G}^i$ is the output of the type-2 fuzzy system for the rule i, and M is the number of rules. As can be seen, the rule structure of type-2 fuzzy logic system is similar to type-1 fuzzy logic system except that type-1 membership functions are replaced by their type-2 counterparts.

(B) Inference engine

In a fuzzy interval type-2 using the minimum or product t-norms operations, the ith activated rule $F^i(x_1, \ldots x_n)$ produces the interval that is determined by two extremes $\underline{f}^i(x_1, \ldots x_n)$ and $\overline{f}^i(x_1, \ldots x_n)$ like written below (Liang and Mendel 2000):

$$F^i(x_1, \ldots x_n) = \left[\underline{f}^i(x_1, \ldots x_n), \underline{f}^i(x_1, \ldots x_n) \right] \stackrel{m}{=} \left[\underline{f}^i, \overline{f}^i \right] \tag{2}$$

Where $\underline{f}^i$ and $\overline{f}^i$ can be defined as follow:

$$\underline{f}^i = \underline{\mu}_{F_1^i}(x_1) * \ldots * \underline{\mu}_{F_n^i}(x_n) \tag{3}$$

$$\overline{f}^i = \bar{\mu}_{F_1^i}(x_1) * \ldots \bar{\mu}_{F_n^i}(x_n) \tag{4}$$

(C)Type reducer

After definition of the rules and executing the inference, the type-2 fuzzy system resulting in type-1 fuzzy system is computed. In this part, the available methods to compute the centroid of type-2 fuzzy system using the extention principle are discussed ((Mendel 2011). The centroid of type-1 fuzzy system A is given by:

$$C_A = \frac{\sum_{i=1}^{n} z_i w_i}{\sum_{i=1}^{n} w_i} \tag{5}$$

Where: n represents the number of discretized domain of A, $z_i \in R$ and $w_i \in [0, 1]$.

If each z_i and w_i is replaced by a type-1 fuzzy system (Z_i and W_i), with associated membership functions of $\mu_Z(z_i)$ and $\mu_W(W_i)$ respectively, and by using the extention principle, the generalized centroid for type-2 fuzzy system $\tilde{A}$ can be expressed by:

$$GC_{\tilde{A}} = \int_{z_1 \in Z_1} \cdots \int_{z_n \in Z_n} \int_{w_1 \in W_1} \cdots \int_{w_n \in W_n} \left[T_{i=1}^{n} \mu_Z(z_i)^* T_{i=1}^{n} \mu_W(z_i) \right] / \frac{\sum_{i=1}^{n} z_i w_i}{\sum_{i=1}^{n} w_i} \tag{6}$$

T is a t-norm and $GC_{\tilde{A}}$ is a type-1 fuzzy system. For an interval type-2 fuzzy system, it can be written:

$$GC_{\tilde{A}} = [y_l(x), y_r(x)]$$

$$= \int_{y^1 \in [y_l^1, y_r^1]} \cdots \int_{y^M \in [y_l^M, y_r^M]} \cdots \int_{f^1 \in [\underline{f}^1 \bar{f}^1]} \cdots \int_{f^M \in [\underline{f}^M \bar{f}^M]} 1 / \frac{\sum_{i=1}^{M} f^i y^i}{\sum_{i=1}^{M} f^i} \tag{7}$$

(D) Defuzzifier

To get a crisp output from a type-1 fuzzy logic system, the type-reduced set must be defuzzified. The most common method to do this is to find the centroid of the type-reduced set. If the type-reduced set Y is discretized to n points, then the following expression gives the centroid of the type-reduced set:

$$y_{output}(x) = \frac{\sum_{i=1}^{n} y^i \mu(y^i)}{\sum_{i=1}^{m} \mu(y^i)} \tag{8}$$

The output can be computed using the iterative Karnik Mendel Algorithms (Castillo and Melin 2012; Juan et al. 2009; Martínez et al. 2009; Wu and Tan 2006). Therefore, the defuzzified output of an interval type-2 FLC is:

$$y_{output}(x) = \frac{y_l(x) + y_r(x)}{2} \tag{9}$$

With:

$$y_l(x) = \frac{\sum_{i=1}^{M} f_l^i y_l^i}{\sum_{i=1}^{M} f_l^i} \quad \& \quad y_r(x) = \frac{\sum_{i=1}^{M} f_r^i y_l^i}{\sum_{i=1}^{M} f_r^i} \tag{10}$$

2.3   Interval Type-2 T-S Fuzzy Logic Controller

Model Rule i

$$\text{If } e_1 \text{ is } \tilde{F}_1^i \text{ and } e_2 \text{ is } \tilde{F}_2^i \text{ and } \ldots e_n \text{ is } \tilde{F}_n^i, \text{Than } \dot{x} = A_i x(t) + B_i u(t), i = 1, 2, \ldots, M$$

Here M is the number of IF-THEN rules. $x(t) \in \mathbb{R}^{n \times 1}$ is the state vector of the system. $u(t) \in \mathbb{R}^{m \times 1}$ is the input vector of the system. A is the state matrix of the system. B is the input matrix of the system.

Control Rule i

$$\text{If } e_1 \text{ is } F_1^i \text{ and } e_2 \text{ is } \tilde{F}_2^i \text{ and } \ldots e_n \text{ is } \tilde{F}_n^i, \text{Than } y^i = \tilde{G}^i \, i = 1, \ldots, M$$

where $\tilde{G}^i \in \mathbb{R}^{1 \times 1}$ is the output of the interval type-2 fuzzy system for the rule i using the centroid defuzzification.

2.4   Passive FTC Design

In order to eliminate the high oscillation, nonlinear system and model uncertainty, a continuous Interval Type-2 T-S Fuzzy logic control (IT2TSFLC) is used to approximate the discontinue control. The proposed control scheme is shown in Fig. 2; it contains conventional PID controller part and IT2TSFLC.

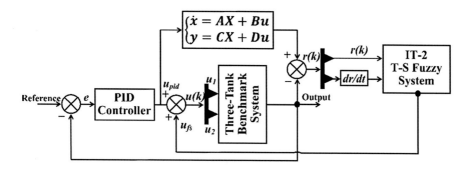

Fig. 2. Block diagram of proposed Passive FTC scheme.

The equivalent control (u_{eq}) is calculated in such a way to have rate of change of $r(k)$ is zero.

$$u_{fs} = IT2TSFLC\big(r(k), r(\dot{k})\big) \tag{11}$$

u_{fs} is the output of the IT2TSFLC, which depends on the normalized $r(k)$ and rate of change of $r(k)$.

All the membership functions of the fuzzy input variable are chosen to be gaussian for all upper and lower membership functions. The uses labels of the fuzzy variable residue and its derivative are: {negative big (NB), negative small (NS), medium (M), Positive small (PS) and positive big (PB)}. Figure 3 presents the type-2 membership functions for the IT2FLC. The corrective control is decomposed into five levels, so total rules can be 25 presented in Table 1.

$$u(k) = u_{fs} + u_{pid} \tag{12}$$

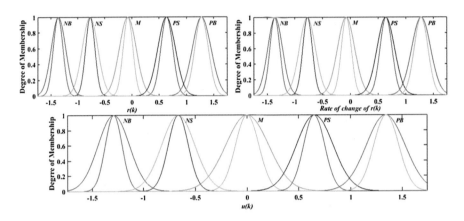

Fig. 3. Membership functions of input variables $(r(k), \dot{r}(k))$ and output $u(k)$.

Table 1. Fuzzy rules for IT2FLC.

u_{fs} $r(k)$	$\dot{r}(k)$				
	NB	NS	M	PS	PB
NB	NB	NS	M	PB	PB
NS	NB	M	M	PB	PB
M	M	M	PS	PS	PB
PS	M	PS	PS	PB	PB
PB	PS	PS	PB	PB	PB

3   Process Description and Mathematical Model

The plant used as the test-bench in this work is the three-tank hybrid system shown in Fig. 4. The system contains three water tanks that can be filled with two independent pumps acting on the outer tanks 1 and 2. The liquid flow rate q_1 and q_2 are the flow delivered by pump-1 and pump-2. And they can be continuously manipulated from a flow of 0 to a maximum flow q_{max}.

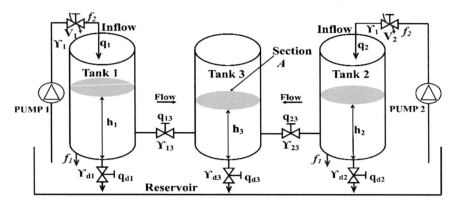

Fig. 4. Three-tank benchmark system prototype.

Interaction in the system is caused by interconnected pipes, which is connected to each other tanks. The flow over these pipes can be interrupted with switching valves V_{13} and V_{23} that can adapt either the completely open or the completely closed situation. The h_1, h_2 and h_3 are the liquid levels of each tank, it can be measured by differential pressure transmitter (DPT) sensor continuously, and this value can be calibrated in terms of water level [in cm]. The nominal outflow from the system are located at the bottom of each tank, having flow coefficients of k_{d1}, k_{d2} and k_{d3} respectively. By changing the valves in the system enables various configurations, which allows us to work with one, two or three tanks, at a time. The system can be configured as an interacting as well as a non-interacting system. (Nandola and Bhartiya 2008) have shown the following dynamic behaviour of the first principles model of the experimental setup:

$$A\frac{dh_1}{dt} = q_1 - q_{13}\Upsilon_{13} - q_{d1} \tag{13}$$

$$A\frac{dh_2}{dt} = q_2 - q_{23}\Upsilon_{23} - q_{d2} \tag{14}$$

$$A\frac{dh_3}{dt} = q_{13}\Upsilon_{13} + q_{23}\Upsilon_{23} - q_{d3} \tag{15}$$

Flows through the solenoid valve can be calculated by the following relations:

$$q_{13} = \Upsilon_{13}sign(h_1 - h_3)\sqrt{2g|h_1 - h_3|} \tag{16}$$

$$q_{23} = \Upsilon_{23}sign(h_2 - h_3)\sqrt{2g|h_2 - h_3|} \tag{17}$$

Outflows through the tanks can be calculated by the following equations:

$$q_{d1} = \Upsilon_{d1}\sqrt{2gh_1} \tag{18}$$

$$q_{d2} = \Upsilon_{d2}\sqrt{2gh_2} \tag{19}$$

$$q_{d3} = \Upsilon_{d3}\sqrt{2gh_3} \tag{20}$$

The range of the flow rate of pump-1 and pump-2 are about 0 to 373 cm^3/sec and 396 cm^3/sec respectively. Three-tank level control system process parameters are given in Table 2.

Table 2. Three-tank level control system process parameters.

Sr. No.	Parameter	Value with unit
1	Inner diameter of all tanks	15 cm
2	Height of tank	100 cm
3	Overflow height of tank	95 cm
4	Inner connecting pipe inner diameter	1.25 cm
5	Orifice diameter of tank	1.25 cm
6	Discharge co-efficient Υ_{13}	0.60 cm^2/sec
7	Discharge co-efficient Υ_{23}	0.72 cm^2/sec
8	Discharge co-efficient Υ_{d1}	0.85 cm^2/sec
9	Discharge co-efficient Υ_{d2}	1.03 cm^2/sec
10	Discharge co-efficient Υ_{d3}	0.1839 cm^2/sec

The output of the system considering $y = [h_1\ 1\ h_2]^T$ and input vector $u = [u_1 u_2]^T$. For the T-S model of the system following matrix is found form the system model:

$$A_1 = \begin{bmatrix} -0.0114 & 0 & 0.0114 \\ 0 & -0.025 & 0.0114 \\ 0.0066 & 0.0066 & -0.0132 \end{bmatrix}, A_2 = \begin{bmatrix} 0.0114 & 0 & 0.0114 \\ 0 & 0.025 & 0.0114 \\ 0.0066 & 0.0066 & 0.0132 \end{bmatrix}$$

$$B_1 = \begin{bmatrix} 64.9351 & 0 \\ 0 & 64.9351 \\ 0 & 0 \end{bmatrix}, B_2 = \begin{bmatrix} 6.4935 & 0 \\ 0 & 64.9351 \\ 0 & 0 \end{bmatrix}$$

4   Simulation Results

This section presents the results from the simulation of Passive FTC using interval type-2 T-S FLC subject to actuator and system component (tank leak) faults.

Mainly to assess the performance of PFTC, following conditions have been taken:

(i) When interaction is effect means while the level of tank $h_1 > h_2$.
(ii) When interaction is in effect and system component (tank-1 and tank-2 bottom leak) faults occur.
(iii) When interaction is in effect and actuator faults occur in V_1 and V_2 control valve.

In this literature two combination is taken for conditions one is (i) & (ii) have been accommodated together and second one is (i) & (iii) have been accommodated together.

For the design and simulation of the control system MATLAB environment has been used, in order to validate the efficiency of the propped FTC. The volumetric inflow rate q_1 and q_2 of both the pumps pump 1 and pump 2 respectively in the three-tank system used as manipulating variable.

Condition (i) and (ii) of investigation has accommodated together in Fig. 5. To ensure the possible interaction level and system component faults (f_1) occur in tank-1 and tank-2 bottom. Here also same reference-point consider for tank-1 50 cm and tank-2 40 cm. Here, first we are letting the controller track the set-point and brings the system in the steady state. After system reaching steady state (at t = 17 s) we are introducing leak in tank bottom. This causes the system to deviate from steady-state and degrade the performance, however PFTC taking the appropriate action to regulate the level by changing the manipulating variable.

Figure 6 shows the simulation results for set-point tracking of levels of tank-1 and tank-2 for the three-tank system for condition-(i & iii). For condition-(i) 50 cm and 40 cm height has been taken as a reference-point for tank-1 and tank-2 respectively, and for condition-(ii) Actuator faults taken in control valve V_1 and V_2 which is controlling the manipulated variable q_1 and q_2 volumetric inflow rate of pump 1 and pump 2 of three-tank system. In this simulation the actuator faults (f_2) behavior is abrupt in nature applied at t = 20 s after steady state height is achieved. Control signal generated by IT2TSFLC and PID controller for changing, manipulating variable (which is inflow rate) has also been plotted in Fig. 6, in order to represent the performance of controller with time.

Form the simulation error value are obtained, as the minimal value of ISE will represent the elimination of large errors quickly and often this leads to faster responses. The IAE and ISE error results for proposed FTC scheme presented in Tables 3 and 4 subject to different conditions and magnitudes.

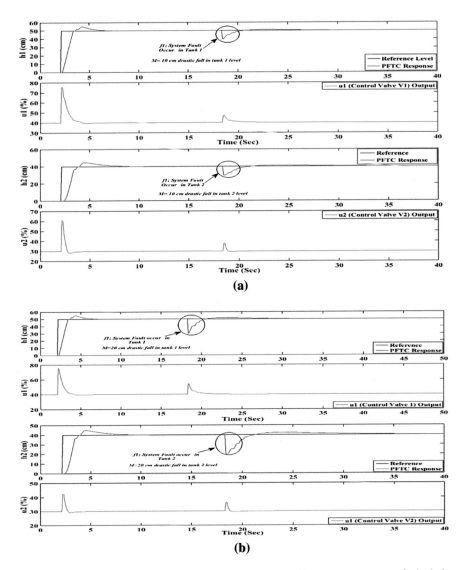

Fig. 5. Step responses for conditions (i and ii) (a) response with system component faults in low magnitude and (b) response with system component faults in high magnitude.

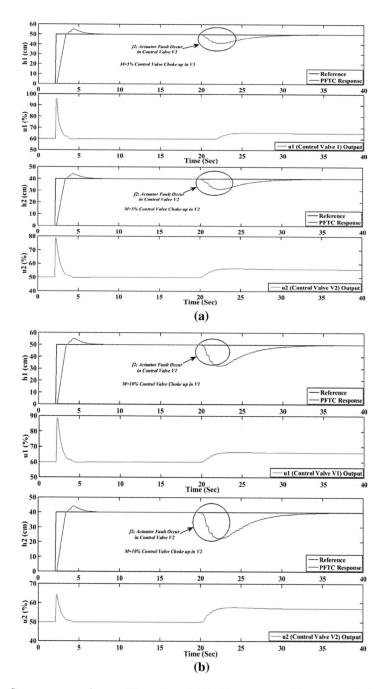

Fig. 6. Step responses for conditions (i and iii) (a) response with actuator faults in low magnitude and (b) response with actuator faults in high magnitude.

Table 3. IAE and ISE error results from simulation with low magnitude of faults.

Levels/conditions	IAE		ISE	
	h_1	h_2	h_1	h_2
Condition i & ii	0.2691	0.3681	4.5814	3.6912
Conditions i & iii	0.3941	0.4375	4.9347	4.1863

Table 4. IAE and ISE error results from simulation with high magnitude of faults.

Levels/conditions	IAE		ISE	
	h_1	h_2	h_1	h_2
Condition i & ii	0.5912	0.5318	10.4857	9.3628
Conditions i & iii	0.6721	0.5825	11.6328	10.9551

5   Conclusion

In this paper, an Interval Type-2 T-S FLC and PID controller has been used to design FTC for a three-tank level control system subject to actuator and system component faults, the simulation results have been produced and efficacy of the proposed FTC investigated using IAE and ISE error. The key benefit of the IT2TSFLC appears to be its expertise to remove persistent fluctuations, model uncertainty and nonlinear behavior of the system. As the IT2FLC can tolerate larger faults and no need to measured accurate faults magnitudes. The prime benefits of the proposed strategy is to tolerate system component and actuator faults and maintain control performance at acceptable level.

References

Boskovic, J.D., Mehra, R.K.: Failure detection, identification and reconfiguration in flight control. In: Fault Diagnosis and Fault Tolerance for Mechatronic Systems. Springer, New York (2002)

Castillo, O., Melin, P.: A review on the design and optimization of interval type-2 fuzzy controllers. Appl. Soft Comput. **12**, 1267–1278 (2012)

Castillo, O., Melin, P.: Type-2 Fuzzy Logic: Theory and Applications. Springer, Heidelberg (2008)

Guerra, P., Puig, V., Witczak, M.: Robust fault detection with unknown input set-membership state estimators and interval models using zonotopes. In: Proceedings of the 6th IFAC Symposium on Fault Detection, Supervision and Safety of Technical Processes, SAFE-PROCESS 2006, Beijing, China, pp. 1303–1308 (2006)

Patel, H.R., Shah, V.A.: Fault tolerant control systems: a passive approaches for single tank level control system. i-manager's J. Instrum. Control Eng. **6**(1), 11–18 (2018a)

Patel, H.R., Shah, V.A.: Fault detection and diagnosis methods in power generation plants-the Indian power generation sector perspective: an introductory review. J. Energy Manag. **02**(02), 31–49 (2018b)

Patel, H.R., Shah, V.A.: Fuzzy logic based passive fault tolerant control strategy for a single-tank system with system fault and process disturbances. In: IEEE 5th International Conference on Electrical and Electronic Engineering (ICEEE 2018c), Istanbul, Turkey, pp. 257–262 (2018c). https://doi.org/10.1109/iceee2.2018.8391342

Patel, H.R., Shah, V.A.: A framework for fault-tolerant control for an interacting and non-interacting level control system using AI. In: Proceedings of the 15th International Conference on Informatics in Control, Automation and Robotics, ICINCO, vol. 1, pp. 180–190 (2018d). ISBN 978-989-758-321-6. https://doi.org/10.5220/0006862001800190

Patel, H.R., Shah, V.A.: A fault-tolerant control strategy for non-linear system: an application to the two tank canonical non-interacting level control system. In: Proceedings of 2nd International Conference on Distributed Computing, VLSI, Electrical Circuits and Robotics (DISCOVER 2018), Mangalore, India, 13–14 August (2018e)

Ouyang, H., Lin, Y.: Adaptive fault-tolerant control for actuator failures: a switching strategy. Automatica **81**, 87–95 (2017)

Yang, H., Jiang, B., Staroswiecki, M., Zhang, Y.: Fault recoverability and fault tolerant control for a class of interconnected nonlinear systems. Automatica **54**, 49–55 (2015)

Yang, H., Staroswiecki, M., Jiang, B., Liu, J.: Fault tolerant cooperative control for a class of nonlinear multi-agent systems. Syst. Control Lett. **60**(4), 271–277 (2011)

Castro, J.R., Castillo, O., Melin, P., Rodríguez-Díaz, A.: A hybrid learning algorithm for a class of interval type-2 fuzzy neural networks. Inf. Sci. **179**, 2175–2193 (2009)

Korbicz, J., Kościelny, J., Kowalczuk, Z., Choleva, W.: Fault Diagnosis: Models, Artificial Intelligence, Applications, p. 920. Springer, Berlin (2004)

Karnik, N., Mendel, J.: Introduction to type-2 fuzzy logic systems. In: Fuzzy Systems Proceedings IEEE World Congress on Computational Intelligence, vol. 2, pp. 915–920 (1998)

Liang, Q., Mendel, J.: Interval type-2 fuzzy logic systems: theory and design. IEEE Trans. Fuzzy Syst. **8**(5), 535–550 (2000)

Mendel, J.: Uncertain Rule-Based Fuzzy Logic Systems: Introduction and New Directions. Prentice-Hall, Upper Saddle River (2011)

Martínez, R., Castillo, O., Aguilar, L.: Optimization of interval type-2 fuzzy logic controllers for a perturbed autonomous wheeled mobile robot using genetic algorithms. Inf. Sci. **179**(13), 2158–2174 (2009)

Noura, H., Theilliol, D., Ponsart, J., Chamssedine, A.: Fault-tolerant control systems: design and practical applications. In: Advances in Industrial Control. Springer, London (2009)

Nandola, N., Bhartiya, S.: A multiple model approach for the predictive control of nonlinear hybrid system. J. Process Control **18**, 131–148 (2008)

Puig, V.: Fault diagnosis and fault tolerant control using set-membership approaches: application to real case studies. Int. J. Appl. Math. Comput. Sci. **20**(4), 619–635 (2010). https://doi.org/10.2478/v10006-010-0046-y

Patton, R.: Fault-tolerant control systems: the 1997 situation. In: Proceedings of the IFAC Symposium on Fault Detection, Supervision and Safety for Technical Processes, Kingston Upon Hull, UK, pp. 1033–1054 (1997)

Sudha, K.R., Santhi, R.V.: Robust decentralized load frequency control of interconnected power system with generation rate constraint using type-2 fuzzy approach. Int. J. Electr. Power Energy Syst. **33**, 699–707 (2011)

Lam, H.K., Seneviratne, L.D.: Stability analysis of interval type- 2 fuzzy-model-based control systems. IEEE Trans. Syst. Man Cybern.-Part B: Cybern. **389**(3), 617–628 (2008)

Wang, P.: Interval type-2 fuzzy T-S modeling for a heat exchange process on CE117 process trainer. In: Proceedings of 2011 International Conference on Modelling, Identification and Control, pp. 457–462 (2011)

Wu, D., Tan, W.: A simplified type-2 fuzzy logic controller for real-time control. ISA Trans. **45**, 503–510 (2006)

Zhang, Y., Jiang, J.: Bibliographical review on reconfigurable fault-tolerant control systems. Annu. Rev. Control **32**(2), 229–252 (2008)

Zadeh, L.A.: The concept of a linguistic variable and its application to approximate reasoning-1. Inf. Sci. **8**, 199–249 (1975)

Yang, Y., Du, J., Liu, H., Guo, C., Abraham, A.: A trajectory tracking robust controller of surface vessels with disturbance uncertainties. IEEE Trans. Control Syst. Technol. **22**(4), 1511–1518 (2014)

A Semi-local Method for Image Retrieval

Hanen Karamti[1,2(✉)]

[1] MIRACL-ISIMS, BP242 City Ons, 3021 Sfax, Tunisia
karamti.hanen@gmail.com
[2] Princess Nourah bint Abdulrahman University,
PO Box 84428, Riyadh, Saudi Arabia

Abstract. The visual content of an image is expressed by global or local features. Global features describe some properties of the image such as color, texture and shape. Local features were successfully used for object category recognition and classification to extract the local information from a set of interest points or regions. In this paper, we propose a semi-local method to extract the features based on the previous features extraction methods. Our technique is called the "Spatial Pyramid Matching: SPM". It works by partitioning the image into increasingly fine sub-regions (or blocs) and computing histograms of global features found inside each bloc.

The results obtained by the proposed method are illustrated through some experiments on Wang and Holidays Dataset. The obtained Results show the simplicity and efficiency of our proposal.

Keywords: Global descriptors · Local descriptors ·
Spatial Pyramid Matching · Features · Image retrieval · Visual content ·
Semi-local

1 Introduction

Content Based Image Retrieval (CBIR) is the application of computer vision techniques to the image retrieval domain to search the digital images in large databases. It has been used in different application domains like health, cultural heritage, etc. Many CBIR systems have been developed for each domain such as (Query By Image Content) of IBM [1] and [2] for publishing, SPIRS [3] and IRMA [4] in medicine, etc. Content-based image retrieval is based on the features extraction to extract the visual content from images. The visual content of an image is expressed by global or local features. Global features describe some properties of the image such as color, texture and shape [5]. Local features were successfully used for object category recognition and classification to extract the local information from a set of interest points or regions [6].

A lot of work has been done on the global descriptors, which is justified by their large number. We can note in the literature: the color descriptors like the Scalable Color Descriptor (SCD) [7] and the Color Layout Descriptor (CLD) [8]. The shape descriptors like the invariant moments [24] and the texture descriptors such as the contour orientation histogram [9], and the homogeneous texture descriptor [10]. Many features have been developed since the 1980's, these features describe the overall

A. Abraham et al. (Eds.): ISDA 2018, AISC 941, pp. 165–172, 2020.
https://doi.org/10.1007/978-3-030-16660-1_16

information of the image. These global descriptors are very varied and their usefulness is highly recommended until today.

In the 90s, local descriptors attracted more attention in the CBIR domain [11]. Many systems, based on the local content of images, have been developed to improve the performance of CBIR, such as the system developed by Lowe [12] that proposes the scale invariants descriptor [13], and the system based on the Entity Transformation Descriptor (EIPD) to capture the local image information. The EIPD function detects the salient regions in each image and describes each local region with a vector of 128 features [12]. One of the main advantages of these descriptors is that the objects are described by their spatial information. So, an image that can be considered as a bag of features (BOF-points) [15] describing each object in image by points. With each BOF representation, the image retrieval is done by comparing the points represented in the query with those of the images on the collection. The similar works are SIFT [16], the GIST [14], the SURF [17], the BOW (Bag-of-Words) [18], etc. Local descriptors are advised in the object recognition domain because they ignore the image background and focus on find more interesting regions in image.

In order to benefit the advantages of local and global descriptors, some researchers use a hybrid method that uses all descriptors to extract the features based on the segmentation. This method can be considered as a semi-local method, image will be segmented on blocks or sub-regions, and then the features are extracted from each block. The extracted blocks can have the same size or different sizes. In fact, the search in blocks can do by global descriptors [19] or local descriptors [20]. The problem with these approaches is that each block must be compared to the different extracted blocks. This type of comparison is not always optimal if we work on large databases. An improvement for the semi-local method is the spatial pyramid (SP), which works on three levels of segmentation and each level has a fixed size. Spatial Pyramid is widely used in the field of pattern recognition to detect objects in an image. Different variants of SP representation have been used, but the original work of SP has been developed by Lazebnik et al. [21]. This technique works by partitioning the image into increasingly fine sub-regions and computing histograms of local features found inside each sub-region. This work shows an improvement compared to other features extraction methods in CBIR. Other extension of SP is developed [22, 23] adding the concept of overlapping spatial blocks. This work is useful for capturing spatial information [21]. In [25], a generalized deformable spatial pyramid (GDSP) is created to compute the similarity between images. This method is a solution to solve the problem of geometric variations of images. GDSP is another extension of DSP (Deformable Space Pyramid) which is used to treat deformations of images.

In this paper, we propose a new semi-local technique to extract the features from image. This technique combines the advantages of the global and local descriptors. Our semi local method, called Spatial Pyramid matching (SPM), segments the image in blocs under 3 levels and then it extracts from each bloc the global descriptors.

The remaining of the paper is organized as follows: we present our SPM method in Sect. 2. Section 3 presents and discusses the results obtained with our technique on two benchmark datasets. Finally, Sect. 4 concludes and presents further research directions.

2   SPM with Global Descriptors

To retrieve a specific region in image, researchers use the local descriptors to extract the features like SIFT [16], Haris Detector [27], etc. or through the segmentation of the image in blocks, namely the technique of Spatial Pyramid Matching (SPM). In this technique, the image is subdivided into blocks with the same size where each subdivision is assimilated to a level. That is, at each level of segmentation (we have 3 levels), the image is segmented into a finite number of fixed-sized blocks. The choice of the size of these blocks will make it possible to have homogeneous blocks which each contain a well-defined region.

The SPM is a partitioning method, the method partitions the image into $2^l * 2^l$ segments or blocks at each level ($l = 0, 1, 2$). Within $2l$ segments, we calculate a histogram for each descriptor, and then we concatenate these histograms to form a vector representation of the image. At level 0 the SPM works with extracted features for the entire image. If we use different descriptors, we will establish for each one a SPM, and then we take advantage of the late fusion according by rank [28] to merge the 4 SPM issues (one SPM for each descriptor).

To build our SPM, we start from the entire image, then we divide it into four rectangular cells and we keep the division until we reach the predefined number of levels of pyramid (3 levels). This is a classical spatial pyramid as described in literature. Indeed, SPM is applied in the literature with local descriptors such as SIFT and word bags. However, our goal is to apply SPM with global descriptors. Figure 1 illustrates the principle of adapting SPM in our CBIR system. In Fig. 1, we note that each block represents part of the image. The number of blocks is fixed by going from one resolution level to another ($N0$, $N1$ and $N2$). For level 0 ($N0$), the number of blocks is 1 because SPM works on the whole image. At level 1 ($N1$) we have 4 blocks and we have 16 blocks at level 2 ($N2$).

There are different methods in the literature for calculating the correspondence between the SPM of which we distinguish two methods: the kernel search method called Pyramid Match Kernels [26] and the Pyramid Spatial Matching method [21]. All of both were used with local descriptors only. Our goal is to apply SPM with global descriptors.

Taking into account the benefits of global descriptors, we tried to apply them to search images by regions. The first tests carried out by Spatial Pyramid with the global descriptors, using the two measurements Pyramid Matching Kernel and Spatial Pyramid Matching did not give good results. The accuracies found were inferior to those by a search with local descriptors. That's why we thought about using a new match function.

To apply the SPM with global descriptors, we have merged the two formulas used respectively for Pyramid Matching Kernel [26] and Spatial Pyramid Matching [21], then we obtain:

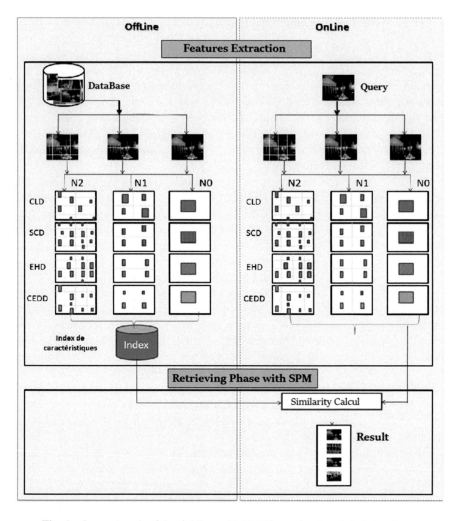

Fig. 1. Our approach of Spatial Pyramid Matching using the global descriptors

$$SPM^L(X, Y) = \sum_{m=1}^{M} SPM^L(X_m, Y_m)$$

$$SPM^L(X, Y) = I^0 + \frac{1}{2^L} \sum_{m=1}^{M} SPM^L(X_m, Y_m)$$

$$= I\left(H_X^0, H_Y^0\right) + \frac{1}{2^L} \sum_{m=1}^{M} SPM^L(X_m, Y_m)$$

$$= \sum_{i=1}^{D} \min\left(H_X^0(i), H_Y^0(i)\right) + \frac{1}{2^L} \sum_{m=1}^{M} SPM^L(X_m, Y_m)$$

Where H_X^0 and H_Y^0 are respectively the histograms of X and Y at the resolution in level 0. $H_X^0(i)$ and $H_Y^0(i)$ are the features of X and Y which correspond to the ith of the grid cell. Then the number of matches at level 0 is given by the intersection function of histograms, where D is the dimension of the descriptor:

$$I\left(H_X^l, H_Y^l\right) = \sum_{i=1}^{D} \min\left(H_X^l(i), H_Y^l(i)\right)$$

The weight associated with level l is set to $\frac{1}{2^l}$, for $l = 0...L - 1$, where L is the number of levels and is equals to 3.

Each channel m gives us two sets of two-dimensional vectors, X_m and Y_m, representing the coordinates of features of type m found in the respective images. All feature vectors are quantized into M (number of features of each descriptor) discrete types, and make the simplifying assumption that only features of the same type can be matched to one another. The final SPM is then the sum of the separate channel.

3 Experimental Comparison

We evaluate the performance of our model on two standard datasets:

- INRIA Holidays[1] dataset contains 1491 vacation snapshots corresponding to 500 groups each having the same scene or object. One image from each group serves as a query. The performance is reported as mean average precision over 500 queries.
- Wang[2] database is a subset of 1000 images of the Corel stock photo database which have been manually selected and forms 10 classes of 100 images each. The 10 classes are used for relevance estimation: given a query image, we assume that the user is searching for images from the same class, and therefore the remaining 99 images from the same class are considered relevant and other images irrelevant.

In this section, we will evaluate our SPM method and we will justify our choice of SPM as a semi-local search method. Table 1 represents a comparison on the Wang dataset between our SPM method, our initial search system (IS) using the global descriptors only (IS), and two other search systems that employ local descriptors (SIFT and CMI). The IS is based on a late fusion [28] of the 4 global descriptors ($CEDD$, SCD, CLD and EHD). In experiments, the CMI (Central Moment Invariant) makes recourse to a vocabulary of 300 words [24]. SPM is very sensitive to the descriptors used.

From Table 1, we notice that the MAP value of SPM (56.43) is better than the initial system (56.35), but it is close to it as well because the SPM is very sensitive to the descriptors used, we just reused the descriptors of our initial system on the blocks.

Comparing this value with $SIFT$ and CMI, we notice that SPM gives higher results than other systems. The results found show the effectiveness of SPM by using global

---

[1] https://lear.inrialpes.fr/~jegou/data.php.
[2] http://wang.ist.psu.edu/docs/related/.

Table 1. Comparison between SPM and other methods using the local descriptors on the Wang dataset. The improvement is significant (*) if p-value < 0.05.

Category	IS	SPM	SIFT	CMI
0	50.5	42.37	36.17	53.73
1	39.32	35.4	28.7	28.85
2	39.15	46.29	23.47	37.07
3	66.18	56.34	76.04	50.08
4	99.34	99.51	77.09	96.91
5	37.63	39.49	44.32	42.04
6	68.17	77.86	68.79	40.28
7	82.06	80.47	56.79	70.15
8	40.81	42.00	23.93	29.58
9	40.37	44.58	35.66	38.85
Average	*56.35*	*56.43(*)*	*47.09*	*48.75*

descriptors versus the use of other local descriptors. This improvement is significant compared to the initial search system because $p\text{-}value = 2.9E - 05 < 0.05$.

We also note that the *SPM* does not have a high exact accuracy for Wang's database. The precision of the *IS* is *56.35* while the SPM is *56.43*. A very small evolution of the values is significant that the *SPM* cannot be adapted to a collection of images having a single object.

Table 2 summarizes the results obtained by our initial system and by the *SPM* on the holiday's dataset. We find that the *MAP* of SPM (*0.73*) is better than the initial system, it is the same case with the median precision (*P-med*), the value of the *SPM* is *0.72* and the *IS* value is *0.6*. Concerning the calculation time, the *SPM* (*2613 ms*) is very slow than the *IS* (*741 ms*). So, this result corresponds the consumed time for the image segmentation and for the features extraction for each block.

Table 2. Comparison between SPM and our initial system on the holidays dataset.

Value of	IS	SPM
MAP	*0.68*	*0.73*
P-med	*0.6*	*0.72*
Standard deviation	*0.26*	*0.26*
Execution time	*741*	*2613*

4 Conclusion

This paper has presented an approach semi-local for image retrieval based on spatial of pyramid match matching. Our method, which works with the global descriptors, divides the images on blocs and computes the histograms of each bloc features for each level where the number of level is three. Our method is simple and achieves consistently an improvement over our initial system. This is not a trivial accomplishment, but our method present good results compared to other techniques that use the local descriptors. Future work focuses on the use of deep-learning technique to build the spatial pyramid matching.

References

1. Flickner, M., Sawhney, H.S., Ashley, J., Huang, Q., Dom, B., Gorkani, M., Hafner, J., Lee, D., Petkovic, D., Steele, D., Yanker, P.: Query by image and video content: the QBIC system. IEEE Comput. **28**, 23–32 (1995)
2. Gony, J., Cord, M., Philipp-Foliguet, S., Gosselin, P.H., Precioso, F., Jordan, M.: RETIN: a smart interactive digital media retrieval system. In: Proceedings of the 6th ACM International Conference on Image and Video Retrieval, pp. 93–96 (2007)
3. Hsu, W., Long, L.R., Antani, S.K.: SPIRS: a framework for content-based image retrieval from large biomedical databases. In: (MEDINFO) - Proceedings of the 12th World Congress on Health (Medical) Informatics - Building Sustainable Health Systems, pp. 188–192 (2007)
4. Deserno, T.M., Guld, M.O., Plodowski, B., Spitzer, K., Wein, B.B., Schubert, H., Ney, H., Seidl, T.: Extended query refinement for medical image retrieval. J. Digit. Imaging **21**, 280–289 (2008)
5. Schettini, R., Ciocca, G., Gagliardi, I.: Feature extraction for content-based image retrieval. In: Encyclopedia of Database Systems, pp. 1115–1119 (2009)
6. Anh, N.D., Bao, P.T., Nam, B.N., Hoang, N.H.: A new CBIR system using SIFT combined with neural network and graph-based segmentation. In: Intelligent Information and Database Systems, Second International Conference, ACIIDS, pp. 294–301 (2010)
7. Fung, Y.-H., Chan, Y.-H.: Producing color-indexed images with scalable color and spatial resolutions. In: Asia-Pacific Signal and Information Processing Association Annual Summit and Conference, Hong Kong, 16–19 December, pp. 8–13 (2015)
8. Imran, M., Hashim, R., Khalid, N.E.A.: Segmentation-based fractal texture analysis and color layout descriptor for content based image retrieval. In: 14th International Conference on Intelligent Systems Design and Applications, ISDA, pp. 30–33 (2014)
9. Dalal, N., Triggs, B.: Histograms of oriented gradients for human detection. In: Proceedings of the 2005 IEEE Computer Society Conference on Computer Vision and Pattern Recognition (CVPR 2005), vol. 1, pp. 886–893 (2005)
10. Ro, Y.M., Kim, M., Kang, H.K., Manjunath, B.S.: MPEG-7 homogeneous texture descriptor. ETRI J. **23**, 41–51 (2001)
11. Lowe, D.G.: Object recognition from local scale-invariant features. In: Proceedings of the International Conference on Computer Vision, vol. 2, p. 1150 (1999)
12. David, L.G.: Distinctive image features from scale-invariant keypoints. Int. J. Comput. Vis. **60**, 91–110 (2004)
13. Roy, S.K., Bhattacharya, N., Chanda, B., Chaudhuri, B.B., Ghosh, D.K.: FWLBP: a scale invariant descriptor for texture classification (2018)
14. Douze, M., Jegou, H., Sandhawalia, H., Amsaleg, L., Schmid, C.: Evaluation of GIST descriptors for web-scale image search. In: Proceedings of the 8th ACM International Conference on Image and Video Retrieval, CIVR, Santorini Island, Greece, 8–10 July, p. 19 (2009)
15. Delaitre, V., Laptev, I., Sivic, J.: Recognizing human actions in still images: a study of bag-of-features and part-based representations. In: Proceedings of the British Machine Vision Conference, pp. 1–11 (2010)

16. Wu, C.: SiftGPU: a GPU implementation of scale invariant feature transform (SIFT) (2007)
17. Bay, H., Tuytelaars, T., Van Gool, L.: Surf: speeded up robust features. In: ECCV, pp. 404–417 (2006)
18. Uijlings, J.R.R., Smeulders, A.W.M.: Visualising bag-of-words. In: Demo at ICCV (2011)
19. Li, X.: Image retrieval based on perceptive weighted color blocks. Pattern Recogn. Lett. **24**, 1935–1941 (2003)
20. Takala, V., Ahonen, T., Pietikäinen, M.: Block-based methods for image retrieval using local binary patterns. In: Proceedings of the 14th Scandinavian Conference on Image Analysis (SCIA), pp. 882–891 (2005)
21. Lazebnik, S., Schmid, C., Ponce, J.: Beyond bags of features: spatial pyramid matching for recognizing natural scene categories. In: Proceedings of the 2006 IEEE Computer Society Conference on Computer Vision and Pattern Recognition, vol. 2, pp. 2169–2178 (2006)
22. Yang, J., Li, Y., Tian, Y., Duan, L., Gao, W.: Group-sensitive multiple kernel learning for object categorization. In: ICCV (2009)
23. Harada, T., Ushiku, Y., Yamashita, Y., Kuniyoshi, Y.: Discriminative spatial pyramid. In: IEEE-CVPR, pp. 1617–1624 (2011)
24. Doretto, G., Yao, Y.: Region moments: fast invariant descriptors for detecting small image structures. In: 2010 IEEE Conference on Computer Vision and Pattern Recognition (CVPR), pp. 3019–3026 (2010)
25. Hur, J., Lim, H., Park, C., Chul Ahn, S.: Generalized deformable spatial pyramid: geometry-preserving dense correspondence estimation. In: IEEE-CVPR, pp. 1392–1400 (2015)
26. Grauman, K., Darrell, T.: The pyramid match kernel: efficient learning with sets of features. J. Mach. Learn. Res. **8**, 725–760 (2007)
27. Harris, C., Stephens, M.: A combined corner and edge detector. In: Proceedings of Fourth, pp. 147–151 (1988)
28. Karamti, H., Tmar, M., Visani, M., Urruty, T., Gargouri, F.: Vector space model adaptation and pseudo relevance feedback for content-based image retrieval. Multimed. Tools Appl. **77**, 5475–5501 (2018)

Physical Modeling of the Tread Robot and Simulated on Even and Uneven Surface

Rashmi Arora[1] and Rajmeet Singh[2]

[1] Chandigarh University, Gharuan, Mohali, Punjab, India
rashmimech.cu@gmail.com
[2] Baba Banda Singh Bahadur Engineering College,
Fatehgarh Sahib, Punjab, India
rajmeet.singh@bbsbec.ac.in

Abstract. Over the years, the researchers have done a lot to developed different surface mobile robot such as wheel, walking and clawing robots. A new method of modeling and simulation of robot based on *SimMechanics* is proposed in this paper. The proposed robot consist of sets of wheel with belt. Due to use of belt the robot have capability of the high gripping with surface. The proposed model of the robot with ground contact model is used for maneuvering on even and uneven surface. The surface with bumps is considered as uneven surface for the simulation purpose. The simulation results for even and uneven surface are compared for tread robot. The simulation result shows the performances of the proposed model of the robot.

Keywords: Tread robots · SimMechanics · Even · Uneven surface · Simulation

1 Introduction

Robotics is an exciting, dynamic interdisciplinary field of study. The fundamental capability of robotic systems can be recognized in [1]

- Mechanical versatility
- Re-programmability

Due to the kinematic and mechanical design of the robot manipulator arm, it is capable of performing a variety of tasks which can be understood as Mechanical versatility of a robot. Because of the capability of the robot's controller and computer facilities, it has the flexibility to perform a variety of task operations which can be termed as re-programmability of a robot. The field of robotics includes all human operated systems and makes improved working conditions in the modern world. With the use of robotics, efficiency of any work is increased up to a great extent with minimum errors. It is basically an automation area integrating various techniques like different mechanisms, electronic sensors and control systems, artificial intelligence and embedded systems [2]. The biped walking robot named as EP-WAR2 (Electro pneumatic Walking Robot) was designed [3] to walk on flat floors and to climb stairs with good performance. The leg mechanism for this biped robot was composed of a pantograph and a double articulated

© Springer Nature Switzerland AG 2020
A. Abraham et al. (Eds.): ISDA 2018, AISC 941, pp. 173–181, 2020.
https://doi.org/10.1007/978-3-030-16660-1_17

parallelogram whereas actuation of each leg is done by pneumatic cylinders. Different types of stair climbing robots have been developed in the past with wheelchair mechanism for elderly and handicapped people [4, 5]. Similarly, crawler robots to cross the obstacles have also been studied [6] for its application in inspection of power transmission line. Even the stair climbing robots with load carrying capacity has also been operated using Arduino [7]. The crawling robots have also been used in pipes in hazarding working conditions and places unapproachable to humans [8]. An integrated jumping-crawling robot was proposed [9–12] which is controlled by trajectory. Here the height of jumping can also be controlled. The modeling of the crawling robot is done in [13] which have capability to move in the tube section.

The *SimMechanics Matlab* toolbox is a graphical representation of system dynamics that are used for development of the robotic arm [14]. It is difficult to mode the complex model of the robotic arm therefore it is directly import from CAD software to *Simulink* model. The CAD model of three Mecanum wheel based mobile robot is developed and simulated into Matlab environment [15]. The *SolidWorks* and *Matlab/SimMechanics* based modeling of multi spindle drilling SCARA robot is developed [16]. The Simmechanics based modeling of the biped robot is proposed in [17]. In this paper, the *SimMechanics Matlab* toolbox is used to develop the dynamic model of the tread robot. The *Simulink* based contact model is also proposed for the robot to move on the even and uneven surface.

The motivation for the work is a physical modeling of the tread robot based on *SolidWorks* and *Matlab/SimMechanics* tools and simulated on uneven surfaces. This is not reported in literature. The robot consist of the 8 wheels (4 small and 4 large) connected by the belt. Each wheel of the robot is controlled by separated DC motor which increases the torque of the robot during maneuvering on uneven surfaces. The contact surface model is proposed for the high gripping capacity of the robot. The proposed model is used to simulate the tread robot performance on even and uneven surface in same path. The novelty of the proposed approach is that it is applicable to the all uneven and rough terrains such as deep valleys, up stairs, down stairs and bumps etc. In this work the approach is applied on the tread robot and simulated for the surface with bumps.

The paper is organized as follow. In Sect. 2, the physical modeling of the tread robot is presented using *Simulink Simscape Toolbox*. The CAD model is developed first and then *Simulink* model is converted from CAD model *xml* file. Also the ground contact model is proposed. In Sect. 3, the simulation results for proposed model for even and uneven surface is presented. Finally, concluding remarks are presented in Sect. 4.

2   Physical Modeling of Tread Robot

2.1   CAD Model of the Robot

A schematic figure of the tread robot is shown in Fig. 1. As shown in Fig. 1, the robot consists of body, 8 set of wheels (4 large and 4 small) and flat belt. The individual motors are used to move the left and right tread of the robot. The both tread is connected to the body of the robot.

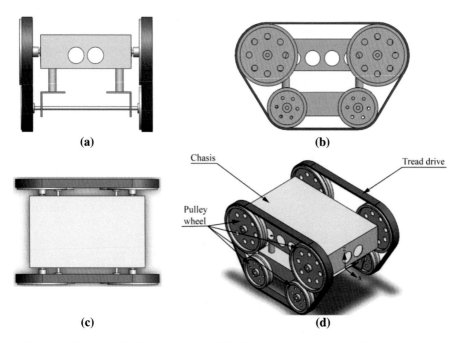

(a)   **(b)**

(c)   **(d)**

Fig. 1. CAD model (a) Front view, (b) Side view, (c) Top view and (d) Isometric view

2.2   Simulink Model of the Tread Robot

In this section the *Simulink* model of the robot. The CAD model is imported to the *SimMechanics Matlab* toolbox. First of all, the *Simulink* model of right tread is developed and then applied for left tread. Figure 2 shows the robot CAD and *Simulink* mechanics explorer environment representation.

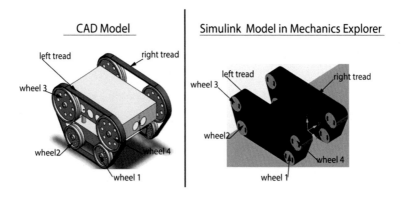

Fig. 2. Representation of tread robot in CAD and *Simulink* environment

Figure 3 shows the schema of the right tread robot. The input to the revolute joints is the angle (θ). The each wheel revolute joints are attached to the coordinate transformation frame (CTF) to the right tread of the robot. Also the revolute joints are attached to the robot body. The complete Simulink model of the tread robot is shown in the Fig. 4. The Solver configuration, world frame and mechanism configuration blocks are used to build the robot model under Simscape environment. The Solver Configuration block specifies the solver parameters that your model needs before you can begin simulation. Each topologically distinct Simscape block diagram requires exactly one Solver Configuration block to be connected to it. This block represents the global reference frame in a model. This frame is inertial and at absolute rest. Rigidly connecting a frame to the World frame makes that frame inertial. Frame axes are orthogonal and arranged according to the right-hand rule. This block provides mechanical and simulation parameters to a mechanism, i.e., a self-contained group of interconnected Simscape™ Multibody™ blocks. Parameters include gravity and a linearization delta for computing numerical partial derivatives during linearization. These parameters apply only to the target mechanism, i.e., the mechanism that the block connects to. In our model the parameters for mechanism configuration is [0, 0, −0.98] m/s2. The signal builder is used to feed the rotational angle (degree) of the DC motor shaft. The tread of the robot is movable on the ground surface.

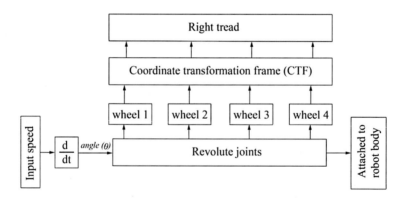

Fig. 3. Schema of *Simulink* model (right tread)

2.2.1   Ground Contact Model

The model of the contact with ground is a vital part of the robot model. When robot tread touches the ground it exerts a certain force. According to Newton's third law, ground applies on the foot a force of equal intensity and opposite direction. The movement of the body on a given surface is achieved due to this action-reaction pair. *Simscape Multibody* Contact Forces Library block *"Sphere to Plane Contact Force (3D)"* and *"Belt to plane Contact Force"* is used to develop the foot ground contact model of robot. Figure 5 shows the detail of the block "*Sphere to Plane Contact Force*". This block implements a contact force between a sphere and a plane. The force is active above and below the plane.

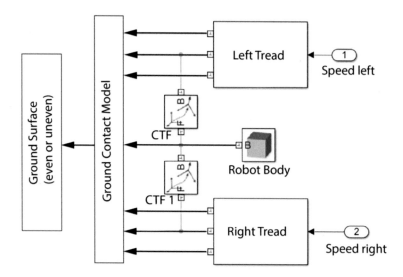

Fig. 4. Complete *Simulink* model of the tread robot

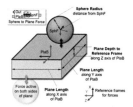

Fig. 5. Ground contact force model (*Simmechanics model*)

3   Simulation Results

In this section, the simulation is done for proposed tread robot on even and uneven surfaces. Figure 6 shows the simulation environment. The uneven surface is created by introducing some bumps on the surface as shown in Fig. 6(b). The bumps may be small or big in size.

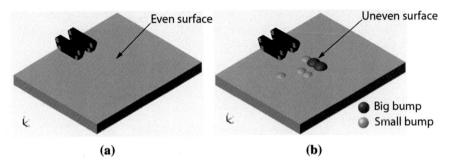

Fig. 6. Simulation environment (a) Even surface (b) Uneven surface

The simulation parameters are shown in the Table 1.

Table 1. Simulation parameters

Parameters	Description	Value
G_x	Ground x dimension	12 (m)
G_y	Ground y dimension	15 (m)
G_z	Ground z dimension	0.6 (m)
C_s	Contact stiffness	50^5 (N/m)
C_d	Contact damping	10^5 (N/(m/s))
T	Simulation time	52 (s)
T_l	Tread length	2.8 (m)
T_k	Co-eff. of kinetic friction	0.8
T_s	Co-eff. of static friction	1

The comparison of the simulation results for even and uneven surface is shown in Fig. 7. Figure 7(a–c) shows the displacement of the tread robot in X, Y and Z direction. The variation in result during uneven surface is depicted in Fig. 7(a–c).

During the motion on the even surface there is no deviation in Z direction but in case of uneven surface the deviation is shown in Fig. 7(c) because of the presence of the bumps. The change in velocity in X direction during even and uneven surface is illustrated in Fig. 7(d). The yaw angle (rad) during movement on even and uneven surface is shown in the Fig. 7(e). Similarly, the path travelled by the robot during maneuvering on even and uneven surface is shown in Fig. 7(f).

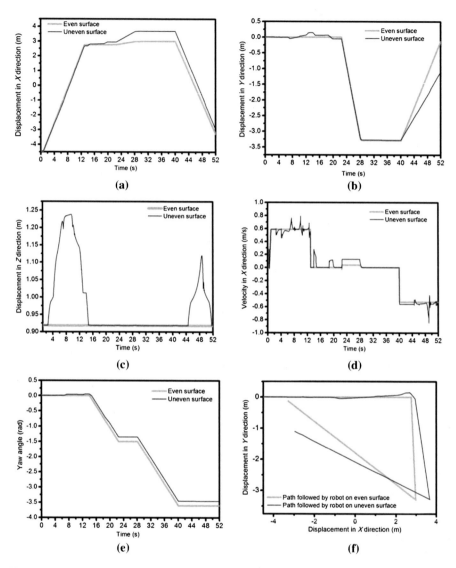

Fig. 7. Simulation results during even and uneven surface (a) Displacement in X direction, (b) Displacement in Y direction, (c) Displacement in Z direction, (d) Velocity in X direction, (e) Yaw angle and (f) Path followed by robot.

4   Conclusions

The physical modeling of the tread robot based on *SimMechanics* approach was proposed in this paper. The CAD model of the robot is developed and import into *SimMechanics Matlab/Simulink* model. The ground contact friction model is used for moving the wheels on the tread part of the robot and also belt on plane ground contact model is used to generate the motion between tread and ground surface. The proposed tread robot model was simulated on even and uneven surface to check the performance of the robot. The simulation results were shown and discussed. The proposed robot model with contact model is suitable for all types of the surfaces. Also in future the model will be tested on stairs (up or down). The applications such as pick and place, obstacle avoidance will be considered in the future scope.

References

1. Cecccarelli, M., Ottaviano, E.: Kinematic design of manipulators (2004). http://cdn. intechopen.com/pdfs-wm/5580.pdf
2. Narendra, K., Gopichand, A., Gopala, M., Gopi Krishna, B.: Design and development of adjustable stair climbing robot. Int. J. Res. Eng. Technol. **2**, 232–267 (2013)
3. Figliolini, G., Ceccarelli, M.: Climbing stairs with EP-WAR2 biped robot. In: IEEE International Conference on Robotics and Automation, Seoul, Korea (2001)
4. Shiatsu, T., Lawn, M.: Modeling of a stair-climbing wheelchair mechanism with high single-step capability. IEEE Trans. Neural Syst. Rehabil. Eng. **11**, 323–332 (2003)
5. Wang, M., Tu, Y.: Design and implementation of a stair-climbing robot. In: IEEE Workshop on Advanced Robotics and its Social Impacts, Taipei, Taiwan (2008)
6. Wang, J., Sun, A., Zheng, C., Wang, J.: Research on a new crawler type inspection robot for power transmission lines. In: 1st International Conference on Applied Robotics for the Power Industry, Montreal, QC, Canada (2010)
7. Jeyabalaji, C., Vimalkhanna, V., Avinashilingam, N., Zeeshan, M., Harish, K.: Design of low cost stair climbing robot using Arduino. Int. J. Eng. Res. Appl. **4**, 15–18 (2014)
8. Singh, P., Ananthasuresh, G.K.: A compact and compliant pipe-crawling robot. IEEE Trans. Robot. **29**, 251–260 (2013)
9. Jung, G., Casarez, C.S., Jung, S.P., Fearing, R.S., Cho, K.: An integrated jumping-crawling robot using height adjustable jumping module. In: International Conference on Robotics and Automation, ICRA, Stockholm, Sweden (2016)
10. Liu, J., Wang, S., Li, B.: Analysis of stairs-climbing ability for a tracked reconfigurable modular robot. In: IEEE International Workshop on Safety, Security and Rescue Robotics, Kobe, Japan, pp. 36–41 (2005)
11. Brynedal, N., Rasmusson, N., Matsson, J.: An overview of legged and wheel robotic locomotion. In: 12th International Conference on Advancement of Robotics and its Mechanism, California, USA, pp. 23–34 (2012)
12. Yoneda, K., Ota, Y., Hirose, S.: Stair climbing robots and high-grip crawler. In: Scorpus Robotics and Terminology for Stair Climbing, Tokyo, Japan, pp. 74–96 (2002)
13. Nakamura, T., Iwanaga, T.: Locomotion strategy for a peristaltic crawling robot in a 2-dimensional space. In: IEEE International Conference on Robotics and Automation, Pasadena, CA, USA, pp. 25–29 (2008)
14. Udai, A.D., Rajeevlochana, C.G., Saha, S.K.: Dynamic simulation of a KUKA KR5 industrial robot using MATLAB SimMechanics. In: 15th National Conference on Machines and Mechanisms, pp. 1–8. IIT Kanpur, India (2015)

15. Kubela, T., Pochyly, A., Singule, V.: Advanced tools for multi-body simulation and design of control structures applied in robotic system development. In: Solid State Phenomena, vol. 164, pp. 387–391 (2010)
16. Mariappan, S.M., Veerabathiran, A.: Modelling and simulation of multi spindle drilling redundant SCARA robot using SolidWorks and MATLAB/SimMechanics. Revista Facultad De Ingenieria, Universidad de Antioquia **81**, 63–72 (2016)
17. Yuan, S., Liu, Z., Li, X.: Modeling and simulation of robot based on Matlab/SimMechanics. In: 27th Chinese Control Conference, Kunming, China, pp. 8–12 (2008)

ipBF: A Fast and Accurate IP Address Lookup Using 3D Bloom Filter

Ripon Patgiri[✉], Samir Kumar Borgohain, and Sabuzima Nayak

National Institute of Technology Silchar, Silchar 788010, Assam, India
ripon@cse.nits.ac.in, samir@nits.ac.in, sabuzimanayak@gmail.com

Abstract. IP address lookup is a crucial part of router in Computer Network. There are millions of IP addresses to be searched per second. Hence, it is immensely necessitated to enhance the performance of the IP address lookup. Therefore, this paper presents a novel approach of IP address lookup using 3D Bloom Filter, called ipBF. ipBF inherits the properties of 3D Bloom Filter. Thus, ipBF features - (a) high accuracy, (b) low memory consumption, and (c) high performance. In addition, ipBF consumes $8 - bits$ per IP address which is very less as compared to its contemporary solution. Besides, ipBF filters the false positive probability in eight layers by deploying eight 3D Bloom Filters. Hence, ipBF is able achieve higher accuracy. We show the accuracy using theoretical calculations.

Keywords: Bloom Filter · IP address lookup · Router · Networking · 3D Bloom Filter · Prefix matching

1 Introduction

Bloom Filter is introduced by Burton Howard Bloom in 1970 [4]. Bloom Filter is extensively experimented and applied in various domain since its inception. The application domains are, namely, Big Data [8], IoT [28], Computer Networking [10,21,27], Network Security and privacy [2,7,11,20,26,29], Biometrics [5,12,25], Bioinformatics [16,22,31], and Metadata server [14,15,30]. Bloom Filter emerges due to diverse applicability in Computer Network. Most of the Bloom Filters are developed for Computer Networking purposes. For instance, counting Bloom Filter. In recent year, Bloom Filter is applied to enhance various Networking Systems. For instance, IP address lookup. Performance enhancement of IP address lookup is a grand challenge in Modern Computer Networking. There are numerous IP address lookup techniques available, however, the space and time complexity become barriers for all those techniques. Therefore, Bloom Filter is deployed in IP lookup for the advantages of both space and time complexity. Moreover, Bloom Filter becomes a crucial part of router in routing table. Mun and Lim [23] presents a new variant of IP address lookup using Bloom Filter. Bloom Filter is used to reduce the number accesses of off-chip memory while membership of an IP address is programmed in on-chip Bloom Filter. Thus,

© Springer Nature Switzerland AG 2020
A. Abraham et al. (Eds.): ISDA 2018, AISC 941, pp. 182–191, 2020.
https://doi.org/10.1007/978-3-030-16660-1_18

IP address lookup is unaffected by false positive. Off-chip IP address is stored using a trie based algorithm. Kwon et al. [17] proposes IP address lookup using Cuckoo Filter. As a consequence, the propose system inherits the disadvantage of Cuckoo Filter, for instance, kick-off issues [9]. Moreover, Byun and Lim [6] presents a vectored Bloom Filter for IP lookup without supporting of off-chip memory in most cases. Bloom Filter is also deployed in IPv6 address lookup by the support of GPU [18,19].

Despite many research works, there is a call for high accuracy Bloom Filter for IP address lookup. There are diverse variants of Bloom Filter which can be deployed in IP address lookup, for instance, counting Bloom Filter. However, false positive key barrier for counting Bloom Filter. For other variants, the space complexity become a key barrier of the IP address lookup system. Also, performance enhancement is an ultimate grand challenge. The performance of Bloom Filter depends on the number complex operators, and hashing algorithm.

Current state-of-the-art IP lookup system uses different variant of Bloom Filters, however, there are trade-offs among space consumption, performance, and accuracy. The IP address lookup system requires a low spaced table, very high performance, and high accuracy to be deployed in network devices. There is a high demand for new variants of Bloom Filter which meets all requirements.

The contribution of this paper is as follows - this paper presents a novel IP address lookup system using Bloom Filter which meets all required parameters, called ipBF. ipBF deploys 3D Bloom Filter (3DBF) [24] for better performance and low space consumption. 3DBF uses a single bit per input item which is the lowest on-chip memory consumption. ipBF deploys eight 3DBFs to implement high accuracy IP address lookup. Moreover, the bitwise operator enhance the performance of ipBF. Also, ipBF is able to stand without support of off-chip memory.

This paper is organized as follows - Sect. 2 presents our proposed system. In this proposed system, 3DBF is exposed and deployed to achieve our target. Section 3 presents a theoretical analysis on the performance, accuracy and space consumption. Finally, the paper is concluded in Sect. 4.

2   ipBF: The Proposed System

2.1   3D Bloom Filter

Let us assume, $\mathbb{B}$ be the 3-dimensional Bloom Filter (3DBF) [24], and $\mathbb{B}_{x,y,z}$ be the 3D array to implement Bloom Filter. The 3DBF is initialized by zero. The x, y and z be the dimension of the filter, which are prime numbers. Let, $\mathbb{B}_{i,j,k}$ be a particular location in $\mathbb{B}$. Let ρ be the bit position of a particular cell $B_{i,j,k}$. The cell size of $B_{i,j,k}$ depends on the memory occupied by the filter for each cell, termed as β, for example, 64-bits. Let, κ be an item to input in 3DBF. Now, hash function $h = \mathcal{H}(\kappa)$ returns a value, where $\mathcal{H}()$ is murmur hashing [3]. Instead of placing κ in various locations, 3DBF performs four modulus operations and places κ in a single bit position. The operations are - $i = h\%x$, $y = h\%y$, $k = h\%z$, and $\rho = h\%\beta$, where $\%$ is a modulus operator and β is the number

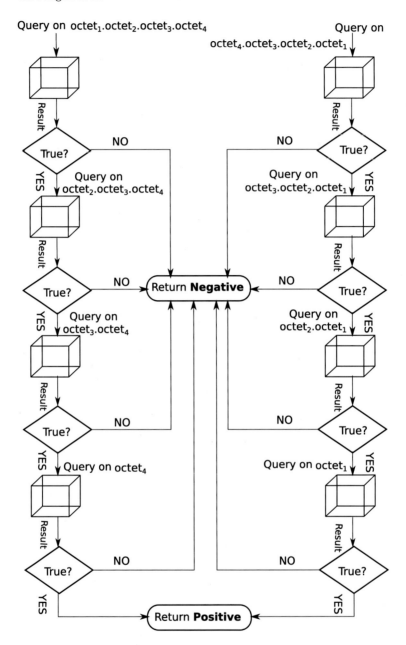

Fig. 1. ipBF architecture.

bits size per cell of the Bloom array. Thus, the item κ is inserted into the 3DBF using Eq. (1). The existence of the item κ is tested using Eq. (2). Therefore, an item requires a single bit in 3DBF to store.

$$\mathbb{B}_{i,j,k} \leftarrow \mathbb{B}_{i,j,k} \; OR \; (1 << \rho) \tag{1}$$

where OR is a bitwise operator. Similarly, Eq. (2) is invoked to test whether a particular bit is set or not.

$$Flag \leftarrow (\mathbb{B}_{i,j,k} \oplus (1 << \rho))AND(1 << \rho) \tag{2}$$

where AND is a bitwise operator. If $Flag$ and $\mathbb{B}_{i,j,k}$ are zero, then the particular bit is set to zero; one otherwise. The Eqs. (1) and (2) is used for set and test a bit respectively.

2.2 Insertion

ipBF inserts an IP address in eight 3DBF. Let, the eight 3DBF be $\mathbb{B}_1$, $\mathbb{B}_1'$, $\mathbb{B}_2$, $\mathbb{B}_2'$, $\mathbb{B}_3$, $\mathbb{B}_3'$, $\mathbb{B}_4$, and $\mathbb{B}_4'$. Let $octect_1.octet_2.octet_3.octet_4$ be the IP address and its reversal is $octect_4.octet_3.octet_2.octet_1$ as depicted in Fig. 1. All four octets of IP address are inserted into the $\mathbb{B}_1$ using Eq. (1), and reversal IP address is inserted into the $\mathbb{B}_1'$ using Eq. (1). For second level insertion, an octet is removed and inserted into $\mathbb{B}_2$ using Eq. (1), i.e., the IP address becomes $octet_2.octet_3.octet_4$ and it is inserted into $\mathbb{B}_2$ using Eq. (1). For another part, remove an octet from the reversal IP address and it is inserted into $\mathbb{B}_2'$ using Eq. (1), i.e., the reversal IP address becomes $octet_3.octet_2.octet_1$ and it is inserted into $\mathbb{B}_2'$ using Eq. (1), and so on. At the final phase, IP address contains a single octet, and the remaining octet is $octet4$. The single octet is inserted into $\mathbb{B}_4$ using Eq. (1). On the other hand, the reversal IP address also contains a single octet and the remaining octet is $octe_1$. This remaining octet is inserted into $\mathbb{B}_4'$ using Eq. (1). Therefore, the insertion of an IP address passes through eight phases. However, each item occupies a single bit in 3DBF, and hence, all eight insertions cause $8 - bits$ consumption of on-chip memory.

2.3 Lookup

An IP address is arranged in two formats, particularly, forward order, and reverse order. Checking of the two formats is serial. Therefore, let us, first, examine the existence of an IP address in forward direction. The lookup query sends an IP address in four octet format, i.e., the IP address is $octect_1.octet_2.octet_3.octet_4$. The lookup query is passed to $\mathbb{B}_1$. Using Eq. (2), if $\mathbb{B}_1$ returns true, then proceeds for next lookup. Otherwise, return false. In the next step, remove an octet from the IP address, and query the resulting IP address to $\mathbb{B}_2$. If $\mathbb{B}_2$ returns true, then proceeds to the next step, otherwise returns false, and so on. Equation (3) define the rules for returning 'true' or 'false'.

$$Flag = (\mathbb{B}_1 = 0?false : (\mathbb{B}_2 = 0?false : (\mathbb{B}_3 = 0?false : (\mathbb{B}_4 = 0?false : true)))) \tag{3}$$

Similarly, the IP address is reversed as $octect_4.octet_3.octet_2.octet_1$. This IP address is passed through $\mathbb{B}'_1$. Using Eq. (2), if $\mathbb{B}'_1$ returns true, then proceeds to the next step, returns negative otherwise. Again, remove an octet and lookup at $\mathbb{B}'_2$ for the reverse IP address. If $\mathbb{B}'_2$ returns true, then proceeds to the next step, otherwise return negative, and so on. Equation (4) defines the rules for returning 'true' or 'false'.

$$Flag' = \left(\mathbb{B}'_1 = 0? false : \left(\mathbb{B}'_2 = 0? false : \left(\mathbb{B}'_3 = 0? false : \left(\mathbb{B}'_4 = 0? false : true\right)\right)\right)\right)$$
(4)

Finally, ipBF returns 'true' if both $Flag$ and $Flag'$ have 'true' value. Otherwise, ipBF returns 'false'. This an eight layered filters to remove the false positive probability.

2.4   Hashtable

The hashtable is programmed in on-chip memory. The hashtable retains the routing information. The addresses are hashed into hashtable. If ipBF returns true, then hashtable is looked-up. Therefore, ipBF enhance in avoiding the unnecessary access to the hashtable. Thus, IP address lookup does not depend on off-chip memory.

3   Analysis

3.1   False Positive Probability

The false positive probability is calculated by Let m be the bit size of 3DBF, n be the total input size of a 3DBF, and thus, the probability of a bit to be '0' is [1]

$$\left(1 - \frac{1}{m}\right)^n$$
(5)

Unlike conventional BF, an item is placed in single location, i.e., a single bit of a 3DBF. Thus, the number of hash function is $h = 1$. Therefore, the probability of total bit to be '1' is

$$\left(1 - \left(1 - \frac{1}{m}\right)^n\right)$$
(6)

Let, X be the random variable of total number of bit filled by '1', then

$$E[X] = m\left(1 - \left(1 - \frac{1}{m}\right)^n\right)$$
(7)

Grandi [13] demonstrates the false positive probability using $\delta-transformation$. Let, X be the random variable to represent the total number of '1' in the BF, then

$$E[X] = m\left(1 - \left(1 - \frac{1}{m}\right)^n\right)$$

The probability of false positive is conditioned to a number by $X = x$, then

$$Pr(FP|X = x) = \left(\frac{x}{m}\right)$$

Therefore, false positive probability is

$$FPP_s = \sum_{x=0}^{m} Pr(FP|X = x)Pr(X = x)$$

$$FPP_s = \sum_{x=0}^{m} \left(\frac{x}{m}\right) f(x)$$

where $f(x)$ is probability mass function of X. Grandi [13] deploys $\delta -$ $transformation$ to calculate $f(x)$ and derives FPP as follows-

$$FPP_s = \sum_{x=0}^{m} \left(\frac{x}{m}\right) \binom{m}{x} \sum_{j=0}^{x} (-1)^j \binom{x}{j} \left(\frac{x-j}{m}\right)^n \qquad (8)$$

Equation 8 is calculation of false positive probability of a 3DBF. ipBF is constructed using eight 3DBF. Therefore, the total false positive probability of ipBF is

$$FPP_{ipBF} = \prod_{p=1}^{8} \left(\sum_{x=0}^{m} \left(\frac{x}{m}\right) \binom{m}{x} \sum_{j=0}^{x} (-1)^j \binom{x}{j} \left(\frac{x-j}{m}\right)^n \right) \qquad (9)$$

$$FPP_{ipBF} = \left(\sum_{x=0}^{m} \left(\frac{x}{m}\right) \binom{m}{x} \sum_{j=0}^{x} (-1)^j \binom{x}{j} \left(\frac{x-j}{m}\right)^n \right)^8 \qquad (10)$$

The false positive probability (FPP) is very low as shown in Eq. (9).

3.2 Memory Consumption

The ipBF consume very low memory per IP address. Figure 1 depicted that ipBF uses eight 3DBF. Thus, the ipBF inherits the properties of 3DBF. 3DBF consumes a single bit per input item. Therefore, ipBF consumes $8 - bits$ per IP address. Conventionally, the size of IPv4 address is 32bytes. For higher accuracy Bloom Filter, $8 - bits$ per IP address is reasonable and justifiable.

3.3 Accuracy

Lookup accuracy is the utmost important part of Bloom Filter. Most of the Bloom Filters suffer from false positive. ipBF reduces the false positive probability by deploying eight 3DBF. Figure 1 depicts the architecture of ipBF. Four 3DBFs are dedicated for the forward direction of input IP address, and other four 3DBFs are dedicated to reverse order of input IP address. Let us analyze the forward direction of IP address first. The IP address consists of four octets. The four octets are queried to $\mathbb{B}_1$. If $\mathbb{B}_1$ returns negative, then ipBF concludes that no such element is found. Otherwise, remove an octet from the input IP address, and the three octet input is queried to next 3DBF $\mathbb{B}_2$. Again, if this 3DBF returns negative, then ipBF concludes that no such element is present in the Bloom Filter. Otherwise, remove an octet from the three octet input, and so on. Finally, the input contains single octet which is queried to 3DBF $\mathbb{B}_4$. If $\mathbb{B}_4$ may return true or false.

On the other hand, the reversal of input IP address is queried to $\mathbb{B}'_1$. If $\mathbb{B}'_1$ returns negative, then ipBF concludes that there is no such kind of element present in the filter. Otherwise, remove an octet from reverse input. The three octet input is queried to $\mathbb{B}'_2$. If $\mathbb{B}'_2$ returns negative, then ipBF concludes that there is no such element present in the filter. Otherwise, remove an octet and query to next 3DBF. Finally, the single octet is queried to $\mathbb{B}'_4$. The $\mathbb{B}'_4$ may return either true or false. If both $\mathbb{B}_4$ and $\mathbb{B}'_4$ returns true, then ipBF concludes that the element is present in the filter. Otherwise, ipBF concludes that there is no such element present in the filter. If both $\mathbb{B}_4$ and $\mathbb{B}'_4$ returns true directly implies that all six 3DBFs have decided that the element is present in the filter.

Suppose, $\mathbb{B}_1$ returns a false positive. Let us also assume, $\mathbb{B}'_1$ returns a false positive. We agree that both can return false positive results. Furthermore, $\mathbb{B}_2$ and $\mathbb{B}'_2$ return a false positive on the same input, and so on. Finally, we also assume that $\mathbb{B}_4$ and $\mathbb{B}'_4$ also return a false positive. But, the last two filters cannot return false positive, since, all filter cannot return false positive. Otherwise, the ipBF is saturated. In the scenario of saturation, ipBF is input without checking its memory capacity. Therefore, we claim that "ipBF returns a false positive" is impractical.

3.4   Performance

ipBF achieves high accuracy without compromising the performance of filters. The performance of ipBF is inherited from 3DBF. Let us analyze the performance of 3DBF. 3DBF deploy a fast string hashing algorithm, murmur hashing [3]. Instead of placing k places in conventional Bloom Filter, 3DBF uses four modulus operations to place an input item in the correct location. A single bit location is selected to place the item. Moreover, 3DBF avoids using complex arithmetic operations, and uses bitwise operator for performance enhancement. The time complexity of murmur hashing is the length of the string. In our case, string length is $O(\mathcal{L})$ where $\mathcal{L}$ is the length of an IP address. The length of an IP address is constant, hence, the time complexity of 3DBF is

$$T_c = O(\mathcal{L}) + \mathcal{C} \tag{11}$$

where $\mathcal{C}$ is other constant operations. Therefore, the total time complexity is

$$T_c = O(\mathcal{L} + \mathcal{C}) \tag{12}$$
$$T_c = O(1) \tag{13}$$

Now, the total time complexity of ipBF is derived from Eq. (12), which is

$$T_{ipBF} = 8 \times O(\mathcal{L} + \mathcal{C}) \tag{14}$$

$$T_{ipBF} = O(1) \tag{15}$$

Since, ipBF uses eight 3DBF. Therefore, ipBF can achieve extremely high performance in lookup of IP addresses. Also, ipBF achieves high accuracy without compromising the performance.

4 Conclusion

This paper has presented a novel variant of Bloom Filter and IP address lookup using 3DBF, called ipBF. ipBF is able to achieve high performance, high accuracy and low space consumption. Also, ipBF uses less number of complex arithmetic operators to improve its performance. In addition, ipBF filters the probability of false positive in eight layers. Therefore, ipBF have less number of false positive probability as compared to contemporary solution. Besides, ipBF consumes space at most $8 - bits$ per IP address of on-chip. Moreover, ipBF is able to stand itself without supporting of the off-chip memory.

Acknowledgement. Authors would like to acknowledge TEQIP-III, NIT Silchar for supporting this research work.

References

1. Almeida, P.S., Baquero, C., Preguica, N., Hutchison, D.: Scalable bloom filters. Inf. Process. Lett. **101**(6), 255–261 (2007)
2. Antikainen, M., Aura, T., Särelä, M.: Denial-of-service attacks in bloom-filter-based forwarding. IEEE/ACM Trans. Netw. **22**(5), 1463–1476 (2014). https://doi.org/10.1109/TNET.2013.2281614
3. Appleby, A.: Murmur hash (2018). https://sites.google.com/site/murmurhash/. Accessed Aug 2018
4. Bloom, B.H.: Space/time trade-offs in hash coding with allowable errors. Commun. ACM **13**(7), 422–426 (1970). https://doi.org/10.1145/362686.362692
5. Bringer, J., Morel, C., Rathgeb, C.: Security analysis and improvement of some biometric protected templates based on bloom filters. Image Vis. Comput. **58**(Supplement C), 239–253 (2017). https://doi.org/10.1016/j.imavis.2016.08.002
6. Byun, H., Lim, H.: IP address lookup algorithm using a vectored bloom filter. Trans. Korean Inst. Electr. Eng. **65**(12), 2061–2068 (2016). https://doi.org/10.5370/kiee.2016.65.12.2061
7. Calderoni, L., Palmieri, P., Maio, D.: Location privacy without mutual trust: the spatial bloom filter. Comput. Commun. **68**(Supplement C), 4–16 (2015). https://doi.org/10.1016/j.comcom.2015.06.011. Security and Privacy in Unified Communications: Challenges and Solutions
8. Chang, F., Dean, J., Ghemawat, S., Hsieh, W.C., Wallach, D.A., Burrows, M., Chandra, T., Fikes, A., Gruber, R.E.: Bigtable: a distributed storage system for structured data. ACM Trans. Comput. Syst. **26**(2), 4:1–4:26 (2008). https://doi.org/10.1145/1365815.1365816

9. Fan, B., Andersen, D.G., Kaminsky, M., Mitzenmacher, M.D.: Cuckoo filter: practically better than bloom. In: Proceedings of the 10th ACM International on Conference on Emerging Networking Experiments and Technologies, pp. 75–88. ACM, New York (2014). https://doi.org/10.1145/2674005.2674994

10. Gao, W., Nguyen, J., Wu, Y., Hatcher, W.G., Yu, W.: A bloom filter-based dual-layer routing scheme in large-scale mobile networks. In: 2017 26th International Conference on Computer Communication and Networks (ICCCN), pp. 1–9 (2017). https://doi.org/10.1109/ICCCN.2017.8038405

11. Geravand, S., Ahmadi, M.: Bloom filter applications in network security: a state-of-the-art survey. Comput. Netw. **57**(18), 4047–4064 (2013). https://doi.org/10.1016/j.comnet.2013.09.003

12. Gomez-Barrero, M., Rathgeb, C., Li, G., Ramachandra, R., Galbally, J., Busch, C.: Multi-biometric template protection based on bloom filters. Inf. Fusion **42**(Supplement C), 37–50 (2018). https://doi.org/10.1016/j.inffus.2017.10.003

13. Grandi, F.: On the analysis of bloom filters. Inf. Process. Lett. **129**, 35–39 (2018). https://doi.org/10.1016/j.ipl.2017.09.004

14. Hua, Y., Zhu, Y., Jiang, H., Feng, D., Tian, L.: Supporting scalable and adaptive metadata management in ultralarge-scale file systems. IEEE Trans. Parallel Distrib. Syst. **22**(4), 580–593 (2011)

15. Huo, Z., Xiao, L., Zhong, Q., Li, S., Li, A., Rua, L., Wang, S., Fu, L.: MBFS: a parallel metadata search method based on bloomfilters using mapreduce for large-scale file systems. J. Supercomput. **72**(8), 1–27 (2015)

16. Jackman, S.D., Vandervalk, B.P., Mohamadi, H., Chu, J., Yeo, S., Hammond, S.A., Jahesh, G., Khan, H., Coombe, L., Warren, R.L., et al.: ABySS 2.0: resource-efficient assembly of large genomes using a bloom filter. Genome Res. **27**(5), 768–777 (2017)

17. Kwon, M., Reviriego, P., Pontarelli, S.: A length-aware cuckoo filter for faster IP lookup. In: 2016 IEEE Conference on Computer Communications Workshops (INFOCOM WKSHPS), pp. 1071–1072 (2016). https://doi.org/10.1109/INFCOMW.2016.7562258

18. Lin, F., Wang, G., Zhou, J., Zhang, S., Yao, X.: High-performance IPv6 address lookup in GPU-accelerated software routers. J. Netw. Comput. Appl. **74**, 1–10 (2016). https://doi.org/10.1016/j.jnca.2016.08.004

19. Lucchesi, A., Drummond, A.C., Teodoro, G.: High-performance IP lookup using Intel Xeon Phi: a bloom filters based approach. J. Internet Serv. Appl. **9**(1), 3 (2018). https://doi.org/10.1186/s13174-017-0075-y

20. Maccari, L., Fantacci, R., Neira, P., Gasca, R.M.: Mesh network firewalling with bloom filters. In: 2007 IEEE International Conference on Communications, ICC 2007, pp. 1546–1551. IEEE (2007)

21. Marandi, A., Braun, T., Salamatian, K., Thomos, N.: BFR: a bloom filter-based routing approach for information-centric networks. arXiv preprint arXiv:1702.00340 (2017)

22. Melsted, P., Pritchard, J.K.: Efficient counting of k-mers in DNA sequences using a bloom filter. BMC Bioinform. **12**(1), 333 (2011). https://doi.org/10.1186/1471-2105-12-333

23. Mun, J.H., Lim, H.: New approach for efficient IP address lookup using a bloom filter in trie-based algorithms. IEEE Trans. Comput. **65**(5), 1558–1565 (2016). https://doi.org/10.1109/TC.2015.2444850

24. Patgiri, R., Nayak, S., Borgohain, S.K.: rDBF: a r-dimensional bloom filter for massive scale membership query. J. Netw. Comput. Appl. (2019). Accepted

25. Sadhya, D., Singh, S.K.: Providing robust security measures to bloom filter based biometric template protection schemes. Comput. Secur. **67**(Supplement C), 59–72 (2017). https://doi.org/10.1016/j.cose.2017.02.013

26. Sarela, M., Rothenberg, C.E., Aura, T., Zahemszky, A., Nikander, P., Ott, J.: Forwarding anomalies in bloom filter-based multicast. In: 2011 Proceedings IEEE INFOCOM, pp. 2399–2407 (2011). https://doi.org/10.1109/INFCOM.2011. 5935060

27. Sasaki, K., Nakao, A.: Packet cache network function for peer-to-peer traffic management with bloom-filter based flow classification. In: 2016 18th Asia-Pacific Network Operations and Management Symposium (APNOMS), pp. 1–6 (2016). https://doi.org/10.1109/APNOMS.2016.7737214

28. Singh, A., Garg, S., Batra, S., Kumar, N., Rodrigues, J.J.: Bloom filter based optimization scheme for massive data handling in IoT environment. Future Gener. Comput. Syst. (2017). https://doi.org/10.1016/j.future.2017.12.016

29. Xiao, P., Li, Z., Qi, H., Qu, W., Yu, H.: An efficient DDoS detection with bloom filter in SDN. In: 2016 IEEE Trustcom/BigDataSE/ISPA, pp. 1–6 (2016). https:// doi.org/10.1109/TrustCom.2016.0038

30. Zhu, Y., Jiang, H., Wang, J., Xian, F.: HBA: distributed metadata management for large cluster-based storage systems. IEEE Trans. Parallel Distrib. Syst. **19**(6), 750–763 (2008)

31. Ziegeldorf, J.H., Pennekamp, J., Hellmanns, D., Schwinger, F., Kunze, I., Henze, M., Hiller, J., Matzutt, R., Wehrle, K.: BLOOM: BLoom filter based oblivious outsourced matchings. BMC Med. Genomics **10**(2), 44 (2017). https://doi.org/10. 1186/s12920-017-0277-y

From Dynamic UML/MARTE Models to Early Schedulability Analysis of RTES with Dependent Tasks

Amina Magdich[1(✉)], Yessine Hadj Kacem[2], Bouthaina Dammak[1],
Adel Mahfoudhi[3], and Mohamed Abid[3]

[1] College of Community, Princess Nourah bint Abdulrahman University,
Riyadh, Saudi Arabia
{ASMagdich,BAdammak}@pnu.edu.sa
[2] College of Computer Science, King Khalid University, Abha, Saudi Arabia
y.hadjkacem@kku.edu.sa
[3] CES Laboratory, National School of Engineers of Sfax (ENIS), Sfax, Tunisia
adel.mahfoudhi@yahoo.fr, mohamed.abid@ceslab.org

Abstract. The process of verifying whether Real-Time Embedded Systems (RTES) meet their temporal requirements is a major step during the system design. This step, called schedulability analysis, must be carried out at early design stages to avoid system failures. Currently, researchers are interested in using high-level techniques to raise the abstraction level and reduce the designers' effort. Nevertheless, only the scheduling approaches that prohibit task migration have been supported. An attempt to consider semi-partitioned and global scheduling approaches, which allow task migration, has been recently proposed. However, it doesn't support dependent tasks. In this context, this paper proposes an automatic process for early schedulability analysis considering dependent tasks and scheduling approaches with task migration. The focus is on the transformation of dynamic models annotated through the Unified Modeling Language (UML) profile for Modeling and Analysis of Real-Time Embedded systems (MARTE) to the model of SimSo tool.

Keywords: Semi-partitioned scheduling · Global scheduling ·
MARTE · MDE · Dynamic models · SimSo ·
Model to Text transformation

1 Introduction

Temporal determinism remains a goal to satisfy while developing Real-Time Embedded Systems (RTES). In fact, the allocation of tasks on the available processors while meeting deadlines, represents an important issue. This step is typically performed using scheduling theory. The latter consists in assigning tasks to be executed on the available processors while respecting the temporal requirements. With this regard, three major scheduling approaches are available

© Springer Nature Switzerland AG 2020
A. Abraham et al. (Eds.): ISDA 2018, AISC 941, pp. 192–201, 2020.
https://doi.org/10.1007/978-3-030-16660-1_19

to schedule multiprocessor systems; the partitioned, semi-partitioned and global scheduling approaches. The partitioned approach assigns each task to be executed on only one processor during its periods. Using this approach comes to apply the uni-processor scheduling since each task can be allocated to only one processor. A downside of the partitioned approach is that it prevents task migration. Consequently, some computing resources may be in an idle state while a ready task is waiting for execution. Using the partitioned approach, the CPU utilization rate may not be optimal. To overcome the shortcomings caused by the use of the partitioned approach, scheduling approaches that enable task migration have been proposed; mainly the semi-partitioned and global scheduling approaches. The global scheduling approach allows tasks to migrate inter-processors; it enables full task migration. The use of this approach avoids idles processors while tasks are waiting to be executed. Therefore, it improves the CPU utilization. As for the semi-partitioned approach, it enables a controlled task migration. This approach assigns tasks to be executed on specific processors like in the partitioned scheduling approach. However, tasks which could not be assigned to any processor are allowed to migrate inter-processors. Compared to the global scheduling approach, the semi-partitioned one offers less task preemption and migration. Accordingly, the migration costs and cache misses may be reduced.

Regardless of the scheduling approach, the dramatically increasing complexity of RTES has made their scheduling a difficult step that needs to be managed using rigorous methodologies to reduce designers' effort and avoid system failures. A prominent effort has been focused on the use of high-level techniques, which automate the development flows, promote model reuse and raise the abstraction level. Basically, earlier results have been focused on the use of the Model Driven Engineering (MDE) to support the scheduling step at a high-level of abstraction. In particular, the UML/MARTE profile is an adequate solution that has been adopted to support the modeling of such systems during the different stages of RTES development. In fact, the use of high-level methodologies helps designers to overcome the growing complexity of RTES and accelerate the scheduling step. However, there are still open issues regarding the temporal constraint verification step called schedulability analysis step. Because of the big variety of schedulability analysis tools, bridging the gap between system design models and models of the used scheduling analysis tools becomes an important issue. Bearing this in mind, some research studies have focused on the automatic transformation of system models to schedulability analysis tool models. Nevertheless, only the partitioned approach has been supported. An attempt to support the semi-partitioned and global scheduling approaches has been recently proposed [6], but only independent tasks have been considered. Accordingly, in this paper we propose a methodology for automatic schedulability analysis using the semi-partitioned and global scheduling approaches while supporting dependent tasks. Our proposal illustrates the process for the transformation of an UML/MARTE model to the meta-model of the schedulability analysis tool SimSo [1].

The remainder of this paper starts with some related works in Sect. 2. In Sect. 3, we compare some existing schedulability analysis tools and we justify the use of SimSo tool. Section 4 illustrates the proposed process for automatic and early schedulability analysis step. To validate our proposal, a case study is given in Sect. 5. Finally, conclusions and future work are discussed in Sect. 6.

2   Motivation and Related Work

Schedulability analysis is a major step that enables the verification of system temporal behavior. It represents an important issue that was widely studied during the last years. However, there are still many deficiencies which need to be carefully studied. In particular, the big variety of schedulability analysis tools represents a major challenge during the temporal verification step.

Each analysis tool is based on an input model which may be defined using different modeling languages such as UML, XML, SysML, AADL, MARTE. Since the input model of analysis tools differ from one tool to another, there is a need to propose approaches which reduce the gap between system models and analysis tool models. In this context, authors of [3] have proposed to build a specific UML/MARTE-based model that represents an input for the MAST tool [2]. This proposal has two major shortcomings; Firstly, it has been used to support only uni-processor systems. Secondly, the construction of input models for analysis tools has been done manually. To overcome these shortfalls, some research studies have been focused on the automatic transformation of system models to models conforming to the analysis tool meta-model to automatically analyze uni-processor or multiprocessor systems. Researchers' attention has been then focused on the transformation of system models to analysis tool meta-models in order to analyze uniprocessor or multiprocessor systems [7,8]. In [7], MARTE models have been transformed to the meta-model of the cheddar tool [9] to analyze the temporal behavior of multiprocessor systems. Nevertheless, only the partitioned approach that prevents task migration is supported. In the same context, an attempt to translate MARTE models to analysis tool meta-models has been highlighted in [8]. In this paper, authors have proposed a M2M transformation to transform activity diagrams annotated using MARTE/SAM (Schedulability Analysis Modeling) to the Colored Petri Net (CPN) tool. A shortcoming of this proposal is that only the multiprocessor partitioned scheduling is supported. In [9], an early schedulability analysis based on Cheddar tool has been performed. In this context, a M2M transformation from AADL models to the meta-model of cheddar has been established. Originally, Cheddar supported only the uni-processor and multiprocessor partitioned scheduling. It has been extended in this proposal to support global scheduling, but it still does not support the semi-partitioned scheduling approach.

In the previous cited research studies for automatic and early schedulability analysis, only uni-processor and multiprocessor partitioned scheduling have been supported. An attempt to overcome this shortcoming has been recently proposed [6]. In this proposal, authors have suggested a process for early and automatic schedulability analysis step while supporting both semi-partitioned and global scheduling approaches. To this end, they have transformed class diagrams annotated through MARTE/GRM to the meta-model of the SimSo tool.

In fact, the use of class diagrams does not allow designers to specify the dependency between tasks (such as the sending of data). Consequently, we propose in this work a process for automatic schedulability analysis of RTES with dependent tasks. Both semi-partitioned and global scheduling approaches are supported in our proposal. High-level modeling languages are also used to raise the abstraction level of the schedulability analysis step.

3   Overview of Schedulability Analysis Tools

Due to the importance of timing requirements during the design of RTES, various schedulability analysis tools have been documented to support the analysis of uni-processor and multiprocessor systems. Each schedulability analysis tool supports specified types of architectures and tasks. As example of tools supporting the schedulability analysis of RTES, we mention RTsim [4], MAST [2] and Cheddar. These scheduling analysis tools may be used to analyze systems, which are scheduled using uni-processor or multiprocessor scheduling approaches. Whereas, considering multiprocessor scheduling approaches, only the partitioned and global scheduling (for Cheddar) approaches are supported by these tools. Moreover, these tools are designed to validate, test and analyze systems without handling direct overheads such as scheduling overheads (context switching and scheduling decisions). To deal with this issue, SimSo tool for Simulation of Multiprocessor Scheduling with Overheads [1] has been proposed. This tool will be used during the proposed process for schedulability analysis of RTES with dependent tasks and regarding semi-partitioned and global scheduling approaches.

4   Proposed Process for Early Schedulability Analysis of RTES with Dependent Tasks

4.1   Overview

The proposed process enables the schedulability analysis of RTES at early design stages while supporting semi-partitioned and global scheduling approaches. A main advantage of this proposal is the support of dependent tasks. The proposed process is based on the use of MDE concepts to raise the abstraction level and ease the temporal verification step. In fact, MDE fosters independence of the development flow regarding analysis tools. Consequently, any schedulability analysis tool may be used. Designers have to only establish some transformation rules to sustain the transformation from the system model to the chosen schedulability analysis tool (SimSo). In case of non schedulability, appropriate feed-back needs to be produced to allow the user to change the temporal properties.

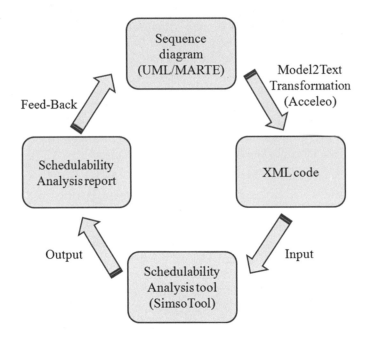

Fig. 1. The process of early schedulability analysis.

4.2   UML/MARTE Model

Figure 1 shows the major steps to be performed for early and automatic schedulability analysis. In fact to decide about the temporal correctness of RTES, SimSo tool which accepts only XML files as input is used in our proposal. Given a system model (sequence diagram) annotated through the MARTE profile and mainly using SAM (Schedulability Analysis Modeling) sub-profile, a Model To Text (M2T) transformation must be performed to translate the system properties from the sequence diagram to the meta-model of SimSo tool. Accordingly, we have implemented an ACCELEO template to support this transformation. This template implements the transformation of MARTE/SAM concepts to SimSo meta-model concepts. The execution of the implemented ACCELEO template allows the generation of an XML file, which contains the system features. This file represents the input of the SimSo tool for schedulability analysis. The sequence diagram displayed in Fig. 2 illustrates the modeling of temporal behavior of RTES. It represents a generic model annotated through MARTE profile and mainly using the sub-profile MARTE/SAM. This model may be reused to model any kind of system allowing the task migration (global scheduling in this case). The stereotypes and attributes used to annotate this model allow the specification of temporal requirements, which will be transformed to the schedulability analysis tool to check the temporal behavior of the studied system. The global scheduling is mainly based on the task set, the execution resources, the global scheduler and the mutual exclusion resources.

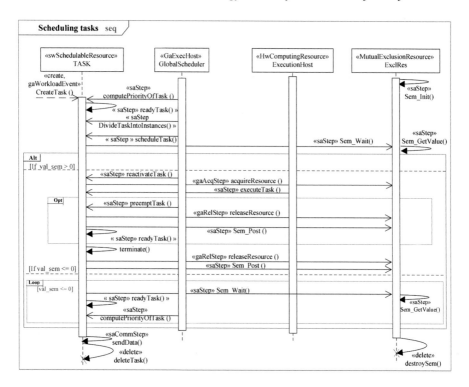

Fig. 2. The dynamic view of the DP-AGS

Following its creation, a task is placed in the global scheduler queue. This placement is based on the chosen scheduling algorithm. The task is waiting for an available processor («readyTask()»). The «sem_Wait()» has to be executed to check the availability of a shared resource. In this context, two cases arise:

- If the result value of the operation «sem_wait()» <0:
  In this case, the task remains in the waiting state until the availability of an execution host (processor). The test of the availability of an execution host is repeated until the satisfaction of the request. This is represented through the fragment «Loop».

- If the result of «sem_wait()» >0:
  In this case, a free processor is available. The scheduler reactivates a task («reactivateTask()») to lock the shared resource («acquireResource()»). The task begins its execution on the corresponding processor («executeTask()»). If the processor is unable to execute the whole task, this later will be subdivided into instances («DivideTaskIntoInstances()») and assigned, not simultaneously, to different execution hosts («scheduleTask()»). Once the task has finished its execution, it releases («releaseResource()») the shared resource that will be locked through another ready task. Consequently, the «Sem_Post()» operation is launched to change the semaphore value and to specify that the shared resource becomes available.

4.3   Simso Tool

SimSo [1] is an open source tool written in Python language. It has been designed to ease the study of the scheduler behavior and enable fast simulations and a rapid prototyping of scheduling policies. SimSo supports aperiodic, sporadic and periodic tasks. Moreover, it encloses a library that contains a big set of uniprocessor and multiprocessor schedulers from partitioning to global ones. SimSo has various advantages such as: supporting partitioned, semi-partitioned and global scheduling approaches, supporting all task types, taking into consideration the properties of the target architecture, taking into consideration the overheads associated with the scheduling, being open source and easy to extend, using a simple language for the input model (XML).

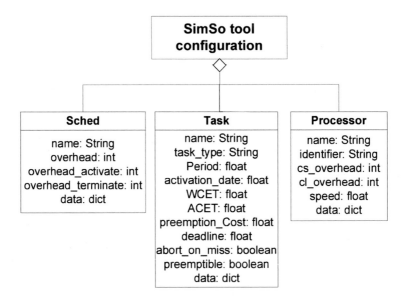

Fig. 3. Meta-model of the Simso tool

As shown in Fig. 3, the architecture of SimSo is based on three main components used to parameter the schedulability analysis; scheduler, task, processor.

- **Task:** it corresponds to a concurrent resource that needs to be executed on a processor.
- **Scheduler:** it represents the scheduling algorithm used to schedule tasks.
- **Processor:** it models the execution resource of the system.

4.4   From Annotated Dynamic Model to SimSo Tool Meta-Model: Model to Text Transformation Rules

This transformation allows the translation of the data stored in a dynamic view to the model of the chosen analysis tool. This dynamic view may be a sequence or activity diagram containing information regarding the dynamic temporal behavior of the studied system. In fact, the transformation of the dynamic view to a textual code is based on some transformation concepts mentioned in Table 1.

Table 1. MARTE to SimSo transformation.

	MARTE	SimSo
Processor stereotypes and annotations	«HwProcessor» «HwComputing Resource»	processor
	«name»	name
	«speedFactor»	speed
Tasks stereotypes and annotations	«SwSchedulable Resource» «Schedulable Resource»	task
	«isPreemptable»	preemptible
	«pattern»	task_type
	«period»	period
	«deadline»	deadline
	«priority»	priority
	«readyT»	activation _date
	«preemptT»	preemption _cost
	«execTime»	WCET
Scheduler stereotypes and annotations	«GaExecHost» «Scheduler»	sched
	«otherSchedPolicy»	class
	«clockOvh»	overhead

The stereotypes «HwProcessor» and «HwComputingResource» annotate a processor. They are transformed to the element *processor*. «name» which is an attribute of the stereotypes «HwProcessor» and «HwComputingResource», is transformed to the element *name*. The attribute «speedFactor» is also an attribute of the stereotypes «HwProcessor» and «HwComputingResource». It is transformed to the element *speed*. The stereotype «SwSchedulableResource » annotates a task. It is transformed to the element *task*. «isPreemptable» is an attribute of the «SwSchedulableResource» stereotype and indicates whether a task may be preemptable. It is transformed to the element *preemptible*. The attribute «pattern» indicates the type of the task/job (periodic, sporadic or aperiodic). It is transformed to the property *type*. «period» gives the value of the task/job priority. It is transformed to the property *period*.

The attribute «deadline» specifies the deadline of a task/job. It is transformed to the property *deadline*. «priority» is an attribute of MARTE and it specifies the value of a priority. It is transformed to *priority*, which is added to the dictionary of SimSo. «readyT» represents the date of activation. It is transformed to the property *activation_date*. «preemptT» is an attribute of MARTE and it specifies the duration of preemption of a task/job. It is transformed to the property *preemption_cost*. The attribute «execTime» represents the execution time(s) of a task. It is transformed to the property *WCET*.

5 Case Study

To validate the proposed methodology for early and automatic schedulability analysis of RTES, we consider the networking router application described in [5]. The software part of the studied system is composed of twenty three periodic and dependent tasks such that each task is characterized by a Worst Case Execution Time (WCET), a priority and a deadline (Table 2).

Table 2. Task parameters

Task	Priority	WCET	Deadline
IF1	1	200	201
IF2	2	100	101
IF3	3	400	401
IF4	4	200	201
IF5	5	200	201
IF6	6	400	401
Mem1	7	100	201
Mem2	8	100	201
Mem3	9	100	101
PktRx1/2	10	1442	1443
Chlk	11	450	451
HdrCal	12	2400	2401
PktRx3	13	1530	1531
PktRx4	14	670	671
Pkt_fwd	15	600	601
Proc1	16	2800	2801
Cnfg	17	1600	1601
Assm	18	800	2401
Recalc1	19	3000	3801
PktRx5/6	20	2100	2101
Enc	21	4900	4901
Recalc2	22	1750	6381
Proc2	23	5600	5601

These tasks are mapped to a preemptive execution platform composed of two embedded processors (CPU1 and CPU2), an accelerator (ASIC), two physical memories (RAM1 and RAM2) and a communication bus. The different features are mentioned in the dynamic view (sequence diagram). This model represents a reuse of the sequence diagram exposed by Fig. 2. It will be then transformed to an XML view using an ACCELEO template. The generated XML file represents the input of the SimSo tool. After checking the properties specified in the sequence diagram, SimSo tool has proven that the studied system is schedulable (Fig. 4).

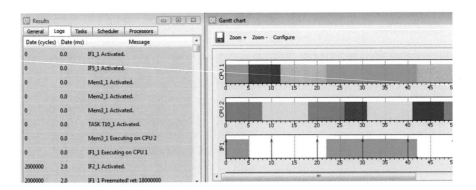

Fig. 4. Screenshot of the results of the schedulability analysis using Simso

6   Conclusion

In this paper we proposed an MDE-based process for an early schedulability analysis of RTES. This methodology strives to automate the transformation of system models to the meta-models of schedulability analysis tools in order to provide an easy and safe way to analyze the temporal behavior of systems as well as to reduce the designers' efforts. To perform this process, we implemented an ACCELEO template, which enables the transformation of dynamic models annotated through MARTE to the model of Simso tool (XML file). In this process both semi-partitioned and global scheduling approaches, which support task migration, are supported. A key feature of this proposal is that it supports the scheduling of systems with dependent tasks.

As future work, we seek to automate the reuse of generic models annotated through MARTE profile in order to facilitate the modeling of systems mainly for designers who do not have enough knowledge about MARTE.

References

1. Chéramy, M., Hladik, P.-E., Déplanche, A.-M.: SimSo: a simulation tool to evaluate real-time multiprocessor scheduling algorithms. In: WATERS 2014 (2014)
2. Gonzalez Harbour, M., Gutierrez Garcia, J.J., Palencia Gutierrez, J.C., Drake Moyano, J.M.: Mast: modeling and analysis suite for real time applications. In: 2001 13th Euromicro Conference on Real-Time Systems, pp. 125–134 (2001)
3. Jensen, K.E.A.: Schedulability analysis of embedded applications modelled using MARTE. Ph.D. thesis, Technical University of Denmark (2009)
4. Manacero Jr., A.: Real time systems (2006). http://www.dcce.ibilce.unesp.br/spd/rtsim/english
5. Madl, G.: Model-based analysis of event-driven distributed real-time embedded systems. Ph.D. thesis, Long Beach, CA, USA (2009)
6. Magdich, A., Kacem, Y.H., Mahfoudhi, A., Abid, M.: From UML/MARTE models of multiprocessor real-time embedded systems to early schedulability analysis based on SimSo tool. In: ICSOFT, Lisbon, Portugal (2016)
7. Medina, J.L., Cuesta, Á.G.: From composable design models to schedulability analysis with UML and the UML profile for MARTE. SIGBED Rev. 8(1), 64–68 (2011)
8. Naija, M., Ahmed, S.B., Bruel, J.-M.: New schedulability analysis for real-time systems based on MDE and petri nets model at early design stages. In: ICSOFT-EA 2015, Colmar, Alsace, France, 20–22 July 2015, pp. 330–338 (2015)
9. Rubini, S., Fotsing, C., Singhoff, F., Tran, H.N., Dissaux, P.: Scheduling analysis from architectural models of embedded multi-processor systems. In: EWiLi Workshop (2013)

Improving Native Language Identification Model with Syntactic Features: Case of Arabic

Seifeddine Mechti[1], Nabil Khoufi[2(✉)], and Lamia Hadrich Belguith[3]

[1] LARODEC Laboratory, ISG of Tunis,
University of Tunis, Tunis, Tunisia
mechtiseif@gmail.com
[2] ANLP Research Group, MIRACL Laboratory, IHE of Sfax,
University of Sfax, Sfax, Tunisia
nabil.khoufi@fsegs.rnu.tn
[3] ANLP Research Group, MIRACL Laboratory, FSEG of Sfax,
University of Sfax, Sfax, Tunisia
l.belguith@fsegs.rnu.tn

Abstract. In this paper, we present a method based on machine learning for Arabic native language identification task. We expose a hybrid method that combines surface analysis in texts with an automatic learning method. Unlike the few techniques found in the state of the art, the features selection phase allowed improving performances. We also show the impact of syntactic features for native language identification task. Therefore, the obtained results outperformed those provided by the best methods used for Arabic native language detection.

Keywords: Arabic native language identification · Machine learning · Syntactic features

1 Introduction

Native language identification (NLI) is typically modelled as a sub-class of text classification, particularly text author profiling which aims to detect some author's features, such as age, gender, educational level and the native language, the focus of our study.

Currently, many research works have focused on the texts classification for learners of English while few studies have dealt with the Arabic language [1].

In this article, we try to find the English, Chinese and French people speaking Arabic. We also detect the Arabs speaking native language (L1). Based on automatic learning, we use the best attributes combination to get better classification results. We also present the contribution of the syntactic features in the classification results.

This paper is organized as follows: Sect. 2 presents the related works. Section 3 describes our methodology. Section 4 shows the experiments and assessment of our findings. Finally, we end up the paper with some concluding remarks.

© Springer Nature Switzerland AG 2020
A. Abraham et al. (Eds.): ISDA 2018, AISC 941, pp. 202–211, 2020.
https://doi.org/10.1007/978-3-030-16660-1_20

2 Related Works

English, the first world language, has been widely studied by many works in SLA and NLP. Thus, since [2], the first NLI study, most of the works carried out in automatic native language prediction has focused on English. In their study [3] used researchers dealt with native language among other dimensions including age, gender and personality in authorship profiling approach. Because of the unavailability of comprehensive corpus, they resorted to the use of separate corpus to accomplish their studies related to ICLE for the sub-task of language dimension identification. Their cross-validation tests were carried on each of the following features types separately (stylistic features only, content features only) and on both. Results showed that content-based feature is slightly useful in gender dimension identification. Indeed, when combined with style-based feature, it gave a classification accuracy of 76% for gender and 77.7 for age. Unlikely, classifiers, learned using only style features, provided an accuracy rate of 63% for personality dimensions.

Concerning language dimension, an outperformance of 82.3% was achieved by using only the content features that show up the preference word usage between speakers of different languages. However, such results were discussed by the researchers themselves who noted that they may be infected by possible topic bias. For this reason, the majority of the following researches avoided the use of content features.

Wong and Dras [4] conducted their experiments on the same data (i.e. ICLE). They integrated three common syntactic types of error made by English learners. These mistakes are related to subject-verb disagreement, noun-number disagreement and misuse of determiners with lexical features, function words, character n-grams and POS n-grams, used in previous approaches. They achieved an accuracy of 73.71% with all features combined. Results showed the usefulness of these error types in NLI task. This fact is more investigated by Kochmar [5] who suggested using character n-grams error. Although most English NLI studies were conducted on (ICLE), other corpora have been used in parallel. Bykh and Meurers [6] conducted a research on the second version of the International Corpus of Learners of English (ICLE) and three other corpora (NOCE, USE and HKUST in this case). They used different Support Vector Machines (SVMs) as classifiers.

The researchers defined features based on recurring n-grams of all accruing lengths. Three classes of recurring n-grams are defined in their work, viz.: one word-based n-gram and two generalizations of the first class (POS- and Open-Class-POS based n-grams). They conducted experiments based on random samples from ICLE in a single-corpus evaluation and a crosscorpus1 evaluation. The word-based n-gram was proved to be the best performing class with a high accuracy reaching 89.71%. The results also demonstrated that the pattern learned on ICLE generalized well across corpora and gave an accuracy of 88%.

In another study, [7] used a large longitudinal data to identify the native language of the learners of second tongue. They focused mainly on corpus collected from Cambridge University to investigate the different proficiency levels. The authors employed accurate learning machine based on Support Vector Machines SVM. Both the syntactic and lexical features were tested separately and combined in this experiment.

Results showed that lexical features outperformed syntactic features when tested individually at all levels. The same findings were achieved when combining the features, especially at the advanced level.

In [8], Nisioi investigated the proficiency of the different features for the task of native language identification benefiting from the EF Cambridge Open Language error annotated Database. Nisioi's objective was firstly to analyze the different features used for automatic text classification in the frame of NLI. Secondly, he tended to highlight the important learner's linguistic background role in the learning process. Then, he used it to distinguish the learners' native country of people who share the same mother tongue. Analyzed features in this study covered topic-independent features (function words, POS n-grams, anaphoric shell nouns and annotated errors) and other characteristics that are dependent on topic (character n-games and positional-token frequencies). His Experiments demonstrated that anaphoric shell nouns and positional token frequencies contributed to achieve the best accuracy. In fact, topic-independent features combination reached an accuracy of 93.75. For the sensitive topic features characters, 4-grams achieved the highest accuracy of about 99% across corpora. He carried out the linguistic background analysis to explain some misclassifications of native country for some native languages.

In [9], the researchers were interested in the identification of authors' native languages based on their writing on the Web. Their proposed method was based on automatic classification using various types of features (namely Lexical, Syntactic, Structural and Content features) collected from Web-based texts written by native and non-native authors as a source corpus. To achieve this goal, the researchers compared three different classification techniques (the C4.5 decision tree, the support vector machine and Naïve Bayes). Their experiments showed the obvious efficiency of the lexical and the content specific features compared with the rest of the features.

Concerning the learning algorithm, SVM outperformed Naïve Bayes and C4.5 significantly with a satisfactory accuracy of 70% to 80%. In recent years, other languages apart from English attracted researches' attention in NLI field.

To the best of our knowledge, [10] and [11] addressed the Chinese language. The first work presented the first expansion of NLI application to non-English data. Based on features set that involves part-of-speech tags, n-grams, function words and context-free grammar production rules. Their system revealed that the use of all the combined features outperformed the use of individual features with an accuracy of 70.61%.

In the second work, Lan and Hayato were the first to use skip-grams as features in the NLI problem combined with the traditional lexical features based on the Jinan Chinese Learner Corpus JCLC. Because the skip-gram feature number or dimension grow enormously, they accept only n-grams that occur in more than ten essays as feature reduction strategy.

Unlike most of the NLI studies which have adopted term frequency (TF) or term frequency–inverse document frequency (TF-IDF). The plus point of this study is that a special attention is paid to assign an effective weight to each feature. They adopted the BM25 term-weighting method [12]. Their proposed system reached a higher performance with 75% accuracy using hierarchical linear SVM classifiers.

Arabic was currently perceived as a critical and strategically useful language. However, Malmasi and Dras work [1] was the unique study dealing with this language in NLI field. An interesting study might be relevant to the present one. Their aim was to investigate the usefulness of the syntactic features, mainly CFG production rules, Arabic function word and Part-of-speech n-grams more detailed in the flowing section. They used a supervised multi-class classification approach. As a result, such experiments proved to be successfully applicable to Arabic NLI. Added to that, it is notable that combining features led to a reasonable accuracy of about 41%, which was 10% lower compared with their previous study of English using the same set of features. This was due to, on the one hand, to the fact that Arabic is significantly different from English morphological and syntactic richness, which gave challenge to the use of syntactic features, and on other hand to the small size of data set used in the learning phase.

3   Proposed Method

In this study, we address the prediction of Arabic learner's native language inspired from the [1] work. To come up with an optimal classification model, our focus is on the features selection step which has not given great importance in most of the previous works. Optimality refers to the reduction of features and performance improvement.

Our proposed method, as shown in Fig. 1, is divided into three steps. Firstly, in the pre-processing step, we prepared the text to be used in the next step. Then, in the features extraction phase, we extracted the features set that seem to be useful to Ll learner background discrimination. Finally, we applied classification algorithm to generate the classification model. Obviously, the two last steps were supported by a sub-step of features selection. After that, we ensured that our system was as optimal enough to be used in new L1 text prediction.

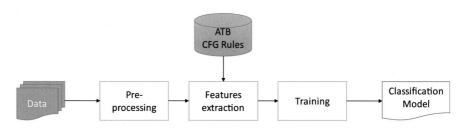

Fig. 1. Proposed method

3.1   Text Pre-processing

This phase includes removing the inappropriate characters, words and marks for example:

إن الأهرامات ضخمة في تكون أحجارها فوزن الحجر الواحد
يبلغ 1035 kg

In this sentence, the non-Arabic author forgot to translate the word “Kg” (basic international weight unit) into Arabic (كغ). Thus, we omitted this unnecessary word.

3.2 Features

In this study, we explored three syntactic feature types, namely ‘function word’, ‘Part of speech n-grams’ and ‘Context-free grammar production rule’. In this way, we generated three sets of features for each text. For each individual feature, we measured the frequency (*TF*) with which it appears.

Function Words. Refer to context and topic-independent words used differently by learners to obtain coherent sentences. Here, we adopted 411 common Arabic function words regrouped into seventeen types (classes). Below are examples of the Arabic function words listed by classes (Table 1).

Table 1. Examples of Arabic function words

Type	Examples
Linking words	علاوة على(furthermore), برغم (despite), حيث أن (whereas), etc.
Conjunctions	أو(or), بل(but/ rather), و(and) etc.
Prepositions	من(from), إلى (to), في (in), على (on), etc.
…	…

Part of Speech n-grams. They are representative features that highlight the words linguistic category. In our experiment, we used POS n-gram with n = 1..3. We applied Alkhalil Arabic-specific morphological analyser [13] to tag word.

Syntactic Features. The aim of the syntactic feature is to verify if the syntactic rules (CFG rules) used in the text are commonly used in Arabic. To do that, we need to compare these syntactic rules to an Arabic syntactic rules database. Figure 2 represents an example of CFG rules extracted from the correspondence parse tree. Texts have been parsed using the Stanford parser [14] in order to extract context-free grammar rules.

```
(S
 (VP (VBP اريد)
  (PP (IN في)
   (NP (DTNN المستقبل)))
   (SBAR (IN ان)
   (S (VP (VBP اكون)
     (NP ((NN اقتصاديا) (JJ ماهرا))))))))
```

```
Production rules:
S→ VP | VB NP
VP→ VBP PP SBAR
PP→ IN NP
NP→ DTNN | NN JJ
SBAR→ IN S
```

Fig. 2. A constituent parse tree along with the extracted context-free grammar production rules

Therefore, we begin by constituting the Arabic syntactic rules database using the PATB corpus. In the following we start by presenting the corpus and then explain the induction process of syntactic rules.

PATB Corpus. In our work, we chose to use the well-known corpus, the Penn Arabic Treebank (PATB). This choice was motivated not only by the richness, the reliability and professionalism with which it was developed but also by the syntactic relevance of its source documents (converted to several other Treebank representations). This is shown by its efficacy in many of research projects in various fields of NLP [14]. The good quality of the text and its annotations is demonstrated by its performance in the creation of other Arabic Treebanks such as the Prague Arabic Dependency Grammar [15] and the Columbia Arabic Treebank [16], which converted the PATB to its syntactic representations in addition to other annotated texts.

Indeed, these annotations were manually elaborated and validated by linguists. Moreover, this treebank is composed of data from linguistic sources written in Modern Standard Arabic. This corpus is also the largest Arabic corpus which integrates syntactic tree files.

The Penn Arabic Treebank (PATB) was developed in the Linguistic Data Consortium (LDC) at the University of Pennsylvania [17].

We used the PATB 3 version 3.2 of this corpus, which consists of 599 files, and includes POS tags, morpho-syntactic structures at many levels and glosses. It comprises 402,291 tokens and 12,624 sentences. It is available in various formats: The "sgm" format refers to source documents. The "pos" format gives information about each token as fields before and after clitic separation. The "xml" format contains the "tree token" annotation after clitic separation. The "penntree" format generates a Penn Treebanking style. And finally, the "integrated" format brings together information about the source tokens, tree tokens, and the mapping between them and the tree structure.

Syntactic Rules Induction. Adeep study of the PATB allows us to identify rules which guide the CFG rules induction process. Indeed, we focused on the morpho-syntactic trees of the PATB and we identified the following rules which guides the induction process:

R1: Tree root → Start symbol
R2: Internal tree node → Non-terminal symbol
R3: Tree word → Terminal symbol
R4: Tree fragment → CFG rules

We present below some statistics about obtained syntactic rules. Table 2 presents the most frequent syntactic rules generated from the PATB corpus and Table 3 presents the overall count of rules (contextual rules and lexical rules).

Table 2. Most frequent rules

Left-hand symbol (LHS)	NP	VP	S	FRAG	ADJP	UCP	PP
Rule count	1821	1311	1154	360	330	196	150

Table 3. Rule count of the induced PCFG

Contextual rules	Lexical rules	Total
5757	38 901	44 658

3.3   Classification Model

In our experiments, we used LIBSVM[1] package as variant of Support Vector Machines (SVMs) which has been widely used and proved to be efficient for text classification areas.

Table 4. L1 distribution by number of words

L1	Words number
Chinese	11073
Urdu[a]	12341
Malay[b]	6686
French	5942
Fulani[c]	5571
English	5774
Yoruba[d]	4794
Total	52181

[a]The national language and lingua franca of Pakistan
[b]The Malaysian language
[c]Non-tonal language spoken in 20 countries of West and Central Africa
[d]One of the five most spoken languages in Nigeria

4   Experimental Setup

4.1   Data Set

We trained our model to the segment of the second version of Arabic learner corpus ALC [18]. The latter was annotated by language within other metadata containing essays written by native and non-native Arabic learners who speak 66 different L1s. We included in this experiment the seven top L1s in term of text numbers with average text length of 166 words.

Table 4 shows the L1 distribution by number of words. More details can be found in [2].

We notice that ALC is not only small, but it is also the only available language annotated dataset. Such issue was also dealt with by [1] who noted that their system performance might be influenced by the small amount of data used in the training

---

[1] http://www.csie.ntu.edu.tw/~cjlin/libsvm.

step. Their hypothesis is supported by [3] finding revealing that the system performance improves as data size increases.

4.2 Results

In order to be consistent with the de facto standard of reporting NLI results, we reported K-fold cross-validation results of our experiment with k = 10.

We conducted serval individual experiments testing our features separately and combined. Then, we represented the result of our features set. Table 2 summarizes the full classification accuracies. For individual features, CFG production rules as well as function words have shown their ability to distinguish L1 learners with 36.5% accuracy for production rule and 31% for function words.

Unsparingly, POS n-grams consistently with those given in earlier studies outperformed the other syntactic features. A best accuracy of 38% was reached with n = 2. We note here, though POS trigrams we yield a reasonable accuracy of 29%, it turned out that combining trigrams POS with other features did not give better results. It rather underperformed the global performance. This can be explained by the fact that these trigrams represent a redundant information compared with the production rules. Thus, we exclude it when we use features together.

We combined 278 features distributed as follows: 16 unigrams, 145 bigrams, 11 classes of function words and 106 production rules. This set allowed achieving better classification result for Arabic NLI of 45%, which highlights the importance of features selection step (Table 5).

Table 5. Number of features and their accuracy

Features	Number of features	Accuracy (%)
CFG rules	106	36.5
Function Words	11	31.0
POS unigrams	16	34.9
POS bigrams	145	38.0
POS trigrams[a]	580	29.0
Combined	278	45.0

[a]POS-Trigrams are excluded from the classification when we combine features.

We noticed that the use of the syntactic features is very useful for a better performance of the model. In fact, we obtained a global accuracy of 43.2% without the use of the syntactic features (CFG rules). Therefore, the uses of syntactic features improved the performance of the model by 1.8%.

Our results can be comparable with Malmasi and Dras, our results are on global well performed than those reported by him, around 5% up in accuracy.

5 Conclusion

We presented an original method for Arabic native language detection. We combined CFG production rules, function words and POS bigrams to perform machine learning process. By inspiring from the study of [1], we obtained the best accuracy (45% vs 41%). The experimental results show that we were able to detect Asiatic learners of Arabic more than European authors. In fact, for Chinese authors, we obtained 80% accuracy rate, while this percentage was 36% for French authors and 30% for English writers.

Consequently, it was proven that Asians have more ability to learn Arabic than European people. This can be explained by the fact that these Semitic languages are more difficult compared to French or English. Furthermore, to allow a better author detection, we think of going beyond the native language and consider the detection of age, gender and geographical background of the author and above all the detection his/her personality.

Currently, our approach is based on a static-learning model where the corpus is not updated, and the already-predicted documents are not included in the training. Therefore, in future work, we plan to address this issue to be more adequate for real time author profiling scenarios.

References

1. Malmasi, S., Dras, M.: Arabic native language identification. In: Proceedings of the Arabic Natural Language Processing Workshop, Doha, Qatar (2014)
2. Koppel, M., Schler, J., Zigdon, K.: Automatically determining an anonymous author's native language. In: International Conference on Intelligence and Security Informatics, pp. 209–217. Springer, Heidelberg (2005)
3. Argamon, S., Koppel, M., Pennebaker, J.W., Schler, J.: Automatically profiling the author of an anonymous text. Commun. ACM **52**(2), 119–123 (2009)
4. Wong, S.M.J., Dras, M.: Contrastive analysis and native language identification. In: Proceedings of the Australasian Language Technology Association Workshop, pp. 53–61 (2009)
5. Kochmar, E.: Identification of a writer's native language by error analysis. Doctoral dissertation, Master's thesis, University of Cambridge (2011)
6. Bykh, S., Meurers, D.: Native language identification using recurring n-grams–investigating abstraction and domain dependence. In: Proceedings of COLING 2012, pp. 425–440 (2012)
7. Ionescu, R.T., Popescu, M., Cahill, A.: Can characters reveal your native language? A language-independent approach to native language identification. In: Proceedings of the 2014 Conference on Empirical Methods in Natural Language Processing (EMNLP), pp. 1363–1373 (2014)
8. Jiang, X., Guo, Y., Geertzen, J., Alexopoulou, D., Sun, L., Korhonen, A.: Native language identification using large, longitudinal data. In: LREC, pp. 3309–3312 (2014)
9. Nisioi, S.: Feature analysis for native language identification. In: International Conference on Intelligent Text Processing and Computational Linguistics, pp. 644–657. Springer, Cham (2015)

10. Malmasi, S., Dras, M., Temnikova, I.: Norwegian native language identification. In: Proceedings of the International Conference Recent Advances in Natural Language Processing, pp. 404–412 (2015)
11. Lan, W., Hayato, Y.: Robust Chinese native language identification with skip-gram. In: DEIM Forum (2016)
12. Boudlal, A., Lakhouaja, A., Mazroui, A., Meziane, A., Bebah, M.O.A.O., Shoul, M.: Alkhalil morpho sys1: a morphosyntactic analysis system for arabic texts. In: International Arab Conference on Information Technology, Benghazi, Libya, pp. 1–6 (2010)
13. Klein, D., Manning, C.D.: Accurate unlexicalized parsing. In: Proceedings of the 41st Annual Meeting on Association for Computational Linguistics, vol. 1, pp. 423–430. Association for Computational Linguistics (2003)
14. Habash, N.Y.: Introduction to Arabic natural language processing. In: Hirst, G. (ed.) Synthesis Lectures on Human Language Technologies, vol. 3, no. 1 (2010)
15. Hajic, J., Vidová-Hladká, B., Pajas, P.: The Prague dependency treebank: annotation structure and support. In: Proceedings of the IRCS Workshop on Linguistic Databases, pp. 105–114 (2001)
16. Habash, N.Y., Roth, R.M.: CATiB: the Columbia Arabic treebank. In: Proceedings of the ACL-IJCNLP 2009 Conference Short Papers, pp. 221–224. Association for Computational Linguistics, Stroudsburg (2009)
17. Maamouri, M., Bies, A., Buckwalter, T., Mekki, W.: The Penn Arabic treebank: building a large-scale annotated Arabic corpus. In: The NEMLAR Conference on Arabic Language Resources and Tools, pp. 102–109 (2004)
18. Alfaifi, A.Y.G., Atwell, E., Hedaya, I.: Arabic learner corpus (ALC) v2: a new written and spoken corpus of Arabic learners. In: Proceedings of Learner Corpus Studies in Asia and the World 2014, vol. 2, pp. 77–89. Kobe International Communication Center (2014)

Comparison of a Backstepping and a Fuzzy Controller for Tracking a Trajectory with a Mobile Robot

Rodrigo Mattos da Silva[1]([⊠]),
Marco Antonio de Souza Leite Cuadros[2],
and Daniel Fernando Tello Gamarra[1]

[1] Universidade Federal de Santa Maria, Santa Maria, RS 97105-900, Brazil
rodrigo-mattos@hotmail.com.br
[2] Instituto Federal do Espirito Santo, Serra, Espirito Santo 29173-087, Brazil

Abstract. This work aims to compare the application of two tracking controllers for a mobile robot in a trajectory tracking task. The first method uses a heuristic approach based on the prior knowledge of the designer, while the second method uses mathematical model based on the robot kinematics. Both systems employ the estimated robot position derived from the encoder sensors using the dead reckoning method. The paper shows experimental results with a real robot following a predefined path to explore the use of these techniques.

Keywords: Mobile robots · Fuzzy control · Dead-reckoning ·
Backstepping controller

1 Introduction

In the last few years, the number of applications involving mobile robots has been escalating and one of the factors responsible for that is the diversity and relevance of the tasks that this category of robot may perform [1]. The referred tasks require that the robot could be able to perform actions such as following a given trajectory and knowing its localization. This work presents a comparative approach of two control systems to resolve the trajectory tracking problem.

There are several approaches to trajectory control, these methods can be classified into two categories, the classic control approach [2–4], and the heuristic approach [5, 6]. This work presents a backstepping controller based on kinematics model and a heuristic control method based on fuzzy logic. This approach was previously explored by Omrane et al. [7] with a fuzzy controller to track trajectory and avoid obstacles using a simulator in Matlab called SIMIAN. Xiong et al. [8] applied a method based on machine vision and fuzzy control for intelligent vehicles driving, the fuzzy controller replaces a traditional PID controller. Saifizi et al. [9] explored a similar approach using fuzzy logic controller and computer vision in order to recognize a circle as a landmark and identify the distance and orientation by knowing the diameter of the circle and the calibration of the camera.

© Springer Nature Switzerland AG 2020
A. Abraham et al. (Eds.): ISDA 2018, AISC 941, pp. 212–221, 2020.
https://doi.org/10.1007/978-3-030-16660-1_21

Kanayama et al. [10] shows a robot-independent tracking control rule for non-holonomic systems that can be applied to mobile robots with a dead reckoning ability and the stability is proved through the use of a Lyapunov function. Fierro et al. [6] proposes a control structure that makes possible the integration of a Backstepping controller and a neural network (NN).

One of the most common procedures to estimate position and orientation of robots is dead-Reckoning. Park et al. [11], for example, presents a navigation system for a mobile robot with a differential traction drive that uses the dead-Reckoning method. The system consists of a gyroscope and encoders attached on the wheels, and, in order to raise its reliability a Kalman filter is applied for data fusion of the sensors. Santana et al. [12] proposed a system to localize a robot using odometry and images of pre-existing lines on the floor, an extended Kalman filter is also used for the fusion of the information of the image processing and the Odometry. Borenstein et al. [13] states that the methods of estimation of position and orientation may be summarily categorized in two groups according to their measurement: relative and absolute.

The article will tackle the problem of following a trajectory with a mobile robot using two controllers, one heuristic controller based on Fuzzy logic and the other one based on backstepping. The present article presents the relative method implementation and it is divided into eight sections, after a brief introduction in the first section, the second section deals with the kinematic model of the robot, highlighting the variables employed on the fuzzy and backstepping controller, the third section approaches the dead-reckoning method, In the fourth section, the backstepping control approach is explained, The fifth section explains the development of the fuzzy controller, the sixth section shows the obtained results, followed by the conclusions in the last section.

2   Kinematic Model of the Robot

The kinematic model of the robot used in this work can be observed in Fig. 1. The robot has two independent wheels of differential traction and a free movement wheel, denominated castor wheel, in order to maintain the balance of the structure.

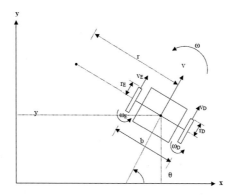

Fig. 1. Kinematic model of a mobile robot with differential traction

Where:

(x, y) = Position of the fixed yardstick on the robot in relation to the fixed yardstick in the workspace.

θ = Robot orientation angle in relation to the fixed yardstick in the workspace.

b = Axis length.

r = Robot turning radius.

$r_R(r_L)$ = Right (and left) wheel radius.

ω = Robot angular speed.

$\omega_R(\omega_L)$ = Angular speed of Right (and left) wheel.

v = Linear speed of the robot.

$v_R(v_L)$ = Linear speed of the right edge (and the left).

The linear and angular speed of the robot may be written in terms of angular speed on the wheels, the axis radius and length of the robot (distance between wheels).

$$\begin{bmatrix} v \\ \omega \end{bmatrix} = \begin{bmatrix} (r_R/2)(r_L/2) \\ (r_R/b) - (r_L/b) \end{bmatrix} * \begin{bmatrix} \omega_R \\ \omega_L \end{bmatrix} \tag{1}$$

It is relevant to observe that, in case the wheels speeds are the same ($\omega_R = \omega_L$), the robot will cover a straight line. If the speeds are different and the right wheel speed is higher than the left ($\omega_R > \omega_L$), the robot will depict a circular trajectory counter-clockwise. Conversely, when the left wheel speed is higher than the right wheel speed ($\omega_R > \omega_L$), the robot will cover a circular trajectory clockwise.

Figure 2 shows the mobile robot used for experimental results.

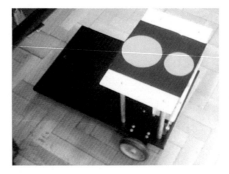

Fig. 2. Mobile robot with differential traction.

3   Dead-Reckoning

One of the primary issues in mobile robotics consists of determining the current localization of the robot. From this assumption, it is possible to displace it to a nearer point and by doing that, to depict the intended trajectory. The dead-reckoning method consists of calculating the linear and angular displacement of the robot from the accumulated angular displacement of the wheels obtained by incremental encoders.

It is possible to find the distance covered by the central point of the robot through the average distance between the two wheels

$$d_{center} = \frac{d_{left} + d_{right}}{2} \qquad (2)$$

Where d_{left} and d_{right} correspond, respectively to the distance covered by the left and right wheels.

The distance covered by each wheel is given by the relation:

$$d = 2 . \pi . R \frac{\Delta tick}{N} \qquad (3)$$

Where:

N = Number of pulses in the encoders for a full turn of the wheel.

Δtick = Difference between the current absolute count of pulses and the count of pulses in the last measurement.

The variation in robot orientation angle, named as phi (ø), is determined with basis on the distances covered by each wheel

$$\emptyset = \frac{d_R - d_L}{b} \qquad (4)$$

given that b is the distance between the wheels (axis length).

Therefore, the current robot orientation angle is incrementally updated using Eq. 5 bellow:

$$\theta' = \theta + \emptyset \qquad (5)$$

Finally, the current position of the robot may be obtained by the relation that follows

$$x' = x + d_{center}\cos(\theta) \qquad (6)$$

$$y' = y + d_{center}\sin(\theta) \qquad (7)$$

It is evidenced that this method consists of counting the pulses obtained from the encoders of each wheel in a given period of time. Therefore, the calculation of the current position depends on the previous position, causing the errors to be accumulated. This method is not indicated for long distances, it must have an additional sensor to correct the accumulated errors.

4 Backstepping Controller

The stable tracking control rule employed in this work was withdrawn from [10] which stability is guaranteed by Lyapunov theory. This method is useful to the class of mobile robots in which reference path specification and current position estimation are given separately.

The tracking control rule purpose is to converge the error posture to zero, making the robot follow a specified path. The reference path is given for an algorithm that generates a reference trajectory points.

Considering reference posture $p_r = (x_r, y_r, \theta_r)$ as a goal posture given by reference and current posture $p_c = (x_c, y_c, \theta_c)$ as instantly posture of the robot estimated through dead reckoning, the error posture $p_e = (x_e, y_e, \theta_e)$ can be represented as follows:

$$p_e = \begin{bmatrix} x_e \\ y_e \\ \theta_e \end{bmatrix} = \begin{bmatrix} \cos\theta_c & \sin\theta_c & 0 \\ -\sin\theta_c & \cos\theta_c & 0 \\ 0 & 0 & 1 \end{bmatrix} \cdot (p_r - p_c) \tag{8}$$

Error posture is a linear transformation of the reference posture frame to current posture frame. The architecture of control system consists in two inputs: reference posture $p_r = (x_r, y_r, \theta_r)$ and reference velocities $q_r = (v_r, w_r)$, reference velocities are respectively the linear and angular velocity that reference is moving. The system has two control action outputs, which are the target velocities $q = (v, w)$ necessary to reach the reference. The control law is depicted in Eq. 9.

$$q = \begin{bmatrix} v \\ w \end{bmatrix} = \begin{bmatrix} v_r \cos\theta_e + K_x x_e \\ w_r + v_r(K_y y_e + K_\theta \sin\theta_e) \end{bmatrix} \tag{9}$$

Where K_x, K_y, and K_θ are positive constants computed in a heuristic approach. This control law approach is useful only for small reference velocities. The constants heuristically computed used in experimental results are present in Table 1.

Table 1. Backstepping controller parameters.

K_x	K_y	K_θ	v_r (m/s)	w_r (rad/s)
32.5	650	32.5	0.025	0.018

In order to test our model, we define that robot could draw a trajectory given by the algebraic curve denominated Bernoulli Lemniscate. The curve can be represented by the following Cartesian equation.

$$\left(x^2 + y^2\right)^2 = 2a^2\left(x^2 - y^2\right) \tag{10}$$

By programming the curve to represent an equation in terms of only one parameter t, it is obtained the following equation:

$$x = \frac{4\cos(t)}{\left(1 + (\sin(t))^2\right)} \tag{11}$$

$$y = \frac{4\cos(t)\sin(t)}{\left(1 + (\sin(t))^2\right)} \tag{12}$$

So, the robot will have to follow the Lemniscate trajectory in order to test the performance of our controllers.

5 Fuzzy Controller

From the information obtained in Sect. 3, the fuzzy controller is elaborated for the robot to move from one point to another. The fuzzy controller aims to minimize the angle γ and the distance between the two straight lines and the fuzzy rules are represented by two inputs and two outputs. The basic model of the controller known in the literature and the fuzzy rules were proposed in [14].

The controller receives a distance (D) and the difference in angle (γ) between the robot and the point of destination returns to the linear and angular speed necessary to minimize, respectively, the distance and the difference in angle. In the figure below, it is possible to observe the graphic representing the center of mass of the mobile robot (x, y) and the point that is intended to achieve (x_r, y_r). Where $\gamma = \alpha - \theta$.

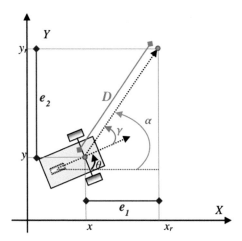

Fig. 3. Distance between the robot center of mass and the destination point (Source: Faria et al.[14]).

The fuzzy variables for the controller are depicted in Fig. 3. If γ is positive, $\alpha > \theta$, the robot will need to turn left to in-crease θ, minimizing the difference between angles and vice-versa. If the angle γ is null, the robot shall cover a straight line. The fuzzy variables and fuzzy sets are shown in Table 2 and the rules are explained in the Table 3:

Table 2. Fuzzy variables.

Variable	Denomination	Depiction
γ	NB	Negative big
	NS	Negative small
	Z	Zero
	PS	Positive small
	PB	Positive big
D and v	VS	Very small
	S	Small
	M	Medium
	B	Big
	VB	Very big
ω	LF	Left fast
	LS	Left slow
	Z	Zero
	RS	Right slow
	RF	Right fast

Table 3. Fuzzy rules.

		γ				
		NB	NS	Z	PS	PB
	VS	VS (RF)	VS (RS)	VS (Z)	VS (LS)	VS (LF)
	S	S (RF)	S (RS)	S (Z)	S (LS)	S (LF)
D	M	M (RF)	M (RS)	M (Z)	M (LS)	M (LF)
	B	B (RF)	B (RS)	B (Z)	B (LS)	B (LF)
	VB	VB (RF)	VB (RS)	VB (Z)	VB (LS)	VB (LF)

From the maximum speed that each wheel may assume, it is possible to calculate the angular speed (ω) and the linear speed (v) of the robot, therefore, vel_e *and* vel_d are respectively the left and the right wheel speed, as follows:

$$v = \frac{r}{2} * (vel_d + vel_e) \tag{13}$$

$$\omega = \frac{r}{b} * (vel_d - vel_e) \tag{14}$$

Table 4 describes the range of values used in the fuzzy variables for our application.

Table 4. Value ranges of the fuzzy variables with dead reckoning.

Distance (D)	$0 \sim 3$ (meters)
Angle difference (γ)	$-3.14 \sim 3.14$
Linear speed of the robot (v)	$0 \sim 7.6$
Angular speed of the robot (ω)	$-15.6 \sim 15.6$

6 Results

Both controllers were implemented in a robot and tested in a real environment. The following results shows how the mobile robot accomplished the predefined Lemniscate trajectory.

6.1 Fuzzy Experimental Results

Figure 4 shows the desired trajectory in blue color and the trajectory followed by the robot with the green color. It can be observed that robot is able to follow the proposed trajectory.

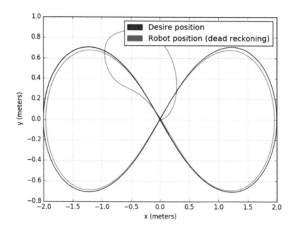

Fig. 4. Experimental results in x and y axis using fuzzy controller.

6.2 Backstepping Experimental Results

Experiments were performed with the control law obtained in the Sect. 4. Figure 5 plots the x and y results together, it is possible to observe the Lemniscate curve.

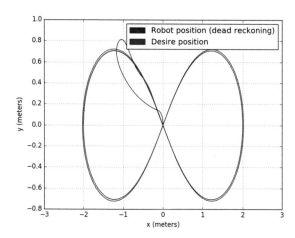

Fig. 5. Experimental results in both x and y axis using backstepping controller.

6.3   Integral Absolute Error (IAE)

Integral absolute error is a method to measure a control system performance, it consists on integrating the absolute error over time. It doesn't add weight to any of the errors in a systems response. In this case, the error is computed as the difference between the goal position and the estimated current position.

Table 5 contains the controllers performance indicators for each variable. These values are the accumulated errors in the x and y variables in a metric scale. It is noted a large error accumulation in the fuzzy controller variables, leading to a worse tracking trajectory as is depicted in Fig. 4. Also, we could observe in Table 5, a subtle improvement of the backstepping controller response relation to the fuzzy controller.

Table 5. Controllers performance indicators for each variable.

Variable	Backstepping	Fuzzy
X	270.09	1666.52
Y	77.92	863.67

7   Conclusion

In this work, two tracking controllers were presented. The designed controllers had as a final objective to make the robot track a predefined trajectory, in order to accomplish this objective, the robot should use the information provided by its encoders using the dead reckoning method. Experimental results in a real robot with the developed controllers validate our approach and it is possible to make a comparison of their performance.

Based on the results obtained, we could observe that the backstepping controller shows just a little better performance compared to the fuzzy controller. This is mainly due to the fact that, a fuzzy system isn't an exact approach, and it relies completely on previous knowledge of the designer about the subject. On other hand, the backstepping is a mathematical model based on robot kinematics, ensuring greater accuracy even if does not include a dynamic model. However, the fuzzy logic option is one of the most convenient ones, as it directs the focus to the resolution of the problem, facilitating and optimizing the implementation of the controller, as it removes the need for expensive mathematical models.

Future works using the same controllers, but employing computer vision are in progress.

References

1. Bezerra, C.G.: Localização de um robô móvel usando odometria e marcos naturais. MS thesis, Universidade Federal do Rio Grande do Norte (2004)
2. Mu, J., Yan, X.G., Spurgeon, S.K., Mao, Z.: Nonlinear sliding mode control of a two-wheeled mobile robot system. Int. J. Model. Ident. Control 27(2), 75–83 (2017)

3. Cuadros, M.A.S.L., De Souza, P.L.S., Almeida, G.M., Passos, R.A., Gamarra, D.F.T.: Development of a mobile robotics platform for navigation tasks using image processing. In: Asia-Pacific Computer Science and Application Conference (CSAC 2014), Shangai, China (2014)
4. Jiang, Z.-P., Nijmeijer, H.: Tracking control of mobile robots: a case study in backstepping. Automatica **33**(7), 1393–1399 (1997)
5. Gamarra, D.F.T., Bastos Filho, T.F., Sarcinelli Filho, M.: Controlling the navigation of a mobile robot in a corridor with redundant controllers. In: Proceedings of the IEEE International Conference on Robotics and Automation (ICRA 2005), Barcelona (2005)
6. Fierro, R., Lewis, F.L.: Control of a nonholonomic mobile robot using neural networks. IEEE Trans. Neural Netw. **9**(4), 589–600 (1998)
7. Omrane, H., Masmoudi, S.M., Masmoudi, M.: Fuzzy logic based controller for autonomous mobile robot navigation. Comput. Intell. Neurosci. **2016**, 10 (2016). Article ID 9548482
8. Xiong, B., Qu, S.R.: Intelligent vehicle's path tracking based on fuzzy controller. J. Transp. Syst. Eng. Inf. **10**(2), 70–75 (2010)
9. Saifizi, M., Hazry, D., Nor, R.M.: Vision based mobile robot navigation system. Int. J. Control Sci. Eng. **2**(4), 83–87 (2012)
10. Kanayama, Y.J., Kimura, Y., Miyazaki, F., Noguchi, T.: A stable tracking control method for an autonomous mobile robot. In: Proceedings of IEEE International Conference on Robotics and Automation (ICRA), pp. 384–389 (1990)
11. Park, K., Chung, H., Lee, J.G.: Dead reckoning navigation for autonomous mobile robots. In: 3rd IFAC Symposium on Intelligent Autonomous Vehicles, Madrid, Spain (1998)
12. Santana, A., Souza, A., Alsina, P., Medeiros, A.: Fusion of odometry and visual datas to localization a mobile robot using extended Kalman filter. In: Thomas, C. (ed.) Sensor Fusion and its Applications, pp. 407–421. InTech (2010). ISBN 978-953-307-101-5
13. Borenstein, J., Everett, H., Feng, L., Wehe, D.: Mobile robot positioning: sensors and techniques. J. Robot. Syst. **14**, 231–249 (1997)
14. Faria, H.G., Pereira, R.P.A., Resende, C.Z., Almeida, G.M., Cuadros, M.A.S.L., Gamarra, D.F.T.: Fuzzy trajectory tracking controller for differential drive robots. In: INDUSCON 2016 is the 12th IEEE/IAS International Conference on Industry Applications, Curitiba (2016)
15. Olson, E.: A primer on odometry and motor control, pp. 1–15 (2004)

Modelling Complex Transport Network with Dynamic Routing: A Queueing Networks Approach

Elmira Yu. Kalimulina$^{(\boxtimes)}$

V. A. Trapeznikov Institute of Control Sciences, Russian Academy of Sciences,
Profsoyuznaya Street 65, Moscow 11799, Russia
elmira.yu.k@gmail.com
http://www.ipu.ru/staff/elmira

Abstract. In this paper we consider a Jackson type queueing network with unreliable nodes. The network consists of $m < \infty$ nodes, each node is a queueing system of M/G/1 type. The input flow is assumed to be the Poisson process with parameter $\Lambda(t)$. The routing matrix $\{r_{ij}\}$ is given, $i, j = 0, 1, ..., m$, $\sum_{i=1}^{m} r_{0i} \leq 1$. The new request is sent to the node i with the probability r_{0i}, where it is processed with the intensity rate $\mu_i(t, n_i(t))$. The intensity of service depends on both time t and the number of requests at the node $n_i(t)$. Nodes in a network may break down and repair with some intensity rates, depending on the number of already broken nodes. Failures and repairs may occur isolated or in groups simultaneously. In this paper we assumed if the node j is unavailable, the request from node i is send to the first available node with minimal distance to j, i.e. the dynamic routing protocol is considered in the case of failure of some nodes. We formulate some results on the bounds of convergence rate for such case.

Keywords: Dynamic routing · Queueing system · Jackson network

1 Introduction

Queueing systems and networks are the most suitable mathematical tools for modelling and performance evaluation of complex systems such as modern computer systems, telecommunication networks, transport, energy and others [1–3]. The reliability is another important factor for quality assessment of these systems, models with unreliable elements are a subject of great interest last years [9–11]. A large number of research papers study queueing systems with unreliable servers [4,15]. The less ones consider queueing networks. In this paper we analyse the performance characteristics of an open queueing network, whose nodes are subject to failure and repair. This assumption is often missed in theoretical papers, but it's essential for applications, for example, for telecommunication, sensor, ad-hoc, mesh and other kind of networks.

© Springer Nature Switzerland AG 2020
A. Abraham et al. (Eds.): ISDA 2018, AISC 941, pp. 222–229, 2020.
https://doi.org/10.1007/978-3-030-16660-1_22

This work is motivated by a practical task of modelling of modern complex networks (telecommunication, transport, distributed computing systems, etc.). We consider the mathematical model of the queueing network as a set of connected nodes that can break down and repair. We propose a modification of the classical model of an open queueing network (see e.g. [16, Chapter 2]), based on the principle of dynamic routing.

A strong mathematical definition of the term "dynamic routing" doesn't exist. It originally appeared in telecommunication industry. Dynamic routing is the technology that enables active network nodes (called routers) to perform many vital functions: detection, maintaining and modification of routes with considering of a network's topology, as well as some functions of routes calculation and their estimations. In difference from static routing technology routes are calculated dynamically using any one of a number of dynamic routing protocols. Dynamic routing is crucial technology for reliable packet transmission in case of failures. Protocols are used by routers to detect nodes availability and find routes available for packets forwarding over network. Figure 1 shows a simple example of six-node inter-networking transmission. The transmission between *Local network 1* and *Local network 4* uses the link *Router D - Router C*. Assume the transmission facility between Gateway *Routers C* and *D* has failed. This renders the link between C and D unusable and the data transmission between *Local network 1* and *2* is impossible (for static routing). The dynamic routing protocols use a route redistribution scheme and dynamically define new paths for transmission between *Router D* and *Router C* via *B* and *A*, or *E*, or central router (dotted lines). Thus network availability can be significantly enhanced through dynamic routing, alternate routes of more length are redundant communications links between nodes D and C.

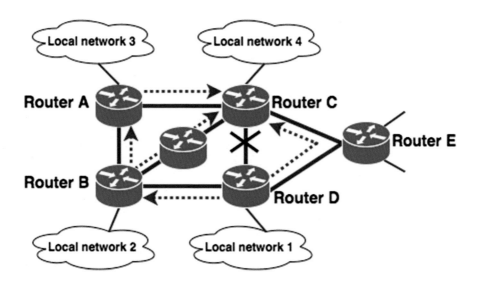

Fig. 1. A six-node inter-network with some route redundancy.

There are some math research papers where queueing networks with dynamic routing were considered.

Definition 1. *A routing matrix $R = (r_{ij})$ of a queueing network is a stochastic matrix, i.e. $\sum_{i=0}^{m} r_{ij} = 1$ $\forall i$, $i, j = 0, \ldots, m$, where r_{ij} is a probability the job from node i will be sent to node j, "0" means an external node.*

Queueing networks with constant routing matrix were considered in papers [5,6], each node was modelled as a multichannel system, principle of dynamic routing was a random selection of a channel at the node. We will adhere to interpret the dynamic routing as it's defined in telecoms and we will understand it as reconstruction of the route depending on the availability/unavailability of a specific node in the network, sending the requests (message) via alternate paths. In terms of queueing networks models it means the change the values in a routing matrix $\{r_{ij}\}$ [9,10,12]. A request will be rerouted to the available node in the case of failure of node j, i.e., this concept is as close as possible to this definition in telecommunication networks.

It is noted that other approaches can be applied to the problem of estimation of networks reliability. For example, the Erdos-Renyi random graphs model can be used to analyse the network connectivity (Erdos-Renyi graph) [17].

But as our task is the performance evaluation and analysis of traffic flows in networks taking into account reliability, we use queueing theory models and a model of open queueing network. There are some researches on unreliable queueing networks. There are several algorithms for modifying the routing matrix $\{r_{ij}\}$ in the case of failures of nodes, the common idea they are based on is the principle of blocking of jobs and repeated service after nodes recovery. The result related to the rate of convergence to the stationary distribution for unreliable network is given in [7,8]. In this paper we give some results for unreliable networks similarly as it was done in [8], but we propose another approach to the modification of the route matrix $\{r_{ij}\}$ and consider a more general model for network nodes. Queueing networks have been traditionally represented as studied as static graphs (fixed routing matrix), in applications of queueing theory to real systems, it has been recognized that network structure should change in time to better reflect the reality of interactions between nodes. There are several main classical problems in the analysis of queueing systems (networks) models that have to been considered: stability, ergodicity and probability of overflow. These problems are strongly connected with the transient behaviour of networks and the speed of convergence to stationary regime, so here we will also consider this problem for unreliable network systems.

2   Process Definition

It is assumed that nodes at the network are unreliable and may break down or repair. Failures can be both individual and in a group (as in models in [7,8]). We will refer to $M_0 = \{0, 1, 2, ..., m\}$ as the set of nodes, where "0" is the "external node" (entry and exit from the network) and to $D \subset M$ as the subset of failed nodes, $I \subset M \backslash D$ the subset of working nodes, nodes from I may break down with the intensity $\alpha_{D \cup I}^{D}(n_i(t))$. Nodes from $H \subset D$ may recover with the intensity $\beta_{D \backslash H}^{D}(n_i(t))$. It is assumed the routing matrix (s_{ij}) is given. Additionally the adjacency matrix for our network (s_{ij}) is considered:

$$s_{ij} = \begin{cases} 1, & \text{if} \quad r_{ij} \neq 0, \\ 0, & \text{if} \quad r_{ij} = 0. \end{cases}$$

Now we can consider all possible paths of the network graph. To find them we need to calculate the following matrix: $(s_{ij})^2$, $(s_{ij})^3$, ..., $(s_{ij})^m$, $m < \infty$, $(s_{ij})^1 = (s_{ij})$. The matrix $(s_{ij})^m$ has the following property: the element in row i and column j is the number of paths from node i in the unit j of length m (including $(m-1)$ transitional nodes).

We take the following routing scheme for network nodes from the subset D (we call it as "dynamic routing without blocking"). Only transitions to $M_0 \setminus D$ are possible for nodes from D:

$$r_{ij}^{D} = \begin{cases} 0, & \text{if} \quad j \in D, i \neq j, \\ r_{ij} + r_{ik}/s_{ik}^{p}, & \text{if} \quad j \notin D, k \in D \\ \exists\, i \to j \to i' \to j' \to \dots \to i'' \to k : \underbrace{s_{ij}^1 * s_{ji'}^1 * s_{i'j'}^1 * \dots * s_{i''k}^1}_{p+1} \neq 0, \\ \text{where} \quad p = \min\{2, 3, ..., m : s_{ik\,k \in D}^{p} \neq 0\}, \\ r_{ii} + \sum_{\substack{k \in D \\ s_{ik}^{p}=0\ \forall\ 1<p\leq m}} r_{ik}, \quad \text{if} \quad i \in M_0 \setminus D, i = j, \end{cases}$$

where s_{ik}^{p} - element of a matrix $(s_{ij})^p$.

The routing matrix is changed according to the same way for the input flow:

$$\Lambda r_{0j}^{D} = \begin{cases} \Lambda r_{0j}, & \text{if} \quad j \in M \setminus D, \\ \Lambda(r_{0j} + r_{0k}/s_{0k}^{p} * \underbrace{(s_{0j}^1 * s_{ji'}^1 * s_{i'j'}^1 * \dots * s_{i''k}^1))}_{p+1}, & \text{if } j \notin D, k \in D \\ 0, & \text{otherwise}. \end{cases}$$

Further we will refer to the modified routing matrix as $R^D = (r_{ij}^D)$, the intensities of failures and recoveries depend on the state of nodes and does not depend on network load and are defined as $\alpha(D, I)$, and $\beta(D, H)$.

A more general model than in [7] is considered for network nodes. It is assumed that each network node is a queueing system type $M/G/1$. The system's dynamic will be described by a continuous in time random process $X(t)$ taking values from the following enlarged state space $\mathbb{E}$:

$$\tilde{\mathbf{n}} = ((n_1, z_1), (n_3, z_2), ..., (n_m, z_m), D) \in \{\mathbb{Z}_+ \times \{R_+ \cup 0\}\}^m \times |D| = \mathbb{E},$$

where n_i is the number of requests at the node i, z_i - time elapsed from the beginning of service for the current request i, $|D|$ - the cardinality of set D.

Intensity rates $\mu_i(n_i, z_i)$ depend on both the number of requests at nodes $n_i(t)$ and time $z_i(t)$, time elapsed from the beginning of service for the current request at time t. The probability of any jump (finishing of service in one of the nodes or new request arriving in a network) is defined similarly:

$$\mu_j(n_j(t), z_j(t))\Delta t \left(1 - \int_t^{t+\Delta t} \left(\Lambda(s) + \sum_{i \neq j}^m \mu_i(n_i(t), z_i(t) + s)\right)ds + O(\Delta t)^2\right) \quad (1)$$

$$\Lambda(t)\Delta t \left(1 - \int_t^{t+\Delta t} \left(\sum_{i=1}^m \mu_i(n_i(t), z_i(t) + s)\right)ds + O(\Delta t)^2\right). \quad (2)$$

The following transitions in a network are possible:

$$T_{ij}\tilde{\mathbf{n}} := (D, n_1, \cdots, n_i - 1, \cdots, n_j + 1 \cdots, n_m),$$
$$T_{0j}\tilde{\mathbf{n}} := (D, n_1, \cdots, n_j + 1, \cdots, n_m),$$
$$T_{i0}\tilde{\mathbf{n}} := (D, n_1, \cdots, n_i - 1, \cdots, n_m),$$
$$T_H\tilde{\mathbf{n}} := (D \setminus H, n_1, \cdots, n_m),$$
$$T^I\tilde{\mathbf{n}} := (D \cup I, n_1, \cdots, n_m).$$

Definition 2. *The Markov process* $\mathbf{X} = (X(t), t \geq 0)$ *is called unreliable queueing network if it's defined by the following infinitesimal generator:*

$$\tilde{\mathbf{Q}}f(\tilde{\mathbf{n}}) = \sum_{j=1}^m [f(T_{0j}\tilde{\mathbf{n}}) - f(\tilde{\mathbf{n}})]\Lambda(t)r_{0j}^D$$

$$+ \sum_{i=1}^m \sum_{j=1}^m [f(T_{ij}\tilde{\mathbf{n}}) - f(\tilde{\mathbf{n}})]\mu_i(n_i(t), z_i(t))r_{ij}^D \quad (3)$$

$$+ \sum_{I \subset M} [f(T^I\tilde{\mathbf{n}}) - f(\tilde{\mathbf{n}})]\alpha(D, I) + \sum_{H \subset M} [f(T^I\tilde{\mathbf{n}}) - f(\tilde{\mathbf{n}})]\beta(D, H)$$

$$+ \sum_{j=1}^m [f(T_{j0}\tilde{\mathbf{n}}) - f(\tilde{\mathbf{n}})]\mu_j(n_i(t), z_i(t))r_{j0}^D.$$

3   Main Results

Like the classical Jackson network the existence of a stationary distribution for an unreliable network with dynamic routing may be proved.

Theorem 1. *It is assumed the following conditions for unreliable network from the Definition 2*

$$(1) \inf_{n_j, t} \mu_j(n_j, z_j) > 0 \quad \forall \, j,$$

(2) time of service and time between new arrivals are independent random variables,

(3) routing matrix R^D is reversible,

then the stationary distribution for unreliable networks is defined by formulae

$$\pi(\tilde{\mathbf{n}}) = \pi(D, n_1, n_2, \cdots, n_m) = \frac{1}{C} \frac{\psi(D)}{\phi(D)} \prod_{i=1}^{m} \frac{1}{C_i} \frac{\lambda_i^{n_i}}{\prod_{k=1}^{n_i} \mu_i(k)}$$

where

$$C_i = \sum_{n=0}^{\infty} \frac{\lambda_i^{n_i}}{\prod_{y=1}^{n} \mu_i(y)}, \quad \lambda_i = \sum_{j=0}^{m} \Lambda * r_{ji}.$$

The main result for the convergence rate is formulated in terms of the spectral gap for unreliable queueing network. The preliminary notations and results on the spectral gap: there is a Markov process $\mathbf{X} = (X_t, t \geq 0)$ with the matrix of transition intensities $Q = [q(\mathbf{e}, \mathbf{e}')]_{e,e' \in \mathbb{E}}$, with stationary distribution π and an infinitesimal generator given by

$$\mathbf{Q}f(\mathbf{e}) = \sum_{\mathbf{e}' \in \mathbb{E}} (f(\mathbf{e}') - f(\mathbf{e}))q(\mathbf{e}, \mathbf{e}').$$

The usual scalar product on $L_2(\mathbb{E}, \pi)$ is defined as

$$\langle f, g \rangle_{pi} = \sum_{\mathbf{e} \in \mathbb{E}} f(\mathbf{e})g(\mathbf{e})\pi(\mathbf{e}). \tag{4}$$

The spectral gap for $\mathbf{X}$ is

$$Gap(\mathbf{Q}) = \inf\{-\langle f, \mathbf{Q}f \rangle_{\pi} : \|f\|_2 = 1, \langle f, \mathbf{1} \rangle_{\pi} = 0\}.$$

The main result for a network is formulated in the following theorems:

Theorem 2. *If $\mathbf{X}$ is a Markov process with infinitesimal generator $\mathbf{Q}$, it is assumed that $\mathbf{Q}$ is bounded, the minimal intensity of service is strictly positive $\inf_{n_j, t} \mu_j(n_j, z_j) > 0$ and the routing matrix (r_{ij}^D) is reversible, then $Gap(\mathbf{Q}) > 0$, if the following condition is true: for any $i = 1, \cdots, m$, for the birth and death process, corresponding to the node i with parameters λ_i and $\mu_i(n_i, z_i)$ the spectral gap is strictly positive $Gap_i(\mathbf{Q}_i) > 0$.*

Theorem 3. *If $\mathbf{X}$ is a Markov process with a bounded infinitesimal generator $\mathbf{Q}$, positive minimal intensity of service $\inf_{n_j, t} \mu_j(n_j, z_j) > 0$ and reversible routing matrix (r_{ij}^D), then $Gap(\mathbf{Q}) > 0$ iff for any $i = 1, \cdots, m$, the distribution $\pi = (\pi_i), i \geq 0$ has light tails, i.e. the following condition $\inf_k \frac{\pi_i(k)}{\sum_{j>k} \pi_i(j)} > 0$.*

Theorem 4. *(Corollary from [14]) If $\mathbf{X}$ is unreliable queueing network with dynamic routing from Definition 2 with infinitesimal generator $\mathbf{Q}$ and transition probabilities matrix P_t. It is assumed that routing matrix (r_{ij}^D) is reversible and $(r_{ij}^D)^k > 0$ dor $k \geq 1$. If the distribution π_i has light tails for any $i = 1, \cdots, m$, then the following conditions are equivalent*

– *for any* $f \in L_2(E, \pi)$

$$\|P_t f - \pi(f)\|_2 \le e^{-Gap(\mathbf{Q})t} \|f - \pi(f)\|_2, t > 0, \tag{5}$$

– *for any* $\mathbf{e} \in \mathbb{E}$ *the constant* $C(\mathbf{e}) > 0$ *exists such, that*

$$|\delta_e - \pi(f)|_{TV} \le C(e) e^{-Gap(Q)t}, t > 0. \tag{6}$$

References

1. Lakatos, L., Szeidl, L., Telek, M.: Introduction to Queueing Systems with Telecommunication Applications, 388 p. Springer, Heidelberg (2012)
2. Daigle, J.: Queueing Theory with Applications to Packet Telecommunication, 316 p. Springer, Heidelberg (2006). Technology & Engineering
3. Thomasian, A.: Analysis of fork/join and related queueing systems. ACM Comput. Surv. **47**(2), 71 p. (2014). https://doi.org/10.1145/2628913. Article 17
4. Jain, M., Sharma, G.C., Sharma, R.: Unreliable server M-G-1 queue with multi-optional services and multi-optional vacations. Int. J. Math. Oper. Res. **5**(2) (2013). https://doi.org/10.1504/IJMOR.2013.052458
5. Vvedenskaya, N.D.: Configuration of overloaded servers with dynamic routing. Probl. Inf. Transm. **47**, 289 (2011). https://doi.org/10.1134/S0032946011030070
6. Sukhov, Yu.M., Vvedenskaya, N.D.: Fast Jackson networks with dynamic routing. Probl. Inf. Transm. **38**, 136 (2002). https://doi.org/10.1023/A:1020010710507
7. Lorek, P., Szekli, R.: Computable bounds on the spectral gap for unreliable Jackson networks. Adv. Appl. Probab. **47**, 402–424 (2015)
8. Lorek, P.: The exact asymptotic for the stationary distribution of some unreliable systems. arXiv:1102.4707 [math.PR] (2011)
9. Kalimulina, E.Yu.: Rate of convergence to stationary distribution for unreliable Jackson-type queueing network with dynamic routing. In: Distributed Computer and Communication Networks. Communications in Computer and Information Science, vol. 678, pp. 253–265 (2017). https://doi.org/10.1007/978-3-319-51917-3_23.
10. Kalimulina, E.Yu.: Queueing system convergence rate. In: Proceedings of the 19th International Conference, Distributed Computer and Communication Networks, DCCN 2016, Moscow, Russia, vol. 3, pp. 203–211. RUDN, Moscow (2016)
11. Kalimulina, E.Yu.: Analysis of system reliability with control, dependent failures, and arbitrary repair times. Int. J. Syst. Assur. Eng. Manag., 1–11 (2016). https://doi.org/10.1007/s13198-016-0520-5(2016)
12. Kalimulina E. Yu.: Analysis of unreliable Jackson-type queueing networks with dynamic routing, December 2016. SSRN: https://doi.org/10.2139/ssrn.2881956
13. Dorogovtsev, S.N., Mendes, J.F.F.: Evolution of Networks: From Biological Nets to the Internet and WWW (Physics). Oxford University Press Inc., New York (2003)
14. Chen, M.F.: Eigenvalues, Inequalities, and Ergodic Theory. Springer, Heidelberg (2005)
15. Klimenok, V., Vishnevsky, V.: Unreliable queueing system with cold redundancy. In: Gaj, P., Kwiecień, A., Stera, P. (eds.) Computer Networks. Communications in Computer and Information Science, vol. 522, pp. 336–346. Springer, Heidelberg (2015). https://doi.org/10.1007/978-3-319-19419-6_32

16. Chen, H., Yao, D.D.: Fundamentals of Queueing Networks: Performance, Asymptotics, and Optimization. Stochastic Modelling and Applied Probability, vol. 46. Springer, Heidelberg (2001). https://doi.org/10.1007/978-1-4757-5301-1
17. Yavuz, F., Zhao, J., Yağan, O., Gligor, V.: Toward k -connectivity of the random graph induced by a pairwise key predistribution scheme with unreliable links. IEEE Trans. Inf. Theory **61**(11), 6251–6271 (2015). https://doi.org/10.1109/TIT.2015.2471295
18. Sauer, C., Daduna, H.: Availability formulas and performance measures for separable degradable networks. Econ. Qual. Control **18**(2), 165–194 (2003)

Math Modeling of the Reliability Control and Monitoring System of Complex Network Platforms

Elmira Yu. Kalimulina[✉]

V. A. Trapeznikov Institute of Control Sciences, Russian Academy of Sciences,
Profsoyuznaya Street 65, Moscow 11799, Russia
elmira.yu.k@gmail.com
http://www.ipu.ru/staff/elmira

Abstract. This paper is concerned with analytical models and methods for reliability planning, optimization, and operation of the control and monitoring system of SMS-network platforms. The difference between classical models of reliability and models developed here is that the last ones take into account the economy is often an important aspect of reliability planning. Network providers operate in a competitive environment, so the main principle of reliability planning is not so much to increase reliability as well (as it's done in fully technical approach) as to maximise the total profit of the network. Thus, new methods must allow to find the combination of factors (costs, investments into reliability, redundancy and increased maintenance, the profit from system operation, penalties for delays and failures, risks defined in a service level agreement, etc.) that maximises the profit. We employ the well-known from an actuarial science a model that considers a network through two cash flows: incoming cash from customers and outgoing claims paid due unreliability under SLA. Also we include into the optimisation model more general assumptions about the reliability of systems: different ways of reliability improvements and redundancy, non-exponentially distributed failures and recoveries, dependent failures, and reliability control.

Keywords: Reliability model · Reliability planning ·
Reliability optimisation · Redundancy · Reservation ·
Dependent failures · Markov models · Cash flows

1 Introduction

The reliability of modern telecommunication networks is an important factor to ensure the quality of IT-services and it is of great importance for customers. Telecommunication networks have a very complex structure, including a lot of components (on a hardware and software level), so there is a huge diversity of ways to improve the reliability and achieve the maximum of it so far as possible. Many authors in the reliability optimization of telecommunication networks take

© Springer Nature Switzerland AG 2020
A. Abraham et al. (Eds.): ISDA 2018, AISC 941, pp. 230–237, 2020.
https://doi.org/10.1007/978-3-030-16660-1_23

into consideration only initial cost for redundancy and doesn't consider the other parameters such as a network profit, costs per maintenance and other expenditures [1,2]. But network providers often operate in a competitive environment and their abilities of achieving a certain level of dependability differ [7]. On the one hand, network provider must satisfy user needs and provide the high reliability of network operation. On the other hand, the costs, including costs for reliability and penalties paid under a service level agreement (SLA) must not exceed the total profit of the network. So, there is a need of establishing a model for optimal reliability and cost planning and allocating in the telecommunication network. This problem was considered in different formulations. The most common problem statement is to find the maximum of the reliability $R(\cdot)$ at a minimum initial cost $C(\cdot)$ or finding the maximum/minimum of the reliability/cost under some restrictions on total cost of a system or reliability:

$$\begin{matrix} \max R(\cdot) \\ \min C(\cdot) \end{matrix} \quad , \quad \begin{matrix} \max R(\cdot) \\ C(\cdot) \leq C_0 \end{matrix} \quad , \quad \begin{matrix} \min C(\cdot) \\ R(\cdot) \geq R_0. \end{matrix} \tag{1}$$

So, the main stress is done on an application of different methods to evaluation of the reliability and performance, and algorithms of optimization in itself. For example, the papers [2] addresses an NP-hard problem, referred to as Network Topology Design with minimum Cost subject to a Reliability constraint, to design a minimal-cost communication network topology that satisfies a pre-defined reliability constraint. This paper describes a dynamic programming scheme to solve the problem. The paper [3] is also about new method and the algorithm to classical problem of redundancy allocation subject to a set of resource constraints.

A good overview of key publications on reliability approaches in various fields of engineering and physical sciences and on models and methods for reliability optimization problems is presented in [4] and [6]. The first paper provides the major areas i.e. past, current and future trends of reliability methods and applications. The second one discusses models and methods for reliability optimisation problems including reliability allocation, redundancy allocation and reliability-redundancy allocation and gives good overview of other research papers on the system reliability. But the general idea of these articles does not consider the system reliability as the key factor for an entire business (it takes into account only costs for redundancy, and no other parameters such as Maintenance, penalties for failures, SLA, profit and etc.).

Our idea in this paper is to consider many factors (investments, increased maintenance, etc.) that maximise the total profit of the network [7]:

$$\max\{TR(total\ revenue) - TC(total\ costs)\},$$

where

$$TC = TFC(total\ fixed\ costs) + TVC(total\ variable\ costs),$$

and TC include the investment costs to achieve a certain level of reliability (redundancy costs and allocation costs), the expected maintenance costs, the costs for an interruption of the network operation, caused by faults.

2   Model of the Reliability Optimisation

We consider a network provider activity experiencing two opposing cash flows (as in an insurance company): incoming cash from a traffic revenue and outgoing cash claims for the traffic disturbance due to failures [9, 10].

Without loss of generality we suppose that a revenue from customers arrives with the constant rate $c > 0$. For a practical usage it may be calculated as: $c = \sum_k V_k * p_k$, where V_k - the average network traffic volume of k-type transmitted per hour and p_k - tariff for transmission of traffic of k-type. In the case of absolutely reliable network a network provider during the time t receives the revenue equal to $c * t$. Denote the reliability function for time t as $R(t)$. It may be considered as an indication of the fraction of time that a system is available in a cycle of system operation t. Thus, we can estimate the real revenue during the time t equal to $c * t * R(t)$. In the classical Cramer-Lundberg model claims arrive according to a Poisson process N_t with intensity λ and are independent and identically distributed non-negative random variables with distribution F and mean α

$$X(t) = x_0 + c * t - \sum_{i=1}^{N_t} \xi_i, \quad t \geq 0.$$

For network provider we replace the sum in this formula by the following expression:

$$\sum_{i=1}^{N(t)} T_i * \xi_i,$$

where failures in a network occurs according to a random process N_t (so, N_t is the number of failures before the time t), T_i - the duration of a network downtime after a failure i, ξ_i - a non-negative random variables which characterises the size of a payout for an hour of a network downtime [10]. Moreover, we take into account initial costs C_0 for achieving the desired level of system reliability $R(t)$. In general case there are many ways of reliability improvements, but we consider two ones: redundancy and reliability allocation. A redundancy means, that we improve a system reliability via addition of $\mathbf{n} = (n_1, \cdots, n_m)$ parallel elements, m - the total number of subsystems in a network, whose reliability can be increase via redundancy. Thus we can estimate the total costs for redundancy as $\sum_{i=1}^{m} n_i * c_i$. For subsystems, whose reliability will not increased via redundancy, we suppose $\mathbf{R} = (R_1(t), \cdots, R_l(t))$ are reliability values, which we want to achieve, l - the total number of subsystems, whose reliability can be improved via allocation. The cost of the reliability improvement to the value $R_i(t)$ is defined by the following function [11, 13]:

$$f(R_i(t)) = a_i \ln \frac{1}{1 - R_i(t)}, \quad i = 1, 2, \cdots, l,$$

where a_i is a scale parameter that reflects the sensitivity of the reliability of a subsystem to changes in the cost.

Thus the model describing a network provider takes a form:

$$X(t) = x_0 + t * R(t) * \sum_k V_k * p_k - \sum_{i=1}^{N(t)} T_i * \xi_i$$

$$- \sum_{i=1}^{m} n_i * c_i - \sum_{i=1}^{l} a_i \ln \frac{1}{1 - R_i(t)}. \tag{2}$$

For simplicity some terms in (2) can be replace by their mean values:

$$X(t) = x_0 + t * R(t) * \sum_k V_k * p_k - \bar{N}(t) * \xi_i$$

$$- \sum_{i=1}^{m} n_i * c_i - \sum_{i=1}^{l} a_i \ln \frac{1}{1 - R_i(t)}, \tag{3}$$

or

$$X(t) = x_0 + t * R(t) * \sum_k V_k * p_k - MTBF * \bar{N}(t) * p$$

$$- \sum_{i=1}^{m} n_i * c_i - \sum_{i=1}^{l} a_i \ln \frac{1}{1 - R_i(t)}, \tag{4}$$

where $MTBF$ is the mean time between failures, $\bar{N}(t)$ - the mean number of failures until the time t, ξ - a non-negative random variable (which characterises the mean size of a payout for an hour of a network downtime), p - the economic valuation of interruption of a network operation per an hour (a non-random variable). $R(t)$, $R_i(t)$, $MTBF$ and $\bar{N}(t)$ depends on reliability characteristics of network subsystems (on parameters of distributions of failure and recovery times), and in general case it may be very complicate and non trivial task. The problem of deriving these analytical expressions should be considered independently.

In this paper we will present some examples of reliability estimation.

3   Reliability Optimisation of a System of Two Independent Elements

Now we illustrate an application of this model with a simple example. Let's consider a system consisting of two independent subsystems operating in series. The times of failure and recovery of subsystems are exponentially distributed with parameters λ_i and μ_i, $i = 1, 2$. The reliability of the single subsystem is defined by the well-known formula:

$$R_i(t) = \frac{\mu_i}{\lambda_i + \mu_i} + \frac{\lambda_i}{\lambda_i + \mu_i} e^{-(\lambda_i + \mu_i)t}. \tag{5}$$

The reliability of the second subsystem is improved via active redundancy, n identical subsystems are working in parallel. so the reliability of a system is defined by the formula:

$$R(t) = \left(\frac{\mu_1}{\lambda_1 + \mu_1} + \frac{\lambda_1}{\lambda_1 + \mu_1} e^{-(\lambda_1+\mu_1)t} \right) \tag{6}$$
$$\left(1 - \left(\frac{\lambda_2}{\lambda_2 + \mu_2} \right)^n (1 - e^{-(\lambda_2+\mu_2)t})^n \right).$$

We can find the probability of failure up to and including time t, the failure density function $f(t)$, and the mean time between failures (MTBF) :

$$Pr(T \le t) = F(t) = 1 - R(t) = \tag{7}$$
$$1 - \left(\frac{\mu_1}{\lambda_1 + \mu_1} + \frac{\lambda_1}{\lambda_1 + \mu_1} e^{-(\lambda_1+\mu_1)t} \right)$$
$$\left(1 - \left(\frac{\lambda_2}{\lambda_2 + \mu_2} \right)^n (1 - e^{-(\lambda_2+\mu_2)t})^n \right)$$

$$f(t) = \frac{n\lambda_2^n e^{-(\lambda_2+\mu_2)t}}{(\lambda_2 + \mu_2)^{n-1}} \left(\frac{\mu_1}{\lambda_1 + \mu_1} + \frac{\lambda_1 e^{-(\lambda_1+\mu_1)t}}{\lambda_1 + \mu_1} \right) * \tag{8}$$
$$\left(1 - e^{-(\lambda_2+\mu_2)t} \right)^{n-1} + \lambda_1 e^{-(\lambda_1+\mu_1)t} -$$
$$\lambda_1 e^{-(\lambda_1+\mu_1)t} \left(1 - \frac{\mu_2}{\lambda_2 + \mu_2} - \frac{\lambda_2 e^{-(\lambda_2+\mu_2)t}}{\lambda_2 + \mu_2} \right)^n,$$

$$\int_0^\infty t f(t) dt = \int_0^\infty t(A e^{Et} + B e^{Et}(1 - e^{Ft})^n +$$
$$C e^{Ft}(1 - e^{Ft})^{n-1} + D e^{Gt}(1 - e^{Ft})^{n-1}) dt =$$
$$= \left(\frac{A(Et - 1)}{E^2} e^{Et} + B \sum_{k=0}^n \binom{n}{k}(-1)^k \frac{(kF + E)t - 1}{(kF + E)^2} e^{(kF+E)t} + \right.$$
$$C \sum_{k=0}^{n-1} \binom{n-1}{k}(-1)^k \frac{(k+1)Ft - 1}{(k+1)^2 F^2} e^{(k+1)Ft} +$$
$$\left. D \sum_{k=0}^{n-1} \binom{n-1}{k}(-1)^k \frac{(kF + G)t - 1}{(kF + G)^2} e^{(kF+G)t} \right) \Big|_0^\infty \Rightarrow$$
$$\Rightarrow \quad MTBF = \frac{A}{E^2} + \sum_{k=0}^n \binom{n}{k}(-1)^k \frac{B}{(kF + E)^2} + \tag{9}$$
$$\sum_{k=0}^{n-1} \binom{n-1}{k}(-1)^k \left(\frac{C}{(k+1)^2 F^2} + \frac{D}{(kF + G)^2} \right),$$

where

$$A = \lambda_1, \quad B = -\lambda_1 \left(\frac{\lambda_2}{\lambda_2 + \mu_2}\right)^n,$$

$$C = \frac{n\mu_1(\lambda_2 + \mu_2)}{\lambda_1 + \mu_1} \left(\frac{\lambda_2}{\lambda_2 + \mu_2}\right)^n, D = \frac{n\lambda_1(\lambda_2 + \mu_2)}{\lambda_1 + \mu_1} \left(\frac{\lambda_2}{\lambda_2 + \mu_2}\right)^n,$$

$$E = -(\lambda_1 + \mu_1), F = -(\lambda_2 + \mu_2), G = E + F,$$

where $F(t)$ is the failure distribution function, T is the failure time, here we used the fact that n is positive finite integer ($n \in \mathbb{N}$, $n << \infty$).

The mean number of failures $N(t_1)$ till time t_1 can be estimated using the following relation:

$$E[N(t_1)] = \int_0^{t_1} \lambda(t)dt, \tag{10}$$

where $\lambda(t)$ is a failure rate, t - time.

Since we consider the simplest case of the exponentially distributed failure and recovery times, $\lambda(t) = \lambda = \frac{1}{MTBF}$. So, using (9), we have

$$E[N(t_1)] = \frac{t_1}{MTBF}. \tag{11}$$

Thus, using (5), (6), (9), (11) we can concretize a function (3) for our example:

$$X(t) = x_0 + t * \left(\frac{\mu_1}{\lambda_1 + \mu_1} + \frac{\lambda_1}{\lambda_1 + \mu_1} e^{-(\lambda_1 + \mu_1)t}\right) \tag{12}$$

$$\left(1 - \left(\frac{\lambda_2}{\lambda_2 + \mu_2}\right)^n (1 - e^{-(\lambda_2 + \mu_2)t})^n\right) * \sum_1^l V_l * p_l -$$

$$t * \xi_i \left(\frac{\lambda_1}{(\lambda_1 + \mu_1)^2} + \frac{\lambda_1 \lambda_2^n}{(\lambda_2 + \mu_2)^n} \sum_{k=0}^n \binom{n}{k} \frac{(-1)^k}{(\lambda_1 + \mu_1 + k\lambda_2 + k\mu_2)^2} + \right.$$

$$\frac{n\lambda_2^n}{(\lambda_1 + \mu_1)(\lambda_2 + \mu_2)^n} \sum_{k=0}^{n-1} \binom{n-1}{k} (-1)^k \left(\frac{\mu_1}{(k+1)^2(\lambda_2 + \mu_2)} + \right.$$

$$\left.\left.\frac{\lambda_1(\lambda_2 + \mu_2)}{((k-1)(\lambda_2 + \mu_2) + \lambda_1 + \mu_1)^2}\right)\right)^{-1}$$

$$-n * c - a_1 \ln \frac{\lambda_1 + \mu_1}{\lambda_1(1 - e^{-(\lambda_1 + \mu_1)t})}.$$

Now we assume that x_0, λ_1, λ_2, μ_2, l, V_l, p_l, the period of planning t, c, a_1 are given and $\xi_i \sim \mathcal{N}(1,1)$. An optimization problem consists of maximizing a real function (12) by choosing the recovery rate for the first subsystem μ_1 ($\mu_1 > 0$) and the number of redundant components n ($n \in \mathbb{N}$) for the second one:

$$\max_{\substack{\lambda_1 > 0 \\ n \in \mathbb{N}}} X(\lambda_1, n). \tag{13}$$

Assume the following initial values: $x_0 = 1000000$, $\lambda_1 = 1/25$, $\lambda_2 = 1/5$, $\mu_2 = 0.75$, $l = 5$, $\mathbf{V} = (10, 20, 11, 18, 15)$, $\mathbf{p} = (2, 2, 4, 1, 7)$, $t = 24 * 365 * 7$, $c_2 = 170000$, $a_1 = 100000$. For the optimization we will use the $\boldsymbol{R}$ language and the package *DEoptim* based on the Differential Evolution algorithm. The critical point is $\mu_1 = 5.7342, n = 3$ and The value of the function at a critical point is a critical value is $1.3671 * 10^7$. As we can see from results the reliability grows closer to unity with an increase of μ_1 and n. The contour plot of the revenue function demonstrates the decrease of the total revenue starting from the certain point, since the cost for reliability grows faster than revenue starting from this point. A real-valued function $X(t)$ has a global maximum point at $\mu_1 = 5.7342, n = 3$, this point is optimal in the sense of balance between revenue and cost. Thus, our results meet the modern principle of reliability planning to maximise the profit. This result reflects the underlying assumption that optimization of reliability with problem statement in one of the forms from (1) is not as efficient as using the objective function in the form (2), (3) or (4).

4   Conclusion

Thus, the formulae to make definitive conclusions about design or existing systems, for example, to plan the load, adjust the performance parameters, make decisions and commissioning or decommissioning of additional capacity. Calculations in the case of these examples show that it is possible to increase the load on the system.

References

1. Lei, L.: Study on reliability optimization problem of computer network. Int. J. Secur. Appl. **9**(4), 161–174 (2015)
2. Elshqeirat, B., Soh, S., Rai, S., Lazarescu, M.: Topology design with minimal cost subject to network reliability constraint. IEEE Trans. Reliab. **64**(1), 118–131 (2015)
3. Caserta, M., Vob, S.: An exact algorithm for the reliability redundancy allocation problem. Eur. J. Oper. Res. **244**(1), 110–116 (2015)
4. Mangey, R.: On system reliability approaches: a brief survey. Int. J. Syst. Assur. Eng. Manag. **4**(2), 101–117 (2013)
5. Moradijoz, M., Moghaddam, M.P., Haghifam, M.R., Alishahi, E.: A multi-objective optimization problem for allocating parking lots in a distribution network. Int. J. Electr. Power Energy Syst. **46**, 115–122 (2013)
6. Soltani, R.: Reliability optimization of binary state non-repairable systems: a state of the art survey. Int. J. Ind. Eng. Comput. **5**(3), 339–364 (2014)
7. ITU-T Recommendations, E.862(06/92): Dependability planning of telecommunication networks, 15 pages. ITU Telecommunication Standardization Sector, Geneva (1992)
8. Kalimulina, E.Y.: Reliability computation for complex systems with parallel structure that are completely repairable during use. Autom. Remote Control **71**(6), 1257–1264 (2010)

9. Kalimulina, E.Yu.: Mathematical model for reliability optimisation of distributed telecommunications networks. In: 2011 International Conference on Computer Science and Network Technology (ICCSNT), 24–26 December 2011, vol. 4, pp. 2847–2853 (2011)
10. Kalimulina, E.Y.: Analysis of system reliability with control, dependent failures, and arbitrary repair times. Int. J. Syst. Assur. Eng. Manag. **8**, 180 (2017). https://doi.org/10.1007/s13198-016-0520-5
11. Kalimulina, E.Y.: A new approach for dependability planning of network systems. Int. J. Syst. Assur. Eng. Manag. **4**, 215 (2013). https://doi.org/10.1007/s13198-013-0185-2
12. Gertsbakh, I.B., Shpungin, Y.: Models of Network Reliability : Analysis, Combinatorics, and Monte Carlo. CRC Press, Boca Raton (2010)
13. Kalimulina, E.: Analytical reliability models and their application for planning and optimisation of telecommunication networks, 11 February 2016. SSRN: https://ssrn.com/abstract=2731170 or https://doi.org/10.2139/ssrn.2731170

Construction and Merging of ACM and ScienceDirect Ontologies

M. Priya[✉] and Ch. Aswani Kumar

School of Information Technology and Engineering,
Vellore Institute of Technology, Vellore, India
sumipriya@gmail.com

Abstract. An Ontology is an absolute formal conceptualization of some realm of significance. Nowadays Ontologies play a vibrant part in Information Architecture, Biomedical Informatics, Electronic commerce, Software Engineering, Semantic Web, Knowledge management, Artificial Intelligence and etc. Huge number of Ontologies and extensive variety of Ontologies are available for every single domain. It creates very difficult to maintain and access all the existing Ontologies. Ontology merging is the solution to overcome this kind of problems. Ontology merging is a procedure of fetching two existing Ontologies as input and obtains a newly merged Ontology as output. The merged Ontology will have common concepts and relationships between two Ontologies. This paper presents how two Ontologies can be constructed and merged using Protege and Conexp tools with an example.

Keywords: ACM · Conexp · Formal concept analysis · Ontology ·
Ontology construction · Ontology merging · Prompt · Protégé · ScienceDirect

1 Introduction

Ontology construction is a process of making individuals, classes, relations, attributes, functions, axioms, events, rules, and restrictions for the specified domain. Individuals are the basic objects. Classes are the set of domain concepts. Attribute describes the properties of the concepts. Relations state the relationship between classes and individuals. Functions are the distinctive way of defining associations in which the nth item of the correlation is uncommon for the $n - 1$ prior items. Restrictions are the set of summarization of what must be valid all together for some declaration to be acknowledged as information. Rules are the form of if-then sentence [1]. Formal Concept Analysis is a mathematical concept which includes a group of concept hierarchies and concepts. Using formal concept analysis we can able to display the connotation among the group of objects and their attributes by examining the given data [2]. Based on Formal concept analysis, formal context can be defined as a cross-table between group of objects (rows) and group of attributes (columns), FC: = (X, Y, Z) where X defines group of objects, Y defines group of properties, and Z defines an incidence among the X and Y. A concept lattice can be generated from the formal context which contains entire formal concepts [2].

© Springer Nature Switzerland AG 2020
A. Abraham et al. (Eds.): ISDA 2018, AISC 941, pp. 238–252, 2020.
https://doi.org/10.1007/978-3-030-16660-1_24

Both ACM and ScienceDirect websites are peer-reviewed which provide us a lot of quality article from various journals, conference proceedings and magazines in the heterogeneous domain. Both aim at providing in-depth information on multiple domains to their users in a different way. Our idea is to merge these two online platforms and offering all the information on the single source itself. To perform the merging process first we need to construct two ontologies either manually or using any tool. Ontologies can also be merged using formal concept analysis in terms of concept lattices. Many Ontology construction tools are available. Among the available tools, Protégé is one of the user-friendly tool. Hence we have used Protégé tool to construct the ontologies. We have developed two Ontologies namely ACM Ontology and ScienceDirect Ontology using Protégé tool. Then merged the process has been performed based on the association between the ontologies. Apart from Protégé we have used Conexp tool to create the concept lattice of ACM and ScienceDirect. Then the merging process can be performed based on the generated concept lattices.

The thought process of the paper is to give an elaborated view of Ontology development and integration with an example using Protégé Prompt plug-in. The content of the paper is structured as follows: Sect. 2 presents Construction of ACM Ontology, Sect. 3 deals with Construction of ScienceDirect Ontology, Sect. 4 describes the Protégé Prompt plug-in, Sect. 5 defines merging of ACM and ScienceDirect Ontologies and Sect. 6 briefs on generation of concept lattice using Conexp tool and finally Sect. 7 concludes the merging process.

2　ACM Ontology Construction

Association for Computing Machinery (ACM) is a global scholarly society for computing. ACM is the world's biggest educational and scientific Computing Classification System. It maintains a complete hierarchy of entire subjects. It provides a cognitive map of the computing space from the most common subject field to most specific subject fields [https://en.wikipedia.org/wiki/Association_for_Computing_Machinery]. Figure 1 shows ACM Ontology which includes 6 main classes as follows.

1. Applied computing
2. Computer systems organization
3. Computing methodologies
4. General and reference
5. Hardware
6. Human-centred computing
7. Information systems
8. Mathematics of computing
9. Networks
10. Security and privacy
11. Social and professional topics
12. Software and its engineering
13. Theory of computation

Fig. 1. ACM ontology

Applied computing class provides information on inter-discipline. It includes eleven subclasses namely Electronic commerce, Law, Physical sciences and engineering, social and behavioral sciences, Arts and humanities, Enterprise computing, Computer forensics, Education, Computers in other domains, Document management and text processing, Life and medical sciences, and Operations research [http://dl.acm.org/ccs/ccs_flat.cfm]. Computer systems organization class provides complete information about system architectures which is having four subclasses namely Embedded and cyber-physical systems, Architectures, Dependable and fault-tolerant systems, and networks and Real-time systems. Figure 2 shows the class hierarchy of applied computing and Computer systems organization. Each subclass gives information on the respective domain.

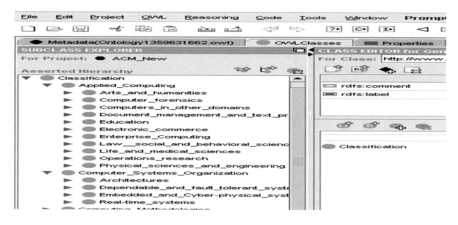

Fig. 2. Class hierarchy of applied computing and computer systems organization

Computing methodologies class deals with various computing techniques and methodologies which contain eight subclasses namely Distributed computing methodologies, Modeling and simulation, Artificial intelligence, Symbolic and algebraic manipulation, parallel computing methodologies, machine learning, concurrent computing methodologies and computer graphics. General and reference class contain complete data on various generic information in terms of general guidelines, tools and documents. It deals with two subclasses namely cross-computing tools and techniques and various document types. Figure 3 shows the class hierarchy of Computing Methodologies and General and References.

Fig. 3. Class hierarchy of computing methodologies and general and references.

Hardware class deals with complete data on hardware and their implementation. There are ten subclasses namely communication hardware, emerging technologies, integrated circuits, electronic design automation, power and energy, hardware test, printed circuit boards, robustness and VLI design. Human-centred computing class provides detailed knowledge of Human computer interaction, Ubiquitous and mobile computing, Accessibility, collaborative and social computing, Interaction design, and visualization. Figure 4 shows the class hierarchy of Hardware and Human-Centred Computing.

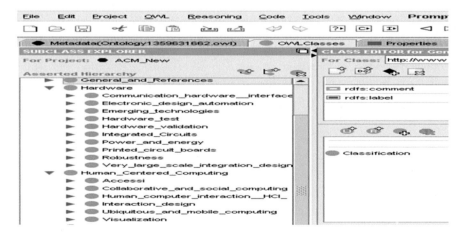

Fig. 4. Class hierarchy of hardware and human-centred computing

Information systems class provides entire knowledge of various databases and information storage and retrieval techniques. It's having five subclasses namely Information storage systems, Information retrieval, World Wide Web, Data management systems, and Information systems applications. Mathematics of computing class has a complete material on various mathematical methods and techniques such as Continuous mathematics, Discrete mathematics, Information theory, Mathematical analysis, Mathematical software and Probability and statistics. Figure 5 shows the class hierarchy of Information System and Mathematics of Computing.

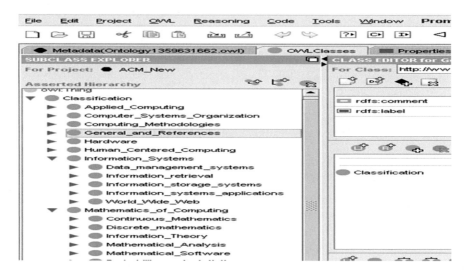

Fig. 5. Class hierarchy of information system and mathematics of computing

Network class elaborates on various network types, techniques, tools, application and so on. There are eight subclasses namely Network components, Network architectures, Network types, Network algorithms, Network performance evaluation, Network properties, Network services, and Network protocols. Security and privacy class elaborate various security methods and services. There are ten subclasses namely Formal methods and theory of security, Intrusion/anomaly detection and malware mitigation, Software and application security Cryptography, Security services, Network security, Database, and storage security, Security in hardware, Human and societal aspects of security and privacy and Systems security. Figure 6 shows the class hierarchy of Networks and Security and Privacy.

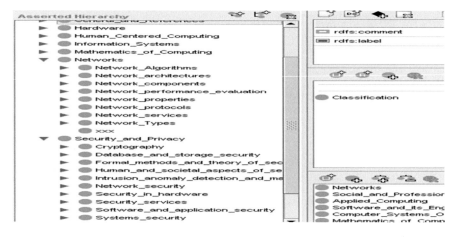

Fig. 6. Class hierarchy of networks and security and privacy

Social and professional topics class shows the material on general topics about professional and social such as User characteristics, Computing/technology policy and Professional topics. Software and its engineering class providing overall information software tools, techniques, and methods. There are three subclasses namely Software creation and management, Software notations and tools, and Software organization and properties. Figure 7 shows the class hierarchy of Social and Professional topics and Software and its engineering.

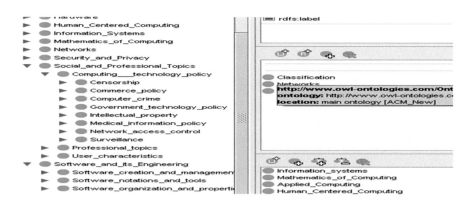

Fig. 7. Class hierarchy of social and professional topics and software and its engineering

Theory of computation is one of the vast areas which provide material on various formal languages and tools. There are eight subclasses namely Computational complexity and cryptography, Design and analysis of algorithms, Logic, Formal languages and automata theory, Models of computation, Randomness, geometry and discrete structures and Theory and algorithms for application domains. Figure 8 shows the class hierarchy of Theory of Computation.

Fig. 8. Class hierarchy of theory of computation

3   ScienceDirect Ontology Construction

The ScienceDirect website offers a huge amount of medical and scientific databases to the user on subscription based. ScienceDirect clouds include over 5 million pieces of information from 34,000 e-books and 3,500 research journals. It has been categorized into four main sections: Health Sciences, Life Sciences, Physical Sciences, and Engineering and Social Sciences and Humanities. The abstracts of all the articles are available at free of cost, but subscription or pay-per-view buy is needed for full-text access [https://en.wikipedia.org/wiki/ScienceDirect].

ScienceDirect Ontology having four main classes as follows:

1. Health sciences
2. Life sciences
3. Physical sciences and engineering
4. Social sciences and humanities.

Figure 9 shows ScienceDirect Ontology with subclasses.

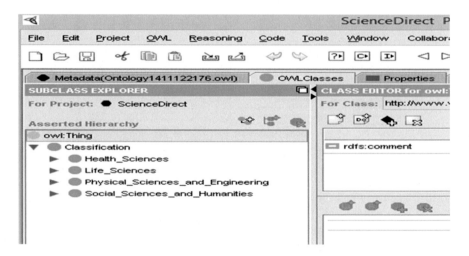

Fig. 9. ScienceDirect ontology

Health sciences provide information about human medical sciences and veterinary science which includes complete information on various human and veterinary diseases, disorders, diagnosis, available treatments, and departments. There are four subclasses namely Pharmacology Toxicology and Pharmaceutical Science, Medicine and Dentistry, and Veterinary science and Veterinary medicine and Nursing and Health professions. Life sciences provide material on a scientific study of living organisms i.e. Human beings, plants, microorganisms, and animals. There are five subclasses namely Immunology and Microbiology, Environmental Science, Neuroscience, Agricultural and Biological Sciences, and Biochemistry Genetics and Molecular Biology. Figure 10 shows the class hierarchy of health sciences and Life sciences.

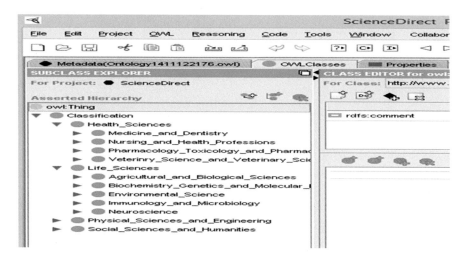

Fig. 10. Class hierarchy of health sciences and life sciences

Physical sciences and engineering class deal with various departments coming under the engineering field and physical sciences. It includes nine subclasses namely Earth and Planetary Sciences, Computer Science, Engineering, Material Science, Chemical engineering, Physics and Astronomy, Mathematics, Chemistry and Energy. Social sciences and humanities provide information on core disciplines of social science and humanity. It includes six subclasses namely Arts and human, Business Management and Accounting, Decision Sciences, Economics Econometrics and Finance, Psychology and Social Sciences. Figure 11 shows the class hierarchy of Physical Sciences and Engineering and Social Sciences and Humanities.

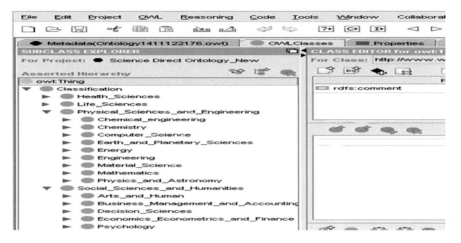

Fig. 11. Class hierarchy of physical sciences and engineering and social sciences and humanities

4 Protégé

Protégé is an open-source tool which is used to build Ontologies with knowledge-based applications and domain models. The Ontology range can fall into any of the four categories; classifications, taxonomies, database schemas, and fully axiomatized theories. Nowadays Ontologies play a vital role in most of the fields such as Semantic Web, electronic commerce, information management, knowledge management, and Web services. Using Protégé Ontologies can be modeled in two methods namely Frame-based modeling and OWL–based modeling. Frame-based modeling uses Protégé-Frames editor and OWL–based modeling uses Protégé-OWL editor for Ontology creation. Protege Frames editor allow the user to construct and populate frame-based Ontologies based on Open Knowledge Base Connectivity Protocol which includes Classes, Slots for properties and relationships and Instances for class. Protege-OWL editor allows the users to construct the Semantic Web Ontology, in specific to OWL which includes Classes, Properties, Instances and reasoning [http://protege.stanford.edu].

4.1 Prompt Plug-In

Prompt is a plug-in available in the Protégé tool. The user can manage multiple Ontologies in Protégé using Prompt Tab. A Prompt plug-in allows the users to perform the following five operations on Ontology [1, 3, 4].

1. **Compare** – The compare option enables the user to compare the various version of the identical Ontology with their current Ontology and build a merged ontology version as a result.
2. **Map** – The map option enables the user to map their existing two Ontologies and make over the data between two Ontologies.

3. **Move** – The move option enables the user to move the frames between the projects.
4. **Merge** – The merge option enables the user to merge two existing Ontologies and create a newly merged Ontology.
5. **Extract** – The extract option enables the user to excerpt a chunk from an existing Ontology and augment the chunk into the recent project.

Prompt offers a semi-automatic method for aligning and merging the Ontologies. Prompt achieves certain actions automatically and provide suggestions to the user in achieving the rest of the action as needed. Prompt can also define potential conflicts in the posture of the Ontology. Based on this, the user can perform the needed actions to overcome the inconsistencies [1, 4]. To merge Ontologies using Prompt the user has to give two Ontologies as input to and will get a merged new Ontology. During the merging process initially Prompt offer suggestions for merging classes to the user. Then the user can either accept the given suggestions or the user can offers their own suggestions for merging.

5   Merging of Ontologies - ACM Ontology and ScienceDirect Ontology

Ontology merging is a process of merging two existing Ontologies and creating a new extended Ontology. The merged Ontology will have the common classes and relationship between two Ontologies. The new Ontology will replace the original source Ontologies [3, 4]. Ontology merging comprises an association of related domains. In Fig. 12, Ontology 1 and Ontology 2 are the source Ontologies with few related concepts and domain values. Merged Ontology is the newly attained Ontology from the source Ontologies. The transition from Ontology 1 and Ontology to merged Ontology is indicated by the arrows. The merging process initiated by combining of entire conceptual based material [5].

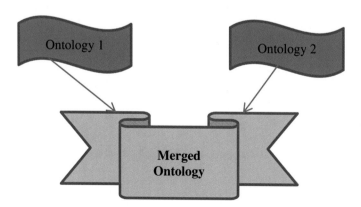

Fig. 12. Ontology merging process

Both ACM and ScienceDirect Ontology provide information in the form of journals, eBooks, conferences and etc. on various domains. Almost both Ontologies provide information about all the domains but in their own ways. It makes very difficult for the user to retrieve the needed information from both of the Website. The merging of these Ontologies provides a better solution to the user. The ultimate aim of the paper is to obtain a newly merged Ontology by merging the ACM Ontology and ScienceDirect Ontology based on the related concepts and relationships.

The Ontology merging process starts by choosing the ACM and ScienceDirect Ontologies as two input Ontologies and by clicking the "merge" option in the Prompt plug-in. Figure 13, shows these steps Then the lexical matching or the initial comparison can be performed. After completing the lexical matching process, the source matching process will start and then name comparison will be done and finally merging process will start by clicking begin merging option [6–8].

Fig. 13. Initial process of ontology merging

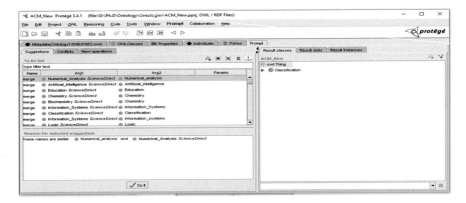

Fig. 14. Suggestions given by Prompt for merging ACM and ScienceDirect ontologies

After that, we have to select the "New Operations" option to choose our own suggestions for merging input Ontologies based on classes, subclasses, and instances [6–8]. Figure 14 depicts this situation. Now choose "Do it" option for finishing the merging process. The newly merged Ontology will have common classes, subclasses, and instances between the input Ontologies. Figure 15 shows the merged output Ontology which contains the similar classes from both input Ontologies ACM and ScienceDirect.

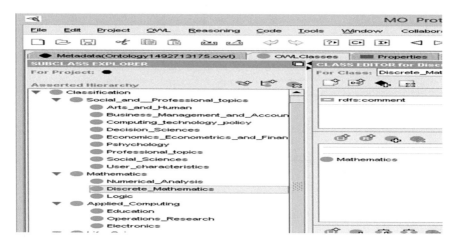

Fig. 15. Class hierarchy of merged ontology

6 Concept Lattice

Concept lattice plays a vital role in formal concept analysis. A Concept Lattice can be represented in terms of the pool of formal concepts of a specified formal context. The concept lattice can also be called as Galois graph [9–27].

Let us consider a formal context (X, Y, Z).

The pool of concept of (X, Y, Z) is represented as £ (X, Y, Z).

Then the concept lattice of £ (X, Y, Z) is represented as (£ (X, Y, Z), $\leq$).

The concept lattice can be generated using many tools. Among the available tool, Conexp is one of the easiest and flexible tool.

6.1 Conexp

Conexp (Concept Explorer) is a universal software in the realm of formal concept analysis which is an open-source tool. Using Conexp we can able to create the formal contexts with a group of objects and group of attributes. From the formal context, the user can generate the concepts and identify the total number of concepts of a specified context. Based on the concepts, concept lattice can be generated. The Conexp tool allows the user to visualize and manage the generated concept lattice. Conexp allows the user to perform the attribute exploration as well as forming association rules based on the identified concepts. Using Conexp we have created two concept lattice based on ACM and ScienceDirect Ontologies. Figures 16 and 17 shows the concept lattices of ACM and ScienceDirect Ontology. From ACM Formal context, we have identified 31 concepts and from ScienceDirect formal context we have identified 15 concepts. Finally, the concept lattices of ACM and ScienDirect has been merged together to form a new concept lattice which contains similar concepts between these two lattices.

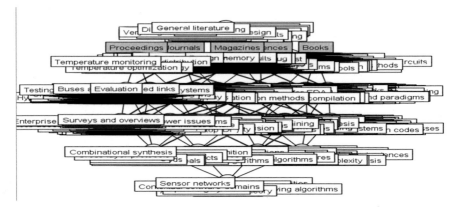

Fig. 16. Concept lattice of ACM

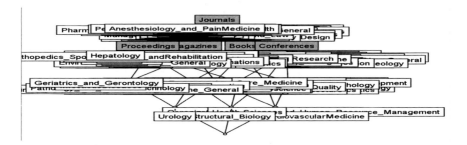

Fig. 17. Concept lattice of ScienceDirect

7 Conclusion

In this paper, we have discussed on construction and merging of ontologies using Protégé tool with an example. The illustration demonstrates the merging of ACM Ontology and ScienceDirect Ontology using Protégé Prompt plug-in. We have also merged the ontologies using Conexp tool by means of concept lattices. This paper may provide a better idea in the Ontology merging process and enable the researchers to proceed their work towards Ontology merging.

References

1. Li, S., Lu, Q., Li, W.: Experiments of ontology construction with formal concept analysis. In: International Joint Conference on Natural Language Processing, pp. 67–75 (2005)
2. Priya, M., Aswani Kumar, Ch.: A survey of state of the art of ontology construction and merging using formal concept analysis. Indian J. Sci. Technol. **8**(24), 2–7 (2015)
3. Lian, Z.: A tool to support ontology creation based on incremental mini-ontology merging. Data Extraction Research Group, Department of Computer Science, Brigham Young University (2008)
4. de Bruijn, J., Ehrig, M., Feier, C., Martin-Recuerda, F., Scharffe, F., Weiten, M.: Ontology mediation, merging, and aligning. In: Davies, J., Studer, R., Warren, P. (eds.) Semantic Web Technologies: Trends and Research in Ontology-Based Systems, pp. 102–104. Wiley, New York (2006)
5. https://protegewiki.stanford.edu/wiki/PROMPT
6. Noy, N.F., Musen, M.A.: PROMPT: algorithm and tool for automated ontology merging and alignment. In: Proceedings of the AAAI 2000, pp. 1–6 (2000)
7. Malik, S.K., Prakash, N., Rizvi, S.A.M.: Ontology merging using prompt plug-in of protégé in semantic web. In: International Conference on Computational Intelligence and Communication Networks, pp. 476–481 (2010)
8. Ostrowski, D.A., Schleis, G.M.: Enterprise ontology merging for the semantic web. In: Proceedings of 2008 International Conference on Semantic Web and Web Services, WorldComp 2008, Las Vegas, Nevada, USA, 7–17 July (2008)
9. http://ksl.stanford.edu/software/chimaera/
10. Huang, Z., Van Harmelen, F., Teije, A.T., Groot, P., Visser, C.: Reasoning with inconsistent ontologies: a general framework. EU-IST Integrated Project (IP) IST-2003-506826 SEKT (2005)
11. Stephens, L.M., Gangam, A., Huhns, M.N.: Constructing consensus ontologies for the semantic web: a conceptual approach. World Wide Web: Internet Web Inf. Syst. **7**, 421–442 (2004)
12. http://protege.stanford.edu/
13. Uschold, M., Gruninger, M.: Ontologies: principles, methods and applications. Knowl. Eng. Rev. **11**(2), 93–136 (1996)
14. Ganter, B., Stumme, G.: Creation and merging of ontology top-levels. In: International Conference on Conceptual Structures, vol. 2746, pp. 131–145, July 2003
15. Fan, Z., Zlatanova, S.: Exploring ontologies for semantic interoperability of data in emergency response. Appl. Geomat. **3**(2), 109–122 (2011)
16. Gómez-Pérez, A., Fernández-López, M.: Ontological Engineering: With Examples from the Areas of Knowledge Management, E-Commerce and the Semantic Web. Springer, London (2010)
17. Li, X., Liu, G., Ling, A., Zhan, J., An, N., Li, L., Sha, Y.: Building a practical ontology for emergency response systems. In: 2008 International Conference on Computer Science and Software Engineering, vol. 4, pp. 222–225 (2008)
18. Aswani Kumar, Ch., Srinivas, S.: Concept lattice reduction using fuzzy k-means clustering. Expert Syst. Appl. **9**(1), 2696–2704 (2010)
19. Aswani Kumar, Ch.: Fuzzy clustering based formal concept analysis for association rule mining. Appl. Artif. Intell. **26**(3), 274–301 (2005)
20. Aswani Kumar, Ch., Dias, S.M., Vieira, N.J.: Knowledge reduction in formal contexts using non-negative matrix factorization. Math. Comput. Simul. **109**, 46–63 (2015)

21. Aswani Kumar, Ch.: Mining association rules using non-negative matrix factorization and formal concept analysis. In: International Conference on Information Processing, vol. 157, no. 1, pp. 31–39 (2011)
22. Mouliswaran, S.C., Aswani Kumar, Ch., Chandrasekar, C.: Modeling Chinese wall access control using formal concept analysis. In: International Conference on Contemporary Computing and Informatics (IC3I) (2017)
23. Aswani Kumar, Ch., Srinivas, S.: Concept lattice reduction using fuzzy K-means clustering. Expert Syst. Appl. **37**(3), 2696–2704 (2010)
24. Prem Kumar, S., Aswani Kumar, Ch.: Bipolar fuzzy graph representation of concept lattice. Inf. Sci. **288**, 437–448 (2014)
25. Chunduri, R.K., Aswani Kumar, Ch.: Scalable formal concept analysis algorithm for large datasets using spark. J. Ambient Intell. Humanized Comput. 1–21 (2018)
26. Chunduri, R.K., Aswani Kumar, Ch.: HaLoop approach for concept generation in formal concept analysis. JIKM **17**(3), 1850029 (2018)
27. Subramanian, C.M., Aswani Kumar, Ch., Chelliah, C.: Role based access control design using three-way formal concept analysis. Int. J. Mach. Learn. Cybern. **9**(11), 1807–1837 (2018)

An Empirical Assessment of Functional Redundancy Semantic Metric

Dalila Amara[1(✉)], Ezzeddine Fatnassi[1,2], and Latifa Rabai[1,3]

[1] SMART Lab, Institut Supérieur De Gestion De Tunis, Université de Tunis,
Tunis, Tunisie
dalilaa.amara@gmail.com,
Ezzeddine.fatnassi@gmail.com, latifa.rabai@gmail.com
[2] Insitut des Hautes Etudes de Tunis, Tunis, Tunisie
[3] College of Business, University of Buraimi,
Al Buraimi 512, Sultanate of Oman

Abstract. Software dependability is a generic concept that reflects the system' trustworthiness by its users. It consists of different quality attributes like reliability and maintainability. To achieve dependable and reliable software systems, different dependability means are defined including fault tolerance. Most of fault tolerance techniques are based on the redundancy concept. To reflect the ability of a program to tolerate faults, the quantitative assessment of the program' redundancy is required. Literature review shows that a set of semantic metrics whose objective is to assess the programs' redundancy and to reflect their potential to tolerate faults is proposed. Despite the importance of the different metrics composing this suite, literature shows that they are manually computed for procedural programs, and only a theoretical basis of them is presented. Consequently, we aim in this paper to propose a way to automatically compute one of these metrics termed functional redundancy for different object oriented java programs. The automatic computing is necessary required for the different metrics to perform their empirical validation as software quality indicators.

Keywords: Software dependability · Fault tolerance · Semantic metrics · Functional redundancy metric

1 Introduction

Improving quality of software systems is the objective of all organizations using software systems. Hence, software quality characteristics have to be analyzed. Dependability is one of the most important characteristics since it gathers different attributes like reliability [1] and based on different means including fault tolerance [2].

In software systems, fault tolerance is the ability of a system to continue its function despite the presence of faults [1]. It is also one of the quantifiable attributes that helps to assess the product reliability [3]. Most of fault tolerance techniques are based on a common concept that is redundancy [4] defined as the use of extra elements like instructions, programs and functions in order to detect, remove and mask errors [5, 6]. Despite the importance of these techniques [5], they still limited in assessing programs'

© Springer Nature Switzerland AG 2020
A. Abraham et al. (Eds.): ISDA 2018, AISC 941, pp. 253–263, 2020.
https://doi.org/10.1007/978-3-030-16660-1_25

redundancy in a quantitative way which is needed to make comparisons and appropriate decisions. Hence, software metrics are proposed as quantitative measures to assess various quality attributes including redundancy [6, 16]. For instance, a suite of semantic metrics is recently proposed by Mili et al. [3] to assess programs' redundancy.

The proposed suite consists of four basic metrics namely state redundancy, functional redundancy, error masking (non-injectivity) and error recovery (non-determinacy). The objective of these metrics is to quantitatively assess the programs' redundancy in order to reflect their ability to tolerate faults [7]. The major problem of these metrics is that they are manually computed [3] and empirical studies focusing on their validation as measures of programs fault tolerance and reliability arc lacking.

Consequently, the main motivation of this work is that despite the importance of these metrics as quantitative measures of programs' ability to tolerate faults, there is a lack of an automated way to compute the cited metrics. Thus, our objectives are:

- Proposing an automated way to compute one of these metrics namely functional redundancy metric.
- Studying the statistical relationship between this metric and the state redundancy one discussed in our previous work [8].

Two major contributions are driven from the automated computing of this metric. Thus, we show that it is possible to automatically compute this metric for object oriented programs which leads us to obtain an empirical data base needed to perform further empirical studies. Moreover, we state that there exist a statistical relationship between this metric and the state redundancy one.

This paper is organized into five sections. In Sect. 2, we present fundamentals of software quality including software dependability, fault tolerance, software redundancy and software metrics use for quality assessment. In Sect. 3, we describe the purpose of functional redundancy metric, its formulation and the way of its automatic computing. Section 4 discusses the correlation between functional redundancy and state redundancy metrics. Finally, Sect. 5 draws the conclusion and perspectives.

2 Fundamentals of Software Quality

This section describes the basic elements required to improve software quality as it is shown in Fig. 1.

Figure 1 shows that to improve software quality, we need to understand the different elements related to this concept. Thus, we need firstly to analyze basic characteristics of software quality. Among these characteristics, software dependability was identified as a communal concept that assembles different attributes [9]. It aims at avoiding threats using different means including fault tolerance (See Fig. 1).

Different techniques are proposed to perform fault tolerant systems and most of them are based on the redundancy concept [5]. To make appropriate decisions and comparisons, quantitative measures are also required. In this context, Fig. 1 shows that programs' redundancy may be assessed through software semantic metrics.

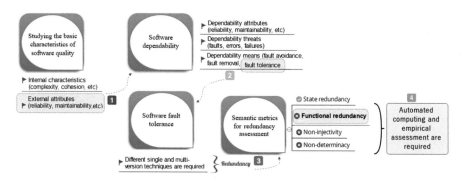

Fig. 1. General purpose of the paper

The different elements of software quality presented in Fig. 1 namely software dependability, software fault tolerance and semantic metrics are described along following subsections:

2.1 Software Dependability

Software dependability is a generic concept that reflects the overall quality of software systems from the user perspective and subsumes other important attributes like reliability and maintainability [10]. It is defined as *"the ability to deliver service that can justifiably be trusted"* [11] and consists of three basic elements:

- Dependability attributes: are quality properties that a system must verify. They include reliability, maintainability, integrity, etc. [9].
- Dependability means: are techniques required to perform dependability attributes and include fault avoidance, fault removal and fault tolerance (See Fig. 1).
- Dependability threats: represent the unwanted behavior of the system like errors, faults and failures [12].

To sum up, fault tolerance is identified as one of the important dependability means since it is difficult to totally avoid the presence of faults and failures [7].

2.2 Software Fault Tolerance

As noted above, fault tolerance is the ability of a system to continue its intended functions in the presence of faults [13, 14]. It is based on three major phases [4, 6, 10]:

- Error detection: is the ease of detecting errors in the state of a program in execution [7]. Two redundancy semantic metrics are proposed for error detection namely state redundancy and functional redundancy [3].
- Error recovery: identifies the erroneous state before its substitution with an error-free one [6]. Mili et al. [3] proposed a semantic metric for error recovery termed error non-determinacy.
- Error compensation: provides fault masking. A semantic metric called error non-injectivity as measure of the program' ability to mask errors is proposed [3].

Among these phases, different techniques are defined in order to detect, correct and mask errors [5, 9, 10]. Most of them are based on redundancy concept [13, 16–18]. However, they do not perform its quantitative assessment. Hence, software metrics are proposed as quantitative measures of various quality characteristics including redundancy [19].

2.3   Software Semantic Metrics for Redundancy Assessment

Different studies related to redundancy use to perform fault tolerant and reliable systems are proposed [9, 15, 19, 20]. In recent years, a suite of four semantic metrics is proposed by Mili et al. [3] to quantitatively assess programs' redundancy based on the entropy concept (the number of bits used by the program' variables [21]). The proposed suite consists of four major metrics described as follows [3, 7]:

- State redundancy: uses the program' redundancy as the number of bits losses which are declared but not used by the program in order to help detecting errors. It subsumes two metrics which are the initial state redundancy (excess data before executing the program) and the final state redundancy (excess data when executing the program) [8].
- Functional redundancy: is also proposed as error detection measure based on the quantitative assessment of programs redundancy. It reflects the excess output data generated by a program function. Its purpose is to verify the proper execution of the function.
- Non-injectivity: is based on the program' redundancy to measure in bits the amount of erroneous information that can be masked by this program.
- Non-determinacy: is proposed to measure the ability of a program to tolerate faults.

To sum up, the different proposed metrics are quantitative measures of programs redundancy. Also, they are important means to achieve fault tolerant systems. However, to the authors' best knowledge, these metrics are theoretically presented and manually computed [3]. Hence, in the following section, we will present how the functional redundancy metric may be automatically computed.

3   Automatic Computing of Functional Redundancy Semantic Metric

This section presents the theoretical basis and the formulation of the functional redundancy metric as well as its automatic computing for object oriented java programs.

3.1   Theoretical Definition and Objective

The functional redundancy metric reflects the excess output data generated by a program' function in order to verify the proper execution of the function. Mathematically speaking, this metric is equated with the non-surjectivity of the program' function which means that not all program' outputs (final states) are mapped to at least one input (initial state) [3]. Hence, the functional redundancy of a program g on a space S is denoted by φ(g) and defined by:

$$\varphi(g) = (H(S) - H(Y))/H(Y) \tag{1}$$

- H(S) is the state space of the program: the maximum value (size in bits) that the declared program variables may take,
- Y is the random variable (range or output space) of g defined as the set of final states of the program's function,
- $H(Y) = H(\sigma f)$, is the entropy of the output produced by g defined as the number of bits required to store the result of the program' execution,
- φ(g) is the functional redundancy of the program g.

The main purpose of this metric is to quantify the excess information when executing the program in order to measure its ability to detect errors and avoiding failures. From the value of this metric, three possible interpretations are drown [3]:

- if $\varphi(g) = 0$: there is no scope for checking any property since all bits are used, the input and output spaces are equal.
- if $0 < \varphi(g) < 1$: we can check part of the result against redundant information,
- if $\varphi(g) > 0$: there are a larger bits of redundancy when executing the program.

3.2 Illustrative Example

Consider two functions F1 = x and F2 = x%4, and B5, the input range that is a set of natural numbers represented in a word of five bits as shown in Fig. 2.

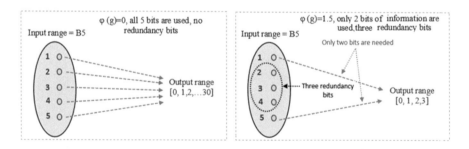

Fig. 2. Illustrative example of the functional redundancy metric

Figure 2 shows that for the first function, the possible output range is [0...31], hence all five bits are used and then there is no redundancy. However, for the second function, the output range is [0, 1, 2, 3]. In other words, only two bits of information are needed, whereas the three other ones are redundancy bits [7].

3.3 Automatic Computing Process

Our experimental study consists of three java programs which are the addition of two integers (Sum), the Power program (Power) and the maximum value program (Max-Value). The computing process for the Sum program is presented in Fig. 3.

```
26    int x= rand.nextInt(Integer.MAX_VALUE) + 1;
27
28    int y= rand.nextInt(Integer.MAX_VALUE) + 1;

36    int values= Integer.MAX_VALUE;
37    int statespace=2*sizeOfBits(values);

41    int initialsizex= sizeOfBits(x);
42    int initialsizey= sizeOfBits(y);
43    int initialstatespace=initialsizex+initialsizey;

60    int finalsizex= sizeOfBits(x);
61    int finalsizey= sizeOfBits(y);
62    int finalstatespace=finalsizex+finalsizey;

78    float functionalredunduncy=(float)(statespace-finalstatespace)/finalstatespace;
```

Fig. 3. A part of functional redundancy computing for the Sum program

This process consists of three major steps:

- First, the state space H(S) of each program as the entropy (bits) of the declared variables is computed. For instance, for the Sum program, the state space is computed as shown in lines 36 and 37 in Fig. 3.
- Second, we compute the final state space H(σf) that is the number of bits (entropy) required to store the result of the program' execution. For instance, for the Sum program, this value is computed as shown in lines 60 to 62 in Fig. 3.
- Third, the functional redundancy metric is deduced using Eq. (1). For instance, for the Sum program, it is computed as it is shown in line 78 of Fig. 3. The same steps are used to compute this metric for the other programs considering the variables used for each one.

A part of the generated data base for the Sum program is shown in Table 1.

Table 1. A part of Sum program functional redundancy values for random generated inputs.

Input values		Functional redundancy
x	y	
9582	5,22E+08	0.33
2884139	3,92E+08	0.2
37069549	1,49E+08	0.17
4,75E+08	2,95E+08	0.09

The state redundancy metric is also computed for the same programs. Its formulation and automatic computing is detailed in our previous work [8]. The objective is to study the statistical relationship between these two metrics as described under here.

4 Correlation Between State Redundancy and Functional Redundancy Metrics

For each of the computed metrics (state [8] and functional redundancy), 10.000 output values are generated from random inputs as shown in Table 1. These data are used to study the correlation between these two metrics in order to identify if one of them may be expressed by the other one. To perform this analysis, the STATA statistical tool is used.

4.1 Data Standardization and Normality Tests

To perform the correlation between the two metrics, two basic steps are required which are data standardization and normality tests:

- Data standardization: this step aims to unify the units of the two metrics. For instance, the state redundancy metrics (initial and final states) are expressed in bits, whereas, the functional redundancy is expressed in dimensionless i.e. the degree of programs functional duplication [3].
- Normality tests: are required to identify which correlation coefficient is the appropriate to use [22]. We use in our study the Jarque-Bera (JB). Under normality, the p-value should be less than 0.05. Moreover, JB test is a function of the measures of skewness S and kurtosis K computed from the sample. These values have to be respectively 0 for S and 3 for K [23]. The different tests that are performed for our data are shown in Table 2.

Table 2. Normality tests for the generated metrics

Metrics	JB test	Normal distribution?	S	Normal distribution?	K	Normal distribution?
Initial state redundancy	0.834	No	1.619	No	0.795	No
Final state redundancy	0.834	No	1.619	No	0.795	No
Functional redundancy	0.834	No	0.757	No	-1.589	No

Table 2 shows that the generated data set is not normal since S and K values are respectively different from 0 and 3. Also JB test' values are large (0.834).

4.2   Correlation Analysis and Results' Interpretation

The obtained results of normality tests presented in Table 2 show that the generated data have not a normal distribution. In this case, we resort to the Spearman correlation test since for normally distributed data, Pearson's correlation coefficient is used, whereas, for non-normal distributed data, the Spearman's correlation coefficient is used [24, 25].

The Spearman correlation test is performed based on the following hypothesis. Results of this correlation are summarized in Table 3.

H0: $\rho = 0$ (null hypothesis) there is no significant correlation between state and functional redundancy metrics.

H1: $\rho \ddagger 0$ (alternative hypothesis) there is a significant correlation between these metrics.

Table 3. Correlation between state and functional redundancy metrics

Programs' functional redundancy	Correlation coefficients			
	Initial state redundancy		Final state redundancy	
Sum	0.987	p-value = 0.000	0.907	p-value = 0.000
Power	0.999	p-value = 0.000	0.901	p-value = 0.000
MaxValue	1	p-value = 0.000	1	p-value = 0.000

From Table 3, we see that there is an evident relationship between state redundancy metric (initial and final) and functional redundancy one. Thus, the functional redundancy is positively correlated with initial and final redundancy ones because all of the correlation coefficient values are positive and large for the different programs. Additionally, their p-values are good (less than 0.05). In short, the null hypothesis H0 of no correlation between these metrics is rejected. We notice that according to the alternative hypothesis H1, there is a positive correlation between the functional redundancy and the state redundancy metrics. This means that if the initial and final state redundancy metrics increase, the functional redundancy increase too and vice versa.

Considering the definition of these metrics explained above, we can explain the obtained correlation as follows: the initial and final state redundancy metrics represent respectively the unused amount of bits at the beginning and at the end of the program. Concerning functional redundancy, it represents the ratio of the unused entropy (bits) by the used one. Consequently, each time the unused entropy deduced by the state redundancy metrics tends to increase, the functional redundancy will also increase and vice versa.

To sum up, the presented study is benefic and presents different contributions compared with the related work [3] as shown in Table 4:

Table 4. Comparison between the performed study and the related work

Related work [3]	Our work
Functional redundancy is manually computed for procedural programs	Functional redundancy metric is computed for different oriented object java programs
The paper do not perform the statistical relationship between the different metrics	In this paper, a study of the statistical relationship between this metric and state redundancy one is performed

As seen in Table 4, two major contributions are driven from this study. First, we show that it is possible to automatically compute functional redundancy metric for different programs. This step is necessary required by developers to reflect the redundancy of their programs and to help detecting errors. Moreover, this metric is positively correlated with state redundancy one. This means that if the program' state redundancy increases, then the functional redundancy will increase too.

5   Conclusion and Perspectives

Software dependability is generally reflected by assessing the different attributes related to it. The literature performed in this paper shows that fault tolerance is one of the quantifiable attributes that helps to assess the dependability attributes. Moreover, Considerable attention has been paid to the use of semantic metrics as quantitative measures of software fault tolerance.

Among these metrics, the functional redundancy semantic metric is defined to show the excess output data generated by a program function. A key limitation of this metric is that it is manually computed for procedural programs and only a theoretical basis was presented for it. To solve this problem, we proposed an automated way to compute it for different java programs.

Different benefits are driven from the automatic calculation of this metric. Thus, for developers, an automatic calculation of this metric is necessary required. Additionally, it is an important step that helps to construct an empirical data base which is useful to perform other experimental studies. Furthermore, it is proved that the functional redundancy metric may be expressed by the state redundancy one.

Even though the efficiency of the automatic computing of the recently proposed semantic metrics (state and functional redundancy), it is possible to further improve this study. Hence, the next step of our research will focus on extending our work to incorporate open source java programs and to study the use of these metrics as measures of different quality attributes like defect density.

References

1. Randell, B., Laprie, J.C., Kopetz, H., Littlewood, B. (eds.): Predictably Dependable Computing Systems. Springer, Heidelberg (2013)
2. Asghari, S.A., Marvasti, M.B., Rahmani, A.M.: Enhancing transient fault tolerance in embedded systems through an OS task level redundancy approach. Future Gener. Comput. Syst. **87**, 58–65 (2018)
3. Mili, A., Jaoua, A., Frias, M., Helali, R.G.M.: Semantic metrics for software products. Innov. Syst. Softw. Eng. **10**(3), 203–217 (2014)
4. Randell, B.: System structure for software fault tolerance. IEEE Trans. Softw. Eng. **2**, 220–232 (1975)
5. Pullum, L.L.: Software Fault Tolerance Techniques and Implementation. Artech House, Norwood (2001)
6. Lyu, M.R.: Handbook of Software Reliability Engineering (1996)
7. Mili, A., Tchier, F.: Software Testing: Concepts and Operations. Wiley, New York (2015)
8. Amara, D., Fatnassi, E., Rabai, L.: An automated support tool to compute state redundancy semantic metric. In: International Conference on Intelligent Systems Design and Applications, pp. 262–272. Springer, Cham, December 2017
9. Avizienis, A., Laprie, J.C., Randell, B., Landwehr, C.: Basic concepts and taxonomy of dependable and secure computing. IEEE Trans. Dependable Secur. Comput. **1**(1), 11–33 (2004)
10. Laprie, J.C.: Dependability: basic concepts and terminology. In: Dependability: Basic Concepts and Terminology, pp. 3–245. Springer, Vienna (1992)
11. Laprie, J.C.: Dependable computing and fault tolerance: concepts and terminology. In: Twenty-Fifth International Symposium on Fault-Tolerant Computing, 1995, p. 2. IEEE, June 1985
12. Isermann, R.: Fault-Diagnosis Systems: An Introduction from Fault Detection to Fault Tolerance. Springer, Heidelberg (2006)
13. Dubrova, E.: Fault-Tolerant Design, pp. 55–65. Springer, New York (2013)
14. Jaoua, A., Mili, A.: The use of executable assertions for error detection and damage assessment. J. Syst. Softw. **12**(1), 15–37 (1990)
15. Carzaniga, A., Mattavelli, A., Pezzè, M.: Measuring software redundancy. In: Proceedings of the 37th International Conference on Software Engineering, vol. 1. pp. 156–166. IEEE Press, May 2015
16. Fenton, N., Bieman, J.: Software Metrics: A Rigorous and Practical Approach. CRC Press, Boca Raton (2014)
17. Hamming, R.W.: Error detecting and error correcting codes. Bell Syst. Tech. J. **29**(2), 147–160 (1950)
18. Shannon, C.E.: A mathematical theory of communication. ACM SIGMOBILE Mob. Comput. Commun. Rev. **5**(1), 3–55 (2001)
19. Lyu, M.R., Huang, Z., Sze, S.K., Cai, X.: An empirical study on testing and fault tolerance for software reliability engineering. In: 14th International Symposium on Software Reliability Engineering, 2003, ISSRE 2003, pp. 119–130. IEEE, November 2003
20. Jiang, L., Su, Z.: Automatic mining of functionally equivalent code fragments via random testing. In: Proceedings of the Eighteenth International Symposium on Software Testing and Analysis, pp. 81–92. ACM, July 2009. https://doi.org/10.1145/1572272.1572283
21. Davis, J.S., LeBlanc, R.J.: A study of the applicability of complexity measures. IEEE Trans. Softw. Eng. **14**(9), 1366 (1988)

22. Yazici, B., Yolacan, S.: A comparison of various tests of normality. J. Stat. Comput. Simul. **77**(2), 175–183 (2007). https://doi.org/10.1080/10629360600678310
23. Thadewald, T., Büning, H.: Jarque-Bera test and its competitors for testing normality–a power comparison. J. Appl. Stat. **34**(1), 87–105 (2007). https://doi.org/10.1080/02664760 600994539
24. Gall, C.S., Lukins, S., Etzkorn, L., Gholston, S., Farrington, P., Utley, D., Virani, S.: Semantic software metrics computed from natural language design specifications. IET Softw. **2**(1), 17–26 (2008). https://doi.org/10.1049/iet-sen:20070109
25. Olague, H.M., Etzkorn, L.H., Gholston, S., Quattlebaum, S.: Empirical validation of three software metrics suites to predict fault-proneness of object-oriented classes developed using highly iterative or agile software development processes. IEEE Trans. Softw. Eng. **33**(6), 402–419 (2007). https://doi.org/10.1109/TSE.2007.1015

An Enhanced Plagiarism Detection Based on Syntactico-Semantic Knowledge

Wafa Wali[(⊠)], Bilel Gargouri[(⊠)], and Abdelmajid Ben Hamadou[(⊠)]

MIRACL Laboratory, Sfax University, Sfax, Tunisia
{wafa.wali,bilel.gargouri}@fsegs.rnu.tn,
abdelmajid.benhamadou@isimsf.rnu.tn

Abstract. The issue of plagiarism in documents has been present for centuries. Yet, the widespread dissemination of information technology, including the internet, made plagiarism much easier. Consequently, methods and systems aiding in the detection of plagiarism have attracted much research within the last two decades. This paper introduces a plagiarism detection technique based on the semantic knowledge, notably semantic class and thematic role. This technique analyzes and compares text based on the semantic allocation for each term in the sentence. Semantic knowledge is superior in semantically generating arguments for each sentence. Weighting for each argument generated by semantic knowledge to study its behavior is also introduced in this paper. It was found that not all arguments affect the plagiarism detection process.
In addition, experimental results on PAN13-14 (http://pan.webis.de/.) data sets revealed significant speed-up, which outperforms the recent methods for plagiarism detection in terms of Recall and Precision measure.

Keywords: Plagiarism detection · Semantic similarity ·
Thematic role · Semantic class · Arguments weight

1 Introduction

Plagiarism, such as can be considered as one of the electronic crimes such as, computer hacking, computer viruses, spamming, phishing, copyrights violation and others crimes. In fact, it is defined as the act of taking or attempting to take or to use (whole or parts) of another person's works without referencing to him as the owner of this work. Recently, there has been much interest in automatic plagiarism. Written-text plagiarism is a wide-spread problem which many organizations have to deal with.

Various methods, such as copy and paste, redrafting or paraphrasing in the text, plagiarism of idea, and plagiarism through translation from one language to another are used in this field.

Actually, several documents are available on the word wide web and are easy to access. Due to this availability, users can easily create a new document by

© Springer Nature Switzerland AG 2020
A. Abraham et al. (Eds.): ISDA 2018, AISC 941, pp. 264–274, 2020.
https://doi.org/10.1007/978-3-030-16660-1_26

copying and pasting from these resources. Sometimes, users can reword the plagiarized part by replacing the word with their synonyms. This kind of plagiarism is difficult to be detected by the traditional plagiarism detection system, such as SCAM (Stanford Copy Analysis Mechanism) [10], CHECK [11], etc.

According to literature review by BAO [2], the current plagiarism detection system is found to be too slow and takes too much time for checking. The matching algorithms are also dependent on the text lexical structure rather than on the semantic structure. Therefore, it becomes difficult to detect semantically the paraphrased text. The big challenge is to provide plagiarism checking with an appropriate algorithm in order to improve the percentage of finding result and time checking. The important question about the plagiarism detection problem in this study is whether it is possible to apply new techniques, such as Semantic arguments, notably semantic class and thematic role to handle plagiarism problems for text documents.

Semantic arguments notably semantic class and thematic role, is one of the Natural Language Processing techniques that are used in many fields, such as text summarization [12], text categorization [9] and sentence similarity [15]. In this paper, we proposed a plagiarism detection tool, built around a content-based method based on a semantic class and a thematic role and an improved similarity measure using argument weighting.

The proposed method can detect, copy and paste plagiarism, rewording or synonym replacement, changing of word structure in the sentences, modifying the sentence from passive voice to active voice and vice versa. These semantic arguments, especially semantic class and thematic role, were used to analyze the sentence semantically. Indeed, WordNet [6] and VerbNet [4] thesaurus were used to extract the concepts or synonymies for each word inside the sentences. Weight score is calculated for each argument to study their behaviour and effect in plagiarism detection.

The rest of this paper is organized as follows. Section 2 presents the related work in plagiarism detection methods. The plagiarism detection method and the similarity measure are described in Sect. 3. Section 4 discusses the experimental results the proposed approach, whereas Sect. 5 draws the conclusions and some future research directions.

2   Related Studies

This section presents the state of the art solutions for the plagiarism detection problem. The proposed methods can be divided into character-based, structural-based, cluster-based, syntax-based, cross language-based and semantic-based methods.

In this section, we limited ourselves to just presenting the different semantic-based methods in order to explore their advantages and limitations.

Tsatsaronis et al. [13] propose a semantic-based approach to text-plagiarism detection which improves the efficiency of traditional keyword matching techniques. The method can detect a larger variety of paraphrases, including the use of synonymous terms, the repositioning of words in the sentences. Plagiarism is detected through two steps: First, each document while candidate for plagiarism is compared to all the original documents in the collection and the suspect documents are selected for the second step. More specifically, the documents are processed in segments so that only the suspect segments inside each suspect document are retrieved. In the second step, the authors compared each suspect segment to its possible source and decided whether it is a plagiarism or not.

On their part, Alzahrani and Salim [1] used fuzzy semantic-based string similarity. This method was developed via four main steps. The first step consists in pre-processing that involves tokenization, stop word removal and stemming. In the second step, a list of candidate documents is retrieved for each suspicious document using Jaccard coefficient and shingling algorithm. Suspicious documents are then compared to sentence-wise with the associated candidate documents. This step entails the computation of a degree of fuzzy similarity that ranges between two edges: 0 for completely different sentences and 1 for identical sentences. Two sentences are marked as similar if they gain a fuzzy similarity score above a certain threshold. The last stage is post-processing in which consecutive sentences are joined to form single paragraphs or sections. Besides, a different method of semantic-based plagiarism detection was proposed by Chow and Salim [5]. The proposed method calculates the similarity between the suspected and the original documents according to the predicates of the sentences. In fact, each sentence predicate is extracted using the Stanford Parser Tree (SPT). The degree of similarity between the extracted predicates is calculated using the WordNet thesaurus. The drawback of this method is that it does not cover all the parts of the sentence, only the subject, the verb and the object.

On the other hand, Osman et al. [7] introduced a plagiarism detection technique based on the Semantic Role Labeling (SRL). This technique analyzes and compares texts based on the semantic allocation for each term inside the sentence. The SRL is superior in semantically generating arguments for each sentence. Weighting for each argument is generated by SRL to study its behaviour.

Paul et al. [8] proposed a new plagiarism detection technique which uses Semantic Role Labeling and Sentence Ranking for plagiarism detection. Sentence ranking gives suspicious and original sentence pairs by vectorizing the document. Then, the proposed method analyses and compares the ranked suspected and the original documents based on the semantic allocation of each term in the sentence using SRL.

Moreover, Vani and Gupta [14] proposed a plagiarism detection method based on syntax-semantic concept extractions with a genetic algorithm in detecting cases of idea plagiarism. The work mainly focuses on idea plagiarism where the source ideas are plagiarized and represented in a summarized form. Plagiarism detection is employed at both the document and passage levels by exploiting the document concepts at various structural levels. Initially, the idea embedded within the given source document is captured using sentence level concept extraction with genetic algorithm. Document level detection is facilitated with word-level concepts where syntactic information is extracted and the non-plagiarized documents are pruned.

Through the literature review, we show in this section that many efforts have been made in the past to detect the similarity between text documents. However, these methods still need to improve the detecting technique to capture plagiarized parts, especially in semantically plagiarized parts.

The main difference between our proposed method and other techniques is represented in three points: first, our proposed method is a comprehensive plagiarism detection technique which focuses on many types of detection, such as detecting copy paste plagiarism, rewording or synonym replacement, changing of word structure in the sentences and modifying the sentence from passive voice to active voice and vice versa. The second point corresponds to the comparison mechanism where our proposed method is taken from the semantic class and thematic role as a means of analysis and comparison to capture a plagiarized meaning of text while the other methods focused on traditional comparison techniques, such as character-based and string matching. The important point of this difference is an improvement of similarity score by our proposed method using the semantic class and thematic role weighting scheme where none of the proposed methods was used before.

3   Plagiarism Detection Process

This paper proposes a new method for plagiarism detection which is very reliable and takes less time for reporting plagiarism in text documents. The proposed method uses the semantic class and the thematic role of each constituted term of a sentence based on its semantic predicate. The plagiarism detection process is based on a sentence level of the semantic parsing which determines the object and subject of a sentence. It depends on the delineation of cases that determines how: "who" did "what" to "whom" at "when" and "where". Therefore, it becomes clear that the main objective of semantic arguments, notably the semantic class and the thematic role is to determine the semantic roles of each term based on the semantic relationship between their predicates and terms. The method also uses a sentence ranking method that keeps enhance the plagiarism detection.

The proposed method has three main steps, as shown in Fig. 1, which are:

– Pre-processing
– Linguistic manipulation
– similarity detection

3.1   Pre-processing

Pre-processing is one of the key steps in Natural Language Processing (NLP). This steps has two sub-steps, which were text segmentation and stop word removal.

The process of dividing a text into meaningful units is called text segmentation. The text can be divided into sentences, words or topics. We used sentence-based text segmentation as a first step in our proposed method, where the original and suspected texts are divided into sentence units. This is because our

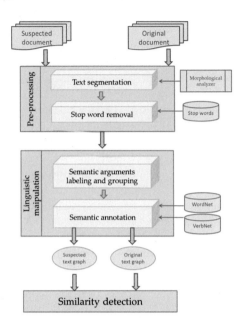

Fig. 1. Proposed method of general architecture.

proposed method aims at comparing a suspected text with the original text based on the sentence matching approach.

In the process of stop word removal, some of the English words that are most frequently used not contribute any meaning to the content such as, articles, pronouns, prepositions, and determiners. The removal of such words can improve accuracy and time required for comparisons by saving memory space and thus increasing the speed of processing.

3.2   Linguistic Manipulation

Linguistic manipulation is the second step in the plagiarism detection method. This step comprise two sub-steps, which were semantic argument labeling and grouping and semantic annotation.

In the following paragraphs, we will explain in detail the two sub-steps.

3.2.1   Semantic Argument Labeling and Grouping

This sub-step represents the sentences in the form of nodes. Semantic arguments, notably semantic class and thematic role are used to extract the arguments and role of each term in the argument from the sentences using WordNet and Verb-Net. The concepts extracted from these arguments are represented as a graph. All the arguments extracted from the text will be grouped in the nodes according to the argument type. Each node type contains similar extracted arguments. Each group is named by the argument name, such as Arg0, Arg1, Verb, Time and Location.

3.2.2 Semantic Annotation

The concepts that were extracted from each term in the argument groups using WordNet and VerbNet thesaurus is called Semantic Term Annotation (STA). The concept extraction is an important step in our proposed method where different concepts having the same meaning can be referred to as one term, which makes the text easier to be checked for plagiarism. In this step, concept extraction is carried out using a WordNet thesaurus by extracting the hypernyms and synonyms for the terms. We believe that by using STA, the difficulty of detecting plagiarized text will be reduced, especially for plagiarism using paraphrasing or rewording. The terms in the original topic signature and the suspected topic signature are compared. When we find two identical terms, we look directly to the argument label group that contains these terms, and then compare the sentences that convey these terms.

This step compares the argument labels of possible sentences that have been plagiarized to the corresponding argument labels in the original sentences. The argument label group guides us to the main argument and each argument inside the group that gives a quick guide to the possible plagiarized sentence. The comparison between the terms must be done in the right way. If we compare the terms in Arg0 (subject) in the suspected text to all the other arguments in the original text to determine the plagiarism ratio, it can be wrong. For instance, it is not fair to compare the subject with the Location argument (Arg-LOC) to the subject with Time argument (Arg-TIM) as shown in Fig. 2.

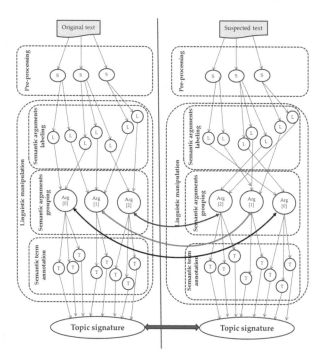

Fig. 2. Similar arguments comparison.

3.3   Similarity Detection

In this stage, sentence-based similarity analyses between the ranked suspected and the original sentences are performed. The sentences in the suspected documents are compared with each sentence in the candidate documents according to the arguments of the sentences. Here, we detect not only the arrangement similarity between sentences, but also possible semantic similarity between two sentences. For this reason, we use Wordnet and VerbNet thesaurus as a core tool for the calculation of similarity values. Wordnet Taxonomy returns a path similarity score denoting how similar two words depend on the shortest path between these two words in the taxonomy. This score ranges from 0 to 1.

For example, consider the sentences given below:

John painted the entire house (Original Sentence). The entire house was painted by john (Suspected Sentence). Figure 3 illustrates the analysis of the suspected sentence and original sentence using semantic arguments, notably, the semantic class and the thematic role in the example given above. Plagiarism detection is based on thee similarity between the arguments of the suspected document and the original document which is calculated according to Jaccard coefficient [3] that can be defined in the following equation:

Output S1:

	Semantic arguments		Stanford parser
	Thematic role	Semantic class	
Jhon	Agent [Arg0]	Human	S ((NP (NNP John)))
painted	Verb: paint		(VP (VBD painted)
the	Object [Arg1]	Inanimate	(NP (DT the)
entire			(JJ entire)
house			(NN house)

Output S2:

	Semantic arguments		Stanford parser
	Thematic role	Semantic class	
The	Object [Arg1]	inanimate	S(NP(DT(The)
entire			(JJ entire)
house			(NN house)
was	Verb: paint		(VP (VBD was)
painted			(VP (VBN painted)
by			(PP (IN by)
jhon	Agent [Arg0]	Human	(NP (NN john))

Fig. 3. Analysis sentences using semantic arguments.

$$SimilarityCi(ArgSj, ArgSk) = \frac{C(ArgSj) \cap C(ArgSk)}{C(ArgSj) \cup C(ArgSk)} \qquad (1)$$

Where C(ArgSj) are Concepts of the argument sentence in the original document and C(ArgSk) are Concepts of the argument sentence in the suspected document. We then compute the overall similarity between the original and suspected documents using the below equation:

$$Totalsimilarity(Doc1, Doc2) = \sum_{i=}^{l} \sum_{j,k=1}^{m,n} SimilarityCi(ArgSj, ArgSk) \qquad (2)$$

where Similarity Ci(ArgSj, ArgSk) is similarity between Arguments sentence, j in suspected document containing concept, i Arguments sentence, k in original document containing concept i, l is the number of concepts, m is number of Arguments sentence in suspected document and n is the number of Arguments sentence in original document.

4 Experimental Results

Our plagiarism detection method is tested on 100 documents. The suspected documents are plagiarized in different ways of plagiarism, such as copy and paste, change some terms with their corresponding synonyms, and modify the structure of the sentences. The experiments are performed on these 100 suspicious files each of which is plagiarized from one or more original documents according to the PAN13-14 corpus.

In this experiment, we looked only at the amount of the detected plagiarized sentences from the original documents. For this reason, sentence-based similarity analysis between the ranked suspected and original documents are performed. Suspicious and original sentence pairs obtained from the sentence ranking are compared according to their arguments.

To evaluate the effectiveness of the proposed method, we use the recall and precision metrics. Which are testing parameters that are commonly used in plagiarism detection.

$$Recall = \frac{Number\ of\ detected\ arguments}{Total\ number\ of\ arguments} \qquad (3)$$

$$Precision = \frac{Number\ of\ detected\ arguments}{Number\ of\ detected\ arguments} \qquad (4)$$

The former is defined as the percentage of detected semantic arguments as plagiarized with respect to the total number of arguments between two documents and the latter is defined as the percentage of plagiarized arguments identified with respect to the total number of detected arguments.

The proposed method is evaluated and compared to existing SRL based method, such as [7] and [8] on the basis of detection accuracy and time complexity.

Table 1. Performance evaluation of the proposed method.

	Recall	Precision	Time complexity
Our proposal	**0.92**	**0.91**	$O(n^2)$
[8]	0.89	0.85	$O(f(n)N^2)$
[7]	0.89	0.90	$O(n^2)$

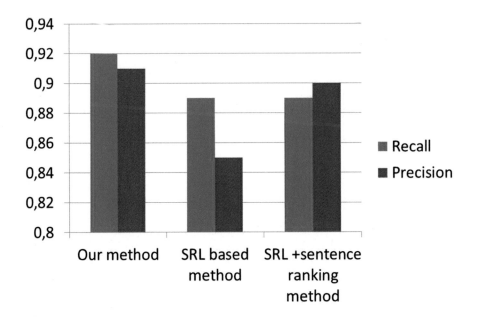

Fig. 4. Comparison results with plagiarism detection techniques.

Indeed, we evaluated our proposed method based on time complexity. We found that our proposed method belongs to the O(n2) class and therefore it is similar to the plagiarism detection algorithm proposed by [7]. However, the plagiarism detection system suggested by [8] has $O(f(n)N^2)$, where N is the size file number of the collection, and f(n) is the time to make the comparison between one pair of files of length n.

The results are shown below in Table 1.

Figure 4 gives a comparison between the proposed method with SRL-based method [7] and SRL+ sentence ranking-based method [8]. From the obtained result itself, it is clear that the proposed method gives better results than the other methods. Moreover, it becomes clear that the proposed method reduces the execution time for checking plagiarism.

In fact, the improvement provided by our proposed method is represented on the selection of the important arguments, notably the semantic class and the thematic role that can affect plagiarism detection. Another improvement represented in the comparison mechanism between two texts should be made by comparing the corresponding arguments. Furthermore, the proposed method can detect copy paste and semantic plagiarism, rewording or synonym replacement, changing of word structure in the sentences, modifying the sentence from passive voice to active voice and vice versa.

5   Conclusion

This paper presents an improved plagiarism detection method with semantic arguments, notably the semantic class and the thematic role. Indeed, the effects of arguments are studied using argument weighting. Through these arguments, we used only the important arguments in the similarity calculation process.

An experimental evaluation is performed on the PAN13-14 standard dataset for plagiarism detection. The results suggest that the measure performs encouragingly well (a recall $= 0.92$ and a precision $= 0.91$) and significantly better than the SRL- based method and SRL+sentence ranking method.

As a perspective of our work, we plan to improve our proposed method by using more documents from the PAN-PC dataset. Furthermore, we wish to investigate the role of the terms using Fuzzy Inference System (FIS) to enhance the similarity score that improved by the argument weighting schema. For instance, we can develop a way to determine all the important arguments that have more contribution than others in a sentence. Finally, we propose to apply our plagiarism detection method to other languages.

References

1. Alzahrani, S., Salim, N.: Fuzzy semantic-based string similarity for extrinsic plagiarism detection. Braschler Harman **1176**, 1–8 (2010)
2. Bao, J.-P., Shen, J.-Y., Liu, X.-D., Song, Q.-B.: A survey on natural language text copy detection. J. Softw. **14**(10), 1753–1760 (2003)
3. Jaccard, P.: Etude comparative de la distribution florale dans une portion des Alpes et du Jura. Impr, Corbaz (1901)
4. Kipper, K., Korhonen, A., Ryant, N., Palmer, M.: A large-scale extension of Verb-Net with novel verb classes. In: Corino, C.O.E., Marello, C. (eds.) Proceedings of the 12th EURALEX International Congress, pp. 173–184, Torino, Italy. Edizioni dell'Orso, September 2006
5. Kent, C.K., Salim, N.: Web based cross language plagiarism detection. In: 2010 Second International Conference on Computational Intelligence, Modelling and Simulation (CIMSiM), pp. 199–204. IEEE (2010)
6. Miller, G.A.: WordNet: a lexical database for English. Commun. ACM **38**(11), 39–41 (1995)
7. Osman, A.H., Salim, N., Binwahlan, M.S., Alteeb, R., Abuobieda, A.: An improved plagiarism detection scheme based on semantic role labeling. Appl. Soft Comput. **12**(5), 1493–1502 (2012)
8. Paul, M., Jamal, S.: An improved SRL based plagiarism detection technique using sentence ranking. Procedia Comput. Sci. **46**, 223–230 (2015)
9. Shehata, S., Karray, F., Kamel, M.S.: An efficient model for enhancing text categorization using sentence semantics. Comput. Intell. **26**(3), 215–231 (2010)

10. Shivakumar, N., Garcia-Molina, H.: SCAM: a copy detection mechanism for digital documents (1995)
11. Si, A., Leong, H.V., Lau, R.W.: Check: a document plagiarism detection system. In: Proceedings of the 1997 ACM Symposium on Applied Computing, pp. 70–77. ACM (1997)
12. Suanrnali, L., Salim, N., Binwahlan, M.S.: Automatic text summarization using feature-based fuzzy extraction. Jurnal Teknologi Maklumat **2**(1), 105–155 (2009)
13. Tsatsaronis, G., Varlamis, I., Giannakoulopoulos, A., Kanellopoulos, N.: Identifying free text plagiarism based on semantic similarity. In: Proceedings of the 4th International Plagiarism Conference. Citeseer (2010)
14. Vani, K., Gupta, D.: Detection of idea plagiarism using syntax-semantic concept extractions with genetic algorithm. Expert. Syst. Appl. **73**, 11–26 (2017)
15. Wali, W., Gargouri, B., Hamadou, A.B.: Sentence similarity computation based on WordNet and VerbNet. Computación y Sistemas **21**(4), 627–635 (2017)

Towards an Upper Ontology and Hybrid Ontology Matching for Pervasive Environments

N. Karthik$^{(\boxtimes)}$ and V. S. Ananthanarayana

Department of Information Technology, National Institute of Technology Karnataka,
Mangalore, India
nkarthikapce@gmail.com, anvs@nitk.ac.in

Abstract. Pervasive environments include sensors, actuators, handheld devices, set of protocols and services. The specialty of this environment is its power to manage with any device at any time anywhere and work autonomously for providing customized services to user. The different entities of pervasive environment collaborate with each other to accomplish an objective by sharing data among them. It raises an interesting problem called semantic heterogeneity. To address this problem, a hybrid ontology matching technique which combines direct and indirect matching techniques is proposed in this paper. To share and integrate data semantically, ontology matching technique establishes a semantic correspondence among various entities of pervasive application ontologies. To find the efficiency of proposed approach, we carried out set of experiments with real world pervasive applications. Experimental results prove that the proposed approach shows excellent performance in hybrid ontology matching. Results also proved that the use of background knowledge has influence over the performance of ontology matching technique.

Keywords: Ontology matching · Pervasive environments ·
Upper ontology

1 Introduction

Pervasive environments are established on acquiring of real time data from various sources and shared ad hoc by different applications to reach its full potential. For example, based on user location, smart home application interacts with traffic system to predict user arrival for controlling thermostat. When entities of different application try to share and exchange data, a semantic heterogeneity problem occurs. Ontologies are used to overcome the heterogeneous data integration problem [1]. Ontology supplies a set of vocabulary that describes the domain of interest and explicitly specifies the meaning of terms used in it [1]. It is also used to represent knowledge of concepts in certain relationship with classes, properties and rules. In this work, we give an attention towards ontology

© Springer Nature Switzerland AG 2020
A. Abraham et al. (Eds.): ISDA 2018, AISC 941, pp. 275–283, 2020.
https://doi.org/10.1007/978-3-030-16660-1_27

as a technique for sharing common knowledge and integration of pervasive applications. The ontology is used to have a common understanding about concepts and serves as a tool for interaction among heterogeneous application. Ontology matching could supply a semantic connection between several ontologies for accessing and exchanging data semantically. Most of the pervasive applications are developed and maintained by different developers who have diverse knowledge background and different terms are used to describe the same concepts in pervasive domain.

Different developers contributed different ontologies for same domain concepts. This leads to semantic heterogeneity problem and limits interaction among entities. Ontology matching is introduced to overcome this heterogeneity problem and interoperability among various pervasive applications [2].

The main contributions of this work include: (i) we propose an upper ontology for pervasive environments with trust mechanism to deal with faulty data, missing data of various entities; (ii) we propose a hybrid ontology matching technique which combines context, instance and upper ontology with trust mechanism for ontology alignment and (iii) we tested our approach with four different ontology matching tasks of pervasive environments.

Rest of the paper is organized as follows: Sect. 2 presents related work in hybrid and instance ontology matching with trust mechanism. The upper ontology for pervasive environments with trust mechanism is introduced in Sect. 3. The proposed hybrid ontology matching is discussed in Sect. 4. Experimental results and analysis are given in Sect. 5. Section 6 conclude our paper.

2   Related Work

In this section, we discuss the recent works of ontology matching techniques. In general, we have three types of ontology matching process: direct matching, indirect matching and hybrid matching. Direct matching process uses multiple ontology architecture to find the set of correspondence among concepts. In indirect matching process, the global shared vocabulary is used as background knowledge for finding semantic correspondence among various concepts. The hybrid matching is the combination of direct and indirect ontology matching for establishing semantic correspondences among similar concepts of various Ontologies [2]. Notable methods of instance based ontology matching techniques are reviewed and their future research directions are highlighted in [3]. It is found that similarity based and machine learning based methods are renowned techniques in instance based ontology matching [3].

A hybrid ontology matching technique is proposed in [4], in which multiple matchers are used to find similarities between elements of ontology. It uses hierarchical information to find weights of similarities. Finding the semantic similarity of Ontologies based on WordNet and structure level is introduced in introduced in [5]. The combination of ontology driven and keyword matching system is introduced as hybrid approach in [6]. Trust mechanism supplies a framework that infers a correct and wrong matching among entities of pervasive environments with trust metrics [7–11]. But none of them consider the trustiness of instances and entities for ontology matching and there is no support for large scale matching [3]. Research has to be carried out for dealing with combination of direct and indirect matching. To overcome these problems, we propose a hybrid ontology matching technique which uses upper ontology as background knowledge for indirect matching and context and instance based matcher with trust mechanism for direct matching.

3   Upper Ontology for Pervasive Environments

In this section, we represent the upper ontology of pervasive environments. In general, upper ontology describes high level concepts which are not restricted to single application or domain. Usually it contains the core concepts and set of requirements which are independent of any pervasive domains. We listed out four main concepts for upper ontology of pervasive environments. They are temporal properties, spatial properties, entities and trust management as shown in Fig. 1. There are five temporal properties of data which are generated from pervasive environments. They are rate of change of data, data generation time, data validity time, sampling time and temporal dimension. Rate of change of data describes about the rate at which dynamic data changes per time. Data generation time describes about the time at which the data is generated from source. Data validity time explains about the validity of data. The rate at which the data is collected or sampled from the environment is called sampling time. The generated data may represent either past, current or future state of environment is referred to temporal dimension.

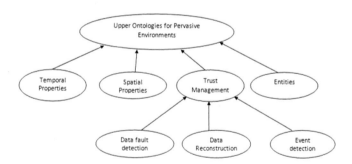

Fig. 1. Upper ontology for pervasive environments

There are two ways to represent spatial information in pervasive environment. They are physical location coordinate representation and symbolic location representation. Physical location coordinates are generated by GPS either in 2D or 3D. The symbolic locations are human friendly location names like conference room and seminar hall. Entities in pervasive environment refer to data sources (sensors, users) and actuators. Trust management in pervasive environment refers to the process of finding the trustiness of data, devices and user. It also includes trust based data fault detection, trust based data reconstruction and trust based event detection [12]. The temporal and spatial properties of data are used in trust based data fault detection, data reconstruction and event detection. We borrow methods from [13–15] to find the trustiness of instance and entities of ontologies.

4   Hybrid Ontology Matching

In this section, the hybrid ontology matching is introduced as shown in Fig. 2. The proposed ontology matcher takes two schemas, namely, source and target ontology as input. Ontology matching is referred to the process of detecting similar entities between source and target Ontologies and establishing communication among them for data sharing and exchange. It produces an ontology alignment as an output. In addition to source and target schemas, the matcher takes some parameters and resources as input to support the process of ontology alignment. Usually, the parameter includes minimum trust values of various entities, instances and some heuristic rules for establishing the communication between various entities. The upper ontology acts as a background ontology resource in ontology matching process. The context based matcher collects information about entity id, entity name, entity neighbors and does the matching operation between source and target Ontologies with respect to collected information. Instance matcher is similar to record linkage technique in databases [3]. It matches the entities with value similarity and heuristic rule based methods [7]. Existing instance does not handle missing instance value in ontology matching process. The main advantage of proposed method is reconstructing the missing instance value and faulty instance value with the help of trust based data reconstruction method [7]. Moreover, the trustiness of instance value is checked before the ontology alignment is made between various entities. If the trust value of instance is below threshold (say 0.3), then mapping is ignored with particular entity in target ontology. The alignment between entities of Ontologies can be

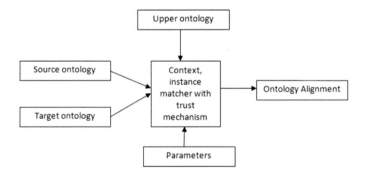

Fig. 2. Hybrid ontology matching

represented in quadruple format <aid, es, et, t> where aid is the alignment id, es and et are entities of source and target schemas and t is the trust value which holds the alignment between entities.

5   Experimental Results

We carried out the set of experiments on pervasive real world examples with three metrics from the field of information retrieval: precision, recall and F-measure. We evaluated the performance of proposed approach against direct and indirect matching. We run our proposed approach three times on each source and target schemas pairs of each application. In the first run, we considered only direct matching. During second run, we did with indirect matching where upper ontology is considered as background knowledge. In the third and final set of experiments, we combined both direct and indirect matching technique together. List of Ontologies used for experiment is shown in Table 1. Table 2 shows the list of applications and reference alignments.

Table 1. List of ontologies used for experiment

Ontologies	Concepts	Properties
Smart home Os	10	18
Healthcare Oh	12	22
Traffic system Ot	13	24
Environment Oe	12	20
Upper ontology Ou	24	68

Table 2. List of applications and reference alignments

Applications	Ontologies	Reference alignments
Thermostat control	Os and Ot	24
Medical event detection	Oh and Os	16
Route design	Ot and Os	18
Lighting system	Oe and Os	20

Let M be the number of total matches found by domain expert. The number of right matches done by proposed approach is represented as C. Let W be the number of wrong matches done by proposed approach. The precision is calculated by $P = (C/(C + W)) * 100$ and recall is calculated by $R = (C/M) * 100$. The F-measure is evaluated by using $F = (2PR/(P + R)) * 100$.

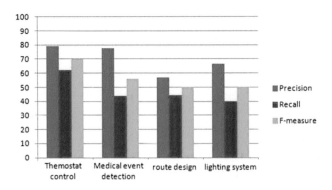

Fig. 3. Performance of direct ontology matching without trust

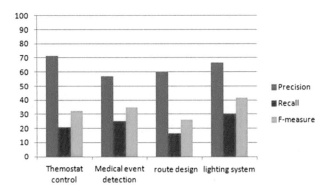

Fig. 4. Performance of indirect ontology matching without trust

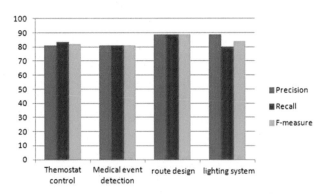

Fig. 5. Performance of hybrid ontology matching without trust

For thermostat control application, smart home ontology Os interacts with traffic system ontology Ot to predict the user arrival for switching on thermostat. In healthcare application, healthcare ontology Oh interacts with smart home

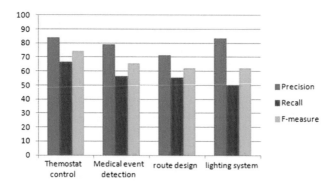

Fig. 6. Performance of direct ontology matching with trust

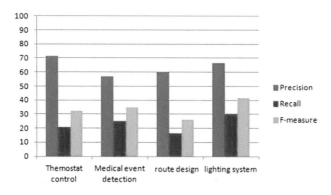

Fig. 7. Performance of indirect ontology matching with trust

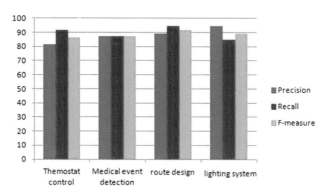

Fig. 8. Performance of hybrid ontology matching with trust

ontology Os to find the activities of patient and medical event detection. For designing pollution and traffic free route, traffic system ontology Ot interacts with environment ontology Oe. To minimize the energy consumption of electrical equipments, smart home ontology interacts with environment ontology.

Based on the results presented in Figs. 3, 4 and 5, we conclude that the hybrid ontology matching achieves reasonable performances in ontology matching when compared to direct and indirect matching techniques. However the performance of matching technique will gradually increase only if we have quality upper ontology as background knowledge. We manually inserted 5% of untrustworthy data and simulated few entities to behave abnormally in all four applications to check the impact of trust management in ontology matching. Based on the results presented in Figs. 6, 7 and 8, we can say that the ontology matching with trust mechanism gives better performance than ontology matching without trust mechanism.

6 Conclusion

We proposed an upper ontology for pervasive environments for hybrid ontology matching. Ontology matching establishes the semantic correspondence between matching entities of various application ontologies for data sharing using background knowledge as upper ontology. Our experimental results show that upper ontology plays a key factor in the performance of ontology matching. Even though experiment is restricted to pervasive environment, we did evaluations with several heterogeneous applications within pervasive environment for ontology matching. In future, we would like to test the performance of proposed approach with benchmarked datasets and also we have plan to compare and validate the proposed upper ontology with existing upper ontologies in pervasive environments.

References

1. Shvaiko, P., Euzenat, J.: Ontology matching: state of the art and future challenges. IEEE Trans. Knowl. Data Eng. **25**(1), 158–176 (2013)
2. Cerdeira, L.O.: Study and application of new methods for ontology matching. Dissertation, Universidade de Vigo (2014)
3. Abubakar, M., et al.: Instance-based ontology matching: a literature review. In: International Conference on Soft Computing and Data Mining. Springer, Cham (2018)
4. Wang, Z., Bie, R., Zhou, M.: Hybrid ontology matching for solving the heterogeneous problem of the IoT. In: IEEE TrustCom (2012)
5. He, W., Yang, X., Huang, D.: A hybrid approach for measuring semantic similarity between ontologies based on WordNet. In: International Conference on Knowledge Science, Engineering and Management. Springer, Heidelberg (2011)
6. Ducatel, G., Cui, Z. and Azvine, B.: Hybrid ontology and keyword matching indexing system. In: Proceedings of IntraWebs Workshop at WWW (2006)
7. Liu, X., Wang, Y., Zhu, S., Lin, H.: Combating web spam through trust-distrust propagation with confidence. Pattern Recognit. Lett. **34**, 1462–1469 (2013)
8. Wang, X., Su, J., Wang, B., Wang, G., Leung, H.F.: Trust description and propagation system: semantics and axiomatization. Knowl.-Based Syst. **90**, 81–91 (2015)

9. Jiang, C., Liu, S., Lin, Z., Zhao, G., Duan, R., Liang, K.: Domain-aware trust network extraction for trust propagation in large-scale heterogeneous trust networks. Knowl.-Based Syst. **111**, 237–247 (2016)
10. Wu, J., Xiong, R., Chiclana, F.: Uninorm trust propagation and aggregation methods for group decision making in social network with four tuple information. Knowl.-Based Syst. **96**, 29–39 (2016)
11. Xiong, F., Liu, Y., Cheng, J.: Modelling and predicting opinion formation with trust propagation in online social networks. Commun. Nonlinear Sci. Numer. Simul. **44**, 513–524 (2017)
12. Karthik, N., Ananthanarayana, V.S.: Data trust model for event detection in wireless sensor networks using data correlation techniques. In: IEEE ICSCN (2017)
13. Karthik, N., Ananthanarayana, V.S.: A hybrid trust management scheme for wireless sensor networks. Wirel. Pers. Commun. **97**(4), 5137–5170 (2017)
14. Karthik, N., Ananthanarayana, V.S.: An ontology based trust framework for sensor-driven pervasive environment. In: 2017 Asia Modelling Symposium (AMS). IEEE (2017)
15. Karthik, N., Ananthanarayana, V.S.: A trust model for lightweight semantic annotation of sensor data in pervasive environment. In: 2018 IEEE/ACIS 17th International Conference on Computer and Information Science (ICIS). IEEE (2018)

Emotion Assessment Based on EEG Brain Signals

Sali Issa[✉], Qinmu Peng, Xinge You, and Wahab Ali Shah

Huazhong University of Science and Technology, Wuhan, China
I201622211@hust.edu.cn

Abstract. This paper presents an emotion assessment method that classifies five emotions (happy, sad, angry, fear, and disgust) using EEG brain signals. Public DEAP database is chosen for the proposed system evaluation, Fz channel electrode is selected for the feature extraction process. Then a Continuous Wavelet Transform (CWT) is used to extract the proposed Standard Deviation Vector (SDV) feature which describes brain voltage variation in both time and frequency domains. Finally, several machine learning classifiers are used for the classification stage. Experiment results show that the proposed SDV feature with SVM Classifier produce robust system with high accuracy result of about 91%.

1 Introduction

Recently, Brain Computer Interface (BCI) technology is the most important and up-to-date science, where it is possible to combine artificial intelligence techniques and human brain information such as Electroencephalogram (EEG) brain signals.

Emotion assessment or prediction is one of the BCI technologies, it covers several applications such as patient monitoring and therapy as well as criminal detection. Hence, it is considered as one of the hot smart life technologies in the modern era.

Actually, emotion recognition researches are not new; the first proposed researches depended on face expressions, body gestures and voice interpretations [1,2], they sometimes have problems because human could change or control expressions and voice tone. As a normal trend, scientists turned to use brain signals, because emotion itself is an electrical brain activity voltages that correspond to human inner feelings and cognition [3,4], and it is impossible to fake brain signals, so the recent science pays more attention to EEG brain signals [5,6].

In this work, a new emotion assessment system is proposed based on the frequency-time analysis of the EEG brain signals. The EEG brain recording signals were taken from the public DEAP emotional stimuli database [7]. Continuous Wavelet Transform (CWT) is applied using just one EEG electrode to get the signal information data in the frequency-time domain, then Standard

© Springer Nature Switzerland AG 2020
A. Abraham et al. (Eds.): ISDA 2018, AISC 941, pp. 284–291, 2020.
https://doi.org/10.1007/978-3-030-16660-1_28

Deviation Vector (SDV) feature is created as a simulation of the CWT function coefficients, later in the classification part, multiple classifiers are applied for evaluation purpose.

This paper is arranged as follows: Sect. 2 provides a brief description of previous works. Section 3 illustrates and discusses the proposed system in details. Section 4 shows the experimental results of the framework, and Sect. 5 lights up conclusion as well as some recommendation for future work.

2 Related Works

In general, emotion classification using EEG brain signals has been started since 2004. Depending on the technical and feature extraction analysis, emotion classification based on EEG brain signals has several approaches: time domain approach such as computing signal Fractal Dimension (FD), Hjorth parameters, and the common Statistical information like minimum and maximum magnitudes, mean, standard deviation and variance [8,9]. Frequency domain approach as calculating Power Spectral Density (PSD), relative power, power band and Fast Fourier Transform (FFT) [10,11].

While the time-frequency analysis such as using Short Time Fourier Transform (STFT), Gabor function, as well as discreet and continuous Wavelet Transform (WT), this kind of analysis gave better results, but unfortunately leads to other problems such as huge calculation space, and time consuming calculations [12,13].

There are many recent researches for emotion classification based on EEG brain signals such as Wang et al. [14] research in 2015, he used large number of EEG electrodes to compute power spectral feature, as well as visual-audio features from the stimuli movies as a privileged information. Feature selection methods were used to generate just one new combination feature, and Support Vector Machine (SVM) was applied to distinguish four emotions with accuracy of 75%.

Atkinson et al. [15] used 14 EEG electrodes to compute several extraction features such as statistical features, band power, Hjorth parameters and fractal dimension. Feature selection methods were used due to the large number of feature extraction, and genetic analysis with Support Vector Machine (SVM) were applied to distinguish two emotions with accuracy of 73%.

Kumar et al. [16] used just two EEG electrodes to calculate the bispectrum feature for theta, alpha, and beta frequency bands using Higher Order Spectrum Analysis (HOSA). Support Vector Machine (SVM) was implemented to classify four emotions with the average accuracy of 63.01%.

Mehmood paper et al. [17] applied Hjorth Parameter for 14 EEG electrodes to calculate feature extraction, then made a comparison between the results of multiple classifiers such as LDA, KNN, SVM, Naïve/Bayes Net, Random Forest, Voting, Boosting, Bagging and Deep Learning, where, Voting and Deep Learning algorithms achieved the highest accuracy of about 76.6%.

According to the new survey paper on emotion classification based on EEG brain signals (2017) [18], it is obvious that this topic still has challenges, actually, no high or robust accuracy has been achieved without certain restrictions. Additionally, most of the previous works and researches were used additional peripheral signals, for example, heart rate, ECG, pressure, temperature, voice signals and face expressions [8, 12, 14].

3   The Proposed Work

The flow chart of emotion system classification based EEG brain signals is shown in Fig. 1. From that figure, signal recognition system consists of four main processes including data recording, pre-processing, feature extraction and classification process.

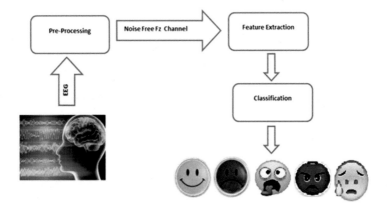

Fig. 1. The proposed emotion classification system processes.

3.1   Database Acquisition and Pre-processing

DEAP database [7] is a multimodal data set used for recording EEG and other peripheral signals of 32 participants (19–37 years old and 50% were females).

During the stimuli experiment each participant watched 40 one-minute long excerpts of music videos, and rated each video in terms of the levels of arousal, valence, like/dislike, dominance and familiarity. For 22 of the 32-participants, frontal face video was also recorded.

During EEG Recording, the international 10/20 System is used to describe the position of electrodes on scalp. It is based on the relation between the electrode location and the underlying area of the cerebral cortex; number 10 and 20 represents the distance between adjacent electrodes which are 10% or 20% of the total front-back or right-left distance of the skull [19] (Fig. 2).

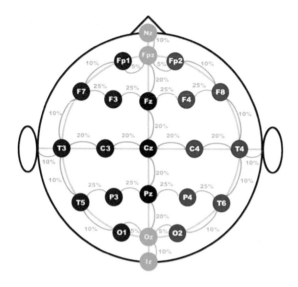

Fig. 2. The EEG 10/20 system electrode distances [19]

In Pre-Processing Stage, dataset is sampled to 128 Hz, EOG artefacts are removed and averaged to the common reference, a bandpass frequency filter (4–45 Hz) is applied.

According to medical and psychological science, the frontal head area is responsible for emotions [20], hence only the middle frontal channel electrode (Fz) is chosen as a dataset entry to the next feature extraction step.

3.2 Feature Extraction and Classification

In the feature extraction process, the selected Fz electrode is used to calculate the Standard Deviation Vector (SDV) feature using Continuous Wavelet Transform (CWT).

For non-stationary signals such as EEG brain signals where the frequency components change within the time, the frequency domain information is not enough and the full knowledge for signal frequency band changes within the time could not be available, so an alternating transform is used such as Continuous Wavelet Transform (CWT).

The Continuous Wavelet Transform (CWT) is defined in the following equation [21]:

$$CWT_x^\psi(\tau, S) = \Psi_x^\psi(\tau, S) = \frac{1}{\sqrt{s}} \int X(t)\Psi^*(\frac{t-\tau}{s}) \tag{1}$$

Depending on the above equation, the transformed signal is a function of two variables; the translation (τ) refers to the location of the window, as the window is shifted through the signal, and scale (s) variable. $\boldsymbol{\Psi}(t)$ represents the transforming function, and it is called the mother wavelet.

Standard Deviation Vector (SDV) feature calculation could be divided into the following steps:

1. Convert the Fz raw data from time domain space to frequency-time domain using Continuous Wavelet Transform (CWT) as mentioned in equation number one. Bump wavelet function is used as the mother wavelet function because it is a good choice when the time domain signal is oscillatory (see Eq. 2)[22]:

$$\Psi(A, h, w, f, t, y, x) = 1 - \frac{(y, f)^2}{h^2} + \frac{(x - t)^2}{w^2} \tag{2}$$

   Where y and x are the time and frequency positions of the adaptation window on the time-frequency map (fixed parameters); f and t are the time and frequency positions of the bump on the time-frequency map; h and w are the height and width of the bump; and A is the function amplitude.

2. Resize the resulted time-frequency voltage values to 45×45 2D matrix by dividing the time domain into 45 blocks and then find the maximum voltage magnitude for each block as shown in Eq. 3.

$$\forall c \in C\{M = \max M(R, c = 1, 2, ...45)\} \tag{3}$$

   Where M is a 2D CWT matrix information; R is the frequency value (ranged from 1 Hz to 45 Hz); C is the time block number (ranged from 1 to 45); and c is the specific time within the current time block C.

3. Compute the standard deviation magnitude for each row in the 2D CWT matrix which corresponds to a specific frequency band (Eq. 4). So a single vector of size 45×1 is produced which is the Standard Deviation Vector (SDV) feature.

$$\forall r \in R\{SDV(r) = \sigma(M(r, \forall c \in C))\} \tag{4}$$

   Where SDV is the standard deviation vector, σ is the standard deviation magnitude; M is the resized 2D CWT matrix; c is the specific time within the overall time column vector C; r is the specific frequency band within the overall frequency row vector R.

In classification process, different machine learning methods are applied to the extracted feature including K-Nearest Neighbors (KNN), Neural Network (NN), Linear Discriminant Analysis (LDA), Support Vector Machine (SVM), and Probabilistic Neural Network (PNN).

4   Results and Discussion

In the experimental part, MATLAB 2017 software was utilized to evaluate the proposed method. All of the experiments were tested and implemented using a laptop equipped with an Inter(R) Core(TM) i5-5200 2.2 GHz CPU.

The participants' ratings and questionnaires were studied because emotion feeling is a person dependent psychological state. For example, a happy video could cause happiness to somebody while it may cause nothing to others. Standard Deviation Vector (SDV) feature was calculated for the database, then, it was randomly divided into training and testing groups. Each category in the training data has 60% of the database and the testing set has the remaining ratio for each emotion.

Five different machine learning methods were applied to the extracted feature including [23]:

1. Support Vector Machine (SVM): the polynomial kernel function is used.
2. K-Nearest Neighbors (KNN): Euclidean distance is used with 'nearest' property for the classification policy, and the number of nearest neighbors is 3.
3. Neural Network (NN): it is a feed forward neural with 10 hidden neurons.
4. Linear Discriminant Analysis (LDA): the chosen discriminant type is pseudoLinear.
5. Probabilistic Neural Network (PNN): the spread of the radial basis function is 0.4.

Table 1 sums up the accuracy results for all the tested classifiers, it is clear that SVM achieved the best accuracy result of about 91%. While Table 2 provides the comparison between our proposed system and other recent related works.

Table 1. The accuracy result of the tested classifiers.

Classifier	Accuracy
KNN	85%
SVM	91%
PNN	60%
LDA	86%
NN	70%

Table 2. Comparison between the proposed method and some existing methods

Research	No. channels	Results
Wang et al. [14]	32 EEG channels other peripheral signals	71.32% 4 emotions
Atkinson et al. [15]	14 EEG channels	61.51% 3 emotions
Kumar et al. [16]	2 EEG channels	63% 4 emotions
Mehmood et al. [17]	All EEG channels	76.6% 4 emotions
Proposed work	1 EEG channel	91% 5 emotions

According to the tested experiments and the previous related works, this paper produces a new feature extraction using just one EEG electrode (Fz electrode) and SVM classifier without neither the slow and complex feature selection methods nor the peripheral signals.

5   Conclusion

The paper presents a new effective emotion classification system using just one frontal EEG electrode and Continuous Wavelet Transform (CWT). The constructed SVM classifier achieved high accuracy of 91% on distinguishing five distinct emotions. In future work, different types of deep learning methods will be used to enhance the accuracy results.

References

1. Busso, C., Deng, Z., Yildirim, S., Bulut, M., Lee, C.M., Kazemzadeh, A., Lee, S., Neumann, U., Narayanan, S.: Analysis of emotion recognition using facial expressions, speech and multimodal information. In: Proceedings of the 6th International Conference on Multimodal Interfaces, pp. 205–211. ICML, State College, Pennsylvania, USA (2004)
2. Emerich, S., Lupu, E., Apatean, A.: Emotions recognition by speechand facial expressions analysis. In: 17th European on Signal Processing Conference, Glasgow, UK, pp. 1617–1621. IEEE (2009)
3. Arnold, M.B.: Emotion and Personality: Psychological Aspects. Columbia University Press, New York (1960)
4. Frijda, N.H.: The Emotions. Cambridge University Press, UK (1986)
5. Canon, W.: The James Lange theory of emotion: a critical examination and an alternative theory. Am. J. Psychol. **39**, 106–124 (1927)
6. LeDoux, J.E.: Brain mechanisms of emotion and emotional learning. Curr. Opin. Neurobiol. **2**(2), 191–197 (1992)
7. Koelstra, S., Muhl, C., Soleymani, M., Lee, J.-S., Yazdani, A., Ebrahimi, T., Pun, T., Nijholt, A., Patras, I.: DEAP: a database for emotion analysis using physiological signals. IEEE Trans. Affect. Comput. **3**(1), 18–31 (2012)
8. Chanel, G., Rebetez, C., Btrancourt, M., Pun, T.: Emotion assessment from physiological signals for adaptation of game difficulty. IEEE Trans. Syst. Man Cybern. Part A: Syst. Hum. **41**(6), 1052–1063 (2011)
9. Atkinson, J., Campos, D.: Improving BCI-based emotion recognition by combining EEG feature selection and kernel classifiers. Expert. Syst. Appl. **47**, 35–41 (2016)
10. Soleymani, M., Lichtenauer, J., Pun, T., Pantic, M.: A multimodal database for affect recognition and implicit tagging. IEEE Trans. Affect. Comput. **3**(1), 42–55 (2012)
11. Stikic, M., Johnson, R.R., Tan, V., Berka, C.: EEG-based classification of positive and negative affective states. Brain-Comput. Interfaces **1**(2), 99–112 (2014)
12. Murugappan, M., Juhari, M.R.B.M., Ramachandran, N., Yaacob, S.: An investigation on visual and audiovisual stimulus based emotion recognition using EEG. Int. J. Med. Eng. Inform. **1**(3), 342–356 (2009)
13. Nasehi, S., Pourghassem, H.: An optimal EEG-based emotion recognition algorithm using Gabor features. WSEAS Trans. Signal Process. **8**(3), 87–99 (2012)

14. Wang, S., Zhu, Y., Yue, L., Ji, Q.: Emotion recognition with the help of privileged information. IEEE Trans. Auton. Ment. Dev. **7**(3), 189–200 (2015)
15. Atkinson, J., Campos, D.: Improving BCI-based emotion recognition by combining EEG feature selection and kernel classifiers. Expert Syst. Appl. **47**(C), 35–41 (2016)
16. Kumar, N., Khaund, K., Hazarika, S.M.: Bispectral analysis of EEG for emotion recognition. In: 7th International conference on Intelligent Human Computer Interaction, IHCI 2015, pp. 31–35, IIIT-Allahabad, India (2015). Procedia Comput. Sci.
17. Mehmood, R.M., Du, R., Lee, H.J.: Optimal feature selection and deep learning ensembles method for emotion recognition from human brain EEG sensors. IEEE Access **5**, 14797–14806 (2017)
18. Alarcao, Soraia M., Fonseca, Manuel J.: Emotions recognition using EEG signals: a survey. IEEE Trans. Affect. Comput. **PP**(99), 1–20 (2017)
19. Trans Cranial Technologies. https://www.trans-cranial.com
20. Newsweek Media Group Magazine. http://www.newsweek.com/sad-brain-happy-brain-88455
21. Polikar, R.: The wavelet tutorial. Rowan University, Glassboro, Camden, Stratford, New Jersey, U.S (1999)
22. Vialatte, F.B., Solé-Casals, J., Dauwels, J., Maurice, M., Cichocki, A.: Bump time frequency toolbox: a toolbox for time-frequency oscillatory bursts extraction in electrophysiological signals. BMC Neurosci. **10**, 1186 (2009)
23. Duda, R.O., Hart, P.E., Stork, D.G.: Pattern Classification. Wiley, New York (2001)

Characterization of Edible Oils Using NIR Spectroscopy and Chemometric Methods

Rishi Ranjan[1,3(✉)], Navjot Kumar[2,3], A. Hepsiba Kiranmayee[1,3],
and P. C. Panchariya[2,3]

[1] CSIR-Central Electronics Engineering Research Institute, Chennai Center,
Chennai, Tamil Nadu, India
rishiranjan@ceeri.res.in
[2] CSIR-Central Electronics Engineering Research Institute,
Pilani, Rajasthan, India
[3] Academy Scientific and Innovative Research, New Delhi, India

Abstract. Authenticity and characterization of edible oils have become necessary to tackle the problems like adulteration and quality assurance of edible oils. The present paper deals with the use of Near Infrared (NIR) spectroscopy combined with exploratory data analysis methods for possible characterization and identification of eight different types of edible oils (sesame oil, safflower oil, mustard oil, palmolein oil, groundnut oil, extra virgin olive oil, canola oil and refined soya oil) used in Indian cuisine. NIR absorbance spectra covering the 1050–2400 nm spectral range of all the samples pertaining to eight different varieties of edible oil were collected. Spectra data has been corrected using iterative restricted least square (IRLS) method of baseline correction. The principal component analysis was used for exploratory analysis of edible oil spectra. Loadings vector of PCA has been used to select the important wavelength regions. Based on the loading vector five wavelength regions were selected. For each wavelength region PCA-DA, PLS-DA, and k-NN classification models were developed. Effect of different data pretreatment methods such as Savitzky-Golay smoothing, standard normal variate (SNV) correction, multiplicative scatter correction (MSC) and extended multiplicative scatter correction (EMSC) on the error rate of classifier has been presented. The result shows that in the wavelength region R4 (2110–2230 nm) all the classifiers have zero error rate with external validation samples. The thence proposed models using specific wavelength bands show good probability for characterization and identification of all the eight varieties of edible oil are presented.

Keywords: NIR spectroscopy · Wavelength selection ·
Principal component analysis · Edible oils

1 Introduction

Edible oil is one of the important ingredients in Indian cuisine which plays a major role in determining the taste, texture, nutrient profile and shelf life of food products. India is one of the largest producers and consumers of edible oil in the world [1]. Some of the many kinds of edible oils include olive oil, palm oil, soybean oil, canola oil, pumpkin

© Springer Nature Switzerland AG 2020
A. Abraham et al. (Eds.): ISDA 2018, AISC 941, pp. 292–300, 2020.
https://doi.org/10.1007/978-3-030-16660-1_29

seed oil, corn oil, sunflower oil, safflower oil, peanut oil, grape seed oil, sesame oil, groundnut oil, argan oil, mustard oil, and rice bran oil. Edible oils can be discriminated by their different botanical origin, manufacturing processes, and their physio-chemical properties [2, 3]. Primary methods of identifying and authenticating different edible oils involve various laboratory methods involving destructive chemical methods. In food control laboratories edible oils are classified by their fat contents, amount of tocopherol content, photometric color index and CIELAB methods [4]. Different varieties of edible oils have been characterized using the fluorescence spectroscopy and PCA and effect of heating on deterioration of edible oils was studied [5]. Principal component analysis and Random forest have been applied to fatty acid profiles for edible oil classification and detection of adulteration [6]. Data mining methods are applied to the chromatograms of various Olive oils mixtures for their classification [7]. Fourier Transform-mid-IR spectral model of physio-chemical properties for the classification of edible oil of different botanical origin has been proposed [8]. LDA based model for the classification of extra virgin olive oil (EVOO) and low-cost oils using the derivative of FTIR spectra for the classification of different types of olive oil samples gathered from Morocco using hierarchical cluster analysis (HCA) and PCA was developed [9].

Near Infrared (NIR) spectroscopy with exploratory multivariate technique gains a lot of popularity in the last twenty-five years due to its wide range of applications and advantages in the food industry [10]. NIR spectroscopy does not require any prior sample preparation. NIR spectra are typically composed of broad overlapping and, thus, ill-defined absorption bands of overtones and combination of the fundamental vibration frequency of absorption for O–H, C–H, C–O, and N–H relating to the chemical and physical information of the sample [11]. NIR spectra are therefore hardly selective. Nevertheless, for qualitative and quantitative analysis i.e., to relate the spectral variables to the chemical and physical properties of the samples analyzed, chemometrics are used to extract the information of interest. NIR spectroscopy with chemometrics is widely used for the quantification of oil parameters such as acidity, refractive index, viscosity, peroxide value [12, 13].

NIR spectra have been used to classify the virgin olive oils of different geographical origin of France and for the estimation of their fatty acid value, partial least square (PLS) regression model has been used [2]. A model for detecting low-cost oil adulteration in olive using NIR spectrum has also been developed [14]. NIR along with chemometric techniques has been used to detect and quantify the lard adulteration in palm oil [15]. Laser NIR spectroscopy along with various spectral treatment methods and support vector machine classification model has been used to classify 11 varieties of edible oils with prediction accuracy of 95% [16]. NIR spectra and Fusion of Multilayer perceptron and SVM classification techniques has been used to classify the five varieties of edible oils with accuracy of 98.2% [17].

The present paper deals with the use of Near Infrared (NIR) spectroscopy combined with chemometric methods for possible characterization and identification of eight different types of edible oils (Sesame oil, Safflower oil, Mustard oil, Palmolein oil, Groundnut oil, Extra Virgin Olive oil, Canola oil and Refined Soya oil) used in Indian cuisine. NIR absorbance spectra covering the 1050–2400 nm spectral range of all the samples pertaining to eight different varieties of edible oil were collected. Data treatment was performed and different models for characterizing and authenticating the different varieties have been presented. Loadings of principal component analysis

(PCA) are used in finding and extracting the specific wavelength bands which are responsible for characterizing and identifying the different samples. PCA, Partial Least Square Discriminant analysis (PLS-DA) and k-NN classifier. Besides, the paper also deals with the different data treatment methods for spectral correction.

2   Materials and Methods

2.1   Samples

Eight varieties of oils namely Sesame oil, safflower oil, Mustard oil, Palmolein oil, Groundnut oil, Extra Virgin Olive Oil and Canola Oil, Refined Soya oil has been used for analysis. All samples were collected from the local supermarket and stored in airtight plastic jars at room temperature.

2.2   Instrument

The near infra-red absorption spectra of edible oil samples are obtained from double beam spectrophotometer UV3600 from SHIMADZU (Kyoto, Japan). It consists of three detectors (photomultiplier tube for ultraviolet and visible regions, InGaAs and cooled PbS detector for near infrared region), and a double monochromator to achieve high resolution (maximum 0.1 nm) in the large wavelength range from 185–3000 nm. For temperature-controlled measurement, TCC-240 accessory from SHIMADZU was used. For instrument control and spectra collection, UVProbe version2.34 software from SHIMADZU was used.

2.3   Spectra Collection

Each sample under investigation was transferred to a 0.35ul, 1 mm path length quartz cuvette from the bottle and placed into the TCC240. Measurement has been performed by collecting the absorption spectra of each of the samples in the wavelength range 1050 nm to 2400 nm with a resolution 1 nm and slit width 5 nm. The spectra of all the samples were recorded at 30 °C and the time for spectra collection was 6 min per spectra. Each sample has been scanned thrice and then averaged out as a single measurement for further analysis. The experimental setup for collecting the spectra of the samples is shown in Fig. 1.

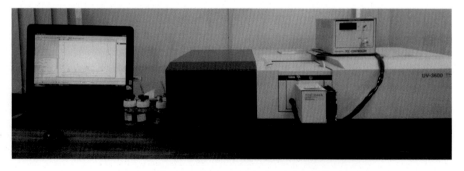

Fig. 1. Experimental setup for NIR spectra collection.

2.4 Spectral Preprocessing and Outlier Detection

Baseline Correction. NIR spectra of oil samples contain baseline noise as well as scatter noise due to instrumental error. The collected spectra are corrected from baseline shift using iterative restricted least square (IRLS) method which uses second derivate constraints for primary smoothing and repeated baseline suppression [18].

Savitzky Golay (S-G) Filtering. NIR spectra have the overlapping bands and have high-frequency spectral noise part. S-G filter is a window-based filter in which for each point in the window is calculated using a polynomial expression of given order which is calculated using least square fit to the window data points [19].

Standard Normal Variate (SNV) Correction. SNV is one of the popular methods used for scatter correction in the NIR spectra. Using SNV corrected spectra is obtained by normalizing the mean subtracted original spectra by the standard deviation of the sample spectrum [20].

Multiplicative Scatter Correction (MSC). This technique is used for removing the additive and multiplicative scatter effect due to stray light. Spectra correction using the MSC is performed by estimating the correction coefficients from the reference spectra and then correction of original spectra using those coefficients [20]. Mean spectra of each type of the edible oils are taken as a reference spectrum for calculating the correction coefficients. And each of the oil's samples is corrected by their respective mean spectrum.

Extended Multiplicative Scatter Correction (EMSC). EMSC is used to correct the wavelength dependent noise along with multiplicative and additive scatter noise. This uses the a priori knowledge of spectra of interest for second order polynomial fitting of reference spectrum for calculating the wavelength dependent light scattering correction coefficients [20].

2.5 Outlier Detection

Most of the multivariate analysis techniques are susceptible to outlier present in the sample which may be due to instrumental reasons or due to human error. The outliers in each variety are separately detected using Hoteling's T2 statistics on PC1 and PC2, the first and the second largest principal components. The points which are outside the 95% confidence region have been considered as an outlier in the measurement and removed from the further analysis step. The score plot of the mustard oil samples with Hotelling's T-square ellipse is shown in Fig. 2. The new data matrix consisting of remaining spectra is formulated by removing these outliers from the data set.

2.6 Wavelength Selection

To find out specific wavelength bands by which all the edible oils can be discriminated or characterized, PCA is applied on the whole data matrix involving all samples of all the eight varieties of edible oil after outlier removal.

Our concern is to find the "true" underlying sources of data variation within samples of all the varieties. Loadings of PCA describe the data structure in terms of *variable contributions and correlations.* Every variable analyzed has a loading on each PC, which reflects how much the individual variable contributes to that PC, and how well the PC considers the variation contained in a variable [21]. The flow diagram of baseline correction, outlier removal, and wavelength selection is shown in Fig. 3.

3   Result and Discussions

The NIR spectra are first corrected using the using iterative restricted least square (IRLS) method and then subjected to outlier detection. PCA is applied to NIR spectra for oil samples individually for each class. The score plot of the mustard oil samples with Hotelling's T-square ellipse is shown in Fig. 3. Samples which falls under the 95% confidence interval in PC1vs PC2 score plot is selected for further analysis and rest are treated as outlier. The new data matrix consisting of remaining spectra is formulated by removing these outliers from the dataset.

For checking the classification ability of NIR spectra for classification of eight different types of edible oils, exploratory chemometric method Principal Component Analysis (PCA) was used. PCA was applied to baseline corrected new dataset created after outlier removal and score plots were examined to assess the classification ability of the spectra. 3D score plot is shown in Fig. 4. In the 3D score plot we can observe that different oils sample falls under different clusters in the plot which shows that the sample spectra have the capability to classify the oil samples.

3.1   Wavelength Selection

Loadings of the first three principal components have been shown in Fig. 5. Based on the variations in loading plots which attributes to the contribution of each variable in the model, five wavelength regions (R1:1150 nm–1245 nm; R2:1380 nm–1480 nm; R3:1640 nm–1820 nm; R4: 2110 nm–2230 nm; R5: 2250 nm–2400 nm) was selected.

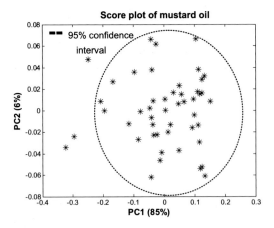

Fig. 2. Score plot of NIR spectra of mustard oil sample with Hotelling's T-square ellipse in 95% confidence interval.

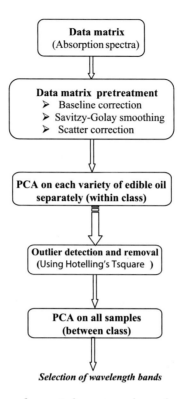

Fig. 3. Flow diagram for spectral pre-processing and wavelength selection

3.2   Classification Results

For developing the classification models, the complete dataset is treated through four types of pretreatment methods namely S-G smoothing, SNV, MSC and EMSC, and four new datasets were created. Each dataset was then divided into five wavelength regions. For each wavelength region data set is divided into a training dataset and test dataset using Kennard Stone sampling algorithm [22]. Using the training data set for each wavelength region and pretreatment method three classification models (PCA-DA, PLS-DA, k-NN) has been developed using the classification toolbox v5.1 [23]. An optimal number of factors for PCA-DA and PLS-DA models has been chosen by cross-validating the models by samples selected using venetian blind sampling method. Each model has been validated using the external samples and error rate for each model has been calculated using the Eqs. 1 and 2.

$$Error\ rate = \frac{sum(true\ positive\ rate)}{total\ number\ of\ samples} \tag{1}$$

$$True\ positive\ rate = \frac{true\ positive}{true\ positive + false\ negative} \tag{2}$$

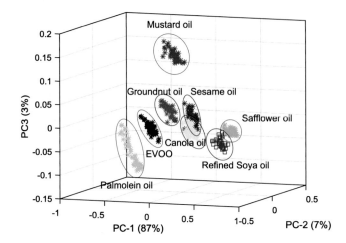

Fig. 4. 3D scatter plot of PC1, PC2, and PC3 by PCA analysis on the full spectrum of oil samples

The error rate for each of the classification models is shown in Table 1. It can be noted that the all the developed classification models (PCA-DA, PLS-DA, and k-NN) with different pretreatment methods gives zero error rate in the selected wavelength region R4 (2110 nm–2230 nm). Also, the PCA-DA model developed using MSC and EMSC corrected dataset of wavelength region R3 and R5 gives the zero-error rate. It is also shown that the region R3 (1640 nm–1820 nm) and R5 (2250 nm–2400 nm) also shown very less error rate in classification results.

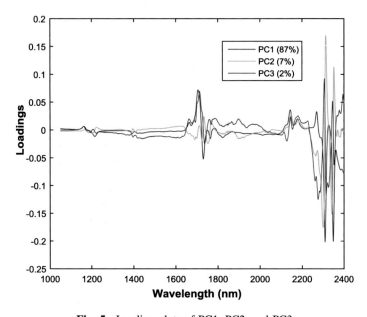

Fig. 5. Loading plots of PC1, PC2, and PC3

Table 1. The error rate of classification models for all wavelength regions using external validation samples.

Wavelength region	Classifier	Error rate			
		S-G smoothing	SNV	MSC	EMSC
R1	PCA-DA	0.018	0	0.025	0
	PLS-DA	0.025	0.02	0.18	0.24
	kNN	0.016	0	0.01	0.03
R2	PCA-DA	0.07	0	0.07	0.01
	PLS-DA	0	0.11	0.12	0.12
	kNN	0.02	0.10	0.05	0.07
R3	PCA-DA	0	0.07	0	0
	PLS-DA	0	0	0	0
	kNN	0.07	0.01	0.07	0
R4	PCA-DA	0	0	0	0
	PLS-DA	0	0	0	0
	kNN	0	0	0	0
R5	PCA-DA	0	0	0	0
	PLS-DA	0.01	0.03	0	0
	kNN	0	0.07	0	0.07

4   Conclusion

This paper summarizes the use of NIR spectroscopy in combination with chemometric techniques to positively characterize and classify the eight different varieties of edible oil and the detection of specific wavelength bands responsible for the classification of the samples. The use of NIR absorption spectra of the samples of eight varieties of edible oil for their classification and characterization has been achieved. Various data treatments and analysis methods are applied to the spectra of the samples to discriminate the samples. The results show that the wavelength centric approach in the classification and characterization of the edible oils. Selected wavelength regions R3, R4, and R5 shows great classification ability using the developed classification models. Models developed using the wavelength region R4 shows zero error rate in all classification models. This analysis leads to a chance for developing a system based on specific wavelength bands sources and detectors for classifying and characterizing the edible oils.

References

1. Edible oilseeds supply and demand scenario in India: Implications for policy. http://www. iari.res.in/files/Edible_Oilseeds_Supply_and_Demand_Scenario_in_India.pdf. Accessed 08 Nov 2018
2. Galtier, O., Dupuy, N., Dreau, Y.L., Ollivier, D., Pinatel, C., Kister, J., Artaud, J.: Geographic origins and compositions of virgin olive oils determinated by chemometric analysis of NIR spectra. Anal. Chim. Acta **595**, 136–144 (2007)
3. Sato, T.: Application of principal-component analysis on near-infrared spectroscopic data of vegetable oils for their classification. J. Am. Oil Chem. Soc. **71**, 293–298 (1994)

4. Giacomelli, L.M., Mattea, M., Ceballos, C.D.: Analysis and characterization of edible oils by chemometric methods. J. Am. Oil Chem. Soc. **83**, 303–308 (2006)
5. Kongbonga, Y.G.M., Ghalila, H., Onana, M.B., Majdi, Y., Lakhdar, Z.B., Mezlini, H., Sevestre-Ghalila, S.: Characterization of vegetable oils by fluorescence spectroscopy. Food Nutr. Sci. **2**, 692–699 (2011)
6. Zhang, L., Li, P., Sun, X., Wang, X., Xu, B., Wang, X., Ma, F., Zhang, Q., Ding, X.: Classification and adulteration detection of vegetable oils based on fatty acid profiles. J. Agric. Food Chem. **62**, 8745–8751 (2014)
7. Ruiz-Samblas, C., Cadenas, J.M., Pelta, D.A., Cuadros-Rodriguez, L.: Application of data mining methods for classification and prediction of olive oil blends with other vegetable oils. Anal. Bioanal. Chem. **406**, 2591–2601 (2014)
8. Luna, A.S., DaSilva, A.P., Ferre, J., Boque, R.: Classification of edible oils and modeling of their physico-chemical properties by chemometric methods using mid-IR spectroscopy. Spectrochim. Acta Part A Mol. Biomol. Spectrosc. **100**, 109–114 (2013)
9. Luca, M.D., Terouzi, W., Ioele, G., Kzaiber, F., Oussama, A., Oliverio, F., Tauler, R., Ragno, G.: Derivative FTIR spectroscopy for cluster analysis and classification of morocco olive oils. Food Chem. **124**, 1113–1118 (2011)
10. Blanco, M., Villarroya, I.: NIR spectroscopy: a rapid-response analytical tool. Trends Anal. Chem. **21**, 240–250 (2002)
11. Siesler, H.W., Ozaki, Y., Kawata, S., Heise, H.M.: Near infrared spectroscopy: Principles, Instruments and Applications, 1st edn. Wiley-CH, Dortmund (2002)
12. Pereira, A.F.C., Pontes, M.J.C., Neto, F.F.G., Santos, S.R.B., Galvao, R.K.H., Araujo, M.C. U.: NIR spectrometric determination of quality parameters in vegetable oils using iPLS and variable selection. Food Res. Int. **41**, 341–348 (2008)
13. Armenta, S., Garrigues, S., De La Guardia, M.: Determination of edible oil parameters by near infrared spectrometry. Anal. Chim. Acta **596**, 330–337 (2007)
14. Chirsty, A., Kasemsumran, S., Du, Y., Ozaki, Y.: The detection and quantification of adulteration in olive oil by near-infrared spectroscopy and chemometrics. Anal. Sci. **20**, 935–940 (2004)
15. Basri, K.N., Hussain, M.N., Bakar, J., Sharif, Z., Khir, M.F.A., Zoolfakar, A.S.: Classification and quantification of palm oil adulteration via portable NIR spectroscopy. Spectrochim. Acta Part A Mol. Biomol. Spectrosc. **173**, 335–342 (2017)
16. Song, Z., Tu, B., Zeng, L., Yin, C., Zhang, H., Zheng, X., He, D., Qi, P.: Classification of edible vegetable oils by laser near-infrared spectra. In. IEEE International Conference on Progress in Informatics and Computing, pp. 123–127. IEEE, Shanghai (2014)
17. Saha, S., Saha, S.: Decision template fusion for classifying Indian edible oils using singular vector decomposition on NIR spectrometry data. In. 2nd International Conference on Research in Computational Intelligence and Communication Networks, pp. 229–234. IEEE, Kolkata (2016)
18. Liland, K.H., Mevik, B.H.: Baseline: baseline correction of spectra. R pacakage v1.1-4. http://CRAN.R-project.org/package=baseline
19. Shafer, R.W.: What is a Savitzky-Golay Filter? IEEE Signal Process. Mag. **28**, 111–117 (2011)
20. Rinnan, A., Berg, F.V.D., Englsen, S.B.: Review of most common preprocessing techniques for near infra-red spectra. Trends Anal. Chem. **28**(10), 1201–1222 (2009)
21. Bro, R., Smilde, A.K.: Principal component analysis. Anal. Methods **6**, 2812–2831 (2014)
22. Wu, W., Walczak, B., Massart, D.L., Heuerding, S., Erni, F., Last, I.R., Prebble, K.A.: Artificial neural network in classification of NIR spectral data: design of training set. Chemom. Intell. Lab. Syst. **33**(1), 35–46 (1996)
23. Ballabio, D., Consonni, V.: Classification tools in chemistry. Part1: linear models. PLS-DA. Anal. Methods **5**, 3790–3798 (2013)

Design and Application of Controller Based on Sine-Cosine Algorithm for Load Frequency Control of Power System

Saswati Mishra, Shubhrata Gupta, and Anamika Yadav[✉]

Department of Electrical Engineering, National Institute of Technology Raipur,
Raipur, India
ayadav.ele@nitrr.ac.in

Abstract. Load frequency control (LFC) has emerged as one of the potential research areas in the field of power system. LFC is a mechanism by which the system frequency is maintained within allowable limits by maintaining equilibrium between generation and load. In this study, design and application of controller based on sine-cosine algorithm (SCA) is utilized for LFC of interconnected power system. A proportional-integral-derivative (PID) controller with a filter consisting of derivative term is used and its parameters are tuned using SCA. The performance criterion chosen for tuning process is the minimization of integral error of variations in frequency and tie-line power. To examine the efficacy of SCA-PIDN controller, its performance is compared with other controllers reported in literature. Further, time-domain simulations are illustrated to support the obtained results. Additionally, the robustness of SCA-PIDN controller is examined with random load perturbations. The outcomes of the test cases reveal the superiority of SCA-PIDN controllers over others.

Keywords: Area control error · Integral error · Load frequency control ·
PID controller · Sine-cosine algorithm

1 Introduction

In last few decades, the increasing power demand has resulted in interconnections of existing power networks [1]. These interconnections have increased the complexities in secure and reliable operation of power system. The prime aim of electric utility is to preserve constant frequency by maintaining equilibrium between generation and load [2]. However, with sudden change in load demand, the system frequency varies. Therefore, it is essentially required to bring the system frequency within specified limits. The mechanism by which this objective is achieved is often regarded as load frequency control (LFC).

In general, LFC utilizes specific type of controllers to achieve the target of maintaining system frequency within targeted limits. Different controllers used for this purpose are reported in literature. These controllers can be broadly classified into two types, namely, conventional and modern controllers. The conventional controllers are proportional (P), proportional-derivative (PD), proportional-integral (PI), and proportional-integral-derivative (PID) controller [3]. Some modern controllers like fuzzy controllers

© Springer Nature Switzerland AG 2020
A. Abraham et al. (Eds.): ISDA 2018, AISC 941, pp. 301–311, 2020.
https://doi.org/10.1007/978-3-030-16660-1_30

[4], ANFIS controller [5] etc. are also reported and found to be more effective than conventional controllers. However, modern controllers are structurally complex than conventional one and their proper design depends upon accurate knowledge of system model. On the contrary, the conventional controllers are structurally simple and require less computational effort for their design. The associated advantages of conventional controllers made them more popular among the research community.

The appropriate design of conventional controller generally depends upon two factors, namely, the proper choice of design objectives and the technique for parameter tuning. Several design objectives (DOs) are reported in literature regarding controller design for LFC. The prime DOs considered are the minimization of area control error (ACE), integral error, settling times (STs), peak overshoots (POs) and maximization of damping ratios. Out of these DOs, the minimization errors is most preferred performance criteria. The integral-time-absolute-error (ITAE) is one of the types of such error and is widely used in literature [6]. The other factor which directs the proper design of controller is the choice of technique for parameter tuning. Several techniques have been reported in literature and can be classified in two categories: conventional and modern techniques. Conventional techniques like Zeigler-Nichols [7] and others, have their own limitations and disadvantages such as slow convergence and local minima trappings. These disadvantages were omitted by the introduction of modern meta-heuristic optimization techniques.

Modern techniques are utilized to a greater extent in solving LFC problem. Some of the relevant techniques are genetic algorithm (GA) [8], particle swarm optimization (PSO) [9], grey wolf optimization (GWO) [10], cuckoo search algorithm (CSA) [11], teacher-learner based optimization (TLBO) [12], ant-lion optimization (ALO) [13], Jaya algorithm (JA) [6], bat algorithm (BA) [14], bacteria foraging algorithm (BFA) [15], non-dominated sorting genetic algorithm II (NSGA-II) [16], differential evolution (DE) [17], flower pollination algorithm (FPA) [18], whale optimization algorithm (WOA) [19], salp swarm optimization (SSO) [20] etc. The mentioned techniques are found to be effective in solving LFC problem. However, there is further scope of utilization and exploitation of recent modern techniques in solving LFC problem.

In the present work, a recently proposed sine-cosine algorithm (SCA) is utilized to design controllers for LFC of interconnected system. SCA is modern optimization technique which does not contain any algorithm-specific parameters to be tuned for specific problem. A conventional PID controller with a derivative term (PIDN) is tuned by SCA for the problem. The minimization of sum of ITAEs of frequency deviations (FDs) of both areas along with tie-line power deviations (TPDs) is chosen as design objective. To examine the performance of SCA based controller, varied test conditions are carried out. A comparative analysis is tabulated to validate superior performance of SCA-PIDN controller against other controllers reported in literature for solving LFC problem. Time-domain simulations (TDS) are illustrated to enhance the outcomes from the comparative results. Robustness of the controller is examined by exposing the system to random load disturbances (RLDs). The major highlights of the paper are outlined below.

- LFC of two-area interconnected power system is carried out by tuning parameters of PIDN controller.
- SCA is utilized to tune the parameters of considered controller.
- Minimization of ITAEs of deviations in frequencies and tie-line power is taken as the objective function.
- The performance SCA based controller is compared to other controllers reported in literation to validate its superiority.
- Robustness of proposed controller is tested with random RLDs.

Rest of the paper is organized as follows. The system and controller model used in this study is discussed in Sect. 2. The problem description is presented in Sect. 3. The basics of SCA and its implementation to the problem is described in Sect. 4. In Sect. 5, the study outcomes and respective discussion are presented. The conclusion is drawn in Sect. 6.

2 Model Description

In this section, the system model and the controller under study is described. A two-area non-reheat thermal-thermal interconnected power system (TNTTIPS) is used for simulation study. The Simulink model of the studied system is shown in Fig. 1. The nominal parameters of the studied system is reported in [15].

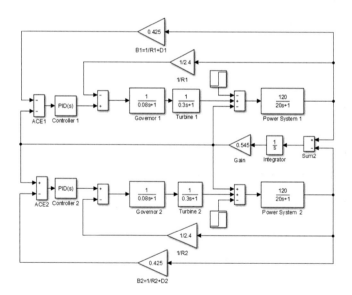

Fig. 1. Simulink model of the studied system for LFC.

A PID controller with a derivative filter (PIDN) is used in this study. Figure 2 shows the block diagram of PIDN controller. Area control error (ACE) of each area serves as input to the controller. ACEs are linear relationship of FD and TPD of the area. ACEs of the respective area are represented as:

$$ACE_1 = B_1 * \Delta f_1 + \Delta P_{tie} \tag{1}$$

$$ACE_2 = B_2 * \Delta f_2 + a_{12} * \Delta P_{tie} \tag{2}$$

where ACE is the ACE; B is the bias-factor used for frequency; Δf is the FD; ΔP_{tie} is TPD; a_{12} is a constant equal to -1 and subscripts 1 and 2 is used for representing the two areas, respectively.

The transfer function PIDN controller is given as

$$TF_{PIDn} = K_p + K_i\left(\frac{1}{s}\right) + K_d\left(\frac{N}{1 + \frac{N}{s}}\right) \tag{3}$$

where K_p, K_i and K_d are different gains and N is the coefficient of the derivative filter to reduce noise.

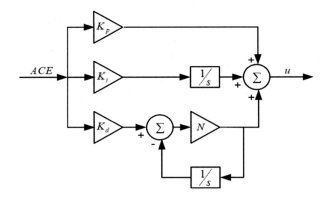

Fig. 2. Block diagram representation of controller.

3   Problem Description

This section formulates DO for LFC problem. The DO considered is the minimization of the sum of ITAE of the Δf_1, Δf_2 and ΔP_{tie}. The corresponding ITAEs are represented as

$$\Delta f_1^{ITAE} = \int_0^{t_{sim}} (|\Delta f_1|) \cdot t \cdot dt$$

$$\Delta f_2^{ITAE} = \int_0^{t_{sim}} (|\Delta f_2|) \cdot t \cdot dt \tag{4}$$

$$\Delta P_{tie}^{ITAE} = \int_0^{t_{sim}} (|\Delta P_{tie}|) \cdot t \cdot dt$$

where t_{sim} is termed as total simulation time. Therefore, the DO, J, is expressed as follows:

$$J = \Delta f_1^{ITAE} + \Delta f_2^{ITAE} + \Delta P_{tie}^{ITAE} \tag{5}$$

The DO, J, is minimized by SCA satisfying different system constraints defined as:

$$K_p \in \left\{ K_p^{min}, K_p^{max} \right\}; \quad K_i \in \left\{ K_i^{min}, K_i^{max} \right\} \tag{6}$$
$$K_d \in \left\{ K_d^{min}, K_d^{max} \right\}; \quad N \in \left\{ N^{min}, N^{max} \right\}$$

4   Sine-Cosine Algorithm

4.1   Sine-Cosine Algorithm (SCA)

This section presents the basic description of SCA with its implementation to LFC problem. SCA is developed to solve mathematical optimization problems [21]. Unlike many recent algorithms, SCA does not possess any algorithm-specific parameters to be tuned in accordance with the type of problem. Similar to other modern optimization algorithms, SCA starts with random generation of initial population and then, updates the solution with knowledge of best solution based on sine or cosine model. Let, there are m number of solutions with n-dimensions. The number of decision variables are represented by n. Thus, the population size becomes $(m \times n)$ where each row is a solution and each column represent a decision variable. The ith solution is represented as $X_k = \left(x_{k,1}, x_{k,2}, \cdots, x_{k,n} \right)$ where $k = 1, 2, \cdots, m$. The mathematical model by which each solution in the population is updated in SCA is represented as:

$$X_{k,j}^{r+1} = \begin{cases} X_{k,j}^r + \phi_1 \sin(\phi_2) \left| \phi_3 \cdot X_{best,j}^r - X_{k,j}^r \right| & \text{if } \phi_4 < 0.5 \\ X_{k,j}^r + \phi_1 \cos(\phi_2) \left| \phi_3 \cdot X_{best,j}^r - X_{k,j}^r \right| & \text{if } \phi_4 \geq 0.5 \end{cases} \tag{7}$$

$$\mu_1 = b - r(b/R) \tag{8}$$

where r is the current iteration, $X_{k,j}^{r+1}$ is the updated value of jth decision variable of kth solution at rth iteration i.e. $X_{k,j}^r$ and $X_{best,j}^r$ is the jth decision variable of best solution at rth iteration. The control parameter ϕ_1 maintains the exploration versus exploitation of the search space region by dictating the next solution's position and decreases linearly from a constant value b to 0 with each iteration following Eq. (8). R is the maximum number of iterations. The ϕ_2, ϕ_3 and ϕ_4 are random numbers where $\phi_1, \phi_2, \phi_3 \in \{0, 1\}$. The pseudo code of SCA is provided in Algorithm 1.

```
Algorithm 1
begin SCA Main
        Random initialization of each solution.
        X_k   (k = 1, 2, ···, m);
        Read total number of iterations R;
        while r < R do for each iteration
                for each solution X_k do
                        Evaluate objective function f(X_k)
                        if f(X_k) < f(X_best) then
                        X_best = X_i
                        end
                end
                Update φ_1.
                Generate randomly new values for φ_2, φ_3 and φ_4.
                for each solution X_k do
                        Update X_k.
                end
        end
        Print X_best.
end
```

4.2   Implementation to LFC Problem

The LFC problem formulated in (5) is minimized using SCA algorithm subject to design constraints

$$0 \leq K_p \leq 2; \quad 0 \leq K_i \leq 2$$
$$0 \leq K_d \leq 2; \quad 100 \leq N \leq 500 \tag{9}$$

The initial population of the design variables is randomly generated in the range given in (9). Steps for implementation of SCA to LFC problem are as follows

Step 1: Define population size i.e. $(m \times n)$ and maximum number of iterations R. Initialize random population X.
Step 2: Evaluate objective function defined in (5) and identify best solution X_{best}.
Step 3: Update each solution of X using (7).
Step 4: Select better among old and updated solutions.
Step 5: Update ϕ_1 using (8).
Step 6: Jump to step 2 and continue till any termination criterion gets satisfied.

5 Case Study and Discussion

The work carried out here shows controller tuning for LFC of TNTTIPS. The parameters of the studied system are taken from [15]. A PIDN controller is utilized to meet the objectives of LFC. It is very obvious in any interconnected system that any change in one area is going to affect other area too. Using SCA, the parameter tuning of the controller for the problem is done. An increased constant load disturbance (CLD) of 10% in area 1 at $t = 0$ s is initiated and area 2 is kept unchanged for simulation and testing capabilities of SCA-PIDN controller. A comparative assessment of SCA-PIDN controller is carried out with controllers reported in literature like GA-PI controller [15], BFOA-PI controller [15], NSGA-II PI controller [16] and NSGA-II PIDN controller [16]. The TDSs are demonstrated to further validate the superior capability of SCA-PIDN controller. A total of 100 iterations and 30 population size is considered for SCA algorithm. MATLAB platform is used for execution of all simulations and implementation of algorithm.

Table 1 lists the obtained results from the simulation carried out under afore mentioned test scenario. The tuned controller parameters are tabulated. From the table, it is found the minimum value of DO, J, i.e. ITAE, equal to **0.0516**, is obtained from SCA based PIDN controller which is better than other controllers. Additionally, the STs of FDs of both areas with TPD are found to be **2.6162** s, **3.5916** s, and **3.4636** s, respectively with the SCA-PIDN controller which is minimum in comparison to other controllers. The TDSs of frequency deviations and TPD are shown in Figs. 3, 4 and 5. The figure illustrations agree with the tabulated results of Table 1. The figures suggest the frequency of area 2 and power flowing through tie-line is getting affected by load disturbance in area 1 which is as per the expectations. From keen observation of the figure, it is easy to identify the SCA based PIDN is better performer in terms of STs than other controllers. From the above discussion, the conclusion can be drawn that SCA based PIDN controller are outperforming other controllers in solving LFC problem.

To further examine the robust behavior of SAC-PIDN, the system is subjected to RLD in the range 0–10% in area 1. The TDS for the system under mentioned scenario is illustrated in Fig. 6. In the figure, the FDs of both areas along with TPD subjected to RLD is shown. Since, area 1 is experiencing RLD but, area 2 is also getting affected by the same. The TPD is under the same disturbance as of FDs. From the figure, it can clearly be seen that the controller is acting very fast in achieving objective of LFC by compelling the FDs to zero values whenever there is some disturbance. The same observations can be seen for the case of TPD. From the above discussion, it can be deduced that SCA-PIDN controller is robust in its performance as it is performing better in case of RLD.

The overall discussion obtained from different test scenarios suggests that SCA-PIDN controller is better performer for LFC problem and is immune to the type of load disturbances.

Table 1. Comparative results for studied case

Controllers	ITAE	Controller parameters				Settling times (s)		
	J	K_d	K_i	K_d	F	Δf_1	Δf_2	ΔP_{tie}
GA-PI [15]	2.7160	−0.2346	0.2662	–	–	8.5119	9.6128	9.6431
BFOA-PI [15]	1.8376	−0.4207	0.2795	–	–	4.1589	6.5579	6.3667
NSGA-II-PI [16]	1.6764	−0.4280	0.2967	–	–	4.5141	6.3509	6.0929
SCA-PI	1.1763	−0.3126	0.4629	–	–	5.2670	6.1586	6.6358
NSGA-II-PIDN [16]	0.6012	0.4262	0.8063	0.6146	384.58	3.6878	5.1291	5.4009
SCA-PIDN	**0.0516**	1.0847	2.0000	0.3849	405.75	**2.6162**	**3.5916**	**3.4636**

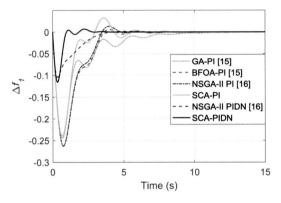

Fig. 3. Frequency deviation of area 1 under studied case.

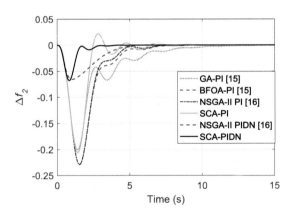

Fig. 4. Frequency deviation of area 2 under studied case.

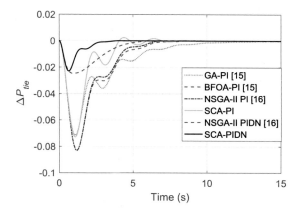

Fig. 5. Tie-line power deviation under studied case.

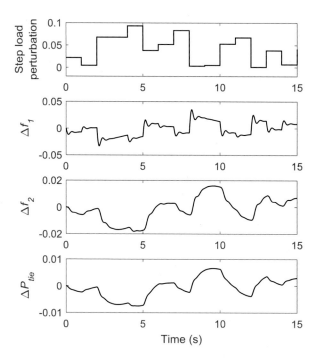

Fig. 6. Time domain simulations under RLD.

6 Conclusion

In this paper, design and application of controller based on sine-cosine algorithm for load frequency control of power system is presented. SCA is applied to obtain the tuned parameters of PIDN controller for the problem. A single DO is formed to minimize ITAE of the STs of the FDs and TPD. To examine the capability of SCA-

PIDN controller, a test scenario of a constant SLP is simulated. A comparative assessment of the results obtained for the system with SCA-PIDN controller is depicted with other controllers reported in literature. The simulation results suggest that SCA-PIDN controller is outperforming other controllers. To further validate the superiority of the SCA-PIDN controller, TDSs of FD of each areas and TPD are illustrated. From the simulations, the superior performance of SCA based controller can be established. The future scope of this work lies in extending it to a large system with inclusion of non-linearity in the system like governor dead-band and other non-linearity.

References

1. Wood, A.J., Wollenberg, B.F.: Power Generation, Operation, and Control. Wiley, Hoboken (2012)
2. Kundur, P., Balu, N.J., Lauby, M.G.: Power System Stability and Control. McGraw-Hill, New York (1994)
3. Shayeghi, H., Shayanfar, H., Jalili, A.: Load frequency control strategies: a state-of-the-art survey for the researcher. Energy Convers. Manag. **50**(2), 344–353 (2009)
4. Çam, E., Kocaarslan, I.: Load frequency control in two area power systems using fuzzy logic controller. Energy Convers. Manag. **46**(2), 233–243 (2005)
5. Khuntia, S.R., Panda, S.: Simulation study for automatic generation control of a multi-area power system by ANFIS approach. Appl. Soft Comput. **12**(1), 333–341 (2012)
6. Singh, S.P., Prakash, T., Singh, V., Babu, M.G.: Analytic hierarchy process based automatic generation control of multi-area interconnected power system using Jaya algorithm. Eng. Appl. Artif. Intell. **60**, 35–44 (2017)
7. Mallesham, G., Mishra, S., Jha, A.: Ziegler-Nichols based controller parameters tuning for load frequency control in a microgrid. In: 2011 International Conference on Energy, Automation, and Signal (ICEAS), pp. 1–8. IEEE (2011)
8. Abdel-Magid, Y., Dawoud, M.: Genetic algorithms applications in load frequency control (1995)
9. Ghoshal, S.P.: Optimizations of PID gains by particle swarm optimizations in fuzzy based automatic generation control. Electr. Power Syst. Res. **72**(3), 203–212 (2004)
10. Guha, D., Roy, P.K., Banerjee, S.: Load frequency control of interconnected power system using grey wolf optimization. Swarm Evol. Comput. **27**, 97–115 (2016)
11. Abdelaziz, A., Ali, E.: Cuckoo search algorithm based load frequency controller design for nonlinear interconnected power system. Int. J. Electr. Power Energy Syst. **73**, 632–643 (2015)
12. Barisal, A.: Comparative performance analysis of teaching learning based optimization for automatic load frequency control of multi-source power systems. Int. J. Electr. Power Energy Syst. **66**, 67–77 (2015)
13. Raju, M., Saikia, L.C., Sinha, N.: Automatic generation control of a multi-area system using ant lion optimizer algorithm based PID plus second order derivative controller. Int. J. Electr. Power Energy Syst. **80**, 52–63 (2016)
14. Abd-Elazim, S., Ali, E.: Load frequency controller design via BAT algorithm for nonlinear interconnected power system. Int. J. Electr. Power Energy Syst. **77**, 166–177 (2016)
15. Ali, E., Abd-Elazim, S.: Bacteria foraging optimization algorithm based load frequency controller for interconnected power system. Int. J. Electr. Power Energy Syst. **33**(3), 633–638 (2011)

16. Panda, S., Yegireddy, N.K.: Automatic generation control of multi-area power system using multi-objective non-dominated sorting genetic algorithm-II. Int. J. Electr. Power Energy Syst. **53**, 54–63 (2013)

17. Mohanty, B., Panda, S., Hota, P.: Controller parameters tuning of differential evolution algorithm and its application to load frequency control of multi-source power system. Int. J. Electr. Power Energy Syst. **54**, 77–85 (2014)

18. Jagatheesan, K., et al.: Application of flower pollination algorithm in load frequency control of multi-area interconnected power system with nonlinearity. Neural Comput. Appl. **28**(1), 475–488 (2017)

19. Simhadri, K.S., Mohanty, B., Panda, S.K.: Comparative performance analysis of 2DOF state feedback controller for automatic generation control using whale optimization algorithm. Optim. Control. Appl. Methods

20. Guha, D., Roy, P., Banerjee, S.: A maiden application of salp swarm algorithm optimized cascade tilt-integral-derivative controller for load frequency control of power systems. IET Gener. Transm. Distrib. (2018)

21. Mirjalili, S.: SCA: a sine cosine algorithm for solving optimization problems. Knowl.-Based Syst. **96**, 120–133 (2016)

A Perusal Analysis on Hybrid Spectrum Handoff Schemes in Cognitive Radio Networks

J. Josephine Dhivya[✉] and M. Ramaswami

Department of Computer Applications, Madurai Kamaraj University,
Madurai, India
josedhivya@gmail.com

Abstract. Spectrum handoff management is an important issue which needs to be addressed in Cognitive Radio Networks (CRN) for interminable connectivity and productive usage of unallocated spectrum for the unlicensed users. Spectrum handoff which comes under the phase of Spectrum mobility in CRN plays a vital role in ensuring seamless connectivity which is quite exigent. Handoff process in general comes under active and proactive types. The intelligent and hybrid handoff methods which combines both these types based on the network conditions proves to be quite satisfactory in the recent works. This paper proposes a hybrid novel method for handling the handoff mechanism based on Fuzzy rough set theory (FRST) with Support Vector machine (SVM), which enables the decision making stage of the handoff process more tenable and productive. The implied method predicts the node wherein handoff is to be initiated in the lead through which the handoff delay time and number of handoffs are minimized. The experimental results are compared with the previously proposed hybrid schemes including Fuzzy genetic algorithm (FGA) based handoff, FGA with cuckoo search (CS) optimization technique, FRS with CS and the findings portray the suggested methodology attains better prediction mechanism with minimal handoffs.

Keywords: Spectrum handoff · Fuzzy rough set · Support Vector Machine

1 Introduction

The prototype behind wireless communications is opportunistic spectrum access, based on which Cognitive Radio Networks (CRN) are deployed. CRN enables secondary user (SU) to make use of unused spectrum, however upon the arrival of the primary user (PU) having the privilege of accessing the spectrum the SU needs to vacate the spectrum, resulting in spectrum handoff [1]. This issue is more prevalent in CRN and hence need to be investigated to ensure seamless connectivity. The SU enters into handoff state based on decision which is inferred by sensing the spectrum [2]. This decision making process should be highly influential to promote the effectiveness of the system. Various hybrid handoff schemes are proposed to enhance the decision process including proactive and reactive strategies [3]. This work illustrates a hybrid fuzzy rough set based strategy with SVM for better optimization. The main aim of incorporating rough set theory is the wide usage of the network parameters which provides

© Springer Nature Switzerland AG 2020
A. Abraham et al. (Eds.): ISDA 2018, AISC 941, pp. 312–321, 2020.
https://doi.org/10.1007/978-3-030-16660-1_31

enormous information for predicting the spectrum sensing phase in CRN. The handoff decision is then made based on the SVM method. At the end of the paper comparison of various hybrid handoff schemes are depicted thereby indicating the effectiveness of the proposed method.

2 Related Work

The process of spectrum handoff decision making in CRN is a challenging task and the concept of hybrid schemes in determining the decision process is fairly explored in the literature. Various works are focused either on active or proactive handoff modes as such. These were analyzed extensively to monitor the pros and cons of the model. At the outset it was standardized that the handoff strategy should make use of the two handoff modes and switch over to each case based on the situations. This resulted in the evolution of hybrid, intelligent models which deploy both handoff modes. A comprehensive classification of various handoff schemes were presented in [4]. The authors in [5] suggested an adaptive or hybrid handoff mechanism is mandatory for an efficient handoff process. An intelligent algorithm was proposed in [6] which compared the handoff process which deployed models such as Analytical hierarchical processing and Fuzzy logic. A hybrid channel selection and spectrum decision function which determines the occurrence of the handoff process was proposed and handoff decision was made based on the probability of PU on a certain channel incorporating Fuzzy logic [7]. The authors in [8] proposed a vertical handoff decision algorithm based on fuzzy systems and RST. An adaptive handoff algorithm based on PU prioritized markov process which captured the interactions between the primary and secondary users was presented in [9]. Our previous work include the spectrum handoff process based on fuzzy genetic with cuckoo search algorithm for optimization [10].

The intelligent hybrid decision making process for spectrum handoff based on FRST with SVM in the proposed method is clearly distinct from the already existing works and shows an overall increase in performance in comparison to the defined methods in the literature.

3 Proposed System

The proposed work for spectrum handoff decision making based on Fuzzy Rough set theory (FRST) enables the handoff process to be conducive in CRN. In this method the network parameters such as mobility, received signal strength (RSS), distance, speed, and bandwidth are fed as inputs to the hybrid FRST model and the CRN is analyzed to determine the handoff decision pattern ahead for enhanced data transmission. This method enables the SU to switch to another channel which is unoccupied in advance thereby minimizing link failures. This hybrid handoff strategy ensures decisions taken to be predictive and accurate because of taking into account all the possible outcomes even though the knowledge about the environment is uncertain. The flow of proposed work is divided into two sections first part comprises of the preprocessing data and training data by FRST.

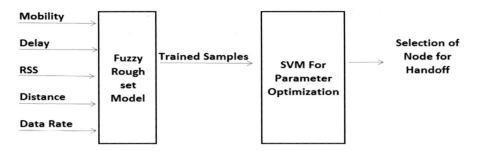

Fig. 1. Flow of the proposed hybrid FRST with SVM handoff strategy

The second part includes the optimization of parameters by SVM based on which node is selected for handoff. Figure 1 illustrates the flow of the proposed hybrid FRST with SVM handoff strategy.

3.1 Fuzzy Rough Set Model

The main objective of using this model is to predict the neighboring nodes that are idle so as to make the spectrum handoff decision. In general a fuzzy rough set is obtained from the approximation of a fuzzy set wherein the values of conditional attribute are crisp and decision attribute values are fuzzy and these sets define the lower and upper approximation [11]. This model is deployed to compute reducts which is mandatory for attribute reduction. The decision rules are generated based on classifying and clustering the nodes. It is essential for the SU to occupy the opportunistic spectrum based on the availability of the channel which is uncertain and hence the FRS model is incorporated. The input parameters that are fed initially into FRS model includes Mobility which refers to the movement of the nodes, Data rate represents the frequency range over which the signal is transmitted, Distance which represents the distance from the node to the base station, Delay refers to the transmission delay, RSS refers to the received signal strength which is the measurement of power present in the radio signal.

The main role of rough set model is the construction of the indiscernability relations which is not possible in fuzzy systems because of complexity in analysis. In this FRS based model there are five conditional attribute and one decision attribute. The indis-cernability matrix is obtained according to the values of the attributes then reducts are computed [12]. The inputs are fuzzified using the membership functions. The membership functions are calculated based on trapezoidal membership functions and the values for Mobility ranges from low, medium and high the corresponding values obtained are 0.625, 0.166 and 0. For the distance variable the membership functions include far, medium, close and the range of values for these are 0.125, 0.50 and 0. In case of speed the membership functions includes fast, average and slow with values ranging between 0 and 1.

Table 1. Rules for handoff decision

Mobility	Distance	Delay	RSS	Data rate	Priority
Low	Far	Fast	Low	Low	Very High
Low	Far	Average	Low	Low	Very High
Low	Far	Slow	Low	Low	High
Low	Medium	Fast	Medium	Medium	Med High
Low	Medium	Average	Medium	Medium	Med High
Low	Medium	Slow	Medium	Medium	Med High
Low	Close	Fast	High	High	Med
Low	Close	Average	High	High	MedLow
Low	Close	Slow	High	High	MedLow
Medium	Far	Fast	Low	Low	Very High
Medium	Far	Average	Low	Low	High
Medium	Far	Slow	Low	Low	Med High
Medium	Medium	Fast	Medium	Medium	Med High
Medium	Medium	Average	Medium	Medium	Med
Medium	Medium	Slow	Medium	Medium	MedLow
Medium	Close	Fast	High	High	MedLow
Medium	Close	Average	High	High	Low
Medium	Close	Slow	High	High	VeryLow
High	Far	Fast	Low	Low	Med High
High	Far	Average	Low	Low	Med High
High	Far	Slow	Low	Low	Med
High	Medium	Fast	Medium	Medium	MedLow
High	Medium	Average	Medium	Medium	MedLow
High	Medium	Slow	Medium	Medium	MedLow
High	Close	Fast	High	High	Low
High	Close	Average	High	High	VeryLow
High	Close	Slow	High	High	VeryLow

Bandwidth and RSS parameters generate membership functions that range from 0.125.0.50, 0 for low, medium and high. The Fuzzy Inference system (FIS) converts the fuzzified input values to rules. The application of reducts in the RS enables the FIS to generate rules thereby achieving increased performance in computation. Table 1 shows the handoff rules generated on fuzzy rough based model and it is inferred that the possibility of occurrence of handoff is at three cases wherein the mobility, bandwidth, RSS ranges between low, medium and speed, distance seems to be fast, far. The rules generated from FIS shows inconsistencies in making decision hence it mandatory to classify the rules and there is a vital need for parameter optimization. For this reason and to show an increase the potent of the system the concept of Support Vector Machine (SVM) is incorporated.

3.2 Optimization by SVM

The underlying fact for posing SVM in this handoff process is for its smooth decision making. The process involves the training and the decision stage [13]. The training stage includes the construction of a hyper plane which separates the mobile nodes based on the availability of the channel for access. Then the parameter optimization is carried out for which kernel function selection is mandatory in this case the Radial basic function (RBF) type of kernel is selected the whole process falls under the SVM classifier [14]. In the decision stage optimization is achieved by specifying a class wherein the nodes are allocated to the channels and then identifying the node from which the handoff process has to occur.

4   Results and Analysis

The proposed work is compared to the various other hybrid handoff schemes which includes fuzzy genetic algorithm (FGA) handoff, Fuzzy based genetic algorithm (FBGA) with cuckoo search(CS) handoff, Fuzzy rough set with CS handoff and from the results it is evident that the proposed hybrid handoff scheme with SVM optimization yields better results. The network simulator (ns-2) tool was deployed to carry out the simulation with mobile nodes set to 20, number of channels set to 10 and simulation time as 200 ms. The handoff performance indicators were monitored based on the simulation carried out. It is based on these qualitative parameters the efficiency of the model is determined. The parameters that are taken into consideration include delay, number of handoffs occurred, number of failed handoffs and throughput.

(i) Delay: This parameter specifies the prolonged time period wherein data is transferred from the source node to destination node [15]. Figure 2 poses the analysis of the delay parameter with respective to time. The transmission time is varied and the delay graph shows a drastic decrease in FRS-SVM in comparison to the other various hybrid handoff schemes. FBGA with CS shows the next minimum value followed by FRS with CS and FGA. The increase in delay in these models is mainly due to the high competition of channel access. The proposed model shows minimum delay because the waiting time is minimized in this case and the secondary users are given quick access to the channels.

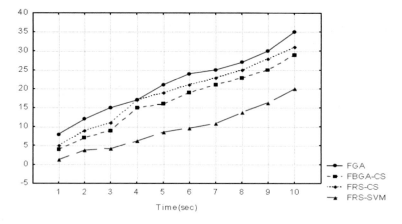

Fig. 2. Delay parameter plotted against time in x- axis.

(ii) Failed Handoffs: The number of unsuccessful transmission occurred in a particular state of a channel refers to the number of failed handoffs [15]. More the number of failed handoffs less is the efficiency of the system. Figure 3 depicts the occurrence of failed handoffs and it is clearly noticeable that the proposed FRS with SVM model has minimal number of failed handoffs. The FGA with CS scheme shows the next minimal states and then comes the FRS with CS and FGA. These models shows increase in failed handoffs due to inaccurate prediction and minimum availability of channels for SU to access. The number of failed handoffs is comparatively minimized in the proposed model because of the prediction of link failure ahead of range failure which results in accurate prediction in terms of handoff occurrence as well as channel availability.

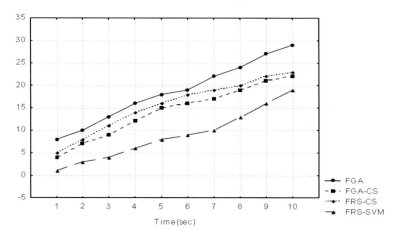

Fig. 3. Failed handoffs represents time plotted in x-axis against the number of failed handoffs plotted in the y-axis.

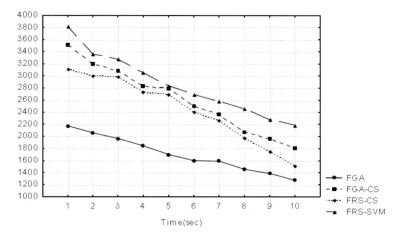

Fig. 4. Throughput represents time plotted in x-axis against the amount of data transferred plotted in the y-axis.

(iii) Throughput: This refers to the total amount of data transmitted for a given period from source node to destination node, data that is transferred must be carried out with minimal packet loss [16] and from Fig. 4 its clear that this value is minimum for FGA because it does not deploy any optimization technique the FRS with SVM strategy shows increase in throughput in comparison to FGA with CS and FRS with CS. The proposed intelligent method makes decision prior, so that collisions are avoided and minimal handoffs occur during the simulation in with increase in data transmission. Also the increase in opportunities for accessing the spectrum, availability of channels for user access promotes the overall throughput of the system.

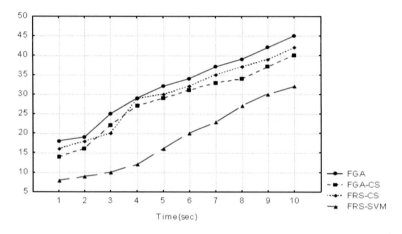

Fig. 5. Handoffs represents time plotted in x-axis against the total number of handoffs occurred plotted in the y-axis.

Table 2. Comparison of handoff schemes

Handoff parameters	Transmission time in sec	Handoff based on DBA	Handoff based on FGA	Handoff based on hybrid FBGA with CS	Handoff in FRS with CS	Handoff in FRS with SVM
Average delay in sec	1	12	8	4	5	1.21
	2	16	12	7	9	3.76
	3	19	15	9	11	4.11
	4	21	17	15	17	6.19
	5	23	21	16	19	8.54
	6	26	24	19	21	9.68
	7	28	25	21	23	10.75
	8	31	27	23	25	13.82
	9	32	30	25	28	16.35
	10	38	35	29	31	19.93

(continued)

Table 2. (*continued*)

Handoff parameters	Transmission time in sec	Handoff based on DBA	Handoff based on FGA	Handoff based on hybrid FBGA with CS	Handoff in FRS with CS	Handoff in FRS with SVM
Average throughput in kbps	1	1892	2174	3512	3112	3819
	2	1750	2058	3201	3001	3358
	3	1627	1964	3085	2985	3275
	4	1582	1846	2840	2740	3057
	5	1488	1702	2792	2692	2840
	6	1201	1600	2502	2402	2701
	7	1057	1592	2359	2259	2589
	8	937	1458	2075	1975	2461
	9	893	1387	1957	1757	2275
	10	742	1272	1810	1510	2187
Number of handoffs	1	24	18	14	16	8
	2	26	19	16	18	9
	3	31	25	22	20	10
	4	34	29	27	29	12
	5	38	32	29	30	16
	6	42	34	31	32	20
	7	44	37	33	35	23
	8	49	39	34	37	27
	9	52	42	37	39	30
	10	55	45	40	42	32
Number of failed handoffs	1	10	7	4	5	1
	2	13	10	7	8	3
	3	15	13	9	11	4
	4	19	16	12	14	6
	5	22	18	15	16	8
	6	26	19	16	18	9
	7	28	22	17	19	10
	8	31	24	19	20	13
	9	34	27	21	22	16
	10	37	29	22	23	19

(iv) Number of Handoffs: The number of change of state in the channels present for seamless data transmission in the nodes refers to the number of handoffs. In general this number must be minimal for unswerving connectivity during data transmission [1, 16]. The proposed FRS with SVM hybrid scheme depicts a significant minimal number of handoffs occurred in comparison to the other hybrid schemes. This minimal value is attained by the optimization of SVM which includes prediction and classification done ahead of decision making in FRS with SVM. Also from Fig. 5 it is clear that the proposed method intelligently predicts the occurrence of handoff depending on the predicted data delivery time of the primary user which minimizes the occurrences of handoff.

Table 2 poses the overall comparison of various handoff schemes analyzed in terms of the analyzed network parameters including delay, failed handoffs, throughput and total occurrences of handoff. The dynamic bandwidth allocation strategy fails due to the active handoff mode chosen and lack of addressing the handoff parameters, this method however works good for LTE networks but not optimum for CRN. The FGA scheme analysis shows it to be a poor handoff strategy due to the lack of optimization scheme. The FRS with CS method is better compared to FGA but not in case of FBGA with CS, FRS with SVM due to training of the input parameters and the deployment of cuckoo search as optimization technique which causes the time complexity issue. On the contradictory cuckoo search serves well for FBGA method because it incorporates the optimization of fitness value obtained by means of genetic algorithm. To encapsulate the FRS with SVM proves to be a reliable handoff scheme because of taking in to account the cons of the other scheme in designing this model.

5   Conclusion

The proposed method is a type of intelligent handoff scheme which combines the features of active and proactive types and makes ingenious efforts to shift to the respective modes based on the network performance indicators. This switching process is further made efficient by the concepts of FRS with SVM. The FRS technique provides a wide usage of attributes for making the handoff decision and SVM acts as an optimization method to ensure accurate prediction. From the overall comparison of the various hybrid handoff schemes and the qualitative analysis of the handoff performance indicators it is vividly implied that the hybrid FRS with SVM strategy reports to be an efficient and conducive handoff scheme which is one of the demanding need in the Spectrum Mobility phase of the Cognitive Radio Networks.

References

1. Kumar, K., Prakash, A., Tripathi, R.: Spectrum handoff scheme with multiple attributes decision making for optimal network selection in cognitive radio networks. Digit. Commun. Netw. 3(4), 164–175 (2017)
2. Bhatia, M., Kumar, K.: Network selection in cognitive radio enabled wireless body area networks. Digit. Commun. Netw. 1–11 (2018)
3. Ujarari, C.S., Kumar, A.: Handoff schemes and its performance analysis of priority within a particular channel in wireless systems. Int. J. Res. Appl. Sci. Eng. Technol. 3(5), 1021–1026 (2015)
4. Kumar, K., Prakash, A., Tripathi, R.: Spectrum handoff in cognitive radio networks: a classification and comprehensive survey. J. Netw. Comput. Appl. 61(C), 161–168 (2016)
5. Christian, I., Moh, S., Chung, I., Lee, J.: Spectrum mobility in cognitive radio networks. IEEE Commun. Mag. 50(6), 114–121 (2012)
6. Salgado, C., Hernandez, C., Molina, V.: Intelligent algorithm for spectrum mobility in cognitive wireless networks. Procedia Comput. Sci. 83, 278–283 (2016)
7. Hernandez, C., Pedraza, E.: Multivariable adaptive handoff spectral model for cognitive radio networks. Contemp. Eng. Sci. 10(2), 39–72 (2017)

8. Yan, S., Yan, X.: Vertical handoff decision algorithm based on predictive RSS and reduced fuzzy system using rough set theory. J. Inf. Comput. Sci. **12**(12), 4677–4688 (2015)
9. Mir, U., Munir, A.: An adaptive handoff strategy for cognitive radio networks. Wirel. Netw. **24**(6), 2077–2092 (2017)
10. Josephine Dhivya, J., Ramaswami, M.: Ingenious method for conducive handoff appliance in cognitive radio networks. Int. J. Electr. Comput. Eng. **8**(2), 5195–5202 (2018)
11. Mardani, A., Nilashi, M., Antucheviciene, J., Tavana, M., Bausys, R., Ibrahim, O.: Recent fuzzy generalizations of rough sets theory: a systematic review and methodological critique of the literature. Complexity **2017**, Article ID 1608147, 1–33 (2017)
12. Kumar, M., Yadav, N.: Fuzzy rough sets and its application in data mining field. Adv. Comput. Sci. Inf. Technol. **2**(3), 237–240 (2015)
13. Cho, M.-Y., Hoan, T.T.: Feature selection and parameter optimization of SVM using particle swarm optimization for fault classification in power distribution systems. Comput. Intell. Neurosci. **1**, 1–9 (2017)
14. Eitrich, T., Lang, B.: Efficient optimization of support vector machine learning parameters for unbalanced data sets. J. Comput. Appl. Math. **196**(2), 425–436 (2006)
15. Josephine Dhivya, J., Ramaswami, M.: A study on quantitative parameters of spectrum handoff in cognitive radio networks. Int. J. Wirel. Mob. Netw. **9**(1), 31–38 (2017)
16. Josephine Dhivya, J., Ramaswami, M.: Analysis of handoff parameters in cognitive radio networks on coadunation of wifi and wimax systems. In: IEEE International Conference on Smart Technologies and Management for Computing, Communication, Controls, Energy and Materials 2017, pp. 190–194. IEEE Xplore Digital Library, Chennai (2017)

On Human Identification Using Running Patterns: A Straightforward Approach

R. Anusha$^{(\boxtimes)}$ and C. D. Jaidhar

Department of Information Technology, National Institute of Technology Karnataka,
Surathkal 575 025, Karnataka, India
it16fv01.anusha@nitk.edu.in, jaidharcd@nitk.edu.in

Abstract. Gait is a promising biometric for which various methods have been developed to recognize individuals by the pattern of their walking. Nevertheless, the possibility of identifying individuals by using their running video remains largely unexplored. This paper proposes a new and simple method that extends the feature based approach to recognize people by the way they run. In this work, 12 features were extracted from each image of a gait cycle. These are statistical, texture based and area based features. The Relief feature selection method is employed to select the most relevant features. These selected features are classified using k-NN (k-Nearest Neighbor) classifier. The experiments are carried out on KTH and Weizmann database. The obtained experimental results demonstrate the efficiency of the proposed method.

Keywords: Classification · Feature extraction · Gait recognition · Human identification

1 Introduction

A significant role in human recognition is played by biometrics in recent times. Many researchers across the world are working on gait recognition in today's society to recognize the individuals by the pattern of their walking [15,17]. Using gait as a biometric is motivated by occlusion of criminals face and that they either walk or run to getaway a scene of the crime. On the other hand, the likelihood of identifying the individuals by the way they run remains predominantly unexplored. Very often, robbers and criminals naturally run to escape, instead of walking [23]. Therefore, it was necessary to come up with a recognition method which distinguishes people by considering their running patterns.

Walking can be defined as the combination of single support and double support phase. It is cyclical in nature. Running is considered as an extension of walking, and it comprises of different joint movements, coordination, and higher velocities. Both the running cycle and walking cycle can be called as gait cycle. The running cycle and walking cycle are not distinguished by velocity, however by whether a person happens to be airborne during movement. The foot contacts the ground in a different way for running and walking. The main

© Springer Nature Switzerland AG 2020
A. Abraham et al. (Eds.): ISDA 2018, AISC 941, pp. 322–331, 2020.
https://doi.org/10.1007/978-3-030-16660-1_32

difference between jogging and running is the traveling speed is prominent during running, and there is also a difference in the distance between the two legs [20].

In this work, we have made an attempt to identify the individuals by using their running patterns. The rest of the paper is organized as follows: Literature review and proposed methodology are described in Sects. 2 and 3 respectively. Experimental results are discussed in Sect. 4. The conclusions are given in Sect. 5.

2   Related Work

Gait identification approaches are divided into two categories. They are model based and appearance based approaches. In model based approaches, a model is built for the movement of the human body, and the gait characteristics are obtained from this model [3, 26]. These methods concentrate on the dynamics of gait and give less importance to the appearance of people. Model based methods do not attain good results while compared to appearance based methods. Besides, these methods are computationally expensive.

Appearance based approaches perform measurements on gait images to obtain the gait characteristics [5, 18, 21]. The features such as shape, step size, stride length, cadence, etc. were extracted directly from the images or the different templates created from these images. These methods are more sensitive to factors such as carrying and clothing conditions and less sensitive to factors such as illness, change in the speed of walking and so on. The appearance based methods include many templates. One of it is Gait Energy Image (GEI) proposed by Bhanu et al. [16] which is computed by an average of all images of a gait cycle. GEI is used extensively in gait recognition methods. Other templates are optical flow image [14], gait entropy image [4], chrono gait image [22], and so on.

Many different methods such as a histogram of oriented gradients [11], local binary patterns [13], deep learning methods [2], etc. have been used in recent times for gait recognition. Some of the gait detection methods show satisfactory performance under controlled set-ups [5, 17, 18], however the application of gait recognition in real life is still limited, predominantly because of the various covariate issues such as change in viewing angle, carrying and clothing conditions, walking speed variations, walking surface conditions which impact the individuals gait and hence make recognition more complicated [20, 21].

On reviewing the literature, it has been discovered that very few attempts have been made for recognizing the individuals by using their running video database. Yam et al. [23] proposed a model based approach on a dynamically coupled oscillator. This model has been used to recognize the people using walking and running dataset. He obtained reasonably good results, but the experimentation was done on the database which consists of only 7 subjects. Yam et al. [24] further proposed a model for lower leg and thigh. The gait features are obtained from the phase-weighted magnitude of the lower order Fourier components of the knee rotation and thigh. The features were extracted from a small database of 5 subjects, and then they were classified to get the recognition results. Keeping in view the above factors, there was a need to propose a robust method for recognition of people using their running and jogging patterns.

3   Proposed Method

Given a running video dataset, human recognition is to identify the person based on the features derived from the gait cycle of that person. Initially, the video is converted into the sequence of images, which are then processed by some steps such as background subtraction, de-noising, post processing and normalization to get silhouette images. The framework of the proposed recognition method is shown in Fig. 1. Figures 2 and 3 show the sample silhouette images of the different movement pattern. i.e., running and jogging.

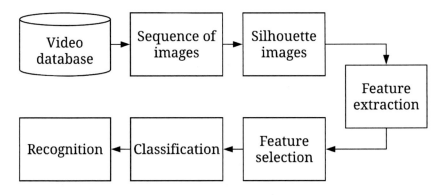

Fig. 1. Structure of the proposed approach

Fig. 2. Sample silhouette images from KTH running database.

The proposed system includes the following steps:

1. Feature extraction: A set of features, which contribute to the identification of an individual are extracted from the silhouette images.
2. Feature selection: A subset of the relevant features are selected.
3. Classification: Decision about the recognition of a person is done using k-NN (k-Nearest Neighbors) classifier with the selected feature set.

Fig. 3. Sample silhouette images from KTH jogging database.

3.1   Feature Extraction and Selection

The features extracted in this work and their description are shown in Tables 1 and 2. It is a combination of statistical, texture based and silhouette region based features. The features from F10 to F12 are the texture features extracted from an image using the SFTA (Segmentation-based Fractal Texture Analysis) algorithm [7]. All these features were extracted from each image of the gait cycle. The final feature vector for each gait cycle is obtained by calculating the mean of a particular feature extracted from all images of a gait cycle. The usage of only mean values for the classification reduces the redundant information and also decreases the size of the feature vector to a large extent. The Relief feature selection algorithm [12] is applied to the extracted feature vectors as it facilitates the improvement of recognition accuracy by reducing the feature space. It removes some of the non-relevant features. The resulting classifier will be simpler and potentially faster. The steps followed to obtain the feature vector is shown in Algorithm 1. The ranking of features obtained by applying Relief algorithm is shown in Fig. 5.

Table 1. Various parameters extracted for gait evaluation

Feature category	Feature name
F1	Area
F2	Perimetre
F3	Convex_hull
F4	Minor_axis_length
F5	Major_axis_length
F6	Contrast
F7	Correlation
F8	Energy
F9	Homogenity
F10	Box_counting
F11	Mean_graylevel
F12	Pixel_count

Table 2. Type of parameters extracted for gait recognition

Features	Description
F1, F2, F3, F4, F5	Area-based features
F6, F7, F8, F9	Statistical features
F10, F11, F12	Texture features

Algorithm 1. The framework followed in the proposed work to extract gait feature vector is as follows.

Input: Number of subjects $S = \{s_1, s_2, ..., s_n\}$.
Begin
for all subjects of S **do**
    for subject s_1 of S **do**
        Identify 4 gait cycles $C = \{c_1, c_2, c_3, c_4\}$ for s_1.
        for all the values C **do**
            for gait cycle c_1 of C **do**
                Convert c_1 into silhouette images $I = \{i_1, i_2, ..., i_m\}$.
                for all silhouette images of I **do**
                    Extract 12 features, $i_k F = \{i_k F_1, i_k F_2, ..., i_k F_{12}\}$.
                end for
                Compute mean, $M = \left\{ \frac{i_1 F_1 + ... + i_m F_1}{m}, ..., \frac{i_m F_{12} + ... + i_m F_{12}}{m} \right\}$.
                Obtain the feature vector of size 12×1 for c_1, $F_{c1} = F_1, F_2, ..., F_{12}$.
            end for
        end for
    end for
end for
Apply Relief feature selection algorithm on F.
Select the top scoring features $F_s = \{F_1, F_3, F_5, F_6, F_{11}, F_{12}\}$ for classification.
End

---

4   Experimental Results and Discussion

All silhouette images used in this work are of size 240×240. As the top scoring six most relevant features were considered for classification, the size of the feature vector obtained for a gait cycle is 6×1. The classifier used is k-NN [8] where $k = 1$. The 5-fold cross validation [9] is applied to the classifier to protect against overfitting by partitioning the dataset into folds and estimating accuracy on each fold. The Correct Classification Rate (CCR) is computed to measure the results which is given by the proportion of the number of correctly classified subjects over the total number of subjects. In this study, the experiments were conducted on two databases. They are

1. KTH database
2. Weizmann database

4.1 Experimental Results on KTH Database

The KTH database is a "video database containing six types of human actions (walking, jogging, running, boxing, hand waving and hand clapping) performed several times by 25 subjects in four different scenarios: outdoors $s1$, outdoors with scale variation $s2$, outdoors with different clothes $s3$ and indoors $s4$. Currently, the database contains 2391 sequences. All sequences were taken over similar backgrounds with a static camera with 25 fps frame rate" [19]. In this work, we considered only KTH running and jogging database. Scenarios $s1$, $s2$, $s3$, and $s4$ were considered for experimentation. The sample images of KTH running and jogging dataset are shown in Fig. 4. Here, we conducted two experiments.

Fig. 4. Sample images from KTH running and jogging database, where $s1$, $s2$, $s3$, $s4$ represent the four scenarios. (a)-(d) represent jogging patterns, and (e)-(h) represent running patterns.

1. Initially, the subjects of each scenario ($s1$, $s2$, $s3$, $s4$) were used separately for classification, and the results are reported in Table 5. For each subject, 4 gait cycles were considered for experimentation. Hence, 100 feature vectors of 25 subjects for each scenario were used for classification. More information about gait of the subject is always acquired for the view angle near $90°$, and it gradually decreases for other views. Therefore, CCR of scenario $s2$ is less compared to scenario $s1$ and $s4$. It is evident from the literature that the clothing condition influences the CCR to a large extent [1,25]. The considerable changes in the subject's silhouettes are caused by clothing condition, and

Table 3. Recognition accuracy of the proposed method on KTH jogging database for different (probe, gallery) dataset combination.

Probe dataset s1	Gallery dataset	s2	s3	s4
		63.50	95.00	98.50
Probe dataset s2	Gallery dataset	s1	s3	s4
		64.50	47.00	62.00
Probe dataset s3	Gallery dataset	s1	s2	s4
		95.50	46.50	94.56
Probe dataset s4	Gallery dataset	s1	s2	s3
		98.00	60.50	95.00

Table 4. Recognition accuracy of the proposed method on KTH running database for different (probe, gallery) dataset combination.

Probe dataset s1	Gallery dataset	s2	s3	s4
		55.50	91.50	95.50
Probe dataset s2	Gallery dataset	s1	s3	s4
		55.00	35.50	50.00
Probe dataset s3	Gallery dataset	s1	s2	s4
		92.00	37.50	88.50
Probe dataset s4	Gallery dataset	s1	s2	s3
		95.00	48.50	90.00

hence, it decreases the accuracy as the appearance based techniques depend on the temporal and spatial variation of silhouette over a gait cycle. So, from the Table 5, it is apparent that less CCR is obtained for scenario $s3$, while compared to $s1$ and $s4$.

2. The experimental results corresponding to probe and gallery dataset for KTH running and jogging database are illustrated in Tables 3 and 4. From Tables 3 and 4, it is noticeable that when the (probe, gallery) pair is $s1$ and $s4$, higher CCR is obtained as the view angle for both the scenarios is same. Due to the same view angle $90°$, the subjects silhouettes do not vary largely. When we consider the probe dataset as $s1$, $s4$ and gallery dataset as $s3$ or the other way around, the value of CCR decreases slightly because of the variations in the subject's silhouettes caused due to clothing conditions. When the probe dataset is $s1$, $s3$, $s4$ and gallery dataset is $s2$ or vice versa, the CCR obtained is very less because of the scale variation of $s2$.

4.2   Experimental Results on Weizmann Database

"The Weizmann database was recorded in 2005. Its background is relatively simple, and only one person is acting in each frame. It contains 10 human actions

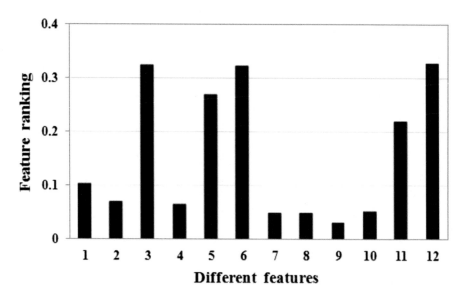

Fig. 5. Different ranking of the features obtained by Relief algorithm. Here, 1, 2,...,12 represents F1, F2, ..., F12.

Table 5. Recognition accuracy of the proposed method on KTH running and jogging database for various scenarios

Sl. no.	Database	s1	s2	s3	s4
1	KTH running	97.66	91.00	94.66	99.00
2	KTH jogging	98.33	92.66	96.00	99.66

(walking, running, jumping, galloping sideways, bending, one-hand waving, two-hands waving, jumping in place, jumping jack, and skipping), each performed by nine people [6]".

The sample images of Weizmann database are shown in Fig. 6. Among all human actions, only running is considered for experimentation in this work. The results are shown in Table 6. The experiments were conducted for 4 gait cycles for each subject. Hence, the total number of gait cycles considered is $9 \times 4 = 36$. As the number of subjects are less, 100% recognition accuracy is obtained for this database.

Table 6. Recognition accuracy of the proposed method on Weizmann database.

No. of subjects	Total no. of gate cycles	CCR (%)
9	36	100

Fig. 6. Example images from Weizmann database.

5   Conclusion

In this work, we have proposed an effective and simple approach for recognition of individuals by using their running patterns. We extracted 12 features from each image of a gait cycle. We computed the mean of each feature for a gait cycle. Further, we used Relief feature selection algorithm to remove the irrelevant features. We got a single feature value by computing the mean of around 21 to 35 values. As a result, high recognition accuracy is obtained. The experimentation of the proposed method is carried out on KTH and Weizmann database. Experimental results demonstrate that the proposed method gives higher recognition accuracy. This method is more feasible for lateral view. For other views, it may give less accuracy, as this method is completely dependent on the shape of the silhouette. Furthermore, experiment on a much larger database needs to be done.

Acknowledgment. We owe our sincere thanks to the team behind KTH action database [19] and Weizmann action database [10], for sharing the database with us, without which the work could not have been done.

References

1. Al-Tayyan, A., Assaleh, K., Shanableh, T.: Decision-level fusion for single-view gait recognition with various carrying and clothing conditions. Image Vis. Comput. **61**, 54–69 (2017)
2. Alotaibi, M., Mahmood, A.: Improved gait recognition based on specialized deep convolutional neural network. Comput. Vis. Image Underst. **164**, 103–110 (2017)
3. Ariyanto, G., Nixon, M.S.: Marionette mass-spring model for 3D gait biometrics. In: 2012 5th IAPR International Conference on Biometrics (ICB), pp. 354–359. IEEE (2012)
4. Bashir, K., Xiang, T., Gong, S., Mary, Q.: Gait representation using flow fields. In: BMVC, pp. 1–11 (2009)
5. Binsaadoon, A.G., El-Alfy, E.S.M.: FLGBP: improved method for gait representation and recognition. In: 2016 World Symposium on Computer Applications & Research (WSCAR), pp. 59–64. IEEE (2016)
6. Chaquet, J.M., Carmona, E.J., Fernández-Caballero, A.: A survey of video datasets for human action and activity recognition. Comput. Vis. Image Underst. **117**(6), 633–659 (2013)
7. Costa, A.F., Humpire-Mamani, G., Traina, A.J.M.: An efficient algorithm for fractal analysis of textures. In: 2012 25th SIBGRAPI Conference on Graphics, Patterns and Images (SIBGRAPI), pp. 39–46. IEEE (2012)

8. Cunningham, P., Delany, S.J.: k-nearest neighbour classifiers. Mult. Classif. Syst. **34**, 1–17 (2007)
9. Fushiki, T.: Estimation of prediction error by using k-fold cross-validation. Stat. Comput. **21**(2), 137–146 (2011)
10. Gorelick, L., Blank, M., Shechtman, E., Irani, M., Basri, R.: Actions as space-time shapes. IEEE Trans. Pattern Anal. Mach. Intell. **29**(12), 2247–2253 (2007)
11. Hofmann, M., Rigoll, G.: Improved gait recognition using gradient histogram energy image. In: 2012 19th IEEE International Conference on Image Processing (ICIP), pp. 1389–1392. IEEE (2012)
12. Kira, K., Rendell, L.A.: The feature selection problem: traditional methods and a new algorithm. In: Aaai, vol. 2, pp. 129–134 (1992)
13. Kumar, H.M., Nagendraswamy, H.: LBP for gait recognition: a symbolic approach based on GEI plus RBL of GEI. In: 2014 International Conference on Electronics and Communication Systems (ICECS), pp. 1–5. IEEE (2014)
14. Lam, T.H., Cheung, K.H., Liu, J.N.: Gait flow image: a silhouette-based gait representation for human identification. Pattern Recognit. **44**(4), 973–987 (2011)
15. Lumini, A., Nanni, L.: Overview of the combination of biometric matchers. Inf. Fusion **33**, 71–85 (2017)
16. Man, J., Bhanu, B.: Individual recognition using gait energy image. IEEE Trans. Pattern Anal. Mach. Intell. **28**(2), 316–322 (2006)
17. Prakash, C., Kumar, R., Mittal, N.: Recent developments in human gait research: parameters, approaches, applications, machine learning techniques, datasets and challenges. Artif. Intell. Rev. **49**(1), 1–40 (2018)
18. Rida, I., Almaadeed, S., Bouridane, A.: Gait recognition based on modified phase-only correlation. Signal Image Video Process. **10**(3), 463–470 (2016)
19. Schuldt, C., Laptev, I., Caputo, B.: Recognizing human actions: a local SVM approach. In: 2004 Proceedings of the 17th International Conference on Pattern Recognition, ICPR 2004, vol. 3, pp. 32–36. IEEE (2004)
20. Semwal, V.B., Raj, M., Nandi, G.C.: Biometric gait identification based on a multilayer perceptron. Robot. Auton. Syst. **65**, 65–75 (2015)
21. Sharma, S., Tiwari, R., Singh, V., et al.: Identification of people using gait biometrics. Int. J. Mach. Learn. Comput. **1**(4), 409 (2011)
22. Wang, C., Zhang, J., Pu, J., Yuan, X., Wang, L.: Chrono-gait image: a novel temporal template for gait recognition. In: European Conference on Computer Vision, pp. 257–270. Springer (2010)
23. Yam, C.Y., Nixon, M.S., Carter, J.N.: Extended model-based automatic gait recognition of walking and running. In: International Conference on Audio-and Video-Based Biometric Person Authentication, pp. 278–283. Springer (2001)
24. Yam, C., Nixon, M.S., Carter, J.N.: Gait recognition by walking and running: a model-based approach (2002)
25. Yu, S., Tan, D., Tan, T.: A framework for evaluating the effect of view angle, clothing and carrying condition on gait recognition. In: 2006 18th International Conference on Pattern Recognition, ICPR 2006, vol. 4, pp. 441–444. IEEE (2006)
26. Zheng, S., Zhang, J., Huang, K., He, R., Tan, T.: Robust view transformation model for gait recognition. In: 2011 18th IEEE International Conference on Image Processing (ICIP), pp. 2073–2076. IEEE (2011)

Analysis of Encoder-Decoder Based Deep Learning Architectures for Semantic Segmentation in Remote Sensing Images

R. Sivagami[✉], J. Srihari, and K. S. Ravichandran

School of Computing, SASTRA Deemed to be University,
Thirumalaisamudiram, Thanjavur 613401, Tamilnadu, India
sivagamiramadass@sastra.ac.in

Abstract. Semantic segmentation in remote sensing images is a very challenging task. Each pixel in a remote sensing image has a semantic meaning to it and automatic annotation of each pixel remains as an open challenge for the research community due to its high spatial resolution. To address this issue deep learning based encoder-decoder architectures like SegNet and ResNet that is widely used for computer vision dataset is adopted for remote sensing images and its performance is analyzed based on the pixel wise classification accuracy. From the experiment conducted it is inferred that SegNet suffers from degradation problem when the depth of the network is increased with an overall accuracy of about 86.086% whereas the Residual network manages to overcome the degradation effect with an overall accuracy of about 87.747%.

Keywords: Semantic segmentation · Deep learning ·
Encoder-Decoder architectures · Remote sensing images

1 Introduction

Recent advancement in on-board sensor technology and data acquisition systems in satellite and aircrafts generates a huge volume of easily available and accessible remote sensing data. Most importantly, these remote sensing data are used to monitor the earth as an entire system and used in numerous applications like Land use and Land cover classification, weather prediction, monitor the expansion of urban area, water resources, deforestation, species identification, pollution monitoring, meteorology, detect and classify objects, and in military applications like surveillance system, anomaly detection, and so on. Abundant information can be inferred from these data and give meaningful information for the end user. Computer-based image analysis of these huge volume of data is ever increasing due to the rapid growth in the advent of more sophisticated and computationally intensive hardware's like GPU and efficient image processing algorithms. Recent research community takes up more challenging task related to the effective use of these data to solve many problems that exist in earth observation. One among the open research problem is image classification.

Per pixel classification or semantic segmentation in a remote sensing image is the task of assigning a class label/category of interest to each pixel in the given image. i.e., mapping pixel values to symbols.

In the field of pattern recognition, computers are trained in such a way to understand the relationship between the raw data and the information classes for effective classification. Automatic annotation of pixels is achieved with machine learning techniques. Accurate classification of remote sensing images using machine learning techniques mainly relies on efficient feature sets and learning techniques, i.e., the attributes of the data say intensity information of that particular pixel, texture features, statistical features, relationship between its neighboring pixels, etc. Though there are a number of different proposed algorithms, each one has its own limitations.

2 Recent Works

In recent works many researchers addressed per pixel labelling of images obtained from UAVs and satellites. Recent success in using deep learning for image net classification with less error rate compared to the existing state of the art models, motivated the researchers in computer vision to adopt deep learning models trained for one application to the other and propose a novel architecture based on encoder and decoder for semantic segmentation task. Semantic segmentation methods can be categorized as traditional machine learning based approaches and more recent deep learning approaches. The success of traditional machine learning approaches mostly relies on the manual feature extraction techniques. Better the features for learning, higher the classification accuracy.

In deep learning era there is no need for manual feature extraction or hand crafted features. The CNN itself will learn the low-level features to high level features directly from the input image. Paisitkriangkrai et al. (2016) combined the features extracted using CNN from an aerial image and hand crafted features with CRF as post processing for label probabilities to delineate the boundary pixels. Audebert et al. (2016a), investigated the issues of deep features extracted from CNN and proposed a method using super pixel segments as pre-processing, feature extraction from theses super pixel segments and final classification using SVM for better classification accuracy. Audebert et al. (2016b), addressed encoder and decoder based semantic segmentation using multiscale and a fusion approach for error correction.

A new CNN based segmentation architecture defined by Long et al. (2015) replace the last fully connected layer with a fully convolutional layer not only to predict the class labels but also to find the location of the class labels, this architecture managed to generate a coarse level semantic segmented image as output. The loss of information from this coarsely predicted image is addressed in Sun et al. (2018) by a maximum fusion strategy. Another notable work by Audebert et al. (2018) used ResNet model and analyzed the performance of early and late fusion of composite image obtained from concatenating the Digital Surface Model (DSM), Normalized DSM, Vegetation index data. Kemker et al. (2018) reviewed different deep learning architecture for multispectral images that has a brief description of the state of the art techniques.

A multi filter CNN proposed by Sun et al. (2018), fuses the Lidar point cloud data and high resolution optical images for multiresolution segmentation.

From the literature it is inferred that efficient feature extraction from a CNN and using CRF for boundary delineation as a post processing step makes the architecture more complex with less or no improvement in accuracy. The deep learning models with deep encoder and decoder architecture performs better by eliminating the need for post processing and achieves significant improvement in overall classification accuracy.

3 Methodology

Symmetrical encoder and decoder architectures like SegNet (Badrinarayanan et al. 2015), DeepLab (Chen et al. 2018), ResNet (He et al. 2015) have proven effective results on computer vision dataset like PASCAL VOC2012 and Microsoft COCO. The main objective is to transpose traditional architectures like SegNet and ResNet used on computer vision dataset to Earth observation data in order to understand a given remote sensing image at its pixel level and analyze the performance of both the SegNet and ResNet deep learning architectures in terms of its accuracy. Computer vision dataset focuses on everyday scenes from human level insight with meaningless background information and few objects of interest. The earth observation data is acquired from birds view point covering a large area and each pixel in a remote sensing image has a meaningful information. As effective localization of features in a satellite image is a crucial task, satellite images that are rich in spatial and structural information are less investigated compared to everyday images in computer vision.

3.1 System Architecture

SegNet, an end to end trained deep symmetrical encoder-decoder architecture consists of a series of convolution, Batch normalization and activation functions followed by pooling layer in the encoder part. Decoder is similar to encoder except that the pooling layer is replaced with up-sampling layers to match the resolution of the learned features to the input image. Finally a softmax layer is used to predict the class probabilities for each pixel based on the learned features. Figure 1 shows the architecture of SegNet inspired from Badrinarayanan et al. (2015).

The deeper networks are often exposed to degradation problem when the network starts to converge during training. If more number of layers are used the error rate may increase if this degradation issue not handled properly. SegNet suffers from this degradation issues which is addressed in Residual network (He et al. 2015, 2016). To overcome this degradation problem and to ease the training of deeper network a residual layer that adds the input for the convolution layer with the learned feature values obtained as output from CNN. ResNet achieved an error rate of only 3.5% for a network of depth up to 152 layers. ResNet trained for image net classification problem is used as encoder and constructed a symmetrical decoder corresponding to the encoder to perform semantic segmentation. In decoder for up sampling the feature maps to match the spatial resolution of the input image Deconvolution operation is used. The basic building block for a ResNet is as shown in Fig. 2 from the paper by He et al. (2015, 2016).

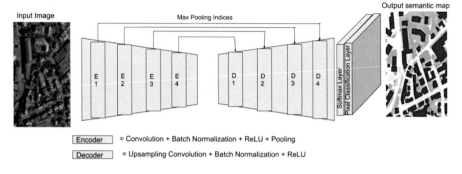

Fig. 1. SegNet architecture adopted for Semantic segmentation

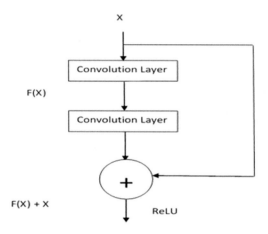

Fig. 2. Residual Connection in a ResNet

4   Experimental Framework

4.1   Implementation

The models used for semantic segmentation is implemented using MATLAB R2018b (2018) and Cuda enabled NVIDIA K2200 with computing capability of 4.0 of 4 GB memory is used to train the architecture.

4.2   Dataset

The input image used to validate the method described above is obtained from http://www2.isprs.org/commissions/comm3/wg4/2d-semlabel-vaihingen.html. The Vaihingen dataset consists of 33 images true orthophoto in 8 bit Tiff format extracted from a larger TOP mosaic with a resolution of about 9 cm per pixel. The approximate size of a single tile image is about 2100 × 2100 pixels. Out of 33 images 16 images has public labelled/ground truth images for training and testing.

Out of this 16 images 12 (75%) images are used as training and 4 (25%) images for inference. The Tiff images has three bands that corresponds to the near infrared, red and green bands (IRRG) and the dataset has Digital Surface Model (DSM) data obtained from Lidar point cloud that gives the height information. The DSM data is used to distinguish trees and low vegetation areas and buildings. A sample input image and its equivalent ground truth image with five classes is shown in Fig. 3.

4.3   Data Preprocessing

Image Patch Extraction
2D semantic labelling high resolution dataset obtained from ISPRS is too large with an approximate size of about 2100 × 2100 × 3 pixels to process through a CNN. If the input size is large the number of parameters to be handled will be large. Due to memory limitations, a number of small image patches and its corresponding ground truth labels are extracted from a large tile. For this experiment a number of image patches of size 128 × 128 × 3 are extracted from a large image tile.

Image Normalization
Image normalization is a linear transformation technique that transforms the input pixel values to follow a similar data distribution. The distribution of data resemble Gaussian curve centered at zero. Training a neural network is solving a non-convex optimization problem using gradient based approach. In backprogation gradients are calculated to update the weights. If the data are not normalized the optimizer has to do a lot of searching to find a good solution. If the image is equal variance and zero mean the optimizer makes it lot easier to find the optimal solution. Image normalization does not change the content of the image, but makes easier for the optimizer to proceed numerically. As the gradients depend on the inputs, zero centering of the input data reduces bias in the gradients. This normalization is purposely done to make convergence faster while training the network.

$$Zero\ centering = \frac{X - \mu(X)}{\sigma(X)} \tag{1}$$

where

X - Input image to be normalized
μ - Mean value calculated from the pixel values of the image to be normalized
σ - Standard deviation calculated from the pixel values of the image to be normalized

4.4   Hyperparameter Initialization

Hyperparameters are the model parameters to be predefined before training a CNN as they cannot be learned directly from the data as in standard machine learning models.

The hyperparameters to be initialized are learning rate, batch size for training, learning algorithm like Stochastic Gradients Descent (SGD), ADAM optimizer, etc. All models for semantic segmentation used for this experiment are trained using SGD optimizer with a learning rate of about 0.001, momentum of about 0.9 and L2 regularization method to avoid over fitting. The value for regularization parameter is set to 0.0005. Due to GPU memory constraints the Mini Batch size is set to 4 and Trained for a Max epoch of about 100. All the hyperparameters used for training is as recommended from Audebert et al. (2017). Initial weights for the learnable filters are assigned based on He initialization technique (He et al. 2015, 2016). These learnable weights are later updated during backpropagation based on the average loss computed.

4.5   Class Balancing

In the given dataset the number of pixels in impervious class labels and building labels are higher than the car classes for learning. The car class covers a maximum of up to 28 pixels in a given large image tile. Here there is an imbalance in class labels. This the dataset is used as such this can be detrimental for learning process as it learns more dominant classes and biases in favor of these dominant classes. In order to improve the learning process, a systematic way to improve low class should be adopted. One such technique is class weighting approach in which more class weights are assigned for labels with less number of total pixels and less weight for labels that cover more pixels (Buda et al. 2018). Class weights are calculated as follows:

$$Image\,frequency = \frac{Number\,of\,pixels\,available\,in\,a\,particular\,class}{Total\,number\,of\,pixels\,in\,all\,classes} \tag{2}$$

$$Class\,weight = \frac{Median\,(Image\,frequency)}{Image\,frequency} \tag{3}$$

5   Results and Discussion

In this experiment we have successfully adopted the deep learning architectures previously used for semantic segmentation of street view and everyday images to Earth observation data. For SegNet architecture the encoder is adopted from VGG16 image net classification model and a symmetrical decoder is constructed for semantic segmentation task. The model is analyzed with different learning rate for encoder and decoder. For ResNet the same input parameters are adopted as in SegNet and analyzed its performance. The inference result for our SegNet model and ResNet model is given in Tables 1 and 2.

The trained model is tested on 4 inference images. It is evident that the performance is better when weight values trained on a specific task is adopted and fine-tuned for earth observation data than training from scratch. The accuracy of the model is assessed in terms of pixel-wise classification accuracy, F1 score and Kappa statistics to measure the closeness of the predicted class labels with the ground truth. Classification accuracy is defined by:

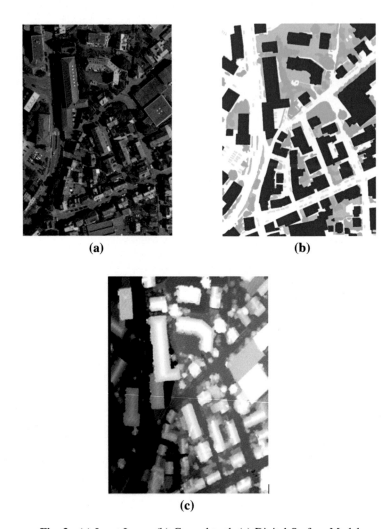

Fig. 3. (a) Input Image (b) Ground truth (c) Digital Surface Model

$$Global\ pixelwise\ Accuracy = \frac{Number\ of\ correctly\ classified\ pixels}{Total\ number\ of\ pixels} \quad (4)$$

$$F1\ score = 2 \times \left[\frac{precision\ \times\ recall}{precision\ +\ recall}\right] \quad (5)$$

$$precision = \frac{Number\ of\ true\ positives}{Number\ of\ False\ positives} \tag{6}$$

$$recall = \frac{Number\ of\ true\ positives}{Number\ of\ true\ positives + Number\ of\ false\ negatives} \tag{7}$$

$$Kappa\ statistics = \frac{observed\ accuracy\ -\ expected\ accuracy}{1 - expected\ accuracy} \tag{8}$$

Table 1. Validation results on 2D semantic labelling Vaihingen dataset by SegNet

Classes	Impervious surface	Buildings	Low vegetation	Trees	Car	Clutter
Impervious surface	2308197	238142	22792	9159	38398	0
Buildings	73699	1176130	5411	3082	12592	0
Low vegetation	38479	18377	244577	57586	3577	0
Trees	35676	10228	61011	487185	2174	0
Car	12759	4344	531	168	139024	0
Clutter	24482	24501	54	17	6548	0

Overall Accuracy = 86.088%
Kappa = 0.787

Table 2. Validation results on 2D semantic labelling Vaihingen dataset by ResNet

Description	Impervious surface	Buildings	Low vegetation	Trees	Car	Clutter
Impervious surface	2252320	85542	19244	9913	24248	5554
Buildings	54995	887082	19488	2879	6924	10381
Low vegetation	64996	6917	205783	42304	1171	168
Trees	39338	2496	38413	436500	725	0
Car	51165	4918	378	779	89161	200
Clutter	26083	14845	826	982	492	0

Overall Accuracy = 87.747%
Kappa = 0.804

The training accuracy of SegNet reached about 97% in some iteration and started to degrade eventually in the next few iterations and obtained 94%. Whereas in ResNet the training accuracy kept on increasing without any degradation and managed to obtain 94% training accuracy in 100 epoch.

6 Conclusion

SegNet a convolutional layer based encoder decoder architecture followed by pixel classification layer is chosen to perform pixel wise labelling in ISPRS 2D semantic labelling dataset over other architectures as it provides a good balance between accuracy and computational cost. In SegNet the encoder part is inspired from VGG16 architecture that accepts input image in three channel format (RGB format). The decoder part which is also based on convolution layers similar to encoder and does up sampling to match the resolution of the feature maps to the resolution of the input image. Overall classification accuracy obtained by SegNet is 86.088%. SegNet suffers from degradation issues during training which overcomed by the ResNet and achieved a classification accuracy of about 87.747%. In deepnets it is believed that if the depth of the architecture is increased the learning is better leading to good classification accuracy. If the depth increases there are more chances that the deep learning architecture suffers from degradation issues which can be overcome by using a residual connection.

Acknowledgement. The authors would like to thank DRDO-ERIPR for their funding under research grant no: ERIP/ER/1203080/M/01/1569. The first author would like to thank CSIR for their funding under grant no: 09/1095(0033)18-EMR-I. The Vaihingen dataset is obtained from German society for photogrammetry, Remote Sensing and Geoformation (DGPF) (Cramer 2010): http://www.ifp.uni-stuttgart.de/dgpf/DKEP-Allg.html. The authors thank ISPRS for making the dataset openly available.

References

Paisitkriangkrai, S., Jamie, S., Janney, P., van den Hengel, A.: Semantic labeling of aerial and satellite imagery. IEEE J. Sel. Top. Appl. Earth Obs. Remote Sens. **9**(7), 2868–2881 (2016)

Audebert, N., Le Saux, B., Lefevre, S.: How useful is region-based classification of remote sensing images in a deep learning framework? In: 2016 IEEE International Geoscience and Remote Sensing Symposium (IGARSS), pp. 5091–5094. IEEE (2016a)

Sun, W., Wang, R.: Fully convolutional networks for semantic segmentation of very high resolution remotely sensed images combined with DSM. IEEE Geosci. Remote Sens. Lett. **15** (3), 474–478 (2018)

Sun, Y., Zhang, X., Xin, Q., Huang, J.: Developing a multi-filter convolutional neural network for semantic segmentation using high-resolution aerial imagery and LiDAR data. ISPRS J. Photogram. Remote Sens. 143, 3–14 (2018)

Kemker, R., Salvaggio, C., Kanan, C.: Algorithms for semantic segmentation of multispectral remote sensing imagery using deep learning. ISPRS J. Photogramm. Remote. Sens. **145**, 60–77 (2018)

He, K., Zhang, X., Ren, S., Sun, J.: Deep residual learning for image recognition. In: Proceedings of the IEEE Conference on Computer Vision and Pattern Recognition, pp. 770–778 (2016)

Badrinarayanan, V., Kendall, A., Cipolla, R.: Segnet: a deep convolutional encoder-decoder architecture for image segmentation. arXiv preprint arXiv:1511.00561 (2015)

Audebert, N., Le Saux, B., Lefèvre, S.: Beyond RGB: very high resolution urban remote sensing with multimodal deep networks. ISPRS J. Photogramm. Remote. Sens. **140**, 20–32 (2018)

Audebert, N., Le Saux, B., Lefèvrey, S.: Fusion of heterogeneous data in convolutional networks for urban semantic labeling. In: Urban Remote Sensing Event (JURSE), 2017 Joint, pp. 1–4. IEEE (2017)

Audebert, N., Le Saux, B., Lefèvre, S.: Semantic segmentation of earth observation data using multimodal and multi-scale deep networks. In: Asian Conference on Computer Vision, pp. 180–196. Springer, Cham (2016b)

Long, J., Evan, S., Darrell, T.: Fully convolutional networks for semantic segmentation. In: Proceedings of the IEEE Conference on Computer Vision and Pattern Recognition, pp. 3431–3440 (2015)

Chen, L.-C., Papandreou, G., Kokkinos, I., Murphy, K., Yuille, A.L.: Deeplab: semantic image segmentation with deep convolutional nets, atrous convolution, and fully connected CRFs. IEEE Trans. Pattern Anal. Mach. Intell. **40**(4), 834–848 (2018)

Buda, M., Maki, A., Mazurowski, M.A.: A systematic study of the class imbalance problem in convolutional neural networks. Neural Netw. **106**, 249–259 (2018)

He, K., Zhang, X., Ren, S., Sun, J.: Delving deep into rectifiers: surpassing human-level performance on imagenet classification. In: Proceedings of the IEEE International Conference on Computer Vision, pp. 1026–1034 (2015)

Cramer, M.: The DGPF-test on digital airborne camera evaluation–overview and test design. Photogrammetrie-Fernerkundung-Geoinformation, **2010**(2), 73–82 (2010)

MATLAB version 9.5.0. Natick, Massachusetts: The MathWorks Inc., September 2018

Predicting Efficiency of Direct Marketing Campaigns for Financial Institutions

Sneh Gajiwala$^{(\boxtimes)}$, Arjav Mehta, and Mitchell D'silva

Dwarkadas J Sanghvi College of Engineering, Mumbai, India
sneh.gajiwala@gmail.com

Abstract. All marketing campaigns are dependent on the data that their clients provide. These datasets include everything from their name, number, their salary, the loans they've already taken, the money they have in their account etc. These datasets are huge and it is impossible for a human to analyze the patterns in the client deposits and whether the campaign will be a success or not. This paper introduces prediction of the success rates of the campaigns with the help of various machine learning predictive algorithms: Random forest, KNN and KNN using tensor flow framework. In the recent years, the success of general campaigns led by financial institutions have been declining and to step up their game and increase campaign effectiveness they need to target the clients very specifically. The predictive results obtained, with the highest accuracy, help in increasing the target audience and ensure that it will be a success for the financial institutions in picking out clients who will subscribe to the different marketing schemes. This paper uses multiple accuracy parameters to predict the potential success a direct marketing campaign will have on specific clients.

Keywords: Direct marketing · KNN · Random Forests · TensorFlow · Sensitivity · Specificity

1 Introduction

Banks maintain a record of all the details of their customers ranging from their marital status to their occupation and the deposits they make or the schemes they have chosen over the years. This data is very useful to the executives who devise marketing campaigns. Nowadays, companies and banks use a strategy called as direct marketing which removes the middle man from the process and the company provides messages directly to the potential client. To make direct marketing cost effective the customer database has to be well managed [2].

When running or devising a marketing campaign it is required to judge how effective the strategy would be in terms of reaching the audience. It is essential to know whether the target audience or the clients will have interest as well as the capacity to accept new schemes. Are the clients responding positively to the campaigns? Are they investing in these schemes? This paper uses various machine learning classification models to predict the performance marketing strategies developed by the banks will have. It also makes their work easier and hopes to advise them to put their marketing expenditure and efforts in different types of campaigns if the results are not satisfactory

© Springer Nature Switzerland AG 2020
A. Abraham et al. (Eds.): ISDA 2018, AISC 941, pp. 342–352, 2020.
https://doi.org/10.1007/978-3-030-16660-1_34

[1]. Having knowledge of and access to customer profiles will enable the banks to have a deeper understanding about potential clients, which will influence their decisions about direct marketing campaigns. This can lead to improved marketing campaigns, targeted sales and better customer service. A clearer view about the customer profile enables companies to send out triggered messages, calls, texts, and emails etc. which is a good way to reinforce the brand and target customers.

There are two types of machine learning models, Regression models and classification models. This paper implements two types of classification models to find the best one. These models help to solve the analytical problems and reach a decision. It enables to determine whether the customers will be interested in the campaign, whether these specific customers should be targeted for direct marketing campaigns or whether the customer has a potential to enroll in the bank schemes.

One of the methods implemented is the K-nearest neighbor (KNN) algorithm. The aim is to primarily find out how the input parameters affect the output i.e. based on the available dataset finding out whether the clients targeted through various campaigns have the potential to sign-up for the campaign or not. This technique eliminates the cost of learning process and complex concepts can be learned by local approximations of simple procedures. In this method, each sample is classified according to its neighbors and if the sample is unknown then the sample is predicted by considering the nearest neighbor. KNN is robust for training a noisy dataset and apt for predicting if the dataset is large.

The second method used is Random forest which is considered to be a general-purpose technique available for working on classification and regression as it is fast and easy to implement [5]. Random forest enables us to capture the variance of the multiple input variables without deletion and at the same time enables a large dataset to participate in prediction [5]. Since the financial institutions have huge number of clients and a lot of variables to be considered in the prediction of the success of the campaigns this method is very useful. It also helps us detect the interactions between various input variables in the predictions.

2   Related Works

Machine Learning based classification algorithms have been used for a long time in various markets such as finance, marketing, healthcare, etc. algorithms such as KNN have been used in healthcare for heart disease prediction, in the weather industry for rainfall and humidity prediction and in the financial industry for stock market prediction. Random forests have been used to classify trending topics and aid in recommendation systems for systems such as online shopping websites, movie recommendation systems.

Such algorithms have been proved to be an important aspect in forming the business strategy and aiding the decision making process of the organization. Given the important part they play in business, neither regression nor classification models have been used to predict the success ration of business strategies or marketing campaigns.

The financial institutions have plentiful opportunities to make improvements with a lot of marketing campaigns being devised every day. The success of these campaigns can be foreseen based on the factors influencing the campaign. Such a system has not been implemented yet and the proposed system is to be used to predict the efficiency of direct marketing campaigns for a financial institution based on their myriad clientele and the probability of success in different client groups.

3 System Implementation

To design the direct marketing campaigns for prospective client's, the financial institutions need to know what their customers consider as important. A predictive algorithm, using machine learning helps solve this problem and predict the success rates by uncovering meaningful patterns amongst the customer profile and transactional data.

Figure 1 illustrates the basic architecture for the implemented system.

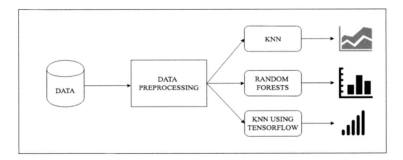

Fig. 1. Basic architecture of the implemented system

3.1 Input to System

The dataset used has been extracted from the UCI Machine Learning Repository. It is a classic marketing bank dataset which has information regarding a direct marketing campaign carried out by the bank or financial institution on their customers. The dataset consists of variety of attributes ranging from the age, education, account balance, loans, campaign duration etc. Various classification models have been applied on this dataset to successfully predict the outcome of the future marketing campaigns and strategies carried out by the financial institutions to maximize and improve their success rates. Table 1 shows the attributes in the dataset [3].

3.2 Data Preprocessing

Data preprocessing is performed so that the data is free of any anomalies and can be efficiently used for prediction. This preprocessed data is then entered as input to the three models and the results are eventually compared.

Data preprocessing is an important step in data analysis and entails the transformation of raw data into understandable and reusable forms. Data preprocessing is necessary because data from the real world is usually incomplete and inconsistent and may lack certain attributes and behaviour.

The dataset consists of 18 columns and over 11,000 epochs. A two tier approach is used for the preprocessing of the selected data set. In the first stage, the data is cleansed of all the columns that are not relevant for prediction. Columns like day, month and contact were stripped off to make the dataset more compact and easy to manage. The second stage involves data transformation wherein relevant columns such as jobs and education that are composed of textual categorical values are transformed to numeric categorical values. Transformation is applied as handling of numerical values speeds up the execution of the classification model and thus improves time complexity as compared to that of textual values. After preprocessing, the attributes considered and converted are shown in Table 1.

Table 1. Attributes after preprocessing

Attributes	Kind
Age	Numerical value
Job	Numerical value
Marital status	Numerical value
Education	Numerical value
Balance	Numerical value
Output	Binary

Thus, the preprocessed dataset of over 11,000 epochs is used as the training set by the models and 20% of this training set is used as the test set to determine and compare their accuracy.

3.3   K - Nearest Neighbour

K - Nearest Neighbor or KNN is a pattern recognition method used for classification and regression. KNN is a non-parametric algorithm, as it is not just based on the parametric distribution of probabilities. KNN works on the simple principle of voting. The class with the highest of votes gets assigned to the test case. In this algorithm, all the neighbours have equal part in the prediction of the output variable irrespective of their relative distance from the query point [5].

K- Nearest Neighbour (or KNN) algorithm uses the property of 'feature similarity' to predict the value of the new data points, i.e. the new data point is assigned its value based on how far it is from the other data points in the training data set. This resemblance or similarity is calculated using Euclidean distance, Manhattan distance or Minkowski distance. KNN is an algorithm that stores all the available data (training set) and classifies new data (test set) based on a distance measure. The distance measure resembles how close the new data point is to the available data points. The output of KNN, when used for classification, is a discrete value. This discrete value is the predicted class and the classification is based on its neighbours. The object is classified to the class that is the nearest amongst its k-neighbours [9].

KNN finds application in classifications well as regression predictive problems. There are three parameters that are generally used to evaluate a prediction technique -

- Ease with which the input is interpreted
- Time taken for calculations
- Accuracy of the prediction

The KNN algorithm excels across all three parameters of consideration, and is therefore widely used. Figure 2 shows the basic architecture of KNN.

The working of KNN can be explained in the following steps [8] -

- Initially, the K data points that are closest to the new data point are selected. K is a user defined value and heavily influences the accuracy of the algorithm. K represents the number of neighbours that are considered to classify the new data point. The most accepted way to determine the value of K is to use the square root of the total number of data points. However, a very large or a very small value of K can result in erroneous predictions. Thus, the value of K must be chosen very carefully.
- The distance between the new data point and all the data points in the training sets is calculated. This paper uses Euclidean distance as a distance measure for implementing KNN. The formula for Euclidean distance is as shown in Eq. 1. It is the square root of the summation of the squared distance between the two points. Here, p_i is the new data point and q_i is the data point in the training population and i is the number of neighbours, i.e., K

$$\sqrt{\sum_{i=1}^{n}(q_i - p_i)^2} \tag{1}$$

- Next, the values predictions are sorted according to their values. For our dataset, the predicted value is a Boolean. So the predictions are sorted into two groups, either the customer enrolls in the marketing scheme or he does not.
- The group with the majority predictions is considered as the final classification and assigned to the testing value.

Figure 2 illustrates the basic architecture of KNN implementation.

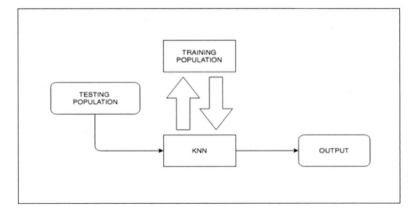

Fig. 2. Basic architecture of KNN

K - Nearest Neighbour has been implemented using the "KNeighborsClassifier" class from the sklearn.neighbors library. The dataset is automatically split into training set and testing set. The StandardScaler is used fit the data and transform it as required. The sklearn.metrics library provides the accuracy score for the prediction.

3.4   Random Forest

Random Forests is a supervised algorithm that is used for classification. Random Forests is an ensemble learning technique that operates by constructing decision trees in the training period and the resultant output of the classification is the mean of the classification output of the individual trees. This method involves splitting the sample dataset or population into multiple homogeneous subsets depending on the most significant variable. Random Forests uses the same parameters as that of a decision tree classifier such as an indexing, maximum features etc. and thus can be considered as an extended implementation of decision tree [8].

Random Forests is a supervised and flexible algorithm that is used for classification. Random Forests also takes care of dimensional reduction, handles missing and outlier values. It builds an ensemble of decision trees that are trained using the 'Bagging' method. The general idea for Random Forests is to combine a group of weak models to form a powerful one to improve the overall accuracy [10].

Random Forests are called as "random" because instead of searching for the most significant feature while splitting a node, it searches for the best feature from the set of random features. This ensures a diversified model that substantially improves the accuracy of the model. Thus, in Random Forests only a random subset of the population's features are considered for splitting the node.

In contrast to the Classification and Regression Trees (CART) model, Random Forest uses multiple trees for classification. To predict the class, each of the trees casts a vote for that class. Ultimately, the forest chooses the class with the majority of votes.

The algorithm can be summed up as follows [10] –

1. Consider that there are N classes in the training set. A sample of the N classes is taken at random with replacement. This sample is the training set used for growing the tree.
2. If there are X input variables, a random number 'x' that is less than X is specified such that at each of the nodes, 'x' variables are selected at random out of the X.
3. The best split on these 'x' values is used to split the node. This 'x' is held constant for the tree to grow
4. Since there is no pruning, each tree grows to its largest extent
5. The new data is predicted by aggregating the majority of the votes for classification.

Figure 3 illustrates the basic architecture of Random Forest implementation.

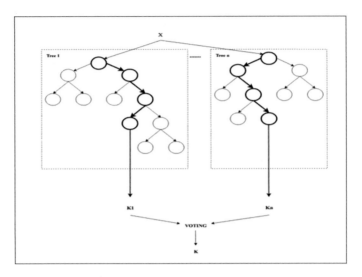

Fig. 3. Basic architecture of Random Forest

Random Forest has been implemented using the RandomForestClassifier class from the sklearn.preprocessing library. The dataset is automatically split into training set and testing set. The StandardScaler is used to fit the data and transform it as needed. The sklearn.metrics library provides the accuracy score for the prediction [7].

3.5   K - Nearest Neighbour Using TensorFlow

TensorFlow is a cross-platform and open source software library provided by Google that is used as a computational framework for developing machine learning algorithms and data-flow graphs. It runs on anything ranging from GPUs and CPUs to mobile embedded systems and even specialized Tensor Processing Units that is hardware for tensor math [11].

TensorFlow can be used for a much broader category of algorithms than just neural networks. Simple algorithms like K - Nearest Neighbour can be implemented very efficiently using TensorFlow libraries.

The TensorFlow implementation is much more efficient and easier that the primitive KNN implementation. TensorFlow uses its own distance formula and automatically chooses a regulated value for K based on the dataset. The KNN algorithm uses a TensorFlow session to store the training set and test set. The session responds with a predicted class for every index in the test set. A TensorFlow method, called as the FeedDictionary is used to get the prediction nearer to the test parameters from the training set [10].

4   Result Analysis

This section discusses the results obtained by the implemented models. Three approaches have been implemented for classification, namely K - Nearest Neighbours, Random Forests, and K - Nearest Neighbours using TensorFlow. The training set comprises of over 11,000 epochs and the testing set is 20% of the test set, that is, approximately 2230 epochs. Figure 4 illustrates the original values for 100 epochs.

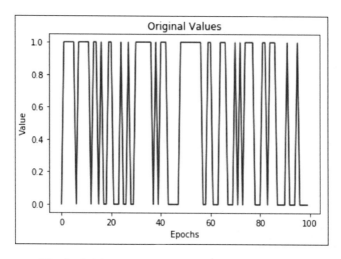

Fig. 4. Original values of 100 epochs from the test set

4.1 K - Nearest Neighbour (KNN)

Figure 5 displays the predicted value of 100 epochs from the test set for K - Nearest Neighbours. The accuracy score obtained is 0.59561. This implies that out of the 2,200 plus epochs tested, approximately 1,330 epochs were correctly classified. The accuracy of this algorithm can thus be considered as approximately 59.56%.

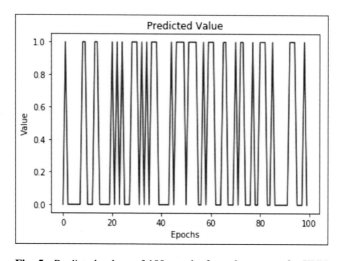

Fig. 5. Predicted values of 100 epochs from the test set for KNN

4.2 Random Forest

Figure 6 displays the predicted value of 100 epochs from the test set for Random Forests. The accuracy score obtained is 0.61307. This implies that out of the 2,200 plus epochs tested, approximately 1,360 epochs were correctly classified. The accuracy of this algorithm can thus be considered as approximately 61.31%.

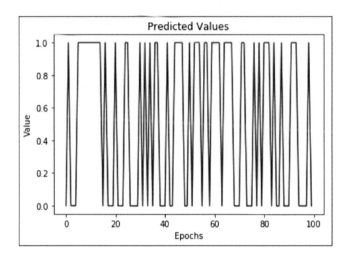

Fig. 6. Predicted values of 100 epochs from the test set for Random Forests

4.3 KNN Using Tensorflow

Figure 7 displays the predicted value of 100 epochs from the test set for K - Nearest Neighbours. The accuracy score obtained is 0.97179. This implies that out of the 2,200 plus epochs tested, approximately 2,170 epochs were correctly classified. The accuracy of this algorithm can thus be considered as approximately 97.18%.

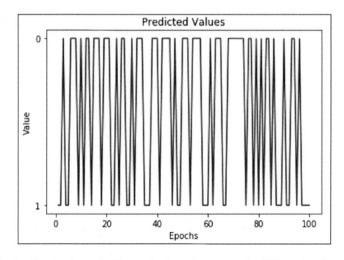

Fig. 7. Predicted values of 100 epochs from the test set for KNN using TensorFlow

4.4   Comparative Analysis

Table 2 illustrates the accuracy parameters for the three models.

Table 2. Properties of models

Property	KNN	Random Forests	KNN using TensorFlow
Positive	1039	1028	1033
Negative	1194	1205	1201
True Positive	567	600	1033
True Negative	735	769	1158
False Positive	459	436	43
False Negative	472	428	0
Accuracy	0.5830	0.6130	0.9807
Error Rate	0.4169	0.3869	0.0192
Sensitivity	0.5457	0.5836	1.0
Specificity	0.6155	0.6381	0.9641

5   Conclusion

The results in Sect. 5 make it abundantly clear that KNN using TensorFlow performs extensively better than the primitive KNN model and the ensemble Random Forests model. The reason is that the neural network architecture implemented in TensorFlow reduces error rate by a very large margin. TensorFlow also beats other well-known frameworks such as Theano and Torch because of its optimization of numerical computations and better data visualization using convolutional filters, images, and graphs.

The results thus obtained assist the decision makers at the financial institution to understand who their potential clientele is. The subsequent marketing of the scheme can be focused on that group of clients to ensure the success of the scheme.

This predictive analysis helps the financial institutions to accurately score and profile the clients on the basis of the output. These models play a crucial role in maximizing the ROI and building & maintaining relationships with the clientele.

References

1. Forbes, Direct Marketing. https://www.forbes.com/sites/forbescommunicationscouncil/2017/09/29/10-metrics-worth-watching-when-judging-the-effectiveness-of-your-next-marketing-campaign/#5cf52a777edc. Accessed 02 Oct 2018
2. Investopedia, Direct Marketing. https://www.investopedia.com/terms/d/direct-marketing.asp. Accessed 04 Oct 2018
3. Elsalamony, H.A.: Bank direct marketing analysis of data mining techniques. Int. J. Comput. Appl. **85**(7), (2014). ISSN 0975–8887. https://research.ijcaonline.org/volume85/number7/pxc3893218.pdf

4. Syarif, S., Anwar, Dewiani: Trending topic prediction by optimizing K-nearest neighbor algorithm. In: 2017 4th International Conference on Computer Applications and Information Processing Technology (CAIPT) (2017). https://doi.org/10.1109/caipt.2017.8320711

5. Imandoust, S.B., Bolandraftar, M.: Application of k-nearest neighbor (KNN) approach for predicting economic events: theoretical background. Int. J. Eng. Res. Appl. 3(5), 605–610 (2013). https://www.ijera.com/papers/Vol3_issue5/DI35605610.pdf

6. Alkhatib, K., Najadat, H., Hmeidi, I., Shatnawi, M.K.A.: Stock price prediction using K-nearest neighbor (KNN) algorithm. Int. J. Bus. Humanit. Technol. 3(3), 32–44 (2013). https://www.ijbhtnet.com/journals/Vol_3_No_3_March_2013/4.pdf

7. Analytics Vidhya, K - Nearest Neighbour. https://www.analyticsvidhya.com/blog/2018/03/introduction-k-neighbours-algorithm-clustering/. Accessed 04 Oct 2018

8. Medium, K - Nearest Neighbour. https://medium.com/@adi.bronshtein/a-quick-introduction-to-k-nearest-neighbors-algorithm-62214cea29c7. Accessed 04 Oct 2018

9. Medium, Random Forest Algorithm. https://medium.com/@williamkoehrsen/random-forest-simple-explanation-377895a60d2d. Accessed 05 Oct 2018

10. Towards Data Science, Random Forest Algorithm. https://towardsdatascience.com/the-random-forest-algorithm-d457d499ffcd. Accessed 05 Oct 2018

11. Tensorflow, TensorFlow Guide. https://www.tensorflow.org/guide/. Accessed 07 Oct 2018

12. TensorFlow, Keras. https://www.tensorflow.org/guide/keras. Accessed 07 Oct 2018

Le Vision: An Assistive Wearable Device for the Visually Challenged

A. Neela Maadhuree[(⊠)], Ruben Sam Mathews, and C. R. Rene Robin

Department of Computer Science and Engineering,
Jerusalem College of Engineering, Chennai, Tamilnadu, India
123maadhuree@gmail.com

Abstract. We are only as blind as we want to be. In this paper, The Lé Vision is one assistive technology product that is focused on helping people with vision loss to be able to read. The device recognizes texts, snaps a picture, and relays the message to the user via an audio outlet device. The device is small, portable, and discreet allowing users to blend in with the crowd. To reduce the creation cost, we used a single-board computer, such as Raspberry Pi. It also provides obstacle detection enhancing the travelling experience of the visually impaired. We attach ultrasonic sensors for measuring the distance, interfaced to the Arduino. The presence of an obstacle is intimated via haptic feedback.

Keywords: Binarized · Thresholding · Analyze · Snapshot · Pixels

1 Introduction

With the advancement in today's technology we can enhance the standard of living. The sophisticated components and concepts available enables us to strive for the development of aiding the people in need which is the basic principle of engineering. The Lé Vision is one assistive technology product that is focused on helping people with vision loss to read. The device recognizes texts, snaps a picture, and relays the message to the user via an audio outlet device. The text recognition is performed by applying the concepts of neural networks and deep learning. The device is small, portable, and discreet allowing users to blend in with the crowd. We have also provided added features such as obstacle detection mechanism which guides the visually challenged to avoid it. The device has an efficiency of 82% which was measured on the basis of the sample set that was used to analyze the output. The individual word efficiency was also analyzed in terms of word score which are in detail explained in the paper. Thus the work was focused to exploit the concepts of neural networks to enable the text recognition that is applied with other basic concepts to develop a product that helps as a daily utility item for the visually challenged to aid them in their hurdles thus contributing towards the welfare of the society.

© Springer Nature Switzerland AG 2020
A. Abraham et al. (Eds.): ISDA 2018, AISC 941, pp. 353–361, 2020.
https://doi.org/10.1007/978-3-030-16660-1_35

2   Related Work

An Artificial Neural Network (ANN) is a computational model that is inspired by the way biological neural networks in the human brain process information. Artificial Neural Networks have generated a lot of excitement in Machine Learning research and industry, thanks to many breakthrough results in speech recognition, computer vision and text processing [7]. Deep Learning is a subfield of machine learning concerned with algorithms inspired by the structure and function of the brain called artificial neural networks. The core of deep learning according to Andrew is that we now have fast enough computers and enough data to actually train large neural networks. as we construct larger neural networks and train them with more and more data, their performance continues to increase. This is generally different to other machine learning techniques that reach a plateau in performance [8]. In [1] a methodology was suggested to enable the visually challenged to read newspaper on the daily basis for which the system was proposed by having a voice or braille output attached to the computer. This is an efficient method for a very specific task of enabling the newspaper narration. However the utility domain is very restrictive and has highly specific location dependency. It also involves a lot of hardware and computation. [2] suggests a technology for assisting the visually challenged to read the labels on product packaging for which it proposes the user to withhold the product in front of the camera and an image is captured which is converted to speech. however the limitation of this system could possibly be the necessity to exactly focus the particular region of the text location on the label which practically for a visually challenged person is hard to determine. Another perspective also suggests that this is only in terms of the printed text thus the computerized fonts are alone expected to be analyzable however the peer advantage of this mechanism is the utilization of motion based method to define a region of interest by which the untidy backgrounds and other surroundings can be isolated from the actual image. They utilized Novel Text Localization algorithm. In the earlier versions of text recognition a paper [3] presents a prototype for extracting text from images using Raspberry Pi. In this the images are captured in a webcam and processed using tools such as the Open CV or OTSU algorithm. These images are converted to gray scale, rescaled and vertical ratios are applied to bring in some form of a transformation. Later which some form of a adaptive thresholding is applied. Some special functions are utilized to draw the boundaries to determine the text. Whose similar text extraction approaches are enhanced in [4]. The ring as stated in [5, 11] is a assistive device that helps the blind read. It is in a ring shape worn on the index finger of the user. The user needs to scroll across the text in the pages. During which a miniature camera mounted on top of the ring captures the data and determines the text. The algorithm in the Ring device is provisioned to determine the end of line positions so that the user can scroll to the next line. The haptic actuators are used to provide the feedback to the user [14]. In [10] the various current trends of text recognition their methodology with limitations and advantages are analyzed in depth.

3   Problem Specification

People with complete blindness or partial vision often have a difficult time self-navigating outside even in well known environments. In fact, physical movement is one of the biggest challenges for blind people, explains World Access for the Blind [6].

Blindness affects a person's ability to perform many duties, which severely limits their employment opportunities, explains the World Health Organization [12]. This may not only affect a person's finances, but also their self-esteem [12].

In addition, they are limited with their activities like reading and need to adopt special mechanisms such as the Braille system which requires more effort, time and cost as special material has to be developed as well the personnel must be trained to use the Braille material.

4   Proposed Framework

Our work aims at developing smart glasses which can be used by the visually challenged people. The glasses include features such as text recognition and obstacle detection with reflexive responses. The glasses contain a small camera for image capturing and sensors to detect the distance [9,13]. The concept of obstacle detection by ultra sonic sensor has been used here. As soon as the obstacle is detected by the sensor, its distance is sent to the processing module. The glasses enable reflexive responses by calculating the distance between the person and the obstacle [15,17]. For the text recognition we rely on the computer vision technology based on Artificial Intelligence to analyse the text, the analyzed text is converted to speech using a speech API and provided to the user. The device is small, portable, and discreet allowing users to blend in with the crowd. The product is inexpensive since it is achieved by using smaller and cheaper components. Thus our product results in the complete replacement of the Braille system saving cost and effort. It also provides guidance to the visually impaired during travelling by providing the obstacle information leading to a safer travel experience. The aim of our work is to aid the visually challenged in their hurdles and thus two of the biggest problems faced by them are considered and a assistive device making the issues simpler for them is created. It also boosts the self-esteem of the visually impaired and brings in a confidence making them more independent.

5   Le Vision System Design

The system is basically distinguished as a two module set as shown in Fig. 1. Where the first module consists of the text recognition and conversion to speech which is achieved using a raspberry pi. The second module consists of the obstacle avoidance by using ultrasonic sensor and processing it in an Arduino kit. The raspberry pi and Arduino variations are provided for fault tolerance and backup system. For the text to speech conversion an image is captured and uploaded to the server. The text recognition is performed by an appropriate algorithm. The text is retrieved as response and converted into speech which is delivered to the user. We utilize the Stroke Width transform (SWT) for the text detection, the scale invariant feature transform (SIFT) is also employed to adhere to the varying input types.

Algorithm

```
predict text (image):
    init_image_stage_1 = sift(image)
    init_image_stage_2 = swft(image)
    sentence        = null
    text_level_1    = ann(init_image_stage_2)

      FOR everyword  IN text_level_1:
          word_1_prob  = calculate_probability (text_level_1)
          text_level_2 = lstm(text_level_2)
          word_2_prob  = calculate_probability (text_level_2)

          IF (word_1_prob  <  word_2_prob):
            final_word    = text_level_2
          ELSE
            final_word    = text_level_1
          END IF
      sentence := sentence + final_word
    END FOR
return sentence
```

5.1   Le Vision Framework

Thus the overall basic design view of the le vision has been shown in Fig. 2. There are much more processing involved in the intricate details of each component. It does not consist of a strict basic structure and can be modified according to the requirements and convenience. The structure proposed here is the one used in the implementation of the working model.

6   Emperical Evaluation

The developed product is tested with different input domains, the corresponding outputs are recorded and the performance is analyzed.

The set of different texts with various alignments, font sizes, font case (uppercase and lowercase), and various spacing's were given as input and the results were observed to be mostly approximate. The handwritten texts of various persons were also given as input. The system executed and provided the appropriate

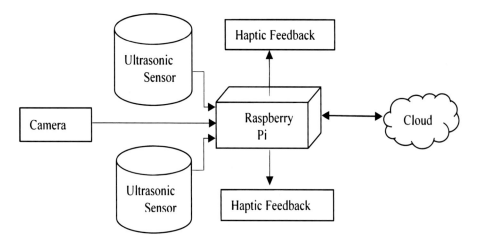

Fig. 1. Le Vision Framework

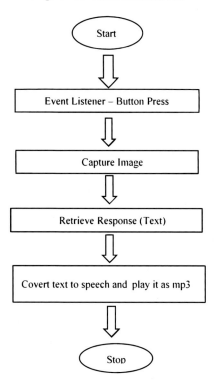

Fig. 2. Le Vision Process

results successfully. One of the test cases given as input is shown in Fig. 4 and its corresponding output in Fig. 5.

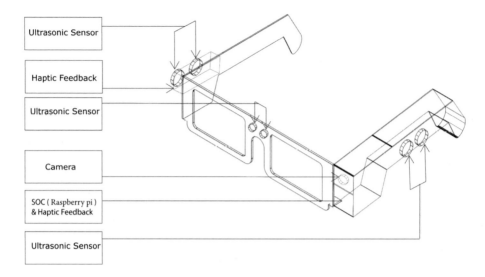

Fig. 3. Product [16]

Fig. 4. Input

The output of the test case was obtained as Output: GRAB IT WHILE IT'S HOT HotMeals Choose from a wide range of Hot Meals from the Spice Jet meal menu and save upto 10%. This corresponds to 100% accuracy but on a series of test cases it was determined that the overall system achieved 80% accuracy. Thus the system is tested and analyzed for its performance. The various stats are recorded. The system is considered to consist of acceptably effective results. A set of numerous test cases have been analyzed, the corresponding input and output are tabulated as in Table 1.

The performance measures are determined as

$$Precision = TP/(TP + FP) \tag{1}$$

$$Recall = TP/(TP + FN). \tag{2}$$

Table 1. Sample Output.

Image Input	Audio Output in Textual Form
	Dreams Re Not Whi You See N Sleep, They Are The One That Do Not Let You Sleep. Dreams 0.93423 Re 0.34533 NO 0.43545 Whi 0.6812 You 0.85678 See 0.87156 N 0. 74562 Sleep 0.6783 ,They 0.76453 Do 0.784753 Not 0.86835 Let 0.89545 You 0.84539 Sleep 0.81579
	Eminem Eminem 0.78233
	WISH YOR TEAM A GREAT SUCCESS. Wish 0.89904 Yor 0.78978 Team 0.56788 A 0.91234 Great 0.76899 Success 0.79350
	CONGRATULATIONS TO STUDENTS Congratulations 0.96341 To 0.77894 Student 0.76578
	You Don't Gt Another Chance, Life s No Nintendo Gam. You 0.78342 Don't 0.87775 Gt 0.67844 Another 0.64735 Chance 0.84234 Life 0.78543 s 0.5456 No 0.76743 Nintendo 0.78453 Gam 0.87349

```
High Performance MPEG 1.0/2.0/2.5 Audio Player for Layer 1, 2, and 3.
Version 0.3.2-1 (2012/03/25). Written and copyrights by Joe Drew,
now maintained by Nanakos Chrysostomos and others.
Uses code from various people. See 'README' for more!
THIS SOFTWARE COMES WITH ABSOLUTELY NO WARRANTY! USE AT YOUR OWN RISK!

Directory: /home/pi/Vision
Playing MPEG stream from cap.mp3 ...
MPEG 2.0 layer III, 32 kbit/s, 24000 Hz mono

[0.01] Decoding of cap.mp3 finished.
LOG :Sending Data
 GRAB IT WHILE IT'S HOT HotMeals Choose from a wide range of Hot Meals from the Spice Jet meal menu and save upto 10%
High Performance MPEG 1.0/2.0/2.5 Audio Player for Layer 1, 2, and 3.
Version 0.3.2-1 (2012/03/25). Written and copyrights by Joe Drew,
now maintained by Nanakos Chrysostomos and others.
Uses code from various people. See 'README' for more!
THIS SOFTWARE COMES WITH ABSOLUTELY NO WARRANTY! USE AT YOUR OWN RISK!

Playing MPEG stream from sp.mp3 ...
MPEG 2.0 layer III, 32 kbit/s, 24000 Hz mono

[0.09] Decoding of sp.mp3 finished.
High Performance MPEG 1.0/2.0/2.5 Audio Player for Layer 1, 2, and 3.
Version 0.3.2-1 (2012/03/25). Written and copyrights by Joe Drew,
now maintained by Nanakos Chrysostomos and others.
Uses code from various people. See 'README' for more!
THIS SOFTWARE COMES WITH ABSOLUTELY NO WARRANTY! USE AT YOUR OWN RISK!

Directory: /home/pi/Vision
Playing MPEG stream from end.mp3 ...
MPEG 2.0 layer III, 32 kbit/s, 24000 Hz mono

[0.01] Decoding of end.mp3 finished.
LOG   Bye Message
```

Fig. 5. Output

where,

TP represents true positive

FP represents false positive

FN represents false negative

According to our above test cases on an overall evaluation we obtain, $TP = 31$ $FP = 7$ $FN = 7$ hence precision is determined as 82% and corresponding recall also stands for 82%.

7   Conclusion

The glasses enable the visually challenged to enhance their standard of living. With its reflexive response it prevents the accidents by collision. With the text recognition the visually impaired people can enjoy the joy of reading the book. The Braille system can be completely replaced [18]. No more will people require to adopt or learn new techniques or will require specialized materials. The device enhances self-confidence among the visually challenged people as well enables them to be more independent. The text conversion is given to the user based on the necessity making it more usable. Voice texting is provided for better user interactivity.

The le vision is seen to be of great usability and can be improvised further by including better components such as webcam and other sophisticated components which are restricted in this paper due to fund constraints. We also planned to take forward and further extend the functionality of the product by introducing the face recognition. We can provide the translation of the text recognized and also work on scene description.

References

1. Electronic Newspapers for the Blind Available (Using Voice or Braille Output Device Attached to Their Computers, Users Can Read the Newspaper at Home), Feliciter (Ottawa), vol. 41, no. 6, 6 January 1995
2. Rajkumar, N., Anand, M.G., Barathiraja, N.: Portable camera-based product label reading for blind people. Int. J. Eng. Trends Technol. (IJETT), **10**(11), 521–524 (2014)
3. Rajesh, M., et al.: Text recognition and face detection aid for visually impaired person using Raspberry PI. In: 2017 International Conference on Circuit, Power and Computing Technologies (ICCPCT). IEEE (2017)
4. Liu, X., Samarabandu, J.: Multiscale edge-based text extraction from complex images. In: 2006 IEEE International Conference on Multimedia and Expo. IEEE (2006)
5. MIT News Magazine, May 2017. www.technologyreview.com/mit-news/2017/05/
6. Csapó, Á., et al.: A survey of assistive technologies and applications for blind users on mobile platforms: a review and foundation for research. J. Multimodal User Interfaces **9**(4), 275–286 (2015)
7. Turki, H., Ben Halima, M., Alimi, A.M.: Text detection based on MSER and CNN features. In: 2017 14th IAPR International Conference on Document Analysis and Recognition (ICDAR), vol. 1. IEEE (2017)
8. Deep Learning, Wikipedia page. https://en.wikipedia.org/wiki/Deep_learning
9. Elmannai, W., Elleithy, K.: Sensor-based assistive devices for visually-impaired people: current status, challenges, and future directions. Sensors **17**(3), 565 (2017)
10. Manwatkar, P.M., Yadav, S.H.: Text recognition from images. In: 2015 International Conference on Innovations in Information, Embedded and Communication Systems (ICIIECS). IEEE (2015)
11. 10 March 2015. http://news.mit.edu/2015/finger-mounted-reading-device-blind-0310
12. World Health Organization news releases. http://www.who.int/
13. Bharathi, S., Ramesh, A., Vivek, S.: Effective navigation for visually impaired by wearable obstacle avoidance system. In: 2012 International Conference on Computing, Electronics and Electrical Technologies (ICCEET). IEEE (2012)
14. Ingber, J.: An update on the finger reader, an on-the-go reading device in development at MIT. Access World Mag. **16**(7) (2015)
15. Kumar, A., Kaushik, A.K., Yadav, R.L.: A robust and fast text extraction in images and video frames. In: Advances in Computing, Communication and Control, pp. 342–348. Springer, Heidelberg (2011)
16. Base template for Fig:3 Created by Rawpixel.com - Freepik.com used under license CC
17. Chen, D., Odobez, J.-M., Bourlard, H.: Text detection and recognition in images and video frames. Pattern Recognit. **37**(3), 595–608 (2004)
18. Karungaru, S., Terada, K., Fukumi, M.: Improving mobility for blind persons using video sunglasses. In: 2011 17th Korea-Japan Joint Workshop on Frontiers of Computer Vision (FCV). IEEE (2011)

Permission-Based Android Malware Application Detection Using Multi-Layer Perceptron

O. S. Jannath Nisha[(⊠)] and S. Mary Saira Bhanu

National Institute of Technology, Tiruchirappalli 620015, Tamilnadu, India
406115006@nitt.edu, msb@nitt.edu

Abstract. With the increasing number of Android malwares, there arises a need to develop a system that automatically detects malware in Android applications (apps). To discriminate malware applications (malapps) from benign apps, researchers have proposed several detection techniques to detect malicious apps automatically. However, most of these techniques depend on hand-crafted features which are very difficult to analyze. In this paper, the proposed Android Malware Detection approach uses MultiLayer Perceptron (MLP) to discriminate malware apps from benign ones. This approach uses permissions as features based on the static analysis techniques from a disassembled APK file. Apps permission features are automatically learned by the neural network and thereby removing the need for hand-crafted features and are trained to discriminate apps. It is computationally feasible with a large number of dataset samples to perform classification. After the training, the MLP network can be executed, allowing a substantial amount of files to be detected and providing high accurate classification rate. The proposed approach is evaluated using the benchmark dataset. The experimental results show that the proposed approach can achieve high accuracy than that of the existing techniques.

Keywords: Android operating system · Benign apps · Malapps ·
Permissions · Neural networks · MLP

1 Introduction

Mobile devices are used in the modern world mainly as a tool to call others and sending messages for communication. Now mobile users can use the device for sharing multimedia contents and also use some software for personal and business purpose with the help of operating systems. Among several Operating Systems such as Apple IOS, Symbian, Blackberry OS, and Windows OS, Android [20] is the world's most widely used operating system.

Due to the open nature of Android [5], an attacker can easily download one benign app, repackage it with malicious code and upload the app to official or third-party markets. Android malware running on mobile devices violates

© Springer Nature Switzerland AG 2020
A. Abraham et al. (Eds.): ISDA 2018, AISC 941, pp. 362–371, 2020.
https://doi.org/10.1007/978-3-030-16660-1_36

security and privacy. For the protection of the information and devices, Android has two security mechanisms [7,11], sandbox environment at the kernel level to prevent access to the file system and an API of permissions that controls the privileges of other resources of the apps. Permissions [3] are the security mechanisms at the apps development level. Researcher community has developed powerful solutions for malware detection based on permissions.

With the increasing popularity of mobile devices and their official or unofficial app markets, it is challenging to examine each app manually to detect the behavior of the app. Traditionally, apps behavior has analyzed manually, but this process cannot interpret a large number of apps. Consequently, automatic malware detection techniques [2,9] have proposed based on Android apps requested permissions. However, the expert knowledge is required to extract the relevant features that are used to perform the classification task.

Deep learning [16] is a part of the machine learning method that reflects the way the human brain acts and has achieved attention in the field of artificial intelligence. It plays a major role in speech recognition and natural language processing. The method derives the essential features from the high level representation of features using a hierarchical of multiple layers. Each layer transforms the input features into more abstract representation. The output layer combines those features to make predictions.

This paper uses an approach of MLP-based Neural Networks for Android malware detection based on apps permissions. The advantage of this approach is that features are learned from the given permissions automatically and hence removes the need to preprocess the features. Furthermore, the accuracy of malware detection will improve by giving more training samples because the MLP has an enormous learning capacity to train the model. Once the model is trained, a large number of malapps can be efficiently detected. A new malapps that appear in the markets at any time can also detected by retraining with different malware samples to accommodate to the growing malware environment using incremental learning algorithm [10].

The outline of this paper is as follows: Sect. 2 explains different Permission-based malware detection techniques. A novel approach is presented in Sect. 3 to discriminate malware apps from benign apps and the experiments and results are discussed in Sect. 4. Finally, Sect. 5 concludes the paper.

2   Related Work

This section explains previous research works of permission-based malware detection method using hand-crafted features [18] applied to static technique. Previous static malware detection approaches used manually extracted features such as permissions, and API calls with various classification algorithms such as SVM [1], Logistic Regression, Decision Tree, Random Forest, Linear Discriminant Analysis and K-Nearest Neighbor [13].

Sahs and Khan [12] proposed a machine learning system to detect Android malware in mobile devices with the One-Class SVM classification algorithm and static analysis as a technique to obtain the information from the apps. During the development, they used the Androguard tool to extract the information of the APK. From the AndroidManifest.xml, they developed a binary vector that contains the information of each permission used for every application. Wei et al. [17] proposed a machine learning method to detect Android malapps. The authors implemented an Androidetect malicious app detection tool to examine the relationship between system functions, sensitive permissions, and app programming interfaces. Aung et al. [2] introduced a framework to built a machine learning detection system to detect malapps and to secure the mobile users. This framework extracts permissions from the android apps, and examines these features by using various classifiers to detect whether the app is malicious or not.

In [8], the authors proposed a security advisory system based on the combinations of permissions and the rating from the app market. They used predefined set of rules for recognizing unusual combinations of permissions and also calculate a risk range based on combination of permissions, the number of downloads and its rating. The resulting accuracy is not acceptable, but the authors showed that there is a relationship between malapps and its statistics from the market that is most malapps found on the market have less number of ratings and downloads.

Artificial intelligence techniques have also been explored for safety in Android. In conventional machine learning methods, the expert analyzes the malapps and extracts the features manually from the raw data. Then a classifier is used to develop a model. A neural network can study the features from the raw data automatically to perform classification. In [6], the authors proposed MLP algorithm to assess the security vulnerabilities presented in permission structure of an Android application and can predict app categories efficiently. But the proposed method is not able to detect whether tha app is malicious or not. This paper uses MLP neural networks to train the model and perform the classification between malware and benign.

3   Approach

This section presents an Android Malware Detection approach using MLP neural networks. This approach has a simple design, where feature extraction and detection are based on MLP neural networks.

3.1   Multi-Layer Perceptron

MLP [19] is a fully connected artificial neural network, and it is composed of more than one perceptron. MLP neural network consists of multiple layers such as the input layer, hidden layers, and the output layer that are shown in Fig. 1. Each layer has many artificial neurons and the neurons in each layer connected to neurons in the previous layer. The input layer receives the input features, an output layer that makes a detection about the input apps, and in between, an arbitrary number of hidden layers that are the computational engine of the MLP. A single neuron can be computed by the following equation:

$$U_k = \sum_{j=1}^{m} W_{kj}.x_j + b_k \qquad (1)$$

Then the sigmoid function is used to calculate the output from a single neuron. The advantage of this MLP network is that it learn the representation automatically from the features to perform detection. This paper takes one more step towards Android malware detection with automatic representation. To attain this, MLP considers the permissions from an Android app file for malware detection. In this work, the proposed approach can be divided into three steps: (1) Permission extraction, (2) Training phase, and (3) Detection phase.

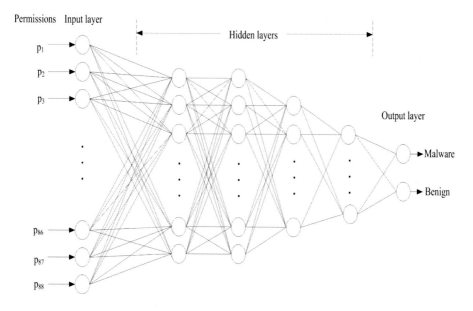

Fig. 1. Overall view of the Android malware detection using MLP neural networks

Permission Extraction. Google Android [14] uses 'all-or-nothing' permission granting policy to prevent access to privileged system resources. App developers have to declare the requested permissions explicitly. When the user grants all the required permissions, then only app is installed successfully. In this way, the permissions can protect the user's private information from unauthorized access. For instance, the INTERNET permission is required to perform network communications. So, establishing a network connection is prevented by the INTERNET permission and the READ CONTACTS permission to read entries in a user's contacts as well. To request permission, the developer declares a

$< uses - permission >$ attribute in the Manifest file. An app can also declare its custom permissions to protect its resources, by declaring a $< permissions/ >$ attribute.

Reverse Engineering process [21] can identify and explain the complete working of an app. Each application [22] is decompiled from an Android Application Package [APK] file, which includes the code of the classes.dex files, and the AndroidManifest.xml file. In that, the manifest file is a resource file which contains all the information needed by the operating system about the application. Permissions are extracted statically using the tools such as Androguard [4] and APKtool [7]. In this work, the benchmark dataset from [15] is used for the evaluation. This dataset consists of 88 permissions that is $P = \{p_1, p_2, p_3 ...... p_{88}\}$ and feature vectors are generated from the datasets.

Training Phase. The feature vector is given to MLP, which consists of four fully-connected hidden layers and a fully-connected output layer. The first two hidden layer contains 60 neurons. Next, the third and fourth layer consists of 40 and 15 neurons. The input of this phase is $(A(n), C(n))$, where $A(n)$ is the vector constructed from the nth app in the training sample, and $C(n)$ is the corresponding label. When $C(n) = 1$, the app is malicious, otherwise $C(n) = 0$ (benign). During the training process, the work is to train the MLP networks so that they can detect malware from benign apps.

Initially, the induced local fields and function signals are computed by proceeding forward through the network. The induced local field for neuron j in layer l is defined as

$$v_j^{(l)}(n) = \sum_i W_{ji}^{(l)}(n) . y_i^{(l-1)}(n) \tag{2}$$

where $W_{ji}^{(l)}(n)$ is the weight of neuron j in layer l that is fed from neuron i in layer $l - 1$ and $y_i^{(l-1)}(n)$ is the output of neuron i in layer $l - 1$ at iteration n. When $i = 0$, $w_{j0}^{(l)}(n) = b_j^{(l)}(n)$ is the bias applied to neuron j in layer l and $y_0^{l-1}(n) = +1$ The output of neuron j in layer l is

$$y_j^{(l)}(n) = (v_j^{(l)}(n)) \tag{3}$$

When $l = 1$, $y_j^{(l)}(n) =$ the jth element of $x(n)$. Next, the neural network depth is denoted by L. When $l = L$, compute the error signal using the following equation:

$$e(n) = d(n) - y_1^{(L)}(n) \tag{4}$$

This method detects malapps based on permissions. Sigmoid function is used in hidden layers and an output layer. This phase used a 2-neuron output layer to represent the 2-D binary vectors is either "1" (malware) or "0".

Detection Phase. After training the network, the MLP makes a permission-based detection when being fed to a testing dataset. This phase contains 20% of apps from the dataset. By using the above approach, the MLP has achieved the capability to detect the app. This step requires the neural network should follow only forward computations. Primarily, feed the testing feature vectors to the trained approach. According to Eqs. 2 and 3, the performance are obtained by calculating the output of each node in the neural network. The value will be in between 0 and 1. Using this value, the networks do the classification. Depending upon the value, the application will be recognized as a malicious or a benign.

4   Experiments and Results

The MLP program developed using the TensorFlow computing environment. During training, the parameters were optimized with a learning rate of 0.2, for 1000 epochs. The MLP weights were randomly initialized using TensorFlow initialization. Nvidia GTX 980 GPU is used for the development of the MLP network and training the network to detect malware which takes around a few minutes for the large dataset. Once the network is trained, the MLP network can classify the malapps approximately.

The benchmark permission vector dataset [15] is used for the experiments and evaluation. The dataset consists of 3,10,926 benign apps and 3207 malicious apps. A subset of the permission vector dataset is used for evaluating the proposed model. Three subsets generated from the permission vector set are named as Permission Vector set 1 (1000 apps), Permission Vector set 2 (2000 apps), and Permission Vector set 3 (7000 apps). The dataset description is given in Table 1.

Experiments were conducted on three different datasets to estimate the performance of the proposed approach. A large number of training samples should be available in the dataset to avoid over-fitting and also achieve high accuracy. From the Fig. 2, it can be observed that the large dataset has higher accuracy than the other two smaller datasets.

Tables 2, 3, and 4 present the values of each PV sets for all the classification results using the accuracy, precision, recall, TPR, F1-score and FPR. In Table 2, the solutions proposed by RF has higher detection performance than the proposed solution. In the case of the solutions proposed by RF has equal performance in Table 3. In Table 4, the results reveal that MLP can predict malapps efficiently and reliably using a large number of training samples.

Table 1. Datasets descriptions

Datasets	Benign	Malware
Permission vector set 1 (PV set 1)	700	300
Permission vector set 2 (PV set 2)	1000	1000
Permission vector set 3 (PV set 3)	4000	3000

Table 2. Classification results of PV set 1

Classifier	Accuracy	Recall	Precicion	TNR	F1 score	FPR
KNN	92.30	0.95	0.94	0.87	0.94	0.12
SVC	87.63	0.98	0.86	0.66	0.92	0.34
LDA	91.97	0.99	0.90	0.76	0.94	0.24
LR	90.97	0.97	0.90	0.77	0.93	0.22
DT	92.64	0.94	0.95	0.90	0.95	0.09
RF	94.31	0.97	0.95	0.89	0.96	0.11
MLP	93.93	0.85	0.94	0.98	0.90	0.02

Table 3. Classification results of PV set 2

Classifier	Accuracy	Recall	Precicion	TNR	F1 score	FPR
KNN	92.32	0.93	0.93	0.92	0.93	0.08
SVC	92.15	0.96	0.89	0.87	0.93	0.13
LDA	93.65	0.97	0.91	0.89	0.94	0.10
LR	94.15	0.96	0.93	0.92	0.96	0.08
DT	93.65	0.92	0.95	0.95	0.94	0.05
RF	95.49	0.96	0.96	0.95	0.96	0.04
MLP	95.17	0.97	0.93	0.93	0.95	0.07

Table 4. Classification results of PV set 3

Classifier	Accuracy	Recall	Precicion	TNR	F1 score	FPR
KNN	84.62	0.90	0.88	0.73	0.89	0.27
SVC	83.61	0.97	0.83	0.55	0.89	0.45
LDA	84.28	0.94	0.85	0.62	0.89	0.38
LR	86.62	0.91	0.90	0.76	0.90	0.24
DT	86.95	0.92	0.89	0.75	0.91	0.25
RF	88.62	0.93	0.91	0.78	0.92	0.22
MLP	96.83	0.96	0.96	0.97	0.96	0.02

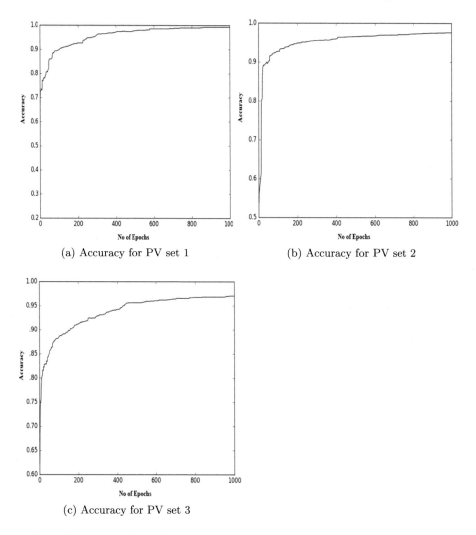

(a) Accuracy for PV set 1

(b) Accuracy for PV set 2

(c) Accuracy for PV set 3

Fig. 2. Comparing the performance of MLP approach for three different datasets

5 Conclusion

The importance of the security of mobile devices has increased throughout the years. In order to solve this, this work introduced a new Android malware detection system based on MLP NNs. The proposed approach shows high classification rate, and also the comparison with K Nearest Neighbor, Support Vector Machine, Linear Discriminant Analysis, Logistic Regression, Decision Tree, and Random Forest shows that MLP offers the best results among the algorithms examined, and validated in Android apps dataset. This approach has the capacity of learning simultaneously to perform feature extraction and malware classification using a large number of app samples with labels. The merits are that it eliminates the necessary for hand-crafted features, and computationally efficient than current malware classification task.

In future, extend the approach to both dynamic and static analysis of malware. For example, the NN could process system calls sequences produced during dynamic analysis of Android apps. Likewise, by adjusting the disassembled preprocessing step, the same network structure could be utilized to analyze malware on various levels of features in Android apps.

References

1. Arp, D., Spreitzenbarth, M., Hubner, M., Gascon, H., Rieck, K., Siemens, C.: Drebin: effective and explainable detection of android malware in your pocket. In: Ndss, vol. 14, pp. 23–26 (2014)
2. Aung, Z., Zaw, W.: Permission-based android malware detection. Int. J. Sci. Technol. Res. **2**(3), 228–234 (2013)
3. Brahler, S.: Analysis of the android architecture. Karlsruhe institute for technology **7**(8) (2010)
4. Desnos, A.: Androguard: Reverse engineering, malware and goodware analysis of android applications... and more (ninja!) (2015)
5. Faruki, P., Bharmal, A., Laxmi, V., Ganmoor, V., Gaur, M.S., Conti, M., Rajarajan, M.: Android security: a survey of issues, malware penetration, and defenses. IEEE Commun. Surv. Tutor. **17**(2), 998–1022 (2015)
6. Ghorbanzadeh, M., Chen, Y., Ma, Z., Clancy, T.C., McGwier, R.: A neural network approach to category validation of android applications. In: 2013 International Conference on Computing, Networking and Communications (ICNC), pp. 740–744. IEEE (2013)
7. Jha, A.K., Lee, W.J.: Analysis of permission-based security in android through policy expert, developer, and end user perspectives. J. UCS **22**(4), 459–474 (2016)
8. Matsudo, T., Kodama, E., Wang, J., Takata, T.: A proposal of security advisory system at the time of the installation of applications on android OS. In: 2012 15th International Conference on Network-Based Information Systems (NBiS), pp. 261–267. IEEE (2012)
9. Moonsamy, V., Rong, J., Liu, S.: Mining permission patterns for contrasting clean and malicious android applications. Future Gener. Comput. Syst. **36**, 122–132 (2014)
10. Polikar, R., Upda, L., Upda, S.S., Honavar, V.: Learn++: an incremental learning algorithm for supervised neural networks. IEEE Trans. Syst. Man Cybern. Part C (Appl. Rev.) **31**(4), 497–508 (2001)
11. Rashidi, B., Fung, C.J.: A survey of android security threats and defenses. JoWUA **6**(3), 3–35 (2015)
12. Sahs, J., Khan, L.: A machine learning approach to android malware detection. In: 2012 European Intelligence and Security Informatics Conference (EISIC), pp. 141–147. IEEE (2012)
13. Sharma, A., Dash, S.K.: Mining API calls and permissions for android malware detection. In: International Conference on Cryptology and Network Security, pp. 191–205. Springer (2014)
14. Tchakounté, F.: Permission-based malware detection mechanisms on android: analysis and perspectives. J. Comput. Sci. **1**(2) (2014)

15. Wang, W., Wang, X., Feng, D., Liu, J., Han, Z., Zhang, X.: Exploring permission-induced risk in android applications for malicious application detection. IEEE Trans. Inf. Forensics Secur. **9**(11), 1869–1882 (2014)
16. Wang, X., Yang, Y., Zeng, Y.: Accurate mobile malware detection and classification in the cloud. SpringerPlus **4**(1), 583 (2015)
17. Wei, L., Luo, W., Weng, J., Zhong, Y., Zhang, X., Yan, Z.: Machine learning-based malicious application detection of android. IEEE Access **5**, 25591–25601 (2017)
18. Wu, D.J., Mao, C.H., Wei, T.E., Lee, H.M., Wu, K.P.: DroidMat: android malware detection through manifest and API calls tracing. In: 2012 Seventh Asia Joint Conference on Information Security (Asia JCIS), pp. 62–69. IEEE (2012)
19. Xiao, X., Wang, Z., Li, Q., Xia, S., Jiang, Y.: Back-propagation neural network on Markov chains from system call sequences: a new approach for detecting android malware with system call sequences. IET Inf. Secur. **11**(1), 8–15 (2016)
20. Zaidi, S.F.A., Shah, M.A., Kamran, M., Javaid, Q., Zhang, S.: A survey on security for smartphone device. Int. J. Adv. Comput. Sci. Appl. **7**(4), 206–219 (2016)
21. Zhou, W., Zhou, Y., Jiang, X., Ning, P.: Detecting repackaged smartphone applications in third-party android marketplaces. In: Proceedings of the Second ACM Conference on Data and Application Security and Privacy, pp. 317–326. ACM (2012)
22. Zhou, Y., Wang, Z., Zhou, W., Jiang, X.: Hey, you, get off of my market: detecting malicious apps in official and alternative android markets. In: NDSS, vol. 25, pp. 50–52 (2012)

Accelerating Image Encryption with AES Using GPU: A Quantitative Analysis

Aryan Saxena⬤, Vatsal Agrawal⬤, Rajdeepa Chakrabarty⬤,
Shubhjeet Singh⬤, and J. Saira Banu$^{(\boxtimes)}$⬤

School of Computer Science and Engineering (SCOPE),
Vellore Institute of Technology, Vellore, India
{aryan.saxena2016,vatsal.agrawal2016,
rajdeepa.chakrabarty2016,
sshubhjeet.singh2016}@vitstudent.ac.in,
jsairabanu@vit.ac.in

Abstract. From military imaging to sharing private pictures, confidentiality, integrity and authentication of images play an important role in the Internet of modern world. AES is currently one of the most famous symmetric cryptographic algorithms. Performing encryption/decryption of high-resolution images with AES is highly computation-intensive and time-consuming due to their large sizes and the underlying complexity of AES algorithm. Earlier increasing cryptographic complexity meant an increase in the processing time for encryption as well as decryption. Now, with the rise of the powerful GPUs containing thousands of high-performance and efficient cores and with the evolvement of GPU computing, the processing time has been reduced to a fraction of the time it used to take earlier. This paper presents a parallel implementation of AES using NVIDIA CUDA and OpenCV to encrypt images rapidly. We achieved an average speed up of four times on GPU as compared to CPU-only.

Keywords: AES · GPU · Image encryption · OpenCV · CUDA · Speedup

1 Introduction

Parallel programming today is being exploited in various arenas of computer science to improve the performance of algorithms. Moreover, with the advent of GPU computing, GPUs are not only restricted to application in graphical computation, they are also being used to perform general purpose computing. Hence, it has given us the opportunity to utilize the power of GPUs on various new applications to reduce the processing time drastically.

Security of data is the prime concern of this era. We need a much higher security in data transmission as compared to previous generations. AES (Advanced Encryption Standard) is one of the most popular and extensively used algorithm in cryptography. It is secured enough to be used by the US government for transmitting their classified information. It consists of different rounds of permutations and substitutions, depending upon the key size used: 10 rounds for 128-bit, 12 rounds for 192-bit and 14

© Springer Nature Switzerland AG 2020
A. Abraham et al. (Eds.): ISDA 2018, AISC 941, pp. 372–380, 2020.
https://doi.org/10.1007/978-3-030-16660-1_37

rounds for 256-bit key. As the key size increases, the resistance against attacks increases but this also leads to an increased complexity and computation intensity. AES is quite efficient when a 128-bit key is used but it is extremely slow and computation-heavy when it comes to using 192 or 256-bit keys. Encrypting a file of as large size as a high-resolution image requires a mechanism which has to be highly efficient to be of any practical use for us. The images can be viewed as a two-dimensional matrix of integers of varying values depending upon the bit depth of the image. Higher resolution images have a large number of pixels and due to this higher size, it takes a long time to process them. To solve this problem, we harness the power of the thousands of cores present inside a GPU. Each core can act independently and parallelly process different parts of the image.

The advantage in our methodology is that we can share the encrypted image in the image form itself. Without knowing the exact key, it is computationally infeasible to know the contents of the image as the image will be just a random pattern of different colours. The next section presents a brief review of the different methods researchers have devised in the past for encryption/decryption of text and images. We then present our proposed method in detail along with the observations and results in the subsequent sections.

2 Literature Review

Nagendra et al. [1] showed how AES implemented on dual processor by OpenMP API takes 40–45% lesser time for performing the encryption and decryption than the sequential implementation. Elkabbany et al. [2] used eleven stages of pipelining in order to exploit the sources of parallelism in both initial and final round. The paper combined pipelining of rounds and parallelization of Mix_Column and Add_Round_Key transformations. It showed that Pipelining improves both encryption and decryption by approximately 95%. Daniel and Stratulat [3] discuss the implementation of plain-text encryption with AES using CUDA. Since data is stored in local memory of the GPU this implementation offers speedups of almost 40 times in comparison to the CPU. It also aimed to implement and test AES 192, AES 256 on GPU, encryption and decryption and OpenSSL adaptation to use these algorithms.

Sowmiya et al. [4] showed that to avoid the data redundancy the pixel and magic square method has been combined to give a new algorithm that is also named as Pan Magic Square method. The plain text is divided into pixels and a total of 64 keys are generated. The encrypted result is very different when compared to the text so can easily be transmitted over the internet.

Feng, Xia and Tian [5] introduce a method which is based on fractional Fourier equations for the purpose of image encryption. Further the magical cube rotation scrambling is also used in this paper. Ren et al. [6] showed in their paper the security concerns of the image encryption using Chaotic maps. It shows how the decryption of the image can be done by the proposed algorithms very easily and hence they proved the vulnerability of the previous algorithms. Lei et al. [7] presented an image encryption algorithm which is derived from SMS4 commercial cipher algorithm. In the proposed algorithm, encryption, decryption and a safe transmission of the image is ensured. The conclusion drawn is that grey level images are more equally distributed. The security of the image is enhanced and gray level pixel produces better results.

Li et al. [8] showed how the AES encryption and decryption make high performance enhancement. Bandwidth of PCI-E bus and page-lock memory allocation costs are vital limitations. Abdelrahman [9] implemented AES on CUDA and derived the observation that workload distribution over threads and thread blocks gave higher performance but some of the optimizations such as parallel granularity tweaking did not have effect on the older platforms. Khan et al. [10] concluded that input plaintext pattern was more random and less repetitive. Speed up of 87 times compared to modern CPU was achieved. Execution time was increased by taking more random input plaintext. Patchappen et al. [11] implemented multi variant AES cipher and found that a higher throughput and a greater performance than multi core CPU are achieved when processing data size larger than 512 MB. Ma, Chen, Xu, Shi [12] concluded that GPU can accelerate the speed of implementation of AES to a large extent when input data is large. They also proposed to optimize the data transferring overhead and to implement other encryption algorithms on GPU.

Subramanyan et al. [13] showed encryption of image using AES algorithm. This method uses a 128-bit key and the key used is different for all the blocks of pixels which are generated independently on both sides i.e. the sender and the receiver. They proposed this method to make the encryption process of images more secure but they failed at achieving time optimized encryption for large sizes of images.

Zhang [14] in his paper carried out the test to verify whether AES algorithm can be used for highly secured and time optimized encryption of images. He finally concluded that AES encryption is a very secure algorithm for image encryption and it also performs pretty fast on general purpose computers. He suggested that the speed of AES encryption should be considered as a benchmarks and algorithms performing encryption at speeds lower than this should be rejected. The only drawback is that his implementation is on a CPU based platform whereas a higher time optimized implementation can be done on a GPU based platform.

So, after reviewing these papers, we found that the best way we can achieve a high speed for encryption/decryption of images without compromising the security is to implement AES in a counter (CTR) mode parallelly on the GPU. CUDA global memory has to be used instead of shared memory because of the large size of images as the total amount of shared memory per block is only 48 KB, whereas the total amount of global memory is 2004 MB on the machine we implemented our code (Table 1). We discuss our proposed implementation in detail in the following section.

3   Proposed Work

The image to be encrypted is taken as one of the input and the other input has to be the private key which the users want to use to encrypt their images. This method utilizes a 256-bit key for a better protection. For this key, users are required to enter a text of their choice of any arbitrary length. This text can be anything - a paragraph, a sentence, a word or even a single character. Hash digest of the entered text is calculated with the hashing algorithm SHA-256 for producing a 256-bit length key.

OpenCV C++ library is used for pixel-wise encryption of images. The input image is read using the `imread()` function into the program in RGB mode so that it will have three channels, one each for red, green and blue colours. Each has its own 8-bit value (0–255). An unsigned char data type has been used to represent each channel component of each pixel. These will be stored in the form of a two-dimensional matrix.

AES is a block cipher which operates upon 128-bit of input at a time. Hence, image needs to be divided into blocks of sixteen pixels each. This is because each pixel is an 8-bit unsigned character and sixteen pixels together will sum to 128-bits. Padding can be used to make sum to 128 bits if the number of pixels is lesser than 16 in the last block.

NVIDIA CUDA provides abstraction to the programmer. It offers a single Grid divided into Blocks containing equal number of Threads. The CUDA platform architecture is greatly simplified and is shown in Fig. 1. Allocating each of these user-defined threads to a GPU core is done automatically by the CUDA.

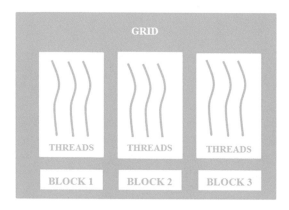

Fig. 1. CUDA Grid, Blocks and Threads

For efficiency, we have taken the number of blocks in grid as the number of processors in the multiprocessor GPU. Our GPU has three processors and hence we defined three blocks in our grid. We divide the number of pixel-blocks of input image by this number of blocks to obtain the number of threads. If the number of threads in a block exceeds 1024, we allocate a new block to maintain the number. Each block of sixteen pixels will then be run on an independent thread for the further rounds of AES which will ensure maximum parallelization. AES-256 algorithm basically has 14 rounds consisting of four operations: substitution, row shifting, column mixing and adding round keys as shown in Fig. 2.

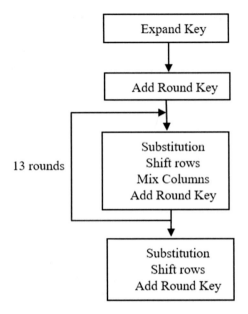

Fig. 2. The rounds in AES-256 which the plaintext undergoes to transform itself into ciphertext

For the purpose of parallel encryption, it is quite evident that we cannot use the CBC (Cipher Block Chain) mode of operation as each thread will require the resultant cipher text of the previous block and this dependency will hinder our performance. We also cannot use the ECB (Electronic Code Book) mode because there is a large amount of redundancy present in the images and since the blocks are encrypted independently, there is a high possibility to guess the pattern present in the image. Hence, we will be using the CTR (counter) mode, which indexes and encrypts an auto-incrementing integer counter with the symmetric key and then performs the XOR operation with the plaintext mode.

After obtaining the cipher-text, we convert the cipher-text into hexadecimal numbers and then represent each encrypted pixel as a combination of these hexadecimal numbers taking two numbers at a time. These pixels can then be displayed in the form of an encrypted image. This image will mostly contain random structures.

The same technique in the reverse order has been used for decrypting the encrypted image back to the original image.

4   Results

We processed various images of different resolutions ranging from a low-resolution image of size 1280 × 720 (0.3 MB) to a high-resolution image of size 7680 × 4320 (11.6 MB) using the same 256-bit key. Figure 3 shows the original image and the encrypted image obtained. After that we compared the time taken for processing between GPU and CPU. The configuration details of the machine upon which the experiments were performed are tabulated in Table 1. Table 2 lists the timing results for images of various sizes. We also depict the time it takes for encryption and decryption of general text files of the same sizes as that of images in Table 3 for the purpose of comparison.

Note that the timings in Table 2 depict the total time taken to read an image, encrypt it, generate and save the encrypted image, then again read the encrypted image, decrypt it and finally output the original image. The formula used for calculating throughput (in Mbps) is given in Eq. 1.

$$Throughput\ (in\ Mbps) = \frac{8 \times file - size\ (MB)}{Time\ taken\ (s)} \tag{1}$$

Table 1. Machine hardware configuration

GPU	CPU
NVIDIA GeForce 940MX	Intel® Core™ i5
2 GB DDR3	7th Generation
3 Multiprocessors	4 cores
128 CUDA Cores/MP	8 GB DDR4 RAM

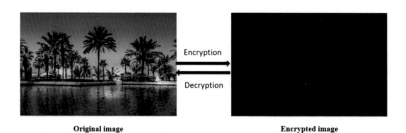

Original image   Encrypted image

Fig. 3. Notice how the encrypted image has randomised pixels and hence the original image cannot be guessed just by looking at it.

Table 2. Timing results for encrypting/decrypting images

File size (MB)	Image resolution	GPU (sec)	CPU (sec)	Speedup	Throughput (Mbps)	
					GPU	CPU
11.6	7680 × 4320	29.52	147.63	5x	3.14	0.63
5.32	5120 × 2880	13.28	64.94	4.9x	3.20	0.65
3.17	3840 × 2160	7.58	36.71	4.8x	3.34	0.69
1.32	2560 × 1440	3.53	16.49	4.7x	2.99	0.64
0.71	1920 × 1080	1.90	9.14	4.8x	2.99	0.62
0.30	1280 × 720	0.91	4.09	4.5x	2.64	0.59

Table 3. Timing results for encrypting/decrypting text files

File size (MB)	GPU (sec)	CPU (sec)	Speedup	Throughput (Mbps)	
				GPU	CPU
11.6	1.33	6.19	4.65x	69.77	14.99
5.32	0.66	2.86	4.33x	64.48	14.88
3.17	0.46	1.66	3.61x	55.13	15.27
1.32	0.21	0.71	3.38x	50.28	14.87
0.71	0.17	0.40	2.35x	33.41	14.20
0.30	0.12	0.16	1.33x	20.00	15.00

We observe that it takes a long time for the CPU to encrypt an image while a GPU can complete similar task in a fraction of time. We can infer from Fig. 4 that there is just a small difference between the execution times on GPU and CPU for low resolution images like 1280 × 720 and 1920 × 1080 images. But as we encrypt/decrypt high resolution images of the order of 5120 × 2880 and 7680 × 4320, the time difference becomes significant. Similar trend can be observed in Fig. 5 for text files. Another observation we infer from Tables 2 and 3 is that the speedup achieved on GPU over CPU increases at a higher rate for text files than for images on increasing the size of files.

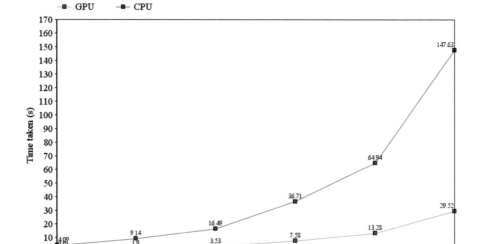

Fig. 4. CPU vs GPU - Time comparison for encryption/decryption of images

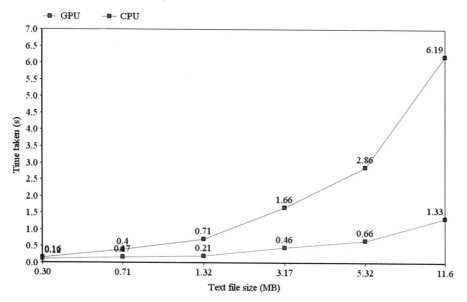

Fig. 5. CPU vs GPU - Time comparison for encryption/decryption of text files

5   Conclusion

We have presented a method to accelerate image encryption/decryption with AES using the GPU. We have also calculated the timing results for images of different resolutions and have compared them with a CPU-only computation. The speed-up achieved on GPU for images is around five times with respect to the CPU. Hence, the user need not compromise the security of the image to encrypt it in a shorter time by taking a smaller key or by using some less secure cryptographic algorithm.

Our future aim is to try the same parallel implementation on other parallel architectures too and compare their performance. We also aim to reduce the space complexity of our implementation. With increasing sharing of multimedia over social networking sites, we also need efficient encryption techniques for secure transfer of media types like audio and videos. Also, the method described in this paper works on GPUs of personal computers. We aim to extend this technique to GPUs of modern smartphones.

References

1. Nagendra, M., Sekhar, M.C.: Performance improvement of advanced encryption algorithm using parallel computation. Int. J. Softw. Eng. Appl. **8**(2), 287–296 (2014)
2. Elkabbany, G.F., Aslan, H.K., Rasslan, M.N.: A design of a fast parallel- pipelined implementation of AES: advanced encryption standard. arXiv preprint arXiv:1501.01427 (2015)

3. Daniel, T.R., Stratulat, M.: AES on GPU using CUDA. In: 2010 European Conference for the Applied Mathematics & Informatics. World Scientific and Engineering Academy and Society Press (2010)

4. Sowmiya, S., Tresa, I.M., Chakkaravarthy, A.P.: Pixel based image encryption using magic square. In: 2017 International Conference on Algorithms, Methodology, Models and Applications in Emerging Technologies (ICAMMAET), pp. 1–4. IEEE (2017)

5. Feng, X., Tian, X., Xia, S.: A novel image encryption algorithm based on fractional fourier transform and magic cube rotation. In: 2012 4th International Congress on Image and Signal Processing (CISP), vol. 2, pp. 1008–1011. IEEE (2012)

6. Ren, S., Gao, C., Dai, Q., Fei, X.: Attack to an image encryption algorithm based on improved chaotic cat maps. In: 2012 3rd International Congress on Image and Signal Processing (CISP), vol. 2, pp. 533–536. IEEE (2010)

7. Lei, Z., Li, L., Xianwei, G.: Design and realization of image encryption system based on SMS4 commercial cipher algorithm. In: 2012 4th International Congress on Image and Signal Processing (CISP), vol. 2, pp. 741–744. IEEE (2012)

8. Li, Q., Zhong, C., Zhao, K., Mei, X., Chu, X.: Implementation and analysis of AES encryption on GPU. In: 2012 IEEE 14th International Conference on High Performance Computing and Communication & 2012 IEEE 9th International Conference on Embedded Software and Systems (HPCC-ICESS), pp. 843–848. IEEE (2012)

9. Abdelrahman, A.A., Fouad, M.M., Dahshan, H., Mousa, A.M.: High performance CUDA AES implementation: a quantitative performance analysis approach. In: Computing Conference 2017, pp. 1077–1085. IEEE (2017)

10. Khan, A.H., Al-Mouhamed, M.A., Almousa, A., Fatayar, A., Ibrahim, A.R., Siddiqui, A.J.: Aes-128 ECB encryption on GPU and effects of input plaintext patterns on performance. In: 2014 15th IEEE/ACIS International Conference on Software Engineering, Artificial Intelligence, Networking and Parallel/Distributed Computing, SNPD, pp. 1–6. IEEE (2014)

11. Patchappen, M., Yassin, Y.M., Karuppiah, E.K.: Batch processing of multi-variant AES cipher with GPU. In: 2015 Second International Conference on Computing Technology and Information Management (ICCTIM), pp. 32–36. IEEE (2015)

12. Ma, J., Chen, X., Xu, R., Shi, J.: Implementation and evaluation of different parallel designs of AES using CUDA. In: 2017 IEEE Second International Conference on Data Science in Cyberspace (DSC), pp. 606–614. IEEE (2017)

13. Subramanyan, B., Chhabria, V.M., Sankar Babu, T.G.: Image encryption based on AES key expansion. In: 2011 Second International Conference on Emerging Applications of Information Technology (2011)

14. Zhang, Y.: Test and verification of AES used for image encryption. Published online: 12 January 2018. 3D Research Center, Kwangwoon University and Springer-Verlag GmbH Germany, part of Springer Nature (2018)

Intelligent Analysis in Question Answering System Based on an Arabic Temporal Resource

Mayssa Mtibaa[✉], Zeineb Neji, Mariem Ellouze,
and Lamia Hadrich Belguith

Faculty of Economics and Management, Computer Department,
Miracl Laboratory, University of Sfax, Sfax, Tunisia
maysamtibaa@gmail.com, zeineb.neji@gmail.com,
mariem.ellouze@planet.tn, l.Belguith@fsegs.rnu.tn

Abstract. In this article, we present the overall structure of our specific system for generating answers of temporal questions based on Arabic temporal resource, called **Ar-TQAS** (**A**rabic **T**emporal **Q**uestion **A**nswering **S**ystem). This system deals with answering temporal questions involving several forms of inference to obtain a relevant answer. The Ar-TQAS system was built and the experiments showed good results with an accuracy of 72%.

Keywords: Question answering system · Arabic language ·
Complex temporal information · Temporal inference · Arabic temporal resource

1 Introduction

A question answering (QA) system aims to generate a precise answer to a given question in natural language. For example, for this question: للكرة القدم ؟متى فازت تونس بكأس إفريقيا/When did Tunisia win the African Cup of Football?, a system must answer by: in 2004. Indeed, many approaches have been developed in this context with various domains, various types of questions and information resources for example: a database, a collection of documents, a knowledge base and a Web [1]. In this paper, we focus on the task of answering to temporal question in Arabic. We propose a new method which allows improving the performance of traditional Arabic question answering systems. This new method is based on the construction of an Arabic temporal resource for the resolution of the problems of temporal ambiguities. Then the automatic processing of complex natural language like Arabic is rather difficult challenge [2]. This complexity is mainly due to the inflectional nature of Arabic. Moreover, in our chosen field, research on temporal entity extraction in English, German, French, or Spanish, uses local grammars, finite state automata [3] and neural networks [4] to detect temporal entities. These techniques do not work well directly for Arabic due mainly to the rich morphology and high ambiguity rate of Arabic temporal information [5]. Thus we propose a solution to infer complex temporal information through an Arabic temporal resource. This article is organized as follows: in the next section, we give some earlier and related researches in Arabic question answering

© Springer Nature Switzerland AG 2020
A. Abraham et al. (Eds.): ISDA 2018, AISC 941, pp. 381–391, 2020.
https://doi.org/10.1007/978-3-030-16660-1_38

systems. Then, we give the motivation of our research work in the third section. We will present the temporal notion in the Sect. 4. In the Sect. 5, we detail our proposed method and its different stages. Finally, we close with a conclusion of this work and some perspectives.

2 Related Work

Question answering systems are less studied in Arabic language in comparison with other languages (e.g.: English, French…). We give in this section an overview on existing Arabic question answering systems.

We start with AQAS is the first question answering system for the Arabic language developed by [6] since the 1993s. It is a QA system that allows returning answers to questions that start with interrogative pronouns (e.g.: Who, What, Where, When…) from structured data and not from a raw text.

QARAB [7] is a question answering system that takes a set of Arabic questions and attempts short answers. The system's source of knowledge is a collection of Arabic newspaper texts taken from the Al-Raya, a newspaper published in Qatar. QARAB uses superficial language understanding to deal with problems and it doesn't attempt to understand the content of the question at a deep semantic level.

ArabiQA is an Arabic question answering system presented in [8] that deals with factoid questions. This system based on a generic architecture made up of three integrated modules; a Named Entity Recognition (NER) module, a passage retrieval system (JIRS) and a module of answer extraction (AE).

QArabPro [9] is a question answering system for Arabic language, deals with complex questions where starting with interrogative pronouns (e.g.: How, Why) to analyze questions and retrieve the answers. Akour et al. present an approach based on a set of rules for each type of WH question.

A work is specialized for the Holy Quran provides Al-Bayan (An Arabic Question Answering System for the Holy Quran) presented in [10]. This system provides a semantic understanding of the Quran to answer question using reliable Quranic resources.

A QA system that deals with definitional questions is DefArabicQA (Arabic definitional Question Answering System) developed by [11]. The system searches for candidate definitions using a set of lexical patterns and categorizes them using heuristic rules. DefArabicQA classifies definitions using a statistical approach.

NArQAS (New Arabic Question Answering System) developed by [12] is a system that answers to questions based on the automatic understanding of Arabic texts to transform them into representational logic. It uses text implication recognition techniques.

3 Motivation

When the question asked by the user refers directly or indirectly to a temporal expression, the answer is expected to validate the temporal constraints. In this case, TPE (Temporal Pattern Extraction) [13] is an Arabic QA system based on temporal inference. It deals with the relations between temporal expressions and events mentioned in the question and to rely on temporal inference to justify the answer. This system consists in answering questions that start with temporal signals (e.g.: متى/When, منذ متى/Since when, كم دام/How Long, في أي عام/In which day...) by an answer pattern. During this work, Omri et al. found problems at the level of the temporal inferences. In the same context, we present a new method based on Arabic temporal resource for the resolution of temporal ambiguities. For example, the expected answer type of question Q1 is a Date:

Q1 : متى تشتدُّ الرياح ؟

MtY t$td AlryAH?

Q1: When is the wind hardening?

The answer to Q1: الليالي السود/Black nights, extracted from the context P1:
تقوى الرياح في الليالي السود/tqwYA lryAH fy <u>AllyAly Alswd</u>/Winds rise in <u>black nights</u>.

In the paragraph P1, the Verb = تقوى/rise which has an argument الرياح/Winds. We need an analysis to get the simple answer like for example (14 جانفي/14 jAnfy/14th January). This intelligent analysis is called Temporal Inference. It is concerned with building systems that automatically answer questions in a natural language by extracting a precise answer from a corpus of documents. But, for the example above, the answer A1 presents complex temporal information which is not extracted; it needs to refer to a temporal resource.

4 Particularity of Arabic Language and Temporal Constraints

Arabic is a spoken language by more than hundreds of millions people in the world and it is also considered one of the six official languages of the United Nations (English, Arabic, Chinese, French, Russian and Spanish). Also, Arabic is a very rich language because of it complex set, its morphology characteristics and its derivation. This richness needs special manipulation which makes regular NLP systems, designed for other languages are unable to manage it. Among the manifestations of the richness of this language is the fact that the temporal notion benefits from a very wide palette of nuances.

So, there are several motivating factors for the choice to use the Arabic temporal notion. Such factors are:

- Arabic is a very rich and complex language,
- Temporal information is an important dimension of any information space,
- Arabic temporal information is an essential component in text comprehension and useful in a large number of automatic language processing applications,
- The temporal entities are expressed in different ways,
- For temporal information, we find several representations. This causes ambiguity.

Example of ambiguity: For the example of حرب لبنان الثانية/Second Lebanon War can be expressed as follows:

12/ 07/2006
12-07-2006
2006 الثاني عشر من جويلية / AlvAny E$ rmn jwylyp 2006.
2006 جويلية 12 / 12 jwylyp 2006.
2006تموز 12 / 12 tmwz2006.
2006 يوليو 12/ 12 ywlyw 2006.
1427 جمادى الأخرة 16 / 16 jmAdY Al|xrp 1427.

For the word "جويية/jwylyp/July», we also find the following words which are equivalent «تموز/tmwz, جمادى الأخرة/jmAdY Al|xrp » and for the year 2010 we can also find the year 1427هجري/1427 hjry/1427 Hijri or 2006 ميلا دي/2006 mylAdy/2006 gregorian.

5   Proposed Approach

The proposed method presented in this section aims at automating the question answering system based on the construction of a temporal resource for the resolution of temporal inference. The proposed method involves three main stages presented in Fig. 1, namely: (1) question Analysis, (2) document processing, finally (3) answer processing.

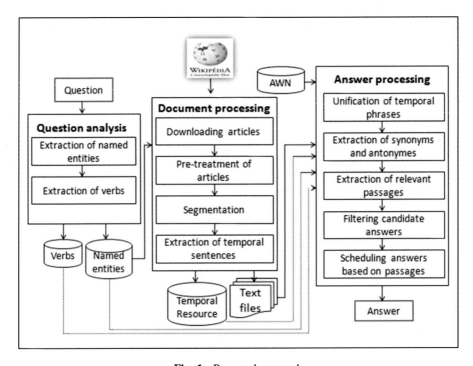

Fig. 1. Proposed approach

In the following subsections, we will detail the different sub-stages used in this proposed method.

5.1 Question Analysis

The objective of question analysis is to understand the asked question. This stage consists of representing the characteristics of each question, it facilitates the treatment. This analysis focuses on the extraction of named entities and verbs. This stage is preceded by a pretreatment phase in which we have checked the temporality of the question.

Indeed, our input is a question dealing with temporal information, i.e. a question that starts with one of the signals presented in Table 1.

Table 1. Temporal signals

Temporal signals in Arabic	Temporal signals in English
متى	When
منذ متى	Since when
كم (دام ـ مكث ـ أقام ـ كان ـ أصبح)	How(long-stay-many-much)
في أي (عام ـ وقت ـ زمن ـ شهر ـ يوم ـ تاريخ)	In Which (year-time-day-date-month-era)
إلى أي (عام ـ وقت ـ زمن ـ حقبة ـ شهر ـ يوم ـ تاريخ)	Until which(year-time-day-date-month-era)

Extraction of Named Entities
The detection of named entities (NE) in Arabic is a potential pretreatment and represents a serious challenge taken into account the specificities of the Arabic language. We provide at this level the extraction of the NE that occurs in each question for the purpose of building a base of NE that will be useful to the following stage.

Extraction of Verbs
The extraction of verbs is based on the decomposition of the question in order to extract the verb that exits. This verb will be stored in a base of verbs that will be useful for the following step of answer processing.

5.2 Document Processing

Downloading Articles
This step consists in downloading articles from the online encyclopedia Wikipedia. In fact, we automatically download the articles which contain the extracted named entities (NE) from the question analyzing and we will retrieve the Infobox, if it exists because

we will use it after for the verification of candidate answers. The structure of Wikipedia articles requires pretreatment, a segmentation that leads to the extraction of relevant sentences.

Pre-treatment of Articles

We keep the textual content of previously downloaded articles. The content of the article requires a pre-treatment like; elimination of parentheses, symbols, spacious characters and words that are not in Arabic as well as links and images.

Segmentation

Segmentation consists of dividing the text into paragraphs and sentences. We applied the approach proposed by [14]. This approach is based on the contextual exploration of markers of punctuations, connective words as well as some particles such as coordination conjunctions.

Extraction of Temporal Sentences

The extraction of relevant sentences consists in the identification of sentences which contain temporal information. We notice that there are temporal sentences that have not been detected by Nooj software [15] for example presented in Fig. 2. So, these sentences contain complex temporal information. It will be identified and unified through a temporal resource.

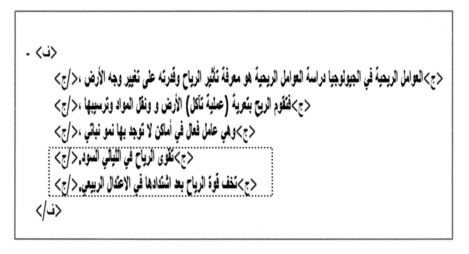

Fig. 2. The not extracted temporal sentences

Those temporal informations الأعتدال الربيعي and الليالي السود present in Fig. 2 are called complex temporal information. They must be identified and added into a temporal resource.

Construction of Arabic Temporal Resource

In this step, we will construct a temporal resource specific to the Arabic language. It allows to identify and unify all the extracted complex temporal information substituted into categories (as illustrated in Table 2), which purpose is to solve all the problems of temporal inferences and to facilitate the acquisition of the relevant answer.

We classified the set of complex temporal notions identified according to 6 categories composing a temporal resource.

Table 2. Categorization of temporal information

CATEGORY
Months/ Al>$hr / الأشهر
Revolutions / AlvwrAt / الثورات
Seasons/ AlmwAsm / المواسم
Feasts / Al>EyAd / الأعياد
Calendar / tqwym / تقويم
Wars / AlHrwb / الحروب

We collected a set of Arabic temporal information of different categories from the Wikipedia and Internet. The purpose is to facilitate the identification of temporal information, to solve all ambiguities and to have a quick answer in a standard format (Table 3).

Table 3. Example of complex temporal information

CATEGORY	EXAMPLE
months / Al>$hr / الأشهر	غشت / g$t
Revolutions / AlvwrAt / الثورات	الثورة الفرنسية /Alvwrp Alfrnsyp
seasons / AlmwAsm / المواسم	قرة العنز / qrp AlEnz
Feasts / Al>EyAd / الأعياد	اليوم العالمي للصحة/ Alywm AlEAlmy llSHp
Calendar / tqwym / تقويم	1440 هجري / 1440 hjry
Wars / AlHrwb / الحروب	حرب لبنان /Hrb lbnAn

5.3 Answer processing

Unification of Temporal Sentences

A task of unification is applied to the extracted temporal sentences whose aim is to normalize all the complex temporal information under the same writing format to facilitate the extraction of the relevant answers. We used for this phase our Arabic temporal resource.

P1: تقوى الرياح في **الأيام السود**.
P1: Winds rise in **black nights**.
P2: تخف قوة الرياح بعد اشتدادها في **الاعتدال الربيعي**.
P2: The strength of the wind is reduced after increasing in the **spring equi-
nox**.

After Unification:

P1: تقوى الرياح في **14 جانفي إلى 2 فيفري**.
P1: Winds rise on 14th January to 02nd February.
P2: تخف قوة الرياح بعد اشتدادها في **20 مارس**.
P2: The strength of the wind is reduced after increasing on 20th March.

Extraction of Synonyms and Antonyms of the Question's Verb

In this step, we will extract the list of synonyms and antonyms for the verbs detected in the previous module of question analyzing. This extraction is done by referring to Arabic WordNet (AWN).

Extraction of Relevant Passages

We have a set of relevant sentences from which we will extract a set of candidate sentences. Indeed, a sentence is considered as a candidate if:

- The response must contain the same named entity as the original question or one of its synonyms belonging to the Arabic WordNet ontology (AWN).
- The answer must contain the same verb as the original question or one of its synonyms or antonyms belonging to the Arabic WordNet ontology (AWN).

Filtering Candidate Answers

The filtering task makes it possible to obtain a set of candidate sentences containing the most maximum of answers likely to be correct. It eliminates answers that do not carry certain conditions and filters for the rest of the candidate answers based on a temporal inference mechanism.

Scheduling Answers based on Passages

We calculated for each answer corresponding to the question the number of occurrences of the extracted named entity. We ranked the relevance of the passages according to the number of occurrences of terms (C_t). C_t varies between 1 and 15 named entities: we fixed this condition according to the results obtained during this step.
If the score

$C_t >= 1$ and $C_t < 4$: the passage is considered to be weakly relevant,
$C_t >= 4$ and $C_t <= 8$: the passage is considered moderately relevant,
$C_t > 8$ and $C_t <= 15$ the passage is considered highly relevant.

6 Evaluation

Evaluation is an essential task in the development of computer applications for the TALN. It aims to analyze the detailed capabilities of our proposed method cited in the previous section. For this evaluation, we collected a corpus in Arabic language composed of a set of 500 temporal questions from the TREC corpus (Text REtrieval Conference) and from a list of questions produced in TERQAS Workshop (Table 4).

Table 4. Experiment results

Number of questions	Number of articles	Temporal Relations
500	25533	12947

In our collection of downloaded articles, almost 0.48% of Wikipedia articles not contain an infobox, 0.30% are empty articles. Indeed, these extracted temporal relations are relevant sentences whose correct answer is in one of them (Table 5).

Subsequently, our temporal resource is composed of 6 categories presented previously in the step of construction of temporal resource.

Table 5. Statistics of RcTA

CATEGORY	NUMBER
months / Al>$hr/ الأشهر	54
Revolutions / AlvwrAt / الثورات	31
Seasons/ AlmwAsm / المواسم	24
Feasts / Al>EyAd / الأعياد	73
Calendar /tqwym/ تقويم	34
Wars/ AlHrwb /الحروب	56

Finally, the performance of an Ar-TQAS system can be measured by the metrics of evaluations introduced by [16] (Table 6).

Table 6. Results of system Ar-TQAS

Recall	Precision	F-measure
0.76	0.70	0.72

7 Conclusion

This paper deals with temporal information involving several forms of inference in Arabic language. We introduced a new method based on the construction of an Arabic temporal resource to solve problems of temporal inference for QA that enables us to enhance the recognition of the exact answers to a variety of questions. The Ar-TQAS system has been developed. This system makes it possible to deals with the complex temporal information and generates an answer from a corpus of texts. The evaluation Ar-TQAS system has shown encouraging results. In the future, we aim to extend the corpus of questions and the resource of the Arabic complex temporal information too.

References

1. Ben-Abacha, A.: Finding specific Answers to Medical Questions: The MEANS Questions and Answers System. Ph.D. thesis. Universite PARIS-SUD 11 LIMSI-CNRS. JUIN 2012
2. Neji, Z., Ellouze, M., Belguith, L.: Question answering Based on temporal inference. Res. Comput. Sci. **117** (2016)
3. Koen, D.B., Bender, W.: Time frames: temporal augmentation of the news. IBM Syst. J. **39**, 597–616 (2000)
4. Li, H., Gao, Y., Shnitko, G., Meyerzon, Y., Mowatt David, D.: Techniques for extracting authorship dates of documents, December 2009
5. Zaraket, F., Makhlouta, J.: Arabic temporal entity extraction using morphological analysis. IJCLA **3**(1), 121–136 (2012)
6. Mohammed, F., Nasser, K., Harb, H.: A knowledge based Arabic question answering system (AQAS). ACM SIGART Bull. **4**, 21–33 (1993)
7. Hammo, B., Ableil, S., Lytinen, S., Evens, M.: Experimenting with a question answering system for the Arabic language. Comput. Humanit. **38**(4), 397–415 (2004)
8. Benajiba, Y., Rosso, P., Lyhyaoui, A.: Implementation of the ArabiQA question answering system's components. In: Proceedings of Workshop on Arabic Natural Language Processing, 2nd Information Communication Technologies International Symposium. ICTIS-2007, Fez, Morroco, 3–5 April 2007
9. Akour, M., Abufardeh, S., Magel, K., Al-Radaideh, A.Q.: QArabPro: a rule based question answering system for reading
10. Abdelnasser, H., Ragab, M., Mohamed, R., Mohamed, A., Farouk, B., El-Makky, N., Torki, M.: Al-Bayan: an Arabic question answering system for the holy Quran. In: Proceedings of the EMNLP 2014 Workshop on Arabic Natural Language Processing (ANLP), pp. 57–64 (2014)
11. Trigui, O., Belguith, L.H., Rosso, P.: Arabic definition question answering system. In: Workshop on Language Resources and Human Language Technologies for Semitic Languages, 7th LREC, Valleta, Malta, pp. 40–45 (2010)
12. Bakari, W., Bellot, P., Neji, M.: A logical representation of Arabic questions toward automatic passage. Int. J. Speech Technol. **2**, 339–353 (2017)
13. Omri, H., Neji, Z., Ellouze, M., Belguith, L.: The role of temporal inferences in understanding Arabic text. In: International Conference on Knowledge Based and Intelligent Information and Engineering Systems, Marseille, France (2017)

14. Belguith, H., Baccour, L., Ghassan, M.: Segmentation de texts arabes basée sur l'analyse contextuelle des signes de ponctuations et de certaines particules. 6–10 juin 2005
15. Mesfar, S.: Analyse lexicale et morphologique de l'arabe standard utilisant la plateforme linguistique Nooj. In: RECITAL 2006, Leuven, 10–13 avril 2006
16. Van Rijsbergen, C.J.: Information Retrieval, 2nd edn. Butterworth & Co (Publishers) Ltd, London (1979)

Towards the Evolution of Graph Oriented Databases

Soumaya Boukettaya[1,2](✉), Ahlem Nabli[1,3], and Faiez Gargouri[1,4]

[1] MIRACL Laboratory, Sfax University, Sfax, Tunisia
[2] Faculty of Economics and Management of Sfax, Sfax University, Sfax, Tunisia
soumayaboukettaya@gmail.com
[3] Faculty of Computer Sciences and Information Technology, Al-Baha University,
Al Bahah, Kingdom of Saudi Arabia
Ahlem.nabli@fss.usf.tn
[4] Institute of Computer Science and Multimedia, Sfax University, Sfax, Tunisia
faiez.gargouri@isims.usf.tn

Abstract. As one of NoSQL data models, graph oriented databases are highly recommended to store and manage interconnected data. Used as back-end for today applications, NoSQL databases come with the challenge of effectively managing data evolution. In fact, NoSQL graph oriented databases offer a great flexibility. Usually such flexibility helps developers to manage huge data quantities with heterogeneous structure. Nevertheless, they may struggle to deal with legacy entities in production. The problem of evolution in NoSQL databases is not well treated. The common procedure is to migrate all data eagerly, but that comes with the cost of the application downtime. So lazy migration strategy may be more cost-efficient, as legacy entities are only migrated in case they are actually accessed by the application. In this paper, we propose an approach to control the evolution of data in the graph oriented databases by highlighting a lazy migration process.

Keywords: Graph oriented databases · Lazy migration ·
Database evolution

1 Introduction

With the rise of the Internet and the emergence of the Big Data era, data is becoming very varied in terms of structure and volume (e.g. data from social media). The graph oriented model is one of the NoSQL models that can efficiently manage highly interconnected data (social networks, traffic graphs, geographic locations, etc.). Thus, many applications are moving towards the use of graph oriented databases as back-end to their codes. However, when it comes to manipulating data in applications deployed continuously on the same database, this flexibility bring forth difficulties in managing the increasing entropy of the data structure and ensuring proper handling of existing data. This problem may induce severe data loss and execution errors.

© Springer Nature Switzerland AG 2020
A. Abraham et al. (Eds.): ISDA 2018, AISC 941, pp. 392–399, 2020.
https://doi.org/10.1007/978-3-030-16660-1_39

Data evolution is not correctly handled in graph oriented databases. As a matter of fact, graph oriented databases can sufficiently handle heterogeneous data structure. However, each new release of the application can induce some changes to the structure of some entities. Usually, that leads to the creation of a totally new type of data. In this case, both, the old and the new data are stored in the database. Thus, special tools are required in order to manage data evolution in graph oriented databases.

2   Related Work

A very wide corpus of literature exists today reflecting the vast work in the field of schema evolution. Even though it has been studied for many decades, data and schema evolution represents a problem that it's difficult to maintain without the appropriate tools. Many works were done over the relational model including those presented in [1,2,9,11]. Also, some recent works such as the one presented in [4], describe a schema evolution language based on schema modification operations (SMO) and adapt a lazy migration strategy.

Data evolution covers other domains like the object-oriented model [7], ontologies [8] and data warehouses [19]. As a recent area of research, the evolution of data and schemas in NoSQL databases is not widely spreading. Similar to the relational model, there are two basic schema evolution strategies in NoSQL databases. The first strategy consists of detecting entities that have undergone changes from different versions of the database depending on schema versioning strategies, such the works in [5,10,13]. The second strategy is to follow the migration of data over time in order to be able to control the implicit evolution of the schema. The most basic data migration techniques are: eager migration and lazy migration. Eager data migration consist of eagerly migrating all legacy data in one go. Popular tools like Flyway and Liquibase (for relational databases), or Mongeez (for MongoDB) follow this approach [16]. Eager migration requires careful deployment, to avoid application down-time, and may be costly due to the need of a high number (and costs) of migration operations. Lazy migration consist of only migrating lagacy entities in production when the updated application accesses an object in the old format. Few works were done in this area starting with object-NoSQL mappers in [12], plug-in like ControVol presented in [14] in its first extension which verifies the type of the object-mapper declaration classes and makes it possible to correct type incompatibilities immediately, already during the development process and its second extension which represented in [3]. That extension brings the possibility for developers to choose their own migration strategy. In fact, all legacy data can be migrated using NotaQL transform scripts or can be migrated, as declared by the object-mapper annotations. Other studies such as [6,15,17] proposed languages and operations for lazy data migration.

All these approaches are done over data in JSON format and in level of key/values entities. Thus they are not adapted for graph oriented databases structure. Taking into consideration the structure of a graph, in this paper, we propose an approach to control the evolution of data in graph oriented databases.

3   Formalization of the Graph-Oriented Model

The facts that graph oriented databases are based on graph theory make it different to manage data evolution and migration over such databases. In this section, we propose a process that can handle lazy data migration over graph oriented databases. But keeping in view the structure of graph, a formalism of a graph oriented database seems to be important. Graph oriented databases present one of the most powerful NoSQL database models that rely on graph theory to manipulate and store data. Designed to explore highly inter-connected data, the graph oriented database structure enables the modeling of complex data in a simple and intuitive way. A graph oriented database is composed mainly of the nodes that represent the different entities and arcs representing the different relations which organize the nodes. Nodes and relationships can have properties formed by a key/value pair. Additionally, nodes can be labeled with one or more labels. A graph oriented database is a couple (N, R), where:

- N: the total set of nodes and which form the entities of the (OGDB).
- R: the set of relations that join the different nodes.

Nodes. Each node n_i is composed of its identifier id_{ni} and possibly a set of properties P_{ni} and eventually a set of labels L_{ni}. It should be noted that the identifier does not contain any semantic information. Semantic is usually expressed through one or more labels. A node can be written as follow:

$$\forall n \in N; n_i = (id_{ni}, P_{ni}, L_{ni}) \tag{1}$$

- id_{ni}: is the unique identifier of each node n.
- P_{ni}: is the set of properties $(p_1, .., p_{ni})$ related to the node. It should be noted that a property is a couple of a (key/value) pair.
- L_{ni}: is the set of labels $(l_1,.., l_{ni})$ attached to the node n.

Relationships. A relation r_i is defined as $(id_{ri}, N_{eri}, N_{sri}, T_{ri}, P_{ri})$ which contains the identifier id_{ri}, the incoming N_{eri}/outgoing nodesN_{sri}, the type of relation T_{ri} and its set of properties P_{ri}. The relation can be presented as:

$$\forall r \in R; r_i = (id_{ri}, N_{eri}, N_{sri}, T_{ri}, P_{ri}) \tag{2}$$

- N_{eri}: the identifier of the incoming node.
- N_{sri}: the identifier of the outgoing node.
- T_{ri}: the type of relationship, it is a string of characters that bears the name of the relationship.
- P_{ri}: a set of properties $(p_1,.. ,p_{ri})$ specific to Tr.

4   Data Migration Process

By definition, a lazy data migration is done each time an updated application access an old object (or legacy entity) in the database. In this section, we present our process to safely manage lazy data migration over the different graph oriented

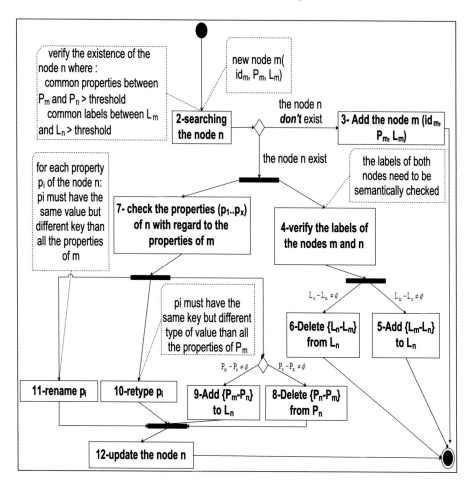

Fig. 1. Data migration process over nodes.

database entities. In a previous work, we presented a set of basic operations that can safely manage data when executing the different CRUD operations over the entities of an graph oriented database [18]. Depending on the nature of the entity that the application accessed, a migration process can be done over nodes or relationships. In the remaining of this section we will describe in details the process we propose in order to safely managing nodes and relationships migration.

4.1 Nodes Migration Process

Generally, nodes adding, nodes updating or nodes deleting, leads to a schema evolution. Controlling such operations in graph oriented databases is crucial in order to maintain a correct schema. In the scope of this paper, we will focus on the node adding operation and describe a migration process that can maintain a correct schema when a new node is presented and ready to be stored in the database. When new node is presented to the database via the application code, it will be processed, so it can finally be safely stored. The lazy adding data migration process over a node is shown in Fig. 1. The first step of this process consists to search the similar node n to the new node m in the database by extracting the appropriate part of the schema (step1). n is considered similar to m only if the common properties and labels between the two nodes are superior or equal to a fixed threshold (step2). If the node doesn't exist, it will be automatically added and stored in the database (step3). If the node exists, an update operation will be performed over the old node. To properly update a node, operations over properties and labels of the node n are done. The decision to add the non-common labels (step5) or deleting them (step6) is taken after semantically comparing the labels of the two nodes m and n (step4). After checking every property of the node n with regard to all the properties of m, three kinds of operations are processed. The first one is to weather add or delete the non-common properties between the two nodes (steps 8 and 9). The second is to check if there is any property that needs to be renamed (the property of n must have the same key but different type of value than all the properties of m) (step 10). The last operation is to retype a property of the node n (the property to retype must have the same value but different key than all the properties of m)(step11).

Table 1 explain the different operations which can be done over the nodes.

Table 1. Lazy migration operations for the nodes.

Operations	Description	Pre-conditions	Post-conditions
Add node	Add a new node to the database	The node must not exist in the database	The added node must not be obsolete
Delete node	Delete an existing node from the database	The node must exist in the database, so its properties and labels	–
Rename property	Rename a property of a specific node	The old property must exist and attached to a specific node	–
Retype property	change the type of a specific node's property	–	–
Add property	Add a new property to a specific node	The property to add should not previously attached to the node	–
Delete property	Delete an existing property from a node	The property to delete must be one of the node properties	–
Add label	Add a new label to a specific node	The label to add should not previously attached to the node	–
Delete label	Delete an existing label from a node	For semantic purposes, the label that will be deleted must not be the only label of the specific node	–

4.2 Relationships Migration Process

Relationships are also one of the important components of a graph oriented database. Thus, safely migrating a relationship, present an important issue to deal with. In order to manage relationship migration, we propose a lazy migration process that is automatically executed whenever the application aims to store a new relationship in the database. Figure 2 highlights the lazy migration process on the relationships. The first step to extract the appropriate part of the schema in order to verify the existence of the relationship to add (step1). In fact a

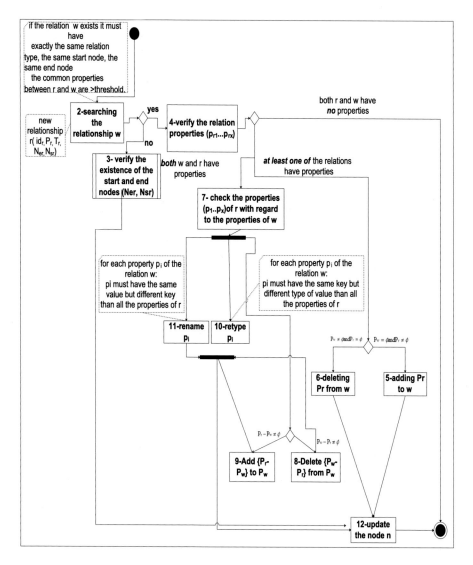

Fig. 2. Data migration process over relationships.

relationship may exist only if it has the exact relation type, start/end nodes, and properties as an existing relationship (step2). On the assumption that the relation doesn't exist in the database, a test to make sure of the existence of the start and end nodes of the relationship is done (step3). In fact, in this step the "node migration process" will be executed. In the other hand, and once the relation proved to be existed, an automatic update will be processed over the old relationship (step4), depending on the checking results of the relation properties (step4). If both of relations don't have any properties no further stages will be accomplished and the process will end. In case that one of the relations has properties, two different tasks will be performed depending on which relationship has properties. If the old relation doesn't have properties and the new one has: the idea is to delete the properties from the old relationship (step6), then, to add the properties of the new relationship to the old one (step5). Supposing that both of the relations have properties, another set of tasks will be accomplished. The first task is to check every property of the relation w with regard to all the properties of r, a specific set of operations will be processed. The first one is to weather add or delete the non common properties between the two relations (steps 8 and 9). The second is to check if there is any property that needs to be renamed (the property of w must have the same key but different type of value than all the properties of r) (step 10). The last operation is to retype a property of the relation w (the property to retype must have the same value but different key than all the properties of r) (step11). Almost the same operations over nodes are done to relationships such adding, deleting and updating the relations by manipulating its properties.

5   Conclusion

This work studies the evolution and the migration of data in graph oriented databases. Usually, as a database evolves, so does its schema. However, graph oriented databases have no schema and are not yet equipped with practical data management tools. In the scope of this article, we feature the basics of the systematic management of data evolution in the context of graph oriented databases. We suggested two different lazy migration processes over nodes and relationships, we also highlighted the different conditions to ensure a fluent data migration in order to prevent errors and data losses of the data in production. In this paper, we highlighted lazy the migration process over adding nodes and relationships. In the future, we aim to put forward, complex data migration operations along with the update and delete migration process.

References

1. Caruccio, L., Polese, G., Tortora, G.: Synchronization of queries and views upon schema evolutions: a survey. ACM Trans. Database Syst. (TODS) **41**(2), 9 (2016)
2. Cleve, A., Gobert, M., Meurice, L., Maes, J., Weber, J.: Understanding database schema evolution: a case study. Sci. Comput. Program. **97**, 113–121 (2015)

3. Haubold, F., Schildgen, J., Scherzinger, S., Deßloch, S.: Controvol flex: flexible schema evolution for NoSQL application development (2017)
4. Herrmann, K., Voigt, H., Rausch, J., Behrend, A., Lehner, W.: Robust and simple database evolution. Inf. Syst. Front. **20**(1), 45–61 (2018)
5. Klettke, M., Störl, U., Scherzinger, S., Regensburg, O.: Schema extraction and structural outlier detection for JSON-based NoSQL data stores. In: BTW, vol. 2105, pp. 425–444 (2015)
6. Klettke, M., Störl, U., Shenavai, M., Scherzinger, S.: NoSQL schema evolution and big data migration at scale. In: 2016 IEEE International Conference on Big Data (Big Data), pp. 2764–2774. IEEE (2016)
7. Li, X.: A survey of schema evolution in object-oriented databases. In: Technology of Object-Oriented Languages and Systems, 1999. TOOLS 31, Proceedings, pp. 362–371. IEEE (1999)
8. Mahfoudh, M.: Adaptation d'ontologies avec les grammaires de graphes typés: évolution et fusion. Ph.D. thesis, Université de Haute Alsace-Mulhouse (2015)
9. Manousis, P., Vassiliadis, P., Zarras, A., Papastefanatos, G.: Schema evolution for databases and data warehouses. In: European Business Intelligence Summer School, pp. 1–31. Springer (2015)
10. Meurice, L., Cleve, A.: Supporting schema evolution in schema-less NoSQL data stores. In: 2017 IEEE 24th International Conference on Software Analysis, Evolution and Reengineering (SANER), pp. 457–461. IEEE (2017)
11. Qiu, D., Li, B., Su, Z.: An empirical analysis of the co-evolution of schema and code in database applications. In: Proceedings of the 2013 9th Joint Meeting on Foundations of Software Engineering, pp. 125–135. ACM (2013)
12. Ringlstetter, A., Scherzinger, S., Bissyandé, T.F.: Data model evolution using object-NoSQL mappers: Folklore or state-of-the-art? In: Proceedings of the 2nd International Workshop on BIG Data Software Engineering, pp. 33–36. ACM (2016)
13. Ruiz, D.S., Morales, S.F., Molina, J.G.: Inferring versioned schemas from NoSQL databases and its applications. In: International Conference on Conceptual Modeling, pp. 467–480. Springer (2015)
14. Scherzinger, S., de Almeida, E.C., Cerqueus, T., de Almeida, L.B., Holanda, P.: Finding and fixing type mismatches in the evolution of object-NoSQL mappings. In: EDBT/ICDT Workshops (2016)
15. Scherzinger, S., Klettke, M., Störl, U.: Managing schema evolution in NoSQL data stores. arXiv preprint arXiv:1308.0514 (2013)
16. Scherzinger, S., Sombach, S., Wiech, K., Klettke, M., Störl, U.: Datalution: a tool for continuous schema evolution in NoSQL-backed web applications. In: Proceedings of the 2nd International Workshop on Quality-Aware DevOps, pp. 38–39. ACM (2016)
17. Scherzinger, S., Störl, U., Klettke, M.: A datalog-based protocol for lazy data migration in agile NoSQL application development. In: Proceedings of the 15th Symposium on Database Programming Languages, pp. 41–44. ACM (2015)
18. Boukettaya, S., Ahlem Nabli, F.G.: Vers l'évolution des bases de données orientées graphes : opérations d'évolution. In: ASD 2018: Big data & Applications, ASD, pp. 557–569, May 2018
19. Subotic, D., Poscic, P., Jovanovic, V.: Data warehouse schema evolution: state of the art. In: Central European Conference on Information and Intelligent Systems. Faculty of Organization and Informatics Varazdin, p. 18 (2014)

Arabic Logic Textual Entailment with Feature Extraction and Combination

Mabrouka Ben-sghaier[1(✉)], Wided Bakari[1,2], and Mahmoud Neji[2]

[1] Faculty of Economics and Management, 3018 Sfax, Tunisia
mabrouka.bensghaier@gmail.com
[2] MIR@CL, Sfax, Tunisia
{wided.bakkari,mahmoud.neji}@fsegs.rnu.tn

Abstract. Determining the textual entailment between texts is important in many NLP tasks, such as, question-answering, summarization, and information retrieval, etc. In question-answering, this technique is frequently used to validate an answer retrieved by a question-answering system. In this paper, we address the problem of textual entailment in Arabic. We employ some features to determine the textual entailment between pairs of logical representations of a question and a passage of text. We have implemented our approach in an Arabic recognizing textual entailment system called Ar-SLoTE (Arabic Semantic Logic based Textual Entailment). The experiments results of the entailments classification achieved a precision and a recall successively equal to 73% and 68% on average.

Keywords: Recognizing textual entailment · Arabic language ·
Question-answering system · Logic representation · Ar-SLoTE

1 Introduction

Textual entailment is a modern field of research in natural language processing (NLP). Its purpose is to federate the NLP researches in order to provide lexical, syntactic and semantic language processing methods independently in a wide range of natural language processing applications, including, question-answering, automatic summary, text generation, machine translation, information retrieval, etc., (Dagan et al. 2013). This task was introduced by Dagan and Glickman in the first evaluation campaign called Recognizing Textual Entailment (RTE).

The purpose of this task is to model the human capacity to know if a hypothesis can be deduced from a text. More specifically, it aims to automatically determine whether a text segment (H) is derived from another text segment (T) (Dagan et al. 2005). Since 2005, RTE has been proposed as a task where its goal is to capture the main semantic inference needs between applications in computational linguistics (Dagan et al. 2009). A new evaluation campaign took place each year with redefinitions of the task (from 2005 to 2013). The RTE is useful for question-answering systems where a text represents an answer to a question if the existential closure of the representation of the question is implied by the representation of this text (Dagan et al. 2006).

© Springer Nature Switzerland AG 2020
A. Abraham et al. (Eds.): ISDA 2018, AISC 941, pp. 400–409, 2020.
https://doi.org/10.1007/978-3-030-16660-1_40

This study explores an approach for determining the textual entailment in Arabic language. The proposed approach is dedicated to a question-answering system in the sense that a text is an answer to a question if a representation of the question is entailed by a representation of this text. This approach is based on the extraction and the combination of features to determine textual entailment between pairs of logical representations of a question and a passage of text.

The remainder of the paper is organized as follows. Section 2 describes the related works, while Sect. 3 details our proposed approach to determine the entailment relation between pairs of logic representations of the text and the hypothesis. Section 4 discusses evaluation results. Finally, Sect. 5 summarizes the conclusion.

2 Related Works

Many experiments have been conducted to detect textual entailment in English language for many NLP applications, including question-answering. The first of these, presented in (Glickman et al. 2006), assumes that, to entail the hypothesis, the text must contain the words of the hypothesis identically or in a variant form. This study looks at the lexical entailments and four phenomena were revealed as well as their frequencies of occurrences. The COGEX system (Tatu and Iles 2006) deals with the validation of English answers by making an entailment on meaning. The system is based on the idea that the text entails the hypothesis if it logically entails its meaning. In this formalism the predicates correspond to verbs, nouns and adjectives. The relations are also obtained by a syntactical analysis. A set of axioms provided by the extended WordNet (Mihalcea and Moldovan 2001) is also considered.

The (Fowler et al. 2005) approach uses the COGEX system (a modified version of the OTTER prover) to recognize the textual entailments. The authors use axioms to express the syntactic equivalences and to weaken the complexity of other logical formulas. Other types of systems apply to English have used WordNet (Fellbaum 1998) to detect the connection between two words. The simplest method, presented in (Pakray et al. 2009) or (Ofoghi 2009), is based on synonymic or hyperonymic relationships. Many systems consider named entities correspondence to detect the validity of answers or recognize textual entailment.

(Ferrández et al. 2009) defines two criteria: the first tests if all the named entities of the hypothesis are found in the text, the second one corresponds to the proportion of named entities of the hypothesis presented in the text. The system of the University of Hagen (Glöckner 2007; Glöckner and Pelzer 2008) also contains a logical proof mechanism but starts with a step to standardize the text by modifying the words so that they are closer to those of the hypothesis. The proof is performed by a recursive relaxation mechanism. First, the system takes all the predicates of the text and the hypothesis and determines if there is an entailment. If it does not, it removes the predicates from the hypothesis. This mechanism is executed until getting entailment. The final decision is then made from the words presented in this new hypothesis.

In comparison with English, the Arabic language has relatively less attention for the recognition of textual entailment, because of the challenges in determining whether one Arabic text is entailed from another. One of these challenges is that the Arabic language has a productive derived morphology, where from one root, several forms could be derived. These derivative words become confused in case the diacritics are missing (Alabbas 2011). Arabic is one of the most difficult languages to deal with because of its morphological richness and its relatively free words order as well as its diglossic nature (where the norm and dialects are confounded in most types of data) (Almarwani and Diab 2017).

Moreover, there are little published researches in the literature of recognition of Arabic textual entailment, which motivated the work on this theme. To our knowledge, the work of (Alabbas 2011) was the first that targeted this question which its purpose is to highlight the Arabic Textual Entailment (ArbTE) system to evaluate existing textual entailment techniques when they are applied to recognize Arabic textual entailment. Indeed, the adopted technique by the author corresponds to pairs of textual hypotheses using the tree editing distance algorithm (TED). Authors in (Alabbas and Ramsay 2013) have proposed the use of extended tree editing distance with sub trees, resulting a more flexible matching algorithm to identify textual entailments in Arabic.

In addition, (AL-Khawaldeh 2015) examined the negation and the polarity as additional features for the recognition of Arabic textual entailment. The author focused on the importance of classifying the contradictions in RTE systems, as the contradiction could reverse the current polarity of the sentence. The words of contradiction are treated as empty words and removed from the text in the pretreatment stage before the entailment stage is completed. (Khader et al. 2016) proposed a textual entailment method that was constructed using the Python language. This method is based on a lexical and semantic combination. It consists of two phases (Word Overlap Calculation and Bigram Matching Verification). Recently, (Almarwani and Diab 2017) posed the problem of textual entailment as a task of binary classification. Without relying on external resources, the authors use both traditional characteristics and distributive representations to identify whether a text T entails a hypothesis H.

It appears that no existing work has addressed the RTE's task for factual question-answering systems in Arabic language. So, we propose in this work a semantic method for the RTE in Arabic which determines the relation of entailment between the logical representations of the question and the answer. The proposed method combines three features namely overlapping predicate-arguments, correspondence between named entities and semantic similarity, in order to extract the entailment relation between the question and its candidate answers.

In our case, textual entailment is considered to be a problem of logical entailment between the meanings of the two sentences (Tatu and Iles 2006). At this level, the entailment between two sentences is accepted if their senses are involved. In other words, the predicate can have various types of arguments (e.g. subject, direct object complement, and indirect object complement). For this, the predicate-argument structure is often used, that is, the text sentences T and the questions considered as hypotheses H are transformed into a set of predicates and arguments through a transformation algorithm to deduce the entailment after being transformed into conceptual graphs. Therefore, to determine whether a text T1 entails a text T2, we translate the semantic representations presented by conceptual graphs into logical formulas. Then, we determine the relation of the logical entailment. Given two texts T1 and T2, T1 entails T2 if and only if (the first-order logic translation of) one of the semantic representations of T1 entails (the first-order logic translation of) one of the semantic representations of T.

To our knowledge, there are some logical approaches to the task of the RTE. In addition, although this task has been defined much less rigorously than logical entailment, Harabagiu and his associates believe that the RTE between a question and a set of candidate answers may allow question-answering systems to identify the correct answers with greater precision than with keywords or model-based techniques (Harabagiu and Hickl 2006).

3 Proposed Approach for Logic RTE in Arabic Language

In our work, we determine the textual entailment between pairs of logical representations of a question (hypothesis) and a passage of text that represents an answer of this question (text). To do this, it is necessary to understand precisely the meaning of the passage and the question. We propose an approach that is essentially based on extracting and combining features to determine the logical implication relationship between each pair of a question and its passage answer.

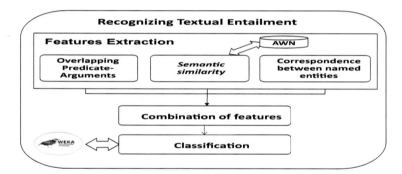

Fig. 1. Proposed approach

As shown in Fig. 1, our proposed approach consists of three main phases in order to recognize textual entailment in Arabic language, namely, features extraction, combination of features and classification. The extracted features are overlapping Predicate-Arguments, semantic similarity and correspondence between named entities.

3.1 Features Extraction and Entailment Determination

In this subsection, we describe in detail the features used to generate a feature vector for each text (T) and hypothesis (H) pair, which would be used for training and decision entailment. We discover three features used to determine textual entailment. Only shallow features are taken into account in our work. These features have been constructed for use in determining and classifying the entailments. The features are based on scores calculated by similarity metrics. For each metric, we use the logical input representations of the T-H pairs respectively FOLT and FOLH; the features are as follows:

(i) Feature 1: Overlapping Predicate-Arguments

To measure the overlap of words between the text and the hypothesis, we assume that we are talking about similar entities. Our general model is a simple word bag model. Therefore, the overlap is generated for all word pairs (w1, w2) where w1 $\in$ FOLT and w2 $\in$ FOLH. A word can be either a unary predicate or a binary predicate. Indeed, a binary predicate consists of a pair of linked arguments that can represent a constant or a variable that refers to a predicate. However, a unary predicate consists of a single argument that is either a constant or a variable.

In the case of a binary predicate, we first take into account the relations (e.g. "objOf (X, تونس)") for the comparison, and then the arguments. Finally, in the case of a unary predicate, if the argument represents a constant (e.g. "تونس"), we take its name for the comparison. On the other hand, if the argument is a variable (e.g. "X"), we take into account its corresponding predicate. The overlap is calculated as follows:

$$\text{Predicat}_{\text{argument}_{\text{overlap}}} = \frac{(\textbf{Number of words in common between text and hypothesis})}{(\textbf{Total number of words in the hypothesis})}$$

(1)

Let's consider the following example to understand the principle of the overlapping predicate-arguments feature.

Question: متى فازت تونس بجائزة نوبل للسلام ؟
"When Tunisia won the Nobel Peace Prize?"

FOLH : $\exists TX\ \exists X\ \exists Y\ \exists Z\ \exists W$: date(TX) Λ فوز(X) Λ TMP(X,TX) Λ Location(تونس) Λ objOf(X,جائزة) Λ (Y)تونس Λ Arg(X,Y) Λ نوبل(Z) Λ is(Y,Z) Λ سلام(W) Λ attributeOf(W,Y)

Text Passage: وفي 2015، اقتسمت أربع منظمات تونسية جائزة نوبل للسلام العربية الخامسة، "لمساهمتها في إنجاح مسار الانتقال الديمقراطي في تونس
"In 2015, four Tunisian organizations shared the fifth Arab Peace Nobel Prize, "for their contribution to the success of the democratic transition process in Tunisia".

FOLT : $\exists X\ \exists Y\ \exists Z\ \exists W\ \exists T\ \exists E\ \exists F\ \exists G\ \exists H\ \exists I\ \exists J\ \exists K\ \exists L\ \exists M\ \exists N$: قسم(X) Λ 2015(Y) Λ isEqual(X,Y) Λ أظما(Z) Λ ربع(W) Λ isEqual(Z,W) Λ agentOf(X,Z) Λ تونس(T) Λ propertyOf(Z,T) Λ جائزة(E) Λ objOf(E,X) Λ نوبل(F) Λ is(E,F) Λ سلام(G) Λ attributeOf (G,E) Λ عرب(H) Λ AdjOf(G,H) Λ سهم(I) Λ is(I,E) Λ مسار(J) Λ جوح(K) Λ is(J,K) Λ مسيرة (L) Λ attributeOf(L,J) Λ ديمقراطي(M) Λ AdjOf(L,M) Λ تونس(N) Λ is(J,N)

In this example, we first calculate the overlap between the pairs of unary predicates. So, it has four pairs overlapping. We calculate their overlap as follows:

$$\text{Unary}_{\text{predicate}_{\text{overlap}}} = \frac{4}{6} = 0.66$$

Second, we calculate the overlap between pairs of binary predicates. We present, first, the set of relations of FOLH by R1 = {"TMP(X,TX)", "objOf(X,تونس)", "Arg(X, Y)", "is(Y,Z)" and "attributeOf(W,Y)"}, the set of relations of FOLT by

R2 = {"isEqual(X,Y)", "isEqual(Z,W)", "agentOf(X,Z)", "propertyOf(Z,T)", "objOf (E,X)", "is(E,F)", "attributeOf(G,E)", "AdjOf(G,H)", "is(I,E)", "is(J,K)", "attributeOf (L,J)", "AdjOf(L,M)" and "is(J,N)"}. We find that the pair "FOLH/FOLT" has two shared relationships, their overlap is equal to:

$$\textbf{Binary}_{\textbf{predicate}_{\text{overlap}}} = \frac{2}{5} = 0.4$$

The measure of overlap "Predicat_argument_overlap" of the unary and binary predicates is equal to:

$$\textbf{Predicat}_{\textbf{argument}_{\text{overlap}}} = \frac{\textbf{Unary}_{\textbf{predicate}_{\text{overlap}}} + \textbf{Binary}_{\textbf{predicate}_{\text{overlap}}}}{2} = 0.53$$

It should be noted that the binary predicates extracted from the logical representation of a sentence P indicate the coded semantic relations between the constituents of P. Thus, the determination of the overlap between these predicates plays an important role in the determination of the implication.

(ii) Feature 2: Correspondence between named entities

For each named entity NE1 in the logical representation of the passage, we look for a named entity NE2 in the hypothesis that it entails. We consider that NE1 of FOLT entails NE2 of FOLH if they have the same type described in the logical forms of FOLT and FOLH, and the text string of NE1 contains the text string of NE2. Specifically, the matching process between named entities follows these steps:

1. Extract the lists of named entities found in FOLT, FOLH and in the ArNER (Zribi et al. 2010) result file.
2. Compare each named entity of FOLT with the named entities of FOLH.
3. Calculate the matching score between FOLH and FOLT. This score is calculated as follows:

$$\textbf{Score}_{\textbf{NE}_{\text{corresp}}} = \frac{\textbf{Number of NEs shared between the text and the hypothesis}}{\textbf{Number of NEs in the hypothesis}} \quad (2)$$

Consider the example studied in the previous feature, we calculate the matching score between FOLT and FOLH. This example has a shared pair of named entities, namely, "Location: تونس". So, the assigned score is determined as follows:

$$\textbf{Score}_{\textbf{NE}_{\text{corresp}}} = \frac{1}{2} = 0.5$$

(iii) Feature 3: Semantic similarity

To determine the semantic similarity between two logical representations, we seek, first, the semantic similarity between each word of the hypothesis with all the words of the text using the Wu-Palmer semantic measurement provided by Arabic Wordnet (Black et al. 2006). Indeed, Wu-Palmer is based on depths and lengths in taxonomy. It takes into account the length between the concepts C1 and C2, the length between the LCS and the root of the classification in which the concepts exist. The similarity is determined as follows:

$$sim_{WP}(C_i, C_j) = \frac{2 * \mathbf{depth}(\mathbf{ICS}(C_i, C_j))}{\mathbf{depth}(C_i) + \mathbf{depth}(C_j)} \quad (3)$$

With:

- Depth (C) is the depth of the synset C using edge counting in the taxonomy.
- LCS (C1, C2) is the smallest sub-segment in common of C1 and C2.
- Depth (LCS (C1, C2)) is the length between LCS of C1 and C2 and the root of the taxonomy.

Second, we calculate the overall similarity between two logical representations using the corresponding average strategy. Given the two logical forms FOLH and FOLT which correspond respectively to the hypothesis and to any answer passage, we denote m for the length of FOLH, n for length of FOLT. To compute the corresponding average, we propose to construct a relative semantic similarity matrix R [m, n] of each pair sense of words of FOLH and FOLT, where R [i, j] is the semantic similarity between the word at the position i of FOLH and the word at position j of FOLT. Thus, the similarity between FOLH and FOLT is reduced to the problem of calculating the maximum total matching weight of a bipartite graph. This is ensured by exploiting the Hungarian algorithm on this graph where X and Y are FOLH and FOLT and the nodes of the graph are the related words (Dao and Simpson 2005). The corresponding average is calculated as follows:

$$\mathbf{Corresponding_{average}} = \frac{2*Match(FOLH, FOLT)}{|FOLH| + |FOLT|} \quad (4)$$

For our example of the FOLT/FOLH couple, we first look for the semantic similarity between the senses of all FOLH words with each word of FOLT using the Wu-Palmer measure. We then construct the relative matrix of the semantic similarity of each pair of meanings of the words. The matching results are then combined into a single similarity value by the average match for the FOLT and FOLH pair which is equal to 0.31.

$$\mathbf{Corresponding_{average}} = 0.31$$

3.2   Features Combination and Entailment Decision

We have chosen only the three features, taking into account the fact that larger sets of features do not necessarily lead to the improvement of the performance of the decision of entailment. In addition, we find in some cases that it is difficult to find a passage which, at the same time, determines all the features, although it may contain an answer to the given question. Therefore, we come to the conclusion that the combination of features is often more efficient than taking individually the best feature.

The final step is to classify the entailments. At a certain level, the problem of entailment is simply considered as a problem of classification. More precisely, it is a problem with two classes, a class "TRUE" and a class "FALSE".

For each hypothesis (question) and their answer passages represented in logical forms, respectively FOLH and FOLTs, we use the features that were evaluated separately first as learning data to form features vectors. Then, we use the WEKA J48 decision tree classifier (Witten and Frank 1999) which classifies these vectors as "TRUE" or "FALSE" entailments.

4   Experiments and Results

To evaluate our approach, we used text passages and questions from the AQA-WebCorp corpus (Bakari et al. 2016). The entailment classification step produces for each question a number of answers that are either (positive "true") or (negative "true"). Indeed, a positive "true" answer is a passage with "true" entailment with the question and which contains the answer to that question. On the other hand, a negative "true" answer is a passage with "true" entailment with the question but which does not answer the question. To evaluate our approach, only the positive "true" answers are counted using the three metrics: precision, recall and F-measure.

(a) Precision:

Precision is the proportion of correct answers among the true returned by the classification step.

$$Precision = \frac{Number\ of\ true\ positive\ answers\ returned}{Number\ of\ "true"\ answers} \tag{5}$$

(b) Recall:

The recall is the proportion of the number of true answers correct returned by the classification step in relation to the number of true expected answers.

$$Recall = \frac{Number\ of\ true\ positive\ answers\ returned}{Number\ of\ positive\ true\ expected\ answers} \tag{6}$$

(c) F-measure:

Precision and Recall can be both combined in a weighted harmonic average called F-measure.

$$F - measure = \frac{2 * Precision * recall}{Precision + recall} \qquad (7)$$

The results of the classification of the entailments show a precision and a recall successively equal to 73% and 68% on average. To our knowledge, there is no existing work that addresses the RTE task for question/answering systems in Arabic.

5 Conclusion

In this paper, we have presented our approach for Arabic recognizing textual entailment. In our work, we determine the textual entailment between pairs of logical representations of the question (hypothesis) and the passage of text that can answer this question (text). In the entailment determination and classification, some features are automatically, extracted, such as, predicate-argument overlap, named entities correspondence and semantic similarity. The textual entailment determination is based on scores calculated by a similarity metric for each feature. Finally, for each metric, we use as input the logical representations of the T/H (text passage/question) pairs named respectively FOLT and FOLH. In the future, we plan to use the inference rules for the determination of textual implications in Arabic language.

Acknowledgements. I give my sincere thanks for my collaborators Professor Mahmoud NEJI and Doctor Wided BAKARI (University of Sfax-Tunisia) that i have benefited greatly by working with them.

References

Dagan, I., Roth, D., Sammons, M., Zanzotto, F.M.: Recognizing textual entailment: models and applications. Synth. Lect. Hum. Lang. Technol. **6**(4), 1–220 (2013)

Dagan, I., Glickman, O., Magnini, B.: The PASCAL recognising textual entailment challenge. In: Machine Learning Challenges Workshop, pp. 177–190 (2005)

Dagan, I., Dolan, B., Magnini, B., Roth, D.: Recognizing textual entailment: rational, evaluation and approaches. Nat. Lang. Eng. **15**(4), i–xvii (2009). Editorial of the special issue on Textual Entailment

Dagan, I., Glickman, O., Magnini, B.: The PASCAL recognising textual entailment challenge. In: Machine Learning Challenges. Evaluating Predictive Uncertainty, Visual Object Classification, and Recognising Tectual Entailment, pp. 177–190. Springer, Heidelberg (2006)

Glickman, O., Dagan, I., Keller, M., Bengio, S., Daelemans, W.: Investigating lexical substitution scoring for subtitle generation. In: Proceedings of the Tenth Conference on Computational Natural Language Learning, pp. 45–52. Association for Computational Linguistics, June 2006

Tatu, M., Iles, B., Moldovan, D.: Automatic answer validation using COGEX. In: Workshop CLEF 2006, Alicante, Spain (2006)

Mihalcea, R., Moldovan, D.I.: Extended WordNet: progress report. In: Proceedings of NAACL Workshop on WordNet and Other Lexical Resources (2001)

Fowler, A., Hauser, B., Hodges, D., Niles, I., Novischi, A., Stephan, J.: Applying COGEX to recognize textual entailment. In: Proceedings of the PASCAL Challenges Workshop on Recognising Textual Entailment, pp. 69–72 (2005)

Fellbaum, C. (ed.): WordNet: An Electronic Lexical Database. MIT Press, Cambridge (1998)

Pakray, P., Bandyopadhyay, S., Gelbukh, A.F.: Lexical based two-way RTE system at RTE-5. In: TAC, November 2009

Ofoghi, B.: Enhancing factoid question answering using frame semantic-based approaches. Doctoral Dissertation, Ph.D. thesis, Université de Ballarat, Ballarat, Australia (2009)

Ferrández, O., Izquierdo, R., Ferrández, S., Vicedo, J.L.: Addressing ontology-based question answering with collections of user queries. IPM 45(2), 175–188 (2009)

Glöckner, I.: Filtering and fusion of question-answering streams by robust textual inference. In: Proceedings of KRAQ, vol. 7, pp. 43–48 (2007)

Glockner, I., Pelzer, B.: Exploring robustness enhancements for logic-based passage filtering. In: Knowledge Based Intelligent Information and Engineering Systems (Proceedings of KES2008, Part I). LNAI 5117, pp. 606–614. Springer, Heidelberg (2008)

Alabbas, M.: ArbTE: Arabic textual entailment. In: Proceedings of the 2nd Student Research Workshop Associated with RANLP, pp. 48–53 (2011)

Almarwani, N., Diab, M.: Arabic textual entailment with word embeddings. In: WANLP 2017 (Co-located with EACL 2017), pp. 185–190 (2017)

Alabbas, M., Ramsay, A.: Natural language inference for Arabic using extended tree edit distance with subtrees. J. Artif. Intell. Res. 48, 1–22 (2013)

AL-Khawaldeh, F.T.: A study of the effect of resolving negation and sentiment analysis in recognizing text entailment for Arabic. World Comput. Sci. Inf. Technol. J. (WCSIT) 5(7), 124–128 (2015)

Khader, M., Awajan, A., Alkouz, A.: Textual entailment for Arabic language based on lexical and semantic matching. Int. J. Comput. Inf. Sci. 12(1), 67 (2016)

Harabagiu, S., Hickl, A.: Methods for using textual entailment in open-domain question answering. In: Proceedings of the 21st International Conference on Computational Linguistics and the 44th annual meeting of the Association for Computational Linguistics, pp. 905–912. Association for Computational Linguistics, July 2006

Zribi, I., Hammami, S.M., Belguith, L.H.: L'apport d'une approche hybride pour la reconnaissance des entités nommées en langue arabe. In: TALN'2010, Montréal, 19–23 juillet 2010, pp. 19–23 (2010)

Dao, T.N., Simpson, T.: Measuring similarity between sentences. WordNet. Net, Technical report (2005)

Witten, I.H., Frank, E.: Data Mining: Practical Machine Learning Tools and Techniques with Java Implementations. Morgan Kaufmann, Burlington (1999)

Bakari, W., Bellot, P., Neji, M.: AQA-WebCorp: web-based factual questions for Arabic. Procedia Comput. Sci. 96, 275–284 (2016). ISO 690

Black, W., Elkateb, S., Rodriguez, H., Alkhalifa, M., Vossen, P., Pease, A., Fellbaum, C.: Introducing the Arabic WordNet project. In: Proceedings of the Third International WordNet Conference, pp. 295–300, January 2006

Transformation of Data Warehouse Schema to NoSQL Graph Data Base

Amal Sellami[1(✉)], Ahlem Nabli[1,2], and Faiez Gargouri[1]

[1] MIRACL Laboratory, University of Sfax, Sfax, Tunisia
sellami.amal91@gmail.com, ahlem.nabli@fss.usf.tn,
faiez.gargouri@isims.usf.tn
[2] Al-Baha University, Al Bahah, Kingdom of Saudi Arabia

Abstract. Big volumes of data cannot be processed by traditional ware-houses and OLAP servers which are based on RDBMS solutions. As an alternative solution, Not only SQL (NoSQL) databases are becoming increasingly popular as they have interesting strengths such as scalability and flexibility for an OLAP system. As NoSQL database offer great flexibility, they can improve the classic solution based on data warehouses (DW). In the recent years, many web applications are moving towards the use of data in the form of graphs. For example, social media and the emergence of Facebook, LinkedIn and Twitter have accelerated the emergence of the NoSQL database and in particular graph-oriented databases that represent the basic format with which data in these media is stored. Based on these findings and in addition to the absence of a clear approach which allows the implementation of data warehouse under NoSQL model, we propose, in this paper, new rules for transforming a multidimensional conceptual model into NoSQL graph-oriented model.

Keywords: NoSQL · Graph-oriented databases · Data warehouse

1 Introduction

Nowadays, data volume analysis is reaching critical sizes challenging traditional data warehouse approaches. Current common solutions are mainly based on relational databases (using R-OLAP approaches) that are unfortunately no longer adapted to these data volumes. With the rise of large Web platforms (e.g. Google, Facebook, Twitter, Amazon, etc.) solutions for "Big Data" management have been developed. These solutions are based on decentralized approaches managing large data amounts and have contributed to developing "Not only SQL" (NoSQL) data management systems.

Indeed, NoSQL solutions allows us to consider new approaches for data warehousing. In this paper, we focus on one class of NoSQL stores, namely graph-oriented systems.

Graph-oriented systems are one of the famous families of NoSQL systems. They are used for managing highly connected data and perform complex queries

© Springer Nature Switzerland AG 2020
A. Abraham et al. (Eds.): ISDA 2018, AISC 941, pp. 410–420, 2020.
https://doi.org/10.1007/978-3-030-16660-1_41

over it. Not only data values but also graph structures are involved in queries. Specifying a pattern and a set of starting points, it is possible to reach an excellent performance for local reads by, first, traversing the graph, then collecting and aggregating information from nodes and edges.

In order to benefit from the NoSQL model advantages, and knowing that there is no clear approach allowing the implementing of the NoSQL DW under graph oriented data base, the objective of this paper is to propose a mapping between the concepts of a multidimensional schema and the concepts of the graph-oriented NoSQL model. We propose than a set of rules able to transform conceptual model to logical graph-oriented models.

This paper is organized as follows. Section 2 represents a state of the art. In Sect. 3, we introduce the formal representation of the multidimensional schema. Our proposal for formal representation of the NoSQL graph-oriented Databases is disclosed in Sect. 4. Details about the transformation rules is given in Sect. 5. In Sect. 6, we conclude this paper by some future work.

2   Related Work on NoSQL Data Warehousing

NoSQL solutions have proven clear advantages compared to relational database management systems (RDBMS). In the literature, several research focused on the translation of data warehousing concepts to NoSQL ones. Indeed, in 2014, a first approach was proposed to couple the graph-oriented model and OLAP [4]. In this approach authors propose to structure the data within the graph-oriented NoSQL system. Authors present then two formalisms to represent the fact and the dimensions at the level of the logical model. These formalisms provide two types of relationships, those linking the fact to dimensions, and those linking the attributes of dimensions to each other. The latter makes possible to preserve the hierarchical relationship. In [9], authors have developed a new benchmark for the columnar NoSQL DW, without giving the formalization for the modeling process. This work is considered as the first work proposing implemented star DW under column oriented NoSQL RDBMS directly from dimensional model. Later, this work was extended by proposing three approaches which allow big data warehouses to be implemented under the column oriented NoSQL model [8]. Each approach differs in terms of structure and the attribute types used when mapping the conceptual model into logical model is performed.

Otherwise, many researches [5,6] have tried to define logical models for NoSQL data stores (oriented columns and oriented documents). They proposed a set of rules to map star schema into two NoSQL models: column-oriented (using HBase) and document-oriented (using MongoDB). In [12], authors move towards a column-oriented NoSQL data model for both reasons; first, to solve the problem of attribute distribution between column families and second to optimize queries and facilitate data management. In [14], authors proposed transformation rules that ensure the successful translation from conceptual DW schema to two logical NoSQL models (column-oriented and document-oriented). Authors proposed then two possible transformations namely: simple and hierarchical transformations. The first one stores the fact and dimensions into one column-family/collection. The second transformation uses different column-families/collections for storing fact and dimensions while explaining hierarchies. In the same context, in [1] authors define UML class diagram translation processes in column oriented NoSQL with HBase. In [3] a new suggestion is presented to transform an RDF to NoSQL graph-oriented model by proposing a set of transformation rules. In [7], the authors analyze several issues including modeling, querying, loading data and OLAP cuboids. Authors compared document-oriented models (with and without normalization) to analogous relational database models.

Other recent works, focused on simplifying the heterogeneous data querying in the graph-oriented NoSQL systems [10]. In [13], authors propose a diagram to address this lack of a generic and comprehensive notation for graph database modeling, called GRAPHED (Graph Description Diagram for Graph Databases).

All the mentioned works present several interesting mining. The majority of them have shown the transformation of the data warehouse schema in the column-oriented or document-oriented NoSQL model. It should be noted that the column and document oriented model share the major disadvantage of the relational database which is the join. The latter must be highlighted during the interrogation which represents a painful work. The major interesting advantage of the graph oriented model is related to the ability for supporting complex queries without using joins. Therefore, we thought it would be interesting to define a process to implement a multidimensional data warehouse based on graph-oriented NoSQL model. We are mainly interested to define rules allowing the representation of a conceptual diagram of DW in graph-oriented NoSQL model.

3   Conceptual Multidimensional Model

Conceptual modeling is considered as a major foundation phase. In this phase, DW are modeled in a multidimensional way. The basic concepts of multidimensional modeling are: facts, measures, dimensions and hierarchies. With reference to [11], we suggest the following formalization:

Multidimensional Schema. A DW is characterized by its multidimensional schema (MS) composed of a fact schema with a single or a constellation of facts and dimensions. Formally, a multidimensional schema MS is defined by (F^{MS}, D^{MS}, Func) where:

- $F^{MS} = \{F_1, ..., F_n\}$ is a set of facts,
- $D^{MS} = \{D_1, ..., D_m\}$ is a set of dimensions,
- Func a function that associates fact of F^{MS} to sets of dimensions along which it can be analyzed (link Fact-Dimension).

Dimension and Hierarchy. The dimensions represent the axes of the multidimensional analysis. They are usually textual and discreet. They are used to restrict the scope of queries to limit the size of the responses. Formally, a dimension, is defined by (N^D, Att^D, H^D) where:

- N^D is the name of the dimension,
- $Att^D = \{A_1, ..., A_m\}$ is the set of strong and weak attributes of a dimension,
- $H^D = \{H_1, ..., H_K\}$ is a set of hierarchies.

The attributes of the dimensions are organized into one or more hierarchies. A hierarchy is composed of several levels, representing different degrees of information accuracy. A hierarchy of the dimension H, is defined by $(N^{Hi}, Param^{Hi},$ Pred, $Weak^{Hi})$ where:

- N^{Hi} is the name of the hierarchy,
- $Param^{Hi}$ is a set of parameters,
- Pred is a function that associates each parameter its predecessor,
- $Weak^{Hi}$ is a function that associates with each parameter zero or more weak attributes.

Fact. A fact represents the analyzed subject. It is composed of measures reflecting the information to be analyzed. The measures of a fact are generally numeric and continuously valued to summarize a large number of records. Formally, a fact, is defined by (N^F, M^F) where:

- N^F is the name of the fact,
- M^F is a set of measures, each associated with an aggregation function.

Example. Based on the above formalization, the multidimensional schema MS illustrated by the Fig. 1 will be defined as following:

- F $= \{F_{lineOrder}\}$, D $= \{D_{Customer}, D_{Product}, D_{Store}, D_{Date}\}$ and Fonc$(F_{lineOrder}) = \{D_{Customer}, D_{Product}, D_{Store}, D_{Date}\}$.

The fact $F_{lineOrder}$ is defined by (LineOrder, {SUM(Quantity), SUM(Discount), SUM(Total)}) and it is analyzed according to four dimensions, each consisting of several hierarchical levels (called detail levels or parameters):

- The Customer dimension $(D_{Customer})$ with parameters Customer (along with the weak attribute Name), City and Region.
- The Product dimension $(D_{Product})$ with parameters Prod_key (with weak attributes Prod_Name), Category, and Type, organized using two hierarchies H_{Type} and H_{Categ}.
- The Date dimension (D_{Date}) with parameters Day, Month (with a weak attribute, Month_name) and Year.
- The Store dimension (D_{Store}) with parameters Store-id (with weak attributes Name), City and Region.

For instance, the dimension $D_{Product}$ is defined by (Product, {Prod_key, Category, Type, ALL }, {H_{Type}, H_{Categ} }) with Prod_key $= id_{Product}$ and:

- $H_{Type} = (H_{Type}, \{$Prod_key, Type, ALL $\},($Prod_key, $\{$Prod_Name $\}))$
  Note that Weak H_{Type} (Prod_key) $=$ Prod_Name
- $H_{Categ} = (H_{Categ}, \{\{$Prod_key, Category, ALL$\}, ($Prod_key, $\{$Prod_Name $\})\})$

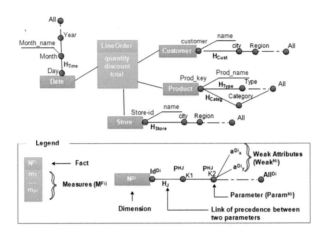

Fig. 1. Multidimensional conceptual schema

4   Formalization of Graph-Oriented Databases

The growing usage of "Not-only-SQL" storage systems, referred as NoSQL, has given ability to efficiently handle large volume of data. Graph-oriented systems are among the most increasingly used NoSQL approaches. A special attention has focused on how to model data in the form of graphs. Graph-oriented databases are based upon graph theory (set of nodes, edges, and properties). It's useful for inter-connected relationship data such as communication patterns, social networks, and biographical interactions. The relational database performed better on executing queries when the amount of data is relatively limited. However, as queries became complex, the graph database outperformed the relational one. For the conceptual modeling, an entity-relationship diagram is readily translated into a Property Graph Model, making a conceptual model for graph databases necessary. It helps to understand which entities can be logically connected to which other entities. Graph databases support only binary relationships. On the other hand, graph modeling is much easier than for a relational data model because real world objects are explicit in terms of connections. The formalization of graph-oriented model consists on the formalization of the following three components: Nodes, Edges, and Properties (Fig. 2).

Graph-Oriented Database. A Graph-oriented databases (G) is composed of a set of nodes, edges, and properties. Formally, a graph databases G, is defined by (V^G, E^G, P^G) where:

- $V^G = \{V_1, ..., V_p\}$ is a set of node that represent the entities,
- $E^G = \{E_1, ..., E_y\}$ is a set of edges that represent the relation between the nodes,
- P is a set of properties attributed to each component of the graph-oriented database. A property is formed by a couple of a key and value pair.

Node. Each node has property and label. Formally, a node, is defined by (id^V, P^V, L^V) where:

- id^V is the identifier of the nodes,
- P^V is a set of properties that describe a node,
- L^V is a set of labels or etiquette attached to the node. In order to express the semantic of the nodes, usually a node can have 0 or more labels written as $L^G = \{L_1, ..., L_q\}$.

Relation. The relations connecting the nodes can eventually have properties. Formally, a relation is defined by $(id^R, V_1{}^R, V_\emptyset{}^R, T^R, P^R)$ where:

- id^R is the identifier of the relation,
- $V_1{}^R$ is the identifier of the incoming node,
- $V_\emptyset{}^R$ is the identifier of the outgoing node,
- T^R is the type of relation that bears the name of the relation,
- P^R is a set of properties of a relation.

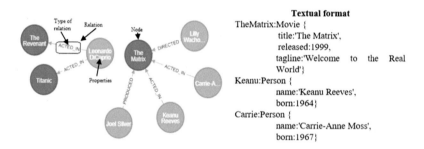

Fig. 2. A graph example

5 Transformation Rules

In this section, we show how the concepts of a Data Warehouse can be modeled based on the graph paradigm. The target NoSQLl model has been formalized in [2]. DW schema consists of fact with measures, as well as a set dimensions with attributes, we map then the dimensions according to its attributes and the fact according to its measures. As several alternatives are possible, we will detail some of these alternatives in order to choose the best one. In this paper special attention is given to two transformations that define the manner how create a DW under NOSQL graph data base.

5.1 First Transformation

In this transformation (Fig. 3) each fact and dimension parameter are transformed into nodes as described by the following rules:

Rule.1- Fact Transformation. Each fact is transformed into a node with the label of the node takes the name of the concept of the multidimensional model which is 'fact' then we add the name of the fact as a second label at the same node. Each measure is transformed by a property of Fact node.

RF.1. Each fact $F \in F^{MS}$ is transformed into a node, defined by V (id^V, P^V, L^V) where:

- Label l_1 is the type of the multidimensional concept: $l_1 = $ 'Fact' / $L^V = \{l_1\}$,
- Label l_2 is the name of the fact: $l_2 = N^F$ / $L^V = L^V \cup \{l_2\}$,
- Each measure $m_i \in M^F$ is represented as a property with p $\leftarrow m_i/P^V = P^V \cup \{p\}$.

Rule.2- Name and Identifier of Dimension Transformation. Each dimension is transformed into a node with the label of the node takes the name of the concept of the multidimensional model (in this case is the dimension). Then, we allow the name of the dimension as a second label at the same node. After, the identifier is transformed into a property in the node. Finally, any weak attribute associated to the identifier is transformed into a property in the same node.

RD.1. Each name and identifier of a dimension is transformed into a node V (id^V, P^V, L^V) where:

- Label l_1 is the type of the multidimensional concept: $l_1 = $ 'Dimension' / $L^V = \{l_1\}$,
- Label l_2 is the name of the dimension: $l_2 = N^D$ / $L^V = L^V \cup \{l_2\}$,
- The identifier a_i modeling by a property p with p $\leftarrow a_i/P^V = P^V \cup \{p\}$,
- Each weak attribute a_w associated to a_i is transformed into a property p with p $\leftarrow a_w/P^V = P^V \cup \{p\}$.

Rule.3- Hierarchies Transformation. A hierarchy consists of a set of parameters and a link of precedence between parameters. Each parameter is transformed into a node with the label of the node takes the name of the concept of the multidimensional model (in this case is the parameter). Then, we allow the name of the parameter as a second label at the same node. After that, each weak attribute is represented in the node in the form of property. Finally, each link of precedence is transformed into a relation.

RH.1. Transformation of Parameter/Modeling the Link of Precedence Between Parameter

RH.1.1 Transformation of Parameter
Each parameter a_i is transformed into a node V (id^V, P^V, L^V) where:

- Label l_1 is the type of the multidimensional concept: $l_1 = $ 'Parameter' / $L^V = \{l_1\}$,
- Label l_2 is the name of the parameter: $l_2 = a_i$ / $L^V = L^V \cup \{l_2\}$,
- Each weak attribute a_w associated to a_i is transformed into a property p with $p \leftarrow a_w / P^V = P^V \cup \{p\}$.

RH.1.2. Transformation of Link of Precedence Between Parameter

Each $a_i \rightarrow a_{i-1} \subset H^D$ is transformed into a relation R, defined by $(id^R, V_1{}^R, V_\emptyset{}^R, T^R, P^R)$ where:

- $V_1{}^R$ is the node represented a_i,
- $V_\emptyset{}^R$ is the node represented a_{i-1},
- The type t_1 is the name of the relation: $t_1 = $ 'Precede'$/T^R = \{t_1\}$.

Rule.4- Transformation of the Link Fact-Dimension.

Each link between fact and dimension is represented as a relation having as node source the node modeling the fact and as node destination the node modeling the dimension. The relation has as name 'link fact-dimension'.

RFD.1.

Each link fact-dimension is transformed into a relation R, defined by $(id^R, V_1{}^R, V_\emptyset{}^R, T^R, P^R)$ where:

- $V_1{}^R$ is the node represented the fact,
- $V_\emptyset{}^R$ is the node represented the dimension,
- The type t_1 is the name of the relation: $t_1 = $ 'link fact-dimension'$/T^R = \{t_1\}$.

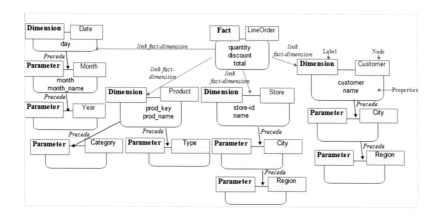

Fig. 3. FT examples

5.2 Second Transformation

This transformation (Fig. 4) has the same rules as the first one but with only one difference at **Rule.4-Transformation of the link fact-dimension**, all

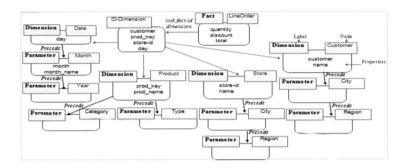

Fig. 4. ST examples

dimension identifier is transformed into one node, as explained by the following rule:

Rule.5- Transformation of the Link Fact-Dimension. A relation was created between the fact and the node created with the first parameters of each dimension. The relation has as name 'link fact-id dimension'.

RFD.2. The link fact-id dimension is transformed into a relation R, defined by $(id^R, V_1^R, V_\emptyset^R, T^R, P^R)$ where:

- V_1^R is the node represented the fact,
- $V_\emptyset^R$ is the node represented the id of each dimension,
- The type t_1 is the name of the relation: $t_1 =$ 'link fact-id dimension'/$T^R=\{t_1\}$.

The reason why graph databases are an interesting category of NoSQL is because, contrary to the other approaches, they actually go the way of increased relational modeling, rather than doing away with relations, that is, one to one, one to many, and many to many structures can easily be modeled in a graph-based way.

In a way, a graph database is a hyper-relational database, where JOIN tables are replaced by more interesting and semantically meaningful relationships that can be navigated (graph traversal) and/or queried, based on graph pattern matching. Here, we're using the Cypher query language, the graph-based query language introduced by Neo4j, one of the most popular graph databases. Let's look at some examples:

- A SQL SELECT query:
  MATCH(p:Product)
  RETURN p;
- Create data relationships:
  MATCH (d:id-dimension), (p:Product)
  WHERE d.prod_key = p.prod_key
  CREATE (d)-[:PART_OF]→(p)
- ORDER BY and LIMIT statements:
  MATCH(p:Product)

RETURN p
ORDER BY p.type **DESC**
LIMIT 10;
- List all Leonardo Dicaprio movies:
  MATCH (Leo:Person{name: "Leonardo   Dicaprio"})-[:ACTED_IN]→
  (LeonardoDicaprioMovies)
  RETURN Leo, LeonardoDicaprioMovies
- Who directed "The matrix":
  MATCH (Thematrix {title: "The matrix"})←[:DIRECTED]-(directors)
  RETURN directors.name
- Find the produce store:
  MATCH (p:Product{category: "Produce"})←(p:Product)←(s:Store)
  RETURN DISTINCT s.name as ProduceStore

6   Conclusion

NoSQL solutions have proven some clear advantages when compared to classical relational database management systems. Nowadays, many researches atten-
tions have moved towards the use of these systems in order to store, manipulate and analyzing "big data". We have proposed two types of transformations
rules ensuring successful passage from conceptual DW schema to NoSQL graph-
oriented models. Both of these transformations are based on a set of rules presented with examples. As future work, we will focus on evaluating our proposed
transformation based on a set Olap query to choose the best transformation.

References

1. Abdelhedi, F., Ait Brahim, A., Atigui, F., Zurfluh, G.: Processus de transformation
   mda d'un schéma conceptuel de données en un schéma logique NoSQL. INFORSID
   (actes électroniques) (2016)
2. Sellami, A., Ahlem Nabli, F.G.: Entrepôt de données nosql orienté graphe:règles
   de modélisation. In: ASD: Big data & Applications. ASD (2018)
3. Bouhali, R., Laurent, A.: Exploiting RDF open data using NoSQL graph databases.
   In: IFIP International Conference on Artificial Intelligence Applications and Innovations, pp. 177–190. Springer (2015)
4. Castelltort, A., Laurent, A.: NoSQL graph-based OLAP analysis. In: KDIR (2014)
5. Chevalier, M., El Malki, M., Kopliku, A., Teste, O., Tournier, R.: Implementing
   multidimensional data warehouses into NoSQL. In: 17th International Conference
   on Enterprise Information Systems, vol. DISI (2015)
6. Chevalier, M., El Malki, M., Kopliku, A., Teste, O., Tournier, R.: Implementation of
   multidimensional databases in column-oriented NoSQL systems. In: East European
   Conference on Advances in Databases and Information Systems. Springer (2015)
7. Chevalier, M., El Malki, M., Kopliku, A., Teste, O., Tournier, R.: Entrepôts de
   données orientés documents: cuboïdes étendus. Document numérique **20**(1) (2017)
8. Dehdouh, K., Bentayeb, F., Boussaid, O., Kabachi, N.: Using the column oriented
   NoSQL model for implementing big data warehouses. In: Proceedings of PDPTAs
   (2015)

9. Dehdouh, K., Boussaid, O., Bentayeb, F.: Columnar NoSQL star schema bench-mark. In: International Conference on Model and Data Engineering. Springer (2014)

10. El Malki, M., Hamadou, H.B., Chevalier, M., Péninou, A., Teste, O.: Querying heterogeneous data in graph-oriented NoSQL systems. In: International Conference on Big Data Analytics and Knowledge Discovery, pp. 289–301. Springer (2018)

11. Nabli, A., Feki, J., Gargouri, F.: Automatic construction of multidimensional schema from OLAP requirements. In: AICCSA, p. 28. IEEE (2005)

12. Scabora, L.C., Brito, J.J., Ciferri, R.R., Ciferri, C.D.d.A., et al.: Physical data warehouse design on NoSQL databases OLAP query processing over hbase. In: International Conference on Enterprise Information Systems, vol. XVIII (2016)

13. Van Erven, G., Silva, W., Carvalho, R., Holanda, M.: Graphed: a graph description diagram for graph databases. In: World Conference on Information Systems and Technologies. Springer (2018)

14. Yangui, R., Nabli, A., Gargouri, F.: Automatic transformation of data warehouse schema to nosql data base: comparative study. Procedia **96**, 255–264 (2016)

Translation of UML Models for Self-adaptive Systems into Event-B Specifications

Marwa Hachicha[(✉)], Riadh Ben Halima, and Ahmed Hadj Kacem

ReDCAD, University of Sfax, B.P. 1173, 3038 Sfax, Tunisia
marwahachicha@gmail.com

Abstract. Engineering self-adaptive systems using a standard or a guide is a challenging task. In this context, we propose using a set of MAPE (Monitor-Analyze-Plan-Execute) patterns for modeling self-adaptive systems. Ensuring that there is no ambiguity, incompleteness and misunderstanding of self-adaptive systems that instantiate MAPE patterns is very crucial. For this purpose, we propose to specify self-adaptive systems instantiating MAPE patterns using the Event-B formal method to provide a precise reference for designer to carry out an effective verification. To this end, we propose to generate Event-B specifications from conceptual modeling of the self-adaptive systems. In this paper, we propose a set of structural transformation rules to transform each element in the model to its corresponding concept in the Event-B method.

1 Introduction

The concept of self-adaptation was introduced so that the software systems can deal with the increasing complexity and dynamism of modern software systems autonomously. In this context, the MAPE control loop is considered as an essential concept for the achievement of self-adaptation. The MAPE control loop is based on the following functions: monitoring the managed system and its environment, analyzing the managed system's behavior, planning adaptation actions when an undesired system's behavior is detected and executing the planned adaptation actions to adapt the managed system.

While the principles behind the MAPE control loop are rather simple, designing and guaranteeing the correctness of the behavior of self-adaptive systems are not easy tasks. The key identified challenge is how to engineer self-adaptive systems using a standard or a guide. In this context, design patterns have been proved as an established and robust solution for designing software systems. They include guides of good practice to abstractly describe the architecture of a software system. In fact, when systems are complex and involve multiple components, it is essential to integrate several MAPE control loops that coordinate their operations to realize the adaptation of the system. In this context, Weyns et al. [2] proposed a set of MAPE patterns for decentralized control in

© Springer Nature Switzerland AG 2020
A. Abraham et al. (Eds.): ISDA 2018, AISC 941, pp. 421–430, 2020.
https://doi.org/10.1007/978-3-030-16660-1_42

self-adaptive systems. Coordinated control and information sharing are based on a fully decentralized approach, while master/slave, regional planning, and hierarchical control, are instead based on a hierarchical distribution model, where higher level MAPE components control subordinate MAPE components. These patterns are presented using a simple notation and do not fully enumerate all the possible decentralization patterns. Therefore we have proposed in previous works [1] a standard notation based on the Unified Modeling Language (UML) to describe and compose the different MAPE patterns proposed by Weyns et al. [2]. Structural features are designed using a UML profile that extends UML component diagram and behavioral features are designed using a UML profile that extends UML activity diagram. Providing evidence that self-adaptive system models respect the expected behaviour and they are correct with respect to given specifications is very crucial. Consequently, it is essential to provide a formal specification model for the designed self-adaptive system. Formal methods provide the means to define precisely the system using mathematical notations. It allows engineer to prove that the specification of the system is realizable and also to prove its properties without running it. Therefore, we propose defining a mapping mechanism for UML system models to a formal model based on the Event-B method for formal verification purposes.

In this paper, we are interested in structural transformation rules to transform each element in the model to its corresponding concept in the Event-B method. These transformation rules are expressed using the XSLT language to transform the UML system modeling into Event-B specifications. These specifications will be then imported into the Rodin theorem prover for verification.

The remainder of this paper is organized as follows. Section 3 presents a set of algorithms to translate the different elements in the system model to Event-B specification using the XSLT language. Section 4 describes a case study to illustrate our approach. Finally, the last section concludes the paper and gives future work directions.

2   Translation of Self-adaptive System UML Model into Event-B Specifications

This section is devoted to present the different transformation rules for mapping structural features of the self-adaptive system model to their respective concepts in the Event-B formal model.

STR1: Pattern Component Transformation Rules. We propose to classify the different components that constitute the architecture of a MAPE pattern into four classes:

- The first class involves the main components (Master, Slave, Regional planner, etc.) that form the architecture of the MAPE pattern.
- The second class involves the different MAPE components: Monitor, Analyzer, Planner and Executor components.

- The third class involves the different MAPE sub-components: Receiver, Processor and Sender.
- The fourth class involves the Probe, Effector and Context components.

STR1.1: Main Components Transformation Rule. This rule specifies formally the main components included in the design of a self-adaptive system instantiating a MAPE pattern, such as managers, managed elements, regional planners. This rule transforms the name of the main component into a constant in the CONSTANTS clause. This is specified in the Algorithm 1 from line 6 to line 27.

In the master/slave pattern, the set of master components is composed of one element. So, this is transformed formally into a set in the AXIOMS clause as presented in the Algorithm 1 (line 31).

For the self-adaptive systems instantiating the other patterns, the set of the main components is composed of all the component names. This is formally transformed into a partition. For example, the set of managed elements is composed of all the managed element names as shown in the Algorithm 1 (from line 51 to line 56). This is means that $ManagedElement = \{MD_1, ..., MD_n\} \land MD_1 \neq MD_2 \land ... \land MD_{n-1} \neq MD_n$

Algorithm 1. Main component transformation rules algorithm

Begin

1: Write (" SETS ")
2: Write ('Master')
3: Write ('Slaves')
4: Write ('RegionalPlanners')
5: ...
6: Write (" CONSTANTS ")
7: **if** exist Master **then**
8:     Write ('Master. Name')
9: **end if**
10: **if** exist Slave **then**
11:     **For each** Slave do
12:     Write(Slave.Name)
13: **end if**
14: **if** exist RegionalPlanner **then**
15:     **For each** RegionalPlanner do
16:     Write(RegionalPlanner.Name)
17:     EndFor
18: **end if**
19: **if** exist MangedElement **then**
20:     **For each** MangedElement do
21:     Write(MangedElement.Name)
22:     EndFor
23: **end if**
24: **if** exist MangerElement **then**
25:     **For each** MangerElement do
26:     Write(MangerElement.Name)

27:     EndFor
28: **end if**
29: Write (" AXIOMS ")
30: **if** exist Master **then**
31:     Write ($'Master - set : Master = \{Master.Name\})'$)
32: **end if**
33: **if** exist Slave **then**
34:     Write ($'Slave - partition : partition(Slaves,'$)
35:     **For each** Slave do
36:     Write(Slave.Name)
37:     EndFor
38: **end if**
39: **if** exist RegionalPlanner **then**
40:     Write ($'RegionalPlanner - partition : partition(RegionalPlanners,'$)
41:     **For each** RegionalPlanner do
42:     Write(RegionalPlanner.Name)
43:     EndFor
44: **end if**
45: **if** exist MangedElement **then**
46:     Write ($'MangedElement - partition : partition(MangedElements,'$)
47:     **For each** MangedElement do
48:     Write(MangedElement.Name)
49:     EndFor
50: **end if**
51: **if** exist MangerElement **then**
52:     Write ($'MangerElement - partition : partition(MangerElements,'$)
53:     **For each** MangerElement do
54:     Write(MangerElement.Name)
55:     EndFor
56: **end if**

End

ST1.2: MAPE Components Transformation Rule. This rule specifies formally the different MAPE components of the MAPE control loop. It transforms the name of the different MAPE components into a constant in the CONSTANTS clause. This is specified in the Algorithm 2 from line 6 to line 26.

The set of monitor (respectively analyzer, planner, executor) components is composed of all monitor (respectively analyzer, planner, executor) component names. This is transformed formally to a partition.

Algorithm 2. MAPE components transformation rule algorithm

Begin

1: Write (" SETS ")
2: Write ("MonitorComponents ")
3: Write (" AnalyzerComponents ")
4: Write (" PlannerComponents ")
5: Write (" ExecutorComponents ")
6: Write (" CONSTANTS ")
7: **if** exist Monitor **then**

```
8:      For each Monitor do
9:          Write(Monitor.Name)
10:         EndFor
11: end if
12: if exist Analyzer then
13:         For each Analyzer do
14:             Write(Analyzer.Name)
15:             EndFor
16: end if
17: if exist Planner then
18:         For each Planner do
19:             Write(Planner.Name)
20:             EndFor
21: end if
22: if exist Executor then
23:         For each Executor do
24:             Write(Executor.Name)
25:             EndFor
26: end if
27: Write (" AXIOMS ")
28: if exist Monitor then
29:         Write ('Monitor − partition : partition(MonitorComponents,')
30:         For each Monitor do
31:             Write(Monitor.Name)
32:             EndFor
33: end if
34: if exist Analyzer then
35:         Write ('Analyzer − partition : partition(AnalyzerComponents,')
36:         For each Analyzer do
37:             Write(Analyzer.Name)
38:             EndFor
39: end if
40: if exist Planner then
41:         Write ('Planner − partition : partition(PlannerComponents,')
42:         For each Planner do
43:             Write(Planner.Name)
44:             EndFor
45: end if
46: if exist Executor then
47:         Write ('Executor − partition : partition(ExecutorComponents,')
48: end if
```

End

STR1.3: MAPE Sub-components Transformation Rule. This rule specifies formally the different MAPE sub-components of the MAPE control loop. In fact, each component in the MAPE control loop is composed of a receiver, a processor and a sender components.

This rule transforms the name of each receiver, processor and sender into a constant in the CONSTANTS clause. This is specified in the Algorithm 3.

The set of monitor sub-components (respectively analyzer, planner, executor) is composed of the sender, processor and receiver names. This is formally transformed into a set.

Algorithm 3. MAPE sub-components transformation rule algorithm

Begin

```
1: Write (" SETS ")
2: Write ("MonitorSubcomponents ")
3: Write (" AnalyzerSubcomponents ")
4: ...
5: Write (" CONSTANTS ")
6: For each Monitor do
7: if exist Receiver then
8:     Write ("Receiver.Name")
9: end if
10: if exist Processor then
11:     Write ("Processor.Name")
12: end if
13: if exist Sender then
14:     Write ("Sender.Name")
15: end if
16: EndFor
17: For each Analyzer do
18: if exist Receiver then
19:     Write ("Receiver.Name")
20: end if
21: if exist Processor then
22:     Write ('Processor.Name')
23: end if
24: if exist Sender then
25:     Write ("Sender.Name")
26: end if
27: EndFor
28: ...
29: Write (" AXIOMS ")
30: if exist Monitor then
31:     Write ('Monitor − Subcomponent − Set : MonitorSubcomponents = '{Monitor.
32:     Receiver.Name, Monitor.Processor.Name, Monitor.Sender.Name}')
33: end if
34: if exist Analyzer then
35:     Write ('Analyzer − Subcomponent − Set : AnalyzerSubComponents =
        '{Analyzer.Receiver.Name, Analyzer.Processor.Name, Analyzer.Sender.Name}')
36: end if
37: if exist Planner then
38:     Write ('Planner − Subcomponent − Set : PlannerSubcomponents = '
        {Planner.
39:     Receiver.Name, Planner.Processor.Name, Planner.Sender.Name}')
40: end if
41: ...
```

End

STR1.4: Probes, Effectors and ContextElement Transformation Rule.
The different actions planned by the autonomic manager are executed through effectors. Besides, the monitor component collects data about a context through probes or sensors. This rule generates the formal specification of the different

probes, effectors and context components. It transforms the name of each probe, effector and context element into a constant in the CONSTANTS clause. This is specified in the Algorithm 4 from line 5 to line 14. The set of probe components (respectively effector and context element) is composed of all the probe (respectively effector and context element) names. This is formally transformed into a partition.

Algorithm 4. Probes, effectors and context transformation rule algorithm

Begin

```
 1: Write (" SETS ")
 2: Write ("Probes ")
 3: Write (" Effectors ")
 4: Write (" ContextElements ")
 5: Write (" CONSTANTS ")
 6: For each Probe do
 7: Write ('Probe.Name')
 8: EndFor
 9: For each Effector do
10: Write ('Effector.Name')
11: EndFor
12: For each ContextElement do
13: Write ('ContextElement.Name')
14: EndFor
15: Write (" AXIOMS ")
16: if exist Probe then
17:     Write ('Probes − partition : partition(Probes,')
18:     For each Probe do
19:     Write(Probe.Name)
20:     EndFor
21: end if
22: if exist Effector then
23:     Write ('Effectors − partition : partition(Effectors,')
24:     For each Effector do
25:     Write(Effector.Name)
26:     EndFor
27: end if
28: if exist ContextElement then
29:     Write ('ContextElements − partition : partition(ContextElements,')
30:     For each ContextElement do
31:     Write(ContextElement.Name)
32:     EndFor
33: end if
```

End

STR2: Connection Transformation Rule. This rule transforms each connection (*ObservedProperty, Symptom, RFC*, etc.) into a constant in the CONSTANTS clause, as shown in the Algorithm 5. Additionally, this rule defines the graphical connection (Required/Provided interface) with an Event-B relation between two entities. For example, the connection *ObservedProperty* is specified using a relation between *ContextElements* and *Probes* as shown in line 23 of the Algorithm 5. The set of connections is composed of all connection names. This is formally transformed into a partition (*ConnectionName − partition*). This rule also generates *Domain* and *Range* axioms for each connection to define its source and its destination.

Algorithm 5. Connection transformation rule algorithm

Begin

```
1:  Write (" CONSTANTS ")
2:  if exist ObservedProperty then
3:      Write ('ObservedProperty')
4:      For each ObservedProperty do
5:      Write(ObservedProperty.Name)
6:      EndFor
7:  end if
8:  if exist Monitoringdata then
9:      Write ('Monitoringdata')
10:     For each Monitoringdata do
11:     Write(Monitoringdata.Name)
12:     EndFor
13: end if
14: if exist Symptom then
15:     Write ('Symptom')
16:     For each Symptom do
17:     Write(Symptom.Name)
18:     EndFor
19: end if
20: ...
21: Write (" AXIOMS ")
22: if exist ObservedProperty then
```
23: $\quad$ Write ('*ObservedProp − Relation : ObservedProperty ∈ ContextElements ↔ Probes*')
24: $\quad$ Write ('*ObservedProperty − partition : partition(ObservedProperty,*')
```
25:     For each ObservedProperty do
26:     Write(ObservedProperty.Name)
27:     EndFor
28:     For each ObservedProperty do
29:     Write ('ObservedProperty.Name-Domain:')
30:     Write (Dom(ObservedProperty.Name)=)
31:     Write (ObservedProperty.Source)
32:     Write ('ObservedProperty.Name-Range:')
33:     Write (Ran(ObservedProperty.Name)=)
34:     Write (ObservedProperty.Destinataire)
35:     EndFor
36: end if
37: ...
```

End

---

3   Illustration: The Forest Fire Detection System (FFDS)

In an efficient forest fire detection system, each sensor node is equipped with a sensing device that measures its energy level. These measurements are sent to an autonomic manager element (it can be a cluster head, a base station, etc.) to be analyzed. When there is a degradation of the battery level, then the autonomic manager element makes rapidly decisions such as reduction of the monitoring frequency or turning off a node in order to save energy. Finally, adaptation actions are executed using actuators or effectors. In this context, we propose instantiating the master/slave pattern to design such self-adaptive system. The FFDS instantiating the master/slave pattern includes a base station which plays the role of the master element. Also, the FFDS includes two sensor nodes which send their energy level to the corresponding cluster head. In the structural modeling phase, we model, the master, the slave, the different managed elements and the different MAPE components. In addition, we model the different connections between the different main components. Formally, the structural modeling phase is transformed into a context entitled *FFDS-Context*.

3.1 Structural Transformation of UML Self-adaptive System Model Instantiating the Master/Slave Pattern

After applying Algorithm 1, we transform the name of the master, slave and managed element components into a constant in the context file *FFDS-Context* of the Event-B specification (Lines 9–13, Listing 1.1). Besides, the set of the different cited components is transformed to a set or a partition in the AXIOMS clause of the context file *FFDS-Context* (Lines 15–17, Listing 1.1). The application of Algorithm 1 produces the Event-B specification presented in Listing 1.1.

Listing 1.1. Specification of the main components of the FFDS instanciating the master/slave pattern

```
1   CONTEXT      FFDS−Context
2   SETS
3   Master
4   Slaves
5   ManagedElements
6   MasterSteps
7   SlavesSteps
8   CONSTANTS
9   Basestation
10  ClusterHead1
11  ClusterHead2
12  Node1
13  Node2
14  AXIOMS
15  Master − set  :  Master = {Basestation}
16  Slave − partition  :  partition(Slaves, {ClusterHead1}, {ClusterHead2})
17  ManagedElment − partition  :  partition(ManagedElements, {Node1}, {Node2})
18  END
```

Formally, the name of each connection in the model is transformed into a constant in the CONSTANTS clause of the context file (Lines 3–16, Listing 1.2). Also, according to Algorithm 4, each connection must be specified using a relation in the AXIOMS clause and the set of connection names is transformed into a partition. Besides, the domain and the range of each connection are specified using an axiom in the AXIOMS clause. For instance, Lines 18–24 show the formal specification of the connection *MonitoringData*. The application of Algorithm 4 produces the Event-B specification presented in Listing 1.2.

Listing 1.2. Specification of the different connections of the FFDS instantiating the master/slave pattern

```
1   CONTEXT       FFDS−Context
2   CONSTANTS
3   MonitoringData
4   Symptom
5   RFC
6   AdaptationActions
7   Commands
8   EnergyMeasurementsN1
9   EnergyMeasurementsN2
10  LowBattery
11  PCBattery
12  Request1
13  ReduceMF
14  TurnOff
15  SetMF
16  SetMode
17  AXIOMS
18  MonitoringData − Relation  :  MonitoringData ∈ ManagedElements ↔ MonitorComponent
19  MonitoringData − partition  :  partition(MonitoringData, {EnergyMeasurementN1},
20  {EnergyMeasurementN2})
21  EnergyMeasurementN1 − Domain  :  dom({EnergyMeasurementN1}) = {Node1}
22  EnergyMeasurementN1 − Range  :  ran({EnergyMeasurementN1}) = {MCH1}
23  EnergyMeasurementN2 − Domain  :  dom({EnergyMeasurementN2}) = {Node2}
24  EnergyMeasurementN2 − Range  :  ran({EnergyMeasurementN2}) = {MCH2}
25  Symptom − Relation  :  Symptom ∈ MonitorComponents ↔ AnalyzerComponents
26  Symptom − partition  :  partition(Symptom, {LowBattery}, {PCBattery})
27  ...
28  END
```

4 Conclusion

In this paper, we presented a set of algorithms to generate Event-B specifications from the conceptual modeling of a self-adaptive system instantiating a MAPE pattern. In particular, we proposed a set of XSLT transformation rules to transform structural features of the system into Event-B specifications. As future work, we plan to translate behavioral features of the self-adaptive system model to Event-B specifications.

References

1. Hachicha, M., Ben Halima, R., Hadj Kacem, A.: Designing compound MAPE patterns for self-adaptive systems. In: Abraham, A., Muhuri, P.K., Muda, A.K., Gandhi, N. (eds.) Intelligent Systems Design and Applications, pp. 92–101. Springer, Cham (2018)
2. Weyns, D., Schmerl, B., Grassi, V., Malek, S., Mirandola, R., Prehofer, C., Wuttke, J., Andersson, J., Giese, H., Goschka, K.M.: On patterns for decentralized control in self-adaptive systems. In: de Lemos, R., Giese, H., Müller, H.A., Shaw, M. (eds.) Software Engineering for Self-Adaptive Systems II: International Seminar, Dagstuhl Castle, Germany, 24–29 October 2010, Revised Selected and Invited Papers, pp. 76–107. Springer, Heidelberg (2013)

Evolutionary Multi-objective Whale Optimization Algorithm

Faisal Ahmed Siddiqi$^{(\boxtimes)}$ and Chowdhury Mofizur Rahman

United International University, Dhaka 1212, Bangladesh
ahmedfaisal.fa21@gmail.com

Abstract. Whale Optimization Algorithm (WOA) is a recently proposed metaheuristic algorithm and achieved much attention of the researchers worldwide for its competitive performance over other popular metaheuristic algorithms. As a metaheuristic algorithm, it mimics the hunting behavior of humpback whale which uses its unique spiral bubble-net feeding maneuver to search and hunt prey. The WOA has been designed to solve mono-objective problems and it shows great performance and even surplus other state of the art metaheuristics in terms of fast convergence and other performance criteria. But this such a distinctive and successful metaheuristic's performance in dealing multi-objective problems especially in dealing with multi-objective benchmark problems has not been studied that much extent. In this paper, we developed a multi-objective version of WOA which incorporates both whale search and evolutionary search strategy. The obtained results are also compared with NSGA-II, NSGA-III, MOEA/D, MOEA/D-DE, MOPSO and d-MOPSO state of art multi-objective evolutionary algorithms.

Keywords: Multi-objective Whale Optimization Algorithm (MOWOA) ·
Non-dominated sorting genetic algorithm (NSGA) · Pareto-optimal set (PS) ·
Pareto-optimal front (PF)

1 Introduction

Whale Optimization Algorithm (WOA) [1] currently is gaining popularity among the researchers for its well-balanced exploration and exploitation characteristics. As this is comparatively a new algorithm, the full potential of this metaheuristic algorithm has not been explored yet. Its ease of application, competitive performance and reasonable computation cost motivated the researchers to use WOA in solving diversified problems in engineering fields [2–5]. The performance of WOA in multi-objective cases so far been studied in limited scale [6, 18, 23] but especially its performance against different multi-objective bench-mark problems have not been studied elaborately in the available literature. For optimal mobile robot path planning, Dao [6] investigated WOA in the multi-objective scenario. The WOA has also been applied to IEEE CEC 2009 [28] bi-objective test function. In [18] non-dominated sorting approach is used along with a multi-objective version of WOA and also applied to some benchmark and engineering problems. Some distinct multi-objective versions of WOA have been applied to engineering problems like image retrieval [30], wind forecasting [31].

© Springer Nature Switzerland AG 2020
A. Abraham et al. (Eds.): ISDA 2018, AISC 941, pp. 431–446, 2020.
https://doi.org/10.1007/978-3-030-16660-1_43

Besides these above-mentioned literatures, numerous works and projects are available which successfully use WOA in various research and engineering fields. In this paper, we propose an evolutionary multi-objective Whale Optimization Algorithm (WOA) which can perform both swarm based whale search and evolutionary search. We also compared its performance for different benchmark problems with some other well-known state of the art (such as NSGA-II [7], NSGA-III [22], MOEA/D [24], MOEA/D-DE [25], MOPSO [26], dMOPSO [27]) multi-objective algorithms.

2   Multi-objective Optimization

Suppose $x = x_1, x_2, \ldots, x_D$ is a vector of decision variables with dimension D and Ω is the search space. Considering only a single objective case, we name our objective function as $f(x)$. Similarly, we can call $f(x^*)$ as the global minimum of a given function $f(x) : x \in \mathbf{R}, D \to \mathbf{R}$, for $x \in \Omega$ if and only if:

$$\forall x \in \Omega : f(x^*) \leq f(x) \tag{2.1}$$

However, when we consider our case is multi-objective where multi-objective optimization deals with optimizing problems and eventually those problems have multiple objectives and often these objectives have conflicting nature. Generally, a multi-objective optimization (MOO) problem can be stated as finding a solution vector $x \in \Omega$ which can minimize the function z with n conflicting objectives:

$$z = f(x) = (f_1(x), f_2(x), \ldots, f_n(x)) \tag{2.2}$$

Subject to k inequality constraint

$$g(x) = (g_1(x), g_2(x), \ldots, g_k(x) \geq 0) \tag{2.3}$$

And m equality constraint

$$h(x) = (h_1(x), h_2(x), \ldots, h_m(x)) = 0 \tag{2.4}$$

The multi-objective optimization (MOO) generally behaves in such a way that the improving performance of each objective cannot be done without sacrificing or degrading the performance of at least another one objective. Therefore the solution to a multi-objective optimization problem (MOO) problem eventually comes into existence in the form of a balance between convergence and diversity [8, 13] generally known as a Pareto optimal set, Pareto dominance, Pareto front [9, 15] as depicted in the following figure (Fig. 1).

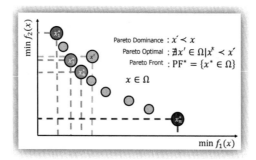

Fig. 1. Pareto dominance, Pareto optimality and Pareto Front in objective space. Each circle denotes the objective value in 2D space and inside each circle x' and x^* symbolizing the corresponding decision vectors.

3   Methods for Developing MOWOA

3.1   Whale Optimization Algorithm

As mentioned in the previous section, WOA is a metaheuristic algorithm [1] which ultimately mimics the foraging behavior of humpback whales. The humpback whales generally hunt krill or small fishes close to the sea surface. They approach their prey by swimming around them and follow a unique trajectory within a shrinking circle and creating bubbles along a circle or '9'-shaped path (see Fig. 2). The whale's encircling prey and spiral bubble-net attacking method are represented as the exploitation phase and in the exploration phase the humpback whale search randomly for its prey.

Fig. 2. Bubble net hunting behavior of humpback whales.

3.2   Exploitation Phase (Includes Encircling Prey and Bubble-Net Attacking Method)

Humpback whales begin hunting a pray with first encircling it and this encircling mechanism can be modeled mathematically by Eqs. (3.1) and (3.2) [1].

$$\vec{D} = \left| \vec{C}.\overrightarrow{X^*}(t) - \vec{X}(t) \right| \tag{3.1}$$

$$\vec{X}(t+1) = \overrightarrow{X^*}(t) - \vec{A}.\vec{D} \tag{3.2}$$

where t is the current iteration, $\overrightarrow{X^*}$ represents the best solution (Position of whale leader) obtained so far, $\vec{X}$ is denoted as the position vector of the selected whale or the selected search agent. Similarly, $\vec{A}$ and $\vec{C}$ [1] are coefficients vectors can be calculated as shown in Eqs. (3.3) and (3.4), respectively:

$$\vec{A} = 2\vec{a}.\vec{r} - \vec{a} \tag{3.3}$$

$$\vec{C} = 2.\vec{a} \tag{3.4}$$

In above two equations here the parameter $\vec{a}$ decreases linearly from 2 to 0 over the course of iterations (in both exploration and exploitation phases) and r is a random vector generated having range [0, 1]. According to Eq. (3.2) in this exploitation phase, the search agents (whales) update their positions over the courses of iteration according to the position of the best- known solution (prey) obtained in the previous iteration. The values of $\vec{A}$ and $\vec{C}$ vectors actually control the areas where a whale can be located in the close proximity of the prey (best solution). This exploitation technique which hump-back whale follows is called the shrinking encircling behavior (See Fig. 3) is achieved by decreasing the value of a (See Eqs. (3.3) and Eq. (3.4)) according to Eq. (3.5).

$$a = 2 - \frac{t}{MaxIter} \tag{3.5}$$

Where t in above equation is the iteration number and *MaxIter* is the maximum number of allowed iterations. Another exploitation technique of humpback whale is a spiral path hunt, in this mechanism the distance between a search agent (X) and the best-known search agent obtained so far (X^*) is calculated (See Fig. 4), then a spiral equation is used to create the position of the neighbor search agent as formulated in Eq. (3.6).

$$\vec{X}(t+1) = D'.e^{bl}.\cos(2\pi l) + \overrightarrow{X^*}(t) \tag{3.6}$$

where $D' = \left| \overrightarrow{X^*}(t) - \vec{X}(t) \right|$ and it indicates the distance of the i-th whale and the prey (best solution obtained so far), similarly b is a constant which defines the shape of the logarithmic spiral, and l is a random number ranging within $[-1, 1]$. Now a procedure is followed to select one exploitation mechanism from the above two exploitation techniques (shrinking encircling search and the spiral-shaped path search) and accordingly a probability of 50% is set to choose between them over the course of iteration as in Eq. (3.7).

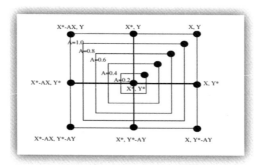

Fig. 3. Shrinking circle exploitation phase in 2D space Eq. (3.2), here *(X*Y*)* denotes the current best position.

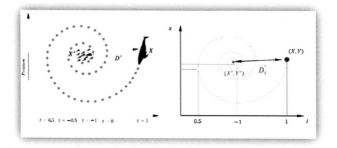

Fig. 4. Spiral exploitation phase of humpback whales, time domain spiral position (left), spiral shrinking calculation (right) in 2D space based on parameter "*l*" Eq. (3.6).

$$\vec{X}(t+1) = \begin{cases} Shrinking\ Encircling\ Eq.\ (3.2)\ if\ p < 0.5 \\ Spiral\ shaped\ path\ Eq.\ (3.6)\ if\ p \geq 0.5 \end{cases} \tag{3.7}$$

Where in above equation p is a random number in [0, 1] associated with the uniform distribution.

3.3 Exploration Phase (Random Search for Prey)

To obtain efficient exploration capability in WOA, a random search agent is selected to guide the other search agents rather than updating the positions of them according to the position of the best one search agent of the current iteration. Accordingly, a vector $\vec{A}$ is considered having random values greater than 1 or less than −1 and this vector is used to force search agent to move far away from the best-known search agent. This mechanism can be mathematically formulated as shown in Eqs. (3.8) and (3.9).

$$\vec{D} = \left| \vec{C} . \overrightarrow{X_{rand}} - \vec{X} \right| \tag{3.8}$$

$$\vec{X}(t+1) = \overrightarrow{X_{rand}} - \vec{A} . \vec{D} \tag{3.9}$$

The visualization of exploration phase of WOA is very much similar as depicted in Fig. 3 except the leader whale (Best position obtain so far) is randomly chosen and $\overrightarrow{X_{rand}}$ denotes the current position of the randomly chosen whale leader. The figure below shows the algorithm executed in WOA [1] (Fig. 5).

Algorithm 1: WOA

Initialization of the whale population $X_i (i = 1, 2, ..., n)$
Calculate the fitness of each whale(search agent)
X*=the best search agent (obtained from the initial population)
while(t < maximum number of iterations)
for each search agent (whale)
Update parameters a, A, C, l, and p
if1 (p<0.5)
   if2 (|A|< 1)
      Update the position of the current search agent by the Eq.(3.2)
   else if2 (|A|=>1)
      Select a random search agent ()
      Update the position of the current search agent by the Eq.(3.9)
   end if2
 elseif1 (p=>0.5)
Update the position of the current search agent by the Eq. (3.6)
 end if1
end for
Check if any search agent goes outside the given search space and amend it [33].
Calculate the fitness of each whale or search agent.
Update X* if there is a better solution found in the current iteration.
t=t+1
end while
return **X***

Fig. 5. Algorithm for WOA.

3.4   Multi-objective Whale Optimization Algorithm

To develop or formulate a multi-objective algorithm practical and efficient it needs to have an archive as well as a balanced search strategy. We adopted both search strategy and archiving strategy for our proposed MOWOA. First, we acquired simple archiving mechanism based on non-dominated sorting and crowing distance [10] and we included grouped local WOA search associated with the evolutionary search using both Simulated Binary Crossover (SBX) [12] and Polynomial mutation[14]. The purpose of the inclusion of evolutionary search is to repair the potential vulnerability of WOA search which may occur in the courses of iteration. The following Fig. 6 shows the basic procedures of proposed MOWOA in brief.

Algorithm 2: MOWOA

Initialize the whales population X, where $X= (x_1, x_2, ... x_n)$
while (t < maximum number of iterations)
   Archive Update (Algorithm 3)
   Perform SBX and Polynomial Mutation on Archive A
   Grouped WOA (Algorithm 4)
t=t+1
end while

Fig. 6. Algorithm for MOWOA

3.5 Archiving Mechanism

Developing an efficient archive is a very important step for MOWOA. In this stage, we adopted a simple archiving mechanism where first, in archive **A** non-dominated solutions [7] with higher crowding-distance [7, 11] values are preserved which are considered to be good representatives of the entire PF (Pareto Front). After the completion of the task of WOA search or evolutionary search, the newly achieved non-dominated solutions are collected into archive **A**. The size of archive can't be infinite, and for a finite archive size, the number of non-dominated solutions may be overloaded, so it is necessary to use an efficient and proper selection criterion for archive update, which can effectively preserve both convergence and diversity and guide the population towards the true PF (Pareto Front). To fulfill the above archiving requirements we selected a well known and popular archive update technique used in [10] which is based on Pareto dominance and crowding distance. The pseudo-code of the archive update mechanism can be briefly described in Fig. 7 [10], where N is the maximum size of archive **A**. We assume newly generated solution set is **S** and the solution set in the external archive is **A**, we also have two functions CheckDominance(x, y) and CrowdingDistanceAssignment (A), the function CheckDominance(x, y) actually returns the pareto dominance relationship between solution x and y. If the function returns 1, it means that x certainly dominates y. Otherwise, when the function returns −1 it means y dominates or is equal to x. Another function CrowdingDistanceAssignment(A) will calculate the crowding distance value [10] for each solution in **A**.

```
                    Algorithm 3: Archive Update
for1 i =1 to | S |
     for2 j =1 to | A|
     State = CheckDominance( S_i , A_j );
        if State = 1
           mark A_j as a dominated solution;
        else
           break;
        end if
     end for2
  delete the marked dominated solutions from A;
     if1 State! = -1
        add S_i to A;
        if2 | A| > N
        CrowdingDistanceAssignment(A);
        delete the most crowded one;
        end if2
     end if1
end for1
```

Fig. 7. Algorithm for archive update.

3.6 Group WOA Local Search

Dealing with multi-objective problems requires preserving both diversity and fast convergence, but as mentioned earlier sections basically, the WOA is designed for mono-objective problems and fast convergence is the primary goal where diversity issue is not involved in those mono-objective cases. Therefore, to guide the whole population towards the complex and irregular pareto-surfaces (Pareto Fronts), it is a good strategy to divide the whole population into certain sub-divisions and perform WOA search (See Algorithm 4) for each sub-division which preserves both diversity and fast convergence as well. Before conducting group local search, we replace 20% to 33% of the current population with archive population and this mechanism also faster the convergence (Fig. 8).

Algorithm 4 : Grouped Local WOA Search

Replace current population with 20% to 33% of archive population.

Divide the population into k sub-population $X = (X_1, X_2, X_3.....X_{k-1}, X_k)$, and $X_K = (x_1, x_2,...x_m)$ where each subpopulation X_k contains m population

for each sub-population
    WOA (Algorithm 1)
end for

Aggregate the population $X = (X_1, X_2, X_3.....X_{K-1}, X_K)$
return X

Fig. 8. Algorithm for grouped local WOA search

4   Experiment and Results

In this section, we performed several experiments to examine the performance of MOWOA for different benchmark problems. Firstly we tuned parameters of MOWOA for simulations then we select the IGD (Inverted Generational Distance) [16] as the performance metric for comparison with other state of the art evolutionary algorithms NSGA-II [7], NSGA-III [22] MOEA/D[24], MOEA/D-DE [25], MOPSO [26], dMOPSO [27]. We have chosen well-known and popular benchmark problems namely ZDT [19], DTLZ [20] BT [21] and CEC-2017 [32] for our experiment.

4.1   Experimental Setup

Most of the parameters of the original WOA remain unchanged throughout our simulation. First, we tune the percentage of population replacement with archive population ranging from 20% to 33% (See Algorithm 4). For Grouped WOA local search we also divide the whole population (Total population is 100 and archive size is also set to 100) into 10 to 20 equal subdivision. Since our Grouped WOA local search mechanism significantly preserve diversity therefore for specific families of benchmark problem like DTLZ [20] and BT [21], we reduced the probability of choosing random agent selection (See Eqs. (3.8) and (3.9)) as best search agent by setting $A \geq 3$ (See Algorithm 1) and this approach makes the convergence faster as well.

The performance metric (IGD) of each algorithm for each of benchmark problems been computed 30 times and the tables below show the averaged IGD computed over 30 individual IGD obtained after completion of given evolutions. To maintain a standard simulation environment we successfully used recently developed MATLAB based evolutionary multi-objective optimization platform "PlatEMO" version 1.3 [17] as our simulation tool (Figs. 9, 10, 11 and 12).

Parameters specified for each algorithm

Global Parameters
Simulated binary crossover and polynomial mutation (Termed as "EAreal" in PlatEMO)
proC = 1 = The probability of doing crossover
disC = 20 = The distribution index of simulated binary crossover
proM = 1 = The expectation of number of bits doing mutation
disM = 20 = The distribution index of polynomial mutation
Differential evolution and polynomial mutation (Termed as "DE" in PlatEMO)
CR = 1 = Parameter CR in differential evolution
F = 0.5 = Parameter F in differential evolution
proM = 1 = The expectation of number of bits doing mutation
disM = 20 = The distribution index of polynomial mutation

Algorithms	Parameters
MOWOA	Gr = No of subdivision = 10 to 20
	AR = Percentage of Archive population replaced to WOA population = 20% to 33%
NSGA-II	As implemented in PlatEMO (v 1.3)
NSGA-III	As implemented in PlatEMO (v 1.3)
MOEA/D	As implemented in PlatEMO (v 1.3)
MOEA/D-DE	As implemented by PlatEMO (v 1.3)
	Some Important Parameters
	delta = 0.9 = The probability of choosing parents locally
	nr = 2 = Maximum number of solutions replaced by each offspring
	operator = Differential Evolution (DE)
MOPSO	As implemented in PlatEMO (v 1.3)
dMOPSO	As implemented in PlatEMO (v 1.3)

Keywords used in Tables
*Gr = Number of sub-division. **AR** = Percentage of Archive population replaced to WOA population. **A** = Control parameter (See Algorithm 1). **M**= Number of Objective.* IGD = Inverted generational distance, *std* = Standard Deviation.
Population=100
1 (One) Iteration = Population × 1 Evolution =100 Evolution
Archive Size=100

Table 1. Performance Table of MOWOA for ZDT benchmark problem (Gr = 10, AR = 20%, A ≥ 1)

Evolution		10,000	10,000	10,000	10,000	10,000
Problem		ZDT1	ZDT2	ZDT3	ZDT4	ZDT6
M		2	2	2	2	2
MOWOA	IGD	5.2523e−3	**5.3208e−3**	**5.9617e−3**	**4.6998e−3**	4.3985e−3
(proposed)	(*std*)	(3.58e−4)	**(2.94e−4)**	**(2.62e−4)**	**(2.06e−4)**	(3.32e−4)
NSGAII	IGD	4.8419e−3	5.6184e−1	1.5447e−1	2.3946e−1	9.4342e−2
	(*std*)	(1.73e−4)	(1.43e−1)	(9.07e−2)	(1.85e−1)	(5.07e−2)
NSGAIII	IGD	3.8891e−3	9.6664e−1	4.3706e−1	8.6091e−1	5.1218e−1
	(*std*)	(1.21e−6)	(2.09e−1)	(2.19e−1)	(3.06e−1)	(2.41e−1)
MOEA/D	IGD	3.9754e−3	3.5017e−1	2.2553e−1	4.8762e−1	8.4125e−2
	(*std*)	(1.34e−5)	(1.99e−1)	(8.04e−2)	(2.26e−1)	(4.60e−2)
MOEA/	IGD	**3.8857e−3**	5.8958e+0	4.6781e+0	3.7492e+0	4.7868e−2
D-DE	(*std*)	**(1.61e−6)**	(3.80e+0)	(2.60e+0)	(1.59e+0)	(1.10e−1)
MOPSO	IGD	4.8611e−3	4.6691e+1	4.3016e+1	1.8361e+1	1.3352e+0
	(*std*)	(1.81e−4)	(1.03e+1)	(9.24e+0)	(5.34e+0)	(2.61e+0)
dMOPSO	IGD	3.9924e−3	4.5278e−1	1.6201e−1	5.3953e−1	**3.5075e−3**
	(*std*)	(1.42e−5)	(2.63e−1)	(2.28e−1)	(2.91e−1)	**(9.25e−4)**

Table 2. Performance Table of MOWOA for BT benchmark problem (Gr = 20, AR = 33%, A ≥ 3)

Evolution		100,000	100,000	100,000	100,000	100,000
Problem		BT1	BT2	BT3	BT4	BT5
M		2	2	2	2	2
MOWOA	IGD	**1.3194e−2**	**2.5934e−2**	1.5762e−2	**9.7273e−3**	**8.5861e−3**
(proposed)	(*std*)	**(1.38e−2)**	**(1.11e−2)**	(7.53e−3)	**(2.92e−3)**	**(3.43e−3)**
NSGAII	IGD	3.4374e−2	1.2035e−1	**8.7933e−3**	1.8769e−2	4.9093e−2
	(*std*)	(5.10e−2)	(2.17e−2)	**(1.55e−3)**	(2.45e−3)	(5.03e−2)
NSGAIII	IGD	1.3840e−1	1.4368e−1	2.1810e−2	2.7727e−2	1.4654e−1
	(*std*)	(1.20e−1)	(2.83e−2)	(6.58e−3)	(4.06e−3)	(1.32e−1)
MOEA/D	IGD	9.5096e−1	1.5558e−1	1.3260e−1	1.2453e−1	1.1738e+0
	(*std*)	(1.80e−1)	(1.81e−2)	(8.98e−2)	(4.36e−2)	(1.54e1)
MOEA/	IGD	1.9866e+0	4.6127e−1	1.1244e+0	9.1710e−1	1.8740e+0
D-DE	(*std*)	(1.76e−1)	(2.55e−2)	(2.58e−1)	(1.87e−1)	(1.95e−1)
MOPSO	IGD	4.3416e+0	2.9276e+0	4.6978e+0	4.4527e+0	4.5570e+0
	(*std*)	(1.67e−1)	(2.68e−1)	(1.95e−1)	(2.60e−1)	(1.81e−1)
dMOPSO	IGD	4.7526e+0	3.2899e+0	4.8622e+0	4.6716e+0	4.7185e+0
	(*std*)	(1.69e−1)	(2.07e−1)	(1.77e−1)	(1.47e−1)	(1.54e−1)

Table 3. Performance Table of MOWOA for DTLZ benchmark problem (Gr = 20, AR = 20%, A ≥ 3)

Evolution		10,000	10,000	10,000	10,000	10,000	10,000	10,000
Problem		DTLZ1	DTLZ2	DTLZ3	DTLZ4	DTLZ5	DTLZ6	DTLZ7
M		3	3	3	3	3	3	3
MOWOA	IGD	**6.7589e−2**	7.4635e−2	**7.495e−1**	**7.0870e−2**	6.460e−3	**6.4807e−3**	8.2654e−2
(proposed)	(std)	**(1.75e−1)**	(3.73e−3)	**(1.67e+0)**	**(2.33e−3)**	(3.13e−4)	**(4.60e−4)**	(5.57e−3)
NSGAII	IGD	3.0522e−1	6.8912e−2	8.5934e+0	1.1785e−1	**6.0290e−3**	1.8141e−2	1.0219e−1
	(std)	(2.42e−1)	(2.87e−3)	(4.04e+0)	(1.34e−1)	**(2.17e−4)**	(6.73e−2)	(4.94e−2)
NSGAIII	IGD	2.4690e−1	5.4980e−2	1.0464e+1	1.5244e−1	1.2347e−2	3.4171e−2	9.9548e−2
	(std)	(2.69e−1)	(2.02e−4)	(4.13e+0)	(1.98e−1)	(1.63e−3)	(7.01e−2)	(9.28e−3)
MOEA/D	IGD	2.4892e−1	**5.4861e−2**	1.2838e+1	4.7370e−1	3.2316e−2	6.9485e−2	1.7481e−1
	(std)	(3.51e−1)	**(1.68e−4)**	(6.99e+0)	(3.41e−1)	(7.01e−4)	(1.63e−1)	(1.21e−1)
MOEA/	IGD	9.8866e−1	7.7737e−2	8.0809e+0	1.8492e−1	1.4304e−2	1.4099e−2	5.0430e−1
D-DE	(std)	(1.45e+0)	(1.43e−3)	(9.90e+0)	(7.62e−2)	(3.22e−4)	(1.91e−4)	(1.35e−1)
MOPSO	IGD	1.1972e+1	5.0051e−1	1.6219e+2	6.0039e−1	9.3583e−3	3.9266e+0	6.2210e+0
	(std)	(7.29e+0)	(1.54e−2)	(5.50e+1)	(1.47e−1)	(1.96e−3)	(1.16e+0)	(1.25e+0)
dMOPSO	IGD	7.1663e+0	1.4707e−1	5.5427e+1	3.2038e−1	4.3117e−2	3.2861e−2	2.2342e−1
	(std)	(6.39e+0)	(9.18e−3)	(5.74e+1)	(2.84e−2)	(7.04e−4)	(2.36e−4)	(2.34e−1)

Table 4. Performance Table of MOWOA for CEC-2017 (F1–F6) benchmark problem (Gr = 5, AR = 20%, A ≥ 3)

Evolution		10,000	10,000	20,000	10,000	10,000	15,000
Problem		F1	F2	F3	F4	F5	F6
M		3	3	3	3	3	3
MOWOA(proposed)	IGD	6.5821e−2	5.2432e−2	**8.9947e−2**	**1.0489e+0**	**3.2150e−1**	5.8995e−3
	(std)	(3.52e−3)	(2.83e−3)	**(3.38e−2)**	**(1.03e+0)**	**(1.23e−2)**	(3.89e−4)
NSGAII	IGD	**5.7571e−2**	4.7103e−2	8.3875e+0	2.1487e+1	3.7097e−1	**5.3377e−3**
	(std)	**(2.60e−3)**	(3.20e−3)	(7.62e+0)	(1.15e+1)	(2.66e−1)	**(2.15e−4)**
NSGAIII	IGD	6.2103e−2	**3.6646e−2**	1.8356e+1	3.9539e+1	5.4475e−1	1.3109e−2
	(std)	(1.27e−3)	**(7.58e−4)**	(2.03e+1)	(1.61e+1)	(5.86e−1)	(1.35e−3)
MOEA/D	IGD	7.0634e−2	4.3855e−2	1.4609e+0	2.4756e+1	1.9085e+0	1.6628e−1
	(std)	(4.23e−4)	(1.96e−3)	(2.00e+0)	(1.89e+1)	(1.67e+0)	(2.00e−1)
MOEA/D-DE	IGD	7.2389e−2	4.6651e−2	3.5714e+2	1.0503e+2	8.0682e−1	1.4331e−2
	(std)	(1.41e−3)	(9.67e−4)	(6.95e+2)	(9.04e+1)	(2.66e−1)	(3.41e−4)
MOPSO	IGD	3.4651e−1	2.4746e−1	4.2321e+4	7.4721e+2	1.9078e+0	1.3515e−1
	(std)	(1.77e−2)	(1.20e−2)	(1.42e+4)	(1.87e+2)	(9.43e−1)	(1.26e−1)
dMOPSO	IGD	1.1532e−1	4.5889e−2	3.3762e+1	2.3132e+2	1.1821e+0	4.3140e−1
	(std)	(6.72e−3)	(1.55e−3)	(1.46e+2)	(2.38e+2)	(1.59e−1)	(1.14e−1)

4.2 Discussion on Obtained Results

From Table 1 we see our proposed MOWOA dominates over ZDT family of benchmark problems. It performs best for ZDT2, ZDT3 and ZDT4. However, MOEA-DE and dMOPSO obtained best results for ZDT1 and ZDT6 respectively.

In Table 2 we find our proposed MOWOA dominates most of the BT (BT1–BT5) benchmark problems, on the other hand, NSGA-II gives its best for BT3. Other remaining algorithms (MOEA/D, MOEA/D-DE, MOPSO and dMOPSO) did not score

Table 5. Performance Table of MOWOA for CEC-2017(F7, F10–F13, F15) benchmark problem (Gr = 5, AR = 20%, A $\geq$ 3)

Evolution		10,000	50,000	10,000	10,000	20,000	10,000
Problem		F7	F10	F11	F12	F13	F15
M		3	3	3	3	3	3
MOWOA(proposed)	IGD	**8.2315e−2**	3.5280e−1	2.0079e−1	2.9095e−1	1.2194e−1	2.1508e−1
	(std)	**(5.90e−3)**	(5.47e−2)	(8.87e−3)	(1.34e−2)	(1.30e−2)	(2.60e−2)
NSGAII	IGD	1.0257e−1	2.5209e−1	**1.9824e−1**	2.8106e−1	1.2124e−1	1.5363e+0
	(std)	(5.09e−2)	(2.66e−2)	**(1.17e−2)**	(1.28e−2)	(1.03e−2)	(4.26e−1)
NSGAIII	IGD	1.3249e−1	**2.1620e−1**	2.0095e−1	**2.4073e−1**	**9.6907e−2**	1.2152e+0
	(std)	(9.49e−2)	**(3.61e−2)**	(3.39e−2)	**(1.09e−2)**	**(9.19e−3)**	(3.63e−1)
MOEA/D	IGD	1.7375e−1	3.6560e−1	1.0726e+0	3.8950e−1	1.2661e−1	4.6742e−1
	(std)	(1.19e−1)	(3.12e−2)	(1.27e−1)	(7.43e−2)	(4.60e−2)	(1.07e−1)
MOEA/D-DE	IGD	5.1620e−1	1.4282e+0	5.0053e−1	3.5500e−1	9.9796e−2	5.5154e−1
	(std)	(1.13e−1)	(9.70e−2)	(6.61e−2)	(2.05e−2)	(1.79e−2)	(1.63e−1)
MOPSO	IGD	6.1968e+0	1.7595e+0	8.4693e−1	2.0671e+0	6.4049e−1	4.9754e+0
	(std)	(1.54e+0)	(7.65e−2)	(3.44e−1)	(2.33e−1)	(8.29e−2)	(3.69e+0)
dMOPSO	IGD	1.9081e−1	1.5363e+0	8.9055e−1	3.5507e−1	1.0292e−1	**1.8096e−1**
	(std)	(1.94e−1)	(4.68e−3)	(1.68e−1)	(1.72e−1)	(1.52e−2)	**(2.72e−2)**

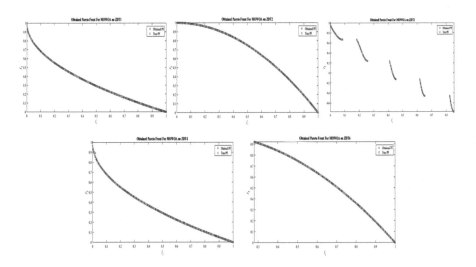

Fig. 9. Obtained Pareto Front for ZDT benchmark problems using proposed MOWOA

a single win. In the case of DTLZ benchmark problems MOWOA also shows its dominating performances (Table 3) for DTLZ1, DTLZ3, DTLZ4, DTLZ6 and DTLZ7. However, MOEA/D and NSGA-II obtained best results for DTLZ2 and DTLZ5 respectively. The other remaining algorithms (NSGA-III, MOEA/D-DE, MOPSO, dMOPSO) could not score the best result for the rest of DTLZ benchmark problems in Table 3. Tables 4 and 5 show that our MOWOA's performance on most of the CEC-2017 [32] benchmark problems and the results are quite significant. MOWOA scored

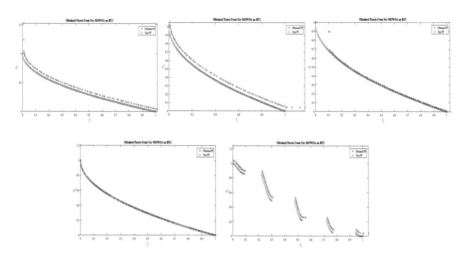

Fig. 10. Obtained Pareto Front for BT benchmark problems using proposed MOWOA

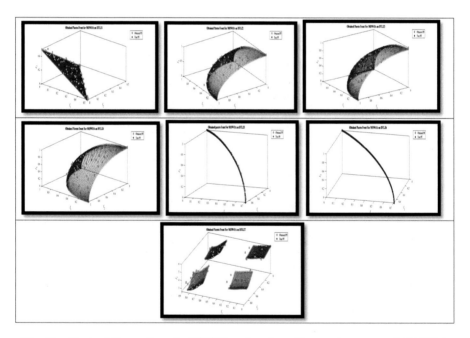

Fig. 11. Obtained Pareto Front for DTLZ benchmark problems using proposed MOWOA

the best result for F3, F4, F5 and F7 while NSGA-II performed best for F1, F6 and F11. The NSGA-III also did best for F2, F10, F12 and F13, however dMOPSO only scored best for F15.

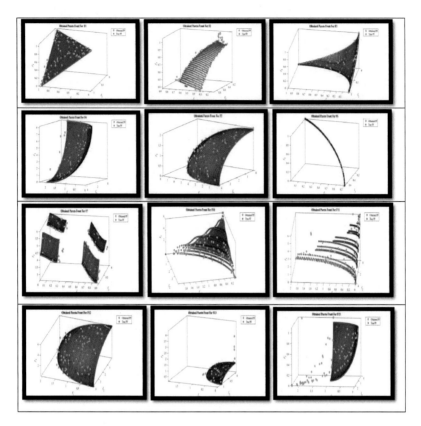

Fig. 12. Obtained Pareto Front for CEC-2017 benchmark problems using proposed MOWOA

The time complexity of our proposed MOWOA can be determined from Fig. 3 (Algorithm 2). For archive update (Algorithm 3) time complexity is $O(N^2)$, for SBX and polynomial mutation, time complexity is $O(N)$ and finally for Group WOA (Algorithm 4), time complexity is $O(N^2)$. In total time complexity comes up with $O(2\,N^2 + N) \sim O(N^2)$ which is reasonable compared to other metaheuristics and evolutionary algorithms.

Our proposed MOWOA performed significantly well for different benchmark problems as discussed above. Inherently swarm based whale search (especially for its unique spiral exploitation mechanism) is faster than PSO [29] and other evolutionary algorithms. Our group WOA search strategy added a balance between diversity and fast convergence. Overall, the strategies and mechanisms we used to design our MOWOA contributed together to perform its best.

5   Conclusion

In this paper, we proposed a multi-objective version of Whale Optimization Algorithm (MOWOA) which was originally proposed to deal with mono-objective problems. In order to transform a mono-objective algorithm into multi-objective one, we have successfully integrated both search strategy of WOA and evolutionary search mechanism using genetic operators like crossover and mutation. We have also included an efficient archive and also a grouped local search strategy which faster the convergence rate and preserve diversity as well.

Proposed MOWOA gives better results on most of four families (ZDT, BT, DTLZ, and CEC-2017) of benchmark problems against the well-known state of the art evolutionary multi-objective algorithms. Our future study will include enabling the MOWOA to deal with more than three objectives with more enhanced performance and we hope application of proposed MOWOA in some specific practical engineering problems will also be studied in our future work.

References

1. Mirjalili, S., Lewis, A.: The whale optimization algorithm. Adv. Eng. Softw. **95**, 51–67 (2016)
2. Prakash, D.B., Lakshminarayana, C.: Optimal siting of capacitors in radial distribution network using whale optimization algorithm. Alex. Eng. J. **56**, 499–509 (2016)
3. Reddy, P.D.P., Reddy, V.C.V., Manohar, T.G.: Whale optimization algorithm for optimal sizing of renewable resources for loss reduction in distribution systems. Renew. Wind Water Sol. **4**(1), 3 (2017)
4. Mafarja, M., Mirjalili, S.: Whale optimization approaches for wrapper feature selection. Appl. Soft Comput. J. **62**(November), 441–453 (2018)
5. Mostafa, A., Hassanien, A.E., Houseni, M., Hefny, H.: Liver segmentation in MRI images based on whale optimization algorithm. Multimed. Tools Appl. **76**(April), 1–24 (2017)
6. Dao, T.K., Pan, T.S., Pan, J.S.: A multi-objective optimal mobile robot path planning based on whale optimization algorithm. In: 2016 IEEE 13th International Conference on Signal Processing, pp. 337–342 (2016)
7. Deb, K., Agrawal, S., Pratap, A., Meyarivan, T.: A fast elitist non-dominated sorting genetic algorithm for multi-objective optimization: NSGA-II. In: Parallel Problem Solving from Nature PPSN VI, pp. 849–858 (2000)
8. Yagyasen, D., Darbari, M., Shukla, P.K., Singh, V.K.: Diversity and convergence issues in evolutionary multiobjective optimization: application to agriculture science. IERI Procedia **5**, 81–86 (2013)
9. Bosman, P.A.N., Thierens, D.: The balance between proximity and diversity in multi – objective evolutionary algorithms. IEEE Trans. Evol. Comput. **7**(2), 174–188 (2003)
10. Lin, Q., Li, J., Du, Z., Chen, J., Ming, Z.: A novel multi-objective particle swarm optimization with multiple search strategies. Eur. J. Oper. Res. **247**(3), 732–744 (2015)
11. Sierra, M.R., Coello, C.A.C.: Improving PSO-based multi-objective optimization using crowding, mutation and ∈-dominance. In: International Conference on Evolutionary Multi-criterion Optimization, pp. 505–519. Springer, Heidelberg, March 2005
12. Deb, K., Agrawal, R.B.: Simulated binary crossover for continuous search space. Complex Syst. **9**, 1–34 (1994)
13. Jiang, S., Yang, S.: Convergence versus diversity in multiobjective optimization. In: International Conference on Parallel Problem Solving from Nature, pp. 984–993. Springer, Cham, September 2016
14. Khare, V.: Performance Scaling Multi-objective Evolutionary Algorithms. School of Computer Science, The University of Birmingham, Birmingham (2002)

15. Ngatchou, P., Zarei, A., El-Sharkawi, A.: Pareto multi objective optimization. In: Proceedings of the 13th International Conference on Intelligent Systems Application to Power Systems 2005, pp. 84–91. IEEE, November 2005

16. Zhou, A., Jin, Y., Zhang, Q., Sendhoff, B., Tsang, E.: Combining model-based and genetics-based offspring generation for multi-objective optimization using a convergence criterion. In: 2006 IEEE International Conference on Evolutionary Computation, pp. 892–899 (2006)

17. Tian, Y., Cheng, R., Zhang, X., Jin, Y.: PlatEMO: a MATLAB platform for evolutionary multi-objective optimization. IEEE Comput. Intell. Mag. **12**, 73–87 (2017)

18. Jangir, P., Jangir, N.: Non-dominated sorting whale optimization algorithm (NSWOA): a multi-objective optimization algorithm for solving engineering design problems. Glob. J. Res. Eng.: F Electr. Electron. Eng. **17**(4) (2017). Version 1.0

19. Zitzler, E., Deb, K., Thiele, L.: Comparison of multiobjective evolutionary algorithms: empirical results. Evol. Comput. **8**, 173–195 (2000)

20. Deb, K., Thiele, L., Laumanns, M., Zitzler, E.: Scalable test problems for evolutionary multi-objective optimization. In: Evolutionary Multiobjective Optimization, Advanced Information and Knowledge Processing, pp. 105–145. Springer, London (2005)

21. Li, H., Zhang, Q., Deng, J.: Biased multiobjective optimization and decomposition algorithm. IEEE Trans. Cybern. **47**, 52–66 (2016)

22. Deb, K., Jain, H.: An evolutionary many- objective optimization algorithm using reference-point-based nondominated sorting approach, part I: solving problems with box constraints. IEEE Trans. Evol. Comput. **18**(4), 577–601 (2014)

23. Kumawat, I.R., Nanda, S.J., Maddila, R.K.: Multi-objective whale optimization. TENCON - IEEE Region 10 Conference, November-2017

24. Zhang, Q., Li, H.: MOEA/D: a multiobjective evolutionary algorithm based on decomposition. IEEE Trans. Evol. Comput. **11**(6), 712–731 (2007)

25. Li, H., Zhang, Q.: Comparison between NSGA-II and MOEA/D on a set of multiobjective optimization problems with complicated pareto sets. IEEE Trans. Evol. Comput. **13**(2), 284–302 (2009)

26. Parsopoulos, K., Vrahatis, M.N.: Particle swarm optimization method in multiobjective problems. In: SAC 2002, Madrid, Spain (2002)

27. Zapotecas Martínez, S., Coello Coello, C.A.: A multi-objective particle swarm optimizer based on decomposition. In: Proceeding of the 13th Annual Conference on Genetic and Evolutionary Computation - GECCO '11, p. 69 (2011)

28. Zhang, Q., Zhou, A., Zhao, S., Suganthan, P.N., Liu, W., Tiwari, S.: Multiobjective optimization test instances for the CEC 2009 special session and competition. University of Essex, Colchester, UK and Nanyang Technological University, Technical report. CES-487, Technical report (2008)

29. Eberhart, R., Kennedy, J.: A new optimizer using particle swarm theory. In: Proceedings of the Sixth International Symposium on Micro Machine and Human Science 1995. MHS'95, pp. 39–43. IEEE, October 1995

30. El Aziz, M.A., Ewees, A.A., Hassanien, A.E.: Multi-objective whale optimization algorithm for content-based image retrieval. Multimed. Tools Appl. **77**, 1–38 (2018)

31. Wang, J., Du, P., Niu, T., Yang, W.: A novel hybrid system based on a new proposed algorithm—multi-objective whale optimization algorithm for wind speed forecasting. Appl. Energy **208**(October), 344–360 (2017)

32. Cheng, R., et al.: Benchmark functions for the CEC 2017 competition on evolutionary many-objective optimization (2017)

33. https://www.mathworks.com/matlabcentral/fileexchange/55667-the-whale-optimization-algorithm

Comparative Performance Analysis of Different Classification Algorithm for the Purpose of Prediction of Lung Cancer

Subrato Bharati[1(✉)], Prajoy Podder[1], Rajib Mondal[1],
Atiq Mahmood[2], and Md. Raihan-Al-Masud[3]

[1] Department of EEE, Ranada Prasad Shaha University,
Narayanganj, Bangladesh
subratobharati1@gmail.com, prajoypodder@gmail.com,
rajibeee06@gmail.com
[2] Department of EEE, World University of Bangladesh, Dhaka, Bangladesh
atiq.mahmood@eee.wub.edu.bd
[3] Department of EEE, Stamford University Bangladesh, Dhaka, Bangladesh
raihan.emasud@gmail.com

Abstract. At present, Lung cancer is the serious and number one cause of cancer deaths in both men and women in worldwide. Cigarette Smoking can be considered as the principle cause for lung cancer. It can arise in any portion of the lung, but the lung cancer 90%–95% are thought to arise from the epithelial cells, this cells lining the bigger and smaller airways (bronchi and bronchioles). Mainly this paper focus on diagnosing the lung cancer disease using various classification algorithm with the help of python based data mining tools. For this purpose, Lung Cancer dataset has been collected from UCI machine learning repository. Three types of pathological cancers have been illustrated in the datasets. In this research paper, the proficiency and potentiality of the classification of Naïve Bayes, Logistic Regression, K-Nearest Neighbors (KNN), Tree, Random Forest, Neural Network in examining the Lung cancer dataset has been investigated to predict the presence of lung cancer with highest accuracy. Performance of the classification algorithms has been compared in terms of classification accuracy, precision, recall, F1 score. Finding out the confusion matrix, Classifier's overall accuracy, user and producer accuracy individually for each classes and value of kappa statistics have been determined in this paper. Area under Receiver Operating Characteristic (ROC) curve and distribution plot of the mentioned classifiers have also been showed in this paper. This paper also implemented Principal component analysis (PCA) and visualized classification tree, Multidimensional scaling (MDS) and Hierarchical Clustering.

Keywords: Naïve Bayes · Logistic regression · Decision tree ·
Random forest · Neural network · AUC · KNN · Confusion matrix · PCA ·
MDS · Hierarchical clustering

© Springer Nature Switzerland AG 2020
A. Abraham et al. (Eds.): ISDA 2018, AISC 941, pp. 447–457, 2020.
https://doi.org/10.1007/978-3-030-16660-1_44

1 Introduction

Lung cancer is a serious disease which is the major cause of cancer deaths in people worldwide. Cigarette Smoking can be considered as the principle cause for lung cancer. Although it is a very dangerous disease but it is curable if the treatment is started at the early stage of the lung cancer. Pair of lungs are the vital organs, which is the function of the respiratory system in human body. Continuing cigarette smoking is due to 85% majority of cases of lung cancer [1]. Approximately 10–15% never smoking who cases lung cancer. Small cell lung cancers (SCLC) and non-small cell lung cancers (NSCLC) are the two types of lung cancer which grow and types are different [2]. Its treatment technique depends on its types and classification of various patients big data. Some several researchers realized the various process to classify health diseases such as lung cancer and survival of the patient prediction. Different performance data mining tools, methods, and techniques provide various results [3].

In this paper, information of a dataset about the patient diagnosis for lung cancer has been collected from UCI machine learning repository database [4]. This data narrate 3 types of pathological lung cancers. and This dataset has been implemented in Orange data mining tool in order to compare the classifiers performance and get highest positive prediction rate. The number of Instances and attributes is 32 and 57 (1 class attribute, 56 predictive) respectively. There are 3 types of classes. 1st, 2nd and 3rd class have 9, 13 and 10 observations respectively. These methods are – Neural Network, Naïve Bayes, Logistic Regression, K-Nearest Neighbors (KNN), Tree, Random Forest algorithm. Fifty-six attributes provided in the data set and we contracted predictions and evaluation results for some cross-validation number. Some cross-validation number has been proved significant in creating the decisions. Implemented data sets and growing scatter plot which is compared to in terms of some classifier. In this paper, refer to a confusion matrix of the different classifier and describe the performance of that classifier. We have also been described the distribution of all classifier with three groups and three folds.

In this Paper, Receiver Operating Characteristics (ROC) Area Curve has also been showed which graphical plot compares the analytical performance of each other. The area under the ROC curve (AUC) is a measure of how well a classifier can separate between two analytical groups (normal/diseased).

2 Related Works

Lynch et al. [5] applied some supervised learning methods including linear regression, Decision Trees, Gradient Boosting Machines, SVM in SEER database in order to classify lung cancer patients in terms of survival. Tazin et al. [6] compared the performance of Naïve Bayes, SVM, KNN and Decision tree classifier using WEKA tool in order to analyze and predict chronic kidney disease. Kirubha et al. [7] compares data mining classification algorithms such as REP tree, Naive Bayes in order to classify the risk factor level of lung cancer affected people. Abdar et al. [8] implemented and compared C5.0, Neural Network, SVM, KNN in order to predict the peril of heart disease.

3 Theoretical Description

In this paper, different classification procedures have been applied to the collected data. Each classifier is mainly some sets of rules and learning methods. Six classifiers have been selected and constructed on the cross-validation number of folds in the dataset. They are Naïve Bayes, Logistic Regression, K-Nearest Neighbors (KNN), Tree, Random Forest, Neural Network.

3.1 Naïve Bayes

Naïve Bayes is a simple machine learning classifier which is a family of probabilistic classifiers built on Bayes' theorem. The rule of Bayes utilizes the statement that attributes are temporarily independent. It works and simple to construct an enormous quantity of data as better. This classification identifies the parameters using training data. It predicts conditional probability model such as classify of a given instance problem. It is correlated to a specific class or not [9]. Bayes theorem gives the posterior probability, $P(c|x)$, from $P(c)$, $P(x)$, and $P(x|c)$.

So,

$$P(c|x) = \frac{P(x|c)P(c)}{P(x)} \tag{1}$$

Here, $P(c|x)$ is the posterior probability of a target class; $P(c)$ is the class prior probability; $P(x|c)$ is the likelihood which is the predictor of probability given class; $P(x)$ is the predictor prior probability.

3.2 Logistic Regression

Logistic regression is estimated the parameters of a logistic model and used for binary classification. It estimated the probability of a binary response established on one or more independent variables. In machine learning, logistics regression is used in several fields such as a medical field [10].

3.3 K-Nearest Neighbors (KNN)

In machine learning, K-nearest-neighbor (kNN) classification is one of the most simple and essential classification techniques. Its classifier performed discriminant analysis for developing when reliable parametric estimates of probability densities are difficult to define [11].

3.4 Tree

In machine learning, a Decision Tree is a fateful model which observed about an item such as represented in the branches to perfection about target class of dataset. It is created a model by a goal that calculates the value of a target class based on some input variables of that class [12].

3.5   Random Forest

Random forest is called random decision forest which is a collaborative learning method for regression, classification, and other tasks, that constructing a multitude of the decision tree is operated at training time. It is outputting the class that means a prediction of the specific trees. The habit of their training set behaves overfitting when Random decision forests accurate for decision trees [13].

3.6   Neural Network

The Neural network is known as multi-layer perceptron (MLP) algorithm with back-propagation of orange data mining tools. Artificial neural networks are unusual to traditional statistical modeling techniques so that it is performing as useful in many scientific disciplines. The neural network is a feed-forward a multi-layer perceptron (MLP) algorithm that is achieved from subbands energy of the wavelet by maps sets of energy [14].

4   Flow Diagram

The work flow diagram in orange tool is shown in Fig. 1.

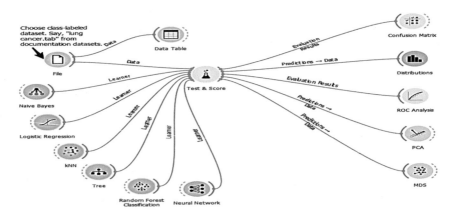

Fig. 1. Work flow diagram in orange data mining environment

5   Simulation Results

5.1   Predictions and Evaluation Results

Area under ROC curve (AUC): From the table it has been observed that Naïve Bayes has comparatively large AUC value and classification accuracy than the other classifiers. But KNN classifier has the largest precision value (0.575) while Naïve Bayes has largest recall value (0.531). KNN has good precision value closest to the highest value. Tree Classifier (0.375) has the smallest precision value. High recall value increases the probabilities of eradicating healthy cells (negative outcome) and rises the chances of eliminating all cancer cells (positive outcome). Precision can be seen as a measure of exactitude or quality on the contrary recall is a measure of completeness or quantity. F1 score is the arithmetic mean of precision and recall. These parameters have been calculated for cross validation number of folds 3 (Table 1).

Table 1. Evaluation results for cross-validation number of folds 3

Classifier name	Area under ROC curve	Classification accuracy	F1 score	Precision	Recall
Naïve Bayes	0.748	0.531	0.503	0.562	0.531
Logistic regression	0.684	0.500	0.502	0.505	0.500
KNN	0.641	0.438	0.411	0.575	0.438
Tree	0.574	0.375	0.370	0.375	0.375
Random forest	0.614	0.469	0.464	0.463	0.469
Neural network	0.645	0.500	0.500	0.500	0.500

5.2 Confusion Matrix

It can be defined as a special kind of contingency table having two dimension namely actual and predicted and identical sets of classes in both dimensions. From the Confusion matrix under Naïve Bayes condition (from Table 2), calculated overall accuracy is 57.047% and kappa statistics is 0.356.

Table 2. Confusion matrix of Naïve Bayes

Classifier results	Truth data			Producer accuracy (Precision)
	Class 1	Class 2	Class 3	
Class 1	77.8%	22.2%	0%	77.778%
Class 2	69.2%	23.1%	7.7%	23.232%
Class 3	0%	30%	70%	70%
Truth overall	146	75	77	
User accuracy (Recall)	52.74%	30.667%	90.909%	

Table 3. Confusion matrix of Logistic Regression

Classifier results	Truth data			Producer accuracy (Precision)
	Class 1	Class 2	Class 3	
Class 1	44.4%	55.6%	0%	44.444%
Class 2	46.2%	23.1%	30.8%	23.232%
Class 3	0%	40%	60%	60%
Truth overall	90	118	90	
User accuracy (Recall)	48.889%	19.492%	66.667%	

From the Confusion matrix under logistic regression classifier (from Table 3), calculated overall accuracy is 42.617% and kappa statistics is 0.139.

Table 4. Confusion matrix of KNN

Classifier results	Truth data			Producer accuracy (Precision)
	Class 1	Class 2	Class 3	
Class 1	77.8%	22.2%	0%	77.778%
Class 2	69.2%	23.1%	7.7%	23.232%
Class 3	20%	40%	40%	40%
Truth overall	166	85	47	
User accuracy (Recall)	46.386%	27.059%	85.106%	

From the Confusion matrix under KNN classifier (from Table 4), calculated overall accuracy is 46.98% and kappa statistics is 0.205.

Table 5. Confusion matrix of Tree

Classifier results	Truth data			Producer accuracy (Precision)
	Class 1	Class 2	Class 3	
Class 1	33.3%	66.7%	0%	33.333%
Class 2	30.8%	23.1%	46.2%	23.232%
Class 3	10%	40%	50%	50%
Truth Overall	73	129	96	
User accuracy (Recall)	45.205%	17.829%	52.083%	

From the Confusion matrix under Tree classifier (from Table 5), calculated overall accuracy is 35.57% and kappa statistics is 0.034.

Table 6. Confusion matrix of Random Forest

Classifier results	Truth data			Producer accuracy (Precision)
	Class 1	Class 2	Class 3	
Class 1	55.6%	44.4%	0%	55.556%
Class 2	38.5%	53.8%	7.7%	54.082%
Class 3	10%	50%	40%	40%
Truth overall	103	147	47	
User accuracy (Recall)	53.398%	36.054%	85.106%	

From the Confusion matrix under Random Forest classifier (from Table 6), calculated overall accuracy is 49.832% and kappa statistics is 0.249.

Table 7. Confusion matrix of Neural Network

Classifier results	Truth data			Producer accuracy (Precision)
	Class 1	Class 2	Class 3	
Class 1	33.3%	66.7%	0%	33.333%
Class 2	38.5%	38.5%	23.1%	38.384%
Class 3	0%	30%	70%	70%
Truth overall	71	134	93	
User accuracy (Recall)	46.479%	28.358%	75.269%	

From the Confusion matrix under Neural Network classifier(from Table 7), calculated overall accuracy is 47.315% and kappa statistics is 0.21.

5.3   Distribution

For discrete attributes, the graphical representation shows how many instances each attribute value appears in the data. If a class variable is contained in the data, class distributions for each of the attribute values will be displayed. In distribution plot (Figs. 2 and 3), x-axis indicates classifier algorithm such as Naïve Bayes, Logistic Regression, KNN, Tree, Random Forest, Neural Network against y-axis indicates frequency.

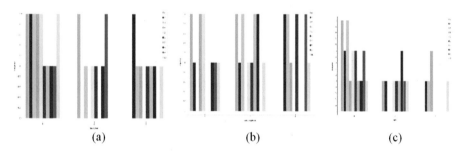

Fig. 2. Distribution of (a) Naïve Bayes (b) Logistic Regression (c) KNN grouped by 'Fold'

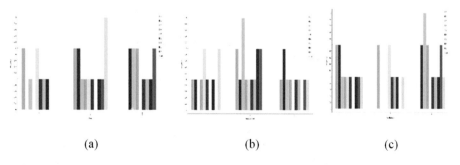

Fig. 3. Distribution of (a) Tree (b) Random Forest (c) Neural Network grouped by 'Fold'

5.4 ROC Analysis

Receiver operating characteristics curve is called as ROC curve. Classification models are a comparison between each other when it assists. a false positive rate of the ROC curve plots on an x-axis (1-specificity; the probability that true value is zero for the target is equal to one) against a true positive rate on a y-axis (sensitivity; the probability that the true value is one when the target is equal to one). Data is separated into three target class and the curve of Naïve Bayes, Logistic Regression, KNN, Tree, Random Forest, Neural Network has been showed. The performance line shows the better performance of the classifier. Black colour curve defined as performance line (Fig. 4).

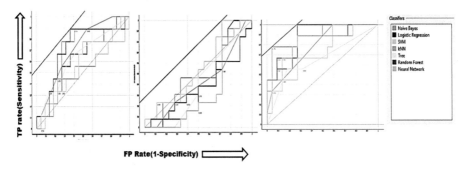

Fig. 4. ROC analysis for target class 1, 2, 3 respectively

6 PCA Analysis

Principal component analysis (PCA) is a statistical process that procedures an orthogonal conversion to transform a set of explanations of probably associated variables into a set of principles of linearly uncorrelated variables known as principal components. PCA is typically cast-off as a tool in experimental data exploration and for creating predictive representations. It is frequently used to imagine genetic distance besides understanding between populations. PCA can be prepared by eigenvalue decomposition of a data covariance matrix otherwise particular value decomposition of a data matrix, PCA can be assumed of as suitable ellipsoid to the data of an n-dimensional, somewhere each axis of the ellipsoid signifies a principal component.

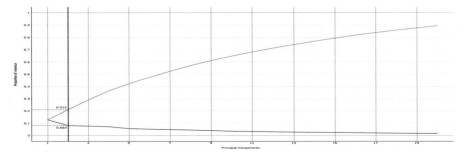

Fig. 5. Principal component analysis for lung cancer dataset

Components selection: Components: 2, Variance covered: 21%.

Principal components diagram, where the red (lower) line is covered the variance of each component and the green (upper) line is covered cumulative variance by components. Figure 5 illustrates the proportion of variance 0.083 and 0.212.

7 Classification Tree

A Classification tree (Fig. 6) can also offer an amount of assurance that the classification is accurate. A Classification tree is constructed through a method recognized as binary recursive separating. This is an iterative method of excruciating the data into dividers, and then excruciating it up added respectively of the outlets.

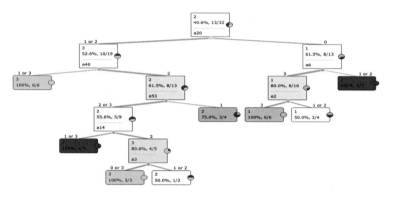

Fig. 6. Lung cancer dataset classification tree viewer

8 MDS

Multidimensional scaling (MDS) (Fig. 7) is a resource of imagining the level of comparison of specific cases of using dataset. It mentions to a set of associated ordination methods cast-off in data visualization, in specific to presentation the info held in a distance matrix. It is a method of non-linear dimensionality decrease. An MDS procedure goals to place separately purpose in N-dimensional position such as the between-object distances are conserved along with possible. For each object is formerly allocated identic in separately of the N dimensions. In the input, the widget desires both a matrix or a dataset of distances. While picturing distances between rows, you can correspondingly synthesize the color of the points, variation their shape, print them, then output them upon collection.

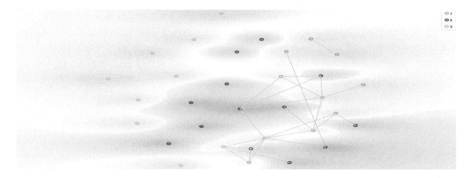

Fig. 7. Lung cancer dataset multidimensional scaling visualization for three types of classes

9   Hierarchical Clustering

In statistics and data mining, hierarchical clustering (also known as hierarchical cluster analysis or HCA) is a technique of cluster exploration which search for construct a hierarchy of clusters. Approaches for hierarchical clustering usually fall into two types such as agglomerative and divisive. Overall, the splits and merges are realized in an avaricious way. The consequences of hierarchical clustering are commonly represented in a dendrogram (Fig. 8).

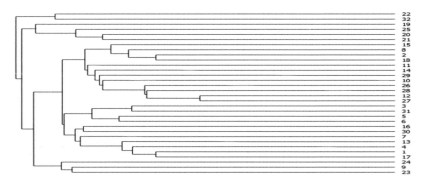

Fig. 8. Hierarchical clustering of lung cancer dataset (Dendrogram representation)

10   Conclusion

Lung Cancer has been analyzed and predicted for different classifiers: Naïve Bayes, K-Nearest Neighbors (KNN), Logistic Regression, Tree, Random Forest, Neural Network. To, corresponding the performance of these classifier algorithms, Orange data mining tool has been used. This paper mainly discusses the performance of various classification algorithm on the basis of distribution plot and ROC analysis. Here the evaluation results for cross-validation number of folds and confusion matrix have also been determined and illustrated. Results of the confusion matrix are observed and Naïve Bayes provides overall highest accuracy is 57.047% and also provides highest kappa statistics value 0.356. The second highest overall accuracy is 49.832% provides for Random Forest. It has also been included that Naïve Bayes has highest percentage (74.8%) of Area under of ROC. This paper is implemented PCA, classification tree, MDS and Hierarchical Clustering.

References

1. Alberg, A.J., Brock, M.V., Samet, J.M.: Epidemiology of lung cancer. In: Murray & Nadel's Textbook of Respiratory Medicine, 6th edn., Chap. 52. Saunders Elsevier (2016)
2. Thun, M.J., Hannan, L.M., Adams-Campbell, L.L., et al.: Lung cancer occurrence in never-smokers: an analysis of 13 cohorts and 22 cancer registry studies. PLoS Med. **5**(9), e185 (2008)
3. Hong, Z.Q., Yang, J.Y.: Optimal discriminant plane for a small number of samples and design method of classifier on the plane. Pattern Recogn. **24**(4), 317–324 (1991)
4. Oh, J.H., Al-Lozi, R., El Naqa, I.: Application of machine learning techniques for prediction of radiation pneumonitis in lung cancer patients. In: 8th International Conference on Machine Learning and Applications, ICMLA 2009, pp. 478–483 (2009)
5. Lynch, C.M., Abdollahi, B., Fuqua, J.D., de Carlo, A.R., et al.: Prediction of lung cancer patient survival via supervised machine learning classification techniques. Int. J. Med. Inform. **108**, 1–8 (2017)
6. Tazin, N., Sabab, S.A., Chowdhury, M.T.: Diagnosis of chronic kidney disease using effective classification and feature selection technique. In: International Conference on Medical Engineering, Health Informatics and Technology (MediTec) (2016)
7. Kirubha, V., Manju Priya, S.: Comparison of classification algorithms in lung cancer risk factor analysis. Int. J. Sci. Res. (IJSR) **6**(2), 1794–1797 (2017)
8. Abdar, M., Kalhori, S.R.N., Sutikno, T., Subroto, I.M.I., Arji, G.: Comparing performance of data mining algorithms in prediction heart diseases. Int. J. Electr. Comput. Eng. (IJECE) **5**(6), 1569–1576 (2015)
9. Hristea, F.T.: The Naïve Bayes Model for Unsupervised Word Sense Disambiguation: Aspects Concerning Feature Selection. Springer, Berlin (2012)
10. Hilbe, J.M.: Logistic Regression Models. CRC Press, Boca Raton (2009)
11. Retmin Raj, C.S., Nehemiah, H.K., Elizabeth, D.S., Kannan, A.: A novel feature-significance based k-nearest neighbour classification approach for computer aided diagnosis of lung disorders. Curr. Med. Imaging Rev. **14**(2), 289–300(12) (2018)
12. Kamiński, B., Jakubczyk, M., Szufel, P.: A framework for sensitivity analysis of decision trees. Central Eur. J. Oper. Res. **26**, 135–159 (2017)
13. Trevor, H., Robert, T., Jerome, F.: The Elements of Statistical Learning, 2nd edn. Springer, Berlin (2008)
14. Tosh, C.R., Ruxton, G.D.: Modelling Perception with Artificial Neural Networks. Cambridge University Press, Cambridge (2010)

Image Encryption Using New Chaotic Map Algorithm

S. Subashanthini[✉], Aswani Kumar Cherukuri, and M. Pounambal

VIT University, Vellore 632 014, Tamil Nadu, India
{subashanthini,mpounambal}@vit.ac.in,
cherukuri@acm.org

Abstract. In this digital era, all the information takes the form of various media during communication. This leads to the need for more concern on the security measures so that the data being transmitted is resilient to various malicious attacks. In this paper, a modified line map algorithm is proposed and it is considered as a chaotic map to scramble the image. The resultant chaotic Line map has been employed to permute the image pixels, whereas diffusion is performed by XOR operation to completely encrypt the image. The robustness of the image against attacks has been increased by repeated permutation-diffusion round for 'n' number of times. Histogram analysis, Number of pixels change rate and correlation analysis has been performed to analyze the robustness of the proposed technique. Simulation results proves that this methodology can be implemented in real time applications.

Keywords: Chaotic map · Cryptography · Image encryption · Line map · Scrambling · UACI

1 Introduction

Cryptography is the widely used technique that hides any data by converting it into an unintelligible form such that it can be deciphered only by the intended receiver who possess the key. This technique has been evolved from the ancient era in order to communicate a secret message between the sender and receiver.

Guomin et al. [1], have proposed a technique that employs skew tent map and the line map for scrambling the image pixels. It uses a key that is a function of left and right map. It is applicable for both color and gray scale images. Analysis results show the attainment of the optimal values for security analysis.

Murillo [2], have proposed a technique which also used the chaotic map for permutation and diffusion process. In addition to that, an additional "Z" value is added after the completion of permutation and diffusion process. It provided added security by compromising the overall time for encryption. NPCR values is closer to 100%.

Feng [3, 4], have proposed a technique in which the image is divided across its diagonal so that the locations of all the image pixels will get re-arranged on both side of the diagonal with the help of the line map algorithm. Encryption is performed by XORing the subsequent pixels to perform permutation. The performance analysis show

© Springer Nature Switzerland AG 2020
A. Abraham et al. (Eds.): ISDA 2018, AISC 941, pp. 458–466, 2020.
https://doi.org/10.1007/978-3-030-16660-1_45

that the correlation is almost equal to zero. It also eliminates the limitation of Baker map algorithm.

Yong Feng et al., have proposed a technique similar to the previous work but with a different kind of mathematical expression. A secret key has been employed to perform encryption. In addition to that, the left and right map procedures are repeated for a predefined number of times. The main advantage of this work is that it can be applied for both square as well as rectangular images. It provide a vast key space which is one of the important feature of the proposed technique. If a minute changes happens on the key value, then the information cannot be decrypted which states the robustness of the techniques against malicious attacks.

A mathematical model for securing images through encryption has been proposed by Wu et al. [5]. They have derived the NPCR values for their methodology and compared these values against various encryption schemes.

Yang Jun et al., discussed how the complexity of the encryption procedure is reduced by replacing the traditional 8 S boxes with a single S-box. In addition to that, it also minimizes the consumption of power for the overall circuit. They have stated that, it results in less hardware overhead, more reliable and it's more economical. But when compared to line map algorithm, it is more vulnerable to attacks, because the secret key used for encryption is not as efficient as other schemes.

Jun et al. [6], emphasis on the security of the algorithm that generates the key is ensured by combining the RSA and Elgamal algorithm. In this method, two keys have been used. Former key is derived by using the prime numbers and the later key is derived by using the Elgamal algorithm. The experimental results show that the computational time is highly reduced compared to the original RSA algorithm. A 256-bit prime number has been used for the key generation, but these public and private key cannot be easily found out because of the application of Elgamal algorithm.

Yajam et al. [7] work mainly focusses on the encryption technique which is appropriate for Anti-forensics applications. It uses a forged private key. A hash function is employed along with the secret key. The major pros of this method is that if the receiver is forced, then the receiver can compromise the private key, Any how it will not reveal the existence of any secret information that is embedded along the code. There is no need to share the secret key between the communicating parties.

Bora et al. [8], provides a technique that aims at a better encryption rate by combining the blowfish algorithm and Cross-Chaos map algorithm. Cross- Chaos map is a type of chaotic map which is mainly used for image pixels randomization by employing various mathematical operations. The experimental results show that the NPCR values attains an optimal value after the completion of double encryption.

Zhang and Ding [9], have implemented the encryption procedure by repeating 'N' iterations with Bye replacement, Line displacement transformation, mixed column transformation and key transformation. The last iteration is differed by excluding the mixed column transformation. The image is decomposed into 4 × 4 matrix for each unit of 8 bits. The results has portrayed a flat histogram and in addition to that the generated key is very sensitive to any changes. The time complexity is very low.

Our proposed technique uses a Line map for scrambling the image pixels. It is a very modest transformation. Efficient implementation using hardware and software is possible using Line map which is a mandatory need for any image encryption in real time applications. It aims at disrupting the location of each element in a sequence. Since there is no sub-block processing, parallel implementation of algorithm is feasible.

This paper is further organized as follows: Sect. 2 details about the idea that acts a backbone for the proposed algorithm. It is followed by Sect. 3 narrates the detailed description of the proposed algorithm using Line map for the image scrambling process. Section 4 gives a detailed analysis on the simulation results. Section 5 concludes with the pros and cons of the proposed algorithm along with the discussion on future work.

2 Background

Encryption is a process which facilitates the transmission of secret messages between two authorized end parties without any kind of interruption. Here the secret message takes the form of image which is scrambled using the modified line map algorithm and the cipher image is transmitted to the receiver. Any encryption scheme uses a pseudo random value as a key to be shared with the receiver for the decryption process.

The original image undergoes pro-processing by getting converted into matrix of pixel values. To this resultant matrix, the state of the art line map algorithm is applied in order to re-locate the pixels. In the traditional Line map algorithm, each pixel value is re-located between two contiguous pixels with respect to the diagonal value of the line map algorithm. This scrambling of pixel is very simple and efficient to be implemented.

3 Proposed Methodology

Given a plain image that can be represented as an $n \times n$ matrix of pixels. The following steps are performed to scramble the image in order to increase the robustness of the image against hackers. The robustness of the confusion process increases with the increased number of rounds. The scrambled image is then encrypted to the get the c image.

Step 1: Pre-processing of image
The plain color image is separated into R, G and B components. The below proposed algorithm is applied to all the three components separately.

Step 2: Image scrambling using new Line-Map Algorithm
The new location of all the pixel values from the pixel matrix of the original image after applying the proposed chaotic-map algorithm is derived using the following steps, for horizontal and vertical mapping respectively.

Step 2.1 Vertical mapping

$$L(c) = \begin{cases} A(i,j) \, \forall i = 0 \, to \, n - 2(iter) - 1, j = 2(iter), \; if \; c \; is \; even \\ \\ A(i,j) \, \forall i = 0 \, to \, n - 2(iter) - 1, j = 2(iter) + 1, \; if \; c \; is \; odd \end{cases} \quad (1)$$

n - order of matrix

c - index value for L array ranging as follows,

$$c = \begin{cases} 0 \, to \, \frac{(n*n)}{4} - 1, & if \; iter = 0 \\ \frac{n*n}{4} + 4(k) \, to \, \frac{n*n}{4} + 4(2*k), & if \; iter = 1, k = 3 \\ \frac{n*n}{4} + 4(k) \, to \, \frac{n*n}{4} + 4(5*k), & if \; iter = 2, k = 2 \\ \frac{n*n}{4} + 4(k) \, to \, \frac{n*n}{4} + 4k + 1, & if \; iter = 3, k = 12 \end{cases} \quad (2)$$

iter - Number of iteration ranging from 0 to ((n/2)−1) (For vertical mapping).

Step 2.2: Horizontal mapping

$$L(c) = \begin{cases} A(i,j) \, \forall i = n - 2 - 2(iter), j = 2(iter) + 2 \, to \, n - 1, if \; c \; is \; even \\ A(i,j) \, \forall i = n - 1 - 2(iter), j = 2(iter) + 2 \, to \, n - 1, if \; c \; is \; odd \end{cases} \quad (3)$$

n - order of matrix

c - index value for L array ranging as follows,

$$c = \begin{cases} \frac{(n*n)}{4} \, to \, \frac{n*n}{4} + (8*k) + 3, & if \; iter = 0, k = 1 \\ \frac{n*n}{4} + 8(k) \, to \, \frac{n*n}{4} + 8(k+1) - 1, & if \; iter = 1, k = 3 \\ \frac{n*n}{4} + 8(k) \, to \, \frac{n*n}{4} + (8*k) + 3, & if \; iter = 2, k = 5 \end{cases} \quad (4)$$

iter - Number of iteration ranging from 0 to ((n/2)−2) (For horizontal mapping).

The resultant values are stored in an 1D array named as L with the index ranging from 0 to 63. These values are replaced in a 2D matrix so that it can be employed for encryption.

Step 3: Encryption

The result of the above scrambled image will act as the input for encryption. The process of the diffusion is carried by using XOR the adjacent pixels, which is a reversible process. The process of encryption is carried out in either row-wise or column wise represented as follows (Figs. 1 and 2).

$$\text{Row wise encryption} : A(i,j) = A(i-1,j) \oplus A(i,j) \tag{5}$$

$$\text{Column wise encryption} : A(i,j) = A(i,j-1) \oplus A(i,j) \tag{6}$$

(a)

11	12	13	14	15	16	17	18
21	22	23	24	25	26	27	28
31	32	33	34	35	36	37	38
41	42	43	44	45	46	47	48
51	52	53	54	55	56	57	58
61	62	63	64	65	66	67	68
71	72	73	74	75	76	77	78
81	82	83	84	85	86	87	88

(b)

11	12	21	22	31	32	41	42
51	52	61	62	71	72	81	82
73	83	74	84	75	85	76	86
77	87	78	88	13	14	23	24
33	34	43	44	53	54	63	64
55	65	56	66	57	67	58	68
15	16	25	26	35	36	45	46
37	47	38	48	17	18	27	28

Fig. 1. (a) Initial and (b) Final values after one round of scrambling

(a)

11	12	21	22	31	32	41	42
51	52	61	62	71	72	81	82
73	83	74	84	75	85	76	86
77	87	78	88	13	14	23	24
33	34	43	44	53	54	63	64
55	65	56	66	57	67	58	68
15	16	25	26	35	36	45	46
37	47	38	48	17	18	27	28

(b)

11	12	51	52	73	83	77	87
33	34	55	65	15	16	37	47
25	38	26	48	35	17	36	18
45	27	46	28	21	22	61	62
74	84	78	88	43	44	56	66
53	57	54	67	63	58	64	68
31	32	71	72	75	85	13	14
76	23	86	24	41	42	81	82

Fig. 2. (a) Initial and (b) Final values after two rounds of scrambling

Step 4: Decryption & Descrambling

The original image is recovered by a decryption of encrypted pixels followed by 'n' rounds of de scrambling. The decryption process follow a reverse procedure of the encryption process.

The steps are as follows:

Step 4.1: The encrypted image is decrypted by reverse XORing either row-wise or column wise as per the encryption.

Step 4.2: The decrypted image which is a 2-d matrix of pixels are converted into a 1-d array.

Step 4.3: The resultant 1-d array will undergo the reverse process to descramble the pixels.

4 Results and Discussion

Standard images like Lena, Baboon, Peppers, Tajmahal and Mahatma Gandhi have been used for the simulation process.

4.1 Histogram

Total distribution of an image can be depicted using Histogram. It is a measure that ensures whether the algorithm employed for encrypting the image satisfy the property of unpredictability so that the histogram attack can be eliminated. The information is un-predictable if the image has a flat-histogram.

Results from Table 4 shows that the encrypted image has the flat histogram which proves that the proposed algorithm has encrypted the image successfully.

Table 3 values shows that the original image correlation is always higher and is almost close to 1 and the correlations value of the encrypted image is much lower nearing to zero. Thus these values proves that the proposed algorithm achieves a good security level.

4.2 NPCR, Number of Pixel Change Rate

The probability of an image to be highly resistance to differential attacks depends on the NPCR score. It is directly proportional to the NPCR value. It is defined as the rate of change of the pixels location from the original image to the encrypted image. It is calculated as overall percentage. An algorithm is coined as an ideal encryption if the NPCR value is $\sim 99\%$.

4.3 UACI, Unified Average Change Intensity

The strength of an encryption algorithm is determined by analyzing the value of NPCR along with UACI value. An ideal encryption should achieve a UACI score of $\sim 33\%$ and it is also calculated as overall percentage.

$$D(i,j) = \begin{cases} 0, \text{ if } C1(i,j) = C2(i,j) \\ 1, \text{ if } C1(i,j) \neq C2(i,j) \end{cases} \tag{7}$$

$$NPCR : N(C1, C2) = \sum_{i,j} \left(\frac{D(i,j)}{m * n} \right) * 100 \tag{8}$$

$$UACI : U(C1, C2) = \sum_{i,j} \frac{|C1(i,j) - C2(i,j)|}{F * m * n} * 100 \tag{9}$$

In (7), (8) and (9) C1 and C2 represents the encrypted images of the original image and later the original image whose one pixel got changed respectively. m * n represents the total number of pixels and the value F represents the largest pixel compatible.

From the Table 2, it can be analyzed that all the images have a NPCR value greater than 99%, which proves that the rate of pixel change is high and therefore the image attains more robustness against various attacks. Moreover the UACI values for each encrypted image lies in the close proximity of 16–18% which also enhances the robustness of the image (Table 1).

Table 1. Original, Encrypted and Decrypted images using New Line Map Algorithm

Image	Original Image	Encrypted Image	Decrypted Image
Baboon			
Lena			
Peppers			
Tajmahal			
Mahatma Gandhi			

Table 2. NPCR and UACI values of the proposed Line Map Algorithm

Image	Proposed line map (average of three planes)	
	NPCR (in terms of %)	UACI (in terms of %)
Baboon	99.6301	18.0086
Lena	99.5995	17.4472
Peppers	99.6104	17.5479
Tajmahal	99.5991	17.7081
Mahatma Gandhi	99.6125	17.6443

Table 3. Correlation values for various images using the Proposed Chaotic Map Algorithm

Image	Correlation for proposed line map algorithm (correlation)	
	Original	Encrypted
Baboon	0.9677	0.0181
Lena	0.9636	0.0098
Peppers	0.9728	0.02208
Tajmahal	0.8786	−0.0020
Mahatma Gandhi	0.9424	0.0338

Table 4. Histogram of the original and encrypted image using proposed Chaotic Map Algorithm

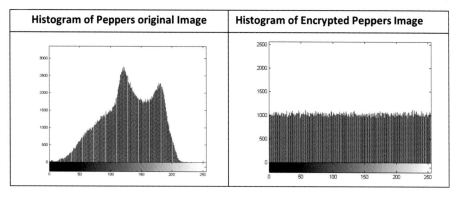

Histogram of Peppers original Image	Histogram of Encrypted Peppers Image

5 Conclusion

The proposed Line map algorithm needs a square image whose width and height should be same. This algorithm is able to encrypt both gray scale and color image. Various measures for security and performance analysis prove that the algorithm is robust against malicious attacks. The robustness can be increased by repeating the confusion process for n- number of rounds and by alternating the Left and Right Map based on a key value.

References

1. Zhou, G., Zhang, D., Liu, Y., Yuan, Y., Liu, Q.: A novel encryption method based on chaos and line map. Neurocomputing **169**, 150–157 (2015)
2. Murillo, Escobar: A RGB image encryption algorithm based on total plain image characteristics and chaos. Signal Process. **109**, 119–131 (2015)
3. Huang, F., Lei, F.: A novel symmetric image encryption approach based on a new invertible two-dimensional map. In: Intelligent Information Hiding and Multimedia Signal Processing (2008)
4. Feng, Y., Li, L., Huang, F.: A symmetric image encryption approach based on line maps. In: Systems and Control in Aerospace and Astronautics 2006. ISSCAA 2006 (2006)
5. Wu, Y., Noonan, J.P., Agaian, S.: NPCR and UACI randomness tests for image encryption. Cyber J.: Multi. J. Sci. Technol., J. Sel. Areas Telecommun. (JSAT), Apr. Ed. (2011)
6. Jun, Y., Na, L., Jun, D.: A design and implementation of high-speed 3DES algorithm system. In: Future Information Technology and Management Engineering 2009. FITME '09 (2009)
7. Yajam, H.A., Ahmadabadi, Y.K., Akhaee, M.: Deniable encryption based on standard RSA with OAEP. In: 2016 8th International Symposium on Telecommunications (IST'2016) (2016)
8. Bora, S., Sen, P., Pradhan, C.: Novel color image encryption technique using blowfish and cross chaos map. In: IEEE ICCSP 2015 Conference (2015)
9. Zhang, Q., Ding, Q.: Digital image encryption based on advanced encryption standard (AES) algorithm. In: Communication and Control (IMCCC) (2015)

Fast Implementation of Tunable ARN Nodes

Shilpa Mayannavar[1]([✉]) [iD] and Uday Wali[2] [iD]

[1] C-Quad Research, Belagavi 590008, India
mayannavar.shilpa@gmail.com
[2] KLE Dr. MSS CET, Belagavi 590008, India
udaywali@gmail.com

Abstract. *Auto Resonance Network* (ARN) is a general purpose *Artificial Neural Network* (ANN) capable of non-linear data classification. Each node in an ARN resonates when it receives a specific set of input values. *Coverage* of an ARN node indicates the spread of values within which the gain is guaranteed to be above half-power point. Tuning of ARN nodes therefore refers to adjusting the coverage of an ARN node. These tuning equations of ARN nodes are complex and hence a fast hardware accelerator needs to be built to reduce performance bottlenecks. The paper discusses issues related to speed of computation of a resonating ARN node and its numerical accuracy.

Keywords: Artificial Intelligence · Auto Resonance Network · Deep learning · Graphic Processor Unit · Hardware accelerators · Tuned Neural Networks

1 Introduction

Artificial Intelligence (AI) research is gaining momentum because of the availability of low cost Graphic Processor Units (GPUs) [1, 2]. Large amounts of data collected at Internet servers has been used to train Deep Learning (DL) systems implemented using various types of Artificial Neural Networks (ANNs). Neurons used in these ANNs need non-linear activation function(s) to compute their outputs. Fast computation of neuronal activation is critical to attain real-time performance. Therefore, it is necessary to implement such activation functions in GPUs or other specialized neural hardware to support massively parallel operations. Google has developed its own processing unit called Tensor Processing Unit (TPU) to accelerate the inference phase of neural networks. It is designed using TensorFlow framework [3] and being used for MultiLayer Perceptron (MLP), Long Short Term Memory (LSTM), Convolutional Neural Network (CNN) and in many Neural Network applications [4]. Cambricon is an Instruction Set Architecture (ISA) specifically designed for neural network accelerators [5]. Cambricon core has been used in recent mobile chip Kirin 970 by Huawei [6]. Other companies like Intel, IBM have also developed their own hardware for Neural Network applications [7–9]. Neural Network Processors should be able to support AI workload with thousands of cores and local memory. There is also a need to look at reconfigurability of the processors to be able to adapt to the needs of applications running on such processors. These demands require a completely new approach to processor design.

© Springer Nature Switzerland AG 2020
A. Abraham et al. (Eds.): ISDA 2018, AISC 941, pp. 467–479, 2020.
https://doi.org/10.1007/978-3-030-16660-1_46

Typical DL networks use several layers of neurons with varying capabilities. It is also possible to define a minimal set of modules necessary to implement all the required functionality [10, 11]. These modules can then be reconfigured to implement various layers of DL ANNs. Essentially, there is also a need to look at re-configurability of the processors to be able to adapt to the needs of applications running on such processors. These demands require a completely new approach to processor design.

Classical processor designs have debated the use of RISC and CISC architectures for improved efficiency, Silicon area, code protection, etc. However, such processor designs are not well suited for deep learning networks. Accelerating developments in AI and General Intelligence (GI) present many design challenges. Some of them are briefly discussed in our other publication [12]. Noise tolerance of ANN systems implies that numerical accuracy is less important compared to the speed efficiency of computation. Therefore there is a complete paradigm shift from classical digital system design to a new representation that needs to be explored at several levels and layers of technology.

Auto Resonance Network was proposed as a generalized data classifier for use in deep learning networks [13]. It has been used in varied applications like robotic path planning [14], predictive analytics, etc. ARN is a dynamic network that accepts multi-dimensional real inputs, with each neuron having an adjustable *acceptance threshold* and a *coverage* that depends on a *tuning parameter* and the *acceptance threshold*. Further details of ARN are available in [13, 14] and elsewhere.

One of the commonly used activation function in ANNs is the sigmoid. Direct implementation of a sigmoid requires computation of an exponent (Taylor series expansion), which is computationally intensive. In our earlier publication [15] we have demonstrated fast implementations of sigmoid function using piece-wise linear and second order interpolation methods, which reduces the complexity to a considerable extent, with a tight control on the computational error. While the sigmoid is a single curve, tunable nodes in ARN require a family of curves to be implemented. The concepts can be easily extended to other activation functions like tanh, ReLu that are functions of a single variable. However, tuning of ARN nodes depends on the input and *a resonance control parameter* (ρ), requiring a family of curves to be approximated. In this paper, we have discussed some implementation issues related to fast implementation of tunable ARN node.

2 Basic ARN Implementation

In this paper we have demonstrated ARN with two real valued inputs. Input nodes of ARN represent the parameters (features) of classification, while output nodes represent each class of data. Each output node corresponds to one unique label/class of data. Each node is tuned to recognize a specific pattern of input vector. Output of any node is limited to a range of values in the range 0 to 1 (scaling is also possible but often not required). Each node also has a label indicating the trained class. On installing an ARN, there are no nodes in output layer. When the first input is applied to ARN, an output node is created by tuning its output value to inputs applied at that time. When second or any other subsequent input is applied, existing nodes compute their outputs and a node with maximum output is selected as winner. If the output of a winning node is above a threshold, then we say that the class of input is belongs to the label of the winner node. On the other hand, if the output of the winner node is less than the threshold, current input is not recognized by the ARN and therefore, another output node will be created such that it is tuned to the applied input. This process continues until the input is recognized by any one of the output nodes in the ARN. It is important to note that, each node has a coverage which allows it to generate an output above threshold when the output is close enough to the tuned value(s). Set of all such inputs that make a node produce the output above a threshold is called *cover* of the node. *Coverage* of a node implies the range of input values for which the output of node is maximum and above threshold. Each output node can recognize the input if it is within the coverage area. The node which recognizes the input is called as the winner node.

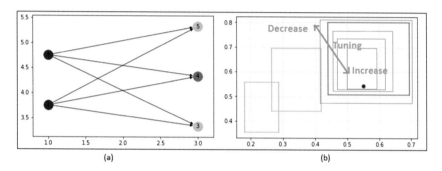

(a)   (b)

Fig. 1. (a) Basic structure of ARN (b) Two-dimensional view of Coverage of ARN nodes

Table 1. Resonant values used to create ARN

Output nodes	Input values $u = \{x, y\}$	
Node 3	0.234	0.456
Node 4	0.567	0.654
Node 5	0.342	0.567
Recognized by node 4	0.55	0.543

The network in Fig. 1(a) is an example of basic Auto Resonance Network obtained from the input values given in Table 1. It has two input nodes and three output nodes that are created during training. Nodes are labeled in an increasing order. Labels 1 and 2 represent input nodes (red color). Labels 3, 4 and 5 represent output nodes, with winner node being represented by green color and other two nodes represented by yellow color. For simplicity, all nodes are set to *threshold* = 0.95.

The *range* of each output node can describe its coverage, where, *range* = {*length, width*} for the given 2-input system as shown in Fig. 1(b). Range will have the same dimensionality as the number of inputs, assuming that all the inputs are independent. In Fig. 1(b), the rectangle in green color indicates the coverage area of winner node and those in yellow color indicate the coverage area of other two nodes. The red dot is the coordinate $u = \{x, y\}$ of the input, which is recognized by the winner node. Input within the rectangle indicates that it is within the *coverage* of a winner node.

3   Tunable ARN Nodes

A brief overview of tunable ARN node is discussed here. Details are available in literature [13]. Success of ARN depends on finding a suitable function that offers good tunability and variable *coverage*. In this paper we have discussed the scaled, shifted sigmoid resonator, which supports input values in any [$\mathfrak{R}$], though a mapping to a smaller range is commonly used.

A resonator can be implemented using several algebraic or transcendental functions. A class of these functions is of the form $y = d(1 - d)$, where $d = f(x)$ and $f(x)$ is a monotonic function with value between 0 and 1.

Sigmoid is one of the activation function defined as in the Eq. (1). Generalized form of Eq. (1) can be written as in Eq. (2), by incorporating a tuning factor   and shifting the input origin to a new mean value x_m. A resonator constructed from the sigmoid is given by Eq. (3).

$$f(x) = \frac{1}{1 + e^{-x}} \tag{1}$$

$$d = \frac{1}{\left(1 + e^{-\rho(x - x_m)}\right)} \tag{2}$$

$$y = d(1 - d) \tag{3}$$

In Eq. (2), x is the current input and x_m is the point of resonance, ρ (rho) is the *resonance control parameter* and y is output value of a node. The nature of scaled shifted sigmoid resonator for $x_m = 0.5$ and $x_m = 0$ are shown in Fig. 2. The sharpness of this curve depends on the value of ρ. Increase in ρ, leads to reduced *coverage* i.e., better is the tuning.

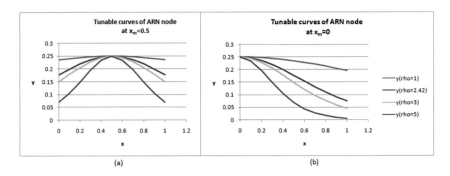

Fig. 2. Tunable curves of ARN node (a) at $x_m = 0.5$ (b) at $x_m = 0$

4   Applications Using Hierarchical ARN

Hierarchical Auto Resonance Network uses multiple layers of ARN nodes. Layers of other types of nodes, e.g., Hebbian Nodes, ReLu, Softmax etc. may also be mixed at higher layers of the hierarchy. However, the first layer of such hierarchical networks will always use ARN nodes. We have been using such networks in various applications in the last few years. Two such applications are discussed here.

Robotic Path Planning is a case of a generalized Mover's problem [16] which belongs to a class of NPHard problems called PSpaceHard. Aparanji et al. [14] have used ARN to solve robotic path planning in presence of dynamically positioned obstacles for multi segmented joint systems with more than six degrees of freedom. The network uses a five layer hierarchical ARN with Hebbian nodes at higher layers. There are two layers of ARN nodes, one of which is generated completely by perturbation of nodes in the layer below. The system learns with very few training inputs using supervised learning. Most of the later learning is by reinforcement learning. The results reported by the authors are very encouraging.

An optical character recognition (OCR) system using two layer ARN has also been developed by the authors, results of which are to be published soon. We have used MNIST data set to train the OCR system. The system was able to learn to classify test data set with high reliability. Both the training data and the test data are taken from MNIST data set [17]. The network was trained with around 10 thousand sample images of hand written characters ranging from 0–9. The accuracy of recognition with no specific optimization was more than 70%. In case of wrong identification, the system will improve the network by adjusting the node statistics or by adding a new node with applied input as a new node.

The network is being used for several types of AI related problems like natural language processing, prediction of radio channel occupancy, classification of human embryo for In-vitro fertilization etc. Therefore we observe that the proposed network is capable of learning in various environments and therefore holds promise as a viable neural network structure. In fact the success of ARN in such diverse applications has prompted us to implement ARN in hardware.

5   Approximation Methods to Implement Tunable ARN Node

Basic sigmoid resonator function given in Eq. (3) involves an exponent function and hence is computationally intensive. Direct hardware implementation of this equation is impractical as it requires large Silicon area. We have used two approximation methods namely Piecewise Linear (PWL) and Second Order Approximation (SOI) to reduce computational complexity with controlled accuracy [15]. We have also introduced a new numerical format for low precision operations and the accuracy up to three fractional digits (0.001) was achieved. This accuracy is sufficient for many applications using ARN and other ANN. By using these approximation methods the computational complexity and the Silicon area can be considerably reduced. In this paper, implementation of tunable curves of ARN node is carried out using the approximation methods and numerical format discussed in [15]. Results of approximating the Eq. (3) for $\rho = 2.42$ are shown in Figs. 3 and 4.

As an illustration, the implementation details of SOI method are described here. Detailed discussion is available in [15]. The idea is to sample a nonlinear curve like the tuning curve of ARN at specific inputs and store the values in a (x, y) table where x is the input variable and y is the computed value on the curve to be interpolated. These values are stored in a ROM in case of an embedded system, or in a predefined array. This stored table can be used to interpolate the values on the curve using these stored points. In case of second order interpolation (SOI) three sequential points viz., (x_1, y_1), (x_2, y_2) and (x_3, y_3) which cover the given input value $x \mid x_1 \leq x \leq x_3$ are used to calculate the output y of a resonator. The coefficients a, b and c are obtained using the Eqs. (4–6). Note that the values of a, b, c are independent of input and output (x, y) and hence can be pre-computed. The output of a resonator is then calculated using the Eq. (7) which involves only two multiplications and two additions [15]. Therefore, the complexity of approximating the transcendental function is reduced to O(2). Note that spacing of sample points x_1, x_2,... need not be uniform, as long as the underlying curve is differentiable.

$$a = \frac{((x_2 - x_1)(y_3 - y_1)) - ((x_3 - x_1)(y_2 - y_1))}{(x_2 - x_1)(x_3 - x_1)(x_3 - x_2)} \tag{4}$$

$$b = \frac{(y_2 - y_1) - a(x_2^2 - x_1^2)}{(x_2 - x_1)} \tag{5}$$

$$c = y_1 - ax_1^2 - bx_1 \tag{6}$$

$$y = (ax + b)x + c \tag{7}$$

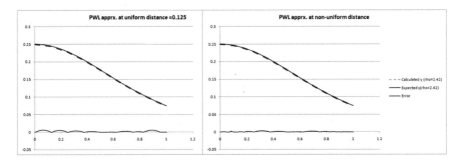

Fig. 3. PWL approximation for uniform and non-uniform distance

The computational accuracy depends on the input spacing and type of approximation. Accuracy can be improved by increasing the number of intervals but it requires larger look-up table (LUT), increasing the cost of implementation. The distance between inputs was uniformly spaced as 0.125 and the error due to PWL interpolation was 0.56%. With non-uniform spacing and increased number of sample, error was further reduced to 0.34%. The Second Order Interpolation (SOI) was introduced to improve the accuracy obtained using PWL method. SOI is three point approximation method, which uses the same number of computations as PWL with added memory storage. With SOI method, the error was further reduced to 0.025%, which is much better compared to as that of PWL (See Fig. 4).

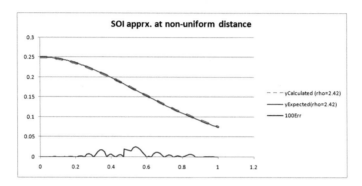

Fig. 4. SOI approximation for Node tuning

All the resonating curves shown in Fig. 2 are approximated using the similar procedure. Each resonating curve will have its own LUT. Each row in the LUT will have the values of input (x) and the corresponding values of coefficients $(a, b$ and $c)$. Output of a node is calculated using $y = (ax + b)x + c$. All the incoming inputs values are assumed to be between 0 and 1, which may be scaled to match required range.

6   Hardware Implementations

6.1   Numerical Representation

To implement neural computations in hardware, the first issue to be considered is number representation. The performance of a design depends on computational speed, numerical accuracy and Silicon area. It is necessary to use optimum number of bits while balancing this trade-off. Lowering the bit width will considerably reduce the Silicon area while compromising for small loss of accuracy. As the ANNs are noise tolerant, computations can be performed using low precision. For all the implementations, we have used our new numerical format [15].

The essential modules required to implement ARN node with N inputs are scaled shifted sigmoid resonator, state transition, multiplier, adder, prioritized bus, look UpFetch and comparator. These are discussed in the following sections. The details of prioritized bus module are available in [18].

6.2   Scaled Shifted Sigmoid Resonator

As it is mentioned in the previous section, *resonance control parameter* (ρ) is inversely proportional to threshold (T). The tuning curves shown in Fig. 2 have four different ρ values viz., 1, 2.42, 3 and 5. All these curves are implemented using verilog Hardware Description Language and the results are simulated using Xilinx ISE simulator. Due to the symmetric nature of a curve as shown in Fig. 2(a), we have considered the curve shown in Fig. 2(b) for implementation.

For simplicity the range of input is considered as (0, 1). Details of LUT for SOI approximation is given in Table 2. The block diagram and Finite State Machine (FSM) used to implement the resonator is shown in Fig. 5.

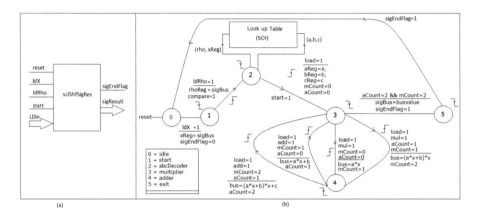

Fig. 5. (a) Block diagram (b) FSM implementation of scaled shifted sigmoid resonator

Table 2. Look up table for SOI approximation

Storage register (2 bytes each)	Number of values	Bytes	Total bytes
Input x	11	2	22
a (For 4 sets of ρ)	10 * 4	2	80
b (For 4 sets of ρ)	10 * 4	2	80
c (For 4 sets of ρ)	10 * 4	2	80
Total	**262 bytes of storage**		

The *table look up* procedure is used to fetch the values of a, b and c for the corresponding values of ρ and x. The *lookUpFetch* module is used to perform the *table look up* procedure.

LookUpFetch

When new input is applied to the network, *lookUpFetch* will select the corresponding a, b and c values from the Look up table. The inputs to this module are ρ, x and output is *abcReg*. Internally, it has comparator module to compare the given input value with the stored values of *x*. The block diagram of *lookUpFetch* is as shown in Fig. 6.

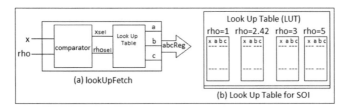

Fig. 6. (a) Block diagram of *lookUpFetch* module (b) LUT for SOI approximation

There are two multipliers and two adders used to calculate the output of Eq. (3). We have implemented this using serial multiplier. The details of serial multiplier will be available in our other publication [12]. The output of *lookUpFetch* module is *abcReg,* which is 48 bit wide and it has the values of $\{a, b, c\}$, where *a*, *b* and *c* are 16-bits each, represented using the number format [15]. It is noticed that maximum of 35 clock cycles are required to compute the output of a resonator. The simulation results for different values of ρ are shown in Figs. 7 and 8.

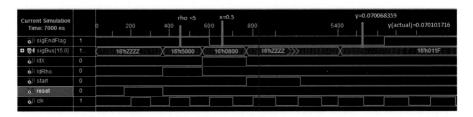

Fig. 7. Simulation result of $d(1 - d)$ for $\rho = 5$

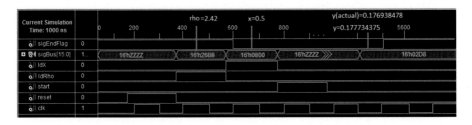

Fig. 8. Simulation result of $d(1 - d)$ for $\rho = 2.42$

6.3   State Transition Module

State transition is one of the important modules in the design of hardware accelerators. The internal structure of state transition is shown in Fig. 9.

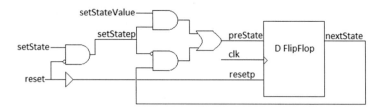

Fig. 9. Internal structure of State transition module

It has four inputs viz., *clk*, *reset*, *setState* and *setStateValue*. Among *reset* and *setState*, *reset* has higher priority. A *setState* command is used to set the state to the value stored in *setStateValue*. The clock input is dual edge triggered. D flipflop is used to store the values of present state (*preState*) and next state (*nextState*). The simulation result for 4-bit state transition module is shown in Fig. 10.

Fig. 10. The simulation of state transition module (4-bit)

6.4 ARN Node

The inputs to ARN can be N- dimensional. Input in each dimension can be a vector of any length. The output node of any layer in ARN can be calculated as weighted sum of resonators. The basic structure of ARN for single node with N inputs and for k-patterns are shown in Fig. 11(a) and (b) respectively. For ARN with k-patterns, the output value, which is maximum and above threshold is selected as a winner node. The equation used to calculate the output value of each node is given in Eq. (7).

$$y_k = \left(\frac{4}{N}\right) \sum d_{ki}(1 - d_{ki}) \tag{7}$$

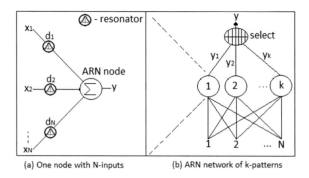

(a) One node with N-inputs   (b) ARN network of k-patterns

Fig. 11. Structure of one ARN node

The block diagram and FSM implementation of ARN node with two inputs is shown in Fig. 12.

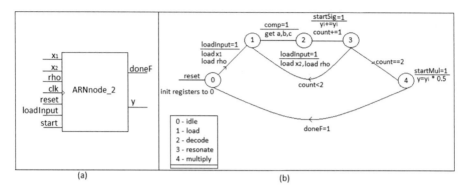

(a)   (b)

Fig. 12. (a) Block diagram (b) FSM of ARN node with N = 2

Number of resonators required is equal to the number of inputs. For two inputs there are two resonators required. All the inputs may have same or different values of ρ. When the new input is applied to the network, depending on the value of ρ and x, corresponding values of a, b and c are fetched and the output value corresponding to the first input is calculated. This result is stored in the accumulator. Same procedure is applied for the second input and this result is now added with the accumulator. The result of accumulator is then multiplied with 0.5 (because N = 2) and the final output value will be available on the output port y with *doneF* raised to 1. It is important to note that all the state transition and data transfer use dual edge triggered clock.

The simulation result of ARN node with N = 2 is shown in Fig. 13 and the computational complexity involved in the implementation of ARN node is summarized in Table 3. As it can be noticed from Table 3, the number of computations using accelerator is much lower than that that of direct computation.

Table 3. Comparison of computational complexity of accelerator vs. direct computation

Operations	Module	No. of clock cycles		
		SOI	PWL	Direct computation
Look Up	Resonator	3	3	0
Compare	Decoder	N	N	0
Multiplication	Resonator	34	7	1105
	ARN node	17	17	17
Addition	Resonator	2	1	6
	ARN node	2(N−1)	2(N−1)	2(N−1)
Subtraction	Resonator	0	1	6
Division	Resonator	0	0	170
Total (for N = 2)		**60**	**33**	**1306**

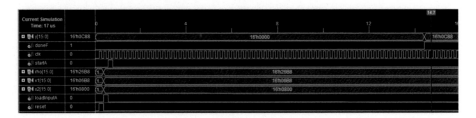

Fig. 13. Simulation result of ARN node with N = 2

7 Conclusion

The paper describes implementation of a hardware accelerator for a typical neural node of an Auto Resonance Network (ARN). ARN has been used for several applications including robotic path planning, optical character recognition, sequence prediction, etc. All essential modules required for an ARN node have been described, implemented, integrated, simulated and tested. At the core of the node is a second order interpolation module for calculating the partial products of a tunable ARN node. Other modules include a prioritized bus access control module, FSM control module for operation of an ARN neuron etc. A generalized description of the integrated module for a two input system is presented in the paper. The system can be made to accept multiple inputs with very little modification. Further work is required to implement large arrays of these modules.

References

1. Lindholm, E., Nickolls, J., Oberman, S., Montrym, J.: NVidia Telsa: a unified graphics and computing architecture. IEEE Micro **28**(2), 39–55 (2008)
2. Halfhill, T.R.: Parallel processing with CUDA. Reed electronics group. http://www.nvidia.in/docs/IO/55972/220401_Reprint.pdf
3. Abadi, M., Agarwal, A., Barham, P., Brevdo, E., Chen, Z., Citro, C., Corrado, G., Davis, A., et al.: TensorFlow: large-scale machine learning on heterogeneous distributed systems. White paper, Google Research, November 2015
4. Jouppi, N.P., Young, C., Patil, N., Patterson, D., Agarwal, G., Bajwa, R., et al.: In-datacenter performance analysis of a tensor processing unit. In: 44th International Symposium on Computer Architecture (ISCA), Toronto, Canada, June 2017
5. Liu, S., Du, Z., Tao, J., Han, D., Luo, T., Xie, Y., et al.: Cambricon: an instruction set architecture for neural networks. In: ACM/IEEE 43rd Annual International Symposium on Computer Architecture (2016)
6. Neural Processor News: Kirin 970's Neural Processing Unit and Cambricon, November 2017. https://npu.ai/2017/11/kirin-970s-neural-processing-unit-and-cambricon/
7. Akopyan, F., Sawada, J., Cassidy, A., Alvarez-Icaza, R., Arthur, J., Merolla, P.: TrueNorth: design and tool flow of a 65 mW 1 million neuron programmable neurosynaptic chip. IEEE Trans. Comput. Aided Des. Integrated Circuits Syst. **34**, 1537–1557 (2015). https://doi.org/10.1109/TCAD.2015.2474396

8. Andres, R., Eden, S., Etay, M., Evarist, F., Young, J.K., Haihao, S., Barukh, Z.: Lower numerical precision Deep Learning inference and training. White paper, Intel, January 2018
9. Koster, U., Webb, T., Wang, X., Nassar, M., Bansal, A., Constable, W., et al.: Flexpoint: an adaptive numerical format for efficient training of Deep Neural Networks. Intel, December 2017
10. Schmidhuber, J.: Deep learning in Neural Networks: An overview. Neural Netw. **61**, 85–117 (2015). Archieves, Cornell university library
11. LeCun, Y., Bengio, Y., Hinton, G.: Deep learning. Nature **521**, 436–444 (2015). https://doi.org/10.1038/nature14539. Review paper
12. Mayannavar, S., Wali, U.: Performance comparison of Serial and parallel multipliers in massively parallel environment. In: ICEECCOT-2018, IEEE Xplore Digital Library (2018). (Accepted for publication)
13. Aparanji, V.M., Wali, U., Aparna, R.: A novel neural network structure for motion control in joints. In: ICEECCOT-2016, IEEE Xplore Digital Library, pp. 227–232, Doc no. 7955220 (2016)
14. Aparanji, V.M., Wali, U., Aparna, R.: Pathnet: a neuronal model for robotic motion planning. In: 3rd International Conference on Cognitive, Computing and Information Processing (CCIP 2017), Springer CCIS, December 2017
15. Mayannavar, S., Wali, U.: Hardware implementation of an activation function for neural network processor. In: IEEE Xplore Digital Library, January 2018. (In press)
16. Reif, J.H.: Complexity of the mover's problem and generalizations. Department of Computer Science, University of Rochester, Research Report TR58, August 1979
17. LeCun, Y., Cortes, C., Burges, C.J.C.: The MNIST database of handwritten digits. http://yann.lecun.com/exdb/mnist/
18. Mayannavar, S., Wali, U.V.: Design of modular processor framework. Int. J. Technol. Sci. (IJTS) **IX**(1), 36–39 (2016). ISSN (Online) 2350-1111, (Print) 2350-1103

Efficient Framework for Detection of Version Number Attack in Internet of Things

Rashmi Sahay$^{(\boxtimes)}$, G. Geethakumari, Barsha Mitra, and Ipsit Sahoo

BITS-Pilani Hyderabad Campus, Hyderabad, India
{p2016009,geetha,barsha.mitra,f2014009}@hyderabad.bits-pilani.ac.in

Abstract. The vision of the Internet of Things (IoT) is to connect minimal embedded devices to the Internet. The constrained nature of these embedded devices makes the use of Internet Protocol impossible in its native form to establish global connectivity. To resolve this, IETF proposed 6LOWPAN, the wireless internet for embedded devices, which makes use of RPL as its routing protocol. RPL organizes low power and lossy networks in the form of one or more Destination Oriented Directed Acyclic Graphs (DODAGs). Each DODAG is assigned a version number. The purpose of the version number is to ensure that there are loop free paths to the root node, the routing table entries of nodes in the DODAG are not obsolete and there is no inconsistency in the DODAG. The root node in a DODAG increments the version number in case of any inconsistency. This calls for a global repair process and the DAG is reconstructed. A malicious node may advertise a false version number in its control message to force a global repair. In this paper, we propose an efficient framework for detecting version number attacks in the IoT. We also present mechanisms to detect the attack and identify the malicious nodes instigating the version number attack.

Keywords: IoT · 6LOWPAN · RPL · Version number attack · Attack detection

1 Introduction

The low power and lossy networks (LLNs) comprising embedded devices like sensors and RFIDs, form the driving force of the Internet of Things (IoT) as they provide global connectivity to devices which were earlier not connected to the Internet. These embedded devices are constrained by low power supply, limited memory space, processing capabilities and radio range. These restrict the size of data link layer frames and long IPv6 addresses do not fit in them. To overcome these problems, Internet Engineering Task Force (IETF) introduced IPv6 over low power and lossy network (6LOWPAN) which acts as an adaptation layer and efficiently encapsulates the IPv6 long headers in packets of size 128 bytes [1].

© Springer Nature Switzerland AG 2020
A. Abraham et al. (Eds.): ISDA 2018, AISC 941, pp. 480–492, 2020.
https://doi.org/10.1007/978-3-030-16660-1_47

IoT applications like smart homes employ large scale deployment of sensors and generate huge volume of data. Hence, routing becomes an essential requirement. IPv6 Routing Protocol over low power and lossy networks (RPL) is the routing protocol used with 6LOWPAN in most of the IoT applications as RPL supports broadcast, unicast and multicast communications [2]. Though RPL has several advantages over other routing protocols, it is prone to several attacks [3–5]. Routing security is essential for uninterrupted data transmission, securing the network traffic and ensuring network performance and resource optimization.

RPL supports three security modes namely, unsecured mode, pre-installed mode and authenticated mode. In unsecured mode, the control messages responsible for topological formation and maintenance is unencrypted and security is taken care by link layer security feature. In pre-installed mode, nodes joining a network must obtain a pre-installed encryption key. Using this key, a node can join the network as a host or as a router. In authenticated mode, a node can join a network as a host using pre-installed key. To become a router, the node should obtain an authentication key from an authentication server. This requires additional protocols defining the process of obtaining key which places additional computational overhead on the constrained nodes. Thus, in most of the IoT applications, incorporating complex security mechanisms at the device level is difficult. This leaves RPL prone to several insider attacks [6]. Insider attacks are those attacks in which a node initiates a malicious activity after it has joined the LLN as a host and router. RPL is a route over routing protocol in which all routing decisions are taken at the network layer unlike mesh under protocols where routing decisions are taken at the adaption or link layer. Hence, it is important that the sink node and the nodes in the LLN have the current view of the topology. In case of inconsistent, it should be instantly addressed to avoid loops and nodes becoming unreachable. To achieve this, RPL uniquely identifies a network topology within an RPL instance using version number. The sink node and every other node in the LNN have the same copy of the version numbers. In case of inconsistency, the sink node calls for a global repair and the entire topology is reorganized. This ensures that the routing entries in the routing table of each router node in the LLN is not obsolete. Ironically, this phenomenon can be used by a malicious node to disrupt normal flow of network traffic by simply advertising a wrong version number and making the neighboring nodes believe that their routing table entries are obsolete. As a result, rather than transmitting data packets, the network is engaged in reorganizing itself. This also results in the unnecessary consumption of node resources. In this paper, we analyze the version number attack and propose an efficient framework to detect the presence of version number attack in RPL-6LOWPAN in IoT. We present mechanisms within the framework which identifies the attack and also identifies the malicious nodes instigating the attack.

The rest of the paper is organized as follows. In Sect. 2, we present a novel threat model of version number attack. In Sect. 3, we review the existing literature and present the motivation of the present work. In Sect. 4, we analyze the performance of LLNs under version number attack. Based on the analysis, we propose an efficient framework for version number attack detection and identification of malicious nodes based on techniques of learning methods and frequency estimation in Sect. 5. In Sect. 6, we demonstrate the results of the attack detection and malicious nodes identification mechanisms. Section 7 concludes the paper.

2   DODAG Version Number Attack (DVA) Threat Model

RPL organizes the LLNs as Directed Acyclic Graphs (DAGs) which are also termed as RPL instances. A DAG can be partitioned into one or more Destination Oriented DAGs (DODAG) where each DODAG has a root (sink) node. Multiple sinks are connected through a backbone network to the Internet or any High Performance Computing environment. To uniquely identify a node in a DODAG, RPL makes use of the triplet [RPL instance id, DODAG id, Rank], where the Rank is determined using an Objective Function [7,8]. Rank describes the relative position of a node with respect to the sink node. Over the period of time as nodes join and leave a DODAG, discrepancies may occur resulting in topological instability. Discrepancies may also occur due to the presence of greedy nodes attempting to increase their parent set by moving deeper in the network resulting in loop formation. In such events, the sink node calls for a global repair mechanism and the DODAG is reorganized. To ensure that all nodes are aware of the new configuration, RPL makes use of DODAG version number. Each time the global repair mechanism is invoked, the version number of the DODAG is incremented. If a node holds an old copy of version number, it means that its routing table entries are obsolete. The current state of a DODAG is uniquely identified by the triplet [RPL instance id, DODAG id, DODAG Version Number]. Hence, version number is an important parameter to maintain loop free and stable topology of the LLNs in the IoT environment.

2.1   Attack Threat Model

Nodes in DODAG solicit information through DODAG Information Solicitation (DIS) messages from neighboring nodes in order to ensure that their routing table entries are not obsolete. If the version number of the response DODAG Information Object (DIO) of the neighboring node does not match the current copy version number with the node, it implies topological inconsistency and that the node with old version number has not migrated to the current version of the DODAG. If multiple nodes in the DODAG have old copies of the version number, the sink calls for a global repair. A malicious node can create a similar situation by falsely advertising an incremented version number at regular intervals. This results in the DODAG being mostly busy in reorganizing itself rather than transmitting data packets. We formally define the version number attack threat model according to the guidelines presented in [9] as following:

1. Definition: DODAG version number attack is an integrity attack where a malicious node advertises an incremented version number in its DIS messages in order to make the neighboring nodes believe their routing table and network configuration data to be obsolete.
2. Threat Source: Version number attack is an insider attack and can be initiated by a node which is part of the DODAG.
3. Adversary Motivation: The goal of the attacker is to force the sink node to call for a global repair and reorganize the DODAG.
4. Adversary Capability: The adversary is capable of depleting the network resources. The transmission links and node's battery power are consumed due to heavy circulation of control messages.
5. Threat Consequence: Disruption of traffic, churning, looping, topological instability, clog and cut.
6. Threat Consequence Zone: Version number attack initiated by one node can affect the entire DODAG.

The following section presents the existing literature on version number attacks.

3   Existing Literature and Motivation

The primary objective of any routing protocol is to ensure seamless transmission of data packets, maintaining the topological and reachability information. DODAG version number attack interrupts all the three objectives of the RPL routing protocol. In [10, 11], authors have studied the performance of LLN under DODAG Version Number attack (DVA) with respect to packet delay, power consumption and control overhead. In the present work, we have analyzed the effect of DVA on all other important link and node parameters like beacon interval and routing metric of nodes in the DAG which are important in framing a detection mechanism for the attack. In [12], authors have proposed a mechanism to prevent the nodes in the DODAG from modifying the version number information in their DIO messages by making use of hash chain and message authentication code. The hash values and the message authentication code are added in the DIS header and each node saves this digital signature upon verification. This process imposes additional overheard on the control messages and computational overhead on the already contained nodes. Also hash chains are themselves vulnerable to several attacks. Hence, this mechanism can not assure that DVA will not take place after applying preventive measures. Therefore, it is important to detect version number attacks and identify the nodes instigating the attack in a timely manner.

To the best of our knowledge, only a limited amount of work has been done for detecting version number attack. In [13], the authors have proposed a DVA detection mechanism based on monitoring nodes distributed over the network. The monitoring nodes check for version number inconsistency in their respective zones and send the inconsistency information along with the address of the suspect node to the sink node. The disadvantage of this mechanism is that it labels non-malicious nodes including the monitoring node as malicious and has high false alarm rate. Monitoring nodes will be of advantage, if they can restrict the malicious node from multicasting false DIO messages. Moreover, only the sink node can take any action against an information of incremented version numbers. Due to the routing procedure followed by RPL protocol, any message (control message/data packet) received by any node is forwarded to the sink via the next hop node. The sink receives the information of increased version number as well as the source of the message. Besides this, the monitoring nodes should be computationally more powerful than regular nodes. In our work, the proposed framework for the version number attack detection can be deployed at the cloud or at the IoT-LLN edge. Also, there is no miss-identification of malicious node.

Before we present our framework and detection mechanism, we describe the analysis of the performance of RPL-6LOWPAN LLNs under the version number attack.

4   Analyzing RPL-6LOWPAN Performance Under Version Number Attack

To understand the characteristics of the DODAG version number attack, we emulated DVA attack scenarios by reproducing RPL instances in Cooja Simulator available in Contiki OS [14].

4.1   Emulation of DODAG Version Number Attack

LLNs in IoT comprises several UDP clients which send packets to the UDP sink via their parent node. RPL-UDP is an implementation of a UDP Server on top of RPL that initializes the RPL DODAG, sets up UDP connections and receives sensor information.

To emulate the version number attack we altered the RPL udp-sender.c file of the malicious node such that the node multicasts an incremented version number in its DIO messages. This process initiates the attack and in no time the sink

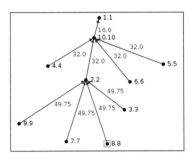

Fig. 1. Simulation setup

Operating System	Contiki OS
Simulator	Cooja Simulator
Adaptation Layer	6LOWPAN
Routing Protocol	RPL
Routing Metric	Expected Transmission Count (ETX)
Mote Type and Radio Range	Sky Mote and 30-50m

Fig. 2. Simulation parameters

node calls for a global repair mechanism and the entire DODAG is reorganized. An example scenario as depicted in Fig. 1 consists of 1 sink node, 9 fair nodes and 1 attacker node. Node 1 is the sink node and node 10 is the attacker node. The details of the simulation parameters is shown in Fig. 2. The moment the attack is initiated, the beacon interval of all the nodes in the DODAG drops to minimum. This is because nodes increase their frequency of broadcasting beacon frames while the DODAG is reorganized. After the global repair is completed, the beacon interval stabilizes to its maximum value. If this interval is unnoticed, the malicious node may continue to instigate the version attack at regular intervals, thus keeping the DODAG busy in reorganizing itself again and again rather than transmitting the data packets. Also, this results in sudden increase in power consumption and the routing metric as shown in Table 1. Hence, timely detection of version number attack is important.

Table 1. Experimental results of version attack simulation

No. of nodes in DODAG	No. of attacker nodes	Increase in avg. power consumption		Increase in avg. routing metric	Drop in beacon interval
		Attacker nodes	Non attacker nodes		
4	1	187.01%	42.74%	74%	94.40%
8	2	250.53%	49.67%	88.22%	90.60%
12	2	251.11%	47.77%	86.59%	86.70%
16	3	178.00%	95.05%	139.40%	84.50%
20	3	230.10%	80.19%	166.03%	84.30%

Table 2. Effect of DODAG version number attack

Observed parameter	Effect	Interpretation
Sensor output (Temperature, Humidity)	No effect	The attacker node changes only the data in DIO messages.
Loss of data packets	None	The malicious node does not block any data packet
Received packets per node	The number of packets received drops	The DIO/DIS radio messages are transmitted at regular intervals since the version attack causes the global repair mechanism. Hence, received packets per node is reduced
Routing metric	Routing metric increases	Possible result of frequent transmission of control messages
Average power and instantaneous power	There is an increase in power consumption by all the nodes in the DODAG. Maximum power is consumed by the attacker node	Reason is due to increased circulation of control messages as a result of frequent global repairs

Based on the various DVA attack simulations as presented in Table 1, we present the observations as depicted in Table 2. Thus, we conclude that the following parameters also termed as Detection Parameters hereafter should be monitored to detect DVA: (a) Beacon Interval, (b) Average/Instantaneous Power Consumption and (c) Routing Metric of Nodes.

5   Proposed Framework for Detection of DODAG Version Number Attacks

To analyze and interpret data and control packets generated in enormous volumes in IoT, IoT applications are integrated with high performance computing capabilities. In the present work, we propose a framework where the analysis of network data for attack detection is pushed to the cloud or edge of IoT-LLN. The proposed framework for the detection of version number attack is shown in Fig. 3. The state of the network is captured through pcap files and mote output files. These files are pumped in the temporary storage area where they are integrated before being sent to the cloud or edge for analysis in order to detect the version number attack. In the proposed framework, the process of attack detection is done at the cloud by a cloud service. The same can also be achieved at the edge (also termed as fog computing). The result of the analysis conducted either at the cloud or at the edge is detection of DVA and identification of the malicious

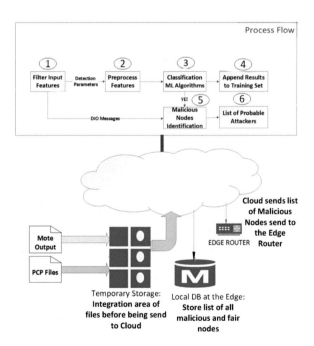

Fig. 3. Proposed framework for DODAG version number attack detection

node instigating DVA. This result is sent to the edge router and stored at the local database at the edge for future reference. The advantage of this framework for attack detection is that it does not place any computational overhead on the constrained nodes in the LLN environment of IoT. Also, at the cloud or edge, as there is no resource constraint, efficient algorithms can be used for the timely detection of attacks and identification of malicious nodes. The framework can be used for detection of other RPL attacks as well. In the following subsection, we present mechanism within the proposed framework for attack detection and identification of malicious nodes.

5.1   Mechanism for Detection of DODAG Version Number Attack

Version number attack is characterized by the rapid increase in version number of the DODAG over a short duration of time. As discussed in Section III, the

Table 3. Sample output of filter input feature module

Time stamp	Node id	Beacon interval(ms)	Routing metric	Power consumption(mW)
507230902	514	16	519	0.708
527769032	2056	1048	512	0.493
568808597	1799	1048	731	0.394
570922271	2056	1048	512	0.416
572405444	2570	131	718	0.598
576220334	1285	131	384	0.572
580537531	514	131	512	0.521
586466267	1028	131	584	0.639
588647352	1542	131	384	0.732
591143355	2313	1048	731	0.409
601258304	771	1048	705	0.511
613457915	**1799**	**16**	**1577**	**0.638**
614692449	514	32	845	2.024
619151651	1542	8	1053	1.696
626862818	2056	32	1428	1.175
640852403	1028	65	790	1.575
648169631	2313	65	1026	1.511
648785082	1285	8	773	0.977
659076260	2570	8	384	1.384
660697574	771	65	912	1.439
669870829	1028	16	964	1.573
673357145	514	32	698	2.194
673380700	2056	16	1223	1.721
674737886	1542	8	860	2.066

phenomena that signals the presence of version number attack is sudden drops in beacon intervals, rapid increase in power consumption and increase in the routing metric of all the nodes in the DODAG. The approach is to detect the stated phenomenon followed by the estimation of the frequency of version numbers changes initiated by each node to identify the malicious nodes.

The contiki OS provides the collect view tool by virtue of which, the sink node receives node information like number of packets transmitted by the node, routing metric, next hop ETX, beacon interval, power consumption and churn count. This mote output is pumped to the file integration area and pushed to the cloud which houses the attack detection service. The attack detection service comprises 6 modules as shown in Fig. 3. Module 1 named as Filter Input Feature extracts the three detection parameters along with the time stamp and node id. A sample output of module 1 is depicted in Table 3. We can observe that for all the nodes, the beacon interval drops whereas routing metric and power consumption increases after 613457915 time stamp. This phenomenon is observed after an interval of 106227013 time stamps from the starting time stamp 507230902.

Based on this observation, we present Algorithm 1. Module 2 named as the preprocess feature makes use of Algorithm 1 to prepare the training data for classification. Algorithm 1 calls the procedures SteepFallBeacon(Beacon) to detect steep fall in beacon interval, SteepRisePower(Power Consumption) to detect rise in power consumption and SteepRiseRM(Routing Metric) to detect rise in routing metric. If a steep fall is found in beacon interval, it labels the beacon values as 1, else they are labeled 0. Similarly, if steep rise is found in the other two detection parameters, their values are labeled as 1, else they are labeled 0. The training data is fed to the module 3 named as the classification ML Algorithms.

Algorithm 1. PREPROCESS TRAINING DATA

Input: Filtered Mote output as depicted in Table 3
Output: Training-data
1 **for** N Sample Data points **do**
2 $\quad$ Sort data by time stamp and Node id;
3 $\quad$ **for** Node i **do**
4 $\quad\quad$ Find max and min of all Detection Parameters;
5 $\quad\quad$ **SteepFallBeacon(Beacon);**
6 $\quad\quad$ **if** Steep_Fall **then**
7 $\quad\quad\quad$ Beacon_Fall[i] = True;
8 $\quad\quad\quad$ Beacon[i] = 1 ;
9 $\quad\quad$ **else** Beacon[i] = 0   ;
10 $\quad\quad$ **SteepRisePower(PowerConsumption);**
11 $\quad\quad$ **if** Power[Current_Sample] >   2.5 * Power[Previous_Sample] **then**
12 $\quad\quad\quad$ Power_Rise[i] = True
13 $\quad\quad\quad$ Power[i] = 1 ;
14 $\quad\quad$ **else** Power[i] = 0   ;
15 $\quad\quad$ **SteepRiseRM(RoutingMetric);**
16 $\quad\quad$ **if** rm[Current_Sample] >   1.2 * rm[Previous_Sample] **then**
17 $\quad\quad\quad$ RM_Rise[i] = True
18 $\quad\quad\quad$ rm[i] = 1 ;
19 $\quad\quad$ **else** rm[i] = 0   ;
20 **return** Training Data;

To detect the presence of version number attack, any standard classification algorithm can be used. The classification algorithm will label the preprocess training data as safe or under attack. The result of the module 3 is saved in module 4 to append the existing training data. In case of attack module 5 identifies the malicious nodes instigating the version number attack as explained in the following subsection. Module 6 maintains the history and details of the identified malicious nodes for future reference.

5.2   Identification of Malicious Nodes

For identification of malicious nodes, we require live capture of LLN data. Contiki OS is popularly used in the sensor devices in the IoT environment. Wireshark is a part of default Contiki OS implementation, but it fails to provide information about dormant nodes or nodes with less memory. Foren6 tool can also be used to capture network data. The LLN data collected in form of pcap files consist of the following types of data: (1) UDP packets: Data packets transmitted by the nodes and (2) icmpv6 packets: Control messages (DIO, DIS, DAO). For detection of version number attack we track DIO messages. The packet header DIO message has Type field value of 155 and Code field value of 2.

Algorithm 2. FIND MALICIOUS NODES

Input: DIO Messages from PCAP files
Output: List of Attacker Nodes
1 **for** N DIO $messages$ **do**
2 $\quad$ Sort Messages by time stamp and Node id;

3 Extract version number at all timestamps;
4 Find Max and Min Version Number;
5 Find variance in Version Number over input DIO messages;
6 **for** $EachChangeInVersionNo.Found$ **do**
7 $\quad$ **if** $Node$ i $initiated$ $change$ **then**
8 $\quad\quad$ $Attacker[j] = nodei$;
9 $\quad\quad$ $No.OfAttacker = No.OfAttacker + 1$;
10 $\quad$ MaxV= Max Version No displayed by node i;
11 $\quad$ MinV= Min Version No displayed by node i;
$\quad\quad$ **FindProbablity(MaxV, MinV, timestamps)**;

12 **return** $List$ of $Attacker$ $Nodes$ $with$ $Probability$;

The frequency of version number change initiated by each node is estimated to find the list of probable attacker node as depicted in Algorithm 2. The nodes with high probabilities are marked as malicious and their information is sent to the edge router and saved in the database. Sink node on receiving the list of probable malicious nodes may black list them. In the following section, we present the results of our proposed detection mechanism.

6   Results and Discussion

To classify the captured network data as safe or under attack we used three different standard classification algorithm as show in Table 4. To evaluate the performance of the classification model, the following metrics are used:

– Confusion Matrix: It visualizes the performance of the classifier in tabular form. The row elements of the matrix represents the instances in a predicted class while column element represents the instances in an actual class as shown in Fig. 4(a).

– Accuracy: It is the number of correct predictions made by the learning method over all the predictions made and is given by Eq. 1.

$$Accuracy = (TP + TN)/(TP + FP + FN + TN) \tag{1}$$

– Precision: It defines the ratio of correct prediction as given by Eq. 2.

$$Precision = TP/(TP + FP) \tag{2}$$

– Recall: It defines how accurately a class is recognized as given by Eq. 3.

$$Recall = TP/(TP + FN) \tag{3}$$

Confusion Matrix	Predicted Normal	Predicted Attack
Actual Normal	True Positive (TP)	False Negative (FN)
Actual Attack	False Positive (FP)	True Negative (TN)

(a)

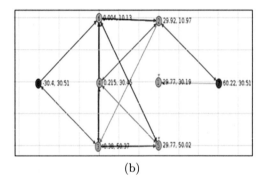

(b)

Fig. 4. (a) Confusion matrix and (b) simulation of version number attack

The classification model detects version number attack. In case, an attack is detected, we identify the attacker node. To illustrate this, we have used the example scenarios as shown in Fig. 4(b) with one and two malicious nodes. The accuracy achieved by various classification models in detecting the presence of version number attack were similar as depicted in Table 4.

In Fig. 4(b), the DODAG consists of 7 fair nodes and one malicious node. Node 6 is the sink node and node 8 is the malicious node. Initially node 8 behaves normally and the LLN is classified as safe. During this period, if we run the algorithm to identify the malicious node, the version number has a variance 0 and the number of times the version number changed is also 0 as depicted in Table 5.

After node 8 initiates the version number attack, node 8 is identified as the malicious with probability of 0.75, the variance in version number and the number of times the version number changed is displayed. Node 6 which is the sink node and node 8 are listed as the nodes which may have incremented the version number with a probability of 0.25 and 0.75 respectively. The algorithm thus, correctly identifies node 8 as the malicious node. At 2000 milliseconds (ms) of simulation time node 7 is also made malicious. As shown in Table 5, the algorithm identifies node 8, node 7 and node 6 as the nodes which may have initiated increment of version number with a probability of 0.5, 0.4 and 0.1 respectively. Thus, node 8 and node 7 can be identified as malicious nodes.

At time 1275.7 ms and 2266 ms, the sink node also appears in the list of attacker nodes with a small probability assigned to it. This happens as the sink node is responsible for upgrading the DODAG which involves version number change.

Table 4. Performance of classification algorithms

Classification algorithm	Accuracy	Precision	Recall	Specificity
Decision tree	0.98	1.00	0.95	1.00
Support vector machine	0.98	1.00	0.94	1.00
Bernoulli RBM and LR	0.98	1.00	0.95	1.00

Table 5. Results of malicious nodes identification algorithm

Simulation time	26.85 ms	715.5 ms	1275.7 ms	2266.37 ms
Variance in version number	0	0	50.35	227.71
No. of version number changes	0	0	2328	8173
Identified nodes responsible for version number change (with probability)	NIL	NIL	1. Node 8 (Probability: 0.75)	1. Node 8 (Probability: 0.50)
			2. Node 6 (Probability: 0.25)	2. Node 7 (Probability: 0.4)
				2. Node 6 (Probability: 0.10)

7 Conclusion

In this work, we have presented an in depth analysis of Version Number Attacks in the IoT environment. Version number attack depletes the network resources and diminishes the network performance. Hence, it is important to detect the attack and take remedial actions. In this paper, we proposed an efficient architecture for analysis and detection of version number attacks. We also presented a mechanism based on the frequency estimations of version number changes to efficiently identify the malicious nodes instigating version number attack. In future, we will test the framework for detection of other RPL attacks like flooding attacks and packet dropping attacks.

References

1. Kushalnagar, N., Montenegro, G., Schumacher, C.: IPv6 over low-power wireless personal area networks (6LoWPANs). https://tools.ietf.org/html/rfc4919. Accessed Nov 2016
2. Winter, T., Thubert, P., Brandt, A., Hui, J., Kelsey, R., Pister, K., Struik, R., Vasseur, J.P., Alexander, R.: RPL: IPv6 routing protocol for low-power and lossy networks. https://tools.ietf.org/html/rfc6550. Accessed Dec 2016
3. Mayzaud, A., Badonnel, R., Chrisment, I.: A taxonomy of attacks in RPL-based internet of things. Int. J. Netw. Secur. **18**(3), 459–473 (2016)
4. Wallgren, L., Raza, S., Voigt, T.: Routing attacks and countermeasures in the RPL-based internet of things. Int. J. Distrib. Sens. Netw. **2013**, 1–11 (2013)
5. Dhumane, A., Prasad, R., Prasad, J.: Routing issues in internet of things: a survey. In: Proceedings of the International Multi Conference of Engineers and Computer Scientists, 16–18 March 2016 (2016)
6. Tsao, T., Alexander, R., Dohler, M., Daza, V., Lozano, A., Richardson, M.: A security threat analysis for the routing protocol for low-power and lossy networks (RPLs). https://tools.ietf.org/html/rfc7416. Accessed April 2017
7. Thubert, P.: Objective function zero for the routing protocol for low-power and lossy networks (RPL). https://tools.ietf.org/html/rfc6552. Accessed April 2017
8. Vasseur, J.P., Kim, M., Pister, K., Dejean, N., Barthel, D.: Routing metrics used for path calculation in low-power and lossy networks. https://tools.ietf.org/html/rfc6551. Accessed April 2017
9. Barbir, A., Murphy, S., Yang, Y.: Generic threats to routing protocols. https://tools.ietf.org/html/rfc4593. Accessed Jan 2018
10. Mayzaud, A., Sehgal, A., Badonnel, R., Chrisment, I., Schönwälder, J.: A study of RPL DODAG version attacks. In: IFIP International Conference on Autonomous Infrastructure, Management and Security, 30 June 2014, pp. 92–104 (2014)
11. Aris, A., Oktug, S.F., Yalcin, S.B.: RPL version number attacks: in-depth study. In: IEEE Symposium in Network Operations and Management Symposium (NOMS), 25 April 2016, pp. 776–779 (2016)
12. Dvir, A., Buttyan, L.: VeRA-version number and rank authentication in RPL. In: IEEE 8th International Conference on Mobile Adhoc and Sensor Systems (MASS), 17 October 2011, pp. 709–714 (2011)
13. Mayzaud, A., Badonnel, R., Chrisment, I.: A distributed monitoring strategy for detecting version number attacks in RPL-based networks. IEEE Trans. Netw. Serv. Manag. **14**(2), 472–86 (2017)
14. Osterlind, F., Dunkels, A., Eriksson, J., Finne, N., Voigt, T.: Cross-level sensor network simulation with COOJA. In: 31st IEEE Conference on Local Computer Networks, 14 November 2006, pp. 641–648 (2006)

Facial Keypoint Detection Using Deep Learning and Computer Vision

Venkata Sai Rishita Middi[1]([✉]), Kevin Job Thomas[2],
and Tanvir Ahmed Harris[3]

[1] RNS Institute of Technology, Bangalore, India
rishimiddi@gmail.com
[2] VIT, Vellore, India
[3] IIT Bombay, Mumbai, India

Abstract. With the advent of Computer Vision, research scientists across the world are working constantly working to expedite the advancement of Facial Landmarking system. It is a paramount step for various Facial processing operations. The applications range from facial recognition to Emotion recognition. These days, we have systems that identify people in images and tag them accordingly. There are mobile applications which identify the emotion of a person in an image and return the appropriate emoticon. The systems are put to use for applications ranging from personal security to national security. In this work, we have agglomerated computer vision techniques and Deep Learning algorithms to develop an end-to-end facial keypoint recognition system. Facial keypoints are discrete points around eyes, nose, mouth on any face. The implementation begins from Investigating OpenCV, pre-processing of images and Detection of faces. Further, a convolutional Neural network is trained for detecting eyes, nose and mouth. Finally, the CV pipeline is completed by the two parts mentioned above.

Keywords: Pipeline · Edges · De-noising · Blurring · Detection · Losses ·
Landmark · Recognition · Robust

1 Introduction

In our day to day lives, facial landmark detection is very pervasive. Its applications include, Expression understanding, face registration, face recognition, building 3D models etc. Top notch companies like Facebook and Google are working towards building advanced facial recognition systems to identify human images effectively through their projects 'DeepFace' and 'FaceNet' respectively. Facial detection is being employed for fraud detection in Passports and Visas. China has begun using this technology in ATM and banks as well, which augmented the security of the card user. In this work we built a facial keypoint detection system which could be used for face tracking and emotion recognition. The ultimate aim of the project is to build a Computer Vision pipeline that detects and localizes certain landmarks on the face. The landmarks used here to characterize are eye corners, nose tip, eyebrow arcs and mouth corners.

© Springer Nature Switzerland AG 2020
A. Abraham et al. (Eds.): ISDA 2018, AISC 941, pp. 493–502, 2020.
https://doi.org/10.1007/978-3-030-16660-1_48

The entire paper is divided into three parts:

- Investigating OpenCV, pre-processing and face detection.
- Training a Convolutional neural network to detect facial keypoints.
- Putting parts 1 and 2 together to identify facial keypoints on any image.

2 Literature Review

Over the years there have been many advances in the field of facial landmark detection. In [1], the authors have proposed a Landmark detection system using an amalgamation of Support Vector Regression and Markov Models which primarily focusses on reducing the time and increasing the robustness of the system. Mapping between appearance of the area and positions of points is learned by regressors. Another system proposed in [2] in which the facial components are found via Subclass Determinant Analysis. Here, multiple models are built for eyes and mouth. In [3], face is detected with skin features and eyes located with Support Vector machines. Belhumer proposed a local detector which employs Bayesian model and SVMs [4]. Sagonas et al. [5] proposed 300 faces with the purpose of evaluation of performance of distinct systems on a newly collected dataset. In [6] the authors have proposed strategies in order to update a model that is trained by a cascade of Regressors. Wu and Ji have done a comprehensive literature survey on various facial landmark detection systems and have effectively compared the performance of each of the discussed strategies [7]. A sparse Bayesian extreme learning machine (SBELM) have been proposed in [8] for real-time face detection.

3 Part 1: Pre-processing and Face Detection

3.1 Step 0: Detection of Faces Using Classifier

We often wonder how the 'photos' application on an iphone segregates the pictures based on the people in it. Certain cameras focus on a specific person's face. These are nothing but the implications of Facial Detection which identifies the human face automatically in pictures. The first and foremost task of Face Detection is to establish a demarcation between the human faces and various other things. Object detection is performed using Haar feature-based classifiers, a method proposed in [9]. Numerous negative and positive images are employed to train a cascade of classifiers, where features are clubbed into stages. Specific features of human face, eyes, nose and mouth are used to distinguish the faces from other objects. The negative samples are all the images that contain the objects that are not to be detected. In OpenCV the file path has to be mentioned for negative samples. Positive samples just need to be converted to binary format. Once the training set is in place, load one of the pre-trained face detectors of OpenCV in a directory.

Import all the required libraries- matplotlib, cv2 etc. Upload and display one test image as well. The default colour channels order, as perceived by OpenCV is the BGR format (Blue, Green, Red). However, a majority of images are available in the RGB format. This calls for swapping the Blue and Red channels. cvtColor function is used for the transformations of formats- Colour-grayscale, RGB- HSV, BGR-RGB. Plot the image using subplots to specify a size and title.

The pattern of pixel intensities is used by the face detector. The color image is converted to grayscale. The trained architecture employed for face detection takes four parameters. They are scaleFactor, minNeighbors, minSize (minimum object size that has to be detected) and maxSize (maximum possible object size). Extract it from the xml file. Make a copy of the original image to draw the face detections in the form of a bounding box. The number of faces detected is also printed. Number of faces detected is 13.

Fig. 1. Detected images indicated by bounding boxes

3.2 Step 2: De-noise an Image for Better Face Detection

It is not always practical to have images free of noise. But for training a deep learning network, it should be free of visual noise. Noise is aberrant pixels, that do not represent the image in the right way. The process of cleaning the image is called pre-processing. This includes cleaning phases like blurring, de-noising, color transformations present in OpenCV. For a comparison, consider the Fig. 1 and add noise to it to analyse the facial detection on a noisy image. Here in this work, we have added noise sampled randomly from a Gaussian distribution. Plot the noisy image and converting it back to uint8 format. Perform face detection on this noisy image by indicating a red bounding box around the detected faces. The number of faces detected in noisy image is 12 (Fig. 2).

The noisy image illustrated above has to be de-noised for accurate detection of faces. OpenCV has a built-in functionality called "fastN1MeansDenoisingColored". The parameters accepted by this function are src (8-bit stream of image), dst (Output

Fig. 2. Noisy image with face detections

image), h_luminance (The strength of the filter is regulated by the h value), photo_render (Parameter that regulates colored noise removal), search_window (window to compute weighted average for given pixel), stream (For asynchronous invocations). The function transforms the given image to CIELAB colorspace. It then individually denoises the L and AB components. L stands for lightness, A stands for green-red and B stands for blue-yellow components. The denoising happens according to the h parameters. Perform face detection on the denoised image (Fig. 3).

Fig. 3. De-noised image with face detection

3.3 Edge Detection

An image many consist of lots of redundant information which may not be necessary for a particular application. Edge detection reduces the amount of data that is needed to be processed. It is an algorithm that has numerous stages. The first stage involves reduction of noise. One could employ the usage of a Gaussian filter. The edge gradient and direction for each pixel can be found out as follows.

$$Edge\_Gradient(G) = (G_x^2 + G_y^2)^{1/2} \tag{1}$$

$$Angle = \tan^{-1}(G_y/G_x) \tag{2}$$

G_x indicates the horizontal direction and G_y indicates the vertical direction. Gradient direction and edges are orthogonal to each other. Every pixel that is encountered is checked to see if it is the local maximum in its surroundings. It is checked for the local maximum if its direction is perpendicular to gradient direction. The next stage of Edge Detection is Hysteresis Thresholding where Minimum and maximum threshold values are determined. If a pixel value is greater than the maximum value, it is an edge. If the pixel value is lesser than the minimum value, then it is not considered to be an edge and therefore it is discarded. Any value within the range is looked up for the connectivity.

Convert an image to Grayscale and perform Canny Edge Detection. Dilate the image to amplify the edges. Plot the edge detected image (Fig. 4).

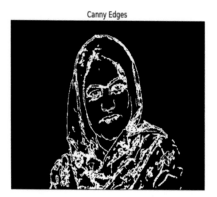

Fig. 4. Edge detected image

It is evident from the figure above that the Edges are not sharp with many local structures thickening the edges. This can be achieved by blurring the original image by the filter2D present in OpenCV. A 5 × 5 matrix with [1] * (1/25) is placed above a pixel. The central pixel is replaced by the average value of 2 pixels below this kernel (Fig. 5).

Canny Edges

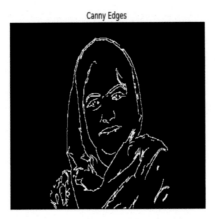

Fig. 5. Canny edge detection on a blurry image

The above mentioned filter itself can be used for hiding the identity of individual. The identity hiding pipeline consists of Face Detection and blurring the region within the bounding box. The blurring happens within the confinements of the parameters of the bounding box.

4   PART 2: Train a CNN to Detect Facial Keypoints

4.1   Step 5: Create a CNN to Identify Facial Keypoints

Initially the Convolutional Neural Network is built for a miniscule dataset of human faces that are cropped. Then the expanse of the network is extended to more generalized images. Human faces have certain characteristics that are specific only to them. These are coined the term "Facial Keypoints". Each Human face has 15 Keypoints. The facial keypoint detection is a Regression problem. For the training purpose, the dataset picked has thousands of 96 × 96 grayscale images commensurately with KeyPoints. The placement of the keypoints is within the x-y place. The ordered pair determines the position. The training data is loaded into the directory (Fig. 6).

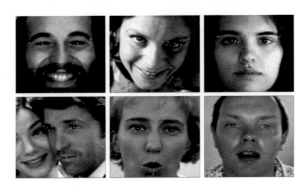

Fig. 6. Visualization of the training set

There were multiple steps involved in reaching an appropriate network. Initially an attempt was made to try a simple three-layer convolutional neural network with decreasing number of nodes at each layer and relu activations. 32, 16 and 8 nodes were given in each of the three convolutional layers. The network was incapable of reaching an optimal value of validation loss. The validation loss kept bouncing around while the training loss gradually decreased. After heuristically finding the best optimizer, the validation loss converged to an optimum value. We have also added a dropout between all the convolutional layers to prevent overfitting. However, this resulted in an increase in the validation loss. When dropout was added to only one layer, there was an improvement in validation loss, suggesting that the extra dropout was not letting the network train to optimum capacity.

As overfitting of network was observed, 2-layer networks and 3-layer networks were experimented with less nodes. Finally, we arrived at the current architecture containing 16, 8 and 4 nodes in each layer (Fig. 7).

```
Layer (type)                 Output Shape              Param #
=================================================================
conv2d_26 (Conv2D)           (None, 94, 94, 16)        160
_____
max_pooling2d_25 (MaxPooling (None, 47, 47, 16)        0
_____
activation_16 (Activation)   (None, 47, 47, 16)        0
_____
dropout_16 (Dropout)         (None, 47, 47, 16)        0
_____
conv2d_27 (Conv2D)           (None, 45, 45, 8)         1160
_____
max_pooling2d_26 (MaxPooling (None, 22, 22, 8)         0
_____
activation_17 (Activation)   (None, 22, 22, 8)         0
_____
conv2d_28 (Conv2D)           (None, 20, 20, 4)         292
_____
max_pooling2d_27 (MaxPooling (None, 10, 10, 4)         0
_____
flatten_11 (Flatten)         (None, 400)               0
_____
dense_11 (Dense)             (None, 30)                12030
=================================================================
Total params: 13,642.0
Trainable params: 13,642.0
Non-trainable params: 0.0
```

Fig. 7. Summary of the model

Compilation of a Keras model requires the optimizer as a parameter. Various optimizers were tested: 'sgd', 'rmsprop', 'adagrad', 'adadelta' etc. They were tested by compiling a model with each optimizer and training it against the full dataset of 10 epochs. The final validation loss generated by each optimizer was compared and the best one was chosen. The optimizer chosen was the Robust 'adadelta' in which the learning continues in spite of many updates having been done (Fig. 8).

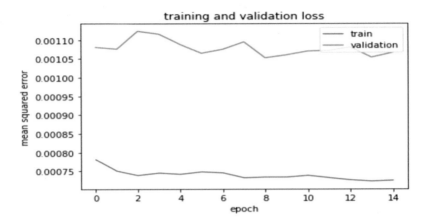

Fig. 8. The plots of training and validation losses

There is clearly some overfitting in the above model as the training loss converges near 0.00072 and the validation loss converges near 0.0011. This was reduced by using dropout. Initially an attempt was made to reduce the number of hidden layers. This led to the validation loss converging at 0.0014. Therefore, we decreased the number of nodes at each layer, thereby reducing the number of parameters to train. A subset of test predictions is given below after running the CNN model (Fig. 9).

Fig. 9. Visualizing the subset of test predictions

5   PART 3: Completing the Pipeline

Complete the pipeline by combining Part 1 and Part 2. If a color image is given, the faces can be detected automatically using OpenCV, predict the facial keypoints in each detected face, indicate the predicted keypoints on each face detected.

The expanse of the keypoint detection is extended to a color image of any size. Here, the grayscale image of arbitrary size has to be normalized during training to the interval $\{-1, 1\}$ before feeding it to the keypoint detector. To be shown correctly, on the original image, the output keypoints from the detector need to be shifted from $\{-1, 1\}$ to the height and width of the detected face.

6 Results

Fig. 10. The final result with keypoint detection

The final result has been obtained after extracting the pre-trained face detector, detecting the faces in the image and indicating facial keypoints on the image (Fig. 10).

7 Conclusion

In this work we have successfully combined Computer vision techniques and Deep learning algorithms to build a Facial Keypoint Detection. First, a human face classifier was built and tested on noisy and noiseless images. An eye detector was also built and indicated by using a bounding box around the eyes. After that, a CNN model was built for keypoint detection. The model with the best validation loss was finalized after varying the nodes and optimizers. At the end, we built completed the pipeline by combining the first two parts. The pipeline was tested on an image with two faces. The Faces were detected and facial keypoints were marked accurately as indicated in the results section above.

Acknowledgment. We would like to thank Dr. Suresh D, Department of Electronics and Communication, RNS Institute of Technology for his technical and writing assistance.

References

1. Valstar, M., Martinez, B., Binefa, X., Pantic, M.: Facial point detection using boosted regression and graph models. In: Proceedings of Conference on Computer Vision and Pattern Recognition, San Francisco, CA, USA, pp. 2729–2736 (2010)
2. Ding, L., Martinez, A.M.: Features versus context: an approach for precise and detailed detection and delineation of faces and facial features. IEEE Trans. Pattern Anal. Mach. Intell. **32**(11), 2022–2038 (2010)
3. Arca, S., Campadelli, P., Lanzarotti, R.: A face recognition system based on automatically determined facial fiducial points. Pattern Recogn. **39**, 432–443 (2006)
4. Belhumeur, P.N., Jacobs, D.W., Kriegman, D.J., Kumar, N.: Localizing parts of faces using a consensus of exemplars. In: Proceedings of Conference on Computer Vision and Pattern Recognition, Providence, RI, USA, pp. 545–552 (2011)
5. Sagonas, C., Tzimiropoulos, G., Zafeiriou, S., Pantic, M.: 300 faces in-the-wild challenge: the first facial landmark localization challenge. In: Proceedings of the IEEE International Conference on Computer Vision Workshops, pp. 397–403 (2013)
6. Asthana, A., Zafeiriou, S., Cheng, S., Pantic, M.: Incremental face alignment in the wild. In: Proceedings of the IEEE Conference on Computer Vision and Pattern Recognition, pp. 1859–1866 (2014)
7. Wu, Y., Ji, Q.: Facial landmark detection: a literature survey. Int. J. Comput. Vis. Received Nov 2016. Accepted Apr 2018
8. Vong, C.M., Tai, K.I., Pun, C.M., Wong, P.K.: Fast and accurate face detection by sparse Bayesian extreme learning machine. Neural Comput. Appl. **26**(5), 1149–1156 (2015)
9. Turk, M.A., Pentland, A.P.: Face recognition using eigenfaces. In: Proceedings of the International Conference on Pattern Recognition, pp. 586–591 (1991)
10. Negi, R.S., Garg, R.: Face recognition using hausdroff distance as a matching algorithm. Int. J. Sci. Res. **4**(9) (2016)
11. Dewan, M.A.A., Qiao, D., Lin, F., Wen, D.: An approach to improving single sample face recognition using high confident tracking trajectories. In: Canadian Conference on Artificial Intelligence, pp. 115–121. Springer, Cham (2016)
12. Shi, X., Wu, J., Ling, X., Zheng, Q., Pan, X., Zhao, Z.: Real-time face recognition method based on the threshold determination of the positive face sequence. In: Proceedings of the 22nd International Conference on Industrial Engineering and Engineering Management 2015, pp. 125–136. Atlantis Press (2016)
13. Goswami, G., Vatsa, M., Singh, R.: Face recognition with RGB-D images using kinect. In: Face Recognition Across the Imaging Spectrum, pp. 281–303. Springer, Cham (2016)
14. Dahal, B., Alsadoon, A., Prasad, P.W.C., Elchouemi, A.: Incorporating skin color for improved face detection and tracking system. In: 2016 IEEE Southwest Symposium on Image Analysis and Interpretation (SSIAI), pp. 173–176. IEEE (2016)
15. Aghaei, M., Dimiccoli, M., Radeva, P.: Multi-face tracking by extended bag-of-tracklets in egocentric photostreams. Comput. Vis. Image Underst. **149**, 146–156 (2016)

Implementation of Harmonic Oscillator Using Xilinx System Generator

Darshana N. Sankhe[✉], Rajendra R. Sawant, and Y. Srinivas Rao

Department of Electronics and Tele-communication Engineering,
Sardar Patel Institute of Technology, Mumbai University,
Andheri, Mumbai 400058, Maharashtra, India
darshana.sankhe@djsce.ac.in, rrsawant@ieee.org,
ysrao@spit.ac.in

Abstract. Advances in technology demands replacement of all analog blocks by its counter digital blocks. Harmonic oscillator (HO) is a mathematical implementation, facilitates generation of sinusoidal waveform, with adjustable frequency, amplitude and harmonics of fundamental frequency with different phase shifts. It is a fundamental block for many communication systems and power control applications.

This paper presents implementation of harmonic oscillator using different discretization techniques to test its stability. Most stable hybrid method is then used for generation of pulse width modulation (PWM) pulses, used in power control applications. Further using HO, an Amplitude Shift Keying (ASK) modulation system is implemented to witness its communication applications. HO simulation model is developed using Xilinx System Generator (XSG) in Matlab Simulink. Its VHDL simulation and synthesis is done to verify the functionality and identify the implementation resources required by various discretization methods. Hardware implementation is tested on Spartun-7 Field Programmable Gate Array (FPGA) platform.

Keywords: Harmonic oscillator · Discretization methods · PWM

1 Introduction

Technologists are working on developing a mathematical model of a sinusoidal wave oscillator, generating required amplitude, frequency and phase signal accurately. HO facilitates generation of sinusoidal waveforms with adjustable amplitude, frequency and phase shift. These sinusoidal are used as signal source in simulation and implementation of digital modulation techniques and in power control applications.

Analogy of HO is derived from pendulum set-up placed in vacuum chamber as shown in Fig. 1. Mathematically simple harmonic motion of this pendulum is represented by a differential equation,

$$\frac{d\theta}{dt} + \frac{g}{L}Sin(\theta) = 0 \tag{1}$$

© Springer Nature Switzerland AG 2020
A. Abraham et al. (Eds.): ISDA 2018, AISC 941, pp. 503–512, 2020.
https://doi.org/10.1007/978-3-030-16660-1_49

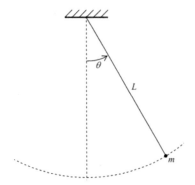

Fig. 1. Simple pendulum motion

Where, θ represents an angular position of a pendulum, L is the length of pendulum and g is the gravitational acceleration. Solution to this differential equation can be obtained by applying an initial force to a pendulum that leads to self-sustainable oscillation, provided the chamber is free from any air friction.

Similar idea of HO is originated from a mass-spring model shown in Fig. 2. Here a block with mass 'm' is tied through a spring having spring constant k, to a rigid support. Assume the spring has zero losses and block is resting on a frictionless surface.

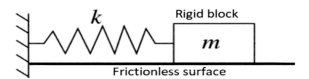

Fig. 2. Mass-spring model

Then any momentarily movement of block in one direction, results into continuous back and forth movement of the block. This change in position and velocity of the block resembles sine and cosine function of time, results into self-sustained oscillations with frequency $f = \frac{1}{2\pi}\sqrt{\frac{k}{m}}$. If no initial force is applied, then system will remain in a rest position. Following this analogy, HO's differential equation is written as:

$$\frac{dx}{dt} = \omega\,Cos(\omega t) \tag{2}$$

$$\frac{dy}{dt} = -\omega\,Sin(\omega t) \tag{3}$$

Where, $x(t)$ and $y(t)$ are function of independent variable time t, ω is angular frequency (radians/seconds). There exists a unique solution for these differential equations as:

$$x = Sin(\omega t) \qquad y = Cos(\omega t) \qquad (4)$$

Where, $x(0) = 0 \, and \, y(0) = 1$, as initial conditions.

Recently, FPGA based systems are becoming popular for digital control and communication applications, due to its high speed, real time processing capabilities, low power consumption and reconfigurable capabilities. Parally, evolution of FPGA hardware and its simulation tools has enabled, exploring different system implementation, on FPGA platform such as phase measurement [1], motor control [2], and impedance measurement [3, 4], and PID controller [5]. HO, being fundamental block can be used in implementation of photovoltaic inverter [6], phase lock loop [7]. Power electronic converters using Xilinx System Generator (XSG) are analyzed in [8, 9].

Main focus of this paper is discrete time implementation of HO, an analog block that generates sinusoidal, on a digital platform. Any analog block implementation on digital platform is challenging task due to its varied nature of hardware resources requirement and speed. Various analog blocks are already available in DSP library. Author aims to bring it on reconfigurable platform, FPGA because discretization errors are not accumulated in FPGA implementation.

This paper is organized as follows: Sect. 2 describes different HO discretization methods and their implementation using XSG. A simulation result defines stability of the system. VHDL simulation and FPGA implementation is described in Sect. 3. Sample HO based applications are elaborated in Sect. 4 while Sect. 5 concludes the paper.

2   Harmonic Oscillator Discretization Methods

Discretization method facilitates conversion of a continuous time system in to its equivalent discrete time. Different discretization techniques discussed in literature are Euler's Explicit and Implicit integration methods, Bilinear transformation, Hybrid method, Trapezoidal method etc.

2.1   Euler's Explicit Integration Method

Discretization of HO's differential Eqs. (2) and (3) using this method results into:

$$x(n+1) = x(n) + h * y(n) \qquad (5)$$

$$y(n+1) = y(n) - h * x(n) \qquad (6)$$

With same initial conditions $x(0) = 0, y(0) = 1$. Here h is a constant, $h = \omega * Ts = 2\pi f * Ts$ and f is required frequency of oscillation, Ts is time interval of discrete time samples. These equations are implemented using XSG in MATLAB Simulink, shown in Fig. 3. Here frequency of sinusoidal output is decided by constant h. Mux block sets initial condition, $y(0) = 1$, to start the iterations. Add, subtract, multiplier and delay blocks are used to generate required sine and cosine waveforms. Simulation results are shown in Fig. 4. This discretization technique produces continuously increasing amplitude oscillations. Hence implementation using this discretization method is unstable, as it doesn't produce self-sustained oscillations. Detail stability analysis of HO implementation is carried out in [10].

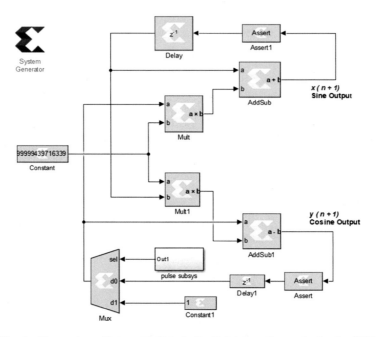

Fig. 3. Harmonic oscillator with Euler's Explicit integration method using XSG

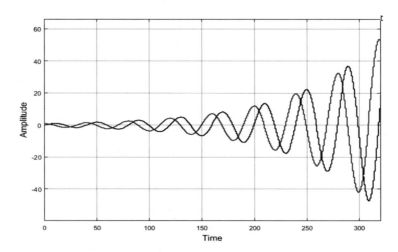

Fig. 4. Simulation results with Euler's Explicit integration method

2.2   Euler's Implicit Integration Method

Differencial Eqs. (2) and (3) can be discretized using above method results into following set of equations:

$$x(n+1) = x(n) + h * y(n+1) \tag{7}$$

$$y(n+1) = y(n) + h * x(n+1) \tag{8}$$

With initial condition $x(0) = 0 \, and \, y(0) = 1$.

Implementation of above equations results into continuously decreasing amplitude oscillations as shown in Fig. 5. Again this system does not produce self-sustainable oscillations, hence unstable.

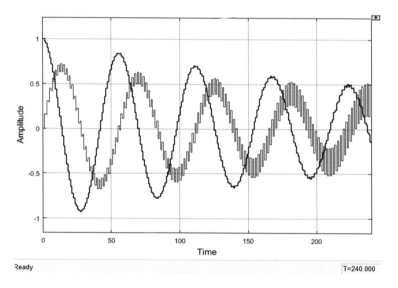

Fig. 5. Simulation results with Euler's Implicit integration method

2.3   Bilinear Transformation Method

This is one of the most popular technique, of converting continuous time signal into discrete time. Here discrete time domain equations of HO are written as,

$$x(n+1) = \frac{1 - \frac{h^2}{4}}{1 + \frac{h^2}{4}} * x(n) + \frac{h}{1 + \frac{h^2}{4}} * y(n) \tag{9}$$

$$y(n+1) = \frac{-h}{1 + \frac{h^2}{4}} * x(n) + \frac{1 - \frac{h^2}{4}}{1 + \frac{h^2}{4}} * y(n) \tag{10}$$

Implementation and simulation of these equations results into self-sustainable oscillations as shown in Fig. 6. Thus it is a stable system, but involves more computations. Here computations can be optimized by pre-calculating the constants and substituting them in Eqs. (9) and (10) respectively.

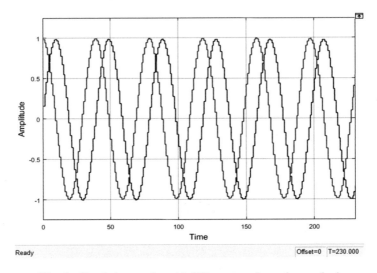

Fig. 6. Simulation results with Bilinear transformation method

2.4   Hybrid Method

This is preferred technique over all discretization methods, due to its less computation and stable output. Equations (11) and (12) represent discrete time form of HO using this approach.

$$x(n+1) = x(n) + h * y(n) \tag{11}$$

$$y(n+1) = y(n) - h * x(n+1) \tag{12}$$

Matrix form representation of these equations is written as:

$$\begin{bmatrix} x(\dot{n}+1) \\ y(\dot{n}+1) \end{bmatrix} = \begin{bmatrix} 1 & h \\ -h & 1-h^2 \end{bmatrix} \cdot \begin{bmatrix} x(n) \\ y(n) \end{bmatrix} \tag{13}$$

Here, A is a system matrix, given by:

$$A = \begin{bmatrix} 1 & h \\ -h & 1-h^2 \end{bmatrix} \tag{14}$$

To evaluate the Eigenvalues of this system, write a characteristic equation as:

$$|Z.I - A| = 0 \tag{15}$$

Where, Z represents discrete-time eigenvalues, I is an identity matrix and A is a system matrix. Evaluating above determinant gives a quadratic equation as:

$$Z^2 + (h^2 - 2) * Z + 1 = 0 \tag{16}$$

Solving above quadratic equation gives two roots of this equation as:

$$Z = \left(1 - \frac{h^2}{2}\right) \pm \sqrt{1 - \frac{(h^2 - 2)^2}{4}} \tag{17}$$

These roots are also the eigenvalues of the hybrid system. Here, it is observed that present system has complex poles, lying on the unit circle in Z-plane. Thus above discrete time system is stable and is capable of producing self-sustained oscillations. Implementation and simulation of these Eqs. (11) and (12) generates self-sustainable oscillations as shown in Fig. 7.

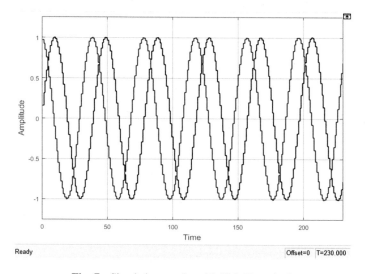

Fig. 7. Simulation results with Hybrid method.

3   VHDL Simulation and FPGA Implementation Results

It is a known fact that programming FPGA is not so easy. Writing Very High Speed Integrated Circuit Hardware Description Language (VHDL) code and debugging for required functionality is time consuming task. Thus XSG using MatLab Simulink is preferred for generation of VHDL codes. It reduces the design time of the system. Here HO implementation using Hybrid method is chosen for VHDL simulation, as both the Euler's methods are unstable while Bilinear Transformation is computationally intense method. VHDL code is generated by setting implementation strategy "Vivado Implementation Defaults". Behavioral simulation results using Vivado Design Suit 2017.4 is shown in Fig. 8.

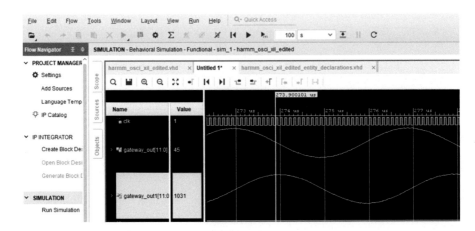

Fig. 8. Simulation of HO using Vivado Design Suit simulator

RTL implementation on XC70Z010 Zinc device and synthesis is done for hybrid method HO implementation and resources used are shown in Table 1. As resources utilized are less, this can be implemented on smaller and cheaper FPGA platform.

Table 1. FPGA resource utilization summary for HO implementation

Resources	Used	Available resources	% Utilization
LUTs	59	17600	0.34
FLIP FLOPS	26	35200	0.07
DSP	2	80	2.5
IO	25	100	25
BUFS	1	32	3.13

4   Applications of Harmonic Oscillator

Harmonic oscillator is a fundamental block, commonly used for generation of 3-phase signals, PWM control pulses, ramp signals, in Phase Lock Loop (PLL) systems and various digital modulation systems. Any arbitrary waveform can be generated using HO as a basic block and different Fourier coefficients a_0, a_n, b_n. Implementation of park's transformation, reference frame transformation, orthogonal transformation and 3-phase PLL all uses HO. PWM generation and ASK modulation system are discussed here as a case study.

Digital PWM is commonly used control technique in power converters, in motor drives and power electronics applications. Figure 9 shows 10% and 40% duty cycle PWM pulses generated using HO subsystem and comparator blocks. ASK digital modulation system is implemented using HO as a basic block and XSG simulation results are shown in Fig. 10.

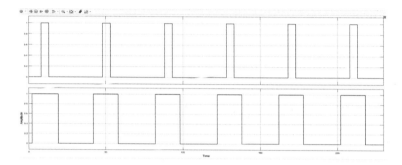

Fig. 9. Center aligned PWM control pulses

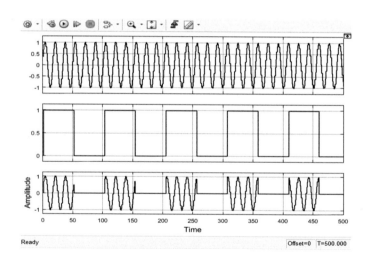

Fig. 10. Simulation results of ASK modulation system

To verify VHDL simulation results experimentally, Spartan7 xc7s100-1fgga676 FPGA hardware platform is chosen. Using Xilinx Vivado simulator, logic synthesis, placement and routing is carried out to generate an implementation file i.e. Bit file. Through JTAG interface this file is downloaded on Spartan7 hardware board. Digital bit pattern is verified on the output port of FPGA for stable HO, PWM and ASK system.

5 Conclusion

Harmonic oscillator is an analog block, typically used as signal source in communication systems and different power control applications. Implementation of HO using different discretization techniques is done and stability is analyzed. It is found that hybrid method is most stable method; both the Euler's methods are unstable while Bilinear transformation is computationally complex. VHDL synthesis and implementation is carried out for hybrid method HO, to verify the functionality and resource utilization. ASK modulation and PWM generation are implemented as an application using HO block. Finally hardware implementation of PWM is tested on Spartan7 xc7s100-1fgga676 FPGA platform.

References

1. Mitra, J., Nayak, T.K.: An FPGA-based phase measurement system. IEEE Trans. Very Large Scale Integr. (VLSI) Syst. **26**(1), 133–142 (2018)
2. Diao, L., Tang, J., Loh, P.C., Yin, S., Wang, L., Liu, Z.: An efficient DSP–FPGA-based implementation of hybrid PWM for electric rail traction induction motor control. IEEE Trans. Power Electron. **33**(4), 3276–3288 (2018)
3. Jiménez, O., Barragán, L.A., Navarro, D., Lucía, O., Artigas, J.I., Urriza, I.: FPGA-based real-time calculation of the harmonic impedance of series resonant inductive loads (2010)
4. Jimenez, O., Lucia, O., Barragan, L.A., Navarro, D., Artigas, J.I., Urriza, I.: FPGA-based test-bench for resonant inverter load characterization. IEEE Trans. Ind. Inform **9**(3), 1645–1654 (2013)
5. Sreenivasappa, B.V., Udaykumar, R.Y.: Analysis and implementation of discrete time PID controllers using FPGA. Int. J. Comput. Eng. **2**(1), 71–82 (2010)
6. Attia, H.A., Ping, H.W., Al-Mashhadany, Y.: Design and analysis for high performance synchronized inverter with PWM power control. In: Proceedings of the IEEE Conference on Clean Energy and Technology (CEAT), Langkawi TBD, Malaysia, pp. 265–270 (2013)
7. Rao, Y.S., Iyer, S.: DSP-FPGA implementation of a phase locked loop for digital power electronics. In: Proceedings of the IEEE Region 8 International Conference on Computational Technologies in Electrical and Electronics Engineering (SIBIRCON), Irkutsk, Russia, pp. 665–670 (2010)
8. Selvamuthukumaran, R., Gupta, R.: Rapid prototyping of power electronics converters for photovoltaic system application using Xilinx system generator. IET Power Electron. **7**, 2269–2278 (2014)
9. Mondragon, M., Calderon, E., Hernandez, M., Resendiz, R.: Implementation of high resolution unipolar PWM inverter using Xilinx system generator. In: IEEE (2016)
10. Sawant, R.R., Chauhan, M., Yerramreddy, S.S., Rao, Y.S.: Harmonic oscillator: a classical fundamental building block for modern electric power control. In: IEEE National Power Electronics Conference (NPEC), Pune India, December 2017

Image Classification Using Deep Learning and Fuzzy Systems

Chandrasekar Ravi[✉]

National Institute of Technology Puducherry, Karaikal, India
chand191987@gmail.com

Abstract. Classification of images is a significant step in pattern recognition and digital image processing. It is applied in various domains for authentication, identification, defense, medical diagnosis and so on. Feature extraction is an important step in image processing which decides the quality of the model to be built for image classification. With the abundant increase in data now-a-days, the traditional feature extraction algorithms are finding difficulty in coping up with extracting quality features in finite time. Also the learning models developed from the extracted features are not so easily interpretable by the humans. So, considering the above mentioned arguments, a novel image classification framework has been proposed. The framework employs a pre-trained convolution neural network for feature extraction. Brain Storm Optimization algorithm is designed to learn the classification rules from the extracted features. Fuzzy rules based classifier is used for classification. The proposed framework is applied on Caltech 101 dataset and evaluated using accuracy of the classifier as the performance metric. The results demonstrate that the proposed framework outperforms the traditional feature extraction based classification techniques by achieving better accuracy of classification.

Keywords: Image classification · Deep learning · Fuzzy systems

1 Introduction

The discipline of Artificial Intelligence that describes the manner in which knowledge from videos or images to computers is termed as Computer Vision. This inter disciplinary branch enables the machines to see the real world just as a human eye can do in co-ordination with the brain [1–3]. It extracts the features from the images, analyses the images using the extracted features and finally derives useful knowledge automatically. This is all totally done with the help of mathematical techniques in the background [4]. The source of information for this computer vision system in often images or videos from single or several cameras. The data could also be multi dimensional like the data from medical imaging devices [5]. These systems finds several applications in various domains like identification (fingerprint, iris, face, voice recognition and so on), manufacturing industry which can use it to detect faults, robotics for building robots for various purposes, surveillance, medical image processing (cancer, neurological disorder detection and so on), unmanned vehicles and so on. The objective of such systems would be to decide whether the target image belongs to a particular class or not. This

A. Abraham et al. (Eds.): ISDA 2018, AISC 941, pp. 513–520, 2020.
https://doi.org/10.1007/978-3-030-16660-1_50

decision is made by using several steps like image acquisition, preprocessing, feature extraction, segmentation, image recognition, image registration and decision making. In this paper, few of the aforementioned steps are used for image classification. The remaining sections of this paper are organized as follows. Section 2 discusses literature review. Section 3 proposes the novel framework. Section 4 discusses the findings and finally conclusion summarizes the entire paper.

2 Literature Review

Table 1, summarizes the literature based on image classification using convolution neural network for feature extraction. The review clearly highlights that the researchers are recently focusing on feature extraction using convolution neural network. The benefit of this approach is that huge volume and wide variety of images can be used to train the convolution neural network to extract the features. Thus the extracted features would be of good quality and thus result in a better classifier model. Also the enormous time and hardware required to learn from the huge volume of images can be drastically reduced by employing a pre-trained convolution neural network for feature extraction. Fuzzy logic based classification is becoming popular recently due to its capacity to handle uncertainty and produce highly interpretable knowledge.

Table 1. Summary of literature.

Journal & Year	Title	Features
Neurocomputing (2018)	Aurora image search with contextual CNN feature [6]	Reduced mis-classification, better precision
IEEE Transactions on Cybernetics (2018)	Feature extraction for classification of hyperspectral and LiDAR data using Patch-to-Patch CNN [7]	Precise representation, multi-scale features between two diverse sources are merged
IEEE Journal of Selected Topics in Applied Earth Observations and Remote Sensing (2018)	Spectral–spatial feature extraction for HSI classification based on supervised hypergraph and sample expanded CNN [8]	Capture complex relationship
IEEE Transactions on Geoscience and Remote Sensing (2018)	Supervised deep feature extraction for hyperspectral image classification [9]	Extract precise characteristic features
IEEE Access (2018)	No-reference stereoscopic image quality assessment using convolutional neural network for adaptive feature extraction [10]	Quality features extracted

(*continued*)

Table 1. (*continued*)

Journal & Year	Title	Features
IEEE Transactions on Cognitive and Developmental Systems (2018)	Zero-shot image classification based on deep feature extraction [11]	Over-fitting is prevented
IEEE Transactions on Image Processing (2018)	Retrieval oriented deep feature learning with complementary supervision mining [12]	Semantic aware feature are extracted.
IEEE Access (2018)	Deep convolution neural network and autoencoders-based unsupervised feature learning of EEG signals [13]	Convergence is fast and training times reduces
IEEE Geoscience and Remote Sensing Letters (2018)	Remote sensing image registration using convolutional neural network features [14]	Extract precise characteristic features
IEEE Access (2018)	Multi-temporal remote sensing image registration using deep convolutional features [15]	Extracts robust features
IEEE Access (2018)	Iris Recognition with off-the-shelf cnn features: a deep learning perspective [16]	Complex characteristics of image is extracted
IEEE Transactions on Smart Grid (2018)	Convolutional neural networks for automatic state-time feature extraction in reinforcement learning applied to residential load control [17]	Hidden discriminative features are extracted
Optik (2018)	Optimized CNN based image recognition through target region selection [18]	Accuracy in increased
Computer Vision and Image Understanding (2018)	Hierarchical semantic image matching using CNN feature pyramid [19]	Better performance with dense images

3 Proposed Framework

The proposed framework is depicted in Fig. 1. The Caltech-101 dataset [20] contains images of 101 categories of objects. ResNet [21] is the Convolutional Neural Networks trained on ImageNet dataset, that has 1000 categories of object and 1.2 million images. When the Caltech-101 dataset is given as input to the Resnet-50, the features are extracted. Then, Caltech-101 is partitioned into training and test datasets which comprises the features extracted using Resnet-50. The training dataset is given as input to the Brain

Storm optimization [22] algorithm which derives the optimal rule base for the image classification. The test set is then classified using the Fuzzy Inference System and the optimal rule base.

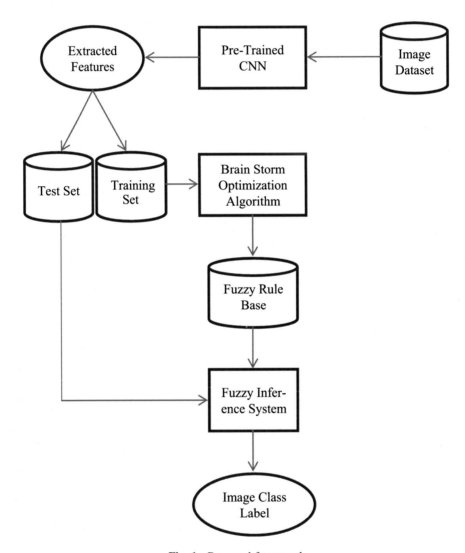

Fig. 1. Proposed framework

The Brain Storm optimization [22] algorithm, described below, is customized according to the proposed framework, to produce optimal rule base. The brainstorm optimization algorithm is preferred than other algorithms like Genetic algorithm, Particle Swarm Optimization and so on. This is because brainstorm optimization algorithm is based on brain storming activity done by humans, whereas other algorithms are based on the social behavior of birds, ants, etc. in Particle Swarm Optimization, Ant Colony Optimization respectively. Since humans are considered to be the superior most in the ecology, the brainstorm optimization algorithm is assumed to produce the best results.

Algorithm : BSO
Input   : Training dataset from Resnet-50, Probability constants (P_{5a}, P_{6b}, P_{6c}), no. of ideas 'n', constants 'e' and 'k', termination criteria (maximum fitness value)
Output   : Optimal rule base for Fuzzy System Procedure:
  Begin
    Initialize 'n' ideas using the proposed representation customized for the proposed frameworkApply clustering
    Repeat
      Evaluate fitness of ideas using the proposed fitness function
      Generate new ideas using P_{5a}, P_{6b}, P_{6c}
      Replace the worst ideas with better ones
    Until termination criteria
    Return the best idea
  End

The individuals in the population of the brainstorm optimization algorithm are called as ideas. The ideas are represented as vectors for easy of computation. The idea vector consists of 'm + 2' elements, where 'm' represents the number of features extracted by Resnet-50, $(m + 1)^{th}$ element represents image class and the $(m + 1)^{th}$ element represents the AND or OR method used by the Fuzzy Inference System. Initial population consists of 'n' ideas.

The fitness function for the brainstorm optimization algorithm to generate optimal rule base is designed based on two factors, namely, length of the rule and adaptiveness of the rules to the training dataset. Adaptiveness of the rules is described based on how well the rules match the training dataset. Generally optimal rules are those which are having small length and more adaptive. Hence the fitness function is inversely proportional to rule length and directly proportional to adaptivity. The below Eq. (1) describes the proposed fitness function.

$$F = (w * m/l) + (w * r/p) \qquad (1)$$

Where, 'w' is a constant deciding the weightage of the length of the rule and adaptiveness factor. Generally 'w' takes the value 0.5 indicating that both the factors are of equal weightage. 'm' represents the number of features extracted by Resnet-50, 'l' represents the length of the rule generated by brain storm optimization algorithm, 'r' represents the number of rules matching the training dataset and 'p' represents the total number of instances in the training dataset.

The 'e' and 'k' constants, which ranges between 0 to 1, are experimentally decided. The input probabilities P_{5a}, P_{6b}, P_{6c} are randomly chosen between 0 to 1. Initially 'n' ideas are generated and clustered into groups. The cluster center is decided based on the fitness value of the ideas in the cluster. New ideas are generated and worst ideas are replaced with them. This is repeated until the maximum fitness value is achieved for the cluster centers. These cluster centers form the optimal rule base for the proposed framework.

4   Results and Discussions

The features are extracted from Caltech-101 dataset using a traditional Local Binary Pattern (LBP) [23] approach and classified using traditional classifiers like support vector machine, naïve bayes, decision tree and k-nearest neighbours. Then the features are extracted from Caltech-101 dataset using Resnet-50 and classified using Fuzzy Inference System.

The 'e' and 'k' values for the brain storm optimization algorithm are experimentally decided. Figure 2 depicts that the average accuracy of classification is constant for 'e' values in the range 0.3 to 0.7 and Fig. 3 depicts that the average accuracy of classification is constant for 'k' values in the range 10 to 30. Thus 'e' and 'k' are chosen in this range.

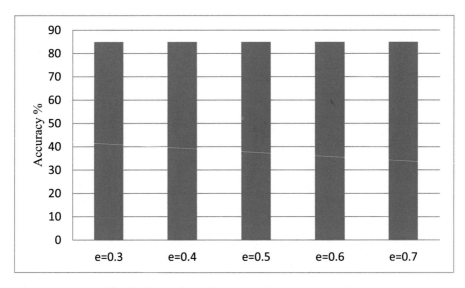

Fig. 2. Comparison of accuracy for various 'e' values

The accuracy of the classifier is defined as the ratio of correctly classified images to total images in the dataset. The k-folds average accuracy of the above combinations of feature extraction and classification in tabulated in Table 2. From the results, it is evident that the proposed framework outperforms the traditional feature extraction techniques based classification.

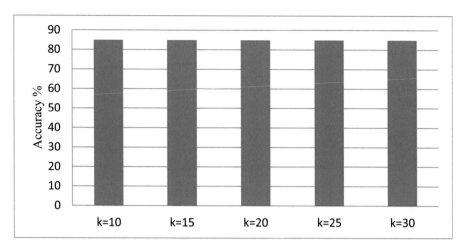

Fig. 3. Comparison of accuracy for various 'k' values

Table 2. Comparison of accuracy.

Classifier	Accuracy	Specificity	Sensitivity
LBP + SVM	61%	65%	81%
LBP + NB	65%	67%	83%
LBP + DT	32%	30%	35%
LBP + KNN	68%	70%	85%
LBP + Fuzzy system	82%	83%	84%
Proposed (Resnet-50 + Fuzzy) system	86%	85%	86%

5 Conclusion

A novel image classification framework has been proposed. The framework employs a pre-trained convolution neural network for feature extraction. Brain Storm Optimization algorithm is designed to learn the classification rules from the extracted features. Fuzzy rules based classifier is used for classification. The proposed framework is applied on Caltech 101 dataset and evaluated using accuracy of the classifier as the performance metric. The results demonstrate that the proposed framework outperforms the traditional feature extraction based classification techniques by achieving better accuracy of classification.

References

1. Ballard, D.H., Brown, C.M.: Computer Vision. Prentice Hall, Upper Saddle River (1982)
2. Huang, T., Vandoni, C.: Computer Vision: Evolution and Promise. 19th CERN School of Computing, pp. 21–25. CERN, Geneva (1996)

3. Sonka, M., Hlavac, V., Boyle, R.: Image Processing, Analysis, and Machine Vision. Thomson, Pacific Grove (2008)
4. http://www.bmva.org/visionoverview
5. Murphy, M.: Star Trek's tricorder medical scanner just got closer to becoming a reality
6. Yang, X., Gao, X., Song, B., Yang, D.: Aurora image search with contextual CNN feature. Neurocomputing **281**, 67–77 (2018)
7. Zhang, M., Li, W., Du, Q., Gao, L., Zhang, B.: Feature extraction for classification of hyperspectral and LiDAR data using patch-to-patch CNN. IEEE Trans. Cybern. (2018). Early Access
8. Kong, Y., Wang, X., Cheng, Y.: Spectral–spatial feature extraction for HSI classification based on supervised hypergraph and sample expanded CNN. IEEE J. Sel. Top. Appl. Earth Obs. Remote Sens. **11**, 4128–4140 (2018). Early Access
9. Liu, B., Yu, X., Zhang, P., Yu, A., Fu, Q., Wei, X.: Supervised deep feature extraction for hyperspectral image classification. IEEE Trans. Geosci. Remote Sens. **56**(4), 1909–1921 (2018)
10. Ding, Y., Deng, R., Xie, X., Xu, X., Zhao, Y., Chen, X., Krylov, A.S.: No-reference stereoscopic image quality assessment using convolutional neural network for adaptive feature extraction. IEEE Access **6**, 37595–37603 (2018)
11. Wang, X., Chen, C., Cheng, Y., Wang, Z.J.: Zero-shot image classification based on deep feature extraction. IEEE Trans Cogn. Dev. Syst **10**(2), 432–444 (2018)
12. Lv, Y., Zhou, W., Tian, Q., Sun, S., Li, H.: Retrieval oriented deep feature learning with complementary supervision mining. IEEE Trans. Image Process. **27**(10), 4945–4957 (2018)
13. Wen, T., Zhang, Z.: Deep convolution neural network and autoencoders-based unsupervised feature learning of EEG signals. IEEE Access **6**, 25399–25410 (2018)
14. Ye, F., Su, Y., Xiao, H., Zhao, X., Min, W.: Remote sensing image registration using convolutional neural network features. IEEE Geosci. Remote Sens. Lett. **15**(2), 232–236 (2018)
15. Yang, Z., Dan, T., Yang, Y.: Multi-temporal remote sensing image registration using deep convolutional features. IEEE Access **6**, 38544–38555 (2018)
16. Nguyen, K., Fookes, C., Ross, A., Sridharan, S.: Iris recognition with off-the-shelf CNN features: a deep learning perspective. IEEE Access **6**, 18848–18855 (2018)
17. Claessens, B.J., Vrancx, P., Ruelens, F.: Convolutional neural networks for automatic state-time feature extraction in reinforcement learning applied to residential load control. IEEE Tran. Smart Grid **9**(4), 3259–3269 (2018)
18. Hao, W., Bie, R., Guo, J., Meng, X., Wang, S.: Optimized CNN based image recognition through target region selection. Optik-Int. J. Light Electron Opt. **156**, 772–777 (2018)
19. Yu, W., Sun, X., Yang, K., Rui, Y., Yao, H.: Hierarchical semantic image matching using CNN feature pyramid. Comput. Vis. Image Underst. **169**, 40–51 (2018)
20. Fei-Fei, L., Fergus, R., Perona, P.: Learning generative visual models from few training examples: an incremental Bayesian approach tested on 101 object categories. In: IEEE CVPR 2004, Workshop on Generative-Model Based Vision (2004)
21. He, K., Zhang, X., Ren, S., Sun, J.: Deep residual learning for image recognition. In: LSVRC 2015 (2015)
22. Ojala, T., Pietikäinen, M., Harwood, D.: Performance evaluation of texture measures with classification based on Kullback discrimination of distributions. In: Proceedings of the 12th IAPR International Conference on Pattern Recognition (ICPR 1994), vol. 1, pp. 582–585 (1994)
23. Shi, Y.: Brain storm optimization algorithm. In: Advances in Swarm Intelligence, LNCS, vol. 6728, pp. 303–309 (2011)

An Evidential Collaborative Filtering Dealing with Sparsity Problem and Data Imperfections

Raoua Abdelkhalek[(✉)], Imen Boukhris[(✉)], and Zied Elouedi[(✉)]

LARODEC, Institut Supérieur de Gestion de Tunis, Université de Tunis,
Tunis, Tunisia
abdelkhalek_raoua@live.fr, imen.boukhris@hotmail.com, zied.elouedi@gmx.fr

Abstract. One of the most promising approaches commonly used in Recommender Systems (RSs) is Collaborative Filtering (CF). It relies on a matrix of user-item ratings and makes use of past users' ratings to generate predictions. Nonetheless, a large amount of ratings in the typical user-item matrix may be unavailable. The insufficiency of available rating data is referred to as the sparsity problem, one of the major issues that limit the quality of recommendations and the applicability of CF. Generally, the final predictions are represented as a certain rating score. This does not reflect the reality which is related to uncertainty and imprecision by nature. Dealing with data imperfections is another fundamental challenge in RSs allowing more reliable and intelligible predictions. Thereupon, we propose in this paper a Collaborative Filtering system that not only tackles the sparsity problem but also deals with data imperfections using the belief function theory.

Keywords: Recommender Systems · Collaborative Filtering ·
User-based · Item-based · Sparsity · Belief function theory ·
Uncertainty

1 Introduction

Over the years, a panoply of RSs [1] has been proposed to help users dealing with the flood of information. Among these approaches, CF has been very promising and widely used in both academia and industry [2]. Neighborhood-based CF approaches represent an important class for CF. They are typically divided into user-based and item-based. To make predictions, user-based method identifies the users sharing the same preferences of the active user. While item-based method tends to find items similar to the item being predicted. Nevertheless, a large amount of ratings corresponding to similar items or similar users may be unavailable due to the sparse characteristic inherent to the rating data. The sparsity problem is known as a major drawback related to CF methods since the users typically rate only a small proportion of the available items. Actually,

© Springer Nature Switzerland AG 2020
A. Abraham et al. (Eds.): ISDA 2018, AISC 941, pp. 521–531, 2020.
https://doi.org/10.1007/978-3-030-16660-1_51

when the ratings matrix is sparse, two users or items are unlikely to have common ratings. Consequently, neighborhood-based CF approaches will make predictions based only on a very limited number of neighbors, which may affect the reliability of the provided predictions.

So far, various recommendation approaches have been proposed aiming to overcome the sparsity problem. Most of the popular ones rely on dimensionality reduction techniques for predicting unprovided ratings. However, the reduction process may lead to a lost of potentially valuable information. Another direction is to add items' contents in the prediction process to alleviate the data sparsity. In such case, additional information regarding items are often required. In contrast to these works, the proposed approach does not reduce the user-item matrix and does not need items' contents to make predictions. It relies on the whole ratings matrix where both items' neighbors and users' neighbors come into play.

Another fundamental and important challenge arising when dealing with RSs is managing the uncertainty pervaded in the final predictions. Providing reliable and intelligible predictions would improve the users' confidence towards the RS and would also ensure their satisfaction. Thereupon, our aim in this paper is not only to deal with the sparsity problem in CF but also to represent the uncertainty of the provided predictions using the belief function theory (BFT) [3–5].

The rest of this paper is organized as follows: Sect. 2 recalls the belief function theory. Section 3 reveals the related works. In Sect. 4, we present our evidential hybrid smoothing-based CF approach. Section 5 discusses the experimental results conducted on a real-world data set. Finally, we conclude the paper in Sect. 6.

2 Belief Function Framework

Under the belief function framework, the frame of discernment Θ represents a finite set of n elementary events E. It contains hypotheses concerning the given problem such that: $\Theta = \{H_1, H_2, \cdots, H_n\}$. The power set of Θ, denoted by 2^{Θ}, is made up of all the subsets of Θ such that: $2^{\Theta} = \{E : E \subseteq \Theta\}$.

In the belief function theory, uncertain knowledge about the elements of the frame of discernment is described by a basic belief assignment (bba). It is defined as a mapping function m from the power set 2^{Θ} to $[0, 1]$ verifying: $\sum_{E \subseteq \Theta} m(E) = 1$.

The value $m(E)$ is called a basic belief mass (bbm). It is considered to be the part of belief exactly assigned to the event E of Θ. The fusion of imperfect data is another crucial task in the BFT owing to its ability to represent more flexible information and to improve decision making. In fact, two bba's m_1 and m_2 can be fused using Dempster's rule of combination which assumes pieces of evidence to be reliable and independent. This rule is defined as follows:

$$(m_1 \oplus m_2)(E) = K. \sum_{F,G \subseteq \Theta : F \cap G = E} m_1(F) \cdot m_2(G) \tag{1}$$

$$where \quad (m_1 \oplus m_2)(\varnothing) = 0 \; and \; K^{-1} = 1 - \sum_{F,G \subseteq \Theta : F \cap G = \varnothing} m_1(F) \cdot m_2(G)$$

To evaluate the reliability of each piece of evidence, a discounting mechanism [5] could be performed on m as follows:

$$m^\delta(E) = (1 - \delta) \cdot m(E), \; for \; E \subset \Theta \; and \; m^\delta(\Theta) \; = \delta + (1 - \delta) \cdot m(\Theta) \quad (2)$$

where the coefficient $\delta \in [0,1]$ reflects the discounting factor.

To make decisions, beliefs held at the credal level may induce at a final stage a pignistic probability measure $BetP(E)$ as follows:

$$BetP(E) = \sum_{F \subseteq \Theta} \frac{|E \cap F|}{|F|} \frac{m(F)}{(1 - m(\varnothing))} \; for \; all \; E \in \Theta \quad (3)$$

3 Related Work

A common problem of CF approaches is their sensitivity to data sparsity. This problem occurs when many users have rated only a few items, or when many items have been rated by a few users. In most cases, data sparsity cannot be avoided. That is why, several attempts have been proposed to deal with the sparsity problem. The most popular approaches that have been used in this context consist in reducing the dimensionality of the user-item ratings matrix. Clustering techniques [6] are among the simpler strategies commonly used to reduce the dimensionality, where the items or the users in the system are first grouped in different clusters. Then, these clusters are considered as a basis of the prediction phase. More advanced approaches rely on statistical techniques such as Principal Component Analysis (PCA) [7] and information retrieval techniques such as Latent Semantic Indexing (LSI) [8]. Based on the reduced matrix, predictions are then performed. The main drawback of these approaches is that useful information can be lost during the reduction process which may yield unreliable predictions. Otherwise, combining CF and content-based techniques is another possible solution to the sparsity problem [9]. It consists on considering also the items' contents in addition to the user-item interaction to generate predictions. Although such hybrid techniques have achieved a considerable success in RSs, they can only be used when information about items' contents is available, which is not always the case. In this work, we are notably interested in neighborhood-based CF where the whole ratings matrix is explored and both users' aspects and items' aspects are taken into account. Furthermore, the uncertain aspect of the final predictions as well as the reliability of the information sources should also be considered. To do so, we opt for the belief function theory, one of the most powerful theories dealing with data imperfections. In fact, item-based CF has been extended under this theory where similar items have been considered as pieces of evidence during the prediction process [10, 11]. An evidential user-based CF has also been proposed where predictions are performed based on user's neighbors [12]. More recently, authors in [13] have proposed to combine the predictions of item-based and user-based using the belief function tools. In such work, both the neighbors of items and users are selected based on the initial user-item matrix to predict ratings. In this paper, we propose a smoothing mechanism which is first adopted based on the ratings predicted by the items' neighbors. The user-side is then considered based on the smoothed user-item matrix and the final predictions are provided to the active user accordingly.

4   Evidential Hybrid Smoothing-Based CF (E-HSBCF)

As previously mentioned, neighborhood-based CF, both user-based and item-based ones, rely only on a very small portion of the ratings which may alter the final results. A large number of ratings in the user-item matrix is generally unavailable which reflects a problem of data sparsity. To tackle this problem, the proposed approach relies first on items' neighbors to fill the sparse ratings matrix. Then, the obtained matrix and the users' neighbors are used to generate predictions. Note that the predictions performed by both similar items and similar users are also combined to derive a more accurate recommendation as illustrated in Fig. 1.

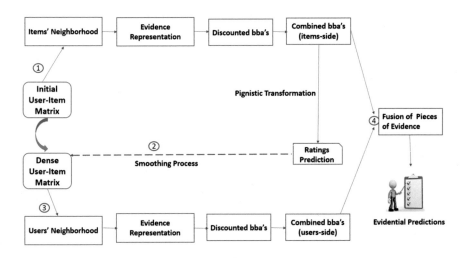

Fig. 1. Evidential Hybrid Smoothing-Based CF (E-HSBCF)

4.1   Notations

Given a set $U = \{U_1, U_2, \cdots, U_M\}$ of M users and a set $I = \{I_1, I_2, \cdots, I_N\}$ of N items, the user-item rating matrix is represented as an $U \times I$ matrix such that: r_{ij} denotes the rating provided by the user U_i on the item I_j where, $i = \overline{1, M}$ and $j = \overline{1, N}$. Our proposed approach relies on both similar items and similar users to provide predictions. Since we embrace the belief function theory to represent the uncertainty pervaded in the ratings predictions, each neighbor's preference rating is transformed into a mass function spanning over the frame of discernment $\Theta_{pref} = \{\theta_1, \theta_2, \cdots, \theta_L\}$, a rank-order set of L preference labels, where $\theta_p < \theta_l$ whenever $p < l$. We denote by I_t the target item for which we would like to predict users' preferences and U_a the active user for whom the prediction is performed.

4.2   Identifying Items' Neighborhhod

Based on the initial user-item matrix, the first step consists on identifying the set of the k-nearest neighbors of the target item I_t. To this end, the distances between I_t and the whole items in the system are computed. Note that the similarity computation strategy is performed based on the formalism proposed in [14] which consists on isolating the co-rated items. That is to say, to compute the similarity between two items, only the ratings of the users who rated both of these items are considered. Formally, the distance between the target item I_t and each item I_j is computed as follows:

$$dis(I_t, I_j) = \frac{1}{|U(t,j)|} \sqrt{\sum_{U \in U(t,j)} (r_{i,t} - r_{i,j})^2} \qquad (4)$$

$|U(t,j)|$ denotes the number of users that rated both the target item I_t and the item I_j where $r_{i,t}$ and $r_{i,j}$ correspond to the ratings of the user U_i for the target item I_t and for the item I_j. Finally, only the k items having the lowest distances are selected leading to the items' neighborhood formation.

4.3   Modeling Items' Neighborhood Ratings

Let Γ_k be the set of the k-nearest neighbors of the target item. The rating of each item I_j belonging to Γ_k is transformed into a *bba* defined as:

$$m_{I_t, I_j}(\{\theta_p\}) = \alpha_0 \exp^{-(\gamma_{\theta_p}^2 \times (dis(I_t, I_j))^2)} \qquad (5)$$

$$m_{I_t, I_j}(\Theta_{pref}) = 1 - \alpha_0 \exp^{-(\gamma_{\theta_p}^2 \times (dis(I_t, I_j))^2)}$$

Following [15], α_0 is set to 0.95 and γ_{θ_p} is defined as the inverse of the mean distance between each pair of items having the same ratings. To evaluate the reliability of each similar item, these *bba's* are then discounted as follows:

$$m_{I_t, I_j}^{\beta_I}(\{\theta_p\}) = (1 - \beta_I) \cdot m_{I_t, I_j}(\{\theta_p\}) \qquad (6)$$

$$m_{I_t, I_j}^{\beta_I}(\Theta_{pref}) = \beta_I + (1 - \beta_I) \cdot m_{I_t, I_j}(\Theta_{pref})$$

Where β_I is a discounting factor depending on the items' distances such as: $\beta_I = \frac{dis(I_t, I_j)}{max(dist)}$ where $max(dist)$ is the maximum value of the computed distances. That is to say, we consider the items within the highest similarities to be the more reliable ones.

4.4   Generating Items' Neighborhood Predictions

Once the evidence of each similar item is discounted, these k $bba's$ are aggregated as follows:

$$m^{\beta_I}(\{\theta_p\}) = \frac{1}{Z}(1 - \prod_{I \in \Gamma_k}(1 - \alpha_{\theta_p})) \cdot \prod_{\theta_p \neq \theta_q} \prod_{I \in \Gamma_k}(1 - \alpha_{\theta_q}) \qquad \forall \theta_p \in \{\theta_1, \cdots, \theta_{Nb}\}$$

(7)

$$m^{\beta_I}(\Theta_{pref}) = \frac{1}{Z} \prod_{p=1}^{Nb}(1 - \prod_{I \in \Gamma_k}(1 - \alpha_{\theta_p}))$$

Note that Nb is the number of the ratings given by the similar items, α_{θ_p} is the belief committed to the rating θ_p, α_{θ_q} is the belief committed to the rating $\theta_q \neq \theta_p$ and Z is a normalized factor defined by [15]:

$$Z = \sum_{p=1}^{Nb}(1 - \prod_{I \in \Gamma_k}(1 - \alpha_{\theta_p}) \prod_{\theta_q \neq \theta_p} \prod_{I \in \Gamma_k}(1 - \alpha_{\theta_q}) + \prod_{p=1}^{Nb}(\prod_{I \in \Gamma_k}(1 - \alpha_{\theta_q})))$$

(8)

4.5   Smoothing Process

Since a large amount of ratings is generally unavailable, we consider first the predictions previously obtained from the items' neighborhood. These predictions, which are represented through $bba's$, are transformed in pignistic probabilities. To predict the rating $r_{i,j}$ corresponding to the user U_i for the item I_j, the value having the highest pignistic probability is selected as follows:

$$\widehat{r_{i,j}} = \widehat{\theta_p} = argmax_E(BetP(E)) \; for \; all \; E \in \Theta_{pref}$$

(9)

The smoothing strategy consists in filling the missing ratings values in the user-item matrix by the ratings already predicted using item-based CF. Consequently, we get a new user-item matrix containing both original ratings and predicted ratings which reduce the sparsity of the ratings data. This matrix will be considered in the next phase as the input data of the evidential user-based CF. Hence, the smoothing process can be considered as a pre-processing step allowing to alleviate the sparsity of data and to provide more appropriate recommendations.

4.6   Identifying Users' Neighborhhod

Until now, we have performed a smoothing of the user-item matrix by the predicted ratings of the similar items. We explore then the new obtained matrix to compute the distances between the active user U_a and the other users as follows:

$$dist(U_a, U_i) = \frac{1}{|I(a,i)|}\sqrt{\sum_{I \in I(a,i)}(r_{a,j} - r_{i,j})^2}$$

(10)

$|I(a,i)|$ denotes the number of items rated by the active user U_a and the user U_i, $r_{a,j}$ and $r_{i,j}$ correspond respectively to the ratings of the user U_a and U_I for the item I_j. Based on the computed distances, we select the Top-k most similar users to be used during the next phase.

4.7 Modeling Users' Neighborhood Ratings

In this phase, the k-similar users are the pieces of evidence considered in the prediction process. Based on the same strategy of item-based, a *bba* is generated over each rating θ_p of the selected neighbors as well as Θ_{pref} in order to model uncertainty. It defined as:

$$m_{U_a,U_i}(\{\theta_p\}) = \alpha_0 \exp^{-(\gamma_{\theta_p}^2 \times (dis(U_a,U_i))^2)} \tag{11}$$

$$m_{U_a,U_i}(\Theta_{pref}) = 1 - \alpha_0 \exp^{-(\gamma_{\theta_p}^2 \times (dis(U_a,U_i))^2)}$$

Similarly to the items' neighborhood evidence, α_0 is fixed to 0.95 and γ_{θ_p} is the inverse of the average distance between each pair of users who provided the same ratings θ_p. Based on the same intuition of item-based, the discounted *bba*'s are represented as follows:

$$m_{U_a,U_i}^{\beta_U}(\{\theta_p\}) = (1 - \beta_U) \cdot m_{U_a,U_i}(\{\theta_p\}) \tag{12}$$

$$m_{U_a,U_i}^{\beta_U}(\Theta_{pref}) = \beta_U + (1 - \beta_U) \cdot m_{U_a,U_i}(\Theta_{pref})$$

Where β_U is a discounting factor defined on the basis of the users' distances. It is defined as: $\beta_U = \frac{dis(U_a,U_i)}{max(dist)}$ where $max(dist)$ is the maximum value of the computed distances.

4.8 Generating Users' Neighborhood Predictions

After discounting the evidence of each similar user, the combination of the $bba's$ is performed as follows:

$$m^{\beta_U}(\{\theta_p\}) = \frac{1}{Y}(1 - \prod_{U \in \chi_k}(1 - \alpha_{\theta_p})) \cdot \prod_{\theta_p \neq \theta_q} \prod_{U \in \chi_k}(1 - \alpha_{\theta_q}) \qquad \forall \theta_p \in \{\theta_1, \cdots, \theta_{Nb}\} \tag{13}$$

$$m^{\beta_U}(\Theta_{pref}) = \frac{1}{Y} \prod_{p=1}^{Nb}(1 - \prod_{U \in \chi_k}(1 - \alpha_{\theta_p}))$$

The number of the ratings given by the similar users is denoted by Nb, χ_k corresponds to the set of the k-nearest neighbors of the active user, α_{θ_p} is the belief committed to the rating θ_p, α_{θ_q} is the belief committed to the rating $\theta_q \neq \theta_p$ and Y is a normalized factor defined by [15]:

$$Y = \sum_{p=1}^{Nb}(1 - \prod_{U \in \chi_k}(1 - \alpha_{\theta_p}) \prod_{\theta_q \neq \theta_p} \prod_{U \in \chi_k}(1 - \alpha_{\theta_q}) + \prod_{p=1}^{Nb}(\prod_{U \in \chi_k}(1 - \alpha_{\theta_q}))) \tag{14}$$

4.9 Final Predictions and Recommendations

In the last phase, we propose a unified evidential framework that incorporates both items' aspects and users' aspects and allows the fusion of these two kinds of information sources. The final prediction is obtained through a combination of the bba's of the k-similar items (m^{β_I}) and the $bba's$ of the k-similar users (m^{β_U}). The aggregation process is performed using Dempster's rule of combination such that:

$$m_{Final} = (m^{\beta_I} \oplus m^{\beta_U}) \tag{15}$$

5 Experimental Study and Analysis

To evaluate the proposed approach, experiments were conducted on Movie-Lens[1] data set which contains in total 100.000 ratings collected from 943 users on 1682 movies. We compare our proposed evidential hybrid smoothing-based CF (E-HSBCF) to three other individual CF approaches namely, the evidential user-based CF (E-UBCF) [11], the evidential item-based CF (E-IBCF) [10] and the discounting-based item-based CF (E-DIBCF) [12], as well as the hybrid neighborhood-based CF approach (E-HNBCF) proposed in [13]. Following [16], movies are ranked based on the number of their given ratings such as:

$$Nb_{user}(movie_1) \geq Nb_{user}(movie_2) \geq \cdots \geq Nb_{user}(movie_{1682})$$

where $Nb_{user}(movie_i)$ is the number of users who rated the $movie_i$. Then, we extract 10 different subsets by progressively increasing the number of the missing rates. Each subset will have a specific number of ratings which leads to different degrees of sparsity. For each subset, 10% of the users were considered as a testing data and the remaining ones were considered as a training data.

5.1 Evaluation Measures

In our experiments, we used two evaluation metrics commonly used in RSs: the *Mean Absolute Error* (MAE) and the *Root Mean Squared Error* (RMSE) defined as follows: $MAE = \frac{\sum_{i,j} |\widehat{r_{i,j}} - r_{i,j}|}{\|r_{i,j}\|}$ and $RMSE = \sqrt{\frac{\sum_{i,j} (\widehat{r_{i,j}} - r_{i,j})^2}{\|r_{i,j}\|}}$. We have also used the *Distance criteron* (Dist_crit) proposed in [17] defined as:

$$Dist\_crit = \frac{\sum_{i,j} (\sum_{j=1}^{l} (BetP(\{r_{i,j}\}) - \sigma_j)^2)}{\|r_{i,j}\|}$$

$r_{i,j}$ corresponds to the real rating given by the user U_i for the item I_j and $\widehat{r_{i,j}}$ is the predicted value of the rating. $\|r_{i,j}\|$ denotes the total number of the predicted ratings over all the users, l is the number of the possible ratings that can be provided in the system and σ_i is equal to 1 if $r_{i,j}$ is equal to $\widehat{r_{i,j}}$ and 0 otherwise. Note that the smaller values of MAE, RMSE and Dist_crit reflect a better prediction accuracy.

---

[1] http://movielens.org.

Table 1. The Comparison results in terms of MAE, RMSE and Dist_crit

Measure	Sparsity	Individual CF			Hybrid CF	Hybrid smoothing-based CF
		E-UBCF	E-IBCF	E-DIBCF	E-HNBCF	E-HSBCF
MAE	53%	0.945	0.663	0.652	0.641	0.714
RMSE		0.949	1.090	0.996	1.530	1.600
Dist_crit		1.672	1.672	1.054	0.946	0.930
MAE	56.83%	0.826	0.875	0.864	0.815	0.687
RMSE		1.28	1.199	1.317	1.240	1.050
Dist_crit		1.452	1.598	0.962	0.933	0.905
MAE	59.8%	1.120	0.890	0.850	0.691	0.645
RMSE		1.369	1.279	1.292	1.110	1.231
Dist_crit		1.160	1.420	0.953	0.940	0.892
MAE	62.7%	0.889	0.741	0.758	0.797	0.744
RMSE		0.886	1.022	1.063	1.120	1.100
Dist_crit		1.043	1.327	0.944	0.932	0.928
MAE	68.72%	0.798	0.854	0.877	0.847	0.750
RMSE		1.147	1.087	1.303	1.130	1.000
Dist_crit		1.131	0.758	0.961	0.961	1.057
MAE	72.5%	0.86	0.886	0.864	0.822	0.741
RMSE		1.154	1.224	1.266	1.147	1.075
Dist_crit		2.000	1.385	0.952	0.936	0.808
MAE	75%	1.000	0.837	0.853	0.881	0.749
RMSE		1.402	1.160	1.339	1.000	1.000
Dist_crit		1.770	1.477	0.907	0.872	1.038
MAE	80.8%	1.000	0.932	0.922	0.539	0.666
RMSE		1.000	1.200	1.230	1.020	0.982
Dist_crit		2.000	1.98	1.206	1.080	0.821
MAE	87.4%	1.000	0.620	0.544	1.000	1.000
RMSE		1.000	0.946	0.916	0.850	0.851
Dist_crit		2.000	1.680	0.818	0.665	0.714
MAE	95.9%	1.000	1.000	1.000	1.000	1.000
RMSE		1.000	3.000	3.000	3.000	3.000
Dist_crit		2.000	2.000	1.900	1.930	1.902
Overall MAE		0.943	0.828	0.818	0.773	**0.769**
Overall RMSE		1.318	1.322	1.375	1.314	**1.288**
Overall Dist_crit		1.594	1.529	1.065	1.01	**0.990**

5.2   Results

Experiments were performed over the 10 extracted subsets characterized by different sparsity degrees. We compute the MAE, the RMSE and the Dist_crit for each subset and we report the obtained results. These results correspond to the average of 10 repetitions performed using 10 different values of neighborhood size. The experimental results are displayed in Table 1. As can be seen, the proposed evidential hybrid smoothing-based CF (E-HSBCF) shows a better performance than the three other individual neighborhood-based approaches. Indeed, the hybrid smoothed-based method achieves the lowest mean values in terms of MAE, RMSE and Dist_crit. For example, the E-HSBCF approach acquires the lowest MAE value corresponding to 0.769 compared to E-HNBCF having a value of 0.773. This emphasizes the fact that the smoothing mechanism as well as the fusion process have a great effect on the prediction performance. The new approach achieves also better results in term of $RMSE$ with a value of 1.288 compared to 1.375 for the E-DIBCF, 1.322 for the E-IBCF and 1.318 for the E-UBCF. Similarly, the Dist_crit results indicate that the proposed approach having a value of 0.990 outperforms E-HNBCF (equal to 1.01), E-UBCF (equal to 1.594), E-IBCF (equal to 1.529) and E-DIBCF (equal to 1.065).

6   Conclusion

In this paper, we have proposed an evidential hybrid CF approach that alleviates the data sparsity while taking into account the uncertainty pervaded in the predictions. The new approach relies first on item-based CF to provide the basis for data smoothing. Predictions are then performed using the obtained matrix and based on both users' neighbors and items' neighbors. The final predictions are represented through basic belief assignment reflecting more credibility and intelligibility. As a future work, we tend to integrate implicit ratings inferred through users' behaviors in order to get more personalized recommendations.

References

1. Ricci, F., Rokach, L., Shapira, B.: Recommender systems: Introduction and challenges. In: Recommender Systems Handbook, pp. 1–34. Springer (2015)
2. Su, X., Khoshgoftaar, T.M.: A survey of collaborative filtering techniques. In: Advances in Artificial Intelligence, pp. 1–19. Hindawi Publishing Corporation (2009)
3. Smets, P.: The transferable belief model for quantified belief representation. In: Quantified Representation of Uncertainty and Imprecision, pp. 267–301. Springer (1998)
4. Dempster, A.P.: A generalization of Bayesian inference. J. Roy. Stat. Soc.: Ser. B (Methodol.) **30**, 205–247 (1968)
5. Shafer, G.: A Mathematical Theory of Evidence, vol. 1. Princeton University Press, Princeton (1976)
6. Xue, G.R., Lin, C., Yang, Q., Xi, W., Zeng, H.J., Yu, Y., Chen, Z.: Scalable collaborative filtering using cluster-based smoothing. In: International ACM SIGIR Conference on Research and Development in Information Retrieval, pp. 114–121. ACM (2005)
7. Goldberg, K., Roeder, T., Gupta, D., Perkins, C.: Eigentaste: a constant time collaborative filtering algorithm. Inf. Retr. **4**, 133–151 (2001)
8. Sarwar, B., Karypis, G., Konstan, J., Riedl, J.: Application of dimensionality reduction in recommender system-a case study. Minnesota Univ Minneapolis (2000)
9. Lian, J., Zhang, F., Xie, X., Sun, G.: CCCFNet: a content-boosted collaborative filtering neural network for cross domain recommender systems. In: International Conference on World Wide Web Companion, pp. 817–818. World Wide Web (2017)

10. Abdelkhalek, R., Boukhris, I., Elouedi, Z.: Evidential item-based collaborative filtering. In: International Conference on Knowledge Science, Engineering and Management, pp. 628–639. Springer (2016)
11. Abdelkhalek, R., Boukhris, I., Elouedi, Z.: Assessing items reliability for collaborative filtering within the belief function framework. In: International Conference on Digital Economy, pp. 208–217. Springer (2017)
12. Abdelkhalek, R., Boukhris, I., Elouedi, Z.: A new user-based collaborative filtering under the belief function theory. In: International Conference on Industrial, Engineering and Other Applications of Applied Intelligent Systems, pp. 315–324. Springer (2017)
13. Abdelkhalek, R., Boukhris, I., Elouedi, Z.: Towards a hybrid user and item-based collaborative filtering under the belief function theory. In: International Conference on Information Processing and Management of Uncertainty in Knowledge-Based Systems, pp. 395–406. Springer (2018)
14. Sarwar, B., Karypis, G., Konstan, J., Riedl, J.: Item-based collaborative filtering recommendation algorithms. In: International Conference on World Wide Web, pp. 285–295. ACM (2001)
15. Denoeux, T.: A K-nearest neighbor classification rule based on Dempster-Shafer theory. IEEE Trans. Syst. Man Cybern. **25**, 804–813 (1995)
16. Su, X., Khoshgoftaar, T.M.: Collaborative filtering for multi-class data using Bayesian networks. Int. J. Artif. Intell. Tools **17**, 71–85 (2008)
17. Elouedi, Z., Mellouli, K., Smets, P.: Assessing sensor reliability for multisensor data fusion within the transferable belief model. IEEE Trans. Syst. Man Cybern. Part B: Cybern. **34**, 782–787 (2004)

Study of E-Learning System Based on Cloud Computing: A Survey

Sameh Azouzi[1,2(✉)], Sonia Ayachi Ghannouchi[2,3], and Zaki Brahmi[1,2]

[1] ISITCom Hammam Sousse, University of Sousse, Sousse, Tunisia
azouzi_sameh@yahoo.fr
[2] Laboratory RIADI-GDL, ENSI, University of Manouba, Manouba, Tunisia
[3] ISG Sousse, University of Sousse, Sousse, Tunisia

Abstract. It is worth to mention that nowadays we observe an evolving interest for the acquisition and the exploitation of new technologies in the context of learning specifically the Internet. They are based on the use of approaches with diverse functionalities (e-mail, Web pages, forums, LMS, and so on) as a support of the process of teaching-learning. E-learning systems usually require many hardware and software resources. There are many education institutions that cannot afford such investments, and cloud computing is the best solution. Cloud computing is the basic environment and platform of the future e-learning. This paper mainly focuses on the application of cloud computing in e-learning environment.

Keywords: Cloud e-learning · Cloud computing · BPMN · E-learning process · Web2.0

1 Introduction

The education sector is confronting an increasing challenge that is to say the emergence of technologies. In fact, various types of equipments (laptop, smart phones, touchpads, etc.) are used by students and teachers. Any delay taken by the education on the rhythm of the technological innovation risks to break even more the contact between the world of the education and the society. More than ever, it is a question of modifying the means of the education to meet the expectations of learners and to confront new challenges of the professional world and the digital society. Therefore, collaborative work has to take place and collaborative tools such as virtual classrooms, discussion forums, sharing online document tools (goolgedrive.com) are becoming a necessity. Institutions will have to invest in the educational space, not only within a class or between the students of the same year, but also between the classes and between the institutions (close or distant) [1]. Students and teachers need a more flexible and adaptive infrastructure. With these characteristics, the Cloud platforms arise as a solution that has a significant impact on teaching and learning. Moreover, cloud computing is a model of deployment of resources and computing capacities which tends to minimize the load of implementation and management for the user organization. Cloud computing offers infrastructures, software, processes and a variety of services which can be consumed by the user

A. Abraham et al. (Eds.): ISDA 2018, AISC 941, pp. 532–544, 2020.
https://doi.org/10.1007/978-3-030-16660-1_52

according to his/her demand via Internet, intranet/extranet by paying only his real consumption. These services offered by the cloud computing are essentially Web applications which supply services when they are ordered. They can be in particular applications of Web 2.0. In fact, web2.0 allows people to create, to exchange, to publish and share information in a new route of communication and collaboration. Thus, to build an environment of successful and effective education, it is necessary to use new technologies such as web2.0 and cloud computing which supplies the collaboration and the interaction in e-learning environment. This new environment supports the creation of new processes of collaborative e-learning which use web2.0 and which can work on a vast range of devices while the data is stored in the cloud. This allows bursting the geographical and organizational structures. To benefit from the advantages of the cloud technology in the educational domain, we need the management of e-learning processes. Indeed, the approach of BPM (Business Process Management) allows a better management of e-learning processes in the sense that it offers a continuous improvement of business processes, increases the productivity and reduces costs. Founded on these ideas, we propose an approach of management of collaborative e-learning processes by using BPM and cloud computing. Our contribution, which is focused on the process of learning, will consist in the construction of a collaborative and reconfigurable learning process, which will be easily adaptable to change and will easily evolve to meet the needs of learners. More precisely, in this paper, we concentrate on the advantages of using the cloud computing and the Web services 2.0 in this context, particularly in terms of collaborative activities. In order to overview all these aspects, this contribution is arranged as follows.

In Sect. 2 we introduce the main concepts on Cloud Computing, including its infrastructure and main layers and we present the features of the e-Learning approach, stressing the advantages of the migration of such a system to the Cloud. Next, in Sect. 3 we show some examples of real applications of this kind. Section 4 presents our approach. Finally, the main concluding remarks are given in Sect. 5.

2 Basic Concepts on Cloud Computing

2.1 Introduction to Cloud Computing

In the literature, there are several definitions of the cloud computing that are more or less vague. However, the definition given by the NIST - National Institute of Standards and Technology is an authoritative one. According to National Institute and Standard of Technology NIST [23, 24], CC is a model that allows easy on-demand network access to shared containers of configurable computing resources (e.g., networks, servers, storage, applications, and services) that can be set up and released quickly with little or no management effort or service provider interaction. CC has emerged to provide computing services based on demand, measurable, 'pay-per-use' and virtual centralized via the internet to improve the company's ability to cope with a flexible and highly competitive business environment. The cloud technology has evolved through combining the advantages of SOA, virtualization, grid computing, and management automation with the following features (8): (1) virtualization, reuses hardware equipment to provide an expandable system environment with extra flexibility such as the use VMware and Xen that act as a demand-based virtualization IT equipment. Users can configure their personal network and system environment through virtualization network known as VPN; (2) service flow and workflow, provides a complete set of service environment as per demand; (3) web service and SOA, through standard of WSDL, SOAP, and UDDI, and other cloud services can be delivered in the web service; (4) web 2.0, can strengthen information share and interactive cooperation of users; (5) large-scale distributed systems, requires large-scale distributed memory system and computing ability to realize the rental of computing resources and memory spaces by users; (6) programming model, allows user to write application program under cloud environment [25].

2.2 Cloud Computing Layers

The view of the NIST is that the cloud computing has three service models, four deployment models and five essential characteristics as shown in Fig. 1 [22]. Three widely referenced service models have evolved [27]:

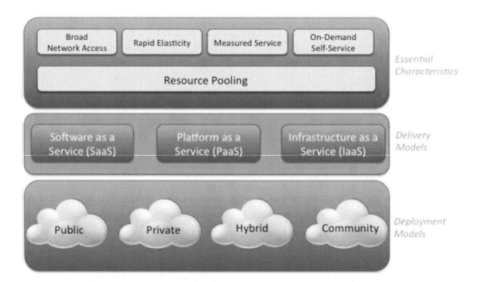

Fig. 1. NIST visual model of cloud computing definition [22].

- Infrastructure as a Service (IaaS): remote management and control of hardware resources provided by a system.
- Platform as a Service (PaaS): offers the cloud platform along with a series of libraries to develop applications in which the distribution of tasks, the persistence and other layers are transparent for the developer.
- Software as a Service (SaaS): consists in offering different applications to be used through the internet as opposed to a local installation.

In addition to the NIST definition, we can find other service models such as:

- Hardware as a Service (HaaS): contrarily to the SaaS and PaaS that provide applications and services to the customers, HaaS offers only the hardware.
- Database as a Service (DaaS): the aim of a DaaS is to offer a database and the services allowing its management to avoid the complexity and running cost of a database if hosted in the own network of a company or organization.
- Business process as a service (BPaas): Above the SaaS layer, was added the BPaaS model which provides the client with a business process environment. In this type of service, cloud providers are not only limited to providing software solutions on behalf of companies, but also participate in the management of the company, to ensure that the objectives are achieved.

In addition to these service models, four deployments have been added:

- Public cloud: The cloud infrastructure is made available to the general public or alarge industry group and is owned by an organization selling cloud services.
- Private cloud: The cloud infrastructure is accessible for an organization only. It may be managed by the organization itself or a third party and can be internal or external.
- Community cloud: A private cloud that is shared by several customers with similar security concerns and the same data and applications sensitivity.
- Hybrid cloud: It merges more than one Cloud Computing model into a single, hybrid model; using a public cloud for hosting sites that must be published publically and containing uncritical data, and using a private cloud for all the other sensitive data or services. This scenario is good for economic and business requirements.

2.3   Impact of Cloud Computing in E-Learning

E-Learning in the Cloud can be viewed as Education Software-as-a-Service. The benefits of "software as a service" in universities may be described by several factors:

- Decrease of the cost: it provides a solution to the problem of licensed software that requires constant updating. The second is that the learning process requires searching and experimentation [27].
- A flexibility, provided by cloud technologies, enables to modify, test and compare different types of software, various forms of use that would be impossible if it was necessary to purchase every time new software and equipment is required [28].
- Elasticity: Institutions using learning process may have trouble with elasticity since their learning process software may only allow coordinating simultaneously a limited number of process instances. Consequently, institutions can be in situations where the demands of learners/teachers increase in a considerable way. In this case, they need to buy additional servers to make sure that the demands of their customers will be satisfied. This causes a problem, because these new servers are rarely used although their purchase and their maintenance can be expensive. With Cloud Computing, institutions can increase or decrease the capacity of their systems depending on their needs. This is called resource allocation at Iaas level.

- Variability and configuration: The requirements for institutions and their learners are in a perpetual change. At the same time they vary from an institution to another. This leads us to return to the cycle of business process and to make such reconfigurations on the concerned process to make it adapted to the needs of establishments and their learners. Cloud computing is a good infrastructure for setting up configurable online learning processes as BPaas and each institution is looking for the configuration it wants. Dealing with the variability of e-learning processes, we have this point in previous work such as [29].
- Re-use: The integration of a learning process in the cloud gives the possibility of reusing it by other institutions/learners in a new context. Thus, we talk about a reusable e-learning process.
- Collaboration: If we talk about a reusable process of learning, then this is a first kind of collaboration because we provide other institutions/learners everywhere with accessible processes and activities. Nevertheless, if we use the web2.0 services and collaborative tools in the process of e-learning, we move then to a second kind of collaboration between various actors involved in the process of e-learning.

3   Related Work

During the last years, we have been able to witness a very considerable use of the applications which use the technique of cloud computing. These applications are numerous such as (Gmail, Hotmail, Google apps, Lotus Live, Web Service Apps). The success of the solution software as a service (Saas) is real and is easily applicable for e-learning. Cloud computing, Saas and e-learning are completely complementary. In this context, several studies have focused on the design and integration of e-learning systems in the cloud. Among these works we can mention those of Ouf et al. [14] that has proposed an e-learning system based on the integration of cloud computing and Web 2.0 technologies to meet the requirements for e-learning environment such as flexibility and compliance towards students' needs and concerns, improve and enhance the efficiency of learning environment. The authors mentioned the most important cloud-based services such as Google App Engine and classified the advantages of implementing cloud-based e-Learning 2.0 applications such as scalability, feasibility and availability. They also emphasized the improvements in cost and risk management.

Aljenaa et al., in [3] introduced the main components of a system of effective e-learning through using cloud computing technology. The authors insist on the importance of the following components:

- "Cloud software platform" which contains the LMS platform and the necessary tools of collaboration and communication to meet the needs of learners throughout the process of learning;
- "Operational and management components" to make the management of the learning process;

The authors in [13] propose an architecture of an e-Learning system in the cloud (e-Learning as a service) for the following purposes:

- Bring together all the institutions to offer them the same e-Learning service and to improve the collaboration aspect.
- Improve learners' knowledge with resource updates.
- Improve the quality of learning.
- Enable distant learners to deliver classes and lectures.

Authors in [15] propose an architecture of a learning system based on mobile agents. These agents act as intermediaries between the cloud and the learner to:

- Provide adaptive learning services according to the preferences and level of the learner.
- Respond to the needs of learners according to the resources available in the cloud to facilitate the learning process.
- Make the selection of services necessary for the learning process to take place.

The following table presents a comparison between his various works cited above (Table 1). It summarizes three aspects related to e-learning and cloud computing and their consideration in the work presented. We consider as follows:

Table 1. Evaluation of previous approaches

Criteria approaches	Cloud computing	BPM	Collaboration
Ouf et al. [14]	+	+	+
Aljenaa et al. [3]	+	+	+
Al Noor et al. [4]	+	−	−
Babu et al. [16]	+	−	−
Our approach	+	+	+

- Cloud computing technology;
- Managing of learning process with BPM approach;
- Synchronous and asynchronous collaboration.

Despite all these initiatives, we still consider that research works concerning the systems of e-learning in the cloud and the modelling of the existing processes of e-learning suffer from some limits namely:

- The absence of a complete initiative or an approach for the construction of a general learning process that meets all the needs of different institutions and learners.
- Lack of collaboration and re-use in learning processes by other universities teachers in spite of the possibilities brought by the cloud computing;
- The absence of an automated mechanism of communication, which informs concerned people at adequate moments when an intervention is required during the execution of the learning process.

In previous work such as [29], we showed the strong similarity between the learning process and business process. In this context, our objective is to make a combination of BPM with cloud computing for the construction of a process of e-learning as a reconfigurable and collaborative service (BPaas).

4 Proposed Model

Using cloud computing technology in the field of e-learning meets some main problems: the absence of an effective management of the learning process that achieves the learning objectives, the absence of the collaboration aspect and re-use of the process of learning. Our objective is the adoption of an approach of continuous improvement of the learning process in the cloud. To achieve this goal, we suggest the following approach:

– The modelling of a general and collaborative learning process: the graphic modelling of a learning process favors its improvement and re-use. BPMN notation is the standard adopted for presenting the process in a graphic and very expressive way. More precisely, for collaborative learning processes, the models must be rich in interactions and collective actions frequently arise. In fact, in an objective-oriented educational approach, acquisition of knowledge in learner's skills development is fundamental;
– Modeling e-learning process variability: modeling of the variability meets adaptability and customization. E-learning and cloud computing become basic needs for institutions and each institution has its own needs. But our objective is to provide a learning process line that meets the needs of all institutions/learners. The SPL approach (Software product Line) enables application development in e-learning to meet the needs of all customers, improve software quality and reduce the cost and development time. Modeling variability is described by a hierarchy of the overall characteristics of the system as model features (functions). Another variability modeling represents the processing characteristics of the system; that is to say, the details of technical implementation of the system (database, content formats, operating system, etc.).
– The migration of this process to the cloud as a Business Process as a service (Bpaas): after the modelling and the deployment of our learning process, our next stage will be the migration of this process to the cloud as a reusable BPaas, useful for several learners, teachers and institutions and discovery of this BPaas in cloud.
– The discovery of this BPaas in cloud and the selection of services allowing the implementation of this process to make it executable: a process of collaborative and interactive learning allows learners and teachers to interact (face-to-face communication, virtual class, etc.). In this way, every learner will participate to the learning process and will not be isolated, but will be located in a wider context, which includes other learners.

To validate the first step in our approach and as we indicated in the work [29], we propose new models of learning process based on learning activities and the interactions between learners and teachers. To select the learning scenarios, we were based on Lebrun [10] and Monnard [11] models. In addition, we made reference to the IMS-

Learning Design standard [26]. The activities of learning are of several types: Scripting, Informing, Interacting, Producing and Summative evaluation (Fig. 2).

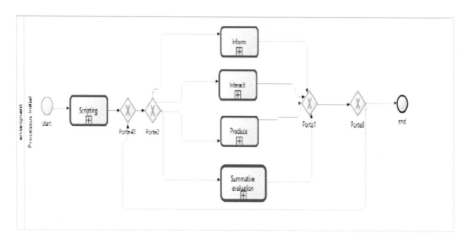

Fig. 2. Global e-learning process

We focus through our work on the aspect of collaboration and communication between the various actors of an activity. For that purpose, and for every process, we propose sub-processes and activities of learning which run in a collaborative context with the use of tools of collaborative work such as docs.google.com, Google drive, discussion forum and webinars. As an illustration of our idea, we choose the process "Interact" that itself is detailed into two sub-processes, which are "Group work" and "thematic Debate" (Fig. 3).

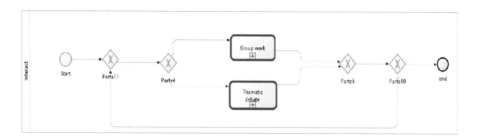

Fig. 3. Sub-process «Interact»

Subsequently, we detail every sub-process through the corresponding tasks dealing with communication between learners and teachers, with the use of collaborative tools (Fig. 4).

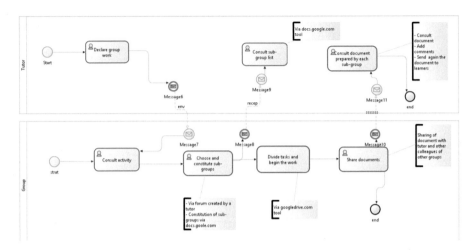

Fig. 4. Sub-process «Group Work»

Group work is a collaborative/cooperative activity which is often used in e-learning. On the one hand, in education, it's required to be capable of working in a group. On the other hand, in socio-constructivism, it's considered that the group work and the confrontation/collaboration with the others are beneficial to the learner himself. The "Group work" activity presented in Fig. 3 appears as a process which is rich in interactions between actors. Furthermore, collaborative tools are given to maintain an important degree of interaction between learners and teachers and between learners. This interactivity between teacher and learner allows adopting a set of actions which offers to learners the possibility of accomplishing their goals. To allow learners to interact in a course, we need to organize a thematic debate on a question/problem. The objective to be achieved is that learners better understand various points of view by identifying the corresponding arguments. Thus, they can improve their capacity of argumentation. Until now we talk about modeling. For showing the feasibility of our e-learning process model, we move to the implementation and enactment of this process with the BPMS engine executions. As it is mentioned in Fig. 1, the process starts with a type of activity "scripting". This activity is mandatory where the teacher defines the proposed course. Figure 5 shows the interface in which the teacher introduces the details related to the course to offer.

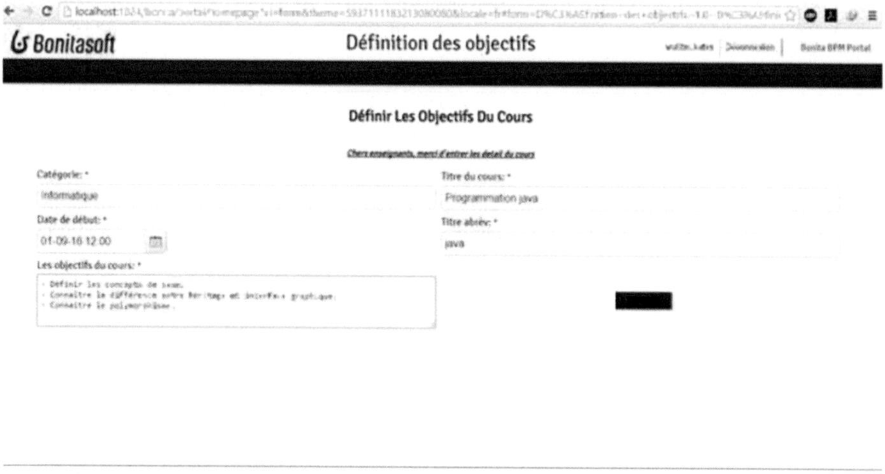

Fig. 5. Definition of course objectives

Thereafter, the learner sees the details of the course proposed by the tutor and gives his opinion. Subsequently, the teacher may decide to choose the next activity (Fig. 6).

Fig. 6. Choice interface

The tutor chooses one type of activity "interact" that is the "group work" so he first sets objectives to the activity (Fig. 7).

Fig. 7. Groupwork activity interface

5   Conclusion

In this work we have exposed the main components of cloud computing and we are focusing on the impact of using cloud computing in e-learning. Then, we have enumerated several approaches that have been already proposed for addressing e-Learning on Cloud Computing, describing these models and how they take advantage of this environment to enhance the features of the educational system. However, we must stress that these are just initial steps towards an open line for research and exploitation of e-learning and cloud computing platforms. Next, we explain the different stages of our proposed solution and next, we proposed the use of BPM approach, cloud computing technologies and web2.0 for the management of e-learning processes. This work aims to overcome the shortcomings of the existing learning systems, mainly by satisfying changing requirements of learners and institutions, providing a collaborative and reconfigurable learning process with a minimum of cost, and benefiting from advantages of BPM and cloud computing. As future work, we intend to:

- Model the variability of our process to make it adaptable to different needs of learners and institutions.
- Improve our e-Learning product line and we are going to investigate the possibilities offered by cloud computing technologies for a better management of our process. This will lead as to consider it as a configurable Business Process as a Service (BPaaS).
- Reflect on an approach to discovering a cloud-based e-learning process and discovering services such as Saas, Paas, and Iaas necessary for performing e-learning activity.

References

1. IBM. Le cloud au service de l'enseignement (2012). www.ibm.com/solution/education/cloudacademy
2. Mircea, M., Andreescu, A.: Education: a strategy to improve agility in the current financial crisis. Commun. IBIMA (2011). https://doi.org/10.5171/2011.875547
3. Aljenaa, E., Al-Anzi, F.S., Alshayeji, M.: Towards an efficient e-learning system based on cloud computing. In: Proceedings of the Second Kuwait Conference on e-Services and e-Systems. ACM (2011)
4. Al Noor, S., Mustafa, G., Chowdhury, S.A., Hossain, Z., Jaigirdar, F.T.: A proposed architecture of cloud computing for education system in Bangladesh and the impact on current education system. IJCSNS Int. J. Comput. Sci. Netw. Secur. 10(10), 7–13 (2010)
5. Samitha, R., Krutika, G., Chandra, S.: A generic agent based cloud computing architecture for e-learning. Springer (2014). https://doi.org/10.1007/978-3-319-03107-1_58
6. Angels, R.G., Miguel-Angel, SU., Garcia, E., Plazuelos, G.M.: Beyond contents and activities: specifying processes in learning technology. In: Current Developments in Technology-Assisted Education (2006)
7. Chua, F., Lee, C.S.: Collaborative learning using service-oriented architecture: a framework design. Knowl.-Based Syst. 22(4), 271–274 (2009)
8. Julien, D.C.: BPMN 2.0 pour la modélisation et l'implémentation de dispositions pédagogiques orientées processus, Mémoire présenté pour l'obtention du master MLALT (2014)
9. Ayodejil, A.: Virtual learning process environment (VLPE): a BPM based Learning process management architecture (2013)
10. Daniel, K.: Les approches scénarisation et la modélisation du workflow pédagogique (2011)
11. Marcel, L.: La formation des enseignants aux TIC: Allier pédagogique et innovation, Institut de pédagogie universitaire et des Multimédias (IPM) (2004)
12. Gérald, C., Sergio, H., François, J., Jacques, M., Hervé, P.: Treize scénarios d'activités de cours avec Moodle, Centre NTE (2013)
13. Méndez, J., Gonzalez, J.: Implementing motivational features in reactive blended learning: application to an introductory control engineering course. IEEE Trans. Educ. 54(4), 619–627 (2011)
14. Ouf, S., Nasr, M., Helmy, Y.: An enhanced e-learning ecosystem based on an integration between cloud computing and Web2.0. In: Signal Processing and Information Technology (ISSPIT), pp. 48–55 (2010)
15. Chandran, D.: Hybrid E-learning platform based on cloud architecture model: a proposal. In: 2010 International Conference on Signal and Image Processing (ICSIP), Chennai, pp. 534–537 (2010)
16. Babu, S.R., Kulkarni, K.G., Sekaran, K.C.: A generic agent based cloud computing architecture for e-learning. In: ICT and Critical Infrastructure: Proceedings of the 48th Annual Convention of Computer Society of India-Vol I. Springer, Cham (2014)
17. Clermont, P., Nelly, A.: La construction de l'intelligence dans l'interaction sociale, ISSN 0721-3700, édition: 5, éditeur: P. Lang. ISBN 3906758230, 9783906758237, vol. 305 (2000)
18. Dessus, P.: Quelles idées sur l'enseignement nous révèlent les modèles d'instrucutional-disign. Revue suisse des sciences de l'éducation, academicPress 28(1), 137–157 (2006)
19. Noel, J.: La gestion des processus métiers. Lulu.com, 372 p. (2007). ISBN 2952826609, 9782952826600

20. Ter Hofstede, A.H.M. et al., (eds.): Modern Business Process Automation: YAWL and Its Support Environment. Springer Science & Business Media (2009)
21. Patil, P.: A study of e-learning in distance education using cloud computing. Int. J. Comput. Sci. Mobile Comput. **5**(8), 110–113 (2016)
22. Van Dar Alast, W.M., TerHofstede, A.H., Weske, M.: Business process management: a survey. In: Business Process Management, pp. 1–12. Springer (2003)
23. Mell, P., Grance, T.: The NIST definition of cloud computing, Recommendations of the National Institute of Standards and Technology (2011)
24. Schulte, S., Janiesch, C., Venugopal, S., Weber, I., Hoenisch, P.: Elastic Business Process Management: state of the art and open challenges for BPM in the cloud. Future Gener. Comput. Syst. **46**, 36–50 (2015)
25. Silver, B.: BPMS watch: ten tips for effective process modeling (2009)
26. Mercia, Gunawan, W., Fajar, A.N., Alianto, H., Inayatulloh: Developing cloud-based Business Process Management (BPM): a survey. In: 2nd International Conference on Computing and Applied Informatics (2017)
27. Pernin, J.-P.: LOM, SCORM et IMS-Learning Design: ressources, activités et scénarios. actes du colloque «L'indexation des ressources pédagogiques numériques», Lyon, vol. 16 (2004)
28. Al Tayeb, A., et al.: The impact of cloud computing technologies in e-learning. Int. J. Emerg. Technol. Learn. (iJET) **8**(2013), 37–43 (2013)
29. Sultan, N.: Cloud computing for education: a new dawn. Int. J. Inf. Manag. **30**, 109–116 (2010). https://doi.org/10.1016/j.ijinfomgt.2009.09.004
30. Azouzi, S., Ghannouchi, S.A., Brahmi, Z.: Software product line to express variability in e-learning process. In: European, Mediterranean, and Middle Eastern Conference on Information Systems, pp. 173–185. Springer, Cham, September 2017

Trusted Friends' Computation Method Considering Social Network Interactions' Time

Mohamed Frikha[1]([⊠]), Houcemeddine Turki[2],
Mohamed Ben Ahmed Mhiri[1], and Faiez Gargouri[1]

[1] MIRACL Laboratory, University of Sfax, Sfax, Tunisia
med.frikha@gmail.com, med.mhiri@gmail.com,
faiez.gargouri@isimsf.rnu.tn
[2] Faculty of Medicine of Sfax, University of Sfax, Sfax, Tunisia
turkiabdelwaheb@hotmail.fr

Abstract. Based on the assumption that users generally tend to use entities proposed by friends rather than strangers and that trust among friends significantly correlates with user's trends, we decided to refer to research conducted on the evolving field of social trust computation. Although many models were proposed to analyze computational trust for various applications of social web, little importance is given to the time factor even by models that represent trust as a source of recommendation measurement. In this paper, we propose to integrate the temporal factor in assessing trust between social network friends. We define a time-aware method for an implicit trust determination between Facebook friends. Then, we integrate it in a mobile app we have developed for identifying trusted friends in social networks. Finally, we assess our method using an objective statistical method and compare it with others methods in the literature to show the importance of the time of interactions between users for a better calculation of social trust and consequently for an enhanced identification of trusted friends in social networks.

Keywords: Trust · Social networks · Users' interactions ·
Recommender systems

1 Introduction

In real daily life, when we think of purchasing a particular and unfamiliar product, we most often tend to seek immediate advice from some of our friends who have run over this product or experienced it. We, similarly, tend to acknowledge and use friends' recommendations since we trust them. Therefore, integrating social networks in recommender systems can result in more accurate recommendations [1]. Indeed, interpersonal influence plays an imperative role in personalized recommender systems. It is observed to be useful in recommending items on social networks and it has as a target relating recommended items to the person's historical behavior and interpersonal relationships. The quality of the recommendation can be ensured based on the support of user's interpersonal interests in a social network. Data acquired about users and their friends makes it useless to look for similar users and to measure their rating likeness.

© Springer Nature Switzerland AG 2020
A. Abraham et al. (Eds.): ISDA 2018, AISC 941, pp. 545–555, 2020.
https://doi.org/10.1007/978-3-030-16660-1_53

Based on the assumption that users generally tend to utilize items recommended by friends rather than strangers and that trust among friends positively correlates with user preference, we decided to refer to research conducted on the evolving field of trust-based recommender system. Trust between two users, therefore, means that a user believes on the quality of the recommendation of a trusted user. This area of study focuses on providing users' personalized item recommendations with reference to the trust relationships among users, which is found to be helpful in solving a significant number of matters associated with traditional systems, such as data sparsity [2] and cold start [1].

Social trust-based recommendation systems are based on giving recommendations using only trust scores' calculation between users' interactions or on a combination of trust and similarity scores while giving suggestions. [3] argues that users prefer to receive recommendations from people they know and that trust-based recommendation approaches perform better than approaches based on only user similarity. If we like to search the most trusted friends in a social network for a $user_x$, we need to compare all interactions between $user_x$ as well as each of his/her friends, taking into consideration the temporal factor of every interaction between them. That is because $user_x$ can trust a friend in the past, but the value of trust can degrade with time. In fact, interactions are not perceived the same way over time because some interactions are more important than others when computing an opinion [4]. That is way, time information can be useful in facilitating tracking the evolution of user interests and improving recommendation accuracy [5].

The rest of the paper is organized as follows. In Sect. 2, we present previous works related to social trust-aware recommender systems. Then, in Sect. 3, we describe a time-sensitive method for calculating trusted friends on social networks, we assess it and we compare it with other works using a Facebook application we have created. Finally, we conclude the paper and present our future work.

2   Related Works

Trust can play an important role across different disciplines and is an important feature of our everyday lives. Additionally, trust is a feature associated with people in the real world and with users in social media [6, 7]. In recommender systems, trust is defined with reference to the other users' ability to provide valuable recommendations [8]. The trust value can be binary or in real numbers (i.e., in the range of [0; 1]). Binary trust is the simplest way of expressing trust. Two users can choose to trust or not trust each other. A more in-depth method is a continuous trust model, which gives real values to the trust relations. In both binary trust and continuous models, 0 and 1 means no trust and full trust, respectively [9].

Nowadays, many trust models were proposed in many social web applications. [10] uses the subjective logic in his attempt to define a trust model considering the local, collective and global trust. Indeed, the subjective logic is found to have a better formal framework for modeling trust. It is also found to be useful in representing the relationship of trust between users through probabilistic opinions. An opinion in the subjective sense is the resulted accumulation of several interactions between the user and the object of opinion. However, during the modeling of opinions, this theory has some disadvantages: It models different views in the same way, that is to say, by the principle of equal probability and it only models singleton opinions [11]. Moreover, this method does not take into consideration the temporal factor in measuring trust between users. Although a simple opinion in subjective logic has a probabilistic structure based on the multiple interactions between users, it is not time-sensitive [4]. This means that all interactions have the same importance. We assume that interactions are not perceived the same way over time. To put it differently, some interactions are more important than others when computing an opinion [5]. Similarly, [12] suggested a general model to measure trust among social network users. They have also explained how a general model is implemented using a specific social network as a medium. In their work, however, [12], also have not taken into consideration the temporal factor in measuring trust between users. In the real world, the value of an interaction between two users is different according to its time of apparition [4].

As shown, most of social trust computation methods do not take into consideration the time of an interaction between two users in the social network. In our recent work, we have provided a method to calculate trusted friends for an ego user with an implicit trust metrics [13]. Implicit trust is generally inferred from user behaviors and trust relationships among social friends. Our method takes into consideration the temporal factor of interactions that exists between social network friends. In this work, we will apply this method through a Facebook application so that it can be assessed and compared to other methods in the literature.

3 Time-Sensitive Method for Calculating Trusted Friends on Social Networks

Research has demonstrated the effectiveness of trust in recommender systems. Lack of explicit trust information in most systems emphasizes the need to propose some trust metric to infer implicit trust from the user and his/her friends' interactions. Our goal is to determine the list of friends trusted by a $user_x$. We begin, first, by all the friends who have made interactions or shared information with the user in a very specific period of time. Interactions among users in a social network are of different types. They can be comments or mentions "Like" on objects in the profile (from the user or from friends). We apply some trust metric to calculate the value of trust between $user_x$ and each of his/her friends. Our goal is to demonstrate how an ego's social activities and his/her friends can be used to calculate a "level of trust" between two friends. Finally, we choose friends who have the highest level of trust. The extraction of the trusted friends' list of a user is performed to determine the preferences and interests of each friend. Indeed, these interests can be useful to know the interests of the *ego user*.

A *social activity is* a social interaction between two directly connected social network friends at a period d. An example of social activity can be the joint tagging of two friends on the same photo. We can define the $PT(user_x)^d$ as a list of friends who are tagged on the same photo with the $user_x$ at a period of time d. This period is precised starting from date d till now. For example $PT(user_x)^d$ can be defined as:

$$PT(user_x)^d = \{user_a, user_b, user_c\} \tag{1}$$

This means that $user_x$ is tagged with $user_a$, $user_b$ and $user_c$ after the fixed date d. Formula 1 does not take into consideration the number of tags between our *ego user* and another tagged user. In fact, in our work, we do not take into consideration only the number of interactions the way [12] did because for any existing interaction between two users after the date d, a weight value is calculated. This value is a real number limited between 0 and 1. It is fixed according to the date of this interaction. That is to say, the more this interaction is recent, the more its value is near to 1 and vice versa. Formula 2 is used to calculate the weight value for any interaction between the $user_x$ and his/her friend $user_y$:

$$W_{int_i}(User_x, User_y) = \frac{100 - (Time(int_i) * 10)}{100} \tag{2}$$

Where *Time (int_i)* is a function that calculates the year duration of the interaction i from the time of its apparition.

This function return a real variable that represents the age of this interaction with year. For example, a 40-day interaction has a value of $W_{int} = 0,99$, because the function **Time (int) = 0,1** (40/365). This means, we take into consideration only the first number after the decimal point. But in another example, the weight of an interaction between $user_x$ and $user_y$ that is two and a half years old (**Time(int) = 2,5**) has a value of $W_{int} = 0,75$ and the weight of an older interaction has a value inferior to this weight, etc. (in our work, we do not take into consideration interactions that are older than ten years old). Then, we calculate the sum of values of all interactions of the same kind between $user_x$ and $user_y$ from this prefixed date d with the following formula. N is a social activity between these users (i.e., Tag, Like, Comment, etc.).

$$N(User_x, User_y)^d = \sum_{i \in \text{interaction kind N}} W_{int_i}(User_x, User_y) \tag{3}$$

Where i is an interaction of kind N. N is a social activity between these users. For example, if N is a Tag, we calculate the sum of weight of all Tags between $user_x$ and $user_y$ that exists after date d. Formula 3 can be applied for any social activity.

Different social networks have different types of social activities among users. To calculate a "level of trust" between two friends in a social network, we have to describe all social activities among these friends taking into consideration the temporal factor. [12] have classified social activities but they did not take into consideration the video interaction between friends in Facebook. In our work, we have added all video interactions between social network friends (friends tagged in the same video, friends who liked or commented on a video). Table 1 describes all the social activities:

Table 1. Lists of $user_x$' facebook friends classified by different social activities on facebook

List label	Description of the list	Weight assigned
$PT(user_x)$	List of friends who are tagged on a same photo with the Facebook $user_x$	W_{PT}
$S(user_x)$	List of friends who write on the Facebook $user_x$'s *Wall*	W_S
$C(user_x)$	List of friends who leave comments on the Facebook $user_x$'s *Wall*	W_C
$L(user_x)$	List of friends who like posts on the Facebook $user_x$'s *Wall*	W_L
$I(user_x)$	List of friends who write to the Facebook $user_x$'s inbox	W_I
$M(user_x)$	List of friends on whose *Walls* the Facebook $user_x$ writes or comments	W_M
$PL(user_x)$	List of friends who like the Facebook $user_x$'s photos	W_{PL}
$PC(user_x)$	List of friends who leave comments on the Facebook $user_x$'s photos	W_{PC}
$VT(user_x)$	List of friends who are tagged on a same video with the Facebook $user_x$	W_{VT}
$VL(user_x)$	List of friends who like the Facebook $user_x$'s video	W_{VL}
$VC(user_x)$	List of friends who leave comments on the Facebook $user_x$'s video	W_{VC}
$F(user_x)$	List of all friends of the Facebook $user_x$	–

Social activities of the $user_x$ are described as the ***social_activity_set(User_x)***. In other words, it is the set of lists where every list describes a different social activity. The lists summarized in Table 1 (except list $F(user_x)$) form a complete set of social activities between the $user_x$ and all of their Facebook friends:

$$Social\_activity\_set(User_x) = \{PT(user_x), S(user_x), C(user_x), L(user_x), I(user_x), M(user_x),$$
$$PL(user_x), PC(user_x), VT(user_x), VL(user_x), VC(user_x)\}$$

$$(4)$$

Every list from the set ***social_activity_set(User_x)*** contains a list of users who have this specific activity with the $user_x$ (Formula 1 is an example of the list of users who are tagged with the $user_x$).

In order to calculate trust from interactions' data contained in the ***social_activity_set(User_x)***, we multiply the specific activity data from every list in the ***social_activity_set(User_x)*** with certain weights peculiar to every interaction type. Formula 5 is used to calculate the *level of trust* between the social network *ego user* ($user_x$) and one of his/her friends ($user_y$):

$$Trust\left(User_x, User_y\right)^d =$$
$$\frac{\sum_{N \in social\_activity\_set(User_x)} W_N \times N\left(User_x, User_y\right)^d}{\sum_{N \in social\_activity\_set(User_x)} W_N}, \forall\, User_y \in F(User_x) \tag{5}$$

Where $user_x$ is the *ego user* and $user_y$ is a friend with this *ego user*. $Trust\left(User_x, User_y\right)^d$ calculates the level of trust between the *ego user* and a direct friend with him/her according to all interactions between these users from a date d. After the calculation of the level of trust between the *ego user* and all his/her friends, we conserve the 10 friends that have the higher level of trust as the *"Trusted Friends"* for this *ego user*.

4 Facebook Application for Identifying Trusted Friends

"Trusted Friends" is the Facebook application responsible for identifying trusted friends for a particular Facebook user. The implemented Facebook application collects activities (i.e., "liking" and commenting posts, inbox information "writing" on the Facebook Wall, etc.) of friends and calculates a trust level between the *ego user* and all of their friends.

The data collection process includes collecting personal and behavioral data of the ego user when connecting to his/her Facebook account, installing the *"Trusted Friends"* application, and allowing the application to access all these information (photos, user's Wall, user's inbox). Then, the system will extract some data to determine the user's trusted friends. We mean by extracted data, the user's list of friends, and those who write and make comments on the ego user's wall. The list also comprises friends tagged with the ego user in their photos, friends who like posts on the ego user's Wall, friends who comment on the ego user's Wall, friends who write to the ego user's inbox, friends on whose Walls the ego user writes or comments, friends who like ego user's photos, and finally, friends who comment on ego user's photos. When we extract any information for the user profile, we extract also its date when it happened. The data processing consists of a grouping of interactions (social activities) according to their type. Then for every friend, we will calculate the sum of their social activities taking into consideration the weight of every interaction to classify user's friends according to the sum of their social activities. As it has been explained, every social activity has a weight that represents its importance for determining trust between users. After making some experimentations and taking into consideration the result found by [12], we have proven the following: When we give more weighting to the lists of friends who write on the Facebook $user_x$'s wall and inbox and the list of friends on whose walls the Facebook $user_x$ writes or comments, we have the closest finding to the manual user's trust determination. In the rest of our work, we will use the following set for weighting every social activity presented in Table 2:

Table 2. Weights assigned to social activities

W_{PT}	W_S	W_C	W_L	W_I	W_M	W_{PL}	W_{PC}	W_{VT}	W_{VL}	W_{VC}
2	4	3	2	5	5	1	1	2	1	1

To test our application, we have taken into consideration only the users who have installed the *"Trusted Friends"* application and allowed the application to access all the *ego user*'s photos, the *ego user*'s Wall, and the *ego user*' inbox.

We have chosen an objective formula to assess our suggested method. By objective formula, we mean a mathematical calculation of the *precision measure* used for the evaluation of recommender systems. Precision refers to the quality of recommendation (In other words, at what point those proposed friends confirm to the friends manually chosen by the user). To calculate the precision of our method, every user who tested our application will be asked to choose his/her trusted friends manually at first. Then we apply our algorithm and calculate its trusted friend automatically with our method. After that, we calculate the precision measure of our algorithm for this *ego user*.

$$Precision = \frac{number\ of\ pertinent\ suggestions}{number\ of\ trusted\ friends\ suggested} \qquad (6)$$

The recall, meanwhile, is the number of recommendations that are relevant (number of pertinent suggestions) divided by the total number of recommendations (number of manually selected friends) that exist in the relevant data set in question. So, the recall highlights the portion of the relevant recommendations that has been returned to the user from the total set of relevant recommendations.

$$Recall = \frac{number\ of\ pertinent\ suggestions}{number\ of\ manually\ selected\ friends} \qquad (7)$$

While the precision measures the proportion of relevant recommendations and the recall measures the proportion of relevant recommendations that appear in the system's recommendations, the F-measure considers both measures simultaneously in order to calculate the weighted score. Thus, the F-measure can be interpreted as a weighted average of accuracy and recall, which has a value between 0 and 1, and which indicates the overall usefulness of the recommendation list. Good accuracy does not necessarily mean a good recall.

The "Trusted Friends" application calculates the user's trust level for each friend, then offers the user friends who have a confidence level above a given threshold. In order to determine the best threshold that must be considered later in our final proto-type, we tested our method on a dynamic threshold. That is to say, we evaluated our method with a threshold varying between 0 and 2 (friends who have a threshold greater than 2 are still considered as trusted friends). In other words, for each user, the pre-cision and the recall for several values of the threshold variable have been calculated in order to be able to determine the optimal threshold to be chosen in the future (when we apply our method of determining trusted friends in a recommendation system).

5 Evaluation and Comparison

An experiment is carried out on 80 users who have tested our application in order to be able to evaluate the accuracy of the method of determination of the trusted friends for a user connected to his Facebook account. To test our application, we only considered users who installed the "Trusted Friends" application and allowed the app to access their profile. Each of the users, who have tested our app, will be asked to choose their trusted friends manually at the beginning. Then we apply our method to automatically determine trusted friends. Then, we calculate the precision of the method for this user. Figure 1 shows the accuracy, recall, and average F-score of all users who tested our application.

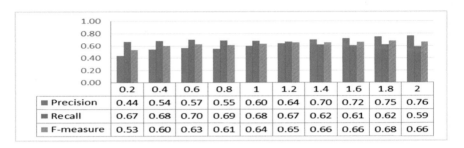

Fig. 1. Assessment of the method "Trusted Friends"

We note in Fig. 1 that the best F-measure obtained (F-measure = 0.68) is with a threshold equal to 1.8. We deduce then that this value (threshold = 1.8) is the optimal value of the threshold that must be applied in the future to set the threshold and automate the task of determining trusted friends.

Our method of calculating confidence is an improvement of the method of [12] by taking into consideration the age of each interaction between users and also adding other types of interactions as inputs. Therefore, we need to compare our method with that of [12]. For this, we will need to evaluate the confidence calculation method proposed by [12]. Indeed, the same users who tested our Android application "Trusted Friends" also tested the "Closest Friends" app. So for the evaluation of the method, we collected the different results given by this application "Closest Friends" in order to

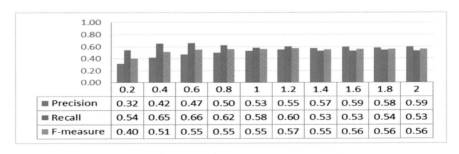

Fig. 2. Assessment of the method "Closest Friends"

compare it with our application "Trusted Friends". Figure 2 shows the accuracy, recall, and average F-score of all users who tested the "Closest Friends" app.

Subsequently, we compared the precision and recall of the two methods "Trusted Friends" and "Closest Friends" to show the importance of the improvements we have made. Figure 3 shows a comparison of the average F-measure (the average of F-measures of all thresholds) for each user who tested the "Closest Friends" application and the "Trusted Friends" application.

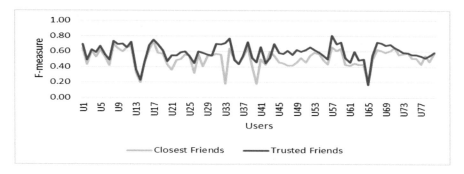

Fig. 3. Comparison of results between "Closest Friends" and "Trusted Friends" algorithms for every user

After having tested the application by several users and on several thresholds, we calculate the average of the precisions for all the thresholds applied. The value of the precision obtained shows the importance of taking into consideration the time factor (time of appearance of each interaction) and also the importance of adding other types of interactions for the calculation of the trust value. In fact, we have taken into consideration the interactions that are added to any video sequence shared by our user. The following figure shows that the mean precision of the *"Trusted Friends"* method is higher than the value found with the *"Closest Friends"* method of [12]. Figure 4 shows the result of the comparison between the two methods.

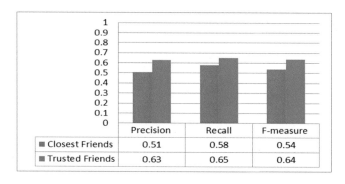

	Precision	Recall	F-measure
■ Closest Friends	0.51	0.58	0.54
■ Trusted Friends	0.63	0.65	0.64

Fig. 4. Comparison between the methods "Closest Friends" and "Trusted Friends"

The average of the precision that we found is superior to the value found by [12]. We can conclude, therefore, that the more interaction types we add (like video interactions between social network friends) the more the accuracy of trust calculation between friends can be improved. Furthermore, when we take into consideration the time duration of every interaction between the ego user and his/her friend and when we give a weight for the interaction taken into consideration its temporal factor, we can improve trusted friends' determination process and give more precise results. As a result, we can say that trust is sensitive to the interaction time between social network users. Many models were proposed to represent computational trust in different applications of social web. However, even models that represent trust as incremental measurement device do not give enough importance to the time axis. Consequently, exploiting temporal context is found to be an effective approach to improve recommendation performance.

6   Conclusion and Future Work

In this research work, we proposed a method for a time-sensitive implicit trust determination between Facebook friends and demonstrated its significantly better efficiency to identify the trusted friends of a user using an objective statistical evaluation method. Trust can play an important role across recommender systems. In our future work, we will be interested in integrating this method in social recommendation mobile applications able to run in smartphones in order to improve recommendation accuracy. Such mobile applications can also be useful to add geographic proximity as a factor in our recommender system through the global positioning systems currently available on most mobile devices.

References

1. Rathod, A., Indiramma, M.: A survey of personalized recommendation system with user interest in social network. Int. J. Comput. Sci. Inf. Technol. **6**, 413–415 (2015)
2. Papagelis, M., Plexousakis, D., Kutsuras, T.: Alleviating the sparsity problem of collaborative filtering using trust inferences. In: Trust Management, pp. 224–239. Springer, Berlin (2005)
3. Golbeck, J.: Trust and nuanced profile similarity in online social networks. ACM Trans. Web (TWEB) **3**, 12 (2009)
4. Haydar, C., Boyer, A., Roussanaly, A.: Time-aware trust model for recommender systems. In: International Symposium on Web Algorithms (2014)
5. Campos, P.G., Díez, F., Cantador, I.: Time-aware recommender systems: a comprehensive survey and analysis of existing evaluation protocols. User Model. User-Adap. Inter. **24**, 67–119 (2014)
6. Lewis, J.D., Weigert, A.: Trust as a social reality. Soc. Forces **63**, 967–985 (1985)
7. Mayer, R.C., Davis, J.H., Schoorman, F.D.: An integrative model of organizational trust. Acad. Manag. Rev. **20**, 709–734 (1995)

8. Guo, G., Zhang, J., Thalmann, D., Basu, A., Yorke-Smith, N.: From ratings to trust: an empirical study of implicit trust in recommender systems. In: Proceedings of the 29th Annual ACM Symposium on Applied Computing, pp. 248–253. ACM (2014)
9. Gupta, S., Nagpal, S.: Trust aware recommender systems: a survey on implicit trust generation techniques. Int. J. Comput. Sci. Inf. Technol. **6**(4), 3594–3599 (2015)
10. Haydar, C.: Les systèmes de recommandation à base de confiance. Université de Lorraine (2014)
11. Selmi, A., Brahmi, Z., Gammoudi, M.M.: Trust-based recommender systems: an overview. In: 27th IBIMA Conference (2016)
12. Podobnik, V., Striga, D., Jandras, A., Lovrek, I.: How to calculate trust between social network users? In: 2012 20th International Conference on Software, Telecommunications and Computer Networks (SoftCOM), pp. 1–6. IEEE (2012)
13. Frikha, M., Mhiri, M., Zarai, M., Gargouri, F.: Time-sensitive trust calculation between social network friends for personalized recommendation. In: Proceedings of the 18th Annual International Conference on Electronic Commerce: e-Commerce in Smart Connected World, p. 36. ACM (2016)

Delay and Quality of Link Aware Routing Protocol Enhancing Video Streaming in Urban VANET

Emna Bouzid Smida[✉], Sonia Gaied Fantar, and Habib Youssef

PRINCE Lab, University of Sousse, Sousse, Tunisia
emna.bouzid@isetso.rnu.tn, soniagaied3@gmail.com,
habib.youssef@fsm.rnu.tn

Abstract. Multimedia communication over VANET let drivers, passengers to capture live scenes, share, and access stored content in order to be informed of a specific event or to be entertained and have an enjoyable trip. The main concern of video streaming transmission in this highly dynamic network is the improvement of the Quality of Experience (QoE). Indeed, one of the critical issues in VANET is the design of suitable routing protocol dealing with the QoE requirements. In this paper, we focus on multi-hop level routing and we propose a Delay and Quality of Link aware cross-layer Routing (DQLR) protocol enhancing the perceived video quality. In the aim to find out a route from communicating vehicles, we define a set of fuzzy rules for simultaneously minimizing the end-to-end delay and the Packet Loss Ratio and maximizing the qualities of links. To demonstrate the efficiency of DQLR for video streaming, we compare it with the AOMDV protocol in terms of QoE and QoS parameters. The simulation results show that our protocol achieves better performance compared to AOMDV in terms of PSNR (Peack Signal to Noise Ratio), MOS (Mean Opinion Score), SSIM (Structural SIMilarity), end-to-end delays and frame loss ratio.

Keywords: VANET · Video streaming · QoE · Fuzzy · Unicasting · Routing · Delay · Quality of link · Packet Loss

1 Introduction

In Vehicular ad hoc network, nodes move with different speeds and directions. As vehicles mobility increases, the VANET topology is changing rapidly. Hence, the dynamic and changeable vehicles moving in opposite directions or changing directions at cross road will result on very little connection time. There are several others factors which cause frequent link quality degradation and or link disconnection such as traffic density, and weather or environmental condition etc. Therefore, VANET is a challenging environment for the video content transmission due to the above mentioned factors, and the video streaming requirements. The design of a feasible solution for the better delivering of video content over VANETs has to be in agreement with constrained characteristics of this network. The heterogeneity of this network in terms of speed and quality of links between vehicles are the crucial factors to be considered in

© Springer Nature Switzerland AG 2020
A. Abraham et al. (Eds.): ISDA 2018, AISC 941, pp. 556–566, 2020.
https://doi.org/10.1007/978-3-030-16660-1_54

the design of routing protocols. Besides, the increasing necessities for media contents bring another important issue to be considered which is the perceived video quality or QoE. In this research, we are interested in constraints imposed by routing and management of QoE necessities. The selection of relaying vehicles is achieved according to a set of fuzzy rules. Hence, the cross-layer routing protocol, implemented in NS3, aims to minimize the distance with the number of hops in order to minimize the delay of the communication between vehicles and selecting the relaying vehicles having the best qualities of links to maintain the communication without breaking and enhance the QoE (Quality of Experience) of received video. The performance evaluation of this proposed protocol is compared to AOMDV (Ad-hoc On-demand Multipath Distance Vector routing in VANET) [1]. The remainder of this paper is structured as follows. We will provide an overview of the related work in the next section. In Sect. 3, we will present DQLR routing protocol with a set of fuzzy rules, modified RREQ fields and the route discovery process. Section 4 will present the performance evaluation of DQLR in terms of perceived video quality and the final section will conclude this paper.

2 Related Work

In this section a synthesis of the most relevant proposals dealing with routing approaches and them enhancing the delivery of video content in VANET is presented. An improved AODV introduced in [2], uses the speed and distance between vehicles to optimize the route discovery and enhance the route selection process. The proposed metric LET (Link Expiration Time) reveals an interesting improvement of the packet delivery ratio and a minimizing of the overhead. But, this metric can not reveal the actual quality of a link. In [3], a Link Quality Prediction metric (LQ) was proposed. This metric is based on the ETX metric and the prediction of the future locations of vehicles. This metric can not usually reflect the link availability always when the vehicle is changing its direction. In [4], the authors proposed the R-AOMDV protocol to reduce the route discovery occurrence by using the multipath routing concept. R-AOMDV combines two cross-layer routing metrics which are the number of hops and retransmission counts. Two different scenarios were considered: the high and low densities. However, the performance evaluation of this protocol suffers from packet loss and delay. An interested VANET-routing protocol called link-state aware geographic routing protocol (LSGR) is proposed in [5]. In LSGR, the authors were based on a routing metric called expected one-transmission advance (EOA) which is a ratio of the distance between two vehicles and the ETX metric. Compared to others geographic routing protocols, their proposed protocol achieves better packet delivery rate and network throughput. However, the proposed metric is not sufficient to reflect the state of a link compared to our main proposed metric which is the link quality based on SNR (Signal to Noise Ratio) value. In the above mentioned works, the video streaming transmission constraints are not considered in their proposed protocols. An interesting research on video transmission over VANET was performed in [6]. The main objective of the authors is to describe in details the process of video dissemination over VANETs. They present a remarkable evaluation of existing solutions like network coding techniques and delay-based approaches. They propose the necessity of a cross-layer design in order to achieve better delivery ratio and minimum end-to-end delay. In [7], the authors propose a geographic backbone-based unicasting routing protocol for video transmission. The next locations of vehicles are predicted with a Bayesian model. One of the main differences between our protocol and the ones mentioned above is our main routing metric which is the Quality of link. Another important difference is at the performance evaluation where we are based on QoE metrics which are neglected in the most works above. Hence, we propose in the following section a cross-layer routing protocol which will improve the perceived video quality according to the proposed routing metric.

3   Delay and Quality of Link Aware Routing Protocol

In this section, we propose a reliable routing protocol which is appropriate for video streaming communication in VANET. By exchanging information of the neighboring vehicles in order to update the "NeighborsTable", the routing process is achieved by minimizing the delay and the packet loss and maximizing the link quality. The proposed protocol consists on a two preliminary phases: the first one is collecting information of neighboring vehicles and the second one is monitoring the vehicles Qualities of links. Based on these above mentioned phases, the selecting of the routing path with a route discovery phase and the route maintenance will be faster achieved. In multi-hop communication network such as in VANET, more than one path can exist to transfer a packet between a source and destination nodes. The adopted routing algorithm is charged of selecting the best path based-on some routing metric. Routing metric is a weight or a cost needed to take decision of the path that will transfer the data such as hop-count or number of hops. The hop-count metric can be coupled to cumulative distances between source and destination in order to minimize the transmission delay. However in wireless networks, wireless links are affected by many factors such as environmental circumstances and or mobility scenarios. This fact motivates us to propose link quality metric for routing protocols. This metric will enhance the quality of communication and especially the perceived video quality.

3.1   Collecting Information of Neighboring Vehicles

The mobility information such as speed, distance between vehicles and direction has a significant impact on the selecting of the routing path. The distance between two vehicles or between a vehicle and an RSU on a road can be calculated with the position coordinates of GPS system. The vehicle gathers information about the neighboring vehicles by listening to beacon messages. Then, the vehicle creates a table, called "NeighborsTable", storing the neighboring vehicles information. For example, a vehicle maintains a list of nearby vehicles where the distance between them is calculated from the position coordinates collected from GPS system. We suppose that each vehicle send to neighboring vehicles a beacon message. This message contains the vehicles positions, speed and direction. The "NeighborsTable" contains the distance between the current vehicle and neighboring vehicles and the calculated SNR of a received beacon message as shown in the Fig. 1. Also this table contains the number of hops (Nh_i) to reach the vehicle, the calculated quality of link (Ql_i) parameter and the Packet Loss Ratio (PL_i). In the routing process, the decision of the next-hop is achieved according to information computed in this table.

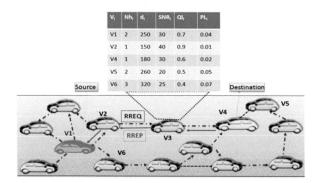

V_i	Nh_i	d_i	SNR_i	Ql_i	PL_i
V1	2	250	30	0.7	0.04
V2	1	150	40	0.9	0.01
V4	1	180	30	0.6	0.02
V5	2	260	20	0.5	0.05
V6	3	320	25	0.4	0.07

Fig. 1. NeighborsTable

3.2   Monitoring the Vehicles Qualities of Links

In order to estimate the best Qualities of links for the forwarding path, we propose a dynamic updating (after each received beacon from a neighbor vehicle) of the Quality of link for each vehicle in the NeighborsTable. The Quality of link Ql_i of each vehicle is updated according to SNR_i (Signal to Noise Ratio) and speed parameters. Therefore, the quality of links between vehicles can be estimated. In order to estimate the link quality, we compute the received power between a vehicle and another vehicle according to the two-ray ground reflection model. P_r is calculated according to Eq. (1):

$$P_r = \frac{P_t * G_t * G_r * h_t^2 * h_r^2}{d_{lv}^4 * l_s} \tag{1}$$

Where: P_t and P_r are the transmitted and received power, G_r and G_t are the antenna gain of receiver and transmitter, h_t and h_r are the height transmitter and receiver antenna, d_{lv} is the distance between two vehicles and l_s is the system loss. Based on the received power, the interference power and the thermal noise power, we admit that the SNR (Signal to Noise Ratio) in decibels is calculated as in Eq. (2):

$$SNR_i = 10\log_{10}\left(\frac{a_f^2 * P_r}{P_{if} * a_f^2 + P_{tn}}\right) \tag{2}$$

Where: P_{if} represents the interference power, a_f is the amplitude of the fading channel, and P_{tn} is the thermal noise power. As the speed and direction alteration of a vehicle has an impact on the quality of communication, we calculate the Quality of link (Ql_i) of a current vehicle j and another vehicle i according to the SNR_i value of the beacon message and the average of speeds (S_{avg}) as given in (3, 4). S_i and S_j are the respective speeds of vehicle i and vehicle j. The parameter p_{dc} is the probability of directions changes of the vehicle i and the vehicle j getting their position in the roadmap, their direction and their closeness to a cross road. p_{dci} and p_{dcj} are the values of probabilities of direction change for vehicle i and vehicle j. The p_{dci} have the value 1 if the vehicle i will change direction and respectively for vehicle j.

$$Ql_i = \frac{SNR_i}{S_{avg}} - p_{dc} \tag{3}$$

$$\text{Where :} \quad \begin{cases} S_{avg} = \frac{Si + Sj}{2} \\ P_{dc} = p_{dci} \cdot p_{dcj} \end{cases} \tag{4}$$

3.3 Route Discovery

Route Request Process

The route discovery of our protocol is achieved by modified RREQ packets designed for AODV protocol [8] where we consider three additional fields which are the "cumulative distances" between relaying vehicles, "Packet Loss" and the "Link Quality" of each intermediate link. The number of hops or hop count is already designed as a field of RREQ packet of AODV. These three added fields must be updated by each intermediate vehicle receiving an RREQ packet. When a vehicle has a video content to send in a streaming mode to a destination vehicle (like in emergency cases), it broadcasts a Route Request (RREQ) for that destination. At each intermediate vehicle, when an RREQ is received, a new entry in its routing table to the source is created. If the receiving vehicle hasn't receive this RREQ before, that it is not the destination and does not have a route to the destination, it rebroadcasts the RREQ. A vehicle receiving rather than one RREQ from other intermediate vehicles will check the adequate fields for decision routing (distance and number of hops as delay metric, Packet Loss Ratio and link Quality). The vehicle records the route of the intermediate vehicles having the best quality of link and the minimum packet loss and delay (minimum distance and hop count) which are calculated according to fuzzy rules (Table 1). This step will ensure that the routing path will be towards vehicles having the best link quality, minimums delay and Packet Loss. The delay D_P of a routing path or a sub-path (at an intermediate vehicle) is calculated as in Eq. (5):

$$D_P = (d_i + Nh_i) + \left(\sum_j^P (d_j + Nh_j) \right) \tag{5}$$

Where $(d_i + Nh_i)$ indicates the delay of a link between the current vehicle and the next-hop vehicle and $\left(\sum_j^P (d_j + Nh_j) \right)$ indicates the sum of the delay of vehicles links of the routing path or the sub-path P. Each intermediate vehicle selects the next-hop vehicle to forward the RREQ which fulfill the minimum delay (Eq. (5)), the minimum packet loss (Eq. (6)) and the maximum quality of link (Eq. (7)), as described by the set of fuzzy rules.

In wireless environments, links may have various Packet Loss ratios. A vehicle may need to retransmit a packet several times on a link having a high packet loss ratio, which will degrade the perceived video quality. The Packet Loss can be calculated during consecutive times by the vehicle in order to be outputted as routing metric. Admitting that two neighbor vehicles are previously communicating between each other, hence each vehicle can compute the Packet Loss Ratio. This calculated ratio will be added to the field Packet Loss of RREQ packet. This field will accumulate all the packet loss of the links travelled by an RREQ packet as in Eq. (6):

$$PL_P = PL_i + \left(\sum_j^P PL_j \right) \tag{6}$$

Where PL_i indicates the Packet Loss Ratio of a link between the current vehicle and the next-hop vehicle and $\left(\sum_j^P PL_j \right)$ indicates the sum of the Packet Loss Ratios of vehicles links of the routing path or the sub-path P.

The field "Link Quality" of modified RREQ packet is calculated according to Eqs. (3, 4 and 7).

$$Ql_P = Ql_i + \left(\sum_j^P Ql_j \right) \tag{7}$$

Where Ql_i indicates the Quality of a link (Eq. 3) between the current vehicle and the next-hop vehicle and $\left(\sum_j^P Ql_j \right)$. indicates the sum of the Qualities of vehicles links of the routing path or the sub-path P.

The destination vehicle, receiving one or more RREQ checks the field "Cumulative Distances", "Packet Loss" and "LinkQuality" by previously calculated distances and Packet Loss Ratios between this vehicle and the vehicle(s) originating of RREQ(s). If multiple RREQs are received by the destination vehicle, the route with the minimum hop count and minimum distances (minimum delay), minimum Packet Loss and the best Link Qualities is chosen according to Fuzzy rules. This process will lead to a minimization of the transmission delay and the Packet Loss Ratio of video streaming and a maximization of the link qualities which will enhance the perceived video quality.

Route Reply Process

If the receiving vehicle of broadcasted RREQ is the destination or an intermediate vehicle having a valid route to the destination, it generates Route Reply (RREP). The packet RREP is transmitted by the same route discovered by RREQ in a hop-by-hop mode to the vehicle source. As the RREP propagates, each intermediate vehicle creates a new entry in its routing table to the destination. When the source vehicle receives the RREP, it records the discovered route to the destination and begins sending content.

3.4   Route Maintenance

The route maintenance of our proposed protocol is processed with the same manner as AODV protocol. If a failure in the route is detected during a data transfer or in a current video streaming, a Route Error (RERR) is sent to the source of the content in a hop-by-hop fashion. As the RERR is transmitted toward the source, each intermediate vehicle invalidates unreachable destination. When the source of the content receives the RERR, it deletes the route and re-initiates another route discovery process.

3.5   Fuzzy Formulation of DQLR

DQLR aims to find the best multi-hop route from the source vehicle to the destination vehicle. The efficiency of a multi-hop route depends on all direct wireless links that constitute the route. With the help of fuzzy logic DQLR estimates whether a Path is good or not by considering multiple metrics which are the Delay (Eq. 5), the Packet Loss (Eq. 6) and Link Qualities (Eq. 7) and uses this approach to select a route in a way that can provide multi-hop reliability and high efficiency.

Table 1. DQLR fuzzy rules

Rules	Delay (D_P)	Packet loss (PL_P)	Links qualities (Ql_P)	Evaluation
1	Low	Low	High	Very good
2	Low	Low	Medium	Good
3	Low	Low	Low	Fair
4	High	High	Low	Very bad
5	High	Medium	Low	Bad
6	Medium	Medium	Medium	Good
7	Medium	Medium	High	Fair
8	Medium	Low	High	Good
9	Low	Medium	High	Good
10	High	Medium	High	Bad

4   Performance Evaluation

To simulate our proposed routing protocol and its impact on the video streaming transmission over VANET, we used a network simulator (NS3) [9]. To inject and evaluate the video streaming over NS3, we used Evalvid [10]. In our simulation of DQLR we used two video sequences with different formats. The first one is the test video sequence called "*akiyo*" having 300 frames with a frame size 352 × 288 pixels (CIF format). The second is called "*tempete*" having 260 frames with a frame size 176 × 144 pixels (QCIF format). The frame rate of the two sequences is 30 frames/second. We had encoded the sequences with H.264/ffmpeg. During our simulation, the channel bandwidth is 10 MHz and the vehicle speeds are 20 m/s and 30 m/s. A routing process is taken place between 8 vehicles placed at an area of 1500 m * 1500 m. The below table summarizes the simulation parameters (Table 2).

Table 2. Simulation parameters

Parameter	Value
Vehicles speeds	20 m/s and 30 m/s
MAC, PHY parameters	IEEE 802.11p
Channel bandwidth	10 MHz
Rate	6 Mbps
Propagation transmission model	Two ray ground
Number of vehicles	8
Videos	tempete (QCIF 176 × 144 pixels) akiyo(CIF 352 × 288 pixels)
Simulation area	1500 m * 1500 m
Distances between moving vehicles	150–180 m

4.1 PSNR

The results in Figs. 2 and 3 demonstrate higher values of DQLR values of PSNR for CIF and QCIF format compared to AOMDV. The Fig. 2 shows an analysis of the video quality PSNR metric (PSNR: Peak Signal to Noise Ratio) of the reconstructed received "akiyo" video sequence between AOMDV protocol and our proposed protocol. With AOMDV, the PSNR of "akiyo" video sequence (CIF) decreases to the value 29 dB at frames [211–213] however with our proposed protocol the minimum value is 30 dB. The average of PSNR values for "akiyo" video is for AOMDV 32.93 dB and for DQLR 34. 87 dB. For the "tempete" video DQLR outperforms AOMDV since the most DQLR's PSNR values are between [17 dB–18.5 dB], however AOMDV values are between [16 dB–17.5 dB]. The average of PSNR values for tempete video is for AOMDV 16.13 dB and for DQLR 17.27 dB. The beginning video frames of "tempete" and "akiyo" are distressed by a quality degradation (low PSNR values) due to frame loss and the higher time range taken by AOMDV in the discovering of the routing path compared to our approach.

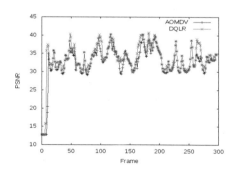

Fig. 2. PSNR of akiyo received frames

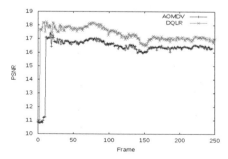

Fig. 3. PSNR of tempete received frames

4.2 MOS and SSIM

The metric MOS, reflecting the user-satisfaction on the perceived video quality, is correlated to PSNR metric since its ranges are dependent on PSNR values. The outperformance of DQLR on the average MOS of "akiyo" and "tempete" videos are shown in Fig. 4. With CIF format (akiyo), the two compared protocols are in "Good" range since PSNR values are between 31 dB and 37 dB. With QCIF format (tempete), the two compared protocols are in "Bad" range since PSNR values are lower than 20 dB, and hence, the MOS is 1 for the two protocols.

The metric SSIM improves the PSNR by measuring the frames based on their structural, luminance and contrast similarity. The Fig. 5 demonstrates the average SSIM, measured by "MSU Video Quality Measurement Tool" [11] of our protocol compared to AOMDV. SSIM results reveal that DQLR transmits video sequences with an enhanced quality level. This means that the decoded videos have an acceptable correlation with the original video flows. For the majority frames, the "tempete" SSIM values for DQLR are greater than the value 0.35, however, the minimum value for AOMDV is 0.27 for frame 150. The pic-value for DQLR is 0.46 for the same video, however, its maximum value with the same metric is 0.36 for frame 17 and 0.35 for frame 223 with AOMDV. The average of SSIM values is for AOMDV 0.308 and for DQLR 0.413. The outperformance of our solution on the average SSIM of akiyo and tempete videos are shown in Fig. 5.

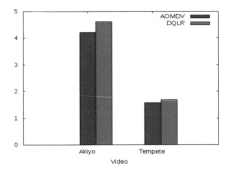

Fig. 4. Average MOS of tested videos

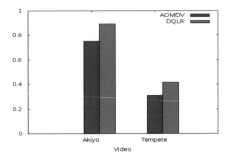

Fig. 5. Average SSIM of tested videos

4.3 End-to-End Delay

Another crucial parameter for video quality is the en-to-end delay (Fig. 6-akiyo video and Fig. 7-tempete video). This score reaches its pic-values 0.20 s for frame 31 and the value 0.22 s for frame 23 of tempete video with AOMDV. For DQLR, the pic-values are: 0.12 s for frames 37 and 38, and 0.17 s for frame 23. Besides, the majority of AOMDV end-to-end delays values are higher than DQLR values.

4.4   Frame Loss

With AOMDV, the overall frame loss for tempete video frames is 6.62% where for I frames the percentage of loss is 9%, and for P frames it is 4.62%. This high loss level degrades the perceived video quality. However, with our proposed protocol, the overall frame loss is 3.41% and the respective Frame loss percentage is over 3.65% (I frames) and 3.16% (P frames). The below table summarizes the simulation results of the two tested video sequences with our approach and AOMDV protocol (Table 3).

Table 3. Video quality parameters

Parameter	Akiyo		Tempete	
	AOMDV	DQLR	AOMDV	DQLR
Average PSNR	32.93	34.87	16.13	17.27
Average SSIM	0.753	0.891	0.308	0.413
Average MOS	4.21	4.62	1.57	1.68
Overall frame loss	3.35%	2.25%	6.62%	3.41%

The lower values of frame loss and end-to-end delays, and the better results of PSNR, MOS and SSIM values compared to AOMDV explain the enhancement of the perceived video quality of our solution as shown in Figs. 8 and 9.

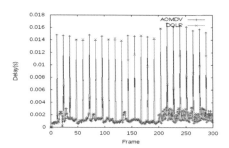

Fig. 6. Akiyo frames end-to-end delays

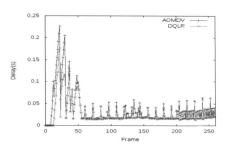

Fig. 7. Tempete frames end-to-end delays

Fig. 8. Tempete with DQLR

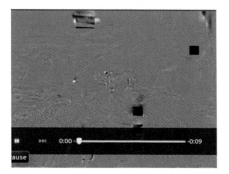

Fig. 9. Tempete with AOMDV

5   Conclusion

In this paper, we have proposed a routing protocol that optimizes the Quality of Experience of video streaming in vehicular network environment by minimizing the delay of transmission, monitoring the vehicles quality of link and taking decision of relaying vehicles according to delay, packet loss and link Quality. Our results demonstrate a performance enhancement in terms of QoE parameters PSNR, MOS and SSIM in comparison with AOMDV protocol. Moreover, the QoS parameters (frames end-to-end delay and frame loss) are also enhanced. In our future work, we will improve this routing protocol by selecting links according to their stabilities against different moving scenarios in urban context.

References

1. Ledy, J., Poussard, A.-M., Vauzelle, R., Hilt, B., Boeglen, H.: AODV enhancements in a realistic context. In: International Conference on Wireless Communications in Unusual and Confined Areas (ICWCUCA), pp. 1–5, August 2012
2. Ding, B., Chen, Z., Wang, Y., Yu, H.: An improved AODV routing protocol for VANETs. In: 5th International Conference on Wireless Communications, Networking and Signal Processing (WCSP), pp. 1–5, November 2011
3. Ngo, C.T., Oh, H.: A link quality prediction metric for location based routing protocols under shadowing and fading effects in vehicular ad hoc networks. In: International Symposium on Emerging Inter-networks, Communication and Mobility (EICM) (2014)
4. Chen, Y., Xiang, Z., Jian, W., Jiang, W.: A cross-layer AOMDV routing protocol for V2V communication in urban VANET. In: Proceedings of IEEE 5th International Conference on Mobile Ad-Hoc Sensor Network (MSN), pp. 353–359, December 2009
5. Li, C., Wang, L., He, Y., Zhao, C., Lin, H., Zhu, L.: A link state aware geographic routing protocol for vehicular ad hoc networks. EURASIP J. Wirel. Commun. Netw. **2014**, 176 (2014)
6. Naeimipoor, F., Rezende, C., Boukerche, A.: Performance evaluation of video dissemination protocols over vehicular networks. In: IEEE 37th Conference on Local Computer Networks Workshops (LCN Workshops), pp. 694–701 (2012)
7. Rezende, C., Ramos, H.S., Pazzi, R.W., Boukerche, A., Frery, A.C., Loureiro, A.A.: VIRTUS: a resilient location-aware video unicast scheme for vehicular networks. In: IEEE International Conference Communications (ICC), pp. 698–702 (2012)
8. https://www.ietf.org/rfc/rfc3561.txt, Experimental [Page 2], RFC 3561 AODV Routing, July 2003
9. Ns3: https://www.nsnam.org/
10. Klaue, J., Rathke, B., Wolisz, A.: EvalVid - a framework for video transmission and quality evaluation. In: 13th International Conference, TOOLS 2003, Urbana, IL, USA, 2–5 September 2003
11. http://www.compression.ru/video/quality_measure/

Incremental Algorithm Based on Split Technique

Chedi Ounali$^{(\boxtimes)}$, Fahmi Ben Rejab$^{(\boxtimes)}$, and Kaouther Nouira Ferchichi$^{(\boxtimes)}$

ISGT, LR99ES04 BESTMOD, Université de Tunis, Le Bardo, Tunisia
chedy.ounelly@gmail.com, fahmi.benrejab@gmail.com,
kaouther.nouira@planet.tn

Abstract. Most clustering algorithms become ineffective when provided with unsuitable parameters or applied to data-sets which are composed of clusters with diverse shapes, sizes, and densities.

In our paper we present a new version of k-means method, that allows adding one new cluster to the k cluster we already had with out retraining from scratch. This method is based on the splitting process, we are looking for the cluster that had the highest score to be split, our score is based on three criteria; SSE, Dispersion-index and the size of cluster. Finally, the split process is performed by using standard K. Experimental results demonstrate the effectiveness of our approach both on simulated and real data-sets.

Keywords: Clustering · Clusters · K-Means · Incremental K-Means ·
SSE · Split · Dispersion-index

1 Introduction

Clustering has an important role in data mining, it aims at grouping N data points into k clusters so observations with in same cluster are similar, while observations in diverse cluster are different from each other.

Clustering is an unsupervised learning method as it classifies data-sets without any a prior knowledge. It has been used in many different fields such as bio-informatics, image processing, genetics, speech recognition, market research, document classification, anomalies detection and weather classification [2].

There are various algorithms for the data clustering. The k-means algorithm is one of the most popular, it is very simple in operation, suitable for unraveling compact clusters and a fast iterative algorithm [10]. k-means algorithm divides N elements from data-set for k clusters that used center-based clustering methods [10]. Consequently, the main challenge for these clustering methods is in determining the number of clusters [2]. In general, the number of clusters has been set by users or archives from knowledge of research [11]. But a bad choice of the number of cluster can lead to a wrong distribution of observations. That's how the term adaptive clustering was born. Their for incremental was always related to observations and which cluster should elements be inserted into [12].

© Springer Nature Switzerland AG 2020
A. Abraham et al. (Eds.): ISDA 2018, AISC 941, pp. 567–576, 2020.
https://doi.org/10.1007/978-3-030-16660-1_55

In this research, we study a new version of k means that allows adding a new cluster from splitting an other one with out retraining from scratch our method is based on the Sum of squared Error (SSE) calculation [17], index of dispersion [1] and Size of clusters. The rest of our paper is organized as follow: Sect. 2 contains the standard version of K-Means. Section 3 talked about the notion of incremental clustering. Section 4 present the proposed approach. Section 5 contains the results of experimentation, and the last section presents the conclusion.

2   K-Means

K-means is an iterative refinement problem that is composed of two stages, the first one is the initialization in which centroid of each cluster is assigned and a second step which is iterative and called Lloyd's algorithm. The Lloyd algorithm consist on three steps, in the first stage each element is assigned to the closest cluster. Then centroids of clusters been recalculated based on the mass of elements that where assigned on the previous step. Finally a stop criterion is reached and the clustering process will be finished at this point [2].

2.1   K-Means Algorithm

K-means algorithm is an iterative algorithm which can be described by the following steps [3].

1. Choose initial centroids $m_{1..k}$ of the clusters $C_{1..k}$.
2. Calculate new cluster membership. A feature vector x_j is assigned to the cluster C_i if and only if:

$$I = arg_{k=1...k}min||x_j - m_k||^2 \qquad (1)$$

3. Recalculate centroids for the cluster according.

$$m_i = \frac{1}{|c_i|}\sum x_j \qquad (2)$$

4. If none of the cluster centroids have changed, finish the algorithm. Otherwise go to Step (2).

The standard k-means always retrain from scratch if their is a modification in the data-set which causes a big loss of time, that's why the use of incremental clustering has been improved last years.

2.2   Distance Measure

There are different methods that the algorithm K-means uses for distance measure between these distance measures there are:

1. Euclidean distance and squared Euclidean distance, are generally calculated from row data and not from standardized data. The advantage of the Euclidean distance is that the addition of a new elements cannot influence the measure of distance between two other elements [4].
2. Manhattan distance, consider that the shortest path between two points in the xy-plain is the hypotenuse which refer to the Euclidean distance [5].
3. The Jacquard distance, is a metric measure that inform how dissimilar tow set are. It represents a complement to the Jacquard index, and it's obtained by subtracting the Jaccard coefficient from one [6].

K-means cant deal with incremental datasets. Whenever new elements added, the algorithm needs to retrain from scratch which causes loss of information and time where time is such challenge for this kind of algorithms. So to deal with this problem incremental k-meas was proposed.

3   Incremental Clustering

A large area of research in clustering has focused on improving the clustering process such that the clusters are not dependent on the initial identification of cluster representation.

Some research have present versions of adaptive clustering that allows the regeneration of clustering procedure from scratch to response to the change of elements but those techniques can produce large deference in the size of clusters and a huge waste of time [13].

Some others researcher have used the split technique in the clustering possess by transforming the data-set to a tree and every time a new element that cannot be inserted into any node the closest one will be split and a new one will be created [14]. Many other research uses the splitting technique to create an incremental clustering procedure, but the problem was always which cluster should split. The following approaches are typically used for the selection of the cluster to split [15, 16]:

1. Complete partition: every cluster is split, so obtaining a complete binary tree.
2. The cluster having the largest number of elements is split.
3. The cluster with the highest variance.
4. The shape of the cluster.

The above criteria are extremely simple. The first criteria split every cluster that provide a complete tree, but it completely ignores the issue of the quality of the clusters. The second one is also very simple: it does not provide a complete tree, but it has the advantage of yielding a "balanced" tree, where the leaves have approximately the same size. The tow last criteria is the most sophisticated in relation with the tow previous, since it is based upon a simple but meaningful property (the "scatter") of a cluster. This is the reason why highest variance criteria is the most commonly used criterion for cluster selection.

Adding to those there are other version of incremental clustering. Li et al. [7] represented a clustering boundary detection method by the transformation of affine space, this method where argued by Tong et al. [8] by claiming that boundary points are essential for clustering due to their representation of the distribution of the dataset. In literature we found that all works either search for a new manner to create a new algorithm that allows the incremental in the clustering level or used the split technique but with a simple criterion in the phase of the choice of the cluster to split. one of the most known algorithm based on k-means and uses the split technique to create new clusters is Bisecting k-means [9]; but bisecting k-means creates a complete tree from the k-cluster that given at the beginning. In our case we are searching for the cluster that had the most spreading out elements and none of these algorithm can respond to the problem.

To cover this issue we propose a new version of k means based on split technique that uses different dividing criterion.

4   Proposed Approach

Our proposed work takes into consideration the adding a new cluster to the actual distribution of cluster without retraining from scratch. In order to showcase this work their is an ultimate challenge witch is; witch cluster should we choose to be split. In literature like it was induced in the previous section their are so many different criteria to chose the cluster to split.

Our proposal work is based on three different steps to get final clusters, they are organized as follow:

1. Calculate Score for all clusters
   – Calculate SSE
   – Calculate Dispersion Index
2. Searching for the highest Score
3. Split the cluster with the highest Score using K-Mean.

For the sake of clarity Fig. 1 shows the main different stages from the choice of the cluster to split until getting the $k+1$ clusters as output.

In our work we use both of Sum of Squared Error (SSE) calculation and Index of dispersion at first level and size of clusters at a second level. Through a way or an other those tow indexes refers to all other split criteria that have been mentioned; the shape, size an the variance of clusters, because it calculate the dispersion of elements from the centroid of the cluster [17].

– **SSE** is the sum of the squared difference between each observation and its group's mean. It can be used as a measure of variation within a cluster. If all cases within a cluster are identical the SSE would then be equal to 0.
  It's formula would be like:

$$SSE = \sum_{1}^{n}(x_i - \overline{x}) \tag{3}$$

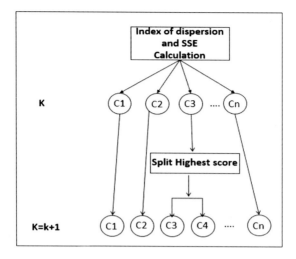

Fig. 1. Structure of the incremental k-Means

– **Index of dispersion** It quantifies if observations in a data set are dispersed or highly related to its centroids. The index of dispersion is the square of the standard deviation divided by the mean of the observation: $\frac{\sigma^2}{\mu}$ where σ calculated like:

$$\sigma = \sqrt{\frac{1}{N} \sum_{i=1}^{n} (x_i - \mu)^2} \tag{4}$$

4.1 Incremental K-Means Algorithm

– Calculate$\mu : \sum \frac{x_i}{n}$.
– Split(cluster): aim to rerun the simple k-means with number of cluster k equal to tow clusters.

– The split process is based on three different split criteria that have been induced from statistic and other used with other machine learning algorithms. The first split criteria is SSE (Sum of Error Square) which used by decision trees to split nodes and used as an evaluation criteria for clustering algorithms, the second one is the index of dispersion, and the last one is the size of clusters. The split process begins by calculating both of SSE and index of dispersion of each cluster. If the tow indexes gives tow different clusters as result then moving to the third criteria which is the size of cluster. in this case we will use the clusters that had the largest number of elements between the tow chosen. Otherwise, the algorithm will split the cluster that had the biggest SEE and Dispersion at the same time with out referring to the size criteria.
– Incremental K-Means then split the chosen cluster using Simple K-means.
– Incremental k-means algorithm stages are divided as follows:

Input: **k clusters**
Output: k+1 clusters
Begin
ID,SSE=0
S,D=First cluster
    for each cluster k **do**
    $Calculate \mu$
    $Calculate \sigma$
    $SSE_c \leftarrow \sum_{i=1}^{n}(x_i - \mu)$
    if $(SSE_c > SSE)$
    $SSE \leftarrow SSE_c$
    $S \leftarrow S_i$
      end if
    $ID_c \leftarrow \frac{\sigma^2}{\mu}$
    if $(IDc > ID)$
    $ID \leftarrow ID_c$
    $D \leftarrow D_i$
      end if
      end for
    if $(clusterS <> clusterD)$
    Split (max (cluster size (S,D))
    else if $(clusterS = clusterD)$
    Split (cluster(S)
      end if
End.

5   Experimentation

5.1   Framework

For the evaluation of our proposal, we test the Incremental-k-means using three real-world data-sets and one simulated dataset taken from UCI machine learning repository [18] and OpenML [19]. They are described in Table 1 as follows.

Table 1. Description of the used datasets

Databases	#Instances	Attributes
Airlines (A)	539382	8
Bank-data (BD)	600	12
3D road network (RD)	434874	4
BNG (vehicle) (VH)	792698	19

5.2   Evaluation Criteria

To test our Incremental k-means, we essentially use three evaluation criteria described as follows.

1. SSE(total) = $\sum(SSE_i)$ It reflect the dispersion of elements within the cluster, it gives the mean of how far elements are from the cluster center.
2. Run-time: The execution time needed to build and get the final model. From choosing the cluster to split until generating the result.
3. Used space: It indicates the amount of used memory (RAM) during the algorithm run.

5.3   Results and Discussion

Table 2. Number of instance when k = 3

Databases	A	BD	RD	VH
Cluster 1	175085	259	208329	232555
Cluster 2	254354	183	87929	256205
Cluster 3	109944	157	138616	303938

Table 3. SSE of clusters when k = 3 (E10)

Databases	A	BD	RD	VH
Cluster 1	102,02	16,51	0.317	3,05
Cluster 2	318,48	22,02	0.134	9,23
Cluster 3	41,22	12,52	0.022	5,15

Table 4. Index of dispersion when k = 3

Databases	A	BD	RD	VH
Cluster 1	43.93	165.2	4.127	8.36
Cluster 2	55.29	189.2	4.208	12.15
Cluster 3	40	170.8	4.224	9.17

By analyzing Tables 3 and 4 we notice that SSE and Index of dispersion gives both the same cluster to be splat every time with different dataset. But arriving to the RD road network dataset there is different between in the cluster choices; SSE shows that the highest dispersion is in the first cluster. Otherwise, the index of dispersion indicate the third cluster as the who contain the most dispersed elements. So the algorithm had to move to third split criteria which is the size of clusters. From this part the cluster who is going to be splat is cluster number one (Tables 2, 5, 6 and 7).

Table 5. Clusters to be split

Databases	A	BD	RD	VH
Cluster 1	175085	259	**208329**	232555
Cluster 2	**254354**	**183**	87929	**256205**
Cluster 3	109944	157	138616	303938

Table 6. Number of instance when k = 4 using dynamic k-means

Databases	A	BD	RD	VH
Cluster 1	175085	259	**99480**	232555
Cluster 2	**165714**	**108**	**108849**	**128110**
Cluster 2	**88640**	**75**	87929	**128095**
Cluster 4	109944	157	138616	303938

Table 7. Number of instance when k = 4 using k-means

Databases	A	BD	RD	VH
Cluster 1	124351	209	199760	140558
Cluster 2	207723	163	59143	139227
Cluster 2	86492	129	44090	150273
Cluster 4	120817	98	131881	147841

Table 8. Average of run time (s)

Databases	Simple k-means	Dynamic K-means
A	427	107
BD	66	46
RD	231	189
VH	1817	845

Table 9. Average of used space (MB)

Databases	Simple k-means	Dynamic K-means
A	377.9	76.6
BD	58.7	40.2
RD	391.6	58.8
VH	1140.7	679.44

Table 10. SSE total (E10)

Databases	Simple k-means	Dynamic K-means
A	608.1	160.8
BD	24.1	12.5
RD	0.017	0.0096
VH	18.3	12.07

Next, we show the distribution of elements between clusters using both incremental k-means and the standard k-means. After that, we are going to do a comparison based on SSE total, Time and used space.

- **Run Time:** Table 8 shows different run time result for simple k-means and our proposed method. Given the results, can we assume that our approach outperformed the k-means algorithm in terms of processing performance as the minimum run-time obtained by all of datasets.
- **Used space:** Table 9 represents the space used by both of dynamic k-means and the k-means algorithm. we can notice that k-means uses much more space than our approach. in term of used space too our approach is better than k-means.
- **SSE:** Table 10 illustrate the result of SSE of each datasets. it contains both results given by k-means and our approach. the dispersion within cluster using dynamic k-means is lower than those calculated by simple k-means.

6   Conclusion

We have proposed a new version of K-Means algorithm, our proposal is an incremental technique based on the split of the cluster that had the most spreading out elements.

Experimental results demonstrate that our approach perform better then standard k-means in term of element distribution between clusters. The split procedure reduces the SSE for each cluster which provide a lower SSE (total) than that given by standard K-Means. Adding to that our proposed method out performs k-means in term of run time especially when it is related to large data sets. In future work, we aim to add other criteria to the split process for better choosing the cluster to be split.

References

1. Clarke, K.R., Chapman, M.G., Somerfield, P.J., Needham, H.R.: Dispersion-based weighting of species counts in assemblage analyses
2. Yadav, A., Dhingra, S.: A review on K-means clustering technique. Int. J. Latest Res. Sci. Technol. **5**(4), 13–16 (2016)

3. Zhou, P.Y., Chan, K.C.C.: A model-based multivariate time series clustering algorithm. In: Peng, W.-C., et al. (eds.) PAKDD 2014 Workshops. LNAI, vol. 8643, pp. 805–817. Springer, Cham (2014)

4. Dalatu, P.I., Fitrianto, A., Mustapha, A.: Hybrid distance functions for K-means clustering algorithms. Stat. J. IAOS **33**, 989–996 (2017)

5. Strauss, T., Von Maltitz, M.J.: Generalising ward's method for use with Manhattan distances. PLoS One **12**(1), e0168288 (2017)

6. Surya Prasath, V.B., Alfeilat, H.A.A., Lasassmeh, O., Hassanat, A.B.A.: Distance and similarity measures effect on the performance of k-nearest neighbor classifier a review. Preprint submitted to Elsevier, 16 August 2017

7. Li, X., Han, Q., Qiu, B.: A clustering algorithm with affine space-based boundary detection. Appl. Intell. **2**, 1–13 (2017)

8. Tong, Q., Li, X., Yuan, B.: A highly scalable clustering scheme using boundary information. Pattern Recognit. Lett. **89**, 1–7 (2017)

9. Patil, R.R., Khan, A.: Bisecting K-means for clustering web log data. Int. J. Comput. Appl. **116**(19), 36–41 (2015)

10. Capó, M., Pérez, A., Lozano, J.A.: An efficient K-means clustering algorithm for massive data. J. Latex Class Files **14**(8) (2015)

11. Han, J., Kamber, M.: Data Mining Concepts and Techniques, 2nd edn. Morgan Kaufmann Publishers, Burlington (2006). Fast kernel classifiers with online and active learning **6**, 1579–1619

12. Mall, R., Ahmad, M., Lamirel, J.: Comportement comparatif des methodes de clustering incrmentales et non incrmentales sur les donnes textuelles htrogenes (2014)

13. Bao, J., Wang, W., Yang, T., Wu, G.: An incremental clustering method based on the boundary profile. PLoS One **13**(4) (2018)

14. Zhang, Y., Li, K., Gu, H., Yang, D.: Adaptive split-and-merge clustering algorithm for wireless sensor networks. In: International Workshop on Information and Electronics Engineering (IWIEE) (2012)

15. Savaresi, M., Boley, D., Bittanti, S., Gazzaniga, G.: Choosing the cluster to split in bisecting divisive clustering algorithms

16. Jain, A.K., Dubes, R.C.: Algorithms for Clustering Data. Prentice-Hall Advance Reference Series. Prentice-Hall, Upper Saddle Rive (1988)

17. Thinsungnoena, T., Kaoungkub, N., Durongdumronchaib, P., Kerdprasopb, K., Kerdprasopb, N.: The clustering validity with silhouette and sum of squared errors. In: Proceedings of the 3rd International Conference on Industrial Application Engineering (2015)

18. Bache, K., Lichman, M.: (UCI) machine learning repository (2013) http://archive.ics.uci.edu/ml

19. Rijn, J.V. (2014). https://www.openml.org/d/268

20. Brahmi, P.I., Ben Yahia, S.: Detection des anomalies base sur le clustering (2014)

Family Coat of Arms and Armorial Achievement Classification

Martin Sustek, Frantisek Vidensky[✉], Frantisek Zboril Jr.,
and Frantisek V. Zboril

FIT, Brno University of Technology, IT4Innovations Centre of Excellence,
Bozetechova 1/2, 612 66 Brno, Czech Republic
{isustek,ividensky}@fit.vutbr.cz

Abstract. This paper presents an approach to classification of family coats of arms and armorial achievement. It is difficult to obtain images with coats of arms because not many of them are publicly available. To the best of our knowledge, there is no dataset. Therefore, we artificially extend our dataset using Neural Style Transfer technique and simple image transformations. We describe our dataset and the division into training and test sets that respects the lack of data examples. We discuss results obtained with both small convolutional neural network (convnet) trained from scratch and modified architectures of various convents pretrained on Imagenet dataset. This paper further focuses on the VGG architecture which produces the best accuracy. We show accuracy progress during training, per-class accuracy and a normalized confusion matrix for VGG16 architecture. We reach top-1 accuracy of nearly 60% and top-5 accuracy of 80%. To the best of our knowledge, this is the first family coats of arms classification work, so we cannot compare our results with others.

Keywords: Coats of arms · Image classification ·
Convolutional neural network · Artificial intelligence · Machine learning

1 Introduction

Family coats of arms were used as a representation of noble families. That still holds for some families nowadays, especially in Europe. Since ancient times, there was a need for distinction of groups of people, individuals or authorities all over the world. That was achieved by means of different symbols. These symbols may have special meaning in some cases; in others, they were chosen according to the environment they originate from. Afterwards, symbol creation respected fixed rules. In Europe, those rules began to arise in the eleventh century. First, symbols were displayed only on shields, later, they started to appear on other parts like helmets, crest or mantling. Although these rules have undergone many changes, they are still used in some countries today. Coats of arms with all components are called an armorial achievement. We do not distinguish between the term *coat of arms* and *armorial achievement* in this text.

M. Sustek and F. Vidensky—Contributed equally to this work.

© Springer Nature Switzerland AG 2020
A. Abraham et al. (Eds.): ISDA 2018, AISC 941, pp. 577–586, 2020.
https://doi.org/10.1007/978-3-030-16660-1_56

Fig. 1. Two coats of arms belonging to the same class. The first image was taken from [1] and the second one was obtained from a website[1].

Despite the fact that computer vision and image processing have been applied in many areas recently, we have not observed any classification of coats of arms.

We focus on a creation of a classifier that will be able to distinguish different classes of coat of arms. Every class represents the same type of coat of arms. Coats of arms of the same family with different components are illustrated in Fig. 1.

There are thousands of family coat of arms classes. More than thousand classes [2,3] could be derived only from the Czech Republic. In order to create a classifier that could distinguish all of them is a tremendously difficult task. Trying to classify only a subset of all classes still remains a complex task. State of the art classification approach is to use one of existing convolutional neural networks architecture [4]. Large dataset is a crucial in order to obtain decent accuracy. Thus, the absence of public dataset makes the task even harder. The most of books containing coat of arms are not digitized and coats of arms are often found only in castles and churches. Therefore, our goal is to classify only 50 family coat of arms types using convnet.

2 Related Work

To the best of our knowledge, this is the first family coats of arms classification work. A recognition system for individual parts of family coats of arms was introduced in [5]. This system was able to detect coat of arms in an image. Afterward, they decomposed the image into individual parts and classified them.

Convnets have many different architectures and we will describe few of the most popular ones. Some architectures were introduced during competitions such as the ImageNet Large Scale Visual Recognition Challenge (ILSVRC)[2]. One of the most common architectures is VGG [6]. VGG iteratively reduces the input

---

[1] http://gis.fsv.cvut.cz/zamky/genealogie/harrach/php/info.php.

[2] http://image-net.org/challenges/LSVRC/.

size and increases the number of feature maps through convolution and max pooling layers. This can be understood as a feature extraction that is followed by fully connected layers. The Inception (GoogLeNet) architecture [7,8] consists of inception modules. These modules have several convolutional blocks that are connected together and each block has different filter size. Each module consists of parallel branches and their outputs are concatenated. Each branch can have multiple sequentially connected blocks. The Inception architecture has far fewer parameters than the VGG architecture, and for this reason, it is faster (even though the sequential blocks slow down the computation). The ResNet architecture allows to train a really deep networks. Residual connections were introduced to deal with a vanishing gradient problem [9]. The Xception [10] network is similar to the Inception architecture with residual connections. Inception modules are replaced by depthwise separable convolutions. Depthwise convolution has fewer parameters than standard convolution because it assumes that the pattern in feature channels is the same at each position.

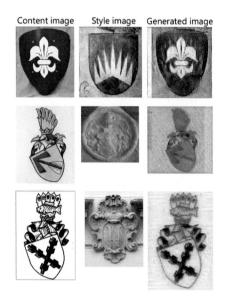

Fig. 2. An example of artificially generated coat of arms. The first column contains coats of arms that are considered to be the content that we want to keep in the generated image. The second column represents the style, the third is the output. The first combination is not far from the painting. The second resembles an old drawing on the wall and the last one almost looks like it is engraved in a concrete wall. Content image and style image on the first row were obtained from websites[3] and other content images were taken from books [1,11].

---

[3] http://www.papirovehelmy.cz/zbozi/339-erb-rodu-bekovt.html
http://www.boskovice.cz/znak-erb-a-prapor-mesta/d-22085.

3   Dataset Description

In order to train convnet, we need to collect a labeled images. We constructed our dataset using books about coat of arms [1,3,11]. Afterward, we explored more sources including websites and literature referenced from [11] to acquire even more data.

Our dataset consists of only 592 images belonging to 50 classes. The image distribution among classes is not balanced. For some classes, we have only 4 images, for others, we have 25 images. Whole distribution is shown in Fig. 3. The need for dataset extension in order to improve classification accuracy, is evident. We decide to use data augmentation to overcome this issue. Specifically, we applied Neural Style Transfer technique introduced in [12] to generate augmented dataset.

In a nutshell, this technique takes two input images and tries to apply style from the first image and content from the second one and combine them into an artificial image. It is not as simple as described, therefore, it does not always work as expected. Extracted style sometimes reflects colors, textures or various patterns in the painting. Few examples are shown in Fig. 2. We extended our dataset by 1282 artificial images.

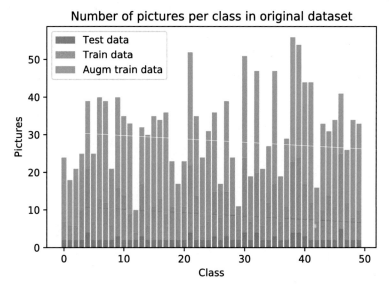

Fig. 3. The number of test, train and augmented train examples for each class in dataset. Some of augmented examples might be excluded as described above.

3.1   Train and Test Data Preparation

The lack of data forced us not to use a validation set. Therefore, to be able to observe partial result, we use a test set. Using a test set as a validation set would result in incorrect outcomes and it would not be a valid approach because the dataset would not be independent. We, therefore, divide dataset into two different distinct (train and test) parts each time we run an experiment. In order to create a balanced dataset, we set a constraint that each class must have at least 2 images in the test set. Moreover, the number of test examples per class for each test set is 20% of total number of examples for classes having more than 10 examples. We created augmented dataset using two images; source and style. Source image is always taken from the dataset. We have used 10 arbitrary chosen style images that are not presented in our dataset and the rest of style images comes from our dataset. In order to prevent leaking test set into train set, we eliminate every artificial image that was generated using style or source image from our test set. Full dataset is visualized in Fig. 3.

Table 1. Experimental result for convnet trained from scratch. The first row shows results when using basic dataset, the third shows dataset extended by images generated by Neural Style Transfer technique. The second row of the results restrict extended dataset to only those images that were generated from style and content images both occurring in training set. As a measurement, we use top-1 and top-5 accuracy on test set.

	No Dropout or Augmentation		Dropout and Augmentation	
	Top-1 ACC	Top-5 ACC	Top-1 ACC	Top-5 ACC
Basic dataset	0.2685	0.4867	**0.3844**	**0.6487**
Partially ext. dataset	0.3140	0.5123	0.3223	0.6115
Extended dataset	0.2809	0.4958	0.3471	0.5537

Table 2. Experimental results for different pretrained architectures after 30 epochs trained on extended dataset.

	30 Epochs		100 Epochs	
	TOP-1 ACC	Top-5 ACC	Top-1 ACC	Top-5 ACC
VGG16	0.5785	**0.8016**	**0.5950**	0.7933
VGG19	0.5371	0.7851	0.5867	0.7685
Xception	0.1404	0.3388	-	-
ResNet50	0.2644	0.5371	-	-
InceptionV3	0.1570	0.3553	-	-

4   Model

We experimented with various convolutional neural networks architecture types. Training a small network from scratch did not produce sufficient results due to

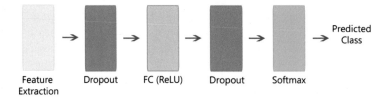

Fig. 4. Schematic illustration of our final architecture.

fast overfitting of a small dataset. We employed dropout and data augmentation [13] to fight overfitting. We augmented our dataset through image transformations. Hyperparameters defining rotation, zoom, width shift, height shift and shear were chosen arbitrarily. We also tried to increase an accuracy through extended dataset. Results after 70 epochs are shown in Table 1. Unfortunately, extended dataset seem to have little effect if any. In order to obtain better results, we decided to use pretrained model. We report results with pretrained model after 30 and 100 epochs in Table 2. VGG16 and its extended version VGG19 [6] show better results than other architectures. Therefore, we further focus on these two architectures.

4.1 Final Architectures

In our final architecture we removed top four layers from the VGG16 architectures and treated the rest of the network as a feature extractor. The feature extraction part is followed by flattening layer, dropout layer with a rate of 0.5, fully connected layer with 512 hidden neurons and ReLU activation function, dropout layer with rate of 0.25 and a softmax layer. Our architecture is visualized in Fig. 4. We used one extra fully connected layers with VGG19 architecture, increased the number of hidden neurons to 1024 and set both dropout rates to 0.5.

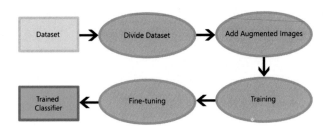

Fig. 5. Training process visualization.

5   Training

We use the categorical cross-entropy as a loss function for our model. Pretrained models were initialized by weights trained on ImageNet[4]. Optimization was performed using RMSProp[5] and our code is written in Keras. We set a minibatch size to 20 images and learning rate to 10^{-4} for learning and 10^{-5} for fine-tuning. Whole training process is illustrated in Fig. 5. After feature extraction phase, we tried to fine-tune the network by unfreezing last 4 layers in the model that were used as a feature extraction (VGG).

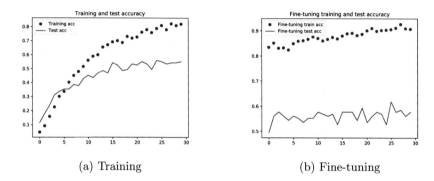

(a) Training   (b) Fine-tuning

Fig. 6. Accuracy during training for VGG16 trained for 30 epochs.

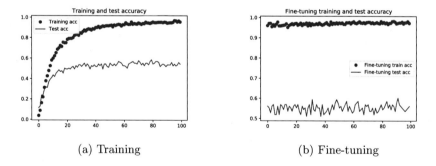

(a) Training   (b) Fine-tuning

Fig. 7. Accuracy during training for VGG16 trained for 100 epochs.

---

[4] http://www.image-net.org/.
[5] http://www.cs.toronto.edu/~tijmen/csc321/slides/lecture_slides_lec6.pdf.

6 Results and Discussion

We have already reported results for pretrained models after 30 epochs in Table 2. Figures 6 and 7 show detailed accuracy progress during training and fine-tuning for 30 and 100 epochs. Training was done on extended dataset with data augmentation, however, images obtained by Neural Style Transfer did not have significant impact on accuracy. Figures suggest that there is no need for 100 epochs of training and it is enough to run fine-tuning just for few epochs. The biggest fine-tuning improvement is typically observed during the first epoch.

We also visualize per-class accuracy on test dataset in Fig. 8. We get 0% accuracy for few classes but it is not crucial since there are only two examples for these classes in test set. For all classes that contain at least three images, model predicted correctly at least one example and these classes also show higher top-5 accuracy than other classes. In order to demonstrate which classes were predicted, we also provide visualization of normalized confusion matrix in Fig. 9.

A typical overfitting pattern can be noticed from the test and train accuracy relation caused by our small dataset. As was mentioned before, our model deals better with classes for which we have more images. Therefore, to obtain better result, it is necessary to collect more data or to try to apply even more sophisticated image augmentation methods than we did. An alternative approach could be to use different architecture.

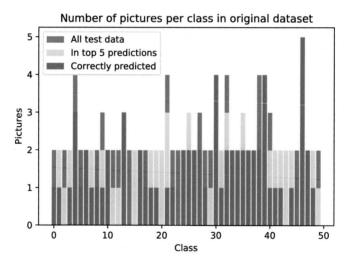

Fig. 8. Per-class accuracy for each class in test set for VGG16 after 100 epochs. Each example is colored according to the model prediction. Green color corresponds to correctly classified image. Examples for which model outputs true label in top five predictions are yellow. Remaining examples are red.

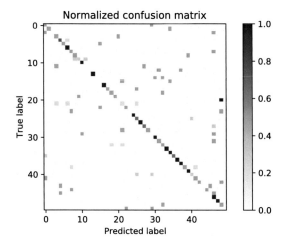

Fig. 9. Normalized confusion matrix for VGG16 after 100 epochs.

7 Conclusion

In this paper, we introduced an architecture for coat of arms classification. We experimented with Neural Style Transfer technique to augment our dataset and described conditions to enforce an independence between augmented training set and test set. Even though some of artificially generated images look realistically, our architecture does not seem to benefit from augmented dataset. We reached top-1 accuracy of nearly 60% and top-5 accuracy of 80%. Trained model could be used to distinct a small number of coat of arms classes. In order to achieve even better accuracy, it is required to collect more data. This is the first step to ultimate classifier that would be able to classify thousands of existing family coat of arms with sufficient accuracy using current machine learning techniques. Our implementation can be useful for historians or researchers.

Acknowledgment. This work was supported by the BUT project FIT-S-17-4014 and the IT4IXS: IT4Innovations Excellence in Science project (LQ1602).

References

1. Halada, J.: Lexikon české šlechty: erby, fakta, osobnosti, sídla a zajímavosti. Akropolis, Praha (1992)
2. Mysliveček, M.: Erbovník 2, aneb, Kniha o znacích i osudech rodů žijících v Čechách a na Moravě podle starých pramenů a dávných ne vždy věrných svědectví. Horizont, Praha (1997)
3. Mysliveček, M.: Erbovník, aneb, Kniha o znacích i osudech rodů žijících v Čechách a na Moravě podle starých pramenů a dávných ne vždy věrných svědectví. Horizont, Praha (1993)
4. Rawat, W., Wang, Z.: Deep convolutional neural networks for image classification: a comprehensive review. Neural Comput. **29**(9), 2352–2449 (2017)

5. Vidensky, F., Zboril, F.: Computer aided recognition and classification of coats of arms. In: International Conference on Intelligent Systems Design and Applications, pp. 63–73. Springer (2017)

6. Simonyan, K., Zisserman, A.: Very deep convolutional networks for large-scale image recognition. arXiv preprint arXiv:1409.1556 (2014)

7. Szegedy, C., Liu, W., Jia, Y., Sermanet, P., Reed, S., Anguelov, D., Erhan, D., Vanhoucke, V., Rabinovich, A.: Going deeper with convolutions. In: Proceedings of the IEEE Conference on Computer Vision and Pattern Recognition, pp. 1–9 (2015)

8. Szegedy, C., Vanhoucke, V., Ioffe, S., Shlens, J., Wojna, Z.: Rethinking the inception architecture for computer vision. In: Proceedings of the IEEE Conference on Computer Vision and Pattern Recognition, pp. 2818–2826 (2016)

9. He, K., Zhang, X., Ren, S., Sun, J.: Deep residual learning for image recognition. In: Proceedings of the IEEE Conference on Computer Vision and Pattern Recognition, pp. 770–778 (2016)

10. Chollet, F.: Xception: Deep learning with depthwise separable convolutions. arXiv preprint arXiv:1610.02357 (2017)

11. Janáček, J., Louda, J.: České erby, Oko edn. Albatros, Praha (1974)

12. Gatys, L.A., Ecker, A.S., Bethge, M.: Image style transfer using convolutional neural networks. In: Proceedings of the IEEE Conference on Computer Vision and Pattern Recognition, pp. 2414–2423 (2016)

13. Perez, L., Wang, J.: The effectiveness of data augmentation in image classification using deep learning. arXiv preprint arXiv:1712.04621 (2017)

FAST Community Detection for Proteins Graph-Based Functional Classification

Arbi Ben Rejab[(⊠)] and Imen Boukhris

LARODEC, Institut Supérieur de Gestion de Tunis, Université de Tunis,
Tunis, Tunisia
Arbibenrejab@gmail.com, imen.boukhris@ensi-uma.tn

Abstract. In this paper we present and evaluate a fast and parallel method that addresses the problem of similarity assessment between node-labeled and edge-weighted graphs which represent the binding pockets of protein. In order to predict the functional family of proteins, graphs can be used to model binding pockets to depict their geometry and physiochemical composition without information loss. To facilitate the measure of similarity on graphs, community detection can be used. Our approach is based on a parallel implementation of community detection algorithm which is an adaptation and extension of Louvain method. Compared to the existing solutions, our method can achieve nearly well-balanced workload among processors and higher accuracy of graph clustering on real-world large graphs.

Keywords: Bioinformatics · Graph-based similarity ·
Community detection · Protein binding sites classification ·
Parallel processing

1 Introduction

In the big data era, comparing graphs becomes an important and challenging task, because of the exponential growth of the size of the graphs. Several applications involve graph comparison such as flow networks, scheduling and planning [22], modeling bonds in chemistry, graph coloring [7], neural networks in artificial intelligence [9], etc. To compare graphs, the notion of similarity should be formalized. Similarity between graphs can be handled using graph matching techniques [15]. Since it is known to be an NP hard problem, many algorithms have been proposed to solve different relaxations [2].

Community detection can be used to match graphs [23]. It has shown its efficiency in partitioning original graph into sub-groups (clusters). Each of them is characterized by high connection into the group and sparse connection between groups.

However, a graph may contain millions of vertices and billions of edges, which make the task of comparison very hard using sequential processing [24]. Parallel

© Springer Nature Switzerland AG 2020
A. Abraham et al. (Eds.): ISDA 2018, AISC 941, pp. 587–596, 2020.
https://doi.org/10.1007/978-3-030-16660-1_57

partitioning is the key to cope with large data. Despite of its importance, there is still a lack of scalable parallel algorithms for large graphs.

In this paper, we investigate graph based functional classification of proteins binding pockets. These latter are represented with large scale graphs which make their comparison hard. Indeed, we propose a method based on parallel community detection that allows to reduce the execution time and achieving good performance. In particular, we implement a parallel modularity-based algorithm using threads. We demonstrate the effectiveness and scalability of our proposed algorithm on real-world large datasets.

The rest of this paper is organized as follows: Sect. 2 gives necessary background information of protein binding sites graphical modeling, graph matching and community detection related work. Section 3 is dedicated to our proposed method for similarity assessment between graphs and proteins functional family prediction based on parallel community detection. Experimental results are covered in Sect. 4. Section 5 concludes the paper and outlines possible directions for future work.

2   Related Work

2.1   Protein Binding Sites Graphical Modeling

Protein binding sites extracted from protein 3D structures can be represented by a simple, connected, undirected, node-labeled and edge-weighted graph G(V, E, lV , lE) [12] which characterized by two sets V and E and two functions lV and lE where:

- $V = \{v_i, i \in [1, \ldots, |V|]\}$: is the set of nodes representing binding atoms, where $|V|$ is the number of nodes of G.
- $E = (v_i, v_j) \subset V \times V$: is the set of edges representing connections between binding atoms that are linked in spatial space, where (v_i, v_j) represent an edge between v_i and v_j, and $|E|$ is the number of edges of G.
- lV $: V \rightarrow N$: is a node labeling function.
- lE $: V \times V \rightarrow R$: is an edge weighting function.

As shown in Fig. 1, physicochemical properties namely, DO (Donor), A (Acceptor), DA (Mixed donor/acceptor), PI ($\pi - \pi$ interaction), AL (hydrophobic aliphatic), IM (Ion metal) and Ar (Aromatic) [18] are interpreted as nodes labels and edges represent links between them.

In order to reduce the complexity of the graph representation, two nodes are only linked when the Euclidean distance separating them is below 11 Ångström[1]. Note that even by ignoring longer edges, graphs remain dense. It was demonstrated in [21] that this representation is sufficient to capture the binding site geometry.

---

[1] Ångström is a unit of length equal to 10^{-10} m.

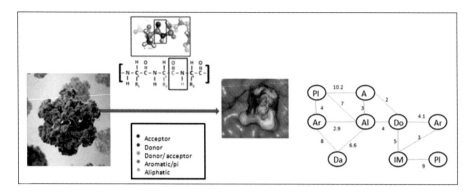

Fig. 1. From protein structure to binding site represented into graph representation

2.2   Graph Matching

– **Maximum Common Subgraph**

A subgraph of a graph G_i is a graph whose vertex set and edge set are subsets of those of G_i [13]. A graph G_{max} is called a maximum common subgraph *(mcs)* of two graphs, G_x and G_y, if there exists no other common subgraph that has more nodes than G_{max}. Finding the maximal common subgraph is an NP-complete problem. Several algorithms based on a backtrack search strategy (e.g., [14,17]) have been proposed to solve this problem. Besides, clique, quasi-clique and community detection adopt different strategies for deriving the maximum common subgraph.

– **Clique**

A clique is a subgraph of a graph in which all the vertices are adjacent to each other (complete subgraph). A maximal clique is a clique that cannot be extended by including one more adjacent vertex.

The problems of finding maximum size clique for the entire graph as well as the maximal size cliques for the individual nodes are NP-hard problems [10]. Accordingly, several exact algorithms have been proposed to solve this problem. However, no polynomial time algorithm is known. For instance, one of the fastest algorithms is the Bron-Kerbosch algorithm [6] which can be used to list all maximal cliques in worst-case optimal time and to list them in polynomial time per clique. Another approach [20], similar to the Bron-kerbosh, consists in a depth-first search algorithm with pruning methods.

– **Quasi-Clique**

The problem of the graph restrictiveness is also solved by replacing clique by a more tolerant concept which is a $\gamma - quasiclique$. It is a complete connected graph where each vertex has to be connected to at least $\gamma * (n - 1)$ other vertices where γ is a relaxation parameter specified by the user that satisfies $0 \prec \gamma \prec 1$ and n is the number of vertices of the $\gamma - clique$. Yet, detection of the maximum $\gamma - clique$ in a graph is an NP-complete problem. Approaches such as [8] mine $\gamma - cliques$ by pruning the search space of the vertex set. Other approaches [5,11] based on mining quasi-cliques have been proposed to classify protein binding sites. These methods have shown their efficiency comparing to those based on cliques but they are very complex.

– **Community Detection**

Another solution that was applied to the problem of *mcs* is community detection. This latter gives a dense subgraph [1]. A community in a graph refers to a group of vertices that are highly connected with each other and are weakly connected to nodes in other groups. The Louvain algorithm [4] is one of the most known method used to detect communities. It is based on greedy approach. The idea behind this algorithm is to put every vertex of a graph in distinct community. Then, for each vertex i, two operations are performed: (1) calculate the modularity gain ΔQ and verify if there is gain when vertice i join any neighbor community j; (2) choose the neighbor j that have the largest gain in ΔQ and join the corresponding community.

Modularity Gain Concept

The modularity variation results from moving an isolated vertex i into a neighboring community C is measured by Modularity gain ΔQ, if i belongs already to a community, it is removed and added to C [19]. It is defined as:

$$\Delta Q = [\frac{\sum_{in} + K_{i,in}}{2m} - (\frac{\sum_{tot} + K_i}{2m})^2] - [\frac{\sum_{in}}{2m} - (\frac{\sum_{tot}}{2m})^2 - (\frac{K_i}{2m})^2]. \quad (1)$$

where $\sum_{in}$ is the number of edges inside the community C, $\sum tot$ is the number of edges incident to community C, k_i is the degree of vertex i, $K_{i,in}$ is the number of edges incident from i to vertices in C and m corresponds to $|V|$.

The first step is repeated until having no more change in communities. As result of this phase, a first level partitions is obtained. Then, reconstruct a new graph where vertices are the communities and an edge exists if at least two vertices of the corresponding communities are connected. Finally, computing the weight of edges between all communities by calculating the sum of weight of all vertices existing in lower level. A super graph is considered as result of this two steps, then repeat those steps on it. The algorithm stops when communities become stable. The flexibility of proteins in the level of 3D make the evaluation of similarity between proteins using community detection approach efficient in term of performance [1]. However, it is not scalable. This have motivated us to develop the FAST Community Detection method extended from Louvain method and based on a parallel implementation.

3   FAST Community Detection for Predicting Proteins Family

Due the huge amount of biological data and the flexible nature of the protein on the 3D level, predicting the functional family of proteins become a challenging task. It has been shown in [16] that the Louvain algorithm gives good results in term of time complexity and high quality of community detection compared to any other sequential algorithms such as clique and $\gamma - clique$ algorithms. However, parallelism is the key to increase the performance of the algorithm. The most computationally intensive parts of the Louvain algorithm are the evaluation of the modularity gain for all vertices and the construction of the next-level supergraph based on the new community structure.

Our proposal Fast Community Detection (FAST-CD) emphasizes on one main aim which is improving the overall modularity and the quality of detected communities in order to extract a dense substructure (maximum common community) in order to evaluate the similarity between a pair of graphs representing proteins binding pockets. Its idea is to compute the modularity gain of vertices in a parallel way using multithreading partitioning of a graph so-called a product graph.

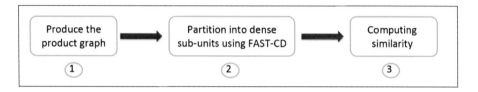

Fig. 2. Three main steps of our approach

As shown in Fig. 2, our approach is made up by three main steps:

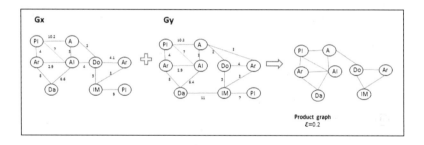

Fig. 3. Product graph

1. The first step is to produce the product graph.
   Let $G_P = (V_P, E_P)$ be a product graph of two graphs $G_X = (V_X, E_X)$ and $G_Y = (V_Y, E_Y)$ it is defined by: A set of nodes $V_P \subseteq V_X \times V_Y$ and a set of edges $E_P \subseteq V_P \times V_P$ where: $V_P = (v_{X_I}, V_{Y_J}) | L_V(v_{X_I}) = L_V(v_{Y_J})$ and $E_P = ((V_{X_i}, V_{Y_j}), (V_{X_k}, V_{Y_l})) | L_E(V_{X_i}, V_{X_k}) = L_E(V_{Y_j}, V_{Y_l})$.
   In other words, a node is inserted in the product graph if it has exactly the same labels in both G_X and G_Y. Similarly, an edge is inserted if it exists and has the same weight between two matched nodes. However, real world data

can be affected with noise and measurement errors which made the possibility to introduce some tolerance in the construction of the product graph G_P by defining the set of edges as:

$$E_P = ((V_{X_i}, V_{Y_j}), (V_{X_k}, V_{Y_l}))|L_E(V_{X_i}, V_{X_k}) - L_E(V_{Y_j}, V_{Y_l}) \prec \epsilon. \quad (2)$$

Where ϵ is a tolerance threshold parameter for edge length differences specified by the user (in our case $\epsilon = 0.2$). Figure 3 depicts an example of a product graph between two protein binding sites graphs where ϵ is equal to 0.2.

2. Then, partition the product graph into dense sub-units without restrictions to the degree of connectedness of vertices using FAST-CD (as explained in Algorithm 1). Our method makes an adaptation and an extension of the original Louvain algorithm.

The original algorithm is permeated by serial dependencies at every level, from vertex to vertex and iteration to iteration. The challenge is the evaluation of the modularity gain ΔQ for all vertices with updated community membership. In other words, each vertex computes the modularity gain for joining its neighbors communities.

The pseudo-code of the FAST-CD algorithm takes as input the product graph G in order to calculate the number of edges inside and outside community for each vertex.

For each vertex, the FAST-CD calculates modularity gain using threads by executing the same code simultaneously. Afterwards, it increments the number of pass done in order to compare it with the maximum number tolerated. The stability in the Community is assessed afterwards. If it is stable, then the algorithm stops otherwise it continues to do another iteration.

3. The last step is computing the similarity assessment [5] between two graphs $G_x(V_x, E_x)$ and $G_y(V_y, E_y)$ representing proteins binding sites. It is computed according to their common maximum common substructure $G_{max}(V_{max}, E_{max})$ as:

$$sim(G_X, G_Y) = \phi * min(\alpha, \beta) + (1 - \phi) * max(\alpha, \beta). \quad (3)$$

where, $0 \prec \phi \prec 1$ denotes a parameter that compromises between an equivalence relation and an inclusion one (in our case $\phi = 0.5$, since it has proved to be a reasonable choice in [4]). $\alpha = \frac{|V_{max}|}{|V_X|}$ expresses the degree to which G_X is a subset of G_Y and $\beta = \frac{|V_{max}|}{|V_Y|}$ is the degree to which G_Y is a sub set of G_X.

4   Experiments

4.1   The Framework

We have tested and applied our new algorithm to real-world datasets (more than 8000 nodes and 25000 edges for dataset1) obtained from the CavBase

Algorithm 1. FAST-CD

INPUTS

(a) Product graph G
(b) The number of pass for level computation nb_{max}.

OUTPUTS

(a) Resulting communities.

FAST-CD Algorithm
begin

                $nb\_pass\_done = 0$
                $stability = false$
     repeat
        $nb\_inside0 = nb\_inside$
        $nb\_outside0 = nb\_outside$
        For each community, compute the number of node inside
        community (nb_inside)
        For each community, compute the number of node between
        nodes and community ($nb\_outside$)
        For $i = 0$ **to** $nb\_vertex - 1$ **do**
            $thread[i].add(vertex[i])$
            $thread.start$ // compute modularity gain
            $thread.join$ // join vertex into the best community
        end for
        $nb\_pass\_done + +$
        if $(nb\_inside = nb\_inside0)$ and $(nb\_outside = nb\_outside0)$
            $stability = true$
        else
            $stability = false$
    until $(nb\_pass\_done <= nb\_max)$ and $(stability = true)$
end

database [3] of two protein binding sites families ATP (Adenosine triphosphate) and NADH (Nicotinamide adenine dinucleotide) in order to predict their protein class. It is a challenging task because ATP is a substructure of NADH so it may possibly bind to the same ligands. In order to confirm the performance of our approach, we have tested the algorithm on larger dataset (dataset2) which contain 40 proteins binding sites more than 24000 nodes and 70000 edges.

Actually, the class determination of the graph G is done by means of computing the distance between G and the rest of instances in the training set. Subsequently, we move to the selection of K closest graphs which are the closest to G. G is classified based on the majority classes in relation to the KNN.

4.2   Evaluation Criteria

We compare our approach with the maximum densest common community (MDCC) method proposed in [16]. Two evaluation criteria are used to test and evaluate the FAST-CD method.

The Percent of Correct Classification (PCC) defined as:

$$PCC = (\frac{number\ of\ well\ classified\ instances}{total\ number\ of\ classified\ instances}) \cdot 100. \qquad (4)$$

It reflects the classification accuracy.

The execution time which reflects the time needed to detect communities.

4.3   Results and Discussion

In this section, we report and detail the results of our new proposal. The results of comparisons between the MDCC method and FAST-CD methods using as evaluation criteria the PCC and the execution time are shown in Tables 1 and 2.

FAST-CD gives better classification than MDCC. For example, for $k = 1$ it reaches 71.42% for the FAST-CD compared to the MDCC with PCC equal to 57.14%. Hence, a great performance, a high degree of robustness and flexibility of the proposed approach are empirically confirmed.

Table 1. Comparison of FAST-CD and MDCC based on PCC

PCC			
Approaches		Dataset1	Dataset2
MDCC	$k = 1$	57.14	52.5
	$k = 3$	57.14	45
	$k = 5$	42.83	47.5
FAST-CD	$k = 1$	71.42	55
	$k = 3$	62.38	52.5
	$k = 5$	62.38	55

Table 2. Comparison of FAST-CD and MDCC basing on execution time

Execution time		
Approaches	Dataset1	Dataset2
MDCC	3220 s	1421 min
FAST-CD	760 s	333 min

Moreover, as seen in Table 2, the MDCC needs much more time to process data compared to our proposal. We can also remark that the proposed FAST-CD method is considerably faster than the MDCC algorithm and their execution time is linear with the dataset size. The execution time of the FAST-CD (780 s) is lower than the execution time of MDCC (3220 s).

We can conclude that the FAST-CD provides final results faster than the MDCC. These results are explained by the use of parallelism concept when detecting communities. Generally, experiments prove that our new approach can effectively minimize the execution time while providing a high classification precision.

5   Conclusion

In this paper, we have highlighted the problem of assessing similarities between node-labeled and edge-weighted graphs that represent protein binding sites by providing a new approach based on parallel community detection FAST-CD applicable on large graphs. The FAST-CD has succeeded to reduce the time allocated to compute modularity by using a model extended from Louvain algorithm and then, making several improvements and updates on it in order to create parallelism. Our proposal is tested using datasets from the PDB (Protein Data Bank). Results of FAST-CD prove the improvement made by this new approach compared to the MDCC. When dealing with classification task with a large amount of data, our experiments validate that the proposed algorithm have higher classification precision and lower time execution.

References

1. Awal, G.K., Bharadwaj, K.: Team formation in social networks based on collective intelligence: an evolutionary approach, pp. 627–648 (2014)
2. Bengoetxea, E.: Inexact graph matching using estimation of distribution algorithms. Ecole Nationale Supérieure des Télécommunications, Paris **2**(4), 49 (2002)
3. Berman, H.M., Westbrook, J., Feng, Z., Gilliland, G., Bhat, T.N., Weissig, H., Bourne, P.E.: The protein data bank. Nucleic Acids Res. **28**, 235–242 (2000)
4. Blondel, V.D., Guillaume, J.L., Lambiotte, R., Lefebvre, E.: Fast unfolding of communities in large networks, P1008 (2008)
5. Boukhris, I., Elouedi, Z., Fober, T., Mernberger, M., Hullermeier, E.: Similarity analysis of protein binding sites: a generalization of the maximum common subgraph measure based on quasi-clique detection. In: ISDA, pp. 1245–1250. IEEE Computer Society (2009)
6. Bron, C., Kerbosch, J.: Algorithm 457: finding all cliques of an undirected graph. Commun. ACM **16**(9), 575–577 (1973)
7. Cohen, J., Castonguay, P.: Efficient graph matching and coloring on the GPU. In: GPU Technology Conference, pp. 1–10 (2012)
8. Daxin, J., Jian, P.: Mining frequent cross-graph quasi-cliques. ACM Trans. Knowl. Discov. **16**(1), 16–42 (2009)
9. Emmert-Streib, F., Dehmer, M., Shi, Y.: Fifty years of graph matching, network alignment and network comparison. Inf. Sci. **346**, 180–197 (2016)
10. Ferrer, M., Valveny, E., Serratosa, F.: Median graph: a new exact algorithm using a distance based on the maximum common subgraph. Pattern Recogn. Lett. **30**(5), 579–588 (2009)

11. Fober, T., Klebe, G., Hullermeier, E.: Local clique merging: an extension of the maximum common subgraph measure with applications in structural bioinformatics. In: Algorithms from and for Nature and Life, pp. 279–286 (2013)
12. Frasconi, P., Passerini, A.: Predicting the geometry of metal binding sites from protein sequence **9**, 203–213 (2012)
13. Harary, F., Norman, R.Z.: Graph theory as a mathematical model in social science, p. 45 (1953)
14. Levi, G.: A note on the derivation of maximal common subgraphs of two directed or undirected graphs. Calcolo **9**(4), 341 (1973)
15. Malewicz, G., Austern, M.H., Bik, A.J., Dehnert, J.C., Horn, I., Leiser, N., Czajkowski, G.: Pregel: a system for large-scale graph processing. In: Proceedings of the 2010 ACM SIGMOD International Conference on Management of Data, pp. 135–146 (2010)
16. Mallek, S., Boukhris, I., Elouedi, Z.: Community detection for graphbased similarity: application to protein binding pockets classification. Pattern Recogn. Lett. **62**, 49–54 (2015)
17. McGregor, J.J.: Backtrack search algorithms and the maximal common subgraph problem. Softw.: Pract. Experience **12**(1), 23–34 (1982)
18. Schmitt, S., Kuhn, D., Klebe, G.: A new method to detect related function among proteins independent of sequence and fold homology. J. Mol. Biol. **323**(2), 387–406 (2002)
19. Shiokawa, H., Fujiwara, Y., Onizuka, M.: Fast algorithm for modularity-based graph clustering. In: AAAI, pp. 1170–1176 (2013)
20. Tomita, E., Tanaka, A., Takahashi, H.: The worst-case time complexity for generating all maximal cliques and computational experiments. Theor. Comput. Sci. **363**(1), 28–42 (2006)
21. Weskamp, N., Hullermeier, E., Kuhn, D., Klebe, G.: Multiple graph alignment for the structural analysis of protein active sites. IEEE/ACM Trans. Comput. Biol. Bioinf. (TCBB) **4**(2), 310–320 (2007)
22. Wu, S.D., Byeon, E.S., Storer, R.: A graph-theoretic decomposition of the job shop scheduling problem to achieve scheduling robustness. Oper. Res. **47**(1), 113–124 (1999)
23. Yang, J., McAuley, J., Leskovec, J.: Community detection in networks with node attributes. In: Data Mining (ICDM), pp. 1151–1156 (2013)
24. Chi, Y., Dai, G., Wang, Y., Sun, G., Li, G., Yang, H.: Nxgraph: an efficient graph processing system on a single machine. In: 2016 IEEE 32nd International Conference on Data Engineering (ICDE), pp. 409-420, May 2016

A Group Recommender System
for Academic Venue Personalization

Abir Zawali[(✉)] and Imen Boukhris

LARODEC, Institut Supérieur de Gestion de Tunis, Université de Tunis,
Tunis, Tunisia
abirzaouali1@gmail.com, imen.boukhris@hotmail.com

Abstract. With the increasing number of academic venues and scientific activities, it is generally difficult for researchers to choose the most appropriate conference or journal to submit their works. A recommender System (RS) may be used to suggest upcoming venues for scientists. Although standard recommender systems have shown their efficiency in supporting individual decisions, they are not appropriate for suggesting items when more than one person is involved in the recommendation process. Since a scientific paper is generally written by a group of researchers, we propose in this paper a new group recommender system that suggests for these researchers personalized conferences that fit their preferences and interests. The main idea is to recommend academic venues for a group of researchers based on the venues attended by not only their co-authors, i.e., the group members, but also on their co-citers. Our recommender system is also able to filter out irrelevant conferences that do not meet the requirements of those researchers, their preferences in terms of conferences location, publisher and ranking. Experimental results demonstrate the efficiency of our new group recommender system.

Keywords: Collaborative filtering · Group recommender system ·
Venue preferences

1 Introduction

Recommender systems are an effective means to filter information in many areas such as books, movies, musics, web pages, etc [13,17]. Recent works [2,4,10,19] are investigating the crucial role of recommender systems to suggest appropriate academic venues to allow researchers to publish their scientific discoveries.

Since the online activities have grown exponentially, for instance a familly is generally looking for a restaurant that satisfy all its members' tastes. Hence, it is important to have a group recommender system that provides relevant suggestions for the group's decision and satisfies the needs of the group [5].

The difference between individuals and groups lies in the influences and interactions among group members. Subsequently, these factors will influence group decision-making.

© Springer Nature Switzerland AG 2020
A. Abraham et al. (Eds.): ISDA 2018, AISC 941, pp. 597–606, 2020.
https://doi.org/10.1007/978-3-030-16660-1_58

Group recommender systems assume that the input of the system include items evaluations given by individuals, namely the group members. The group final recommendations are obtained by aggregating individual recommendations or preferences. Actually, doing this way does not take into account the interactions between the users. Indeed, ignoring how individuals or subgroups influence and interact with each other can lead to an inappropriate recommendations for the whole group.

Accordingly, we present in this paper a new method allowing the recommendation of upcoming conferences to a group of researchers in the field of computer science. This approach considers the interactions, i.e., the past collaborations between (sub)groups' members. Note that a scientist do not need to provide any additional infirmation.

The rest of the document is organized as follows: Sect. 2 discusses related work. The proposed group recommender engine is explained in Sect. 3. In Sect. 4, we present the experimental results and discuss them. Section 5 concludes the paper.

2   Related Work

With the flood of information, intelligent systems, namely recommender systems, are needed to filter out irrelevant items.

There are two basic group recommendation strategies as shown in Fig. 1, namely preference aggregation strategy [21] and recommendation aggregation strategy [13]. The first strategy takes as input each group member's preferences, aggregate them and treat them as a single user. The second strategy generates for each group's member his own recommendation using an individual recommender system, then aggregate the recommendation lists of each member to get at the end a recommendation list for the whole group.

The are several approaches in ways to produce recommendations. The content-based filtering [14] is one of the most common approches. It allows to recommend items similar to those previously liked by the user by comparing the user's profile with the item's profile, i.e., attributes or products characteristics. The collaborative filtering approach [3,11] is based on a user-item matrix. It provides recommendations for a particular user based on similarity measurements between users and/or items. It is the most widely used approach because of its simplicity and the serendipity of its recommended items. Hybrid approaches [12] combines several recommendation methods using different hybridization strategies that can be categorized into three according to the used architecture namely, Monolithic, Parallelized, and Pipelined.

Recommender systems should be addressed to individuals, e.g., a single user that wants to buy a cell phone, as well as for a group, e.g., friends who will go to travel together and will choose a destination.

The most popular recommendation system approaches used in the academic venue recommendation context are those based on authors and those based on paper contents. Works based on authors [4,8,19] are presented in the collaborative filtering context. They use researcher's past publications to suggest scientific

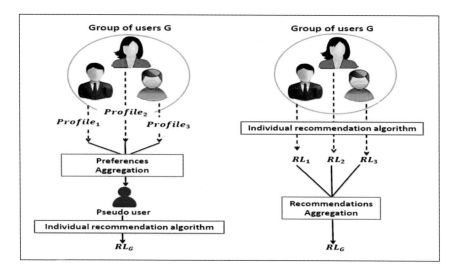

Fig. 1. Group recommendation strategies

conferences. These works have shown their efficiency according to the serendipity of the suggestions. Recommendations based on papers content [16] are proposed in the content-based context. Indeed, information such as the title, the subject, the keywords were used. The change of interest is taken into account in [7,9,18]. Personal bibliographies and researchers citations have been also used [7] to obtain appropriate recommendations.

While scientific papers are often written by several authors, existing academic venues recommendations were proposed in the context of individual recommendations. This have motivated us to propose a new academic venues recommender engine for a group of researchers, i.e., computer scientists able to suggest venues even if the computer scientists are currently working on, even if they are working on different areas or if some of them are young researchers and they do not have published papers before.

3   Academic Venues Group Recommendations

The proposed group recommendation methodology aims to find the most suitable academic venues for a computer research group, based on their interests and needs while putting action on the possible interactions of group members. Indeed, to extract a list of personalized conferences to a target group, we will use a new group recommendation system based on collaborative filtering.

The group denoted by G is composed by the list of authors for whom we would like to suggest academic venues. Each group member is denoted by u_i and each subgroup by g_i.

Figure 2 shows how subgroups are composed reflecting how group's members can interact together.

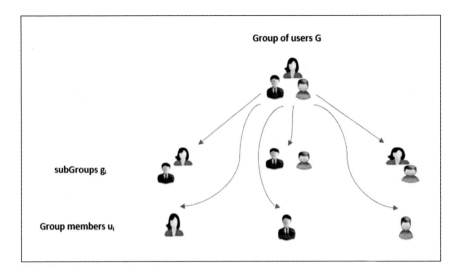

Fig. 2. Possible group's interaction

Our approach is based on the past publications of the authors, co-authors and co-citers, of each group member, each subgroup and the whole group. As shown in Fig. 3, our proposed recommender engine is mainly composed of four phases. The first three ones are respectively dedicated to authors, co-authors and co-citers while the last one allows to hybridize the previous phases then refine the list of academic venues according to researchers' preferences.

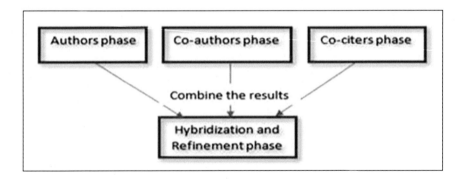

Fig. 3. Group academic venues recommendation steps

3.1 Authors Phase

The authors phase is composed of two steps. As explained in Fig. 4, it takes from the publications database as input a user-item matrix representing the number of published papers in an academic venue v_j, j ∈ J, where J is the set of conferences. For each group, subgroups and researcher, the number of published papers in each venue v_j are denoted respectively by P_{u_i,v_j}, P_{g_i,v_j} and P_{G,v_j}.

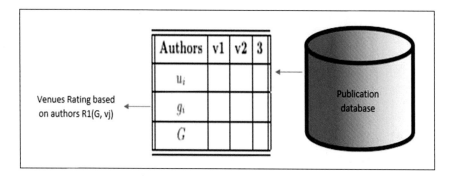

Fig. 4. Authors phase

Since there are some venues that took place each year while others each two years, then to avoid to skew the results, this information is taken into consideration by multiplying the number of publications in conferences occuring each two years by two.

1. **Similarity calculation:** In this step, we calculate the similarity between two academic venues v_1 and v_2 using the cosine similarity [1] as follows:

$$\cos(\overrightarrow{v_1}, \overrightarrow{v_2}) = \frac{\overrightarrow{v_1} \bullet \overrightarrow{v_2}}{\|\overrightarrow{v_1}\| * \|\overrightarrow{v_2}\|}. \tag{1}$$

2. **Prediction author's ratings:** The group's rating denoted by $R_1(G, v_j)$ is computed for each venue v_j taking into consideration the level of similarity as follows:

$$R_1(G, v_j) = \frac{\sum_{j\in J} w_{j,j+1} P_{G,v_j}}{\sum_{j\in J} w_{j,j+1}}. \tag{2}$$

where $w_{j,j+1}$ represents the similarity between two academic venues v_j and v_{j+1}.

3.2   Co-authors Phase

Let us denote by u_{q,u_i} each co-author of a group's member, by u_{q,g_i} each co-author of a subgroup and by $u_{q,G}$ each co-author of the group.

Similarly, the co-authors phase is composed of two steps. As explained in Fig. 5, it takes as input, for each group, subgroup and researcher, the number

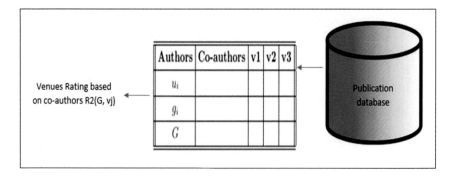

Fig. 5. Co-authors phase

of published papers of their co-authors in each venue v_j denoted respectively by P_{u_{q,u_i},v_j}, P_{u_{q,g_i},v_j} and $P_{u_{q,G},v_j}$.

The co-author's rating is denoted by $R_2(G, v_j)$ is computed ,for each v_j, as follows:

$$R_2(G, v_j) = \frac{\sum_{j \in J} w_{j,j+1} P_{u_{q,G},v_j}}{\sum_{j \in J} w_{j,j+1}}. \tag{3}$$

3.3   Co-citers Phase

Let us denote by u_{t,u_i} each co-citer of a group's member, by u_{t,g_i} each co-citer of a sub group and by $u_{t,G}$ each co-citer of the group.

The co-citers phase is composed of tree steps. As explained in Fig. 6, it takes as input, for each group, subgroup and researcher, the number of published papers of their co-citers in each venue v_j denoted respectively by P_{u_{t,u_i},v_j}, P_{u_{t,g_i},v_j} and $P_{u_{t,G},v_j}$.

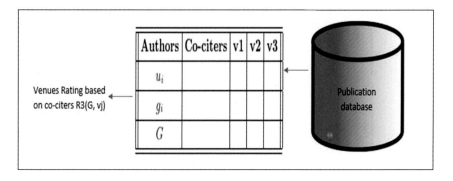

Fig. 6. Co-citers phase

Since there are some venues that took place each year while others each two years, then to avoid to skew the results, this information is taken into consideration by multiplying the number of publications in conferences occuring each two years by two.

1. **Normalisation:** To avoid that co-citers who publish a large number of papers skew the results, we take into account the total number of co-citer's published papers denoted by N_{u_t,v_j} as follows:

$$P_{u_t,j} = \frac{P_{u_t,v_j}}{N_{u_t,v_j}}. \tag{4}$$

2. **Discounting:** In this step, the reliability given to a co-citer is taken into consideration. The frequency is calculated using the number of co-citer's published papers taking as references the target u_i, g_i, or G works according to the total number of u_i, g_i or G published papers denoted by T_{u_i}, T_{g_i}, or T_G. The idea is to weight most heavily v_j. It is defined by: $P_{u_t,u_i,v_j} = P_{u_t,u_i,v_j} \cdot F_{u_i,u_t}$, $P_{u_t,g_i,v_j} = P_{u_t,g_i,v_j} \cdot F_{g_i,u_t}$ and $P_{u_t,G,v_j} = P_{u_t,G,v_j} \cdot F_{G,u_t}$.

   Where:
   $F_{u_i,u_t} = \frac{\beta}{T_{u_i}}$, $F_{g_i,u_t} = \frac{\beta}{T_{g_i}}$, $F_{G,u_t} = \frac{\beta}{T_G}$.
   And β is the number of co-citer's published papers taking as references the target u_i, g_i, or G works.

3. **Prediction co-citer's ratings:** Calculate the rating denoted by $R_3(G,v_j)$ by summing up P_{u_t} for each academic venue.

$$R_3(G,v_j) = \sum_{t=1}^{n} P_{u_t,v_j}. \tag{5}$$

3.4 Hybridization Phase

The final prediction rating denoted by $R^f(G,v_j)$ considers the score from authors $R_1(G,v_j)$, co-authors $R_2(G,v_j)$ and co-citers $R_3(G,v_j)$. It is computed as:

$$R^f(G,v_j) = R_1(G,v_j) + R_2(G,v_j) + R_3(G,v_j). \tag{6}$$

To refine the list of recommended academic venues, the location, ranking and publishers are used according to the needs and preferences indicated by the group. Noted that in the default case, all locations, rankings and editors are preferred by the group. The refinement process is performed by calculating the utility of each upcoming academic venue, denoted by $U(v_j)$, according to a utility value, denoted by $utility(v_j)$, describing the combination of preferences already indicated by the group. The inappropriate venues based on group preferences will be rejected.

$$U(v_j) = utility(v_j) * R^f(G,v_j). \tag{7}$$

Finally, the set of conferences will be sorted according to their utility $U(v_j)$. Hence, a list of the most appropriate venues, i.e, with the highest $U(v_j)$, will be recommended to the group G.

4 Experiments

4.1 Dataset

To evaluate our proposed recommendation approach, we resort to one of the widely used recommendation dataset "DBLP citation dataset" [15] which is publicly available on the aminer website[1]. The DBLP Dataset consists of bibliographic data augmented by the citation relations between papers. It contains information about 3,079,007 papers and 25,166,994 citation relationships published until october 2017. We have adapted 17% of the DBLP data and used it as a training set.

Information about call of papers locations are extracted from WikiCFP[2]. Information about computer science venues, i.e, venue's publisher and previous venue ranking, are extracted from the CORE Conference Portal[3]. We can distinguish six ranking categories ranging from A* to unranked ones.

In order to evaluate our proposed group recommendation approach, we compared the recommendations results addressed to a single researcher with those proposed by the new group recommender personalized academic venue method.

4.2 Evaluation Metrics

Precision and Recall [6] measure the frequency with which a RS makes correct or incorrect suggestions about whether an item is good. Therefore, they allow to assess how much the RS can recommend potentially pertinent items for the user. Those latters are defined as follows:

$$Precision = \frac{N_{rs}}{N_s}. \tag{8}$$

$$Recall = \frac{N_{rs}}{N_r}. \tag{9}$$

Where N_{rs} is the number of relevant items selected by an active user, N_s is the total number of items selected by an active user and N_r is the total number of relevant items available.

F-measure [20] is a harmonic mean of the precision and recall values, defined as follows:

$$F - mesure = \frac{2 * Precision * Recall}{Precision + recall}. \tag{10}$$

4.3 Results and Discussion

In this section, we will compare the recommendation for a single user and the recommendations obtained with our proposed engine. In order to highlight the

---

[1] https://aminer.org/citation.

[2] http://www.wikicfp.com/cfp/.

[3] http://103.1.187.206/core/.

extent to which group recommendation can improve the suggested academic venues, we will compare the results addressed to a single researcher to those of our group recommendation.

Table 1. Comparison of recommendations in term of precision, recall and F-mesure

	Recommendation for a single user	Recommendation for group
Precision	0.45	0.58
Recall	0.71	0.75
F-mesure	0.54	0.67

From Table 1, it is noticeable that our proposed method performs better than a single-user recommendation method with an improve of 13% in terms of precision, 4% in terms of recall and 13% in terms of F-measure. Indeed, using a group recommender system and considering the interactions between the authors improves the quality of recommendations.

5   Conclusion

In this paper, we have presented group recommendations of academic venues for a group of researchers that will submit their works to a scientific conference. Our recommender engine exploits not only the past publications of the group but also publications of each subgroup and each individual researcher as well as those of their co-authors and co-citers. Personalized recommendation are refined by considering the preferred location, publisher and ranking of the academic venues.

With regard to this work, the results obtained show that the proposed method provides high quality group recommendations and satisfactory for group decisions. As future work, we plan to investigate the scalability of the proposed method by suggesting a method dealing with big data.

References

1. Adomavicius, G., Tuzhilin, A.: Toward the next generation of recommender systems: a survey of the state-of-the-art and possible extensions. IEEE Trans. Knowl. Data Eng. **6**, 734–749 (2005)
2. Alhoori, H., Furuta, R.: Recommendation of scholarly venues based on dynamic user interests. J. Informetrics **11**(2), 553–563 (2017)
3. Azizi, M., Do, H.: A collaborative filtering recommender system for test case prioritization in web applications. In: Proceedings of the 33rd Annual ACM Symposium on Applied Computing, pp. 1560–1567 (2018)
4. Boukhris, I., Ayachi, R.: A novel personalized academic venue hybrid recommender. In: 2014 IEEE 15th International Symposium on Computational Intelligence and Informatics (CINTI), pp. 465–470. IEEE (2014)

5. Chen, Y.L., Cheng, L.C., Chuang, C.N.: A group recommendation system with consideration of interactions among group members. Expert Syst. Appl. **34**(3), 2082–2090 (2008)
6. Herlocker, J.L., Konstan, J.A., Terveen, L.G., Riedl, J.T.: Evaluating collaborative filtering recommender systems. ACM Trans. Inf. Syst. (TOIS) **22**(1), 5–53 (2004)
7. Kang, N., Doornenbal, M.A., Schijvenaars, R.J.: Elsevier journal finder: recommending journals for your paper. In: Proceedings of the 9th ACM Conference on Recommender Systems, pp. 261–264. ACM (2015)
8. Klamma, R., Cuong, P.M., Cao, Y.: You never walk alone: recommending academic events based on social network analysis. In: International Conference on Complex Sciences, pp. 657–670. Springer, Berlin (2009)
9. Küçöktunç, O., Saule, E., Kaya, K., Çatalyürek, Ü.V.: TheAdvisor: a webservice for academic recommendation. In: Proceedings of the 13th ACM/IEEE-CS Joint Conference on Digital Libraries, pp. 433–434. ACM (2013)
10. Mhirsi, N., Boukhris, I.: Exploring location and ranking for academic venue recommendation. In: International Conference on Intelligent Systems Design and Applications, pp. 83–91. Springer, Cham (2017)
11. Nilashi, M., Ibrahim, O., Bagherifard, K.: A recommender system based on collaborative filtering using ontology and dimensionality reduction techniques. Expert Syst. Appl. **92**, 507–520 (2018)
12. Parekh, P., Mishra, I., Alva, A., Singh, V.: Web Based Hybrid Book Recommender System Using Genetic Algorithm (2018)
13. Qin, S., Menezes, R., Silaghi, M.: A recommender system for Youtube based on its network of reviewers. In: 2010 IEEE Second International Conference on Social Computing (SocialCom), pp. 323–328. IEEE (2010)
14. Ricci, F., Rokach, L., Shapira, B., Kantor, P.B.: Recommender Systems Handbook (2010)
15. Sinha, A., Shen, Z., Song, Y., Ma, H., Eide, D., Hsu, B.J.P., Wang, K.: An overview of microsoft academic service (MAS) and applications. In: Proceedings of the 24th International Conference on World Wide Web, pp. 243–246. ACM (2015)
16. Wei, F.A.N.G.: Recommending publication venue in context using abstract. DEStech Trans. Comput. Sci. Eng. (MMSTA) (2017)
17. Yang, X., Guo, Y., Liu, Y., Steck, H.: A survey of collaborative filtering based social recommender systems. Comput. Commun. **41**, 1–10 (2014)
18. Yang, Z., Davison, B.D.: Venue recommendation: submitting your paper with style. In: 2012 11th International Conference on Machine Learning and Applications (ICMLA), vol. 1, pp. 681–686. IEEE (2012)
19. Yu, S., Liu, J., Yang, Z., Chen, Z., Jiang, H., Tolba, A., Xia, F.: PAVE: personalized academic venue recommendation exploiting co-publication networks. J. Netw. Comput. Appl. **104**, 38–47 (2018)
20. Zhang, F., Gong, T., Lee, V.E., Zhao, G., Rong, C., Qu, G.: Fast algorithms to evaluate collaborative filtering recommender systems. Knowl.-Based Syst. **96**, 96–103 (2016)
21. Zhao, Z.D., Shang, M.S.: User-based collaborative-filtering recommendation algorithms on hadoop. In: 2010 Third International Conference on Knowledge Discovery and Data Mining, WKDD 2010, pp. 478–481. IEEE (2010)

Imprecise Label Aggregation Approach Under the Belief Function Theory

Lina Abassi$^{(\boxtimes)}$ and Imen Boukhris

LARODEC Laboratory, Institut Supérieur de Gestion de Tunis,
University of Tunis, Tunis, Tunisia
lina.abassi@gmail.com, imen.boukhris@hotmail.com

Abstract. Crowdsourcing has become a phenomenon of increasing interest in several research fields such as artificial intelligence. It typically uses human cognitive ability in order to effectively solve tasks that can hardly be addressed by automated computation. The major problem however is that so far studies could not completely control the quality of obtained data since contributors are uncertainly reliable. In this work, we propose an approach that aggregates labels using the belief function theory under the assumption that these labels could be partial hence imprecise. Simulated data demonstrate that our method produces more reliable aggregation results.

Keywords: Crowdsourcing · Label aggregation ·
Belief function theory · Precision · Exactitude

1 Introduction

Over the last decade, Crowdsourcing have been proposed as a form of collective intelligence aiming to solve problems that are often intractable for computers such as image or video classification or tagging [2], sentiment analysis [1,3] as well as tasks that require creativity or content creation. It enables the democratization of participation. In other words, rather than recruiting experts to accomplish work, crowdsourcing offers work as an open call letting a wide range of anonymous individuals submit contributions.

This concept has been used massively in the artificial intelligence field particularly in machine learning domain for the sake of academic research. In fact, microtasking, a particular type of crowdsourcing known as a process that calls workers to carry out elementary tasks, produces processed data that can be used as training data for the supervised learning algorithms. This new way of getting human-labeled data has become more popular thanks to the availability of internet platforms such as Amazon Mechanical Turk. These platforms put together tasks and individuals willing to work in exchange of a small financial reward. It is indeed a great and cheap way to get work done but it has challenges to overcome in order to achieve the most of its potential. For one, workers can

© Springer Nature Switzerland AG 2020
A. Abraham et al. (Eds.): ISDA 2018, AISC 941, pp. 607–616, 2020.
https://doi.org/10.1007/978-3-030-16660-1_59

deliver imperfect labels for different reasons either intentionally or not. There-fore, microtasking platforms had to implement some techniques to prevent bad quality data such as redundancy [4] which means matching a task to many work-ers. The submissions are then combined using an aggregation technique. If this latter is effective enough, obtained results can be equally accurate as experts ones [3].

In this paper, we propose a new approach of label aggregation under the assumption that workers can give partial answers when they do not know the correct label with certainty. Hence, we resort to the Belief function theory [5,6], a rich framework that allows to manage such imperfections and even combine pieces of information.

Our approach proposes a strong groundwork under the belief function theory that represents workers' labels whatever their imperfections and estimates their levels of expertise through two degrees namely the exactitude and precision degrees. Labels are finally aggregated regarding these expertise levels.

The remainder of the paper is organized as follows: In Sect. 2, we give an overview of existing label aggregation methods. Section 3, presents the basic concepts of the belief function theory. Then in Sect. 4 our proposed approach is detailed. The experimental evaluation is presented in Sect. 5. Finally, Sect. 6 concludes our work and presents some future works.

2   Related Work

For a crowdsourcing system, the main challenge remains in finding the most effective way to obtain accurate results from humans that have different knowl-edge levels, dedication and evaluation criteria. As the solution of assigning tasks to more than one worker was proposed, this required a method to combine the collected answers efficiently. Therefore, research in this field was able to propose various types of algorithms that can be classified into two categories according to their basic assumption. The first category assumes that gold data (i.e true answers provided by experts) are available such as the Expert Label Injected Crowd Estimation (ELICE) approach [12] that uses gold data to estimate par-ticipants expertise and question difficulty or methods like the (GS-BLA) [18] and (CGS-BLA) [20] where gold data are also used to estimate workers qualities before integrating them in the aggregation of labels. As for the second category, it includes methods that have only noisy labels as input. Among these methods we find principally the Majority Decision method (MD) [22] which result gives the label with maximum votes and methods based on the Expectation-Maximization (EM) algorithm [15] which contributes iteratively in the estimation of latent vari-ables. Works inspired from this latter [10,11,16] infer the unknown true labels in the E step and then estimate a set of parameters (e.g. worker expertise, task difficulty) in the M step to be integrated in the next E step.

In this paper, we assume that no gold data is available. We propose a method that does not belong to the probabilistic models but based on the belief function theory. This latter has the asset to be a generalization of the probabilistic theory hence it is more flexible and plenary. A group of methods actually adopted the belief function theory to integrate labels. The Belief Label Aggregation (BLA) [17] proposed to infer worker capabilities using MD and to join them to the aggregation process which is performed using different tools of the BFT. As for [19] (I-BLA), it is also founded on the BFT and inspired from the EM algorithm and it infers the set of true labels, labeling qualities and problem difficulty parameters.

In this paper, unlike the mentioned methods, we assume that a worker is allowed to respond with more than one label in case he ignores the exact answer. This will leave us with imprecise answers. Therefore, we have recourse to the belief function theory since it allows to deal with partial and total ignorance. Our proposed approach, the Imprecise Belief Label Aggregation (IMP-BLA) aims to aggregate such imperfect data besides of taking into account the expertise of participants computed with two complementary degrees notably the exactitude and precision degrees. These latter are used effectively in [14] to identify different types of workers in crowdsourcing systems also under the belief function framework.

3   Background on the Belief Function Theory

The belief function theory [5,6] is considered as one of the main frameworks for incertainty modeling that offers various tools to represent different types of imperfection, manage and combine information. In this work, we rely on the Transferable Belief Model (TBM) [13] as it is one of the most common interpretations of the belief function theory.

Let $\Theta = \{\theta_1, \theta_2,..., \theta_k\}$ denotes a finite set including all the events that can possibly occur in a given problem case. Θ is called frame of discernment.

The information given by a source about an event can be represented by a basic belief assignement (*bba*) (a mass function) denoted by m. It is defined from the power set of Θ, 2^Θ to $[0, 1]$ such that $\sum_{E \subseteq \Theta} m(E) = 1$.

A subset E of Θ is called a focal element when $m(E) > 0$. Some special *bbas* are presented in Table 1.

Table 1. Special *bbas*

bba	Definition
Certain	$m(E) = 1$, $E \in \Theta$ (E is a singleton and it is the only focal element)
Vacuous	$m(\Theta) = 1$ (Θ is the focal element)
Categorical	$m(E) = 1$, $E \subseteq \Theta$ (E is the only focal element)
Simple support function	$m(E) = 1 - \theta$, $m(\Theta) = \theta$ and $\theta \in [0, 1]$
Normal	$m(\emptyset) = 0$
Conflictual	$m(\emptyset) > 0$

The discounting operation [5] allows to take into consideration the reliability of a source providing a piece of information represented by a mass function m. Let $\alpha \in [0, 1]$ be the discounting rate, the discounted mass function m^α is defined by:

$$\begin{cases} m^\alpha(E) = (1 - \alpha)m(E), & \forall\, E \subset \Theta, \\ m^\alpha(\Theta) = (1 - \alpha)m(\Theta) + \alpha. \end{cases} \tag{1}$$

In order to enable the decision process, the belief function framework offers different combination rules.

Two distinct and reliable *bbas* m_1 and m_2 belonging to the same frame of discernment Θ can be combined using:

– Conjunctive rule of combination proposed by Smets [9] and defined by:

$$m_1 \bigcirc\!\!\!\!\!\wedge m_2(E) = \sum_{F \cap G = E} m_1(F)m_2(G) \tag{2}$$

– Dempster rule of combination introduced by Dempster [6]. It is similar to the conjunctive rule except that it does not tolerate a mass on the empty set leading us to have a normal *bba*. It is defined by:

$$m_1 \oplus m_2(G) = \begin{cases} \dfrac{m_1 \bigcirc\!\!\!\!\!\wedge m_2(G)}{1 - m_1 \bigcirc\!\!\!\!\!\wedge m_2(\emptyset)} & \text{if } \emptyset \neq G \subseteq \Omega \\ 0 & \text{otherwise.} \end{cases} \tag{3}$$

– The Combination With Adapted Conflict rule (CWAC) [8] defined by:

$$m_{\bigcirc\!\!\!\!\!\leftrightarrow}(E) = (\bigcirc\!\!\!\!\!\leftrightarrow m_i)(E) = Dm_{\bigcirc\!\!\!\!\!\wedge}(E) + (1 - D)m_\oplus(E) \tag{4}$$

It acts either like the conjunctive rule or like the dempster rule depending on the maximum Jousselme distance [7] between two *bbas* defined as $D = \max [\mathrm{d}(m_i, m_j)]$ such that:

$$d(m_1, m_2) = \sqrt{\frac{1}{2}(m_1 - m_2)^t \mathrm{D}(m_1 - m_2)}, \tag{5}$$

with D is called the Jaccard index defined as:

$$\mathrm{D}(E, F) = \begin{cases} 0 & \text{if } E = F = \emptyset, \\ \dfrac{|E \cap F|}{|E \cup F|} & \forall\, E, F \in 2^\Theta. \end{cases} \tag{6}$$

The principle of decision making in the Transferable Belief Model is based on the pignistic probability (*BetP*) that consists in finding out the most likely hypothesis to happen. It is defined as follows:

$$BetP(\omega_i) = \sum_{E \subseteq \Omega} \frac{|E \cap \omega_i|}{|E|} \cdot \frac{m(E)}{(1 - m(\emptyset))} \quad \forall\, \omega_i \in \Omega \tag{7}$$

4 IMP-BLA: Imprecise Belief Label Aggregation

In this work, we propose a new label aggregation approach that deals with imprecise answers. In fact, in a common single-class labelling scenario, workers are asked to give one answer for each question. However, in our study, we provide a solution in the case where workers are allowed to respond by multiple labels if they are not completely sure about what the correct label is. Since probabilitic methods could not model such scenario, we adopted the belief function theory to handle imprecise labels. Figure 1 presents the different stages of our proposed approach. We go through each stage in next subsections.

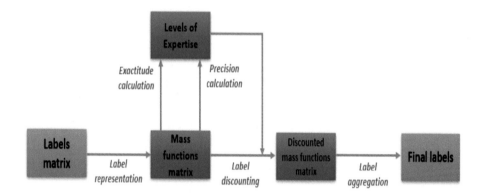

Fig. 1. IMP-BLA

4.1 Step 1: Mass Functions Generation

We suppose that a set of N questions are given to R different workers. Each worker W_j gives an answer A_{ij} to each question Q_i, belongs to subsets in $\{1, 2, ..., K\}$ where K is the number of possible classes. For the sake of simplicity, we suppose that $K = 3$ which means that answers can take values in $\{1, 2, 3, (1, 2), (1, 3), (2, 3), (1, 2, 3)\}$. The final aggregated label $\hat{A}_i$ exists in $\{1, 2, 3\}$.

This step represents labels under the belief function theory because of its effectiveness in modelling imperfect data. As a result each label is transformed into a *bba* m_{ij}^{Θ} with $\Theta = \{\theta_1, \ldots, \theta_n\}$. We consider $\Theta = \{1, 2, 3\}$ in this work.

Example 1. Let us consider the labels matrix presented in Table 2. The transformation of the labels matrix into *bba* matrix is illustrated in Table 3.

As we notice, we obtained 3 types of mass functions:
- Certain *bba*; it means that the worker gives an exact answer (e.g. $m_{11}(\{2\}) = 1$)
- Categorical *bba*; means that the worker is not completly certain about his answer so he gives an imprecise answer (e.g. $m_{13}(\{1, 3\}) = 1$)
- Vacuous *bba*; means that the worker is totally ignorant about what the answer is so he selects all the given answers (e.g. $m_{23}(\{\Theta\}) = 1$)

Table 2. Labels matrix

	Q_1	Q_2	Q_3
W_1	2	1	3
W_2	(2, 3)	1	(1, 3)
W_3	(1, 3)	(1, 2, 3)	3
W_4	(1, 2, 3)	(2, 3)	(1, 2)

Table 3. Label representation

	Q_1	Q_2	Q_3
W_1	$m_{11}(\{2\})=1$	$m_{21}(\{1\})=1$	$m_{31}(\{3\})=1$
W_2	$m_{12}(\{2,3\})=1$	$m_{22}(\{1\})=1$	$m_{32}(\{1,3\})=1$
W_3	$m_{13}(\{1,3\})=1$	$m_{23}(\{\Theta\})=1$	$m_{33}(\{3\})=1$
W_4	$m_{14}(\{\Theta\})=1$	$m_{24}(\{2,3\})=1$	$m_{34}(\{1,2\})=1$

4.2   Step 2: Expertise Estimation

For this step, we propose to compute an expertise degree for each worker that we note x_j. As explained in [14], this degree is estimated after combining two different degrees namely exactitude and precision degree.

Exactitude Degree. The exactitude degree noted e_j is basically the average of the distance between the answer given by the worker m_{W_j} and all the responses of the other workers $m_{W_{\Sigma_{R-1}}}$. This latter is the average of answers proposed by the R − 1 participants for each question, such as:

$$m_{W_{\Sigma_{R-1}}}(X) = \frac{1}{R-1} \sum_{j=1}^{R-1}(X) \tag{8}$$

The distance is then calculated by the distance of Jousselme [7]: $d_j(m_{W_j}, m_{W_{\Sigma_{R-1}}})$. According to this distance, the exactitude degree for each participant is calculated as follows:

$$e_j = 1 - \frac{1}{A_j} \sum_{i=1}^{N} d_j \tag{9}$$

Precision Degree. The precision degree noted p_j consists in calculating the average of the specificity degree [21] of each answer. The specificity degree is computed as follows:

$$\delta_j = 1 - \sum_{X \in 2^\Theta} m_j(X) \frac{log_2(|X|)}{log_2(|\Theta|)} \tag{10}$$

Then the precision degree is obtained by averaging the specificity degrees of all the N questions such as:

$$p_j = \frac{1}{A_j} \sum_{i=1}^{N} \delta_j \tag{11}$$

Expertise Degree. Once the exactitude and the precision degrees are calculated, we combine both of them to obtain the expertise degree as follows:

$$x_j = \beta_j e_j + (1 - \beta_j) p_j \tag{12}$$

4.3   Step 3: Mass Functions Discounting

This step consists in integrating the expertise degree of each worker in the reponses he provided. Therefore, each mass function will be weakened using the discounting operation Eq. 1 such as $\alpha_j = 1 - e_j$. Consequently, we will obtain simple support functions.

Example 2. Let us consider the labels transformed in Table 3 in particular the *bba* of the second worker W_2 for the first question and we suppose that his relative expertise degree is equal to 0.6. Accordingly, the mass function is changed as follows:
$m_{12}(\{2, 3\}) = (1 - 0.4) * 1 = 0.6;$
$m_{12}(\Theta) = 0.4 + (1 - 0.4) * 0 = 0.4$

4.4   Step 4: Mass Functions Aggregation

As a final step, we aggregate for each worker all his discounted mass functions using the CWAC combination rule (Eq. 4) as it is effective in handling the conflict that can be produced. We obtain then a final *bba* for each question that we transform into a decision by applying the pignistic probability (Eq. 7).

Example 3. We consider question 1 in Table 3, after applying the CWAC rule to aggregate all the workers discounted *bbas*, we obtain this final *bba*:
$m_1(\emptyset) = 0.1254;$ $m_1(\{2\}) = 0.5425;$ $m_1(\{3\}) = 0.0598$ $m_1(\{1,3\}) = 0.0399;$
$m_1(\{2,3\}) = 0.1395; m_1(\Theta) = 0.093$

The corresponding pignistic probability is as follows:
$Betp(\{1\}) = 0.0582; Betp(\{2\}) = 0.7354; Betp(\{3\}) = 0.2063$
Accordingly the final label $\hat{A}_1$ is 2 as it obtains the highest pignistic probability.

5   Experimentation

In order to evaluate our proposed approach (IMP-BLA), we conducted experiments on a dataset (R = 50, N = 100) where we randomly generated NxR answers regarding two types of workers namely experts and bad. Experts are those who have more than 80% of exact and precise answers hence a high level of expertise, whereas the bad ones have more than 70% of inexact andnor imprecise answers.

Therefore, in our experiment we vary the ratio of bad workers and observe the accuracy measure calculated as follows:

$$Accuracy = \frac{Number\ of\ correctly\ labeled\ questions}{Number\ of\ total\ questions} \tag{13}$$

Note that, as recommended in [14], we fixed the weight β to 0.5 when computing the expertise levels of workers, since the exactitude and precision degrees have both the same importance.

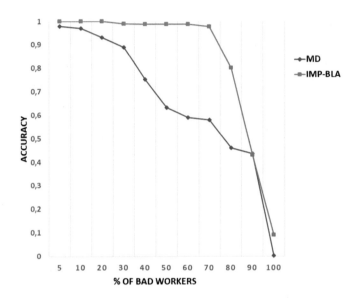

Fig. 2. Accuracy according to different bad labelers ratio

Table 4. Average accuracies of MD and IMP-BLA when varying % of bad workers

Bad workers	[0–30%]	[30–70%]	[70–100%]
MD	0.94	0.63	0.29
IMP-BLA	1	0.98	0.44

Figure 2 displays results comparing our method to majority decision (MD) in terms of accuracy when increasing the ratio of bad workers. We observe that IMP-BLA outperforms the MD when the percentage of workers surpasses 10%

and stays consistant up to 70% with 0.98 accuracy. The average accuracies corresponding to different intervals are reported in Table 4. Even when low quality labelers exceeds 70%, IMP-BLA records an average accuracy of 0.44 overtaking the majority method.

6   Conclusion and Future Work

In this paper, we proposed a new label aggregation approach based on the belief function theory. The IMP-BLA has indeed been reliable in the case of combining imprecise answers provided by a crowd of different levels of expertise. Compared to the majority method, our approach has much better results in terms of accuracy even when a large amount of bad workers is included. As future work, we intend to tackle more scenarii in the crowdsourcing systems such as multichoice labelling.

References

1. Zheng, Y., Wang, J., Li, G., Feng, J.: QASCA: a quality-aware task assignment system for crowdsourcing applications. In: International Conference on Management of Data, pp. 1031–1046 (2015)
2. Yan, T., Kumar, V., Ganesan, D.: Designing games with a purpose. Commun. ACM **51**(8), 58–67 (2008)
3. Snow, R., O'Connor, B., Jurafsky, D., Ng, A.Y.: Cheap and fast but is it good? Evaluation non-expert annotations for natural language tasks. In: The Conference on Empirical Methods in Natural Languages Processing, pp. 254–263 (2008)
4. Sheng, V.S., Provost, F., Ipeirotis, P.G.: Get another label? Improving data quality and data mining using multiple, noisy labelers. In: International Conference on Knowledge Discovery and Data Mining, pp. 614–622 (2008)
5. Shafer, G.: A Mathematical Theory of Evidence, vol. 1. Princeton University Press, Princeton (1976)
6. Dempster, A.P.: Upper and lower probabilities induced by a multivalued mapping. In: The Annals of Mathematical Statistics, pp. 325–339 (1967)
7. Jousselme, A.-L., Grenier, D., Bossé, É.: A new distance between two bodies of evidence. In: Information Fusion, pp. 91–101 (2001)
8. Lefèvre, E., Elouedi, Z.: How to preserve the conflict as an alarm in the combination of belief functions? Decis. Support Syst. **56**, 326–333 (2013)
9. Smets, P.: The combination of evidence in the transferable belief model. IEEE Trans. Pattern Anal. Mach. Intell. **12**(5), 447–458 (1990)
10. Raykar, V.C., Yu, S.: Eliminating spammers and ranking annotators for crowdsourced labeling tasks. J. Mach. Learn. Res. **13**, 491–518 (2012)
11. Dawid, A.P., Skene, A.M.: Maximum likelihood estimation of observer error-rates using the EM algorithm. Appl. Stat. **28**, 20–28 (2010)
12. Khattak, F.K., Salleb, A.: Quality control of crowd labeling through expert evaluation. In: The Neural Information Processing Systems 2nd Workshop on Computational Social Science and the Wisdom of Crowds, pp. 27–29 (2011)
13. Smets, P., Mamdani, A., Dubois, D., Prade, H.: Non Standard Logics for Automated Reasoning, pp. 253–286. Academic Press, London (1988)

14. Ben Rjab, A., Kharoune, M., Miklos, Z., Martin, A.: Characterization of experts in crowdsourcing platforms. In: International Conference on BELIEF 2016, pp. 97–104 (2016)
15. Watanabe, M., Yamaguchi, K.: The EM Algorithm and Related Statistical Models, 250 p. CRC Press, Boca Raton (2003)
16. Whitehill, J., Wu, T., Bergsma, J., Movellan, J.R., Ruvolo, P.L.: Whose vote should count more: optimal integration of labels from labelers of unknown expertise. In: Neural Information Processing Systems, pp. 2035–2043 (2009)
17. Abassi, L., Boukhris, I.: Crowd label aggregation under a belief function framework. In: International Conference on Knowledge Science, Engineering and Management, pp. 185–196. Springer (2016)
18. Abassi, L., Boukhris, I.: A gold standards-based crowd label aggregation within the belief function theory. In: International Conference on Industrial, Engineering and Other Applications of Applied Intelligent Systems, pp. 97–106. Springer (2017)
19. Abassi, L., Boukhris, I.: Iterative aggregation of crowdsourced tasks within the belief function theory. In: European Conference on Symbolic and Quantitative Approaches to Reasoning and Uncertainty, pp. 159–168. Springer (2017)
20. Abassi, L., Boukhris, I.: A worker clustering-based approach of label aggregation under the belief function theory. In: Applied Intelligence, pp. 1573–7497 (2018)
21. Florentin, S., Arnaud, M., Christophe, O.: Contradiction measures and specificity degrees of basic belief assignments. In: 14th International Conference on Information Fusion, pp. 1–8 (2011)
22. Kuncheva, L., et al.: Limits on the majority vote accuracy in classifier fusion. Pattern Anal. Appl. **6**, 22–31 (2003)

An Augmented Algorithm for Energy Efficient Clustering

Ushus Elizebeth Zachariah$^{(\boxtimes)}$ and Lakshmanan Kuppusamy

School of Computer Science and Engineering, Vellore Institute of Technology,
Vellore 632014, Tamil Nadu, India
ushus@vit.ac.in

Abstract. In recent years, Heterogeneous Wireless Sensor Networks are widely used to perform a variety of real-time applications. Since sensor nodes are battery powered, developing an energy efficient algorithm to increase the network life time is challenging. This paper proposes a cluster based mechanism, namely, *Augmented EDEEC*, to improve the network life time. This algorithm is obtained by modifying the distance metric and level of heterogeneity in the existing Protocol. This new protocol exhibits 1.25 times more network lifetime than the existing Distributed Energy Efficient Clustering algorithms. The results are benchmarked with the energy efficient clustering protocols like DEEC, DDEEC and EDEEC.

Keywords: WSN · Energy efficiency · Clustering · Distance metric

1 Introduction

Wireless Sensor Networks (WSN) are collection of very minute sensors usually called nodes, with capabilities of sensing, computing and communication. With the advancement of Micro Electro Mechanical Systems technology (MEMS), WSNs are widely used in various applications like environmental monitoring, health monitoring, precision agriculture, military surveillance etc. [1]. In each of these applications, sensor nodes are expected to run for a longer periods of time, ranging from months to years. The sensor nodes are expected to collect the data periodically from the sensing environment and send back to the target area (also called base station). These nodes work with the help of irreplaceable power source which have limited energy. To improve the network life time under the limited battery life, there is need for development of energy efficient protocols [2, 3].

Developing an Energy efficient mechanism needs specific set of requirements that varies from one application to another. The major WSN requirements include scalability, coverage, latency, Quality of Service, Security, mobility and robustness. These application requirement and energy efficient protocols are interdependent [1].

Researchers have developed various energy efficient mechanisms based on data reduction techniques, optimizing the radio module, Energy efficient routing and so on. This paper addresses the problem of energy consumption in WSN using clustered approach. In Clustered architecture, the nodes are organized in to groups known as clusters. Each cluster have a Cluster Head (CH). CH is responsible for coordinating and

A. Abraham et al. (Eds.): ISDA 2018, AISC 941, pp. 617–626, 2020.
https://doi.org/10.1007/978-3-030-16660-1_60

transmitting the data from nodes to the base station. By balancing the energy consumption (through rotating the CH), turning off certain nodes (while CH transmits data to BS) and improving network scalability, the specific advantages of clustered approach are attained [1, 4, 5].

Clustered Networks can be of homogenous or heterogeneous. The existing routing protocols for homogeneous environment such as LEACH, TEEN, APTEEN, HEED, PEGASIS [6, 13–16] are not well suited for heterogeneous environment. As the nodes of the heterogeneous WSN are provided with different energy levels, the above mentioned protocols are not suitable as these protocols cannot discriminate the nodes in terms of energy [7–12, 15]. In [8], the authors developed SEP protocol, which considers two kinds of node namely normal and advanced. The CH is elected based on the initial energy of the node. Later, DEEC Protocol was proposed by [9] which also had two levels of heterogeneity. Here the CH is selected based on the probability of residual energy and average energy of the network. Usually the node with higher energy level gets the higher chance of becoming the Cluster head.

The DDEEC protocol developed [10] selects the CH based on the residual energy alone. Hence during the initial rounds, the advance node has the highest probability to become the Cluster head. In later stages, when the energy is dissipated, the advanced node will have the same probability as the normal node to become the CH. Saini and Sharma [11] introduced EDEEC protocol which has three levels of heterogeneity normal, advanced and super nodes, which makes the systems more complex, but it prolongs the network life time and stability period. BEENISH [17] is an energy aware protocol considered four levels of heterogeneity, selection of CH is based on residual and average energy of the network. In [18] the author has developed LE-MHR protocol it considers k levels horizontal energy heterogeneity, and it introduces the number of alive nodes in the current round for the CH selection. Similarly Singh [19] introduced n level heterogeneity based on basic protocol as HEED and extended up to MLHEED-n they have taken residual energy and node density as parameter element to choose the cluster head.

In this paper, we propose a protocol named AEDEEC (Augmented Enhanced Distributed Energy Efficient Clustering Algorithm) which is an extension to E-DEEC protocol scheme. In the proposed AEDDEC, the network is deployed initially with four levels of heterogeneity, namely, normal nodes, advanced nodes, super nodes and hyper nodes. While, the normal nodes have the least initial energy, the hyper nodes have the highest. The proposed AEDEEC initially elects the cluster head (CH) based on the energy level, and alters the CH at each epoch to reduce the pace of energy drain. Usually the nodes with higher initial energy are given priority to be elected as the CH. At the end of each round, a different CH is elected based on the residual energy of the node and average energy of the network. The neighboring nodes for each cluster head are elected separately using Manhattan distance/Euclidean distance. The result shows that the proposed AEDEEC has higher network lifetime and higher stability than DEEC, DDEEC and EDEEC.

The paper is organized as follows. Section 2 describes the proposed protocol and system model for AEDEEC. Section 3 discusses simulation results used for analyzing the performance of AEDEEC. Finally concludes the paper with some future work.

2 Proposed Protocol

In this section, we discuss the details of our proposed protocol, which is an extension to E-DEEC protocol by modifying the distance metric and the level of heterogeneity

The network model consist of N number of nodes randomly distributed in an $M * M$ region with Base Station at center. In our proposed protocol we considered four level heterogeneity where the initial energy of all the four levels are different. Based on the energy difference, the heterogeneous nodes are named as normal nodes, advanced nodes, super nodes and hyper nodes. The normal nodes have the least initial energy and the hyper nodes have the highest initial energy. If 'N' is the total number of sensor nodes in the WSN, then 'm', 'm_0' and m_1' represents the percentage of nodes which are advanced, super and hyper nodes, These nodes having 'a', 'b' and 'c' times more energy than the normal nodes.

Similar to the existing protocol [6], the first order radio model have been used.

The model describes about the energy dissipation of a data over a distance (d). The Energy dissipated for transmitting and receiving j-bit message is given by Eqs. 1 and 2 respectively.

$$E_{tx}(j, d) = \begin{cases} J * E_{elect} + J * \epsilon_{fs} d^2 & \text{If } d \leq D_0 \\ J * E_{elect} + J * \varepsilon_{mp} d^4 & \text{If } d > D_0 \end{cases} \tag{1}$$

$$E_{rx}(j, d) = J * E_{elect} \tag{2}$$

Here E_{elect} is the energy required to run the transmitter and receiver circuit. ϵ_{fs} and ε_{mp} represents the free space and multi path models and d is the distance between the sender and receiver. The threshold Distance D_0 is evaluated by $D_0 = \sqrt{\epsilon_{fs}/\varepsilon_{mp}}$.

2.1 Cluster Head Selection and Election Methodology

AEDEEC protocol uses the same strategy of EDEEC [11] for the cluster head selection. The algorithm works in rounds. At the beginning of each round, a node which has highest energy has the chance to become a cluster head using the following threshold equation [1], the node is elected as CH.

$$T(s) = \begin{cases} \frac{p}{1-p*\left(r mod_p^1\right)} & \text{if } n \in G \\ 0 & \text{otherwise} \end{cases} \tag{3}$$

Where p represents the desired percentage of cluster heads, r denotes the current round. G is the set of nodes that have not been the cluster heads for last 1/p rounds. The Probabilities of the four nodes are given by Eq. 4, Threshold for the CH selection is calculated by substituting these equation in Eq. 3.

$$P(i) = \begin{cases} p * E_i(r)/k & \text{for normal nodes} \\ p * E_i(r) * (1+a)/k & \text{for advanced nodes} \\ p * E_i(r) * (1+b)/k & \text{for super nodes} \\ p * E_i(r) * (1+c)/k & \text{for hyper nodes} \end{cases} \tag{4}$$

Where $k = 1 + (m * a + m_0 * b + m_1 * c) * Ea$ and Ei (r) is the energy available and Ea is the average energy. The average energy is calculated by,

$$E_a = \frac{E_t * (1 - r/r_{max})}{N} \tag{5}$$

Where Et is the total energy, r_{max} denotes the total number of rounds of the network life time and N is the total number o nodes. The r_{max} is given by,

$$r_{max} = E_t/E_r \tag{6}$$

Where Er is the total amount of energy which is dissipated in the network during each round and could be evaluated by,

$$E_r = j(2 * N * E_{tx} + N * E_{DA} + K * \epsilon_{mp} * d_1^4 + N * \epsilon_{fs} * d_2^2 \tag{7}$$

Where EDA is the data aggregation cost, d_1 is the average distance between the cluster head and sink and d_2 is the average distance between cluster members and cluster head. The distance d_1, d_2 and K could be calculated by,

$$d_1 = 0.765 * x/2 \tag{8}$$

$$d_2 = x/\sqrt{2 * pi * K} \tag{9}$$

$$K = \sqrt{N/\pi * D_0} * x/d_1^2 \tag{10}$$

Now the round is started and at the end of each round, the energy dissipated is reduced from the initial energy of the node by using the following equation, separately for cluster head and Normal nodes.

$$E_{disch} = \begin{cases} j * \left(E_{tx} + E_{DA} + (\epsilon_{mp} * j * d^4)\right) & \text{if } d_{optBS} > D_0 \\ j * \left(E_{tx} + E_{DA} + (\epsilon_{fs} * j * d^2)\right) & \text{if } d_{optBS} \leq D_0 \end{cases} \tag{11}$$

$$E_{disnn} = \begin{cases} j * \left(E_{tx} + E_{DA} + (\epsilon_{mp} * j * d^4)\right) & \text{if } d_{optCH} > D_0 \\ j * \left(E_{tx} + E_{DA} + (\epsilon_{fs} * j * d^2)\right) & \text{if } d_{optCH} \leq D_0 \end{cases} \tag{12}$$

In the above discussed energy dissipation evaluation d_{optBS} and d_{optCH} are calculated using the Euclidean in all the available methodologies as given in equations.

$$d_{optBS} = \sqrt{(x_i - x_b)^2 + (y_i - y_b)^2} \qquad (13)$$

Where (x_i, y_i) and (x_b, y_b) are the positions of cluster head and base station.

$$d_{optch} = \sqrt{(x_i - x_c)^2 + (y_i - y_c)^2} \qquad (14)$$

Where (x_i, y_i) and (x_c, y_c) are the positions of cluster members and cluster head. In our proposed protocol, the parameter calculation of d_{optBS} and d_{optCH} is modified by using the Manhattan Distance as shown in the equation below. The reason behind using this distance metric is to help the users use the proposed protocol in real time environment deployment as well.

$$d_{optBS} = |(x_i - x_b + y_i - y_b)| \qquad (15)$$

$$d_{optCH} = |(x_i - x_C + y_i - y_C)| \qquad (16)$$

3 Results and Discussions

3.1 Simulation Parameters and Scenarios Created

We have simulated the proposed four level heterogeneous WSN in MATLAB 2017b. To evaluate the performance of the proposed protocol, it is compared with DEEC, DDEEC and EDEEC, with respect to both Euclidean and Manhattan distance.

The radio parameters used for the simulations are shown in Table 1. For the analysis of the proposed work, we have considered network life time and stability as the metric. Whilst the life time of the network field is defined as the last node alive in the network, stability is measured as the period from the death of the *First Node* to the alive *Last Node*.

Table 1. Simulation parameters

Parameters	Value
Area of interest	(100, 100)
Total number of nodes	100
E_0(initial energy)	0.5 J
Size of message(j)	4000 bits
E_{tx}	50 nJ/bit
E_{rx}	50 nJ/bit
ε_{fs}	10 nJ/bit/m^2
ε_{mp}	0.0013/pJ/bit/m^4
E_{DA}	5 nJ/bit/signal
P	0.1

Simulations are performed in two different scenarios which is presented in Table 2. The deployment network is tested under different cases of heterogeneity by varying the percentage of energy at different levels.

Table 2. Different cases for simulation

Cases	Network field	No of nodes	Energy levels
1.	100 * 100	100	**0.25, 0.5, 0.75, 1**
2.	100 * 100	100	**0.5, 1.5, 2.5, 3.5**

We have evaluated the performance of the network in both Euclidean and Manhattan distance by considering 0.25 difference in energy levels. The parameters taken for the analysis as follows $m = 0.4$, $m0 = 0.3$ and $m1 = 0.3$, which containing 31 normal nodes with 0.25 J energy, 28 advanced nodes with 0.5 J energy, 8 super nodes with 0.75 J energy and 3 nodes with 1 J energy.

3.2   Simulation Results

By analyzing the projected network lifetime for the given energy levels in Case-1 (Figs. 1 and 2), it is inferred that all the algorithms (DEEC, DDEEC, EDEEC and AEDEEC) had no nodes left live after 4000 rounds. A closer bifurcation of the general inference with respect to the chosen distance metric revealed that with Euclidean distance criteria, the alive nodes had a steep drop from the initial value of 100 live nodes to zero live nodes with in the first 2000 rounds. In fact, whilst the AEDEEC algorithm had the live nodes in the network until 1700 rounds (through gradual decrement), all the other algorithms (DEEC, DDEEC and EDEEC) had all the nodes dead much earlier at 1100 rounds. Hence, AEDEEC has a modest gain over the existing algorithms by nearly 600 rounds (Fig. 1).

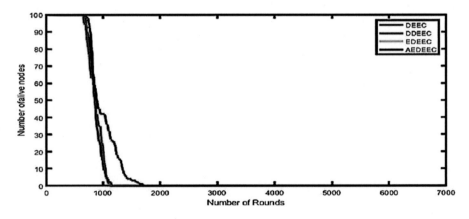

Fig. 1. Alive Nodes during Network life time using Euclidean Distance

Much more promising results were gained through Manhattan distance criterion. Comparing with EDEEC, until 1500 rounds, though the AEDEEC experienced a sharp decline in the number of live nodes (100 nodes to 10 nodes), it could sustain the network life until 4100 rounds - added the network life by 2600 rounds (Fig. 2). But all the other algorithms exhibited almost similar trend in shedding the number of live nodes – all the nodes were dead at 2100 rounds. Hence, AEDEEC has been proved to remain live and successful until 4000 rounds (Fig. 2).

In both Euclidean and Manhattan distance criteria, the first dead node was spotted after 500 rounds.

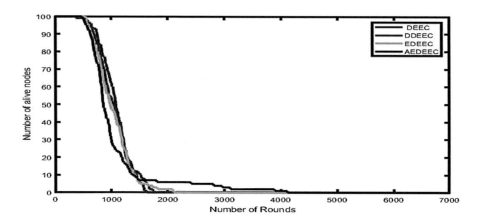

Fig. 2. Alive nodes during Network life time using Manhattan distance

By increasing the energy factor by 1 J (as given in case-2), the network life evaluated by Euclidean distance criterion resulted in exhibiting explicit difference (Fig. 3). With DEEC and DDEEC, whilst the network merely managed to stay alive until 3700 rounds, the EDEED and AEDEEC could sustain the network life with modest number of live nodes (average 20 numbers) until 7000 rounds (Fig. 3).

In between 900–2200 rounds, though the AEDEEC experienced a steep decline in the number of live nodes than its competing algorithms, under the Manhattan distance criterion, the network was live even after 7000 rounds (Fig. 4). Hence, in the combination of both the cases, the AEDEEC was found to be superior than DEEC, DDEEC, EDEEC.

The life time of network and stability is increased when the energy difference are small in manhattan, but in case of eucedian if the energy difference is larger it performs well.

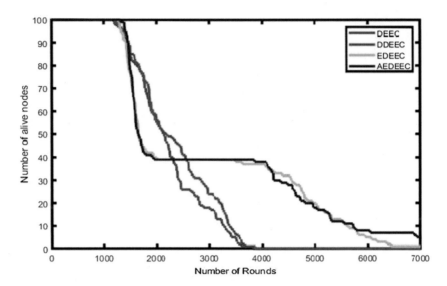

Fig. 3. Alive nodes during Network life time using Euclidean distance

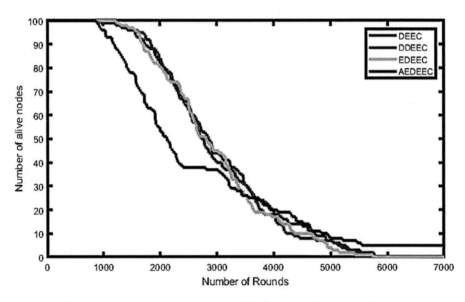

Fig. 4. Alive nodes during Network life time using Manhattan distance

4 Conclusion

In this paper, we have introduced a new protocol called AEDEEC for energy aware hierarchical clustering scheme that can be applied to Heterogeneous Wireless Sensor Networks. The heterogeneity is with respect to the energy levels used in four types of nodes. The CH selection was based on the initial energy at the starting stage and the election was as per the residual energy and average energy. The residual energy was calculated based on the real-time distance metric called Manhattan distance as one of the parameter. The performance of the protocol was analysed using various parameters such as number of alive nodes and stability period. From simulation experiments, we have found that AEDEEC increases the stability period of four level heterogeneous WSN by 1.25 times, the network life time and a few alive nodes, even after 7000 rounds compared to DEEC, DDEEC and E-DEEC.

The future focus that could be of interest is to use some bio-inspired methodology which could optimize the cluster head selection mechanism and further extends to multi-path scenario for increasing the fault tolerance of the wireless heterogeneous network.

References

1. Rault, T., Bouabdallah, A., Challal, Y.: Energy efficiency in wireless sensor networks: a top-down survey. Comput. Netw. **4**(67), 104–122 (2014)
2. Akyildiz, I.F., Su, W., Sankarasubramaniam, Y., Cayirci, E.: Wireless sensor networks: a survey. Comput. Netw. **38**(4), 393–422 (2002)
3. Nack, F.: An overview on wireless sensor networks, Institute of Computer Science (ICS), Freie Universität Berlin (2010)
4. Liu, X.: A survey on clustering routing protocols in wireless sensor networks. Sensors **12**, 11113–11153 (2012)
5. Al-Karaki, N., Kamal, A.E.: Routing techniques in wireless sensor networks: a survey. IEEE Wirel. Commun. (2004). https://doi.org/10.1109/mwc.2004.1368893
6. Heinzelman, W.B., Chandrakasan, A.P., Balakrishnan, H.: Application-specific protocol architecture for wireless microsensor networks. IEEE Trans. Wirel. Commun. **1**, 660–670 (2002)
7. Tanwar, S., Kumar, N., Rodrigues, J.J.: A systematic review on heterogeneous routing protocols for wireless sensor network. J. Netw. Comput. Appl. (2015). https://doi.org/10.1016/j.jnca.2015.03.004
8. Smaragdakis, G., Matta, I., Bestavros, A.: SEP: a stable election protocol for clustered heterogeneous wireless sensor networks. In: Proceedings of the Second International Workshop on Sensor and Actor Network Protocols and Applications, Boston, MA, pp. 1–11 (2004)
9. Qing, L., Zhu, Q., Wang, M.: Design of a distributed energy-efficient clustering algorithm for heterogeneous wireless sensor networks. Comput. Commun. (2006). https://doi.org/10.1016/j.comcom.2006.02.017
10. Elbhiri, B., Saadane, R., Aboutajdine, D.: Developed Distributed Energy-Efficient Clustering (DDEEC) for heterogeneous wireless sensor networks. In: Proceedings of 5th International Symposium On I/V Communications and Mobile Network, pp. 1–4 (2010)

11. Saini, P., Sharma, A.K.: E-DEEC-enhanced distributed energy efficient clustering scheme for heterogeneous WSN. In: Proceedings of the 1st International Conference on Parallel, Distributed and Grid Computing, Solan, pp. 205–210 (2010)
12. Pantazis, N.A., Nikolidakis, S.A., Vergados, D.D.: Energy-efficient routing protocols in wireless sensor networks: a survey. IEEE Commun. Surv. Tutor. (2013). https://doi.org/10.1109/surv.2012.062612.00084
13. Manjeshwar, A., Agrawal, D.P.: TEEN: a routing protocol for enhanced efficiency in wireless sensor networks. In: Proceedings of the 15th International Parallel and Distributed Processing Symposium, San Francisco, CA, USA, pp. 2009–2015 (2001)
14. Manjeshwar, A., Agrawal, D.P.: APTEEN: a hybrid protocol for efficient routing and comprehensive information retrieval in wireless. In: Proceedings of the 16th International Parallel and Distributed Processing Symposium, Ft. Lauderdale, FL, pp. 195–202 (2002)
15. Younis, O., Fahmy, S.: HEED: a hybrid, energy-efficient, distributed clustering approach for ad hoc sensor networks. IEEE Trans. Mobile Comput. (2004). https://doi.org/10.1109/tmc.2004.41
16. Lindsey, S., Raghavendra, C.S.: PEGASIS: power-efficient gathering in sensor information systems. In: Proceedings of the Aerospace Conference, Big Sky, MT, USA, vol. 3, pp. 1125–1130 (2002)
17. Qureshi, T.N., Javaid, N., Khan, A.H., Iqbal, A., Akhtar, E., Ishfaq, M.: BEENISH: balanced energy efficient network integrated super heterogeneous protocol for wireless sensor networks. arXiv preprint arXiv:1303.5285 (2013)
18. Tyagi, S., Tanwar, S., Gupta, S.K., Kumar, N., Rodrigues, J.J.: A lifetime extended multi-levels heterogeneous routing protocol for wireless sensor networks. Telecommun. Syst. **59**, 43–62 (2015)
19. Singh, S.: Energy efficient multilevel network model for heterogeneous WSNs. Eng. Sci. Technol. Int. J. **20**(1), 105–115 (2017)

Analysis of Left Main Coronary Bifurcation Angle to Detect Stenosis

S. Jevitha[1]([✉]), M. Dhanalakshmi[1], and Pradeep G. Nayar[2]

[1] Department of Biomedical Engineering, SSN College of Engineering, Chennai, India
jeevithasankar1294@gmail.com,
dhanalakshmim@ssn.edu.in
[2] Department of Cardiology, MIOT Hospital, Chennai, India
pg_nayar@rediffmail.com

Abstract. Narrowing of blood vessel due to plaque deposition is known as stenosis that acts as prime indicator of the coronary artery diseases (CAD). Coronary cine angiography (CCA) is a digital imaging modality used for the assessment of severity of coronary bifurcation lesions in the coronary arteries. Angiography based stenosis diagnosis is done as subjective analysis by the clinicians that results in overestimations or underestimations of detected stenosis. In stenosis, mostly the plaque deposition occurs at the left main coronary artery (LMCA) branch. Bifurcation angle at site of LMCA act as significant indicators of presences of stenosis. The proposed work involves segmentation of LMCA using various segmentation techniques such as Morphological based segmentation, Hessian detection and Active contour segmentation. Active contour segmentation provides clear visualization of LMCA structure when compare to all other segmentation methods. Then, computation of automatic bifurcation angle measurement at bifurcating regions of LMCA such as left anterior descending (LAD) and left circumflex (LCx) in both normal and stenotic images of CCA is performed. The diagnostic performance of stenosis yields a detection accuracy of 92%. The outcome of proposed work is found to be quantitative tool for the clinicians in accurate analysis of prediction of stenosis and also helpful during stent replacement surgical procedure in percutaneous coronary interventions.

Keywords: Left main coronary artery · Coronary Cine Angiogram ·
Active contour segmentation · Bifurcation angle · Stenosis

1 Introduction

Coronary artery diseases (CAD) are the most common and leading cause of deaths all over the world. It comes under the group of cardiovascular diseases. Coronary arteries are the blood vessel that provides oxygen and nutrient-rich blood supply to the heart. There are mainly two coronary arteries that right and left coronary arteries are present in the heart. Mostly plaque gets deposited at LCA due to anatomical nature of its structure rather than RCA.

© Springer Nature Switzerland AG 2020
A. Abraham et al. (Eds.): ISDA 2018, AISC 941, pp. 627–639, 2020.
https://doi.org/10.1007/978-3-030-16660-1_61

Stenosis is the most common and adverse disease condition of CAD that leads to various cardio-vascular diseases. The accumulation of fat and cholesterol inside the coronary arteries is termed as plaque. This tends to the deprived blood flow to the heart that incurs the existence of stenosis in coronary artery. As this plaque deposits grows, coronary arteries become weak, hardened and narrowed that weaken heart muscles that leads to chest pain, shortness of breath, heart attack, heart failure, etc. the majority of people who gets affected by CAD is in the age group of 60 or older (Fig. 1).

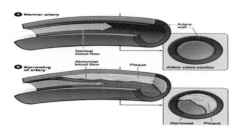

Fig. 1. Normal artery vs Stenotic artery [https://sites.google.com/site/cardiovascularbiotech/coronary-artery-disease-essay]

Stenosis can be diagnosed by three ways. They are Stenosis geometry and plaque composition, cardiac function and vessel function [1]. In this paper, the severity of stenosis is estimated using Stenosis geometry and plaque composition since the geometrical parameters such as diameter and cross sectional area of the artery and bifurcation angle at the site of LCA provides better quantitative assessment of severity of the Stenosis. Bifurcation angle acts as a prime indicator of stenosis as plaque deposition at the site of LCA tend to reduce the blood flow to both LAD and LCx [2, 3]. This bifurcation blockage is more malignant than the single branch blockage.

Coronary cine angiography or x-ray cine angiography is a primary imaging modality used for visualization and diagnosis of coronary artery diseases. It is a 2D projection of 3D vessel. There are several approaches on determination of presence of stenosis as follows. Sun et al. [4] analyzed the calcification of coronary plaques in the left coronary artery based on the measurement of plaque length. Yue Cui et al. [5] presented a method using Virtual Intravascular Endoscopy (VIE) to demonstrate the intraluminal changes of different types of plaque and it is associated with subsequent wider angulation of stenotic artery. Givehcl et al. [6] worked in Multiplanar reformation (MPR) and volume rendering technique (VRT) to measure bifurcation angle between LAD and LCx. Mohan et al. [7], presented a method to predict the presence of stenosis is based on the diameter of blood vessel. In the exiting methods, there were some limitations that include, lower specificity in detection of stenosis and also reliability is find to be less on the basis of accuracy of stenosis prediction.

The aim of this paper is to evaluate the diagnostic performance of the left main coronary artery using bifurcation angle for the detection of stenosis. Quantitative analysis of left main coronary arteries in coronary cine angiogram will guide the clinician in accurate diagnosis of stenosis. The analysis of bifurcation angle provides quantitative and qualitative prediction of stenosis.

The paper is organized as follows. In Sect. 2, briefly explained about overview of the proposed work. The results of stenosis detection on the test images (CCA) are shown in Sect. 3 and conclusion of the proposed work is presented in Sect. 4.

2   Materials and Methods

In the following subsection, we discussed about techniques used to segment the left main blood vessel and then bifurcation angle measurement at the site of stenosed region.

2.1   Materials

The real time coronary cine angiogram dataset were collected from the cardiology department of CHETTINAD HOSPITAL, Chennai that is acquired from 50 patients of 50–75 years of age. The dataset contain both normal and abnormal coronary images. Each individual data has around 20 frames for each sequence and each frame has a resolution of 512 × 512 pixels.

2.2   Methods

The framework of proposed work is shown in the Fig. 2.

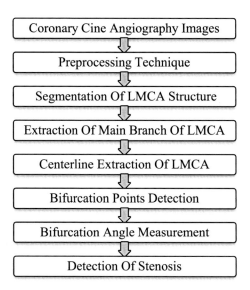

Fig. 2. Flow chart of the proposed work

2.2.1   Pre-processing Technique

The objective of the preprocessing phase is to improve the visual quality of the CCA frames by eliminating the non-uniform illumination and global motion artifact.

Median filter is a sliding window filter used to reduce the noises and also smooth the CCA images. Here the noise from CCA images is reduced by replacing the center value by the median of neighborhood pixels in 3 × 3 window of all the pixel values [8]. It also preserves the edges in the CCA frames. The median filtering output is defined as,

$$f(x, y) = \underset{(s,t) \in S_{xy}}{median} \{g(s, t)\} \tag{1}$$

2.2.2   Segmentation Techniques

The approach behind the segmentation process is to partitioning of image into a set of pixels based on the region of interest. In this paper, the process of segmentation is employed to fragment LMCA from preprocessed image using different segmentation techniques as follows.

2.2.2.1(a) Hessian Matrix Based Segmentation. Hessian-matrix based segmentation method is used in vaious medical images to find tubular structure in an image. Hessian matrix is a 2 × 2 square matrix composed of second-order partial derivatives of the input image. This second derivatives provides the estimation of direction of coronary artery structure in 2D angiographic images. The 2D hessian matrix is defined as

$$H = \begin{bmatrix} Ixx & Ixy \\ Iyx & Iyy \end{bmatrix}$$

Where H and I denote Hessian matrix and original image and Ixx, Iyy, Ixy and Iyx denotes second order derivative of image information, respectively.

From the obtained Hessian matrix its eigenvalues λi and eigenvectors are calculated. Eigenvector decomposition extracts an orthonormal coordinate system that is aligned with the second order structure of the image [9, 10].

2.2.2.1(b) Morphological Based Segmentation. Morphological operations deal with the shape and structure of the objects. Morphological techniques validate the image with a small template called structuring element [11]. Here, erosion and dilation is performed on the preprocessed image to segment the left main coronary artery branch structure.

2.2.2.1(c) Active Contour Segmentation. This method provides sub-regions with continuous boundaries in the given image. In this paper, boundaries of blood vessels are represented as contours (closed curve). The contours deform dynamically in order to match to the shape of blood vessel. Contour deformation is starts by applying shrinking operation iteratively for about 100 iteration. The mask is used to specify the initial state of the active contour. Boundaries of the blood vessels in mask define the initial contour position used for contour evolution to segment the blood vessels from the preprocessed image [12, 13]. It preserves the edges of blood vessel without the discontinuities in the image.

2.2.2.2 Extraction of Main Branch of LMCA. X-ray coronary angiogram images are segmented using various segmentation techniques to remove unwanted structures such as ribs, muscle, etc., while preserving only the left coronary artery tree structure. Active contour segmentation method is found to be best in clear visualization of entire left main coronary artery structure when compared to other segmentation methods.

After the segmentation of left coronary artery tree structure, main branch of LMCA is extracted using Frangi Filter in order to enhance the coronary blood vessel and to reduce noise illumination from the image.

2.2.2.2(a) Frangi's Filter. Frangi filter is also known as vessel enhancement vessel as it mainly designed to increase enhancement of vessel contrast along with suppression of background noises in the image. It works based on hessian matrix. Hessian matrix based vessel extraction filter can be defined as

$$F(x) = max_\sigma f(x, \sigma) \qquad (2)$$

Where x is a position of pixel in an image; f is filter used for vessel extraction and σ is the standard deviation for calculating Gaussian image derivative [14].

2.2.2.3 Centerline Detection. Following the extraction of LMCA using Frangi filter, skeletonization operation is performed to tract the centerline of LMCA. Skeletonization is a process of binary morphological operation, involves in eroding the boundary of pixel in the foreground region without destroying the connectivity of the original region of the image using structuring element [15]. It preserves the topology of the shape of the pattern in the image without any interrupt. Therefore, remaining pixels makes the skeleton of foreground region as thin line in the image.

2.2.2.4 Bifurcation Points Detection. The bifurcation points act as important role in the analysis of coronary vascular diseases. The locations were main blood vessel branch subdivided into two daughter blood vessel branches are known as bifurcation points. To detect bifurcation points in left coronary artery tree structure, a mask is applied on the centerline of skeleton of LMCA image [16]. A 3×3 window that has eight neighborhood pixels to the central pixel is used to pick the potential bifurcation points in the LMCA images. The mask is used to move to right or to down one pixel at each step and all pixels are checked. If the center pixel that lie on the blood vessel that has 3 neighborhoods as shown in the Fig. 3 then it is detected as the bifurcation point. Likewise window is applied to all the image pixels and bifurcation points are detected.

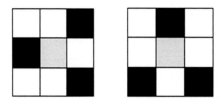

Fig. 3. Bifurcation point template (Grey: Center pixel, Black: Neighborhood pixel)

2.2.2.5 Bifurcation Angle Measurement. The angle between the axis of the main vessel and the axis of the side-branch at its origin is known as bifurcation angle [17]. Natural distribution of four main coronary artery bifurcation angles are [LAD and LCx], [LAD and Diag 1], [LCx and OM1], [PDA and Rpld] [1]. The left main coronary bifurcation angle is mainly formed by two coronary arteries that LAD and LCx at the site of LMCA.

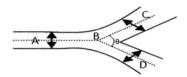

Fig. 4. Schematic diagram of bifurcation angle measurement

In this paper, from the Fig. 4 the bifurcation occurs at the point B. The vessel AB is known as LMA. To the right of B, it is divided between the two vessels BC and BD, i.e. LAD and LCx. The left coronary bifurcation angle (θ) is measured at site of B that between LAD and LCx using the Eq. 1 and 2. Variation in the angle values for normal and stenotic coronary arteries were measured and analyzed.

$$\theta = \cos^{-1}\left(\frac{v1. \,*\, v2}{||v1 \,*\, v2||}\right) \tag{3}$$

$$\text{Angle in degrees} = \left(\theta \,*\, \left(\frac{180}{\pi}\right)\right) \tag{4}$$

3   Results and Discussion

The results of proposed work of various methods and techniques are discussed in the following subsections.

3.1   Preprocessing the Coronary Cine Angiogram Images

For pre-processing the input image, i.e., in order to remove the noise from image and to smoothening the image median filter is used. The pre-processing result is shown in Fig. 5(b).

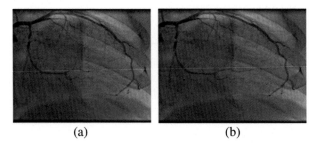

(a)   (b)

Fig. 5. (a) Original image (b) Filtered image using median filter

3.2 Segmentation Techniques

After pre-processing the input image (coronary cine angiogram image), various segmentation techniques are applied on pre-processed image in order to segment coronary blood vessel from it.

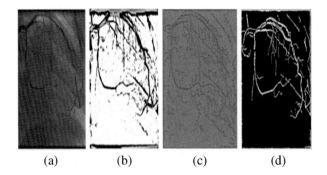

(a)   (b)   (c)   (d)

Fig. 6. (a) Original image, (b) Morphological segmentation, (c) Hessian based segmentation, (d) Active contour segmentation

The morphology operations such as dilation, erosion, filling holes etc., are done to detect the edges of angiogram images and to extract the features in images. The results of morphological operations are shown in the Fig. 6(b). In this method, only the initial results of identified coronary arterial tree are provided while edges of coronary blood vessels are not clearly detected in the given images. Hessian detection is done on the pre-processed image, that detects the coronary arteries structure based on the eigenvalues of Hessian matrix. The results of hessian detection segmentation are shown in Fig. 6(c). In this method, boundaries of coronary arteries are detected clearly while edges of coronary arteries are not segmented properly. Active contour method is performed on pre-processed image in order to segment the blood vessel from an image. It detects the inner boundaries by closed curve evolution. The results of active contour segmentation are shown in Fig. 6(d). This segmentation provides clear and accurate segmentation of coronary blood vessel tree structure from the given images.

3.3   Frangi Filter

From the various segmentation methods, active contour segmentation result is obtained as final output segmented image.

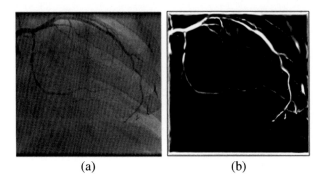

(a)   (b)

Fig. 7. (a) Original image, (b) Extraction of LMCA using Frangi filter

Following this, Frangi filter is applied on the segmented image to extract only the main branch of left coronary artery by suppressing non-vascular structure of short blood vessels and contrast also enhanced as shown in the Fig. 7(b).

3.4   Centerline Detection

Following the extraction of left main coronary blood vessel, skeletonization operation is performed to detect the centerline of the vascular structure. To detect the centerline, first it is converted into binary image and then the contour of vascular structure is eroded until it reaches medial one pixel width.

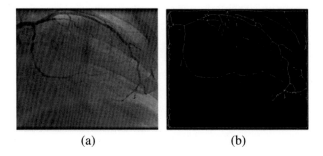

(a)   (b)

Fig. 8. (a) Original image, (b) Extraction of centerline using skeletonization

From the Fig. 8(b), it can be seen that the topology of shape of left main coronary arteries structure is preserved along with connectivity of original image.

3.5   Bifurcation Points Detection

A size of 3 × 3 window as mask on the centerline image is applied in order to detect bifurcation points. In the 3 × 3 window, center pixel has 8 neighborhood pixels, where if the center pixel is 1 and has exactly 3 one-valued neighborhoods, then centerline pixel corresponding to the center of mask is detected as bifurcation points as shown in the Fig. 9(b).

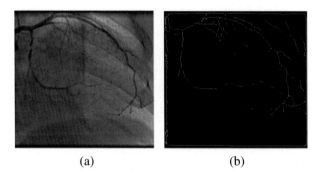

(a)   (b)

Fig. 9. (a) Original image, (b) Image after bifurcation points detection in LMCA

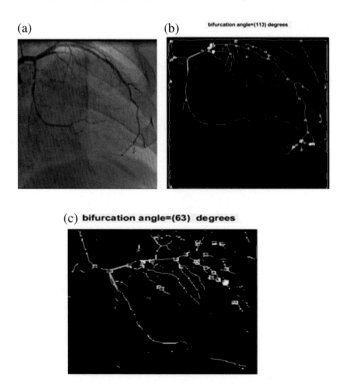

Fig. 10. (a) Original image, (b) Image of bifurcation angle measurement (dark blue) at LAD-LCx for Stenotic image, (c) Image of bifurcation angle measurement (dark blue) at LAD-LCx for normal image

3.6   Bifurcation Angle Measurement

Following the bifurcation point detection from the centerline image, bifurcation angle is measured at the site of LAD and LCx.

The Fig. 10(b), it is the image of 63-year – old male who had significant stenosis of LAD-LCx. The bifurcation angle between LAD and LCx (dark blue arrow) measures 113°. A Fig. 10(c), image of 54-year-old male who had normal artery (not affected by stenosis). The bifurcation angle between LAD and LCx (dark blue arrow) measures 63°.

Table 1. Evaluation of bifurcation angle

Total number of subjects	Age	Mean bifurcation angle	
		Stenotic artery	Normal artery
50	45–75	$101.3^0 \pm 8.23$	$66.1^0 \pm 4.01$

The Table 1 shows that the mean left coronary bifurcation angle(LCBA) for subject with stenosis is 101.30 ± 8.23 and subject without stenosis is 66.10 ± 4.01. From the obtained result, it can be seen that angulation gets wider in the stenotic coronary artery than normal coronary artery.

3.7   Confusion Matrix

A confusion matrix provides information about actual and predicted cases produced by LCBA measurement. Table 2 shows the number of correctly and incorrectly LCBA measurement in the images.

Table 2. Confusion matrix

Actual N = 50	Predicted	
	Stenotic artery	Normal artery
Steontic artery	23(TP)	2(FP)
Normal artery	1(FN)	19(TN)

3.8   Performance Measure

The efficiency of LCBA measurement is evaluated using performance metrics such as sensitivity, specificity and accuracy. A Table 3 shows the performance metrics of LCBA measurement. These metrics can be computed using confusion matrix.

Table 3. Performance metrics

Test case (normal & stenotic arteries)	Sensitivity	Specificity	Accuracy
50	94%	90%	92%

The outcome of obtained results shows that overall accuracy is about 92% in predicting the presence of stenosis in LMCA among 50 subjects of x-ray coronary angiograms.

Table 4. Comparison on performance between existing method and proposed method

Method	Accuracy (%)
Method in [7]	86.67
Proposed method	92

From the Table 4, it can be seen that accuracy in stenosis detection of proposed method is higher when compare with existing method of [7]. The performance of proposed method based on bifurcation angle is finding to be better in predicting presence of stenosis than the other existing methods.

4 Conclusion

In the diagnosis of CAD, coronary cine angiogram is the most commonly used image modality were cardiologist makes decision about stenosis prevalence in the coronary arteries based on the visual inspection of anatomical characteristics of coronary arteries that may vary depend on the experience of clinicians. There is no quantitative tool for the accurate analysis of degree of stenosis. In this paper, a new method for the measurement of bifurcation angle at left main coronary vascular branch of coronary cine angiography images is used. The locations of bifurcation points are very important in terms of finding branch angle. A mask is used on the centerline images to find the location of bifurcation points. Then using location of bifurcation points, bifurcation angle is measured at LAD-LCx. The experimental results shows that the mean left coronary bifurcation angle (LCBA) for 25 subjects with stenosis is $101.3° \pm 8.23$ and 25 subjects without stenosis is $66.1° \pm 4.01$. Based on the analysis of metrics, the performance of proposed method for bifurcation angle measurement achieves accuracy of 92% in detection of stenosis. The bifurcation angle was found to be useful for the quantitative analysis of both diagnosis of CAD and percutaneous coronary intervention. The future work is to find the geometric relations according to hemodynamic properties of blood flow and percentage of plaque burden that can be evaluated in relation with bifurcation measurement. The outcome of obtained results is found to be helpful for clinicians in the assessment of stenosis detection.

Acknowledgment. We thank for the support of 'Chettinad Hospital, Chennai' for providing both normal and stenotic arteries of Coronary Cine Angiograms, which made us to do this project.

References

1. Cui, Y., Zeng, W., Yu, J., Lu, J., Hu, Y., Diao, N., Shi, H.: Quantification of left coronary bifurcation angles and plaques by coronary computed tomography angiography for prediction of significant coronary stenosis: a preliminary study with dual-source CT. PLoS One **12**(3), e0174352 (2017)
2. Juan, Y.H., Tsay, P.K., Shen, W.C., Yeh, C.S., Wen, M.S., Wan, Y.L.: Comparison of the left main coronary bifurcating angle among patients with normal, non-significantly and significantly stenosed left coronary arteries. Sci. Rep. **7**(1), 1515 (2017)
3. Sun, Z., Lee, S.: Diagnostic value of coronary CT angiography with use of left coronary bifurcation angle in coronary artery disease. Heart Res. Open J. **3**(1), 19–25 (2016)
4. Sun, Z., Chaichana, T.: An investigation of correlation between left coronary bifurcation angle and hemodynamic changes in coronary stenosis by coronary computed tomography angiography-derived computational fluid dynamics. Quant. Imaging Med. Surg. **7**(5), 537 (2017)
5. Fatemi, M.R., Mirhassani, S.M., Yousefi, B.: Vessel segmentation in X-ray angiographic images using Hessian based vesselness filter and wavelet based image fusion. In: 10th IEEE International Conference on Information Technology and Applications in Biomedicine (ITAB), pp. 1–5, November 2010
6. Givehchi, S., Safari, M.J., Tan, S.K., Shah, M.N.B.M., Sani, F.B.M., Azman, R.R., Wong, J. H.D.: Measurement of coronary bifurcation angle with coronary CT angiography: a phantom study. Phys. Med. **45**, 198–204 (2018)
7. Mohan, N., Vishnukumar, S.: Detection and localization of coronary artery stenotic segments using image processing. In: International Conference on Emerging Technological Trends (ICETT), pp. 1–5, October 2016
8. Mahmood, N.H., Razif, M.R., Gany, M.T.: Comparison between median, unsharp and wiener filter and its effect on ultrasound stomach tissue image segmentation for pyloric stenosis. Int. J. Appl. Sci. Technol. **1**(5) (2011)
9. Ersoy, I., Bunyak, F., Mackey, M.A., Palaniappan, K.: Cell segmentation using Hessian-based detection and contour evolution with directional derivatives. In: 15th IEEE International Conference on Image Processing ICIP, pp. 1804–1807, October 2008
10. Frangi, A.F., Niessen, W.J., Vincken, K.L., Viergever, M.A.: Multiscale vessel enhancement filtering. In: International Conference on Medical Image Computing and Computer-Assisted Intervention. Springer, Heidelberg, pp. 130–137, October 1998
11. Hassan, G., El-Bendary, N., Hassanien, A.E., Fahmy, A., Snasel, V.: Retinal blood vessel segmentation approach based on mathematical morphology. Procedia Comput. Sci. **65**, 612–622 (2015)
12. Kass, M., Witkin, A., Terzopoulos, D.: Snakes: active contour models. Int. J. Comput. Vision **1**(4), 321–331 (1988)
13. Airouche, M., Bentabet, L., Zelmat, M.: Image segmentation using active contour model and level set method applied to detect oil spills. In: Proceedings of the World Congress on Engineering. Lecture Notes in Engineering and Computer Science, vol. 1, no. 1, pp. 1–3, July 2009
14. Khan, K.B., Khaliq, A.A., Jalil, A.: Shahid, M: A robust technique based on VLM and Frangi filter for retinal vessel extraction and denoising. PloS One **13**(2), e0192203 (2018)
15. Khan, S.A., Hassan, A., Rashid, S.: Blood vessel segmentation and centerline extraction based on multilayered thresholding in CT images. In: The 2nd International Conference on Intelligent Systems and Image Processing (ICISIP 2014), September 2014

16. Bhuiyan, A., Nath, B., Ramamohanarao, K.: Detection and classification of bifurcation and branch points on retinal vascular network. In: International Conference on Digital Image Computing Techniques and Applications (DICTA), pp. 1–8, December 2012

17. Cao, Y., Liu, C., Jin, Q., Chen, Y., Yin, Q., Li, J., Zhao, W.: Automatic Bifurcation angle calculation in intravascular optical coherence tomography images. In: 2nd International Conference on Image, Vision and Computing (ICIVC), pp. 650–654, June 2017

A Crisp-Based Approach for Representing and Reasoning on Imprecise Time Intervals in OWL 2

Fatma Ghorbel[1,2(✉)], Elisabeth Métais[1], and Fayçal Hamdi[1]

[1] CEDRIC Laboratory, Conservatoire National des Arts et Métiers (CNAM), Paris, France
fatmaghorbel6@gmail.com,
{metais, faycal.hamdi}@cnam.fr
[2] MIRACL Laboratory, University of Sfax, Sfax, Tunisia

Abstract. In the Semantic Web field, representing and reasoning on imprecise temporal data is a common requirement. Several works exist to represent and reason on crisp temporal data in ontology; however, to the best of our knowledge, there is no work devoted to handle imprecise temporal data. Representing and reasoning on imprecise time intervals in OWL 2, is the problem this work is dealing with. Our approach is based only on crisp environment. First, we extend the 4D-fluents approach to represent imprecise time intervals and crisp temporal interval relations. Second, we extend the Allen's interval algebra to propose crisp temporal relations between imprecise time intervals. We show that, unlike most related work, our temporal interval relations preserve many of the properties of the Allen's interval algebra. Furthermore, we show how they can be used for temporal reasoning by means of a transitivity table. We experimentally infer the resulting temporal relations from the introduced imprecise time intervals via a set of SWRL rules. Finally, we illustrate the usefulness of our work in the context of the Captain Memo memory prosthesis.

Keywords: Imprecise time interval · Temporal representation ·
4D-fluents approach · Temporal reasoning · Allen's interval algebra · OWL 2

1 Introduction

Temporal data given by the user is often imprecise. For instance, if they give the information "John was married to Maria from early 80 to by 1995", two measures of imprecision are involved. On the one hand, the information "early 80" is imprecise in the sense that it could be 1980 or 1981 or 1982 or 1983 or 1984; on the other hand, the information "by 1995" is imprecise in the sense that it could be 1994 or 1995 or 1996.

In the Artificial Intelligence field, several works exist to handle imprecise temporal data. On the contrary, in the Semantic Web field, many works have been proposed to represent and reason only on crisp temporal data; to the best of our knowledge, there is no work devoted to handle imprecise temporal data. In this paper, we introduce a

© Springer Nature Switzerland AG 2020
A. Abraham et al. (Eds.): ISDA 2018, AISC 941, pp. 640–649, 2020.
https://doi.org/10.1007/978-3-030-16660-1_62

crisp-based approach to represent and reason on imprecise time intervals in OWL 2. Imprecise time intervals are classical time intervals characterized by imprecise beginnings or/and endings.

Our first contribution focuses on representing imprecise time intervals in OWL 2. Earlier work by [1] showed how crisp quantitative temporal data and the evolution of concepts in time can be modeled in OWL using the so called 4D-fluents approach. In the present work, this approach is extended in two ways. (1) It is enhanced with new components to be able to represent imprecise time intervals. (2) It is enhanced with qualitative temporal expressions (e.g., After and During) representing relations between imprecise time intervals.

Our second contribution focuses on reasoning on imprecise time intervals in OWL 2. The Allen's interval algebra [2] is the most used and known formalism for reasoning about time intervals. However, it is not designed to handle situations in which time intervals are imprecise. A number of approaches have been extended the Allen's work to handle imprecision. They propose a fuzzy redefinition of Allen's interval relations i.e., the resulting relations are fuzzy and hold only to a given degree. Hence, they cannot be represented in OWL 2. We extend the Allen's interval algebra to propose crisp temporal relations between imprecise time intervals. Unlike most related work, our interval relations preserve many of the properties of the Allen's interval algebra. Furthermore, they can be used for temporal reasoning by means of a transitivity table. Finally, we infer the resulting relations via a set of SWRL rules.

The current paper is organized as follows. Section 2 is devoted to present some preliminary concepts and related work. Sections 3 and 4 focus, respectively, on representing and reasoning on imprecise time intervals in OWL 2. Section 5 illustrates an application of our work. Section 6 draws conclusions and future research directions.

2   Preliminaries and Related Work

In this section, we introduce some preliminary concepts and related work in the field of temporal data representation in OWL and reasoning on time intervals.

2.1   Representing Temporal Data in OWL

Five main approaches have been proposed to represent temporal data in OWL: Temporal Description Logics [3], Reification [4], Versioning [5], N-ary relations [6] and 4D-fluents. All these approaches represent only crisp temporal data in OWL.

Temporal Description Logics extend the standard description logics with additional temporal constructs e.g., "sometime in the future". The Reification is "a general purpose technique for representing N-ary relations using a language such as OWL that permits only binary relations" [7]. The Versioning approach is described as "the ability to handle changes in ontologies by creating and managing different variants of it" [5]. When an ontology is modified, a new version is created to represent the temporal evolution of the ontology. The N-ary relations approach represents an N-ary relation using an additional object. The N-ary relation is represented as two properties each related with the new object. The two objects are related to each other with an Nary relation. The 4D-fluents approach represents temporal data and the evolution of the last ones in OWL. Concepts varying in time are represented as 4-dimensional objects with the 4th dimension being the temporal dimension.

Based on the present related work, we choose the 4D-fluents approach. Indeed, it minimizes the problem of data redundancy as the changes occur only on the temporal parts and keeping therefore the static part unchanged. It also maintains full OWL expressiveness and reasoning support [7]. We extend this approach in two ways. (1) It is extended with crisp components to represent imprecise time intervals and crisp interval relations in OWL 2.

2.2 Allen's Interval Algebra

Allen has been proposed 13 mutually exclusive primitive relations that may hold between two crisp time intervals. Their semantics is illustrated in Table 1.

Table 1. Allen's temporal relations between the two crisp time intervals A = $[A^-, A^+]$ (━━) and B = $[B^-, B^+]$ (══).

Relation (A, B)	Inverse (A, B)	Definition	Illustration
Before	After	$A^+ < B^-$	
Meets	Met-by	$A^+ = B^-$	
Overlaps	Overlapped-by	$(A^- < B^-) \wedge (A^+ > B^-) \wedge (A^+ < B^+)$	
Starts	Started-by	$(A^- = B^-) \wedge (A^+ < B^+)$	
During	Contains	$(A^- > B^-) \wedge (A^+ < B^+)$	
Ends	Ended-by	$(A^- > B^-) \wedge (A^+ = B^+)$	
Equal	Equal	$(A^- = B^-) \wedge (A^+ = B^+)$	

A number of approaches have been extended the Allen's interval algebra. We classify them into (1) approaches proposing temporal relations between crisp time intervals (i.e., [8, 9] and [10]) and (2) approaches proposing temporal relations between imprecise time intervals (i.e., [11–13] and [14]).

Nagypál and Motik [11] propose a temporal model based on fuzzy sets to extend the Allen's relations. The authors represent an imprecise time interval as a fuzzy set. For example, "the period from the late 20 to the early 30" is represented as a fuzzy set using the following semantics (1928, 1933, 2, 2). They introduce a set of auxiliary operators on time intervals and define fuzzy counterparts of these operators. However, many of the properties of the original Allen's interval algebra are lost. For instance, the relation Equals is not reflexive. Thus, the compositions of the resulting relations cannot be studied by the authors.

Ohlbach [12] extends the Allen's interval algebra based on fuzzy sets. This approach proposes some gradual temporal relations as "more or less finishes". It does not preserve many of the properties of the original Allen's interval algebra. Therefore, it is not suitable for temporal reasoning.

Schockaert and Cock [13] propose a generalization of the Allen's interval algebra. This approach allows handling classical temporal relations, as well as some other gradual relations. It preserves many of the properties of the Allen's interval algebra. The resulting relations are used for fuzzy temporal reasoning by means of a transitivity table.

Gammoudi et al. [14] generalize the 13 Allen's relations to make them applicable to imprecise time intervals in conjunctive and disjunctive ways. The compositions of the resulting relations are not studied by the authors.

All the mentioned approaches propose fuzzy temporal relations between imprecise time intervals. Hence, the resulting relations cannot be represented in OWL 2. In Sect. 4, we extend the Allen's interval algebra to propose crisp interval relations that may be represented in OWL 2.

3 Our Approach for Representing Imprecise Time Intervals in OWL 2

Let $I = [I^-, I^+]$ be an imprecise time interval. The imprecision is represented by a disjunction of mutually exclusive elements; one of them is the true value [9]. We represent the imprecise beginning bound I^- as a disjunctive ascending set $\{I^{-(1)}... I^{-(B)}\}$ and the imprecise ending bound I^+ as a disjunctive ascending set $\{I^{+(1)}... I^{+(E)}\}$. For instance, if we have the information "Alexandre was started his PhD study in 1975 and he was graduated around 1980". The imprecise time interval representing this period is [1975, {1978 ... 1982}]. This means that his PhD studies end in 1978 or 1979 or 1980 or 1981 or 1982. We extend the 4D-fluents approach to represent the imprecise time intervals and the associated crisp temporal relations that may hold between them in OWL 2.

3.1 The Classical 4D-Fluents Approach

The 4D-fluents approach introduces two crisp classes "TimeSlice" and "TimeInterval" and four crisp properties "tsTimeSliceOf", "tsTimeInterval", "hasBegining" and "hasEnd". The class "TimeSlice" is the domain class for entities representing temporal parts (i.e., "time slices"). The property "tsTimeSliceOf" connects an instance of class "TimeSlice" with an entity. The property "tsTimeInterval" connects an instance of class "TimeSlice" with an instance of class "TimeInterval". The instance of class "TimeInterval" is related with two temporal instants that specify its starting and ending points using, respectively, the "hasBegining" and "hasEnd" properties. Properties having a time dimension are called fluent properties. Figure 1 illustrates the use of the classical 4D-fluents approach to represent the following example: "John was started his PhD study in 1975 and he was graduated in 1980".

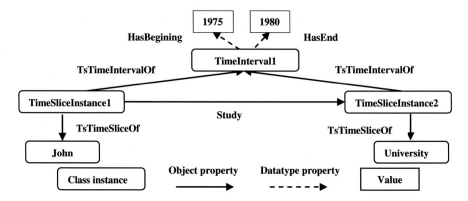

Fig. 1. An instantiation of the classic 4D-fluents approach.

3.2   Our Extension of the 4D-Fluents Approach

Let $I = [I^-, I^+]$ be an imprecise time interval; where $I^- = \{I^{-(1)}... I^{-(B)}\}$ and $I^+ = \{I^{+(1)}... I^{+(E)}\}$. We extend the original 4D-fluents approach to represent imprecise intervals in OWL 2. We add four crisp datatype properties "HasBeginningFrom", "HasBeginningTo", "HasEndFrom", and "HasEndTo" to the class "TimeInterval". "HasBeginningFrom" has the range $I^{-(1)}$. "HasBeginningTo" has the range $I^{-(B)}$. "HasEndFrom" has the range $I^{+(1)}$. "HasEndTo" has the range $I^{+(E)}$.

The 4D-fluents approach is also enhanced with crisp temporal relations that may hold between imprecise time intervals. This is implemented by introducing a temporal relation, called "RelationIntervals", as an object property between two instances of the class "TimeInterval". Figure 2 represents the extended 4D-fluents approach.

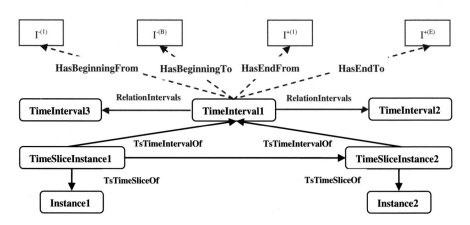

Fig. 2. The extended 4D-fluents approach.

4   Our Approach for Reasoning on Imprecise Time Intervals in OWL 2

To reason on imprecise time intervals, we extend the Allen's interval algebra in a crisp way. We infer the resulting relations in OWL 2 using a set of SWRL rules.

4.1   Our Extension of the Allen's Interval Algebra

We extend the Allen's interval algebra to propose crisp temporal relations between imprecise time intervals. We introduce our transitivity table to reason on the resulting relations.

Our Crisp Temporal Relations Between Imprecise Time Intervals. We redefine the 13 Allen's relations to propose crisp temporal relations between imprecise time intervals. We are based on the Vilain and Kautz's point algebra [15]. For instance, the relation Before between the two imprecise time intervals $I = [I^-, I^+]$ and $J = [J^-, J^+]$ is redefined as:

$$\forall I^{+(i)} \in I^+, \forall J^{-(j)} \in J^- / Precedes\left(I^{+(i)}, J^{-(j)}\right) \tag{1}$$

This means that the most recent time point of I^+ ($I^{+(E)}$) ought to precede the oldest time point of J^- ($J^{-(1)}$):

$$Precedes\left(I^{+(E)}, J^{-(1)}\right) \tag{2}$$

In a similar way, we redefine the other Allen's relations, as shown in Table 2.

Table 2. Our crisp temporal relations between the imprecise time intervals I and J.

Relation (I, J)	Definition	Inverse (I, J)
Before	Precedes $(I^{+(E)}, J^{-(1)})$	After
Meets	Min (Same $(I^{+(1)}, J^{-(1)}) \wedge$ Same $(I^{+(E)}, J^{-(B)}))$	Met-by
Overlaps	Min (Precedes $(I^{-(B)}, J^{-(1)}) \wedge$ Precedes $(J^{-(B)}, I^{+(1)}) \wedge$ Precedes $(I^{+(E)}, J^{+(1)}))$	Overlapped-by
Starts	Min (Same $(I^{-(1)}, J^{-(1)}) \wedge$ Same $(I^{-(B)}, J^{-(B)}) \wedge$ Precedes $(I^{+(E)}, J^{+(1)}))$	Started-by
During	Min (Precedes $(J^{-(B)}, I^{-(1)}) \wedge$ Precedes $(I^{+(E)}, J^{+(1)}))$	Contains
Ends	Min (Precedes $(J^{-(B)}, I^{-(1)}) \wedge$ Same $(I^{+(1)}, J^{+(1)}) \wedge$ Same $(I^{+(E)}, J^{+(E)}))$	Ended-by
Equals	Min (Same $(I^{-(1)}, J^{-(1)}) \wedge$ Same $(I^{-(B)}, J^{-(B)}) \wedge$ Same $(I^{+(1)}, J^{+(1)}) \wedge$ Same $(I^{+(E)}, J^{+(E)}))$	Equals

Example: Let $I_1 = [I_1^-, I_1^+]$, $I_2 = [I_2^-, I_2^+]$ and $I_3 = [I_3^-, I_3^+]$ be imprecise time intervals; where $I_1^- = \{1960 \ldots 1963\}$; $I_1^+ = \{1971 \ldots 1974\}$; $I_2^- = \{1940 \ldots 1943\}$; $I_2^+ = \{1971 \ldots 1974\}$; $I_3^- = \{2017 \ldots 2020\}$ and $I_3^+ = \{2030 \ldots 2033\}$. We obtain:

$$\text{Ends}\,(I_1, I_2) = 1 \qquad \text{Before}\,(I_2, I_3) = 1$$

Properties. Our temporal relations preserve many properties of the Allen's algebra. We obtain generalizations of the reflexivity/irreflexivity, symmetry/asymmetry and transitivity properties. Let $I = [I^-, I^+]$, $J = [J^-, J^+]$ and $K = [K^-, K^+]$ be imprecise time intervals.

Reflexivity/Irreflexivity: The temporal relations {Before, After, Meets, Met-by, Overlaps, Overlapped-by, Starts, Started-by, During, Contains, Ends and Ended-by} are irreflexive, i.e., let R be one of the aforementioned relations. It holds that

$$R\,(I, I) = 0 \tag{3}$$

Furthermore, the relation Equals is reflexive. It holds that

$$Equals\,(I, I) = 1 \tag{4}$$

Symmetry/Asymmetry: The temporal relations {Before, After, Meets, Met-by, Overlaps, Overlapped-by, Starts, Started-by, During, Contains, Ends and Ended-by} are asymmetric, i.e., let R be one of the aforementioned relations. It holds that

$$R\,(I, J)\,and\,R\,(J, I) \Rightarrow I = J \tag{5}$$

Furthermore, the relation Equals is symmetric. It holds that

$$Equals\,(I, J) = Equals\,(J, I) \tag{6}$$

Transitivity: The temporal relations {Before, After, Overlaps, Overlapped-by, Starts, Started-by, During, Contains and Equals} are transitive, i.e., let R be one of the aforementioned relations. It holds that

$$R\,(I, J)\,and\,R\,(J, K) \Rightarrow R\,(I, K) \tag{7}$$

Transitivity Table. The crux of the Allen's interval algebra is the transitivity table. This table lets us reason from R_1 (A, B) and R_2 (B, C) to R_3 (A, C), where $[A^-, A^+]$, $B = [B^-, B^+]$ and $C = [C^-, C^+]$ are crisp time intervals and R_1, R_2 and R_3 are Allen's interval relations. For instance, using the Allen's original definitions, we can deduce from During (A, B) and Meet (B, C) that Before (A, C) holds. Indeed by During (A, B), we have $(A^- > B^-)$ and $(A^+ < B^+)$, and by Meet (B, C), we have $B^+ = C^-$. From $(A^+ < B^+)$ and $(B^+ = C^-)$, we conclude Before (A, C).

We generalize such deductions using the three imprecise time intervals $I = [I^-, I^+]$, $J = [J^-, J^+]$ and $K = [K^-, K^+]$. Based on Table 2, we can deduce from During (I, J) and Meet (J, K) that Before (I, K) holds. Indeed by During (I, J), we have Min (Precedes

$(J^{-(B)}, I^{-(1)}) \wedge$ Precedes $(I^{+(E)}, J^{+(1)}))$, and by Meet (J, K), we have Min(Same $(J^{+(1)}, K^{-(1)}) \wedge$ Same $(J^{+(E)}, K^{-(B)}))$. From Precedes $(I^{+(E)}, J^{+(1)})$ and Same $(J^{+(1)}, K^{-(1)})$, we conclude Before (I, K). Our transitivity table coincides with the Allen's one.

4.2   Inferring Our Crisp Temporal Interval Relations Using SWRL Rules

Let $I = [I^-, I^+]$, $J = [J^-, J^+]$ and $K = [K^-, K^+]$ be imprecise time intervals. Based on Table 2, we propose a set of SWRL rules that allow inferring our crisp temporal relations from the imprecise time intervals entered by the user. For instance, the rule to infer the "Meet (I, J)" is the following:

$$TimeInterval\,(I) \,\wedge\, TimeInterval\,(J) \,\wedge\, HasEndFrom\left(I, I^{+(1)}\right)$$

$$\wedge\, HasBeginningFrom\left(J, J^{-(1)}\right) \,\wedge\, Equals\left(I^{+(1)}, J^{-(1)}\right) \,\wedge\, HasEndTo\left(I, I^{+(E)}\right)$$

$$\wedge\, HasBeginningTo\left(J, J^{-(B)}\right) \,\wedge\, Equals\left(I^{+(E)}, J^{-(B)}\right) \;\rightarrow\; Meet\,(I, J)$$

Based on the transitivity table, we associate for each transitivity relation a SWRL rule. For instance, we can infer "Before (I, K)" as the following:

$$During\,(I, J) \,\wedge\, Meet\,(J, K) \;\rightarrow\; Before\,(I, K)$$

5   Application to the Captain Memo Memory Prosthesis

In the context of the VIVA project[1], we are suggesting the Captain Memo memory prosthesis [16], to assist Alzheimer's patients in overcoming mnesic problems. It supplies a set of services. Among these services, one is devoted to "remember thing(s) about people" i.e., it helps the Alzheimer's patient to remember their convivial surroundings and relatives. It is based on PersonLink [17] which is a multicultural and multilingual OWL 2 ontology for storing, modeling and reasoning on family relations. Interpersonal relations change over time. However, imprecise temporal data inputs are especially numerous when given by an Alzheimer patient. We integrated our work in the PersonLink ontology to handle imprecise time intervals. For instance, we consider the following information: "John was married to Béatrice since about 10 years. John was married to Maria just after he was graduated with a PhD and it lasts 15 years. John was graduated with a PhD in 1980". Let $I = [I^-, I^+]$ and $J = [J^-, J^+]$ be two imprecise time intervals representing, respectively, the duration of the marriage of John and Béatrice and the one of John and Maria. Assume that $I^- = \{2007 \dots 2009\}$, $I^+ = 2018$, $J^- = \{1980 \dots 1983\}$ and $J^+ = \{1995 \dots 1998\}$. Figure 3 illustrates a part of PersonLink ontology which represents this example.

---

[1] http://viva.cnam.fr/.

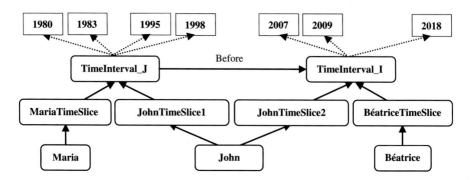

Fig. 3. An instantiation of the extended 4D-fluents approach in OWL 2.

We note that dates are often given in reference to other dates or events. An interesting point in this work is to deal with a personalized slicing of the person's life in order to sort the different events. For each user, we define their slices of life. They serve as reference intervals. For instance, the reference interval which represents the period of living in Paris is [2008, Now] and the interval which represents the period of living in Nantes is [1971, 2008].

For instance, in response to the following question: "When did I marry to Béatrice?". Our system compares the temporal data already entered (i.e., "John was married to Béatrice since about 10 years") to all entered temporal data and all references interval. For instance, we obtain that the patient married to Béatrice at the same time when he moved to Paris and he moved from Nantes.

6 Conclusion

In this paper, we proposed a crisp-based approach to represent and reason on imprecise time intervals in OWL 2. It is based only on crisp environment. First, we extended the 4D-fluents approach to represent imprecise time intervals and crisp interval relations. To reason on imprecise time intervals, we extended the Allen's interval algebra in a crisp way. The resulting relations are crisp. Unlike most previous approaches, generalizations of important properties of Allen's interval relations are valid, in particular those related to reflexivity/irreflexivity, symmetry/asymmetry and transitivity. We introduce a transitivity table to reason about the resulting temporal relations. Inferences are done via a set of SWRL rules.

The main interest of our approach is that it can be implemented with classical crisp tools and that the programmers are not obliged to learn technologies related to fuzzy ontology. Considering that crisp tools and models are more mature and better support scaling than the fuzzy tools, our approach is suitable for marketed products.

The Allen's interval algebra is not intended to relate an interval with a time point or even two time points. We plan to extend our approach to represent and reason about imprecise temporal data i.e., time points and intervals. We will propose temporal relations that may hold between two imprecise time points or an imprecise time interval and an imprecise time point. We will define the composition of the resulting relations.

References

1. Welty, C., Fikes, R., Makarios, S.: A reusable ontology for fluents in OWL. In: Frontiers in Artificial Intelligence and Applications, pp. 226–236 (2006)
2. Allen, J.F.: Maintaining knowledge about temporal intervals. Commun. ACM **26**, 832–843 (1983)
3. Artale, A., Franconi, E.: A survey of temporal extensions of description logics. Ann. Math. Artif. Intell. **30**, 171–210 (2000)
4. Gomez, C., Lopez, J.R., Olive, A.: Evolving temporal conceptual schemas: the reification case. In: International Workshop on Principles of Software Evolution, pp. 78–82 (2000)
5. Klein, M., Fensel, D.: Ontology versioning on the semantic web. In: International Semantic Web Working Symposium (SWWS), pp. 75–91 (2001)
6. Hayes, P., Welty, C.: Defining N-ary relations on the semantic web. In: W3C Working Group Note (2006)
7. Batsakis, S., Petrakis, E.G.: Representing and reasoning over spatio-temporal information in OWL 2.0. In: Workshop on Semantic Web Applications and Perspectives (2010)
8. Guesgen, H.W., Hertzberg, J., Philpott, A.: Towards implementing fuzzy Allen relations. In: Workshop on Spatial and Temporal Reasoning, pp. 49–55 (1994)
9. Dubois, D., Prade, H.: Processing fuzzy temporal knowledge. IEEE Trans. Syst. Man Cybern. **19**, 729–744 (1989)
10. Badaloni, S., Giacomin, M.: The algebra IA^{fuz}: a framework for qualitative fuzzy temporal. J. Artif. Intell. **170**(10), 872–902 (2006)
11. Nagypal, G., Motik, B.: A fuzzy model for representing uncertain, subjective, and vague temporal knowledge in ontologies. In: OTM Confederated International Conferences on the Move to Meaningful Internet Systems, pp. 906–923. Springer, Heidelberg (2003)
12. Ohlbach, H.: Relations between fuzzy time intervals. In: International Symposium on Temporal Representation and Reasoning, pp. 44–51. IEEE (2004)
13. Schockaert, S., De Cock, M., Kerre, E.E.: Fuzzifying Allen's temporal interval relations. IEEE Trans. Fuzzy Syst. **16**, 517–533 (2008)
14. Gammoudi, A., Hadjali, A., Ben Yaghlane, B.: Fuzz-TIME: an intelligent system for fluent managing fuzzy temporal information. Int. J. Intell. Comput. Cybern. **10**(2), 200–222 (2017)
15. Vilain, M.B., Kautz, H.: Constraint propagation algorithms for temporal reasoning. In: Readings in Qualitative Reasoning About Physical Systems, pp. 377–382 (1986)
16. Métais, E., Ghorbel, F., Herradi, N., Hamdi, F., Lammari, N., Nakache, D., Ellouze, N., Gargouri, F., Soukane, A.: Memory prosthesis. Non-Pharmacol. Ther. Dement. (2015)
17. Herradi, N., Hamdi, F., Metais, E., Ghorbel, F., Soukane, A.: PersonLink: an ontology representing family relationships for the CAPTAIN MEMO memory prosthesis. In: International Conference on Conceptual Modeling, pp. 3–13. Springer (2015)

Algorithmic Creation of Genealogical Models

Frantisek Zboril[(⊠)], Jaroslav Rozman, and Radek Koci

FIT, IT4Innovations Centre of Excellence, Brno University of Technology,
Bozetechova 1/2, 612 66 Brno, Czech Republic
{zborilf,rozmanj,koci}@fit.vutbr.cz

Abstract. Need for automatic creation of genealogical models rises as the amount of digitalized and transcribed record from historical sources. Usually it is necessary to distinguish what is genealogical model by an axiomatic system. Automatically created models need not to fulfil the axioms when record is wrongly interconnected with other records. If we are capable to decide whether a model is a genealogical model, we can better produce models automatically from a set of transcribed parish registers record. In this paper we introduce an algorithmic approach to creation of possible models of a parish society in the middle Europe from a set of transcribed baptism, marriage and burial records. It is shown that probabilistic estimation of person interconnection from different records could be helpful for reconstruction of a societies through several centuries.

Keywords: Genealogical models · State space searching method ·
Probabilistic record linkage

1 Introduction

Genealogy as an auxiliary science of history which is not often addressed by researchers from the areas of information technology or computer science. As interest in making family trees grows a need for computer aid grows too in the community of genealogists and some particular IT relating results starts to appear. For example an effort to make population reconstruction for some areas by computer is described in [1] or [2]. A tool that assists genealogist to create family trees is described in [3].

However, there have been just a very little of the matrices transcribed to digital format. This limits our effort to automatize generation of social models based on such data. We believe that in near future the amount of this kind of data will grow and then a system that make correct social models will be required.

Probabilistic models are being intensively studied and Bayesian networks and Probabilistic relational models belongs among very popular tools for stochastic system designs. These types of models have been used for theory of record linkage on which stays most of recent research in population reconstruction. Essential work [4] shows how to estimate identical persons in various records. We studied the problem of record linkage in [11]. Probabilistic estimation of candidates is also used in [5] where a population with over 100000 persons were matter of reconstruction. Some probabilistic and soft computing approaches to this problem can be found in [6, 7] and [8]. All these

© Springer Nature Switzerland AG 2020
A. Abraham et al. (Eds.): ISDA 2018, AISC 941, pp. 650–658, 2020.
https://doi.org/10.1007/978-3-030-16660-1_63

works deal with a problem of estimation whether two records are identical when they have some attributes.

In contrast to these we suppose that we are capable to find a match of persons in records with the same attributes, to be concrete with the same name and surname. But even these persons have the same attributes they need not to be the same. Then we try to find the correct one or ones and merge them together with the original one. Also on the contrary to the above mentioned works we provide an algorithmic approach that shows how to proceed when a genealogical model should be constructed.

We start quite informally with introduction of a format of the data which we have on input. There are three sorts of records formats that depends on whether a record describes birth, marriage or death of a person. Then we have a record with one of the following structures:

A **child** was born to a **father** and **mother**, additionally there may be mentioned persons like godparents, grandparents, cooperators etc.

A **groom** and a bride were **married**, also some testes and relatives could be mentioned there

A **deceased**, was buried. The deceased could be a man, child of a man, wife of a man or a virgin as was usual to distinguish in such records in the time periods of our interest.

In our system we consider three persons that take part in baptism, two in marriage and one in burial. We understand that one record means a relation among several (three or two) roles in a time moment at a given place. So we know that there was such a relation each time a baptism or marriage occurred. But what we do not know is how relate persons that are mention in different records in different roles. Basic idea for our methodology is that if we have b baptism records, m marriage records and r burial records, then we have at most $3 * b + 2 * m + r$ persons. But obviously a person that lived all her or his life in a given location is often mentioned more than just in one record. A girl is born, then she marriages somebody, gives birth to several children, possibly marries again as a widow and then dies. A boy or man can be mentioned in several of records as well. If we want to have complex social model then we have to merge all the corresponding persons together.

For this we try to find out all the candidates for merging. In extreme we may suspect all the mentioned persons anywhere in the record set but our algorithm selects only those that are reasonable for merging. It means that a bride can be only a woman that was adult in the time of marriage etc. We will discuss this in more details later. Also reasoning which of the person, if any, should be merged with another deserves closer look. Probabilistic approach that we adopted for this sort of problem allows heuristic estimation how to interconnect particular records to create a model but not every such a model can be considered to be a genealogical model. As it is impossible that a woman becomes bride before she was born we can find more rules that cannot be broken if we suppose that we have a genealogical model. These rules, or axioms, are verified in the phase that follows interconnection of individual roles in the records. Finally, if there is detected a problem with the structure regarding the axioms then the model should be replaced by another and checked again.

2   Genealogical Models

Before we provide algorithms regarding our methodology we introduce some terms and entities that we will use further in this text. We know that record is a structure that has been transcribed from a parish book and describes an event of baptism, marriage or burial. In fact, a record is an element of a relation that characterize a genealogic model. Semantically it interconnects some persons that took part in an historical event. From above we know that these persons may take roles of a child, a father, a mother, a groom, a bride or a deceased. Each record then establishes relation among these roles and in this work we consider them to be trustworthy. It means that we do not admit any errors and mistakes there. On the other hand, as one person may appear in more records we need to find a way how to merge the roles from particular events together. By clustering roles of hypothetically identical person we obtain a class of roles that we call identity.

The following paragraphs formally introduce our genealogical models and the objects which they contain.

Definition 1: Genealogical model is a tuple $Mg = \{Ro, E, RE, ID\}$ where Ro is a set of roles, E is a set of events, RE and ID are relations over Ro.

Both the relations are relations of equivalence. RE represents relation among all the roles that take part in any event. ID is then a relation that groups together all the roles that refer to the same person.

Algorithm that we are going to present now takes Ro and RE as its input and computes the relation ID as an output. To demonstrate how the algorithm works we need to extend the model with some data that the records provide.

Definition 2: Role annotation is a structure which is assigned to each role and it is denoted as r[an], where $r \in Ro$.

Typically, the annotations contain date of the event, name and surname of a person who plays the role there. Additionally, there can be information about location, age, social state etc. Using these annotations we define a function that for each pair of role maps a number that denotes how likely both the roles are played by the same person. The value is nonzero only when the first role precedes the second. In the following definition we relax the annotation as we consider them to be part of the role element implicitly.

Definition 3: Evaluation function assigns a value to a pair of roles

$$ev : RxR \rightarrow \ <0, 100>$$

Finally the decision functions selects just one of a the preceding role to another role and this is done by using the evaluation function. Formally we define it as:

Definition 4: Decision is a function $dec : R \rightarrow R$ which for a role selects another role from the role set.

After there is a decision function then we may group roles together and create identities.

Definition 5: Identity function IDi constructs a set of roles which are mapped from a role

$$IDi : R \rightarrow 2^R \text{ and it is defined as}$$
$$IDi(ri) = \{ri\} \cup \{rj, \exists k(rj \in IDi(rk) \land dec(rk) = ri)\}$$

The role for which an identity is created is called to be a core role. Then identity contains the core role plus it contains all the roles from the identities whose core role points to the core role of this identity. For example if $dec(ra) = rb$ and $dec(rc) = rb$, then $IDi(rc) = \{rc\}$, $IDi(rb) = \{rb, rc\}$ and $IDi(ra) = \{ra, rb, rc\}$. The identity $IDi(r)$ is a core identity if $dec(r) = r$. Then we can see that the roles from such an identity are not propagated to another identity because the decision of the core role does not point to another role but itself.

Core identity is crucial for creating the relation ID from Definition 1. We try to find all the roles one person plays by finding the very first occurrence of the person in records and then to map all the corresponding roles via a role tree created by the identities. How to do this shows the next section in which we introduce the algorithm.

3 Algorithm for Genealogical Model Creation

The algorithm is in quite abstract way presented in the following steps and it has two phases.

Phase 1: Creation of an initial model

1. create roles from the records (role set RS)
2. for every role assign matching roles which precedes it in reasonable time interval
3. evaluate every such assignment
4. create identities

Phase 2: Model modification

5. For every identity use semantic rules to check its consistency
6. If an identity is not consistent, try to rearrange it to a consistent form
7. Repeat until all the identities are consistent or a limit is reached

3.1 Role Set Creation

The first part of the algorithm is simple. Individual roles from every record are added to RS. Then the set contain every role that can be found in the records and these roles are now ready for evaluation and clustering.

3.2 Assignment of Previous Matching Roles

The matching roles are searched in an overestimated interval. It means that for every role the algorithm looks back in some interval (up to 100 years when corresponding children are searched for deceased, 80 when brides, mothers, fathers and grooms are searched for deceased etc.) and stores all the roles which matches, usually when names and surnames are the same.

3.3 Evaluation of Assigned Roles

In this step the algorithm uses the evaluation function for estimating how probable the matches are true matches. We use a Gaussian distribution with some parameters of mean and variance and using this we assign for every role and every its matching role a value that represents the estimation. In some cases, it happened that the writer noticed an age of groom, bride or deceased, but fathers and mothers usually had no age mentioned in the records. Furthermore, even when we see an age in the record, we cannot rely on that. Back in the XVII. to XIX. centuries people had no knowledge about their real age and because of this the writer just estimated it. This is why we use the Gaussian distribution with some mean and dispersion to distribute probability over possibly several candidates for a preceding role. In Table 1 there are shown the parameters for a pair of role types. Each tuple m, v represents hyperparameters of the function.

Table 1. Parameters for matches estimations

	Child	Mother	Father	Groom	Gr (age)	Bride	Bd (age)
Mother	30;5	2;1	–	–	–	6;1.5	30-Ag;5
Father	35;7	–	2;1	10;2	5-Ag;3	–	–
Bride (age)	Age;2,5	Age-22;3	–	–	–	Age-22;3	difAg;1,5
Bride	22;5	6;1.5	–	–	–	6,3	22-Age;3
Groom (age)	Age;2,5	–	35-Ag;3	25-Age;3	difAg;1,5	–	–
Groom	25,7	–	10;2	6,3	25-Age;3	–	–
Deceased (age)	Age;1	A-30;5	Age-35;5	Age-25;5	difAg;3	Age-22;5	difAg;3
Deceased (child)	6;2.5	–	–	–	–	–	–
Deceased (others)	6;2.5	15;4	10;4	20;5	45-Age;3	23;5	45-Age;4

Average age of brides, grooms, mothers, fathers, deceased and deceased children were estimated to 22, 25, 30, 35, 45 and 6 years. Also the parameters were estimated due to experiences of genealogists. difAg here represents difference of the roles ages.

As we can count the function values to any previous matching role we should count such a value also for the case that there is no previous role where the person acts and this is her or his first occurrence in the records. For children this value is 100. But it may happen also for the non-child roles when a person is mentioned in a record that had not been transcribed yet or she or he simply had moved to the location from somewhere else.

After a migration rate *mr%* and a percentage of transcribed records is rw% are determined, then the value for the possibility that a role *ri* is the core role is *s(ri) = unknown/average_age*. To prove this we show that by *mr* and *rw* we can compute which part of person roles is known, it means that we have the records transcribed to the record set, and which part is unknown. Also we need to include the migration rate *mr* to the known *rw*. By the rate we mean which percentage of people in the area had come there from elsewhere. Then *rwm = (100 − mr) * rw*. Consequently, we may count how many positive matches for a person have been found in the records during previous 'average_age' years where 'average_age' is an estimated life expectancy in that times. Then we may expect that there are other *unknown = known * (100 − rwm)/rwm* non-transcribed person's roles in the 'average_age' range. And in total the estimated value which represents that the referenced one is not mentioned in the records before is *unknown/average_age*.

In this approach we use the values not in the sense of theory of probability but just for the ordering of the matching roles. This will enable us to create the decision function from the Definition 4.

3.4 Identity Creation

The way how the identities are created follows from the above definitions, especially from Definition 5. In this step we create a decision function in that a way that we take a matching identity with the biggest evaluation between referencing role and the matching role. After this there are some core identities as some representatives of the persons. Consequently, the algorithm moves forward to its second phase.

3.5 Semantic Check of the Model

As we declared on the beginning of this text, we may verify whether a model is a genealogical model or not by using some axioms. In our work we used the following ones:

1. Two consequent births regarding one pair cannot be in less than 250 days
2. Death must be the last role of every identity
3. A mother must be younger than 63 years
4. A father must be younger than 83 years
5. When partner one of a pair starts to give births with somebody else, then his or her previous partner must be already dead

The last rule reflects the fact that during the previous centuries for which we make the genealogical models it was almost impossible to divorce.

Algorithm checks identity by identity for every such rules and if any of them is broken then it tries to solve the inconsistence by model reconfiguration.

3.6 Model Reconfiguration

As we have evaluated every match of the roles, we may try to change the decision when an identity is inconsistent. Inconsistency means that at least one decision has been done wrongly. For searching of a better decision we use Best first search algorithm [9]. The state cost function by which we determine which state is the 'cheapest' to follow is built as a sum of evaluations differences of the new and the original roles assignment. It is illustrated in the Fig. 1.

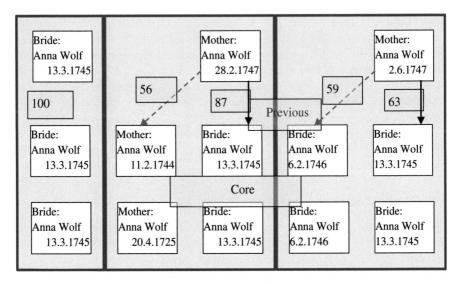

Fig. 1. Reconfiguration of a model example

We see that there are two roles composing an identity of Anna Wolf. But such identity is not consistent as there are two births in the interval of three months. This is not possible, and the identity should be reconfigured. In this case we may reassign both the birth roles to another preceding roles. If we change decision for the first role then the value decreases by 31 and if we reconsider the second decision then the difference is 4. If we change both then the total value change would be 35. We also see that by both re-assignments corresponding core role changes too and because of this the births change the identities to which they belong. First of all, the reconfiguration algorithm tries to change the second decision, then the first decision and finally both the decisions. In this example the identity become consistent right after the first reconfiguration and then the algorithm finishes.

However, there is a state explosion when there are more roles forming an identity (for instance 10) with more possibilities of reconfigurations (5 and more). Because of this we must restrict the number of steps to a limit and when no consistent variant of an identity is found up to the limit then we let the identity inconsistent and proceed forward to the next identity.

4   Experiments and Results

We have implemented the methodology in the SWI Prolog language [10]. Evaluation of our methodology has been limited by the fact that we had just quite small set of genealogical data. Concretely, we had about eleven thousands of records transcribed from the parish books of Jevisovka (former Frelichov) village from the south Moravia region. Because at the moment when we were performing the experiments we had not any genealogical model made by genealogist we just tested the semantic rules on the results. The experiment showed that there were identified about 3000 families - identities that contains at least two births of children (that means that the person had at least two children) and about 600 of them were inconsistent. After performing the BestFS repairment with the limit of 100 steps the number of inconsistent families was reduced to 300.

The second dataset regards village Bukovinka, also at south Moravia, for which we had a referenced genealogical model which had been created 'by hand'. Thus we can evaluate our results comparing them with the original model. In this case we had 1862 birth records containing 5586 roles. We processed these data by the algorithm and reached the following results. Success ratio was evaluated as a 100 - *fails* where fails is a percentage of roles which had been assigned wrongly. Number of incorrectly assigned roles is the lowest number of roles that need to be reassigned for finding a bijection between computed identities and identities in the referenced model. 'Rules broken' is the number of rules which were not satisfied by the identities and 'inconsistent id.' is a number of identities which broke at least one rule (Table 2).

Table 2. Evaluation of genealogical model creation for Bukovinka

	0 steps	150 steps	300 steps	450 steps	600 steps
Score	84,973%	85,155%	85,373%	85,355%	85,574%
Rules broken	162	131	116	115	114
No. of families	590	598	589	589	588
Inconsistent idt	101	81	73	75	75

5   Conclusion

Automatic creation of genealogical models, to the best of our knowledge, has not been matter of research yet. In this paper we tried to prove that our concept of such model creation is acceptable. We realized that the way how we make an estimation of a model from parish records is quite good as we classified over 85% of roles inside the record correctly, that means that we were able to identify most of the persons that acts in that records. On the other hand, the reconfiguration of that model for avoiding inconsistences does not lead to remarkable improvement. In the future there should be improved likelihood estimation functions, concretely parameters of the particular Gaussian distributions. For the semantic checks and repairments there should be used rather semantic approach that enhances the state space searching algorithm with some heuristic. Overall our methodology is suitable for genealogical model creation and it will be a part of a global genealogical system that we are developing for the purposes of genealogical population reconstructions.

Acknowledgement. This work was supported by TACR No. TL01000130, by BUT project FIT-S-17-4014 and the IT4IXS: IT4Innovations Excellence in Science project (LQ1602).

References

1. Dintelman, S., Maness, T.: Reconstituting the population of a small European town using probabilistic record linking: a case study, family history technology workshop, BYU (2009)
2. Milani, G., Masciullo, C., et al.: Computer-based genealogy reconstruction in founder populations. J. Biomed. Inform. **44**, 9971003 (2011)
3. Malmi, E., Rasa, M., Gionis, A.: AncestryAI: a tool for exploring computationally inferred family trees. In: Proceedings of the 26th International Conference on World Wide Web Companion (2017)
4. Fellegi, I.P., Sunter, A.B.: A theory for record linkage. J. Am. Stat. Assoc. **64**, 1183–1210 (1969)
5. Malmi, E., Gionis, A., Solin, A.: Mating computationally inferred genealogical networks uncover long-term trends in assortative mating. In: Proceedings of the 2018 World Wide Web Conference, pp. 883–892 (2018)
6. Efremova, J., Ranjbar-Sahraei, B., Oliehoek, F.A.A., Calders, T., Tuyls, K.: Baseline method for genealogical entity resolution. In: Workshop on Population Reconstruction (2014)
7. Gottapu, R.D., Dagli, C., Ali, B.: Entity resolution using convolutional neural network. Procedia Comput. Sci. **95**, 153–158 (2016)
8. Efremova, J., Garcia, A.M., Zhang, J., Calders, T.: Towards population reconstruction: extraction of family relationships from historical documents. In: First International Workshop on Population Informatics for Big Data (21th ACM-SIGKDD PopInfo 2015), pp. 1–9 (2015)
9. Russel, S., Norvig, P.: Artificial Intelligence: A Modern Approach. Pearson, London (2009)
10. Wielemaker, J.: An overview of the SWI-Prolog programming environment. In: Proceedings of the 13th International Workshop on Logic Programming Environments (2003)
11. Rozman, J., Zboril, F.: Persons linking in baptism records, accepted for PAOS (2018)

Android Malicious Application Classification Using Clustering

Hemant Rathore$^{(\boxtimes)}$, Sanjay K. Sahay$^{(\boxtimes)}$, Palash Chaturvedi$^{(\boxtimes)}$,
and Mohit Sewak$^{(\boxtimes)}$

Department of CS and IS, BITS, Pilani, Goa Campus, Sancoale, India
{hemantr,ssahay,f20150395,p20150023}@goa.bits-pilani.ac.in

Abstract. Android malware have been growing at an exponential pace and becomes a serious threat to mobile users. It appears that most of the anti-malware still relies on the signature-based detection system which is generally slow and often not able to detect advanced obfuscated malware. Hence time-to-time various authors have proposed different machine learning solutions to identify sophisticated malware. However, it appears that detection accuracy can be improved by using the clustering method. Therefore in this paper, we propose a novel scalable and effective clustering method to improve the detection accuracy of the malicious android application and obtained a better overall accuracy (98.34%) by random forest classifier compared to regular method, i.e., taking the data altogether to detect the malware. However, as far as true positive and true negative are concerned, by clustering method, true positive is best obtained by decision tree (97.59%) and true negative by support vector machine (99.96%) which is the almost same result obtained by the random forest true positive (97.30%) and true negative (99.38%) respectively. The reason that overall accuracy of random forest is high because the true positive of support vector machine and true negative of the decision tree is significantly less than the random forest.

Keywords: Android · Classification · Clustering · Malware detection · Static analysis

1 Introduction

The term Malware is derived from **Mal**icious Soft**ware**, and initially, malware was developed to show one's technical skills, but now it has become a profit-driven industry. Over the last decade, android popularity has grown immensely, and its current mobile market share is more than 75% [2]. The popularity of android Operating System (OS) is because of its open source platform and the availability of a large number of feature-rich applications for it. These applications should ideally be benign, but because of malicious intent of the adversaries, it can be made to perform some unwanted activities in the devices, e.g., stealing the personal information, sending short message service to a premium number, spying on the user, etc.

© Springer Nature Switzerland AG 2020
A. Abraham et al. (Eds.): ISDA 2018, AISC 941, pp. 659–667, 2020.
https://doi.org/10.1007/978-3-030-16660-1_64

According to G DATA Security, 3809 new malware samples were detected in 2011 [1]. Since then there has been a rapid rise in the number of android malware, and in 2017 more than three million new malware samples have been identified. In this, nearly seven million applications (70% higher as compared to 2016) were removed from Google Play which was either malware or had some unacceptable content [5]. A report published by McAfee in 2018 shows that there has been an increase of 36% in AdClick Frauds, 23% spyware and 12% Banking Trojan from the previous year [7]. Another McAfee Threat Report suggests that by the end of 2017 there were more than twenty million malware samples on the android platform out of which around ~2.5 millions new samples were detected in quarter-3 itself [6]. For detection of android malware, most of the anti-malware rely on the signature-based detection techniques. But generating and maintaining signature of such a huge number malware is a herculean task. Thus in the last couple of years, researchers have actively started to explore different ways to detect the malware effectively and efficiently. Therefore recently many authors [16,20,21,26,28] have proposed machine learning as a useful technique to counter the malware, and in this field, the research can be broadly divided into two parts: the feature extraction/selection and methods for the classifications.

Feature extraction is an important component of machine learning, and for the effective and efficient classification, features can be extracted without executing the application (static) or during the execution (dynamic) for the detection of malware. Although static analysis gives a better code coverage, it has other limitations. However, with the static feature, several authors have proposed various malware detection models, e.g., Au et al. [12] in PScout extracted permissions used in 4 different android OS and found that there are over 75 unique permissions out of which 22% of the non-system permissions are unnecessary. Lindorfer et al. proposed a fully automated ANDRUBIS system to analyze the android applications from which they found that malicious application on an average request for 13 permission while for good applications the number is just 4.5 [20]. Sharma et al. used a combination of static opcode frequency and dangerous permissions for malware classification [25]. Puerta et al. with the Genome dataset found that opcode frequency is better feature vector than the permissions [21].

Malware can also be analyzed while executing them in a controlled environment to find the features, e.g., network traffic, application programming interface, system calls, information flow for the classifications. In this, Enck et al. built TaintDroid to track the runtime information flow and found many instances of private data being misuse in different applications [16]. Tam et al. developed CopperDroid to model well know process-OS interaction and also inter-process and intra-process communications for effective malware detection [26]. You et al. in 2015 did a comprehensive analysis using Dalvik opcodes as features to find the potential threats [28].

Finally, the features obtained by the static or dynamic approach is an important ingredient for the ***classification***. In this, Puerta et al. proposed a single class classifier for malware detection based on opcode occurrence and achieved an accuracy of 85% [21]. Sharma et al. used opcode occurrence on five different classifiers and obtained the highest accuracy of 79.27% with the functional tree [24]. Feizollah et al. proposed AndroDialysis, which used intent and permission separately and then combine them to make an extensive feature vector [17]. Their analysis shows that independently intent and permission achieved a detection rate of 91% and 83% respectively. However, combing both the features shows an increase in the detection rate (95.5%). Wu et al. proposed DroidMat, which collects permission, intent, API Call and applied various machine learning algorithms like k-nearest neighbor and naive bayes for malware detection, and achieved the highest accuracy of 97.87% [27]. Arp et al. used permission, intent, API call, network address in the Drebin dataset of more than 5000 malware samples and utilized support vector machine for the classification, and achieved an accuracy of 94% [11].

Recently Sharma et al. used opcode frequency as a feature and grouped them based on permission to achieve a detection accuracy of 97.15% [25]. Li et al. used permission as the feature vector and used three level pruning for identification of significant permission for effective detection of malware and benign [19]. They have used 22-most significant permissions and achieved a detection rate of 93.62%. Rana et al. use different tree based classifiers like a decision tree, random forest, gradient boosting and extremely randomized tree with n-gram approach and shown that with 3-gram and random forest classifier one can achieve an accuracy up to 97.24% [22]. Chen et al. used n-gram opcode sequence with exemplar feature selection method and random forest classifier to detect the malware only with 95.6% correctly with 4-gram approach [15].

The above-proposed methods using different machine learning seem to be not sufficient to identify sophisticated advanced malware. However, it appears that detection accuracy can be improved by using the clustering method. Therefore in this paper, we propose a novel scalable and effective clustering method to improve the detection accuracy of the android malware. Hence, the rest of the paper is organized as follows: Sect. 2 introduces the dataset and feature extraction. Section 3 contains the experimental analysis and results. Finally, Sect. 4 concludes the paper.

2   Dataset and Feature Extraction

In the area of machine learning, it is very important that how good one can make/train a model to identify the target in new/test data, and in turn, the quality of the trained model depends on the dataset and how the features are extracted/selected from it. Understanding the fact we used Drebin dataset [11] which is one of the largest benchmark malware samples, which consists of 5550 malware samples from more than 20 different families. It also includes all the malware samples from the Android Malware Genome Project [18].

For the benign file, we have collected 8500 android applications between 2016–17 from various sources (Google Play [4], Third-party app stores, alternate marketplaces, etc.). To test the downloaded files is benign or not, we verified them using VirusTotal [10] (it is a subsidy of Alphabet which is an aggregator of 40–60 antivirus including AVG, Bitdefender, F-Secure, Kaspersky Lab, McAfee, Norton, Panda Security, etc. and provide various API's to check whether the application is malware/benign) services, and we declare an application as malicious if one or more antivirus from virustotal.com categories it as malware. Thus after verifying all the downloaded samples from VirusTotal, we were left with 5720 samples for the experimental analysis.

For the analysis, we disassemble all the sample file by APKtool [3] to extract all the opcodes for the clustering and classifications. For this purpose, we have taken the frequency of the opcode as a feature vector and generated the feature vector for the complete dataset as 11266 × 256 matrix, where rows represent different files, and a column represents the frequency of the opcode in that particular file.

3   Experimental Analysis and Result

A schematic of the experimental analysis of our novel proposed approach/method is depicted in Fig. 1. We first experimented by the regular method given by the various authors [11,17,18,21,27], i.e., without clustering/grouping the data for comparison of accuracy with our novel method (first we find the optimal number of cluster and then identify the malicious android applications), and we used the scikit-learn [9] machine learning library for all the clustering and classifications. Also Numpy python library has been used to handle large multidimensional arrays and matrices [8].

3.1   Malware Detection Without Clustering

To understand the performance of the different type of classifiers by regular used method (i.e., taking the whole data altogether), we used Logistic Regression (LR), Naive Bayes (NB), Support Vector Machine (SVM), Decision Tree (DT), and Random Forest (RF). For the classification, we used all 256 opcodes as features and ten-fold cross-validation (it divides the data into ten equal parts out of which nine parts are used for training the model, and one part is used for testing. This exercise is completed ten times with different combinations of training and testing set and at last average result is taken into account) and the result obtained is shown in Fig. 2. From the analysis, we found that after three-fold cross-validation the accuracy of classifiers is more or less same. The detailed results of the experiment conducted by all the five classifiers in terms of Accuracy, Recall, and Specificity are given in the Table 1, where

$$\text{Accuracy} = \frac{\text{Number of Malware and Benign correctly classified}}{\text{Total Number of files in the dataset}}$$

$$\text{True Postive Rate (TPR)} = \frac{\text{Number of Malware correctly classified}}{\text{Number of Malware in the dataset}}$$

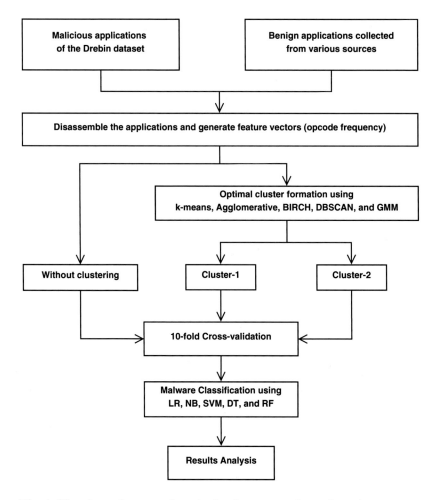

Fig. 1. Flowchart of our novel method to improve malware detection accuracy.

Table 1. Accruacy, TPR and TNR obtained without clustering the data.

	Logistic Regression	Naive Bayes	Support Vector Machines	Decision Trees	Random Forest
Accuracy	87.96	76.66	84.68	93.70	95.82
Recall/TPR	92.44	94.89	69.38	95.27	95.40
Specificity/TNR	84.16	58.49	99.96	92.65	95.73

$$\text{True Negative Rate (TNR)} = \frac{\text{Number of Benign correctly classified}}{\text{Number of Benign in the dataset}}$$

The analysis also shows that the tree-based RF classifier outperformed the other classifiers with the best accuracy of 95.82%.

3.2   Clustering Based Malware Detection

To improve the detection accuracy of malicious applications, we first studied the
five different clustering algorithms (k-means, Agglomerative, BIRCH, DBSCAN,

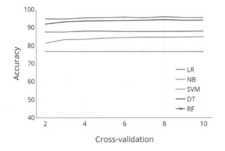

Fig. 2. Classifier accuracy with the k-fold cross-validation.

Fig. 3. Number of clusters with SSE by k-means algorithm.

Table 2. Number of clusters with different clustering algorithm and its Calinski Harabaz and Silhouette score.

	No of clusters	Calinski Harabaz score	Silhouette score
k-means clustering	**2**	**28568.28**	**0.7377**
	3	25493.58	0.7005
	4	22109.38	0.6167
	5	22971.34	0.6013
Agglomerative clustering	2	24706.58	0.7256
	3	21245.43	0.7069
	4	20046.19	0.6043
	5	20939.81	0.6052
BIRCH clustering	2	24735.95	0.7258
	3	20844.19	0.7080
	4	18494.73	0.6656
	5	21261.10	0.6056
Gaussian mixture model clustering	2	8398.36	0.4716
	3	2738.43	0.1967
	4	1640.20	0.1010
	5	4813.65	0.2519
DBSCAN clustering	eps = 5000	1380.75	0.1539
	eps = 10000	3469.39	0.5316
	eps = 15000	3743.52	0.5349
	eps = 20000	4858.33	0.6753

and GMM) to find the clusters in the dataset. The analysis shows (Table 2) that the k-means algorithm forms the best cluster (Calinski-Harabasz [13] and Silhouette score [23] of the k-means is best among the selected five clustering algorithm). Further, we observed that the two clusters formed by the k-means would be best for the classification of malicious applications (Fig. 3) and in the obtained cluster, we find that cluster-1 contains total 8760 files and is dominated by malware files whereas the cluster-2 contains total 2780 files dominated by benign files. However, for the analysis, we have balanced the dataset using SMOTE [14]. Now for the classification, we use the same set of classifiers that are used for the classification without clustering and found that the overall accuracy, TPR and TNR obtained by clustering the data is significantly more than without clustering the data and the results obtained are given in Table 3.

Table 3. Accuracy, TPR and TNR after clustering the data.

	Logistic Regression	Naive Bayes	Support Vector Machines	Decision Trees	Random Forest
Accuracy	92.23	81.01	92.29	97.92	98.34
Recall/TPR	91.49	96.46	84.59	97.59	97.30
Specificity/TNR	92.98	65.57	99.96	98.25	99.38

4   Conclusion

To detect the malware, generally signature-based anti-malware are used which is not good enough to detect the advanced obfuscated malware. Hence time-to-time to detect the advanced malware different machine learning solutions are proposed by various authors. In this, we proposed a novel scalable and effective clustering based android malware detection system. The analysis shows an overall improvement of accuracy with RF (98.34%) by our proposed clustering method. The accuracy achieved by our approach outperformed the recent accuracy obtained by the authors viz. Ashu et al. (97.15%), Li et al. (93.62%) and Rana et al. (97.24%). Also, the experimental analysis shows that despite the TNR of SVM and TPR of DT are marginally better than the RF, the overall accuracy of RF is best among the tested classifiers, this is because the TPR of SVM and TNR of DT is far below the RF classifier. As the results are significant, therefore we are developing an API for the identification of the android malicious apps, which we will be free to the research community.

References

1. G DATA Mobile Internet Security. Technical report, G DATA (2017). https://www.gdatasoftware.com/mobile-internet-security-android. Accessed 02 Oct 2018
2. Smartphone OS Market Share. Technical report, ITC (2017). https://www.idc.com/promo/smartphone-market-share/os. Accessed 02 Oct 2018
3. APKTOOL. Technical report, Apache (2018). https://ibotpeaches.github.io/Apktool/documentation/. Accessed 02 Oct 2018
4. Google Play. Technical report, Google (2018). https://play.google.com/store?hl=en. Accessed 02 Oct 2018
5. How we fought bad apps and malicious developers in 2017. Technical report, Android Developers Blog (2018). https://android-developers.googleblog.com/2018/01/how-we-fought-bad-apps-and-malicious.html. Accessed 02 Oct 2018
6. McAfee Mobile Threat Report December 2017. Technical report, McAfee (2018). https://www.mcafee.com/content/dam/enterprise/en-us/assets/reports/rp-quarterly-threats-dec-2017.pdf. Accessed 02 Oct 2018
7. McAfee Mobile Threat Report Q1, 2018. Technical report, McAfee (2018). https://www.mcafee.com/enterprise/en-us/assets/reports/rp-mobile-threat-report-2018.pdf. Accessed 02 Oct 2018
8. NumPy. Technical report (2018). http://www.numpy.org/. Accessed 02 Oct 2018
9. Scikit-learn. Technical report (2018). http://scikit-learn.org/stable#. Accessed 02 Oct 2018
10. VirusTotal. Technical report, Google (2018). https://www.virustotal.com. Accessed 02 Oct 2018
11. Arp, D., Spreitzenbarth, M., Hubner, M., Gascon, H., Rieck, K., Siemens, C.: DREBIN: effective and explainable detection of android malware in your pocket. In: NDSS, vol. 14, pp. 23–26 (2014)
12. Au, K.W.Y., Zhou, Y.F., Huang, Z., Lie, D.: PScout: analyzing the android permission specification. In: Proceedings of the 2012 ACM Conference on Computer and Communications Security, pp. 217–228. ACM (2012)
13. Caliński, T., Harabasz, J.: A dendrite method for cluster analysis. Commun. Stat.-Theory Methods **3**(1), 1–27 (1974)
14. Chawla, N.V., Bowyer, K.W., Hall, L.O., Kegelmeyer, W.P.: SMOTE: synthetic minority over-sampling technique. J. Artif. Intell. Res. **16**, 321–357 (2002)
15. Chen, T., Mao, Q., Yang, Y., Lv, M., Zhu, J.: TinyDroid: a lightweight and efficient model for android malware detection and classification. Mob. Inf. Syst. **2018**, 9 (2018)
16. Enck, W., Gilbert, P., Han, S., Tendulkar, V., Chun, B.G., Cox, L.P., Jung, J., McDaniel, P., Sheth, A.N.: TaintDroid: an information-flow tracking system for realtime privacy monitoring on smartphones. ACM Trans. Comput. Syst. (TOCS) **32**(2), 5 (2014)
17. Feizollah, A., Anuar, N.B., Salleh, R., Suarez-Tangil, G., Furnell, S.: AndroDialysis: analysis of android intent effectiveness in malware detection. Comput. Secur. **65**, 121–134 (2017)
18. Jiang, X., Zhou, Y.: Dissecting android malware: characterization and evolution. In: 2012 IEEE Symposium on Security and Privacy, pp. 95–109. IEEE (2012)
19. Li, J., Sun, L., Yan, Q., Li, Z., Srisa-an, W., Ye, H.: Significant permission identification for machine learning based android malware detection. IEEE Trans. Ind. Inform. **14**, 3216–3225 (2018)

20. Lindorfer, M., Neugschwandtner, M., Weichselbaum, L., Fratantonio, Y., Van Der Veen, V., Platzer, C.: ANDRUBIS–1,000,000 apps later: a view on current android malware behaviors. In: 2014 Third International Workshop on Building Analysis Datasets and Gathering Experience Returns for Security (BADGERS), pp. 3–17. IEEE (2014)
21. de la Puerta, J.G., Sanz, B., Santos, I., Bringas, P.G.: Using dalvik opcodes for malware detection on android. In: International Conference on Hybrid Artificial Intelligence Systems, pp. 416–426. Springer (2015)
22. Rana, M.S., Rahman, S.S.M.M., Sung, A.H.: Evaluation of tree based machine learning classifiers for android malware detection. In: International Conference on Computational Collective Intelligence, pp. 377–385. Springer (2018)
23. Rousseeuw, P.J.: Silhouettes: a graphical aid to the interpretation and validation of cluster analysis. J. Comput. Appl. Math. **20**, 53–65 (1987)
24. Sharma, A., Sahay, S.K.: An investigation of the classifiers to detect android malicious apps. In: Information and Communication Technology, pp. 207–217. Springer (2018)
25. Sharma, A., Sahay, S.: Group-wise classification approach to improve android malicious apps detection accuracy. Int. J. Netw. Secur. **21**(3), 409–417 (2019)
26. Tam, K., Khan, S.J., Fattori, A., Cavallaro, L.: CopperDroid: automatic reconstruction of android malware behaviors. In: NDSS (2015)
27. Wu, D.J., Mao, C.H., Wei, T.E., Lee, H.M., Wu, K.P.: DroidMat: android malware detection through manifest and API calls tracing. In: 2012 Seventh Asia Joint Conference on Information Security (Asia JCIS), pp. 62–69. IEEE (2012)
28. You, W., Liang, B., Li, J., Shi, W., Zhang, X.: Android implicit information flow demystified. In: Proceedings of the 10th ACM Symposium on Information, Computer and Communications Security, pp. 585–590. ACM (2015)

Performance Evaluation of Data Stream Mining Algorithm with Shared Density Graph for Micro and Macro Clustering

S. Gopinathan$^{(\boxtimes)}$ and L. Ramesh

Department of Computer Science, University of Madras, Chennai 600005, India
gnathans2002@gmail.com, rameshnethaji2012@gmail.com

Abstract. **We propose** to solve the problem of micro clustering using the integration of data stream mining algorithms. Streaming data are potentially infinite sequence of incoming data at very high speed and may evolve over the time. This causes several challenges in mining large scale high speed data streams in real time. This paper discusses various challenges associated with mining data streams. Several algorithm such as data stream mining algorithms of accuracy and micro clustering are specified along with their key features and significance. Also, the significant performance evaluation of micro and macro clustering relevant in streaming data of shared density graph and clustering are explained and their comparative significance is discussed. The paper illustrates various streaming data computation platforms that are developed and discusses each of them chronologically along with their major capabilities. The performance and analysis are different radius activation functions and various number of radius applied to an data stream mining algorithm with shared density graph for micro clustering and macro clustering.

Keywords: Stream mining · Micro clustering · Shared density graph · Macro clustering

1 Introduction

The data streams are in the real time. They are very high speed and very evolve over the time [1]. Mining of these large scale data streams to perform some kind of machine learning or futuristic predictions regarding data instances have drawn a significant attention of researchers in couple of previous years. The data streams resemble the real time incoming data sequence very well. The source of these data streams can be collected from various sensors situated in medical domain to monitor health conditions of patients, in industrial domain to monitor manufactured products and other sources are to collect from streams on social networking.

E-commerce sites twitter posts, various blogs, web logs, and many more [2, 3]. The above mentioned sources not only produce data streams, but they produce them in huge amount (of scale of tera bytes to peta bytes) and at rapid speed. Now, mining huge data in real time raises various challenges and has become the hot area of research recently.

© Springer Nature Switzerland AG 2020
A. Abraham et al. (Eds.): ISDA 2018, AISC 941, pp. 668–676, 2020.
https://doi.org/10.1007/978-3-030-16660-1_65

These challenges include memory limitation, faster computing requirement *etc.* Apart from these challenges, streaming data has inherent nature of evolution that means that concepts that are being mined evolve and change over the time [4, 5]. It makes the traditional data mining algorithms and techniques incapable of appropriately handling data streams and yields the requirement of algorithms suitable for streaming data mining. This may be achieved in two ways; either modify the existing data mining algorithms to make them suitable for stream mining or create new streaming data mining algorithms right from the scratch. Another aspect of this field is the evaluation of the performance of the stream data mining algorithms. Since the performance evaluation is done continuously throughout the mining task and on partial read data streams, it becomes critical to use suitable performance measures in reference to streaming data mining. Various new performance evaluators have been devised specifically for this purpose [6–8].

2 Related Work

Prasad and Agarwal et al. to discuss the stream data mining, Agarwal challenges associated in mining potentially infinite data streams along with various stream mining algorithms for classification and clustering [3].

Sibson to discuss The Single-Link method is a commonly used hierarchical clustering method starting with the clustering obtained by placing every object in a unique cluster, in every step the two closest clusters in the current clustering are merged until all points are in one cluster. Other algorithms which in principle produce the same hierarchical structure have also been suggestion [19].

Hinneburg, Keim et al. to discuss the density-based algorithm DenClue is proposed. This algorithm uses a grid but is very efficient because it only keeps information about grid cells that do actually contain data points and manages these cells in a tree-based access structure. This algorithm generalizes some other clustering approach'es which, however, results in a large number of input parameters. Also the density- and grid-based clustering technique CLIQUE [20].

Jain, Dubes et al. to discuss the common way to find regions of high-density in the data space is based on grid cell densities. A histogram is constructed by partitioning the data space into a number of non-over lapping regions or cells. Cells containing a relatively large number of objects are potential cluster centers and the boundaries between clusters fall in the "valleys" of the histogram. The success of this method depends on the size of the cells which must be specified by the user. Cells of small volume will give a very "noisy" estimate of the density, whereas large cells tend to overly smooth the density estimate [21].

Schikuta et al. to discuss the Hierarchical clustering is based on the clustering properties of spatial index structures. The GRID and the BANG clustering apply the same basic algorithm to the data pages of different spatial index structures. A clustering is generated by a clever arrangement of the data pages with respect to their point density. This approach is not well suited for high-dimensional data sets because it is based on the affectivity of these structures as spatial access methods [22].

Brian, Datar and Motwani et al. to discuss the algorithm of Hoeffding tree. Hoeffding trees a decision tree learning method. Hoeffding trees can be learned in constant time per example while being nearly identical to the trees a conventional batch learner would produce [10].

Prasad, Agarwal et al. to discuss the Fast decision tree (FDTA) which is an improvement classification method based on Hoeffding tree. The t is an algorithm parameter has shown an effect on a decision regarding making a split process on the tree or not. So t is generated in order fashion in domain (0–1) instead of treated as a fixed value as known in traditional Hoeffding tree (t known as 0.05). According to the obtained results above, it shows that FDTA is gained highest accuracy than Hoeffding tree, the same things regarding memory space and execution time [15].

Cao, Ester et al. to discuss the DenStream, an effective and efficient method for clustering an evolving data stream. The method can discover clusters of arbitrary shape in data streams, and it is insensitive to noise. The structures of p-micro-clusters and o-microclusters maintain sufficient information for clustering, and a novel pruning strategy is designed to limit the memory consumption with precision guarantee [17].

Prasad et al. to discuss the specifies the need of new algorithms and evaluation measures relevant to this field and mentioned some of them used in stream mining scenario. The various available tools or platform to provide the appropriate frame work to deal with large scale data streams along with their key features have also been described in chronical order that helped in understanding the evolvement of the streaming data computing and mining platforms [18].

3   Block Diagram of Proposed System

The Block diagram explains our proposed method for the Research work. In this frame work for Performance Evaluation of Data Stream Mining Algorithm with Shared density graph for Micro and macro clustering (Fig. 1).

In this framework, the input dataset is feed into the data stream mining algorithm where the DBSTREAM, CluStream, DenStream, D-Stream, D-Stream + Attraction is done according to the given dataset, and it will takes the method which gives the different radius like radius level r = 0.01, 0.02, 0.03, 0.04, 0.05 to compare the Number of micro clusters and Number of macro clusters and using shared density level is 5 cm.

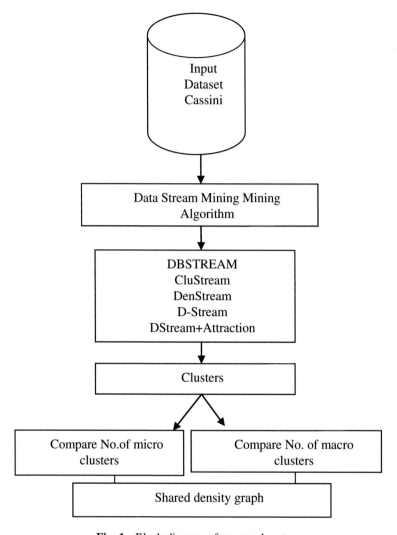

Fig. 1. Block diagram of proposed system

4 Algorithm for Proposed System

Step 1: Initialization
Input: Read the dataset
Begin
Step 2: Data stream mining algorithm are DBStream, Clustream,
Denstream, D-Stream
Step 3: Radius fixation like r = 0.01, 0.02, 0.03, 0.04, 0.05 Set radius of current
data stream mining algorithm

Step 4: Update on All the data stream mining algorithm compare Number of Micro clusters and Number of macro clusters
Step 5: plot for micro and macro clustering of graph presentation
Step 6: Shared density graph Using Macro cluster and macro clustering
Step 7: Calculate the different running time with the data stream mining algorithm
End begin

5   Experiments and Results

In the experiment part the performance of the proposed work implemented with data stream mining algorithm publically available R extension called stream.

Our Research work used real data set called CASSINI developed number of data stream mining algorithm and MOA (Massive Online Analysis) used. Our proposed Algorithm work with best result compare with different method. It showed in Tables (Fig. 2).

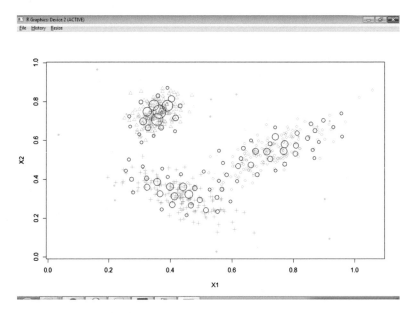

Fig. 2. Cassini data set for micro clustering

DBSTREAM Stream Mining Algorithm

See Table 1 and Fig. 3.

Table 1. DBSTREAM micro and macro cluster with r = 0.01, 0.02, 0.03, 0.04, 0.05.

Sl. no.	Radius level (r)	No. of micro clusters	No. of macro clusters
1	0.01	**52**	22
2	0.02	117	9
3	0.03	95	5
4	0.04	83	**2**
5	0.05	68	**2**

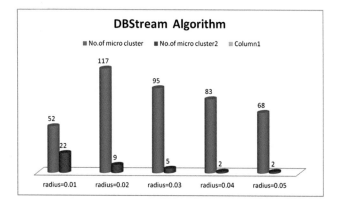

Fig. 3. DBSTREAM micro and macro cluster with r = 0.01, 0.02, 0.03, 0.04, 0.05.

DENSTREAM Stream Mining Algorithm

See Table 2 and Fig. 4.

Table 2. DEN Stream micro and macro cluster with r = 0.01, 0.02, 0.03, 0.04, 0.05.

Sl. no.	Radius level (r)	No. of micro clusters	No. of macro clusters
1	0.01	**26**	2
2	0.02	28	3
3	0.03	37	**1**
4	0.04	60	**1**
5	0.05	47	**1**

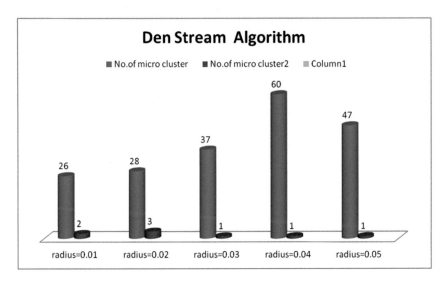

Fig. 4. DEN Stream micro and macro cluster with r = 0.01, 0.02, 0.03, 0.04, 0.05.

Clu STREAM - Stream Mining Algorithm
See Table 3 and Fig. 5.

Table 3. Clu STREAM micro and macro cluster with r = 0.01, 0.02, 0.03, 0.04, 0.05.

Sl. no.	Radius level (r)	No. of micro clusters	No. of macro clusters
1	0.01	**38**	**1**
2	0.02	43	2
3	0.03	49	1
4	0.04	72	2
5	0.05	50	3

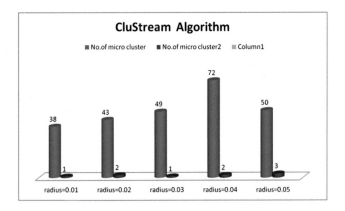

Fig. 5. Clu STREAM micro and macro cluster with r = 0.01, 0.02, 0.03, 0.04, 0.05

D-STREAM - Stream Mining Algorithm
See Table 4 and Fig. 6.

Table 4. D-STREAM micro and macro cluster with r = 0.01, 0.02, 0.03, 0.04, 0.05.

Sl. no.	Radius level (r)	No. of micro clusters	No. of macro clusters
1	0.01	**33**	4
2	0.02	41	5
3	0.03	43	**3**
4	0.04	39	4
5	0.05	45	6

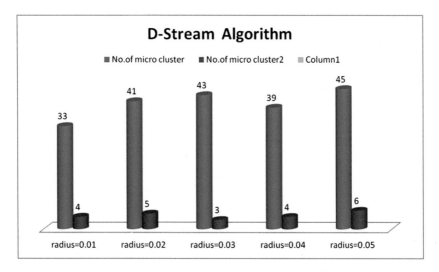

Fig. 6. D-STREAM micro and macro cluster with r = 0.01, 0.02, 0.03, 0.04, and 0.05.

6 Conclusion

In the research paper we study the problem of micro clustering using the integration of data stream mining and algorithm with shared density graph for cm = 5 five different radius functions like r = 0.01, 0.02, 0.03, 0.04, 0.05 to Cassini dataset. It is found after Number of micro and macro cluster when compared with the shared density graph functions. Further, we conclude the best micro and macro clustering in all the data stream mining algorithm mentioned bold.

References

1. Han, J., Kamber, M., Pei, J.: Data Mining: Concepts and Techniques, 3rd edn. Morgan Kaufmann, San Francisco (2011)
2. Gama, J.: Knowledge Discovery from Data Streams. Chapman & Hall CRC, Atlanta (2010)
3. Agarwal, S., Prasad, B.R.: High speed streaming data analysis of web generated log streams. In: 2015 IEEE 10th International Conference on Industrial and Information Systems (ICIIS), pp. 413–418. IEEE (2015)
4. Kifer, D., David, S.B., Gehrke, J.: Detecting change in data streams. In: VLDB Conference (2004)
5. Kranen, P., Kremer, H., Jansen, T., Seidl, T., Bifet, A., Holmes, G., Pfahringer, B.: Clustering performance on evolving data streams: assessing algorithms and evaluation measures within MOA. In: IEEE International Conference on Data Mining - ICDM, pp. 1400–1403 (2010)
6. Bifet, A.: Pitfalls in benchmarking data stream classification and how to avoid them. In: Machine Learning and Knowledge Discovery in Databases, pp. 465–479. Springer, Heidelberg (2013)
7. Song, M.J., Zhang, L.: Comparison of cluster representations from partial second-to full fourth-order cross moments for data stream clustering. In: ICDM, pp. 560–569 (2008)
8. Philipp, K.: Clustering performance on evolving data streams: assessing algorithms and evaluation measures within MOA. In: 2010 IEEE International Conference on Data Mining Workshops (ICDMW). IEEE (2010)
9. Daniel, J.A.: Aurora: a new model and architecture for data stream management. VLDB Int. J. Very Large Data Bases 12(2), 120–139 (2003)
10. Brian, B., Datar, M., Motwani, R.: Load shedding for aggregation queries over data streams. In: 2004 Proceedings of 20th International Conference on Data Engineering. IEEE (2004)
11. Abadi, D.J.: The design of the borealis stream processing engine. In: Proceedings of CIDR (2005)
12. Vowpal Wabbit (2007). http://hunch.net/vw
13. Neumeyer, L., Robbins, B., Nair, A., Kesari, A.: S4: distributed stream computing platform. In: Proceedings of ICDMW, pp. 170–177. IEEE Press (2010)
14. Storm (2011). http://storm-project.net
15. Prasad, B.R., Agarwal, S.: Handling big data stream analytics using SAMOA framework - a practical experience. Int. J. Database Theory Appl. 7(4), 197–208 (2014)
16. Bifet, A.: Mining big data in real time. Informatica 37, 15–20 (2013)
17. Cao, F., Ester, M., Qian, W., Zhou, A.: Density-based clustering over an evolving data stream with noise. In: SDM, vol. 6, pp. 328–339 (2006)
18. Prasad, B.R., Agarwal, S.: Stream data mining: platforms, algorithms, performance evaluators and research trends. Int. J. Database Theory Appl. 9, 201–218 (2016)
19. Sibson, R.: SLINK: an optimally efficient algorithm for the single-link cluster method. Comput. J. 16(1), 30–34 (1973)
20. Hinneburg, A., Keim, D.: An efficient approach to clustering in large multimedia databases with noise. In: Proceedings of 4th International Conference on Knowledge Discovery & Data Mining, New York City, NY (1998)
21. Jain, A.K., Dubes, R.C.: Algorithms for Clustering Data. Prentice-Hall, Inc., Upper Saddle River (1988)
22. Schikuta, E.: Grid clustering: an efficient hierarchical clustering method for very large data sets. In: Proceedings of 13th International Conference on Pattern Recognition, Vol. 2, pp. 101–105 (1996)

Sentiment Analysis for Scraping of Product Reviews from Multiple Web Pages Using Machine Learning Algorithms

E. Suganya$^{(\boxtimes)}$ and S. Vijayarani$^{(\boxtimes)}$

Department of Computer Science, Bharathiar University,
Coimbatore 641046, India
elasugan1992@gmail.com

Abstract. Sentiment analysis is the computational task of automatically determining what feelings a writer is expressed in text. Sentiment analysis is gaining much attention in recent years. It is often framed as a binary distinction, i.e. positive vs. negative, but it can also be a more fine-grained, like identifying the specific emotion an author is expressing like fear, joy or anger. Globally, business enterprises can leverage opinion polarity and sentiment, topic recognition to gain deeper understanding of the drivers and the overall scope. Subsequently, these insights can advance competitive intelligence and improve customer service, thereby creating a better brand image and providing a competitive edge. To extract the content from e-commerce website using web scraping technique. It will be looping through then number of pages or so of comments for each of the products. In this work, online product reviews are collected using web scraping technique. The collected online product reviews are analyzed using opinion or sentiment analysis using classification models such as KNN, SVM, Random Forest, CNN (Convolutional Neural Network) and proposed hybrid SVM-CNN. Experiments for the classification models are performed with promising outcomes.

Keywords: Web scraping · Sentiment analysis · KNN · Random Forest · SVM · CNN

1 Introduction

Sentiment is an attitude, thought, or judgment prompted by feeling. Sentiment analysis, which is also known as opinion mining refers to the use of natural language processing (NLP), text analysis and computational linguistics to identify and extract subjective information from the source materials. It aims to determine the attitude of a writer with respect to a specific topic or the overall contextual polarity of a document [9]. The internet is a resourceful place with respect to sentiment information. From a user's perspective, people are able to post their own content through various social media, such as forums, micro-blogs, or online social networking sites. From a researcher's perspective, many social media sites release their application programming interfaces (APIs), prompting data collection and analysis by researchers and developers [3]. Hence, sentiment analysis seems to have a strong fundament with the support of

© Springer Nature Switzerland AG 2020
A. Abraham et al. (Eds.): ISDA 2018, AISC 941, pp. 677–685, 2020.
https://doi.org/10.1007/978-3-030-16660-1_66

massive online data. However, those types of online data have several flaws that potentially hinder the process of sentiment analysis. The first flaw is that since people can freely post their own content, the quality of their opinions cannot be guaranteed. For example, instead of sharing topic-related opinions, online spammers post spam on the forums. Some spam is meaningless at all, while others have irrelevant opinions also known as fake opinions [11–13]. The second flaw is that ground truth of such online data is not always available. A ground truth is more like a tag of a certain opinion, indicating whether the opinion is positive, negative, or neutral.

Web scraping is about downloading structured data from the web, selecting some of that data, and passing along what the user selected to another process [14]. Web Scraping is known by many other names, depending on how a company likes to call it, Screen Scraping, Web Data Extraction, Web Harvesting and more, is a technique employed to extract large amounts of data from websites. The data are extracted from various websites and repositories and are saved locally for instantaneous use or analysis that is to be performed later on. Data is saved to a local file system or database tables, as per the structure of the data extracted. Most websites, that view regularly, allow us only to see the contents and do not generally allow a copy or download facility. Manually copying the data is as good as cutting newspapers and can take days and weeks. Web Scraping is the technique of automating this process so that an intelligent script can help the user extract data from web pages of your choice and save them in a structured format.

2 Methods

The web pages are scraped using web scraping technique, then preprocessing techniques is performed for removing punctuation, stop words, stemming and identified term frequency. After that the sentiment analysis process has been done using machine learning algorithms such as KNN, SVM, Random Forest, CNN and Hybrid SVM-CNN.

2.1 KNN

It can be used for both classification and regression problems. However, it is more widely used in classification problems in the industry. K nearest neighbors are a simple algorithm that stores all available cases and classifies new cases by a majority vote of its k neighbors. The case being assigned to the class is most common amongst its K nearest neighbors measured by a distance function [7].

These distance functions can be Euclidean, Manhattan, Minkowski and Hamming distance. First three functions are used for continuous function and fourth one (Hamming) for categorical variables. If K = 1, then the case is simply assigned to the class of its nearest neighbor. At times, choosing K turns out to be a challenge while performing KNN modeling. KNN can easily be mapped to our real lives [10, 15].

2.2   SVM

It is a classification method. In this algorithm, plot each data item as a point in n-dimensional space (where n is the number of features the users have) with the value of each feature being the value of a particular coordinate.

For example, if only had two features like Height and Hair length of an individual, first plot these two variables in two dimensional space where each point has two coordinates (these co-ordinates are known as Support Vectors) [5]. Now find some line that splits the data between the two differently classified groups of data. This will be the line such that the distances from the closest point in each of the two groups will be farthest away [10].

a. **Linear Kernel SVM**

The dot-product is used which is called the kernel and it will be written as:

$$K(x, x_i) = sum(x * x_i)$$

Here k is the kernel that defines the similarity or a distance measure between new data and the support vectors. The dot product is the similarity measure used for linear kernel because the distance is a linear combination of the inputs [10].

b. **Radial Kernel SVM**

The Radial kernel is more complex to linear kernel. For example:

$$K(x, x_i) = exp(-gamma * sum((x - x_i^\wedge 2))$$

Where gamma is a parameter that must be used in support vector machine learning algorithm. Gamma 0.1 is a good default value, where gamma is between 0 to 1. The radial kernel can create complex regions within the feature space and transform low dimensional space to high dimensional space [10].

c. **Polynomial Kernel SVM**

Polynomial kernel is the one of the kernel function in support vector machine learning algorithm.

$$K(x, x_i) = 1 + sum(x * x_i)^\wedge d$$

Where d is the degree of the polynomial must be specified by hand to the learning algorithm. Polynomial kernel has an interactive feature that is the polynomial not only determine the similarity measures, but also it uses regression analysis to find out the relationship so, the polynomial kernel is equivalent to polynomial regression [4, 10].

2.3   Random Forest

Random Forest is a trademark term for an ensemble of decision trees. In Random Forest, we've collection of decision trees (so known as "Forest"). To classify a new object based on attributes, each tree gives a classification and say the tree "votes" for that class. The forest chooses the classification having the most votes (over all the trees in the forest) [6].

Each tree is planted & grown as follows:

1. If the number of cases in the training set is N, then a sample of N cases is taken at random but *with replacement*. This sample will be the training set for growing the tree.
2. If there are M input variables, a number m ≪ M is specified such that at each node, m variables are selected at random out of the M and the best split on these m is used to split the node. The value of m is held constant during the forest growing.
3. Each tree is grown to the largest extent possible. There is no pruning [10].

2.4 CNN

The neural network is an information-processing machine and can be viewed as analogous to human nervous system. Just like the human nervous system, which is made up of interconnected neurons, a neural network is made up of interconnected information processing units. The information processing units do not work in a linear manner [9]. In fact, neural network draws its strength from parallel processing of information, which allows it to deal with non-linearity. Neural network becomes handy to infer meaning and detect patterns from complex data sets. The neural network is considered as one of the most useful technique in the world of data analytics. However, it is complex and is often regarded as a black box, i.e. users view the input and output of a neural network but remain clueless about the knowledge generating process [1, 10].

Three convolutional layers and 3 pooling layers are used. Here, convolutional layers are used such as pooling layers, Parametric Rectified Linear Unit (PReLU) layers and dropout layers in CNN. In the architecture of CNN, the most time of training the neural network is spent in the convolution. Meanwhile, the full-connected layer takes up most of the parameters of the network. The main aim of convolution is to extract the input feature, and pooling is to sample the convolution matrix [2].

2.5 Hybrid SVM-CNN

CNNs are efficient at learning invariant features from web pages, but do not always produce optimal classification results. Conversely, SVMs with their fixed kernel function cannot learn complicated invariances, but produce good decision surfaces by maximizing margins using soft-margin approaches [16]. In this context, the proposed algorithm focus for investigating a hybrid system, in which the CNN is trained to learn features that are relatively invariant to irrelevant variations of the input. In this way, an SVM with a non-linear kernel can provide an optimal solution for separating the classes in the learned feature space. The output layer of the CNN is replaced by SVM i.e. the fully connected layer of the CNN acts as an input to the SVM. Observing the similarities in among CNNs, MLPs and SVMs, the decision function f(x) in MLPs (including CNNs) and SVMs can be written in its general form as $f(x) = (w \cdot \varphi(x) + b)$, where w represents the vector of weights, b is a bias, and all parameters are included in φ. For φ-machines and SVMs, φ is an arbitrary function.

Pseudocode for Hybrid SVM-CNN

Initialize: w such that, $\|w\| \leq (1/\sqrt{\lambda})$

for i = {1, 2, ...I} do

Swarm gets n samples: $\{A_t(i)|\ i = 1, 2, ..., n\}$

$A_i^+ = \{(x, y) \in A_i : 1 - y(w_i \cdot x) < 1\}$ and w_{i+1}

$$\eta_i = \frac{1}{\lambda i} w_{i+1/2}$$

$$w_{i+1} = minimize \left\{ 1, \frac{\frac{1}{\sqrt{\lambda}}}{\left\| w_{i+\frac{1}{2}} \right\|} \right\} w_{i+\frac{1}{2}}$$

end

Output: w_{i+1}

---

3   Result and Discussion

a. Dataset Description

The product review comments of online shopping websites are scraped. In Table 1 shows that the dataset description, i.e. online shopping websites, number of products, number of visited web pages and number of scraped web pages.

Table 1. Dataset description

S. No	Websites	No. of products	No. of visited pages	No. of scraped pages
1	Amazon	15	140	140
2	Flipcart	15	130	130
3	Snapdeal	15	110	110

In Table 2 shows the accuracy of scraped web pages of online shopping websites such as Amazon, Flip Cart and Snap Deal. The online shopping websites are randomly selected. Table 3 represents the description of web scraping technique. Each web page contains more number of product review comments which are scraped or extracted using web scraping technique.

Table 2. Scraping accuracy

Websites	Scraped accuracy (%)
Amazon	100
Flipcart	100
Snapdeal	100

Table 3. Scraping technique

Websites	Scraped web pages	No. of reviews
Amazon	140	16040
Flipcart	130	15269
Snapdeal	110	12468
Total	**380**	**43777**

Table 4 and Fig. 1 depicted as the sentiment analysis of the product review comments. Here the review comments are classified into the possible ways i.e. positive, negative and neutral based on the word frequency. The review comments are pre-processed using stop word removal, stemming, term frequency identification and remove punctuation. After performing the preprocessing technique the review comments are classified.

Table 4. Sentiment analysis

Websites	No. of reviews	Positive comments	Negative comments	Neutral comments
Amazon	16040	6864	5478	3698
Flipcart	15269	5576	5123	4570
Snapdeal	12468	5311	4584	2573

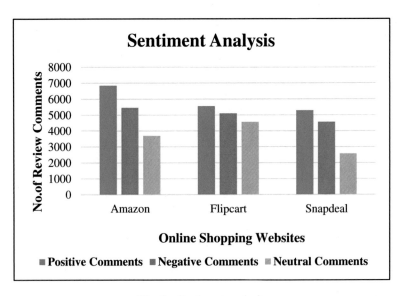

Fig. 1. Sentiment analysis

Table 5 and Fig. 2 shows the performance measures for machine learning algorithms. The performance factors are precision, recall, F-score and accuracy. Here KNN, SVM, Random Forest, CNN machine learning algorithms are considered for classifying the review comments. Hybrid SVM-CNN algorithm has proposed for high accuracy rate. Based on the performance factors hybrid algorithm outperforms well than other algorithms.

$$Precision = \frac{The\ no.\ of\ correctly\ classified\ samples\ of\ this\ type\ of\ polarity}{The\ no.\ of\ marked\ samples\ of\ this\ type\ of\ polarity}$$

$$Recall = \frac{The\ no.\ of\ correctly\ classified\ samples\ of\ this\ type\ of\ polarity}{The\ no.\ of\ this\ samples\ of\ this\ polarity}$$

$$F-Score = 2 * \frac{Precision * Recall}{Precision + Recall}$$

Table 5. Performance measures

Machine learning algorithms	Precision (%)	Recall (%)	F-Score (%)	Accuracy (%)
KNN	86.4	84.1	82.7	84.4
SVM	91.6	89.4	87.6	89.5
Random Forest	88.3	85.8	82.2	85.4
CNN	91.5	90.3	89.8	90.5
Hybrid SVM-CNN	94.2	92.4	90.7	92.4

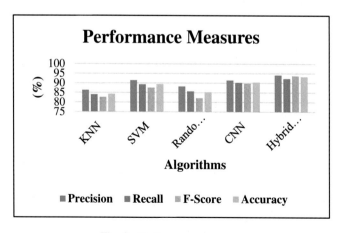

Fig. 2. Performance measures

Table 6 and Fig. 3 shows the execution time of each algorithms in milliseconds.

Table 6. Time taken

Machine learning algorithms	Time taken (milli seconds)
KNN	1400
SVM	1317
Random Forest	1429
CNN	1284
Hybrid SVM-CNN	1102

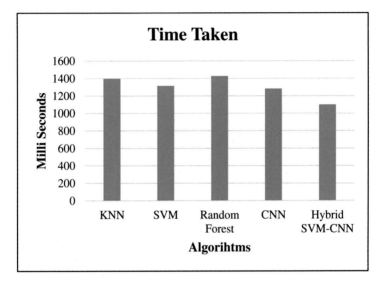

Fig. 3. Time taken

4   Conclusion

A web scraping will automatically load multiple web pages one by one, and extract data, as per requirements. In this paper a hybrid combination of SVM and CNN, machine learning algorithm for sentiment classification is presented. In the proposed algorithm, the classifier performance and accuracy are adopted as heuristic information. Experimental results demonstrate competitive performance. Proposed SVM- CNN algorithm is compared with other algorithms such as KNN, SVM, random forest and CNN for text sentiment classification. In order to evaluate the performance of the proposed algorithm, experiments were carried out on the product review comments og online shopping websites i.e. Amazon, Flipcart and Snapdeal.

References

1. Sonagi, A., Gore, D.: Efficient sentiment analysis using hybrid PSO-GA approach. Int. J. Innov. Res. Comput. Commun. Eng. **5**(6), 11910–11916 (2017)
2. Nirmala Devi, K., Jayanthi, P.: Sentiment Classification Using SVM And PSO. Int. J. Adv. Eng. Technol. **VII**(II), 411–413 (2016)
3. Kumar, A., Khorwal, R., Chaudhary, S.: A survey on sentiment analysis using swarm intelligence. Indian J. Sci. Technol. **9**(39), 1–7 (2016)
4. Redhu, S., Srivastav, S., Bansal, B., Gupta, G.: Sentiment analysis using text mining: a review. Int. J. Data Sci. Technol. **4**(2), 49 (2018)
5. http://amazonreviewscraping.blogspot.com/2014/12/scrape-web-data-using-r.html
6. http://sci-hub.tw/, https://ieeexplore.ieee.org/document/8321910
7. https://www.datacamp.com/community/tutorials/r-web-scraping-rvest
8. https://www.tidytextmining.com/sentiment.html
9. https://www.evoketechnologies.com/blog/sentiment-analysis-r-language/
10. https://www.analyticsvidhya.com/blog/2017/09/common-machine-learning-algorithms/
11. http://content26.com/blog/bing-liu-the-science-of-detecting-fake-reviews/
12. Jindal, N., Liu, B.: Opinion spam and analysis. In: Proceedings of the 2008 International Conference on, Web Search and Data Mining, WSDM 2008, pp. 219–230. ACM, New York (2008)
13. Mukherjee, A., Liu, B., Glance, N.: Spotting fake reviewer groups in consumer reviews. In: Proceedings of the 21st, International Conference on World Wide Web, WWW 2012, pp. 191–200. ACM, New York (2012)
14. https://realpython.com/tutorials/web-scraping/
15. http://dataaspirant.com/2016/12/23/k-nearest-neighbor-classifier-intro/
16. Huang, F.-J., et al.: Large-scale learning with SVM and convolutional nets for generic object categorization. In: Proceedings of the IEEE International Conference on CVPR (2006)

Interval Chi-Square Score (ICSS): Feature Selection of Interval Valued Data

D. S. Guru and N. Vinay Kumar[(⊠)]

Department of Studies in Computer Science, University of Mysore,
Manasagangotri, Mysuru 570006, India
dsg@compsci.uni-mysore.ac.in,
vinaykumar.natraj@gmail.com

Abstract. In this paper, a novel feature ranking criterion suitable for interval valued feature selection is proposed. The proposed criterion simulates the characteristics of the well known statistical criterion - chi-square in selecting the interval valued features effectively and hence called as Interval Chi-Square Score. Moreover, the paper also highlights the alternative approach proposed for computing the frequency of the distribution of interval valued data. For experimentation purpose, two standard benchmarking interval valued datasets are used with a suitable symbolic classifier for classification. The performance of the proposed ranking criterion is evaluated in terms of accuracy and the results are comparatively better than the contemporary interval valued feature selection methods.

Keywords: Interval valued data · Symbolic feature selection ·
Chi-square score

1 Introduction

In designing any pattern recognition or machine learning models, feature selection has become a major step in selecting a feature subset [3]. Currently, it has become a very interesting topic among the community of the researchers due to its vast applications in the field of classification, clustering, and regression analysis [3]. Due to its simplicity and efficiency, the feature selection technique is widely accepted in many applications [3]. Basically, the feature selection techniques are broadly classified into: filter, wrapper, and embedded methods [4].

Generally, the existing conventional feature selection methods [4] fail to perform feature selection on unconventional data like interval, multi-valued, modal, and categorical data. These data are also called in general symbolic data. The notion of symbolic data was emerged in the early 2000, which mainly concentrates in handling very realistic type of data for effective classification, clustering, and even regression for that matter [1]. As it is a powerful tool in solving problems in a natural way, we thought of developing a feature selection model for any one of the modalities. In this regard, we have chosen with an interval valued data, due its nature of preserving the continuous streaming data in discrete form [1]. Thus, in this paper, a novel feature ranking criterion suitable for supervised interval valued data is introduced.

© Springer Nature Switzerland AG 2020
A. Abraham et al. (Eds.): ISDA 2018, AISC 941, pp. 686–698, 2020.
https://doi.org/10.1007/978-3-030-16660-1_67

Initially, Ichino [13] provided the theoretical interpretation of feature selection on interval valued data. The method works based on the pretended simplicity algorithm handled in Cartesian space. Later, there are couple of attempts found on feature selection on mixed type data (i.e., interval, multi-valued, and qualitative). Kiranagi et al., [14] proposed a two stage feature selection algorithm which can handle both interval as well multi-valued data using Mutual Similarity Value proximity measure for un-supervised data. Liu et al., [15] proposed an approach based on information theory which selects an optimal feature subset by computing modified heuristic mutual information. This approach handles both numeric and interval valued features. Hedjazi et al., [11] proposed a feature selection model which handles heterogeneous type data viz., interval, quantitative, and qualitative. The proposed feature selection model makes use of similarity margin and weighting scheme for selecting features. In addition to this the model converts different types of data into a common type and further a common weighting scheme is employed on it. Hedjazi et al., [10] present feature selection of interval valued data based on the concept of similarity margin computed between an interval sample and a class prototype. The similarity margin is computed using a symbolic similarity measure. The authors have constructed basis for the similarity margin and then they worked on the multi-variate weighting scheme. The weight corresponding to each feature decides the relevancy of that feature. Hence, they considered Lagrange Multiplier for optimizing the weights which results with the optimal set of features. The experimentation is done on three standard benchmarking interval dataset and validated using LAMDA classifier. Hsiao et al., [12] have proposed a model which makes use of [10] for selecting the features. The authors have used the robust Gaussian kernel for similarity margin computation. Experimentation and a comparative analysis have been made on only one interval dataset. Dai et al., [2] proposed a heuristic approach for attribute reduction. This approach makes use of rough set and information theory for attribute reduction. In the theory of rough sets, if the efficiency of the optimal subset equals to the efficiency of original feature set then such process is termed as attribute reduction (instead of feature selection). Guru and Vinay [9] proposed a feature selection model based on two novel feature ranking criteria for interval valued data. This model makes use of vertex transformation technique before computing the rank of the features. The ranked features are then sorted based on their relevancy before get selected through experimentation. The limitation of this model is the computation of vertex transformation for higher dimensional data which leads to exponential time complexity. Guru and Vinay [6] proposed a feature selection model based on class dependent feature clustering. Unfortunately, the model turns out be very complex in handling very large dimensional interval valued data. In the similar line, Guru et al., [8] generalized the work of [6] by building up the framework for interval valued feature selection. The framework is built to emphasize the generosity of two ways of clustering the features viz., class independent and class dependent feature clustering. Though, this model works fine for large dimensional but it is very hard to handle relatively larger dimensional data. Recently, Vinay and Guru [18] proposed a novel feature selection model based on class co-variance score [9] computed on two limits of an interval valued feature. The lower limit and upper limit of the interval valued feature are handled independently and the class co-variance scores of the respective limits are further averaged to arrive at a single score corresponding to the interval valued feature. The main limitation of [18] is the sample distribution associated with each class; as the proposed model demands for a uniform sample distribution and also it operates on two independent crisp feature matrices rather on interval feature matrix.

From the literature, very few works found on interval valued feature selection through clustering approaches [6, 8]. There are no works on the filter based feature selection for supervised interval valued data which operates directly on interval valued data rather on transformation of interval valued data [9, 18]. The filter approach basically selects the optimal feature subset independent of learning algorithm. In addition to this, it is very simplistic and versatile in nature, as it solves the curse of dimensionality problem very efficiently [4]. With this background, here in this paper, a novel uni-variate filter based ranking criterion suitable for supervised interval valued feature selection is introduced.

The introduced feature ranking criterion operates on uni-variate interval valued features. Thus each feature is evaluated independently and assigned with a relevancy score. The relevancy score defines the degree of the class discriminating ability that a feature has acquired. The features are then sorted based on their relevancy scores. Further, the top d' features are selected and preserved in the knowledge base during training phase. During testing, an unknown interval sample (with only d' features) is classified with the aid of a suitable symbolic classifier.

The major contributions of this paper are as follows:

1. Proposal of a novel feature ranking criterion suitable for interval valued data.
2. Conduction of extensive experiments on two interval datasets.
3. Comparative study of the proposed model against the state-of-the art models.

The organization of the paper is as follows: The proposed model is described in Sect. 2. Section 3 describes the details of experimental setup, datasets and results. Section 4 presents a comparative analysis. Finally, Sect. 5 concludes the paper.

2 Proposed Model

The proposed model has majorly two steps viz., designing a novel filter based ranking criterion for interval valued data, followed by selecting the optimal subset of features using a suitable learning algorithm. The general architecture of the proposed model is given in Fig. 1.

2.1 ICSS (Interval Chi-Square Score): A Feature Ranking Criterion

A feature ranking criterion called interval chi-square score (ICSS) for ranking the interval valued features is introduced. The ranking criterion evaluates each feature and computes the degree of its relevancy in a supervised environment as explained below.

Basically, Chi-square [16] is used to assess three types of comparison: tests of goodness of fit, tests of homogeneity and tests of independence. In feature selection it is used as a test of independence to assess whether the class label is independent of a particular feature. Here, the interval Chi-square Score has been introduced to test the

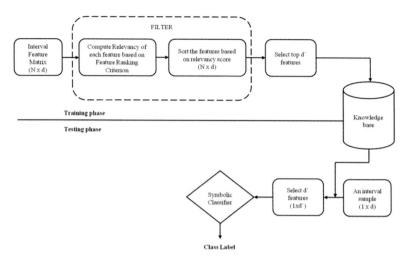

Fig. 1. General architecture of the proposed feature selection model.

degree of independence between the class label and an interval feature. If an interval feature is said to be highly discriminating, then the degree of independence between the class label and an interval feature should be relatively high, otherwise the feature is said to be less discriminating feature.

The interval chi-square for an interval feature with u unique intervals and m different classes is defined as:

$$I\chi^2 = \sum_{i=1}^{u} \sum_{j=1}^{m} \frac{\left(In_{ij} - I\mu_{ij}\right)^2}{I\mu_{ij}}$$

Where In_{ij} is the frequency of samples with i^{th} interval value.

$$I\mu_{ij} = \frac{In_{*j} \, In_{i*}}{In}$$

Where, In_{i*} is the number of samples with the ith interval value for the particular feature, In_{*j} is the number of interval samples in class j, and In is the number of interval samples.

Here, the frequency of the interval value within a particular class or in the samples across the classes is not counted based on the presence of exact interval value, instead it is counted based on the interval similarity kernel.

Let In_{ij}^q and In_{ij}^r be the ith unique interval feature value reference interval value respectively described as,

$$[q_i^-, q_i^+]; \; where, \; q_i^- \leq q_i^+$$
$$[r_i^-, r_i^+]; \; where, \; r_i^- \leq r_i^+$$

The frequency of the unique interval value is counted against the reference interval value, only if $In_{ij}^q = In_{ij}^r$, otherwise the frequency is not counted. This kind of frequency counting introduced here is inspired by the work [14], where they describe the similarity between two intervals based on their spatial relationship rather conventional exact interval matching strategy. The similarity constraints between two intervals are given by:

$$if\ sim(In_{ij}^q, In_{ij}^r) = 1,\ then\ In_{ij}^q = In_{ij}^r$$
$$if\ sim(In_{ij}^q, In_{ij}^r) = 0,\ then\ In_{ij}^q \neq In_{ij}^r$$

$$where,\ \ sim(In_{ij}^q, In_{ij}^r) = \begin{cases} 1 & if\ q_i^- \leq r_i^-\ and\ q_i^+ \geq r_i^+ \\ 0 & Otherwise \end{cases}$$

The example below illustrates the significance of the proposed interval chi-square score which works based on the similarity based matching rather than the conventional exact interval matching based.

Let us consider a sample supervised interval valued dataset (Table 1) consisting of 15 samples and 2 interval valued features spread across two different classes. Now if we compute the parameters of interval chi-square score using equation with the conventional exact interval match and the proposed similarity match, then the obtained values are tabulated in the tables from Tables 2 3, 4 and 5.

From Table 6, one can observe that the relevancy scores corresponding to interval features IF1 and IF2 have same values with respect to conventional interval matching, whereas they are different with respect to the proposed similarity match. Thus results with the different ranking sequence for the given two interval features. It is also shown from the Table 6 that the conventional matching results with the confusion in ranking

Table 1. Sample interval valued dataset.

	IF1	IF2
Class-1	[1.5, 3]	[1.3, 2.5]
	[1.6, 4.5]	[1.2, 2.6]
	[1.6, 4.5]	[1.1, 2.7]
	[4.2, 5]	[1.2, 2.6]
	[1.5, 3]	[1.9, 3.6]
	[4, 5]	[1.9, 3.6]
	[1, 2]	[1.6, 3.7]
Class-2	[9, 10]	[9.2, 11.6]
	[11, 12]	[9.2, 11.6]
	[11.5, 13]	[8.8, 11.8]
	[9.5, 10]	[8.5, 12]
	[11.8, 12.5]	[8.8, 11.8]
	[14, 16]	[9.5, 13]
	[12, 14]	[9.6, 13.5]
	[9.2, 10]	[9.2, 11.6]

Table 2. Interval chi-square parameter values based on exact match for first interval feature (IF1) of sample supervised interval valued dataset

Unique interval values	Conventional exact match for IF1								
	In_{i1}	In_{i2}	In_{i*}	$I\mu_{i1}$	$I\mu_{i2}$	In_{*1}	In_{*2}	In	$I\chi^2$
[1, 2]	1	0	1	0.47	0.53	7	8	15	15
[1.5, 3]	2	0	2	0.93	1.07				
[1.6, 4.5]	2	0	2	0.93	1.07				
[4, 5]	1	0	1	0.47	0.53				
[4.2, 5]	1	0	1	0.47	0.53				
[9, 10]	0	1	1	0.47	0.53				
[9.2, 10]	0	1	1	0.47	0.53				
[9.5, 10]	0	1	1	0.47	0.53				
[11, 12]	0	1	1	0.47	0.53				
[11.5, 13]	0	1	1	0.47	0.53				
[11.8, 12.5]	0	1	1	0.47	0.53				
[12, 14]	0	1	1	0.47	0.53				
[14, 16]	0	1	1	0.47	0.53				

Table 3. Interval chi-square parameter values based on proposed similarity match for first interval feature (IF1) of sample supervised interval valued dataset

Unique interval values	Proposed similarity match for IF1								
	In_{i1}	In_{i2}	In_{i*}	$I\mu_{i1}$	$I\mu_{i2}$	In_{*1}	In_{*2}	In	$I\chi^2$
[1, 2]	1	0	1	0.47	0.53	7	8	15	19.64
[1.5, 3]	2	0	2	0.93	1.07				
[1.6, 4.5]	2	0	2	0.93	1.07				
[4, 5]	2	0	2	0.93	1.07				
[4.2, 5]	1	0	1	0.47	0.53				
[9, 10]	0	3	3	1.40	1.60				
[9.2, 10]	0	2	2	0.93	1.07				
[9.5, 10]	0	1	1	0.47	0.53				
[11, 12]	0	1	1	0.47	0.53				
[11.5, 13]	0	2	2	0.93	1.07				
[11.8, 12.5]	0	1	1	0.47	0.53				
[12, 14]	0	1	1	0.47	0.53				
[14, 16]	0	1	1	0.47	0.53				

the features, whereas the proposed will overcome the confusion occurred while ranking. Hence, the proposed ICSS with similarity matching is better compared to the conventional exact matching.

The Fig. 2 shows the graphical illustration of IF1 and IF2 features. From the figure, one can observe that the degree of separability between the samples of two classes is high with respect to IF2 than IF1. This indicates that the feature IF2 is more capable of

Table 4. Interval chi-square parameter values based on exact match for second interval feature (IF2) of sample supervised interval valued dataset

Unique interval values	Conventional similarity match for IF2								
	In_{i1}	In_{i2}	In_{i*}	$I\mu_{i1}$	$I\mu_{i2}$	In_{*1}	In_{*2}	In	$I\chi^2$
[1.1, 2.7]	1	0	1	0.47	0.53	7	8	15	15.00
[1.2, 2.6]	2	0	2	0.93	1.07				
[1.3, 2.5]	1	0	1	0.47	0.53				
[1.6, 3.7]	1	0	1	0.47	0.53				
[1.9, 3.6]	2	0	2	0.93	1.07				
[8.5, 12]	0	1	1	0.47	0.53				
[8.8, 11.8]	0	2	2	0.93	1.07				
[9.2, 11.6]	0	3	3	1.4	1.6				
[9.5, 13]	0	1	1	0.47	0.53				
[9.6, 13.5]	0	1	1	0.47	0.53				

Table 5. Interval chi-square parameter values based on exact match for second interval feature (IF2) of sample supervised interval valued dataset

Unique interval values	Proposed similarity match for IF2								
	In_{i1}	In_{i2}	In_{i*}	$I\mu_{i1}$	$I\mu_{i2}$	In_{*1}	In_{*2}	In	$I\chi^2$
[1.1, 2.7]	4	0	4	1.87	2.13	7	8	15	28.86
[1.2, 2.6]	3	0	3	1.40	1.60				
[1.3, 2.5]	1	0	1	0.47	0.53				
[1.6, 3.7]	3	0	3	1.40	1.60				
[1.9, 3.6]	2	0	2	0.93	1.07				
[8.5, 12]	0	6	6	2.80	3.20				
[8.8, 11.8]	0	5	5	2.33	2.67				
[9.2, 11.6]	0	3	3	1.40	1.60				
[9.5, 13]	0	1	1	0.47	0.53				
[9.6, 13.5]	0	1	1	0.47	0.53				

Table 6. Ranks and relevancy scores based on conventional and proposed matching for the sample supervised interval valued dataset

		IF1	IF2
Relevancy score	Conventional	15	15
	Proposed	19.64	28.86
Rank	Conventional	IF1	IF2
	Proposed	IF2	IF1

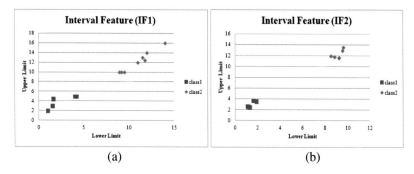

Fig. 2. Graphical illustration of distribution of interval features, (a) IF1, (b) IF2

discriminating the two classes (without class information) than the feature IF1. The same thing has been brought out by the proposed interval chi-square score criterion.

2.2 Feature Selection

After computing the relevancy scores to all interval features, they are sorted based on their associated relevancy score. Further, the top d' (d' << d) features are selected and preserved in the knowledge base for further consideration during testing for classification of an unknown test sample.

2.3 Symbolic Classifier

In this work, though the type of data handled is of type interval; it is difficult to compatible with conventional classifiers such as K-NN, SVM, Random Forest, Naïve Bayesian etc., [3]. Hence, a symbolic classifier [5] which suits well for interval type data for classification is recommended. Guru and Nagendraswamy [5], propose a symbolic classifier which is based on nearest neighbour approach in classifying an unknown sample to a known class.

During testing, the symbolic classifier (SC) is used to classify an unknown interval test sample to a known class with only d' interval features (where d' feature indices are selected from the knowledge base).

3 Experimentation and Results

3.1 Datasets

The proposed feature selection model is validated on two standard benchmarking supervised interval valued datasets viz., Water and Flower datasets. The information about these datasets is detailed below.

The Water dataset [17] is a supervised interval valued dataset consists of 316 samples with 48 interval features spread over two classes. Each class consists of 223 and 93 samples respectively.

The original flower dataset [7] consists of conventional crisp type features. It consists of 3000 samples with 501 features spread across 30 different classes of flower species and each class consisting of 100 samples. In this work, the proposed model demands for interval valued data. Hence, the samples of the dataset in each class are grouped into 20 sub-classes based on [1]. Further, the intra-class variations are preserved using mean-standard deviation interval representation. Thus, resulting with 600 samples with 501 interval features spread across 30 classes and each class consisting of 20 interval samples.

3.2 Experimental Setup

The details of experimentation conducted on the two supervised interval valued datasets for selecting the top d' features using the proposed feature selection model are given in this section.

During experimentation, the interval dataset is divided into training and testing set of samples. The percentage of training-testing samples are varied from 30%–70% to 80%–20% (in steps of 10%) respectively. During training phase, feature relevancy score for each interval feature is computed as discussed in Sect. 2.1. Later on, the features are sorted based on their relevancy scores. Here, the relevancy score possess low score for high discriminating features and it possesses high score for less discriminating features. Hence, the features are sorted in ascending order. Further, the top d' features are preserved in the knowledge base. During testing phase, with the support of the symbolic classifier discussed in Sect. 2.3, an unknown interval test sample is classified to a known class. The β value in the symbolic classifier (SC) is set to 1 throughout the experimentation.

3.3 Results

To evaluate the performance of the proposed model, a performance measure - classification accuracy is recommended. It is defined as the ratio of the number of correctly classified samples to the total number of samples.

To select the top d' features corresponding to a train-test percentage, the experimentation is conducted for 20 trials. In each trial, the feature subset which gives maximum classification accuracy is obtained and from all trials, 20 such accuracies along with associated feature subsets are obtained. Using this, the minimum, maximum and average accuracies along with top d' subsets of features are tabulated. To show the robustness of feature sub-setting, the classification accuracy obtained considering all features are also tabulated. Similarly, the same procedure is followed for all remaining train-test percentage of samples.

Tables 7 and 8 show the minimum, maximum, and average classification accuracies along with associated top d' features obtained under varying train-test percentage of samples for water and flower datasets respectively. They also show the minimum, maximum, and average classification accuracies obtained for all features in consideration under varying train-test percentage of samples for the same datasets. In Tables, (d') denote the number of features selected corresponding to the obtained accuracy.

Table 7. Minimum, maximum, and average classification accuracies obtained from 20 trails under different train-test percentage of samples for Water dataset.

Train-test	With Feature Selection (WFS)			Without Feature Selection (WoFS)		
	Min	Max	Avg	Min	Max	Avg
30–70	72.40 (21)	78.73 (31)	75.34 (24)	68.33 (48)	75.57 (48)	71.76 (48)
40–60	71.28 (07)	80.32 (13)	75.66 (27)	63.83 (48)	76.06 (48)	71.73 (48)
50–50	73.89 (25)	81.53 (24)	77.48 (31)	66.88 (48)	77.71 (48)	73.06 (48)
60–40	73.02 (19)	82.54 (22)	77.74 (28)	65.87 (48)	79.37 (48)	73.06 (48)
70–30	76.34 (39)	86.02 (40)	79.78 (28)	68.82 (48)	82.80 (48)	74.84 (48)
80–20	74.19 (35)	84 (21)	78.87 (22)	66.13 (48)	80.65 (48)	73.95 (48)

Table 8. Minimum, maximum, and average classification accuracies obtained from 20 trails under different train-test percentage of samples for Flower dataset.

Train-Test	With Feature Selection (WFS)			Without Feature Selection (WoFS)		
	Min	Max	Avg	Min	Max	Avg
30–70	90.24 (60)	94.76 (480)	92.48 (298)	88.10 (501)	94.05 (501)	91.12 (501)
40–60	93.33 (340)	96.39 (460)	95.04 (355)	91.11 (501)	96.11 (501)	94.18 (501)
50–50	93.67 (400)	97.67 (440)	95.85 (283)	93.00 (501)	97.33 (501)	95.12 (501)
60–40	95.00 (320)	99.17 (500)	96.94 (308)	93.75 (501)	99.17 (501)	96.08 (501)
70–30	96.11 (140)	99 (420)	97.67 (295)	93.89 (501)	98.89 (501)	97.03 (501)
80–20	95.83 (280)	100 (480)	98.08 (230)	94.17 (501)	100 (501)	97.25 (501)

From Tables 7 and 8, it is very clear that the classification accuracies obtained with feature sub-setting are higher than the classification accuracies obtained without feature sub-setting.

4   Comparative Analysis

To demonstrate the superiority of the proposed model, a comparative analysis has been brought on the state-of-the-art methods. The comparative analysis is given only on the two discussed datasets. In literature, couple of interval valued feature selection methods operated on water dataset are found [8, 10, 12, 15, 18]. But, only a work found on the feature selection with respect to the flower dataset [18]. In addition to this a comparison is also given against the state-of-the-art classification method which does not use any feature selection method during classification [7]. Tables 9 and 10 respectively show the comparative analysis on Water and Flower datasets against the proposed model.

Table 9. Comparative analysis of the accuracies obtained from the proposed feature selection method against the state-of-the-art methods on Water dataset.

Methods		Classifier	Accuracy (feature subset)
Proposed model		SC	86.02 (40)
Method [18]		SC	84.95 (09)
Method [8]	Model-1	C-1	74.19 (2)
		C-2	77.42 (15)
		C-3	73.12 (11)
	Model-2	C-1	79.03 (4)
		C-2	79.57 (10)
		C-3	75.81 (2)
Method [10]		LAMDA	77 (14)
Method [12]		LAMDA	78.66 (11)
Method [15]		KNN	78 (27)

Table 10. Comparative analysis of the classification accuracy of the proposed feature selection method against the without feature selection method on Flower dataset.

Methods	Classifier	Accuracy (feature subset)
Proposed model	SC	100 (480)
Method [18]	SC	100 (40)
Method [7]	PNN	84 (501)

From Tables 9 and 10, it is clear that the proposed model outperforms the state-art-of-the-art methods in terms of classification accuracy.

5   Conclusion

In this paper, a novel interval valued feature ranking criterion for supervised interval valued feature selection is introduced. The introduced feature ranking criterion evaluates each feature and computes an interval chi-square score based on similarity based comparison of two intervals. Further, the features are sorted based on the feature relevancy. Empirically, the top d' features are selected and performed classification with the help of a symbolic classifier. To test the effectiveness of the proposed model, two supervised interval valued datasets are used. The experimental results show the superiority of the proposed model against the state-of-the-art methods in terms of accuracy.

Acknowledgement. The author N Vinay Kumar acknowledges the Department of Science & Technology, Govt. of India for their financial support rendered in terms of DST-INSPIRE fellowship.

References

1. Billard, L., Diday, E.: Symbolic Data Analysis: Conceptual Statistics and Data Mining. Wiley, Hoboken (2007)
2. Dai, J.H., Hu, H., Zheng, G.J., Hu, Q.H., Han, H.F., Shi, H.: Attribute reduction in interval-valued information systems based on information entropies. Front. Inf. Technol. Electron. Eng. **17**(9), 919–928 (2016)
3. Duda, O.R., Hart, E.P., Stork, G.D.: Pattern Classification, 2nd edn. Wiley-Interscience (2000)
4. Ferreira, A.J., Figueiredo, M.A.T.: Efficient feature selection filters for high-dimensional data. Pattern Recogn. Lett. **33**, 1794–1804 (2012)
5. Guru, D.S., Nagendraswamy, H.S.: Symbolic representation and classification of two-dimensional shapes. In: Proceedings of the 3rd Workshop on Computer Vision, Graphics, and Image Processing (WCVGIP), pp. 19–24 (2006)
6. Guru, D.S. Vinay Kumar, N.: Class specific feature selection for interval valued data through interval K-means clustering, RTIP2R 2016, CCIS, vol. 709, pp. 228–239. Springer (2017)
7. Guru, D.S., Sharath, Y.H., Manjunath, S.: Textural features in flower classification. Math. Comput. Model. **54**(3–4), 1030–1036 (2011)
8. Guru, D.S., Vinay Kumar, N., Suhil, M.: Feature selection of interval valued data through interval feature selection. Int. J. Comput. Vis. Image Process. **7**(2), 64–80 (2017)
9. Guru, D.S., Vinay Kumar, N.: Novel feature ranking criteria for interval valued feature selection. In: Proceedings of the IEEE International Conference on Advances in Computing, Communications and Informatics, pp. 149–155 (2016)
10. Hedjazi, L., Martin, A.J., Lann, M.V.L.: Similarity-margin based feature selection for symbolic interval data. Pattern Recogn. Lett. **32**, 578–585 (2011)
11. Hedjazi, L., Martin, J.A., Lann, M.V.L., Hamon, T.K.: Membership-margin based feature selection for mixed type and high-dimensional data: theory and applications. Inf. Sci. **322**, 174–196 (2015)
12. Hsiao, C.C., Chuang, C.C., Su, S.F.: Robust Gaussian Kernel based approach for feature selection. In: Advances in Intelligent Systems and Computing, vol. 268, pp. 25–33 (2014)
13. Ichino, M.: Feature selection for symbolic data classification. In: New Approaches in Classification and Data Analysis, pp. 423–429. Springer (1994). section 2
14. Kiranagi, B.B., Guru D.S., Ichino, M.: Exploitation of multivalued type proximity for symbolic feature selection. In: Proceeding of the Internal Conference on Computing: Theory and Applications, pp. 320–324. IEEE (2007)

15. Liu, Q., Wang, J., Xiao, J., Zhu, H.: Mutual information based feature selection for symbolic interval data. In: Proceedings of International Conference on Software Intelligence, Technologies and Applications, pp. 62–69 (2014)
16. Liu, H., Setiono, R.: Chi2: feature selection and discretization of numeric attributes. In: Proceedings of 7th IEEE International Conference on Tools with Artificial Intelligence, Technologies and Applications, pp. 388–391 (1995)
17. Quevedo, J., Puig, V., Cembrano, G., Blanch, J., Aguilar, J., Saporta, D., Benito, G., Hedo, M., Molina, A.: Validation and reconstruction of flow meter data in the Barcelona water distribution network. J. Control Eng. Pract. **18**, 640–651 (2010)
18. Vinay Kumar, N., Guru, D.S.: A novel feature ranking criterion for supervised interval valued feature selection for classification. In: Proceedings of the 14th IAPR International Conference on Document Analysis and Recognition (ICDAR), pp. 71–76 (2017)

Association Rule Hiding Using Firefly Optimization Algorithm

S. Sharmila$^{(\boxtimes)}$ and S. Vijayarani

Department of Computer Science, Bharathiar University,
Coimbatore, India
sharmilasathyanathan@gmail.com,
vijimohan_2000@yahoo.com

Abstract. Privacy preserving data mining is an important research area which protects the private information and reduces the information loss during the data mining process. There are many data mining techniques whereas Association rule mining is one of the data mining technique which finds existing correlations between data items. Privacy Preserving Association Rule Mining is one of the techniques in this field, which aims to hide sensitive association rules. Many different algorithms with particular approaches have been developed to protect the private information. In this paper, a new approach has been introduced using firefly optimization algorithm for hiding the sensitive association rules. To hide the sensitive rules distortion technique was used. Further in this work fitness function was defined to achieve the optimal solution with fewest side effects. The efficiency of proposed algorithm was evaluated with different databases. The results of the execution of the proposed algorithm and existing algorithm tabu search on different databases indicates that firefly algorithm has better performance compared to other algorithm.

Keywords: Data mining · Privacy preserving data mining ·
Association rule hiding · Firefly optimization algorithm

1 Introduction

The privacy preserving data mining contains two steps: First, sensitive raw data should be modified from the original database. Second, sensitive knowledge which can be mined from database using data mining algorithms should also be excluded. The main objective of privacy preserving data mining is to develop algorithms for modifying the original data so that the private data and knowledge remain private even after the mining process. Privacy of the user data is preserved, and, at the same time, the mining models can be reconstructed from the modified data with reasonable accuracy [5, 7].

Association rule hiding has been widely researched along two principal directions: The first includes approaches that aim at hiding specific association rules among those mined from the original database. The second includes approaches that hide specific frequent item sets from the frequently found item sets while mining original database [2]. Association Rule hiding is the process of hiding strong association rules and creating sanitized database from the novel database to prevent unauthorized party from

© Springer Nature Switzerland AG 2020
A. Abraham et al. (Eds.): ISDA 2018, AISC 941, pp. 699–708, 2020.
https://doi.org/10.1007/978-3-030-16660-1_68

generating frequent patterns [3, 8]. Several association rule hiding algorithms exists namely Genetic, Tabu, Cuckoo, Artificial bee colony and Ant colony Optimization, by using these algorithms sensitive association rules are hidden.

Association rule mining is mostly focused on finding frequent co-occurring associations among a collection of items. It is also defined as Market Basket Analysis. Association rule mining was first introduced by Agarwal [1]. Many association rule mining algorithms are available like Apriori, AprioriTID, H-mine, Eclat, Dclat, RELIM, FP-growth and FIN algorithm, etc. [4]. Association rule finds the frequently occurred patterns, associations and correlation between item sets relational databases [1, 6].

The paper consists of the following sections: Sect. 2 contains literature survey it describes what has been done in the privacy preserving of association rules mining. In Sect. 3, explains problem statement of association rule hiding Sect. 4 gives a brief review of existing algorithm. Section 4.2 explains the concept of proposed optimization algorithms. Section 5 is about experimental results. Section 6 describes conclusion.

2 Literature Survey

Wang et al. [9] had introduced ISL and DSR algorithms. The distortion technique was used in both the algorithms. ISL algorithm hides sensitive rules by reducing the confidence of sensitive rules to less than minimum confidence threshold, through increasing their support for the left hand side element. DSR algorithm hides the sensitive rules by decreasing their confidence to less than minimum confidence threshold, through decreasing their support for the right-hand side element. ISL algorithm is weak in the number of hiding failures and DSR algorithm is weak in the number of lost rules.

Afshari et al. [10] proposed Cuckoo optimization algorithm for hiding sensitive association rules. Hiding process was performed through the data-distortion method. In addition, in this method, three fitness functions are used to achieve an optimal solution with the minimum side effects. In this algorithm, a pre-processing operation is performed on the original database, so that only transactions with sensitive items can be involved in the hiding process.

Dehkordi et al. [11] used a genetic algorithm to provide a new method for sensitive association rules hiding. In addition to sensitive rules hiding, the main goal of this algorithm was to reduce the changes in the database. In order to achieve this goal, the algorithm defines a pre-processing stage in which only sensitive transactions of the database are selected and changes are only made on sensitive transactions. Also, in this algorithm, four different strategies for fitness functions were proposed to hide sensitive rules and sensitive items with minimum side effects.

Khan et al. [14] had defined a new fitness function to improve the method which was proposed by the author. In this method, to reduce the failure in hiding, lost rules and ghost rules, another criterion called the information loss, was also considered.

The proposed fitness function calculates fitness value of each transaction that modifies some transactions in original database. Author had proved that proposed technique produces better results comparing to other traditional algorithms.

Le et al. [15] proposed a heuristic algorithm based on the intersection lattice of frequent itemsets for hiding sensitive rules. The algorithm first determines the target item such that modifying this item causes the least impact on the set of frequent item sets. Then, the minimum number of transactions that need to be modified is specified. After that, the target item was removed from the specified transactions and the data set is sanitized.

Oliveira et al. [16] had discussed about issue of multiple hiding of sensitive association rules. The proposed algorithms include Naïve, IGA, MinFIA and MaxFIA, which perform the hiding process through scanning the database twice. In the first time, this algorithm identifies sensitive transactions and indexes them in order to increase the speed of sensitive transactions recognition. In the second time, after removing the selected element, sanitization is performed with a minimum number of elements.

Telikani et al. [20] have examined many scientific algorithms and analyzed them in terms of four features including hiding strategy, sanitization technique, sanitization approach, and selection method. In terms of results and findings, this review showed that in comparison to other aspects of sanitization algorithms, the transaction and item selection methods more significantly influence the optimality of hiding process and blocking technique increases the disclosure risk while distortion technique is better in knowledge protection field, and transaction deletion/insertion technique is a new direction.

3   Problem Definition

The problem of associated rules hiding can be defined as the conversion of the original database into a sanitization database so that data mining techniques are not capable of mining the sensitive rules of the database, while all non-sensitive rules can be extracted The main objective of this research work is to hide sensitive association rule by using firefly optimization algorithm with less side effects.

4   Methodology

4.1   Tabu Search Algorithm

Tabu search is a Meta heuristic algorithm it is used for solving optimization problems, such as the traveling salesman problem (TSP) [17]. Local search method was used in Tabu search optimization algorithm, it is used to move iteratively from a solution k to a solution k' in the neighborhood of k, until all the conditions are satisfied. To discover regions of the search space that would be left unexplored by the local search procedure,

Tabu search modifies the neighborhood structure of each solution. The memory structures determines solution to new neighborhood. In this research work, the sensitive data items were modified using Tabu search algorithm.

In the first step, all the sensitive items, sensitive transactions and number of modifications required for the sensitive items are initialized. In the second step, the cost i.e. distance of each sensitive transaction is calculated. Based on this cost value the transactions are selected for sensitive item modification [10]. If the particular sensitive item is found in the transaction then modify it as 1 to 0. These modified values are moved into tabu list. This process is repeated until the number of modifications becomes 0. Last step is termination process, before terminating the algorithm ensures that all the sensitive items are modified and the number of modification is 0 then the algorithm is terminated.

4.2   Firefly Optimization Algorithm

Xin-She Yang was first person to developed Firefly Algorithm (FA) in late 2007 and 2008 at Cambridge University [17, 18], it was based on the flashing patterns and behavior of fireflies. FA uses three rules: Fireflies are unisex so that one firefly will be attracted to other fireflies regardless of their sex. The attractiveness is related to the brightness, and they both decrease as their distance increases. Thus any two flashing fireflies, the less bright one will move towards the brighter one. The firefly moves randomly if there is no brighter one. To determine the brightness of a firefly objective function are used [18].

In the proposed method, the algorithm of Firefly optimization is used for database sanitization. Also, to hide the sensitive association rules, the data distortion technique is used. In order to hide the sensitive association rules, the method of support or confidence reduction is used through the removal of elements [19]. The reduction in support or confidence level of the sensitive association rules is accompanied by the reduction in the support level for the set of right-hand side elements of sensitive association rules in transactions which fully support the rules. In the proposed method, fitness functions are defined which lead to the solutions with the less side effects.

In the proposed Firefly Optimization Algorithm first step, initializes the population, here firefly are considered as sensitive items. In the Second step objective function it is used to determine the number of sensitive items in each transaction. Whereas next step is to arrange the transaction in ascending order. Next step calculates the attractiveness of firefly, in ARH it finds the number of modification to be done in each item and modify the sensitive item in the transaction. Finally in this step firefly moves from less brightness to more brightness but in the proposed algorithm sensitive items are modified, the sensitive items marked as absent until all the modifications are done to get an optimal solution. Final step is to update the new modification in the original database.

Input:
      Initialize the parameters I,J, TN_t, NM_m, NE_e // Initialize the population –number of transactions

Procedure:
      I-Total no of transactions
      J- Total no of items in the transaction
      TN_t_No of transactions with sensitive items
      NM_m_No of modifications
      NE_e_No of elimination

Output: Sanitized database

Step 1: Objective Function
      To identify the more number of sensitive items in transactions a formula is used
      Sen_{item}=No of non-sensitive items /number of items.
      Apply this formula in all the transactions to find the highest number of sensitive items in the transactions.

Step 2: Light Intensity
      Arrange the transactions in the ascending order

Step 3: Calculate the number of modification in sensitive item
      $NMm=N_{oc} - 1$ //where m is the sensitive item which determines number of modification to be done
          //Move results to NMm in the forward direction to hide the sensitive items. Sen_{item} modification is done in transaction TN which has sensitive items.

Step 4: Repeat the steps 1.3 to 4 until all the modification becomes 0

Step 5: Elimination
      Ensure all the sensitive items are modified
      Number of modification becomes 0 then the process Completed

Step 6: Exit

---

5 Result and Discussion

5.1 Datasets

In proposed algorithm, different databases have been used. Therefore, for experiments, two datasets i.e mushroom and chess are taken from FIMI repository. The features of the database are shown in Table 1.

Table 1. Features of dataset

Dataset	No of transaction	No of items	Avg. length of items
Mushroom	8124	119	23
Chess	3196	75	18

5.2   Performance Measures

To examine the quality of the hiding, its side effects must be calculated on the database. The hiding process has calculated four performance measures i.e. Hiding Failure, Missing cost, Execution time and memory space were described below.

5.2.1   Hiding Failure (HF)

This measure represents the number of **sensitive** rules that the sanitization algorithm failed to hide, and they can still be extracted from the sanitization database D'. The value of HF is calculated by the following relation: In this relation, |Rs(D')| indicates the number of sensitive rules extracted from the sanitization database D' and |Rs(D)| indicates the number of sensitive rules extracted from the original database D. Table 2 and Fig. 1 shows the analysis of Hiding Failure.

Table 2. Comparative evaluation of hiding failure

Threshold	Hiding failure			
	Tabu		Firefly	
	Mushroom	Connect	Mushroom	Connect
s-20 c-40	0.4	0.35	0.3	0.2
s-30 c-50	0.3	0.3	0.3	0.25
s-40 c-60	0.2	0.1	0.1	0.05
s-50 c-70	0.1	0.1	0.05	0.1

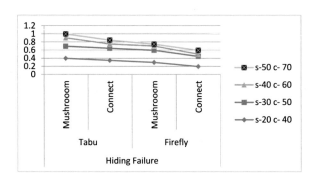

Fig. 1. Analysis of hiding failure

5.2.2   Missed Cost

This performance feature is used to measure the percentage of the nonrestrictive patterns that are hidden as a side-effect of the modification process. It is computed as follows: where RP (D) is the set of all non-sensitive rules in the original database D and RP (D') is the set of all non-sensitive rules in the modified database D. Table 3 and Fig. 2 shows the performance of Missing cost.

Table 3. Performance evaluation of missing cost

Threshold	Missing cost			
	Tabu		Firefly	
	Mushroom	Connect	Mushroom	Connect
s-20 c-40	10	12	8	11
s-30 c-50	7.5	8.9	7	8.3
s-40 c-60	8	5.3	4.6	4.9
s-50 c-70	4.5	4.2	2	4

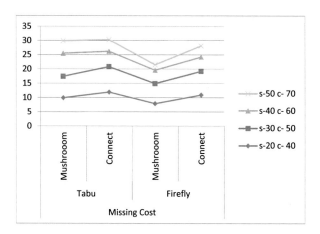

Fig. 2. Analysis of missing cost

5.2.3 Runtime

The length of time the algorithm spends to reach the optimal solution. In this work, the efficiency of the algorithm is calculated by using the CPU time. Table 4 and Fig. 3 illustrates the comparison of Execution Time

Table 4. Comparison of execution time

Threshold	execution time			
	Tabu		Firefly	
	Mushroom	Connect	Mushroom	Connect
s-20 c-40	1573	983	1265	892
s-30 c-50	1126	836	1032	793
s-40 c-60	983	645	956	601
s-50 c-70	765	612	701	712

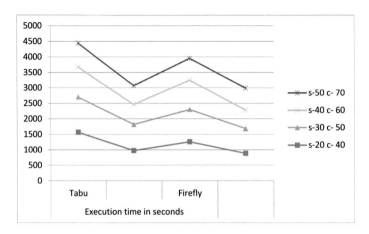

Fig. 3. Analysis of execution time

5.2.4 Memory

The amount of the storage space occupied by then Existing and proposed algorithms. Table 5 and Fig. 4 describes the comparison of memory space, from the analysis it has proved that proposed algorithm produces better result than existing algorithm.

Table 5. Comparison of memory space

Threshold	Missing cost			
	Tabu		Firefly	
	Mushroom	Connect	Mushroom	Connect
s-20 c-40	57	69	51	57
s-30 c-50	43	60	39	53
s-40 c-60	32	51	21	47
s-50 c-70	27	47	21	40

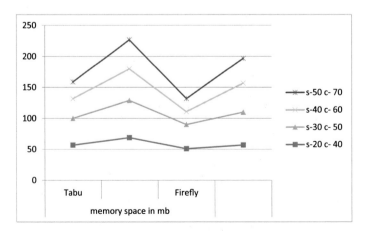

Fig. 4. Analysis of memory space

6 Conclusion and Future Works

The research work has proposed a heuristic method for hiding sensitive association rules using optimization algorithm. The proposed algorithm is capable of hiding sensitive items. In this study, the proposed method hides sensitive rules using Firefly optimization algorithm. Several sensitive items are hided in the method. The proposed method also has fewer missing cost than other algorithm. Fitness functions are proposed to find the solution with minimum side effects. Fitness functions are able to determine the number of hiding failures for each solution without data mining. After calculating the fitness for each solution, the fitness values of the solutions are compared with all fitness values of other solutions in order to select the best solution. The method is implemented on two datasets. The proposed approach was compared to existing algorithms Tabu Search optimization algorithm. From the analysis firefly Optimization Algorithm (FOA) is best comparing with Existing algorithms.

References

1. Agarwal, C.C., Yu, P.S. (eds.): Privacy-Preserving Data Mining: Modeland Algorithms (2008). ISBN 0-387-70991-8
2. Jain, Y.K.: An efficient association rule hiding algorithm for privacy-preserving data mining. Int. J. Comput. Sci. Eng. 3(7), 2792–2798 (2011)
3. Nayak, G., Devi, S.: A survey on privacy preserving data mining: approaches and techniques. Int. J. Eng. Sci. Technol. 3(3), 2127–2133 (2011)
4. Patel Tushar, S., Mayur, P., Dhara, L., Jahnvi, K., Piyusha, D., Ashish, P., Reecha, P.: Association an analytical study of various frequent itemset mining algorithms. Res. J. Comput. Inf. Technol. Sci. 1(1), 6–9 (2013). February Res. J. Computer & IT Sci. International Science Congress
5. Han, J., Kamber, M., Pei, J.: Data Mining: Concepts and Techniques. Morgan Kaufmann, San Mateo (2006)
6. Saygin, Y., Verykios, V.S., Elmagarmid, A.K.: Privacy preserving association rule mining. In: Proceedings of the 2002 International (2002)
7. Schuster, A., Wolff, R., Gilburd, B.: Privacy preserving data mining on data grids in the presence of malicious participants. In: IEEE International Symposium on High Performance Distributed Computing - HPDC (2004)
8. Zhang, N., Wang, S., Zhao, W.: A new scheme on privacy preserving association rule mining. In: Principles of Data Mining and Knowledge Discovery – PKDD, vol. 3202, pp. 484–495 (2004)
9. Otey, M.E., Wang, C., Parthasarathy, S., Veloso, A., Meria, W.: Mining frequent itemsets in distributed and dynamic databases. In: IEEE International Conference on Data Mining (2003)
10. Afshari, M.H., Dehkordi, M.N., Akbari, M.: Association rule hiding using cuckoo optimization algorithm. Expert Syst. Appl. 64, 340–351 (2016)
11. Dehkordi, M.N., Badie, K., Zadeh, A.K.: A novel method for privacy preserving in association rule mining based on genetic algorithms. J. Softw. 4(6), 555–562 (2009)
12. Yang, X.S.: A discrete firefly algorithm for the multi-objective hybrid flow shop scheduling problems. IEEE Trans. Evol. Comput. 18(2), 301–305 (2014)

13. Jia, D., Duan, X., Khan, M.K.: Binary artificial bee colony optimization using bitwise operation. Comput. Ind. Eng. **76**, 360–365 (2014)
14. Khan, A., Qureshi, M.S., Hussain, A.: Improved genetic algorithm approach for sensitive association rules hiding. World Appl. Sci. J. **31**(12), 2087–2092 (2014)
15. Le, H.Q., Arch-Int, S., Nguyen, H.X., Arch-Int, N.: Association rule hiding in risk management for retail supply chain collaboration. Comput. Ind. **64**(7), 776–784 (2013)
16. Oliveira, S.R.M., Zaiane, O.R.: Privacy preserving frequent itemset mining. In: Proceedings of the IEEE International Conference on Privacy, Security and Data Mining, vol. 14, pp. 43–54 (2002)
17. Vijayarani, S., Tamilarasi, A., SeethaLakshmi, R.: Tabu search based association rule hiding. Int. J. Comput. Appl. **19**(1), 0975–8887 (2011)
18. Yuan, F., Chen, S., Liu, H.: Association rules mining on heart failure differential treatment based on the improved firefly algorithm. J. Comput. **9**(4), 822–830 (2014)
19. Neelima, S., Sathyanarayan, N., Murthy, P.K.: A novel multi-objective firefly algorithm for optimization of association rule mining (2017)
20. Telikani, A., Shahbahrami, A.: Data sanitization in association rule mining: an analytical review. Expert Syst. **96**, 406–426 (2017)

Understanding Learner Engagement in a Virtual Learning Environment

Fedia Hlioui[✉], Nadia Aloui, and Faiez Gargouri

Multimedia Information System and Advanced Computing Laboratory,
University of Sfax, Sfax, Tunisia
fediahlioui@gmail.com, alouinadia@gmail.com,
faiez.gargouri@isims.usf.tn

Abstract. The past few years has seen the rapid growth of educational data mining approaches for the analysis of data obtained from the virtual learning environments (VLE). However, due to the open and online characteristics of VLEs, vast majority of learners may enroll and drop a course freely, resulting in high dropout rates problem. One of the key elements in reducing dropout rates is the accurate and prompt identification of learners' engagement level and providing individualized assistance. In this respect, this paper proposes a survival modeling technique to study various factors' impact on attrition over the Open University in UK. We aim to perceive the learning from a psychological engagement perspective, which is necessary to gain a better understanding of learner motivation and subsequent knowledge and skill acquisition. In this way, we provide an innovative process that may help the tutor to interfere with weak learner at the appropriate time, such as dialog prompts, or learning resources to enhance the learning efficiency. It can help developers to evaluate the VLE effectively and expand system function for future development trend.

Keywords: Educational Data Mining · Learner behavior ·
Predicting engagement · Virtual learning environment

1 Introduction

It has been a decade since the emergence and development of the Information and Communication Technologies pushed the online education to a new height. The popularity of VLEs is credited to their ability to provide an easy access to the learning objects, a high quality and a low-cost content on a large scale more efficiently than the traditional learning [1]. Despite all the VLEs' benefits, one problem persists, which is the high learners' dropout rates. Its puts the efficacy of the online learning technology into question. The completion rate is less than 7%, which is lower than that of the tradition learning [2]. For instance, Coursera[1] is one of the leading VLE, which attracts over tens of millions of learners [2]. While recent studies like in [3] proved that in general, the completion rate in Coursera does not exceed even 7%–9%.

---

[1] https://www.coursera.org/.

© Springer Nature Switzerland AG 2020
A. Abraham et al. (Eds.): ISDA 2018, AISC 941, pp. 709–719, 2020.
https://doi.org/10.1007/978-3-030-16660-1_69

In literature, many researchers focused on the factors of learners' dropout. Even with a small number of learners, the instructor is faced with the difficulty of interpreting and evaluating the learning and quality of the learners' contributions due to the distant nature and the sheer size of an online learning object [4]. Secondly, learners have different preferences, learning styles, goals and intentions that interact and change overtime, and because of the low cost of entry and exit for VLEs, the decision to leave can easily be triggered [5]. In this context, we proved in our previous work [6] that the learning scenarios' modeling would be a big challenge for courses' designers by considering all these learners' individual differences. Romero et al. [7] highlighted likewise that the unfamiliarity with the computer use is one of the prominent factors of dropout. Other factors like outside life commitments can cause learners' dropout such as professional, health, academic, family and personal reasons. These external factors are practically impossible to intervene upon, and most are virtually impossible to detect purely through the digital traces of learner behavior on the platform [5].

To overcome these issues, we propose a survival modeling technique to study various factors' impact on attrition over a VLE, which enable the instructor to understand the trend of the learner behavior, improve the curriculum and the quality of learning. On one hand, we present how to observe the learners' interactions and the problems linked to observation through traces (like the collection and anonymization process, the factors affecting learner retention, etc.) and, on the other hand, how to obtain relevant indicators allowing decisions to be made before the dropout behavior happens. A solution to this problem is the use of Data Mining approaches for extracting information from a dataset and to transform it into an understandable structure for further use [8] and is known as Educational Data Mining approaches (EDM) when it apply the unique kinds of data that come from an educational setting [7].

In the remainder of the paper, we begin by describing a comprehensive literature review of exiting efforts along with their strength and weakness. Next, the proposed approach is presented together with the materials and methods used. Finally, Sect. 4 concludes with the main findings and future works of this study

2 Literature Review

Decreasing the learner dropout is arguably the biggest challenge for researchers. In current literature, according to the differences in datasets and prediction purposes adopted by different papers, the confusion of learner dropout's definition persists. Some researchers treat a learner last event within a VLE as the dropout date, where event could be submitting an assessment [9], or future response in quiz [10], watching videos [11], completion courses [12], posting in a discussion forum [4], or any other event whatsoever. Others define dropout as not earning a certificate within a course [13], or achieving the course [7, 14–16]. Li et al. [16] pointed that the grade prediction is similar to drop out prediction to some extent. Both dropout and grade prediction need to analyses learners' course performances. In other context, Wang et al. [17] high-lighted that the dropout is not having any activity records in a specific period. Studies like [18] and [19] considered that the learner withdrawals if he does not have any learning activities in ten continuous days. As we can see, there are heterogeneous definitions of the dropout. Nevertheless, all these studies have one objective, which is analyzing the learners' online behavior in an attempt to identify how they engage with the course materials, and how this engagement can help to receive an accomplishment statement. Table 1 summarizes the main traits of the EDM approaches that depict the learner dropouts.

Table 1. Main traits of the educational datamining approaches that depicts learner dropout

EDM approach	Learning objective	VLE	Collection type	Behavioral features
[13]	Predicting certification	Edx	Server	Learning /Interactive /Test /Access behavior
[7]	Predicting final mark	Moodle	Server	Messages; threads; score; centrality; prestige; etc.
[9]	Predicting submission assignments	Moodle	Server	Number of activities/assignment in time segment; etc.
[20]	Prediction engagement in next/final week	Edx, Coursera	Server	Lecture view/download; Quiz; Forum view/thread/post/comment
[18]	Predicting dropout in next ten days	Edx	Server	Totals clicks; number_dropout; average respond time/by category, etc.
[10]	Predicting future responses	Simulated data, Khan academy data, assignment benchmark data	Server	
[15]	Predicting post test scores	High school computer science	Server	Course; Forum; Coaching environment; Topic of posts
[16]	Predicting dropout	KDD cup 2015 dataset	Server	Viewing assignments/video; accessing wiki/forum/etc.
[4]	Predicting engagement	Coursera	Server	Structural; Temporal; Linguistic; Behavioral Features
[11]	Understanding video dropout	Edx	Server	Video dropout rates; interaction peak profiles
[21]	Predicting performance	Unspecified	Server/client	Statistical features; Attention patterns features; etc.

As shown in Table 1, most of these approaches are experimented by using a public platforms (e.g. Coursera, Moodle, Edx) or a freely datasets (e.g. KDD cup dataset, Assignments benchmark dataset). We remark that most of studies does not have data collection tools from the client side. For instance, Kardan and Conati [21] adopted an eye tracking technology for monitoring the learner's eye gaze. Shareghi Najar et al. [22] stated that this process often contain noise, which can be caused by different factors such as participant's head movements, wearing contact lenses and glasses, or inaccurate calibration. Moreover, the learner perceives that the use of their personal data relates to many privacy issues. May et al. [23] proved that it is not always straightforward or simple to promise absolute privacy, confidentiality and anonymity when using open VLE. In fact, we aim to use a dataset, which the privacy levels are clearly identified, and their protection measures allow us to set rules and policies in terms of learner tracking. Another studies like in [10] turned to simulation dataset in the experiment step. Simulated dataset is a partial solution of the issues presented previously. Nevertheless, using a real and concrete data would provide a more robust and reliable validation.

Due to the complexity and the heterogeneity of learners' individual differences, researchers only use data from online discussion forum to correlate with the learner performance in courses [7], or only focus on in-video dropout rates, interaction peak profiles, and learner activity categorization around peaks [11, 14], or only interpreted the learner's assignments submission [9], or only demographic and performance data are predicting learner's future responses [10]. Careful selection of learning attributes and machine learning techniques is critical. Inappropriate application of the rules or the machine learning techniques may cause degraded performance. Each study looked upon the learner dropout's problem differently. Fei et al. [20] treated dropout prediction as a sequence classification problem and proposed temporal models. Prior work [17] and [3] viewed dropout prediction as a binary classification problem to make distinction between dropout and retention learners. In terms of the machine learning models used, some studies adapted one [10, 15] or more techniques [7, 13, 24] to the experimentation step. However, studies like in [3], applied a multi-view semi-supervised learning method to make use of a large number of unlabeled data. Therefore, it is necessary to establish a more specific effect prediction model combining the completely online learning process and the learning analytics process. In this work, we identify behavioral indicators, which model the learner engagement through online courses and analyze how these engagement metrics are related to the learner's performance. We propose a new survival modeling technique to explore the relationship between performance and learning by coding observed leaners from the Open University.

3   Research Methodology

This work presents a survival modeling technique by using the Open University Learning Analytics Dataset[2] (OULAD) and the data mining techniques for understanding the learner engagement in a VLE. The proposed approach contains the same four steps in the general Educational Data Mining process (see. Fig. 1). Firstly, we

---

[2] https://analyse.kmi.open.ac.uk/open_dataset.

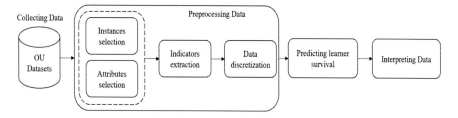

Fig. 1. Our educational data mining architecture for predicting the learner engagement

analyze and explore the collected data, which is gathered between 2013 and 2014. Next, an instance and attribute selection process are applied in order to select relevant records.

In this stage, the log files obtained as an output of the traces' collection step are not always directly exploited, and it is necessary to get through more transformation to extract a meta-knowledge (i.e. the behavioral indicators) of the traces observations. Thus, the dataset will be transformed into a proper format to be mined. Then, Data Mining techniques are applied to get a clear and comprehensible survival model. Finally, the obtained engagement decisions rules are discussed.

3.1   Collecting Data: OU Dataset

The Open University (OU) in UK is one of the largest distant learning universities worldwide. The OULAD [25] contains the log files of the OU learners, which includes

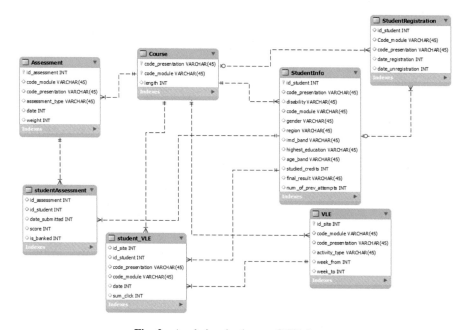

Fig. 2. A relational schema of OU dataset

demographic data, assignments results and interaction data within the VLE. It is freely available as a set of CSV files, in the web under a CC-BY license. We propose the relational schema (see. Fig. 2) to show the dataset in depth.

The OULAD had undergone an anonymization process. It was performed in a series of steps, using the ARX tool [26]. For example, temporal information has been expressed in relative terms with respect to the presentation start and all numeric identifiers (i.e., student_id, code_module, etc.) have been reassigned and completely randomized. The Age and IMD band of the place where the learner lived were generalized. The anonymization process removed 20% of all record entries. It reduced the number of learners from 38,239 to 32,593 learners. The OULAD covers seven modules from different domains, and their 22 presentations. Each presentation contains different materials (wiki, forum, resources, glossary, OUelluminate[3], quiz, shared subpage, urls, etc.) and three assessments (Tutor Marked Assessment, Computer Marked Assessment and Final Exam). The data captures daily the number of clicks that correspond to the learners' interaction with learning material on each day (10,655,280 interactions) [25]. The dropout rate of OULAD converges to more than 52%. The completion rate did not depend only from the assessments' submissions but from the learners' interactions. More than 40% of learners submitted all the assessments, but they did not success the course.

3.2   Pre-processing Data

The pre-processing is a major step for preparing and filtering data before applying the Data Mining techniques. In our case, four main pre-processing tasks have been applied: instance selection, attribute selection, indicators extraction and discretization data.

Instance selection: it is a data reduction task by choosing a subset of relevant instances. By exploiting the dataset in depth, we remark that there are some meaningless instances. For example, there are learners who have passed the course, although they did not submit any assessment. Moreover, learners who has completed the course and submitted all the assessments, but they have a withdrawn status as an achievable result.

Attribute selection: This is an attribute selection task for reducing the data dimensionality by selecting a subset of relevant attributes. In our case, some of them may be irrelevant have no effect on the engagement prediction. For example, we remove the attributes related to the place where the learner live like *region* and the *IMD Band* from the StudentInfo table (See. Fig. 2). The temporal attributes (*date_submitted, date, week_from/to*) will not be considered in this work.

Behavioral indicator extraction: The raw data records in OULAD cannot be directly used as inputs to our survival model. As we mentioned previously that, it is necessary to get through more transformation to extract a meta-knowledge (i.e. indicator) of the observations traces. Throughout our study, we have classified the indicators into two class: demographic and behavioral indicators.

- **Demographic indicators:** they are extracted from the *student_info* table, like *gender, age, highest_education_num_previous_attempts, studied_credit, disability,* etc.
- **Behavioral indicators:** based on the previous research works, we retained four indicators, which cover as much as possible the learners' behaviors in a VLE.
  - *The autonomy indicator:* it expresses the learner's ability to find the information he needs [27]. Thus, the navigation frequency within the VLE is calculated by the number of consulted activities (site) whatever their type (quiz, forum, resources, etc.).

$$Autonomy_L = \sum_{i=1}^{k} nbConsultation\_L(site_i) \Big/ nbConsultations$$

Where:

k: the total sites visited by the learner L whatever the type of the activity;

$nbConsultation\_L(Sitek)$: the number of the consultations of a given site i made by the learner L

$nbConsultations$: the total number of consultation for all sites.

- *The engagement indicator:* It represents thus the ability of the learner to absorb and to use the information that he was provided with. It is calculated by the sum of clicks (interaction) made in each site according to the type of engagement activity.

$$Engagement_L = \sum_{i=1}^{n} \alpha_i \sum_{j=1}^{m} sumclick(site_{i,j})$$

$$\alpha_i = \frac{\sum_{T=1}^{s} sumclick(site_{i,T})}{Total\ of\ learners\ intractions}$$

Where:

n: total number of engagement activities (i.e. type of activities (e.g. quiz, forum, resource, wiki, ouelluminate, etc.));

m: total number of sites per engagement activity;

$sum\_click(sitei, j)$: sum of clicks made by the learner L (attribute *sum_clicks* from table *Student_VLE*) on the site j, which is belonging to an engagement activity i; α_i: weight of the engagement activity presents the consultation frequency with respect to the entire module-presentation.

- *The sharability indicator:* Bouzayane et al. [27] defined the shariabaility or collaboration indicator as an attitude that expresses the willingness of a transmitter to share her own knowledge with the other actors in the VLE. It is calculated by the sum of clicks made in each site of collaborative activities (e.i. forum, wiki, Elluminate).

$$Shariability_L = \sum_{i=1}^{v} \sum_{j=1}^{w} sumclick(site_{i,j})$$

Where:

  v: total number of collaborative activities

  w: total number of sites per collaborative activity;

  sum_click(sitei,j): number of time the learner L interacted with the site *j*, which is belonging to a collaborative activity *i;*

- *The performance indicator*: it is calculated by the sum of the assessments' scores achieved by the learner.

$$Performance_L = \sum_{i=1}^{p} weight_i \times score$$

Where:

  p: total number of assessments

  Weighti: the weight of the assessments; Its value varies from the type of the module as well as the presentation.

Discretization Data: Discretization divides the numerical data into categorical classes that are easier to understand for the instructor [28]. Additionally, some Data Mining techniques required nominal values of the attributes in classification task. All numerical values of the behavioral indicators have been discretized except for the module, presentation, assessment, and VLE site identification number. For instance, the learner performance value is divided with three interval (Fail if the value is lower than 0.4; Pass if the value is higher or equal to 0.4 and lower than 0.8; Distinction if the value is higher or equal to 0.8). The discretization process is applied to the entire proposed behavioral indicator and some attributes like the age.

3.3   Predicting Model

The essence of this study is the construction of predicting model in which the samples are categorized into four classes, namely *withdrawn, fail, pass and distinction*. The variables of a single sample are co posed of two parts, i.e., the input attribution (i.e. behavioral indicators) X = (x1, x2, …, xn) and the category attribution Y; the process for constructing a predicting model is the establishment of a mapping function y = f (X) in which the function can be used to determine the category attribution Y of a sample according to the sample's input attribution X. We choose the JRip as a machine learning technique. It belongs to the rules-based algorithms, which implements a propositional rule learner as optimized version of IREP algorithm [29]. We used 10 cross fold-validation that means that each dataset is randomly divided into 10 disjointed subsets of equal size in a stratified way. JRip is executed 10 times and in each repetition, one of the 10 subsets is used as the test set and the other 9 subsets are combined to form the training set [7]. We used two well-known measures, namely Accuracy and F-measure to calculate the classification performance. We used the Weka[4] tool, which

---

[4] https://www.cs.waikato.ac.nz/ ~ ml/weka/.

is open source machine-learning software that provided us with all the data mining algorithms we used in our execution of the JRip algorithm.

3.4 Interpreting the Result

The empirical results shows that the Cohen's kappa and the Accuracy are acquired by JRip algorithm, with values 0.620 and 0.813 respectively. The model discovered by the JRip algorithms lead us to predict the success or failure of new learner, and thus, they are very useful, for example, to detect learner that are likely to fail, so that the instructor can offer them personalized help in an attempt to avoid failure before the course's end. Thus, it is very important that the model should be comprehensible for instructors. Fig. 3 presents a subset of rules generated by JRip algorithm of the OU learners.

```
(highest_education = Lower Than A Level) and (studied_credits >= 130) and
(num_of_prev_attempts <= 0) => final_result=Withdrawn (230.0/10.0)
(studied_credits >= 80) and (highest_education = Lower Than A Level) and (disability = Y)
=> final_result=Withdrawn (352.0/13.0)
```

Fig. 3. Subset of rules generated by JRip algorithm

As we can see, the rule set produced by JRip is an IF-ELSE-THEN structure in which the THEN operator is indicated by the symbol " =>", and the numbers between brackets at the end of each rule show the number of instances associated with this rule (coverage) and the number of instances incorrectly classified (error) according to the rule. The first number describes the number of instances/students considered in each rule, whereas the second number shows the number of misclassified instances. For example, the meaning of this first rule is that if the highest education of the learner is lower than A level, the studied credits is higher than 130 and the number of previous attempts is lower or equal to 0, then the student is predicted to *withdrawn* the course.

4 Conclusions and Future Work

In this paper, we propose a survival modeling technique to study various factors' impact on attrition over an Open University datasets, which enable the instructor to understand the trend of the learner behavior, improve the curriculum and the quality of learning. Our objective is to investigate to observe the learners' interactions, the problems linked to observation through traces, and how to obtain relevant indicators allowing decisions to be made. We used JRip as a machine learning technique for constructing a prediction model to understand the learner behavior and identify the potential dropouts before it happens. The simulation of our experiment show an acceptable result by using this method. As a future work, we point to validate our survival model with others machine learning techniques to get better performance.

Additionally, we aim to integrate more behavioral indicators as much as possible, in order to enhance the performance of the profiling rules and to cover all the learners' individual differences. Moreover, the contention of our work is to consider the learners' emotional features.

References

1. Saadatdoost, R., Sim, A.T.H., Jafarkarimi, H., Mei Hee, J.: Exploring MOOC from education and information systems perspectives: a short literature review. Educ. Rev. **67**(4), 505–518 (2015)
2. Jiang, S., Kotzias, D.: Assessing the use of social media in massive open online courses arXiv preprint arXiv:1608.05668 (2016)
3. Li, W., Gao, M., Li, H., Xiong, Q., Wen, J., Wu, Z.: Dropout prediction in MOOCs using behavior features and multi-view semi-supervised learning. In: International Conference on Neural Networks (2016)
4. Ramesh, A., Goldwasser, D., Huang, B., Hal Daume, I.I.I., Getoor, L.: Modeling learner engagement in MOOCs using probabilistic soft logic. In: NIPS Workshop on Data Driven Education (2013)
5. Halawa, S., Greene, D., Mitchell, J.: Dropout prediction in MOOCs using learner activity features. In: Proceedings of the Second European MOOC Stakeholder Summit (2014)
6. Hlioui, F., Aloui, N., Gargouri, F.: Automatic deduction of learners' profiling rules based on behavioral analysis. In: Conference on Computational Collective Intelligence Technologies and Applications (2017)
7. Romero, C., Lopez, M.-I., Luna, J.-M., Ventura, S.: Predicting students' final performance from participation in on-line discussion forums. Comput. Educ. **68**, 458–472 (2013)
8. Klosgen, W., Zytkow, J.M.: The knowledge discovery process. In: Handbook of Data Mining and Knowledge Discovery (2002)
9. Druagulescu, B., Bucos, M., Vasiu, R.: Predicting assignment submissions in a multi-class classification problem. TEM J. **4**(13), 244 (2015)
10. Piech, C., Bassen, J., Huang, J., Ganguli, S., Sahami, M., Guibas, L.J., Sohl-Dickstein, J.: Deep knowledge tracing. In: Advances in Neural Information Processing Systems (2015)
11. Kim, J., Guo, P.J., Seaton, D.T., Mitros, P., Gajos, K.Z., Miller, R.C.: Understanding in-video dropouts and interaction peaks in online lecture videos. In: Learning Scale Conference (2014)
12. Tan, M., Shao, P.: Prediction of student dropout in e-learning program through the use of machine learning method. Int. J. Emerg. Technol. Learn. **10**(1), 11–17 (2015)
13. Zhao, C., Yang, J., Liang, J., Li, C.: Discover learning behavior patterns to predict certification. In: 11th International Conference on Computer Science and Education (2016)
14. Yang, T.-Y., Brinton, C.G., Joe-Wong, C., Chiang, M.: Behavior-based grade prediction for MOOCs via time series neural networks. IEEE J. Signal Process. **11**(5), 716–728 (2017)
15. Tomkins, S., Ramesh, A., Getoor, L.: Predicting post-test performance from online student behavior: a high school MOOC case study. In: International Conference on Educational Data Mining (2016)
16. Li, X., Xie, L., Wang, H.: Grade prediction in MOOCs. In: IEEE International Conference on Computational Science and Engineering (2016)
17. Wang, W., Yu, H., Miao, C.: Deep model for dropout prediction in MOOCs. In: Proceedings of the 2nd International Conference on Crowd Science and Engineering (2017)

18. Liang, J., Li, C., Zheng, L.: Machine learning application in MOOCs: dropout prediction. In: 11th International Conference on Computer Science and Education (2016)
19. Whitehill, J., Williams, J., Lopez, G., Coleman, C., Reich, J.: Beyond prediction: first steps toward automatic intervention in MOOC student stopout (2015)
20. Fei, M., Yeung, D.-Y.: Temporal models for predicting student dropout in massive open online courses. In: IEEE International Conference on Data Mining Workshop (2015)
21. Kardan, S., Conati, C.: Comparing and combining eye gaze and interface actions for determining user learning with an interactive simulation. In: International Conference on User Modeling, Adaptation, and Personalization (2013)
22. Shareghi Najar, A., Mitrovic, A., Neshatian, K.: Eye tracking and studying examples: how novices and advanced learners study SQL examples. J. Comput. Inf. Technol. 23(12), 171–190 (2015)
23. May, M., Iksal, S., Usener, C.A.: The side effect of learning analytics: an empirical study on e-learning technologies and user privacy. In: International Conference on Computer Supported Education (2016)
24. Ren, Z., Rangwala, H., Johri, A.: Predicting performance on MOOC assessments using multi-regression models arXiv preprint arXiv:1605.02269 (2016)
25. Kuzilek, J., Hlosta, M., Zdrahal, Z.: Open university learning analytics dataset (2016)
26. Prasser, F., Kohlmayer, F., Lautenschlager, R., Kuhn, K.A.: Arx-a comprehensive tool for anonymizing biomedical data. In: AMIA Annual Symposium Proceedings (2014)
27. Bouzayane, S., Saad, I.: A preference ordered classification to leader learners identification in a MOOC. J. Decis. Syst. 26(2), 189–202 (2017)
28. Romero, C., Ventura, S., Garcia, E.: Data mining in course management systems: moodle case study and tutorial. Comput. Educ. 51(1), 368–384 (2008)
29. Cohen, W.W.: Fast effective rule induction. In: Machine Learning Proceedings (1995)

Efficient Personal Identification Intra-modal System by Fusing Left and Right Palms

Raouia Mokni[1][✉] and Monji Kherallah[2]

[1] Faculty of Economics and Management of Sfax, University of Sfax, Sfax, Tunisia
raouia.mokni@gmail.com
[2] Faculty of Sciences of Sfax, University of Sfax, Sfax, Tunisia
monji.kherallah@fss.usf.tn

Abstract. Palmprint biometric modality is one of the most challenging and active areas of research in the computer vision field due to its high accuracy, high user-friendliness, low cost, high level of security and usability. However, an automatic system designation is a challenging task due to the different variability produced in an unconstrained environment. Since the concept of fusion is robust against these variability, we propose in this paper a intra-modal palmprint fusion based on the features fusion extracted from left and right palms using the SIFT and HOG descriptors. This fusion scheme is performed at both the feature and score levels applying the Canonical Correlation Analysis and Geometric means metrics, respectively. Then, a Random Forest algorithm used in order generate the identity of a person. Practical experiments are performed over CASIA-Palmprint and IIT-Delhi datasets, report encouraging performances which are competitive to other well known palmprint identification approaches.

Keywords: Left and Right palms fusion · SIFT · HOG · Multi-instance information fusion · CCA

1 Introduction

The safety and security of persons, information, and properties requires a beefy guarantee in our society particularly, with the spread of terrorism nowadays throughout the world. In this context, biometrics is a potentially powerful technology for the identification of a person using his physiological traits (palmprint, iris, face, etc.) and/or behavioral traits (voice, signature, etc.). Nowadays, palmprint biometric is a physiologic trait of a person's body parts, which represents one of the privileged techniques in forensic applications thanks to its reliability, stability of its features over time, usability, difficulty to forge and user friendly interface, etc compared to other biometric technologies. Palmprint is a biometric technology that require only a low resolution (less than 100 dpi) which is suitable for civil and commercial applications.

© Springer Nature Switzerland AG 2020
A. Abraham et al. (Eds.): ISDA 2018, AISC 941, pp. 720–729, 2020.
https://doi.org/10.1007/978-3-030-16660-1_70

In literature, palmprint identification systems have attracted more attention among research teams. In fact, an overview of the related work methods about palmprint identification applications is presented. We distinguish two existing approaches which are categorized the palmprint representation features extractors, which are: Structural and Global approaches. Regarding the Structural ones, they analyze the structure of the palmprint, such as the principal lines (cf., [7]), the wrinkles (cf., [4]), the ridges and minutiae (cf., [6]) for represented the palmprint pattern. Although this type of approach relies on taking into account the structural peculiarity of the palmprint representations since it is stable over time, Unluckily, those representations alone cannot provide satisfactory information to identify the person efficiently. Whereas the Global ones, they use the whole area of the palmprint as input to their recognition algorithm. Several descriptors have been used to analyze the palmprint texture based mainly on the Gabor filters (GF) (cf., [18]), Blanket or Fractal dimension (cf., [8]), Wavelets (cf., [10]), Histograms of Oriented Gradients (HOG) (cf., [2]), Scale-Invariant Feature Transform (SIFT) (cf., [1]) and Gray Level Co-occurrence Matrix (GLCM) (cf., [12]), etc.

In the biometric system, the palmprint image acquisition may be conducted in two ways. Firstly, it is performed through contact, which means that the users put their hands directly on the same device surface using the guidance pegs to constraint their hands. Nevertheless, this way is not acceptable by the users due to health reasons. Secondly, image acquisition is conducted without contact way (contactless acquisition system) which leads to more comfortable and hygienic (cf., [13,15]). However, the freedom of the palmprint acquisition means to several challenging problems related to the texture pattern, which include, for instance, translation, texture deformation, scale, position, direction, and noise. These variations lead to the intra-class variations and inter-class similarity of the texture skin pattern. Any variability can be a major source of difficulty of person's identification in a real application. To deal with some of these problems and increase the level of security, a Intra-Modal fusion is developed based on the fusion of multi-sources may allow a more precise identification of the identity.

Multi-source biometric fusion can increase the performance of biometric identification system, can improve the accuracy, less susceptibility to spoof attacks and higher robustness. We distinguish five kinds of Multi-source biometric fusion as follows: Multi-Sensors: it uses several sensors to acquire the same biometric modality; Multi-Algorithms: it is the fusion of multiple descriptors or algorithms to extract the features from a single representation of the same modality; Multi-Representations: it focuses on the fusion of the Multi-Representations extracted from the single modality which are treated through several algorithms or descriptors; Multi-Instances: it associates and combines several instances of the same biometric modality. For example, two irises (left and right iris), two palmprints (left and right hand); and Multi-Modals: it focuses on the fusion of multiple biometric modalities.

Moreover, the fusion procedure can happen at different levels: Fusion at data or sensor level (cf., [20]), Fusion at the feature level (cf., [17]), Fusion at the score level (*cf.*, [22]), Fusion at the rank level [3] or Fusion at the decision level [11]. Several researchers claimed that the fusion at the feature level is considered to be more efficient than it is at the other levels because it focuses on the fusion of multiple sources of information to obtain a single feature set that contains more effective discriminant and richer information than the score, rank and the output decision (*cf.*, [9]). Thus, the fusion at feature level requires that the feature vectors be compatible and focus on the use of several fusion type, such as the Serial feature fusion (it uses a simple concatenation of multiple feature sets into a single feature set) and the Canonical Correlation Analysis (CCA) feature fusion (it incorporates the correlation concept between the features and describes the intrinsic relationship between these features using either the summation or the concatenation).

Thus, in this paper, the above discussion indices and encourages us to propose the multi-instance contactless palmprint fusion, fusing the left and the right palms performed at feature level incorporating the correlation concept between the features. The proposed approach adopts SIFT and HOG descriptors in order to analyze the texture features, for efficient palmprint identification. The main objective of these descriptors is their invariance to different transformations, challenging and variations concerning the texture deformation, lighting conditions, the rotation, position, translation, scale of palmprint texture image in unconstrained environments.

The rest of this paper is structured as follows: In Sect. 2, The Intra-Modal approach related to a multi-instance Fusion is presented. Section 3 elaborates the experiments and results of the proposed approach. Lastly, Sect. 4 presents the concluding remarks and future directions.

2   The Intra-modal Approach Related to a Multi-instance Fusion

The proposed Intra-Modal palmprint approach related to a multi-instance Fusion includes the following modules: (1) Hand Palmprint Acquisition (2) Palmprint Preprocessing (3) Feature Extraction (4) Left and Right palms fusion-based Multi-Instance (5) Palmprint Classification and (6) Decision making.

2.1   Palmprint Preprocessing

The prepossessing module aims to extract the Region Of Interest (ROI) of the hand following three principal phases such as: (1) Hand contour detection (2) Coordinate System Stabilization and (3) ROI preprocessing, as illustrated in Fig. 1. More details of these phases are described in our previous work [19].

2.2   Feature Extraction

Scale Invariant Feature Transform Descriptor. SIFT is a feature extractor, introduced by Lowe in 1999 [14]. This descriptor detects local features from

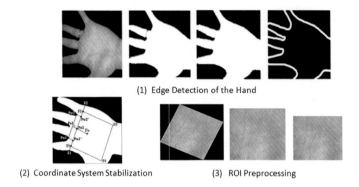

(1) Edge Detection of the Hand

(2) Coordinate System Stabilization   (3) ROI Preprocessing

Fig. 1. The ROI extraction

the images and its extracted keypoints are invariant to different transformations and variations such as translation, rotation, illumination and scale changes. The SIFT descriptor may be detailed briefly as follows:

1. Scale-space extrema detection: The palmprint image $I(x, y)$ is convolved with a set of Gaussian kernels at different scales to construct a continuous function of Gaussian scale space using the following equation:

$$L(x, y, \sigma) = G(x, y, \sigma) \times I(x, y) \tag{1}$$

where $L(x, y, \sigma)$ represents the Gaussian kernel in scale σ), $I(x, y)$ is the segmented image (where x and y are pixel coordinates) and $G(x, y, \sigma)$ represents a variable-scale function defined as:

$$G(x, y, \sigma) = \frac{1}{2\pi\sigma^2} \times e^{-(x^2+y^2)/2\sigma^2} \tag{2}$$

2. Keypoint localization: Detect maxima and minima of difference-of-Gaussian in scale Localization of the keypoints is conducted by using the Gaussian scale transform detecting the maxima and minima of difference-of-Gaussian in scale.
3. Keypoint description: the description of the keypoints is conducted using their gradient magnitude and their orientation. Each keypoint has a location, a scale and an orientation.

In our experiments, the keypoint is described by 16 sub-blocks (4×4 arrays). 8 bin histogram of gradient orientation is calculated for each sub-block, 128 features for one keypoint are obtained. Therefore, this descriptor focuses of construct a vector (V_{SIFT}) of 128 elements representing neighborhood intensity changes of current keypoint.

Histograms of Oriented Gradients Descriptor. HOG descriptor is invented by Dalal and Triggs [5] to extract the local appearance and shape of objects (human body). This descriptor is also considered as one of the efficient descriptors used to extract the relevant texture features of the image in literature. The main objective of these extracted features is their invariance to different object transformations and variations (*e.g.*, texture deformation, lighting conditions, etc). In our proposed system, in order to detail the process of calculating the Histogram of Oriented Gradient, we are initially described as the distribution or histograms of the orientation of gradients which means to the distribution of the local intensity gradients or edge directions of an image being calculated from small connected regions. To this end, practically, the image window is divided into small connected spatial regions named cells, for each cell assembling a local 1-D histogram of gradient directions or edge orientations (1-D gradient filter mask $[-1\ 0\ 1]$) over the pixels of the cell. The combined histogram entries present the representation. The gradients of an image are sensitive to overall lighting. In fact, for better invariance to illumination, shadowing, the local histograms were contrast-normalized by calculating a measure of the local histogram "energy" over a larger region of the image, called blocks, and hence utilizing this value to normalize all cells inside the block. This process aims to normalize the histogram so it is not affected by lighting variations. Note that these blocks typically overlap. This normalized descriptor can be referred as Histogram of Oriented Gradient (HOG) descriptors. Four types of normalization are explored: $L2-norm$, $L2-Hys$, $L1-sqrt$, and $L1-norm$. In our experiment the $L2-norm$ scheme was employed thanks to its more used factor in pattern recognition application. There exist two main blocks: Rectangular $R-HOG$ blocks and Circular $C-HOG$ blocks. In our experiments, the HOG descriptor is calculated over $R-HOG$ blocks that are the square grids. During the calculation of the HOG vector features, it can be noted that each pixel within the image aids in the magnitude of its gradient to a histogram orientation bin which compatible to the orientation of its gradient. The gradient orientations are accumulated into 9 orientation bins. The descriptor of the object of interest would then be represented by the histograms. The feature descriptor is measured by the combination of the histograms from each block. Therefore, the feature vector dimension of the HOG descriptor is computed although the multiplication between the selected numbers of the blocks and the orientation bins. Practically, we have divided the image into 10×10 rectangular blocks and 9 bin histograms per blocks. The 10×10 blocks with nine bins were then concatenated to make a 900-dimensional feature vector: $(V_{HOG} = (10 \times 10 \times 9) = 900)$.

2.3   Left and Right Palms Fusion-Based Multi-instance

The fusion of multi-sources in the biometric system has gained a great interest in several researchers' teams. This fusion can occur in different ways for a recognition system, such as: the data level, feature level, score level, rank level and decision level as already described in Sect. 2.

In our experiments, the Multi-Instance fusion is based on the fusion of both the left and right palms in order to enhance the system performance. This fusion is performed using two fusion manners such as the score level using Geometric mean metric and the feature level using the Canonical Correlation Analysis (CCA) method. This method is accrued to extract the feature pairs describing the intrinsic relationship between multiple feature sets and offer the correlation concept across these independent features and consequently increase the accuracy and decrease the complexity of the proposed identification system. It is a statistical method which uses the correlation criterion to extract canonical correlation features so as to form a single new feature vector, which is more discriminating than any of the original feature vectors [21]. CCA method is based on the examination of the correlation between two feature vectors to obtain the transformed ones that have the maximum correlation between them, however, they are not correlated in every feature vector. CCA technique firstly implements a function of correlation criterion between multiple sets of feature vectors and secondly calculates the projection vector set of these multiple sets depending on the criterion. Finally, the fused canonical correlation features are assumed using either the summation or the concatenation of these projection feature sets. The background and the description of the CCA technique will be detailed in our previous works [16,17]. The CCA technique has some drawbacks such as the Small Sample Size (SSS) issue (dimensionality problem) and the neglecting the class concept issue (lack of the concept of class structure among the samples). In order to surmount this SSS problem and also achieve the integration of the class structure, it is necessary to integrate the supervised dimensionality reduction technique, such as Linear Discriminant Analysis (LDA) based necessary on incorporating the class structure concept and reduce the dimension of such feature sets. In our empirical experiment, we have noticed that the dimension of the obtained HOG feature vector (V_{HOG}) is very large, which implies dimensionality problems. To overcome this problem, we try again to present the performance of our proposed approach applying the LDA feature selection technique in order to pick up the pertinent features, reduce the length of the feature vectors and their redundancies.

2.4 Palmprint Classification

For the classification phase, we used the Multi-class version of the Random Forest (RF) classifier, which was defined as a meta-learner comprised of many individual trees. It was designed to quickly operate over large datasets and more importantly to be diverse by using random samples to build each tree in the forest. Diversity is obtained by randomly choosing attributes at each node of the tree and then using the attribute that provides the highest level of learning. Once trained, Random Forest classifies a new person from an input feature vector by putting it down each of the trees in the forest. Each tree gives a classification decision by voting for this class. Then, the forest chooses the classification having the most votes (over all the trees in the forest).

3   Experiments and Results

3.1 Palmprint Databases

CASIA-Palmprint-Database. This dataset contains 5502 palmprint images captured from 312 subjects. This dataset is taken from Chinese Academy of

Sciences of institution who Automation. For each subject, 8 palmprint images have been collected from both left and right palms. The capture system captures the all palmprint images using a CMOS camera fixed on the top of the device. Since the hand of each person is not touching any platform plate, a presence of distance variation between the hand and the capture device which implies scale variations.

IIT-Delhi-Palmprint Database. The IIT-New Delhi India campus provides a IIT-Delhi palmprint database that includes 2300 images acquired from 230 persons aged from 12–57 years. For each person, five samples have been collected from both left and right palms. These palmprint samples are captured in unconstrained environment with several challenges and variation in terms of position, scale, direction and texture deformation.

In these datasets, the users are not used the guidance pegs to constraint their hands. Therefore, these hand images are captured with different directions and postures, which brings on contracting and stretching the texture skin pattern at the region of interest of the palmprint. These datasets have many variations which make the identification of the persons more difficult.

3.2   Protocol Setting and Empirical Results

All the experiments were performed with MATLAB 2010 and asses on a desktop PC with a 3 GHz Core 2 Duo processor system with 3 GB memory.

So as to evaluate the accuracy of our proposed approach, an empirical experiment is performed over two benchmark databases described above. For the CASIA database, left and right palm images are employed. In fact, we used eight left and eight right palms from 312 users and we indiscriminately selected five samples of each palm as a gallery set and three samples were singled as the probe set for both left and right palms. For the IIT-Delhi database, we used five left and five right palms from 230 users. In fact, we arbitrarily singled three samples of each palm as gallery set and the remainder of samples as the probe set for both left and right palms. In our experiments, Correct Recognition Rate (CRR) is calculated to assess the accuracy of the proposed approach and the Weka Multi-class implementation of Random Forest algorithm is used for the palmprint classification by considering 250 trees.

Table 1. The different results of our proposed Multi-Instance fusion approach over CASIA and IIT-Delhi databases.

Instance	CASIA		IIT-Delhi	
	SIFT	HOG	SIFT	HOG
Left	96.30%	97.81%	95.86%	97.50%
Right	96.35%	97.87%	95.92%	97.65%
Left+Right at feature level (CCA)	98.05%	**99.93%**	98.35%	**99.65%**
Left+Right at score level	97.52%	99.03%	97.42%	98.81%

We revealed the experiment results of the proposed multi-instance fusion at feature level using CCA technique and Score level using Geometric metric. The extracted features of left and right palmprints are performed by both the SIFT and HOG extractors. These experiments are conducted over CASIA and IIT-Delhi palmprint-datasets, as described in Table 1. It can be inferred that the experiments conducted over both datasets applying HOG descriptor surpass the SIFT descriptor using uniformly the left palm, the right palm and the fusion between them (multi-instances). This Table reveals CRRs of each instance alone as well as CRRs of their fusion at feature and score levels. Obviously, the performance of CRR using fusion at feature level outperforms the CRR using score level. Furthermore, the experiment exhibit that the right hand and the left hand gives different performances. This means that both right and left palmprints have different texture patterns. Thus, it is interesting to fuse the two-instance information (Left and right palmprint information) in order to improve the performance of our approach. Consequently, results report that fusion of the two palmprint instances achieves 99.93% and 99.65% of CRR at the feature level (CCA) for CASIA and IIT-Delhi datasets, respectively.

Table 2. Comparison of the results between several Multi-Instance palmprint over contactless datasets. Protocols Description: Gallery Set (G), Probe Set (P).

Proposal	Feature extractors	Dataset-subject	Fusion level	CRRs (%)
Xu et al. [22]	SIFT + Orthogonal line ordinal features	IIT-Delhi-230-6G,4P	Score level	99.57
Leng et al. [13]	Two-dimensional discrete cosine transform (2DDCT)	CASIA-101-10G,6P	Feature level	99.70
Charfi et al. [3]	SIFT + Sparse representation	CASIA-240-6G,4P	Feature level	99.94
The proposed approach	HOG+LDA	CASIA-312-10G,6P	Feature level-CCA	99.93
		IIT-Delhi-230-6G,4P	Feature level-CCA	99.65

Table 2 reveals a comparison between the proposed approach and recent existing approaches [3,13,22] over two contactless datasets such as CASIA and IIT-Delhi. Notice that the presented protocols lie in the sample number during their classification between the gallery and probe data are almost the same over all proposals, whereas a salient difference between the used protocols lies in the subject number. As illustrated in this table, although our chosen subject number used for evaluating is bigger compared to the famous approaches using CASIA dataset, where nearly 101 subjects [13] and 204 subjects [3] are considered, the suggested approach gives significant performances. In fact, the proposed approach is competitive by reaching CRR = 99.93% at feature level fusion using CCA technique, for all subjects in CASIA dataset (312 subjects), compared to the approach of Leng et al. [13], which reaches CRR = 99.70% using only 101 subjects. Furthermore, it can be noted that we have obtained almost the same result to that of Charfi et al. [3]. These authors yielded an improvement of recognition rate of about 0.01% than us, since they considered only 240 subjects and six samples for the gallery set and four samples for the probe set, while we conducted our experiments using all the users on CASIA dataset (312 users) with all samples provided in this dataset (ten samples for the gallery set and six samples for the probe set). The use of a small dataset user number leaves uncertainty about the statistical significance of their performance on large and important challenging dataset. In addition, we compare our obtained experimental result over IIT-Delhi dataset with the works in [22]. Obviously, the performance of our approach demonstrates likewise a better CRR compared to this work, which follows the same protocol as ours. Consequently, this result achieves an improvement rate of about 0.18%.

Although our approach was assessed on large databases with high user number, different challenges and variations, it reached an encouraging and favorably comparable accuracy that proves the efficiency of our proposed approach than the state of the art approaches.

4   Conclusion and Future Directions

In this paper, we investigate a multi-instance contactless palmprint identification approach combining left and right palmprints at feature level using the Canonical correlation analysis method integrating the correlation concept between the features, which leads to the improvement of performance. SIFT and HOG descriptors were adopted in order to analyze the texture features of each instance of palmprint. The experimental results evaluated over "CASIA-Palmprint" and "IIT-Delhi-Palmprint" databases show promising recognition performance of about 99.93% and 99.65%, respectively and prove that the suggested approach is competitive to other multi-instance palmprint approaches. As future work, we plan to integrate another biometric modality, such as the palm veins, using the same developed methods and fuse it with our proposed approach.

References

1. Abeysundera, H.P., Eskil, M.T.: Palmprint verification using sift majority voting. In: Computer and Information Sciences II, pp. 291–297. Springer, London (2011)
2. Arunkumar, M., Valarmathy, S.: Palm print identification using improved histogram of oriented lines. Circ. Syst. 7(08), 1665 (2016)
3. Charfi, N., Trichili, H., Alimi, A.M., Solaiman, B.: Local invariant representation for multi-instance toucheless palmprint identification. In: 2016 IEEE International Conference on Systems, Man, and Cybernetics (SMC), pp. 003522–003527. IEEE (2016)
4. Chen, J., Zhang, C., Rong, G.: Palmprint recognition using crease. In: Proceedings 2001 International Conference on Image Processing, vol. 3, pp. 234–237. IEEE (2001)
5. Dalal, N., Triggs, B.: Histograms of oriented gradients for human detection. In: Proceedings of the 2005 IEEE Computer Society Conference on Computer Vision and Pattern Recognition (CVPR'05) - Volume 1, vol. 01, pp. 886–893. IEEE Computer Society (2005)

6. Duta, N., Jain, A.K., Mardia, K.V.: Matching of palmprints. Pattern Recogn. Lett. **23**(4), 477–485 (2002)
7. Han, C.C., Cheng, H.L., Lin, C.L., Fan, K.C.: Personal authentication using palmprint features. Pattern Recogn. **36**(2), 371–381 (2003)
8. Hong, D., Pan, Z., Wu, X.: Improved differential box counting with multi-scale and multi-direction: a new palmprint recognition method. Optik-Int. J. Light Electron Opt. **125**(15), 4154–4160 (2014)
9. Jagadiswary, D., Saraswady, D.: Biometric authentication using fused multimodal biometric. Procedia Comput. Sci. **85**, 109–116 (2016)
10. Kekre, H.B., Sarode, K., Tirodkar, A.A.: A study of the efficacy of using wavelet transforms for palm print recognition. In: 2012 International Conference on Computing, Communication and Applications (ICCCA), pp. 1–6. IEEE (2012)
11. Kumar, A., Zhang, D.: Personal authentication using multiple palmprint representation. Pattern Recogn. **38**(10), 1695–1704 (2005)
12. Latha, Y.L.M., Prasad, M.V.N.K.: GLCM based texture features for palmprint identification system. In: Computational Intelligence in Data Mining, vol. 1, pp. 155–163. Springer, New Delhi (2015)
13. Leng, L., Li, M., Kim, C., Bi, X.: Dual-source discrimination power analysis for multi-instance contactless palmprint recognition. Multimedia Tools Appl. **76**(1), 333–354 (2017)
14. Lowe, D.G.: Object recognition from local scale-invariant features. In: The Proceedings of the Seventh IEEE International Conference on Computer vision, 1999, vol. 2, pp. 1150–1157. IEEE (1999)
15. Michael, G.K.O., Connie, T., Teoh, A.B.J.: A contactless biometric system using multiple hand features. J. Vis. Commun. Image Representat. **23**(7), 1068–1084 (2012)
16. Mokni, R., Drira, H., Kherallah, M.: Fusing multi-techniques based on LDA-CCA and their application in palmprint identification system. In: 2017 IEEE/ACS 14th International Conference on Computer Systems and Applications (AICCSA), pp. 350–357. IEEE (2017)
17. Mokni, R., Drira, H., Kherallah, M.: Multiset canonical correlation analysis: texture feature level fusion of multiple descriptors for intra-modal palmprint biometric recognition. In: Pacific-Rim Symposium on Image and Video Technology, pp. 3–16. Springer, Cham (2017)
18. Mokni, R., Elleuch, M., Kherallah, M.: Biometric palmprint identification via efficient texture features fusion. In: International Joint Conference on Neural Networks, pp. 4857–4864 (2016)
19. Mokni, R., Zouari, R., Kherallah, M.: Pre-processing and extraction of the ROIs steps for palmprints recognition system. In: International Conference on Intelligent Systems Design and Applications, pp. 380–385 (2015)
20. Raghavendra, R., Busch, C.: Novel image fusion scheme based on dependency measure for robust multispectral palmprint recognition. Pattern Recogn. **47**(6), 2205–2221 (2014)
21. Sun, Q.S., Zeng, S.G., Liu, Y., Heng, P.A., Xia, D.S.: A new method of feature fusion and its application in image recognition. Pattern Recogn. **38**(12), 2437–2448 (2005)
22. Xu, Y., Fei, L., Zhang, D.: Combining left and right palmprint images for more accurate personal identification. IEEE Trans. Image Process. **24**(2), 549–559 (2015)

A Novel Air Gesture Based Wheelchair Control and Home Automation System

Sudhir Rao Rupanagudi[1(✉)], Varsha G. Bhat[1], Rupanagudi Nehitha[1],
G. C. Jeevitha[2], K. Kaushik[2], K. H. Pravallika Reddy[2], M. C. Priya[2],
N. G. Raagashree[2], M. Harshitha[3], Soumya S. Sheelavant[3],
Sourabha S. Darshan[3], G. Vinutha[3], and V. Megha[4]

[1] WorldServe Education, Bengaluru, India
sudhir@worldserve.in
[2] Sir MVIT, Bengaluru, India
[3] YDIT, Bengaluru, India
[4] Bangalore Institute of Technology, Bengaluru, India

Abstract. With almost 1 in 10 people across the world requiring the need of assistance for mobility, the wheelchair finds itself to be the most widely used device for commuting in the world. This has led to a lot of research being carried out in the past few years to ease the way an individual can locomote with the help of the same. From eye gaze to brain waves - a lot of research has been carried out in this regard. In this paper, one such method to control the wheelchair utilizing air gestures has been described. The methodology not only paves the way for easy independent control but also proves to be highly cost effective as well. Further to this, the system is also capable of wirelessly performing various activities such as switching on lights and even opening the front door - all with the help of simple air gestures. The video processing algorithm for identifying the air gestures has been designed and developed using MATLAB 2013a and implemented on a Raspberry Pi 3 Model B. The algorithm was found to be 1.92 times faster than its predecessor algorithm.

Keywords: Air gesture · Android pattern · Arduino UNO · Dilation ·
Erosion · Heaviside · Histogram · Home automation · Image processing ·
Otsu algorithm · Raspberry Pi · Video processing · Wheelchair

1 Introduction

Along with the various changing trends in the world which have prevailed in the 21st Century, there has been an emergence of novel tendencies in the field of science and technology as well. Most of these innovations tend towards improvising human life and easing the way day to day activities are performed. Innovations such as smart trolley systems in supermarkets [1] or cost effective touch screens at restaurants [2] are fine examples of this catching trend. Along with normal humans, inventions which assist the differently abled are also seeing the light of day. Systems such as those which assist the paralyzed to communicate [3] or devices which help the blind to access ATM systems [4] are gaining popularity in the modern world. On similar lines, this paper too

© Springer Nature Switzerland AG 2020
A. Abraham et al. (Eds.): ISDA 2018, AISC 941, pp. 730–739, 2020.
https://doi.org/10.1007/978-3-030-16660-1_71

focuses on a novel application which could assist the wheelchair bound to locomote with ease and also with less dependency on others.

Easing the way differently abled individuals move around in a wheelchair has been a long thought about and intriguing problem on the platter of researchers over the past few decades. The earliest known research in this regard was by [5] in 1985, where the authors described a way to dynamically model an electric wheelchair which can be controlled with the help of a joystick. Since then, a large number of engineering organizations have come up with various designs to assist the immobile in independent wheelchair movement. Though utilizing a joystick eases the task, it proves futile for those having difficulties or disorders for precise movements in their hands, and also for those suffering from arthritis since the repetitive action of pushing the joystick can lead to hand strain or injury. In order to overcome these issues, the authors in [6, 7] came up with a novel method to control wheelchairs with the help of eye movement and blinks. Though advantageous for completely paralyzed patients, this method would result in eye stress and could cause eye pain and tiredness. On similar lines, the authors in [8] suggest the use of face movement and expressions for controlling a wheelchair. The main disadvantage of such a system would be an accidental movement of the chair in case any expressions are made while naturally conversing and hence an additional system is required to confirm the user input. This would potentially cause a slowdown in system performance and also prove unsafe in emergency situations. Similar issues could also be faced in wheelchair control systems which utilize voice control [9] and also those which use brain waves as elaborated in [10]. Hence it can be concluded, that the most safest and viable option would be to use hand control wherever possible and this can be made simpler for the user by using gestures rather than an actual physical device such as the joystick.

Utilizing hand gestures dates back to 1990 where [11] describes the use of a data glove to measure finger flexes. This was later on utilized for identifying American Sign Language. A similar type of device was subsequently used by [12] as well. Wearables such as wristbands and armbands were used by [13] for the same purpose. Though accurate, these methodologies prove to be bulky and also impede free hand movement for the user [14]. In order to overcome these issues researchers in [15] utilized WiFi and ultrasound respectively to identify hand gestures. Though these methods overcome the disadvantage of the previous techniques, they face issues of interference in the form of external noise or by other electrical appliances in the vicinity, and need perfect line of sight at all interval of times with no obstructions. Hence, considering all these disadvantages, it can be concluded that the best methodology to detect hand gestures would be to use image acquisition devices. Researchers in [16] utilized depth sensors as acquiring devices and [17] utilized the same for wheelchair control as well. However, as stated in [18], utilizing such devices could be a costly affair. Therefore in the research conducted in this paper, we have utilized a low cost webcamera for our experiments. Though a similar approach was utilized by [19], care has been taken in this research for a faster processing time and also to incorporate several other features to assist the wheelchair bound.

In the next section, the setup has been described in detail. Section 3 elaborates the algorithm and the results obtained at every stage can be seen in Sect. 4. Conclusion and possible future scope has been explained in Sect. 5.

2   The Setup

As explained in the previous section, the main aim of the research in this paper is to design and develop a high speed and cost effective methodology utilizing air gestures to assist wheelchair bound individuals to locomote. The setup for the same has been elaborated in this section.

Since the user should control the wheelchair utilizing air gestures, the main cause for concern is to intimate the user as to where he or she must show the gestures. In order to simplify this task, we utilized a box created out of thermacole for this purpose and fastened the same to the handle of the chair as shown in Fig. 1.

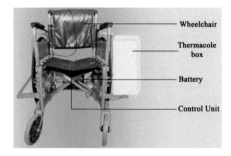

Fig. 1. Image of the actual prototype with the thermacole box attached to the arm of the wheelchair

A camera is fixed at one end of the box while the other end is left open for the user to make gestures. The total dimension of the box depends on the focal length of the camera and in our case was found to be 1 foot in depth. Also, in order to make it easy for the user to access the system, instead of having to show intricate gestures such as sign language, a provision has been made such that the user can perform an action by just placing his finger at a particular position rather than moving the whole hand. In order to achieve this, the complete open end of the box has been divided into several imaginary squares, where each square corresponds to a particular action. As explained in the previous section, since the aim of this research is not only to move the wheelchair but also to perform certain important day to day activities by the user, a few of the imaginary squares have been dedicated for such applications as well. In this project, a few of these applications include switching on or off the lights and fans of the house. Another interesting feature included is the facility to open the front door of user's house as well. Since this is a highly secure task, instead of providing a single imaginary square to open the door, the user is now required to draw an imaginary pattern quite similar to the patterns drawn to unlock a smart phone. The imaginary squares which are used to move the wheelchair can now double up as an area where the user can show the pattern to unlock the door, hence reducing the need of additional space in the open area of the box. Figure 2 shows the demarcations created along with each of the imaginary squares labeled.

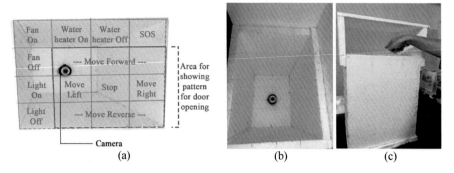

Fan On	Water heater On	Water heater Off	SOS
Fan Off	--- Move Forward ---		
Light On	Move Left	Stop	Move Right
Light Off	--- Move Reverse ---		

Camera

(a)

Area for showing pattern for door opening

(b)   (c)

Fig. 2. (a) Overhead view of the box with each imaginary square labeled with its respective function. (b) White threads crisscrossing the opening of the box and (c) a roof added for extra security

In order to assist the user such that he or she knows the boundaries of each imaginary square, thin white threads are used to earmark the same. This can be seen in Fig. 2b. Since the pattern to unlock the door requires security so that people in the surroundings cannot see the same, the open area of the box is enclosed on three sides along with a roof. This additional provision can be seen in Fig. 2c as well.

The setup along with all components included can be seen in Fig. 3. It can be observed that the camera which acquires the images of the hand gestures is connected to a Raspberry Pi (RPi). This in turn houses the novel video processing algorithm which identifies the hand gestures shown. Based on the hand gesture input from the user, the direction of the wheelchair motion is controlled with the help of a master Arduino UNO microcontroller which is connected to the motors of the wheelchair and receives instructions from the RPi. In the case the user decides to control any home appliance or the door, the same Arduino UNO sends instructions via Bluetooth to a slave Arduino UNO which is connected to these appliances via relays.

The next section describes the video processing algorithm in detail which was first prototyped and designed on MATLAB before being deployed in python on the RPi.

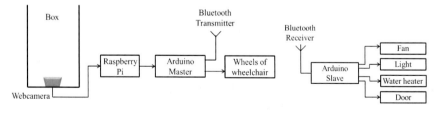

Fig. 3. Complete setup diagram showing both transmitter and receiver

3 Proposed Algorithm

As mentioned in the previous section, the algorithm to identify the hand gesture has been elaborated in this section. Also, as cited earlier, since the application to be performed is decided based on the placement or movement of the finger, there is only a necessity to find the position of the finger rather than the whole hand. The steps involved in the complete algorithm can be seen in Fig. 4 and each step has been explained in detail below.

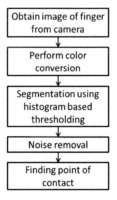

Fig. 4. Flowchart showing steps involved in the proposed algorithm

3.1 Obtaining the Image of the Finger from the Camera

As mentioned in the previous section, a web camera is installed in the box which is used to capture images of the finger at various positions in the open end of the box. The images are captured at a rate of 30 frames per second and are passed on to the next stage of the video processing algorithm in order to extract the same.

3.2 Color Conversion

The image of the hand obtained from the camera in the previous stage is around 800×600 pixels in size. It is a well known fact that every pixel in any color image comprises of red, green and blue pixels which are represented by 8 bits each. In other words a total of 24 bits per pixel are required to represent the image, which in turn would increase the memory required and also affect the speed of execution of the video processing algorithm. Hence in order to reduce this overhead and also to easily distinguish the hand from the background, a process of color conversion is performed.

From various experiments conducted, it was found that the best color model wherein the finger and the roof of the box could be easily differentiated was the chroma-red (Cr) color model. The conversion of the color image to the Cr color model can be represented mathematically by (1) [20].

$$Cr = 0.5 * R - 0.418688 * G - 0.081312 * B \tag{1}$$

Where Cr is the chroma red color component and R, G and B are the Red, Green and Blue pixels of the image respectively.

Upon performing color conversion to the Cr color model, the hand appears much brighter in comparison with the roof. The image thus obtained is then segmented using the procedure elaborated next.

3.3   Segmentation Using Histogram Based Thresholding

Though many segmentation algorithms exist in literature, the main cause for concern in our prototype was to increase the speed of execution. Algorithms such as Otsu's algorithm [19] or background subtraction [21] though accurate, are extremely time consuming and are not a viable solution to be used. Hence for the experiments in this research, histogram based thresholding was adopted. This was mainly possible due to the fact that the hand always appears brighter than its background counterpart and that the background will never change.

Once the image is obtained after color conversion, the histogram of the same is plotted. This can be seen in Fig. 5. As mentioned before, since the finger is the brightest part of the image, the peak encircled and present in the right most part of the histogram belongs to that of the finger. The corresponding intensity range values of the peak are then utilized in a thresholding algorithm which is mathematically represented by (2).

$$Seg_{(x,y)} = \begin{cases} 1 & Cr_{(x,y)} > \theta \\ 0 & otherwise \end{cases} \tag{2}$$

Where Seg is the segmented output, θ is the threshold obtained from the histogram and x and y are the row and column respectively.

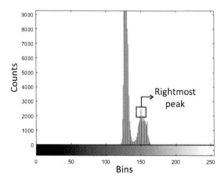

Fig. 5. The histogram of the Cr image with the rightmost peak corresponding to the finger encircled

The outcome of this algorithm is a black and white binary image where the region of interest i.e. the finger is now white while the background is converted to black. However, since a few pixels of the background might have the same intensity as that of the finger due to external interferences, noise removal must be performed and this has been explained in the next section.

3.4   Noise Removal

As mentioned previously, the process of segmentation converts the region of interest into white and the background to black. However, there is a possibility of a few pixels of the background to be accidentally detected as a part of the finger. Hence in order to remove these erroneously identified pixels, a morphological noise removal operation of erosion is performed.

In the process of erosion, a window of n x n pixels is selected which is also known as the structuring element. This window is then made to slide over all the pixels of the image and the center pixel of the part of the image underneath the window is checked to see if it is white or black. In case the pixel is white and any of the surrounding pixels ((n x n)−1) are black, the center pixel in a new image is made black. All other pixels are copied to the new image as it is. This process can be represented by (3).

$$E = B \, \Theta \, S = \cap_{s \in S} B_{-s} \qquad (3)$$

Where E is the eroded image, B is the segmented image and S is the structuring element.

Though erosion is successful in removing stray pixels and pixels which do not belong to the region of interest (ROI), the process also shrinks the edges of the ROI as well. Hence, in order to retrieve these pixels back, a reverse process of dilation is performed. This can be mathematically represented by (4).

$$D = E \oplus S = \cap_{s \in S} E_s \qquad (4)$$

Where D is the dilated image, E is the Eroded image and S is the structuring element.

With the help of the twin morphological operations, in most of the cases all noise is removed and the ROI, in this case the finger, is retained. The next step would be to find the exact position of the finger.

3.5   Finding the Point of Contact

Once the finger is extracted from the background, the next step in the process would be to find the exact point of contact. Usually this procedure is performed by finding the centroid of the ROI. However, since along with the finger, the hand is also extracted in the previous stage, the centroid of the ROI obtained would give the center co-ordinates of the hand rather than that of the finger. Hence in order to avoid this issue, a novel approach is adopted to find the position of the finger and the mathematical equations to calculate the coordinates of the same can be seen in (5) and (6).

$$a = \sum_{x=1}^{r} \sum_{y=1}^{c} (g(x,y) * x * H[-a] + a) \tag{5}$$

$$b = \sum_{x=1}^{r} \sum_{y=1}^{c} (g(x,y) * y * H[-b] + b) \tag{6}$$

$$H[-z] = \begin{cases} 0, \forall z > 0 \\ 1, \forall z \leq 0 \end{cases} \tag{7}$$

Where 'x' and 'y' represent the rows and columns of the picture, 'g' the image obtained after dilation, 'H' the Heaviside equation represented by (7) and 'a' and 'b' are the coordinates of the position of the finger. Once the coordinates are obtained, the corresponding instructions are sent to the Arduino microcontroller and the respective operation is performed.

The results obtained at every stage of the algorithm explained in this section can be seen next.

4 Results

All the steps involved in the video processing algorithm elaborated in the previous section were designed and developed using the MATLAB 2013a software and the results obtained at each step have been shown below.

Figure 6a shows the image obtained from the camera. The image obtained after color conversion can be seen in Fig. 6b wherein it can be clearly seen that the finger appears brighter than the background in the chroma red (Cr) component. The hand image extracted from the background can be seen in Fig. 6c and the images after performing morphological operations of erosion and dilation can be seen in Fig. 6d and e. A close up of the finger and the coordinates obtained can be seen in Fig. 7.

Further to this, in order to perform a speed comparison with an existing algorithm, the speed of execution of the algorithm described in this paper was compared to a similar approach adopted by [19]. The values obtained from both are tabulated in Table 1 and it can be seen that our algorithm is 1.92 times faster than the existing approach.

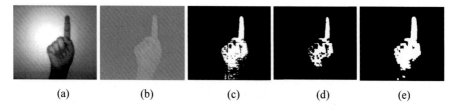

| (a) | (b) | (c) | (d) | (e) |

Fig. 6. Images obtained (a) from the camera (b) after color conversion (c) after segmentation (d) after erosion and (e) dilation

Fig. 7. MATLAB output showing the co-ordinates of the point of contact

Table 1. Comparison of speed of execution for different methods of hand extraction

Method of hand extraction	Speed of execution (s)
Otsu's algorithm [19]	2.27
Proposed method	1.18

5 Conclusion

This paper presents a novel methodology to assist individuals to not only move in a wheelchair but also control various aspects of their day to day life such as switching on lights and also opening doors, utilizing air gestures. The main advantage of the method presented is the ease of use and also cost effectiveness of the whole setup which was found to be less than $150. Also, since the speed efficiency of the algorithm described is 1.92 times faster than popular methods of hand gesture recognition, it provides a far superior user experience and a quicker response time. Overall, the technology discussed in this paper, if put to use on a commercial basis, can improvise wheelchair commute in an inexpensive and efficient manner, in turn changing the life of wheelchair bound individuals radically across the world.

References

1. Rupanagudi, S.R. et al.: A novel video processing based cost effective smart trolley system for supermarkets using FPGA. In: 2015 International Conference on Communication, Information and Computing Technology (ICCICT), pp. 1–6. Mumbai (2015)
2. Ravoor, P., Rupanagudi, S., Bs, R.: Novel algorithm for finger tip blob detection using image processing. In: 2012 4th International Conference on Electronics Computer Technology, ICECT (2012)
3. Rupanagudi, S.R., et al.: An optimized video oculographic approach to assist patients with motor neuron disease to communicate. In: 2017 International Conference on Robotics, Automation and Sciences (ICORAS), pp. 1–5. Melaka (2017)
4. Rupanagudi, S.R., et al.: A high speed algorithm for identifying hand gestures for an ATM input system for the blind. In: 2015 IEEE Bombay Section Symposium (IBSS), pp. 1–6. Mumbai (2015)
5. Johnson, B.W., Aylor, J.H.: Dynamic modeling of an electric wheelchair. IEEE Trans. Ind. Appl. **IA-21**(5), 1284–1293 (1985)

6. Rupanagudi, S.R., Bhat, V.G., Karthik, R., Roopa, P., Manjunath, M., Glenn, E., Shashank, S., Pandith, H., Nitesh, R., Shandilya, A., Ravithej, P.: Design and implementation of a novel eye gaze recognition system based on scleral area for MND patients using video processing. In: 2014 International Conference on Advances in Computing, Communications and Informatics (ICACCI), vol. 320, pp. 569–579, 24–27 September 2014

7. Desai, J.K., Mclauchlan, L.: Controlling a wheelchair by gesture movements and wearable technology. In: 2017 IEEE International Conference on Consumer Electronics (ICCE), pp. 402–403, Las Vegas (2017)

8. Pinheiro, P.G., Pinheiro, C.G., Cardozo, E.: The wheelie—a facial expression controlled wheelchair using 3D technology. In: 2017 26th IEEE International Symposium on Robot and Human Interactive Communication (RO-MAN), pp. 271–276, Lisbon, (2017)

9. Raiyan, Z., Nawaz, M.S., Adnan, A.K.M.A., Imam, M.H.: Design of an arduino based voice-controlled automated wheelchair. In: 2017 IEEE Region 10 Humanitarian Technology Conference (R10-HTC), pp. 267–270, Dhaka (2017)

10. Li, Y., Yang, J.: Intelligent wheelchair based on brainwave. In: 2018 International Conference on Intelligent Transportation, Big Data and Smart City (ICITBS), pp. 93–96, Xiamen (2018)

11. Quam, D.L.: Gesture recognition with a DataGlove. In: IEEE Conference on Aerospace and Electronics, vol. 2, pp. 755–760, Dayton (1990)

12. Ge, Y., Li, B., Yan, W., Zhao, Y.: A real-time gesture prediction system using neural networks and multimodal fusion based on data glove. In: 2018 Tenth International Conference on Advanced Computational Intelligence (ICACI), pp. 625–630. Xiamen (2018)

13. Lee, D., You, W.: Recognition of complex static hand gestures by using the wristband-based contour features. In: IET Image Processing, vol. 12, no. 1, pp. 80–87 (2018)

14. Sang, Y., Shi, L., Liu, Y.: Micro hand gesture recognition system using ultrasonic active sensing. IEEE Access **6**, 49339–49347 (2018)

15. Tian, Z., Wang, J., Yang, X., Zhou, M.: WiCatch: a Wi-Fi based hand gesture recognition system. IEEE Access **6**, 16911–16923 (2018)

16. Liao, B., Li, J., Ju, Z., Ouyang, G.: Hand gesture recognition with generalized hough transform and DC-CNN using realsense. In: 2018 Eighth International Conference on Information Science and Technology (ICIST), pp. 84–90, Cordoba (2018)

17. Yashoda, H.G.M.T., et al.: Design and development of a smart wheelchair with multiple control interfaces. In: 2018 Moratuwa Engineering Research Conference (MERCon), pp. 324–329, Moratuwa (2018)

18. Kakkoth, S.S., Gharge, S.: Survey on real time hand gesture recognition. In: 2017 International Conference on Current Trends in Computer, Electrical, Electronics and Communication (CTCEEC), pp. 948–954, Mysore (2017)

19. Chowdhury, S.S., Hyder, R., Shahanaz, C., Fattah, S.A.: Robust single finger movement detection scheme for real time wheelchair control by physically challenged people. In: 2017 IEEE Region 10 Humanitarian Technology Conference (R10-HTC), pp. 773–777, Dhaka (2017)

20. Rupanagudi, S.R., et al.: A further simplified algorithm for blink recognition using video oculography for communicating. In: 2015 IEEE Bombay Section Symposium (IBSS), pp. 1–6, Mumbai (2015)

21. Mesbahi, S.C., Mahraz, M.A., Riffi, J., Tairi, H.: Hand gesture recognition based on convexity approach and background subtraction. In: 2018 International Conference on Intelligent Systems and Computer Vision (ISCV), pp. 1–5, Fez (2018)

A Comparative Study of the 3D Quality Metrics: Application to Masking Database

Nessrine Elloumi[1,2(✉)], Habiba Loukil Hadj Kacem[1,3], and Med Salim Bouhlel[1]

[1] Research Unit: Sciences of Electronics,
Technologies of Image and Telecommunications,
Higher Institute of Biotechnology, University of Sfax, Sfax, Tunisia
ellouminessrine@gmail.com
[2] Higher Institute of Computer Science and Telecom Hammam Sousse
(ISITCom), University of Sousse, Sousse, Tunisia
[3] Higher Institute of Industrial Management, University of Sfax, Sfax, Tunisia

Abstract. High definition and 3D telemedicine offer a compelling mechanism to achieve a sense of immersion and contribute to an enhanced quality of use. 3D mesh perceptual quality is crucial for many applications. Although there exist some objective metrics for measuring distances between meshes, they do not integrate the characteristics of the human visual system and thus are unable to predict the visual quality.

Keywords: Perceptual quality · Static metrics 3D · 3D meshes · Objective metrics · Quality assessment · 3D triangle mesh · Human visual system · Statistical modeling

1 Introduction

Many applications require a specified level of detail, 3D meshes and 3D optimized models such as in the medical applications that dedicated to surgery. 3D models are widely used including networked 3D games, 3D virtual and immersive worlds and telemedicine 3D information technology to facilitate certain procedures [1]. Telemedicine cover different medical practices, such as viewing or sharing remote data (medical imaging, patient records, etc.) [2, 3]. Telemedicine applies to all areas of medicine, specialized or not [4]. For example Teleconsultation, Tele radiology, and medical hotline (see Fig. 1).

Today, this innovative approach grows and opens new perspectives in the organization of care. It has several advantages: helps develop home care, to improve monitoring of patients and prevent complications, limits the movements (especially for elderly or disabled patients), facilitates access to care in areas of difficult access, and facilitates consultation between general practitioners and specialists.

Furthermore, 3D models are appear as a new media trend, such as 3D screen and 3D devices gaming, these engines open up new perspectives in terms of interaction with the 3D world.

© Springer Nature Switzerland AG 2020
A. Abraham et al. (Eds.): ISDA 2018, AISC 941, pp. 740–748, 2020.
https://doi.org/10.1007/978-3-030-16660-1_72

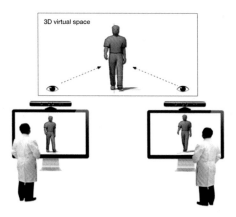

Fig. 1. 3D technology for advanced telemedicine

Many applications required rendering and/or streaming in real time 3D mesh models, these objects are generally composed of a set of vertices, which are, connect to form meshes. This high number of vertices/faces represents the 3D model in a detailed way to increase the visual quality. In addition, 3D applications require a high level-of-detail (LOD) when they process 3D models for optimization and fast rendering. These treatment (Watermarking, simplification or compression) apply on 3D objects requires quality assessments because they cause certain distortion on the 3D models (see Fig. 2).

Fig. 2. Bunny model with two type of distortion (simplification and smoothing)

In the literature, many researchers conducted to extend the 2D objective quality metrics to integrated 3D properties [5, 6]. 3D objects can be viewed from different types of screens.

In addition, these models can be used in different application, which makes the appearance of 3D models dependent properties of the material, texture and lighting used by the application [7, 8].

Moreover, many operations that applied on the 3D model, such as simplification, makes handle changes when they reduce the number of vertices this treatment can delete for example many faces, which represent an important level of details.

In this context, it is important to evaluate the visual quality introduced by the operations performed on the mesh based on the metrics for measuring quality 3D meshes. The current work presents a comparative study between existing approaches for assessing quality assurance. Then their application on a database that contains 26 models distorted with a masking effect.

We introduce our work by describing multiple features approach for assessing quality assurance. The database used for assessment efficiency of measures and experimental results in Sect. 3. Finally, the conclusion represented in Sect. 4.

2   Related Work

2.1   Geometric-Distance-Based Metrics

These metrics based on the calculation of the distance between the two vertices. The simplest estimation of similarity provided by the root mean square Root Mean Square (RMS).

$$RMS(A, B) = \sqrt{\sum_{i=1}^{n} ||ai| - bi||^2} \tag{1}$$

This metric calculates the Euclidean distance between two point A and B, which share the same connectivity ai and bi [9]. Hausdorff Distance (Hd) is the most popular and metric which compare objects with different connectivity. It calculates the similarity between two vertices with the calculation of the distance. The distance between A and B computed as:

$$dist(a, B) = min_{b \in B}(||a - b||)$$
$$D(A, B) = max_{a \in A}(dist(a, B)) \tag{2}$$

If the result is not symmetric, the Hd distance is computed by taking the maximum of the distance between A and B and the distance between B and A. (see Fig. 3).

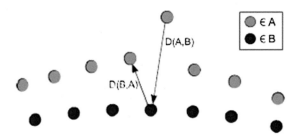

Fig. 3. The Hausdorff distance.

The Hausdorff distance can be defined as:

$$H(A; B) = \max(D(A; B); D(B; A)) \tag{3}$$

These metrics fail to correlate with the human visual system because they compute a geometric distance between two faces [10].

To estimate the perceived quality of 3D objects several metrics was proposed. We can categorize these measurement as roughness-based and structure-based [11]. These metrics integrated different properties of perception to estimate the perceived quality.

2.2 Roughness-Based Metrics

Many metrics developed to measure the quality of 3D shapes by calculate the difference between the original objects and its distorted version based on roughness faces.

These solutions include the properties of the human visual system to calculate the noise related to the roughness or smoothing of the details of a surface.

One of the important perceptual property is the roughness. We can detect the distortion on smooth surfaces easily but it's difficult to determine the distortion on the rough region of the 3D model [12]. This effect related to the masking effect, which can hide some detail (see Fig. 4).

Fig. 4. Roughness map of LION.

Karni propose a new compression approach [13]. To evaluate his roughness-based metrics approach. This metric calculates the geometric laplacian of a vertices v_i.

$$GL(vi) = vi - \frac{\sum_{j \in n(i)} l_{ij}^{-1} vj}{\sum_{j \in n(i)} l_{ij}^{-1}} \tag{4}$$

Where n (i) is the set of neighbors of vertices i, and Lij is the geometric distance between two vertices i and j. The norm of laplician difference between two models M1 and M2 is combined with the norm of the geometric distance between two models where v is the vertices set of M.

$$\left|\left|M^1 - M^2\right|\right| = \frac{1}{2n}\left(\left|\left|v^1 - v^2\right|\right| + \left|\left|GL(v^1) - GL(v^2)\right|\right|\right) \tag{5}$$

This metric has limitation is that the compared models as the RMS approach. Also, Wu and al propose a simplification algorithm to calculate the dihedral angles of the adjacent faces of the 3D models, where they have a greater dihedral angles they find the rough surface [14]. This type of metric have well result with the watermarked models.

Recently, watermarking of 3D objects the attracted attention of researchers owing to the increased diffusion of such objects in several areas of applications, such as in medical, mechanical engineer, design, and entertainment. The 3D Watermarking Perception Metric (3DWPM) is employed to predict the quality of watermarked 3D mesh as perceived by human subjects. Corsini et al. [15] have developed a new quality metric entitled the 3DWPM based on the calculation of the difference in roughness between two 3D meshes. This 3DWPM distance measured between two vertices M1 and M2 are meshes defined by:

$$3DWPM(M_1, M_2) = \log\left(\frac{\rho(M_2) - \rho(M_1)}{\rho(M_1)} + k\right) - \log(k) \tag{6}$$

where, $\rho(M1)$ and $\rho(M2)$ measure the overall roughness of the two meshes, and k is a constant numerical stability. Two variants of 3DWPM were developed using two different roughness descriptors. 3DWPM1 is the first descriptor roughness is inspired by Wu et al. [17]. The roughness value is calculated through the measurement of the dihedral angles between the normal of the facets in a neighborhood. Normal facets on a smooth surface do not strongly vary. However, on textured areas (roughness), normal rough varies more meaningful [18]. A Multi-scale analysis of these entities is considered in [19] to evaluate dihedral angles using the direct vicinity (1 ring) and the extended neighborhood (1 ring, 2 rings, etc.). The second roughness measurement adopted by Corsini et al. for 3DWPM2 is based on estimating the roughness of surfaces [20]. This approach was based on the comparison of a mesh and smoothed version of the same mesh. Smooth regions correspond to small differences, while the rough areas have more significant differences.

The roughness-based metrics correlate very well with the Human visual system when we have a watermarking distortion. Lavoué develop a new approach called local roughness measure. This metric is able to calculate the difference between different parts in a rough or smooth object. His approach based on the curvature analysis of local windows of 3D meshes in depending of the connectivity between faces [21]. This measure provides a local roughness estimation, it can be used to design a future quality metric or hide artifacts.

2.3   Structural Distortion-Based Metrics

The Human visual system is very good at extracting the structural information. Lavoué et al. propose a structural measure (MSDM) [22].

This approach uses the curvature analysis of 3D mesh [23]. It calculates the distortion of 3D mesh by two local windows x and y. Is measured as:

$$LMSDM(x,y) = (\alpha * L(x,y)^a + \beta * C(x,y)^a + \gamma * S(x,y)^a)^{\frac{1}{a}} \tag{7}$$

Where, α, β are defined as 0.4, and 0.2, respectively, L: represent the curvature comparison, C: contrast comparison, and S: structure comparison which are computed as:

$$L(x,y) = \frac{||\mu_x - \mu_y||}{MAX(\mu_x, \mu_y)} \tag{8}$$

$$C(x,y) = \frac{||\sigma_x - \sigma_y||}{MAX(\sigma_x, \sigma_y)} \tag{9}$$

$$S(x,y) = \frac{||\sigma_x\sigma_y - \sigma_x\sigma_y||}{\sigma_x\sigma_y} \tag{10}$$

Where μx is the mean, σx is the standard deviation, and σxy is the covariance of the curvature on local windows x and y.

Then the MSDM is measured as:

$$MSDM(x,y) = (\frac{1}{n_w} \sum_{i=1}^{n_w} LMSDM(x_i, y_i)^a)^{\frac{1}{a}} \in [0, 1] \tag{11}$$

Where X and Y are the compared meshes, xi and y_i are the corresponding local windows of the meshes, and n_w is the number of local windows. a is selected as 3 by the authors, for Eqs. 7 and 11. This metric correlates very well with the Human visual system [24].

3 Experimental Results and Discussion

For evaluating the performance of an objective metric we need to calculate, the correlation between the objective metric and the mean opinion scores (MOS).

There are two coefficients fluently used: the Pearson linear correlation coefficient (rp) that used to measure the accuracy of the prediction (measures the linear dependence between the objective measurement and subjective scores) and the Spearman rank-order correlation (rs) used to measure the prediction monotony as it measures how the relationship between objective and subjective scores can be described by a monotonic function. In order to evaluate the quality assessment of 3D meshes, we compared existing metrics detailed in Sect. 2 using the LIRIS Masking Database. This database contains 26 models. (4 reference meshes: Armadillo, Bimba, Dyno and Lion, and 24 distorted models). The local noise addition is the only type of distortion applied. The specific objective of this database is to test the capability of mesh visual quality metrics in capturing the visual masking effect. Eleven observers participated in the subjective evaluation. This database was created at the University of Lyon, France. Figure 5 shows some models from the LIRIS Masking Database and their distorted versions.

Fig. 5. LIRIS masking database and their distorted versions. (a) Reference meshes. (b) Distorted meshes with noise addition.

Table 1. Correlation coefficients r_p and r_s (%) of different objective metrics on LIRIS masking database.

Metrics	Models							
	Armadillo		Lion		Bimba		Dyno	
	r_s	r_p	r_s	r_p	r_s	r_p	r_s	r_p
HD	48.6	37.7	71.4	25.1	25.7	7.5	48.6	31.1
RMS	65.6	44.6	71.4	23.8	71.4	21.8	71.4	50.3
3DWPM1	58.0	41.8	20.0	9.7	20.0	8.4	66.7	45.3
3DWPM2	48.6	37.9	38.3	22.0	37.1	14.4	71.4	50.1
GL1	65.7	44.4	37.1	22.4	20.0	19.8	71.4	50.0
GL2	65.7	44.2	20.0	21.6	20.0	18.0	60.0	49.8
MSDM	**88.6**	**72.2**	**94.3**	**78.0**	42.9	33.9	**100.0**	**91.7**
MSDM2	81.1	70.6	91.8	100	95.6	100	93.5	90.3

Regarding the LIRIS masking database (Table 1), MSDM produce the highest score for Dyno model (rs = 100.0%, rp = 91.7%) and also for Lion model (rs = 94.3%, rp = 78.0%), in addition MSDM prove a good results for Armadillo model (rs = 88.6%, rp = 72.2%) and fair scores for Bimba model comparing to the other metrics cited in Table 1. Moreover, we can conclude that MSDM is a good detector of visual masking effect as proven by the results.

Generally, the results obtained in Table 1 prove that the metric based on curvature amplitude provides superior results compared to those based on surface roughness. The comparison results obtained from metrics prove that the MSDM metric is closer to the subjective results because it does not take into account the connectivity constraint between two 3D meshes.

Consequently, the experimental comparative study establishes that the results of the MSDM metric are quite effective and provide noble performance to predict the results of which are well correlated with the subjective scores. However, this metric is still limited due to its needs to an implicit correspondence between the tops of two 3D meshes. Thus, proposing an innovative quality assessment metric using perceptually 3D mesh surface attributes can be considered in the future work.

4 Conclusion

The rise of the use of 3D graphic models highlight the importance of assessing the visual quality of 3D objects. In this context, we can evaluate the visual quality of the 3D models by using subjective measurements or an objective evaluation that done through the metrics. The different treatments applied to 3D models also affect the visual quality of 3D models. Given these distortions, it is necessary to develop objective measures that reflect perceptual quality. The measures that integrate the properties of the human visual system that are possibly the masking effect and the contrast sensitivity function gives good results for the measurement of the perceived quality.

References

1. Cooperstock, J.: Multimodal telepresence systems. IEEE Sig. Process. Mag. 28, 77–86 (2011)
2. Triki, N., Kallel, M., Bouhlel, M.S.: Imaging and HMI, foundations and complementarities. In: The 6th International Conferences: Sciences of Electronics, Technologies of Information and Telecommunications 2012, pp. 25–29, Sousse (2012). https://doi.org/10.1109/setit.2012.6481884
3. Aribi, W., Khalfallah, A., Bouhlel, M.S., Elkadri, N.: Evaluation of image fusion techniques in nuclear medicine. In: The 6th International Conferences: Sciences of Electronics, Technologies of Information and Telecommunications 2012, pp. 875–880, Sousse (2012). https://doi.org/10.1109/setit.2012.6482030
4. Daly, L., Brutzman, D.: X3D: extensible 3D graphics standard. IEEE Sig. Process. Mag. 24, 130–135 (2007)
5. Daly, S.: Digital images and human vision. In: Watson, A.B. (ed.) MIT Press, Cambridge (1993). Ch. The visible differences predictor: an algorithm for the assessment of image fidelity, pp. 179–206
6. Masmoudi, A., Bouhlel, M. S., Puech, W.: Image encryption using chaotic standard map and engle continued fractions map. In: The 6th International Conferences: Sciences of Electronics, Technologies of Information and Telecommunications, pp. 474–480, Sousse (2012). https://doi.org/10.1109/setit.2012.6481959

7. Myszkowski, K., Tawara, T., Akamine, H., Seidel, H.-P.: Perception guided global illumination solution for animation rendering. In: Proceedings of the 28th Annual Conference on Computer Graphics and Interactive Techniques, SIGGRAPH 2001, pp. 221–230. ACM, New York (2001)

8. Abdmouleh, M.K., Khalfallah, A., Bouhlel, M.S.: Image encryption with dynamic chaotic look-up table. In: The 6th International Conferences: Sciences of Electronics, Technologies of Information and Telecommunications, pp. 331–337, Sousse (2012). https://doi.org/10.1109/setit.2012.6481937

9. Cignoni, P., Rocchini, C., Scopigno, R.: Metro: measuring error on simplified surfaces. In: Computer Graphics Forum, vol. 17, no. 2, pp. 167–174 (1998)

10. Luebke, D., Watson, B., Cohen, J.D., Reddy, M., Varshney, A.: Level of Detail for 3D Graphics. Elsevier Science Inc., New York (2002)

11. Aspert, N., Santa-cruz, D., Ebrahimi, T.: Mesh: measuring errors between surfaces using the hausdorff distance. In: The International Conference on Multimedia and Expo, ICME 2002, vol. 1, pp. 705–708 (2002)

12. Lavoué, G., Corsini, M.: A comparison of perceptually-based metrics for objective evaluation of geometry processing. IEEE Trans. Multimedia $12(7)$, 636–649 (2010)

13. Karni, Z., Gotsman, C.: Spectral compression of mesh geometry. In: The 27th Annual Conference on Computer Graphics and Interactive Techniques, SIGGRAPH 2000, pp. 279–286. ACM Press/Addison-Wesley Publishing Co., New York (2000)

14. Wu, J.-H., Hu, S.-M., Sun, J.-G., Tai, C.-L.: An effective feature preserving mesh simplification scheme based on face constriction. In: The 9th Pacific Conference on Computer Graphics and Applications, PG 2001, pp. 12–21. IEEE Computer Society (2001)

15. Drelie Gelasca, E., Ebrahimi, T., Corsini, M., Barni, M.: Objective evaluation of the perceptual quality of 3D watermarking. In: The IEEE International Conference on Image Processing, ICIP (2005)

16. Corsini, M., Drelie Gelasca, E., Ebrahimi, T., Barni, M.: Watermarked 3D mesh quality assessment. IEEE Trans. Multimedia $9(2)$, 247–256 (2007)

17. Corsini, M., Larabi, M.C., Lavoué, G., Petřík, O., Váša, L., Wang, K.: Perceptual metrics for static and dynamic triangle meshes. In: Computer Graphics Forum (2013)

18. Torkhani, F., Wang, K., Chassery, J.M.: A curvature-tensor-based perceptual quality metric for 3D triangular meshes. Mach. Graph. Vis. J. (2013)

19. Chowdhuri, S., Roy, P., Goswami, S., Azar, A.T., Dey, N.: Rough set based adhoc network. Int. J. Serv. Sci. Manag. Eng. Technol. (IJSSMET) $3(4)$, 66–76 (2014)

20. Wu, J.H., Hu, S.M., Tai, C.L., Sun, J.G.: An effective feature-preserving mesh simplification scheme based on face constriction. In: Pacific Conference on Computer Graphics and Applications (2001)

21. Abouelaziz, I., Omari, M., Hassouni, M., Cherifi, H.: Reduced reference 3D mesh quality assessment based on statistical models. In: The 11th International Conference on Signal-Image Technology and Internet-Based Systems (SITIS) (2015)

22. Lavoué, G.: A local roughness measure for 3D meshes and its application to visual masking. ACM Trans. Appl. Percept. 5, 1–21:23 (2009)

23. Lavoué, G., Gelasca, E.D., Dupont, F., Baskurt, A., Ebrahimi, T.: Perceptually driven 3D distance metrics with application to watermarking. In: The SPIE Applications of Digital Image Processing XXIX, vol. 6312 (2006)

24. Wang, Z., Bovik, A., Sheikh, H., Simoncelli, E.: Image quality assessment: from error visibility to structural similarity. IEEE Trans. Image Process. $13(4)$, 600–612 (2004)

A Single Ended Fuzzy Based Directional Relaying Scheme for Transmission Line Compensated by Fixed Series Capacitor

Praveen Kumar Mishra[1]($\boxtimes$) and Anamika Yadav[2]

[1] Department of Electrical and Electronics Engineering,
Government Engineering College, Raipur, Chhattisgarh, India
praveenmishraeee@gmail.com
[2] Department of Electrical Engineering, National Institute of Technology,
Raipur, Chhattisgarh, India

Abstract. Protection of the series capacitor compensated transmission line (SCCTL) is a very challenging task due to the abrupt change in apparent impedance seen by the relay, voltage inversion and current inversion phenomena. This paper presents a combined discrete Fourier Transform and fuzzy system based solution to protection issues of the SCCTL. Fuzzy inference system (FIS) have been considered for detection of the existence of fault in a SCCTL and also to recognize its direction whether forward or reverse fault. The presented fuzzy based direction relaying scheme avoid the requirement of a communication link as it uses only single end measurements. The presented scheme has been corroborated for all 10 types of shunt faults that may occur in a transmission line under different fault inception angle (FIA), fault resistances (FR) and fault location (FL). Feasibility of proposed scheme is evaluated in a 735 kV; 60 Hz transmission system with midpoint series capacitor compensation using MATLAB Simulink platform. The presented technique needs current and voltage signals obtainable at one end of transmission line. The presented fuzzy-based relaying algorithm can identify the existence of fault in forward or reverse path in ≤ 8.87 ms time and the test results corroborate its reliability, accuracy and security.

Keywords: Distance relay · Voltage inversion · Current inversion · Fuzzy logic · Series compensation

1 Introduction

Rapid growth in the power demand requires raising power transfer competency of the obtainable transmission line up to thermal boundary. It is practicable by adoption of series capacitor compensation as an alternative and economically worthwhile resolution over adding additional line in parallel [1]. Series capacitor compensation results in reduced transmission line inductive reactance, reduced losses, improved power transfer proficiency as well as it improves the steady-state and transient stability of the network [1]. Nevertheless, the insertion of a series capacitor in the SCCTL requires deviations in standing protection ideas owed to the abrupt alteration in apparent impedance

© Springer Nature Switzerland AG 2020
A. Abraham et al. (Eds.): ISDA 2018, AISC 941, pp. 749–759, 2020.
https://doi.org/10.1007/978-3-030-16660-1_73

comprehended by the relay, voltage inversion, current inversion and also sub-synchronous resonance [2, 3]. SCs and their overvoltage protection devices (typically metal oxide variastor, MOVs and air gaps), when installed on transmission line, create several problem for protective relay and fault locators. As a result of which, direction, distance and fault location algorithms for a series compensated transmission network are affected and leads to malfunction of relays. Fast and reliable fault detection and fault classification technique is an important requirement in power transmission systems to maintain continuous power flow. In recent consequence payable to growing complication of the SCCTL, it is a significant operational obligation to design precise, fast and trustworthy protective relaying scheme [4]. If the transmission line is series compensated then the power system network becomes more complicated.

Recently a comprehensive investigation of different protection techniques conveyed till 2014 has been reported in [1]. During the past two decades, various schemes have been discussed for fault identification and classification problem for SCCTL [5–19]. Some of the recent papers have used the averages of voltage and current [5], Neuro-Fuzzy (ANFIS) [6], support vector machine (SVM) [7–9], fuzzy logic (FL) [10–12], probabilistic neural network (PNN) [13]. In the artificial intelligence based schemes; various signal processing algorithms have been applied such as wavelet transform (WT) [9, 12], DFT [9, 10], S-transform (ST) [13] is applied for detection, identification and classification of faults. In [7–9], SVM have been utilized in conjunction with WT for section identification and fault classification. In [10, 11] fuzzy logic technique is used for detection, classification and location of the fault in uncompensated transmission line. In [12], joint wavelet-fuzzy methodology has been presented for fault classification in series compensated line. In [13], a PNN based technique has been reported for both section identification and fault classification using ST. Combined Wavelet Transform and Extreme Learning Machine (WT-ELM) technique for fault section identification, classification and location in a series compensated transmission line has been presented in [14]. But it requires detecting the fault by some other means and the algorithm does not deal with direction relaying function. An algorithm worked on the concept of equal transfer process of transmission lines (ETPTL) [15] has been proposed for SCCTL to recognize the relative position of the fault w.r.t. the SC and conclude the fault location. Further fault detection in series-compensated using phase angle (PA) of differential impedance of both ends of the line has been presented in [16]. An algorithm for impedance measurement and for fault direction discrimination in series compensated lines has been presented in [17] and adaptive distance relay setting in [18]. The protection schemes discussed above for series compensated lines are non-directional.

In this article, FL based technique is proposed as directional relaying scheme for protection of the series compensated transmission lines. Among all these schemes, FL based scheme is more practicable for online protective relaying schemes beacuse it doesnot requires to solve multifaceted equations to arrive to a trip conclusion and it comprises of the simple configuration of IF-THEN rule base for decision making. Also, it is fast and independent to set the optimal classifier parameters. The presented algorithm is deliberate for the finding of the presense of a fault in a SCCTL and also to recognize its path whether it is forward or reverse fault. Feasibility of presented algorithm is assessed on a 735 kV; 60 Hz transmission network with midpoint SC utilizing MATLAB Simulink platform [19]. In this study, the fuzzy logic input data established on the symmetrical sequence components and fundamental components obtained through data acquisition system is used by considering sampling frequency of 1.2 kHz.

2   Mid-point Series Compensated Transmission System

In this work, a 735 kV, 60 Hz three phase series compensated power system network is considered for the study as shown in Fig. 1. Source-1 at bus-B1 exemplifies a power plant entailing of six 13.8 kV, 350 MVA generators which is associated to step up transformer −1 of 13.8/735 kV, 6 * 350 MVA, 60 Hz. The power is transferred through transmission Line-1 between bus-B3 and B4 and another Line-2 between buses B4–B2 which is series compensated at mid-point, to an equivalent grid signified by Source-2 of 735 kV, 60 Hz. The distance of Line-1 is 100 km and Line-2 is 300 km having distributed parameter line model. The series capacitor is designed to provide 40% reactance compensation to 300 km long transmission line between bus B4 and B2. The positive sequence reactance of line-2 is 105.6 Ω (0.352 Ω/km × 300 km) and fixed capacitor capacity required for 40% compensation is calculated as 62.8 µF. Series capacitor is protected by MOV from overvoltage and in this study, the voltage and energy rating of the MOV are deliberated as 298.7 kV and 30 MJ respectively. Load-1 at Bus B1 is of 100 MW, at bus B2 Load-2 of 132 MW and Load-3 of 332MVAR are connected. Load 4 of 250 MW is connected to load side of Bus B2 through Transformer 2 of 300 MVA, 60 Hz, 735/230 kV. Herein, a directional relay established on fuzzy system is presented, thus the relaying point is taken as bus-B4 where the three phase current and voltage signals are measured and Line-2 is forward line section to be protected and Line-1 is reverse line section.

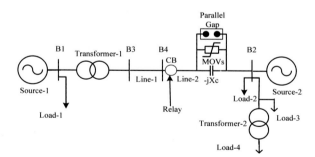

Fig. 1. Considered series compensated test system

3   Proposed Directional Relay Scheme

The flow chart of the presented fuzzy-based directional relay or fuzzy fault direction detector (FDD) is displayed in Fig. 2 where Fault Direction Detector (FDD) is designed for fault detection and direction estimation which defines the existence of fault and discriminates among forward and reverse fault. In Fig. 2, the input variables (crisp) are presented and in this article, the phase angle of positive sequence current (PAPSC) has been used as input to fuzzy system to determine the presence and direction of fault. If the output of FDD is zero, then it is no fault or if it is +1 then it is forward fault and trip signal is issued to the circuit breaker, else if output is −1, then it is a reverse fault.

The basic steps in designing a FIS for protective relaying task involves three steps: fuzzification, forming rule base and defuzzification. As the input variable is crisp in nature, it needs to be first converted into it's corresponding fuzzy variable by fuzzification process before application to the fuzzy inference system. After fuzzification, the fuzzified input is specified to the FIS, which, succeeding the given fuzzy rule base, contributes the direction of fault as response. For executing the fuzzy inference engine, the "min" operator for connecting multiple antecedents in a rule, the "min" implication operator, and the "max" aggregation operator have been utilized. To regulate the crisp output correctly, the fuzzy output necessity to be defuzzified. The centroid defuzzification algorithm has been utilized for this determination in this paper and, subsequently, the presence of fault and its direction is specified. The fuzzy system for fault detection has been designed by utilizing the fuzzy-logic toolbox in the MATLAB platform. FIS utilized here is based on 'Mamdani' type with a triangular shape membership function.

In order to fuzzify the input variable i.e. PAPSC, the PA is divided into three ranges: IpLOW, IpMID, and IpHIGH.

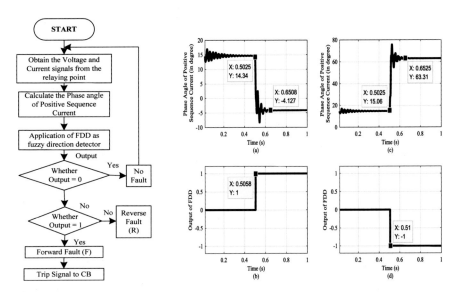

Fig. 2. Flow diagram of the proposed fuzzy-based directional relay.

Fig. 3. (a) Input to FDD during forward fault, (b) Output of FDD during forward fault, (c) Input to FDD during fault and (d) Output of FDD during reverse fault.

The fuzzy response is also divided into three ranges based on the existence of fault in forward direction by trip forward TF (1) or reverse direction by trip reverse TR (−1) or no fault condition by trip not TN (0) using triangular membership function. To determine the presence and direction of fault, we divide the output into three categories as follows:-

FDD output = 0; TN- Trip Not - No Fault
FDD output = −1; TR- Trip Reverse - Reverse Fault
FDD output = 1; TF- Trip Forward - Forward Fault
Fuzzy rule base of FDD is as follows-

1. If PA is IpLOW then (trip is TF)
2. If PA is IpMID then (trip is TN)
3. If PA is IpHIGH then (trip is TR).

4 Results and Discussion

The presented fuzzy based direction relaying technique is tested by numerous fault case studies. For the analysis of the performance of the fuzzy director detector (FDD) we have simulated numerous fault cases by varying fault type, fault location, fault inception angle (FIA) and fault resistance. Both forward side and reverse side fault cases have been deliberated. Test cases have been considered by random distribution of the following parameters:

- 40 and 20 fault locations for forward and reverse faults respectively;
- 8 FIAs (between $0° - 360°$ in step of $45°$);
- 6 fault resistances (0.001 Ω, 10 Ω, 20 Ω, 50 Ω, 100 Ω and 200 Ω) for ground faults and 2 fault resistances (0.001 Ω, 10 Ω,) for phase faults;
- 10 fault types (AG, BG, CG, AB, BC, CA, ABG, BCG, CAG and ABC).

When the system is operating in normal condition, the proposed fuzzy based directional relaying scheme output must be always zero (0−Trip not (TN)) but when a fault occurs in the transmission line it will be either +1−trip forward (TF) or −1−trip reverse (TR) as per occurrence of a fault. To evaluate the effectiveness of the presented algorithm, consider a single phase to ground fault has occurred in "B" phase of the three phase transmission line in forward direction at 50 km from bus B4 at 0.5025 s, FIA = 45° and in another case a reverse fault BG fault has occurred in reverse direction from bus B4 at −50 km, at 0.5025 s, FIA = 45°. The waveform of the input feature to the proposed fuzzy based direction detector i.e. the PAPSC and respective output of FDD in these two fault cases are shown in Fig. 3(a–d). Figure 3(a) presents the PAPSC during forward fault at 50 km at 0.5025 s, FIA = 45°; which decreases after the incpetion of fault and becomes negative (e.g. −4.145°). In this case, the output of the FDD is presented in Fig. 3(b), which is zero before the fault initiation and after 0.5058 s the output rises to +1 depcitng that it is a forward fault. The response time of the proposed scheme is 5.8 ms. Similarly a reverse direction BG fault is simulated at -50 km from the relaying point at 0.5025 s, in this fault condition, the PAPSC component increases after inception of fault and reach to 63.31° as shown in Fig. 3(c). The output of FDD in this reverse fault case as depcited in Fig. 3(d) becomes '−1' after 0.51 s confirming the fault is in reverse direction (R) in 7.5 ms. Detection time for FDD can be calculated as:

$$T_d = T_o - T_i \tag{1}$$

Where T_d, T_o and T_i are detection time, operating instant of relay and fault inception instant respectively. Furthermore above consideration for system conditions, in order to more validate the proposed technique, the fault simulation studies have been carried out under wide variation of fault location, fault resistances, fault inception angles, fault types and compensation level (Xc) and results obtained are discussed in detail hereunder.

4.1   Performance in Case of Voltage Inversion

A voltage inversion will occur for a fault near to the SC when impedance between the source bus to the fault point is inductive but simultaneously impedance from the relay location to the fault is capacitive rather than the inductive as normally encountered in an uncompensated system. Consequently, the voltage at the relay location will be shifted by 90° or more [1]. Conventional diatnce relay overreach, if fault includes capacitor and under reach, if fault occurs just after capacitor [1]. Since the distance relays are deliberate to work properly on an inductive system, the voltage inversion will have an effect on the enactment of the relay. For the mathematical formulation of voltage inversion case consider Xs, Xt, X_{L1} and X_{L2} are inductive reactances of source, transformer and transmission line 1 & 2 respectively. For voltage inversion: Line 1 is 100 km, Line 2 is 300 km, $\%X_c/X_{L2} = 60\%$ and AG fault is created at 120 km from relay bus.

Fault current through relay bus will be:

$$I_{B4} = \frac{Vs}{j(Xs + Xt + X_{L1} + X_{L1} - Xc)} \tag{2}$$

$$I_{B4} = \frac{Vs}{j(X)} \tag{3}$$

From Eq. (3), the relay current I_{B4} lags or leads the voltage by 90° depending upon the sign of the X (where $X = Xs + Xt + X_{L1} + X_{L1} - Xc$).

Voltage at the relay bus can be expressed as

$$V_{B4} = jI_{B4}(X_{L2} - X_C) = \frac{V_S}{X}(X_{L2} - X_C) \tag{4}$$

Voltage at the relay bus as expressed in (4) will be inverted when the following

$$X_C > X_{L2} \text{ and } X_{L1} + X_{L2} + X_S + X_t > X_C \tag{5}$$

So as to, check the enactment of the proposed technique during voltage inversion, a 735 kV, 60 Hz 3-phase series compensated power system network as displayed in Fig. 1 is deliberated. In the network, Line-1 and Line-2 are of 100 km and 300 km length respectively and Line-2 is 60% compensated and the capacitor is located near to the bus B4/relay end. A single phase to ground fault (SLGF) in phase "a" is created at 120 km from the bus B4 with FIA = 0° and R_f = 20 Ω and consequent the three phase voltage waveform in this situation is depicted in Fig. 4(a). From this Fig. 4, it can be visualized that, the magnitude of voltage of faulted phase "a" is reduced due to ag fault at 0.5 s and also the PA is shifted by more than 90° and it is approximately in phase with phase-b. So in this case voltage inversion is taking place. The performance of the proposed FDD is evaluated in voltage inversion case and output result is shown in Fig. 4(b) wherein after 0.51 s the fault has been detected as forward fault. Thus the fault detection time is 10 ms in this case.

4.2 Performance in Case of Current Inversion

The fault current in the SCCTLs network depends on various parameters such as FL, FT, R_f, compensation level, power angle and MOV etc. Fault current in case of faults before series capacitor is similar to fault current of uncompensated lines. But when fault occurs after the series capacitor, the fault current is sensitive to source reactance, faulted line reactance, fault resistance and series capacitor reactance. A current inversion occurs only when current at the relay point leads the source voltage by more or equal to 90° due to large capacitive reactance in the fault loop and concurrently relay point voltage is in phase with the source voltage [1]. The condition for current inversion is given as

$$X_C > X_t + X_{L1} + X_{L2} + X_S \tag{6}$$

To test the proposed scheme for the current inversion case, the same power system network is considered but with few modifications such as: length of Line-1 is 25 km, Line-2 is 300 km, %Xc/X_{L2} = 60% and ag type fault is created at 100 km from the relay bus. Figure 5(a) shows the instantaneous three phase currents (Iabc) wherein the magnitude of current of faulted phase-a increases due to fault and also it is shifted by more than 90°. Hence in this case current inversion has occurred. Figure 5(b) shows the output of fault direction detector FDD during current inversion. Fault is created at 0.5 s and FDD gives active high signal at 0.5158 s. So, fault detection time is 15.8 ms, which is less than one cycle time. This it can be concluded that, the proposed scheme is not affected by current or voltage inversion problems associated with series compensated lines.

4.3   Performance in Case of Varying Fault Location

Effect of fault location have been investigated by applying different types of fault at different locations in the power system network and by keeping FIA = 0° and outcomes of the presented scheme is depicted in Table 1. As shown in Table 1 the proposed method successfully detects the fault direction in less than a ½ cycle time (both forward and reverse faults).

4.4   Influence of Varying Fault Resistance

In order to investigate the performance of the proposed scheme during varying fault resistances; simulation results of various SLGF with varying fault resistances say 0, 20, 40, 60, 80, and 100 Ω have been presented in Table 2. From the results exemplified in Table 2, we can conclude that, with the increase of FR, the detection time of the presented relay also increases. Figure 6 depicts the performance of the proposed scheme during double line to ground fault abg at 205 km from the relaying point at 0.5025 s with FIA = 45° and R_f = 100 Ω. The 3-phase voltage and current signal are shown in Fig. 6(a), (b) and the output of proposed FDD w.r.t time is shown in Fig. 6(c) which detects the fault as forward fault in 13.3 ms. The effectiveness of the proposed algorithm has been equated with conventional mho relay for different fault resistance R_f = 0.001, 50 and 100 Ω at 100 km from relay bus B4. The results on the performance of mho elements after analysis for AG-fault with different value of fault resistance are shown in Fig. 7 which indicates that, when the FR is 0.001 Ω, the impedance trajectory is in zone 1, but when the fault resistance is increased to 50 Ω then it is in zone 2 and while for 100 Ω the trajectory lies outside the three protection zones. Oppositely, the presented algorithm works perfectly with varying FR between 0–100 Ω as exemplified in Table 2. The performance of the recommended technique has been equated with conventional mho relay for dissimilar FR R_f = 0.001, 50 and 100 Ω at 100 km from relay bus B4. The results on the performance of mho elements after analysis for AG-fault with different value of fault resistance are shown in Fig. 7 which indicates that, when the fault resistance is 0.001 Ω, the impedance trajectory is in zone 1, but when the fault resistance

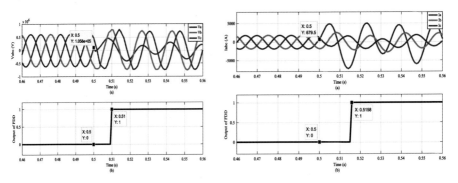

Fig. 4. During voltage inversion (a) 3-phase voltage signal and (b) Response of FDD.

Fig. 5. During current inversion (a) 3-phase current signal and (b) Response of FDD

is increased to 50 Ω then it is in zone 2 and while for 100 Ω the trajectory lies outside the three protection zones. Oppositely, the presented technique works perfectly with varying FR between 0–100 Ω as exemplified in Table 2.

Table 1. Response of proposed scheme for different fault locations

FT	FL (km)	R_f (Ω)	T_d (in ms)	Direction
abg	−95	0.001	6.646	R
ab	**135**	**10**	**5.813**	**F**
Ab	175	10	6.665	F
Ab	−20	10	4.98	R
abg	10	50	5.813	F
abg	95	50	6.646	F
abg	−40	50	4.98	R
abc	120	10	4.98	F
abc	−10	10	4.147	R

Table 2. Response of proposed scheme for different fault resistance

FT	FL (km)	R_f (Ω)	T_d (in ms)	Direction
abg	−55	0.001	2.917	R
abg	−85	60	3.717	R
abg	55	20	4.517	F
abg	25	0.001	3.717	F
abg	95	80	5.417	F
abg	205	40	6.217	F
abg	245	100	13.717	F

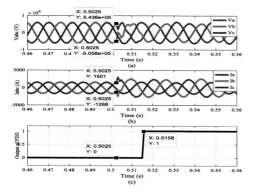

Fig. 6. Effect of fault resistance during double line to ground ABG fault at 245 km, FIA = 45° and Rf = 100 Ω : (a) 3-phase voltage signal, (b) 3-phase current signal and (c) Response of FDD.

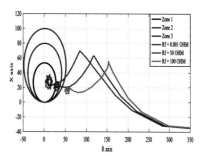

Fig. 7. Impedance trajectory for single line to ground faults with varying fault resistance.

5   Conclusion

In this article, a fuzzy-based direction relaying algorithm for SCCTL has been presented. The fundamental components of the three phase current and voltage signals retrieved at one end of the transmission lines are calculated using discrete Fourier transform. These are then fed to the fuzzy inference system which detects the presence of the fault and its direction. The scheme has been verified on a SCCTL network considering wide variation in FT, fault location, FR, and FIA. The presented method is also robust to different level of compensation and also the problem of voltage and current inversion prevailing in series compensated transmission line has been discussed. It is observed that the presented fuzzy based relaying technique can issue the tripping signal for effective protection measure within ½ cycle time in most of the cases. The outcomes based on extensive work specify that the suggested technique can reliably protect the SCCTL against different fault situations and thus, is a potential contestant for protective relaying.

References

1. Vyas, B., Maheshwari, R.P., Das, B.: Protection of series compensated transmission line: Issues and state of art. Electr. Power Syst. Res. **107**, 93–108 (2014)
2. IEEE Standard C37.116: IEEE Guide for Protective Relay Application to Transmission-Line Series Capacitor Banks (2007)
3. Elmore, W.A.: Line and circuit protection. In: Protective Relaying Theory and Applications, 2nd edn. New York, Marcel Dekker (2003)
4. Phadke, A.G.: Computer Relaying for Power Systems. Wiley, New York (1988)
5. Hashemi, S.M., Hagh, M.T., Seyedi, H.: High-speed relaying scheme for protection of transmission lines in the presence of thyristor-controlled series capacitor. IET Gener. Transm. Distrib. **8**(12), 2083–2091 (2014)
6. Dash, P.K., Pradhan, A.K., Panda, G.: A novel fuzzy neural network based distance relaying scheme. IEEE Trans. Power Deliv. **15**(3), 902–907 (2000)
7. Dash, P.K., Samantaray, S.R., Panda, G.: Fault classification and section identification of an advanced series-compensated transmission line using support vector machine. IEEE Trans. Power Deliv. **22**, 67–73 (2007)
8. Parikh, U.B., Das, B., Maheswari, R.P.: Fault classification technique for series compensated transmission line using support vector machine. Electr. Power Energy Syst. **32**, 629–636 (2010)
9. Parikh, U.B., Das, B., Maheswari, R.P.: Combined wavelet-SVM technique for fault zone detection in a series compensated transmission line. IEEE Trans. Power Deliv. **23**, 1789–1794 (2008)
10. Anamika, Y., Aleena, S.: Enhancing the performance of transmission line directional relaying, fault classification and fault location schemes using fuzzy inference system. IET Gener. Trans. Distrib. **9**(6), 580–591 (2015)
11. Swetapadma, A., Yadav, A.: Fuzzy inference system approach for locating series, shunt, and simultaneous series-shunt faults in double circuit transmission lines. Comput. Intell. Neurosci. **2015**(620360), 12 (2015)
12. Pradhan, A.K., Routray, A., Pati, S., Pradhan, D.K.: Wavelet fuzzy combined approach for fault classification of a series-compensated transmission line. IEEE Trans. Power Deliv. **19** (4), 1612–1618 (2004)
13. Samantaray, S.R., Dash, P.K.: Pattern recognition based digital relaying for advanced series compensated line. Int. J. Electr. Power Energy Syst. **30**, 102–112 (2008)
14. Malathi, V., Marimuthu, N.S., Baskar, S., Ramar, K.: Application of extreme learning machine for series compensated transmission line protection. Eng. Appl. Artif. Intell. **24**, 880–887 (2011)
15. Qi, X., Wen, M., Yin, X., Zhang, Z., Tang, J., Cai, F.: A novel fast distance relay for series compensated transmission lines. Int. J. Electr. Power Energy Syst. **64**, 1–8 (2015)

16. Jena, M.K., Samantaray, S.R.: Intelligent relaying scheme for series-compensated double circuit lines using phase angle of differential impedance. Int. J. Electr. Power Energy Syst. **70**, 17–26 (2015)
17. Saha, M.M., Rosolowski, E., Izykowski, J., Pierz, P.: Evaluation of relaying impedance algorithms for series-compensated line. Electr. Power Syst. Res. **138**, 106–112 (2016)
18. Sivov, O., Abdelsalam, H., Makram, E.: Adaptive setting of distance relay for MOV-protected series compensated line considering wind power. Electr. Power Syst. Res. **137**, 142–154 (2016)
19. MATLAB, Version R2013a: The MathWorks Inc., Natick, MA, USA

Hybrid of Intelligent Minority Oversampling and PSO-Based Intelligent Majority Undersampling for Learning from Imbalanced Datasets

Seba Susan[✉] ⓘ and Amitesh Kumar

Department of Information Technology, Delhi Technological University,
Bawana Road, Delhi 110042, India
seba_406@yahoo.in

Abstract. Learning from imbalanced datasets poses a major research challenge today due to the imbalanced nature of real-world datasets where samples of some entities are few in number, while some other entities have thousands of samples available. A novel hybrid scheme of intelligently oversampling the minority class followed by subsequent intelligent undersampling of the majority class, is proposed in this paper for learning from imbalanced datasets. Different oversampling techniques: SMOTE and the intelligent oversampling versions of Borderline-SMOTE, Adaptive Synthetic Sampling (ADASYN) and MWMOTE, are considered in combination with Sample Subset Optimization (SSO) that is an intelligent majority undersampling technique based on the evolutionary optimization algorithm of Particle Swarm Optimization (PSO). The datasets after balance-correction are applied to the decision tree classifier. Experiments on benchmark datasets from the UCI repository prove the efficiency of our method due to the higher classification accuracies obtained as compared to the baseline methods.

Keywords: Imbalanced learning · Imbalanced datasets · Oversampling · Undersampling

1 Introduction

Learning from imbalanced data is a matter of concern for researchers in the field of data mining and pattern recognition. The disadvantage of applying imbalanced data directly to machine learning is incorrect classification, with the higher accuracies favoring the majority class [1]. The inability of the minority class to contribute in decision making is summarized as the imbalanced learning problem [2]. Examples of real-world datasets in which this problem is prevalent is the LFW face database [28], that gave lower accuracies in experiments in [26], due to the bias towards the majority classes comprising of the faces of popular celebrities. Various classifiers such as neural networks, Bayesian classifiers, support vector machines and reinforcement learning fail to address this issue and easily fall prey to this problem. Several techniques have been proposed for solving the imbalance in real-world datasets. These techniques can be classified into

© Springer Nature Switzerland AG 2020
A. Abraham et al. (Eds.): ISDA 2018, AISC 941, pp. 760–769, 2020.
https://doi.org/10.1007/978-3-030-16660-1_74

sampling methods and the cost sensitive methods [20]. The sampling methods involve the addition or deletion of samples. Cost sensitive methods are defined by cost metrics that induce mis-classification penalty. Some popular cost metrics are Gini index, entropy, DKM etc. that impose cost for misclassifying the data. Cost sensitive neural networks also provide penalty on misclassification [21].

In this paper, we introduce a hybrid of intelligent techniques, for minority over-sampling and majority under-sampling, for learning from imbalanced datasets. The organization of this paper is as follows: Sect. 2 reviews some of the related works and Sect. 3 gets the reader familiarized with the sampling concepts used in this paper. Section 4 introduces the proposed hybrid learning techniques. Section 5 discusses the experimentation and the results, and Sect. 6 draws the final conclusions.

2 Related Work

Since the past few years, many authors have concentrated exclusively on the class-imbalance problem. In 2002, the Synthetic Minority Oversampling (SMOTE) technique [16] was introduced by Chawla *et al*. This oversampling technique is applied on the minority class to generate synthetic samples along line segments joining k-nearest minority neighbors. Improvised versions of SMOTE such as CORE [3] have been introduced since then. Borderline SMOTE [11] and Adaptive Synthetic Sampling [10] are other variants of the SMOTE technique. In borderline SMOTE, only the minority samples close to the boundary of the classifier are oversampled. While in adaptive SMOTE, the gap between minority and majority classes is addressed and filled by the synthetic samples generated from SMOTE.

The One-sided selection (OSS) method [17] is based on data cleansing. Here, from the majority class, a subset of samples (E) is selected which combines with the minority class sample set (S_{min}) to form a set called preliminary set N, where $N = \{E \cup S_{min}\}$. Data cleaning is performed in the next step. The k-Nearest Neighbor (kNN) approach [14] was introduced by Zhang and Mani that is based on under-sampling. Four different under-sampling schemes were introduced that were based on the distances between the nearest majority and minority samples. Cluster-based sampling method [12] addresses the within-class imbalance problem. In cluster-based sampling, all sub classes are inflated equally except major sub-class and finally the minority clusters are oversampled.

Evolutionary algorithms like Particle Swarm Optimization (PSO) and Genetic Algorithm (GA) were used for optimizing the sample subset space using cross-validation procedure in [5]. Genetic Algorithm has been tried before for optimizing the population in [27]. The technique in [5] called Sample Subset Optimization (SSO) proposed by Yang and Yoo is an instance of intelligent majority class under-sampling since it selects the majority samples that when combined with the minority samples gives high classification accuracies. The classifiers used are KNN and J48 decision trees.

SMOTEBoost [15] is another technique for ensemble learning. Here, the samples are selected in such a way that are hard to learn from both classes. Synthetic dataset is created i.e., $E_{syn} = |S| * error(t)$ where, E_{syn} is the synthetic data samples generated for both major and minor classes. $|S|$ are samples selected from both the classes and *error*(t) is the error rate of the current classifier. JOUS-Boost [9] is another ensemble technique. It is based on sampling (either oversampling or undersampling) with jittering. RUSBoost [22] integrates SMOTE with AdaBoost. In the real-world scenario, where data of different classes are overlapping, due to ambiguity in the boundary separating the two classes, the class-imbalance problem is further intensified. In such cases it is desirable to remove data samples which lies near the decision boundary and which are less helpful in classification. Such an approach is observed in the case of Tomek Links [19], where majority samples are selectively removed.

3 Sampling Concepts Used

3.1 SMOTE [16]

SMOTE stands for Synthetic Minority Oversampling Technique. It is an oversampling approach to solving the class-imbalance problem, introduced by S. Chawla in 2002 [16]. A synthetic sample is generated by SMOTE in the following manner. Consider the k-nearest neighbors of a minority sample $x_i \in S_{min}$. One of these k-neighbors $\hat{x}_i$ is randomly selected and the difference between the two vectors is multiplied by a random weight in the range [0,1], and this weighted difference is added to the minority sample x_i to generate the new synthetic sample.

$$x_{new} = x_i + (\hat{x}_i - x_i) * \delta \tag{1}$$

The new instance lies on the line joining the two vectors. The disadvantages of SMOTE include over-generalization and variance.

3.2 SSO-PSO [5]

SSO stands for Sample Subset Optimization for determining the optimal balanced sample subset. The cost function to be minimized is an error term given by

$$\hat{\varepsilon} = \frac{1}{n} \sum_{i=1}^{f} \sum_{j=1}^{n/f} \left| y_j^{(i)} - p(t_j^{(i)} | \Theta^{(i)}, x_j^{(i)}) \right| \tag{2}$$

The various notations in (2) stand for:
F = *number of folds*
N = *number of samples*
$y_j^{(i)}$ = *The given class label of the sample j from the test fold i*
$p(t_j^{(i)} | \Theta^{(i)}, x_j^{(i)})$ = *Prediction of the jth sample from the test fold i*
$x_j^{(i)}$ = *feature vector*
$\Theta^{(i)}$ = *model parameters learned during training*
$t_j^{(i)}$ = *predicted value of the jth sample*

The error in (2) is the f-fold cross-validation error and the subset that minimizes this error is the solution to the optimization problem. Finding the optimum sample subspace containing equal members of majority and minority samples is thus translated as an optimization problem. Evolutionary algorithms are used in [5] to optimize the sample subspace to a global minimum solution for the cost function. A globally optimum solution is targeted by involving the Particle Swarm Optimization as the search heuristic.

Particle Swarm Optimization (PSO) is an evolutionary algorithm proposed by Eberhart and Kennedy [18]. PSO based learning techniques have been observed before in [4, 8, 25] and it is one of the most popular evolutionary optimization techniques due to its ease of convergence. It serves to find the optimum particle position x_{id} through the velocity update equation [18] given by

$$v_{i,d} \leftarrow \omega v_{i,d} + \phi_p r_p \left(p_{i,d} - x_{i,d} \right) + \phi_g r_g \left(g_d - x_{i,d} \right) \tag{3}$$

3.3 Intelligent Oversampling Techniques

Borderline SMOTE [11] is a variation of SMOTE. In Borderline SMOTE, only those samples are considered which are nearer to the decision boundary. The minority samples $x_i \in S_{min}$ selected for over-sampling in Borderline-SMOTE satisfy the constraint:

$$\frac{m}{2} \leq \left| S_{i:m-NN} \cap S_{maj} \right| < m \tag{4}$$

The set of m-nearest neighbors of x_i is denoted by $S_{i:m-NN}$. The intersection of this subset with the majority class should contain a substantial amount of majority class neighbors as evident from (4). The minority samples that satisfy this constraint constitute the borderline samples crucial for decision-making. ADASYN (Adaptive Synthetic Sampling) [10] is an intelligent oversampling technique where the hard to learn minority samples are synthetically generated while oversampling. The number of minority samples that are generated as per Eq. (1), by ADASYN, is determined by the formula:

$$\boldsymbol{number}(x_i) = \boldsymbol{Norm} \left(\frac{\boldsymbol{S}_{i:m-NN} \cap \boldsymbol{S}_{maj}}{K} \right) \times \left| \boldsymbol{S}_{maj} - \boldsymbol{S}_{min} \right| \times \beta, \beta \in [0, 1] \tag{5}$$

where K is the number of nearest neighbors for particular data sample x_i of minority class. MWMOTE [24] executes the same intelligent approach for oversampling of minority samples, however, it generates samples by hierarchical clustering such that the generated samples lie inside a minority class cluster.

4 Proposed Oversampling-Undersampling Hybrid Methods

In this Section, we introduce the proposed hybrid techniques of *SMOTE-SSO, Borderline SMOTE-SSO, ADASYN-SSO* and *MWMOTE-SSO*. The motivation of creating such hybrids can be summarized in the following two points:

a. Minority samples, when oversampled intelligently, present a better comprehensive match during the subsequent undersampling of the majority class
b. Intelligent undersampling of the majority class with respect to the already over-sampled minority class will present a higher accuracy than random oversampling techniques. Intelligent agents such as Particle swarm is used to isolate the optimal subspace using the cross-validation procedure in SSO.

4.1   SMOTE-SSO Hybrid

SMOTE-SSO is a hybrid implementation SSO-PSO technique [5] and SMOTE [16]. It is a combination of oversampling and intelligent evolutionary-based undersampling. The procedure of undersampling is same as that in SSO-PSO [5]. The oversampling technique is applied on minority class samples for each fold and just before SSO-PSO process. The intention of doing this is to balance the class data by identifying other possible data samples that can be useful for classification. We have assumed the number of k-nearest neighbors as nine (i.e. k = 9) and number of synthetic data samples to be generated are either 25% or 85%. The proposed algorithm for SMOTE-SSO is as follows.

Step 1: Divide the dataset in two folds. For each minority class data sample x_i:

For each fold, the fold under consideration will act as training set and the other folds will act as testing set. For each minority class example x_i of training set determine its distance with other class samples N. The distance is calculated as:

$$Distance_{ij} = \sqrt{\left(x_i - x_j\right)_1^2 + \left(x_i - x_j\right)_2^2 + \ldots + \left(x_i - x_j\right)_A^2} \qquad (6)$$

Where, $Distance_{ij}$ is Euclidean distance between x_i and x_j. x_i is selected minority class sample for consideration. x_j is the other sample from minority class of N + 1 data samples. A is the attribute number. Sort the distances and select the k-nearest neighbors samples for the data sample x_i. From the k−nearest neighbors, randomly choose any one neighbor for synthetic data sample generation. Use Eq. (7) to calculate $Difference_{iA}$.

$$Difference_{iA} = \left(x_{jA} - x_{iA}\right) \qquad (7)$$

where, $Difference_{iA}$ is the difference of x_i and x_j on attribute A. x_{iA} and x_{jA} are the values of x_i and x_j respectively on corresponding attribute A.

The new data sample will be calculated as:

$$Synthetic\_Data_{iA} = \left(x_{iA} + gap_{iA} * Difference_{iA}\right) \qquad (8)$$

Where, $Synthetic\_Data_{iA}$ is the attribute of synthetic data sample generated for data sample x_i. gap_{iA} is a random number. $Difference_{iA}$ is calculated from Eq. (7). x_{iA} is the value of x_i corresponding to an attribute A.

Step 2: Training fold data along with synthetic samples are fed to SSO-PSO algorithm for selecting important sample-subsets from the majority class.

4.2   Combining Intelligent Variants of SMOTE with SSO-PSO

Borderline SMOTE-SSO is the hybrid implementation of Borderline SMOTE oversampling [11] and SSO-PSO undersampling [5]. It is thus a combination of intelligent oversampling and intelligent undersampling. The oversampling is applied on the minority class samples for each fold just before the SSO-PSO process. The intention of doing this is to identify minority class samples which are close to boundary line and are preferred for generating the synthetic samples. The proposed hybridization algorithm is as follows:

Step 1: Divide the dataset in two folds. For each minority class data sample x_i:

For each fold, the fold under consideration will act as training set and the other folds will act as testing set. For each minority class example x_i of training set determine its distance with other class samples N of minority class. The distance is determined by Eq. (6). Sort the distance and select k-nearest neighbors for the data sample x_i. Select those samples x_i that satisfy the criteria of Eq. (4). The collection of such sets is called the DANGER set. This DANGER set is fed to SMOTE algorithm for generating the synthetic data samples.

Step 2: Training fold data along with the synthetic samples are fed to SSO-PSO algorithm for selecting important sample-subsets from majority class.

On similar lines, ADASYN-SSO and MWMOTE-SSO are implemented. Both ADASYN and MWMOTE perform intelligent oversampling of selected minority samples, as discussed in Sect. 3. The number of minority samples for the ADASYN and MWMOTE are set as discussed in previous section. This stage is followed by the intelligent undersampling by SSO-PSO.

5   Experimental Results and Discussions

5.1   System Specifications

RAM: 4 GB (3.88 GB usable);   Processor: Intel Core i3-4005U CPU @ 1.70GHz

Programming Language: Java, R;   IDE used: Eclipse Oxygen, R interpreter (32 bit)

Data Modelling Tool: Weka 3.8 data mining tool

5.2   Discussion on Results

The experiments are performed on six datasets of the UCI repository [23] namely-*Diabetes, Hacide, EEG Eye, Banknote* and *Ionosphere* datasets. The details of these datasets including their original state of imbalance is outlined in Table 1.

Table 1. Datasets used for the experimentation

Datasets	Majority samples	Minority samples
Ionosphere	127	48
Banknote	381	305
EEG eye	4128	3362
Diabetes	400	214
Hacide	980	20

The figures showing the balancing action for the proposed hybrid is shown step by step in Fig. 1.

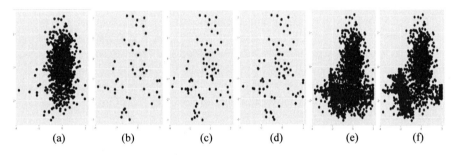

| (a) | (b) | (c) | (d) | (e) | (f) |

Fig. 1. The *Hacide* data distribution before and after balancing (RED: majority sample, BLUE: minority samples) (a) original distribution (b) SMOTE-SSO (25% oversampling) (c) SMOTE-SSO (85% oversampling) (d) borderline SMOTE-SSO (e) ADASYN-SSO (f) MWMOTE-SSO

The classification results after balancing by different techniques is shown in Table 2. As observed, the proposed hybrids give the highest accuracies as compared to all other methods. The following observations were noted from Table 2:

(1) *In case of severely unbalanced datasets like Hacide and Ionosphere, the 85% SMOTE-SSO technique gave the best results*

(2) *In case of less severe and more balanced cases like Diabetes, Banknote and EEG Eye, the 25% SMOTE-SSO technique gave the best results*

(3) *Borderline SMOTE-SSO and MWMOTE-SSO gave consistently high results especially for the severly unbalanced cases*

(4) *All the hybrid methods gave overall higher performance as compared to the baseline SSO-PSO Majority undersampling [5] method and ADASYN and MWMOTE oversampling methods [10, 24].*

Table 3 shows the balance ratio attained between majority and minority classes of SSO and proposed hybrids for *Hacide* dataset. Motivated by the improved classification performance in our experiments, we will be further exploring more variants of hybridization of intelligent oversampling and undersampling techniques as an extension of this work.

Table 2. Classification by the proposed hybrid and baseline methods for datasets of UCI repository (AUC: Area Under the Curve of the Receiver Operating Characteristics curve).

Methods	Ionosphere		Diabetes		Hacide		Banknote		EEG Eye	
	Fold 1 AUC	Fold 2 AUC	Fold 1 AUC	Fold 2 AUC	Fold 1 AUC	Fold 2 AUC	Fold 1 AUC	Fold 2 AUC	Fold 1 AUC	Fold 2 AUC
SSO-PSO [5]	0.8869	0.9065	0.7807	0.8214	0.932	0.9425	0.968	0.9768	0.8096	0.8143
ADASYN [10]	0.86	0.8978	0.782	0.786	0.967	0.976	0.999	0.994	0.7804	0.8277
MWMOTE [24]	0.8821	0.7738	0.7473	0.724	0.7644	0.9345	0.9395	0.9600	0.786	0.823
SMOTE-SSO (25%)	0.8875	0.9226	0.7843	0.8227	0.9425	0.8794	0.9654	0.9717	0.8157	0.8155
SMOTE-SSO (85%)	0.8902	0.9513	0.7706	0.8117	0.9269	0.9855	0.9786	0.9725	0.8061	0.8116
Borderline SMOTE-SSO	0.901	0.913	0.7654	0.7899	0.8977	0.9418	0.9671	0.9755	0.808	0.81
ADASYN-SSO	0.909	0.9325	0.758	0.7862	0.9363	0.9871	0.957	0.969	0.8104	0.8157
MWMOTE-SSO	0.8945	0.9057	0.7689	0.8122	0.9452	0.9955	0.9681	0.9766	0.81	0.81

Table 3. Balance ratio for *Hacide* dataset achieved by different methods

Methods	Majority samples	Minority samples
Original unbalanced dataset	980	20
SSO-PSO [5]	20	20
SMOTE-SSO 25%	24	24
SMOTE-SSO 85%	36	36
Borderline SMOTE-SSO	35	35
ADASYN-SSO	971	971
MWMOTE-SSO	973	973

6 Conclusions

We have proposed hybrids of SMOTE and intelligent oversampling variants with the intelligent undersampling technique of SSO-PSO, in the order of oversampling followed by the undersampling. The proposed techniques have shown relevance through a sign of improvement in ROC performance for various benchmark datasets. We will next shift our focus on improving such hybrid techniques using more variants of intelligent oversampling and undersampling techniques.

References

1. Yan, B., Han, G., Sun, M., Ye, S.: A novel region adaptive SMOTE algorithm for intrusion detection on imbalanced problem. In: 2017 3rd IEEE International Conference on Computer and Communications (ICCC), pp. 1281–1286. IEEE (2017)
2. Zhang, W., Kobeissi, S., Tomko, S., Challis, C.: Adaptive sampling scheme for learning in severely imbalanced large scale data. In: Asian Conference on Machine Learning, pp. 240–247 (2017)
3. Bunkhumpornpat, C., Sinapiromsaran, K.: CORE: core-based synthetic minority over-sampling and borderline majority under-sampling technique. Int. J. Data Mining Bioinf. **12**(1), 44–58 (2015)
4. Huang, C.-L.: A particle-based simplified swarm optimization algorithm for reliability redundancy allocation problems. Reliab. Eng. Syst. Saf. **142**, 221–230 (2015)
5. Yang, P., Yoo, P.D., Fernando, J., Zhou, B.B., Zhang, Z., Zomaya, A.Y.: Sample subset optimization techniques for imbalanced and ensemble learning problems in bioinformatics applications. IEEE Trans. Cybern. **44**(3), 445–455 (2014)
6. Lunardon, N., Menardi, G., Torelli, N.: ROSE: a package for binary imbalanced learning. R J. **6**(1) (2014)
7. Peng, Y., Yao, J.: AdaOUBoost: adaptive over-sampling and under-sampling to boost the concept learning in large scale imbalanced datasets. In: Proceedings of the International Conference on Multimedia Information Retrieval, pp. 111–118. ACM (2010)
8. Del Valle, Y., Venayagamoorthy, G.K., Mohagheghi, S., Hernandez, J.-C., Harley, R.G.: Particle swarm optimization: basic concepts, variants and applications in power systems. IEEE Trans. Evol. Comput. **12**(2), 171–195 (2008)
9. Mease, D., Wyner, A.J., Buja, A.: Boosted classification trees and class probability/quantile estimation. J. Mach. Learn. Res. **8**(Mar), 409–439 (2007)
10. He, H., Bai, Y., Garcia, E.A., Li, S.: ADASYN: adaptive synthetic sampling approach for imbalanced learning. In: IEEE International Joint Conference on Neural Networks, 2008, IJCNN 2008, (IEEE World Congress on Computational Intelligence), pp. 1322–1328. IEEE (2008)
11. Han, H., Wang, W.-Y., Mao, B.-H.: Borderline-SMOTE: a new over-sampling method in imbalanced datasets learning. In: International Conference on Intelligent Computing, pp. 878–887. Springer, Heidelberg (2005)
12. Jo, T., Japkowicz, N.: Class imbalances versus small disjuncts. ACM Sigkdd Explor. Newsl. **6**(1), 40–49 (2004)
13. Haykin, S., Network, N.: A comprehensive foundation. Neural Netw. **2**(2004), 41 (2004)
14. Mani, I., Zhang, I.: kNN approach to unbalanced data distributions: a case study involving information extraction. In: Proceedings of Workshop on Learning from Imbalanced Datasets, vol. 126 (2003)
15. Chawla, N.V., Lazarevic, A., Hall, L.O., Bowyer, K.W.: SMOTEBoost: improving prediction of the minority class in boosting. In: European Conference on Principles of Data Mining and Knowledge Discovery, pp. 107–119. Springer, Heidelberg (2003)
16. Chawla, N.V., Bowyer, K.W., Hall, L.O., Kegelmeyer, W.P.: SMOTE: synthetic minority over-sampling technique. J. Artif. Intell. Res. **16**, 321–357 (2002)
17. Kubat, M., Matwin, S.: Addressing the curse of imbalanced training sets: one-sided selection. In: ICML, vol. 97, pp. 179–186 (1997)
18. Eberhart, R., Kennedy, J.: A new optimizer using particle swarm theory. In: 1995 Proceedings of the Sixth International Symposium on Micro Machine and Human Science, MHS 1995, pp. 39–43. IEEE (1995)

19. Tomek, I.: Two modifications of CNN. IEEE Trans. Syst. Man Cybern. **6**, 769–772 (1976)
20. Chawla, N.V., Japkowicz, N., Kotcz, A.: Special issue on learning from imbalanced datasets. ACM Sigkdd Explor. Newsl. **6**(1), 1–6 (2004)
21. Zhou, Z.-H., Liu, X.-Y.: Training cost-sensitive neural networks with methods addressing the class imbalance problem. IEEE Trans. Knowl. Data Eng. **18**(1), 63–77 (2006)
22. Seiffert, C., Khoshgoftaar, T.M., Van Hulse, J., Napolitano, A.: RUSBoost: improving classification performance when training data is skewed. In: 19th International Conference on Pattern Recognition, 2008, ICPR 2008. pp. 1–4. IEEE (2008)
23. Asuncion, A., Newman, D.: UCI machine learning repository (2007)
24. Barua, S., Islam, M.M., Yao, X., Murase, K.: MWMOTE–majority weighted minority oversampling technique for imbalanced dataset learning. IEEE Trans. Knowl. Data Eng. **26** (2), 405–425 (2014)
25. Susan, S., Ranjan, R., Taluja, U., Rai, S., Agarwal, P.: Neural net optimization by weight-entropy monitoring. In: Computational Intelligence: Theories, Applications and Future Directions, vol. II, pp. 201–213. Springer, Singapore (2019)
26. Susan, S., Jain, A., Sharma, A., Verma, S., Jain, S.: Fuzzy match index for scale-invariant feature transform (SIFT) features with application to face recognition with weak supervision. IET Image Proc. **9**(11), 951–958 (2015)
27. Susan, S., Sharawat, P., Singh, S., Meena, R., Verma, A., Kumar, M.: Fuzzy C-means with non-extensive entropy regularization. In: IEEE International Conference on Signal Processing, Informatics, Communication and Energy Systems (SPICES), 2015, pp. 1–5. IEEE (2015)
28. Huang, G.B., Mattar, M., Berg, T., Learned-Miller, E.: Labeled faces in the wild: a database for studying face recognition in unconstrained environments. In: Workshop on Faces in Real-Life Images: Detection, Alignment, and Recognition (2008)

Data Mining with Association Rules for Scheduling Open Elective Courses Using Optimization Algorithms

Seba Susan[✉][iD] and Aparna Bhutani

Department of Information Technology, Delhi Technological University,
Bawana Road, Delhi 110042, India
seba_406@yahoo.in

Abstract. A new course scheduling based on mining for students' preferences for Open Elective courses is proposed in this paper that makes use of optimization algorithms for automated timetable generation and optimization. The Open Elective courses currently running in an actual university system is used for the experiments. Hard and soft constraints are designed based on the timing and classroom constraints and minimization of clashes between teacher schedules. Two different optimization techniques of Genetic Algorithm (GA) and Simulated Annealing (SA) are utilized for our purpose. The generated timetables are analyzed with respect to the timing efficiency and cost function optimization. The results highlight the efficacy of our approach and the generated course schedules are found at par with the manually compiled timetable running in the university.

Keywords: Timetable scheduling · Genetic Algorithm · Simulated Annealing · Data mining · Students preferences

1 Introduction

Timetable scheduling is a time-consuming task in universities and schools and is also challenging due to the variety of optional electives courses offered in the present educational scenario. Automated timetable generation and optimization has been of great interest to the research community due to the variety of constraints the problem imposes under the limited resources available [1]. Resources refer to classrooms available and the number of teachers. Most of the research works for automated timetable generation have concentrated on scheduling fixed courses with no provision whatsoever to student's preferences for optional elective courses [2–4]. Optimization tools such as Genetic Algorithm (GA) in [2, 5] and Simulated Annealing (SA) in [4, 6, 7] have been utilized for scheduling courses that are fixed in the perspective of the student community. SA particularly, is a very popular means of optimizing schedules due to its fast execution and the optimized schedules it generates [8]. A pioneering work for automated scheduling of elective courses has been observed in [9] that initially needs a timetable template satisfying the constraints and then uses GA for optimizing this timetable. Association rules are used with students' preferences for the

© Springer Nature Switzerland AG 2020
A. Abraham et al. (Eds.): ISDA 2018, AISC 941, pp. 770–778, 2020.
https://doi.org/10.1007/978-3-030-16660-1_75

courses, along with teachers' preferences and habits obtained from historical data [17]. Another important and recent work is [15] that uses graphical models and distance-based local search method for minimizing student conflicts in multiple courses. In this paper, we present a fully automated timetable scheduling using both Genetic Algorithm (GA) and Simulated Annealing (SA) optimization algorithms, that mines for association patterns among students' preferences in the Elective courses. The data is mined from an existing University database for the Elective courses it is currently running. The resources available and courses imposed are also as per the University norms. We compare the timetable generated through our approach using different optimization algorithms, with the manually compiled timetable currently running in the university. The paper is organized as follows. The optimization algorithms of GA and SA are reviewed in Sect. 2. The proposed method and its steps are outlined in Sect. 3. The experimental setup and the results are discussed in Sect. 4. The overall conclusions are drawn in Sect. 5.

2 Optimization Algorithms

In this section, we review the optimization algorithms Genetic Algorithm (GA) and Simulated Annealing (SA) that we use for timetable optimization.

2.1 Optimization by SA

Simulated Annealing [14] comes from the concept of "annealing" in metallurgy. In annealing, the metal is heated to a certain temperature and then cooled under controlled conditions to produce a new metal. This new metal has low energy so it is less prone to defects and is much stronger than the previous metal.

Simulated Annealing is useful for optimization problems. It helps us achieve an optimum value for a given problem. Simulated Annealing uses temperature parameter to control the search. Temperature gives us the degree of randomness of the solution. Lower temperature behaves like hill climbing i.e. if it sees a better move, it'll make it. However, very high temperature behaves like random walk i.e. it will evaluate value of new and current node to decide to make the move or not. If we want to explore more then we keep the temperature high. We usually start with high temperature. At each iteration we have a current node and we select a random neighboring node. We evaluate ΔE which is the difference between energy levels of new node and current node.

$$\Delta E = E(new) - E(current) \tag{1}$$

where $E(new)$ = Energy level for new node

$\quad\quad E(current)$ = Energy level for current node

if $\Delta E > 0$ then we make the move. If $\Delta E < 0$ we still allow that move but with a lower probability. The temperature is slowly lowered/cooled at each iteration. Hence, the algorithm initially, at high temperatures, explores more which prevents it to be stuck at local optima.

2.2   Optimization by GA

Genetic Algorithm is an evolutionary algorithm and works on the theory of natural selection i.e. survival of the fittest. It is used in optimization problems [2, 5, 9, 12, 16] where we try to find optimum values to system parameters with the given values of input. Genetic algorithm is a randomized algorithm i.e. it starts with random number of solutions and tries to reach towards the optimum solution for the given problem.

Steps of Genetic Algorithms include:

1. *Initialization:*
   Initialization includes defining basic elements in a genetic algorithm such as chromosomes, genes and population.
   Chromosome is one individual or one solution for the problem which is being solved. It is usually represented in the form of a binary string. Chromosomes are usually randomly generated. Gene is defined as a single element of a chromosome. If the chromosome is a binary string then gene represents each bit in the string. Population is defined as the collection of individuals or chromosomes.
   Thus in initialization, we define a pool of possible chromosomes/solutions for the given problem.
2. *Fitness function:*
   This is the function which we are trying to optimize. It is also known as the objective function. It gives us a quantitative measure of how fit a chromosome is. Better the fitness value, better is the chromosome. Fitness function also help us direct the simulation towards optimal chromosomes.
3. *Selection:*
   In selection operation, chromosomes with high fitness value are chosen over chromosomes with low fitness value so as to direct the simulation towards obtaining better solution. In selection, the fitness value of each chromosome is calculated. The chromosomes are then arranged in a descending order of fitness values and the most fit individuals are selected.
4. *Crossover:*
   Crossover is a genetic operation which involves creation of a new chromosome by randomly pairing up existing chromosomes and mixing up there genes. The new chromosome is said to have a higher fitness value than that of parent chromosomes. Crossover is like a convergence operation. It directs the population towards the local optima.
5. *Mutation:*
   Mutation is one operation in genetic algorithm which introduces diversity from one generation to the other. In mutation, the new chromosome generated is completely different from the previous chromosome. Mutation prevents the genetic algorithm to be stuck at a local optima by preventing the chromosomes of a population to be too similar to each other. Mutation is a rare event. It is like a divergence operation and usually affects only a few chromosomes in a population for a generation.

3 Proposed Methodology for Timetable Scheduling

In this section, we begin by first presenting the general assumptions and the hard and soft constraints that define our scheduling problem. Hard constraints are ones that cannot be violated. Soft constraints are the ones when violated, induce a penalty cost. Association rules [10, 13] are mined from students' preferences list as shown in Fig. 1. We follow the university system and Core and Elective courses at the university's online database at [11] for the third year undergraduate students batch of Spring 2018.

3.1 Step Wise Procedure for Timetable Generation Using GA Starting from Assumptions, Initial Parameters

Assumptions:

- Number of electives = 2
- Number of professors = 3 (one for Elective 1, two for Elective 2)
- Number of classrooms = 3
- Elective 1-
  – Theory (3 h)
  – Tutorial (1 h for each group)
- Elective 2-
  – Theory (3 h)
  – Tutorial (1 h for each group)

$$\text{Fitness function} = 1/(1 + n_c) \tag{2}$$

where n_c = number of clashes

Initial Parameters for GA:

- Mutation Rate = 0.01
- Population size = 100
- Crossover Rate = 0.9

Hard Constraints:

- Courses are arranged to be taught between certain time periods (08:00 am–12:00 pm).
- There should be exactly two electives.
- All students of the department enroll for these two electives.

Soft Constraints

- Classroom capacity should be greater than the size of the class.
- No room should be assigned to a course or teacher if its already assigned to another course or teacher for that particular time period.
- Every course should have a professor to teach it.
- Each student is assigned to only one classroom at a certain time period.

SUBJECTS	SUPCOUNT	SUPRANK
ML	216	1
CFCC	184	2
ADBMS	48	3
RTS	41	4
ON	12	5
HSN	11	6
MSD	8	7

ASSOCIATION_ID	ASSOCIATIONS	ASSOCIATION_COUNT	RANK
1	ML, CFCC	168	1
2	ML, ADBMS	22	2
3	ML, RTS	20	3
4	CFCC, ADBMS	16	4
5	RTS, ADBMS	10	5
6	RTS, HSN	8	6
7	ON, ML	6	7
8	ON, RTS	3	8
9	ON, HSN	3	9

SUPCOUNT=support count, SUPRANK= support rank

Optional Electives:
ML= Machine Learning
CFCC= Cyber Forensics and Cyber Crime
ADBMS= Advanced Database Management System
MSD= Multimedia System Design

RTS= Real Time Systems
ON= Optical networks
HSN=High speed Networks

Fig. 1. Association rule mining from students' preferences in University website in [11] for the combination of two University Open electives

Procedure:
Stage 1 Use Data Mining to find electives *(Refer to Fig. 1 for the mining of Association patterns)*

- Take preference from each student for the 2 electives they want to study.
- Calculate support count for each subject.
- Decide a threshold value (support value), subjects with support count < threshold are discarded. If course reaches the threshold value it is said to be strongly associated.
- Find associations between courses using FP growth algorithm. Create table for associated courses selected by students.
- Count the associations between courses and arrange them.
- Associations with maximum count are selected as the two electives.
- Selected electives are used in stage 2 to create timetable.

Stage 2: Applying GA on selected electives to create optimized timetable

- Initialize a timetable with given number of professors, time slots, classes and electives.
- Define population size, mutation rate, crossover rate.
- Form one chromosome, calculate number of clashes as per constraints not satisfied.
- Calculate fitness of a chromosome using fitness function = 1/1 + (no of clashes)
- Repeat for all generations
  - Crossover
  - Mutation
  - Selection
  - Compute new fitness value for new chromosome obtained
- Stop when all generations are over or when number of clashes = 0 and fitness function = 1.
- Compare factors- generation number, makespan time with simulated annealing.

3.2   Step Wise Procedure for Timetable Generation Using SA Starting from Assumptions, Initial Parameters

Initial Parameters:
- Initial Temperature = 1000
- Cooling Rate = 0.05

Constraints:

Hard Constraints:
- Courses are arranged to be taught between certain time periods (08:00 am–12:00 pm).
- There should be exactly two electives.
- All students of the department enroll for these two electives.

Soft Constraints
- Classroom capacity should be greater than the size of the class.
- No room should be assigned to a course or teacher if its already assigned to another course or teacher for that particular time period.
- Every course should have a professor to teach it.
- Each student is assigned to only one classroom at a certain time period.

Procedure:
Stage 1 Use Data Mining to find electives *(Refer to Fig. 1 for the mining of Association patterns)*

- Take preference from each student for the 2 electives they want to study.
- Calculate support count for each subject.
- Decide a threshold value (support value), subjects with support count < threshold are discarded. If course reaches the threshold value it is said to be strongly associated.
- Find associations between courses using FP growth algorithm. Create table for associated courses selected by students.
- Count the associations between courses and arrange them.
- Associations with maximum count are selected as the two electives.
- Selected electives are used in stage 2 to create timetable.

Stage 2: Applying SA on selected electives to create optimized timetable

- Initialize a timetable (number of professors, time slots, classes and subjects).
- Define temperature, cooling rate.
- Take initial temperature to be very high (T = 1000).
- Repeat till (Temperature > 1)
- Define current node C, Initialize random neighbor N.
- Evaluate $\Delta E = \text{eval}(N) - \text{eval}(C)$.
- eval (N) = number of clashes (calculated for soft constraint violation) in new node

- eval (C) = number of clashes (calculated for soft constraint violation) in current node
  - If eval(N) < eval(C) i.e. $\Delta E < 0$ choose N, probability = 1.
  - If eval(N) > eval(C) i.e. $\Delta E > 0$ choose N, probability P = $-\Delta E/T$.
  - Fitness Function (cost function) = 1/1 + x
  - Here, x = $-\Delta E/T$
  - Repeat till clashes = 0 and fitness value = 1.
  - If fitness value = 1 stop algorithm
- Temp = temperature * coolingRate.
- Stop when Temperature < 1
- Compare parameters: Iteration at which solution obtained, makespan time with Genetic Algorithm.

4 Discussion on Results

The comparison of the course schedules generated using SA and GA is shown in Table 1. The timetables generated by GA and SA makes sure that no soft or hard constraint are violated. The timetable is randomly generated by both genetic algorithm and simulated annealing. We compare these on the basis of two parameters:

Execution time (makespan time): This is the time taken to generate a timetable by genetic algorithm and simulated annealing respectively as shown in Table 1.

Iteration/generation number at which solution is obtained: For genetic algorithm, it gives the generation and for simulated annealing it gives the iteration number at which the number of clashes are minimum (i.e. 0) and the fitness function is maximized to be 1. From the results in Table 1 we observe that:

- *Simulated annealing takes lesser time to generate timetable and hence gives a faster result than genetic algorithm. Both genetic algorithm and simulated annealing take much lesser time than manual timetable scheduling*
- *Iteration/Generation number at which solution is obtained: Simulated annealing requires lesser number of iterations than genetic algorithm. The timetables generated by genetic algorithm and simulated annealing make sure that no constraints are violated whereas manual timetable scheduling can be prone to manual errors and might cause violation of constraints (like multiple assignment of classes for same time slot). Manual timetable scheduling takes much more time to generate timetable than both GA and SA.*

Table 1. Performance analysis for SA and GA for timetable scheduling

Best case makespan		Average case makespan		Worst case makespan	
Generation no.	Time (in ms)	Generation no.	Time (in ms)	Generation no.	Time (in ms)
1	170	7	265	13	413
Simulated annealing					
Iteration No.	Time (in ms)	Iteration no.	Time (in ms)	Iteration no.	Time (in ms)
1	143	5	220	10	311

For the time-tables generated, we have the following general observations:

ELECTIVE: ML- Machine Learning
   Theory - 3 h (for all students)
   Tutorial - 1 h (for groups G1, G2, G3, G4)

We have taken 4 groups as we are dividing all students of IT.

In ML.
Total no of students = 256
Students in G1 = 64
Students in G2 = 64
Students in G3 = 64
Students in G4 = 64
Capacity of Room TW3TF3 = 300, so all ML classes are conducted here.
Capacity of Room TW1TF3 = 200, so CFCC conducted here.
Capacity of Room TW2GF2 = 200, so CFCC conducted here.

Also, students in ML might be same as in CFCC so those classes cannot coincide so we need more slots.

ELECTIVE: CFCC-Cyber Forensics and Cyber Crime
Theory: Total no of students = 256
      Divided into 2 groups:
Theory - A (128 students) taught by Dr. Kapil Sharma
Theory - B (128 students) taught by Ms. Namita Jain
Tutorial: Divided into 2 subgroups of A and B
Tutorial - A - Group 1 consists of 64 students (taught by Dr. Kapil Sharma)
Tutorial - A - Group 2 consists of 64 students (taught by Dr. Kapil Sharma)
Tutorial - B - Group 1 consists of 64 students (taught by Dr. Kapil Sharma)
Tutorial - B - Group 2 consists of 64 students (taught by Dr. Kapil Sharma)

The Timetables generated by SA and GA are fully automated from scratch with no manual compilation required. There were no clashes reported between the teachers and the classrooms. We also observed that the timetable generated by SA is more spread out over slots while the schedules generated by GA is more concentrated in clusters of nearby slots. Overall SA is more efficient than GA in scheduling problems both with respect to the schedule and the computational efficiency, as we proved from our experiments. We propose to further extend this work by mining for feedback from teachers for their timing preferences and suggested amendments in order to optimize the timetable in a subsequent second phase of optimization.

5 Conclusion

We investigate the optimization algorithms Genetic Algorithm (GA) and Simulated Annealing (SA) for fully automated timetable scheduling in this paper. The generated timetables are found optimized and data mining patterns with association rules are used for enclosing students' preferences for the Elective courses. The SA is found faster as compared to GA for the scheduling task.

References

1. Schaerf, A.: A survey of automated timetabling. Artif. Intell. Rev. **13**(2), 87–127 (1999)
2. Yu, E., Sung, K.-S.: A genetic algorithm for a university weekly courses timetabling problem. Int. Trans. Oper. Res. **9**(6), 703–717 (2002)
3. Yazdani, M., Naderi, B., Zeinali, E.: Algorithms for university course scheduling problems. Tehnicki Vjesnik-Technical Gazette **24**, 241–247 (2017)
4. Duong, T.-A., Lam, K.-H.:. Combining constraint programming and simulated annealing on university exam timetabling. In: RIVF, pp. 205–210 (2004)
5. Rozaimee, A., Shafee, A.N., Hadi, N.A.A., Mohamed, M.A.: A framework for university's final exam timetable allocation using genetic algorithm. World Appl. Sci. J. **35**(7), 1210–1215 (2017)
6. Thompson, J., Dowsland, K.A.: General cooling schedules for a simulated annealing based timetabling system. In: International Conference on the Practice and Theory of Automated Timetabling, pp. 345–363. Springer, Heidelberg (1995)
7. Zheng, S., Wang, L., Liu, Y., Zhang, R.: A simulated annealing algorithm for university course timetabling considering travelling distances. Int. J. Comput. Sci. Math. **6**(2), 139–151 (2015)
8. Brusco, M.J., Jacobs, L.W.: A simulated annealing approach to the cyclic staff-scheduling problem. Nav. Res. Logist. (NRL) **40**(1), 69–84 (1993)
9. Wang, Y.-T., Cheng, Y.-H., Chang, T.-C., Jen, S.M.: On the application of data mining technique and genetic algorithm to an automatic course scheduling system. In: 2008 IEEE Conference on Cybernetics and Intelligent Systems, pp. 400–405. IEEE (2008)
10. Agrawal, R., Srikant, R.: Fast algorithms for mining association rules. In: Proceedings of 20th International Conference Very Large Data Bases, VLDB, vol. 1215, pp. 487–499 (1994)
11. Delhi Technological University, New Delhi, India. reg.exam.dtu.ac.in/register_all.php
12. Kumara, S., Goldberg, D.E., Kendall, G.: Genetic algorithms. In: Search Methodologies, pp. 93–117. Springer, Boston (2014)
13. Savasere, A., Omiecinski, E.R., Navathe, S.B.: An efficient algorithm for mining association rules in large databases. Georgia Institute of Technology (1995)
14. Brusco, M.J., Jacobs, L.W.: A simulated annealing approach to the cyclic staff-scheduling problem. Nav. Res. Logist. (NRL) **40**(1), 69–84 (1993)
15. Müller, T., Rudová, H.: Real-life curriculum-based timetabling with elective courses and course sections. Ann. Oper. Res. **239**(1), 153–170 (2016)
16. Susan, S., Sharawat, P., Singh, S., Meena, R., Verma, A., Kumar, M.: Fuzzy C-means with non-extensive entropy regularization. In: 2015 IEEE International Conference on Signal Processing, Informatics, Communication and Energy Systems (SPICES), pp. 1–5. IEEE (2015)
17. Taha, M., Nassar, H., Gharib, T., Abraham, A.: An efficient algorithm for incremental mining of temporal association rules. Data Knowl. Eng. **69**, 800–815 (2010)

Compressed Sensing in Imaging and Reconstruction - An Insight Review

K. Sreekala[1(✉)] and E. Krishna Kumar[2]

[1] Visvesvaraya Technological University, Belgaum, India
kannothsree@gmail.com
[2] Electronics and Communications Engineering,
EPCET (Visvesvaraya Technological University), Bangalore, India

Abstract. Compressed sensing is a modern approach for signal sensing and sensor design which helps to reduce sampling and computation cost while acquiring signals with sparse or compressible representation. Nyquist-Shannon theorem speaks about the number of samples required to represent a bandlimited signal, but the number of samples required to represent a signal can be extensively reduced if it is sparse in a known basis. Sparsity is an essential property of signals which helps in storing the signals using few samples and recover accurately with sparse recovery techniques. Many signals are sparse in reality or when they are represented in an appropriate transform domain. Many algorithms exists in the literature for sparse signal recovery. Through this paper we make an attempt to look into the idea of compressed sensing, applications in image processing and computer vision, different basic sparse recovery algorithms and recent developments in the application of Compressed sensing in image enhancement.

Keywords: Compressive sampling · Sparse signals ·
Sparse recovery algorithms · Transform domain

1 Introduction

Compressed sensing (CS) [1,2] is an exhilarating, expeditiously emerging field, and has attained significant recognition in various fields of science and engineering. It is a novel technique of taking samples at sub-Nyquist rate. According to Nyquist sampling theory, an analog signal can be restored without distortion if the minimum sampling frequency is equal to twice the maximum frequency exist in the signal, known as Nyquist rate [6]. CS is a technique which allows the perfect recovery of signal even if the signal is sampled at a rate far below the Nyquist rate. This is feasible by relying on two principles - sparsity and incoherence [7].

CS works with sparse or compressible signals, that is the signal has most of the coefficients zero when it is represented in a transform domain. CS uses the fact that most of the natural signals are sparse or compressible when transformed

© Springer Nature Switzerland AG 2020
A. Abraham et al. (Eds.): ISDA 2018, AISC 941, pp. 779–791, 2020.
https://doi.org/10.1007/978-3-030-16660-1_76

to a proper basis, ψ. Incoherence says that signals with sparse representation in ψ will be spread out in the domain which they are captured.

In this paper, a brief overview of CS, its applications in image processing and computer vision, the different algorithms available to recover the sparse signals, and our observations from recent literature are presented. Figure 1 shows the organized structure of this literature survey. Figure 2 shows the details of the CS, based on applications and algorithms. More details of this is given later.

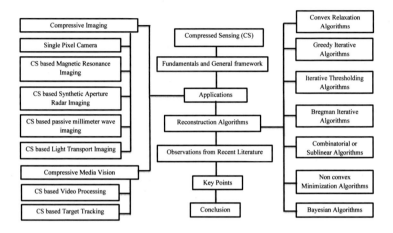

Fig. 1. Organised structure of survey on CS applications, algorithms and trends

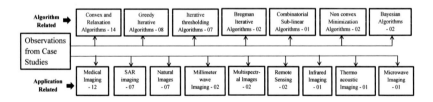

Fig. 2. Details of CS algorithms and applications from recent literature

The flow of the paper is given as follows. Section 2 gives a glance of CS fundamentals, Sect. 3 briefs applications in imaging and media vision, Sect. 4 is the overview of recovery algorithms, Sect. 5 is our observations from the recent literature on the use of CS in different imaging systems, Sect. 6 gives the important observations from this survey and Sect. 7 is the conclusion.

2   Fundamentals and Preliminary Concepts

CS theory speaks about concurrent sensing and compression, where the signal can be represented with minimum number of measurements while making the measurements sufficient to approximate the original signal. In such cases, the signals can be restored from lesser samples by using appropriate recovery algorithms. Consider a vector x of length n. In CS theory, we measure m $(m < n)$ linear measurements of x in the form $y = Ax$, where A is a mXn measurement matrix which meets the Restricted Isometry Property (RIP) [43–45]. Here the matrix is modeled in such a way that it takes minimum number of measurements possible while enabling the reconstruction of x from its measurement vector y. The most important concept used in CS theory is Sparsity [3–5]. According to this, x has got very less number of non-zero elements if x is sparse. A signal is known as k-sparse, if the number of nonzero elements in it is k. So if x is k-sparse, then x can be reconstructed from $y = Ax$ utilizing the m number of measured coefficients. Candes, Tao, Rombergand Donoho have shown that sparse signals can be restored precisely form a linear, non adaptive set of measurements [1,8–13].

Most of the natural signals are compressible when they are transformed to a proper domain and signal representation will be more concise when it is transformed to a new domain or basis. Compression generally helps to reduce the data storage and data transmission [15]. It is possible to achieve high rate of compression using sparse signal models. In signal processing theory, sparsity has been used for compression [15] and denoising [16]. For natural signals the multiscale wavelet domain [14] representation provides approximate sparse representation and hence the sparsity property is used in image processing applications also. Other domains such as DFT, DCT, Curvelet transform etc can be used for sparsifying signals.

The main challenge faced by the existing system is the effective transmission and memory utilization when dealing with the signals from high resolution systems, which results in high data rate. Hence CS technique can be used as an improved alternate solution with a sampling model which replaces the concept of bandlimited signals with sparse signals.

3 Applications

3.1 Compressive Imaging

CS techniques are used in image processing due to the sparsity property on images. The important applications of CS in imaging includes:

Single Pixel Camera: These cameras were proposed based on digital micromirror device(DMD) architecture with CS background [17,18]. It uses only a single light sensor. They are simpler, smaller and cost effective cameras.

CS Based Magnetic Resonance Imaging: MRI scanner can be seen as a system that computes the frequency(Fourier) domain details about an object [19]. Measured data here is directly proportional to the scan time and this creates discomfort for patients. Hence CS techniques are used in MRI system [21–23], so that a high quality image can be generated from less number of measurements.

CS Based Synthetic Aperture Radar Imaging: SAR image gives the details about spatial distribution of the reflectivity function of stationary targets and terrain [26–29]. The majority of these images are sparse in wavelet or complex wavelet basis. CS concepts allows reconstruction of such sparse signal from randomly taken measurements using a proper nonlinear reconstruction method [1, 11, 23].

CS Based Passive Millimeter Wave Imaging: In this, single beam scanning is done in azimuth and elevation to help the image formation. For each sample, total scan duration depends on positioning and process times. In order to decrease the number of samples, CS methods are used [30–35].

CS Based Light Transport Imaging: It is a technique based on CS to capture the light transport data of a scene so that the acquisition time and storage can be reduced. The non-adaptive illumination patterns helps in reducing measurement noise and improving the restoration performance [36].

3.2   Compressive Media Vision

CS techniques are used in media vision as well for video processing and target tracking applications.

CS Based Target Tracking: CS methods can solve the challenges in tracking systems by using CS cameras in place of traditional cameras or by using CS methods in data processing [74]. The methods which use the CS theory to perform tracking tasks are background subtraction [37], signal tracking [38], visual tracking [39], multiview tracking [40], particle filtering [41] etc.

CS Based Video Processing: There are lot of advancements happened in CS videos and periodic signal processing. A CS based method was suggested for checking periodic activities [42], where the sparsity property of periodic signals are used to reconstruct the phenomena at a higher rate.

4   CS Reconstruction Algorithms

Algorithms that can be used for CS based signal reconstruction are presented here. The detailed classification of CS recovery algorithms are given in Fig. 3.

Convex Relaxation Algorithms: In these algorithms, the number of measurements necessary for signal recovery is less, but they are computationally complex [46, 47]. Two important convex relaxation methods are Basis Pursuit [48] and Gradient Descent [49]. In Basis Pursuit the signal recovery is done by finding the smallest l_1 norm by using convex optimization. In Gradient Descent, the smallest l_1 norm calculation is done in an iterative way. Basis Pursuit De-Noising(BPDN) [48], Modified BPDN [50], Least Absolute Shrinkage and Selection Operator(LASSO) [51] and Least Angle Regression(LARS) [52] are the other variation of this algorithm.

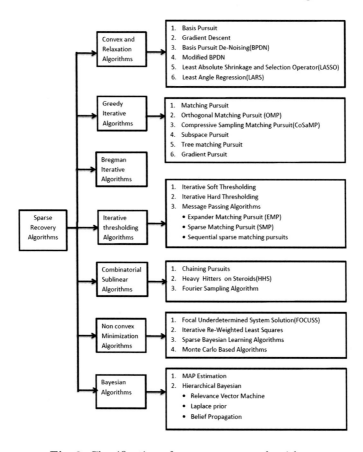

Fig. 3. Classification of sparse recovery algorithms

Greedy Iterative Algorithms: Greedy pursuit algorithms consists of two steps: element selection and coefficient update. In these methods initial estimation is taken as zero. Then the residual error and support set is calculated and updated in each iteration by adding extra elements to support set and hence updating the signal estimate and decreasing the residual error. Different greedy algorithms are Matching Pursuit [53], Orthogonal Matching Pursuit(OMP) [54], The CS Matching Pursuit [CoSaMP] [44] and the Subspace Pursuit [55].

Iterative Thresholding Algorithms: Thresholding type algorithms are easy to implement, fast and have very good performance. In these kind of algorithms, recovery of correct measurements can be done using iterative hard thresholding and iterative soft thresholding [16,56]. Modification on these iterative thresholding algorithms resulted in message passing algorithms, such as Expanded Matching Pursuit (EMP) [57], Sparse Matching Pursuit (SMP) [58], and Sequential sparse matching pursuits [59].

Bregman Iterative Algorithms: Bregman iterative algorithms [60] gives accurate solution for the basis pursuit problem in fixed number of steps. In CS problems this iterative approach attains reconstruction in maximum six steps with better computational speed.

Combinatorial or Sublinear Algorithms: These algorithms are very quick and powerful with the measurements being sparse. Here the sparse signal recovery is done through group testing. The main algorithms under this category are Chaining Pursuits [61,72], Heavy Hitters on Steroids(HHS) [62], Fourier Sampling Algorithm [63] etc.

Non Convex Minimization Algorithms: Non convex Minimization Algorithms can be used for recovering CS signals from minimum number of measurements [64]. The algorithms in this category uses the techniques like Focal Underdetermined System Solution(FOCUSS) [66,73], Iterative Re-Weighted Least Squares [65], Sparse Bayesian Learning Algorithms [67], and Monte Carlo Based Algorithms [68].

Bayesian Algorithms: These are based on statistical assumptions. These algorithms either uses a Maximum Aposteriori (MAP) estimation technique or hierarchical Bayesian technique. Examples for Bayesian network includes Relevance Vector Machine [69], Laplace prior [70] and Belief Propagation [71].

5   Observations from the Survey

For having a better understanding on recent developments for applications of CS in the area of image enhancement, we have done a survey on selected relevant papers from recent literature. Summary of this is given in Fig. 2. The details are given below.

Papers [76,77,79,82,86,87,89–91,94,96,103] describes the application of CS and sparse recovery in medical images such as tomography, microwave breast cancer imaging, MR imaging, ultrasound imaging and EEG signals. [76] uses the sparsity property of tumors. Since tumors are small, the number of pixels required to represent it is less. They use convex optimization technique and LASSO for reconstruction. [77] introduces CS based tomography technique. These signals are sparse when distributed within the array antenna's field of view and iterative Kaczmarz algorithm is used for the signal recovery. [91] and [89] introduces a compressive deconvolution technique for reconstructing ultrasound images. These signals are sparse in a transform domain and uses algorithm based on the alternating direction method of multipliers (ADMM) and simultaneous direction method of multipliers (SDMM) for reconstruction. [82] and [94] also speaks about CS based ultrasound imaging techniques. They use convex optimization technique and algorithm based on Bayesian techniques for the signal recovery. [87,90,96,103] all speak about CS based MR imaging systems. MR images are sparse when transformed to an appropriate domain. Non-linear conjugate gradient method (NLCG) and FOCUSS, pseudo polar fast Fourier transform and Fast Weighted iterative shrinkage/thresholding algorithm (FWISTA) are used for signal reconstruction. It results in good quality image less artifacts and helps in retaining the edge details. [79] proposes a method for medical image sampling, compression, encryption and confidentially homomorphic aggregation. Images are converted to sparse using a proper basis and Orthogonal matching pursuit algorithm used for sparse signal recovery. [86] explains the use of CS techniques in EEG signal recovery. EEG signals have got weak sparsity and they use a sparsifying basis to increase the sparsity.

Papers [80, 84, 95, 97, 101, 104, 105] proposes CS based SAR imaging techniques. It uses the sparsity property of SAR image in transform domain. Use of CS in SAR imaging allows generalized subsampling data acquiring methods with prior information. In 3D SAR imaging cross track signals are sparse in the object domain. In Polarimetric SAR Tomography of Forested Areas, vertical structures and polarimetric signatures are sparse in a transform domain. Algorithms such as Back projection and iterative hard thresholding with partial known support (PKS) and GLRT, perturbed auto focus SAR which is an orthogonal matching pursuit based greedy technique, Bayesian CS algorithm, Truncated SVD and other variations of convex optimization and greedy algorithms are used for the sparse signal reconstruction. These techniques results in a good quality image with less noise and low computation time.

Papers [88, 92, 98–100, 102, 106] in general speak about the use of CS techniques in natural image processing applications. All these papers describes how to obtain a high resolution image from the samples taken from a low resolution sparse image. [88] and [92] uses Two-step iterative shrinkage/thresholding (TwIST) algorithm for the signal recovery. [98–100, 102, 106] uses Split Bregman technique, Total Variation and Directional Nonlocal Means regularization, iterative thresholding, iterative algorithm under majorization–minimization framework and optimization and gradient descent techniques respectively for the signal recovery. They generate High resolution images with less error and noise.

In [75], authors presents a CS based 3D thermo acoustic imaging system. Here Sparsity property of the thermo-acoustic signals helps in decreasing the number of sensors. The algorithm used for sparse signal reconstruction is Gradient projection for sparse reconstruction(GPSR) optimization algorithm. In [78] CS techniques are used for the reconstruction and noise removal from multispectral images because of the sparsity property of these images in transform domain. Bregman split method principle component analysis is used for the signal recovery. [93] uses CS techniques in hyperspectral imaging systems, using the sparsity property of stars throughout the space. Here Block OMP algorithm is used for the signal restoration. [83] and [85] uses CS techniques in millimeter wave phased array imaging systems. Usage of Far-field focusing method helps in getting sparse measurements. Split augmented lagrangian shrinkage algorithm and BPDN are used for signal recovery. [107] and [108] uses single or low pixel per sensor imaging system for remote sensing applications. In both the cases iterative thresholding algorithms are used for signal recovery. An infrared imaging system based on sparse signals is described in [81]. Convolutional sparse coding helps in modeling each individual color channel as a sparse sum of image patches. Algorithm based on ADMM is used for signal recovery.

6 Key Points from the Review

Following are the Key Points emerging from this Review.

- Image Processing applications are extensively referred to in the context of CS or Sparse Systems due to the inherent characteristics of sparsity existing in them.
- Though the images have inherent property of sparsity in them, majority of the CS related research has happened in the field of enhancement because image enhancement problem can be treated as a underdetermined case.
- More research has happened in the area of Medical imaging, SAR imaging, and natural image processing applications. Comparatively less CS related work has happened in the other areas such as thermal acoustic imaging, Remote sensing, and millimeter wave imaging.
- Majority of the applications described in the survey have used Convex Optimization, Greedy iterative and Iterative Thresholding algorithms because in general they are fast and generate good visual output.
- Most of the medical imaging applications uses Convex Optimization and Greedy Iterative algorithms since they results in output with less artifacts and minimum error by preserving the edge details.
- Majority of the SAR imaging applications uses Greedy Iterative, Iterative Thresholding and Convex Optimization algorithms since they reconstruct scattering more accurately with minimum noise, reconstruct scene with lower side lobes, withstand strong clutter noise and simplifies the design.
- For the exact signal reconstruction, the number of measurements required depends on the sparsity and number of iterations, and this decide the complexity of algorithms.
- Algorithms under Greedy category perform much faster than other algorithms by selecting best possible solution at each step.
- Compared to Matching Pursuit, OMP algorithms are simple faster and generate better visual output. In OMP, after each step, all the extracted coefficients are updated and they can be converged in limited number of iterations.
- All Greedy Iterative algorithms except Matching Pursuit and Gradient Pursuit, require a measurement matrix which satisfies the RIP, hence the design of measurement matrix easier for Matching Pursuit and Gradient Pursuit.
- Algorithms under Convex and Relaxation category gives better solution to sparse recovery problems with minimum error. These algorithms are computationally complex and it consists of more computational steps. But the signal recovery can be done from less number of measurements.
- Bayesian algorithms have recovery time and recovery error in between Greedy category and Convex and Relaxation category, but require prior knowledge of sparse signal distribution for reconstruction. It is a trade off between both time and error. These are used in images with continuous structures and helps in suppressing isolated components.
- Bregman algorithms are used in recovering noisy images. For natural images with noise it provides a stable reconstruction performance.

Authors have used metrics like Mean Square Error (MSE) and Peak Signal to Noise Ratio (PSNR)to measure the enhancement technique performance and shown that CS techniques are very efficient in reconstructing signals.

7   Conclusion

This survey on CS applications and algorithms gives an outline of theoretical foundation of CS technology and different algorithms used for sparse signal recovery. We have also stated the application of this technology in different fields of image and video processing. Although the progress is significant in all these fields, there are lot of challenges related to the effectiveness of sparse representation and CS. In addition to this a review is done from the recent literature on application of CS in different imaging fields like Thermal acoustic imaging, Microwave imaging, mm-wave imaging, SAR imaging, Infrared imaging, Medical imaging, Multispectral imaging, Remote sensing and Natural Imaging. The research can be extended further for systems with low energy source such as:

- MRI imaging with magnetic field less than 0.5 Tesla.
- Tomography imaging using low-energy x-rays.
- SAR imaging system with low gain or less transmitter power.
- Natural dim light images where the light source intensity is less.

Noise removal and enhancement are indeed challenging tasks to be considered in these images.

References

1. Donoho, D.: Compressed sensing. IEEE Trans. Inf. Theory **52**(4), 1289–1306 (2006)
2. Tsai, Y., et al.: Extensions of compressed sensing. Sig. Process. **86**(3), 549–571 (2006)
3. Bruckstein, A., et al.: From sparse solutions of systems of equations to sparse modeling of signals and images. SIAM Rev. **51**(1), 34–81 (2009)
4. DeVor, R.: Nonlinear approximation. Acta Numer. **7**, 51–150 (1998)
5. Elad, M.: Sparse and Redundant Representations: From Theory to Applications in Signal and Image Processing. Springer, New York (2010)
6. Shannon, C.E.: Communication in the presence of noise. Proc. IRE **37**, 10–21 (1949)
7. Candes, E.J., et al.: An introduction to CS. IEEE Sig. Proc. Mag. **25**, 21–30 (2008)
8. Baraniuk, R.: Compressive sensing. IEEE Sig. Proc. Mag. **24**(4), 118–121 (2007)
9. Candès, E.: Compressive sampling. In: Proceedings of the International Congress of Mathematicians, Madrid, Spain (2006)
10. Candès, E., Romberg, J.: Quantitative robust uncertainty principles and optimally sparse decompositions. Found Comput. Math. **6**(2), 227–254 (2006)
11. Candès, E., et al.: Robust uncertainty principles: exact signal reconstruction from highly incomplete frequency information. IEEE Trans. Inf. Theory **52**(2), 489–492 (2006)

12. Candès, E., et al.: Stable signal recovery from incomplete and inaccurate measurements. Commun. Pure Appl. Math. **59**(8), 1207–1223 (2006)
13. Candès, E., Tao, T.: Near optimal signal recovery from random projections: universal encoding strategies. IEEE Trans. Inf. Theory **52**, 5406–5425 (2006)
14. Mallat, S.: A Wavelet Tour of Signal Processing. Academic Press, San Diego (1999)
15. Pennebaker, W., et al.: JPEG Still Image Data Compression Standard. Van Nostrand Reinhold, New York (1993)
16. Donoho, D.: Denoising by soft-thresholding. IEEE Trans. Inf. Theory **41**(3), 613–627 (1995)
17. Duarte, M.F., et al.: Single-pixel imaging via CS. IEEE SP Mag. **25** (2008)
18. Shin, J., et al.: Single-pixel imaging using compressed sensing and wavelength-dependent scattering. Opt. Lett. **41**, 886–889 (2016)
19. Wright, G.A.: MR imaging. IEEE Sig. Proc. Mag. **14**(1), 56–66 (1997)
20. Roohi, S., et al.: Super-resolution MRI images using compressive sensing. In: 20th Iranian Conference on Electrical Engineering, Tehran (2012)
21. Lustig, M., et al.: Sparse MRI: the application of compressed sensing for rapid MR imaging. Magn. Reson. Med. **58**(6), 1182–1195 (2007)
22. Lustig, M., et al.: Compressed sensing MRI. IEEE Sig. Proc. Mag. **25**(2), 72–82 (2008)
23. Patel, V.M., et al.: Gradient-based image recovery methods from incomplete fourier measurements. IEEE Tran. Image Process. **21**(1), 94–105 (2012)
24. Chan, T.F., et al.: Recent Developments in Total Variation Image Restoration. Springer, Berlin (2005)
25. Wang, Y., et al.: A new alternating minimization algorithm for total variation image reconstruction. SIAM J. Imaging Sci. **1**(3), 248–272 (2008)
26. Carrara, W.G., et al.: Spotlight SAR: Signal Processing Algorithms. Artech House, Norwood (1995)
27. Chen, V.C., Ling, H.: Time-Frequency Transforms for Radar Imaging and Signal Analysis. Artech House, Norwood (2002)
28. Soumekh, M.: SAR Signal Processing with Matlab Algorithms. Wiley, New York (1999)
29. Cumming, I.G., et al.: Digital Processing of SAR Data. Artech House, Norwood (2005)
30. Patel, V.M., et al.: Compressive passive millimeter-wave imaging with extended depth of field. Opt. Eng. **51**(9), 091610 (2012)
31. Babacan, S.D., et al.: Compressive passive mm-wave imaging. In: IEEE ICIP (2011)
32. Christy, F., et al.: Millimeter-wave compressive holography. Appl. Opt. **49**(19), E67–E82 (2010)
33. Fernandez, C.A., et al.: Sparse fourier sampling in millimeter-wave compressive holography. In: Digital Holography and 3 Dimensional Imaging (2010)
34. Gopalsami, N., et al.: CS in passive mm-wave imaging. In: Proceedings of SPIE, vol. 8022 (2011)
35. Noor, I., et al.: CS for a sub-millimeter wave single pixel images. In: Proceedings of SPIE, vol. 8022 (2011)
36. Peers, P., et al.: Compressive light transport sensing. ACM Trans. Graph **28**, 3 (2009)
37. Cevher, V., et al.: CS for background subtraction. In: ECCV (2008)
38. Vaswani, N.: Kalman filtered CS. In: IEEE International Conference on Image Processing (2008)

39. Cossalter, M., et al.: Joint compressive video coding and analysis. IEEE Trans. Multimedia **12**, 168–183 (2010)

40. Reddy, D., et al.: Compressed sensing for multi-view tracking and 3-D voxel reconstruction. In: IEEE International Conference on Image Processing (2008)

41. Wang, E., et al.: Compressive particle filtering for target tracking. In: IEEE Workshop on Statistical Signal Processing (2009)

42. Veeraraghavan, A., et al.: Coded strobing photography: CS of high speed periodic videos. IEEE Trans. Pattern Anal. Mach. Intell. **33**(4), 671–686 (2011)

43. Candès, E., et al.: Decoding by linear programming

44. Needell, D., Tropp, J.: CoSaMP: iterative signal recovery from incomplete and inaccurate samples. Appl. Comput. Harmon. Anal. **26**(3), 301–321 (2009)

45. Foucart, S.: Sparse recovery algorithms: sufficient conditions in terms of RIC. In: Approximation Theory XIII: San An tonio. Springer, New York (2012)

46. Tropp, J.A.: Algorithms for simultaneous sparse approximation. Part II: convex relaxation. Sig. Process. **86**, 589–602 (2006)

47. Boyd, S., et al.: Convex Optimization. Cambridge University Press, Cambridge (2009)

48. Chen, S.S., et al.: Atomic decomposition by basis pursuit. SIAM Rev. **43**(1), 129–159 (2001)

49. Garg, R., et al.: Gradient descent with sparsification: an iterative algorithm for sparse recovery with RIP. In: Proceedings of the 26th Annual International Conference on Machine Learning. ACM (2009)

50. Lu, W., et al.: Modified basis pursuit denoising for noisy compressive sensing with partially known support. In: Proceedings of the ICASSP (2010)

51. Tibshirani, R.: Regression shrinkage and selection via the lasso. J. Roy. Stat. Soc. Ser. B (Methodol.) **58**, 267–288 (1996)

52. Efron, B., et al.: Least angle regression. Ann. Stat. **32**, 407–499 (2004)

53. Mallat, S., et al.: Matching pursuits with time-frequency dictionaries. IEEE Trans. Sig. Process. **41**, 3397–3415 (1993)

54. Tropp, J., Gilbert, A.: Signal recovery from random measurements via orthogonal matching pursuit. IEEE Trans. Inf. Theory **53**, 4655–4666 (2007)

55. Dai, W., et al.: Subspace pursuit for CS signal reconstruction. IEEE Trans. Inf. Theory **55**, 2230–2249 (2009)

56. Blumensath, T., et al.: Iterative hard thresholding for compressed sensing. Appl. Comput. Harmon. Anal. **27**, 265–274 (2009)

57. Indyket, P., et al.: Near-optimal sparse recovery in L1 norm. In: Proceedings of IEEE FOCS (2008)

58. Berinde, R., et al.: Practical near-optimal sparse recovery in the L1 norm. In: Proceedings of Allerton Conference on Communication, Control, Computing (2009)

59. Berinde, R., et al.: Sequential sparse matching pursuit. In: Proceedings of the Allerton Conference on Communication, Control, and Computing (2010)

60. Yin, W., et al.: Bregman iterative algorithms for L1-minimization with applications to compressed sensing. SIAM J. Imaging Sci. **1**, 143–168 (2008)

61. Gilbert, A., et al.: Algorithmic linear dimension reduction in the L1 norm for sparse vectors. arXiv preprint arXiv:cs/0608079 (2006)

62. Gilbert, A., et al.: One sketch for all: fast algorithms for compressed sensing. In: Proceedings of the ACM Symposium on Theory of Computing (2007)

63. Gilbert, A., et al.: Improved time bounds for near-optimal sparse Fourier representations. In: Proceedings of SPIE, p. 59141 A (2005)

64. Chartrand, R.: Exact reconstruction of sparse signals via nonconvex minimization. IEEE Sig. Process. Lett. **14**, 707–710 (2007)

65. Chartrand, R., et al.: Iteratively reweighted algos for CS. In: Proceedings of IEEE ICASSP (2008)
66. Murray, J., et al.: An improved FOCUSS-based learning algorithm for solving sparse linear inverse problems. In: Proceedings of Asilomar Conference on Signals, Systems and Computers, vol. 1 (2001)
67. Wipf, D., et al.: Sparse Bayesian learning for basis selection. IEEE Trans. Sig. Process. **52**, 2153–2164 (2004)
68. Godsill, S., et al.: Bayesian computational methods for sparse audio and music processing. In: Proceedings of EURASIP Conference on Signal Processing (2007)
69. Ji, S., et al.: Bayesian compressive sensing. IEEE Trans. Sig. Process. **56**, 2346–2356 (2008)
70. Babacan, S.D., et al.: Bayesian CS using laplace priors. IEEE Trans. Image Process. **19**(1), 53–63 (2010)
71. Baron, D., et al.: Bayesian compressive sensing via belief propagation. IEEE Trans. Sig. Process **58**(1), 269–280 (2010)
72. Gilbert, A., et al.: Sublinear approximation of signals. In: Proceedings of SPIE - The International Society for Optical Engineering, vol. 6232
73. Gorodnitsky, I.F., et al.: Sparse signal reconstruction from limited data using FOCUSS: a re-weighted minimum norm algorithm. Trans. Sig. Proc. **45**, 600–616 (1997)
74. Zhang, K., et al.: Real-time compressive tracking. In: Conference Proceedings Computer Vision ECCV 2012. Springer, Heidelberg (2012)
75. Wang, B., et al.: 3D thermoacoustic imaging based on compressive sensing. In: International Workshop on Antenna Technology (iWAT) (2018)
76. Zarnaghi Naghsh, N., et al.: CS for microwave breast cancer imaging. IET Sig. Process. **12**, 242–246 (2018)
77. Ross, D., et al.: Compressive k-space tomography. J. Lightwave Technol. **36**, 4478–4485 (2018)
78. Liu, P., et al.: Compressive sensing of noisy multispectral images. IEEE Geosci. Remote Sens. Lett. **11**, 1931–1935 (2014)
79. Wang, L., et al.: Compressive sensing of medical images with confidentially homomorphic aggregations. IEEE Internet Things J
80. Wei, Z., et al.: Wide angle SAR subaperture imaging based on modified compressive sensing. IEEE Sens. J. **18**, 5439–5444 (2018)
81. Hu, X., et al.: Convolutional sparse coding for RGB+NIR imaging. IEEE Trans. Image Process. **27**, 1611–1625 (2018)
82. Besson, A., et al.: Ultrafast ultrasound imaging as an inverse problem: matrix-free sparse image reconstruction. IEEE Trans. Ultrason. Ferroelectr. Freq. Control **65**, 339–355 (2018)
83. Cheng, Q., et al.: Near-field millimeter-wave phased array imaging with CS. IEEE Access **5**, 18975–18986 (2017)
84. Camlica, S., et al.: Autofocused spotlight SAR image reconstruction of off-grid sparse scenes. IEEE Trans. Aerosp. Electron. Syst. **53**, 1880–1892 (2017)
85. Cheng, Q., et al.: Compressive millimeter-wave phased array imaging. IEEE Access **4**, 9580–9588 (2016)
86. Djelouat, H., et al.: Joint sparsity recovery for CS based EEG system. In: IEEE 17th International Conference on Ubiquitous Wireless Broadband, Salamanca (2017)
87. Yang, Y., et al.: Pseudo-polar fourier transform-based compressed sensing MRI. IEEE Trans. Biomed. Eng. **64**, 816–825 (2017)

88. Sun, Y., et al.: Super-resolution imaging using CS and binary pure-phase annular filter. IEEE Photonics J. **9**, 1–10 (2017)

89. Chen, Z., et al.: Reconstruction of enhanced ultrasound images from compressed measurements using simultaneous direction method of multipliers. IEEE Trans. Ultrason. Ferroelectr. Freq. Control **63**, 1525–1534 (2016)

90. Kustner, T., et al.: MR image reconstruction using a combination of CS and partial fourier acquisition: ESPReSSo. IEEE Trans. Med. Imaging **35**, 2447–2458 (2016)

91. Chen, Z., et al.: Compressive deconvolution in medical ultrasound imaging. IEEE Trans. Med. Imaging **35**, 728–737 (2016)

92. Sun, Y., et al.: Compressive superresolution imaging based on local and nonlocal regularizations. IEEE Photonics J. **8**, 1–12 (2016)

93. Fickus, M., et al.: Compressive hyperspectral imaging for stellar spectroscopy. IEEE Sig. Process. Lett. **22**, 1829–1833 (2015)

94. Gifani, P., et al.: Temporal super resolution enhancement of echocardiographic images based on sparse representation. IEEE Trans. Ultrason. Ferroelectr. Freq. Control **63**, 6–19 (2016)

95. Wu, Q., et al.: High-resolution passive SAR imaging exploiting structured bayesian CS. IEEE J. Sel. Top. Sig. Process. **9**, 1484–1497 (2015)

96. Yang, Y., et al.: Compressed sensing MRI via two-stage reconstruction. IEEE Trans. Biomed. Eng. **62**, 110–118 (2015)

97. Zhang, S., et al.: Truncated SVD-based CS for downward-looking 3-D SAR imaging with uniform/nonuniform linear array. IEEE Geosci. Remote Sens. Lett. **12**, 1–5 (2015)

98. Zhang, J., et al.: Group-based sparse representation for image restoration. IEEE Trans. Image Process. **23**, 3336–3351 (2014)

99. Li, X., et al.: Single image superresolution via directional group sparsity and directional features. IEEE Trans. Image Process. **24**, 2874–2888 (2015)

100. Bourquard, A., et al.: Binary compressed imaging. IEEE Trans. Image Process. **22**, 1042–1055 (2013)

101. Aguilera, E., et al.: A data-adaptive CS approach to polarimetric SAR tomography of forested areas. IEEE Geosci. Remote Sens. Lett. **10**, 543–547 (2013)

102. Pham, D., et al.: Improved image recovery from compressed data contaminated with impulsive noise. IEEE Trans. Image Process. **21**, 397–405 (2012)

103. Guerquin-Ker, M., et al.: A fast wavelet-based reconstruction method for MR imaging. IEEE Trans. Med. Imaging **30**, 1649–1660 (2011)

104. Wang, H., et al.: ISAR imaging via sparse probing frequencies. IEEE Geosci. Remote Sens. Lett. **8**, 451–455 (2011)

105. Zhang, L., et al.: Resolution enhancement for inversed SAR imaging under low SNR via improved CS. IEEE Trans. Geosci. Remote Sens. **48**, 3824–3838 (2010)

106. Yang, J., et al.: Image super-resolution via sparse representation. IEEE Trans. Image Process. **19**, 2861–2873 (2010)

107. Ma, J., et al.: Deblurring from highly incomplete measurements for remote sensing. IEEE Trans. Geosci. Remote Sens. **47**, 792–802 (2009)

108. Ma, J.: A single-pixel imaging system for remote sensing by two-step iterative curvelet thresholding. IEEE Geosci. Remote Sens. Lett. **6**, 676–680 (2009)

Gender Identification: A Comparative Study of Deep Learning Architectures

Bsir Bassem[✉] and Mounir Zrigui

LATICE Laboratory Research Department of Computer Science,
University of Monastir, Monastir, Tunisia
Bsir.bassem@yahoo.fr, mounir.zrigui@fsm.rnu.tn

Abstract. Author profiling, dating back to the earliest attempts at of analyzing quantitative text documents, is an extensivel-studied problem among NLP researchers. Because of its utility in crime, marketing and business. In this paper, three deep learning methods were evaluated for author profiling using tweets in Arabic language. The first method is based on a Convolutional Neural Network (CNN) model, while the second and third technique belongs to the family of Recurrent Neural Networks (RNN). The appropriate choice of some parameters, such as the number of amount of filters, training epochs, batch size, dropout and learning rate of Adam optimizer used in a RNN model is crucial in obtaining reliable results. The experimental findings of our comparative evaluation study demonstrate that GRU model outperforms LSTM and CNN models.

Keywords: Gated recurrent units · Epochs · Batch size · Dropout · GRU · LSTM neural network · CNN · Author profiling · Gender identification · Deep learning

1 Introduction

The purpose of author profiling is to response some questions like: how to find the differences in the writing style, on social networks, between men and women, age groups, location or psychological profiles? By answering such questions, we can solve current issues of the social network era as fake news, plagiarism and theft identification.

Gender detection is one of the most popular sub-task in Author Profiling (AP). However, most of the research works performed in AP were devoted to using texts correctly categorize the author's profile.

Deep learning models have recently achieved remarkable results in computer vision (Szegedy et al. 2016), speech recognition (Amodei et al. 2016) and text classification (Howard et al. 2018).

Within natural language processing, much of the studies based on deep learning methods involve learning word vector representations through neural language models and performing composition over the learned word vectors for classification. (Miao et al. 2016). These approaches have an internal memory that keeps track of the examples they have seen so far in the current sequence. In this context, text is one of the clear use cases for RNNs because of its sequential nature.

© Springer Nature Switzerland AG 2020
A. Abraham et al. (Eds.): ISDA 2018, AISC 941, pp. 792–800, 2020.
https://doi.org/10.1007/978-3-030-16660-1_77

The aim of our approach is to investigate whether reasonable results for gender identification could be obtained by applying CNN, GRU and LSTM models, combined with word embed dings input representations. We analyse its effectiveness on a dataset consisting of twitter texts presented by the PAN Lab at CLEF 2018 in terms of accuracy (Stamatatos et al. 2018). Furthermore, we investigate techniques for decreased training times and compare different neural network architectures.

2 State of the Art

Gender is perhaps the most studied social factor in many disciplines examining the relation between language use and social environment. Indeed, (Sap et al. 2014) created publicly available lexica (words and weights) for age and gender using regression and classification models over language usage in social media. Evaluation of the lexica over Facebook yielded accuracies age ($r = 0.831$) and gender (91.9% accuracy) prediction.

In 2015, (Werlen et al. 2015) analysed features obtained from LIWC dictionaries, these are frequencies of use words by categories, which gives a general view about how the author writes and what he/she is talking about. According the experimental results, those are significant features to differentiate gender, age group and personality. Their study showed that the writing style, word choice and grammar rule are solely dependent on the topic of interest and the difference were found with topic variations. It is obvious that the gender specific topic has an impact on their writing styles. Based on the analysed texts, it was clear that female authors tend to write more about wedding styles and fashions; whereas male bloggers focus more on technology and politics.

(Clauset et al. 2008) used LIWC to analyze 46 million words produced by 11,609 participants. The studied texts include written texts composed as part of psychology experiments carried out by universities in the US, New Zealand and England. Full texts of fictional novels, essays written for university evaluation and transcribed free conversations from research interviews were examined in these studies. Authors concluded that some of their LIWC variables were statistically significant to identify the gender of author.

Recently, few architectures and models have been introduced for authorship attribution employing Deep learning frameworks including LSTM, CNN and Recursive Neural Network (Ruder et al. 2016) (Bsir and Zrigui 2018). Models, in this paradigm, can take the advantage of the general learning procedures relying on back-propagation, 'deep learning', a variety of efficient algorithms and some other tricks to further improve the training procedure.

For instance, (Savoy et al. 2018) evaluated two neural models for gender profiling on the PAN@CLEF 2018 tweet collection. The first model is a character-based Convolutional Neural Network (CNN), while the second is an Echo State Network-based (ESN) recurrent neural network with various features. A lot of features based on words, characters and grammar were suggested to identify the characteristics of authors. There has been a recent revival of interest in using deep learning methods to solve various machine learning problems and NLP issues for the ultimate purpose of learning more robust features using easily available unlabelled data.

(Malmasi et al. 2016) employed deep learning techniques for author profiling. They showed a considerable difference between traditional machine learning models and deep learning models in the participant teams evaluated in the Third Workshop on NLP for Similar Languages, Varieties and Dialects.

In 2018, (Bsir and Zrigui 2018) implemented a bi-directional Recurrent Neural Network with a Gated Recurrent Unit (GRU) combined with stylometric features for author profiling detection. In this work, researchers achieved an accuracy of 79%, in gender classification.

(Stout et al. 2018) identified the gender of authors based on written texts and shared images. They proposed a way to combine multiple predictions on shared content into a single prediction on user level. Their system compare Naive Bayes model and a RNN. Authors got accuracy scores varying between 62.3% and 78.8% depending on the language and whether we used models that classify based on text or on images.

The approach of (Sierra et al. 2018) consisted of evaluating gender by using multi-modal information (texts and images). They learned their multimodal representation by employing GMUs. They obtained 0.80, 0.74 and 0.81 of accuracy rate in the multi-modal scenario for the test partition for English, Spanish and Arabic respectively.

(Veenhoven et al. 2018), in PAN@clef 2018, created an additional training data by translating the data from one language to another. They explored the different directions in which the translations could be made. Their approach based on bi-LSTM model, and their official test accuracy scores are 79.3, 80.4 and 74.9 for English, Spanish and Arabic respectively.

3   The Proposed Approach

It seems that these deep learning models are not explored to their full potential in the author profiling task. For this reason, we focus, in this paper, on implementing these three methods for this task.

In the following section, we propose three neural network models for author gender identification. The first model is based on a Convolutional Neural Network (CNN), while the second and third models rely on Gated Recurrent Unit (GRU) model and bidirectional Long Short-Term Memory network (LSTM), respectively.

3.1   Quantifying Words and Sequences

Word2Vec model pre-trained word vectors are used since Word2Vec captures the word context information (such as word similarity). In this research work, we apply Recurrent Neural Network (RNN) on Word2Vec word vectors as RNN can also capture the word/sentence sequence information, which allows better classifying the authorship of tweets. The experimental results confirm this assumption. The Word2Vec model was formed by the corpus of Arabic Wikipedia with 4 million tweets extracted in order to enrich the vocabulary list with words that do not exist in Wikipedia. For training, we used the skip-gram neural network model with a window of size 5 (1 center word + 2 words before and 2 words after), a minimum frequency of 15 and a dimension equal to 300.

Instead, we used 10-fold cross-validation. The dataset was divided at the note level. We separated out 10% of the training set to form the validation set.

3.2 Dataset

The dataset is part of author profiling task of PAN@CLEF 2018. It was collected from Twitter. For each tweet collections, Arabic texts are composed of tweets written by 2400 authors; 100 tweets per authors. For the Arabic language, four varieties were used in this corpus: Egypt, Gulf, Levantine and Maghrebi.

3.3 CNN Model

The model architecture, shown in Fig. 1, is a slight variant of the CNN architecture of (Collobert et al. 2011). In Our architecture, there are four layers of convolutions with nonlinear activation functions ReLU applied to the results. Instead, we started with an embedding layer with a size equal to the vocabulary size and a dimension of 300 for each word. In the second layer, we created a feature map by using a filter matrix and three different convolutional layers with kernel sizes of 2, 3 and 4 followed by a ReLU nonlinearity.

The third layer is composed of three max pooling layers, one for each preceding convolutional layer.

The final layer is a linear input of size 2 one output per class followed by a Softmax so that the outputs will be the probability for each class. The probability that the user is male and the probability that the user is female.

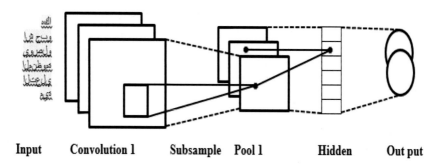

| Input | Convolution 1 | Subsample | Pool 1 | Hidden | Out put |

Fig. 1. The architecture of a CNN model.

3.4 LSTM Model

LSTM model is an extension of the RNNs structure. It is a bi-directional recurrent neural network (BRNN) introduced in (Zhu et al. 2015) where the conventional neuron replaced with a so-called memory cell controlled by input, output and forget gates in order to overcome the vanishing gradient problem of traditional RNNs. The BRNN can be trained without the limitation of utilizing input information just up to a present future frame, which was simultaneously accomplished by being trained in positive and negative time directions.

LSTM networks introduce a new structure called a memory cell where each memory cell is made of two memory blocks and an output layer, as demonstrated in Fig. 2.

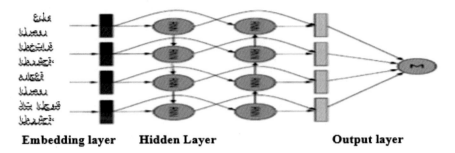

Embedding layer   **Hidden Layer**   **Output layer**

Fig. 2. The architecture of a LSTM model.

Each LSTM cell computed its internal state by applying the following iterative process and for multiple blocks; the calculations were randomly repeated for each block:

$$i_t = \sigma(W_{hi}h_{t-1} + W_{xi}xt + W_{ci}c_{t-1} + b_i) \tag{1}$$

$$f_t = \sigma(W_{hf}h_{t-1} + W_{xf}xt + W_{cf}c_{t-1} + b_f) \tag{2}$$

$$o_t = \sigma(Woh_{t-1} + W_{xo}xt + Woc_t + b_o) \tag{3}$$

$$c_t = f_t \odot c_{t-1} + i_t \odot \tanh(W_{xc}x_t + W_{hc}h_{t-1} + b_c) \tag{4}$$

$$h_t = o_t \odot \tanh(c_t) \tag{5}$$

LSTM contains a hidden state and three gates: forget, input and an output gate:

- Embedding Layer: This layer transforms each word into embedded features. In our LSTM model, we used first the same embedding layer utilized in CNN model.
- Hidden Layer: The hidden layer consists of a Bi-Directional RNN where artificial neurons take in a set of weighted inputs and produce an output through an activation function
- Output Layer: In the output layer, the representation learned from the RNN is passed through a fully connected neural network with a sigmoid output node that classifies the sentence as male or female.

In our LSTM, we employed Adam optimizer in combination with a learning rate of 0.0001, Sigmoid as an activation unit type, 100 nodes, a dropout rate of 0.5 and a batch size of 32.

3.5 GRU Model

Gated Recurrent Units (GRUs) use both past and future information stored by both the forward and backward networks, regardless of the type of network. Indeed, this bi-directional model employs the activation functions of Softmax to calculate its output. It first computes the hidden state forward, by applying the function and the hidden state backwards of an entity in the sequence, as demonstrated in Fig. 3.

The GRU cell has two gates: an update gate (z) and a reset gate (r). It reduces the three gates defined in the LSTM networks (the input, forget and output gates).

The following equations represent the gating mechanism in a GRU:

$$z^{(t)} = \sigma\left(W_z h^{(t-1)} + U_z x^{(t)} + b_z\right) \tag{6}$$

$$r^{(t)} = \sigma\left(W_r h^{(t-1)} + U_r x^{(t)} + b_r\right) \tag{7}$$

$$\widehat{h}^{(t)} = \tanh\left(r^{(t)} \odot W_{\widehat{h}} h^{(t-1)} + U_{\widehat{h}} x^{(t)} + b_{\widehat{h}}\right) \tag{8}$$

Gated recurrent units are designed to provide more persistent memory, making it easier for RNNs to capture long-term dependencies.

In this paper, we applied the same parameters employed with LSTM model.

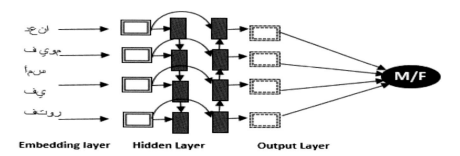

Fig. 3. The architecture of a GRU model.

4 Experimentation

We first evaluate the performance of different RNN models using word embeddings. Table 1 shows the results obtained by various RNN models applied on different metrics (specifically Accuracy, Precision, Recall and F1) after 10-fold cross validation. We observe that GRU architecture slightly outperformed other models.

This validation set was employed to evaluate our bi-directional RNN model. We also utilized a maximum of 100 epochs to train our model on an Intel core i7 machine with 16 GB memory. The accuracy results provided by our method for gender iden-tification is 79.62% for test data, as shown in Table 1. Obviously, our results are very encouraging for two reasons: (i) two model in this work (LSTM and GRU)

Table 1. Performance of various RNN architectures after 10-fold cross validation.

Model	Accuracy	Recall	F1-Score
CNN	0,631	0,652	0,662
LSTM	0,792	0,781	0,795
GRU	0,796	0,786	0,789

outperformed machine learning methods and (ii) the gap between our accuracy is the shortest, compared to the best results obtained in PAN@CLEF 2018 (81%).

From the obtained results, we notice that using BRNN-LSTM model with word embeddings may provide encouraging results and prove that bi-directional deep networks is crucial in Arabic author profiling, especially for gender identification task.

Table 2 shows some examples (1)–(3) in which GRU and LSTM predicts correctly while CNN predicts falsely.

Table 2. Examples (1)–(3) in which CNN, LSTM and GRU predicts gender of author.

Gold	CNN	LSTM	GRU	Examples
M	**M**	**M**	**M**	لماذا لا يقتبسوا من تركيا الإنتاج الهادف من أمثال أرطغرل و عبد الحميد
M	**F**	**M**	**M**	إن حاربوه اشتد وإن تركوه امتد.
F	**F**	**M**	**F**	بعدين شفتولي ممثل معارض؟

5 Conclusions

This paper proposes three deep learning models with words embeddings to predict the gender of Twitter Arabic authors.

We also compared CNN, LSTM and GRU models, finding the GRU models are better suited for this task, with official final scores of 79.62 for Arabic Tweets.

The RNN models were unable to capture stylometric differences due to grammatical structures, or grammatical structures between genders of authors in tweets. One possible future model inspired form this observation would train separate units for each possible grammatical substructure (Noun, verb, adjective, etc.).

References

Ayadi, R., Maraoui, M., Zrigui, M.: Intertextual distance for Arabic texts classification. In: International Conference for Internet Technology and Secured Transactions, ICITST 2009, pp. 1–6. IEEE, November 2009

Amodei, D., Ananthanarayanan, S., Anubhai, R., Bai, J., Battenberg, E., Case, C., Chen, J.: Deep speech 2: end-to-end speech recognition in English and Mandarin. In: International Conference on Machine Learning, pp. 173–182, June 2016

Bsir, B., Zrigui, M.: Bidirectional LSTM for author gender identification. In: Nguyen, N., Pimenidis, E., Khan, Z., Trawiński, B. (eds.) Computational Collective Intelligence, ICCCI 2018. Lecture Notes in Computer Science, vol. 11055. Springer, Cham (2018)

Bsir, B., Zrigui, M.: Enhancing deep learning gender identification with gated recurrent units architecture in social text. Computación y Sistemas **22**(3), 2018 (2018)

Clauset, A., Moore, C., Newman, M.E.: Hierarchical structure and the prediction of missing links in networks. Nature **453**(7191), 98 (2008)

Hochreiter, S., Schmidhuber, J.: Long short-term memory. Neural Comput. **9**(8), 1735–1780 (1997)

Howard, J., Ruder, S.: Finetuned language models for text classification. CoRR, abs/1801.06146 (2018)

Kalchbrenner, N., Grefenstette, E., Blunsom, P.: A convolutional neural network for modelling sentences. arXiv preprint arXiv:1404.2188 (2014)

Kodiyan, D., et al.: Author profiling with bidirectional RNNs using attention with GRUs: notebook for PAN at CLEF 2017. In: CLEF 2017 Evaluation Labs and Workshop–Working Notes Papers, Dublin, Ireland, 11–14 September 2017 (2017)

Werlen, L.M.: Statistical learning methods for profiling analysis. In: Proceedings of CLEF 2015 Evaluation Labs (2015)

Mahmoud, A., Zrigui, M.: Semantic similarity analysis for paraphrase identification in Arabic texts. In: Proceedings of the 31st Pacific Asia Conference on Language, Information and Computation, pp. 274–281 (2017)

Maraoui, M., Antoniadis, G., Zrigui, M.: Un système de génération automatique de dictionnaires étiquetés de l'arabe. CITALA **2007**(18–19), 2007 (2007)

Miao, Y., Yu, L., Blunsom, P.: Neural variational inference for text processing. In: International Conference on Machine Learning, pp. 1727–1736, June 2016

Rangel, F., Rosso, P., Montes-y-Gómez, M., Potthast, M., Stein, B.: Overview of the 6th author profiling task at pan 2018: multimodal gender identification in Twitter. Working Notes Papers of the CLEF (2018)

Ruder, S., Ghaffari, P., Breslin, J.G.: Character-level and multi-channel convolutional neural networks for large-scale authorship attribution. arXiv preprint arXiv:1609.06686 (2016)

Sap, M., Park, G., Eichstaedt, J., Kern, M., Stillwell, D., Kosinski, M., Ungar, L., Schwartz, H. A.: Developing age and gender predictive lexica over social media. In: Proceedings of the 2014 Conference on Empirical Methods in Natural Language Processing (EMNLP), pp. 1146–1151 (2014)

Schaetti, N., Savoy, J.: Comparison of neural models for gender profiling (2018)

Malmasi, S., et al.: Discriminating between similar languages and arabic dialect identification: a report on the third DSL shared task. VarDial 3 (2016)

Socher, R., Perelygin, A., Wu, J., Chuang, J., Manning, C.D., Ng, A., Potts, C.: Recursive deep models for semantic compositionality over a sentiment treebank. In: Proceedings of the 2013 Conference on Empirical Methods in Natural Language Processing, pp. 1631–1642 (2013)

Stamatatos, E., Rangel, F., Tschuggnall, M., Stein, B., Kestemont, M., Rosso, P., Potthast, M.: Overview of PAN 2018. In: International Conference of the Cross-Language Evaluation Forum for European Languages, pp. 267–285. Springer, Cham, September 2018

Stout, L., Musters, R., Pool, C.: Author profiling based on text and images. In: Experimental IR Meets Multilinguality, Multimodality, and Interaction. Proceedings of the Ninth International Conference of the CLEF Association (CLEF 2018), September 2018

Szegedy, C., Vanhoucke, V., Ioffe, S., Shlens, J., Wojna, Z.: Rethinking the inception architecture for computer vision. In: Proceedings of the IEEE Conference on Computer Vision and Pattern Recognition, pp. 2818–2826 (2016)

Veenhoven, R., Snijders, S., van der Hall, D., van Noord, R.: Using translated data to improve deep learning author profiling models. In: Experimental IR Meets Multilinguality, Multimodality, and Interaction. Proceedings of the Ninth International Conference of the CLEF Association (CLEF 2018), September 2018

Zhu, X., Sobihani, P., Guo, H.: Long short-term memory over recursive structures. In: International Conference on Machine Learning, pp. 1604–1612, June 2015

Zrigui, M., Ayadi, R., Mars, M., Maraoui, M.: Arabic text classification framework based on latent dirichlet allocation. J. Comput. Inf. Technol. **20**(2), 125–140 (2012)

A Novel Decision Tree Algorithm for Fault Location Assessment in Dual-Circuit Transmission Line Based on DCT-BDT Approach

V. Ashok$^{(\boxtimes)}$ and Anamika Yadav

Department of Electrical Engineering, NIT Raipur, Raipur, Chhattisgarh, India
ashokjntuk@gmail.com, ayadav.ele@nitrr.ac.in

Abstract. This paper exemplified a novel decision tree algorithm for fault location estimation/assessment based on DCT-BDT approach for a 400 kV dual-circuit transmission line of Chhattisgarh state power transmission system. In this proposed algorithm, a well-known DCT has been used to pre-process the signals, the standard deviation value of 3-cycle data of three- phase current signals (two circuits) & voltage signals from sending end bus are used as an input to erudite the decision tree modules and the percentage of error attained in fault location assessment is within the acceptable limit. A wide-ranging simulation studies have been executed in MATLAB/Simulink software for all types of common shunt faults and the performance of DCT-BDT based fault location estimation is appraised at diverse situations of power system by changing fault type and varying different fault parameters such as fault location, fault resistance, fault inception angle with existence of mutual coupling. The simulation results confide the efficacy of proposed fault location algorithm at widespread fault scenarios.

Keywords: Distance relaying · DCT-Discrete Cosine Transform ·
BDT-Bagged Decision Tree · Dual-circuit transmission line

1 Introduction

The detection and location of faults in long transmission line having significant role to make power transmission system more reliable. Fault location estimation is one of the important features in distance relays which are employed in transmission system. Through proper estimation of fault location, the amount of time required to attend the fault location by line patrolling people will be reduced to minimum. An improper location of faults leads to reliability and stability issues in transmission system. In this context many researchers and practicing engineers have come-up with numerous algorithms to estimate fault location in transmission lines since many years. These algorithms can be classified based on the parameters which have been used as a feature; (a) based on impedance measurements at relay location [1, 2] (b) based on differential components measured at both the end of the transmission line [3] and (c) the traveling wave theory which uses recorded signals of either end of the transmission line [4, 5].

© Springer Nature Switzerland AG 2020
A. Abraham et al. (Eds.): ISDA 2018, AISC 941, pp. 801–809, 2020.
https://doi.org/10.1007/978-3-030-16660-1_78

Many of these location algorithms designed which are using line reactance are vacillating with under reach problem owing to high impedance fault condition and over reach problem owing to DC offset component. Travelling wave theory based fault location schemes have glitches at close-in fault condition and also shunt fault with zero (approximate) fault inception angle.

In the perspective of machine learning models such as tree-based ensemble techniques are commonly used for prediction and classification/regression purposes in many research fields such as protection of transmission lines for instance. One such tree-based ensemble techniques is Bagged Decision Tree (BDT) which builds, by randomization, numerous decision trees and then aggregates their predictions. From an ensemble of trees, one can derive an importance score for each variable of the problem that assesses its relevance for predicting the output. Every tree in the ensemble is grown on an independently drawn bootstrap replica of input data [6]. Observations not included in this replica are "out of bag" for this tree [7]. The bagging is nothing but bootstrap aggregation of set of decision trees. Individual decision trees tend to over-fit. The bagged decision tree (Bootstrap-aggregated) combines the outcomes of several decision trees, thereby overwhelming the effects of over-fitting and advances generalization. This bagging tree grows with the decision trees in the ensemble using bootstrap samples of the data. The expected out-of-bag error discrepancies can be reduced by adjusting a more stable misclassification cost matrix or a less uneven preceding likelihood vector. BDT produces in-bag samples by oversampling classes with large misclassification costs and under sampling classes with small misclassification costs. Thus, out-of-bag samples have less observation from the classes with large misclassification costs and more observations from the classes with lesser misclassification costs. To erudite a BDT ensemble classifier/regression using a small data set, an extremely skewed cost matrix is considered, and then the number of out-of-bag observations per class might be very low. In both bagging and random forest, many individual decision trees are built on bootstrapped version of the original dataset and are ensemble together. The difference is that, in random forest, a random subset of variables are considered during node split while in bagging of decision trees, all the variables are considered in a node split [8]. These bagged decision tree models mainly explores two different functionalities such as classification tree and regression tree. Classification tree analysis is used when the predicted outcome is a class/label to which the data belongs. Regression tree analysis is used when the predicted outcome can be considered a real number.

This paper presents a novel decision tree algorithm based on DCT-BDT approach to estimate/assess fault location in a practical dual-circuit transmission line of Chhattisgarh state transmission system using only one-end data with existence of mutual coupling. The results attained through simulations demonstrated that all the shunt faults are located appropriately within acceptable error. The algorithm is impenetrable to mutual coupling and diverse system circumstances. The proposed fault location algorithm will not entail any communication platform to record neither remote-end nor zero sequence data of neighboring lines. Since, it has not been described previously for estimation of fault location in existing power system network.

2 Practical Power Transmission Network Under Study

A 400 kV, 50 Hz India power transmission network of Chhattisgarh state has been described in Fig. 1 and appropriate network data is taken from [9]. The network consists of two power stations at bus-4 (KSTPS/NTPC), Station-I with 4 × 500 MW and Station-II with 3 × 210 MW, four dual-circuit transmission lines (towards Vindyachal with 215 km, Raipur/PGCIL-B with 220 km, towards Birsinghpur with 231 km and Bhilai/Khedamara with 198 km) and three single-circuit transmission lines (towards Korba-West with 17 km, Bhatpara with 100 km, from Sipat with 60 km). At bus-3 (Bhilai/Khedamara) one triple circuit line (from Raipur/Raita with 65.68 km), three feeders connected to Bhilai 220 kV sub-station through step down transformer, one dual-circuit line (from KSTPS/NTPC 198 km) and seven single-circuit lines (from Korba-West with 212 km, Bhatpara with 90 km, towards Raipur/PGCIL-A with 20 km, Seoni with 250 km, Koradi with 272 km, Bhadrawati with 322 km and from Marwa with 170 km) are connected. Here the double circuit transmission line connected between bus-4&3 (KSTPS/NTPC & Bhilai/Khedamara) of 198 km length is considered. This existing 400 kV power transmission network of Chhattisgarh state has been modeled in MATLAB/Simulink software to conduct simulation studies for various kinds of shunt faults. The proposed fault location algorithm has been applied at bus-4 (KSTPS/NTPC), by simulating shunt faults at a wide range of fault parameters.

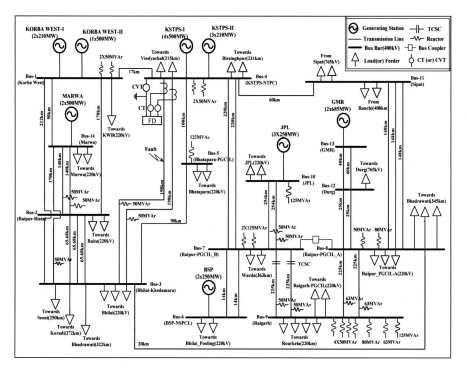

Fig. 1. Schematic view of 400 kV transmission network of Chhattisgarh state

3 Fault Location Algorithm Based on DCT-BDT Approach

The proposed fault location algorithm deals with all types of shunt faults which occur in dual-circuit transmission lines. Herein four fault locator modules (DCT-BDT module) are designed individually to locate all 10 types of shunt faults. With the help of these DCT-BDT modules, location of all kinds of shunt faults can be done by using only one-end data and flow chart of proposed fault location algorithm has been shown in Fig. 2. The simulation studies have been performed in MATLAB/Simulink environment at different fault scenarios to demonstrate the robustness of the proposed fault location algorithm.

3.1 Design of Input/Target Data Set to Erudite DCT-BDT Modules

Designing of DCT-BDT modules are more concerned to input/target dataset of faulty-signals which was recorded at relaying point for different fault scenarios. When the fault occurs, some miscellaneous frequency components will appear along with a DC offset component. Consequently, a number of non-fundamental frequency components transformed at different fault locations. As the performance of decision tree modules will be determined by the input & output topographies hence, it is pre-requisite to pre-process and extract the expedient topographies from recorded data to erudite the DCT-BDT module for precise fault location estimation. A well-known signal processing technique DCT has been used to preprocess fault signals. The Discrete Cosine Transform (DCT) which is analogous to Discrete Fourier Transform though it contemplates symmetry of real/even coefficients to cope up various vacillating frequencies. The standard deviation of 3-cycle data (1-cycle of pre-fault & 2- cycle of post-fault) of voltage and current signals of dual-circuit transmission line are used to design an input data set to erudite DCT-BDT modules. The parameters which have been varied to create input/target data set to erudite DCT-BDT modules are reported in Table 1. Therefore, the total 10 number of inputs are considered to each DCT-BDT module as presented in "(1)". There is one target which is analogous to the input data set of DCT-BDT module. Therefore, the fault location can be estimated as shown in "(2)".

Table 1. Parameters are used to generate dataset for erudition of DCT-BDT module

Parameter	Training	Testing
Fault type	R1G, Y1G, B1G, R1Y1G, Y1B1G, R1B1G, R1Y1, Y1B1, R1B1, R1Y1B1	R1G R1Y1G R1Y1 R1Y1B1
Fault location (L_f)	(2–196) km line in steps of 2 km	(1–197) km line in steps of 2 km
Fault inception angle (Φ_f)	0°, 90° and 270°	0°, 90° and 270°
Fault resistance (R_f)	0 Ω, 50 Ω and 100 Ω	0 Ω, 50 Ω and 100 Ω
No. of fault cases	LG: 2646 LLG: 2646 LL: 2646 LLL: 0882	LG: 891 LLG: 891 LL: 891 LLL: 891

$$P = [Ir1, Iy1, Ib1, Vr1, Vy1, Vb1, Ig1, Ir2, Iy2, Ib2] \qquad (1)$$

$$Y = [L] \qquad (2)$$

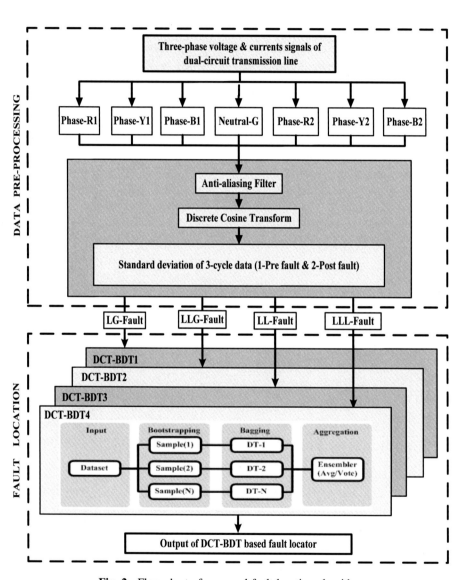

Fig. 2. Flow chart of proposed fault location algorithm

3.2   Training of DCT-BDT Modules for Fault Location Estimation

The erudition of DCT-BDT module has been done by considering a regression tree analysis instead of classification tree analysis. Because the regression tree analysis is most suitable when the predicted outcomes can be expected as a real number unlike classification tree analysis can predicts outcomes as classes/labels [10]. This bagged decision tree sum-up a set of trained weak learners and the data on which these learners are trained. It can predict ensemble response for new data by aggregating predictions from its weak learners. The average of all the predictions from different regression trees are used (in bagged decision tree model) which is more robust than a single decision tree. Herein this work, the total four DCT-BDT modules are designed/erudite to estimate fault location for all common shunt faults; DCT-BDT1 for SLG (R1G, Y1G, B1G) faults, DCT-BDT2 for LLG (R1Y1G, R1B1G, Y1B1G) faults, DCT-BDT3 for LL (R1Y1, R1B1, Y1B1) faults and DCT-BDT4 for LLL (R1Y1B1/R1Y1B1G) faults. All 10 types of shunt faults including LG, LLG, LL, and LLLG are simulated in the three phase dual-circuit transmission line at different fault locations with change in fault resistance (0, 50, 100) Ω and fault inception angles (0, 90, 270)° to confirm efficacy of fault location algorithm. The different DCT-BDT modules and their training outcomes such as mean square error (MSE), mean absolute error (MAE), root mean square error (RMSE), R-Squared (R^2) and training time etc., have been reported in Table 2.

Table 2. Training parameters of different DCT-BDT modules

Module	Size of input dataset	Min. leaf size/no. of learners	MSE/MAE	RMSE/R^2	Training time (sec)
DCT-BDT1	11 × 2646	1/124	0.43/0.48	0.66/1	1.8336
DCT-BDT2	11 × 2646	1/184	0.27/0.33	0.52/1	2.5376
DCT-BDT3	11 × 2646	1/64	0.21/0.31	0.45/1	1.0132
DCT-BDT4	11 × 0882	1/145	0.35/0.36	1.59/1	1.0831

4   Test Results of Proposed Fault Location Algorithm

The fault location assessment based on DCT-BDT approach has been done with widespread data sets consisting of various fault scenarios at different fault locations which have not been considered in erudition of modules. To study the impact of fault parameters on performance of the proposed fault location algorithm, it has been tested with separate data set which is epitomizes wide range of fault circumstances. The trained DCT-BDT modules have been tested and performance was corroborated by simulating different fault cases with change in fault locations $L_f = 1$–197 km in steps of 2 km), fault resistances ($R_f = 0\ \Omega$, 50 Ω, 100 Ω) and fault inception angles ($\Phi_f = 0°$, 90°, 270°). From Table 3, test results of DCT-BDT modules at different fault locations are estimated/assessed properly and their mean error for 99 different locations along the transmission line were summarized. The estimated mean error "E" in percentile for fault location assessment is computed using (3).

$$\%\text{Error} = \left[\left(L_{f\,(\text{Actual})} - L_{f\,(\text{Estimated})} \right) / L_{t\,(\text{Line Length})} \right] \times 100 \qquad (3)$$

Table 3. Test results of different fault location modules (DCT-BDT)

DCT-BDT module	Type of fault	Fault resistance (Ω)	Fault inception angle (°)	Mean error (%) (99-Locations)
DCT-BDT1	R1G	0	0	0.1185
		0	90	−0.1545
		0	270	−0.1546
		50	0	−1.4367
		50	90	−1.2711
		50	270	−1.2710
		100	0	−1.2953
		100	90	−0.9496
		100	270	0.9489
DCT-BDT2	R1Y1G	0	0	−0.0309
		0	90	0.05517
		0	270	0.5518
		50	0	−0.4723
		50	90	−0.1061
		50	270	−0.1069
		100	0	−1.0085
		100	90	−1.1811
		100	270	−1.1811
DCT-BDT3	R1Y1	0	0	0.8921
		0	90	0.0288
		0	270	0.0039
		50	0	0.1988
		50	90	−0.4858
		50	270	−0.5135
		100	0	0.2369
		100	90	−0.9670
		100	270	−0.9097
DCT-BDT4	R1Y1B1	0	0	2.062
		0	90	2.0187
		0	270	2.1864
		50	0	0.0900
		50	90	0.4290
		50	270	0.4335
		100	0	0.2043
		100	90	0.3232
		100	270	0.3239

5 Comparative Assessment of Proposed Fault Location Algorithm

The comparative assessment has been done thereby applying different decision tree algorithms like Bagged Decision Tree (BDT), Boosted Decision Tree (BSDT) which are used based on regression tree analysis. Although DCT-BDT and DCT-BSDT are showing comparable performance, from Table 4 it has been confirmed that DCT-BDT is outperforming to DCT-BSDT with the mean error as minimum as. The main objective is to estimate the fault location accurately so that it can helps to improve power system reliability. It is also worth to mention computer system configuration when dealing with machine learning techniques such as decision tree algorithm, SVM, ANN etc., since the CPU speed and its internal memory capacity plays an vital role while pre-processing of raw data, training & testing as well. This research work has been carried out on: Intel® Xeon® CPU E3-1225 v5 @ 3.30 GHz processor with 8 GB RAM 64-bit Windows 7 operating system.

Table 4. Comparative assessment of different decision tree algorithms

Type of fault	Mean error (%)	
	DCT-BDT	DCT-BSDT
LG	−0.6072	−1.1158
LLG	−0.3866	−0.6628
LL	−0.1947	−0.1152
LLL	0.8967	1.2068

6 Conclusion

In this paper a novel decision tree algorithm for fault location assessment based on DCT-BDT approach has been reported for all 10 types of shunt faults; Standard deviation of 3-cycle data of three phase currents and voltages of dual-circuit transmission line has been given as input to the DCT-BDT modules which are erudite separately for location of LG faults, LLG faults, LL faults and LLL faults. The performance of proposed fault location assessment algorithm has been examined at wide range of fault situations in presence of mutual coupling. The comparative assessment has been done by applying different decision tree algorithms and test results are proved that DCT-BDT is outperforming to DCT-BSDT with minimum error in fault location estimation. The fault location estimation/assessment can be done using this algorithm more accurately which can improve reliability of power transmission system by reducing outage duration.

Acknowledgments. The authors acknowledge the financial support of Central Power Research Institute, Bangalore for funding the project. no. RSOP/2016/TR/1/22032016, dated: 19.07.2016. The authors are grateful to the Head of the institution as well as Head of the Department of Electrical Engineering, National Institute of Technology, Raipur, for providing the research amenities to carry this work. The authors are indebted to the local power utility (Chhattisgarh State Power Transmission Company Limited) for their assistance in data collection.

References

1. Sachdev, M.S., Agarwal, R.: A technique for estimating transmission line fault locations from digital impedance relay measurement. IEEE Trans. Power Deliv. **3**(1), 121–129 (1988)
2. Mazon, A.J., et al.: New method of fault location on double-circuit two-terminal transmission lines. Electr. Power Syst. Res. J. **35**(3), 213–219 (1995)
3. García-Gracia, M., Osal, W., Comech, M.P.: Line protection based on the differential equation algorithm using mutual coupling. Electr. Power Syst. Res. J. **77**(5–6), 566–573 (2007)
4. Thomas, D.W.P., Carvalho, R.J.O., Pereira, E.T.: Fault location in distribution systems based on traveling waves. In: Proceedings of IEEE PowerTech Conference, vol. 2, pp. 1–5 (2003)
5. Thomas, D.W.P., Christopoulos, C., Tang, Y., Gale, P., Stokoe, J.: Single ended traveling wave fault location scheme based on wavelet analysis. In: Proceedings of IEE International conference on Development in Power System Protection, vol. 1, pp. 196–199 (2004)
6. Breiman, L.: Bagging predictors. Mach. Learn. **24**(02), 123–140 (1996)
7. Polikar, R.: Ensemble based systems in decision making. IEEE Circuits Syst. Magzine **06** (03), 21–45 (2006)
8. TreeBagger class. https://in.mathworks.com/help/stats/treebagger-class.html#bvfstrb
9. Ashok, V., Yadav, A., Antony, C.C., Yadav, K.K., Yadav, U.K., Sahu, S.K.: An intelligent fault locator for 400-kv double-circuit line of Chhattisgarh state: a comparative study. In: Soft Computing in Data Analytics. Advances in Intelligent Systems and Computing, vol 758. Springer, Singapore (2018)
10. Loh, W.-Y.: Classification and regression trees. WIREs Data Min. Knowl. Discov. **1**, 14–23 (2011)

Sizing and Placement of DG and UPQC for Improving the Profitability of Distribution System Using Multi-objective WOA

Hossein Shayeghi[1(✉)], M. Alilou[2], B. Tousi[2],
and R. Dadkhah Doltabad[3]

[1] Department of Electrical Engineering,
University of Mohaghegh Ardabili, Ardabil, Iran
hshayeghi@gmail.com
[2] Department of Electrical Engineering, Urmia University, Urmia, Iran
masoud.alilou@yahoo.com, b.tousi@urmia.ac.ir
[3] Department of Electrical Engineering, Technical and Vocational University,
Technical and Vocational Institute Razi, Ardabil, Iran

Abstract. Distributed Flexible AC Transmission System (D-FACTS) devices improve the performance of distributed Generation (DG) units in the distribution system. Optimizing the location and size of DGs and Unified Power Quality Conditioner (UPQC) is done in the distribution system with hourly load model in this study. The model of network load is considered as the hourly variation of consumption. The technical and economic indices are optimized during the proposed method so that the profitability of network increases after locating the devices. The combination of multi-objective whale optimization algorithm and fuzzy decision-making method is utilized to find the optimal location and capacity of DGs and UPQC. Finally, the proposed method is applied to the 34-bus standard distribution system. The results show that the technical performance and the profitability of distribution system are improved considerably after operating the devices in the best location with optimal capacity.

Keywords: Distributed generation · UPQC · Multi-objective optimization · Whale optimization algorithm

1 Introduction

DG unit is an electric power source connected directly to the distribution network or on the customer site of the matter. Nowadays, with respect to the technical developments, enormous benefits can be achieved from DG in economical, technical, and environmental fields. Those advantages could be earned by optimal selection, sizing, and placement of DGs in power electrical systems [1]. On the other hand, UPQC is the practical type of D-FACTS that it is connected to the system in series-shunt. During the voltage sag and over loading, the load voltage of the bus in which the UPQC is connected can be regulated by injection of compensating current into the system. Therefore, it can properly compensate the active and reactive power and also voltage of distribution system [2].

© Springer Nature Switzerland AG 2020
A. Abraham et al. (Eds.): ISDA 2018, AISC 941, pp. 810–820, 2020.
https://doi.org/10.1007/978-3-030-16660-1_79

In the last years, some researchers studied DG units and UPQC in the distribution system. The references [3–8] are sample of these studies. In these articles, the optimization of location and capacity of DG units and UPQC were studied in the different conditions. In Ref [3], optimization of location and size of DG was studied in the distribution system in the presence of capacitor bank; the loss index and the voltage index of distribution system are improved during the optimization. In references of [4, 5], finding the best location and size of distributed generation units was done based on technical indices of distribution system. In other study, locating and sizing of FACTS devices in a power system was developed by using dedicated improved particle swarm optimization algorithm for decreasing the overall costs of power generation and maximizing of profit [6]. Sarker and Goswami utilized cuckoo optimization algorithm for allocating the unified power quality conditioner in the distribution system [7]. Totally, improving the efficiency of distribution system is the main purpose of all studies; of course, the objective functions and proposed procedures of them are different.

In this paper, simultaneous optimization of DG and UPQC is done in the distribution system with nonlinear load model. The load model of distribution system is considered as the hourly variation of consumption. The reduction of active and reactive loss and improvement of voltage profile are considered as the technical indices of problem. Of course, the economic index is also considered in this study. The multi-objective whale optimization algorithm is used to optimize the objective functions and create the Pareto fronts. After multi-objective optimizing the objective functions, fuzzy decision-making method is employed to extract one of the Pareto-optimal solutions as the best compromise one. Finally, the proposed method is evaluated using the 34-bus test distribution system. The results show that the technical performance and the profitability of distribution system are improved considerably after operating the devices in the best location with optimal capacity.

2 Problem Definition

In this paper, the simultaneous optimization of DG units and UPQC is done in the distribution system with hourly load model. The considered model of distributed generation, UPQC and load model of distribution system are explained as follow.

2.1 Load Model

In the practical situation of operating, the amount of consumption changes during the 24-hour. Therefore in this study, the hourly load model is considered for evaluating the proposed method in a more realistic condition of operating [9]. The load model of the distribution system is changed based on customers' daily load patterns [9].

2.2 Distributed Generation

Distributed generation uses different technologies to produce electrical energy. Liquid fuels, natural gas, wind power and solar energy are some of the initial sources that they are changed to electrical energy by distributed generation units. In this study, it is considered that DG unit can produces active and reactive power, simultaneously. Therefore, in the load flow equations, distributed generation unit is considered as a PQ bus which provides active and reactive power; the operating power factor of DG is set at 0.85 [1].

2.3   UPQC

Unified power quality conditioner (UPQC) is integrated of series and shunt compensators. Therefore, it is connected to the distribution system in series-shunt. The sample diagram of distribution system with UPQC is shown in Fig. 1. Totally, the UPQC consists of two voltage source converters connected in cascade through a common DC link capacitor. Each converter can operate with variable pulse pattern; therefore operation of the two converters is completely independent and not linked by the DC capacitor voltage. In this study, the steady state mode of UPQC is considered for improving the efficiency of distribution system [10].

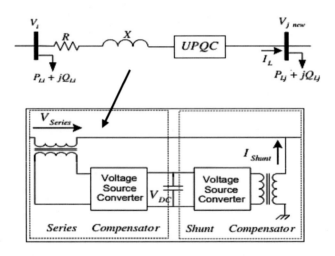

Fig. 1. Single line diagram of system with UPQC [10]

3   Objective Functions

The simultaneous optimization of location and capacity of DG units and UPQC is done as the multi-objective. The active-reactive loss index, voltage profile index and the economic index are considered as the goals of optimization in this study. Mathematically, the main objective function is formulated by (1).

$$objective\ function : \min\{I_L, I_V, I_E\} \tag{1}$$

Where, I_L, I_V and I_E are the loss index, voltage index and economic index, respectively. In the following, the considered objective functions are explained.

3.1   Loss Index

Loss index as the most important technical index is defined with combination active and reactive power loss as:

$$I_L = C_p L_A + C_q L_R \tag{2}$$

$$L_A = P_l / P_{l0} \tag{3}$$

$$L_R = Q_l / Q_{0l} \tag{4}$$

Where, P_l and Q_l are the active and reactive loss after installation while P_{0l} and Q_{0l} are the active and reactive losses before installation, respectively. Active and reactive losses are calculated by (5) and (6). Where, R_i and X_i are the resistance and reactance of branch i, respectively. The parameter I_i is the current of branch i and N_{br} is the branch number.

$$Active_L = \text{sum}_{h=1}^{24} \left\{ \sum_{i=1}^{N_{br}} R_i |I_{hi}|^2 \right\} \tag{5}$$

$$Reactive_L = \text{sum}_{h=1}^{24} \left\{ \sum_{i=1}^{N_{br}} X_i |I_{hi}|^2 \right\} \tag{6}$$

3.2   Voltage Index

In this study, voltage profile index is considered as the voltage index. Therefore, this index is expresses by (7).

$$I_V = VP / VP_0 \tag{7}$$

Where, VP_0 and VP are the voltage profile index of the distribution system before and after installation of devices, respectively. This index indicates bus voltage deviation from nominal voltage. Hence, the network performance will be better when the amount of this index is closer to zero.

$$V_p = \text{sum}_{h=1}^{24} \{V\_profile_h\} \tag{8}$$

$$V\_profile_h = \sum_{i=1}^{n} (V_{ih} - V_b)^2 \tag{9}$$

3.3   Economic Index

Economic issues are an integral part of decision-making in all daily activities. The power grid is also not exempt from this. For this reason, the economic index is defined according to (10).

$$I_E = (I_R - I_C)/(I_{R0} - I_{C0}) \tag{10}$$

Where, I_{R0} and I_{C0} are the initial revenue and cost of distribution system while I_R and I_C are the ultimate revenue and cost of network, respectively. The initial and ultimate revenues of distribution system are equal to obtained income from selling energy to customers (Eq. 11).

$$I_{R0} = I_R = \sum_{i=1}^{n_b} P_{l\_i} \times C_{MR} \times 24 \tag{11}$$

In this equation, C_{MR} is the energy market price. On the other hand, the initial and ultimate costs of distribution system are calculated by (12 and 13).

$$I_{C0} = \left[\sum_{i=1}^{n_b} P_{l\_i} \times C_A \times 24 \right] + \left[\sum_{h=1}^{24} P_{l_h} \times C_A \right] + \left[\sum_{h=1}^{24} Q_{l_h} \times C_R \right] \tag{12}$$

$$I_C = \left[\left(\sum_{i=1}^{n_b} P_{l_i} - \sum_{i=tech} \sum_{j=1}^{n_{DG}} P_{DG_{ij}} \right) \times C_A \times 24 \right] + \left[\sum_{h=1}^{24} P_{l_h} \times C_A \right] +$$
$$\left[\sum_{h=1}^{24} Q_{l_h} \times C_R \right] + \left[\sum_{i=tech} \sum_{j=1}^{n_{DG\_i}} P_{DG_{ij}} C_{DG_i} \right] + \left[\sum_{i=1}^{n_c} Q_{UPQC_i} \times C_i \right] \tag{13}$$

In these equations, the initial cost of network consists of the cost of demand and the cost of losses while the ultimate cost consists of the cost of demand, the cost of losses and the cost of installed devices.

4 Proposed Method

In this study, the multi-objective whale optimization algorithm (MOWOA) is performed for simultaneous optimizing of location and size of DG and UPQC. After multi-objective optimization, the fuzzy decision-making method is employed to extract one of the Pareto optimal solutions as the best compromise one.

In the MOWOA, particles update based on 'Shrinking encircling mechanism' and 'Spiral updating position' during the optimization [11].

Shrinking encircling mechanism: Humpback whales can recognize the location of prey and encircle them. In the WOA, the current best candidate solution is assumed as the target prey. Then, the other search agents will try to update their positions towards the best search agent. This behavior is represented by (14) [11].

$$\vec{X}(t+1) = \overrightarrow{X^*}(t) - \vec{A}.\vec{D} \tag{14}$$

Spiral updating position: in the MOWOA, after calculating the distance between the whale located at (X, Y) and prey located at (X^*, Y^*), a spiral equation is created between the position of whale and prey to mimic the helix-shaped movement of humpback whales (Eqs. 15 and 16) [11].

$$\vec{X}(t+1) = \vec{D}. e^{bl}. \cos(2\pi l) + \overrightarrow{X^*}(t) \qquad (15)$$

$$\vec{D} = \left| \overrightarrow{X^*}(t) - \vec{X}(t) \right| \qquad (16)$$

In the MOWOA, it is assumed that there is a probability of 50% to choose between either the shrinking encircling mechanism or the spiral model to update the position of whales during optimization [11].

In the fuzzy method, the membership function is defined by (17). Moreover, for each member of the non-dominated set, the normalized membership value is calculated using the (18). The maximum value of the membership μ^k is chosen as the best compromise solution [12].

$$\mu_i^k = \begin{cases} 1 & F_i^k \leq F_i^{min} \\ \frac{F_i^{max} - F_i^k}{F_i^{max} - F_i^{min}} & F_i^{min} < F_i^k < F_i^{max} \\ 0 & F_i^{max} \leq F_i^k \end{cases} \qquad (17)$$

$$\mu^k = \frac{\sum_{i=1}^{NO} \mu_i^k}{\sum_{k=1}^{NK} \sum_{i=1}^{NO} \mu_i^k} \qquad (18)$$

Totally, to achieve the best result; firstly, objective functions is optimized using the MOWOA as multi-objective. After applying the intelligent algorithm and creating the Pareto fronts, the fuzzy decision-making method is utilized to select the best particle as the best location and capacity of DG and UPQC in the distribution system. Using the mentioned method, the complete algorithm for simultaneous placement of devices in the distribution system is shown in Fig. 2.

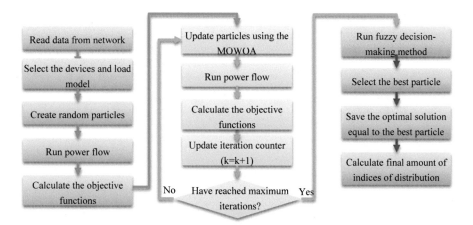

Fig. 2. Flowchart of the proposed method for placement of DG and UPQC

5 Numerical Results and Discussion

In this section, the proposed algorithm for simultaneous placement of DG and UPQC is applied on IEEE 34-bus distribution system. It is presumed that the maximum available capacities of DG and UPQC are 2.5 MW and 2.5 Mvar, respectively. The commercial information is presented in Table 1.

Table 1. The commercial information of network and devices

Parameter	Unit	Value	Market price		
			Load level	Period	$C_{MP}(\$/MW\,h)$
C_A	$\$/MW\,h$	14.192	Light	$(23<h<7)$	35
C_R	$\$/MVar\,h$	3.978	Medium	$(7<h<19)$	49
C_{DG}	$\$/MW\,h$	0.576	Peak	$(19<h<23)$	70
C_{UPQC}	$\$/Mvar\,h$	0.043			

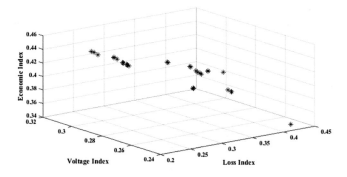

Fig. 3. Pareto front after applying the NSFA

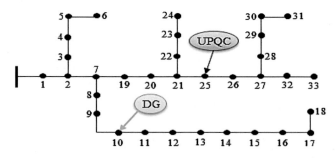

Fig. 4. The single diagram of 34-bus system with DG and UPQC

As mentioned above, the MOWOA is applied to improve the indices (as given by Eq. (1)) of 34-bus distribution system. Figure 3 shows a Pareto front after applying the multi-objective intelligent algorithm. After optimizing the objective functions, the fuzzy method is utilized to select the best location and size of devices. The best location of DG and UPQC in the 34-bus network is shown in Fig. 4. The capacity of DG unit is 1.9723 MW while the capacity of UPQC is 0.8346 Mvar.

Table 2. The values of active and reactive losses before/after optimizing

Load model	Initial		Proposed method	
	$Active_L$ (MW)	$Reactive_L$ (Mvar)	$Active_L$ (MW)	$Reactive_L$ (Mvar)
Constant	1.9019	0.5684	0.5831	0.1346
Industrial	1.5147	0.4527	0.4937	0.1124
Commercial	0.9174	0.2743	0.4612	0.1114
Residential	1.0642	0.3181	0.4511	0.1059

Table 3. The values of voltage index in the different load models

Load model	Initial	Proposed method
Constant	0.32815	0.05159
Industrial	0.26131	0.03061
Commercial	0.15825	0.01429
Residential	0.18357	0.01553

In the following, the performance of the proposed method is evaluated based on the amount of indices of distribution system. The values of active and reactive losses of distribution system with different load models before and after locating of devices are presented in Table 2. The loss index of 34-bus distribution network is improved about 60–70% after operating of DG and UPQC. The percentage of improvement of loss index in the various tests is shown in Fig. 5. The effect of locating of devices on the considered voltage index of 34-bus system is shown in Table 3.

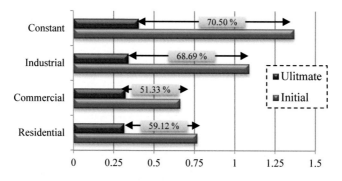

Fig. 5. The improvement of loss index by proposed method

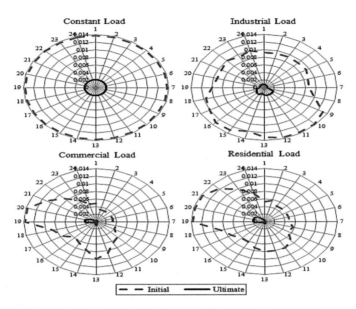

Fig. 6. The hourly variation of voltage profile of 34-bus distribution system

Table 4. The amounts of economic indices of 34-bus system

($)		Load model			
		Constant	Industrial	Commercial	Residential
Initial	Income	3699.05	3301.12	2624.08	2845.35
	Cost	1104.98	988.41	761.98	825.65
	Profit	2594.06	2312.71	1862.09	2019.70
Ultimate	Income	3699.05	3301.12	2624.08	2845.35
	Cost	450.22	333.64	107.22	170.89
	Profit	3248.83	2967.48	2516.86	2674.46

The evaluation of the information of this table indicates that the index of voltage profile of test system is improved about 85% after installing of DG and UPQC. Therefore, the stability of voltage is increased by applying the proposed method. For better showing the performance of the proposed method, the hourly variation of voltage profile of 34-bus distribution system is shown in Fig. 6. As can be seen, the proposed method has proper performance during the day based on the voltage index so that the voltage profile of network is decreased considerably after operating of DG units and UPQC. Therefore, the proposed method has good performance based on the considered technical indices. Operating of distributed generation and UPQC in the best location with optimal capacity increases the profitability of 34-bus distribution system, too. The amounts of economic parameters of standard system are presented in Table 4.

According to this Table, the profitability of distribution system increases about 660 $ (25%) after applying the proposed method. Of course, this amount of reduction is achieved every day; therefore, the improvement of profit of distribution system will be considerable during one year.

Totally, it can be said that the operation of DG and UPQC in the best location with optimal capacity improves the technical indices of distribution system and causes to increase the profitability of power electrical network.

6 Conclusion

In this paper, simultaneous placement of Distributed Generation and Unified Power Quality Conditioner was studied in the IEEE 34-bus distribution system. The combination of MOWOA and fuzzy decision-making method was used for optimizing the proposed technical (the indices of loss and voltage) and economic (the profitability of distribution system) functions. The hourly variation of consumption was considered during the optimization for evaluating the proposed method in the more realistic conditions of operation.

The results show that the proposed algorithm has practical performance in the nonlinear load models and it affects the amount of loss and voltage indices so that their variations become more linear during the change of load model. This substantial reduction causes that if load model of some buses changes to another model, network efficiency will not change much. Distributed generation has the useful effect on the amount of technical indices. Of course, the technology of UPQC improves the performance of DG unit in the distribution system. On the other hand, operating of devices by proposed method increases the profitability of distribution system. The MOWOA has an accurate performance in the proposed problem of simultaneous optimization of location and size of devices. Moreover, the best location and capacity of the DG and UPQC can be selected properly by applying the fuzzy decision-making method to the Pareto fronts. Totally, the results indicate the high performance of the proposed method in improving the considered indices of distribution system in the nonlinear load models.

References

1. Theo, W., Lim, J., Ho, W., Hashim, H., Lee, C.: Review of distributed generation (DG) system planning and optimisation techniques: comparison of numerical and mathematical modelling methods. Renew. Sustain. Energy Rev. **67**, 531–573 (2017)
2. Padiyar, K.R.: FACTS Controllers in Power Transmission and Distribution. New Age International Limited Publishers (2007)
3. Shayeghi, H., Alilou, M.: Application of multi objective HFAPSO algorithm for simultaneous placement of DG, capacitor and protective device in radial distribution network. J. Oper. Autom. Power Eng. **3**, 131–146 (2015)
4. ChithraDevi, S., Lakshminarasimman, L., Balamurugan, R.: Stud Krill Herd algorithm for multiple DG placement and sizing in a radial distribution system. Eng. Sci. Technol. Int. J. **20**, 748–759 (2017)

5. Shahmohammadi, A., Ameli, M.: Proper sizing and placement of distributed power generation aids the intentional islanding process. Electr. Power Syst. Res. **106**, 73–85 (2014)
6. Shayeghi, H., Ghasemi, A.: FACTS devices allocation using a novel dedicated improved PSO for optimal operation of power system. J. Oper. Autom. Power Eng. **1**, 124–135 (2013)
7. Sarker, J., Goswami, S.: Optimal location of unified power quality conditioner in distribution system for power quality improvement. Int. J. Electr. Power Energy Syst. **83**, 309–324 (2016)
8. Khadem, S., Basu, M., Conlon, M.: A comparative analysis of placement and control of UPQC in DG integrated grid connected network. Sustain. Energy Grids Netw. **6**, 46–57 (2016)
9. Ebrahimi, R., Ehsan, M., Nouri, H.: A profit-centric strategy for distributed generation planning considering time varying voltage dependent load demand. Electr. Power Energy Syst. **44**, 168–178 (2013)
10. Hosseini, M., Shayanfar, H.A., Fotuhi-Firuzabad, M.: Modeling of unified power quality conditioner (UPQC) in distribution systems load flow. Energy Convers. Manag. **50**, 1578–1585 (2009)
11. Mirjalili, S., Lewis, A.: The whale optimization algorithm. Adv. Eng. Softw. **95**, 51–67 (2016)
12. Shayanfar, H.A., Shayeghi, H., Alilou, M.: Multi-objective allocation of DG simultaneous with capacitor and protective device including load model. In: The 19th International Conference on Artificial Intelligence, USA (2017)

Opposition Based Salp Swarm Algorithm for Numerical Optimization

Divya Bairathi$^{(\boxtimes)}$ and Dinesh Gopalani

Malaviya National Institute of Technology Jaipur, Jaipur, India
divyabairathijain@yahoo.co.in,dgopalani.cse@mnit.ac.in
http://www.mnit.ac.in

Abstract. In this paper an improved optimization algorithm called Opposition Based Salp Swarm Algorithm (OSSA) is proposed. This is improved version of recently proposed Salp Swarm Algorithm (SSA), which mimics swarming acts of salps when foraging and navigating in oceans. To improve the performance of SSA, Opposition based learning (OBL) is introduced in Salp Swarm Algorithm. The algorithm is evaluated on several numerical standard functions and is compared with some well known optimization algorithms.

Keywords: Optimization · Metaheuristics · Salp Swarm Algorithm · Opposition based learning · Opposition based Salp Swarm Algorithm

1 Introduction

Over the past two decades, metaheuristics have become very popular. Metaheuristics are general purpose stochastic heuristics, that are able to sovle almost all kind of optimization problems [1]. Metaheuristics are flexible, derivative free, stochastic mechanism which are able to avoid local optima entrapment. As these are stochastic techniques, they benefit from randomness in algorithm. This randomness assists in avoiding local solutions and local optima. Nowadays, the uses of meta-heuristics can be seen in different branches of science, engineering and industry.

Metaheuristics can be classified into four main branches: swarm-based, physics-based, evolution-based and Human based methods. Swarm based methods imitates the foraging and social behaviour of animals, birds or insects. Some of the popular algorithm are Particle Swarm Optimization (PSO) [2], Grey Wolf Optimization (GWO) [3], Ant Colony Optimization (ACO) [4], Artificial Bee Colony Optimization (ABC) [5], Salp Swarm Algorithm (SSA) [6], Polar Bear Optimization algorithm [7] etc. Evolution-based methods mimics the laws of natural evolution. Some of the well known evolution based techniques are Genetic Algorithms (GA) [8], Differential Evolution (DE) [9], Genetic Programming (GP) [10], Evolution Strategy (ES) [11] etc. Physics-based methods are motivated by the physical rules in the universe. Some of favoured algorithms

© Springer Nature Switzerland AG 2020
A. Abraham et al. (Eds.): ISDA 2018, AISC 941, pp. 821–831, 2020.
https://doi.org/10.1007/978-3-030-16660-1_80

are Simulated Annealing (SA) [12], Gravitational Search Algorithm (GSA) [13], Charged System Search (CSS) [14], Multi verse optimization (MVO) [15] etc. Human based methods imitate the behaviour of humans, for example Harmony Search (HS), Tabu Search (TS)[16,17], Group Search Optimizer (GSO) [18,19], Firework Algorithm (FA) [20] etc.

In this paper, a population based metaheuristic Opposition based Salp Swarm Algorithm (OSSA) is Proposed. OSSA is opposition learning based version of SSA [6]. Algorithm OSSA is tested for several numerical benchmark functions. Results are compared with some popular metaheuristics.

2   Opposition Based Salp Swarm Algorithm

In this section the base algorithm Salp swarm Algorithm, Opposition based learning (OBL) and finally Opposition based Salp Swarm Algorithm are discussed.

2.1   Salp Swarm Algorithm

Salp Swarm Algorithm is based on the social behaviour of salps. In swarming behavior, Salps Form salpchain. In this salpchain, the salps are connected with each other. The mathematical model of the salp chains concerns of two groups of salps: leader and followers. The first salp of the chain is contemplated as leader and remaining salps in chain are considered as followers. Equations 1 and 2 models position update equations of leader and followers respectively.

$$X_j^1 = \begin{cases} F_j + a_1 \left((UB_j - LB_j) a_2 + LB_j \right) & a3 \geq 0 \\ F_j - a_1 \left((UB_j - LB_j) a_2 + LB_j \right) & a3 < 0 \end{cases} \tag{1}$$

$$X_j^i = \left(X_j^i + X_j^{i-1} \right) / 2 \tag{2}$$

Here X_j^1 and X_j^i show the position of the leader salp (first salp) and i^{th} follower salp in the jth dimension. F_j is the position of the food source in the jth dimension. LB_j and UB_j indicate the lower bound and upper bound of jth dimension respectively. a_2 and a_3 are uniform random numbers in range $[0, 1]$. Value of a_1 is computed by following equation.

$$a_1 = 2e^{-(4l/L)^2} \tag{3}$$

Here l represents current iteration number and L is max number of iterations. The flow chart of SSA is shown in Fig. 1.

2.2   Opposition Based Learning

Opposition-based learning (OBL) is a method introduced in [31] for machine intelligence. The concept is based on opposite numbers. If upper and lower bounds of a range are defined by A and B respectively, then opposite of a real number Z is given by following equation.

$$\bar{Z} = A + B - Z \tag{4}$$

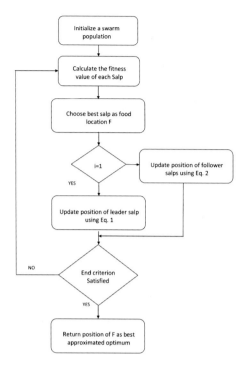

Fig. 1. Flow chart for SSA

Opposition Based Learning: Let Z be real vector in range [A, B], $\bar{Z}$ be opposite of Z, F(Z) be a function of concern and G(Z) be a evaluation function (which calculates the objective value of F(Z)), then according to opposition-based learning, learning continues with Z if $G(F(Z))$ is better than $G(F(\bar{Z}))$, else with $\bar{Z}$.

Opposition based learning is widely used in metaheuristics. OBL increases the convergence speed of optimization algorithm. OBL has inherent selection mechanism which chooses better solution for next iteration. It also doubles population in every iteration and the selection procedure again select the population of original size. It helps to achieve better initial population. Also in each iteration a better solution is obtained. These all points make the optimization algorithm faster than the original algorithm most of the time.

2.3   Opposition Based Salp Swarm Algorithm

OSSA introduces opposition based learning in SSA. The salps are initialized using uniform randomization compel by the lower and upper bounds of the each dimension. Then objective Value of each salp is computed to find the best salp. This best salp's location represents the food location. Using this food location new position is calculated for leader salp and follower salp using Eqs. 1 and 2.

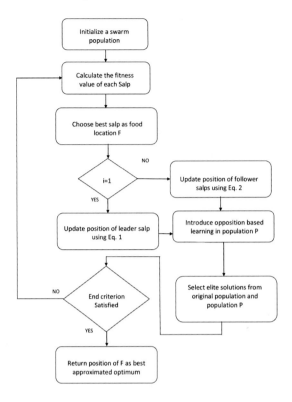

Fig. 2. Flow chart for OSSA

Now a new population of same size as of original population size is created using OBL. Objective values of these solutions is also calculated. Now elite solutions are selected using calculated objective values to form a solution array of size equal to original population size and these array is used for next iteration. Flow chart of OSSA is shown in Fig. 2. The extra population created in each iteration increases exploration of OSSA. Exploration is an important component of a optimization algorithm. During exploration whole search space is looked briefly to find new areas which have better solutions and may contain optima. Efficient exploration power helps the algorithm to search new areas, hence reduces probability of local optimal results. In initial iterations value of a_1 is very higher than in later iterations. Due to this the agents moves abruptly in initial iterations. Further the double population in each iteration increase the exploration (searching) of the search area. In each iteration elite solutions are picked, hence the convergence speed of OSSA is far better than SSA. In later iterations the agents moves slowly and represent the elite promising area. This slow movement, double population and elite solution selection make the algorithm OSCA to finely search through the promising areas. In this way the later iterations are exploitation intensive. As mentioned before, During exploration whole search space is looked briefly to find new areas whereas in exploitation these areas (found during exploration phase) are watched thoroughly [28–30].

3 Experiments and Results

For experiment and analysis purpose, twenty-five benchmark functions are selected from [32,33]. The algorithm OSSA is tested for these benchmark functions and compared with PSO, DE, GWO, WOA, SSA, GSA, ACO, ABC and GA. The benchmark numerical functions are given in Table 1. Here functions F_1–F_{11} are uni-modal functions. These function contain single maxima/minima. Uni-modal function are used for evaluating the exploitation power of optimization algorithms. Functions F_{12}–F_{25} are multi-modal functions. These functions have multiple maxima/minima and useful to evaluate the exploitation power of optimization algorithms. For implementation purpose population size taken is 20 and maximum iteration count is 500. Each algorithm is run 20 times.

In Fig. 3, the behaviour of algorithm OSSA is depicted. Space digram, Search history (only for two dimensions), trajectory of first salp, average fitness of salps and convergence curve for some of the benchmark functions are shown in the figure. From search history, it can be seen that the area nearby the optima is exploited powerfully. It depict fine exploitation provided by OSSA. Also whole search space is explored. This presents the expert exploration power of OSSA. It also shows the strong balance between exploration and exploitation gained by OSSA. The trajectories (of first salp) show that OSSA is able to converge towards optima. It represents high effectiveness and efficiency of the algorithm. Further the average fitness of salps depict that all salps gradually shift towards optimum solution and eventually converge near the optimum. The convergence curve represents the fitness of best position after the iterations. The convergence curve shows that the Best fitness value decreases monotonically with increment in iterations.

Average of best solutions found so far (AVG) and standard deviation of these best solutions (STD) of the functions are shown in Tables 2 and 3. Table 2 shows results for uni-modal functions. As stated before, uni-modal problems are helpful to test exploitation power of optimization algorithms. Table 3 presents multi-modal function results. Multi-modal problems are used to test exploration power of optimization algorithms. The results represent that algorithm OSSA obtains best solutions for most of the functions among the ten algorithms. This shows that OSSA gains high exploration and exploitation power and fine balance between exploration and exploitation. Adaptive changes in value of a_1, double population and elite solution selection make the algorithm OSSA to finely search through the search area. The value of a_1 decreases in each iteration gradually. On account of this, value of a_1 is very higher in initial iterations than in later iterations. High value of a_1 causes food location to move very fast. Owing to this and double population, OSSA is capable of look through whole search space. This

Table 1. Benchmark functions

Test problem	Function	Dim	Range	F_{min}
Sphere	$F_1 = \sum_{i=1}^n x_i^2$	30	[-100, 100]	0
Schwefel 2.22	$F_2 = \sum_{i=1}^n |x_i| + \prod_i^n |x_i|$	30	[-10, 10]	0
Schwefel 1.2	$F_3 = \sum_{i=1}^n \left(\sum_{j=1}^i x_i\right)^2$	30	[-100,100]	0
Max dimension	$F_4 = Max_i\{|x_i|, 1 \le i \le n\}$	30	[-100,100]	0
Rosenbrok	$F_5 = \sum_{i=1}^n [100(2x_{i+1} - x_i^2)^2 + (x_i - 1)^2]$	30	[-30, 30]	0
Shifted sphere	$F_6 = \sum_{i=1}^n (x_i + 0.5)^2$	30	[-100, 100]	0
De jong 4	$F_7 = \sum_{i=1}^n i \cdot (x_i)^4$	30	[-5.12, 5.12]	0
Axis parallel hyper ellipsoid	$F_8 = \sum_{i=1}^n i \cdot x_i^2$	30	[-5.12, 5.12]	0
Exponential	$F_9 = -(\exp(-0.5 \sum_{i=1}^n x_i^2)) + 1$	30	[-1, 1]	0
Sum of different power	$F_{10} = \sum_{i=1}^n |x_i|^{i+1}$	30	[-1, 1]	0
Step	$F_{11} = \sum_{i=1}^n (\lfloor x_i + 0.5 \rfloor)^2$	30	[-100, 100]	0
quartic(noise)	$F_{12} = \sum_{i=1}^n i x_i^4 + random[0, 1)$	30	[-1.28, 1.28]	0
Schwefel	$F_{13} = \sum_{i=1}^n -x_i \sin(\sqrt{|x_i|})$	30	[-500, 500]	-Dim* 418.9829
Rastrigin	$F_{14} = \sum_{i=1}^n [x_i^2 - 10\cos(2\pi x_i) + 10]$	30	[-5.12, 5.12]	0
Ackley	$F_{15} = -20\exp(-0.2\sqrt{\frac{1}{n}\sum_{i=1}^n x_i^2}) - \exp(\frac{1}{n}\sum_{i=1}^n \cos(2\pi x_i)) + 20 + e$	30	[-32, 32]	0
Griewank	$F_{16} = \frac{1}{4000}\sum_{i=1}^n x_i^2 - \prod_i^n \cos(\frac{x_i}{\sqrt{i}}) + 1$	30	[-600, 600]	0
Alpine	$F_{17} = \sum_{i=1}^n |x_i \sin(x_i) + 0.1 x_i|$	30	[-10, 10]	0
levy	$F_{18} = 0.1\sin^2(3\pi x_1) + \sum_{i=1}^n (x_i - 1)^2[1 + \sin^2(3\pi x_i + 1)] + (x_n - 1)^2[1 + \sin^2(3\pi x_n)]\} + \sum_{i=1}^n u(x_i, 5, 100, 4)$	30	[-50, 50]	0
Foxholes	$F_{19} = \left(\frac{1}{500} + \sum_{j=1}^{25} \frac{1}{j + \sum_{i=1}^2 (x_i - a_{ij})^6}\right)^{-1}$	2	[-65, 65]	1
Kowalik	$F_{20} = \sum_{i=1}^{11} \left[a_i - \frac{x_1(b_i^2 + b_i x_2)}{b_i^2 + b_i x_3 + x_4}\right]^2$	4	[-5, 5]	0.00030
Six hump camel back	$F_{21} = 4x_1^2 - 2.1x_1^4 + \frac{1}{3}x_1^6 + x_1 x_2 - 4x_2^2 + 4x_2^4$	2	[-5, 5]	-1.0316
Branin	$F_{22} = \left(x_2 - \frac{5.1}{4\pi^2}x_1^2 + \frac{5}{\pi}x_1 - 6\right)^2 + 10\left(1 - \frac{1}{8\pi}\right)\cos x_1 + 10$	2	[-5, 5]	0.398
goldstein price	$F_{23} = \left[1 + (x_1 + x_2 + 1)^2(19 - 14 x_1 + 3x_1^2 - 14 x_2 + 16 x_1 x_2 + 3x_2^2)\right] \times \left[30 + (2 x_1 - 3 x_2)^2 \times (18 - 32 x_1 + 12x_1^2 + 48 x_2 - 36 x_1 x_2 + 27x_2^2)\right]$	2	[-2, 2]	3
Hartmann-3	$F_{24} = -\sum_{i=1}^4 c_i \exp(-\sum_{j=1}^3 a_{ij}(x_j - p_{ij})^2)$	3	[1, 3]	-3.86
Hartmann-6	$F_{25} = -\sum_{i=1}^4 c_i \exp(-\sum_{j=1}^6 a_{ij}(x_j - p_{ij})^2)$	6	[0, 1]	-3.32

grows exploration component for OSSA. In later iterations, value of a_1 becomes very low. Low value of a_1 causes food location to move slowly. Owing to this and double population, OSSA searches the area around food location rigorously. This proffers exploitation component for OSSA. Further the elite solution selection

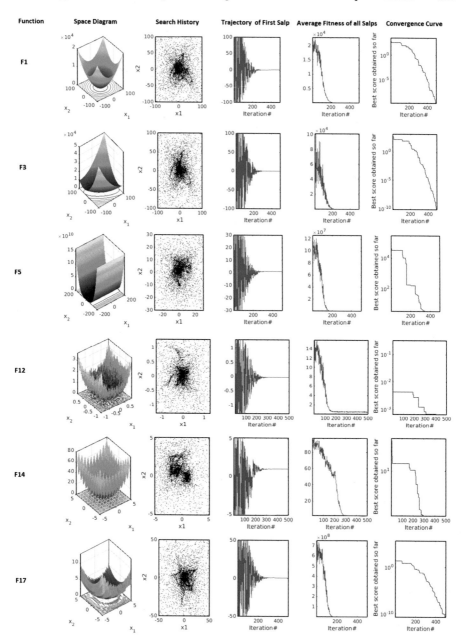

Fig. 3. Behaviour of OSSA

Table 2. Experimental results

Function	OSSA		PSO		DE		GWO		WOA	
	AVG	STD	AVG	STD	AVG	STD	AVG	STD	AVG	STD
F_1	0.0000	0.0000	0.0001	0.0001	0.0004	0.0002	0.0000	0.0000	0.0000	0.0000
F_2	0.0000	0.0000	0.0378	0.0532	0.0019	0.0006	0.0000	0.0000	0.0000	0.0000
F_3	0.0000	0.0000	54.9748	34.9732	3480.340	459.3974	0.0000	0.0000	4878.784	1873.384
F_4	0.0000	0.0000	1.1237	0.2768	15.0000	2.3569	0.0000	0.0000	51.6375	16.4859
F_5	3.2634	4.8971	140.2671	84.0108	162.0895	67.9874	26.3847	0.8931	22.7437	0.2367
F_6	0.0000	0.0000	0.0003	0.0002	0.0006	0.0003	0.3468	0.0455	0.7006	0.3220
F_7	0.0000	0.0000	0.0000	0.0000	0.0000	0.0000	0.0000	0.0000	0.0000	0.0000
F_8	0.0000	0.0000	0.0013	0.0008	0.0000	0.0011	0.0000	0.0000	0.0000	0.0000
F_9	-1	0.0000	-0.9998	0.0001	-1	0.0000	-1	0.0000	-1	0.0000
F_{10}	0.0000	0.0000	0.0000	0.0000	0.0000	0.0000	0.0001	0.0000	0.0000	0.0000
F_{11}	0.0000	0.0000	0.2000	0.4000	0.0000	0.0000	0.0030	0.0012	0.3133	0.1234
Function	SSA		GSA		ACO		ABC		GA	
	AVG	STD	AVG	STD	AVG	STD	AVG	STD	AVG	STD
F_1	0.0000	0.0000	0.0001	0.0001	0.0009	0.0001	0.0003	0.0002	0.0005	0.0002
F_2	0.0000	0.0000	0.0000	0.0000	0.2350	0.0857	0.0088	0.0045	0.0934	0.0263
F_3	8.3847	4.2478	59.0098	41.0083	2243.0034	535.3948	486.374	334.9583	2384.1002	884.3431
F_4	3.0731	2.0138	1.8127	0.9217	23.13523	12.2173	4.2874	3.3497	13.3249	9.3244
F_5	29.0042	0.1466	45.5656	1.4589	173.5669	104.5666	86.4563	23.5665	127.8567	135.7675
F_6	0.0676	0.0456	0.0546	0.0466	0.5677	0.4355	0.6574	0.4556	0.0565	0.0378
F_7	0.0000	0.0000	0.0000	0.0000	0.0001	0.0000	0.0000	0.0000	0.0000	0.0001
F_8	0.0000	0.0000	0.0001	0.0001	0.0001	0.0001	0.0001	0.0000	0.0000	0.0000
F_9	-1	0.0067	-1	0.0029	-0.9994	0.0056	-1	0.0000	-1	0.0001
F_{10}	0.0000	0.0000	0.0000	0.0000	0.0000	0.0000	0.0000	0.0000	0.0000	0.0000
F_{11}	0.0001	0.0000	0.0000	0.0000	0.0086	0.0045	0.1354	0.0574	0.0008	0.0001

in each iteration assist OSSA to converge fast and make balance between exploration and exploitation. In this way the OSSA provide powerful exploration and exploitation of search area as well as balance between the exploration and exploitation. On the basis of results and discussion, OSSA can be presented as one of the competitor among existing efficient and effective optimization techniques.

Results depict that all ten algorithms Support No Free Lunch Theorem [34]. No Free Lunch theorem says that not such a optimization algorithm exist which is best suited for all kind of problems. From results we can see that for different functions, different algorithms are giving best results.

Table 3. Experimental results

Function	OSSA AVG	OSSA STD	PSO AVG	PSO STD	DE AVG	DE STD	GWO AVG	GWO STD	WOA AVG	WOA STD
F_{12}	0.0007	0.0004	0.1899	0.0698	0.1634	0.0287	0.0016	0.0007	0.0031	0.0025
F_{13}	−1858.872	634.3480	−4872.520	7345.1453	−10387.938	2987.7463	−3467.039	4587.4860	−11498.500	473.6430
F_{14}	1.9899	0.2478	49.8374	31.8374	87.6237	45.2938	2.8378	4.8272	0.0000	0.0000
F_{15}	0.0000	0.0000	0.4782	0.2348	0.0033	0.0054	0.0000	0.0000	0.0000	0.0000
F_{16}	0.0033	0.0016	0.0074	0.0059	0.0045	0.0056	0.0030	0.0024	0.0000	0.0000
F_{17}	0.0000	0.0000	0.0438	0.0453	0.0565	0.0337	0.0004	0.0006	0.0000	0.0000
F_{18}	0.0001	0.0000	0.0034	0.0045	0.0004	0.0003	0.4534	0.4556	0.6576	0.5766
F_{19}	1.0067	0.0067	3.6786	3.7867	1.4344	0.5653	4.5634	5.5646	2.5464	3.4566
F_{20}	0.0004	0.0002	0.0009	0.0006	0.0008	0.0006	0.0067	0.0063	0.0013	0.0024
F_{21}	−1.0316	0.0000	−1.0316	0.0000	−1.0316	0.0000	−1.0316	0.0000	−1.0316	0.0000
F_{22}	0.3981	0.0007	0.3982	0.0000	0.3982	0.0000	0.3981	0.0001	0.3981	0.0000
F_{23}	3	0.0000	5.3	4.8973	3	0.0000	3	0.0000	3	0.0000
F_{24}	−3.8598	0.0002	−3.8528	0.0047	−3.8578	0.0034	−3.8590	0.0020	−3.8585	0.0043
F_{25}	−3.319	0.0000	−3.2454	0.0582	−3.3010	0.0003	−3.2435	0.0745	−3.2534	0.1453

Function	SSA AVG	SSA STD	GSA AVG	GSA STD	ACO AVG	ACO STD	ABC AVG	ABC STD	GA AVG	GA STD
F_{12}	0.0002	0.0020	0.0274	0.0345	0.0000	0.0000	0.0290	0.0378	0.0965	0.0674
F_{13}	−12495.100	0.3567	−10665.97	4765.656	−8231.430	7343.055	−9993.121	2381.845	−7786.676	7688.7865
F_{14}	0.6577	0.0876	0.0000	0.0000	45.2874	7.3847	35.4820	5.2194	12.5943	5.5466
F_{15}	0.0000	0.0000	0.0000	0.0000	1.9482	0.0334	0.4533	0.5383	0.7543	0.4543
F_{16}	0.0035	0.0032	0.0034	0.0024	0.0086	0.0082	0.0012	0.0021	0.0034	0.0118
F_{17}	0.0000	0.0000	0.0000	0.0000	0.03435	0.04342	0.0000	0.0000	0.0073	0.0034
F_{18}	0.0025	0.0005	0.0046	0.0046	0.0761	0.0056	0.0033	0.0062	0.0167	0.0054
F_{19}	1.6742	0.5743	1.5674	0.4743	5.6745	5.8798	2.9675	1.8643	3.8964	0.4568
F_{20}	0.0005	0.0003	0.0005	0.0005	0.0054	0.0071	0.0043	0.0067	0.0065	0.0043
F_{21}	−1.0316	0.0000	−1.0316	0.0000	−1.0316	0.0000	−1.0316	0.0000	−1.0316	0.0000
F_{22}	0.3981	0.0006	0.3982	0.0001	0.3990	0.0021	0.3985	0.0004	0.3983	0.0027
F_{23}	3	0.0000	3	0.0000	6.2	2.4644	3.5645	0.0021	3.7645	.3445
F_{24}	−3.8567	0.0019	−3.8594	0.0007	−3.8543	0.0566	−3.8519	0.0029	−3.8515	0.0068
F_{25}	−3.2505	0.1145	−3.2754	0.064	−3.232	0.0271	−3.267	0.0082	−3.2545	0.1896

4   Conclusion

In this paper, Opposition based salp swarm algorithm (OSSA) is proposed. This algorithm Combine opposition based learning and Salp swarm algorithm to provide an improved version OSSA. The algorithm is tested for 25 benchmark functions. It is also compared with other conventional optimization algorithms PSO, DE, GWO, WOA, SSA, GSA, ACO, ABC and GA. Space digram, Search history (only for two dimensions), trajectory of first salp, average fitness of salps and convergence curve are also shown and explained. Results shows that Opposition based Salp Swarm algorithm Provides Efficient exploration and exploitation of search area. It is able to circumvent local optima and find global optima most of the time. Adaptive changes in value of a_1, double population and elite solution selection make OSSA as one of the efficient and effective optimization methods.

For future work, OSSA can be extended for multi-objective, binary and constrained problems. The algorithm can also be applied for real word problems.

References

1. Glover, F.W., Kochenberger, G.A. (eds.): Handbook of Metaheuristics, vol. 57. Springer, Berlin (2006)
2. Eberhart, R., Kennedy, J.: A new optimizer using particle swarm theory. In: Proceedings of the Sixth International Symposium on Micro Machine and Human Science 1995. MHS'95, pp. 39–43. IEEE (1995)
3. Mirjalili, S., Mirjalili, S.M., Lewis, A.: Grey wolf optimizer. Adv. Eng. Softw. **69**, 46–61 (2014)
4. Dorigo, M., Birattari, M., Stutzle, T.: Ant colony optimization. IEEE Comput. Intell. Mag. **1**(4), 28–39 (2006)
5. Karaboga, D.: An idea based on honey bee swarm for numerical optimization, vol. 200. Technical report-tr06, Erciyes University, Engineering Faculty, Computer Engineering Department (2005)
6. Mirjalili, S., Gandomi, A.H., Mirjalili, S.Z., Saremi, S., Faris, H., Mirjalili, S.M.: Salp swarm algorithm: a bio-inspired optimizer for engineering design problems. Adv. Eng. Softw. **114**, 163–191 (2017)
7. Połap, D.: Polar bear optimization algorithm: meta-heuristic with fast population movement and dynamic birth and death mechanism. Symmetry **9**(10), 203 (2017)
8. Holland, J.H.: Genetic algorithms. Sci. Am. **267**(1), 66–72 (1992)
9. Storn, R., Price, K.: Differential evolution-a simple and efficient heuristic for global optimization over continuous spaces. J. Glob. Optim. **11**(4), 341–359 (1997)
10. Koza, J.R.: Genetic Programming. The MIT Press, Cambridge (1992)
11. Rechenberg, I.: Evolution strategy: nature's way of optimization. In: Optimization: Methods and Applications, Possibilities and Limitations, pp. 106–126. Springer, Berlin (1989)
12. Kirkpatrick, S., Gelatt, C.D., Vecchi, M.P.: Optimization by simulated annealing. Science **220**(4598), 671–680 (1983)
13. Rashedi, E., Nezamabadi-Pour, H., Saryazdi, S.: GSA: a gravitational search algorithm. Inf. Sci. **179**(13), 2232–2248 (2009)
14. Kaveh, A., Talatahari, S.: A novel heuristic optimization method: charged system search. Acta Mech. **213**, 267–289 (2010)

15. Mirjalili, S., Mirjalili, S.M., Hatamlou, A.: Multi-verse optimizer: a nature-inspired algorithm for global optimization. Neural Comput. Appl. **27**(2), 495–513 (2016)
16. Glover, F.: Tabu search - Part I. ORSA J. Comput. **1**(3), 190–206 (1989)
17. Glover, F.: Tabu search - Part II. ORSA J. Comput. **2**, 4–32 (1990)
18. He, S., Wu, Q., Saunders, J.: A novel group search optimizer inspired by animal behavioural ecology. In: Proceedings of the 2006 IEEE Congress on Evolutionary Computation, CEC, pp. 1272–1278 (2006)
19. He, S., Wu, Q.H., Saunders, J.: Group search optimizer: an optimization algorithm inspired by animal searching behavior. IEEE Trans. Evol. Comput. **13**, 973–990 (2009)
20. Tan, Y., Zhu, Y.: Fireworks algorithm for optimization. In: Advances in Swarm Intelligence, pp. 355–364. Springer, Heidelberg (2010)
21. Hertz, J.: Introduction to the Theory of Neural Computation, vol. 1. Addison Wesley, Boston (1991)
22. Rumelhart, D.E., Williams, R.J., Hinton, G.E.: Learning internal representations by error propagation. In: Parallel Distributed Processing: Explorations in the Microstructure of Cognition, vol. 1, pp. 318–362 (1986)
23. Mendes, R., Cortez, P., Rocha, M., Neves, J.: Particle swarm for feedforward neural network training. In: Proceedings of the International Joint Conference on Neural Networks, vol. 2, pp. 1895–1899 (2002)
24. Meissner, M., Schmuker, M., Schneider, G.: Optimized particle swarm optimization (OPSO) and its application to artificial neural network training. BMC Bioinform. **7**, 125 (2006)
25. Fan, H., Lampinen, J.: A trigonometric mutation operation to differential evolution. J. Glob. Optim. **27**, 105–129 (2003)
26. Slowik, A., Bialko, M.: Training of artificial neural networks using differential evolution algorithm. In: Human System Interactions, pp. 60–65 (2008)
27. Gao, Q., Qi, K., Lei, Y., He, Z.: An improved genetic algorithm and its application in artificial neural network. In: 2005 Fifth International Conference on Information, Communications and Signal Processing, 06–09 December 2005, pp. 357–360 (2005)
28. Olorunda, O., Engelbrecht, A.P.: Measuring exploration/exploitation in particle swarms using swarm diversity. In: 2008 IEEE Congress on Evolutionary Computation. CEC 2008 (IEEE World Congress on Computational Intelligence), pp. 1128–1134. IEEE (2008)
29. Alba, E., Dorronsoro, B.: The exploration/exploitation tradeoff in dynamic cellular genetic algorithms. IEEE Trans. Evol. Comput. **9**(2), 126–142 (2005)
30. Crepinsek, M., Liu, S.H., Mernik, M.: Exploration and exploitation in evolutionary algorithms: a survey. ACM Comput. Surv. (CSUR) **45**(3), 35 (2013)
31. Tizhoosh, H.R.: Opposition-based learning: a new scheme for machine intelligence. In: 2005 International Conference on Computational Intelligence for Modelling, Control and Automation, and International Conference on Intelligent Agents, Web Technologies and Internet Commerce, vol. 1, pp. 695–701. IEEE, November 2005
32. Ali, M.M., Khompatraporn, C., Zabinsky, Z.B.: A numerical evaluation of several stochastic algorithms on selected continuous global optimization test problems. J. Glob. Optim. **31**(4), 635–672 (2005)
33. Bansal, J.C., Sharma, H., Nagar, A., Arya, K.V.: Balanced artificial bee colony algorithm. Int. J. Artif. Intell. Soft Comput. **3**(3), 222–243 (2013)
34. Wolpert, D.H., Macready, W.G.: No free lunch theorems for optimization. IEEE Trans. Evol. Comput. **1**(1), 67–82 (1997)

A Novel Swarm Intelligence Based Optimization Method: Harris' Hawk Optimization

Divya Bairathi[✉] and Dinesh Gopalani

Malaviya National Institute of Technology Jaipur, Jaipur, India
divyabairathijain@yahoo.co.in, dgopalani.cse@mnit.ac.in
http://www.mnit.ac.in

Abstract. Swarm intelligence is a modern optimization technique, and one of the most promising techniques for solving optimization problems. In this paper, a new swarm intelligence based algorithm namely, Harris' Hawk Optimizer (HHO) is proposed. The algorithm mimics the cooperative hunting behaviour of Harris' hawks. The algorithm is analysed for twenty five well known benchmark functions. Performance of HHO is compared with Particle Swarm Optimization (PSO), Differential Evolution (DE), Grey Wolf Optimizer (GWO) and The Whale Optimization Algorithm (WOA). HHO is implemented and results present HHO as one of the efficient optimization methods.

Keywords: Optimization · Swarm intelligence · Cooperative hunting · Harris' hawk optimization

1 Introduction

Optimization refers to the process of finding optimum solution from a given set of available and feasible alternatives for a given problem. Optimization is used in large variety of fields such as NP-hard combinatorial optimization, scheduling, machine learning, robotics, power systems, molecular optimization problems, electronics and electromagnetic engineering etc. A number of optimization techniques are available in literature. Optimization techniques are broadly classified into traditional approaches and modern approaches [1]. Traditional approaches includes Linear Programming, Non-linear Programming, Dynamic programming, Geometric programming, Integer programming, Stochastic programming, discrete programming etc. The major drawback of traditional approaches is very less exploration of search space, which causes local optima entrapment. Modern approaches alleviate this drawback by using stochastic components. Exploration can be further increased by using multiple agents. These multiple agents search multiple points in parallel in the search space. Agents interact, guide and assist each other to avoid local optima and converge at global optima. Modern approaches are flexible and can be used for different kind of problems (linear,

© Springer Nature Switzerland AG 2020
A. Abraham et al. (Eds.): ISDA 2018, AISC 941, pp. 832–842, 2020.
https://doi.org/10.1007/978-3-030-16660-1_81

Table 1. Optimization algorithms

Algorithm	Inspiration	Year
Genetic algorithm [2]	Darwin's evolution	1975
Simulated annealing [3]	Annealing process	1983
Evolutionary strategies [4]	Evolution	1989
Tabu search [5]	Human behavior	1989
Particle swarm optimization [6]	Bird flock	1995
Differential evolution [7]	Evolution	1997
Artificial bee colony optimization [8]	Bee colony	2005
Ant colony optimization [9]	Ant colony	2006
Gravitational search algorithm [10]	Gravitation	2009
Grey wolf optimization [11]	Grey wolf pack	2014
Whale optimization [12]	Humpback whales	2016

non-linear, non-convex, combinatorial, integer, discrete, real valued etc.) with slight variations in the algorithm. As these methods are derivation free, these are also applicable for non-continuous and non-differentiable functions. Due to these reasons modern approaches are becoming very popular. Some well-known algorithms are listed in Table 1.

Swarm intelligence (SI) is one of the modern approach, based on the cooperative behaviour of social insects, birds and animals. These simple creatures behave intelligently when work in cooperative manner. SI techniques mathematically model this collective behaviour of these creatures. This paper aims to propose a new swarm intelligence based algorithm Harris' Hawk Optimizer (HHO), inspired by hunting behaviour of Harris' hawk. The performance of algorithm is compared with PSO, GWO, DE and WOA.

2   Harris' Hawk Cooperative Hunting

Harris' Hawk is only raptor, known for cooperative hunting in packs [13]. Harris' Hawks are social birds and live in relatively stable group of 2–7 birds. There exists a dominance hierarchy among them. At the top of hierarchy is a mature female bird, followed by a male bird and then other birds of the pack. Harris' hawks live in sparse woodland, marshes and semi-desert. Their diet consist of rats, squirrels, medium-sized birds, rabbits, mammals, lizards etc.

Harris' hawks hunt in packs. This cooperative hunting makes them able to feed in the harsh dessert, where the prey is scarce. It also allows them to hunt bigger prey. The members of pack take turns for scanning the surroundings in search of prey and for attacking the prey. It make them able to hunt for longer duration. Searching is done by perching on the top of power poles, standing dead trees, saguaros and spanning large area around by highly efficient vision. As the prey is spotted, other members are informed through visual displays or

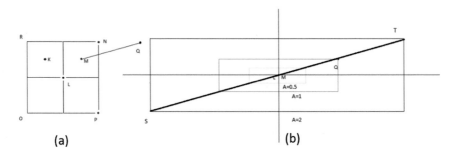

Fig. 1. 2D postion vectors and next possible position of ith hawk

vocalization. In one hunting technique [14,15], hawks fly around the prey. The encircling shadows confuse the prey. One member dives to catch prey and if it misses then another member tries, while the first one gets back in the line. It is continued until prey is caught and shared.

3   Mathematical Model for Harris' Hawk Cooperative Hunting Behaviour

In this section, mathematical model for prey position and hunting is given.

3.1   Mathematical Model

Harris' hawks hunt in packs. The members of pack take turns for hunting. In this way a member can be in hunting phase or non-hunting phase. Hunting is further divided into two phases- search phase and global attack phase. During search phase the prey is tracked and in global attack phase prey is attacked by the pack. In the following equations X_i^{t+1} and X_i^t are position vectors of ith hawk in $t+1^{th}$ and t^{th} iterations respectively and $1 \leq i \leq n$, that is n hawks are involved in group hunting. A and C_1 are scalar coefficient. C_2 is coefficient vector. Values of C_1 and C_2 are given by following equations.

$$C_1 = 2 \bullet A \bullet rand_1 - A \qquad (1)$$

$$C_2 = 2 \bullet rand_2 \qquad (2)$$

Here $rand_1$ is random value and $rand_2$ is random vector in range $[0, 1]$. Value of A linearly decreases from 2 to 0. Range of C_1 and C_2 become $[-A, A]$ and $[0, 2]$ respectively.

Non-hunting phase - In non-hunting phase of hunting, hawks wander at random positions and learn the structure of search space. The promising regions found in this phase are stored in personal best position $X_{pbest,i}^t$ of hawk or in global best position X_{gbest}^t. These positions are used for the exploitation of surrounding area of the positions. Position of ith hawk around a randomly chosen hawk X_{rand}^t is given by the following equation.

$$X_i^{t+1} = X_{rand}^t - C_1 \bullet \left| C_2 \bullet X_{rand}^t - X_i^t \right| \tag{3}$$

Search phase - Hawk search for prey by scanning the area from height. For this they perch on top of power poles, standing dead trees, saguaros etc, so that maximum area can be covered by their sight. This height is represented by personal best position $X_{pbest,i}^t$. $X_{pbest,i}^t$ is the best value of X_i, with optimum value of objective function found so far by i^{th} hawk. Searching around this personal best position is given by the following equation.

$$X_i^{t+1} = X_{pbest,i}^t - C_1 \bullet \left| C_2 \bullet X_{pbest,i}^t - X_i^t \right| \tag{4}$$

Global attack phase - This phase represents the phase when the prey is spotted and hawks get around the prey to attack it. The position of prey is realized by global best X_{gbest}^t. X_{gbest}^t is the best position with optimum value of objective function found so far by all hawks. The dives and attacks of hawks on this position are given by following equation.

$$X_i^{t+1} = X_{gbest}^t - C_1 \bullet \left| C_2 \bullet X_{gbest}^t - X_i^t \right| \tag{5}$$

3.2   Exploration and Exploitation

Modern algorithms generally consist of two phases-exploration and exploitation [16–18]. In exploration whole search space is glanced to find the promising area. In exploitation the promising areas are watched thoroughly to get best solution. We have further divided the exploitation in local exploitation and global exploitation. Exploration and exploitation are incorporated in HHO algorithm as follows.

Exploration - Equation (1) provides exploration of the search space. As shown above that for Eq. (1) $|C_1| >= A_1$, so comparatively a large space around randomly selected hawk is explored.

Local exploitation - Equation (2) searches space around the personal best found so far. In multi-modal problems (containing multiple optima), local exploitation allows to search through different valleys to find new promising areas. In this way, it reduces the chances of stagnation, the case when solutions got entrapped at local optima.

Global exploitation - Equation (3) exploits the nearby space around global best found so far. As $|C_1| < A_2$, comparatively small space around global best is searched. So the global optimization provides exploitation of promising areas in order to find better solutions.

In the proposed algorithm exploration and exploitation are achieved using three controlling variables A_1, A_2 and p. Now position of ith hawk X_i^{t+1} in next iteration is given by-

$$\begin{cases} X_{rand}^t - C1\bullet \left|C2\bullet X_{rand}^t - X_i^t\right| & \text{rand}\,(0,1) < p \ \ \text{and} \ \ |C1| \geq A1 \\ X_{pbest,i}^t - C1\bullet \left|C2\bullet X_{pbest,i}^t - X_i^t\right| & \text{rand}\,(0,1) \geq p \ \text{or A1} > \ |C1| \geq A2 \\ X_{gbest}^t - C1\bullet \left|C2\bullet X_{gbest}^t - X_i^t\right| & |C1| < A2 \end{cases} \quad (6)$$

Here (initial value of A) $> A_1 > A_2$.

The above equations for finding the position of ith hawk are based on following equation.

$$X_i^{t+1} = X - C_1\bullet \left|C_2\bullet X - X_i^t\right| \quad (7)$$

Here X represents X_{rand}^t, $X_{pbest,i}^t$ and X_{gbest}^t for Non-hunting, search and global attack phase phase respectively. Position update for two dimensional problem using Eq. (7) is explained in Fig. 1. In Fig. 1(a) O, L and Q represents origin, X and X_i^{t+1} respectively. Vector $C_2\bullet X$ will lie in rectangle OPNR. Different position corresponding to different values of C_2 are shown in Fig. 1(a). For instance, Positions P, K, M and N corresponds to $C_2 = [0, 2]$, $[1/2, 3/2]$, $[3/2, 3/2]$ and $[2, 2]$ respectively. Similarly, vector $C_2\bullet X - X_i^t$ can be find. The vector $C_2\bullet X - X_i^t$ for $C_2 = [3/2, 3/2]$ is shown by QM in Fig(a). According to Eq. (7), the position of ith hawk around X will depend on C_1 and from Eq. (1) value of C_1 depends on A and $rand_1$. Value of A decides range for X_i^{t+1} around X and $rand_1$ gives the exact position of X_i^{t+1} inside the range. For $C_2\bullet X - X_i^t = QM$, the range on line ST corresponding to A equals to 0.5, 1 and 2 are shown in Fig. 1(b).

Based on these concepts the pseudo code of the algorithm is given as follows.

HHO Algorithm

Create random solutions X_i (i=1,2,......,n)
Initialize A, A_1, A_2 and p
While the end criterion is not satisfied
    Evaluate fitness values $f(X_i)$ for each solution
    Update global best and personal best
    Update value of A (decrease)
    For each solution indexed by i
        Update C_1 and C_2
        If (rand(0,1) < p && $|C_1| \geq A_1$)
            $X_i^{t+1} = X_{rand}^t - C_1\bullet \left|C_2\bullet X_{rand}^t - X_i^t\right|$
        Elseif ($|C_1| \geq A_2$)
            $X_i^{t+1} = X_{pbest,i}^t - C_1\bullet \left|C_2\bullet X_{pbest,i}^t - X_i^t\right|$
        Else
            $X_i^{t+1} = X_{gbest}^t - C_1\bullet \left|C_2\bullet X_{gbest}^t - X_i^t\right|$
        End If
    End For
End While

HHO provides good exploration as well as exploitation. It is achieved by concurrent exploration and exploitation. In commencing iterations some hawks use exploration, some use local exploitation and others use global exploitation. In this way exploration and exploitation occurs simultaneously. As iteration count increases, value of C_1 decreases linearly. It forces reduction of exploration and increase of exploitation of promising area found so far. The advantage of this approach is that the promising areas found during initial iterations are scanned simultaneously for better solutions. In this way, search spaces selected for later iterations, are more promising and exploitation (which consist major part of later iterations) of these spaces gives better solutions.

4   Results and Discussion

For analysis purpose, 25 benchmark functions are selected from [11,19,20] and proposed algorithm is tested for these benchmark functions. These functions are given in Table 2. Here functions F_1–F_6 and F_8–F_{12} are uni-modal functions. Uni-modal function contains single optima. These function are very helpful for evaluating the exploitation power of optimization algorithms. Functions F_{13}–F_{25} are multi-modal functions. Multi-modal functions have multiple optima and allow to evaluate the exploitation power of optimization algorithms. F_7 is noise function, which is highly multi-modal function containing many spikes. The algorithm is also compared with PSO, DE, GWO and WOA. For implementation population size taken is 30 and maximum iteration count is 500. Each algorithm is run 30 times. The parameters for the algorithm HHO are $p = 0.5$, $A = 2$, $A_1 = 1$ and $A_2 = 0.5$. Average of best solution found so far (AVG) and standard deviation of these solutions (STD) of the functions are presented in Table 3.

As shown in Table 3, HHO provides best results (among the five algorithms) for uni-modal functions F_1, F_2, F_4 and F_8–F_{12}. This shows that HHO has good exploitation-capacity. It is also most optimizer (among the five algorithms) for multi-modal functions F_7, F_{13}–F_{17} and F_{20}, which presents high exploration capacity of HHO.

In Fig. 2 convergence curves of HHO, PSO, DE, GWO and WOA are compared for some of the benchmark functions. Here average best obtained so far presents average of best solutions found upto corresponding iteration over 30 runs. From the figure, three kind of convergence behaviour can be seen for HHO. In first behaviour, the HHO algorithm converges to optima in final iterations. This is due to continues exploration and exploitation of new promising areas in order to find better solutions and the found promising areas are exploited in later iterations. Convergence curves for functions F_1, F_4 and F_7 show this behaviour. In second behaviour, HHO converges in middle iterations. For HHO the initial iterations are exploration extensive. As the iteration count increases, number of agents for exploration decreases and for exploitation increases. In second kind of behaviour, the agents of exploration phase are able find promising regions in initial iterations. These promising regions are then exploited in middle iterations. Convergence curves for functions F_{14}, F_{16} and F_{19} show this behaviour.

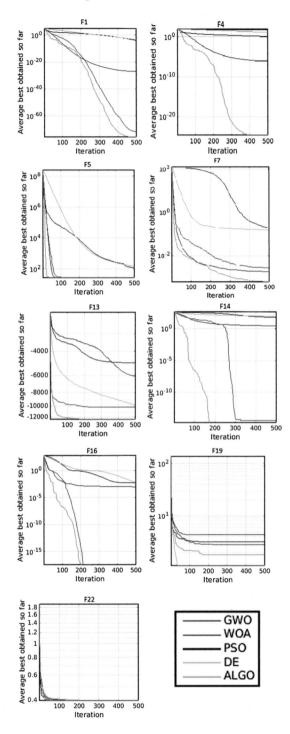

Fig. 2. Comparison of convergence curves for some benchmark problems

Table 2. Benchmark functions

Test problem	Function	Dim	Range	F_{min}
Sphere	$F_1 = \sum_{i=1}^{n} x_i^2$	30	[-100, 100]	0
Schwefel 2.22	$F_2 = \sum_{i=1}^{n} |x_i| + \prod_{i}^{n} |x_i|$	30	[-10, 10]	0
Schwefel 1.2	$F_3 = \sum_{i=1}^{n} (\sum_{j=1}^{i} x_i)^2$	30	[-100,100]	0
Max dimension	$F_4 = Max_i\{|x_i|, 1 \leq i \leq n\}$	30	[-100,100]	0
Rosenbrok	$F_5 = \sum_{i=1}^{n} [100(x_{i+1} - x_i^2)^2 + (x_i - 1)^2]$	30	[-30, 30]	0
Shifted sphere	$F_6 = \sum_{i=1}^{n} (x_i + 0.5)^2$	30	[-100, 100]	0
quartic(noise)	$F_7 = \sum_{i=1}^{n} ix_i^4 + random[0, 1)$	30	[-1.28, 1.28]	0
De jong 4	$F_8 = \sum_{i=1}^{n} i.(x_i)^4$	30	[-5.12, 5.12]	0
Axis parallel hyper ellipsoid	$F_9 = \sum_{i=1}^{n} i.x_i^2$	30	[-5.12, 5.12]	0
Exponential	$F_{10} = -(exp(-0.5 \sum_{i=1}^{n} x_i^2)) + 1$	30	[-1, 1]	0
Sum of different power	$F_{11} = \sum_{i=1}^{n} |x_i|^{i+1}$	30	[-1, 1]	0
Step	$F_{12} = \sum_{i=1}^{n} (\lfloor x_i + 0.5 \rfloor)^2$	30	[-100, 100]	0
Schwefel	$F_{13} = \sum_{i=1}^{n} -x_i \sin(\sqrt{|x_i|})$	30	[-500, 500]	-Dim* 418.9829
Rastrigin	$F_{14} = \sum_{i=1}^{n} [x_i^2 - 10\cos(2\pi x_i) + 10]$	30	[-5.12, 5.12]	0
Ackley	$F_{15} = -20exp(-0.2\sqrt{\frac{1}{n}\sum_{i=1}^{n} x_i^2}) - exp(\frac{1}{n}\sum_{i=1}^{n} \cos(2\pi x_i)) + 20 + e$	30	[-32, 32]	0
Griewank	$F_{16} = \frac{1}{4000}\sum_{i=1}^{n} x_i^2 - \prod_{i}^{n} \cos(\frac{x_i}{\sqrt{i}}) + 1$	30	[-600, 600]	0
Alpine	$F_{17} = \sum_{i=1}^{n} |x_i \sin(x_i) + 0.1x_i|$	30	[-10, 10]	0
levy	$F_{18} = 0.1\sin^2(3\pi x_1) + \sum_{i=1}^{n} (x_i - 1)^2[1 + \sin^2(3\pi x_i + 1)] + (x_n - 1)^2[1 + \sin^2(3\pi x_n)]\} + \sum_{i=1}^{n} u(x_i, 5, 100, 4)$	30	[-50, 50]	0
Foxholes	$F_{19} = \left(\frac{1}{500} + \sum_{j=1}^{25} \frac{1}{j + \sum_{i=1}^{2} (x_i - a_{ij})^6}\right)^{-1}$	2	[-65, 65]	1
Kowalik	$F_{20} = \sum_{i=1}^{11} \left[a_i - \frac{x_1(b_i^2 + b_i x_2)}{b_i^2 + b_i x_3 + x_4}\right]^2$	4	[-5, 5]	0.00030
Six hump camel back	$F_{21} = 4x_1^2 - 2.1x_1^4 + \frac{1}{3}x_1^6 + x_1 x_2 - 4x_2^2 + 4x_2^4$	2	[-5, 5]	-1.0316
Branin	$F_{22} = (x_2 - \frac{5.1}{4\pi^2}x_1^2 + \frac{5}{\pi}x_1 - 6)^2 + 10(1 - \frac{1}{8\pi})\cos x_1 + 10$	2	[-5, 5]	0.398
goldstein price	$F_{23} = [1 + (x_1 + x_2 + 1)^2(19 - 14x_1 + 3x_1^2 - 14x_2 + 16x_1 x_2 + 3x_2^2)] \times [30 + (2x_1 - 3x_2)^2 \times (18 - 32x_1 + 12x_1^2 + 48x_2 - 36x_1 x_2 + 27x_2^2)]$	2	[-2, 2]	3
Hartmann-3	$F_{24} = -\sum_{i=1}^{4} c_i exp(-\sum_{j=1}^{3} a_{ij}(x_j - p_{ij})^2)$	3	[1, 3]	-3.86
Hartmann-6	$F_{25} = -\sum_{i=1}^{4} c_i exp(-\sum_{j=1}^{6} a_{ij}(x_j - p_{ij})^2)$	6	[0, 1]	-3.32

In third behaviour, HHO converges in initial iterations. It is due to concurrent exploration and exploitation of HHO. In commencing iterations some agents perform exploration while others perform local exploitation and global exploitation. Due to concurrent exploration and exploitation, the promising regions found in

Table 3. Experimantal results

Function	HHO		PSO		DE		GWO		WOA	
	AVG	STD	AVG	STD	AVG	STD	AVG	STD	AVG	STD
F_1	0	0	0.000153	0.000167	0.000485	0.00014	1.2976e-27	2.1088e-17	0	0
F_2	0	0	0.03651	0.050355	.0022951	.00048376	1.0196e-16	6.7506e-17	0	0
F_3	7.4292	4.6294	86.4278	40.3963	3296.900	463.6651	9.535e-06	1.7203-05	4258.104	1493.342
F_4	9.0205e-34	3.7843e-33	1.1274	0.2579	14.0159	2.2695	7.6793e-07	7.8401e-07	50.6375	28.3478
F_5	28.6842	0.11541	112.5634	115.1983	142.279	45.9184	26.8943	0.7821	27.9265	0.43023
F_6	0.78378	0.84951	0.000216	0.000240	0.000522	0.000205	0.73982	0.39014	0.35683	0.21451
F_7	0.000152	0.000209	0.19627	0.075373	0.15281	0.02936	0.001902	0.000763	0.002746	0.002161
F_8	0	0	7.0738e-05	0.000114	4.1855e-09	7.0734e-05	0	0	0	0
F_9	0	0	0.001460	0.002341	1.5955e-05	0.001444	3.4654e-29	5.1539e-29	0	0
F_{10}	-1	2.4262e-18	-0.99994	0.000119	-1	5.7829e-05	-1	2.0869e-16	-1	9.0649e-17
F_{11}	0	0	1.6811e-05	7.99e-05	3.6899e-23	1.6811e-05	0	0	0	0
F_{12}	0	0	0.2	0.4	0	0	0	0	0.33333	0.17951
F_{13}	-12297.1	424.408	-4867.62	7826.129	-9832.28	2768.034	-6068.70	6587.486	-10178.5	2856.606
F_{14}	0	0	59.9147	61.5878	87.2869	87.8021	2.7568	4.5781	3.7896e-15	2.0756e-14
F_{15}	8.8818e-16	0	0.32528	0.6111	0.005757	0.005820	1.0332e-13	1.0471e-13	4.4645e-15	4.4645e-15
F_{16}	0	0	0.005512	0.008755	0.006215	0.009402	0.001040	0.004066	0	0
F_{17}	0	0	0.023242	0.02562	0.03375	0.037668	0.000436	0.000567	0	0
F_{18}	0.57664	0.62117	0.003827	0.006437	0.000317	0.000346	0.65927	0.72527	0.51501	0.58853
F_{19}	1.3952	0.76792	3.2287	3.6367	1.1304	0.44336	4.3562	5.475	2.8306	3.6914
F_{20}	0.000605	0.000456	0.000854	0.000575	0.000775	0.000568	0.003715	0.008191	0.001220	0.002783
F_{21}	-1.0316	8.7778e-08	-1.0316	6.1481e-16	-1.0316	6.6613e-16	-1.0316	3.445e-08	-1.0316	4.1705e-10
F_{22}	0.39793	0.000735	0.39789	0	0.39789	0	0.39789	0.000144	0.39789	3.435e-05
F_{23}	3	1.3914e-05	5.7	14.2765	3	1.2665e-15	3	1.5533e-15	3	3.435e-05
F_{24}	-3.8626	0.000190	-3.8628	2.5958e-15	-3.8628	2.6645e-15	-3.8619	0.002031	-3.8585	0.004338
F_{25}	-3.2505	0.11456	-3.2784	0.058245	-3.322	0.000291	-3.275	0.07072	-3.2587	0.13462

commencing iteration are exploited in parallel and best solution is found in early iterations. Convergence curves for functions F_5, F_{13} and F_{22} show this behaviour.

Strength of an optimization algorithm lies in its exploration and exploitation power. Balance between these two is equally important. An optimization algorithm is called successful if it provides good exploration, good exploitation and a fair balance between these two. Results shows that HHO is able to achieve these qualities. The exploration phase provide high exploration, whereas global exploitation and local exploitation provides high exploitation. As the two phases run in parallel, balance between exploration and exploitation is also maintained.

The algorithms also follow No Free Lunch Theorem [21]. According to No Free Lunch theorem, there is no optimization algorithm which is best suited for all kind of problems. So for some functions HHO is giving better results and for some functions other algorithms are providing better results.

5 Conclusion

In this paper, a new SI algorithm (named as HHO, Harris' hawk optimizer) based on cooperative hunting behaviour of Harris' hawks, is proposed. HHO consist of three phases - non-hunting phase, search phase and global attack phase. Non-hunting phase provides exploration, whereas search phase and global attack phase performs exploitation of promising areas. The algorithm is tested for 25 benchmark functions and also compared with other state-of-art modern optimization algorithms PSO, DE, GWO and WOA. Convergence behaviour of HHO is also explained. Results represent that HHO provides good exploration and exploitation, which is key factor for the success of any SI technique. For future work, HHO can be extended for binary, multi-objective and constrained problems.

References

1. Rao, S.S.: Engineering Optimization: Theory and Practice. Wiley, Hoboken (2009)
2. Holland, J.H.: Genetic algorithms. Sci. Am. **267**(1), 66–72 (1992)
3. Kirkpatrick, S., Gelatt, C.D., Vecchi, M.P.: Optimization by simulated annealing. Science **220**(4598), 671–680 (1983)
4. Rechenberg, I.: Evolution strategy: nature's way of optimization. In: Optimization: Methods and applications, possibilities and limitations, pp. 106–126. Springer, Heidelberg (1989)
5. Glover, F.: Tabu search—part I. ORSA J. Comput. **1**(3), 190–206 (1989)
6. Eberhart, R., Kennedy, J.: A new optimizer using particle swarm theory. In: Proceedings of the Sixth International Symposium on Micro Machine and Human Science, 1995, MHS 1995, pp. 39–43. IEEE (1995)
7. Storn, R., Price, K.: Differential evolution-a simple and efficient heuristic for global optimization over continuous spaces. J. Global Optim. **11**(4), 341–359 (1997)
8. Karaboga, D.: An idea based on honey bee swarm for numerical optimization, Technical report-tr06, Erciyes university, engineering faculty, computer engineering department, vol. 200 (2005)
9. Dorigo, M., Birattari, M., Stutzle, T.: Ant colony optimization. IEEE Comput. Intell. Mag. **1**(4), 28–39 (2006)
10. Rashedi, E., Nezamabadi-Pour, H., Saryazdi, S.: GSA: a gravitational search algorithm. Inf. Sci. **179**(13), 2232–2248 (2009)
11. Mirjalili, S., Mirjalili, S.M., Lewis, A.: Grey wolf optimizer. Adv. Eng. Softw. **69**, 46–61 (2014)
12. Mirjalili, S., Lewis, A.: The whale optimization algorithm. Adv. Eng. Softw. **95**, 51–67 (2016)
13. Zoologger: the only raptor known to hunt in cooperative packs, New Scientist. https://www.newscientist.com

14. Coulson, J.O., Coulson, T.D.: Group hunting by Harris' hawks in Texas. J. Raptor Res. **29**(4), 265–267 (1995)
15. Bednarz, J.C.: Cooperative hunting in Harris' hawks (Parabuteo unicinctus). Science **239**(4847), 1525 (1988)
16. Olorunda, O., Engelbrecht, A.P.: Measuring exploration/exploitation in particle swarms using swarm diversity. In: IEEE Congress on Evolutionary Computation, 2008, CEC 2008, (IEEE World Congress on Computational Intelligence), pp. 1128–1134, IEEE (2008)
17. Alba, E., Dorronsoro, B.: The exploration/exploitation tradeoff in dynamic cellular genetic algorithms. IEEE Trans. Evol. Comput. **9**(2), 126–142 (2005)
18. Crepinsek, M., Liu, S.H., Mernik, M.: Exploration and exploitation in evolutionary algorithms: a survey. ACM Comput. Surv. (CSUR) **45**(3), 35 (2013)
19. Ali, M.M., Khompatraporn, C., Zabinsky, Z.B.: A numerical evaluation of several stochastic algorithms on selected continuous global optimization test problems. J. Global Optim. **31**(4), 635–672 (2005)
20. Bansal, J.C., Sharma, H., Nagar, A., Arya, K.V.: Balanced artificial bee colony algorithm. Int. J. Artif. Intell. Soft Comput. **3**(3), 222–243 (2013)
21. Wolpert, D.H., Macready, W.G.: No free lunch theorems for optimization. IEEE Trans. Evol. Comput. **1**(1), 67–82 (1997)

An Improved Opposition Based Grasshopper Optimisation Algorithm for Numerical Optimization

Divya Bairathi[✉] and Dinesh Gopalani

Malaviya National Institute of Technology Jaipur, Jaipur, India
divyabairathijain@yahoo.co.in, dgopalani.cse@mnit.ac.in
http://www.mnit.ac.in

Abstract. In this paper an improved optimization algorithm called Opposition Based Grasshopper Optimisation Algorithm (OGOA) is proposed. This is improved version of recently proposed Grasshopper Optimisation Algorithm (GOA), which mimics swarming behavior of grasshoppers in the living world. To improve the performance of GOA, Opposition based learning (OBL) is introduced in Grasshopper Optimisation Algorithm. The algorithm is tested on several numerical benchmark functions and is compared with some well known optimization algorithms.

Keywords: Optimization · Metaheuristics ·
Grasshopper Optimisation Algorithm · Opposition based learning ·
Opposition based Grasshopper Optimisation Algorithm

1 Introduction

Over the past two decades, metaheuristics have become very popular. Metaheuristics are general purpose stochastic heuristics, that are able to sovle almost all kind of optimization problems [1]. Metaheuristics are flexible, derivative free, stochastic mechanism which are able to avoid local optima entrapment. As these are stochastic techniques, they benefit from randomness in algorithm. This randomness assists in avoiding local solutions and local optima. Nowadays, the employment of meta-heuristics can be seen in different branches of science, engineering and industry.

Some of the popular meta-heuristics are Particle Swarm Optimization (PSO) [2], Grey Wolf Optimization (GWO) [3], Ant Colony Optimization (ACO) [4], Artificial Bee Colony Optimization (ABC) [5], Genetic Algorithms (GA) [6], Genetic Programming (GP) [7], Evolution Strategy (ES) [8], Differential Evolution (DE) [9], Simulated Annealing (SA) [10], Gravitational Search Algorithm (GSA) [11], Tabu Search (TS) [12,13], Whale Optimization Algorithm (WOA) [14], Salp Swarm Algorithm (SSA) [15], Spider Monkey Optimization (SMO) [16] etc.

© Springer Nature Switzerland AG 2020
A. Abraham et al. (Eds.): ISDA 2018, AISC 941, pp. 843–851, 2020.
https://doi.org/10.1007/978-3-030-16660-1_82

In this paper, a population based metaheuristic Opposition based Grasshopper Optimisation Algorithm (OGOA) is Proposed. OGOA is opposition learning based version of GOA [17]. Opposition-based learning (OBL) is a method introduced in [21] for machine intelligence. Algorithm OSSA is tested for several numerical benchmark functions. Results are compared with some popular metaheuristics.

2   Literature Study

Grasshopper Optimisation Algorithm [17] is recently proposed metaheuristic, which is being used for many numerical and other challenging optimization problems. Opposition-based learning (OBL) is a machine learning method [21] and has been applied for improving many existing metaheuristics, reinforcement learning methods etc..

2.1   Grasshopper Optimisation Algorithm

Grasshopper Optimisation Algorithm [17] is based on the social and swarming behaviour of grasshoppers. The mathematical model of the Grasshopper movement based on its current position, the position of all other grasshoppers and the position of the target is given by Eq. 1.

$$X_i^D = C \left(\sum_{j=1, j \neq i}^{N} C \frac{UB_D - LB_D}{2} S \left(\left| x_j^D - x_i^D \right| \right) \frac{x_j - x_i}{D_{ij}} \right) + \hat{T}_d \tag{1}$$

Here X_i and X_j show the position of the ith and j^{th} grasshoppers respectively. LB_D and UB_D are lower bound and upper bound in the Dth dimension. $S(r) = fe^l - e^{-r}$, where f shows the intensity of attraction and l indicates the attractive length scale. $\hat{T}_d$ is target position (best position found so far). C is adaptive coefficient. Value of C is computed as shown in Eq. 2.

$$C = C_{max} - t \frac{C_{max} - C_{min}}{T} \tag{2}$$

Here t represents current iteration number and T is max number of iterations. The flow chart of GOA is presented by Fig. 1.

2.2   Opposition Based Learning

Opposition-based learning (OBL) is a method introduced in [21] for machine intelligence. The concept is based on opposite numbers. Opposite of a real number Z in range [A, B] is given by following equation.

$$\bar{Z} = A + B - Z \tag{3}$$

Opposition Based Learning: Let Z be real vector in range [A, B], $\bar{Z}$ be opposite of Z, F(Z) be a function of concern and G(Z) be a evaluation function (which calculates the objective value of F(Z)), then according to opposition-based learning, learning continues with Z if G(F(Z)) is better than G(F($\bar{Z}$)), else with $\bar{Z}$.

3   Opposition Based Grasshopper Optimisation Algorithm

OGOA introduces opposition based learning in GOA. The grasshopper are initialized using uniform randomization constrained by the upper and lower bounds of the each dimension. Then objective Value of each grasshopper is calculated to find the best grasshopper. This best grasshopper's location represents the target location. Using this target location new position is calculated for grasshoppers using Eq. 1. Now a new population of same size as of original population size is created using OBL. Objective values of these solutions is also calculated. Now elite solutions are selected using calculated objective values to form a solution array of size equal to original population size and these array is used for next iteration. Flow chart of GOA and OGOA is given in Fig. 1.

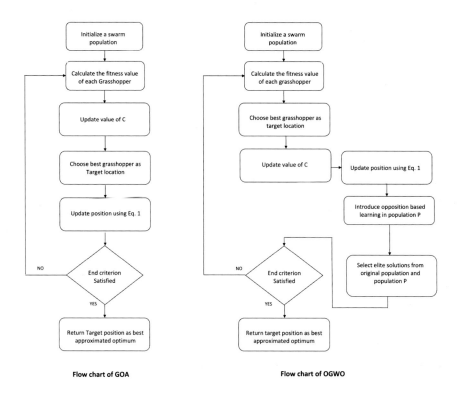

Fig. 1.

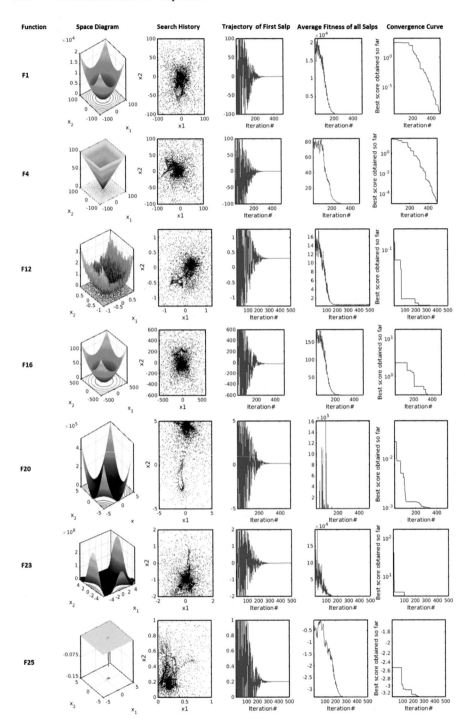

Fig. 2. Behaviour of OGAO

The extra population created in each iteration increases exploration of OGOA. Increment in exploration reduces probability of local optimal results. The selection procedure select elite solutions. This speeds up the convergence of algorithm OGAO. Further adaptive value of C is helpful for exploration, exploitation and balance between exploration and exploitation.

4   Experiments and Results

4.1   Experimental Set-Up and Implementation

For analysis purpose, twenty-five benchmark functions are selected from [22,23]. The algorithm OGOA is tested for these benchmark functions and compared with PSO, DE, GOA and GA. The benchmark numerical functions are given in Table 1. Here functions F_1–F_{11} are uni-modal functions. These function contain single maxima/minima. Uni-modal function are used for evaluating the exploitation power of optimization algorithms. Functions F_{12}–F_{25} are multi-modal functions. These functions have multiple maxima/minima and useful to evaluate the exploitation power of optimization algorithms.

For experimentation purpose, population size is kept 20 agents and maximum iteration count is 500. Each algorithm is executed 20 times. Value of C_{max} and C_{min} is taken as 1 and 0.00001 respectively. MATLAB R2010b is used to carry through the proposed work.

4.2   Qualitative Results

In Fig. 2, the behaviour of algorithm OGOA is depicted. Space digram, Search history (only for two dimensions), trajectory of first grasshopper, average fitness of grasshoppers and convergence curve for some of the benchmark functions are shown in the figure. Search history shows the locations traversed by grasshoppers. From search history, it can be seen that the area nearby the optima is exploited powerfully. It depict fine exploitation provided by OGOA. Also whole search space is explored. This presents the expert exploration power of OGOA. It also shows the strong balance between exploration and exploitation gained by OGOA. The trajectories (of first grasshopper) represent the vale of first grasshopper in first dimension. In the figure, trajectories show that OGOA is able to converge towards optima. It represents high effectiveness and efficiency of the algorithm. The average fitness represents the average fitness of all grasshoppers in each iteration. The average fitness of grasshoppers in figure depict that all grasshoppers gradually shift towards optimum solution and eventually converge near the optimum. The convergence curve represents the fitness of best position after the iterations. The convergence curve shows that the Best fitness value decreases monotonically with increment in iterations.

Table 1. Benchmark functions

Test problem	Function	Dim	Range	F_{min}
Sphere	$F_1 = \sum_{i=1}^{n} x_i^2$	30	[-100, 100]	0
Schwefel 2.22	$F_2 = \sum_{i=1}^{n} |x_i| + \prod_i^n |x_i|$	30	[-10, 10]	0
Schwefel 1.2	$F_3 = \sum_{i=1}^{n} \left(\sum_{j=1}^{i} x_i \right)^2$	30	[-100,100]	0
Max dimension	$F_4 = \text{Max}_i \{|x_i|, 1 \leq i \leq n\}$	30	[-100,100]	0
Rosenbrok	$F_5 = \sum_{i=1}^{n} [100(2x_{i+1} - x_i^2)^2 + (x_i - 1)^2]$	30	[-30, 30]	0
Shifted sphere	$F_6 = \sum_{i=1}^{n} (x_i + 0.5)^2$	30	[-100, 100]	0
De jong 4	$F_7 = \sum_{i=1}^{n} i.(x_i)^4$	30	[-5.12, 5.12]	0
Axis parallel hyper ellipsoid	$F_8 = \sum_{i=1}^{n} i.x_i^2$	30	[-5.12, 5.12]	0
Exponential	$F_9 = -(\exp(-0.5 \sum_{i=1}^{n} x_i^2)) + 1$	30	[-1, 1]	0
Sum of different power	$F_{10} = \sum_{i=1}^{n} |x_i|^{i+1}$	30	[-1, 1]	0
Step	$F_{11} = \sum_{i=1}^{n} (\lfloor x_i + 0.5 \rfloor)^2$	30	[-100, 100]	0
quartic(noise)	$F_{12} = \sum_{i=1}^{n} i x_i^4 + \text{random}[0, 1)$	30	[-1.28, 1.28]	0
Schwefel	$F_{13} = \sum_{i=1}^{n} -x_i \sin(\sqrt{|x_i|})$	30	[-500, 500]	-Dim* 418.9829
Rastrigin	$F_{14} = \sum_{i=1}^{n} [x_i^2 - 10\cos(2\pi x_i) + 10]$	30	[-5.12, 5.12]	0
Ackley	$F_{15} = -20\exp(-0.2 \sqrt{\frac{1}{n} \sum_{i=1}^{n} x_i^2}) - \exp(\frac{1}{n} \sum_{i=1}^{n} \cos(2\pi x_i)) + 20 + e$	30	[-32, 32]	0
Griewank	$F_{16} = \frac{1}{4000} \sum_{i=1}^{n} x_i^2 - \prod_i^n \cos(\frac{x_i}{\sqrt{i}}) + 1$	30	[-600, 600]	0
Alpine	$F_{17} = \sum_{i=1}^{n} |x_i \sin(x_i) + 0.1x_i|$	30	[-10, 10]	0
levy	$F_{18} = 0.1\sin^2(3\pi x_1) + \sum_{i=1}^{n} (x_i - 1)^2 [1 + \sin^2(3\pi x_i + 1)] + (x_n - 1)^2 [1 + \sin^2(3\pi x_n)]] + \sum_{i=1}^{n} u(x_i, 5, 100, 4)$	30	[-50, 50]	0
Foxholes	$F_{19} = \left(\frac{1}{500} + \sum_{j=1}^{25} \frac{1}{j + \sum_{i=1}^{2} (x_i - a_{ij})^6} \right)^{-1}$	2	[-65, 65]	1
Kowalik	$F_{20} = \sum_{i=1}^{11} \left[a_i - \frac{x_1(b_i^2 + b_i x_2)}{b_i^2 + b_i x_3 + x_4} \right]^2$	4	[-5, 5]	0.00030
Six hump camel back	$F_{21} = 4x_1^2 - 2.1x_1^4 + \frac{1}{3}x_1^6 + x_1 x_2 - 4x_2^2 + 4x_2^4$	2	[-5, 5]	-1.0316
Branin	$F_{22} = \left(x_2 - \frac{5.1}{4\pi^2} x_1^2 + \frac{5}{\pi} x_1 - 6 \right)^2 + 10 \left(1 - \frac{1}{8\pi} \right) \cos x_1 + 10$	2	[-5, 5]	0.398
goldstein price	$F_{23} = [1 + (x_1 + x_2 + 1)^2 (19 - 14 x_1 + 3x_1^2 - 14 x_2 + 16 x_1 x_2 + 3x_2^2)] \times [30 + (2 x_1 - 3 x_2)^2 \times (18 - 32 x_1 + 12x_1^2 + 48 x_2 - 36 x_1 x_2 + 27x_2^2)]$	2	[-2, 2]	3
Hartmann-3	$F_{24} = -\sum_{i=1}^{4} c_i \exp(-\sum_{j=1}^{3} a_{ij}(x_j - p_{ij})^2)$	3	[1, 3]	-3.86
Hartmann-6	$F_{25} = -\sum_{i=1}^{4} c_i \exp(-\sum_{j=1}^{6} a_{ij}(x_j - p_{ij})^2)$	6	[0, 1]	-3.32

4.3 Quantitative Results

Average of best solutions obtained so far (Avg.) and standard deviation of these best solutions (Std.) of the functions are shown in Table 2. The results represent that algorithm OGOA obtains best solutions for most of the functions among the five algorithms. This shows that OGOA gains high exploration and exploitation power and fine balance between exploration and exploitation. Adaptive changes

Table 2. Experimental results

Function	OGOA		GOA		PSO		DE		GA	
	Avg.	Std.	Avg.	Std.	Avg.	Std.	Avg.	Std.	Avg.	Std.
F_1	0.0000	0.0000	0.0000	0.0000	0.0023	0.0002	0.0041	0.0009	0.00400	0.0011
F_2	0.0000	0.0000	0.0021	0.0012	0.0101	0.0026	0.0189	0.0064	0.0099	0.0045
F_3	0.0000	0.0000	0.0013	0.0023	0.2598	0.1433	1.6782	0.9276	0.2009	0.1748
F_4	0.0000	0.0000	0.0000	0.0000	0.5111	0.1231	0.7564	0.6576	0.7187	0.4859
F_5	0.0000	0.0000	0.0000	0.0000	0.0405	0.03477	0.3847	0.1931	0.2437	0.1367
F_6	0.0000	0.0000	0.0000	0.0000	0.7867	0.0435	0.3468	0.0455	0.2005	0.1220
F_7	0.0000	0.0000	0.0000	0.0000	0.0000	0.0000	0.0000	0.0000	0.0000	0.0000
F_8	0.0000	0.0000	0.0013	0.0008	0.0000	0.0011	0.0000	0.0000	0.0000	0.0000
F_9	-1	0.0000	-0.9998	0.0001	-1	0.0000	-1	0.0000	-1	0.0000
F_{10}	0.0000	0.0000	0.0000	0.0000	0.0000	0.0000	0.0001	0.0000	0.0000	0.0000
F_{11}	0.0000	0.0000	0.2000	0.4000	0.0000	0.0000	0.0030	0.0012	0.3133	0.1234
F_{12}	0.0104	0.0056	0.0000	0.0000	0.1350	0.0675	0.0290	0.0178	0.0965	0.0674
F_{13}	-12495.100	0.3567	-10615.97	4735.656	-8455.430	5643.055	-9453.121	2351.856	-7564.646	7454.7775
F_{14}	0.000	0.0000	0.0000	0.0000	0.6574	0.3655	0.4820	0.2194	1.5653	0.4566
F_{15}	0.0000	0.0000	0.0000	0.0000	0.8200	0.4532	0.4533	0.5383	0.8754	0.4543
F_{16}	0.0000	0.0000	0.0034	0.0024	0.0086	0.0082	0.4012	0.2321	0.2013	0.0348
F_{17}	0.0000	0.0000	0.0000	0.0000	0.0435	0.0432	0.0000	0.0000	0.0073	0.0034
F_{18}	0.0025	0.0005	0.0044	0.0046	0.0161	0.0056	0.0033	0.0062	0.2043	0.0044
F_{19}	1.1342	0.5435	1.4534	0.4523	5.5543	5.5453	2.9435	1.3453	3.7884	0.6768
F_{20}	0.0007	0.0003	0.0005	0.0005	0.0053	0.0051	0.0043	0.0067	0.0065	0.0043
F_{21}	-1.0316	0.0000	-1.0316	0.0000	-1.0316	0.0000	-1.0316	0.0000	-1.0316	0.0000
F_{22}	0.3981	0.0006	0.3982	0.0001	0.3988	0.0021	0.3986	0.0004	0.3985	0.0025
F_{23}	3	0.0000	3	0.0000	5.2	2.2755	3.6745	0.0022	3.7667	.3444
F_{24}	-3.8567	0.0019	-3.8594	0.0007	-3.8543	0.0566	-3.8519	0.0029	-3.8515	0.0068
F_{25}	-3.3220	0.1455	-3.1698	0.0754	-3.2344	0.0243	-3.229	0.0345	-3.2453	0.1675

in value of C, double population and elite solution selection make the algorithm OGOA to finely search through the search area. The value of C decreases in each iteration gradually. On account of this, value of C is very higher in initial iterations than in later iterations. High value of C causes grasshoppers to move very fast and abruptly. Owing to this and double population, OGOA is capable of look through whole search space. This grows exploration component for OGOA. In later iterations, value of C becomes very low. Low value of C causes target location to move slowly. Owing to this and double population, OGOA searches the area around food location rigorously. This proffers exploitation component for OGOA. Further the elite solution selection in each iteration assist OGAO to converge fast and make balance between exploration and exploitation. In this way the OGAO provide powerful exploration and exploitation of search area as well as balance between the exploration and exploitation. On the basis of results and discussion, OGAO can be presented as one of the competitor among existing efficient and effective optimization techniques.

Results depict that all five algorithms Support No Free Lunch Theorem [24]. No Free Lunch theorem says that there is no optimization algorithm which is best suited for all kind of problems.

5   Conclusion

In this paper, Opposition based Grasshopper Optimisation algorithm (OGOA) is proposed. This algorithm Combine opposition based learning and Grasshopper Optimisation algorithm to provide an improved version OGOA. The algorithm is tested for 25 benchmark functions. It is also compared with other conventional optimization algorithms. Behaviour of OGAO is also shown and explained. Adaptive changes in value of C, double population and elite solution selection make OGOA as one of the efficient and effective optimization methods, able to avoid local optima and find global optima most of the time. For future work, OGOA can be extended for multi-objective, binary and constrained problems. The algorithm can also be applied for real word problems.

References

1. Glover, F.W., Kochenberger, G.A. (eds.): Handbook of Metaheuristics. vol. 57. Springer (2006)
2. Eberhart, R., Kennedy, J.: A new optimizer using particle swarm theory. In: Proceedings of the Sixth International Symposium on Micro Machine and Human Science, 1995, MHS 1995, pp. 39–43, IEEE (1995)
3. Mirjalili, S., Mirjalili, S.M., Lewis, A.: Grey wolf optimizer. Adv. Eng. Softw. **69**, 46–61 (2014)
4. Dorigo, M., Birattari, M., Stutzle, T.: Ant colony optimization. IEEE Comput. Intell. Mag. **1**(4), 28–39 (2006)
5. Karaboga, D.: An idea based on honey bee swarm for numerical optimization. vol. 200, Technical report-tr06, Erciyes university, engineering faculty, computer engineering department (2005)
6. Holland, J.H.: Genetic algorithms. Sci. Am. **267**(1), 66–72 (1992)
7. Koza, J.R.: Genetic programming (1992)
8. Rechenberg, I.: Evolution strategy: nature's way of optimization. In: Optimization: Methods and Applications, possibilities and Limitations, pp. 106–126. Springer, Heidelberg (1989)
9. Storn, R., Price, K.: Differential evolution-a simple and efficient heuristic for global optimization over continuous spaces. J. Global optim. **11**(4), 341–359 (1997)
10. Kirkpatrick, S., Gelatt, C.D., Vecchi, M.P.: Optimization by simulated annealing. Science **220**(4598), 671–680 (1983)
11. Rashedi, E., Nezamabadi-Pour, H., Saryazdi, S.: GSA: a gravitational search algorithm. Inf. Sci. **179**(13), 2232–2248 (2009)
12. Glover, F.: Tabu search – Part I. ORSA J. Comput. **1**(3), 190–206 (1989)

13. Glover, F.: Tabu search – Part II. ORSA J. Comput. **2**, 4–32 (1990)
14. Mirjalili, S., Lewis, A.: The whale optimization algorithm. Adv. Eng. Softw. **95**, 51–67 (2016)
15. Mirjalili, S., Gandomi, A.H., Mirjalili, S.Z., Saremi, S., Faris, H., Mirjalili, S.M.: Salp swarm algorithm: a bio-inspired optimizer for engineering design problems. Adv. Eng. Softw. **114**, 163–191 (2017)
16. Bansal, J.C., Sharma, H., Jadon, S.S., Clerc, M.: Spider monkey optimization algorithm for numerical optimization. Memetic Comput. **6**(1), 31–47 (2014)
17. Saremi, S., Mirjalili, S., Lewis, A.: Grasshopper optimisation algorithm: theory and application. Adv. Eng. Softw. **105**, 30–47 (2017)
18. Olorunda, O., Engelbrecht, A.P.: Measuring exploration/exploitation in particle swarms using swarm diversity. In: IEEE Congress on Evolutionary Computation, 2008, CEC 2008, (IEEE World Congress on Computational Intelligence), pp. 1128–1134, IEEE (2008)
19. Alba, E., Dorronsoro, B.: The exploration/exploitation tradeoff in dynamic cellular genetic algorithms. IEEE Trans. Evol. Comput. **9**(2), 126–142 (2005)
20. Crepinsek, M., Liu, S.H., Mernik, M.: Exploration and exploitation in evolutionary algorithms: a survey. ACM Comput. Surv. (CSUR) **45**(3), 35 (2013)
21. Tizhoosh, H.R.: Opposition-based learning: a new scheme for machine intelligence. In: International Conference on Intelligent Agents, Web Technologies and Internet Commerce, International Conference on Computational Intelligence for Modelling, Control and Automation, 2005, vol. 1, pp. 695–701. IEEE, November 2005
22. Ali, M.M., Khompatraporn, C., Zabinsky, Z.B.: A numerical evaluation of several stochastic algorithms on selected continuous global optimization test problems. J. Global Optim. **31**(4), 635–672 (2005)
23. Bansal, J.C., Sharma, H., Nagar, A., Arya, K.V.: Balanced artificial bee colony algorithm. Int. J. Artif. Intell. Soft Comput. **3**(3), 222–243 (2013)
24. Wolpert, D.H., Macready, W.G.: No free lunch theorems for optimization. IEEE Trans. Evol. Comput. **1**(1), 67–82 (1997)

Classification of Hyper Spectral Remote Sensing Imagery Using Intrinsic Parameter Estimation

L. N. P. Boggavarapu[1,2] and Prabukumar Manoharan[1(✉)]

[1] School of Information Technology Engineering (SITE),
VIT University, Vellore 632-014, India
mprabukumar@vit.ac.in
[2] Department of Information Technology,
V. R. Siddhartha Engineering College, Vijayawada, India

Abstract. A Hyperspectral remote sensing image (HSI) composed of various intrinsic components such as shading, albedo, noise and continuous narrow bands in different wavelengths. The classification of the HSI image is one of the challenging tasks in the area of Remote Sensing as it has numerous applications on environment, mineral exploration, target detection and anomaly detection. The present paper identifies a novel approach in classifying the image by incorporating the albedo intrinsic component retrieved from the image on principal components and factor analysis obtained through the dimensionality reduction. The obtained results are classified via Support Vector Machine classifier. The proposed algorithm tested on the benchmark datasets available worldwide such as Indian Pines, University of Pavia and Salinas. The extraction of albedo intrinsic components helps in effective classification of HSI image and outperforms the results with state of the art techniques, achieved the overall accuracy (OA) on these datasets.

Keywords: Classification · Hyperspectral · Dimensionality reduction · Support vector machine

1 Introduction

An Hyperspectral image is composed with the number of spectral bands with narrow wavelengths. All the bands are continuous, contain abundant information in them and can be represented as a hyperspectral cube in which each pixel is composed of mixture of materials. Its applications are widespread and includes forestry (Tree species classification, vegetation), as it can be seen that Forest and tree cover of India constitutes nearly 789,164 sq. km, which is 24.01% of the geographical area of the country [1], atmosphere, surveillance, identification of target minerals, soil identification, oceanography [2–4]. In general, every image consists of intrinsic and extrinsic parameters [5] so also in each band of the HSI. These parameters include shapes, locations and photometric properties [6]. Incorporation of intrinsic parameters like albedo component in hyper spectral remote sensing image is employed in [7].

© Springer Nature Switzerland AG 2020
A. Abraham et al. (Eds.): ISDA 2018, AISC 941, pp. 852–862, 2020.
https://doi.org/10.1007/978-3-030-16660-1_83

The classification of hyper spectral image is one of the challenging tasks in the area of remote sensing [8]. The basic limitation to classify an HIS image is the number of spectral bands and huge number of pixels in it. Hence, there is a need to reduce the dimensionality of spectral bands and not to suffer from Hughes effect [9].

This paper makes use of Principal Component Analysis and also factor analysis to reduce the dimensionality of the given image with huge number of spectral bands. Then extracts the intrinsic parameter, albedo component as given in [7] Finally, employed support vector machine classifier to classify the image. The methods described in the paper are tested on the bench mark datasets. [5] Discussed elaborately on the various types of feature extraction techniques that can be useful for HSI Images. Table 1 describes various methods used in the literature for the classification of Hyper spectral Remote Sensing Images.

Table 1. Methods used to classify HSI images

Classifier(s)	Description	Remarks
Support vector machine Vs. random forest Vs. neural network Classifiers [10]	Evaluated the three algorithms on APEX hyper spectral data collected in Poland Used dimensionality reduction before applying the classifier Employed the random selection of training and testing datasets	SVM Outperforms Random Forest and ANN Outperforms over RF and SVM Need to find an optimal algorithm for optimal number of bands to be selected
Random forest [11]	Applied variation in Random Forest with variable importance, conditional inference and quantile forest Datasets employed are AVIRIS Indian Pine and selected 1000 random samples	Robust in yielding high accuracy by tuning parameters with tradeoff in computation time Needs to reduce the computation time
Maximum likelihood estimator [12]	It's an algorithm for multitemporal hyper spectral imagery Implemented the classifier on two images taken on two different dates from Airborne Prism Experiment Uses Maximum likely hood classifier Used random stratified sampling	Implemented the ML algorithm on temporal data Implemented on 3 m X 3 m and needs to implement on 1 m resolution
Random forest and SVM [13]	Implemented discriment analysis based on Partial Least Squares These results are tested on Random Forest And SVM Applied Stratified random Sampling technique for sample selection	Implemented competitive adaptive reweighted sampling Provides good results over the dataset
GMM vs. RF, SVM, KNN [14]	Makes use of simple random sampling Smoothing technique is also employed on the data set. Implemented on multitemporal Formosat-2 imagery	Temporal data can be useful for tree special classification
Random forest and multi class classifier [15]	Implemented band reduction through Minimum Noise Fraction to reduce bands to 20 from 116 in the HyMap dataset is used Makes use of Lidar data for height information	Tree species are identified at an high accuracy in urban and semi urban areas

The rest of the paper is organized as follows: Sect. 2 describes the proposed method with its architecture diagram, Sect. 3 dataset description, Sect. 4 describes the results and observations on the three dataset and finally Sect. 5 describes the conclusion.

2   Proposed Approach

This section describes the procedure employed in the paper to classify the given hyper spectral image. The flowchart is shown in the Fig. 1.

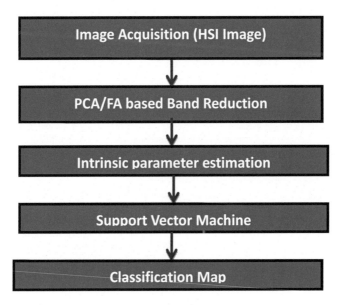

Fig. 1. HSI classification procedure

2.1   Image Acquisition

The input image is acquired from various sensors described in Sect. 3. In general, a hyper spectral image is represented in the vector format as shown in (1) and its class labels are given in (2)

$$X = \left\{ x_m \in R^b, m = 1, 2, \ldots \ldots L \right\}, \tag{1}$$

where L is the number of pixels and b is the number of bands in the image.

$$y \equiv (y_1, y_2, \ldots \ldots, y_n) \in R^n \tag{2}$$

The data is collected from three bench mark datasets available. The reduction of number of bands from the given input HSI image is employed in Sect. 2.2.

2.2 Band Reduction Using Principal Component Analysis/Factor Analysis

This section describes the two methodologies namely Principal Component Analysis and Factor Analysis used for feature selection in spectral bands. The feature selection is one of the important tasks in the processing of data [22, 23] The Principal Component Analysis (PCA) is one of the unsupervised learning algorithms to extract features from the given input image and also to reduce the number of features as well [18]. PCA will extract the uncorrelated data by removing the correlated data from the data. PCA [21] is a two step process in which the covariance matrix C, diagonal matrix D and orthogonal matrix is computed in the first phase as shown in Eqs. (3) and (4) respectively and in the second phase, identify the principal components (uncorrelated data) for the given data on the descending order of Eigen values of the matrix. These can be computed as follows:

If X_n is any spectral vector of a pixel in the data set, then the covariance matrix is computed as

$$C = E[(x_n - E(x_n))(x_n - E(x_n))^T]$$ (3)

where E is the expected value. The diagonal and orthogonal matrix are computed by de composing C as

$$C = EDE^T$$ (4)

where E is the orthogonal matrix computed through eigen vectors and D is the diagonal matrix computed via eigen values.

The Factor Analysis (FA) proposed in [16] can be used for dimensionality reduction. It comprises of the two steps in which examines the latent structure of the bands and interpret interms of the factor scores. FA estimates the decomposition of the reduced correlation matrix C-D where D is the diagonal matrix.

2.3 Intrinsic Parameter Estimation – Edge Preserving and Albedo Estimation

Filtering is one of the important techniques in the image processing and remote sensing applications. This paper employs the Domain transform recursive edge-preserving filter proposed in [17]. Also, an image is the combination of shading and albedo components. This is represented as Image = Albedo component + Shading Component. The albedo component of the given pixel of HSI is extracted from [7]. Here the hyperspectral image is represented as

$$I \in R^{mxnxd} \tag{5}$$

where m, n and d are the rows, columns and the number of bands in the image. Each pixel in (5) can be represented as the combination of albedo A_p and shading component H_p as shown in (6) [17].

$$I_p = A_p H_p \tag{6}$$

The albedo is estimated with the Affinity matrix by computing the similarity between the pixels p and q using Gaussian function [20] as shown in (7)

$$a_{pq} = \begin{cases} \exp\left(-\dfrac{\|p - q\|_2^2}{2\sigma_s^2} + \dfrac{\|I_p - I_q\|_2^2}{2\sigma_r^2}\right), & \text{if } q \in N(p) \\ 0, & \text{otherwise} \end{cases} \tag{7}$$

This helps to find out the albedo components of the image. The paper implemented the albedo on each of the bands obtained after PCA and as well as FA. The albedo of the 2nd band on the three datasets is shown in Fig. 2.

2.4   Support Vector Machine Classifier

Support vector machine (SVM) [19] is employed on the obtained output for classifying the given input images to various classes. Here as proposed in [17], radial basis function kernel with five fold cross validation is employed. The results obtained on these data sets are shown in Sect. 3.

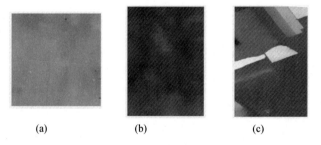

(a)   (b)   (c)

Fig. 2. Albedo component of 2nd band (a) Indian Pines (b) Pavia (c) Salinas

The brief description of the algorithm is specified in Algorithm 1.

Algorithm 1.

Input: Hyperspectral image Data (Hyperspectral Cube which contains m x n x b)
   X: mXn pixel vectors, b: Number of bands, K: Number of class labels
Output: Determine the classification map
Procedure:
Initial Phase:
Dimensionality reduction using Principal Component Analysis or Factor Analysis
Principal Component Analysis
   a) Computation of Covariance Matrix , Orthogonal Matrix and Diagonal Matrix
   b) Computation of Principle Components (uncorrelated) with the orthogonal matrix
Factor Analysis Compute latent variable
Middle Phase:
Domain transform recursive edge-preserving filter to extract intrinsic parameters [7, 17]
Final Phase:
Training on the selected samples using Support Vector Machine

---

3   Data Sets and Experimental Setup

This paper uses the three bench mark data sets obtained through Airborne Visible/Infrared Imaging Spectrometer (AVIRIS) sensor on Indian Pines and Salinas Valley which consists of 220 bands and 16 classes each, through ROSIS Sensor on University of Pavia which consists of 103 spectral bands and 9 classes.

The work is executed on Intel 7200 Work Station with Intel Xcon X3440 Processor, 8 GB DDR RAM and MATLAB R2016b.

4   Results and Observations

Table 2 represents the performance measures on the standard datasets using two different feature extraction techniques.

The ground truth, classification maps of the images using the two feature extraction techniques is shown in Figs. 3, 4 and 5.

The overall accuracy of the SVM, PCA with SVM and FA with SVM are shown in Figs. 6 and 7. The accuracy is increased in both the cases where intrinsic parameter-albedo presented in the process. From Figs. 6 and 7, it is clear that the accuracy of the HSI Classification is improved while applying the intrinsic component using Factor analysis and Principal Component analysis.

Table 2. (a) OA, AA and Kappa coefficient values using PCA. (b) OA, AA and Kappa coefficient values using Factor Analysis

	Overall accuracy (OA)	Average accuracy (AA)	Kappa coefficient
(a)			
Indian Pines	93.08	90.20	92.08
Pavia University	98.21	96.80	97.60
Salinas	94.69	92.84	94.10
(b)			
Indian Pines	94.29	92.83	93.47
Pavia University	98.2	97.52	97.64
Salinas	95.33	93.56	94.81

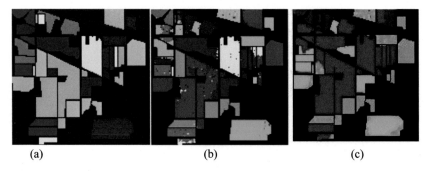

(a)   (b)   (c)

Fig. 3. Results on Indian Pines dataset (a) Ground truth (b) Classification Map using PCA (c) Classification Map using FA

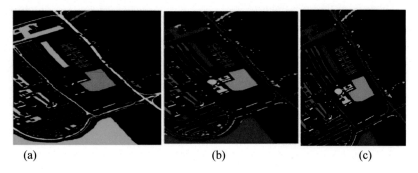

(a)   (b)   (c)

Fig. 4. Results on Pavia dataset (a) Ground truth (b) Classification Map using PCA (c) Classification Map using FA

Also, as in the case of [7], the proposed comparison is also tested on the various number of training sample given for the training in Support vector machine with radial basis kernel function (Factor Analysis is used) and is given in Table 3. In all the cases, the Overall and Average accuracies is improved except on the Salinas dataset. It is slightly reduced when the training samples increased from 7% to 9%.

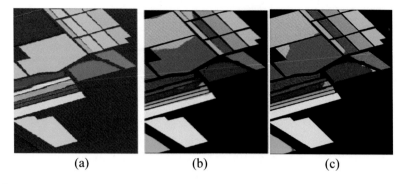

(a)   (b)   (c)

Fig. 5. Results on Salinas dataset (a) Ground truth (b) Classification Map using PCA (c) Classification Map using FA

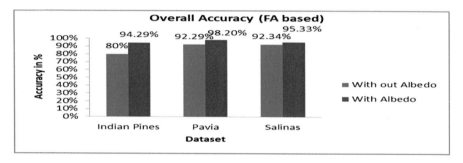

Fig. 6. Overall accuracy with and without intrinsic parameter using FA

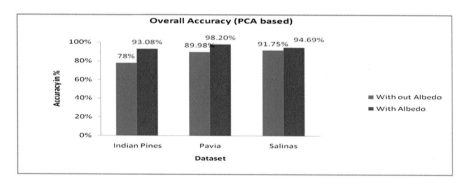

Fig. 7. Overall accuracy with and without intrinsic parameter using PCA

Figure 8 represents the performance analysis on the three datasets employed only SVM [19], PCA and SVM with intrinsic description and Factor Analysis FA and SVM with intrinsic description. The method FA with the intrinsic parameters has high overall accuracy among with the other two methods.

Table 3. OA, AA, Kappa coefficient and time for execution on different training samples

Dataset	No. of samples (%)	Overall accuracy (OA) (%)	Average accuracy (AA) (%)	Kappa coefficient (%)	Time in seconds
Indian Pines	4	92.94	92.48	91.93	156.7
	5	96.4	94.4	95.9	163.2
	9	98.09	97.2	97.8	171.4
Pavia University	5	98.29	99.79	99.7	598.3
	7	99.3	98.75	99.09	646.3
	9	99.47	99.16	99.26	731.2
Salinas	5	99.84	99.74	99.82	786.5
	7	99.95	99.86	99.94	874.5
	9	98.78	98.04	98.32	891.5

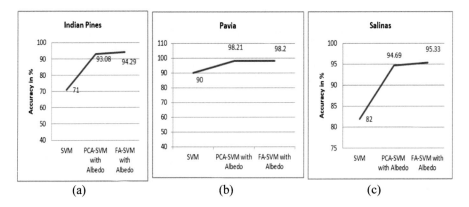

Fig. 8. Comparison of overall accuracy on (a) Indian Pines (b) Pavia (c) Salinas

5   Conclusion and Future Work

The paper presents the comparison of two efficient feature selection techniques, namely, Principal Component Analysis and Factor Analysis to reduce the number of spectral bands to reduce the correlated data between them. Prior to the Support Vector machine classifier with radial basis function kernel, an intrinsic parameter estimation is computed for each band by making use of edge preserving filtering. The overall accuracy with Factor Analysis provides increase in the accuracy: 71% to 94.3%, 90% to 98.2% and 82% to 95.33% on the three standard data sets of Indian Pines, University of Pavia and Salinas respectively. Also, it is evident from the results that among the two feature extraction methods PCA Vs FA, Factor Analysis is a good method for feature selection. The feature selection will help various vegetation indices of remote sensing images.

References

1. State of the forest report 2013: Forest Survey of India, Ministry of Environment and Forests, Government of India, Dehradun (2013)
2. Olmanson, L.G., Brezonik, P.L., Bauer, M.E.: Airborne hyperspectral remote sensing to assess spatial distribution of water quality characteristics in large rivers: the Mississippi River and its tributaries in Minnesota. Remote Sens. Environ. **130**, 254–265 (2013)
3. Zarco-Tejada, P.J., Berjón, A., López-Lozano, R., Miller, J.R., Martín, P., Cachorro, V., González, M.R., de Frutos, A.: Assessing vineyard condition with hyperspectral indices: leaf and canopy reflectance simulation in a row-structured discontinuous canopy. Remote Sens. Environ. **99**, 271–287 (2005)
4. Pascucci, S., Belviso, C., Cavalli, R.M., Laneve, G., Misurovic, A., Perrino, C., Pignatti, S.: Red mud soil contamination near an urban settlement analyzed by airborne hyperspectral remote sensing. In: IEEE International Geoscience and Remote Sensing Symposium (IGARSS 2009), pp. IV-893, IV-896, 12–17 July 2009 (2009)
5. Vaddi, R., Prabukumar, M.: Comparative study of feature extraction techniques for hyper spectral remote sensing image classification: a survey. In: International Conference on Intelligent Computing and Control Systems, 15–16 June 2017. Vaigai College Engineering (VCE), Madurai, India (2017)
6. Barrow, H., Tenenbaum, J.: Computer Vision Systems: Recovering Intrinsic Scene Characteristics from Images. Academic Press, New York (1978)
7. Zhan, K., Wang, H., Xie, Y., Zhang, C., Min, Y.: Albedo recovery for hyperspectral image classification. J. Electron. Imaging **26**(4), 043010 (2017). https://doi.org/10.1117/1.JEI.26.4.043010
8. Chutia, D., Bhattacharyya, D.K., Sarma, K.K., Kalita, R., Sudhakar, S.: Hyperspectral remote sensing Classifications: a perspective survey. Trans. GIS **20**(4), 463–490 (2016)
9. Hughes, G.F.: On the mean accuracy of statistical pattern recognizers. IEEE Trans. Inf. Theory **IT-14**(1), 55–63 (1968)
10. Raczko, E., Zagajewski, B.: Comparison of support vector machine, random forest and neural network classifiers for tree species classification on airborne hyperspectral APEX images. Eur. J. Remote Sens. **50**(1), 144–154 (2017)
11. Nandhini, K., Porkodi, R.: Spatial classification and prediction in hyperspectral remote sensing data using random forest by tuning parameters. Int. J. Adv. Res. Comput. Sci. **8**(3), 259–266 (2017)
12. Tagliabue, G., et al.: Forest species mapping using airborne hyperspectral APEX data. Misc. Geogr. – Reg. Stud. Dev. **20**(1), 28–33 (2016)
13. Richter, R., et al.: The use of airborne hyperspectral data for tree species classification in a species-rich Central European forest area. Int. J. Appl. Earth Obs. Geoinf. **52**, 464–474 (2016)
14. Sheeren, D., et al.: Tree species classification in temperate forests using Formosat-2 satellite image time series. Remote Sens. **8**, 1–29 (2016)
15. Zhang, Z., et al.: Object-based tree species classification in urban ecosystems using LiDAR and hyperspectral data. Forests **7**(6), 1–16 (2016)
16. Rummel, J.R.: Applied factor analysis. Library of congress catalog, card no, pp. 73–78327, United States of America (1988)
17. Gastal, E.S.L., Oliveira, M.M.: Domain transform for edge-aware image and video processing. ACM Trans. Graph. **30**(4), 69:1–69:12 (2011)
18. Abdi, H., Williams, L.J.: Principal component analysis. Wiley Interdiscip. Rev.: Comput. Stat. (2010). https://doi.org/10.1002/wics.101

19. Melgani, F., Bruzzone, L.: Classification of hyperspectral remote sensing images with support vector machines. IEEE Trans. Geosci. Remote Sens. **42**(8), 1778–1790 (2004)

20. Shi, J., Malik, J.: Normalized cuts and image segmentation. IEEE Trans. Pattern Anal. Mach. Intell. **22**(8), 888–905 (2000)

21. Kumar, C.A.: Analysis of unsupervised dimensionality reduction techniques. Comput. Sci. Inf. Syst. **6**(2), 217–227 (2009)

22. Prabukumar, M., Sawant, S., Samiappan, S., Agilandeeswari, L.: Three-dimensional discrete cosine transform-based feature extraction for hyperspectral image classification. J. Appl. Remote Sens. **12**(4), 046010 (2018)

23. Prabukumar, M., Shrutika, S.: Band clustering using expectation–maximization algorithm and weighted average fusion-based feature extraction for hyperspectral image classification. J. Appl. Remote Sens. **12**(4), 046015 (2018)

Probabilistic PCA Based Hyper Spectral Image Classification for Remote Sensing Applications

Radhesyam Vaddi[1,2] and Prabukumar Manoharan[1(✉)]

[1] School of Information Technology Engineering (SITE), VIT University,
Vellore 632-014, India
mprabukumar@vit.ac.in
[2] Department of Information Technology,
V. R. Siddhartha Engineering College, Vijayawada, India

Abstract. Hyper spectral image (HSI) Classification has become important research areas of remote sensing which can be used in many practical applications, including precision agriculture, Land cover mapping, environmental monitoring etc. HSI Classification includes various steps like Noise removal, dimensionality reduction, and classification. In this work, we adopted structure-preserving recursive filter (SPRF) to noise removal and Probabilistic based principal component analysis (PPCA) is applied to reduce dimensionality. Finally classification is performed using multi class large marginal distribution machine (LDM). The proposed (HSI) Classification method is carried out and results are validated across the three widely used standard datasets like Indian Pines, University of Pavia and Salinas. The obtained results show that the proposed method provides results on par with similar type of methods from literature.

Keywords: Hyper spectral image · Principal component analysis · Agriculture

1 Introduction

Classification of Hyper spectral image (HSI) (image which contains hundreds of contiguous narrow spectral bands) is a challenging task which has received significant attention in the remote sensing research community. This is primarily due to impact in various applications like agriculture, Environmental Monitoring, forestry and defense etc. In this work we are interested in identification of application of hyper spectral imaging technologies in agriculture, and in particular, precision agriculture [13]. Example farming applications which use satellite data include monitoring of plant diseases, the estimation of crop yield and the classification of the crops.

The only two major technical aspects required in achieving above discussed farming applications are Dimensionality reduction and classification. Dimensionality reduction or discovering the useful features (feature extraction) from raw HSI data is important step preprocessing step for classification. In fact it directly influences the classification accuracy [4]. Dimensionality reduction is based on transforming the pixel vectors into a new set of coordinates. Principal Components Analysis (PCA) is one of the most generally used methods of this era [6]. Independent Component Analysis

© Springer Nature Switzerland AG 2020
A. Abraham et al. (Eds.): ISDA 2018, AISC 941, pp. 863–869, 2020.
https://doi.org/10.1007/978-3-030-16660-1_84

(ICA) can also be used like PCA for dimensionality reduction and in particular for hyper spectral images [10]. In PCA the basis you want to find is the one that best explains the variability of your data. The first vector of the PCA basis is the one that best explains the variability of your data, the second vector is must be orthogonal to the first one. In ICA the basis you want to find is the one in which each vector is an independent component of your data [18]. The other techniques include image fusion based feature extraction [8], using texture [9] and methods based on deep learning [7]. In this work, a Probabilistic based principal component analysis (PPCA) [5] is applied to reduce HSI image dimensionality [16, 17].

There exist several classification (either supervised or unsupervised) methods useful for hyper spectral images. These include random forest, neural networks and SVM [11]. Large marginal distribution machine (LDM) [3] is applied to HSI classification in this paper. This gives superior results than SVM since it optimizes the margin distribution to obtain improved performance in generalization.

The paper is organized as follows. The glimpse of the HSI classification procedure is presented in Sect. 2. Results and observations in Sect. 3 include analysis of the proposed approach with standard datasets. Scope for Crop type classification and yield estimation is shown in Sect. 4. Section 5 is presented with conclusion and future work.

2 Proposed Approach

The procedure adopted for HSI classification in this article can be illustrated by in Fig. 1. In this Section, noise removal using Structure-Preserving Recursive Filter [1], Probabilistic PCA to reduce the dimension of hyper spectral data and classification of hyper spectral image by using large margin distribution machine are introduced very briefly.

Noise Removal by Structure-Preserving Recursive Filter: An HSI Image data of current interest has rxs pixels with d number of bands. The number of bands is reduced to k by the application of Probabilistic principal components analysis. Noise removal is performed by filtering each band I using structure-preserving recursive filter (SPRF). The mechanism is followed from [2] Robust structure-preserving filtering of HIS images can be done by SPRF. This is due to processing of output signal in such a way that on the same side of edge, neighboring pixels need to maintain similar feature values.

Dimensionality Reduction by Probabilistic PCA: Probabilistic principal components analysis (PPCA) is used to reduce dimensionality [19]. These treats the data by means of a lower dimensional latent space [6]. It is often used for multidimensional scaling like HSI data. In contrast to the traditional PCA it uses a probability model. The maximum-likelihood estimation of parameters in a latent variable model is used to find principal axes of a set of observed data vectors. Further properties of the associated likelihood function are applied iteratively to estimate the principal subspace using EM algorithm [20].

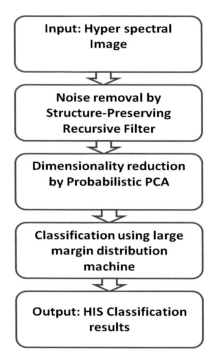

Fig. 1. HSI Classification procedure

Classification Using Multi Class Large Margin Distribution Machine: Large margin distribution machine (LDM) is a binary classifier. But our current Interest is on classification of hyper spectral image data which is of multiclass scenario. So from a set of binary LDM classifiers we have developed multiclass LDM by make use of one-against-one (OAO) (also known as pair wise coupling) strategy [12]. LDMs are so trained to discriminate the samples of one class from the samples of another class for a problem with n classes, n (n − 1)/2. In general, classification of an unknown pattern is done like maximum voting. Here each LDM votes for only one class.

3   Results and Observations

In the present work, three extensively used hyper spectral data sets (Indian Pines, University of Pavia and Salinas) were used to validate the proposed methodology. The complete description of the data sets along with area of interest is shown in Table 1.

The HSI Classification procedure is executed for the above said three datasets. Accuracy also measured in terms of three standard quality metrics, overall accuracy (OA), Average accuracy (AA) and kappa Coefficient. These values corresponds to the each dataset are (for two methods) presented in Table 2.

The HSI Classification results for the proposed approach on three data sets are presented in Fig. 2. The LDM classification uses parameters which are considered for

Table 1. Description of HSI datasets

S no	Name of the HIS data set	No of classes	Pixel size and wavelength range	No of spectral bands	Area coverage
1	Indian Pines	16	145 × 145 0.4–2.5 µm	200	2/3rd portion agriculture 1/3rd portion forest/other natural vegetation
2	Pavia University	9	610 × 340 0.43–0.86 µm	115	Water, trees, meadows and bare soil
3	Salinas	16	512 × 217	200	Vegetation, bare soils, and vineyard fields

Table 2. OA, AA and Kappa coefficient values for (A) (PPCA+SPRF+Multi class LDM) and (B) (FA+SPRF+Multi class LDM) on three standard data sets

	Method	OA (Overall accuracy)	AA (Average accuracy)	Kappa coefficient
Indian Pines data set	A	98.31	98.70	98.05
	B	98.55	96.64	98.33
Pavia University data set	A	98.39	97.10	97.85
	B	95.45	99.11	99.27
Salinas data set	A	99.82	99.66	99.80
	B	99.7	99.58	99.66

the current work as k = 40, sigma_s = 200, sigma_r = 0.3, lambda1 = 1.25e5, lambda2 = 5e4 and c = 1e6. The parameter values and the procedure adopted in this paper is from the motivation of the work done by [1], used Factor analysis (FA) to reduce dimensionality (FA+SPRF+Multi class LDM). As compared with these results, the proposed method results got 2.13% more accuracy for Indian Pines data set in calculating AA. The 0.12%, 0.08% and 0.14% improvement of accuracy obtained for Salinas's data set in terms of OA, AA and Kappa Coefficient respectively.

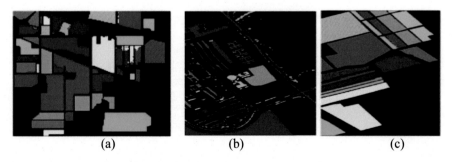

(a)   (b)   (c)

Fig. 2. HSI Classification results (PPCA+SPRF+Multi class LDM) for (a) Indian Pines data set (b) Pavia data set (c) Salinas's data set

Table 3. OA, AA and Kappa coefficient values for proposed HIS Classification (PCA+SPRF +Multi class LDM) on three standard data sets

	OA (Overall accuracy)	AA (Average accuracy)	Kappa coefficient
Indian Pines data set	98.38	96.55	98.14
Pavia University data set	99.70	99.72	99.66
Salinas data set	99.82	99.66	99.80

In order to compare the results further, we have performed HSI classification for the same data sets using the procedure PCA+SPRF+Multi class LDM. The accuracy values are shown in Table 3 and Fig. 3 represents the Classification results.

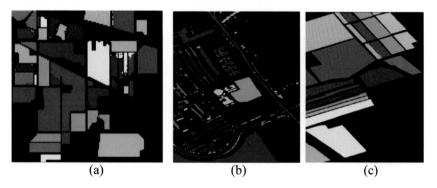

(a)   (b)   (c)

Fig. 3. HSI Classification results (PCA+SPRF+Multi class LDM) for (a) Indian Pines data set (b) Pavia data set (c) Salinas's data set

4   Crop Type Classification and Yield Estimation Using Hyper Spectral Imaginary

The Input images for this project are considered as hyper spectral remote sensing images. Corresponding noise removal using SPRF is to be considered as a part of preprocessing. The immediate processes start with extraction of features using Probabilistic PCA. Classification of crops by taking feature band set is the consequent major step. This is done using multi class LDM. The data sets we use in this project also include more area of agriculture and vegetation so this classification results can be easily validate on crop classification (corn, wheat, soybean, trees etc.).

Yield estimation is also application of the work which depends on vegetation index parameters [14]. Vegetation Index (VI) values represents vegetation condition variations. Spectral reflectance of the EM spectrum represents these variations. These values can be computed for a HSI image, For example, green vegetation (Fig. 4a) absorbs most of the incident VI'S light and reflects a large portion of the NIR light. In contrast, unhealthy vegetation (Fig. 4b) reflects more VIS light and less NIR light [15]. The mathematical calculations are represented in Eqs. 1, 2 and 3.

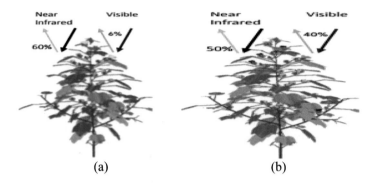

(a)   (b)

Fig. 4. Yield estimation using Vegetation Index example

$$NDVI = \frac{NIR - \mathrm{Re}d}{NIR + \mathrm{Re}d} \tag{1}$$

NDVI values calculated for Fig. 4(a) and for Fig. 4(b) are shown below

$$NDVI = \frac{0.60 - 0.06}{0.60 + 0.06} = 0.82 \tag{2}$$

$$NDVI = \frac{0.50 - 0.40}{0.50 + 0.40} = 0.11 \tag{3}$$

5   Conclusion and Future Work

In this work, Structure-preserving recursive filter (SPRF) is adopted to remove noise. Since HSI Image consists of several narrow bands noise removal is the necessary step. Probabilistic based principal component analysis (PPCA) is applied for dimensionality reduction. This has given a prominent result as compared with simple PCA. Finally classification is performed using multi class large marginal distribution machine (LDM). This outperforms SVM based methods. The proposed (HSI) Classification method is carried out and results are validated across the three widely used standard datasets like Indian Pines, University of Pavia and Salinas. Also results are done by replacing PCA in PPCA. In recent times, Deep learning has become very interested area of research and achieved promising results especially convolutional neural networks (CNNs) have fascinated a lot of interest due to have good accuracy results in remote sensing. Our future work includes incorporating CNN for HSI classification along with PPCA and SPRF.

References

1. Zhan, K., Wang, H., Huang, H., Xie, Y.: Large margin distribution machine for hyper spectral image classification. J. Electron. Imaging **25**(6), 063024 (2016). https://doi.org/10.1117/1.JEI.25.6.063024

2. Gastal, E.S., Oliveira, M.M.: Domain transform for edge-aware image and video processing. ACM Trans. Graph. **30**(4), 69 (2011)

3. Zhang, T., Zhou, Z.-H.: Large margin distribution machine. In: Proceedings of SIGKDD, International Conference on Knowledge Discovery and Data Mining, vol. 20, pp. 313–322 (2014)

4. van der Maaten, L.J.P., Postma, E.O., van den Herik, H.J.: Dimensionality reduction: a comparative review. Tilburg University Technical report, TiCC-TR 2009-005, 2009

5. Tipping, M.E., Bishop, C.M.: Probabilistic principal component analysis. J. R. Stat. Soc.: Ser. B **61**(3), 611–622 (1999)

6. Rodarmel, C., Shan, J.: Principal component analysis for hyper spectral image classification. Purdue University, West Lafayette, 47907-1284, U.S.A (2002)

7. Chen, Y., Lin, Z., Zhao, X.: Deep learning-based classification of hyperspectral data. IEEE J. **7**(6), 2094–2107 (2014)

8. Kang, X., Li, S.: Feature extraction of hyperspectral images with image fusion and recursive filtering. IEEE Trans. Geosci. Remote Sens. **52**, 3742–3752 (2014)

9. Kumar, B., Dikshit, O.: Texture based hyperspectral image classification. In: The International Archives of the Photogrammetry (2014)

10. Diwaker, M.K., Chaudhary, P.T., Bhatt, A., Saxena, A.: A comparative performance analysis of feature extraction techniques for hyperspectral image classification. Int. J. Softw. Eng. Appl. **10**(12), 179–188 (2016)

11. Zhan, K., Wang, H., Xie, Y., Zhang, C., Min, Y.: Albedo recovery for hyperspectral image classification. J. Electron. Imaging **26**, 043010 (2017)

12. Liu, Y., Wang, R., Zeng, Y.: An improvement of one-against-one method for multi-class support vector machine. In: 2007 International Conference on Machine Learning and Cybernetics, Hong Kong, pp. 2915–2920 (2007)

13. Zhang, N., Wang, M., Wang, N.: Precision agriculture-a worldwide overview. Comput. Electron. Agric. **36**, 113–132 (2002)

14. Yang, C., Everitt, J.H., Bradford, J.M.: Airborne hyperspectral imagery and yield monitor data for estimating grain sorghum yield variability. Trans. ASAE **47**(3), 915–924 (2004)

15. Yang, C.: Airborne hyperspectral imagery for mapping crop yield variability. Geogr. Compass **3**(5), 1717–1731 (2009)

16. Boggavarapu, L.N.P., Prabukumar, M.: Survey on classification méthodes for hyper spectral remote sensing imagery. In: International Conference on Intelligent Computing and Control Systems (2017)

17. Boggavarapu, L.N.P., Prabukumar, M.: Robust classification of hyperspectral remote sensing images combined with multihypothesis prediction and 3 dimensional discrete wavelet transform. Int. J. Pure Appl. Math. **117**(17), 115–120 (2017)

18. Vaddi, R., Prabukumar, M.: Comparative study of feature extraction techniques for hyper spectral remote sensing image classification: a survey. In: 2017 International Conference on Intelligent Computing and Control Systems (ICICCS) (2017)

19. Kumar, C.A.: Analysis of unsupervised dimensionality reduction techniques. Comput. Sci. Inf. Syst. **6**(2), 217–227 (2009)

20. Prabukumar, M., Sawant, S., Samiappan, S., Agilandeeswari, L.: Three-dimensional discrete cosine transform-based feature extraction for hyperspectral image classification. J. Appl. Remote Sens. **12**(4), 046010 (2018)

Automatic Determination Number of Cluster for Multi Kernel NMKFCM Algorithm on Image Segmentation

Pradip M. Paithane[1]([✉]) and S. N. Kakarwal[2]

[1] Dr. BAMU, Aurangabad, India
paithanepradip@gmail.com
[2] PES COE, Aurangabad, India
s_kakarwal@yahoo.com

Abstract. In image analysis, image segmentation performed an essential role to get detail information about image. Image segmentation is suitable in many applications like medicinal, face recognition, pattern recognition, machine vision, computer vision, video surveillance, crop infection detection and geographical entity detection in map. FCM is famous method used in fuzzy clustering to improve result of image segmentation. FCM doesn't work properly in noisy and nonlinear separable image, to overcome this drawback, Multi kernel function is used to convert nonlinear separable data into linear separable data and high dimensional data and then apply FCM on this data. NMKFCM method incorporates neighborhood pixel information into objective function and improves result of image segmentation. New proposed method used RBF kernel function into objective function. RBF function is used for similarity measure. New proposed algorithm is effective and efficient than other fuzzy clustering algorithms and it has better performance in noisy and noiseless images. In noisy image, find automatically required number of cluster with the help of Hill-climbing algorithm.

Keywords: Component: clustering · Fuzzy clustering · FCM ·
Hill-climbing algorithm · KFCM · NMKFCM · NMRBKFCM

1 Introduction

Image segmentation is a foremost topic for many image processing research. Image segmentation is acute and vital component of image examination system. It is method of subdividing image keen on different segment (collection of pixel). Segment consist set of similar pixel by using different properties of pixel like quality, shade, color, intensity, character text etc. The goal of method is to make simpler or variation of the presentation of an input image into more expressive and easier image to analyze. This method is performed using four approaches like Clustering, Thresholding, Region Extraction and Edge Detection. Clustering is approach to perform image segmentation on image. It is method of subdividing image into number of group and every image entity assign to group. Image entity allocate such that same entity fit to same group. Clustering perform by using two main approaches like crisp clustering and fuzzy clustering [2]. Crisp clustering is to process in which finding boundary between clusters. In this object belong to only one cluster. Fuzzy clustering has improved solution for this problem which is object fit to many groups. Fuzzy C-MEANS (FCM) has been used commonly as clustering technique for image segmentation. FCM be present method of clustering to which allow one object fits to two or more clusters. FCM is introducing fuzziness with degree of membership function of every object and range of membership function between 0 and 1 [3]. Aim of FCM is to minimize value of objective function and perform partition on dataset into n number of clusters. FCM provide better accuracy result than HCM in noiseless image. FCM is not working properly in noisy image and failed in nonlinear separable data, to overwhelm this drawback Kernel FCM (KFCM) has applied. Role of kernel is convert nonlinear separable data into linear separable data and low dimension into high dimensional feature space [4]. Propose Novel Kernel Fuzzy C-means (NMKFCM) algorithm which is to assimilate neighbor term in objective function and amend result over KFCM and FCM in noisy and noiseless image [5]. NMKFCM is very beneficial and useful method for image segmentation.

2   Clustering Algorithm

2.1   Fuzzy(Soft) Clustering Algorithm (FCM)

Aim of FCM has to curtail objective function [6]. The FCM algorithm is improved outcome over k-mean. In this method feature vector of dataset is subdivided into hard clusters. The feature vector is perfectly a participant of one group only. As an alternative of this method, the FCM has modified the situation. This method is allowed to feature vector for multiple membership positions to several group. FCM smartly handles with such problems by dealing with degree of membership function of feature vector which varies from 0 to 1. FCM is iterative clustering processes that generate ideal c partition by using abate weight inside the group, following formula for sum of square off error objective function Jm_{obj}.

$$J_{m_{obj}} = \sum_{x=1}^{P} \sum_{y=1}^{Q} U_{xy}^m d_{xy}^2 \tag{1}$$

Where: Q: The number of forms in Z, P: The number of cluster, Uxy = The degree of association Z_x in the yth cluster; d_{xy} = Distance unit between object Z_x and cluster center, W_y = The model of the center of cluster y, m: The weighting exponent on each fuzzy relationship.

The FCM emphases over decreasing objective function Jm_{obj} value and focus on the below constraints on U:

$$U_{xy} \in [0, 1], \; x = 1, 2, 3, 4. \ldots P, \text{ number of forms.}$$
$$y = 1, 2, 3, \ldots Q, \text{ number of cluster}$$

$$\sum_{x=1}^{P} U_{xy} = 1, x = 1, 2, 3, 4 \ldots P, \quad 0 < \sum_{x=1}^{P} U_{xy} < 1, y = 1, 2, 3, \ldots Q$$

Objective function J_{mobj} defines a constrained optimization problem. Lagrange multiplier technique has applied for conversion of constrained to unconstrained problem. By using this calculates membership function and update cluster center separately.

$$U_{xy} = \frac{1}{\sum_{x=1}^{P} \left(\frac{d_{xy}}{d_{xl}}\right)^{\frac{2}{(m-1)}}} \tag{2}$$

$$x = 1, 2, \ldots P \text{ and } y, l = 1, 2, \ldots Q$$

If $d_{xy} = 0$ then $U_{xy} = 1$ and $U_{xy} = 0$ for $x \neq y$
And calculate cluster center using following steps

$$W_y = \frac{\sum_{x=1}^{P} (U_{xy})^m Z_x}{\sum_{x=1}^{P} (U_{xy})^m} \tag{3}$$

2.2 Kernel Method

Kernel method is calculating the distance between two data points. The data points are plotted into a high dimensional space. They are openly separable using distance metric [7]. The existing work plans a way of incremental the correctness of the FCM by developing a kernel function in computing the distance of data point from the cluster centers. Radial basis kernel is used in distance to plot feature vector from the input space to a high dimensional space. The kernel function can be used in those algorithm which are exclusively be influenced by the dot product between two vectors. Linear algorithms are converted into decision-boundary algorithms when kernel function used. Those decision-boundary methods are corresponding to linear patterns functioning on the series of a feature space Ψ. Kernels are applied, the Ψ function does not essential to continually clearly computed. Kernel trick is general idea of the distance parameter that calculates distance between two data points. The pattern points are plotted into a high dimensional spaces in which pattern are clearly separable. Known unlabeled data set $Z = \{z_1, z2, z3 \ldots, z_n\}$ in the d-dimensional space R^d, let Ψ be a decision boundary mapping function from this response space to high dimensional feature space $H : \Psi : R^d \rightarrow H_d, z \rightarrow \Psi(z)$. Dot product in the high dimensional feature pattern can be determined with the help kernel trick function $K_T(Z_x, Z_y)$ in the response space R^d.

$$K_T(Z_x, Z_y) = \Psi(Z_x).\Psi(Z_y)$$

Consider the subsequent example. For $d = 2$ also the mapping function Ψ,

$$\Psi : R^2 \rightarrow H_d = R^3(Z_{x1}, Z_{y2}) \rightarrow (Z_{x1}^2, Z_{x2}^2, \sqrt{2Z_{x1}}, Z_{x2})$$

Then the dot product in the feature space H is measured as

$$\Psi(X_i)\Psi(X_j)$$
$$= (Z_{x1}^2, Z_{x2}^2, \sqrt{2Z_{x1}Z_{x2}}) \cdot (Z_{y1}^2, Z_{y2}^2, \sqrt{2Z_{y1}Z_{y2}})$$
$$= ((Z_{y1}^2, Z_{y2}^2) \cdot (Z_{y1}^2, Z_{y2}^2))^2$$
$$= (Z_x, Z_y)^2$$
$$= K_T(Z_x, Z_y)$$

K-trick is the square of the dot product in the response phase. From the observation notice that this illustration used kernel idea function which determine the value of dot product in the feature pattern H_d without remarkably manipulative mapping function Ψ. Following are models of kernel idea function:

1. Polynomial Based K Trick-Function

$$K_T(Z_x, Z_y) = (Z_x Z_y + L)^2, \text{ where } L \geq 0, d \in Q$$

2. Gaussian Based K Trick Function

$$K_T(Z_x, Z_y) = \exp(-\frac{\|Z_x - Z_y\|^2}{2\sigma^2}), \text{ where } \sigma > 0$$

3. Radial Base K-Function

$$K_T(Z_x, Z_y) = \exp\left(-\frac{\sum \left|Z_x^v - Z_y^v\right|^w}{\sigma^2}\right), \text{ where } \sigma, v, w > 0$$

4. Tangent K Trick-Function

$$K_T(Z_x, Z_y) = 1 - \tanh\left(-\frac{\|Z_x - Z_y\|^2}{\sigma^2}\right), \text{ where } \sigma > 0$$

2.3 Novel Modified Kernel FCM Algorithm (NMKFCM)

Novel modified kernel fuzzy method is assimilating neighborhood pixel value in objective function. NMKFCM algorithm is modified version of KFCM. NMKFCM which incorporate neighborhood pixel value using 3×3 or 5×5 window and introduce this value in objective function [1, 8]. In this '$\propto$' parameter is trained for control weight of neighbor's term which is achieved greater value with intense of image noise. Range of $\propto$ value lies within 0 to 1, if percentage of noise is low then choose value of $\propto$ between 0 and 0.5 and percentage of noise is higher then choose value of $\propto$ 0.5 and 1.0. NMKFCM is an iterative process which minimizes value of objective function with neighborhood term [9]. In this objective function introduce window around pixel and $\propto$ parameter.

$$J_{NMKFCM_{obj}}(U, W) = \sum_{y=1}^{Q} \sum_{x=1}^{P} U_{xy}^m (1 - K_T(Z_x, W_y)) \left(\frac{N_R - \alpha \sum_{k \in N_i} U_{yk}}{N_R} \right) \quad (4)$$

Where: NR: The cardinality, Ni: Collection of neighbors of pixel Zi, K_T is Gaussian k trick-Function. Objective function Jnmkm$_{obj}$ describes a constrained optimization problem. Objective function is going to change constrained optimization problem into decision-boundary optimize problematic by applying Lagrange multiplier method.

Update membership function U_{ij}:

$$U_{xy} = \left(\frac{((1 - K_T(Z_x, W_y))(\frac{N_R - \alpha \sum_{l \in N_i} U_{yl}}{N_R}))^{-\frac{1}{m-1}}}{\sum ((1 - K_T(Z_x, W_y))(\frac{N_R - \alpha \sum_{l \in N_l} U_{kl}}{N_R}))^{-\frac{1}{m-1}}} \right) \quad (5)$$

Update Cluster center W_j:

$$W_y = \frac{\sum_{y=1}^{Q} U_{xy}^m K_T(Z_x, W_y) Z_x}{\sum_{x=1}^{P} U_{xy}^m (Z_x, W_y)} \quad (6)$$

NMKFCM work very well in neighborhood pixel information.

2.4   Novel Modified Radial Base Kernel Fuzzy C-Means Algorithm

Novel modified Radial Base kernel fuzzy method is assimilating hesitation degree and RBF kernel function into FCM clustering algorithm. Novel modified kernel FCM algorithm which incorporate neighborhood value into objective function. To determine correct value of α is experimentally hard and range of $\propto$ value differs image by image. NMRBKFCM is worked on similarity of dataset value so accuracy increases as compare to NMKFCM approach. Similarity measure is applied to measure the degree of matching between two objects. In this objective function, introduce hesitation function and RBF:

$$J_{RBNKFCM_{obj}}(\mathbf{U}, \mathbf{W}) = 2 \sum_{x=1}^{Q} \sum_{y=1}^{P} U_{xy}^{m}(1 - \exp(-\frac{\sum |Z_x^v - Z_y^v|^w}{\sigma^2})) \tag{7}$$

Where v, w is greater than 0, It is nominated as kernel width and It has integer positive number, U_{xy}^m is membership function, $\prod_x$ is hesitancy degree.

$\prod_x$ Defined as: $\prod_x = \frac{1}{N} \sum_{K=1}^{N} \prod_{xk}, k \in N[1, N]$

Where we have to calculate $\prod_{xk}$:

$$\prod_{xk} = 1 - U_{xk} - W_y \tag{8}$$

This value is introduced into objective minimize function to maximize the dataset pattern in the class. This method used for decrease entropy of histogram of image. Objective function $J_{RBNKFCM}$ describes a constrained optimization problem. Objective function is going to change constrained optimization problem into decision-boundary optimize problematic by applying Lagrange multiplier method.

Update membership function U_{xy}:

$$U_{xy} = \frac{(\frac{1}{(1-K_T(Z_x, W_y))})^{-\frac{1}{m-1}}}{\sum_{y=1}^{Q} (\frac{1}{(1-K_T(Z_x, W_y))})^{-\frac{1}{m-1}}} \tag{9}$$

Update Cluster center W_j:

$$W_y = \frac{\sum_{x=1}^{P} U_{xy}^m \cdot K_T(Z_x, W_y) \cdot Z_x}{\sum_{x=1}^{P} U_{xy}^m \cdot K_T(Z_x, W_y)} \tag{10}$$

Algorithm

$J_{RBNMKFCMOBJ}$ can be obtaining through an iterative process, which is achieved by following steps:

Input

1. $Z = \{Z_1, Z_2 \dots Z_N\}$, Object Pattern,
2. Q, $2 \leq Q \leq y$, y is numeral of cluster,
3. Set value of ε, it is stopping criteria parameter,
4. Initialize membership function U_{xy} using data set and cluster,
5. Calculate initial cluster center W0 = (w01, w02 ... w0Q).

Output

$W_y = \{W_0, W_1, W_2 \ldots W_y\}$, Final center of clusters.

STEP:

1. Set loop counter s = 0
2. Calculate C cluster center W_y^s with U^s by using Eq. 10
3. Calculate membership function U^{s+1} by using Eq. 9
4. If $\{U^s - U^{s+1}\} < \varepsilon$ then stop, then fix r = r + 1 and move to step number 4.

3   Hill–Climb Algorithm

Hill-climb algorithm is used to determine number of cluster for image segmentation. Clustering approach is used unsupervised model for number of cluster. Hill-Climb algorithm is used to identify required number of cluster for image segmentation. Recognition of salient image region is beneficial for many applications like histogram and segmentation of image, image retrieval. This dilemma is attempted by plotting pixels into various feature patterns. Saliency is calculated as the local contrast of image area with respect to neighborhood of many scales [10]. In this evaluate space among the average feature vectors of pixel with respect to image sub region and average feature vector of neighborhood pixel. At a known scale, divergence based saliency value $c_{i,j}$ for a pixel at position (i, j) in image is calculated as the space S between regular vectors of pixel features of inner region R1 and outer region R2.

$$C_{ij} = S[(\frac{1}{r_1}\sum_{x=1}^{r_1} V_p), (\frac{1}{r_2}\sum_{y=1}^{r_2} V_q)] \tag{11}$$

Where r_1 and r_2: number of pixels in *R1* and *R2* respectively, v: vector of feature components correlated to a pixel. S: Euclidean space if v is a vector of uncorrelated feature elements, Mahalanobis space: elements of the vector are associated.

In experiment, CIELab color distance is used RGB images for generate feature vectors like color and luminance. Meanwhile accepted variances in CIELab color distance are nearly Euclidian, s in Eq. (12).

$$C_{ij} = \|V1 - V2\| \tag{12}$$

Where v1 = [L1; x1; y1]T and v2 = [L2; x2; y2]T: average vectors for regions R1 and R2. Add saliency values across the scales to determined final saliency plan as discussed below:

$$m_{ij} = \sum_s C_{ij} \tag{13}$$

Hill-climbing method is understood as search window run across the space of the s-dimensional histogram to discover the leading bin from window.

4 Experimental Result

Fuzzy clustering methods are performing segmentation on real image, medical image and synthetical image. In medical image open source dataset is available which can used for experiment performance. Real image is captured image by various devices. In this work only .BMP and .jpeg format is used for execution purpose.

NMKFCM method is improved result of image segmentation as compare to FCM and KFCM. NMRBKFCM is also improved segmentation result over other fuzzy clustering algorithm. The performance of NMKFCM method is evaluated by using parameter like CAR, Runtime and Number of iteration. H/W requirement for this experiment is Intel Celeron processor M 1.7 GHz, OS is Microsoft Windows XP, 512-MB memory and the platform is MATLAB 6.5. Clustering Accuracy Rate (CAR) is used to evaluate performance of clustering method. CAR defined in below formula [11]

$$CAR = \frac{\left| A_{xy} \cap A_{ref} \right|}{\left| A_{xy} \cup A_{ref} \right|} \tag{14}$$

Where A_{xy}: set of pixel fitting y^{th} cluster found by x^{th} method and A_{refj}: set of pixel fitting to the jth cluster in the reference segmented image. By using this formula calculate accuracy rate.

Medical Image:
We apply fuzzy clustering algorithms to medical image and Add 2%, 5% and 10% salt and pepper noise into real image. Hill-climbing algorithm automatically determines cluster number 17 is used in image segmentation. In NMKFCM choose $\propto$ value from 0 to 0.5 for less noisy image and for noisier image choose $\propto$ value from 0.5 to 1, $N_R = 8$ (Figs. 1 and 2).

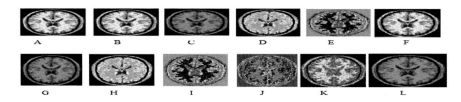

Fig. 1. Medical image

Input image (A) FCM (B) KFCM (C) NMKFCM (D) NMRBKFCM (E) 2% Noise FCM (F) 2% Noise KFCM (G) 2% Noise NMKFCM (H) 2% Noise NMRBKFCM (I) 5% Noise FCM (J) 5% Noise KFCM (K) 5% Noise NMKFCM (L) 5% Noise NMRBKFCM (Table 2).

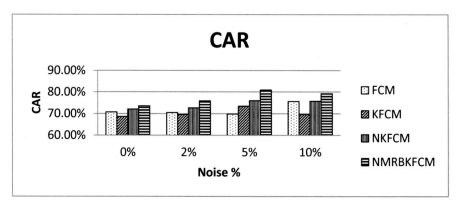

Fig. 2. Comparison between fuzzy clustering algorithms using CAR value from Table 1

Table 1. Comparison with fuzzy clustering algorithms using Cluster Accuracy Rate

NOISE%	FCM	KFCM	NKFCM	NMRBKFCM
0%	70.7439%	68.6309%	72.0572%	73.5222%
2%	70.4721%	69.5362%	72.6506%	75.8815%
5%	69.8048%	73.467%	76.0044%	80.9542%
10%	75.6375%	69.6031%	75.7221%	79.2545%

Table 2. Comparison between FCM, KFCM, NMKFCM, NMRBKFCM using runtime period for 0% noise

Method	Runtime in second
FCM	4.2755 s
KFCM	5.2075 s
NMKFCM	4.2084 s
NMRBKFCM	4.1123 s

5   Conclusion

Proposed algorithm gives efficient image segmentation than FCM and KFCM fuzzy clustering algorithms. Proposed method improves the segmentation performance by incorporating the effect radial base kernel function and hesitation degree. Proposed algorithm has determined automatically required cluster number for image segmentation. There are several things that could be done in the future as the continuation of this work. At firstly, in noisy image propose algorithm which can be determined automatically cluster number but this number is not useful for image segmentation because proposed algorithm has been generated cluster for noisy pixel so image segmentation could not be effective as compare to noiseless pixel. Secondly choosing of optimal factor α is still main issue, proposed algorithm assign value of optimal factor α randomly. In future work, add other kernel function into FCM objective function.

References

1. Yu, C.-Y., Li, Y., Liu, A.L., Liu, J.H.: A novel modified kernel fuzzy c-means clustering algorithms on image segmentation. In: 14th IEEE International Conference (2011). ISSN 978-0-7695-4477

2. Chan, S., Zhang, D.: Robust image segmentation using FCM with spatial constraints based on new kernel - induced distance measure. IEEE Trans. Syst. Man Cybern.-Part B: Cybern. **34**(4), 1907–1916 (2004)

3. Zanaty, E., Aljahdali, S.: Improving fuzzy algorithms for automatic magnetic resonance image segmentation. Int. Arab J. Inf. Technol. **7**(3), 271–279 (2009)

4. Kaur, P., Gupta, P., Sharma, P.: Review and comparison of kernel based fuzzy image segmentation techniques. Int. J. Intell. Syst. Appl. **7**, 50–60 (2012)

5. Islam, S., Ahmed, M.: Implementation of image segmentation for natural images using clustering methods. IJETAE **3**(3), 175–180 (2013). ISSN 2250-2459, ISO 9001:2008 Certified Journal

6. Cannon, R.L., Dave, J.V., Bezdek, J.C.: Efficient implementation of the fuzzy C –means clustering algorithms. IEEE Trans. Pattern Anal. Mach. Intell. **PAMI-8**(2), 248–255 (1986)

7. Hofman, M.: Support vector Machines-Kernel and the Kernel Trick, pp. 1–16 (2006)

8. Zang, D., Chen, S.: A novel kernalized fuzzy C-means algorithm with application in medical image segmentation. Artif. Intell. Med. **32**, 37–50 (2004)

9. Zanaty, E.A., Aljahdli, S., Debnath, N.: A kernalized fuzzy C-means algorithm for automatic magnetic resonance image segmentation. J. Comput. Methods Sci. Eng. Arch. **9**(1, 2S2), 123–136 (2009)

10. Kochra, S., Joshi, S.: Study on hill-climbing algorithm for image segmentation technology. Int. J. Eng. Res. Appl. (IJERA) **2**(3), 2171–2174 (2012). ISSN 2248-9622

11. Paithane, P.M., Kinariwala, S.A.: Automatic determination number of cluster for NMKFC-means algorithms on image segmentation. IOSR J. Comput. Eng. (IOSR-JCE) **17**, 12–19 (2015)

Characteristics of Alpha/Numeric Shape Microstrip Patch Antenna for Multiband Applications

R. Thandaiah Prabu[1,2(✉)], M. Benisha[1,2], and V. Thulasi Bai[3]

[1] Anna University, Chennai, India
thandaiah@gmail.com, benishaxavier@gmail.com
[2] Jeppiaar Institute of Technology, Sriperumbudur, India
[3] KCG College of Technology, Chennai, India
thulasi_bai@yahoo.com

Abstract. In this contribution, alphabets and numeric together called Alpha/Numeric microstrip patch antenna (MPA) is fabricated and tested for 2.4 GHz resonant frequency for Wi-Fi applications. The MPA is preferred for this design because it has the advantage that cut resonant slot inside the patch of different geometry. The Antenna is designed with Flame Retardant 4 (FR4) substrate material with a relative permittivity of 4.4 and with the suitable dimensions of substrate thickness of 1.5 mm, layout layer thickness of 70 μm, height 70 mm and width 60 mm. The above-designed antenna is fabricated (Alpha/Numeric MPA) and tested by network analyzer (E5062A ENA Series). The different characteristics such as return loss and VSWR were analyzed and discussed here.

Keywords: Wi-Fi · FR4 · Microstrip patch antenna (MPA) ·
Network analyzer · Return loss · VSWR

1 Introduction

The antenna can be characteristics neither impedance matching nor radiation pattern. The proposed fabricated antenna impedance matching characteristics like return loss and VSWR were analyzed by using E5062A ENA Series network analyzer. In general, Poor impedance matching may result in standing waves formation. Different ways of representing impedance matching are, Reflection coefficient (ρ); $0 \leq |\rho| \leq 1$, Return loss (RL); $0 \leq RL \leq \infty$ and VSWR (Γ); $1 \leq \Gamma \leq \infty$. For good impedance matching reflection coefficient or VSWR values are low or return loss must be high. Ideal values for good matching are, $|\rho| = 0$; $RL = \infty$ & $\Gamma = 1$. By Industrial standards: $|\rho| \leq -10$ dB & $\Gamma \leq 2$.

The Alpha/Numeric Microstrip Patch Antenna was proposed in [1], which is found to the Wi-Fi applications. All the antennas include 26 shapes for alphabet and 10 shapes in numeric shapes were operated at 2.4 GHz resonant frequency and parameters like return loss and Gain of the antenna were discussed. For simulation and

A. Abraham et al. (Eds.): ISDA 2018, AISC 941, pp. 880–895, 2020.
https://doi.org/10.1007/978-3-030-16660-1_86

optimization ADS 2009 industry standard software is used which works on Method of Momentum (MoM).

In the proposed method, all the simulated antennas were fabricated with FR–4 substrate material thickness of 1.5 mm, layout layer thickness of 70 μm, height 70 mm and width 60 mm. same values were considered during simulation with 50 Ω microstrip line feed. But after the fabrication process coaxial feeding techniques were used, for that feeding point was drilled with 1.4 mm for the standard co-axial connector and soldered. This may produce the losses like connector loss, feeder loss, soldering loss and may produce the effect in resonant frequency.

The various shape of the antenna such as A Shaped, C Shaped, H Shaped, E shaped, L Shaped, I Shaped, T Shaped, U Shaped and monopole antennas were proposed for different applications [2–16], this proposed method has proven that microstrip patch antenna can be any shape but want to satisfy the equivalent values parameter. The proposed antennas have the same dimension for all the shape design and ground height 80 mm and width 70 mm.

2 Fabricated Alphabet Antenna

The following section shows the graphical representation of the various fabricated MPA's based on alphabetical shapes (Figs. 1, 2, 3, 4, 5, 6, 7, 8, 9, 10, 11, 12, 13, 14, 15, 16, 17, 18, 19, 20, 21, 22, 23, 24, 25 and 26). Each alphabet shaped MPA's are validated using their return loss curves and VSWR using Network Analyzer. Resonant frequencies for fabricated antenna were shifted from the simulated frequency based on the feeding methods and also few antennas were produced multiband frequencies.

Table 1 shows the various applications for the different frequencies. Fabricated antennas were tested using Network Analyzer E5062A ENA Series that can measure from 300 MHz to 3 GHz frequency range (Figs. 27, 28, 29, 30, 31, 32, 33, 34, 35 and 36).

Table 1. Various frequency band applications

Frequency	Applications
100–300 MHz	VHF Land Mobile, Fixed, Satellite, Unlicensed Multi Use Radio Service (MURS), FM, Railways, DTV channels, Professional Wireless Microphones, Amateur radio, Military Aircraft Radio
400 MHz	UHF Mobile Antenna
535 kHz and 1.7 MHz	AM Radio Broadcasts
540–890 MHz	Digital Television Antenna Device
694–790 MHz, 700 MHz	4G LTE, Point to Multipoint applications, Public Safety Applications, DVB-T Technology, Terrestrial Television Broadcasting
800 & 900 MHz	GSM Lower Band, LTE, ISM Band, Wireless LAN, Wireless Video Applications & RFID Applications
860 & 960 MHz	Wearable Passive Ultra High Frequency Radio, RFID Tag Antennas

(*continued*)

Table 1. (*continued*)

Frequency	Applications
1 GHz–5 GHz	Ultra High Frequency (UHF) Communication System
1.227 GHz to 1.575 GHz	GPS L1 Band & L2 Band
1.3 GHz	Wireless Infrastructure Receivers, Point To Point Microwave Links, High Dynamic Range Down mixer Applications
1.7 GHz	Cellular Applications, PCS Applications, Indoor Wireless, Multipoint Applications
1.8 GHz	GSM, LTE, DCS, AWS, Point to Point Systems & Point to Multi Point Systems
1.9 GHz	GSM Upper Band, Personal Communication Services, LTE
2.1 GHz	4G/LTE/CDMA/UMTS, Fixed Service, Space Research Service
2.3 GHz	LTE
2.4 GHz	ISM Band, WiFi, IEEE 802.11b, 802.11 g and 802.11n and Bluetooth Applications
2.5–2.7 GHz	LTE, MMDS Band, IEEE 802.16 and 802.20 WiMAX Technology, Sprint 4G WiMax Non Direct Line of Sight (NLOS), Wireless Internet Service Providers (WISP), Sprint 4G WiMax, Point to Point & Point to Multi Point System, Clear wire & wireless internet application [17]

Fig. 1. A shape antenna - network analyzer return loss and VSWR output

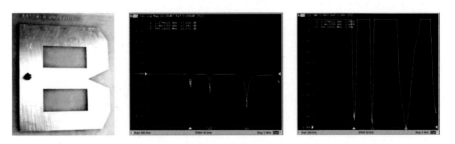

Fig. 2. B shape antenna - network analyzer return loss and VSWR output

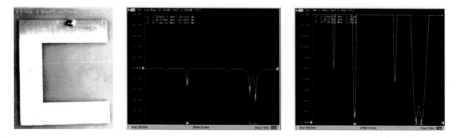

Fig. 3. C shape antenna - network analyzer return loss and VSWR output

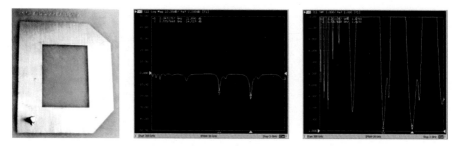

Fig. 4. D shape antenna - network analyzer return loss and VSWR output

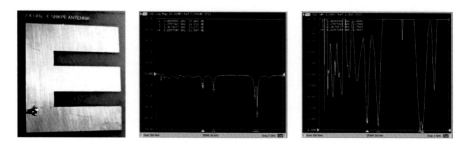

Fig. 5. E shape antenna - network analyzer return loss and VSWR output

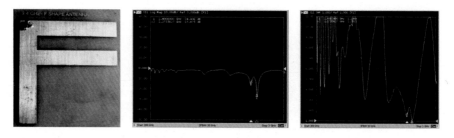

Fig. 6. F shape antenna - network analyzer return loss and VSWR output

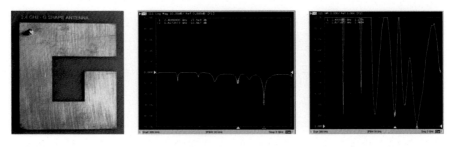

Fig. 7. G shape antenna - network analyzer return loss and VSWR output

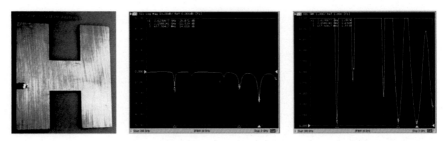

Fig. 8. H shape antenna - network analyzer return loss and VSWR output

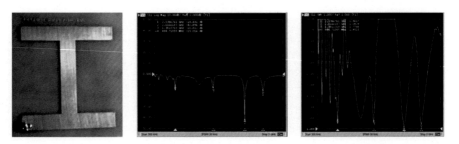

Fig. 9. I shape antenna – network analyzer return loss and VSWR output

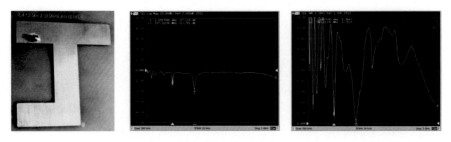

Fig. 10. J shape antenna - network analyzer return loss and VSWR output

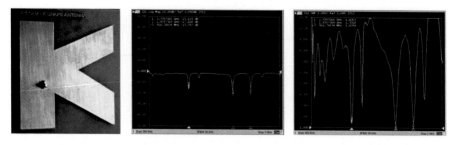

Fig. 11. K shape antenna - network analyzer return loss and VSWR output

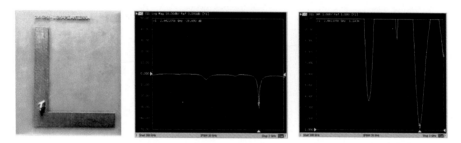

Fig. 12. L shape antenna – network analyzer return loss and VSWR output

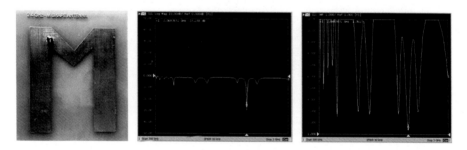

Fig. 13. M shape antenna - network analyzer return loss and VSWR output

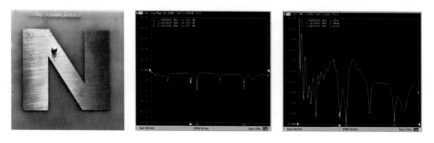

Fig. 14. N shape antenna - network analyzer return loss and VSWR output

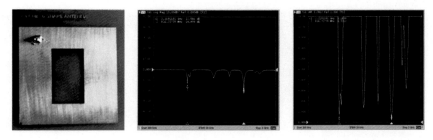

Fig. 15. O shape antenna – network analyzer return loss and VSWR Output

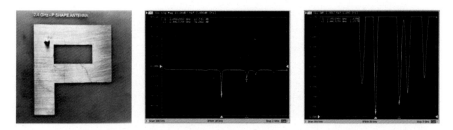

Fig. 16. P shape antenna - network analyzer return loss and VSWR output

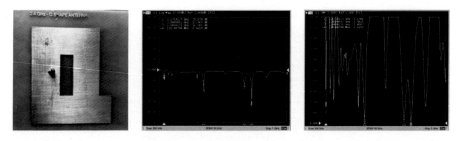

Fig. 17. Q shape antenna - network analyzer return loss and VSWR output

Fig. 18. R shape antenna – network analyzer return loss and VSWR output

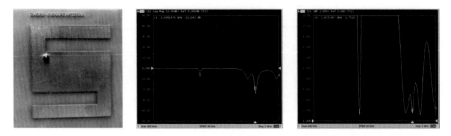

Fig. 19. S shape antenna - network analyzer return loss and VSWR output

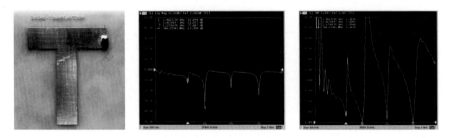

Fig. 20. T shape antenna - network analyzer return loss and VSWR output

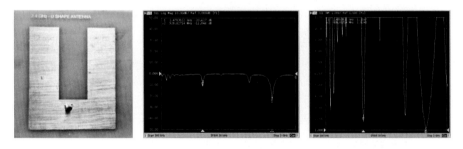

Fig. 21. U shape antenna - network analyzer return loss and VSWR output

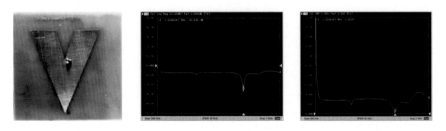

Fig. 22. V shape antenna - network analyzer return loss and VSWR output

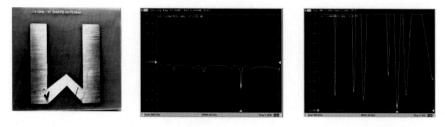

Fig. 23. W shape antenna - network analyzer return loss and VSWR output

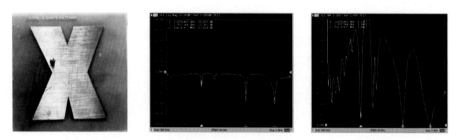

Fig. 24. X shape antenna – network analyzer return loss and VSWR output

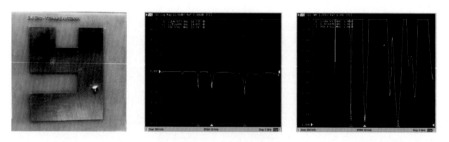

Fig. 25. Y shape antenna - network analyzer return loss and VSWR output

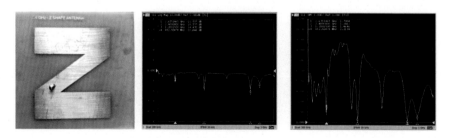

Fig. 26. Z shape antenna - network analyzer return loss and VSWR output

Table 2. Comparison of simulated and fabricated alphabet shape antenna

Shape	Simulated antenna frequency [1]	Simulated antenna return loss (dB) [1]	Simulated antenna gain (dB) [1]	Fabricated antenna frequency	Fabricated antenna return loss (dB)	Fabricated antenna VSWR
A	2.4	−17	5.13	1.30 GHz	−11.85	1.65
				2.15 GHz	−19.97	1.22
B	2.39	−31	6.11	995.15 MHz	−10.53	1.84
				1.44 GHz	−13.04	1.66
				2.27 GHz	−27.15	1.08
C	2.48	−17	6.03	977.06 MHz	−15.59	1.52
				2.41 GHz	**−16.08**	**1.39**
				2.52 GHz	−24.02	1.31
D	2.39	−34	3.02	1.53 GHz	−14.51	1.46
				2.24 GHz	−21.89	1.17
E	2.39	−13	5.57	1.10 GHz	−12.96	1.58
				1.36 GHz	−11.07	1.78
				2.33 GHz	−16.82	1.33
				2.40 GHz	**−33.46**	**1.04**
F	2.43	−20	5.84	2.27 GHz	−15.87	1.39
				2.40 GHz	**−24.40**	**1.14**
G	2.39	−12	5.30	1.82 GHz	−10.46	1.94
				2.40 GHz	**−25.56**	**1.13**
H	2.45	−12	4.13	657.50 MHz	−14.01	1.55
				2.15 GHz	−11.52	1.65
				2.62 GHz	−28.87	1.09
I	2.48	−13	4.28	494.71 MHz	−15.01	1.45
				1.38 GHz	−11.81	1.77
				2.11 GHz	−40.86	1.09
				2.53 GHz	−14.49	1.50
J	2.39	−13	5.57	597.21 MHz	−13.78	1.71
				1.10 GHz	−17.11	1.36
K	2.4	−11.5	4.23	916.76 MHz	−15.78	1.43
				1.93 GHz	−17.10	1.33
				2.33 GHz	−15.12	1.43
L	2.4	−12	3.08	**2.44 GHz**	**−28.49**	**1.14**
M	2.38	−11	4.93	2.06 GHz	−27.25	1.36
N	2.42	−22	3.23	1.01 GHz	−12.59	1.67
				1.18 GHz	−14.29	1.44
				2.40 GHz	**−12.31**	**1.65**
O	2.38	−18	6.15	826.32 MHz	−14.49	1.53
				2.21 GHz	−22.58	1.18
P	2.46	−15	5.77	1.44 GHz	−31.06	1.25

(*continued*)

Table 2. (*continued*)

Shape	Simulated antenna frequency [1]	Simulated antenna return loss (dB) [1]	Simulated antenna gain (dB) [1]	Fabricated antenna frequency	Fabricated antenna return loss (dB)	Fabricated antenna VSWR
Q	2.45	−12	5.55	1.04 GHz	−16.58	1.40
				2.00 GHz	−12.96	1.58
				2.17 GHz	−29.92	1.13
R	2.41	−35	5.96	898.68 MHz	−19.30	1.51
				989.12 MHz	−11.25	1.83
				2.40 GHz	**−18.21**	**1.85**
				2.61 GHz	−18.23	1.29
S	2.39	−44	4.95	**2.43 GHz**	**−21.04**	**1.75**
T	2.47	−13	4.56	766.03 MHz	−13.30	1.56
				1.17 GHz	−33.88	1.05
				1.80 GHz	−11.95	1.69
				2.46 GHz	**−18.89**	**1.26**
U	2.39	−12	5.71	928.82 MHz	−11.04	1.82
				2.47 GHz	**−20.62**	**1.21**
V	2.39	−15	5.59	2.11 GHz	−26.15	1.10
W	2.37	−11.21	4.96	2.068 GHz	−27.23	1.25
X	2.38	−11	4.93	856.47 MHz	−13.42	1.64
				1.93 GHz	−23.30	1.14
				2.63 GHz	−25.03	1.13
Y	2.4	−18	5.26	958.97 MHz	−10.74	1.86
				1.30 GHz	−14.85	1.44
				2.14 GHz	−16.73	1.40
Z	2.4	−22	3.24	1.20 GHz	−14.43	1.46
				2.40 GHz	**−18.53**	**1.28**
				2.63 GHz	−11.53	1.70

Table 2 shows the antenna resonant frequency, return loss, gain and VSWR values of the corresponding MPA's. Simulated results and Fabricated MPA's results were compared. Amongst the 26 fabricated antennas considered, the MPA's fabricated based on alphabet shapes C, E, F, G, L, N, R, S, T, U and Z gives multiband with less return loss and VSWR value. But during the simulation alphabet shapes B, R and S produced high gain with less return loss.

3 Fabricated Numeric Antenna

The seven segment pattern based antenna was proposed in [18], in proposed method the numerical shape based MPA's were fabricated and their return loss curves are shown in the following section.

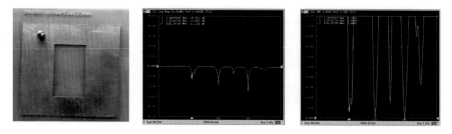

Fig. 27. 0 shape antenna - network analyzer return loss and VSWR output

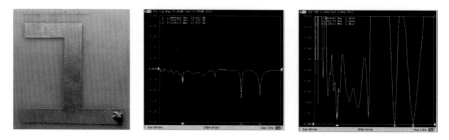

Fig. 28. 1 shape antenna - network analyzer return loss and VSWR output

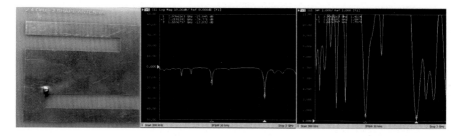

Fig. 29. 2 shape antenna - network analyzer return loss and VSWR output

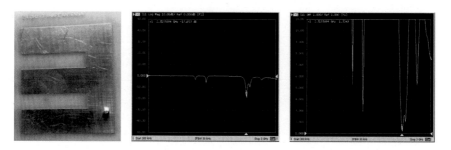

Fig. 30. 3 shape antenna - network analyzer return loss and VSWR output

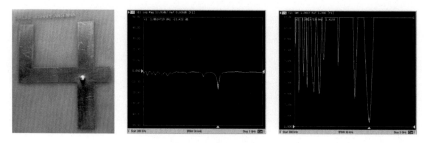

Fig. 31. 4 shape antenna - network analyzer return loss and VSWR output

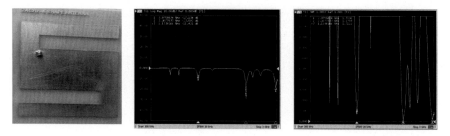

Fig. 32. 5 shape antenna - network analyzer return loss and VSWR output

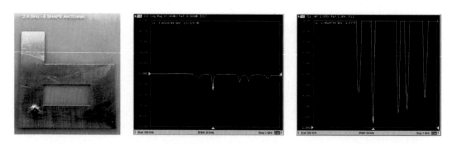

Fig. 33. 6 shape antenna – network analyzer return loss and VSWR output

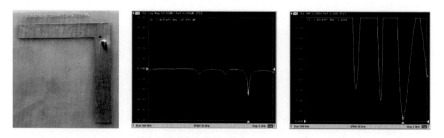

Fig. 34. 7 shape antenna - network analyzer return loss and VSWR output

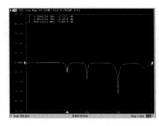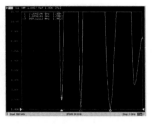

Fig. 35. 8 shape antenna - network analyzer return loss and VSWR output

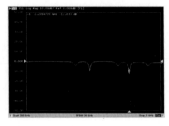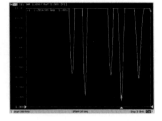

Fig. 36. 9 shape antenna – network analyzer return loss and VSWR output

Table 3. Comparison of simulated and fabricated numeric shape antennas

Shape	Simulated antenna frequency [1]	Simulated antenna return loss (dB) [1]	Simulated gain [1]	Fabricated antenna frequency	Fabricated antenna return loss (dB)	Fabricated antenna VSWR
0	2.38 GHz	−18	6.15	2.20 GHz	−19.30	1.32
				1.47 GHz	−13.00	1.49
				820.29 MHz	−12.23	1.68
1	2.47 GHz	−13	4.56	**2.49 GHz**	−19.49	**1.18**
				2.02 GHz	−23.20	1.14
				573.09 MHz	−12.03	1.58
2	2.39 GHz	−44	4.95	2.93 GHz	−15.84	1.41
				2.21 GHz	−29.37	1.09
				1.10 GHz	−12.03	1.70
3	2.39 GHz	−13	5.57	2.31 GHz	−17.65	1.32
4	2.4 GHz	−11.5	4.23	1.88 GHz	−15.42	1.41
5	2.39 GHz	−44	4.95	2.97 GHz	−17.13	1.32
				2.26 GHz	−23.09	1.17
				1.13 GHz	−11.42	1.71
6	2.46 GHz	−15	5.77	1.45 GHz	−15.32	1.53
7	2.4 GHz	−12	3.08	**2.41 GHz**	−25.59	**1.11**
8	2.39 GHz	−31	6.11	2.20 GHz	−27.27	1.09
				1.45 GHz	−12.83	1.59
				989.12 MHz	−9.89	1.94
9	2.46 GHz	−15	5.77	2.29 GHz	−12.03	1.70

From the Table 3, we infer that the MPA's designed based on numeric shape 7 provides the low return loss and VSWR. These characteristics are obtained using a coaxial feeding technique but during the simulation microstrip line feeding were used. Even though both feeding methods have the same equivalent circuit, resonant frequency was shifted or not produced similarly to simulation results.

4 Conclusion

A new Alpha-Numeric shape MPA is designed fabricated and tested with Network Analyzer E5062A ENA Series which is proposed for Wi-Fi applications. Simulation results were attained using ADS 2009 software, During the Simulation process, all the 36 antennas were produced low return loss with good gain for the resonant frequency 2.4 GHz. Same values were considered for the fabrication process but resonant frequency was shifted, multi band were produced or didn't produce the required frequency. Losses like a material loss, connector loss, feeder loss and soldering loss need to be considered. These proposed antennas were characterized by using return loss and VSWR values. Other characterises like current distribution, impedance matching, outdoor testing, indoor testing and radiation effects need to be analysed. Based on the results filters and amplifiers need to be added along with antenna design.

References

1. Prabu, T. et al.: Design of alpha/numeric microstrip patch antenna for Wi-Fi applications. In: Data Engineering and Intelligent Computing, Advances in Intelligent Systems and Computing, vol. 542 (2017). https://doi.org/10.1007/978-981-10-3223-3_3. ISBN: 978-981-10-3222-6
2. Tripathy, M.R. et al.: Studies on A-shaped microstrip patch antenna for wideband applications. Int. J. Inf. Comput. Technol. 3(11), 1145–1148 (2013). ISSN 0974-2239
3. Srivastava, D.K., et al.: Design and analysis of extended C-shaped microstrip patch antenna for wideband application. In: Advances in Communication and Control Systems. Atlantis Press (2013)
4. Prabu, T., et al.: Design and implementation of E- shaped antenna for GSM/3G applications. IJISET 2(4) (2015)
5. Ramesh, B., et al.: Design of E-shaped triple band microstrip patch antenna. IJERA 3(4), 1972–1974 (2013)
6. Gupta, S., et al.: A novel miniaturized, multiband f shape fractal antenna. IJARCSSE 5(8) (2015)
7. Sun, J.S., Huang, S.Y.: A small 3-D multi-band antenna of F shape for portable phones' applications. Prog. Electromag. Res. Lett. 9, 183–192 (2009)
8. Majumdar, A.: Design of an H shaped microstrip patch antenna for Bluetooth applications. IJIAS 3(4), 987–994 (2013)
9. Trivedi, N., et al.: Design and simulation of novel I shape fractal antenna. IJEST 4(11) (2012)
10. Wu, G.-L., Mu, W., Zhao, G., Jiao, Y.-C.: A novel design of dual circularly polarized antenna feed by L-strip. Prog. Electromag. Res. 79, 39–46 (2008)

11. Arora, H., Jain, K., Rastogi, S.: Review of performance of different shapes (E, S, U) in micro-strip patch antenna. IJSRET **4**(3), 178–181 (2015)
12. Aneesh, M., et al.: Analysis of S-shape microstrip patch antenna for Bluetooth application. Int. J. Sci. Res. Publ. **3**(11), 3–6 (2013). ISSN 2250-3153
13. Kamble, R., Yadav, P., Umbarkar, V., Trikolikar, A.A.: Design of U-shape microstrip patch antenna at 2.4 GHz **4**(3) (2015)
14. Kannadhasan, S., et.al.: Design and analysis of U-shaped micro strip patch antenna. IEEE Dig. Lib. https://doi.org/10.1109/aeeicb.2017.7972333. ISBN: 978-1-5090-5434-3
15. Samsuzzaman, M., et al.: Dual band X shape microstrip patch antenna for satellite applications. ELSEVIER Proc. Technol. **11**, 1223–1228 (2013)
16. Singh, S., Rani, S.: Design and development of Y shape power divider using improved DGS. IJESTTCS **2**(6) (2013)
17. http://www.zdacomm.com
18. Kriti, V., Abishek, A., Ray, T.B.: Development and design of compact antenna on seven segment pattern. IJERGS **3**(3) (2015)

Performance Analysis of Psychological Disorders for a Clinical Decision Support System

Krishnanjan Bhattacharjee[1], S. Shivakarthik[1], Swati Mehta[1], Ajai Kumar[1], Anil Kamath[2(✉)], Nirav Raje[2], Saishashank Konduri[2], Hardik Shah[2], and Varsha Naik[2]

[1] Applied AI Group, Centre for Development of Advanced Computing, Pune, India
{krishnanjanb,shivakarthiks,swatim,ajai}@cdac.in
[2] Department of Information Technology, Maharashtra Institute of Technology, Pune, India
anilkamath21@gmail.com, rajenirav@gmail.com,
kondurisaishashank@gmail.com, hardikshah1796@gmail.com,
varsha.powar@mitpune.edu.in

Abstract. In the domain of psychological practice, experts follow different methodologies for the diagnosis of psychological disorders and might change their line of treatment based on their observations from previous sessions. In such a scenario, a standardized clinical decision support system based on big data and machine learning techniques can immensely help professionals in the process of diagnosis as well as improve patient care. The technology proposed in this paper, attempts to understand psychological case studies by identifying the psychological disorder they represent along with the severity of that particular case, with the help of a Multinomial Naive Bayes model for disorder identification and a regular expression based severity processing algorithm. A knowledge base is created based on the knowledge of human experts of psychology. Psychological disorders however need not possess distinct symptoms to easily differentiate between them. Some are very closely connected with a variety of overlapping symptoms between them. Our work, in this paper, focuses on analyzing the performance of such psychological disorders represented in the form of case studies in a decision support system, with an aim of understanding this gray area of psychology.

Keywords: Standardized clinical decision support system ·
Multinomial Naive Bayes · Psychological disorder ·
Overlapping psychology symptoms

1 Introduction

The world of Artificial Intelligence is trying to overcome the primary inferiority which computational machines possess against humans - understanding.

© Springer Nature Switzerland AG 2020
A. Abraham et al. (Eds.): ISDA 2018, AISC 941, pp. 896–906, 2020.
https://doi.org/10.1007/978-3-030-16660-1_87

The brain's job of understanding, learning and evolving is catered to by deep learning/machine-learning. Machine Learning corroborated by Big Data analytics is turning to be a revolutionary technological force.

With the current computational power, humongous amounts of data can be processed to obtain highly informative analytical insights. These analytical insights can be utilized in clinical decision support systems to assist as well as speed up the process of medical diagnosis. In the domain of mental health as well, such a decision support system would be of great assistance to psychiatrists by providing an efficient and standardized diagnosis platform.

Typically, psychiatrists alter their line of treatment and diagnosis based on their observations and new revelations from a patient in their series of sessions. To better treat the patient by early detection of a psychological disorder and to ensure an accurate diagnosis of the patient, a decision support system can be developed based on a knowledge base which is created by human experts in the field of psychology.

In this paper, we propose developing a decision support system which would aim for early detection of a psychological disorder from the various session transcripts maintained by psychiatrists. Another form of these documents could be patient-authored texts or self-portrayals which a psychiatrist may ask a patient to prepare or share.

We focus our analysis on the psychological text available in these documents for the purpose of disorder identification and severity detection. Psychology being a sensitive area, our research has been comprehensively validated by experts in the disciplines of clinical psychiatry as well as psychological research.

2   Related Works

There is a growing body of related work analyzing psychology-related discourse and language usage in clinical and social media text to better discover and understand mental health related concerns.

In [13], the authors have described the development and evaluation of a decision support system for the diagnosis of schizophrenia spectrum disorders (SAD-DESQ). They created their knowledge base from open interviews with experts in psychology to understand their decision making process. This knowledge base, however, was created by a single expert and hence the decision support model for schizophrenia developed in this process has low generalizability.

A psychiatry database of 400 patients was used to create and analyze a multi-model decision support system for identifying psychiatry problems of depression or anxiety in [15]. An accuracy of 98.75% was achieved by extracting 44 patient features and combining support vector machines, backpropagation neural networks, and radial basis function neural network models to design the system.

[7] used description logic reasoning and ontologies, to diagnose obsessive-compulsive personality disorder, histrionic personality disorder and paranoid personality disorder. The system was evaluated in 2 studies, one using historical data and another performing analysis along with experts.

[12] discusses early diagnosis of mental health problems among children. It compares the diagnosis to different mental health problems among children using 8 machine learning techniques which include AODEsr, Multilayer Perceptron, RBF Network, IB1, KStar, Multiclass Classifier, FT and LAD Tree.

In [10], the authors employed artificial intelligence techniques in mental health expert system. The authors used three AI techniques in this system which were rule based, fuzzy logic and fuzzy genetic logic in this system. The rule-based imitates an expert's reasoning in diagnosis and coming up with a conclusion. The fuzzy logic enhance the conclusion and the fuzzy genetic logic determines several options of suitable set of treatments.

In our research, the knowledge base for the model was developed by a panel of certified psychology experts. Moreover, the phrases which contributed to disorder severity were identified and scored by the experts without any knowledge of the same done by their peers.

3   Methodology

3.1   Dataset

In this study, short case reports collected from standard journals [2,4], books of psychology [11] and biomedical publications [3], have been taken as idealized representations of health records. This dataset includes psychological documents such as:

- Psychiatrist's Session Summary Reports
- Patient Self Portrayals
- Counseling Session Transcripts

which are stored in the form of text files.

The scope of our research is limited to 3 psychological disorders, namely bipolar disorder, schizophrenia and paranoid personality disorder (PPD). The input to the algorithm are text-based documents in English language.

Along with the text documents, a list of phrases is maintained which contribute towards the disorder severity. They have been extracted from these text documents by the psychological experts assisting us in this research and appropriate weights have been assigned to these phrases by them.

3.2   Disorder Identification

Disorder identification is carried out by Apache Mahout [1], which is a platform provided by the Apache Software Foundation to implement algorithms in classification and clustering. Currently Mahout has two flavors of Naive Bayes. The first is standard Multinomial Naive Bayes (MNB). The second is an implementation of Transformed Weight-normalized Complement Naive Bayes (TWCNB) as introduced in [14]. We refer to the former as Bayes and the latter as CBayes.

In our research, the standard Multinomial Naive Bayes is used. An MNB classifier model, created via Mahout, which has been trained on the sample documents is used to identify the disorder. A Java program is used to access the model and calculate scores for the disorders using TF-IDF method. Whichever disorder gets the highest score is the disorder identified by the system.

MNB Implementation in Mahout Distribution. In this section we provide an explanation of Mahout's Naive Bayes classification algorithms presented in [9] and [14]. To understand Mahout's implementation of MNB, consider a corpus of m documents and a vocabulary of size n.

Let $D \triangleq \{d^{(i)} \in \mathbb{R}^n : i = 1, 2, \dots m\}$, where $d_j^{(i)}$ is the number of occurrences of term j in document i.

Let $f = \{f_1 \, f_2 \cdots f_n\}$ be a vector of "document frequency" counts for each term j. f_j is the count of documents in which the term j occurs: $f_j \triangleq |\{d^{(i)} \in D : d_j^{(i)} > 0\}|$.

Let $y \in \{1, 2, \dots \ell\}^m$ be a vector of ordinal document class labels; y_i is the label of the vectorized document $d^{(i)}$.

Let $c \in \{1, 2, \dots \ell\}$ be the classes.

Let α be the smoothing factor.

a. Vectorization

As noted in [14], Term Frequency (TF) and Inverse Document Frequency (IDF) transformations have proven to boost Naive Bayes classification performance.

The TF transformation serves to reduce the weight that a repeated term has on a given document; the IDF transformation of a term frequency for a given corpus gives less weight to a term the more often it appears in other documents in that corpus. The converse is true as well i.e a term is given more weight, if there are fewer documents in which the term appears.

For each $d^{(i)} \in D$ we compute the TF-IDF transformation as:

$$d'^{(i)}_j = \sqrt{d_j^{(i)}} \left(1 + \log \frac{m}{f_j + 1} \right) \tag{1}$$

b. Aggregation by class

Let $C \in \{1, 2, \dots \ell\}$ by a set of ordinal class labels. Let the matrix $N \in \mathbb{R}^{\ell \times n}$ be defined as the aggregate sum per class of each document vector $d^{(i)} \in D$ for each class $c \in C$. N_{cj} is the sum of the (TF-IDF transformed) frequencies of term j, in all documents i, labeled as class c in the training set.

$$N_{c*} \triangleq \sum_i^{y_i = c} D_{i*}, \forall c \tag{2}$$

Class aggregate sum TF-IDF vector can be calculated as:

$$s_c \triangleq \sum_{j=1}^{n} N_{cj} \tag{3}$$

where $s \in \mathbb{R}^{\ell}$ is the total sum of the (IDF transformed) frequencies of all terms used in all documents of each class.

c. Training Multinomial Naive Bayes

The purpose of building an MNB model is to classify a document into one of ℓ classes, c, where $c \in C$. The uncertainty parameters θ_{cj}, are estimated by the algorithm, one parameter for each class-term combination c, j.

A smoothed version of the Maximum Likelihood estimator is used to calculate each parameter estimate $\hat{\theta}_{cj}$.

$$\hat{\theta}_{cj} = \frac{N_{cj} + \alpha}{s_c + n\alpha} \tag{4}$$

Here the default α is the Laplace smoothing factor where $\alpha = 1$. Log parameter estimates are used for decision weights for each term, j, for each class $c \in C$.

$$\hat{w}_{cj} = \log(\hat{\theta}_{cj}) \tag{5}$$

d. Testing the MNB model

The standard MNB predicts by adopting the following rule: it assigns a class, k : $k \in C$, the class to which the document has the largest posterior probability of belonging. Assignment of a label to a vectorized document can be done using a classifier function. As noted in [8], we use the minimum-error classification rule (not considering prior class probabilities):

$$l(t) \triangleq \arg\max_{c} \sum_{j=1}^{n} t_j \hat{w}_{cj} \tag{6}$$

3.3   Gradient Processing Module

In our research, a gradient processing algorithm (Algorithm 1) was developed by us for the computation of the gradient scores of the phrases contributing to the severity of a psychological disorder. Gradient score for the document is calculated using: gradients of all phrases matched, frequency of occurrence of each phrase and exact or partial match of phrase. The algorithm makes use of regular expressions for locating these phrases in any of their possible variations and for negation handling.

Preprocessing. Preprocessing involves: noise removal, POS tagging, stop word removal, negation handling, lemmatization, stemming and token generation. The preprocessed case report and the phrases associated with the identified disorder are given as input to the analytical model which carries out gradient processing of the text.

```
input  : CaseText = Preprocessed Document
         AllPhrases = Phrases of Disorder Identified
output: Disorder Severity Level
begin
    TotalGradient ⟵ 0;
    Count ⟵ 0;
    for i ← MaxLengthOfPhrase to 2 do
        foreach Phrase ∈ AllPhrases do
            Create Regex with length i for Phrase;
            Pass Regex Through CaseText;
            if Regex Matched then
                TotalGradient ⟵ TotalGradient + PhraseGradient;
                Count ⟵ Count + 1;
                Create Regex for Negation Handling;
                Pass Regex Through CaseText;
                if Regex Returns 'not' then
                    TotalGradient ⟵ TotalGradient − PhraseGradient;
                    Count ⟵ Count − 1;
                end
            end
        end
    end
    Result ⟵ TotalGradient ÷ Count;
end
```

Algorithm 1. Gradient Processing Module

Phrase Matching. The phrase list of the matched disorder is iterated through and a regular expression is generated on the fly for each phrase. This regular expression consists of a slope, which is the maximum words allowed between two words of the phrase. On a successful search of the regular expression, the entire matched sentence is replaced with 'XX' followed by a counter, so that the same sentence does not match multiple times.

Sentence: Spent the majority of his time alone making him extremely frustrated and bored.

Sentence after preprocessing: spend major time alon make extrem frustrat bore.

Phrase: extremely frustrated and bored.

Phrase after preprocessing: extrem frustrat bore.

Regex for Exact match:

\b(extrem|frustrat|bore)\W+(\w+\W+){0,slope}?(extrem|
    frustrat|bore)\W+(\w+\W+){0,slope}?(extrem|frustrat|
    bore)\b

Regex for Partial match:

\b(extrem | frustrat | bore)\W+(\w+\W+){0,slope}?(extrem |
    frustrat | bore)\b

The exact match regular expression will get a match and replace 'extrem frustrat bore' with say 'XX02'.

Sentence after phrase matched: spend major time alon make XX02.

Negation Handling. Regular expressions are use to find "not" in the 2 words before the phrase and on finding a match the algorithm replaces the "not" with a 'YY'+counter to prevent further matching and does not consider that phrase.

For example:

Sentence: She didn't feel depressed for the next few weeks.

Sentence after preprocessing: not feel depress next week.

Phrase to be matched: feel depressed.

Phrase after preprocessing: feel depress.

Sentence after phrase matching: not XX03 next week.

Regex Created:

('([^ \r\n]+)\W+(\w+\W+)'+XX03)

Sentence after phrase matched: YY03 XX03 next week.

After the phrase matching process is completed, final gradient score is calculated (Table 1).

4   Results and Observations

The disorder identification step was carried out in pairs on the 3 psychological disorders considered and then on all 3 using the standard MNB model.

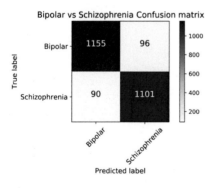

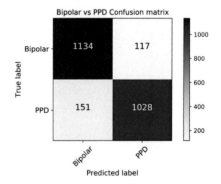

Fig. 1. Bipolar vs Schizophrenia confusion matrix

Fig. 2. Bipolar vs PPD confusion matrix

Table 1. Performance measures

Performance measures	(1)[a]	(2)[b]	(3)[c]	(4)[d]	(5)[e]
Accuracy	92.38%	58.02%	88.97%	69.10%	88.01%
Weighted precision	0.9238	0.5801	0.8899	0.6917	0.8858
Weighted recall	0.9238	0.5802	0.8897	0.6910	0.8801
Weighted F1 score	0.9238	0.5801	0.8896	0.6913	0.8815

[a]1: Bipolar vs Schizophrenia,
[b]2: PPD vs Schizophrenia,
[c]3: Bipolar vs PPD
[d]4: Bipolar vs PPD vs Schizophrenia
[e]5: Gradient Processing for Bipolar

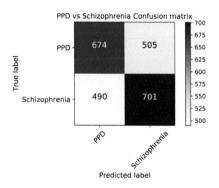

Fig. 3. PPD vs Schizophrenia confusion matrix

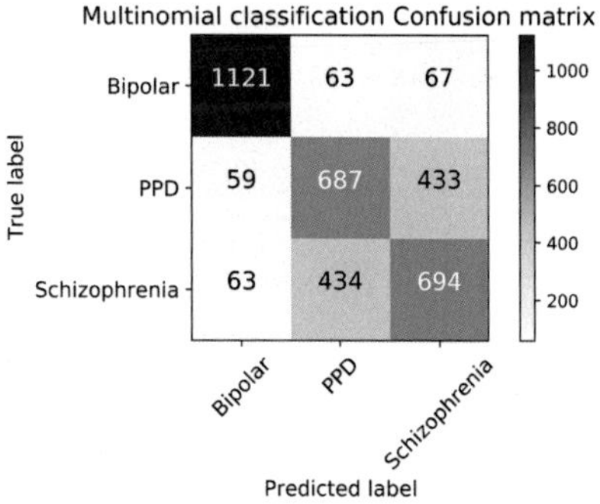

Fig. 4. Multinomial classification confusion matrix

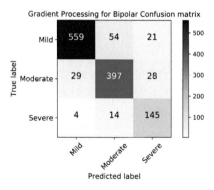

Fig. 5. Gradient processing for bipolar confusion matrix

Figures 1, 2, 3 and 4 show the confusion matrices for each consideration mentioned above. It can be observed from these results that bipolar disorder is easily distinguishable from schizophrenia and paranoid personality disorder(PPD). From Fig. 4, it can be seen that nearly 89.60% of the bipolar disorder documents have been correctly identified by the MNB model. However, clear segregation between schizophrenia and PPD has not been observed. Figure 2 showing the classifier performance for Schizophrenia and PPD. Accuracy for this classification was 58.01% which characterizes either significant misclassifications or overlap due to further types of each disorder.

Our discussions with psychological experts revealed that in the practice of psychological diagnosis as well, the diagnosis usually approaches a gray area of psychology characterized by many overlapping symptoms of schizophrenia and PPD, for which a nuanced knowledge of psychology, as that possessed by a psychiatrist, is required. This further suggests that a clinical decision support system carrying out a uniform diagnosis performs with satisfactory accuracy on a broad scale, however the diagnosis is unclear when the disorders approach a closer connection between symptoms. According to DSM-5 [6], one of the diagnostic criteria for PPD is that it "Does not occur exclusively during the course of schizophrenia, a bipolar disorder or depressive disorder with psychotic features, or another psychotic disorder and is not attributable to the physiological effects of another medical condition." Also as stated in [5], paranoia as a symptom itself involves 3 types of disorders: paranoid personality disorder (PPD), delusional disorder and paranoid schizophrenia. From the discussions we had with psychiatrists, we were apprised that this particular study involved a substantial number of paranoid schizophrenia cases - one of the reasons why the disorder identification module was unable to distinguish satisfactorily, schizophrenia and PPD. There is always a chance in majority of cases to present overlapping symptoms which is why most patients or clients remain at different points on the spectrum of the same condition.

5   Conclusion

In this research, we implemented psychological disorder identification using the Multinomial Naive Bayes Classifier of Apache Mahout on 3 psychological disorders, namely - schizophrenia, bipolar disorder and paranoid personality disorder. The performance results shown by the MNB model indicate that the classifier could identify satisfactorily a case of bipolar disorder from our psychological text documents dataset, however fails to distinguish PPD from schizophrenia due to the heavily overlapping symptoms between the 2 disorders. We hence implemented the severity detection module on bipolar disorder only, with the process of assigning a gradient score to each phrase contributing towards the severity of the case. We remark that the gradient scores attributed to certain phrases for the severity diagnosis of bipolar disorder have been validated by the psychological experts assisting us in our research. We also remark that a real world clinical decision support system which utilizes this approach would require similar validation from a large panel of certified psychological experts.

In the domain of psychological practice, there exist no standardized process of treatment which is followed by all professionals on a global scale. Practitioners often change their line of treatment based on their observations. A clinical decision support system, would thus, not only expedite the process of psychiatric diagnosis but also offer a standardized platform for diagnosis with the help of Big Data analytics and predictive learning techniques. Although, in this work we targeted only 3 psychological disorders and analyzed how they perform, this work can be extended to cover all psychological disorders along with the subtypes of each of these disorders thereby improving the knowledge base and taking a step towards a computer attempting to gain a nuanced knowledge of the psychological world, as possessed by a human expert in this field.

Our work focuses on an intelligent digital assistant to a psychiatrist with a decision support module as a part of it, validated to perform well. There are further features which such an assistant could incorporate such as high-risk patient alerts and generation of a patient profile which graphically demonstrates the progress of a patient during their treatment. In the future, such a system could also involve integration with a web-crawler to provide to the psychiatrist customized search results with the expected relevancy. An additional application of this technology could be a digital assistant to human moderators, often non-specialists of psychological websites which involve forums or open confessions. The technology thus proves beneficial to patients, the psychological profession as well as society as a whole.

References

1. Apache mahout: Scalable machine learning and data mining. https://mahout.apache.org/. Accessed 1 May 2018
2. JMIR mental health. https://mental.jmir.org/. Accessed 1 May 2018
3. Living with schizophrenia. https://www.livingwithschizophreniauk.org/. Accessed 1 May 2018
4. OMICS International. https://www.omicsonline.org/. Accessed 1 May 2018
5. Paranoia - better health channel. https://www.betterhealth.vic.gov.au/health/conditionsandtreatments/paranoia. Accessed 1 May 2018
6. American Psychiatric Association, et al.: Diagnostic and statistical manual of mental disorders (DSM-5®). American Psychiatric Pub (2013)
7. Casado-Lumbreras, C., Rodríguez-González, A., Álvarez-Rodríguez, J.M., Colomo-Palacios, R.: PsyDis: towards a diagnosis support system for psychological disorders. Expert Syst. Appl. **39**(13), 11391–11403 (2012)
8. Duda, R.O., Hart, P.E.: Pattern Classification and Scene Analysis, vol. 3. Wiley, New York (1973)
9. Lyubimov, D., Palumbo, A.: Apache Mahout: Beyond MapReduce. CreateSpace Independent Publishing Platform (2016)
10. Masri, R.Y., Jani, H.M.: Employing artificial intelligence techniques in mental health diagnostic expert system. In: 2012 International Conference on Computer and Information Science (ICCIS), vol. 1, pp. 495–499. IEEE (2012)
11. Miklowitz, D.J., Gitlin, M.J.: Clinician's Guide to Bipolar Disorder: Integrating Pharmacology and Psychotherapy. Guilford Publications (2014)
12. Sumathi, M.R., Poorna, B.: Prediction of mental health problems among children using machine learning techniques. Int. J. Adv. Comput. Sci. Appl. **7**(1), 552–557 (2016)

13. Razzouk, D., Mari, J.J., Shirakawa, I., Wainer, J., Sigulem, D.: Decision support system for the diagnosis of schizophrenia disorders. Braz. J. Med. Biol. Res. **39**(1), 119–128 (2006)
14. Rennie, J.D., Shih, L., Teevan, J., Karger, D.R.: Tackling the poor assumptions of Naive Bayes text classifiers. In: Proceedings of the 20th International Conference on Machine Learning (Icml-03), pp. 616–623 (2003)
15. Suhasini, A., Palanivel, S., Ramalingam, V.: Multimodel decision support system for psychiatry problem. Expert Syst. Appl. **38**(5), 4990–4997 (2011)

Qualitative Collaborative Sensing in Smart Phone Based Wireless Sensor Networks

Wilson Thomas[1] and E. Madhusudhana Reddy[2(✉)]

[1] Research and Development Center, Bharathiar University, Coimbatore, India
[2] Department of CSE, Guru Nanak Institutions Technical Campus,
Ibrahimpatnam, Telangana, India
e_mreddy@yahoo.com

Abstract. Collaborative sensing has become a novel approach for smart phone based data collection. In this process individuals contributes to the participatory data collection by sharing the data collected using their smart phone sensors. Since the data is gathered by human participants it is difficult to guarantee the Quality of the data received. Mobility of the participant and accuracy of the sensor also matters for the quality of data shared in such environment. If the data shared by such participants are of low quality the purpose of collaborative sensing fails. So there must be approach to gather good quality of data from participants. In this paper we propose a Truth Estimation Algorithm (TEA) to identify the truth value of the data received and filter out anomalous data items to improve the quality of data. To encourage the participants to share quality information we also propose an Incentive Allocation Algorithm (IAA) for qualitative data collection.

Keywords: Collaborative sensing · Truth value · Smart phones · Truth discovery · Incentive based approach

1 Introduction

Smart phones are unprecedented tool for collecting empirical data. Modern smart phones are equipped with accelerometer, gyroscope, GPS, and high resolution camera. They are capable of sensing, mobile computing, and data sharing information with the support of wide mobile networks and social networking sites. Since these devices travel with people they can be used to collect more precise and real-time data in space and time. Collaborative sensing [1] is the process where individuals and communities use their smart phones and internet to collect and analyze real-time data on the fly. This technique is useful for developing applications for social well being like Pollution detection, health care applications, security and surveillance etc. The major component of this architecture is the participants with smart phones who collect and share sensor data such as images, voice records, accelerometer data etc. The collaborative sensing application can be distributed among participants through application store such as Google play or App store. In collaborative sensing using smart phones, data gathering can be triggered by participants themselves or through campaigning in which individuals shares sensor data in response to queries of the organizer (Fig. 1).

© Springer Nature Switzerland AG 2020
A. Abraham et al. (Eds.): ISDA 2018, AISC 941, pp. 907–914, 2020.
https://doi.org/10.1007/978-3-030-16660-1_88

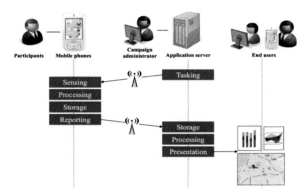

Fig. 1. Collaborative sensing & data analysis

Data collected through the campaign need to be analyzed by application servers to extract useful information. However it is difficult to ensure quality of the data collected by these smart phones as the participant behave unreliable and maliciously. Hence the major challenge in the use of social sensing application is the noisy nature of data received. In this context the algorithms that can estimate the data quality of the data and underlying truth have drawn significant attention. There are many Truth estimation Algorithms [12] available in different research areas. Truth estimation approaches are generally not supervised, hence the originality of the source can be trusted based on the quality of data. High trust values will be assigned to the sources which give reliable data. These data collected from such reliable sources will be considered as truths.

Theseus [2] is a payment based approach that deals with workers' deliberate behavior, and reward good quality data sharing from participants. In Bayesian Nash Equilibrium of the non-cooperative game induced by Teseus, all participants will best work on sensing, which improves their data quality. Hence the final results calculated afterward by truth discovery algorithms based on participant's data will be highly accurate.

Incentives based approach in collaborative sensing focus on inspiring the participants to share useful data by giving monitory reimbursement. Mobility based incentivizing the participants are an approach for quality data collection in collaborative sensing. In this approach people are encouraged to move to the unknown regions for data collection. With this approach the participant gets incentivized and organizer gain more accurate data.

2   Related Work

There has been many research conducted on addressing the problem of data accuracy through data aggregation, trust value calculation, malicious behavior of participants. These people can disrupt the sensing purpose by supplying low quality information.

They can be spamming or misleading other participants. "Privacy and Quality Preserving Multimedia Data Aggregation" [3] Propose an approach for preserving privacy and quality of data in collaborative data sharing with Quality-aware Attribute-based Filtering technique. This approach propose SLICER algorithm, which combines data coding technique and message transfer strategies, to achieve strong protection of participants' privacy and achieve quality data. "Anonymization Techniques for Preserving Data Quality in Participatory Sensing" [4] discuss optimization techniques to preserver data quality. "Context-aware data quality estimation in mobile crowdsensing "[5] Explains data quality estimation in crowed sensing platforms based on context. "A Quality-Based Surplus Sharing Method for Participatory Sensing" [6] Proposes techniques for achieves higher decision accuracy of sensor nodes. It combines quality estimation with incentive technique and show case quality-based data sharing approach is based on unsupervised learning to estimate the value of user data quality and reputations to remove anomalous data from it. "A Survey on Truth Discovery" [7] Suggests quality estimation and monetary incentive, for participatory sensing.

3   Qualitative Data Collection

Truth Discovery Techniques
We have proposed a series of approaches for Truth estimation of the source of the data.

a. Truth Estimation Algorithm (TEA)
Proposed Algorithm below estimates the quality of information gathered considering the probability of truth value of the source of information. Research on trust estimation [8] considers the fact that data gathered about the same event by two or more participants are likely to conflict. This can be due to many reasons like imperfect data collection, network delays, bad weather etc. The following notations are used in the proposed TEA algorithm.

- Object: An object $i \in G$ is the object to be monitored for data gathering.
- Smart Phone: A Smart Phone $s \varepsilon S$ is the individual participant in the event. $\{X_{i,s}\}$ *denote object i is monitored by smart phone s.*
- Source-claim matrix: This matrix is of the order SXG. Each entry S_iG_j says whether a smart phone s has claimed the ownership of entry j.
- Quality: Q_s express the quality metric for the probability that smart phone s provide reliable data.
- Truth: This value for an object i, considered as as T_j, is the most reliable data collected from a set of observations.

Problem Statement: *Let S be a set of smart phones and G be the set of events being monitored. For each smart phone $s \in S$, estimate an optimal truth value $\{T_s\}$ where $s \in S$, from a set of observations $\{X_{i,s}\}$ where $i \in G$, $s \in S$ received from different sources.*

Algorithm1: Framework for Truth value calculation
Input: Observations from sources $\{X_{i,s}\}$ _where i ∈G, s ∈S_
 Output: Optimal truths $\{T_j\}_{i \in G}$ and the estimated source probability $\{P_s\}$ _where s ∈S_
 Initialize: Quality estimation $\{Q_i\}_{i \in S}$
 while optimization not achieved **do**
    for each i ∈ G **do**
        Truth value finding: infer this value for object i
        Using the recent finding of quality
        estimation Q_i
    end
    Quality estimation: update quality value $\{Q_i\}_{i \in S}$ using the latest calculated truth value.
end
Return: $\{T_i\}_{i \in G}$, $\{Q_i\}_{i \in S}$

One of the pioneer approach in truth finding in truth finding [11, 13] decides reputation of each source based on probability that it sends the correct data. It then averages the reputation value to calculate the trustworthiness of the source.

$$Q_i^j = \frac{\sum_{j \in SiGj} x_{i,j}^{k-1}}{|\{SiGj\}j \in E\,|} \tag{1}$$

"Truth Discovery in Crowdsourced Detection of Spatial Events" [9] it estimates the truthiness of a claim by Truth calculates the "probability" of a claim by considering that reliability of data source is the probability that it sends correct data. It takes the average of reliability values to calculate truthiness for every claim from participants.

Incentive Based Approach
In this approach we encourage the participants to contribute valuable data by providing incentives. This algorithm takes set of target events E and participants P and sensor data s, and $\{a_p, b_p\}|\{p \in P\}$ where a_p and b_p are quality factors related to the participants. The calculation of incentive to any participant is based on peer prediction method [10] which decides the incentives based on the data collected by a participant and the difference of it with a reference participant data.

b. Incentive Allocation Algorithm (IAA)

Input: T, P, P', s, $\{a_p, b_p\}|p \in P\}$;
Output: $\{Q_p \mid p \in P\}$;
foreach participant $p \in P$ **do**
  if $p \in P'$ **then**
    Pickup another participant $w \in P'$ in random;
    $Q_s \leftarrow b_p - a_p \frac{1}{T} \sum_{t=1}^{T}(s_t^p - s_t^w)$;
  else
    $Q_s \leftarrow 0$;
  return $\{Q_p \mid p \in P\}$;

As per the above IAA algorithm if the participant p take part in data collection algorithm picks another random participant. The payment Q_s for the participant is calculated as

$$Q_s \leftarrow b_p - a_p \frac{1}{T} \sum_{t=1}^{T} \left(s_t^p - s_t^w \right)$$

If the participant's data shows more similarity to the reference data higher the payment Q_s will be. If the participant moves out from the task the payment Q_s is set to 0 for him. At the end the algorithm return incentives allocated for all the participants.

4 Performance Evaluations

Here, we analyze the performance of our approaches for qualitative data collection, Truth Estimation Algorithm (TEA) and Incentive Allocation Algorithm (IAA). We compare the performance of our IAA algorithm with the well known research work in this area called MSensing [14] auction technique.

A. *Simulation Setup*

For simulation set up we deploy the smart-sense application in an area of 200 m × 200 m. This square region is again partitioned into 100 small squares of 20 m × 20 m area. We assume that 400 sensing tasks and 300 participants are distributed randomly the area.

We consider below parameter for analysis.

- **Completion Ratio** is the ratio of data collection task finished by the individuals.
- **Allocation ratio** is the ration of participants winning the auction being allocated incentives.
- **Social Welfare** is completed data collection jobs minus total data collection cost.

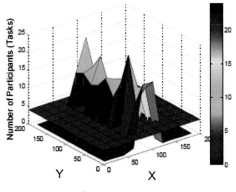

(a) Simulation Model

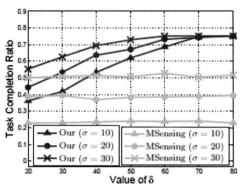

(a) Completion Ratio

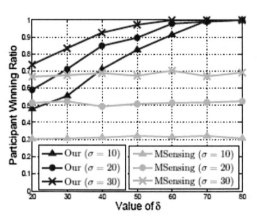

(b) Allocation Ratio

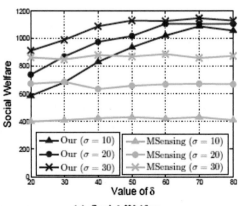

(c) Social Welfare

B. *Performance Comparison*

Simulation results shows that our approach outperforms MSensing with respect to the performance parameters (a) task completion ratio, (b) allocation ratio and (c) social welfare. It performs better when the number of participants are high. When n = 150 and σ = 10, with respect to MSensing, our approach improves parameter (a) by 100%, parameter (b) by 66.6%, and parameter (c) by 63%. Now consider σ = 10 and n = 300, with respect to MSensing, our approach improves the parameter (a) by 145.4%, parameter(b) by 143.3%, parameter (c) by 146.7%. The reason for this is, in MSesnsing mechanism the participants are grouped in popular regions and there is a chance of losing the auction. But incentive based approach encourage the individual to change location to unknown regions and finish data collection.

5   Conclusion

The TEA algorithm helps to calculate truthiness of the collaborative data received. In IAA approach we have proposed a technique to incentivize participants of collaborative data collection based on mobility where participants are motivated for data collection from unpopular areas. Based on theoretical analysis it is clear that the proposed approach holds true for truthfulness and efficiency in qualitative data collection. Hence the incentive based mechanism outperforms the traditional campaigning approach for collaborative data collection.

References

1. Sheng, X., Tang, J., Zhang, W.: Energy-efficient collaborative sensing with mobile phones. In: 2012 Proceedings IEEE INFOCOM, Orlando, FL, pp. 1916–1924 (2012)
2. Jin, H., Su, L.: Theseus: Incentivizing Truth Discovery in Mobile Crowd Sensing Systems. https://arxiv.org/pdf/1705.04387.pdf
3. Qiu, F., Wu, F., Chen, G.: Privacy and quality preserving multimedia data aggregation for participatory sensing systems. IEEE Trans. Mob. Comput. **14**(6), 1287–1300 (2015)
4. Sabrina, T., Murshed, M., Iqbal, A.: Anonymization techniques for preserving data quality in participatory sensing. In: 2016 IEEE 41st Conference on Local Computer Networks (LCN), Dubai, pp. 607–610 (2016)
5. Liu, S., Zheng, Z., Wu, F., Tang, S., Chen, G.: Context-aware data quality estimation in mobile crowdsensing. In: IEEE INFOCOM 2017 - IEEE Conference on Computer Communications, Atlanta, GA,, pp. 1–9 (2017)
6. Yang, S., Wu, F., Tang, S., Gao, X., Yang, B., Chen, G.: Good work deserves good pay: a quality-based surplus sharing method for participatory sensing. In: 2015 44th International Conference on Parallel Processing, Beijing, pp. 380–389 (2015)
7. Li, Y., Gao, J., Meng, C., Li, Q., Su, L., Zhao, B., Fan, W., Han, J.: A survey on truth discovery. SIGKDD Explor. Newslett. **17**(2), 1–16 (2016b)
8. Yin, X., Han, J., Yu, P.S.: Truth discovery with multiple conflicting information providers on the web. IEEE Trans. Knowl. Data Eng. **20**(6), 796–808 (2008). https://doi.org/10.1109/TKDE.2007.190745

9. Ouyang, R.W., Srivastava, M., Toniolo, A., Norman, T.J.: Truth discovery in crowdsourced detection of spatial events. IEEE Trans. Knowl. Data Eng. **28**(4), 1047–1060 (2016)

10. Miller, N., Resnick, P., Zeckhauser, R.: Eliciting informative feedback: peer-prediction method. In: Management Science (2005)

11. Sun, Y., Luo, H., Das, S.K.: A trust-based framework for fault-tolerant data aggregation in wireless multimedia sensor networks. IEEE Trans. Dependable Secur. Comput. **9**(6), 785–797 (2012). https://doi.org/10.1109/TDSC.2012.68

12. Li, X., Zhou, F., Du, J.: LDTS: a lightweight and dependable trust system for clustered wireless sensor networks. IEEE Trans. Inform. Forensics Secur. **8**(6), 924–935 (2013). https://doi.org/10.1109/TIFS.2013.2240299

13. Talasila, M., Curtmola, R., Borcea, C.: Alien vs. mobile user game: fast and efficient area coverage in crowdsensing. In: IEEE MobiCASE (2014)

14. Xue, G., Fang, X., Tang, J.: Crowdsourcing to smartphones: incentive mechanism design for mobile phone sensing. In: ACM Mobi-Com (2012)

Phylogenetic Tree Construction Using Chemical Reaction Optimization

Avijit Bhattacharjee$^{(\boxtimes)}$, S. K. Rahad Mannan, and Md. Rafiqul Islam

Computer Science and Engineering Discipline, Khulna University,
Khulna 9208, Bangladesh
{avijit1412,rahad1414}@cseku.ac.bd, dmri1978@gmail.com

Abstract. Phylogenetic tree construction (PT) problem is a well-known NP-hard optimization problem that finds most accurate tree representing evolutionary relationships among species. Different criteria are used to measure the quality of a phylogeny tree by analyzing their relationships and nucleotide sequences. With increasing number of species, solution space of phylogenetic tree construction problem grows exponentially. In this paper, we have implemented Chemical Reaction Optimization algorithm to solve phylogeny construction problem for multiple datasets. For exploring both local and global search space, we have redesigned four elementary operators of CRO to solve phylogeny construction problem. One correction method has been designed for finding good combination of species according to maximum parsimony criterion. The experimental results show that for maximum parsimony criterion our implemented algorithm gives better results for three real datasets and same for one dataset.

Keywords: Phylogenetic tree · Maximum parsimony ·
Metaheuristics · Chemical Reaction Optimization

1 Introduction

Phylogeny is a visual demonstration showing the genetic affiliation or evolutionary relationship among distinct taxa. A phylogenetic tree is the study of genealogical links between interrelated organisms in evolutionary history [1]. In a computational scenario, phylogenies are binary unrooted trees. Branching patterns of phylogenies represent the evolution of species, which form successions of common ancestors. The aligned entities are located at the tips where common ancestors represent the inner nodes. The correlation between taxa denoted as an edge or branch. The length of a branch represents the mutation time [1]. Phylogenetic tree enriches our understanding capability of how genes, genomes, species (molecular sequences more generally) are evolved. Phylogenies give us an insight of how species change in history and how the species will change in future. It is important in forensics, getting assess of DNA evidence, to identify the pathogen origin, in conservation policy and in Bioinformatics [2]. Phylogeny

© Springer Nature Switzerland AG 2020
A. Abraham et al. (Eds.): ISDA 2018, AISC 941, pp. 915–924, 2020.
https://doi.org/10.1007/978-3-030-16660-1_89

approaches can also be used to know about a new pathogen outbreak. It discovers how species are related to and subsequently the likely source of transmission. This can lead to a new suggestion for public health policy [2]. Phylogenies can be applied in forensics (dental practice in HIV transmission). Other applications of phylogenies include multiple sequence alignment, protein structure prediction, gene and protein function prediction, epidemiology, drug development and drug design [2]. Phylogenetic tree construction methods are divided into two categories. They are algorithmic methods and optimality criterion methods. Algorithmic methods reach to the final solution without judging all solutions in the search space and these are the quick producer of solutions. Algorithmic methods construct a tree using distance base matrix. Examples for Algorithmic methods are Neighbor-Joining (NJ), Unweighted Pair Group Method for Arithmetic Mean (UPGMA). Another type is optimality criterion based, which works by optimizing an objective function and deploying a search technique. In this process, objective function calculates the score of each tree in search space and find the best-scored solution. Example for optimality criterion methods are maximum parsimony and maximum likelihood. In order to get better maximum parsimony score trees e.g. PHYLIP and PAUP are used and for better maximum likelihood score trees e.g. fastDNAml, PHYML, RAxML are used. Phylogenetic tree construction is very important in the study evolutionary process. The good evolutionary tree is necessary in biological studies. Exact algorithms take a large amount of time and space. For better performance, we have looked for a meta-heuristic algorithm. Chemical Reaction Optimization (CRO) is an efficient meta-heuristic algorithm. The main power of CRO algorithm is searching both local and global solution space. We have redesigned four basic operators of CRO and one additional operator has been designed to find a good phylogenetic tree. In recent years, CRO has successfully solved many optimization problems such as quadratic assignment problem, resource-constrained project scheduling problem, channel assignment problem in wireless mesh networks, grid scheduling problem, stock portfolio selection problem, artificial neural network training, network coding optimization problem [3]. Therefore, recent successes of CRO have motivated us for applying it in phylogenetic tree construction problem.

2   Basic Concepts of Phylogeny Construction

The phylogenetic tree represents the evolutionary history of ethnological relationships among linked entities. By statistical and algorithmic procedures, phylogenetic methods have found the most correct assumption about the evolution of taxa. We consider the input is an alignment taken by n sequences of m characters. For DNA sequences, A, C, G, T which represents Adenine, Cytosine, Guanine, and Thymine respectively define site values. The sites represent the nitrogenous base. The output of phylogeny inference is a tree shape structure. Let $T = (V, E)$ be a phylogenetic tree generated from a set of n sequences of m nucleotides, $(u, v) \in V$ two related nodes, $(u, v) \in E$ the ancestral connection between both nodes and u_i, v_i the state at the i-th site on the sequence for u

and v. Construction of phylogenies is a very complex computation task because for 50 species there exists 2.68×10^{78} number of topologies, which is almost near the number of atoms in the universe [2]. For greater values of n (e.g. organism), no polynomial time solution exists. In order to assess the complexity of these problem, two things need to be tackled. Firstly, if the number of sites increases, then the computational cost of objective function will increase. Secondly, if the number of taxa increases in a dataset, then the space of all possible trees grows exponentially. For an unrooted tree, if we have n taxa, then (1) tells the possible number of phylogenetic topologies [1].

$$\frac{(2n-5)!}{(n-3)!2^{n-3}} \tag{1}$$

2.1   Maximum Parsimony Criterion

Maximum parsimony is the process that minimizes total number of character state changes and mutation facts. It is the simplest process to construct the good phylogenetic tree. Let $T = (V, E)$ be a phylogenetic tree generated from a set of n sequences of m nucleotides. Here $(u, v) \in V$ two related nodes and $(u, v) \in E$ the ancestral connection between both nodes. In (2), u_i, v_i the state at the i-th site on the sequence for u and v. The parsimony score of T [4,5]:

$$P(t) = \sum_{i=1}^{m} \sum_{(u,v) \in E} C_i(u, v) \tag{2}$$

where the cost value $C_i(u, v)$ is defined by comparing the character states u_i and v_i as follows: If $u_i \neq v_i$, then $C_i(u, v)$ will be one, otherwise, $C_i(u, v)$ will be zero. In this process, a bifurcating tree is constructed with an additive process. This process gradually takes species and groups them together. In the first step, trees have to be traversed from tips to root. A tip node moves forward to its intermediate node that is linked with another tip node. If two tips node have same state, then intermediate node takes common state. If tips node have different state, then intermediate node takes both state with logical "or". MP calculation algorithm reaches to the root node following above procedure. In the tree of Fig. 1, node 7 has the state C or G because node 6 and node 5 have different states. Similarly, node 8 has A or G state. However, Node 9 has state G as node 7, 8 have common state G. Next, node 10 has state G or C because node 2 and node 9 have different states. By this procedure, MP calculation have reached to root node 1. Now, MP calculation starts from root to tips traversal for counting mutations. In this traversal, MP solves the contradictions created by tips to root traversal. If parent node's state and child node's state differ, then anyone of the "or" state can be selected. Otherwise, common state between parent and child node is placed as child nodes state. Root node 1 and its child node 10 has no common states, so contradiction at node 10 is solved by putting anyone of the "or" state. In this case, node 10 takes state G. The branch connecting node 1 and node 10 has change in state. So, this branch has one parsimony

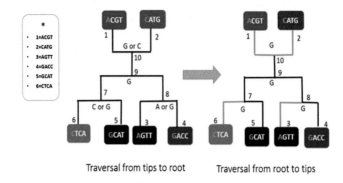

Fig. 1. Maximum parsimony

score. Similarly, any change in site from parent node to child node will increment parsimony score.

3   Related Work

PHYLIP is an integrated program of different construction algorithms. For DNA dataset, PHYLIP has DNAPARS [6] algorithm to get a phylogenetic tree according to maximum parsimony criterion. DNAPARS works using stepwise addition and hill climbing method. It treats gaps as fifth nucleotide sequence. Bifurcating and multifurcating both trees are allowed. However, when the number of sequences increase, its performance degrades in the same proportion. TNT [7] is a culminative program of different parsimony algorithms. Basic TNT method for maximum parsimony performs stepwise addition especially random addition sequences (RAS) and branch-swapping technique such as NNI, TBR. MOABC [1] is a multiobjective swarm intelligence algorithm to get quality evolutionary trees in both maximum parsimony (mp) and maximum likelihood (ml) criteria. 1000 initial solution of mp (500) and ml (500) are generated and then randomly 500 of them are assigned as employee bee. Employee bees search for new nearby solutions by using NNI move. After that, good solutions found by employee bees are assigned to on-looker bees and they explore further. Finally, scout bees deploy PPN move to search in the global optimum. At last, final results are compared with existing mp, ml, and multiobjective algorithms.

4   Proposed Method Using Chemical Reaction Optimization

Chemical reaction optimization is a population-based meta-heuristic algorithm developed by Albert Y.S. Lam [3]. It follows basic events of chemical reaction to optimize a given problem. There are mainly four operators in CRO. They are on-wall ineffective collision, decomposition, inter-molecular ineffective collision and synthesis. These molecules are actually solutions of an optimization problem. Each molecule has the molecular structure (ω), potential energy (PE), kinetic energy (KE) and number of hits. The molecular structure can be a variable, a vector, a matrix or any complex data structure to represent a solution. CRO has two variables α and β for controlling ratio of intensification and diversification.

4.1   Solution Generation

In order to get faster convergence, literature suggest to provide good initial solutions. Therefore, 30 trees from BIONJ [8] algorithm is used as initial solution for each dataset.

4.2   Reaction Operators

We have redesigned four reaction operators of CRO and one additional operator has been designed for our proposed method which are described below.

On-Wall Ineffective Collision. In on-wall collision, we have adopted NNI (Nearest Neighbor Interchange) operator from [9]. Like NNI, our on-wall operator first selects two nearest neighbor nodes of a tree whose grandparents are same. Then the positions of two selected nodes in the tree are swapped. Figure 2 visualizes on-wall ineffective collision. It can be observed that solution m has a tree of four species P, Q, R and S. On-wall ineffective operator selects the nodes of Q and R species. Q and R have the same grandparent. So, the positions of node Q and R are exchanged in solution m. In this way, new solution m' is formed.

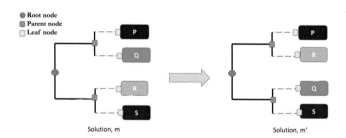

Fig. 2. On-wall ineffective collision

Decomposition. Decomposition operator takes a solution and creates two solutions, which contain significant subtree from parent solution. To explain this, let us assume solution m has a tree of seven species P, Q, R, S, X, Y, Z. Solution m contains two subtrees ((P, Q), (R, S)) and (X, (Y, Z)). In Fig. 3, solution m_1 contains a subtree of solution m that are (X, (Y, Z)) and a random subtree generated from species P, Q, R, S. Similarly, m_2 is formed from m. Therefore, solution m is decomposed into two new solutions m_1 and m_2.

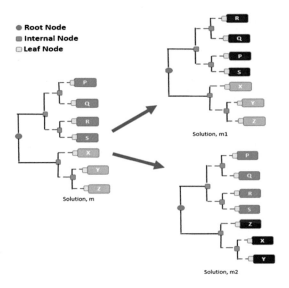

Fig. 3. Decomposition

Inter-molecular Ineffective Collision. This operator works on two solutions m_1 and m_2 by modifying them to some extent. Therefore, the number of solutions in population remains the same in this operator. Our inter-molecular ineffective collision implements subtree pruning and regrafting (SPR) [10]. For example, solutions m_1 and m_2 are selected randomly. Then Subtree pruning and regrafting (SPR) technique discussed above is applied on solution m_1 and m_2. So, two new solutions m_1' and m_2' are found. In Fig. 4, solution m_1 and m_2 contain five species each. Initially, solution m_1 contains a tree ((P, Q), (R, (S, T))). After doing SPR operation on the subtree, (S, T) changes its position and becomes a new solution m_1'. Similarly, solution m_2' is formed from solution m_2.

Synthesis. The operational approach for designing synthesis operator is a random selection of two solutions and making a new single solution. For this work, from two solutions two subtrees are merged together and formed a new solution. Figure 5 shows two solutions m_1 and m_2. Solution m_1 and m_2 have tree structure (((P, Q), (R, X)), (Y, (Z, S))) and (((R, Q), (P, S)), (X, (Y, Z))) respectively. The solution m_1 has two subtrees. They are b1 which has tree structure ((P, Q), (R, X)) and b2 which has tree structure (Y, (Z, S)). Moreover, m_2 has also two subtrees. They are b3 which has tree structure ((R, Q), (P, S)) and b4 which has tree structure (X, (Y, Z)). For example, we select subtree b1 and subtree b4 from solution m_1 and m_2 consecutively. Two subtrees b1 and b4 are added as child of a root node and new solution m is formed.

Reform. During synthesis operation, it is possible that duplicate species remain in solution and some species are absent. Therefore, the reform operator is

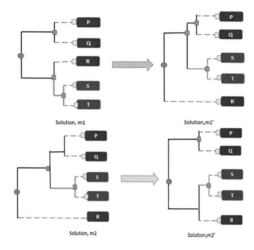

Fig. 4. Intermolecular ineffective collision

designed so that duplicate species are deleted and absent species are added to make the solution valid. In Fig. 6, solution m has duplicated species X and species S is absent. Therefore, Reform deletes species X and adds species S to complete solution. As solution in synthesis operation is created by randomly picking two subtree from two solution, so there is no surety of getting all species in new tree. In order to get valid solution, duplicate species should be removed

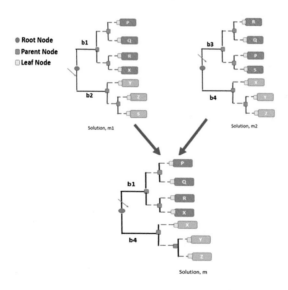

Fig. 5. Synthesis

and absent species should be inserted. Reform operator repairs final solution of synthesis operator to return valid solution tree.

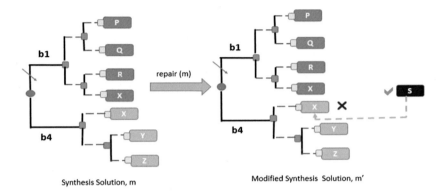

Fig. 6. Repeated species deletion and absent species insertion

4.3   CRO Algorithm for Phylogenetic Tree Construction(PT_CRO)

For each iteration, only one operator is executed on one or two molecules. First of all, CRO decides whether to select one or two molecules. Then, if one molecule is selected and decomposition criteria met, decomposition operator will be called otherwise on-wall operator is called. Secondly, if two molecules are selected and synthesis criteria met, synthesis operator will be called else inter-molecular ineffective operator is called. When Synthesis operator is executed, in order to validate the resultant solution reform operator is also called. At first, we initialized the parameters hit, alpha, beta, and kinetic energy. The values for these parameters are assumed initially as follows: $Hit = 0, Alpha(\alpha) = 40, Beta(\beta) = 20, KineticEnergy(KE) = 100, KELossRate = 0.2, MoleColl = 0.2, Buffer = 0, Iteration = 1000, Pop\_Size = 30$. Bio++ [11] is an object oriented library which includes basic sequence and phylogenetic analysis methods. PT_CRO have adopted this library for implementation.

5   Experimental Results

The four real datasets, rbcl_55 (55 sequences, each sequence has 1314 sites), mtDNA_186 (186 sequences, each sequence has 16,608 sites), RDPII_218 (218 sequences, each sequence has 4182 sites), ZILLA_500 (500 sequences, each sequence has 759 sites) were used to judge the performance of phylogenetic tree construction methods (Table 1).

   These four datasets were collected from the website (http://khaos.uma.es/mophylogenetics/datasets.jsp). Maximum parsimony indicates less genetic

Table 1. Comparison between existing algorithms and our proposed algorithm for Maximum Parsimony (MP) criterion

Datasets	MOABC	TNT	DNAPARS	PT_CRO
rbcl_55	4874	4874	4874	4675
mtDNA_186	2431	2431	2431	2158
RDPII_218	41448	41448	41587	34278
ZILLA_500	16218	16218	16224	16218

change throughout evolution. So, fewer MP(maximum parsimony) score means better phylogeny tree. PT_CRO has found scores better than MOABC [1], TNT [7], DNAPARS [6] in three datasets. On the other hand, for ZILLA_500 MP score of PT_CRO is equal to the good score. In Fig. 7, visual comparison of four datasets with PT_CRO and MOABC [1], TNT [7], DNAPARS [6] are presented. PT_CRO shows significant improvement in rbcl_55, mtDNA_186, RDPII_218 datasets.

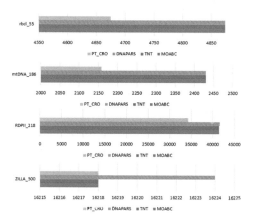

Fig. 7. Comparison of the experimental results of the proposed method with DNA-PARS, TNT, MOABC.

6 Conclusions

The phylogenetic tree construction problem is one of the standard problems in molecular evolution biology. This NP-hard problem has been widely used in biological as well as educational activities that draws attention of the researchers to solve this problem. In this paper, we have proposed an algorithm to solve phylogenetic tree construction problem using the concept of a population based metaheuristic chemical reaction optimization algorithm. We have redesigned elementary operators of CRO and designed one repair method to solve this problem. The experimental results of PT_CRO have been compared with other metaheuristic algorithms such as artificial bee colony, genetic algorithm and metapopulation genetic algorithm for maximum parsimony criterion. The results of PT_CRO is better in maximum parsimony for three datasets and same for one dataset.

References

1. Santander-Jiménez, S., Vega-Rodríguez, M.A.: Applying a multiobjective meta-heuristic inspired by honey bees to phylogenetic inference. Biosystems **114**(1), 39–55 (2013)
2. Stamatakis, A.: Phylogenetics: applications, software and challenges. Cancer Genomics-Proteomics **2**(5), 301–305 (2005)
3. Lam, A.Y., Li, V.O.: Chemical reaction optimization: a tutorial. Memetic Comput. **4**(1), 3–17 (2012)
4. Goeffon, A., Richer, J.-M., Hao, J.-K.: Progressive tree neighborhood applied to the maximum parsimony problem. IEEE/ACM Trans. Comput. Biol. Bioinform. **5**(1), 136–145 (2008)
5. Qin, L., et al.: A novel approach to phylogenetic tree construction using stochastic optimization and clustering. BMC Bioinform. **7**(4), S24 (2006)
6. Felsenstein, J.: PHYLIP (phylogeny inference package), version 3.5 c (1993)
7. Goloboff, P.A., Farris, J.S., Nixon, K.C.: TNT, a free program for phylogenetic analysis. Cladistics **24**(5), 774–786 (2008)
8. Gascuel, O.: BIONJ: an improved version of the NJ algorithm based on a simple model of sequence data. Mol. Biol. Evol. **14**(7), 685–695 (1997)
9. Lemmon, A.R., Milinkovitch, M.C.: The metapopulation genetic algorithm: an efficient solution for the problem of large phylogeny estimation. Proc. Nat. Acad. Sci. **99**(16), 10516–10521 (2002)
10. Zwickl, D.J.: Genetic algorithm approaches for the phylogenetic analysis of large biological sequence datasets under the maximum likelihood criterion (2006)
11. Dutheil, J., et al.: Bio++: a set of C++ libraries for sequence analysis, phylogenetics, molecular evolution and population genetics. BMC Bioinform. **7**(1), 188 (2006)

Hybrid Segmentation of Malaria-Infected Cells in Thin Blood Slide Images

Sayantan Bhattacharya[1], Anupama Bhan[1(✉)], and Ayush Goyal[2]

[1] Department of Electronics and Communication Engineering,
Amity University, Noida, Uttar Pradesh, India
abhan@amity.edu
[2] Frank H. Dotterweich College of Engineering, Texas A&M University,
Kingsville, USA

Abstract. Malaria is a hazardous disease responsible for nearly 400 to 1000 deaths annually in India. The conventional technique to diagnose malaria is through microscopy. It takes a few hours by for an expert to examine and diagnose malarial parasites in the blood smear. The diagnosis report may vary when the blood smears are analyzed by different experts. In proposed work, an image processing based robust algorithm is designed to diagnose malarial parasites with minimal intervention of an expert. Initially, the images are enhanced by extracting hue, saturation and intensity planes followed by histogram equalization. After preprocessing, a median filter is employed to eliminate the noise from the images. After the preprocessing, segmentation of malaria parasite is achieved using k-means clustering to get the clear vision of the region of interest. The clustering is followed by region growing area extraction to remove the unwanted area from the segmented image. The second part deals with counting the number of infected RBCs. The method uses roundness detection for the calculation of the infected RBCs. The experiments give encouraging results for the saturation plane of HSI color space and segmentation accuracy by up to 94%.

Keywords: Malaria · Thin blood smears · Saturation plane clustering ·
Region growing segmentation

1 Introduction

As per the latest report of WHO-UNICEF India is third among 15 countries having the highest cases of malaria and deaths due to the disease [1]. Due to the growing need for correct diagnosis of diseases without human error, it is detrimental to identify the medical images accurately. In case of faulty analysis, any error caused can be life-threatening. Hence authentic detection of medical images is required for proper treatment of the patient. The proposed work highlights a method of accurate identification of malaria-infected cells. Malaria is one of the fatal diseases all across the tropical and sub-tropical regions of the world. It is transmitted to a healthy human body through the bite of female Anopheles mosquito belonging to Plasmodium parasite. In general, there are four categories of Plasmodium parasite-Plasmodium falciparum, Plasmodium

© Springer Nature Switzerland AG 2020
A. Abraham et al. (Eds.): ISDA 2018, AISC 941, pp. 925–934, 2020.
https://doi.org/10.1007/978-3-030-16660-1_90

Malariae, Plasmodium Vivax and Plasmodium Ovale. Out of these parasites, Plasmodium Vivax and falciparum are most commonly witnessed infections in human beings. The life stage of the malaria parasite is generally divided into three stages; Ring-shape trophozoites (rings), Schizont, Gametocytes.

Malaria can be cured effectively if a correct examination is performed efficiently. Practically, the pathologists with their expertise conduct the blood test to identify the infection in the erythrocytes. Low performance of pathologists might lead to a defective examination which will lead to morbidity [2]. Various methods of research have been adopted so far for the development of automatic identification of malaria parasites. The features on basis of color, texture, and arrangement of the cells and parasites are produced. Also, the features relatable to the categorization problem used by human technicians are generated [3].

In the past years, numerous techniques have been implemented for erythrocytes segmentation. Another framework implemented to achieve the identical result is an edge detection technique. Edge detection and edge linking have been carried out in order to obtain the erythrocytes from the background. But edge detection is not useful in the condition wherein the cells are slightly stained. Clump splitting technique has also been implemented to distinguish the overlapping regions, but the technique only works with slightly overlapping regions [4]. The least complex technique for Image segmentation is the thresholding strategy. Otsu's thresholding method has been implemented to apply on a grayscale enhanced image [5]. Walliander has applied the technique on the green channel of an image for the purpose of segmentation process [6]. The proposed paper has been implemented using clustering in saturation plane followed by region growing for segmentation of malaria-infected cells. The prime motivation behind the proposed paper is to form an automatic system for effective and robust detection of stages of the malaria parasite.

The paper is organized in the following manner: Sect. 2 of the paper is defines the proposed methodology which is further divided into Image Acquisition, Pre-processing and Image Segmentation and Sect. 3 of the paper comprises of the experimental results. Section 4 gives the conclusion and Sect. 4 provides the future works.

2   The Proposed Methodology

The algorithm developed for the detection of malaria parasite from the thin blood smear images consists, image acquisition, pre-processing, image segmentation by region growing followed by k-means clustering approach. Figure 1 shows the proposed methodology for the work.

The proposed method of malaria parasite detection is divided into the following three parts:

A. *Image Acquisition.*
B. *Preprocessing.*
C. *Image Segmentation.*

Figure 1 explains the flow of the proposed algorithm method for malaria cell detection.

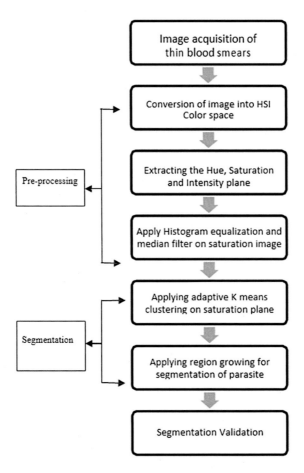

Fig. 1. Proposed methodology for malaria parasite detection.

2.1   Image Acquisition

The traditional method to test malaria is by observing the thin or thick blood smears under a microscope. In the proposed paper, the thin blood smear images are used for the detection of parasites of malaria. About 110 thin blood films images were acquired from the CDC-DPDx-malaria index (Centers for disease control and prevention) [1]. The images acquired are of different life stages such as the ring stage, trophozoite, and gametocyte stage of different Plasmodium parasites with the help of Giesma stained thin blood slides. As observed amid the life cycle stages in blood, there are three morphologically differentiated stages in the of the Plasmodium species. However, failure to effectively separate the Plasmodium species, necessities of costly hardware, time-consuming steps and the high cost restrain the utilization of visual investigation. Hence, a complete diagnostic method equipped with operations to perform: image acquisition, pre-processing, segmentation tasks is advocated. Figure 2 shows different stages of parasites of Malaria.

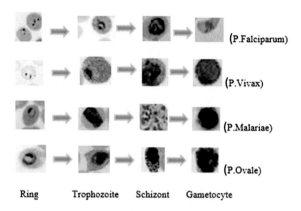

(P.Falciparum)

(P.Vivax)

(P.Malariae)

(P.Ovale)

Ring   Trophozoite   Schizont   Gametocyte

Fig. 2. Different stages of malarial parasites.

2.2   Preprocessing

The images collected are utilized for pre-processing. Pre-processing is an important step before segmentation and is used to enhance the quality of the original image and to eliminate noise present in the image. We have converted convert RGB image to grayscale by working on the Hue Saturation and Intensity planes of the image adopted in order to have a better contrast and enhance the image. Median filter of size 5 was used to remove unwanted noise. Figure 3 shows the histogram of the hue image. Figure 4 shows Saturation Image with Image Histogram. Figure 5 shows Intensity Image with Image Histogram. Figure 6 shows HSI image. Figure 7 shows some of the results attained on preprocessing. Segmentation on malaria parasites images is performed on HIS color space human can interpret better color representation compared to RGB colorspace. The conversion from RGB to HSI color model is computed with the standard conversion formulas. Conversion of the image into HSI and extracting the H, S and I plane: For conversion of the image into HSI the R, G and B plane of the image are extracted. The hue plane is calculated using the Algorithm 1.

Algorithm 1.

Step (a): th=a cos((0.5*((r-g)+(r- b)))./((sqrt((r-g).^2+(r- b).*(g-b)))+eps));
Step (b): H=th;
Step (c): H (b>g) = 2*pi-H(b>g);
Step (d): H=H / (2*pi);

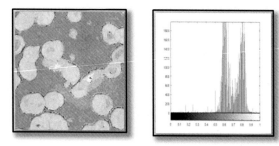

Fig. 3. Histogram of the Hue image.

Then after calculating the H plane the saturation plane is determined using the following Algorithm 2:

Algorithm 2.

Step (a): $S = 1-3.*(min(min(r,g),b))./(r+g+b+eps);$

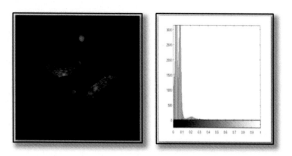

Fig. 4. Saturation image with image histogram.

And then finally the intensity plane is calculated as per the Algorithm 3:

Algorithm 3.

Step (a): $I = (R+G+B)/3;);$

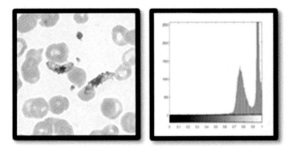

Fig. 5. Intensity image with image histogram.

The above figures have indicated the Preprocessing of Blood Smear Images of Gametocytes of P, Falciparum in thin blood smear.

2.3   Image Segmentation

Segmentation of images is the first step in image analysis. Segmentation refers to subdividing an image into its constituent parts or segments. This is done on the basis of similarities or differences. It is used to obtain the object of interest. It is useful in partitioning an image into multiple regions such that each pixel has similar properties on the basis of texture, color, and intensity. In the proposed paper, K-means clustering is a method for cluster investigation which expects to segment and observations into k groups in which every observation has a place with the group with the closest mean. The K-Means calculation is utilized to discover normal clusters inside given information in light of shifting information parameters.

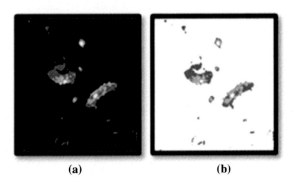

(a)   (b)

Fig. 6. Preprocessing of blood smear images. (a) K-means clustering on S-plane (b) Segmented malaria parasite using region growing.

The strategy tries to create the k-implies calculation to acquire superior and effectiveness. Clusters can be framed for images in view of pixel intensity, location, color or texture or some blend of these. The region is iteratively grown by comparing neighbouring pixels to the region (Figs. 8, 9, 10 and 11).

3   Experimental Results

Experiments have been performed on the total of 110 malarial infected blood samples images and an accuracy of 94% was witnessed. The following figures demonstrate the original and the final image of some samples wherein the malaria-infected red blood cell is successfully segmented from the image. The accuracy of accurate segmentation of malaria-infected cells is given by the equation given below (1):-

$$\frac{\text{Number of Segmented Malaria Cells}}{\text{Total number of Malaria infection cells}} \times 100 \tag{1}$$

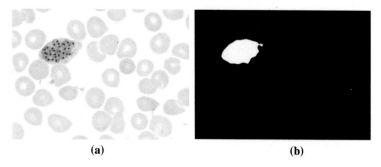

(a)   (b)

Fig. 7. (a) Malaria-infected blood cell sample: Schizont with approximately 30 merozoites and pigment (b) Segmented malarial para

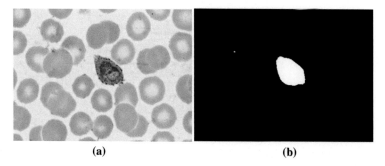

(a)   (b)

Fig. 8. (a) Malaria-infected blood cell sample: Coarse James dots in P. ovale (b) Segmented malarial parasite.

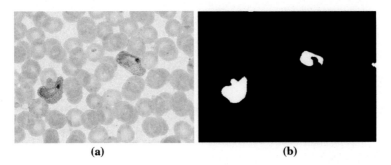

$$\text{(a)} \qquad\qquad\qquad \text{(b)}$$

Fig. 9. (a) Malaria infected blood cell sample: Trophozoites (vacuolated) (b) Segmented malarial parasite.

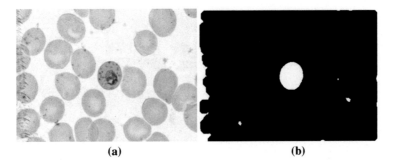

$$\text{(a)} \qquad\qquad\qquad \text{(b)}$$

Fig. 10. (a) Zieman's dots in P. Malariae (b) Segmented malarial parasite.

Using the techniques of image processing the malaria parasite has been detected. For the validation of the proposed work the image processing algorithm was applied to 100 images. The paper presents two work systems, the first one explains the detection of the malaria parasite.

Table 1 shows the classification of the Malaria Parasite which is shown below.

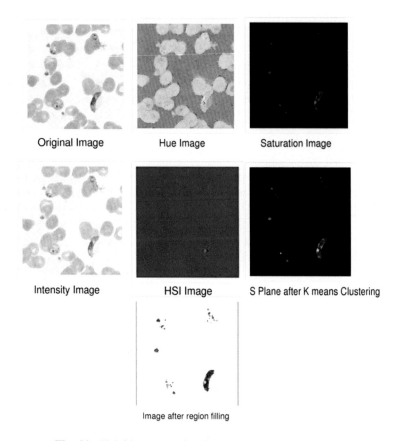

Original Image   Hue Image   Saturation Image

Intensity Image   HSI Image   S Plane after K means Clustering

Image after region filling

Fig. 11. Hybrid segmentation for Malaria parasite detection.

The algorithms were developed in MATLAB R2017 (MathWorks) software on a CPU@ 2.3 GHz, 4 GB RAM, 64-bit operating system.

Table 1. Classification of malaria parasite.

Stage	Total no. of samples	Segmented malarial cells	Accuracy (%)
Plasmodium Falciparum	20	19	95
Plasmodium Knowlesi	20	18	90
Plasmodium Malariae	20	19	95
Plasmodium Ovale	20	19	95
Plasmodium Vivax	20	19	95

4 Conclusion and Future Work

The proposed method of segmentation for extracting malaria-infected cells has been successfully implemented. The processing involves image enhancement, background subtraction, and image segmentation. Segmentation of malaria-infected cells is achieved with the help of a hybrid approach of original images into HSI planes where the saturation plane gives the best segmentation result followed by region growing to remove the unwanted areas. The method proposed facilitates the segmentation of malarial parasites and their life cycle stages with an accuracy of 94%. In this study of work, it is shown that segmentation using the hybrid for different stages of malaria parasite images. The diagnostic process is reliable and effective with a framework involving automated segmentation as compared to the manual process.

The future work will be extended to the research for images wherein overlapping of malaria-infected cells takes place. Multiple overlapping infected cells for segmentation is a tedious task and involves a lot of time. Subsequent work has to be done for real-time applications in the medical industry.

References

1. WHO Malaria Report (2015)
2. Nugroho, H.A., Akbar, S.A., Murhandarwati, E.E.H.: Feature extraction and classification for detection malaria parasites in thin blood smear. In: IEEE 2nd International Conference on Information Technology, Computer and Electrical Engineering (ICITACEE), October 2015
3. Mohammed, H.A., Abdelrahman, I.A.M.: Detection and classification of malaria in thin blood slide images. In: IEEE International Conference on Communication, Control, Computing, and Electronics Engineering (ICCCCEE) (2017)
4. Nanoti, A., Jain, S., Gupta, C., Vyas, G.: Detection of malaria parasite species and life cycle stages using microscopic images of thin blood smear. In: IEEE International Conference on Inventive Computation Technologies (ICICT), August 2016
5. Savkare, S.S., Narote, S.P.: Automatic system for classification of erythrocytes infected with malaria and identification of parasite's life stage. Proc. Technol. **6**, 405–410 (2012)
6. Walliander, M., et al.: Automated segmentation of blood cells in Giemsa stained digitized thin blood films. Diagn. Pathol. **8**, S37 (2012)
7. Khan, W.: Image segmentation techniques: a survey. J. Image Graph. **2**(1), 6–9 (2013)
8. Verma, A., Scholar, M.T., Lal, C., Kumar, S.: Image segmentation: review paper. Int. J. Educ. Sci. Res. Rev. **3**(2) (2016)
9. Das, D.K., Maiti, A.K., Chakraborty, C.: Automated system for characterization and classification of malaria-infected stages using light microscopic images of thin blood smears. J. Microsc. **257**(3), 238–252 (2015)

Application of Artificial Neural Networks and Genetic Algorithm for the Prediction of Forest Fire Danger in Kerala

Maya L. Pai$^{(\boxtimes)}$, K. S. Varsha, and R. Arya

Department of Computer Science and IT, Amrita School of Arts and Sciences,
Kochi, Amrita Vishwa Vidyapeetham, Kochi, India
mayalpai@gmail.com, varshaks123456@gmail.com,
arya.arya.88@gmail.com

Abstract. Forest fire prediction is the most significant component of forest fire management. It is necessary because it plays an important role in resource management and recovery efforts. So, in order to model and predict such a calamity, advanced computing technologies are needed. This paper describes a detailed analysis of forest fire prediction methods based on Artificial Neural Network and Genetic algorithm (GA). The objective is to analyse forest fire prediction in Kerala, India. This paper describes Feed Forward–Back Propagation (FFBP) algorithm to train the networks; to get optimised results, GA is used. Promising results are obtained for GA approach than ANN alone.

Keywords: Kerala forest data · ANN · Back propagation · Genetic algorithm

1 Introduction

Forests are the lungs of the Earth. The rate of occurrence of forest fire might increase with the present rise in temperature. Being earth's reservoir of oxygen, forests play a vital role in the regulation of temperature. Forests cover over 31% of the Earth's total land. Considering the fact that human population grows rapidly, the conservation of forests is of utmost importance.

Forest fire is of two types in terms of occurrence – natural and 'human-made'. Natural fire is caused by the effect of natural phenomenon such as heavy lightning and wind. Human-made fire happens due to various reasons. Some changes in climate and topography can be the reasons for forest fire. Wind and humidity play a vital role in determining the nature of fire. Weather changes can adversely affect the direction and behaviour of fire. Wildfire fuel includes anything that can burn, for instance, grass burns faster while timber at a slower pace. The next factor responsible for forest fire is topography. The major topographical factors include the direction of the Sun, amount of energy received from the Sun, and the slope of forest area. When taken into consideration the working of all these factors in combination results in forest fire. Forest fire overview is depicted in Fig. 1.

The human activities which result in deforestation can be mitigated with proper education and implementation of guidelines by the competent authorities. In order to

© Springer Nature Switzerland AG 2020
A. Abraham et al. (Eds.): ISDA 2018, AISC 941, pp. 935–942, 2020.
https://doi.org/10.1007/978-3-030-16660-1_91

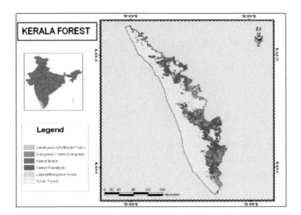

Fig. 1. Kerala forest overview

decrease the impact of natural deforestation, advanced technologies can be used to make the forecast of forest fire more precise; this can reduce the instances of forest fire. Kerala, also known as the 'God's Own Country', is a state with abundant forest cover and is blessed with 44 rivers.

The overarching tourism theme of Green Kerala makes it of utmost importance to maintain its present forest cover and reduce any unforeseeable events which may degrade its pristine quality. Our proposed model aims to provide precise prediction of occurrences and moderation of forest fire. The goal of 'Fire Prediction Modelling' is to predict the effects of fire - ecologically and hydrological. Factors such as consumption of fuel, and rate of burn produced, also come under the purview of the study. The vast majority of forest fires are man-made. Some forest fire may occur due to careless behaviour of human beings such as trekking, camping and carelessness.

2 Literature Review

Forest fires are one of the major natural hazard that leads to the physical, biological, ecological, and environmental consequences. It naturally occurs when the temperature increases which ends in several vulnerability issues [1]. Forest fire results in the widespread devastation of the ecosystem and endangers living beings. This also contributes to the atmospheric pollution [2].

These forest fires release a large amount of gaseous pollutants into the atmosphere directly [3]. Through the destroyed vegetation, the forest fire recharges the quantity of nutrients that is present in the soil then thereby invigorating the sprouting of certain types of seeds [4]. Topography of the area, magnitude of the vegetation and climatic reasons has a significant impact on the intensity of the forest fire [5]. Forest fire, which is a natural phenomenon, is one of the major reasons for the degradation of Indian Forests [6]. The probability of forest fires in Northern Iran was identified using hybrid model of Geographic Information System (GIS) techniques and Dong model. The parameters including but not limited to slope, vegetation type, distance from road, vegetation density, aspect, elevation, and, distance from settlement, and distance from farmland was used in this study [7].

A fire prediction model to identify fire burned areas in Bandipur National Parkand the data and techniques which were used in this study are LISS III data, and RS & GIS techniques. The ecological factors which were considered as parameters for this study are vegetation type, historical data, slope [8–11]. Using RS and GIS techniques, developed a model for accurate and reliable prediction of forest fire zones in Peppara Wildlife Sanctuary, Kerala. A model for identifying forest fire prone areas in Reserve Forests situated in Trivandrum district with the help of GIS techniques was introduced [12]. A proposed architecture based on Adaptive Neuro-Fuzzy Inference System (ANFIS) for forest fire prediction using forest fire hotspot data was also formed and implemented [13–15], and [16]. Forest fire prediction methods based on artificial intelligence are assisted with Support Vector Machine which is used for fire prediction [17]. It developed a novel approach for forest fire prediction. Thereby Artificial Neural Network Based approach is used to solve the problem. The proposed architecture is a multi-layer perception [18, 19]. In [20], the author followed a hybrid model which implements an intelligent sensory system used for detection of component gas. And there are several other similar models like in one research, gas detection is treated and solved using genetic algorithm which proves that ANN-GA gives beneficial results [21]. Moreover, in another paper, this study the author developed an artificial intelligent system based on genetic algorithm for the identification of burned areas in forest [22, 23]. The meteorological data and forest data are used for forest fire prediction.

The objective of this study is to examine forest fire data of Kerala from 1956–2016 and introduce a model for efficient and accurate predict of future forest fire. This paper describes empirical method techniques belonging classification and optimization approach. In this study ANNs are used to implement classification methods. The inputs used are RH, Temperature, Wind, Rain of India Meteorology and the output is burned area of Kerala Forest.

3   Data and Methods

3.1   Data

Kerala forest data of temperature, rainfall, humidity and wind speed collected from Indian Meteorological Department (IMD) [www.imdpune.gov.in] site and its associated institutions such as the National Satellite Data Centre are in charge of providing the weather data.

3.2   Methods

BCNN-Classification Approach
The algorithm indicates two passes through the layers one is forward pass and another is backward pass. In feed forward neural network (FFNN), the data moves only in

forward direction. Each layer consists of different neurons and when each layer receives data, makes separate calculations and forwarded to another layer. The network weights are fixed in forward pass. In NN, there are three set layers; input node is first node and the final node is output node, and both of these layers are interconnected by hidden nodes.

Activation functions are implemented on each node. The sigmoid function is a logistic function which has a sigmoid curve. This can be implemented in hidden layers and output layers and can be calculated using this Eq. 1.

$$S = \frac{1}{1 + e^{-x}} \tag{1}$$

Throughout the rearward pass, the weights are adjusted in accordance with an errors rectification approach. An error signal is calculated by real data of the network is subtracted from predicted data. This signal is then actions backward through the layers of network. So, this algorithm is also known as the Error-back propagation. To make the real data comes closer to the predicted one, connection weights are to be adjusted. The precision of the model can be measured using the Root Mean Squared error (RMSE) function given in Eq. 2.

$$RMSE = \sqrt{\frac{[\sum_{i=1}^{N}(Predicted_N - Actual_N)^2]}{N}} \tag{2}$$

GA-Optimization Approach

GA and ANNs, both are learning and optimization algorithms which centre from biological systems. For training ANN, BPNN is used and GA is used for weight optimization in the network layers. Also, before implementing GA, each individual from the entire population is allotted with a formulated fitness function and parents are nominated for reproduction and crossover to produce new ones. BP learning methods explains the process of moving back from the hidden layers to the input layer of the network depending on the weight modifications. GA is an iterative process that consists of a chromosome, each unique represented by a set of symbols, called the genome, which enhances the given problem. In general, GA is a theoretical method which effectively allows extracting large datasets with BPNN for determining an optimized set of weights.

4   Methodology and Results

The dataset consists of 12 parameters out of which 4 are found to be most influential for the prediction of forest fire danger. In this study, we are using Wrapper method in Weka tool for the parameters selection.

In this study considering the nature of input and output data Multilayer perception (MLP) is used for learning the architecture. The total area of the forest burned in each example is represented as a single number that lie within the output signal. The whole data is divided into three sets, 70% used for training (1956–1996), 15% for validation (1997–2006), and for 15% (2007–2016) is tested. Figure 2 explains the proposed architecture.

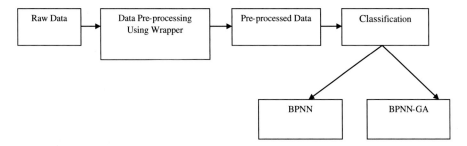

Fig. 2. Proposed architecture

The proposed model is developed using C++ language under a Linux environment. It contains mainly two main parts, the first of which deals with the learning process using the BPNN algorithm. The screenshot of the software used for model building is shown in Fig. 3.

```
ı.lı Result

$g++ -o main *.cpp
$main
Enter the link of file containg the data  : /home/maya/textdata.txt
File opend sucessfully.....
The file system contain 201 rows and 13 coloumn
Enter the number of layers : 3
Enter the numbr of neurons in the layer N 0 :12
Enter the numbr of neurons in the layer N 1 :36
Enter the numbr of neurons in the layer N 2 :1
The leraning rate : 0.1
Network reation Scucessd
The program will uploaded and saved
File opend sucessfully.....
The test process will start soon
Press any key to Start....
Error : 26.0814
Error rate(%)  :9
```

Fig. 3. Input specifications for the learning step

Then input parameters are fed to ANN model and burned area of each section is classified. Sigmoid activation function is applied on each network layer. Figure 4 shows the diagram corresponding to BPNN.

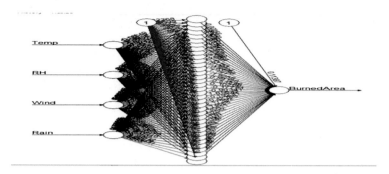

Fig. 4. BPNN architecture

Table 1 describes the accuracy table for the two methods namely BPNN and BPNN-GA. The accuracy for test data is found to be 91.42. To improve the accuracy, the weights of ANN are optimized using GA and the network is learned by BPNN-GA, the combined model gives better accuracy (96.21) than the first model (BPNN). It is evident from the table gives promising results that BPNN-GA is better shown in bold and therefore the model can be used for burned area prediction of other states also. Initially we have used twelve parameters, which give the accuracy of 90.42 and 94.01 for BPNN and BPNN-GA respectively. Later on, we have tried with different number of parameters; the promising results are observed for four parameters which give better accuracy. The graphical representations of both methods are shown in Fig. 5.

Table 1. Accuracy table

Parameter	Hidden layer	No. of neurons	BPNN	BPNN-GA
4	**1**	**36**	**91.42**	**96.21**
4	2	18	87.42	92.21
12	2	18	90.42	94.01

Figure 5 explains pictorial representation of proposed architecture of two methods namely BPNN, BPNN-GA for burned area prediction. From this, it is evident that BPNN-GA gives better results than BPNN alone.

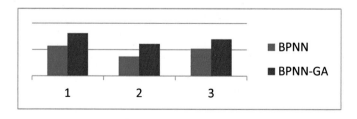

Fig. 5. Comparison graph

5   Conclusion

Forest fire happens as a result of various factors. The phenomenon has got ecological – social – economic implications. Many advanced decision making techniques are required to reduce forest fire. Climatic changes are getting more unpredictable these days, accurate prediction of baseline data is a herculean task. This model describes the need of precise baseline data and to predict forest fire in Kerala.

Forest Fire prediction is very much significant in today's modern world. In this study, we are introducing a hybrid combination of neuro-genetic model for the weight optimization of network layers of ANN. By using the four climatic parameters such as temperature, rainfall, humidity and wind speed of Kerala region to predict forest fire. Results show that the proposed model gives better prediction accuracy for BPNN-GA approach than BPNN alone and hence can be used to predict forest fire prediction.

References

1. Somashekar, R.K., Ravikumar, P., Mohan Kumar, C.N., Prakash, K.L., Nagaraja, B.C.: Burnt area mapping of Bandipur National Park, India using IRS 1C/1D LISS III data. J. Indian Soc. Rem. Sens. **37**(1), 37–50 (2009)
2. Tatli, H., Türkeş, M.: Climatological evaluation of H aines forest fire weather index over the Mediterranean B asin. Meteorol. Appl. **21**(3), 545–552 (2014)
3. Lazaridis, M., Latos, M., Aleksandropoulou, V., Hov, Ø., Papayannis, A., Tørseth, K.: Contribution of forest fire emissions to atmospheric pollution in Greece. Air Qual. Atmos. Health **1**, 14–158 (2008)
4. Dawson, T.P., Butt, N., Miller, F.: The ecology of forest fires. ASEAN Biodiv. **1**(3), 18–21 (2001)
5. Podur, J.J., Martell, D.L.: The influence of weather and fuel type on the fuel composition of the area burned by forest fires in Ontario. Ecol. Appl. **19**, 1246–1252 (2006)
6. Roy, P.S: Forest fire and degradation assessment using satellite remote sensing and geographic information system. Satell. Rem. Sen. GIS Appl. Agric. Meteorol. 361–400 (2003). World Meteorological Organization, Switzerland
7. Eskandari, S., Ghadikolaei, J.O., Jalilvand, H., Saradjian, M.R.: Detection of fire high-risk areas in northern forests of Iran using Dong model. World Appl. Sci. J. **27**(6), 770–773 (2013)
8. Ajin, R.S., Ciobotaru, A., Vinod, P.G., Jacob, M.K.: Forest and wildland fire risk assessment using geospatial techniques: a case study of Nemmara forest division, Kerala, India. J. Wetlands Biodivers. **5**, 29–37 (2015)
9. Veeraanarayanaa, B., Ravikumar, S.K.: Assessing fire risk in forest ranges of Guntur District, Andhra Pradesh: using integrated remote sensing and GIS. Int. J. Sci. Res. **3**(6), 1328–1332 (2014)
10. Rajabi, M., Alesheikh, A., Chehreghan, A., Gazmeh, H.: An innovative method for forest fire risk zoning map using fuzzy inference system and GIS. Int. J. Sci. Technol. Res. **2**(12), 57–64 (2013)
11. Singh, R.P., Ajay, K.: Fire risk assessment in Chitrakoot area, Satna MP, India. Res. J. Agric. Forest. Sci. **1**(5), 1–4 (2013)
12. Gangapriya, P., Indulekha, K.P.: Development of GIS based disaster risk information system for decision making. Int. J. Innov. Res. Sci. Eng. Technol. **2**(Suppl. 1), 140–148(2013)

13. Rothermel, P., Richard, C.: A Mathematical Model for Predicting Fire Spread in Wild Land Fires. USDA Forest Service Research Paper INT – 115, Ogden, Utah, USA (1972)
14. Sowmya, S.V., Somashekar, R.K.: Application of remote sensing and geographical information system in mapping forest fire risk zone at Bhadra Wildlife Sanctuary, India. J. Environ. Biol. **31**(6), 969–974 (2010)
15. Mahdavi, A., Shamsi, S.R.F., Nazari, R.: Forests and rangelands' wildfire risk zoning using GIS and AHP techniques. Caspian J. Environ. Sci. **10**(1), 43–52 (2012)
16. Wijayanto, A.K., Sani, O., Kartika, N.D., Herdiyeni, Y.: Classification model for forest fire hotspot occurrences prediction using ANFIS algorithm. In: IOP Conference on Series: Earth and Environmental Science, vol. 54, p. 012059 (2017)
17. Sakr, E., Elhajj, I.H., Mitri, G., Wejinya, U.C.: Artificial intelligence for forest fire prediction. In: International Conference on Advanced Intelligent Mechatronics Montréal, Canada, 6–9 July 2010
18. Safi, Y., Bouroumi, A.: Prediction of forest fires using artificial neural networks. Appl. Math. Sci. **7**(6), 271–286 (2013)
19. Assaker, A., Darwish, T., Faour, G., Noun, M.: Use of remote sensing and GIS to assess the anthropogenic impact on forest fires in Nahr Ibrahim Watershed, Lebanon. Lebanese Sci. J. **13**(1), 15–28 (2012)
20. Ojha, V.K., Dutta, P., Saha, H.: Performance analysis of neuro genetic algorithm applied on detecting proportion of components in manhole gas mixture. Int. J. Artif. Intell. Appl. **3**(4), 83–98 (2012)
21. Haykin, S.S.: Neural Network a Comprehensive Foundation, 2nd edn. Pearson Prentice Hall, Upper Saddle River (2005)
22. Castelli, M., Vanneschi, L., Popovič, A.: Predicting burned areas of forest fires: an artificial intelligence approach. Fire Ecol. **11**(1), 106–118 (2015)
23. Chavan, M.E., Das, K.K., Suryawanshi, R.S.: Forest fire risk zonation using remote sensing and GIS in Huynial watershed, Tehri Garhwal District, UA. Int. J. Basic Appl. Res. **2**, 6–12 (2012)

A Hybrid Bat Algorithm for Community Detection in Social Networks

Seema Rani$^{(\boxtimes)}$ and Monica Mehrotra

Department of Computer Science, Jamia Millia Islamia, New Delhi, India
seema7519@yahoo.com, drmehrotra2000@gmail.com

Abstract. In this work, a hybrid optimization method is proposed for dealing with the community discovery problem in social networks relying on the bat algorithm. The proposed method hybrids discrete bat algorithm with Tabu search for enhancing the quality of solution in contrast to discrete bat algorithm. The Tabu search is a neighborhood search based method. The local search capability of bat algorithm is improved by introducing the Tabu search strategy. The recommended hybrid approach is tested on a few real-world networks and synthetic benchmark network. The obtained results are very promising and comparable as well. The results are compared with existing algorithms which demonstrate that the proposed method enhances the quality of the obtained solution.

Keywords: Bat algorithm · Tabu search · Community detection · Social networks

1 Introduction

Many complex systems in different domains are characterized in the form of a network structure. The central elements of any network are nodes and edges, which refer to members of the network and their relations. Community detection in networks is largely used in multidiscipline viz physics, biology, computer science etc. In broad-spectrum, the quality of community structure is estimated by a quantitative metric, namely modularity [1, 2]. In case the ground truth partition of the network is known, the Normalized Mutual Information score (NMI) [3] is adopted to evaluate and compare the result with a known community structure. Modularity maximization is an NP-hard problem [4] and falls into the category of optimization problems. Frequently applied optimization algorithms to solve the optimization problems are Genetic Algorithm (GA) [5], Particle Swarm Optimization (PSO) [6, 7], and Simulated Annealing (SA) [8] etc. Recently, new meta-heuristic optimization algorithms inspired by nature are introduced by various researchers viz. Bat algorithm (BA) [9], Grey Wolf Optimizer (GWO) [10], Dragonfly Algorithm (DA) [11], and Ant Lion Optimizer algorithm (ALO) [12] etc. Progressively, researchers' attention is focusing on exploring nature-inspired algorithms and new hybrid approaches.

Yang [9], announced a new nature-inspired optimization algorithm named as bat algorithm, inspired by echolocation behavior of bats. The echolocation behavior of bats is used for finding their target, food/prey. The bats transmit a sound pulse and receive

© Springer Nature Switzerland AG 2020
A. Abraham et al. (Eds.): ISDA 2018, AISC 941, pp. 943–954, 2020.
https://doi.org/10.1007/978-3-030-16660-1_92

the echo that returns from the neighboring objects. The quality of the solution is characterized by loudness and pulse emission rate. However, the original algorithm was for continuous optimization problems and could not be applied equally to discrete or binary optimization problems. With progressive research in the domain of nature-inspired optimization algorithms, a new version of bat algorithm was introduced referred as binary bat algorithm (BBA) for binary optimization problems [13]. In BBA, bats move and search for their target food/prey in the binary search space. The binary version of bat algorithm also supported the fact that it shows better performance than binary PSO (BPSO) [14] and GA [15]. Song *et al.* [16] proposed a Discrete Bat Algorithm (DBA) for solving the community detection problem. BA shows its superiority as it combines the advantage of PSO and SA, both [16].

To an extent, BA is characterized as well adjusted amalgamation of basic PSO and profound local search mechanism [17]. Bat Algorithm is applied for solving different category problems and shown competitive results. In [18], authors gave a model for the detection of electrocardiogram myocardial infarction (ECG MI) using the improved bat algorithm. The improved bat algorithm extracts the best features from each cardiac beat. The generated best features set named as optimized features are applied on the neural net classifier. The result shows that the performance of classifier and accuracy for the detection of MI is upgraded. In the present time, information technology has established its root in the domain of biology and medical science. Researchers are operational on the way of early diagnosis of disease as it leads to a huge impact on the due course of treatment. In [19], a BBA based on Feed-forward Neural Network (FNN) is suggested and shows improved performance in classifying the data into benign and malignant classes. Though, the obtained result accuracy is low with a reduced mean squared error (MSE) and reduction in time consumption. The suggested method BBA shows its usefulness for solving binary classification problem. In [20], an improved binary bat algorithm is offered which outperforms over the BBA [13] and BPSO [14]. The suggested method is tested on standard benchmark functions and zero-one knapsack problem. The result on employed problems shows high accuracy than the BBA and BPSO.

Glover [21] recommended a Tabu Search method which was a neighborhood-based search method. It could be well considered as an exhaustive search method. Tabu search method can be applied to any system whose operation generates a set of moves. These moves offer an instance of a solution to a changed solution. This method offers the best solution from the neighborhood of the current solution.

In the presented work, a hybrid discrete bat algorithm is recommended to deal with the community discovery problem in social networks. Hybridization combines discrete bat algorithm with Tabu search technique. Embedding Tabu search in discrete bat algorithm enhances the local search capability of bats. This hybrid approach is tested on a few real-world data sets. The experimental results and observation show improved performance of proposed hybrid bat algorithm in comparison to Discrete Bat Algorithm (DBA) [16], Discrete Particle Swarm Optimization [22], Fast Newman (FN) [23] and Spectral clustering algorithm [24].

The draft of the paper is organized as follows. Section 2 describes the discrete bat algorithm for community detection. Section 3 discusses the Tabu search procedure in respect of community detection problem. Section 4 discusses the recommended hybrid discrete bat algorithm referred as DBA-tabu along with its pseudo code. In continuation, Sect. 5 discusses the datasets used in the experimentation, initial parameter settings, results and comparison with existing approaches. Finally, Sect. 6 presents the comprehended summary of the work illustrated in the paper.

2   Discrete Bat Algorithm

Yang [9], proposed BA based on the echolocation behavior of bats. The original BA is proposed for continuous optimization problem and the community detection is a discrete optimization problem. Song *et al.* [16] recommended DBA for solving the community detection problem. This section briefs the movement of the bat in order to get their food/prey. Discrete formulas [16] given in DBA are as below:

Pulse frequency: $f_i = f_{min} + (f_{max} - f_{min}) \times \beta$ (1)

Velocity Update: $V_i^t = V_i^{t-1} + (x_i^t \oplus x_i^*) \times f_i$ (2)

with $sig(V_i^t) = \frac{1}{(1+e^{-V_i^t})}$ and $\begin{cases} V_{id}^t = 1 \ if \ rand() < sig(V_i^t) \\ V_{id}^t = 0 \ if \ rand() \geq sig(V_i^t) \end{cases}$, $rand()$ is a random number in range of 0 and 1.

Position update: $X_i^t = \begin{cases} X_i^t & if \ V_{id}^t = 0 \\ X_i^t & if \ V_{id}^t = 1, rand() \geq r_i \\ X_{inew}^t & if \ V_{id}^t = 1, rand() < r_i \end{cases}$ (3)

Local Search: For each node $v_i, 0 < i < n$, if $V_{id}^t = 1$ then $Conn_j = f(v_i, C_j)$ (4)

In Eq. (1), $\beta \in [0, 1]$ is a random number; $f_{min} = minimum frequency$ and $f_{max} = maximum frequency$. In Eq. (2), $\oplus$ is an XOR operator; sig is a Sigmoid function; x_i^t is current position and x_i^* is current best solution. In Eq. (3), r_i is the current emission rate and $rand()$ gives a random number in the range of 0 to 1. In Eq. (4), $C_j \in C, 0 < j < k$, here C is the result of community detection and k is number of communities; $f()$ determines the edges between node v_i and C_j.

3   Tabu Search

To boost the local search ability of bat algorithm, Tabu search scheme is adopted. Tabu search begins with an initial solution and explores the neighborhood search space, in order to find a better solution. The neighborhood of the current solution is determined by taking the union of move operations on all the nodes. The move operation is defined by transferring a node to its adjacent communities. In every iteration, the best eligible neighboring solution is selected. The performance of the search is enhanced by maintaining the record of earlier examined solutions in the tabu list. A solution selected from the neighborhood set is qualified if it is not forbidden by the tabu list or if it is better than the previously encountered solution. A tabu move can be overridden under the condition, named as the aspiration condition [25].

The methodology of the tabu search method for finding the best solution from the current solution enhances the local search ability of bat algorithm. The workflow diagram of a tabu search procedure to generate a new solution around the current best solution in a social network is presented in Fig. 1.

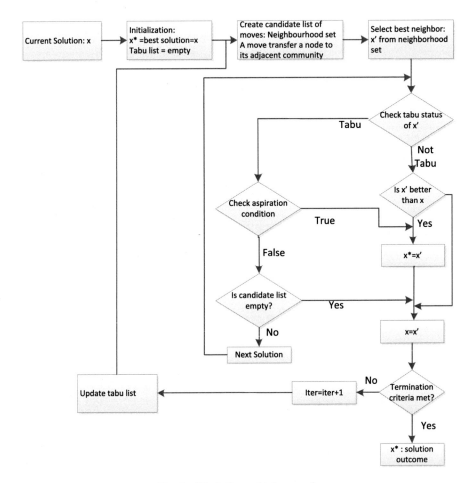

Fig. 1. Work flow of tabu search.

4 Proposed Hybrid Approach

In the proposed method, the discrete bat algorithm is hybridized with Tabu search mechanism using the modularity as the objective function. Bats emit a sound pulse in order to search their food/target which is associated to change in pulse frequency, velocity, and the position. Discrete bat position vector is written as $X = \{x_1, x_2, \ldots, x_n\}$

such that $x_i \in [1, n]$ and velocity vector is a binary vector $V = \{v_1, v_2, \ldots, v_n\}$ such that v_i will be either 0 or 1 only. In the proposed approach, the, position update rule is redefined and local search capability of the bat algorithm is upgraded by applying the Tabu search method. Tabu search is the neighborhood-based search method. It selects the best individual from the neighborhood set around the current best solution.

Movement of the discrete bat in the proposed hybrid approach is simulated as follows:

Pulse frequency: It is expressed as

$$v_i = v_{min} + (v_{max} - v_{min}) \times \beta \tag{5}$$

In Eq. (5), v_{min} = minimum frequency, v_{max} = maximum frequency, β = a random number in range of 0 and 1.

Velocity update rule: In t^{th} iteration, change in velocity is governed by

$$V_i^t = sig\left(V_i^{t-1} + \left(x_i^{t-1} \otimes x_{gbest}\right) \times v_i\right) \tag{6}$$

In Eq. (6), $\otimes$ = XOR operator, x_{gbest} = current best solution and sigmoid function is written as $sig\left(V_i^t\right) = \frac{1}{\left(1+e^{-V_i^t}\right)}$ and $\begin{cases} V_i^t = 1 \, if \, R(0,1) < sig\left(V_i^t\right) \\ V_i^t = 0 \, if \, R(0,1) \geq sig\left(V_i^t\right) \end{cases}$ with $R(0,1)$ = a random number in the range of 0 and 1.

Position update rule: In t^{th} iteration, change in position is defined by

$$x_i^t = x_i^{t-1} \otimes V_i^t \text{ with } \begin{cases} x_i' = x_i \, if \, v_i = 0 \\ x_i' = C(i) otherwise \end{cases} \tag{7}$$

In Eq. (7), $C(i) = argmax_r \sum_{j \in N(i)} \phi(x_j, r)$ in which $\phi(r)$ is an integer value r that maximizes $\phi(r)$. This is the community identifier held by a maximum neighbor of ith node.

Local search: A new solution from the current best solution is generated by

$$x_{new} = TabuSearch\left(x_{gbest}\right) \tag{8}$$

In Eq. (8), x_{gbest} = current best solution and x_{new} = best solution in the neighborhood.

Loudness update: $A_i^{t+1} = \propto A^t$ $\tag{9}$

Pulse emission rate update: $r_i^{t+1} = r_i^0(1 - e^{-\gamma t})$ $\tag{10}$

In Eq. (9), $\propto$ is a constant in the range $0 < \propto < 1$ such that $A_i^t \to 0$ as $t \to \infty$ and in Eq. (10), $\gamma > 0$ is a constant such that $r_i^{t+1} \to r_i^0$ as $t \to \infty$ with r_i^0 as the initial pulse emission rate.

The pseudocode of a hybrid bat algorithm is as below.

ALGORITHM. DBA-tabu

Input: Graph $G(V, E)$

Output: approximate best solution and its modularity value

1. Initialize population size, maximum number of iteration, initial pulse rate, initial loudness, minimum frequency and maximum frequency and the population.
2. f =evaluate the fitness of population (modularity value)
3. Find best solution from all bats (x_{gbest})
4. while termination condition not fulfilled do
  for i=1 to number of bats do
      compute frequency, update velocity and position vector by Eq. (5), Eq.(6) and Eq. (7)
      if $R(0, 1) > r_i$ then
        Apply tabu search around x_{gbest} using Eq. (8)
      end if
      f_{new} = compute modularity of new solution (fitness function)
      if $f_{new} > f_i$ and $R(0, 1) < A_i$ then
       Accepts new solution
       Increase pulse emission rate (r) by Eq. (10)
       Decrease loudness (A) by Eq. (9)
      end if
  end for
   update the best solution from present population
end while

---

5   Experimental Results

The proposed hybrid approach is implemented on desktop PC with Intel Core i7-3770 CPU @ 3.40 GHz processor and 4 GB of RAM with Windows 7 OS. The algorithm code is written in python 3.4.3 with network and community package. The initial input parameters value is based on prior experimentation. The initial frequency values are set to $v_{min} = 0$ and $v_{max} = 1$. A bat pulse emission rate and loudness lie in the range of 0 and 1. So, for simplicity, an average of 0 and 1 i.e. 0.5 is assigned as the initial value of pulse emission rate and loudness, both. The initial value of $\propto = 0.03$ and $\gamma = 0.98$ are chosen in a way that pulse emission rate and loudness constraints are satisfied. In the execution process, number of bats i.e. population size is set to 100. Local search operation by tabu search supports in converging the solution in la esser number of iteration. The obtained optimum value is attained for all the datasets around 15 to 20 iterations. So, the maximum number of iteration is set to 30 and the outcome reveals that the convergence rate is faster with a hybridized approach.

A hybrid algorithm DBA-tabu is implemented for real-world datasets listed in Table 1.

Table 1. Real-world data sets.

Network name	# of nodes	# of edges
Zachary Karate Club	34	78
American Football	115	613
US Political Books	105	441

Zachary Karate Club [26]:- The club network represents the connections among its 34 members.

American Football [1]:- It is a network among teams of American football games during Season Fall 2000.

US Political Books [27]:- A network presents associations among books frequently co-purchased by the same person, sold by the online bookseller Amazon.com.

The execution of DBA-tabu is carried out 10 times. The modularity value (Q) for the algorithms DBA, DPSO, FN and spectral are reported from the literature [16]. The result reported in Table 2 has maximum modularity, average modularity over 10 runs and number of communities at maximum modularity value. Figure 2 presents the best community structure obtained from 10 runs of DBA-tabu.

Table 2. Experimental values.

Dataset name	Algorithm name	Max (Q)	Average (Q)	# of communities
Zachary Karate Club	DBA-tabu	**0.4156**	**0.3926**	4
	DBA	0.392	0.362	2
	DPSO	0.369	0.345	2
	FN	0.376	0.376	2
	Spectral	0.351	0.332	2
American Football	DBA-tabu	**0.6046**	**0.5972**	10
	DBA	0.596	0.531	12
	DPSO	0.563	0.532	10
	FN	0.543	0.543	9
	Spectral	0.531	0.447	12
US Political Books	DBA-tabu	**0.5269**	**0.5251**	4
	DBA	0.479	0.463	3
	DPSO	0.470	0.445	2
	FN	0.467	0.467	4
	Spectral	0.439	0.394	5

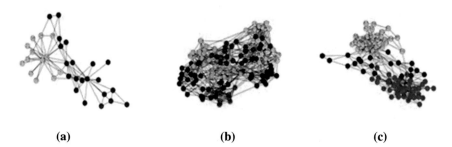

(a)   (b)   (c)

Fig. 2. Community structure at maximum modularity (Q) by DBA-tabu (a) Zachary Karate Club (b) American Football (c) US Political Books.

The visual analysis of results confirms that the quality of community structure obtained by DBA-tabu is enhanced. In Fig. 3(a) comparison of DBA-tabu with existing algorithms at maximum modularity value over 10 runs clearly tells the advantage of using the DBA-tabu. Figure 3(b) shows a comparison of average modularity value over 10 independent runs of DBA-tabu with existing algorithms. Here also, DBA-tabu performance is better than the existing algorithms on all the datasets under test.

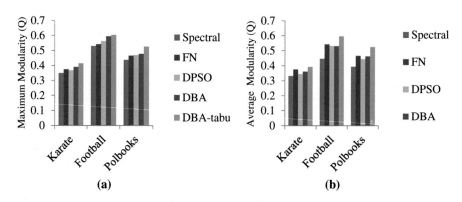

(a)   (b)

Fig. 3. Visual comparison (a) Maximum modularity (b) Average modularity.

Box-plot is straightforward and influential visualization tool to enhance the understanding of results and make comparison across results obtained in ten independent runs by the proposed algorithm. This visualization method is a more communicative way to show the variation in obtained modularity value during independent runs. Figure 4 shows the box-plot diagram of the proposed hybrid approach over 10 runs.

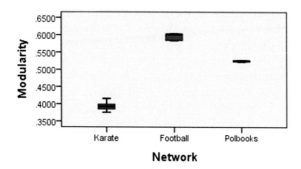

Fig. 4. Box plot diagram of DBA-tabu.

Experimental results on real-world networks and their visual analysis show the benefit of recommended hybrid DBA-tabu. All networks considered under test are small size networks. The largest network used during experimentation is American football network with 115 vertices and 613 edges.

So, a Girvan Newman [28] benchmark network with 128 nodes and 1024 edges is considered. This synthetic network has an average degree of each node as 16 and the network is divided into four equal sized communities with 0.1 as a mixing parameter. The mixing parameter value provides the average percentage of edges that connect with a group of nodes in different communities. DBA-tabu is applied over this synthetic network. For the synthetic network, DBA-tabu experimental value is in line with ground truth community structure of the synthetic network. The community structure of the synthetic network with ground truth and obtained by DBA-tabu is shown in Fig. 5. The NMI value attained for true and predicted communities for the synthetic network is one.

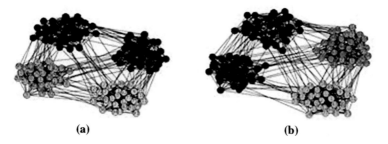

(a)   (b)

Fig. 5. Girvan-Newman network community structure (a) True (b) Predicted by DBA-tabu.

Girvan-Newman benchmark network possesses equal size communities. In a real-world network, it is hard to view that network will have equal size communities. Lancichinetti-Fortunato-Radicchi (LFR) benchmark networks are principally accepted synthetic networks with the varied size of communities and mixed degree of nodes. The assorted characteristics of LFR benchmark networks commonly owned by real-world networks make them stupendous benchmarks and accepted to verify the results of the community detection algorithm. In this work, generated a LFR benchmark network with number of nodes, n = 200; average degree, k = 14.6; maximum degree, maxk = 48 and mixing parameter, mu = 0.1 (./benchmark -N 200 -k 14.6 -maxk 48 - mu 0.1) [28]. The network generated has 1380 edges. The generated network follows the power law distribution with exponent for the degree distribution as 2 and exponent for the community size distribution as 1. DBA-tabu applied over LFR generated benchmark network, the correctness of the proposed approach on a moderately large size network is verified by comparing the obtained result with ground truth result. Figure 6 shows the community partition of LFR network with ground truth and by DBA-tabu. Both have the same number of communities and NMI value for true and predicted communities is one.

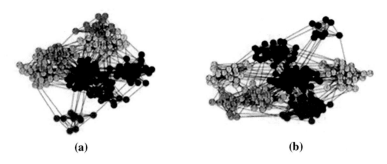

(a)   (b)

Fig. 6. LFR benchmark network community structure (a) True (b) Predicted by DBA-tabu.

The results recommend DBA-tabu has the constructive performance for reasonably large size networks also.

6   Conclusion and Future Work

The work described in this paper is summarized as hybridizing discrete bat algorithm with Tabu search for community detection in social networks. DBA-tabu is applied over real-world networks, Girvan-Newman benchmark network, and LFR benchmark network. The improvement in the quality of community structure as an outcome of an experiment conducted over real-world networks conclude that a hybrid bat algorithm with Tabu search has better performance in contrast to existing algorithms. Further, the correctness of the proposed method is verified by applying DBA-tabu on Girvan-Newman and LFR synthetic benchmark network. In future, the hybrid algorithm can be tested on weighted networks and large-scale networks. Moreover, it can further be explored to identify the overlapping communities as real-world networks have some members that share the membership with more than one community.

References

1. Girvan, M., Newman, M.E.J.: Community structure in social and biological networks. Proc. Natl. Acad. Sci. **99**(12), 7821–7826 (2002)
2. Newman, M.E.J.: Detecting community structure in networks. Eur. Phys. J. B **38**(2), 321–330 (2004)
3. Danon, L., Díaz-Guilera, A., Duch, J., Diaz-Guilera, A., Arenas, A.: Comparing community structure identification. J. Stat. **09008**, 10 (2005)
4. Brandes, U., et al.: On modularity clustering. IEEE Trans. Knowl. Data Eng. **20**(2), 172–188 (2008)
5. Tasgin, M., Bingol, H.: Community Detection in Complex Networks using Genetic Algorithm, arXiv Prepr, p. 6 (2006)
6. Shi, Z., Liu, Y., Liang, J.: PSO-based community detection in complex networks. In: 2009 Second International Symposium on Knowledge Acquisition and Modeling, pp. 114–119 (2009)
7. Chen, Y., Qiu, X.: Detecting community structures in social networks with particle swarm optimization. In: 2nd Communications in Computer and Information Science, vol. 401, pp. 266–275 (2013)
8. Guimerà, R., Nunes Amaral, L.A.: Functional cartography of complex metabolic networks. Nature **433**(7028), 895–900 (2005)
9. Yang, X.S.: A new metaheuristic bat-inspired algorithm. Stud. Comput. Intell. **284**, 65–74 (2010)
10. Mirjalili, S., Mirjalili, S.M., Lewis, A.: Grey wolf optimizer. Adv. Eng. Softw. **69**, 46–61 (2014)
11. Mirjalili, S.: Dragonfly algorithm: a new meta-heuristic optimization technique for solving single-objective, discrete, and multi-objective problems. Neural Comput. Appl. **27**(4), 1053–1073 (2016)
12. Mirjalili, S.: The ant lion optimizer. Adv. Eng. Softw. **83**, 80–98 (2015)
13. Mirjalili, S., Mirjalili, S.M., Yang, X.S.: Binary bat algorithm. Neural Comput. Appl. **25**(3–4), 663–681 (2013)
14. Kennedy, J., Eberhart, R.C.: A discrete binary version of the particle swarm algorithm. In: 1997 IEEE International Conference on Systems, Man, and Cybernetics. Computational Cybernetics and Simulation, vol. 5, pp. 4104–4108 (1997)
15. Holland, J.: Adaptation in natural and artificial systems: an introductory analysis with application to biology. Control Artif. Intell. (1975)
16. Song, A., Li, M., Ding, X., Cao, W., Pu, K.: Community detection using discrete bat algorithm. Int. J. Comput. Sci. **43**(1), 37–43 (2016)
17. Yang, X.S.: Bat algorithm and cuckoo search: a tutorial. Stud. Comput. Intell. **427**, 421–434 (2013)
18. Kora, P., Kalva, S.R.: Improved bat algorithm for the detection of myocardial infarction. Springerplus **4**(1), 666 (2015)
19. Salma, U.M.: A binary bat inspired algorithm for the classification of breast cancer data. Int. J. Soft Comput. Artif. Intell. Appl. **53**(2), 1–21 (2016)
20. Huang, X., Zeng, X., Han, R.: Dynamic inertia weight binary bat algorithm with neighborhood search. Comput. Intell. Neurosci. **2017** (2017). https://doi.org/10.1155/2017/3235720
21. Glover, F.: Tabu search: a tutorial. Interfaces **20**(4), 74–94 (1990)
22. Cai, Q., Gong, M., Shen, B., Ma, L., Jiao, L.: Discrete particle swarm optimization for identifying community structures in signed social networks. Neural Netw. **58**, 4–13 (2014)

23. Newman, M.E.J.: Fast algorithm for detecting community structure in networks. Phys. Rev. E **69**(6), 066133 (2004)
24. Newman, M.E.J.: Community detection and graph partitioning, no. 2 (2013)
25. Mamun-Ur-Rashid Khan, M., Asadujjaman, M.: A tabu search approximation for finding the shortest distance using traveling salesman problem. IOSR J. Math. **12**(05), 80–84 (2016)
26. Zachary, W.W.: An information flow model for conflict and fission in small groups. J. Anthropol. Res. **33**(4), 452–473 (1977). http://www.jstor.org/stable/3629752
27. Newman, M.E.J.: Modularity and community structure in networks. Proc. Natl. Acad. Sci. **103**(23), 8577–8582 (2006)
28. Lancichinetti, A., Fortunato, S., Radicchi, F.: Benchmark graphs for testing community detection algorithms. Phys. Rev. E **78**(4), 46110 (2008)

Design of Effective Algorithm for EMG Artifact Removal from Multichannel EEG Data Using ICA and Wavelet Method

Rupal Kashid[✉] and K. P. Paradeshi

PVPIT, Budhgaon, Maharashtra, India
rupalkashid@gmail.com

Abstract. EMG artifacts are the muscular artifacts, that get introduced in EEG signal during EEG recording because of facial muscle movements. These facial muscle movement includes teeth squeezing, jaw clenching, forehead movement etc. These artifacts surely get embedded into EEG signals because EEG recordings lasts about upto 30 to 35 min. Thus these artifacts create a difficulty in diagnosis of diseases related to brain. So it is necessary to remove EMG artifacts. This paper discusses a technique of Electromyographic i.e. EMG artifact removal from multichannel (16 channel) EEG data by combining Independent Component Analysis and wavelet-based technique for noise reduction. 16 channel EEG data is acquired from 5 subjects by instructing the subject to encounter EMG artifacts that includes teeth squeeze, jaw clench and forehead stretch. ICA is used for computation of independent components. SWT Wavelet decomposition is performed by using Symlet wavelet and Hard thresholding. This approach works satisfactory for EMG artifacts such as teeth squeezing, jaw clenching and forehead movement.

Keywords: EEG · EMG · WICA · Symlet · SWT

1 Introduction

The Electroencephalogram (EEG) is used to measure electrical activity inside the brain. While recording EEG, sensors are mounted on scalp. International 10–20 system is used to decide the placement of these electrodes. Such EEG recordings are affected by various artifacts muscular activities (teeth squeezing, forehead movement, jaw clenching). These muscular activities generate an electrical signals that have large amplitude called as EMG (Electromyogram). EEG recordings lasts upto 30 to 40 min and thus EMG artifacts get embedded in EEG recording. So it is necessary to remove the EMG artifacts.

EMG artifacts are the major contaminated sources to the electroencephalogram (EEG) signals. Research groups considers that the EMG contaminates EEG activity at frequencies greater than 15 Hz. EEG signal is a real time and a non-stationary signal. The frequency of EEG is divided into four sub-bands and is delta – less than 4 Hz, theta – between 4–8 Hz, alpha – between 8–13 Hz, and greater than 13 Hz is beta. Low amplitudes are below 20 μV, amplitudes between 20–50 μV are considered as

© Springer Nature Switzerland AG 2020
A. Abraham et al. (Eds.): ISDA 2018, AISC 941, pp. 955–964, 2020.
https://doi.org/10.1007/978-3-030-16660-1_93

medium ones, and amplitudes greater than 50 μV are assumed as high amplitudes. It is necessary to obtain clean EEG before it is being used.

Raw 16 channel EEG data is shown in Fig. 1.

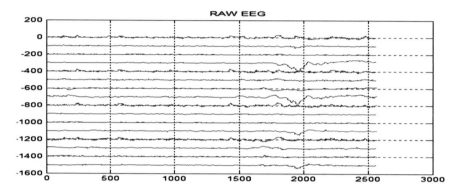

Fig. 1. 16 channel raw EEG signal.

1.1   Background of EEG Artifact Removal Methods

ICA resembles to BSS. Here recorded EEG data signals are divided to get the independent components. BSS estimates sources and parameters of a mixing system. It is based on learning algorithms. In ICA, number of unknown sources must have to be less than or equal to number of recorded signals. For execution of ICA following assumptions are made: linear mixing, square mixing, and stationary mixing are made [4].

The method implemented by Azzerboni et al. explains that DWT approach removes ECG artifact present in few EMG recordings. DWT along with ICA removes artifacts that are present in a surface EMG signals. DWT with ICA incorporates the advantages of DWT as well as ICA. The Wavelet based ICA i.e. (WICA) approach can do extended artifacts removal in medical applications as compared, when DWT and ICA applied separately [5].

Mukul and Matsuno demonstrated wavelet based method using DWT for EEG signal de-noising. It is stronger than Blind source separation (BSS) implemented with Independent component analysis (ICA). DWT separates input signal into its components. Wavelet analysis provides flexibility to localize neuro-electric components, events in time, space and scale. Wavelet transform is able to accurately settle on EEG into specific time and frequency components [6].

Jirayucharoensak and Israsena elaborated on the ICA using LWT (Lifting Wavelet Transform) that extracts the useful EEG data from the artifactual ones. The LWT is absolutely necessary so as to decompose each and every independent component.

LWT then identifies and removes OA (ocular artifacts) and also EMG artifacts (muscular artifacts). LWT ensures that important neural brain activities are retained. The ICA-LWT method gives more proficient method which removes artifacts in raw EEG than that of conventional ICA methods [7].

Chen et al. demonstrates Fast-ICA algorithm. Fast-ICA uses, the sq. root of the typical matrix method so as to achieve symmetric orthogonal and then estimates the source. Fast-ICA avoids the Gram-Schmidt orthogonal method which in turn avoids problem of error accumulation. Fast-ICA eliminates Ocular artifacts (OA) such as eye movement in addition to eye blinking along with ECG and EMG artifacts. FastICA algorithm has fast convergence as well as good separation efficiency [9].

ICA is a technique used for signal processing. It recides on high order statistics, and it separates Independent components from observed multichannel signals. ICA stands on statistical independency principle. Wavelet Transform (WT), is evolved in the mid 1980's. It has wide applications in signal processing, speech recognition as well as image processing. Wavelet analysis has outstanding time and also frequency domain localization properties. Multiresolution analysis is a property of wavelet transform. WT threshold de-noising is developed upon this property. A set of threshold values is selected which in turn is applied on to the raw signal at every scale. Output of this step is denoised signal. After that, to reconstruct denoised signal inverse WT is used [11].

This paper proposes a method for EMG artifact removal by using Independent Component Analysis in addition with wavelet-based (using SWT) signal decomposition for noise reduction.

2 Methodology: SWT-ICA for Artifact Removal

2.1 Independent Component Analysis (ICA)

ICA broadly analyzes and decomposes the raw multichannel EEG signals. If we know a signal then unknown signals information can be calculated by convolving known, unknown signals. In ICA observed raw data is transformed linearly into components that are assumed to be maximum independent from each other. ICA decomposes observed raw multichannel signals into its components that are statistically independent. If there are N observed EEG signals $X_n(t)$ then ICA generates M independent component $S_m(t)$.

n = (1, 2, ... N) and m = (1, 2, ... M)

$$X = As \qquad (1)$$

Where A = mixing matrix and Aij is the mixing coefficient where I varies from 1, 2, ..., M and j varies from 1, 2, ... N. More commonly, A and M are unknown. ICA identifies the M unknown source signals S(t). ICA introduces, inverse of A, i.e. W, a new unmixing matrix:

$$Z = Wx \qquad (2)$$

properties of ICA are Stationary Signal and the Statistical independence. Independent Component can only have nongaussian distribution. Kurtosis, Negentropy, Mutual

Information is calculated to evaluate non-gaussianity. It is vital to carry out preprocessing such as Centering and Whitening, before applying ICA.

2.2 Stationary Wavelet Transform (SWT)

Wavelet transform concatenates features of signal and also of image in wavelet coefficients of large-value. These large value coefficients are small in number. Wavelet coefficients which have small value are noise. Removing these small value coefficients will not disturb image or signal quality. After thresholding, reconstruction of data (image or signal) using the inverse wavelet transform is performed.

While using wavelets for a signal to remove noise requires identifying noisy components, and after that reconstructing the signal excluding those components.

Wavelet Transform is specifically a high quality method used to find quasi harmonic components present in a signal. Classical DWT has drawback that it is not a time-invariant. SWT is designed to retain this property.

The SWT algorithm is very close to DWT. In general, at any step j, approximation coefficients for level $j - 1$ are convolved with suitable novel filters (these are the upsampled filters unlike DWT), this in turn generates detail and approximation coefficients for level j. If N is signal length then the size of approximation and detail coefficients is retained to N because of upsampling.

In our proposed work, SWT is applied with Symlet wavelet throughout the Signal. Symlet is a wavelet family. It is a customized version of Daubechies wavelet. Thresholding is performed by using hard thresholding.

3 Proposed Method

16 channel EEG signals are obtained with artifacts by instructing the subject to do muscle movements such as Jaw clench, Teeth squeeze and forehead movement, during the time of recording. The recorded artifactual EEG data is imported in MATLAB with sampling rate 140 Hz.

Five subjects, 2 males along with 3 females having age between 19–74 years are instructed so as to collect raw EEG data. The electrode placement is according to International 10–20 system.

Wavelet decomposition is performed by using SWT with *symlet2* wavelet. level 2 decomposition and hard thresholding is used. For thresholding, a global threshold of a constant value for all levels is used. Then the decomposed signal is a input for the ICA in matrix. Then ICA is applied to the decomposed signal to find out A and W (the mixing and unmixing matrix respectively) in conjunction with the independent components. Next to this, desired sources of interest are selected. Multiplication of the source of interest with mixing matrix A, gives signal in form of wavelet components. Now, to recover the signal, wavelet reconstruction is done using inverse SWT.

Before applying SWT and ICA, preprocessing is done so as to remove the baseline noise by using zero phase low pass filter. Statistical parameters such as RMSE, PSNR, SNR, PSD are calculated for every artifact free clean EEG signal.

4 Results and Discussion

The proposed technique is implemented using MATLAB. EEG signals with Teeth squeezing, Jaw Clench and forehead artifacts for 5 different subjects are used. The raw EEG signal is processed using wICA method.

Figures 2, 3 and 4 shows artifactual Raw EEG and their results by wICA for three subjects with teeth squeezing artifact.

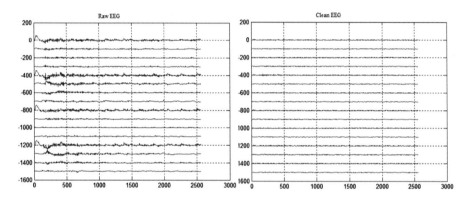

Fig. 2. Raw EEG and clean EEG for subject1 with teeth squeeze artifact

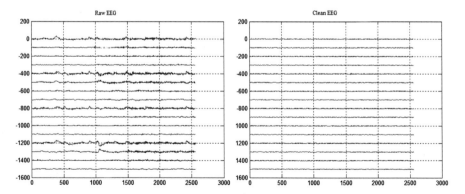

Fig. 3. Raw EEG and clean EEG for subject2 with teeth squeeze artifact

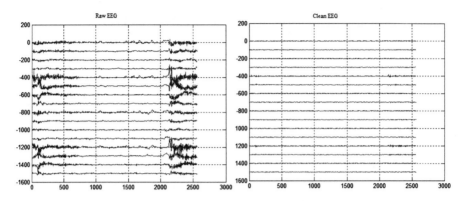

Fig. 4. Raw EEG and clean EEG for subject3 with teeth squeeze artifact

Figures 5, 6 and 7 shows artifactual Raw EEG and their results by wICA for three subjects with jaw clench artifact.

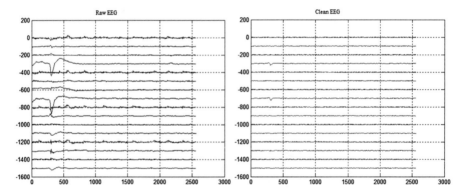

Fig. 5. Raw EEG and clean EEG for subject1 with jaw clench artifact

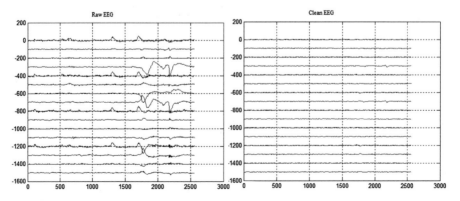

Fig. 6. Raw EEG and clean EEG for subject2 with jaw clench artifact

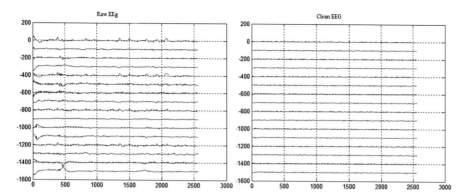

Fig. 7. Raw EEG and clean EEG for subject3 with jaw clench artifact

Figures 8, 9 and 10 shows Raw EEG and their results by wICA for three subjects with forehead artifacts respectively.

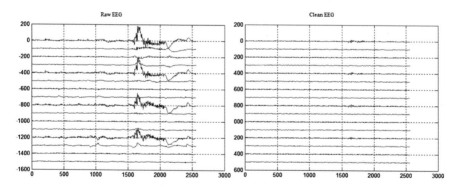

Fig. 8. Raw EEG and clean EEG for subject1 with forehead artifact

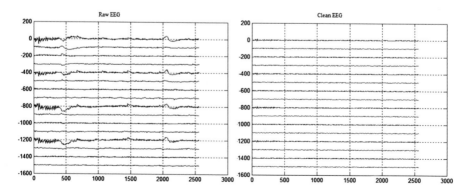

Fig. 9. Raw EEG and clean EEG for subject2 with forehead artifact

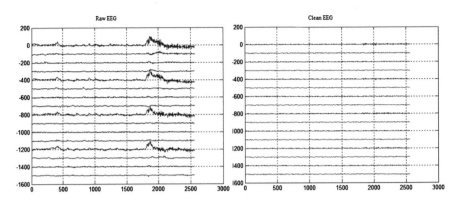

Fig. 10. Raw EEG and clean EEG for subject3 with forehead artifact

Table 1 shows the Statistical Parameters of EEG signal for three subjects with teeth squeeze artifact.

Table 1. Statistical parameters for teeth squeeze artifact

Parameter	Subject1	Subject2	Subject3
RMSE	0.4973	0.4993	0.4915
PSNR	15.2931	15.7491	14.9762
SNR	14.688	14.8327	14.1943
PSD	73.3781	73.8736	72.7936

Table 2 shows the Statistical Parameters of EEG signal for three subjects with Jaw Clench artifact.

Table 2. Statistical parameters for jaw clench artifact

Parameter	Subject1	Subject2	Subject3
RMSE	0.5061	0.4917	0.4891
PSNR	11.6876	11.3741	11.2982
SNR	10.4037	10.3523	10.119
PSD	73.3962	72.2831	72.1974

Table 3 shows the Statistical Parameters of EEG signal for three subjects with forehead artifact.

Table 3. Statistical parameters for forehead artifact

Parameter	Subject1	Subject2	Subject3
RMSE	0.5052	0.4823	0.4902
PSNR	15.9185	14.6759	15.7051
SNR	14.9578	13.4532	15.1587
PSD	73.6279	72.6289	73.1256

5 Conclusion

The proposed method, ICA along with SWT identifies and removes various EMG artifacts which includes Teeth squeeze, Jaw clench, Forehead movement from raw EEG signal. Though the proposed method achieves good performance, some useful data is lost because of hard thresholding. This is the limitation of proposed method.

References

1. Chen, X., Chiang, J., Wang, Z.J., McKeown, M.J., Ward, R.K.: Removing muscle artifacts from EEG data: multichannel or single-channel techniques? IEEE Sens. J. 16(7)
2. Maddirala, A.K., Shaik, R.A.: Removal of EMG artifacts from single channel EEG signal using singular spectrum analysis. In: 2015 IEEE International Circuits and Systems Symposium (ICSyS) (2015)
3. Siamaknejad, H., Loo, C.K., Liew, W.S.: Fractal dimension methods to determine optimum EEG electrode placement for concentration estimation. In: SCIS and ISIS 2014, Kitakyushu, Japan, 3–6 December 2014
4. Nair, K.N., Unnikrishnan, A., Lethakumary, B.: Demonstration of an observation tool to evaluate the performance of ICA technique. In: 2012 IEEE Conference Publications, pp. 1–6 (2012)
5. Azzerboni, B., Carpentieri, M., La Foresta, F., Morabito, F.C.: Neural-ICA and wavelet transform for artifacts removal in surface EMG. In: IEEE International Joint Conference 2004, vol. 4, pp. 3223–3228 (2004)
6. Mukul, M.K., Matsuno, F.: EEG de-noising based on wavelet-transforms and extraction of sub-band components related to movement imagination. In: ICROS-SICE International Joint Conference 2009, 18–21 August 2009. Fukuoka International Congress Center, Japan (2009)
7. Jirayucharoensak, S., Israsena, P.: Automatic removal of EEG artifacts using ICA and lifting wavelet transform. In: 2013 International Computer Science and Engineering Conference (ICSEC), ICSEC 2013 (2013)
8. Mahajan, R., Morshed, B.I.: Sample entropy enhanced ICA denoising technique for eye blink artifact removal from scalp EEG dataset. In: 6th Annual International IEEE EMBS Conference on Neural Engineering San Diego, California, 6–8 November 2013
9. Chen, X., Wang, L., Xu, Y.: A symmetric orthogonal FastICA algorithm and applications in EEG. In: 2009 Fifth International Conference on Natural Computation (2009)
10. Ferdousy, R., Choudhory, A.I., Islam, Md.S., Rab, Md.A., Chowdhory, Md.E.H.: Electrooculography and electromyographic artifacts removal from EEG. In: 2nd International Conference on Chemical, Biological and Environmental Engineering (ICBEE 2010) (2010)

11. Zhou, W., Gotman, J.: Removal of EMG and ECG artifacts from EEG based on wavelet transform and ICA. In: Proceedings of the 26th Annual International Conference of the IEEE EMBS, San Francisco, CA, USA, 1–5 September 2004

12. Mehrkanoon, S., Moghavvemi, M., Fariborzi, H.: Real time ocular and facial muscle artifacts removal from EEG signals using LMS adaptive algorithm. In: International Conference on Intelligent and Advanced Systems (2007)

13. Kachenoura, A., Gauvrit, H., Senhadji, L.: Extraction and separation of eyes movements and the muscular tonus from a restricted number of electrodes using the independent component analysis. In: Proceedings of the 25th Annual International Conference of IEEE 2003, vol. 3, pp. 2359–2362 (2003)

Soft-Margin SVM Incorporating Feature Selection Using Improved Elitist GA for Arrhythmia Classification

Vinod J. Kadam$^{(\boxtimes)}$, Samir S. Yadav, and Shivajirao M. Jadhav

Department of Information Technology,
Dr. Babasaheb Ambedkar Technological University,
Lonere, Maharashtra, India
{vjkadam,ssyadav,smjadhav}@dbatu.ac.in,
http://www.dbatu.ac.in

Abstract. Cardiac arrhythmia is one of the serious heart disorders. In many cases; it may lead to stroke and heart failure. Therefore timely and accurate diagnosis is very necessary. In this paper, we proposed a novel ECG Arrhythmia classification approach which includes an Elitist-population based Genetic Algorithm to optimally select the important features and the Soft-Margin SVM as a base classifier to diagnose arrhythmia by classifying it into normal and abnormal classes. Our improved GA employs the classification error obtained by 10 fold cross-validated SVM classification model as a fitness value. The aim of the Genetic Algorithm is therefore to minimize this value. To show the effectiveness of the proposed method, the UCI ECG arrhythmia dataset was used. Performance of base classifier soft-margin SVM was analyzed with different values of the penalty parameter C. Proposed feature selection method significantly enhances the accuracy and generates fewer and relevant input features for the classifier. With the introduced model, we obtained a promising classification accuracy value. The result of the study proves that the model is also comparable with the existing methods available in the literature. The simulation results and statistical analyses are also showing that the proposed model is truly beneficial and efficient model for cardiac ECG Arrhythmia classification.

Keywords: Soft-Margin SVM · Feature Selection ·
Elitist Genetic Algorithm · Arrhythmia classification · ECG

1 Introduction

Due to advancement in the computer-based technology, new tools for various diseases identification and detection become possible. Many machine learning, expert systems, soft computing and deep learning methods are proposed by various scholars in the almost every sphere of the medical system [1]. Cardiac arrhythmia (cardiac dysrhythmia) is one of the serious heart disorder in which

© Springer Nature Switzerland AG 2020
A. Abraham et al. (Eds.): ISDA 2018, AISC 941, pp. 965–976, 2020.
https://doi.org/10.1007/978-3-030-16660-1_94

the human heartbeats become either too fast (tachycardia) or too slow (brady-cardia). Many times, this disorder becomes life-threatening, therefore prompt and correct diagnosis is really necessary [2]. ECG has important data related to various internal physiological processes; therefore, analysis of ECG signals provides a way to diagnose diseases associated with internal organs. The electro-cardiogram (ECG) is most widely used a test to record or measures the electrical activity of hearts and to detect disease symptoms. The electrocardiogram tests are comparatively easy, inexpensive, non-invasive and very informative [3]. We have proposed the classification model in order to classify the normal and abnor-mal cases of arrhythmia using available ECG dataset. The simulations are carried out on the UCI arrhythmia database [4,5] which is a reasonably difficult dataset with missing, incomplete and ambiguous instances. In past, Many researchers had proposed various approaches to deal with this classification problem and reported classification accuracy ranging between 70% and 85%.

Guvenir et al. applied VF15 supervised learning algorithm on UCI arrhyth-mia database. The VF15 algorithm (Voting Feature Intervals) is based on fea-ture projection and a majority voting among the output class predictions pro-duced by each attribute individually. To improved performance, they also applied the Genetic Algorithm to learn the weights of features in the same study and obtained 68% accuracy with 10 fold cross-validation method [4]. Gao et al. applied the Bayesian artificial neural network model. They used Logistic regres-sion and the back-propagation algorithm to build the model. With 80% training and 20% testing data, they obtained 75% accuracy [6]. Polat et al. proposed the artificial immune recognition system (AIRS) with fuzzy weighted preprocessing. They got 80.7% true accuracy [7]. Lee et al. detected ECG arrhythmias using the neural network with weighted fuzzy membership functions. With 80% training and 20% testing data, this study obtained the accuracy rate 81.21% [8]. Uyar and Gurgen suggested a serial fusion system of Support vector machines and Logistic Regression for ECG arrhythmia detection [9]. Elsayad carried out a compara-tive study of six independent LVQ Learning algorithms: LVQ1, LVQ2.1, LVQ3, OLVQ1, OLVQ3 and HLVQ for ECG arrhythmia classification [10]. Oveisi et al. proposed an efficient tree-based method for Feature extraction using mutual information to improve the performance of classification. They applied this app-roach to eight standard datasets including UCI Arrhythmia dataset to illustrate the performance [11]. Jadhav et al. designed three Artificial Neural network mod-els Multilayer Perceptron, Generalized Feedforward Neural Network (GFFNN) model and Modular Neural Network and obtained 86.67%, 82.35% and 82.22 respectively [12]. Khare et al. applied the SVM classifier with the Spearman Rank correlation for Feature Selection, principal component analysis for Feature Extraction. They obtained 85.98% best accuracy at Gaussian kernel parame-ter sigma = 0.002 and penalty parameter Cost = 0.1 [13]. Yilmaz applied fisher Score technique to enhance the performance of the Least Squares-SVM classifier and obtained the best accuracy of 82.09% [14]. Jadhav et al. proposed feature elimination (FE) based ensembles learning method. They used PART as base classifiers. With 90% training and 10% testing data, they obtained best 91.11%

accuracy [15]. Shensheng Xu et al. investigated feature selection methods like Fisher discriminant ratio and PCA and used SVMs and DNNs for constructing classifiers to classify the selected features. With features selection method based on Fisher discriminant ratio and Deep learning classifier, they achieved the best performance of 80.64% accuracy [16]. Han applied cascade architectures of fuzzy neural networks. They used GA and gradient descent approaches for the optimization of the input subspace and the structure of the cascade architectures [17]. Ayar et al. applied GA to optimally select important the features and then Decision Tree for classification. They Reported 86.96% and 85.04% highest and mean accuracy respectively [18].

2   Proposed Model

We proposed a novel ECG Arrhythmia classification approach which includes an elitist-population based genetic algorithm [19,20] to optimally select the important features and the Soft-Margin SVM as a base classifier [21–24] to diagnose arrhythmia by classifying it into normal and abnormal classes. Figure 1 Shows the proposed model.

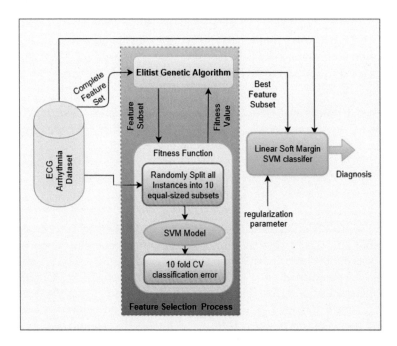

Fig. 1. Proposed model

2.1   Improved Elitist Genetic Algorithm for Feature Selection

Our proposed feature selection approach uses GA assisted by SVM. While the Genetic algorithm generates a trial population of feature subset, the support vector machine calculates the fitness count of each candidate feature subset (chromosome) generated. Each chromosome is represented by a binary string of F bits where F is the number of features present in the dataset. In the chromosome, individual bit value can be either zero or one. '0' means the corresponding feature is absent and '1' means it is present in the candidate feature set. First, an initial population is generated randomly and evaluated using a SVM fitness function. Each individual chromosome (candidate feature subset) is sent into Support Vector Machine to perform a 10-fold cross-validation. The classification error got from this manner is then fixed to each individual chromosome (candidate feature subset) as an indication of its fitness value. The interaction between Genetic algorithm and the SVM is shown in Fig. 1. The all candidate chromosomes are then ranked and based on their rankings; the top m fittest chromosomes (Elitism of size m) are chosen to survive to the subsequent generation. We used m = 10% of total population in this study (for UCI arrhythmia dataset). After these elite chromosomes are pushed to the next generation, the remaining chromosomes in the current population are employed to generate the rest children of the next generation through crossover and mutation operations. In this manner, the SVM based fitness function helps to pick chromosomes for reproduction. The Elitism based GA-SVM based evolutionary procedure runs many generations until the termination state is reached. Best feature subset obtained using this approach is used to classify the ECG arrhythmia dataset using Soft-margin SVM classifier.

Improved Fitness Function. The fitness value is determined using the classification error; the aim of the Genetic algorithm is therefore to minimize this value. 10 fold cross validation fitness value is given by

$$F = \frac{1}{10} \times \sum_{i=1}^{10} E(T_i, \overline{T_i}) \tag{1}$$

$E(T_i, \overline{T_i})$ is classification error of T_i using the model derived by training on $\overline{T_i}$.

　　To compute the fitness value of each individual candidate feature set, the dataset was divided into 10 equal sized partitions (T_1 to T_{10}). The First SVM was trained using complement of T1 ($\overline{T_1}$) partitions and Tested using T1 partition to compute classification error. The second SVM is trained using the complement of T2 ($\overline{T_2}$) partitions and tested using T2 partition to compute classification error and so on. In this manner, total 10 SVMs are trained and tested. The final fitness value is the average of all these individual classification errors.

2.2 Liner Soft-Margin SVM Classifier for Overlapping Classes

SVM classifies instances by finding the optimum or best possible hyper-plane that optimally separates the data into two categories. The data points from both classes closest to the separating hyperplane are called Support vectors.

Consider the problem of binary data classification with given training data points X_i (for some dimension d, the $X_i \in \mathbb{R}^d$ and $i = 1...n$) along with their labels y_i (and $y_i = \pm 1$). The equation of a separating hyperplane is given by

$$f(x) = X^T.W + b = 0 \tag{2}$$

Where W ($W \in \mathbb{R}^d$) is weight vector of the separating hyperplane and b is a real number. The best separating hyperplane (i.e., the decision boundary) can be obtained by solving the optimization problem given in following equation.

$$minimize(\frac{1}{2}\|W\|^2) \tag{3}$$

The optimization problem is feasible when the data is linearly separable. For non-linearly separable cases, the soft margin concept is introduced in SVM.Soft margin means a hyperplane that separates many, but not all data points. It is the extended version of Hard-margin support vector machine.In the formulation of Soft-margin SVM,Slack variables ξ_i and penalty parameter C are added. With the soft margin concept, the optimization problem is defined as

$$minimize(\frac{1}{2}\|W\|^2 + C\sum_{i=1}^{n}\xi_i) \tag{4}$$

Subjected to $\xi_i \geq 0$ and $y_i(X^T.W + b) \geq 1 - \xi_i$, i = 1...n.

This is called 1-norm soft margin problem. Here C is a regularization parameter that controls the trade-off between maximizing the margin and minimizing the training error [21–24].

3 Experiments and Results

3.1 Dataset and Preprocessing

The ECG dataset for this study was obtained from the UC Irvine ML Repository [4]. There are 279 attributes (Linear Valued: 206 Nominal: 73) in the dataset. It is the standard 12 lead ECG signal recordings data. It is well-maintained and benchmark dataset but contains some missing (About 0.33%), incomplete and ambiguous instances. This makes the dataset more similar to the real-world situation. There are 245 normal instances,189 various Arrhythmia instances and 18 unclassified instances in the benchmark UCI dataset. In this study, for Binary classification, we divided dataset instances into two classes (0 and 1) only. Class 0 consists of all normal cases. Class 1 consists of all abnormal cases and unclassified instances. Missing values for attributes in the dataset was handled by replacing them with the average column value.

Table 1. Configuration for the Elitism based Genetic Algorithm

GA parameter	Value	GA parameter	Value
Population size	500	Genomelength	279
Population type	Bitstrings	Number of generations	100
Crossover	Arithmetic Crossover	Crossover probability	0.8
Mutation	Uniform Mutation	Mutation probability	0.2
Selection scheme	Tournament of size 2	Elite count	50

3.2 Experimentation

In order to implement Elitism based GA, its fitness function and soft margin
SVM based classification model, MATLAB software has been used. Table 1 shows
configuration for Elitism based Genetic Algorithm used in the study. Elitism
involves copying some proportion of the fittest candidates, unchanged, into the
next generation. We used 10% of the total population as elite kids for the next
generation. Elitism guarantees that the solution quality obtained by the GA will
not decrease from one generation to the next. Fitness function was implemented
using the functions fitcsvm (default parameters were used except standardization of the predictor was enabled), crossval (to cross-validate the classifier) and
kFoldLoss (to estimate classification error) from Statistics and Machine Learning
toolbox of Matlab. We obtained 92 optimized features (best chromosome) after
the last generation (generation number 100). These optimized 92 features used
for base classifier soft-margin SVM. The soft-margin support vector machine was
also coded using the function fitcsvm (we used default parameters but enabled
standardization of the predictor and the 'KernelScale' flag set to automatic).
Setting the 'KernelScale' flag (a scaling factor by which all inputs are divided)
to 'automatic' employs the heuristic procedure to select the kernel scale value.
We evaluated the true accuracy of the proposed soft-margin classifier for several values (0.1, 0.5, 1, 2, 3, 4, 5, 6, 7, 8, 9 and 10) of 'Box Constraint' flag C
(penalty term). 10-fold CV was chosen to evaluate the accuracy of the classifier.
Figure 2 shows data classification accuracy, Sensitivity, and Specificity, per different values of penalty term C with proposed soft-margin SVM classifier. The
experimental results from Fig. 2 indicate that the maximum amount of classification accuracy, 87.83%, has been achieved with penalty term C = 9. With
penalty term C = 9, we obtained Sensitivity and Specificity 87.83% and 80.19%
respectively (Fig. 3 shows ROC curve, AUC = 0.90). Performance indices used
for comparison are as follows:

$$Accuracy = \frac{TP + TN}{TP + FP + TN + FN} \times 100\% \tag{5}$$

$$Sensitivity = \frac{TP}{TP + FN} \times 100\% \tag{6}$$

$$Specificity = \frac{TN}{FP + TN} \times 100\% \tag{7}$$

These terms are defined using the elements of the confusion matrix given in Table 2. Table 3 gives comparisons between the proposed approach and other methods available in the literature.

Table 2. Confusion matrix of classification

	Prediction as arrhythmia	Prediction as normal
Actual arrhythmia	TP (true positive)	FN (false negative)
Actual normal	FP (false positive)	TN (true negative)

Discussion. In this work, with 92 optimized features obtained using Improved Elitist GA, the highest true 10 fold classification rate is reached when $C = 9$. To make a comparison, true classification accuracies of the different methods present in the literature and the proposed model are provided in Table 3. Our proposed model obtained a remarkable true 10 fold classification accuracy rate of 87.83% and it outperforms other state-of-the-art methods present in the literature except for ensembles learning proposed by Jadhav et al. [15]. It shows that our model produced the comparable and satisfactory result using proper tuning parameter settings.

Table 3. Comparison with other methods

Study	Method	Test method	Accuracy%
Guvenir et al. [4]	VF15 algorithm GA-VF15 algorithm	10-fold-CV	62 65
Gao et al. [6]	Bayesian artificial neural network (ANN) classifier used a logistic regression model and the back-propagation algo	Hold-out	75.00
Polat et al. [7]	Artificial immune recognition system (AIRS) with fuzzy weighted pre-processing	Hold-out 10-fold-CV	80.7 76.2

(*continued*)

Table 3. (*continued*)

Study	Method	Test method	Accuracy%
Lee et al. [8]	neural network with weighted fuzzy membership functions	Hold-out	81.32
Uyar and Gurgen [9]	SVM with Gaussian kernel	Hold-out	76.1
Elsayad [10]	Hierarchical LVQ	Hold-out	76.92
Oveisi et al. [11]	Tree-Structured Feature Extraction Using Mutual Information (TMI-FX)	Hold-out	68.5
Jadhav et al. [12]	Modular neural network generalized feedforward neural network multilayer perceptron model	Hold-out	82.22 82.35 86.67
Khare et al. [13]	Spearman Rank Correlation+PCA+SVM	Hold-out	85.98
Yilmaz [14]	Fisher Score and Least Squares-SVM	10 fold-CV	82.09
Jadhav et al. [15]	Feature elimination based random subspace ensembles learning	Hold-out	91.11
Shensheng Xu et al. [16]	SVM FDR+SVM PCA+SVM Deep Neural Network FDR+Deep Neural Network PCA+Deep Neural Network	10 fold CV	77.77 78.23 76.97 79.18 80.64 73.65
Han [17]	Cascade architectures of fuzzy neural net After gradient decent method	Hold-out	82.5
Ayar et al. [18]	Genetic Algorithm and Decision Tree classifier (c4.5)	Hold-out	86.96
This study	Soft-Margin SVM, Feature Selection using Improved Elitist GA and 10 fold CV SVM fitness function	10 fold CV	87.83

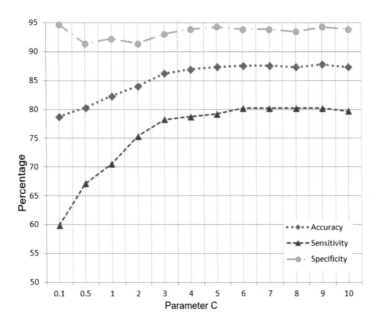

Fig. 2. A comparison of classification performances for different Parameter C

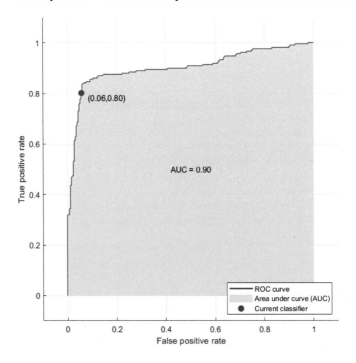

Fig. 3. ROC curve for soft margin SVM classifier (C = 9) used in this study

4   Conclusion

In this paper, we proposed a novel ECG Arrhythmia classification approach which includes an elitist-population based genetic algorithm to optimally select the important features and the Soft-Margin SVM as a base classifier to diagnose arrhythmia by classifying it into normal and abnormal classes. The soft margin SVM is an important and powerful supervised machine learning algorithm due to its generalization ability. The system's performance is assessed using a real Arrhythmia data set with respect to true classification accuracy, sensitivity, and specificity with 10-fold CV. The proposed model achieved a remarkable true 10 fold CV classification accuracy 87.83% and it outperforms many other state-of-the-art methods present in the literature. The experimental results show that the elitist based strategy guarantees that the solution quality obtained will not decline from one generation to the next generation and improve the performance of GA in the dynamic environment. The empirical results show that 92 features are satisfactory for the proposed soft-margin SVM based system to work significantly well in the arrhythmia classification problem. According to experimental findings and results, it is concluded that the proposed soft margin support vector machine classifier (with elitist-population based genetic algorithm based selected features) can aid clinicians to make the reliable diagnosis of ECG arrhythmia.

References

1. Kadam, V.J., Jadhav, S.M.: Feature ensemble learning based on sparse autoencoders for diagnosis of Parkinson's Disease. In: Iyer, B., Nalbalwar, S., Pathak, N. (eds.) Computing, Communication and Signal Processing. Advances in Intelligent Systems and Computing, vol. 810. Springer, Singapore (2019)
2. Why Arrhythmia Matters. Heart.org (2017). http://www.heart.org/HEARTORG/Conditions/Arrhythmia/WhyArrhythmiaMatters/Why-Arrhythmia-Matters_UCM_002023_Article.jsp. Accessed 08 Feb 2018
3. Leng, S., San Tan, R., Chai, K.T.C., Wang, C., Ghista, D., Zhong, L.: The electronic stethoscope. Biomed. Eng. OnLine **14**(1), 66 (2015). https://doi.org/10.1186/s12938-015-0056-y
4. Guvenir, H.A., Acar, B., Demiroz, G., Cekin, A.: A supervised machine learning algorithm for arrhythmia analysis. In: Computers in Cardiology 1997, Lund, Sweden, pp. 433–436 (1997)
5. Dua, D., Taniskidou, E.K.: UCI Machine Learning Repository. School of Information and Computer Science, University of California, Irvine, CA (2017). http://archive.ics.uci.edu/ml
6. Gao, D., Madden, M., Schukat, M., Chambers, D., Lyons, G.: Arrhythmia identification from ECG signals with a neural network classifier based on a Bayesian framework. In: 24th SGAI International Conference on Innovative Techniques and Applications of Artificial Intelligence, December 2004
7. Polat, K., Sahan, S., Günes, S.: A new method to medical diagnosis: artificial immune recognition system (AIRS) with fuzzy weighted pre-processing and application to ECG arrhythmia. Expert Syst. Appl. **31**, 264–269 (2006)

8. Lee, S.-H., Uhm, J.-K., Lim, J.S.: Extracting input features and fuzzy rules for detecting ECG arrhythmia based on NEWFM. In: 2007 International Conference on Intelligent and Advanced Systems, Kuala Lumpur, pp. 22–25 (2007). https://doi.org/10.1109/ICIAS.2007.4658341

9. Uyar, A., Gurgen, F.: Arrhythmia classification using serial fusion of support vector machines and logistic regression. In: 2007 4th IEEE Workshop on Intelligent Data Acquisition and Advanced Computing Systems: Technology and Applications, Dortmund (2007)

10. Elsayad, A.M.: Classification of ECG arrhythmia using learning vector quantization neural networks. In: 2009 International Conference on Computer Engineering & Systems, Cairo (2009)

11. Oveisi, F., Oveisi, S., Erfanian, A., Patras, I.: Tree-structured feature extraction using mutual information. IEEE Trans. Neural Netw. Learn. Syst. **23**(1), 127–137 (2012). https://doi.org/10.1109/TNNLS.2011.2178447

12. Jadhav, S.M., Nalbalwar, S.L., Ghatol, A.A.: Artificial neural network models based cardiac arrhythmia disease diagnosis from ECG signal data. Int. J. Comput. Appl. **44**, 8–13 (2012)

13. Khare, S., Bhandari, A., Singh, S., Arora, A.: ECG arrhythmia classification using spearman rank correlation and support vector machine. In: Deep, K., Nagar, A., Pant, M., Bansal, J. (eds.) Proceedings of the International Conference on Soft Computing for Problem Solving (SocProS 2011), 20–22 December 2011. Advances in Intelligent and Soft Computing, vol. 131. Springer, India (2012)

14. Yılmaz, E.: An expert system based on fisher score and LS-SVM for cardiac arrhythmia diagnosis. Comput. Math. Methods Med. **2013**, Article ID 849674, 6 p. (2013)

15. Jadhav, S., Nalbalwar, S., Ghatol, A.: Feature elimination based random subspace ensembles learning for ECG arrhythmia diagnosis. Soft Comput. **18**(3), 579–587 (2014). https://doi.org/10.1007/s00500-013-1079-6

16. shensheng Xu, S., Mak, M.W., Cheung, C.C.: Deep neural networks versus support vector machines for ECG arrhythmia classification. In: 2017 IEEE International Conference on Multimedia & Expo Workshops (ICMEW), Hong Kong, pp. 127–132 (2017)

17. Han, C.-W.: Detecting an ECG arrhythmia using cascade architectures of fuzzy neural networks. In: Advanced Science and Technology Letters, (ASP 2017), vol. 143, pp. 272–275 (2017)

18. Ayar, M., Sabamoniri, S.: An ECG-based feature selection and heartbeat classification model using a hybrid heuristic algorithm. Inf. Med. Unlocked **13**, 167–175 (2018)

19. Majumdar, J., Bhunia, A.K.: Elitist genetic algorithm for assignment problem with imprecise goal. Eur. J. Oper. Res. **177**(2), 684–692 (2007)

20. Mishra, J., Bagga, J., Choubey, S., Gupta, I.K.: Energy optimized routing for wireless sensor network using elitist genetic algorithm. In: 2017 8th International Conference on Computing, Communication and Networking Technologies (ICC-CNT), pp. 1–5. IEEE, July 2017

21. Tjandrasa, H., Djanali, S.: Classification of P300 event-related potentials using wavelet transform, MLP, and soft margin SVM. In: 2018 Tenth International Conference on Advanced Computational Intelligence (ICACI), pp. 343–347. IEEE, March 2018

22. Nguyen, H.D., Jones, A.T., McLachlan, G.J.: Jpn. J. Stat. Data Sci. **1**, 81 (2018). https://doi.org/10.1007/s42081-018-0001-y
23. Norton, M., Mafusalov, A., Uryasev, S.: Soft margin support vector classification as buffered probability minimization. J. Mach. Learn. Res. **18**, 1–43 (2017)
24. Merker, J.: On sparsity of soft margin support vector machines. J. Adv. Appl. Math. **2**(3) (2017)

Distributed Scheduling with Effective Holdoff Algorithm in Wireless Mesh Networks

K. S. Mathad$^{(\boxtimes)}$ and S. R. Mangalwede

KLS Gogte Institute of Technology, Belagavi, Karnataka, India
ksmathad@gmail.com

Abstract. Wireless IEEE 802.16 provides high speed internet to mobile users. Resource allocation is required that is robust and efficient which provides mobile users with high throughput. Distributed scheduling directs the packets using election algorithm. This algorithm establishes the transmission slots between the competing nodes and uses holdoff algorithm to compute node claimed slots. Mesh node does data transmission without conflicts in multihop networks using control messages. Distributed scheduling with effective holdoff method is proposed to improve throughput. Usual holdoff algorithm has low performance whereas in the proposed algorithm it considers number of adjoining nodes and state of a node for scheduling to improve the throughput.

Keywords: Scheduling · Distributed scheduling · Holdoff · Competing nodes · Throughput

1 Introduction

There are two types of scheduling algorithms namely centralized and distributed scheduling. Base station (BS) schedules the subscriber stations (SS) in centralized scheduling. In this centralized scheduling the performance decreases with the increase in the distance between the subscriber stations. Distributed scheduling provides free access to the control slots without collision. A speculative model gives computation of time interval between consecutive retrieves to the control frames, this provides analysis of performance of distributed scheduling. There are snags in centralized scheduling, nodes using this scheduling visualize needless routes. All packets are forwarded through base station. Contiguous reuse in efficient way cannot be done in this concept of scheduling. Distributed scheduling provides proficient paths to all the subscriber stations in the network. With the increase in the contiguous reuse, the capacity of the network increases. The bandwidth of the network is efficiently used by allocating the minislots to the subscriber stations as their requirement.

Distributed scheduling coordinates by providing schedules to control messages and data packets. It uses Election Algorithm and Reservation Based Distributed Scheduling. A node makes resistless search over all channels to get connected to the network [1]. The limitations in the coordinated distributed scheduling results in reduction in network throughput and increase in end to end delay. Every subscriber station is involved in delay during transmissions initiating from present transmission time. The calculation of holdoff time also involves faulty transmission slots, unused control, unused data slots and unusual use of a node for control signal sending. Combining of pseudo random numbers that are generated by a generator is given as input to identification of slots. It also identifies the nodes for transmission among competing nodes which are subscriber stations. The identification by competing subscriber stations is not based on priority which results in decrease in throughput. To avoid such problems a new algorithm is suggested namely dynamically synchronized distributed scheduling which reduces the delay. The better performance can be achieved with the usage of more number of channels in Wireless Mesh Networks [2]. In traditional coordinated distributed scheduling algorithm (CDS) every subscriber station holdoff for 16 transmission slots. Holdoff time is generated with holdoff defender result which varies from 0–7. In the proposed algorithm the holdoff value is computed based on the count of neighbors for every subscriber station. Dynamically each node can be given priority based on some communication messages. Each SS node has to exchange request, grant and confirm messages before transferring data. This results in lesser number of unused slots. It also results in nearly equal number of transmission possibilities to all the nodes. Some researchers have worked on this scheduling concept by accounting delay right from generation of packet. Delay keeps on getting added till it reaches the destination. In the suggested algorithm the holdoff time is generated by taking the difference between slot of requested node and the time of transmission of contributor node. This value is communicated with the contributor node. Two values help in computing the performance of CDS algorithm. The values are holdoff defender value and the next transmission maximum value. Each SS communicates their next transmission duration which is calculated using above mentioned values. If the slots are unused because of faulty computation due at competitor node. Mesh routers may allow shcedulling to be done locally on the clients and later does scheduling globally by forwarding the flog to another mesh router [3].

2   Working of CDS Algorithm

In CDS algorithm subscriber station behaves like a node in adhoc networks. SS collects the information from all neighbors that are nearby in 2-hop or 3-hop. The control subframe has two parts, control subframe associated with network and control subframe associated with scheduling. Control subframe has 16 possibilities Tpo and data subframe has 256 minislots. The slot of a control subframe consists of 7 Orthogonal Frequency Division Multiplexing OFDM symbols and 4 OFDM symbols. Each node sends the information about entering in the network by sending them messages. It sends two messages network entry message (MSH-NENT) and network configuration message (MSH-NCFG). SS then conveys information either in centrally coordinated mode or distributed coordinated mode. In distributed mode the control information is sent by sending distributed scheduling messages (MSH-DSCH). Mesh distributed number (MSH-DSCH-NUM) is the slot that is available to send distributed control messages. The Control messages are sent at regular intervals and the precomputed time slots are used for sending [4].

The important problem in wireless mesh networks is to increase the state of transfer of packets among the nodes. Resources are allocated in multihop wireless mesh networks and this can be thoroughly inspected in the environment [5]. Every node avoids the conflicts by understanding the information of its neighbors to access control or data slots. The control messages are sent and after that the data minislots can be sent repeatedly from time to time then the scheduling messages are transferred. Network performance can be improved by providing less number of free mini slots which is performed by the node providing the grants [6]. CDS is operated in every SS by getting the required information from the nodes involved in coordination. Ideal performance can be provided in multihop networks by using better routing algorithm at network layer [7]. SS uses holdoff for mentioned time and then continues with the transmission. Transmission time duration from the next time is computed using the formulas mentioned below.

$$\text{HoldoffTime} = 2^{\text{holdoffdefender}+4} \tag{1}$$

$$
\begin{aligned}
2^{\text{holdoffdefender}+4} \times \text{NextTransMax} &< \\
\text{NextimeInterval} &\leq \\
2^{\text{holdoffdefender}+4} \times (\text{NextTransMax} + 1)
\end{aligned}
\tag{2}
$$

3 Effective Holdoff Algorithm

In the proposed algorithm every SS checks for possibilities of transmission by gathering the information from nearby nodes. After collecting the information of possibility of transmission it delays the transmission for a number of slots. It counts the number of nearby nodes to decide the delay of transmission. The nearby nodes of a particular node do not use Tpo. This gives the information that those particular nodes are not lively. The nodes can be grouped as lively and idle. A node becomes lively by transmitting the control signals. Base station and Subscriber stations should synchronize their transmissions with neighbors in multi-hop network with distributed mode [8]. In the traditional algorithms the holdoff time calculation is being done in a way which results in collision between the subscriber stations. Most of the research is focused on achieving ideal performance and maintaining network stability by using ideal throughput scheduling [7]. Current usage of transmission slot recognizes the status of a node. A node is considered lively by collecting the information of usage of transmission slots recently. Recent successive timer and MSH-NCFG are used to determine the status of a node. Every SS notes the information for each instance. The scheduling with control messages identifies the location of resources for sending of scheduling messages [9]. Network performance is evaluated using throughput. Optimizing of the algorithm results in improvement in the performance of a network. Three kinds of messages are used for control of coordination namely request, grant and confirm. When a node J is having a nearby node K and node K is interested to send data to J. J sends a grant message to node K. When there is one more node L then two messages grant and send are sent by node J. In this case P node is utilized which is NULL nearby node.

Algorithm1 Effective Holdoff Algorithm

Require: node lively or not using the value of N and node_lively(i)

Ensure: returning the status of a node

--begin--
for i=1 to N do
        for i=1 to N do
                if(node_lively(i)==0)
                        Node_status(i)=0
                else
                        Node_status(i)=1
                end if
        end for
end for
if(N<=15)
        $H=2^{logw+2cn+2}$
else
        $H=2^{logw+2cn+1}$
end if
if(min(H) && (node_status(i)==1))
        Allocate_slot →Node(i)
else
        i++
end if
--end--

The formula that is used for calculating the throughput is

$$T = D/(C+S+H) \tag{3}$$

Where

T is throughput
C is computation time for using channel for sending data
S is the duration in which node transfers data
H is holdoff time after sending the data
D the amount of data receiver to receive

In the proposed algorithm two concepts are used. First is number of nearby competing nodes increase for a given node then the node has larger holdoff time. In case the mentioned condition is not true the concerned node requires shorter holdoff time. Second is state of a node. Node is busy which is mentioned by the amount of data the node has to send. In this case it has shorter holdoff time. In case the mentioned condition is not true it has longer holdoff time to remove the contention among nearby nodes. Here holdoff time is computed by using the formula.

$$H = 2^{\log w + 2cn + 2} \tag{4}$$

$$H = 2^{\log w + 2cn + 1} \tag{5}$$

Where

H is holdoff time after transmission of data
w is the element which signifies that node is busy
cn is count of contending nodes

In the proposed approach three cases are considered. Case (i) number of nearby nodes are larger it will get larger holdoff time. Case (ii) number of nearby nodes are same the busy node gets shorter holdoff time. Case (iii) The nodes will maintain a considerable balance between the nodes that are competing and states of the nodes. Maximum and minimum decency improves the lowest portion that a move can provide [10].

4   System Model

Entire Wireless Mesh Network is defined as graph G(V, E) where V and E are mentioning nodes and subscriber stations. Edges are connected with each other and are represented in the form of mesh structure. The delay involved while transmission of packet among nodes is the average time involved in reaching destination from source [11]. NS3 is used for simulation. Here the structure is representing distributed model where each mode competes with the other nodes to achieve the target. Each SS holds the information of nearby nodes. SS's are placed at the distance of d. Each node transfers the packet with minimum 2(n − 1) connections which are interconnected. The proposed algorithm obviously improves the throughput when the number of nearby nodes which are competing are less. It increases the usage of channel that is used to control to a greater level. This is done by minimizing the duration of frame among two successive MSH-DSCH messages, thereby increasing the throughput. When the number of nodes competing increases the transmission possibilities decrease. The use of control channel steels as the count of nodes is greater than 9. It results in more competitions with failure. This results in lesser performance. The proposed algorithm groups nodes based on their involvement in transferring of data and list out the hold off time correctly. The performance of the proposed algorithm is better than CDS algorithm. The result of increasing equity results in significant increase in the performance and throughput [6]. The network has the potential to achieve a much higher throughput [12].

Figure 1 shows Comparibility of average throughput with packet length of 1000 bytes against the number of nodes. The proposed approach provides better results than CDS algorithm. The nodes are grouped here on the basis of transferring of data by the nodes and calculating the holdoff time in a correct way.

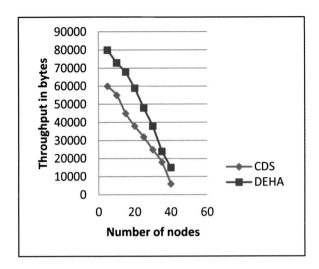

Fig. 1. Comparibility of average throughput with packet length of 1000 bytes against the number of nodes for CDS and DEHA

Figure 2 shows Comparibility of average throughput with packet length of 2000 bytes against the number of nodes. The proposed approach provides better results than CDS algorithm.

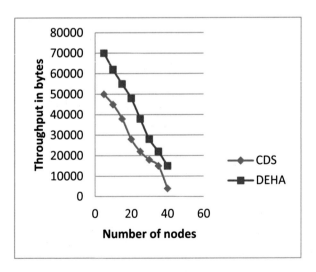

Fig. 2. Comparison of average throughput with 2000 bytes packet size against the number of nodes for CDS and DEHA

Figure 3 Comparibility of average throughput with packet length of 3000 bytes against the number of nodes. The proposed approach provides better results than CDS algorithm.

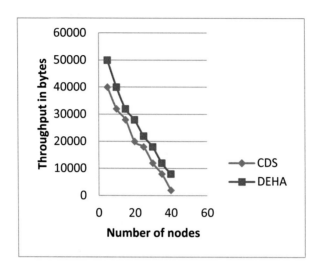

Fig. 3. Comparison of average throughput with 3000 bytes packet size against the number of nodes for CDS and DEHA

5   Conclusion

In this paper, the proposed Distributed Scheduling Effective Holdoff algorithm considers the number of nodes that are competing for transfer of data and the state of the nodes to achieve better results. The CDS algorithm considers the information that is transferred by the nodes immediately after entering the network. The holdoff time in CDS algorithm is static. The proposed Distributed Scheduling Effective Holdoff algorithm calculates the holdoff time based on the network condition. It also considers the busy status of a node during the calculation of holdoff. The results obtained reveal that the proposed algorithm achieves better throughput as compared to CDS algorithm. Since it considers busyness of a node along with the competitiveness among the nodes the performance of DEHA becomes more effective than the traditional CDS algorithm.

References

1. Lee, W.-Y., Hwang, K.-I.: Distributed fast beacon scheduling for mesh networks. In: 2011 Eighth IEEE International Conference on Mobile Ad-Hoc and Sensor Systems, pp. 727–732 (2011)
2. Vallati, C., Mingozzi, E.: Efficient design of wireless mesh networks with robust dynamic frequency selection capability. J. Comput. Netw. **83**, 15–29 (2015)

3. Kim, E.-S., Glass, C.A.: Perfect periodic scheduling for binary tree routing in wireless networks. J. Eur. J. Oper. Res. **247**, 389–400 (2015)
4. Cano, C., Malone, D.: A learning approach to decentralised beacon scheduling. J. Adhoc Netw. **49**, 58–69 (2015)
5. Facchi, N., Gringoli, F., Malone, D., Patras, P.: Imola: a decentralised learning-driven protocol for multi-hop White-Fi. J. Comput. Commun. **105**, 157–168 (2016)
6. Pradhan, S.C., Mallik, K.K.: Minimization of overhead using minislot allocation algorithm in IEEE 802.16 mesh network. In: Fifth International Conference on Eco-Friendly Computing and Communication Systems (ICECCS 2016), pp. 68–72 (2016)
7. Cao, X., Liu, L., Shen, W., Cheng, Y.: Distributed scheduling and delay-aware routing in multihop MR-MC wireless networks. IEEE Trans. Veh. Technol. **65**(8), 6330–6342 (2016)
8. Sabbah, A., Samhat, A.E.: Distributed scheduling in wireless mesh networks using smart antenna techniques. In: International Conference on Parallel and Distributed Processing Techniques and Applications, pp. 55–61 (2015)
9. Park, D.C., Ren, Y., Kim, S.C.: Novel request algorithm for distributed scheduling in wireless mesh networks. In: IEEE 12th Consumer Communication and Networking Conference, pp. 922–924 (2015)
10. Chakraborty, S., Nandi, S.: Distributed service level flow control and fairness in wireless mesh networks. IEEE Trans. Mob. Comput. **14**(11), 2229–2243 (2015)
11. Chattopadhyay, S., Chakraborty, S., Nandi, S.: Leveraging the trade-off between spatial reuse and channel contention in wireless mesh networks. In: 2016 8th International Conference on Communication Systems and Networks (COMSNETS), pp. 1–8 (2016)
12. Deng, X., Luo, J., He, L., Liu, Q., Li, X., Cai, L.: Cooperative channel allocation and scheduling in multi-interface wireless mesh networks. Springer Science and Business Media, LLC, part of Springer Nature (2017)

Continuous Cartesian Genetic Programming with Particle Swarm Optimization

Jaroslav Loebl$^{(\boxtimes)}$ and Viera Rozinajová

Faculty of Informatics and Information Technologies,
Slovak University of Technology in Bratislava,
Ilkovičova 2, 842 16 Bratislava 4, Slovakia
jaroslavloebl@gmail.com, viera.rozinajova@stuba.sk
https://www.fiit.stuba.sk

Abstract. Cartesian Genetic Programming (CGP) is a type of Genetic Programming, which uses a sequence of integers to represent an executable graph structure. The most common way of optimizing the CGP is to use a simple evolutionary strategy with mutations, which randomly changes the integer values of integer sequence. We propose an alternative genotype-phenotype mapping procedure for CGP allowing usage of real-valued numbers in genotype. Novel representation allows continuous transition between various functions and inputs of each given node (hence the name, Continuous CGP), which means, that the optimization of CGP individual is transformed from combinatorial optimization problem to continuous optimization problem. This allows leveraging various metaheuristic optimization algorithms. In this paper, we present results obtained by Particle Swarm Optimization algorithm, showing that continuous representation is able to outperform classic CGP in some benchmarks and provides competitive results with one of the best performing symbolic regression systems in literature.

Keywords: Cartesian Genetic Programming ·
Particle Swarm Optimization · Symbolic regression ·
Evolutionary algorithms

1 Introduction

Evolutionary algorithms have been around here for several decades and they have been generally considered as an optimization method for hard optimization tasks, where other algorithms fail. Flexibility of individual representation, especially in case of Genetic Algorithm (GA), presented by Holland [8], allowed application in various tasks. This flexibility was fully utilized by Koza in his seminal work [14], where Genetic Programming (GP) was introduced. In GP, individual is represented by syntax tree data structure, holding functions as a

© Springer Nature Switzerland AG 2020
A. Abraham et al. (Eds.): ISDA 2018, AISC 941, pp. 985–995, 2020.
https://doi.org/10.1007/978-3-030-16660-1_96

nodes of a tree and inputs (and constants) as a leaves of the tree. GP individual takes an input data and transforms it to output, according to syntax tree GP individual represents. This fundamentally changes the aim of the algorithm, shifting it from the field of optimization, into the field of supervised learning, with symbolic regression (i.e. finding the symbolic prescription of a function from data, not just its parameters) as its prominent task. Though not fully embraced in mainstream practice, GP has become interesting alternative to other machine learning algorithms in many tasks, such as classification [2], feature selection [21], image processing [1], evolution of neural networks [13], data clustering [15] or automatic design of electronic circuits [4]. Koza's standard GP suffers from several issues, most notably a bloat problem (sudden growth of individuals in terms of size, slowing down the evolutionary process), which aggravate the utilization of the framework. Gradually, many techniques were proposed to tackle this problem, one of them is covariant parsimony pressure [22].

Cartesian Genetic Programming (CGP) is a specific kind of Linear Genetic Programming branch of GP. The term CGP was coined in Miller's seminal paper [18], originally as a form of automatic design of digital circuits, but it was later extended to other domains, such as symbolic regression or program synthesis in [19]. In CGP, the phenotype of individual is a directed graph rather than a tree and is encoded in sequence of integer numbers (genotype). Genetic operators are executed on genotype and changes are reflected by explicit genotype-phenotype one-way mapping. CGP holds several advantages, to name the biggest one is the overcoming of the bloat problem. Initially, there have been experiments using genetic algorithm and probabilistic hillclimber [18] as tools to optimize CGP genotype, but during time it has become somewhat standard to optimize the CGP system using simple evolutionary strategy with point mutation and $(1 + \lambda)$ scheme, where λ usually equals to 4. Crossover, while believed as a cornerstone of genetic algorithm, showed only detrimental effects. A few works attempted to design a prosperous crossover operator [6,10,11,25], nevertheless, mutation remained dominant genetic operator. Furthermore, authors of [7] proposed new mutation operators, which outperformed the original point mutation.

In this we work, we propose an entirely different and new method to optimize CGP genotype, aimed at providing a transition between gene values. It is based on novel genotype-phenotype mapping, which allows real-valued, continuous genotype, instead of integer based genotype of classic CGP. We then proceed to optimize such genotype using Particle Swarm Optimization algorithm [12].

The rest of the paper is organized as follows: in Sect. 2 we provide a description of classic CGP, survey of related work is provided in Sect. 3, in Sect. 4 we describe our genotype-phenotype mapping function in detail, in Sect. 5 we describe the setup of our experiments, compare our results with baseline performance of classical CGP using single mutation introduced in [7] and P-tree programming [20] (hereafter PTP) and discuss the results. Final remarks are stated in Sect. 6.

2   Cartesian Genetic Programming

While the foundation ideas for CGP were laid out in paper [17], the first actual appearance of term CGP was in paper [18], where it was aimed solely on boolean functions. In [19] was extended also to non-boolean problems.

CGP lies in Linear Genetic Programming category. This means, that subject of evolution is not syntax tree, or other tree based data structure, but rather linear sequence, be it sequence of instructions, numbers or other symbols. In case of CGP it is a sequence of integers (genotype), which encodes a directed acyclic graph (phenotype). Explicit one-way mapping between genotype and phenotype is therefore required. This sums up the main differences between classic GP as proposed by Koza and CGP: use of directed acyclic graph instead of trees and explicit one-way many-to-one genotype-phenotype mapping procedure.

Phenotype of individual in CGP therefore consists of graph nodes, which are arranged in two-dimensional grid (hence the name, Cartesian), with directed edges among them. Such graph is completely encoded in genotype (linear sequence of integers), which takes the following form:

$$F_0, C_{0,0}, \ldots, C_{0,a}; \ldots; F_{(c+1)r-1}, C_{(c+1)r-1,0}, \ldots, C_{(c+1)r-1,a}; O_0, \ldots, O_m \quad (1)$$

where sequence of genes, starting with F_i gene, followed by sequence of $C_{i,j}$ genes encode a single phenotype node, called function node. F gene is an index of function from function set and C genes represent the indices of nodes, from which current node takes its input. Input nodes are part of phenotype as an initial column, but are not part of genotype. Function nodes from first column can connect only to this input column. Number of input nodes is equal to number of attributes in dataset, plus one node holding the constant value of 1 (in case constant approximation is wanted). Amount of C genes equals to maximum arity of functions in given function set. If F points to a function, with arity lesser than maximum arity, remaining C genes are simply unused (non-coding genes). O genes denotes output of CGP graph and they simply point to a node, from which the resulting value should be taken. m is the number of output nodes, c is the number of columns in grid and r is the number of rows in grid, all are user-specified. Note, that in classical CGP only feed-forward graphs are allowed and also no connection between nodes in same columns are allowed (since that could lead to graph, which would not use any input). This representation has several interesting consequences:

- genotype has fixed length, which poses an upper limit on the size of graph,
- nodes can occur, which do not participate in resulting graph. These are called inactive nodes (and their respecting genes non-coding). This means, that actual graph tends to be much smaller, than upper limit of graph provided by the genotype length,
- we need to evaluate only active nodes of the graph, leading to faster evaluation and it takes only one backward pass to check which nodes are active and which are inactive,

– mutation can produce an individual, which will have identical phenotype as its parent - in this case, we do not need to evaluate the individual again, since it produces same result and thus same fitness.

3   Related Work

There are various alternations of representation in CGP published in literature. However, to our best knowledge, there is only one which uses real-valued genotype [6], but it was designed with different motivation in mind: to leverage arithmetic crossover. Genotype consists of sequence of real-valued numbers from given interval (typically from 0 to 1), which is split into equally sized bins (according to the number of possible gene values), each representing given integer gene value. When producing a phenotype, real-valued genotype is mapped back to integer-based genotype and then to the phenotype using the same mapping as in classic CGP. Several papers further elaborated this idea. In [5] Estimation of Distribution Algorithm was employed, to dynamically scale the size of the bins of each interval, altering the probability of choosing particular gene value. In [16] a forking operator was designed, which decides, whether an individual should be treated as a point in genotype space, or rather a multivariate Gaussian distribution, with gene values as mean of each distribution, from which genotypes are sampled. Authors of [9] leveraged the statistics of population to maintain healthy diversity in population, by adjusting the probabilities of given genetic operators.

Social programming methodology for creating Complex Adaptive Functional Networks is proposed in [24] and is based on ideas of CGP and discrete PSO. While the representation of CGP individual remains standard, the use of swarm intelligence algorithm for optimizing is unique.

On the contrary, we present the modification of CGP individual representation, which alters the way the phenotype is produced (and also what phenotypes can be produced) and allows us to leverage a whole range of known real-valued optimization algorithms.

4   Continuous Cartesian Genetic Programming

Our idea of Cartesian Genetic Programming using real-valued genotype is based on simple linear combination of neighboring genes. To prevent confusion with existing real-valued CGP representation, we named our approach Continuous CGP (hereafter CCGP). For case with maximum arity of 2 we consider following mapping function for triplet of genes (first one being function gene, second and third input genes) representing single node:

$$f(g_0, g_1, g_2) = (1 - c_0) fn_{l_0} (input_1, input_2) + c_0 fn_{u_0} (input_1, input_2) \quad (2)$$

where:

$$
\begin{aligned}
input_i &= (1 - c_i)x_{l_i} + c_i x_{u_i} \\
c_i &= g_i - \lfloor g_i \rfloor \\
l_i &= \lfloor g_i \rfloor \\
u_i &= \lceil g_i \rceil
\end{aligned}
\quad (3)
$$

fn_i denotes function from function table with index i, x_i denotes node under index i, from which input is taken. c_i are weights for linear combinations of functions and inputs for given function.

Table 1. Example function set

Index	Function
0	$+$
1	$-$
2	$\times$

As an example, consider a function set as shown in Table 1, node with genes $1.3, 2.6, 0.1$ and four nodes before our current node (meaning that our node can take input from nodes with index ranging from 0 to 3). By applying mapping function defined in Eq. 2 we get:

$$f(1.3, 2.6, 0.1) = 0.7 fn_1 (input_1, input_2) + 0.3 fn_2 (input_1, input_2) \qquad (4)$$

and for inputs:

$$\begin{aligned} input_1 &= 0.4x_2 + 0.6x_3 \\ input_2 &= 0.9x_0 + 0.1x_1 \end{aligned} \qquad (5)$$

Combining Eqs. 4 and 5 and replacing fn with actual function symbols from Table 1, we get the resulting expression:

$$f(1.3, 2.6, 0.1) = \qquad (6)$$
$$0.7 \left((0.4x_2 + 0.6x_3) - (0.9x_0 + 0.1x_1) \right)$$
$$+ 0.3 \left((0.4x_2 + 0.6x_3) * (0.9x_0 + 0.1x_1) \right)$$

Important property of this mapping is, that it provides a smooth transition between integer gene values: as value of first (function) gene increases, weights align, so that the upper function (fn_2) gains more influence and lower function less. The same applies for input nodes. Also, we do not lose the original integer-based mapping, so if the genes for the same node would be $1, 3, 0$, by applying the same procedure we get:

$$f(1, 3, 0) = x_3 - x_0 \qquad (7)$$

which is exactly the same as in classic integer-based CGP.

Same principle of linear combination applies to output genes, e.g. if output gene has value 3.4, the actual output would be:

$$f_{og}(3.4) = 0.6x_3 + 0.4x_4 \qquad (8)$$

where f_{og} denotes the mapping function for output genes. This representation provides a transition between given functions and inputs and gives us a way to guide search in hyperspace defined by genotype.

5   Evaluation

To a large extend, we adopted the evaluation framework from PTP paper [20], which uses well-known symbolic regression benchmark problems from [26] and [23]. We used the same benchmark functions with same method of sampling of training dataset. Chosen benchmark functions along with their function set and training data are listed in Table 2. For training data, $U[a, b, c]$ denotes c number of points drawn from uniform distribution ranging from a to b, $E[a, b, c]$ denotes c number of equidistant points ranging from a to b. Note, that in this evaluation we omitted the test sets.

For comparison we set up a classic CGP system as our baseline and for further comparison we consider the results from PTP paper [20].

Table 2. Benchmark functions used in evaluation.

Name	Function	Function set	Training set
Ng.4	$x^6 + x^5 + x^4 + x^3 + x^2 + x^1 + x$	$+, -, *, /, \sin, \cos, \exp, \log$	$U[-1, 1, 20]$
Ng.7	$\log(x + 1) + \log(x^2 + 1)$	$+, -, *, /, \sin, \cos, \exp, \log$	$U[-1, 1, 20]$
Pag.1	$\frac{1}{1+x^{-4}} + \frac{1}{1+y^{-4}}$	$+, -, *, /, \sin, \cos, \exp, \log$	$E[-5, 5, 0.4]$
Keij.6	$\sum_i^x \frac{1}{i}$	$+, *, \sin, \cos, \sqrt{}, \log, {}^{-1}$	$E[1, 50, 1]$
Kor.12	$2 - 2.1\cos(9.8x)\sin(1.3w)$	$+, -, *i, /, \sin, \cos, \tan, \tanh, \sqrt{}, \exp, \log, {}^2, {}^3$	$U[-50, 50, 10000]$
Vlad.12	$\frac{10}{5+\sum_{i=1}^5 (x_i - 3)^2}$	$+, -, *i, /, pow, \sin, \cos, \sqrt{}, \exp, \log, {}^{-1}$	$U[0.05, 6.05, 1024]$

5.1   Experiments Setup

As our baseline, we set up a classic CGP system, using

- 1 row and 50 columns,
- simple evolutionary strategy with $(1 + 4)$ scheme,
- single mutation as presented in [7] (mutate, until single active gene is changed),
- input for a given node can be from at most 20 nodes back,
- maximum number of cost function evaluation is 100 000.

For each problem, we add additional input node, with constant value of 1 (for constants approximation). For further comparison, we picked PTP, which significantly outperformed standard GP [20]. Note, that for comparison we consider only results of PTP after 100 000 cost function evaluations.

Our CCGP method, with novel genotype-phenotype mapping (as described in Sect. 4), uses the same settings as baseline CGP, with exception of simple evolutionary strategy and single mutation. It is optimized using PSO algorithm instead. We used PSO implemented in Pygmo library [3], with its default parameters. No hyperparameter tuning was conducted. Our CCGP implementation using Python 3.6 is available on GitHub[1].

---

[1] https://github.com/Jarino/tengp.

For each benchmark function, CGP and CCGP were run 100 times. Mean squared error was used as an objective function for optimization and also as criterion of evaluation. Note, that we did not execute any runs of PTP and we only use numbers published in [20]. Scripts for experiments presented here are available at GitHub[2].

5.2   Results and Discussion

Performance of methods in terms of simple statistics can be seen in Table 3. For each method there is a best result achieved among all 100 runs and median value of best results of all 100 runs. From Table 3 we can see, that in terms of median, our CCGP method outperformed classic CGP in 5 out of 6 benchmarks (CGP showed better performance at Pagie-1), in 4 of them by order of magnitude. In terms of best result achieved, the results are more ambiguous. The minimal error achieved by our method was lesser in one half of benchmarks (Nguyen-7, Keijzer-6 and Vladislasleva-4) and same in one benchmark (Korns-12)

Distributions of best scores from 100 runs are shown in Fig. 1. Clearly, in 4 out of 6 problems, CCGP system showed more stable performance. To prove statistical significance, we used Wilcoxon signed-rank test to determine, whether samples of results from CGP and CCGP come from same distribution. Obtained p-values are shown in title of each plot.

Comparison with PTP is more ambiguous. In terms of median, we outperformed PTP only on Nguyen-7 benchmark, though medians of Korns-12 and Vladislasleva-4 are in the same order of magnitude. When it comes to best result achieved, we achieved better result only on Vladislasleva-4 benchmark.

On Fig. 2 we can see the average best fitness value during evolution. We can see, that CCGP steadily improved even by the end of evolution, while classic CGP seemed to stop improving in Nguyen-7 and Keijzer-6 benchmarks. This mimics the advantageous behavior of PTP, which also continued to improve the best solution regardless of phase of evolution. On contrary, on Pagie-1 and Korns-12 it seems that CCGP got very early stuck at local optima, from which it

Table 3. Mean squared error of algorithms on benchmarks.

Name	CGP		CCGP		PTP	
	Best	Median	Best	Median	Best	Median
Ng.4	5.674e−6	1.314e−3	3.188e−7	3.947e−4	1.17e−32	6.71e−5
Ng.7	1.811e−6	4.352e−4	7.036e−9	1.519e−6	0	5.24e−6
Pag.1	1.188e−32	4.183e−2	1.763e−2	1.126e−1	7.69e−3	2.19e−2
Kei.6	1.719e−6	8.940e−4	4.048e−8	3.539e−6	6.16e−10	1.47e−7
Kor.12	1.117	1.121	1.117	1.119	1.08	1.11
Vlad.12	5.624e−3	2.551e−2	8.968e−3	2.187e−2	1.44e−2	1.53e−2

[2] https://github.com/Jarino/cgp-optimization.

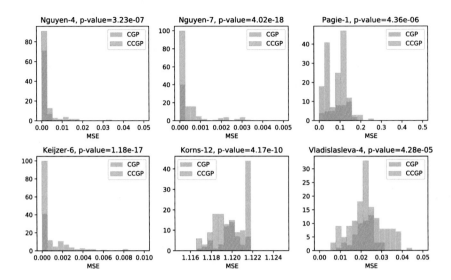

Fig. 1. Distributions of achieved mean squared errors.

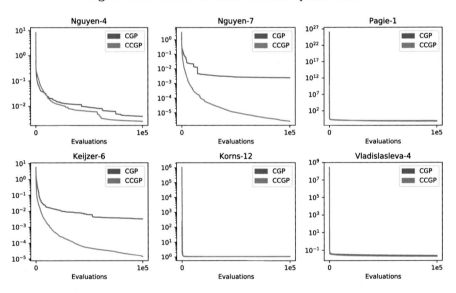

Fig. 2. Evolution of best fitness (MSE) during the run. Value plotted is the average of best fitness at given number of evaluations.

could not escape. Indeed in case of Korns-12, if we were to plot the fitness values of random walk, starting from the optimal solution, we would see, that the MSE quickly increases (by orders of magnitude) and then forms a large plateau. This points to the better ability of overcoming fitness plateaus of classic CGP (also referred to as neutral shift).

6   Conclusion and Future Work

We proposed a novel, real-valued genotype-phenotype mapping for Cartesian Genetic Programming, which allows for continuous transition between functions and inputs of given nodes, aiming to provide a mean to guide search in hyperspace defined by genotype.

We evaluated the proposed method on six benchmark functions, on which it performed significantly better in 4 out of 6 cases, comparing to baseline classic CGP system. Furthermore, we showed that performance is competitive with recent PTP method challenging the genetic programming systems for automatic program synthesis.

Results shown here are preliminary at best: there is a vast space for improvement and new ideas. Here, we outline few directions, which in our opinion could further increase performance:

- our choice of PSO algorithm was nothing more, than sort of educated guess. There is a huge amount of various metaheuristic algorithms for real-valued optimization and there is a possibility, that some other algorithm suits the needs of optimization problem better,
- constant approximation utilized in this paper is the basic used also in classic CGP. Real-valued representation however allows for more precise constant representation, e.g. one can add another gene to each node, representing a constant value, which could be used instead of value computed by inputs (or use it as a simple weights of given node),
- one can imagine, that fitness landscape produced by our system is highly multimodal in high-dimensional space. Genes (in a sense of decision variables) are strongly dependent on each other, forming an optimization problem with high epistasis. Furthermore, ordering of functions in function set could potentially also affect the results - single function might produce error in MSE so high, that the search will simply "not jump" over it. Alternations for the mapping, as well as more problem specific search heuristic could be devised, to target these problems,
- classic CGP holds several advantageous properties, such as low number of active nodes, or ability to skip evaluation of individual, when it has the same phenotype (which is possible thanks to many-to-one mapping between genotype and phenotype). We did not cover this topic in this paper, it would be interesting to see how these properties stand using real-valued genotype-phenotype mapping.

We consider our method as very promising, since achieving really competitive results was matter of relatively simple implementation of relatively simple ideas. We believe, that further research could potentially bring additional improvements.

Acknowledgment. This work was partially supported by the Slovak Research and Development Agency under the contract APVV-16-0213 and by the Operational Programme Research & Innovation, funded by the ERDF, project No. ITMS 26240120039.

References

1. Al-Sahaf, H., Zhang, M., Al-Sahaf, A., Johnston, M.: Keypoints detection and feature extraction: a dynamic genetic programming approach for evolving rotation-invariant texture image descriptors. IEEE Trans. Evol. Comput. **21**(6), 825–844 (2017)
2. Al-Sahaf, H., Zhang, M., Johnston, M.: Binary image classification: a genetic programming approach to the problem of limited training instances. Evol. Comput. **24**(1), 143–182 (2016)
3. Biscani, F., Izzo, D.: esa/pagmo2: pagmo 2.9, August 2018. https://doi.org/10.5281/zenodo.1406840
4. Castejón, F., Carmona, E.J.: Automatic design of analog electronic circuits using grammatical evolution. Appl. Soft Comput. **62**, 1003–1018 (2018)
5. Clegg, J.: Combining cartesian genetic programming with an estimation of distribution algorithm. In: Proceedings of the 10th Annual Conference on Genetic and Evolutionary Computation, GECCO 2008, pp. 1333–1334. ACM, New York (2008). https://doi.org/10.1145/1389095.1389350
6. Clegg, J., Walker, J.A., Miller, J.F.: A new crossover technique for cartesian genetic programming. In: Proceedings of the 9th Annual Conference on Genetic and Evolutionary Computation, GECCO 2007, pp. 1580–1587. ACM, New York (2007). https://doi.org/10.1145/1276958.1277276
7. Goldman, B.W., Punch, W.F.: Analysis of cartesian genetic programming's evolutionary mechanisms. IEEE Trans. Evol. Comput. **19**(3), 359–373 (2015). https://doi.org/10.1109/TEVC.2014.2324539
8. Holland, J.H.: Adaptation in Natural and Artificial Systems: An Introductory Analysis with Applications to Biology, Control, and Artificial Intelligence. MIT press, Cambridge (1992)
9. Kalkreuth, R., Rudolph, G., Krone, J.: Improving convergence in cartesian genetic programming using adaptive crossover, mutation and selection. In: 2015 IEEE Symposium Series on Computational Intelligence, pp. 1415–1422, December 2015. https://doi.org/10.1109/SSCI.2015.201
10. Kalkreuth, R., Rudolph, G., Droschinsky, A.: A new subgraph crossover for cartesian genetic programming. In: McDermott, J., Castelli, M., Sekanina, L., Haasdijk, E., García-Sánchez, P. (eds.) Genetic Programming, pp. 294–310. Springer International Publishing, Cham (2017)
11. Kaufmann, P., Platzner, M.: Advanced techniques for the creation and propagation of modules in cartesian genetic programming. In: Proceedings of the 10th Annual Conference on Genetic and Evolutionary Computation, GECCO 2008, pp. 1219–1226. ACM, New York (2008). https://doi.org/10.1145/1389095.1389334
12. Kennedy, J.: Particle Swarm Optimization, pp. 760–766. Springer, Boston (2010). https://doi.org/10.1007/978-0-387-30164-8_630
13. Khan, M.M., Ahmad, A.M., Khan, G.M., Miller, J.F.: Fast learning neural networks using cartesian genetic programming. Neurocomputing **121**, 274–289 (2013)
14. Koza, J.R.: Genetic programming as a means for programming computers by natural selection. Stat. Comput. 4(2), 87–112 (1994)
15. Lensen, A., Xue, B., Zhang, M.: GPGC: genetic programming for automatic clustering using a flexible non-hyper-spherical graph-based approach. In: Proceedings of the Genetic and Evolutionary Computation Conference, pp. 449–456. ACM (2017)

16. Meier, A., Gonter, M., Kruse, R.: Accelerating convergence in cartesian genetic programming by using a new genetic operator. In: Proceedings of the 15th Annual Conference on Genetic and Evolutionary Computation, GECCO 2013, pp. 981–988. ACM, New York (2013). https://doi.org/10.1145/2463372.2463481

17. Miller, J.F., Thomson, P., Fogarty, T.: Designing electronic circuits using evolutionary algorithms. arithmetic circuits: a case study (1997)

18. Miller, J.F.: An empirical study of the efficiency of learning boolean functions using a cartesian genetic programming approach. In: Proceedings of the 1st Annual Conference on Genetic and Evolutionary Computation - Volume 2, GECCO 1999, pp. 1135–1142. Morgan Kaufmann Publishers Inc., San Francisco (1999)

19. Miller, J.F., Thomson, P.: Cartesian genetic programming. In: Poli, R., Banzhaf, W., Langdon, W.B., Miller, J., Nordin, P., Fogarty, T.C. (eds.) Genetic Programming, pp. 121–132. Springer, Heidelberg (2000)

20. Oesch, C.: P-tree programming. In: 2017 IEEE Symposium Series on Computational Intelligence (SSCI), pp. 1–7, November 2017. https://doi.org/10.1109/SSCI.2017.8280849

21. Papa, J.P., Rosa, G.H., Papa, L.P.: A binary-constrained geometric semantic genetic programming for feature selection purposes. Pattern Recognit. Lett. **100**, 59–66 (2017)

22. Poli, R., McPhee, N.F.: Covariant parsimony pressure in genetic programming. Technical report, Citeseer (2008)

23. Uy, N.Q., Hoai, N.X., O'Neill, M., McKay, R.I., Galván-López, E.: Semantically-based crossover in genetic programming: application to real-valued symbolic regression. Genet. Program. Evolvable Mach. **12**(2), 91–119 (2011). https://doi.org/10.1007/s10710-010-9121-2

24. Voss, M.S.: Social programming using functional swarm optimization. In: Proceedings of the 2003 IEEE Swarm Intelligence Symposium, SIS 2003 (Cat. No.03EX706), pp. 103–109, April 2003. https://doi.org/10.1109/SIS.2003.1202254

25. Walker, J.A., Miller, J.F., Cavill, R.: A multi-chromosome approach to standard and embedded cartesian genetic programming. In: Proceedings of the 8th Annual Conference on Genetic and Evolutionary Computation, GECCO 2006, pp. 903–910. ACM, New York (2006). https://doi.org/10.1145/1143997.1144153

26. White, D.R., McDermott, J., Castelli, M., Manzoni, L., Goldman, B.W., Kronberger, G., Jaśkowski, W., O'Reilly, U.M., Luke, S.: Better GP benchmarks: community survey results and proposals. Genet. Program. Evolvable Mach. **14**(1), 3–29 (2013). https://doi.org/10.1007/s10710-012-9177-2

Detecting Sarcasm in Text

Sakshi Thakur$^{(\boxtimes)}$, Sarbjeet Singh, and Makhan Singh

University Institute of Engineering and Technology, Panjab University,
Chandigarh 160014, India
sakshithakur18@gmail.com,
{sarbjeet, singhmakhan}@pu.ac.in

Abstract. Sarcasm is a nuanced form of speech extensively employed in various online platforms such as social networks, micro-blogs etc. and sarcasm detection refers to predicting whether the text is sarcastic or not. Detecting sarcasm in text is among the major issues facing sentiment analysis. In the last decade, researchers have been working rigorously on sarcasm detection so as to amend the performance of automatic sentiment analysis of data. In this paper, a supervised machine learning (ML) approach, which learns from different categories of features and their combinations, is presented. These feature sets are employed to classify instances as sarcastic and not-sarcastic using different classifiers. In particular, the impact of sarcastic patterns based on POS tags has been investigated and the results show that they are not useful as a feature set for detecting sarcasm as compared to content words and function words. Also, the Naïve Bayes classifier outperforms all other classifiers used.

Keywords: Sentiment analysis · Sarcasm · Feature extraction

1 Introduction

With an advent of the web and explosive outgrowth of social media, there has been an enormous increase in the amount of opinionated content online. Today, anyone with basic Internet access can publish their thoughts and spread their ideas through online platforms. May it be box office predictions [1], business analytics [2], recommender systems [3] or, election results prediction [4], an appropriate understanding of what and how people feel about a particular event is quite valuable. Therefore, there has been a strong demand for tools that can accurately analyze or make predictions from this abundant subjective data produced online.

Sentiment analysis is an ongoing field of research that deals with this natural language text, analyzes public opinions and determines the polarity of the text. Misinterpreting sarcasm in this field formulates a big challenge as it reverses the polarity of the text, which eventually worsens the results of the task performed.

Sarcasm is a form of figurative language which is being employed by people in their day-to-day life for the purpose of being funny, to show anger, to criticize someone or, to avoid giving a clear answer. The Macmillan dictionary defines sarcasm as *"the activity of saying or writing the opposite of what you mean, or of speaking in a way intended to make someone else feel stupid or show them that you are angry"*. In simple

A. Abraham et al. (Eds.): ISDA 2018, AISC 941, pp. 996–1005, 2020.
https://doi.org/10.1007/978-3-030-16660-1_97

words, it can be said that sarcasm is a kind of sentiment where individuals use positive words to express their negative feelings and vice-versa. For example,

"I just love it when I've to stay late at the office for work."

While speaking, this sentence can easily be revealed as sarcastic, but in written form, due to the absence of facial expressions, tonal and gestural cues, it becomes difficult to detect whether the sentence is sarcastic or not. A naïve sentiment analysis system that is not capable of detecting sarcasm would inaccurately classify this tweet as positive which is not accurate as sarcasm has altered the sentiment of the sentence. Therefore, correctly identifying sarcasm in the text is a critical step in predicting the actual sentiment of the text as it can highly affect the overall performance of the system.

In this work, a set of supervised learning experiments for detecting sarcasm in text have been presented. The work draws on the Sarcasm Corpus V2 [5] which is a subset of IAC (Internet Augment Corpus). It is a publicly available corpus of online discussion forum, containing 3260 annotated instances balanced between sarcastic and not-sarcastic. During feature extraction, a feature set consisting of textual and syntactic features and, their different combinations has been considered.

The rest of this paper is arranged as follows: Sect. 2 provides a summarized overview of the work carried out in the field of sarcasm detection. Section 3 describes the proposed approach for sarcasm detection. In Sect. 4, the results obtained from the experiments have been presented and discussed. Finally, this work is concluded in Sect. 5.

2   Related Work

In this section, some state-of-the-art work associated with the field of sarcasm detection has been described.

Tsur et al. [6] presented SASI (Semi-Supervised Algorithm for Sarcasm Identification) to detect sarcasm in product reviews. The algorithm has two stages – (I) Semi-supervised pattern acquisition to identify sarcastic patterns, and (II) Sarcasm classification. They employed two basic feature sets: features based on patterns and features based on punctuation (that included length of sentence in words and number of exclamations '!', capitalized words, question marks '?', quotes in the sentence) and carried out experiments. The proposed approach achieved a precision and F-score of 76.6% and 78.8%, respectively.

Gonzalez et al. [7] proposed an approach to differentiate sarcasm from polar tweets. They employed two lexical features: unigrams; and dictionary-based and three pragmatic features: positive emoticons; negative emoticons; and ToUser (that tells if tweet is a reply to some other tweet). After extracting the features, SVM was employed to classify the tweets as sarcastic, positive and negative, and an accuracy of approximately 75% was achieved.

Riloff et al. [8] presented a bootstrapping algorithm for identifying sarcasm arising from the contrast between a positive sentiment and a negative situation. The bootstrapping process started with a seed word "love" and a set of sarcastic tweets. It then learned negative situation phrases (following a positive sentiment) and positive sentiment phrases (occurring near negative situation phrases). The learning process included: collecting candidate phrases; scoring and then selecting the best candidates. Using this algorithm, an F-score of 51% was achieved.

Liebrecht et al. [9] used Netherland's e-science centre's database containing Dutch tweets and extracted all the tweets with hashtag '#sarcasme', approximately 78,000 tweets. They employed n-grams, specifically 1-, 2- and 3-grams as features and used a balanced Winnow classifier [10] for classifying the tweets.

Justo et al. [11] proposed a set of set of supervised learning approaches to detect sarcasm and nastiness in dialogic language on the web. They extracted a number of feature sets using different criterion: Mechanical Turk Cues, selected by human annotators; Statistical Cues, consisting of n-grams; Linguistic information, which employed part-of-speech (POS) labels; Semantic information; Length information; and Concept and Polarity information. For selecting the most eminent features, Chi-square feature selection method was utilized. For detecting sarcastic or nasty posts, they integrated the above mentioned feature sets to train a Naïve Bayes (NB) classifier and achieved an accuracy of 68.7% and 78.6% respectively. In future, the proposed approach can be improvised by exploring alternative ways of combining relevant knowledge.

Maynard and Greenwood [12] considered the effect of sarcasm contained in hashtags and developed a hashtag tokenizer (an algorithm) to extract individual tokens from the concatenated hashtags. For investigating these hashtags, so as to understand the scope of sarcasm, they developed a set of rules, such as, if the sentiment expressed in the text/tweet is positive and there exists a hashtag containing sarcasm then flip its polarity to negative, and many more. They experimented on a set of general tweets annotated manually and achieved a precision and F-score of 91.03%.

Joshi et al. [13] proposed a rule-based approach for sarcasm detection, taking into consideration an author's historical tweets. The approach they have used comprises three components: Contrast-based predictor, which identifies sarcasm with the help of a sentiment contrast (similar to the one in Riloff et al. [8]); Historical tweet-based predictor, which used the target tweet and author's name to determine whether the opinion conveyed by the tweet is different from the historical sentiment, and; Integrator, that integrates the predictions from both contrast-based and historical tweet-based predictors. They experimented with two lexicons: L1 and L2 that is a list of positive and negative word from [23] and [24], respectively and achieved precision and F-score of 88.0% and 88.2% respectively.

Rajadesingan et al. [14] proposed a behavioral modeling framework named SCUBA (Sarcasm Classification Using a Behavioral modeling Approach) to detect sarcasm on twitter. In this paper, along with analyzing the content of tweets, the behavioral characteristics of the concerned users (based on their past activities) are taken into consideration. They evaluated the performance of SCUBA on a tweets dataset and achieved an accuracy of 83.05%.

Bouazizi and Otsuki [15] used tweets for their work. They proposed a supervised technique which learns sarcastic patterns using various categories of features: features related to sentiments and punctuations, syntactic and semantic features and, features related to patterns. For classification, they employed random forest classifier, SVMs, k-

NN and Maximum Entropy and obtained an F-score of 81.3%, 33.8%, 79.6% and 74.8% respectively.

Mukherjee et al. [16] proposed that for effectively identifying sarcasm, both content words and authorial style play an important role. They employed a number of features: Content word, Function word, POS tags, POS n-gram, and their different combinations. After feature extraction, the Naïve Bayes classification algorithm was applied for classifying the tweets and achieved an accuracy and F-score of 65% and 75% respectively. It was observed that the results improved with an increase in the size of dataset. Therefore, in future, the proposed approach can be further improved by using larger datasets.

Broadly, for supervised approaches, the feature sets that have commonly been used in the literature are: statistical features, syntactic features, pragmatic features and features based on sarcastic patterns. This work employs a supervised learning approach which learns from sarcastic patterns based on POS tags and a combination of content and function words.

3   Methodology

Sarcasm corpus V2 [5], which is a publicly available corpus of online discussion forums, has been used in this work. It consists of 3260 instances balanced between sarcastic and not-sarcastic labels. For training and testing of classifiers, the dataset is randomly split into two sets in the ratio 3:1. In literature, this ratio has been most commonly employed [17].

To avoid the problem of overfitting, 10-fold cross-validation has been performed on the training set. In 10-fold cross-validation, the data is divided into 10 segments, where one of the segments is kept for testing and others for training. The entire algorithm is repeated 10 times such that each segment is kept for testing exactly once. The k-fold cross-validation is performed to avoid the problem of over-fitting and to maximize generalization accuracy [18].

Different features, which are explained in the subsequent section, were extracted from the training set and then tested for performance on the test set using classification methods.

3.1   Feature Extraction

Feature extraction is a technique in which the raw input data is converted into a set of features that represent the data. It forms a key factor in determining the performance of sarcasm detection system. In this work, an extensive list of features has been used for classification purpose. Using more than one feature gives us the chance to analyze results obtained from different features and their combinations. The following categories of features have been used in the proposed system as described below:

POS tags - POS (Part-of-speech) tagging a procedure wherein a word in a corpus is marked with its corresponding part-of-speech based on its relation with neighboring words in an utterance. In the proposed work, POS tags have been used as features for the training set. For example, *absolutely amazing start by heels* gets converted to absolutely/RB, amazing/JJ, start/NN, by/IN, Heels/NNS [19]. Here, RB means absolutely is an adverb, NN means start is a noun, JJ means amazing is an adjective, IN means by is a preposition and NNS means Heels is a noun (plural form).

n-gram POS patterns - In this feature set, part-of-speech tags corresponding to each word in a tweet are used. Broadly, these tags are associated with 8 lexical categories – nouns, pronouns, adjectives, verbs, adverbs, prepositions, conjunctions and interjections. Bigram patterns, trigram patterns & higher-gram patterns were considered and it was found that with higher n-grams, performance doesn't improve. Thus, in this work, a set of POS bigram patterns has been used.

Content words - Content words are the lexical words, having an independent definition i.e. even if it is used outside any sentence, it holds a meaning. Broadly, four classes of content words are there – nouns (words like boy, silver etc.), verbs (words like accept, read etc.), adjectives (words like tall, beautiful etc.) and adverbs (words like easily, slowly etc.) [20].

Function words - Function words are the grammatical words whose purpose is to contribute to the syntax of a sentence or phrase rather than its meaning. These are used to create a structural relation between content words and have little meaning on their own. For example, 'do' in 'I do not like you'. Some of the main classes of function parts-of-speech are pronouns, prepositions, conjunctions, determiners and auxiliaries [20].

Also, various combinations of the above-mentioned features are employed for the purpose of classification.

Not all features are equally important to detect sarcasm. Hence, feature extraction techniques have been employed to find out the relevant ones. In this work, chi-square and extra-trees method has been used for feature selection. Feature selection refers to a process in which the features that contribute the most to the required output are automatically selected. The advantage of using feature selection is that it lessens overfitting because less redundant data means less chances of making decisions on the basis of noise and also, it lessens the computational time required to train the system for classification without altering the performance.

3.2   Classification Method

The above extracted features and the available ground truth are then used to train the model for classifying the text into required categorization i.e. sarcastic and not-sarcastic. On the basis of this training, the classifier assigns label to the text that do not have any label. The classifiers we have employed for evaluating the utility of the above-mentioned features in detecting whether an instance is sarcastic or not are:

Naïve Bayes classifier - Naïve Bayes [21] is a probabilistic model based on the Bayes theorem of conditional probability. Here, the conditional probability is the probability of occurrence of an event that will occur given that another event has already occurred and it is defined as given below:

$$P(Ev_1|Ev_2) = \frac{\{P(Ev_2|Ev_1) \times P(Ev_1)\}}{P(Ev_2)}$$

where $P(Ev_2|Ev_1)$ is the probability of evidence given that the $Event_1$ is true, $P(Ev_1)$ is the probability of $Event_1$ being true, $P(Ev_2)$ is the probability of occurrence of $Event_2$.

Support Vector Machine (SVM) - SVM [22] is a supervised ML algorithm whose main goal is designing a hyper plane that classifies all training vectors in two classes. After this, the hyper-plane which separates two classes very well is used to perform classification. The proposed work makes use of linear kernel as it is faster and performs well for linearly separable data. And also, while training a SVM with linear kernel, only C regularization parameter needs to be optimized, which reduces time.

Random Forest classifier - Random Forest [23] is based on the basic structural principle of Decision Trees but it works by constructing a multitude of Decision Trees at training time and outputting the class that is the mean or mode of the individual trees. To measure the quality of this split, Gini index or information gain is used. Thus in this way it forms an ensemble learning method which works by generating a multitude of decision trees while training. In our work, n_estimators = 100 has been considered.

k-Nearest Neighbor classifier - k-NN [24] is a non-parametric lazy learning algorithm, by this we mean that it does not presume anything on the basis of the training data. It classifies test instances based on computed distances measures to labeled training instances. These distances reveal a set of nearest neighbours (k) which are used to vote on the predicted class. The distance metric that is usually used is Minkowski with p = 2 i.e. the standard Euclidean metric. In the proposed work, a 10-fold cross-validation is being employed to find the most favorable value for k.

The above-mentioned classifiers have extensively been used in the literature, therefore they have been chosen for this work. At the outset of the experiment, it is difficult to say which algorithm will outperform the others. Using multiple algorithms for performing classification is useful in order to select the specific algorithm that maximizes performance. In next section, the performance of the system on the basis of various evaluation metrics is discussed.

4 Experimental Evaluation

The performance of the classification algorithms on the basis of various categories of features was measured by using different evaluation metrics i.e. AUC, F-measure, precision and recall. AUC stands for Area Under the Curve (where curve refers to the ROC curve). More the area covered by the ROC curve, better the classifier is. F-measure is a metric which combines precision and recall through their harmonic mean. It is evaluated using the following formula:

$$F - measure = 2 * \frac{precision * recall}{precision + recall}$$

In this work, precision can be defined as the number of successfully classified sarcastic instances out of total instances being classified as sarcastic and, recall refers to the number of instances correctly classified as sarcastic out of the overall sarcastic instances.

On the basis of the evaluation metrics, it has been found that for detecting sarcasm, combination of content words (CW) and function words (FW) performed better than sarcastic patterns based on POS tags with a precision and AUC of 80% and 81% respectively (see Table 1). However, recall value is higher for sarcastic patterns but for our problem, precision is a more accurate metric for evaluation than recall as precisely identifying the sarcastic instances is of more importance for us. Hence, after evaluating all the metrics, it is found that sarcastic patterns as a feature set does not perform well as compared to others.

Table 1. Comparison of CW + FW and sarcastic patterns based on evaluation metrics used.

		AUC	F1-score	Precision	Recall	Accuracy
Naïve Bayes	CW + FW	**0.81**	0.68	**0.80**	0.57	**0.71**
	Patterns	0.68	0.64	0.64	0.64	0.63
SVM	CW + FW	0.74	0.70	0.66	0.74	0.68
	Patterns	0.69	0.69	0.63	0.78	0.65
Random Forest	CW + FW	0.78	**0.74**	0.68	0.80	**0.71**
	Patterns	0.70	0.65	0.65	0.64	0.64
k-NN	CW + FW	0.68	0.71	0.58	0.92	0.62
	Patterns	0.68	0.72	0.60	**0.91**	0.64

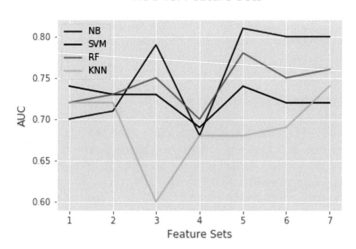

Fig. 1. AUC vs. Feature sets across dataset

In Fig. 1, on x-axis i.e. features, 1, 2, 3, 4, 5, 6 and 7 stands for different features sets namely POS tags, Function Words, Content Words, n-gram POS patterns, Combination-1 i.e. Function Words and Content Words, Combination-2 i.e. Content Words, Function Words and n-gram POS patterns and Combination-3 containing all features, respectively. From Fig. 1, it can be seen that of all the features we have used, CW and FW give the best results for almost all the classifiers.

Figure 2 gives a visualization of the performance of all the classifiers using the finest feature type i.e. a combination of content words and function words. As it is known that more the area under the ROC curve, better the classifier is in terms of performance. From Fig. 2, it can be observed that Naïve Bayes outperforms all the other classifiers with an AUC of 0.81 which shows that the classifier performs well for the sarcasm detection problem.

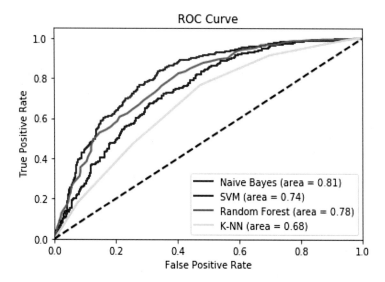

Fig. 2. Performance of all classifiers on the basis of ROC curve

5 Conclusion

The issue of sarcasm is a matter of concern in the field of sentiment analysis. Its presence can completely reverse the meaning of an utterance, so successfully detecting sarcasm is of great importance in this field. In this work, an approach for detecting sarcasm on account of POS-tag based sarcastic patterns has been proposed. But through the experiments, it has been found that they are not as useful as content-words and functions words together.

In future, it is aimed to perk up the performance of the system and reduce the computational time. Also, we hope to work on different feature sets such as pragmatic markers as according to literature, it is believed that they can add-on to the performance of the system when combined with other feature sets.

References

1. Du, J., Xu, H., Huang, X.: Box office prediction based on microblog. Expert Syst. Appl. **41** (4), 1680–1689 (2014)
2. Coussement, K., Van den Poel, D.: Improving customer attrition prediction by integrating emotions from client/company interaction emails and evaluating multiple classifiers. Expert Syst. Appl. **36**(3), 6127–6134 (2009)
3. Li, Y.-M., Shiu, Y.-L.: A diffusion mechanism for social advertising over microblogs. Decis. Support Syst. **54**(1), 9–22 (2012)
4. Ramteke, J., Shah, S., Godhia, D., Shaikh, A.: Election result prediction using Twitter sentiment analysis. In: International Conference on Inventive Computation Technologies (ICICT), vol. 1, pp. 1–5. IEEE (2016)
5. Oraby, S., Harrison, V., Reed, L., Hernandez, E., Riloff, E., Walker, M.: Creating and characterizing a diverse corpus of sarcasm in dialogue. arXiv preprint arXiv:1709.05404 (2017)
6. Tsur, O., Davidov, D., Rappoport, A.: ICWSM-a great catchy name: semi-supervised recognition of sarcastic sentences in online product reviews. In: ICWSM, pp. 162–169 (2010)
7. González-Ibánez, R., Muresan, S., Wacholder, N.: Identifying sarcasm in Twitter: a closer look. In: Proceedings of the 49th Annual Meeting of the Association for Computational Linguistics: Human Language Technologies: Short Papers-Volume 2, pp. 581–586. Association for Computational Linguistics (2011)
8. Riloff, E., Qadir, A., Surve, P., De Silva, L., Gilbert, N., Huang, R.: Sarcasm as contrast between a positive sentiment and negative situation. In: Proceedings of the 2013 Conference on Empirical Methods in Natural Language Processing, pp. 704–714 (2013)
9. Liebrecht, C.C., Kunneman, F.A., van Den Bosch, A.P.J.: The perfect solution for detecting sarcasm in tweets# not (2013)
10. Littlestone, N.: Learning quickly when irrelevant attributes abound: a new linear-threshold algorithm. Mach. Learn. **2**(4), 285–318 (1988)
11. Justo, R., Corcoran, T., Lukin, S.M., Walker, M., Torres, M.I.: Extracting relevant knowledge for the detection of sarcasm and nastiness in the social web. Knowl.-Based Syst. **69**, 124–133 (2014)
12. Maynard, D.G., Greenwood, M.A.: Who cares about sarcastic tweets? Investigating the impact of sarcasm on sentiment analysis. In: LREC 2014 Proceedings. ELRA (2014)
13. Khattri, A., Joshi, A., Bhattacharyya, P., Carman, M.: Your sentiment precedes you: using an author's historical tweets to predict sarcasm. In: Proceedings of the 6th Workshop on Computational Approaches to Subjectivity, Sentiment and Social Media Analysis, pp. 25–30 (2015)
14. Rajadesingan, A., Zafarani, R., Liu, H.: Sarcasm detection on twitter: a behavioral modeling approach. In: Proceedings of the Eighth ACM International Conference on Web Search and Data Mining, pp. 97–106. ACM (2015)
15. Bouazizi, M., Ohtsuki, T.O.: A pattern-based approach for sarcasm detection on Twitter. IEEE Access **4**, 5477–5488 (2016)
16. Mukherjee, S., Bala, P.K.: Sarcasm detection in microblogs using Naïve Bayes and fuzzy clustering. Technol. Soc. **48**, 19–27 (2017)
17. Schürer, S.C., Muskal, S.M.: Kinome-wide activity modeling from diverse public high-quality data sets. J. Chem. Inf. Model. **53**(1), 27–38 (2013)
18. Kohavi, R.: A study of cross-validation and bootstrap for accuracy estimation and model selection. In: IJCAI, vol. 14, no. 2, pp. 1137–1145 (1995)

19. Church, K.W.: A stochastic parts program and noun phrase parser for unrestricted text. In: Proceedings of the Second Conference on Applied Natural Language Processing, pp. 136–143. Association for Computational Linguistics (1988)
20. Biber, D., Johansson, S., Leech, G., Conrad, S., Finegan, E.: Longman grammar of spoken and written English, pp. 89–110 (1999)
21. Murphy, K.P.: Naive Bayes classifiers. University of British Columbia, vol. 18 (2006)
22. Hsu, C.-W., Chang, C.-C., Lin, C.-J.: A practical guide to support vector classification, pp. 1–16 (2003)
23. Breiman, L.: Random forests. Mach. Learn. **45**(1), 5–32 (2001)
24. Islam, M.J., Wu, Q.J., Ahmadi, M., Sid-Ahmed, M.A.: Investigating the performance of naive-bayes classifiers and k-nearest neighbor classifiers. In: International Conference on Convergence Information Technology, pp. 1541–1546. IEEE (2007)

List-Based Task Scheduling Algorithm for Distributed Computing System Using Artificial Intelligence

Akanksha$^{(\boxtimes)}$

Department of Computer Science and Technology,
University Institute of Engineering and Technology (UIET),
Punjab University (PU), Chandigarh, India
akankshasarangal007@gmail.com

Abstract. Job scheduling in DAG (Directed Acyclic Graph) workflow have been a challenging task for the last couple of years. In DAG there is no miner so the time required to search task according to CPU is less. If an appropriate scheduling technique is not selected it may result in an increase in task execution time which may further negatively affect the energy consumption. Energy Prevention is one of the hottest issues in present era which is affecting the global environment. The problem of this research work is to propose a scheduling algorithm in such a manner that the consumption of energy for a DAG G (a, b), on the completion of all jobs is least. In this paper, an energy optimization model with the concept of task scheduling in cloud computing is proposed. List based HEFT (Heterogeneous Earliest Finish Time) algorithm is used to minimize the cost and energy consumption rate. On the basis of total execution time at every processor, the jobs are prioritized. On the basis of job priorities, neural network is trained. The neural network is used to classify the jobs on the basis of energy consumption. The jobs are assigning to the processor that consume less energy. At last the computed parameters such as energy consumption, SLR (Schedule length ratio) and CCR (Computation Cost Ratio) are measured.

Keywords: Task scheduling · HEFT · DAG · Neural network

1 Introduction

With the arrival of distributed heterogeneous systems, like grids and the need to execute complex request for example workflows, issues of selecting a robust schedule is becoming more important [1]. Instead of this situation that depends upon deterministic, static well-known, well-designed scheduling time to execute several tasks for a particular application might be proven to be inefficient [2]. However, the presence of a good schedule is an essential constituent influencing the overall performance of an application. Therefore, to lessen the influence of uncertainties, it is essential to adopt a schedule that ensures robustness, which is, a program that is influenced as short as feasible by numerous run-time changes [3].

A distributed computing system (DCS) consists of a collection of processors that are interlinked with each other through a high-speed network and helps to maintain the

© Springer Nature Switzerland AG 2020
A. Abraham et al. (Eds.): ISDA 2018, AISC 941, pp. 1006–1014, 2020.
https://doi.org/10.1007/978-3-030-16660-1_98

performance of the parallel application. The effectiveness of accomplishing parallel applications on DCSs depends on the procedure which is employed to schedule the tasks of the parallel application onto the possible processors [4]. In DCS the problem of inter-processor communication occurs due to the tasks assigned to various processor exchange data which interrupt the execution of the parallel application [5]. Therefore the need of executing tasks with high quality becomes a critical problem in case of parallel job execution which is executed on DCS using heterogeneous processors. Instead of using parallel job execution the speed is increased but the problem of inter-processor communication overhead exists. In this research work, a HEFT algorithm has been used in which the tasks are executed at different time on different processors [6].

A wrong scheduling decision in heterogeneous DCS might degrade the system's performance by means of the slowest processors. Mainly task scheduling algorithm used for DCS is categorized into two types namely static and dynamic [7]. Static scheduling algorithms: This type of task scheduling algorithm required the information like parallel application model, time of execution of every task along with the cost of communication among the tasks. The static algorithm takes the decision to execute the task before the working of parallel application [8]. Dynamic scheduling algorithm: In this task algorithm the priority to the jobs are assigned during the run time of parallel applications. In the proposed research work, static task scheduling algorithm has been used which is again classified into two types, heuristic and guided random search based. In the heuristic algorithm, an appropriate path to execute the tasks is found out using the heuristic approach and ignoring the other available route. The example of heuristic algorithms is list scheduling, clustering and duplication algorithms. In a guided random approach of arbitrary choices to direct the tasks by itself through the problem space [9].

Scheduling in the cloud is used for selecting the best possible resource for completing the task in less time.

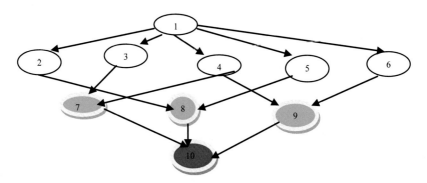

Fig. 1. DAG with Vertices, Edges (10, 14)

Thus for achieving better results the task-scheduling dilemma can be resolved in an efficient manner. In the proposed work workflow scheduling is used in which the tasks are reliant on each other. Dependency means there are priority orders for the tasks, means a task cannot initialize till every of its parent are finished. Workflows are illustrated as DAG. DAG mainly comprises of Vertices and Edges [10].

An example of DAG is shown in Fig. 1. In DAG we are considering mainly two types of costs named as Communication cost and Computation cost. Communication costs apply when the processors change from one to another. The computation costs apply to each node of each processor. The only concern is that the costs should be the lowest. By taking into consideration the cost value, another algorithm named as HEFT is used in this research [11].

HEFT Inspired by HEAG (Heterogeneous Earliest Completion Time) focuses on two aspects, earliest start time (EST) and earliest completion time (EFT). EST is the starting time of the job whereas EFT is the finishing time of the job. The order of job execution is a very important factor in job scheduling. As shown in Fig. 1, consider node 10 for an instance, it has three parent node denoted by 7, 8, 9 nodes. Node 10 cannot be executed until node 7, node 8 and node 9 provide information to node 10 [1].

The main purpose of the research is to propose a scheduling algorithm in such a manner that the consumption of energy for a DAG G(V, E), on the completion of all jobs is least. Thus, the aim of this research work includes the enhancement of HEFT algorithm with a neural network in order to attain minimum consumption of energy and maximum efficiency with minimum SLR [12].

2   Related Work

Arabnejad et al. [13] presented a PEFT (Predict Earliest Finish Time) which is a list based scheduling algorithm used for the heterogeneous computing system. The proposed algorithm has been depending upon an OCT (optimistic cost table) which utilized a rank task for every processor. The proposed algorithm performs well with the minimum quadratic time complexity. Bittencourt et al. [14] proposed an improved HEFT scheduling algorithm that does not consider only an individual task but also considered the children of tasks being assigned by using the concept of lookahead. The concept of using look ahead helps to decrease the schedule makespan. Canon et al. [15] proposed a DAG (directed acyclic graph) scheduling heuristic algorithm to examine the robustness of 20 static makespans for an outsized number of measurements. From the experiment, it has been analyzed that the performance of the static order scheduling algorithm is better than dynamic algorithm. Static algorithms are more robust than a dynamic algorithm and at last, the heuristic algorithms like HEFT perform well for makespan as well as for robustness. Daoud and Kharma [16] presented an LDCP (Longest Dynamic Critical Path) algorithm for heterogeneous DCS along with a restricted number of processors. The LDCP is a "List based scheduling" technique which utilized a novel quality to choose task effectively for scheduling in DCS. Authors have provided the comparison between HEFT and DLS algorithm in terms of SLA and speed. From the experiment, it has been concluded that the LDCP algorithms performs well for the high communication costs in DCS. Zheng and Sakellariou [17] presented a BHEFT cost optimization algorithm based on the "DAG scheduling heuristic HEFT" algorithm. The proposed technique perform better in case of scalability and execution time cost. Munir et al. [18] proposed a SD based task scheduling algorithm named as SDBATS. This algorithm utilized the standard deviation of the expected execution time of a particular task on the obtainable resources in the heterogeneous computing situation in order to allocate task as per priority. Comparison of the proposed algorithm with different scheduling algorithms such as HEFT, PETS, CPOP and DLS has been examined. The proposed algorithm performs well in case of schedule length and speed.

3 Methodology

The steps that are used to execute the proposed work in MATLAB software are described below (Fig. 2):

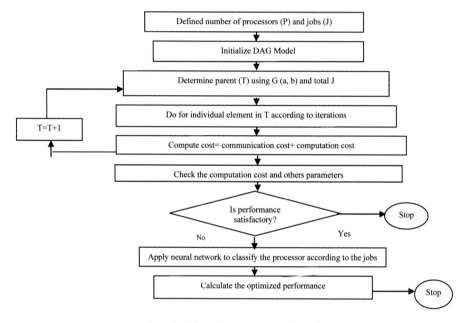

Fig. 2. Flowchart of proposed work

Step 1. Initialize a number of processors along with the number of jobs to simulate the proposed list-based task scheduling algorithm for distributed computing system using artificial intelligence.
Step 2. Defined DAG model along with the vertices a and edges b.
Step 3. Determine the parent and child node and calculate the communication and computation cost for each node which are participating in communication for task scheduling procedure.

Step 4. The process is repeated until the desired cost is achieved. This is possible by comparing the communication cost of each and every node.

Step 5. If the performance of the system is degraded then apply neural network in order to determine an appropriate task according to training of network which helps to find out better task according to processor.

Step 6. Calculate the performance matrices such as SLR, CCR and energy consumption to validate the proposed work and compare the results with existing work that used PHEFT as a task scheduling algorithm.

3.1   Artificial Neural Network

In this research work, the ANN is used to classify the processors as per the jobs. The working of ANN is described below. The Fig. 3 defines the framework of ANN that comprises of four panels named as neural network, algorithm, progress and plots. In a Neural network, three layers are used named as input layer, hidden layer and output layer. The hidden layer consists of 10 neurons from which 10 neurons appear at the output layer.

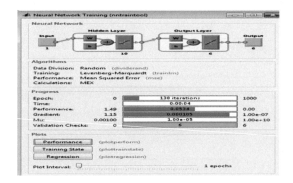

Fig. 3. Structure of ANN

The training of the neural network is completed for 138 iterations. From the above graph, it is observed that there are four lines represented by different colors such as blue, green, red and dotted line and each line signifies the training, validation, test value and best value of ANN respectively. The best value is achieved at 138th iteration with MSE less than 0.1. Lower the value of MSE means better is the training of the neural network. Neural network consists of three inputs namely Training data, group and neurons. On the basis of category NN provide a trained structure on the basis of which we can classify the jobs. The performance of the system trained by NN depends upon the performance parameters such as mutation, MSE, validation and gradient.

Algorithm: Neural network
Input: Training data, group, neurons, number of processors, jobs
Output: Classify the processor as per jobs
Initialize ANN with   -Training data=processors
                             -group=possible jobs
                             -Training algorithm=LM
                             -Neurons=10
-Transfer function=T sigmoid
-Iteration=138
Set performance parameters-MSE
                             -Gradient
                             -Mutation
                             -validation choice
Generate a structure of ANN
Net=newff (training data, group, neurons, processors, jobs)
Net=Train (net, training data, group, processors, jobs)
Count of jobs=sim (net, jobs)
If classify as processor
Consider as job
Else
Not consider as a job
End if
Return; classified jobs
End (function)

4   Experiment Results

The experiment has been performed in MATLAB simulator tool for 'n' number of jobs, single entry points, three processors and 40 exit points. The performance parameters that are observed are shown in the Fig. 4:

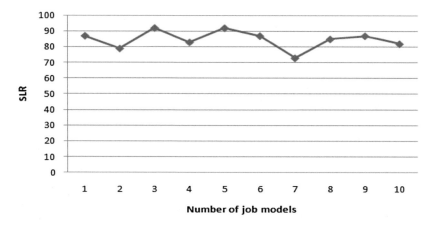

Fig. 4. SLR vs number of jobs

The above figure represents the SLR values obtained the 10 number of job models. From the above graph, it is concluded that as the count of job models increases the value of SLR also increases.

The Fig. 5 represents the energy consumption ratio vs numbers of job models. The average value obtained for the energy consumption ratio is approximately is 0.606.

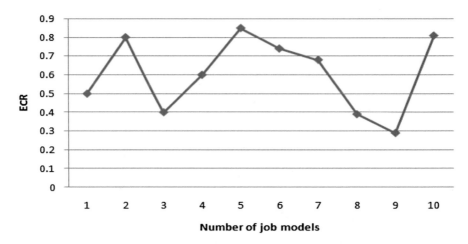

Fig. 5. Energy consumption

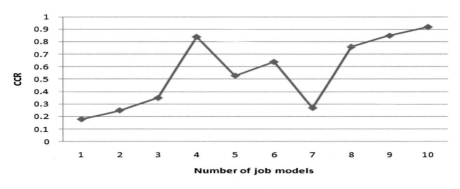

Fig. 6. CCR

The above Fig. 6 represents the computation cost ratio w r t the number of job model. The x-axis represents the node model whereas the y-axis represents the computation Cost ratio. The average value obtained for the CCR is approximately 0.559.

5   Comparison of Proposed Work with Existing Work

In this section, the comparison of SLR has been discussed with the proposed work.

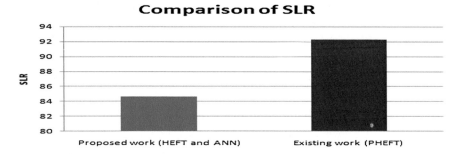

Fig. 7. Comparison of proposed work with existing work

In the existing work, the task scheduling algorithm PHEFT has been used to determine the SLR value. In the proposed work, to increase the efficiency of the proposed work we are using HEFT algorithm along with neural network.

The Fig. 7 represents the comparison between the proposed work and existing work which is compared on the basis of the SLR parameter. Red bar and blue bar represents the average values of existing work and proposed work respectively. From the above figure, it is clear that the value of SLR decreases when the HEFT algorithm along with the neural network has been used. The SLR value is reduced by 8.26% as compared to the PHEFT algorithm used by author Arabnejad and Barbosa [13].

6 Conclusion

In this paper, a novel list aware scheduling algorithm "HEFT" along with neural network as a classification algorithm has been proposed. The proposed algorithm enhances the task scheduling as compared to the existing PHEFT algorithm. The experiment has been performed on 10 numbers of jobs models each comprises 10 jobs and three processors DAG graph which is comprised of V vertex and E edges is taken to find the parent processor. In this research work, we presented two algorithms named as HEFT algorithm and HEFT algorithm with a neural network which is used for scheduling the parallel applications into the system of heterogeneous processors. The proposed algorithm decreases the Scheduling length ratio, energy consumption, as well as the CCR.

References

1. Hu, F., Quan, X., Lu, C.: A schedule method for parallel applications on heterogeneous distributed systems with energy consumption constraint. In: Proceedings of the 3rd International Conference on Multimedia Systems and Signal Processing, pp. 134–141. ACM, April 2018
2. Zhou, N., Li, F., Xu, K., Qi, D.: Concurrent workflow budget-and deadline-constrained scheduling in heterogeneous distributed environments. Soft. Comput. **22**, 1–14 (2018)

3. Aba, M.A., Zaourar, L., Munier, A.: An approximation algorithm for scheduling applications on hybrid multi-core machines with communications delays. In: 2018 IEEE International Parallel and Distributed Processing Symposium Workshops (IPDPSW), pp. 36–45. IEEE, May 2018

4. He, K., Meng, X., Pan, Z., Yuan, L., Zhou, P.: A novel task-duplication based DAG scheduling algorithm for heterogeneous environments. IEEE Trans. Parallel Distrib. Syst. **30**, 2–14 (2018)

5. Maurya, A.K., Tripathi, A.K.: On benchmarking task scheduling algorithms for heterogeneous computing systems. J. Supercomput. 1–32 (2018)

6. Sukhoroslov, O., Nazarenko, A., Aleksandrov, R.: An experimental study of scheduling algorithms for many-task applications. J. Supercomput. 1–15 (2018)

7. Chen, Y., Xie, G., Li, R.: Reducing energy consumption with cost budget using available budget preassignment in heterogeneous cloud computing systems. IEEE Access **6**, 20572–20583 (2018)

8. Padole, M., Shah, A.: Comparative study of scheduling algorithms in heterogeneous distributed computing systems. In: Advanced Computing and Communication Technologies, pp. 111–122. Springer, Singapore (2018)

9. Qin, L., Ouyang, F., Xiong, G.: Dependent task scheduling algorithm in distributed system. In: 2018 4th International Conference on Computer and Technology Applications (ICCTA). IEEE, May 2018

10. Marrakchi, S., Jemni, M.: A parallel scheduling algorithm to solve triangular band systems on multicore machine. Parallel Comput. Everywhere **32**, 127 (2018)

11. AlEbrahim, S., Ahmad, I.: Task scheduling for heterogeneous computing systems. J. Supercomput. **73**(6), 2313–2338 (2017)

12. Zhou, N., Qi, D., Wang, X., Zheng, Z.: A static task scheduling algorithm for heterogeneous systems based on merging tasks and critical tasks. J. Comput. Methods Sci. Eng. (Preprint), pp. 1–18 (2017)

13. Arabnejad, H., Barbosa, J.G.: List scheduling algorithm for heterogeneous systems by an optimistic cost table. IEEE Trans. Parallel Distrib. Syst. **25**(3), 682–694 (2014)

14. Bittencourt, L.F., Sakellariou, R., Madeira, E.R.: Dag scheduling using a lookahead variant of the heterogeneous earliest finish time algorithm. In: 2010 18th Euromicro International Conference on Parallel, Distributed and Network-Based Processing (PDP), pp. 27–34. IEEE, February 2010

15. Canon, L.C., Jeannot, E., Sakellariou, R., Zheng, W.: Comparative evaluation of the robustness of dag scheduling heuristics. In: Grid Computing, pp. 73–84. Springer, Boston (2008)

16. Daoud, M.I., Kharma, N.: A high performance algorithm for static task scheduling in heterogeneous distributed computing systems. J. Parallel Distrib. Comput. **68**(4), 399–409 (2008)

17. Zheng, W., Sakellariou, R.: Budget-deadline constrained workflow planning for admission control. J. Grid Comput. **11**(4), 633–651 (2013)

18. Munir, E.U., Mohsin, S., Hussain, A., Nisar, M.W., Ali, S.: SDBATS: a novel algorithm for task scheduling in heterogeneous computing systems. In: 2013 IEEE 27th International Parallel and Distributed Processing Symposium Workshops & Ph.D. Forum (IPDPSW), pp. 43–53. IEEE, May 2013

Location-Allocation Problem: A Methodology with VNS Metaheuristic

M. Beatriz Bernábe-Loranca[1]([⊠]), Martin Estrada-Analco[1],
Rogelio González-Velázquez[1], Gerardo Martíne-Guzman[1],
and Ruiz-Vanoye[2]

[1] Facultad de Ciencias de la Computación,
Benemérita Universidad Autónoma de Puebla, Puebla, Puebla, Mexico
beatriz.bernabe@gmail.com
[2] Universidad Politécnica de Pachuca, Zempoala, Mexico

Abstract. In this work, we present beginnings of a methodology that allows the establishment of relationships between the location of the facilities and the clients' allocation with a dense demand. The use of this application lets us know the optimal location of production facilities, warehouses or distribution centers in a geographical space. We also solve the customers' dense demand for goods or services; this is, finding the proper location of the facilities in a populated geographic territory, where the population has a demand for services in a constant basis. Finding the location means obtaining the decimal geographical coordinates where the facility should be located, such that the transportation of products or services costs the least. The implications and practical benefits of the results of this work have allowed an enterprise to design an efficient logistics plan in benefit of its supply chain. Firstly, the territory must be partitioned by a heuristic method, due do the combinatory nature of the partitioning. After this process, the best partition is selected with the application of factorial experiment design and the surface response methodology. Once the territory has been partitioned into k zones, where the center of each zone is the distribution center, we apply the continuous dense demand function and solve the location-allocation problem for an area where the population has a dense demand for services.

Keywords: Dense demand · Location-allocation · Methodology ·
Response surface

1 Introduction

In a broad sense, it is understood that logistics is a part of the supply chain, and, in general the logistical networks in the supply chain are a system that manage the merchandise network and its physical flow among the members of the supply chain, influenced by the territorial distribution and by the transportation systems, to reduce the logistical expenses and coordinate the production-distribution activities. The supply chain may be regarded as a set of enterprises: providers, manufacturers, distributors and sellers (wholesale or retail), efficiently coordinated by means of collaborative relationships between their key procedures in order to place the inputs or product

© Springer Nature Switzerland AG 2020
A. Abraham et al. (Eds.): ISDA 2018, AISC 941, pp. 1015–1024, 2020.
https://doi.org/10.1007/978-3-030-16660-1_99

requirements for each link of the chain at the right time and at the lowest cost, looking for the biggest impact on the value chains of the members, in order to satisfy the final consumers' requirements.

The goal of this work is presenting a set of steps towards a methodology to support the strategic decision-making process of an enterprise, to locate the facilities when planning a logistical network. We have made emphasized the case of georeferenced zones and dense demands, which is a territorial design problem combined with a location-allocation problem.

The methodology aims at finding the proper location of the facilities in a populated geographic territory, where the population continuously demands services. Obtaining the locations means finding the longitude and latitude coordinates of each location point in such a way that the transportation of products and services has a minimum cost.

The problem implies solving the territorial design partition. This begins with the selection of a partitioning method, according to the results of an experiment factor and response surface (design of statistical experiments). Then the Weber function that has a demand function multiplied by the Euclidean distances as weights is minimized. The demand function represents the population's demand in every territory, whereas the Euclidean distance is calculated between the potential location points and the demand points.

2 Statistical Methodology for Partitioning Design

The P-median problem considers a territory as made up by geographic zones. The objective is finding N distribution centers to take care of the nearest customers by minimizing the distances between centroids (the distribution centers), and other geographic objects (customers). This geographic partitioning problem is of high computational complexity, so a metaheuristic method has to be included, and in this case Variable Neighborhood Search (VNS) was chosen, and it is here that the methodology here proposed applies: the establishment of a series of steps and surface responses to calibrate a metaheuristic's parameters and reach a parameter combination that leads to the optimum. Lastly, the dense demand problem is set up with an additional methodology that is established in Sect. 3.

The proposed methodology suggests partitioning the territory using a territorial design method that generates compact groups or clusters [1], which have to be obtained first in order to solve the Location-Allocation Problem (LAP) for a Territorial Design Problem (TDP) with dense demand. The Distribution Centers (DC) will provide services to a group of communities that are found in every geographic area, which will be represented by each area's centroid. The location should be the one that minimizes the travelling expenses by finding the geographical coordinates of the center of these centroids. The populations from these communities represent the potential clients of the DC, and the demand is modeled with continuous demand functions with two variables based on the population density of every group [2].

Due to the numeric nature of the solutions obtained, this problem addresses a continuous case of the LAP. Additionally with the associated mathematical approach, we use a geographical information system (GIS) to create maps of the designed territories [3].

There have been many efforts to solve problems related to the location-allocation of services, and the state of the art for it deserves a work of its own [1–5]. Our contribution lies on presenting a methodology that begins with a territory partitioning process that employs a P-median method, and partitioning restrictions to find p-centers by incorporating a metaheuristic [2]. Due to the fact that the population's demand for services is dense, once the distribution centers have been determined, their geographical location (x = longitude, y = latitude) in R2 must be found such that the services or products transportation has a minimum cost, using the Weber function.

For example, let's assume that we wish to know where a healthcare center is to be located (assuming as well that all the associated conditions are met). Then for this particular application we could locate clinics at the centroids of every geographic unit and at the center-most point of all the communities, locating a general hospital as a DC such that transferring patients requires a minimum amount of time.

First we have obtained the territory partition, where the cluster formation is based on the geometric compactness of territorial design and minimum distances between centroids [4], that is, the first part of the methodology consists of the selection of the partition and the heuristic method to continue the statistical experiment (Design of Experiments) which will indicate the number of suitable partitions. Design of Experiments allows the analysis of data using statistical models in order to observe the interaction between the independent variables and their effect on the dependent variable. This requires that experiments be replicated, while data is randomized between replications. Replication yields an estimate of experimental error, and it is expected that the higher the number of replicas is, the experimental error should be lower when all the experiments are given in the same initial conditions. Randomization during the experiment is essential to avoid dependence between samples and ensures results are actually caused by the dependent variables and not by the experimenter.

It is in this scenario that we present a way to develop computational experiments in order to obtain tests in pursuit of the optimum that, in our experience produces good results. Good experimental design is a must for the methodology of experiments to work. An experiment may be performed to: (a) develop the main causes of response variation, (b) find the experimental conditions that yield an extreme value in the variable of interest or response, (c) compare the responses on different observation levels for the controlled variables.

The use of experimental design models is based upon the experimentation and analysis of the results coming from a well-planned experiment. It is seldom possible to use these methods for available or historical data, and that is one of the first errors: assuming data to be correct. One of the main objectives for statistical models, and factorial experiment design models in particular, is the control of variability in a random process, which may have different sources [6].

Experimentation is a natural component of most scientific and engineering research; the results being influenced by different factors whose relevance may be hidden by the very variability in sample results. Therefore, it is necessary to know which factors are actually affecting results, and to estimate said influence. In order to do that, it is recommended that experiments be performed, varying the conditions that affect the experimental units, and observing the response variable. From the analysis and study of the gathered information conclusions are derived [7].

Experimentation has traditionally been used studying factors one by one, varying a chosen factor while the rest remain fixed, with the weakness of manual error. Stages that must be followed to properly plan an experiment have been devised, and must be performed in sequence (given that some basic concepts in the study of models for the design of experiments are known):

1. Defining the experimental objective.
2. Identifying all possible sources of variation.
3. Choosing a rule to assign the experimental units to the conditions of the study (treatments).
4. Specification of the measurements to be performed (response), the experimental procedure and anticipation of possible difficulties.
5. Execution of a pilot experiment.
6. Specification of the model.
7. Schematization of the statistical analysis' steps.
8. Determination of sample size.
9. Review of the mentioned decisions before.

In planning an experiment, three main principles exist: randomization principle, blocking and design factorization.

Lastly, we understand that experimental design is a rule that determines the assignment of experimental units to treatments, although experiments differ greatly in many other aspects. Some of the most widespread are: completely randomized design, block design or factored block, designs with two or more factor blocks, design with nested or hierarchical blocks, designs with two or more factors, two-level factorial designs, etc. [6].

In our case, given the problem conditions, the designs used are Box Bhenken and response surfaces, besides compound central design [6].

The first part of the methodology is shown. (Selection of the partition using design of experiments):

1. Select the partition method
2. Select the candidates of heuristic methods
   2.1 Develop a design of experiment to select the best heuristic method (the one that reaches the best cost function).
       2.1.1. Estimate effects of the factors (heuristic's parameters)
       2.1.2. Form an initial model
       2.1.3. Develop test statistics
       2.1.4. Redesign the model
       2.1.5. Analyze the residuals
       2.1.6. Interpretation of results
3. Determination of the parameters in the selected metaheuristics by identifying the number of partitions (groups)
   3.1. Select the initial model (Box Bhenken, Central Composite, etc.)
   3.2. Development of experimental tests

    3.3. Analysis of the regression model to verify the existence of statistical evidence
       for the reliability of the experiment.
       3.3.1. Verification of the experimental model
       3.3.2. Validation of the parameters, and
4. Selection of the partition that best optimizes the cost function.

2.1 Experiment Design for the Variable Neighborhood Search VNS Metaheuristic for the P-Median Problem

This section briefly presents the use of the experimental design methodology for the P-median problem using VNS to obtain approximate solutions in order to evaluate the quality of these generated solutions [8]. Central Composite and Response Surface have been used to find a balance and adequacy of the parameter values in obtaining good solutions [6].

A. Defining the experimental objectives: The P-median problem is a high computational complexity problem that requires approximate methods. VNS has been chosen as a metaheuristic method. Given the P-median specifications, P = 8 (8 medians to be understood as 8 groups) was chosen [1, 4]. The objective is to apply a factorial analysis to calibrate the VNS parameters that ensure the best performance: response time must considerably reduce computing cost, and the quality of solutions has to be very close to the optimum.

B. Identifying all possible sources of variation, in this case, two parameters define VNS: Local Search (LS) and Neighborhood Structure (NS) [1, 4]. An instance of 463 objects with 8 blocks as a fixed parameter was determined [3].

C. Choosing a rule to assign experimental units to the conditions of study (treatments): Considering the number of parameters Local Search, Neighborhood Structure (LS, NS) a central composite design and response surfaces were chosen.

D. Specification of the measurements to be performed (response), the experimental procedure and anticipation of possible difficulties: Values are chosen for the parameters and the response variable, which is the target function. Experimental levels for LS are between [1031 – 1370], and for NS are between [1718 – 1365] (from previous experiments) [1, 3–5]. The execution platform and the developed program to solve P-median with VNS are also specified.

E. Execution of a pilot experiment: A set of tests was developed from Central Composite Design, whereby 14 parameter combinations were obtained, and the cost function with the respective computing time was registered.

F. Model specification: Once the tests are run, it is verified that the data behave according to a normal distribution, that there are no effects from one test to another and that the second order regression model is adequate.

G. Schematization of statistical analysis' steps: All statistical calculations, variable correlation tests, analysis of variance and hypotheses tests are performed.

H. Determination of sample size: From the prior results it is possible to modify simple size if good results are not obtained.

3 Demand Dense Methodology

Diverse territorial design applications are very useful to solve location problems for services and sales points [2, 9]. For this work, the logistical network design problem with dense demand over a geographical region implies defining (finding) market or services areas, this is an application of Territorial Design (TD).

The partition of one territory into zones generates k zones; this is understood of course as geographical zones grouping. Then, from a logistical point of view, we have available zones to locate facilities that provide services to satisfy the clients' demand. We have chosen a partition of 5 zones for our case study with the goal of allocating Demand Density Functions that we'll denote as DDF [2, 9], where DL1,..,DL6 are linear functions and as DNL1 as DNL2 are non-linear functions. These are shown in the following table:

Table 1. Demand Density Functions (DDF).

DDF	$D(x, y)$
DL1	$100 + 7.5x + 7.5y$
DL2	$100 + 10x + 5y$
DL3	$100 + (100/7)x + (5/7)y$
DL4	$600 + (10/3)x + (5/3)y$
DL5	$600 + 2.5x + 2.5y$
DL6	$100 + (100/21)x + (5/21)y$
DNL1	$100 + (9/80)x^2 + (9/80)y^2$
DNL2	$100 + (3/1.6 \times 10^5)x^4 + (1.6 \times 10^5)y^4$

A well-known problem in integral calculus (Stewart 1999) is the determination of the center of mass of a sheet (such as a metallic plate) with density function $\rho(x, y)$ that occupies a certain region (see Fig. 1).

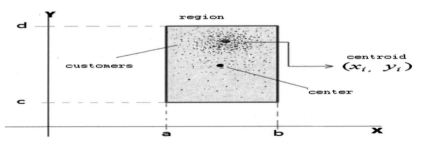

Fig. 1. Center and centroid of a rectangular region in R^2 based on the density of the points.

It's possible to assume that a geographical region in the plane is somewhat like a sheet and its center of mass or centroid may be obtained using Eq. (1), where the density function is a population density function. Said point has geographical coordinates for rectangular regions, but in this project we are applying it to irregular regions. The centroids can be interpreted as location points for facilities because the point is located in a place with high population density [2].

$$x = \frac{\iint x\rho(x,y)dxdy}{\iint \rho(x,y)} \quad ; \; y = \frac{\iint y\rho(x,y)dxdy}{\iint \rho(x,y)} \tag{1}$$

The formulas to obtain the longitude and latitude coordinates of the location point of every geographical unit require the use of double integrals over a region R, with density functions depending on two variables $\rho(x,y)\rho(x,y)$, where R is a geographical region and the density function can be any of the functions from Table 1. The integration limits of every geographical unit have to be obtained by means of a geographical information system (GIS) [2, 9]. The Romberg method has been chosen for double integration. The support software we've employed is free and is known as X numbers, which is an Excel add-on. In this way the coordinates of every centroid for every geographical unit, of territory is obtained.

From a logistical view, each territory that has a population possesses a demand for goods and services that differ per zones due to multiple factors. Let's assume that the demand can be modeled by a DDF that associates a demand volume for a certain service to every geographic point. The demand density term is associated to the population density in the zone. Each of the five groups of geographical unit will be associated with a DDF from Table 1 to integrate them to the location model that is described below as a minimization problem.

Let's consider that a centroid of a geographical unit is a point with geographical coordinates whose location depends on the density of population; we can say that $c_j = (x_j, y_j)$ is representative of the geographical unit.

The solution consists in finding the coordinates $(x,y)(x,y)$ of the point $q \in Gi$ $q \in Gi$ such that the transportation cost from each community to the central facility is minimized.

The mathematical model that represents the conditions mentioned above is written in the following way:

$$\underset{(x,y)\,\in\,G_i}{Minimize} \quad \sum_{j=1}^{|Gi|}|D(x,y)|\sqrt{\left(x - x_j\right)^2 + \left(y - y_j\right)^2} \tag{2}$$

The objective function represented in (2) is the total transportation cost TC, known as the Weber function.

From what we have said, the final model and contribution of our problem is expressed as follows:

Given a set of customers distributed within a territory $T \subseteq R^2$ and $P = \{G_1, G_2, \ldots, G_k, \ldots G_p\}$ a partition of T into p clusters, each G_k is a cluster of geographical units for $k = 1, 2, \ldots, p$. Each geographical object has a representative point called centroid $c_j = (x_j, y_j,)$ from which each community is served. Each point $q = (x,y) \in T$, has a density of demand given by $D(q) = D(x,y)$. Let $d(q, c_j)$, be the Euclidean distance from any point to the centroid. The cost of transportation from a point q to the centroid cj is defined as $D(q)d(q, c_j)$.

The solution consists in finding the coordinates from a point $(x, y) \in G_i$ such that the transportation cost is minimized from each community to a central facility (as in Eq. 2). The mathematical model that represents the mentioned conditions is:

$$\underset{(x, y) \in G_i}{\text{Minimize}} \quad Z = \sum_{j=1}^{Gi} |D(x, y)| \sqrt{(x - x_j)^2 + (y - y_j)^2} \tag{3}$$

$$\text{Subject to} \quad \bigcup_{i=1}^{p} G_i = T \tag{4}$$

$$\text{and} \quad \bigcup_{j=1}^{|G_i|} A_j = G_i \quad \forall\, i = 1, 2, \ldots, p \tag{5}$$

The objective function Z represented in (3) is the total transportation cost and (4) and (5), are the constraints of the partitioning of the territory T and the sub territories Gi. The sequence of necessary steps to obtain the coordinates of the central facility in a cluster is as follows:

1. Define the parameters to partition the territory T.
2. Generate the partition with the VNS metaheuristic.
3. With the file obtained generate a map inside a map with a GIS.
4. Associate with the chosen cluster $P_i, i = 1, 2, \ldots, p$ $P_i, i = 1, 2, \ldots, p$, a demand density function $D(x, y).D(x, y)$.
5. Calculate the centroids of each geographical objet, using (4).
6. Apply (3) for the chosen cluster.

$$x_i = \frac{\int x \rho(x, y) dx dy}{\int \rho(x, y) dx dy} \quad \text{and} \quad y_i = \frac{\int y \rho(x, y) dx dy}{\int \rho(x, y) dx dy} \tag{6}$$

Equation (6) were rewritten to have an order in the methodology. This equation is also (1) and the equations above are the classical formulas of calculus used to calculate the centroid of a metallic plate with density ρ. In this paper we take the GU as metallic plates and a population density given by $\rho(x, y)$.

4 Results

We have applied this methodology to a territory and we can establish that the solutions obtained for the LAP model for TDP with dense demand are consistent with the geographic location of the region. We obtained the coordinates of eight possible location points, depending on the demand density function, which are associated to the geographical units belonging to the cluster under study, selected from the partition of the territory. Figure 2 shows the location of nine points. Therefore, we can select any point as the location point of the cluster; we can also state that the demand density function does not influence the location of the centers of the centroids. The ninth point, with coordinates $P_c = (x_{cc}, y_{cc})$ is calculated as the point whose distance to any center is the minimum, and we can consider it as the "center of centers" and as the best point of location for the entire cluster of geographical units, without losing generality. In a more general way, the location point can be any point within the circle $(x - x_{cc})^2 + (y - y_{cc})^2 = r^2$, where $r = max\{d(P_c, c_j) | j = 1, 2, 3, \ldots, n\}$. This article's proposal provides a structure to solve location allocation models based on geographic information systems as presented in any study case.

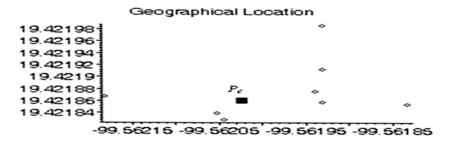

Fig. 2. Location that minimizes the objective function

The methodology was tested in demand regions of irregular shape in comparison with previous papers where the regions are rectangular or convex polygons. According to the analysis of the obtained results, the inclusion of the territorial design aspects with the use of density functions in location-allocation models gives a greater range of possible applications to real problems, for example in the design of a supply chain, among others. The integration of diverse tools such as metaheuristics, geographic information systems and mathematical models provides a strong methodology in visual environments such as maps. Another contribution of this paper is the consideration of three relevant aspects: territorial design, location-allocation, and dense demand.

In general, the proposal here presented contributes to the decision-making process in logistical problems when the population's demand is implicit. As a case study we chose a metropolitan area, however an important advantage of our methodology lies on the fact that it can handle other kinds of geographical data such as blocks, districts or states.

References

1. Bernábe-Loranca, M.B., González, R.: An approximation method for the P-median problem: a bioinspired tabu search and variable neighborhood search partitioning approach. Int. J. Hybrid Intell. Syst. **13**(2), 87–98 (2016)
2. Bernábe-Loranca, M.B., Gonzalez, R., Martínez, J.L., Olivarez, E.: A location allocation model for a territorial design problem with dense demand. Int. J. Appl. Logist. (IJAL) **6**(1), 1–14 (2016)

3. Bernábe-Loranca, M.B., González-Velázquez, R., Estrada-Analco, M., Bustillo-Díaz, M., Martínez-Gúzman, G., Sánchez-López, A.: Experiment design for the location-allocation problem. Appl. Math. **5**(14), 2168–2183 (2014)
4. Bernábe-Loranca, M.B., Rodríguez-F, M.A., González-Velázquez, R., Estrada-Analco, M.: Una propuesta Bioinspirada basada en vecindades para particionamiento. Revista Matemática, Teoría y Aplicaciones **23**(1), 221–239 (2016)
5. Bernábe-Loranca, M.B., González-Velázquez, M., Estrada-Analco, M., Ruíz-Vanoye, J., Fuentes-Penna, A., Sánchez, A.: Bioinspired tabu search for geographic partitioning. In: Advances in Intelligent Systems and Computing, Proceedings of the 7th World Congress on Nature and Biologically Inspired Computing (NaBIC 2015), pp. 189–200, Pietermaritzburg, South Africa (2015)
6. Montgomery, D.: Design and Analysis of Experiments, 2nd edn. Wiley, Hoboken (1991)
7. Barr, R.S., Golden, B.L., Kelly, J., Stewart, W.R., Resende, M.G.C.: Guidelines for designing and reporting on computational experiments with heuristic methods (2001)
8. Hansen, P., Mladenovic, N., Moreno, P.J.: A variable neighborhood search: methods and applications. Ann. Oper. Res. **175**, 367–407 (2010)
9. Love, R.F.: A computational procedure for optimally locating a facility with respect to several rectangular regions. J. Reg. Sci. **18**(2), 233–242 (1972)

Artificial Neural Networks: The Missing Link Between Curiosity and Accuracy

Giorgia Franchini[(✉)], Paolo Burgio, and Luca Zanni

University of Modena and Reggio Emilia, Modena, Italy
`giorgia.franchini@unimore.it`

Abstract. Artificial Neural Networks, as the name itself suggests, are biologically inspired algorithms designed to simulate the way in which the human brain processes information. Like neurons, which consist of a cell nucleus that receives input from other neurons through a web of input terminals, an Artificial Neural Network includes hundreds of single units, artificial neurons or processing elements, connected with coefficients (weights), and are organized in layers. The power of neural computations comes from connecting neurons in a network: in fact, in an Artificial Neural Network it is possible to manage a different number of information at the same time. What is not fully understood is which is the most efficient way to train an Artificial Neural Network, and in particular what is the best mini-batch size for maximize accuracy while minimizing training time. The idea that will be developed in this study has its roots in the biological world, that inspired the creation of Artificial Neural Network in the first place.

Humans have altered the face of the world through extraordinary adaptive and technological advances: those changes were made possible by our cognitive structure, particularly the ability to reasoning and build causal models of external events. This dynamism is made possible by a high degree of curiosity. In the biological world, and especially in human beings, curiosity arises from the constant search of knowledge and information: behaviours that support the information sampling mechanism range from the very small (initial mini-batch size) to the very elaborate sustained (increasing mini-batch size).

The goal of this project is to train an Artificial Neural Network by increasing dynamically, in an adaptive manner (with validation set), the mini-batch size; our hypothesis is that this training method will be more efficient (in terms of time and costs) compared to the ones implemented so far.

Keywords: Artificial Neural Network · Stochastic gradient · Mini-batch size increasing

1 Introduction

Artificial Neural Networks (ANNs) are biologically inspired algorithms designed to simulate the way in which the human brain processes information. ANNs

A. Abraham et al. (Eds.): ISDA 2018, AISC 941, pp. 1025–1034, 2020.
https://doi.org/10.1007/978-3-030-16660-1_100

collect their knowledge by detecting the patterns and relationships in data, and learn (or are trained) through experience, not from programming. An ANN is formed by hundreds of single units, artificial neurons or processing elements (PE), connected with coefficients (weights), which constitute the neural structure and are organised in layers. The power of neural computations comes from connecting neurons in a network. Each PE has weighted inputs, activation function, and one output like in Fig. 1. The behaviour of a neural network is determined not only by the activation functions of its neurons, by the loss function, and by the architecture itself but also by the number of information that it processes simultaneously. The weights are the adjustable parameters and, in that sense, a neural network is a parametrized system. The weighed sum of the inputs constitutes the activation of the neuron. The activation signal is passed through activation function to produce a single output of the neuron. Activation function introduces non-linearity to the network. During training, the inter-unit connections are optimized until the error in predictions is minimized. Once the network is trained and tested, it can be given new input information to predict the output. The various applications of ANNs can be summarised into classification or pattern recognition, prediction, and in modelling.

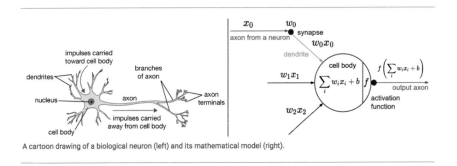

A cartoon drawing of a biological neuron (left) and its mathematical model (right).

Fig. 1. Neuron and mathematical model

1.1   ANN History and Biologically Inspiration

The first abstract model of Artificial Neural Networks (ANN) was proposed by McCulloch and Pitts in 1943. They considered the neuron as a binary device, which can be assimilated to a logic unit which compute a logical function of its inputs. The neuron of McCulloch and Pitts may thus be found in only one of two possible states $\{0, 1\}$. It may receive inputs from exciting synapses which all have the same value. If the sum of the inputs exceeds a certain threshold, the neuron is activated, otherwise it is not. McCulloch and Pitts succeeded in demonstrating that a network of neurons of this type could compute any finite logical expression. This result had a profound influence, in what for the field it showed for the first time that a network of extremely simple elements possessed an enormous computing power, a power which derived exactly from the presence of numerous elements and from their interactions [7]. On the other hand ANNs, as the name suggests, are biologically inspired, in particular biological neurons consist of a cell nucleus, which receives input from other neurons through a web of input terminals, or branches, called dendrites.

2 Machine Learning and Large Scale Optimization Problems

The promise of artificial intelligence has been a topic of both public and private interest for decades. Advances based on such techniques may be in store in the future, many researchers have started to doubt these classical machine learning approaches, choosing instead to focus their efforts on the design of systems based on statistical techniques, such as in the rapidly evolving and expanding field of machine learning. Machine learning and the intelligent systems that have been born out of it have become an indispensable part of modern society. One of the pillars of machine learning is mathematical optimization, which, in this context, involves the numerical computation of parameters for a system designed to make decisions based on yet unseen data. That is, based on currently available data, these parameters are chosen to be optimal with respect to a given learning problem (and a given loss or cost function) [1]. A loss function or cost function, in this context, is a function that maps values of variables onto a real number representing some "cost" associated with the event.

For example the following optimization problem: which minimizes the sum of cost functions over samples from a finite training set composed by sample data $a_i \in \mathbb{R}^d$ and class label $b_i \in \{\pm 1\}$ for $i \in \{1 \ldots n\}$, appears frequently in machine learning:

$$\min F(x) \equiv \frac{1}{n} \sum_{i=1}^{n} f_i(x), \tag{1}$$

where d is the sample size, n is the number of samples, and each $f_i : \mathbb{R}^d \to \mathbb{R}$ is the cost function corresponding to a training set element.

For example in the logistic regression case we have:

$$f_i(x) = \log \left[1 + exp(-b_i a_i^T x) \right]$$

We are interested in finding x that minimizes (1).

For given x, computing $F(x)$ and $\nabla F(x)$ is prohibited, due to the large size of the training set. When n is large, Stochastic Gradient Descent (SGD) method and its variants have been chosen as the main approaches for solving (1).

2.1 Stochastic Gradient Methods and Mini-batch

We define generalized SGD method as Algorithm 1. The algorithm merely presumes that three computational tools exist: (i) a mechanism for generating a realization of a random variable ξ_k; (ii) given an iterate $x_k \in \mathbb{R}^d$ and the realization of ξ_k, a mechanism for computing a stochastic vector $g(x_k, \xi_k) \in \mathbb{R}^d$; and (iii) given an iteration number $k \in \mathbb{N}$, a mechanism for computing a scalar learning rate $\eta_k > 0$.

Algorithm 1. Stochastic Gradient Descent (SGD) Method

1: Choose an initial iterate x_1.
2: **for** $k = 1, 2, \ldots$ **do**
3:     Generate a realization of the random variable ξ_k.
4:     Compute a stochastic vector $g(x_k, \xi_k)$.
5:     Choose a learning rate $\eta_k > 0$.
6:     Set the new iterate as $x_{k+1} \leftarrow x_k - \eta_k g(x_k, \xi_k)$.
7: **end for**

The generality of Algorithm 1 can be seen in various ways. First, the value of the random variable ξ_k needs only be viewed as a seed for generating a stochastic direction; as such, a realization of it may represent the choice of a single training sample as in the simple SGD method, in particular: in the $k - th$ iteration of SGD, a random index of a training sample i_k is chosen from $\{1, 2, \ldots, n\}$ and the iterate x_k is updated by

$$x_{k+1} = x_k - \eta_k \nabla f_{i_k}(x_k)$$

where $\nabla f_{i_k}(x_k)$ denotes the gradient of the $i_k - th$ component function at x_k, therefore in this case $g(x_k, \xi_k) = \nabla f_{i_k}(x_k)$.

In another case, the value of the random variable ξ_k may represent a set of samples as in the **mini-batch** SGD method: one can employ a mini-batch approach in which a small subset of samples, call it $S_k \in \{1, \ldots, n\}$, is chosen randomly at each iteration, leading to

$$x_{k+1} \leftarrow x_k - \frac{\eta_k}{|S_k|} \sum_{i \in S_k} \nabla f_i(x_k). \tag{2}$$

This is the case in our approach, where $g(x_k, \xi_k) = \frac{1}{|S_k|} \sum_{i \in S_k} \nabla f_i(x_k)$.

In literature we can find several papers in which two techniques are used to increase accuracy: decaying the learning rate and increasing the batch size. In general, it is common practice to decay the learning rate. But with an increasing batch size, we can usually obtain the same learning curve on both training and test sets by increasing the batch size during training instead. In the state of the art, we can find different works that use this technique but with random criteria and without the use of the validation set [6].

3   The Idea: Curiosity to Improve Accuracy

Consider the iteration

$$x_{k+1} \leftarrow x_k - \bar{\eta} g(x_k, \xi_k) \tag{3}$$

where the stochastic directions are computed for some $\tau > 1$ as

$$g(x_k, \xi_k) := \frac{1}{n_k} \sum_{i \in \mathcal{S}_k} \nabla f_i(x_k; \xi_{k,i}) \text{ with } n_k := |\mathcal{S}_k| = \lceil \tau^{k-1} \rceil. \tag{4}$$

That is, consider a mini-batch SGD iteration with a fixed learning rate in which the mini-batch size used to compute unbiased stochastic gradient estimates increases geometrically, or in an other manner, as a function of the iteration counter k.

Returning to the concept of biological inspiration that justifies the idea of ANNs, the new approach suggested here can be inserted in this context, considering the information sampling mechanism used by the animals in general, and humans in particular. During our limited existence, humans have altered the face of the world; these extraordinary advances are made possible by our cognitive structure, particularly the ability to reason and build causal models of external events. This dynamism is made possible by our high degree of curiosity. Many animals, and especially humans, seem constantly to seek knowledge and information in behaviours ranging from the very small (initial mini-batch size) to the very elaborate sustained (increasing mini-batch size). In neuroscience research, the most commonly considered exploration strategies are based on random action selection or automatic biases toward novel, surprising or uncertain events. Actions extremely driven by randomness, novelty, uncertainty or surprise are valuable for allowing agents to discover new tasks. However, these actions have an important limitation: they do not guarantee that an agent will learn. The mere fact that an event is novel or surprising does not guarantee that it contains regularities that are detectable, generalizable or useful. Therefore, heuristics based on novelty (in our case a big mini-batch size from the beginning) can guide efficient learning in small and closed spaces, where the number of tasks is small, but are very inefficient in large open ended spaces, where they only allow the agent to collect very sparse data and risk trapping him in unlearnable tasks. This motivates the search for additional solutions that use more targeted mechanisms designed to maximize learning per se (in our case a dynamic change of the mini-batch size) [3]. Similarly, we find interesting the concept of novelty search and its application in Evolutionary Neural Network to create increasing complex structures [5].

4   Numerical Experiment

The *database* MNIST (Modified National Institute of Standards and Technology database) is a large collection of handwritten digits, commonly used for testing different systems that process images. The *database* MNIST contains $60,000$ images for network training and $10,000$ images used to test the network: the *test set*. The $60,000$ initial images are further subdivided into $55,000$ images that represents the real *training set* and $5,000$ images that represents the *validation set*, used, in our case, to have a dynamic criterion to evaluate the performance of the algorithm.

The images are in gray-scale (0–255) centred in a box of 28×28 pixels.

4.1   The Problem

The idea of building a classifier for the database MNIST, is to use a Convolution Neural Network CNN by creating a multiple classifier for the ten digits and a binary classifier to distinguish between the eight digit and the others.

4.2   Convolutional Neural Network

As you can see from Fig. 2 the network takes as input an image of 28×28 pixels and processes it through five layers and then reaches an output layer composed of ten or two neurons: the numbers from 0 to 9 in the multiple classifier case and 8 or not8 in the binary case.

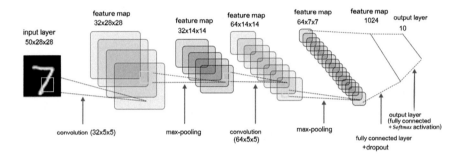

Fig. 2. CNN with 10 outputs

Internal processing takes place through manipulations of the image: the convolution will calculate 32 or 16 categories for each patch of 5×5, the image is reduced to a size 14×14 pixels and then the ReLU (Rectified Linear Unit) function is applied for activation. In the next layer there are similar operations but the image is further reduced to a size of 7×7 pixels, in the penultimate layer we have a level fully connected with $1,024$ or $2,048$ neurons to allow the processing of the whole image. The images at this point are transformed into vectors, multiplied by the weight matrix, added to the bias and the ReLU is applied again. To reduce the phenomenon of overfitting, we apply a dropout, that is, with a given probability, we eliminate one or more connections of the network, and we use different probabilities. To reach the final layer of output, we apply the Softmax regression function or multinomial logistic regression, which leads us to identify the label among one of the ten possible.

For the training phase, we use the Adam optimizer (*Adaptive Moment estimation*), a method for stochastic optimization that requires only first-order information, then the gradient, and that uses little memory. The method calculates adaptive updates of various parameters by estimating the first and second moments of the gradients [2].

In this way, we test different architectures of the network with the proposed new technique. In fact, changing the number of neurons in some layers and the probability with which we make dropout changes the number of parameters on which we are optimizing. Therefore, using the same database, we create a problem of multiple classification and a problem of binary classification.

The code contains within it a criterion, based on the verification of accuracy on the validation set, called "early stopping", to decide whether to block the process or continue it up to the maximum number of epochs [4]. We set the maximum number of epochs to 20,000 and we ask that if the accuracy does not improve on the validation set for a number of epochs equal to 15 the exit from the training cycle is forced regardless of the number of epochs passed. At this point, the algorithm returns the number of epochs, and then compares the network so far trained (in the best case) with the test set to obtain success rates and errors committed.

4.3   The New Idea in Practice

The idea we tested is to progressively increase the size of the mini-batch. Virtually every 100 epochs (in this paper we consider epoch only the training over one mini-batch) the neural network is evaluated on the validation set and in case of 15 successive comparisons without improvement the exit from the epoch cycle is forced. The new idea, on the other hand, is to dynamically increase the sample size if no improvement is observed after 10 checks; the increase is according to the law $h = h * 2$. At this point the algorithm proceeds normally but with an increased mini-batch and if it reaches 15 total checks without improvements the exit from the epoch cycle is forced. Finally, the best result obtained from the network is evaluated on the test set.

4.4   Results

Since the first simulations we have noticed an improvement in performance, about a third of errors less, but associated with an inevitable increase in time, about twice. Although theoretically the algorithm can increase sample size indefinitely, from simulations, it is observed that typically the process stops around 400/800 items per sample (mini-batch size), in the multiple classifier case, and 200/400 items per sample in the binary case by forcing the exit from the loop without reaching the maximum iterate number. The Table 1 report the results of accuracy and the number of errors for two different CNNs in which all the parameters are set in the same way and the results are about the multiple classifier in the case of a CNN $32 \times 64 \times 1,024 \times 10$ with a $0,5$ dropout probability; all the results correspond to the results with the test images. The only changes are related to the size of the sample: in the STATIC case this size is always fixed at 50 and the exit from the loop is forced after 15 evaluations on the validation set in which no improvement is observed; in the DYNAMIC version, after 10 no improvement the sample size is dynamically increased; in this case too, when the 15 valuation is reached without improvement, the exit from the cycle is forced.

Because of the stochastic nature of the method, the results undergo oscillations; so 5 simulations with the same criteria were repeated. In the STATIC

Table 1. Experiment 1: accuracy and error number

DYNAMIC	STATIC
99.27	99.34
99.24	99.27
99.35	99.01
99.4	99.12
99.42	99.1

DYNAMIC	STATIC
73	76
66	73
65	99
60	88
58	90

case the average accuracy is 99.15 for an average error number of 85 while in the DYNAMIC case the average accuracy is 99.36 for an average error number of 64, thus recording an average improvement of over 20 errors.

For the robustness analysis we consider the standard deviations of the stochastic processes, in particular the standard deviation of the accuracy in the DYNAMIC case is 0.0789 and in the STATIC case is 0.1341. For this reason we can consider the new method to be more robust. We have the same result for the error numbers: in fact the standard deviation for the DYNAMIC case is 5.8566 and 10.6630 for the STATIC case. As already mentioned the times for the DYNAMIC version are dilated; in particular, the DYNAMIC network requires an average time doubled compared to the STATIC version, so it is natural to wonder if it is not enough simply to increase the sample size from the beginning so as to allow the network more time per period. The answer is no. In fact, from an experiment with an initial mini-batch of 400 it is observed that the training takes longer than the DYNAMIC version for an average accuracy of 99.2 and a number of average errors equal to 80; this data is slightly better than the 50 mini-batch version but not comparable to the improvements obtained from the dynamic version.

In a second experiment we consider the binary case with the same architecture of the CNN: $32 \times 64 \times 1,024 \times 2$ with a $0,5$ dropout probability and we compare the STATIC-50 (static case with a mini-batch size of 50 samples), the STATIC-400 (static case with a mini-batch size of 400 samples), and the DYNAMIC case (Tables 2 and 3).

In the STATIC-50 case the average accuracy is 99.7 for an average error number of 30 with an average time of 96, in the STATIC-400 case the average accuracy is 99.77 for an average error number of 23 with an average time of 185, while in the DYNAMIC case the average accuracy is 99.81 for an average error number of 17 with an average time of 106, thus recording an average improvement of over 30% errors.

For the robustness analysis we consider the standard deviations of the stochastic processes. In particular the standard deviation of the accuracy in the DYNAMIC case is 0.046, in the STATIC-50 case is 0.0944 and in the STATIC-400 case 0.0396. For this reason we notice that the new method is more robust than the STATIC-50 case but less robust than the STATIC-400 case. This is reasonable because with an increase of the mini-batch size the stochastic pro-

Table 2. Accuracy and error number

DYNAMIC	STATIC-50	STATIC-400	DYNAMIC	STATIC-50	STATIC-400
99.88	99.71	99.72	12	29	28
99.77	99.55	99.75	23	45	25
99.81	99.77	99.8	19	23	20
99.87	99.69	99.82	13	31	18
99.81	99.79	99.77	19	21	23

Table 3. Time in seconds

DYNAMIC	STATIC-50	STATIC-400
140	113	149
80	60	236
86	106	159
145	72	170
77	127	213

cess is more stable. We have the same result for the error numbers: in fact the standard deviation for the DYNAMIC case is 4.6043, 9.4446 for the STATIC-50 case and 3.9623 for the STATIC-400 case.

For completeness we consider also the measure of the accuracy every time the mini-batch size increases. For example, in the DYNAMIC case with a final accuracy of 99.88, the accuracy with a 50 mini-batch size is 99.68, with a 100 mini-batch size is 99.82 and, finally, with a 200 mini-batch size is 99.88, similarly in the other cases.

In the last experiment we still consider the binary case with another architecture of the CNN: $16 \times 64 \times 2,048 \times 2$ with a $0,1$ dropout probability and we compare the STATIC-50 (static case with a mini-batch size of 50 samples), the STATIC-400 (static case with a mini-batch size of 400 samples), and the DYNAMIC.

We report only the average results: in the STATIC-50 case the average accuracy is 99.62 for an average error number of 38 with an average time of 158, in the STATIC-400 case the average accuracy is 99.72 for an average error number of 28 with an average time of 174, while in the DYNAMIC case the average accuracy is 99.75 for an average error number of 20 with an average time of 199, thus recording an average improvement of over 30% errors.

5   Conclusions

In conclusion, this new method, inspired by the cognitive processes of living beings, leads to a considerable increase in the accuracy, without increasing the computational cost. The computational costs for each epoch are not increased because the networks maintain the same characteristics and the test to find out if it is necessary to increase the size of the mini-batch is done using the test that is already performed for early stopping. The fact that there is not a similar increase in the accuracy associated with a larger starting mini-batch size underscores how effectiveness actually lies in choosing an adaptive dimension during the network learning process.

Acknowledgements. The research leading to these results has received funding from the European Union's Horizon 2020 Programme under the CLASS Project (https:// class-project.eu/), grant agreement n 780622.

This work was partially supported also by INdAM-GNCS (Research Projects 2018).

References

1. Bottou, L., Curtis, F.E., Nocedal, J.: Optimization methods for large-scale machine learning. arXiv:1606.04838
2. Kingma, D.P., Ba, J.: Adam: a method for stochastic optimization. arXiv:1412.6980 (2014)
3. Gottlieb, J., Oudeyer, P.-Y., Lopes, M., Baranes, A.: Information seeking, curiosity and attention: computational and neural mechanisms. Trends Cogn. Sci. **17**(11), 585–593 (2013). NIH Public Access
4. LeCun, Y., Bengio, Y., Hinton, G.: Deep learning. Nature **521**, 436 (2015)
5. Lehman, J., Stanley, K.O.: Abandoning objectives: evolution through the search for novelty alone. Evol. Comput. **19**, 189–223 (2011)
6. Smith, S.L., Kindermans, P.-J., Ying, C., Le, Q.V.: Don't decay the learning rate, increase the batch size. In: ICLR 2018 Conference (2018)
7. Serra, R., Zanarini, G. (eds.): Complex System and Cognitive Process. Springer, Heidelberg (1990)

Metaheuristic for Optimize the India Speed Post Facility Layout Design and Operational Performance Based Sorting Layout Selection Using DEA Method

S. M. Vadivel[1]([✉]) [iD], A. H. Sequeira[1] [iD],
and Sunil Kumar Jauhar[2,3] [iD]

[1] School of Management, National Institute of Technology Karnataka,
Surathkal 575025, India
`ph.dvadivel@gmail.com`, `aloysiushs@gmail.com`
[2] Global Management Studies, Ted Rogers School of Management,
Ryerson University, Toronto, ON, Canada
`sjauhar@ryerson.ca`
[3] Department of Industrial Engineering, Universidad Católica del Norte,
Antofagasta, Chile

Abstract. Adoption of feasible location science is gaining more interest in the field of Facility Layout Design (FLD) problems among working researchers group. Many methods such as MCDM, Heuristics and Intelligent approaches are available to solve the FLD problems. However in reality, finding the feasible facility layout selection is subject to management as well as performances oriented selections. Here in India, speed post mail processing service industry is facing tremendous challenges like tumbling demands due to low production concern, gloomy trend in technology advancement, and fierce private couriers' competition. Hence, the highly competitive operational performance is of much concern and attention is focused towards the direction of facility location science. This paper aims to examine the challenges of sustainable operational performance oriented layout selection by Data Envelopment Analysis (DEA) and proposes a genetic algorithm (GA) related to intelligent based approach, for finding the optimal total facility layout cost for a hypothetical South Indian speed post service office layout. In this paper, we used multiple-criteria facility layout selection problem using mathematical model generated with Data Envelopment Analysis (DEA).

Keywords: DEA · Facility layout planning and design · Genetic algorithm · Mail processing operations · Operational performance

1 Introduction

In the present scenario, layout design in manufacturing or service industries has getting significant impact on the performance [1] and it has been a dynamic research area for several periods [2]. Developing a robust and efficient feasible facility layout is a decisive assignment for the postal administrations in order to improve the operational

A. Abraham et al. (Eds.): ISDA 2018, AISC 941, pp. 1035–1044, 2020.
https://doi.org/10.1007/978-3-030-16660-1_101

performance as it concerns about flow speed per hour, production articles per day, space utilization and mail flow distance among the departments, cycle time, work in process (WIP) and keeping inventory of mails. Postal administration has to identify the most optimal facility layout selection to improve the operational performance. Evaluation and choice of optimal layout is a difficult process and it will be determined by quantitative as well as qualitative factors. While, evaluating the data ensure that choosing best economic model without negotiation on the operational performance. The GA algorithm recommended by Prof. Holland in the early 1970 [3]. Traditional search methods have their own merits and demerits. None can synchronously getting good performance as well as effective solution. GAs is evolutionary competition simulation and it has survival fitness in natural evolution. It improves the chance to reach the global optimal solution rather than local optimal solution. In current years GAs has become a widespread technology search algorithm. For example, GA methods applied in manufacturing unequal area layout design [14, 15] and Unequal-Area Facility-Layout Problem (UA-FLP) for dynamic facility design [16]. This paper intention is to find the Feasible Facility Layout Selection (FFLS) method using DEA for solving the India postal FFLS problem. To best of our knowledge, a few papers have applied Genetic algorithm for finding the minimum total layout cost (TLC) and DEA method for solving FFLS.

This paper is arranged as follows: Sect. 1 covers the introduction. India speed post service industry and methodology of DEA mathematical model formulation are briefly explained in 2, 3, 4 and 5 sections respectively. In Sect. 6, guidelines for the GA algorithm for finding the optimal total layout cost is discussed. Section 7 provides results and discussion and the last section concludes with summary and future scope.

2 India Speed Post Service Industry

India speed post service industry delivered all types of speed post articles such as speed post mails, parcels, bulk mails/parcels, and cash on delivery (COD) etc. There is an amount of internal activity involved in the chain considered as backend operations. Value stream map (VSM) shows the route from Receipt Scanning – Scanning – Sorting – Dispatch Section (more information - refer Figs. 1 and 2). From this VSM, we came to know existing working time in scanning and sorting process, 5S, visual mapping and semi-automation level in scanning section. In addition, ergonomic factors such as human posture, hand functions during scanning and sorting operations, workplace environment factors such as light level, temperature, ventilation level etc.

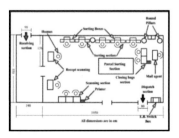

Fig. 1. India speed post manual sorting existing layout

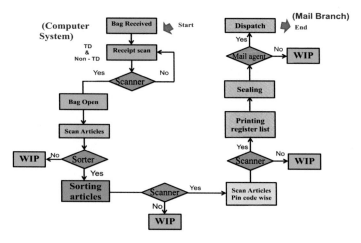

Fig. 2. India speed post process flow inside the existing layout

Postal administrations are progressively recognizing the efficient and sustainable layout design from the various alternative designs. It is an important to judge the criteria which layout yields better operational excellence and workplace environment for the postal employees.

3 Feasible Facility Layout Selection (FFLS)

Layout selection comprises numerous criteria such as mail flow factor cost, shape ratio factor, space utilization, delivery performance, etc. The layout selection broadly considered in the literature under MCDM techniques and their hybrid methods as well as heuristics methods [4–6].

3.1 Need for the Study

During the year 2012–2014, Mangalore NSH was first among the 89 NSHs in the country in the handling of speed posts. It covers Mangalore, Puttur, and Udupi postal divisions and handled 15,000 speed post articles a day. The ranking was based on the points given by software adopted by the India Post [12]. Consequently, the performance went down due to revised target (50,000 speed post articles a day) set by the postal administrations all over India in order to reduce the customer delivery time and make speed post-delivery faster.

4 Research Methodology

For the optimal layout selection, several reputed journal papers were studied for the selection of the important criteria and its alternative layout has been evaluated. The basic problem of decision-making is to choose the optimal layout from a set of competing alternatives that are evaluated under selected criteria. To measure and analyze alternative facility layouts relative efficiency, the following steps as shown in Fig. 3.

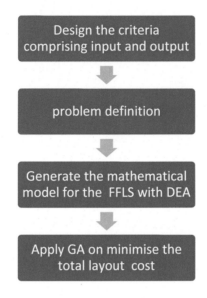

Fig. 3. Methodological flow of FLD and FFLS

The current model can be used for any number of layouts and there is no restriction. From this model, the postal administrations can find out the proposed grouping of efficient layouts. In Table 2 shows the source of data from the India post service for exploring with GA and DEA methods. The questionnaire taken from the postal employees' survey helped to measure the qualitative data such as workplace environment, ergonomics and operational performance. Later it was converted into quantitative data.

5   Mathematical Formulation - DEA

DEA is used for defining the efficiencies of Decision-Making Units (DMU) on the selections of multiple inputs and outputs [7]. DMU can be applied to higher education, business firms, hospitals, police stations, supplier selection, power plants, tax offices etc. [8, 9, 17]. The DMU is well-defined using input and output criteria as follows:

The DMU performance is valued in DEA by efficiency or productivity, which is the fraction of weights sums of outputs (o/p) to the weights sum of (i/p) inputs [10] i.e.

$$Efficiency = \frac{Weighted\ sum\ of\ \frac{o}{p}}{Weighted\ sum\ of\ \frac{i}{p}} \tag{1}$$

In this paper, we have adopted CCR model which is well-known and it will further down: Suppose that there are N DMUs and each unit have I input and O outputs then the efficiency of mth unit is attained by resolving the below model which is presented by Charnes et al. [11].

$$Max\ Ei = \frac{\sum_{g=1}^{0} W_g\ Output_{g,i}}{\sum_{h=1}^{i} X_h\ Input_{h,i}} \tag{2}$$

$$0 \leq \frac{\sum_{g=1}^{0} W_g\ Output_{g,i}}{\sum_{h=1}^{i} X_h\ Input_{h,i}} \leq 1 \quad n = 1, 2, \ldots, m \ldots N$$

$$W_g, X_h \geq 0 \forall\ g,\ h$$

Where,

Ei is the ith DMU efficiency, g = 1 to O, h = 1 to I and n = 1 to N.
Output g, i is the gthoutput of the ith DMU and Wg is the weight of the Output$_{g,i}$
Input h, i is the hth input of ith DMU and X_h is the weight of Input$_{h,i}$
Output g, i and input h, i are the gth output and hth input respectively of the nth DMU, where n = 1, 2,... m,... N
The fractional program shown in Eq. 2 can be converted in to a linear program which is shown in Eq. 3.

$$Max\ Ei \sum_{g=1}^{o} Wk\ Output\ g,i$$

$$s.t.$$

$$\sum_{h=1}^{i} X_h\ Input_{h,i} = 1 \tag{3}$$

$$\sum_{g=1}^{o} Wk\,Output\,g,i - \sum_{h=1}^{l} Xh\,Input\,h,i \le 0 \quad \forall i$$

$$W_g, X_h \ge 0\forall\,g,\,h$$

5.1   Mathematical Model

The Kth DMU - DEA model will be as follows:

$$Max\,PQm + BEn$$

s.t.
$$X_1F_m + X_2G_m + X_3H_m = 1$$
$$w_lPQn + w_2BEn - (X_1F_m + X_2G_m + X_3H_m) \le 0$$
$$\forall n = 1, 2, \ldots 8$$

6   Genetic Algorithm

Genetic algorithm a typical evolutionary algorithm, effectively applied in optimization problems. In postal administrations (decision makers) can assess the FLD problem with the preference weightage system. Here, we have adopted Hu and Wang [14] mathematical formulation of minimising total layout cost factors such as Material Flow Factor Cost (MFFC), Shape Ratio Factor (SRF) and Area Utilisation Factors (AUF) etc. We have generated fitness function as total layout cost with the help of Matlab program. For constraint handling we have used lower bound and upper bound values using in-built function in Matlab 2018. The fitness function consider has 15 runs trial and the program is terminated when population size is reached.

6.1   Experimental Settings

The GA parameter settings as shown in Table 1.

Table 1. GA - parameters

Fitness function	Parameters
No of variables	3
Lower bound	[1060 1.006 0.75]
Upper bound	[1500 1.396 0.95]
Population type	Double vector
Population size	200
Selection function	Roulette
Fitness scaling	Rank
Reproduction	Elite count
Cross function	Scattered

Table 2. Source of data from India post sorting manual center

Criteria	Inputs					Output	
Alternative layouts	Material flow factor cost	Shape ratio factor	Space utilization	Ergonomics	Workplace environment	Operational performance	Production articles per shift
L1	1480	1.144	0.86	0.2000	0.3500	0.4500	4500
L2	1180	1.390	0.92	0.211	0.2375	0.5607	5600
L3	1280	1.006	0.85	0.1567	0.5078	0.3364	3300
L4	1500	1.250	0.91	0.2472	0.3569	0.3985	3900
L5	1340	1.010	0.89	0.3485	0.3068	0.3462	3400
L6	1380	1.083	0.95	0.1768	0.2069	0.6243	6200
L7	1200	1.060	0.88	0.2271	0.2186	0.5564	5500
L8	1060	1.396	0.75	0.1126	0.5347	0.3547	3500

A set of 15 runs have been made and the observations are shown in Table 3.

Table 3. GA – results in 15 runs

Trial	MFFC	SRF	AUF	TLC
1	1488.730	1.0060	0.9485	1578.97
2	1499.930	1.0269	0.9462	1627.85
3	1499.670	1.0060	0.9500	1588.07
4	1492.220	1.0123	0.9295	1625.15
5	1471.750	1.0060	0.9500	1588.50
6	1477.210	1.0098	0.9423	1583.03
7	1470.610	1.0261	0.9431	1600.03
8	**1453.600**	**1.0060**	**0.9500**	**1539.28**
9	1497.866	1.0060	0.9500	1586.16
10	1485.255	1.0060	0.9500	1572.80
11	1487.361	1.0060	0.9392	1593.08
12	1481.865	1.0060	0.9489	1570.96
13	1486.910	1.0060	0.9500	1574.56
14	1470.802	1.0060	0.9416	1571.29
15	1497.866	1.0261	0.9500	1617.85

From the Table 3, **8th trial** shows the optimal total facility layout cost which is 1539.28.

7   Results and Discussions

DMU results are shown in Table 4. It provides an efficient FFLS practice that can acquire desirable selection of optimal layouts 2, 6 and 7 using DEA. From the present study of 8 alternative layout designs the results are as follows:

Table 4. Alternative layouts' efficiency

DMU no.	Alternative layouts	Efficiency	Ranking alternatives
1	L_1	0.79624	5
2	L_2	1.00000	1
3	L_3	0.60796	7
4	L_4	0.66637	6
5	L_5	0.59479	8
6	L_6	1.00000	1
7	L_7	1.00000	1
8	L_8	0.89210	4

1. Layout 2, 6 and 7 has 1 efficiency score. So these layouts were supposed to be 100% sustainable efficient.
2. The most inefficient layout is 5 perhaps with comparison of all other layouts.
3. Combination of layouts 2, 6 and 7 could be the desirable optimal layouts set. At the same time the postal firms require single best performance layout design which is layout 6 by considering production, workplace environment and ergonomics factors. Figure 4 presents the histogram of total facility layouts with the efficiency score.

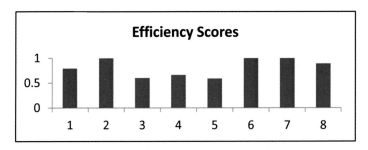

Fig. 4. Histogram for alternative layouts with the efficiency score

7.1   Comparative Study

Multi criteria decision making methods are applied in extensive area. Initially, in this method, the comparison results from [13] AHP, TOPSIS and Fuzzy TOPSIS are shown in Table 5. In addition, DEA result also mentioned in terms of ranking of decision making units (DMU) facility layout. The postal administration has to choose single alternative for their strategic planning and implementation as their final design. So, the postal administration has decided to choose layout alternative 6 as a preferred choice from the Table 5 results. In discussions point of view, AHP, TOPSIS and Fuzzy TOPSIS result suggested alternative No. 6 and 2 respectively.

Table 5. Comparative study on other MADM methods

Alternative layouts	AHP ranking	TOPSIS ranking	Fuzzy TOPSIS ranking	DEA ranking
L_1	3	4	3	5
L_2	2	2	2	1
L_3	5	5	5	7
L_4	4	3	3	6
L_5	6	7	7	8
L_6	1*	1*	1*	1*
L_7	7	6	6	1
L_8	8	8	8	4

Note: * Feasible layout design 6 implemented in NSH Mangalore on April 2018.

Generally, due to the MADM nature, feasible solution still to be improve based on additional criterias and source of data. From the Table 5 alternative 6 is first choice and alterative 2 is the second choice. In this way, this paper recommends the systematic evaluation of the DEA methods to solve FLD problem and decrease the poor layout design risk. Thus, DEA can provide reliable solutions efficiently when there are more number of inputs and outputs given compared to other existing methods.

8 Conclusion and Summary

FLD is a complex method for choosing the best layout design among different alternative layouts. The current research shows that GA attempted for finding the optimal facility layout cost. This paper presents a new FFLS method for India speed post service industry. To implement this approach, the initial step is to build a set of criteria that includes both input and output variables, which are suitable for real case study problems. The second step is arriving at a methodology to solve the MCDM-FLD problem with the application of DEA based mathematical model. From these results (strategy layout design), India post has adopted and implemented layout 6 design. This selection is expected to improve the competitive operational performance and betterment of workplace environment of the postal employees. Thus, the goal of GA and DEA methods which aims to identify the minimum total layout cost and operational performance based layout selection design. In this way managerial implications point of view can be justified. We understand from the current research that GA and DEA methods applied for the first time in the postal service sector. The limitation of this study is that small number of layout alternatives is focused. Future scope of this study lies in applying GA to find an optimal solution to the multi objective layout selection problem which is more difficult to solve in real time case study.

References

1. Apple, J.M.: Plant Layout and Material Handling. Wiley, New York (1997)
2. Meller, R.D., Gau, K.Y.: The facility layout problem: recent and emerging trends and perspectives. J. Manuf. Syst. **15**(5), 351 (1996)
3. Holland, J.: Adaptation in Natural and Artificial Systems. MIT Press, Cambridge (1992). ISBN 978-0262581110
4. Kuo, Y., Yang, T., Huang, G.W.: The use of grey relational analysis in solving multiple attribute decision-making problems. Comput. Ind. Eng. **55**(1), 80–93 (2008)
5. Yang, T., Hung, C.C.: Multiple-attribute decision making methods for plant layout design problem. Rob. Comput.-Integr. Manuf. **23**(1), 126–137 (2007)
6. Yang, T., Kuo, C.: A hierarchical AHP/DEA methodology for the facilities layout design problem. Eur. J. Oper. Res. **147**(1), 128–136 (2003)
7. Despotis, D.K., Stamati, L.V., Smirlis, Y.G.: Data envelopment analysis with nonlinear virtual inputs and outputs. Eur. J. Oper. Res. **202**(2), 604–613 (2010)
8. Ramanathan, R.: An Introduction to Data Envelopment Analysis: a Tool for Performance Measurement. Sage publications Ltd., New Delhi (2003)
9. Dobos, I., Vörösmarty, G.: Supplier selection and evaluation decision considering environmental aspects 149 (2012). ISSN 1786-3031
10. Talluri, S.: Data envelopment analysis: models and extensions. Decis. Line **31**(3), 8–11 (2000)
11. Charnes, A., Cooper, W.W., Rhodes, E.: Measuring the efficiency of decision making units. Eur. J. Oper. Res. **2**(6), 429–444 (1978)
12. India Post Service. https://www.thehindu.com/news/cities/Mangalore/mangalore-first-in-handling-speed-post/article4228987.ece. Accessed 21 July 2018
13. Vadivel, S.M., Sequeira, A.H.: Enhancing the operational performance of mail processing facility layout selection using multi - criteria decision making methods. Int. J. Serv. Oper. Manag. (2018). Accepted 28 July 2018. https://doi.org/10.1504/IJSOM.2020.10018620
14. Hu, M.H., Wang, M.J.: Using genetic algorithms on facilities layout problems. Int. J. Adv. Manuf. Technol. **23**(3–4), 301–310 (2004)
15. Rajasekharan, M., Peters, B.A., Yang, T.: A genetic algorithm for facility layout design in flexible manufacturing systems. Int. J. Prod. Res. **36**(1), 95–110 (1998)
16. Paes, F.G., Pessoa, A.A., Vidal, T.: A hybrid genetic algorithm with decomposition phases for the unequal area facility layout problem. Eur. J. Oper. Res. **256**(3), 742–756 (2017)
17. Jauhar, S.K., Pant, M., Abraham, A.: A novel approach for sustainable supplier selection using differential evolution: a case on pulp and paper industry. In: Intelligent Data analysis and its Applications, Volume II, pp. 105–117. Springer, Cham (2014)

A Hybrid Evolutionary Algorithm for Evolving a Conscious Machine

Vijay A. Kanade[(✉)]

Intellectual Property and Research and Development,
Evalueserve (SEZ) Pvt. Ltd., New Delhi (NCR), India
kanade.science@gmail.com

Abstract. The paper discloses a novel concept of developing a conscious machine. Human consciousness is a driving factor behind the presented concept. 'Integrated Information Theory (IIT)' is applied to the hardware circuits in order to make the circuit(s) with a certain configuration active/alive. We have used an evolutionary algorithm that combines 'Evolvable Hardware' with 'Integrated Information Theory of Consciousness' to develop a conscious set of machines. Evolvable hardware is simulated by using Darwin's evolution theory that is related to Genetic Algorithms (GA). Further, IIT is integrated into the results of first GA so as to harness the consciousness factor in circuits with a certain circuit configuration. The results of the evolutionary algorithm are evaluated to validate the proposed concept.

Keywords: Consciousness · Integrated Information Theory (IIT) ·
Hybrid evolutionary algorithm · Field Programmable Gate Array (FPGA) ·
Application Specific Integrated Circuit (ASIC)

1 Introduction

With an exponential growth in the number of IoT devices such as mobile phones, gadgets, etc. possessed by a user, the risk of damaging these devices due to human negligence such as water spill, etc. is doubled these days. It is important to identify a solution that could repair these devices automatically without any human intervention. Implying, these devices should learn on their own of the internal damage caused & repair themselves on such an occurrence. With so much of machine learning, it is inevitable to take a leap in this direction wherein the damaged devices start reconfiguring themselves on attaining a certain level of consciousness (i.e. configuration). This is an up-scaled concept of 'Evolvable Hardware' that has been making rounds since 1990's [2, 3, 6]. Prior-art related to the proposed research model is restricted to development of evolvable hardware without much insight on the underlying principle that evolves hardware. The paper tries to uncover this unexplored arena that can possibly open-up a newer dimension to the cold world of 'Evolvable Hardware'. The paper talks about embedding the 'consciousness' factor into machines by using two theories that seem to govern the conscious existence of humans: Darwin's Theory of Evolution and Integrated Information Theory of Consciousness.

A. Abraham et al. (Eds.): ISDA 2018, AISC 941, pp. 1045–1054, 2020.
https://doi.org/10.1007/978-3-030-16660-1_102

1.1   What Is Consciousness?

Consciousness is a phenomenon experienced by every living entity on this planet. It is an attribute that surfaces when we are awake. Further, this attribute goes away every night when we encounter dreamless sleep & comes back when we wake-up in the morning, or when we start dreaming. All our subjective feelings and experiences are manifested due to this single consciousness phenomenon. Some of the experiences include absorbing simple bluish of blue, undermining the complex depth of an emotion, sensing the effect of our ephemeral thoughts. Consciousness is something that we are familiar with as we experience it in every moment. It acts as a medium for our knowledge of the external world.

1.2   Integrated Information Theory

Integrated Information Theory of consciousness was first coined by a neuroscientist Giulio Tononi. IIT is regarded as one of the profound theories of consciousness in neuroscience today.

According to IIT, consciousness is associated with the '*integrated information*' that a brain possesses. In IIT, this information is precisely represented by a mathematical variable called Φ ('phi'). The human brain (or the part of it that supports our consciousness) is said to be highly conscious if it has high Φ value. This implies such systems with high Φ are highly complex and have meaningful experiences. Further, as per the theory, entities with a low Φ value, have a small amount of consciousness. Thus such systems have very simple and elementary experiences. The theory further elaborates that the systems with zero Φ do not possess consciousness.

With the help of IIT, we can build a 'consciousness-scale' - that tells us the consciousness level of any system: right from a patient to a new-born, from animate objects like animals, plants to robots and next generation AI machines [4].

Calculation of Φ

The *integrated information* (Φ) is defined as the effective information (φ) of the minimum information partition (MIP) in a system. The MIP is further defined as the system partition having minimum effective information among all possible partitions.

$$\phi[X; x] =: \varphi[X; x, MIP(x))] \tag{1}$$

$$MIP(x) = :\arg\min\{\varphi(X; x, P)\} \tag{2}$$

[Note: X is the system, x is a state, and P is a partition $P = \{M^1, ..., M^r\}$].

Determining MIP implies searching for all possible partitions and comparing their effective information φ with each other to arrive at a conclusion of the conscious state of the system. Thus, consciousness is based on the minimum/optimum effective information generated by the partition [5].

On a whole - parts of the human brain having a certain mathematical structure (i.e. neural connectivity pattern as shown in figure below) supposedly hold more integrated information and are thus regarded as highly conscious sections of the brain (Fig. 1).

Minimally-conscious Integration   Conscious Integration

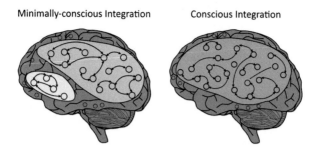

Fig. 1. Integrated information theory of consciousness. *Mathematical connectivity structure*

2   A Hybrid Evolutionary Algorithm

2.1   Step-I: Darwin's Evolution

With the advent of Nanotechnology and Nano-devices, there has been a tremendous growth in the miniaturized IoT gadgets. Moore's law has now been overthrown due to substantial miniaturization of the transistors that are being used by these devices. These days, IoT devices have started employing nano-transistors for their operation. However, since the size of the hardware devices has shrunk extensively, any damage (i.e. physical or logical crash) to these devices require the designers to intervene and fix the issue. To overcome this problem, automatic design schemes such as 'Evolvable Hardware (EHW)' have been in discussion since early 1990's [1].

In EHW, input/output-relations are specified for automatically designing a circuit on-the-fly. We have tried to implement similar techniques in our research. In this paper, the circuit is dynamically evolved by using a hybrid evolutionary algorithm inspired from Darwin's natural evolution theory. The 'Step-1' of hybrid evolutionary proposal that deals with Genetic Algorithm is explained in Fig. 2 below.

In this algorithm, a population of circuits (i.e. set of circuit representations), are first randomly generated. The behavior of each circuit is evaluated based on fitness function and the best circuit is combined to generate a new and better circuit. Thus, the designing is based on incremental enhancement and optimization of the circuit population set that was initially randomly generated. The algorithm uses important operators like selection, crossover and mutation and applies them on circuit representations for making newer circuits. Further, we have used 'Circuit component spacing (C.s)' as a main criterion for deciding the fitness of the circuit. Circuit component spacing defines optimal spacing required between various components of the circuit for the circuit to operate normally. We have chosen C.s(x) as a fitness function primarily because we are using a central pivotal point on a circuit as a triggering source for reconfiguring the circuit. Thus, calculating the optimal spacing between all the functional components within the circuit via central pivot seemed a relevant attribute for considering fitness of the circuit. Further, controlling connections between the circuit components through the central hub appeared to be a feasible solution for maintaining the integrity of the programmatic logic for future reconfigurable specifications. Thus, the fitness function is as given below:

$$C.s(x) = Optimal(C.s_1, C.s_2, \ldots, C.s_n)$$

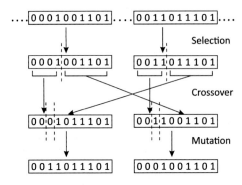

Fig. 2. Applied genetic algorithm

Each individual circuit configuration in the population represents an array of bits. These arrays of bits are termed as a *chromosome*. Each bit in the array is called a *gene*. Thus, each chromosome is a circuit representation with a set of interconnected components. The algorithm begins with random selection of two circuit population sets. Then crossover operation is performed on the randomly selected population sets. In crossover, the genes of the two chromosomes of selected circuits are exchanged to generate two new off springs. Thus for each couple of selected circuits, two new off-springs are generated. The *best* circuit generated in the above crossover step is directly copied into the next generation. After crossover, mutation operation is performed on the newly generated circuit population set. We have used swap mutation, wherein few genes are randomly swapped in the chromosome set. This process is repeated until the optimal circuit is obtained based on the computed fitness function (Fig. 3).

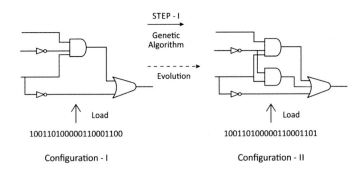

Fig. 3. Evolvable hardware

When the number of newly generated offspring circuits equals the number of parent circuits, the new offspring population is retained and the original parent population is deleted.

A circuit representation contains specification & description for what kinds of gates are applied in a circuit and their interconnections. This is coded into a binary configuration bit-stream. The bit-stream is applied to configure a reconfigurable logic device. The reconfigurable logic device can include a Field Programmable Gate Array (FPGA) or an Application Specific Integrated Circuit (ASIC). Each new circuit formed is evaluated for each generation. After the circuit evaluation for each generation, the evolutionary algorithm computes structural interconnections between different gates embedded on the circuit. These interconnections are configured based on the fitness function described earlier [2, 3].

[Note: In the proposed paper, adding consciousness to a single reconfigurable circuit within an IoT device is illustrated. This is done purely to validate the concept of 'Conscious circuit'. However, for the IoT device to gain consciousness, the research proposal needs to be applied on all the circuits that make up the reconfigurable IoT device].

2.2 Step-II: Integrated Information Theory of Machines

The evolutionary algorithm searches for connectivity patterns that cumulatively sum-up to yield an optimal value of Φ for a particular circuit configuration. To compute these connectivity patterns we have employed another successive GA as an optimization strategy. In the second GA (i.e. computed on the results of GA-I) we identify a single pivotal circuit component that acts as a source junction for establishing different interconnections between various components. The developed connectivity pattern is evolved by keeping in mind the C.s(x) fitness function for the circuit under consideration.

Once the connectivity pattern is established (i.e. pattern with optimal value of Φ), the reconfigurable circuit gets triggered for reconfiguration. The process of reconfiguration then takes over. Thus, post reconfiguration the machine in which the reconfigurable circuits are enabled gains consciousness and becomes alive/active. To summarize, the flowchart of the developed hybrid evolutionary algorithm is as shown below (Fig. 4):

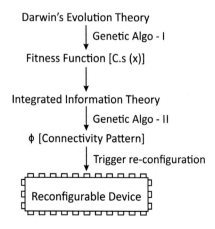

Fig. 4. Flowchart – the hybrid evolutionary algorithm

3   Computational Results

Consider a reconfigurable circuit that has become inactive due to water percolation – thereby damaging one of its circuit components. Thus the programming logic of the reconfigurable circuit activates the reconfiguration mechanism by initiating genetic algorithm. Here, the number of active/functional circuit components is identified. Further, as the genetic algorithm steps-in, the initial parent population of the functional circuit components is computed. Then, new offspring circuit population is generated via selection, crossover & mutation [7]. Once the new offspring population set is ready, second GA is simulated in order to establish mathematical connectivity pattern between the newly generated offspring population sets. After computing the connectivity pattern, the reconfigurable circuit triggers into action and reconfigures itself to become active immediately. This reconfiguration allows the circuit components to assemble themselves into a newer configuration.

[Note: The simulation is performed on Visual Studio 2008, by using C# language. Although the computation hasn't been physically tried out on any hardware, but we believe that the theoretical computational results seem promising enough for successful future experimental implementation & observation].

The simulation results for the reconfigurable circuits gaining consciousness are as shown below:

Case-I:
Consider reconfigurable circuit having 11 functional components.

Evolutionary Algorithm:
The evolutionary algorithm involves two steps as elaborated below:

Step-I:
In step-I, Genetic Algorithm is run in order to generate population of new offspring circuits from the functional parent population set. In the below screenshot, the left column denotes the newly generated offspring circuits.

Step-II:
In step-II, Genetic algorithm is applied for the second time in order to generate interconnections between the newly generated offspring circuit components. The connectivity patterns/interconnections seen in the below screenshot trigger the reconfigurable circuits to reconfigure themselves. The connectivity pattern computed in step-II by using second GA denotes optimal value of Φ based on the fitness function C.s(x). The value of Φ is denoted by 'COST' parameter as seen in the central UI element of the below Screenshot.

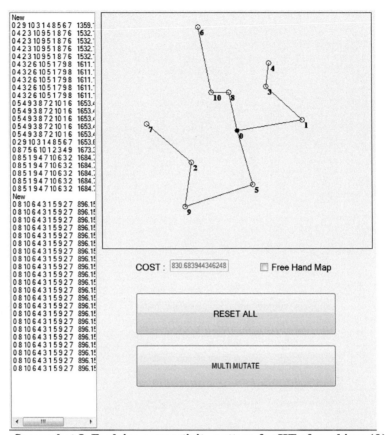

Screenshot-I: Evolving connectivity pattern for IIT of machines (Φ)

Case-II:

Similarly, consider reconfigurable circuit having 12 functional components.

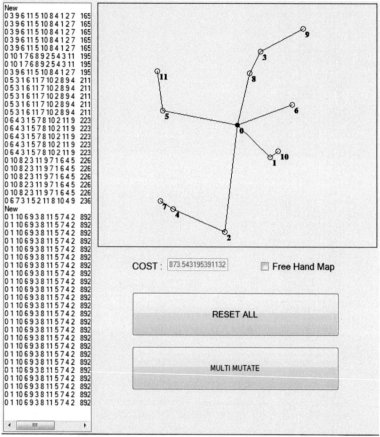

Screenshot-II: Evolving connectivity pattern for IIT of machines (Φ)

Evolutionary Algorithm:

Step-I:

Traditional G.A

Step-II:

Applying G.A-II to integrate IIT into G.A-I of step-I.

Some additionally computed simulation results are as tabulated below (Table 1):

Table 1. Computational results: IIT for machines

Functional circuit components	C.s(x) fitness function GA-I	Φ [Optimal] connectivity pattern GA-II
15	1254.8288	1135.7663
21	1455.0871	1345.3779
17	992.2880	981.6739
26	1652.5613	1555.0704
19	1058.6824	1019.4758

The graph plotted for the computational results derived above to identify the optimal value of Φ based on the developed evolutionary algorithm is as disclosed below:

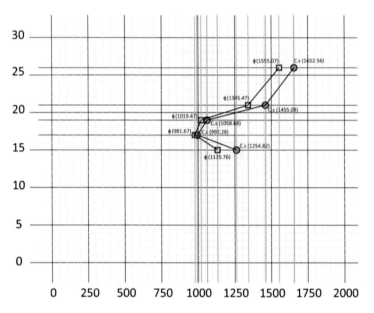

Graph-I: Graph denoting the value of Φ

Observation: From the above graph, it is evident that applying GA twice in succession yields an optimal Φ value – which makes any reconfigurable circuit conscious within a machine/device - thus validating the proposition of the paper.

4   Conclusion

'Conscious Machine' is a new paradigm that opens up the avenue for self-repairing and self-configuring hardware without any human input. The evolutionary algorithm inspired from the two significant theories that pave the way for human existence: 'Darwin's Evolution Theory' & 'Integrated Information Theory of Consciousness' seem apt for the cause of imbibing life into machines. The results drawn by calculating the value of Φ validate the proposed concept of machines with significant amount of consciousness.

Acknowledgement. I would like to extend my sincere gratitude to Dr. A. S. Kanade for his relentless support during my research work.

References

1. Torresen, J.: An evolvable hardware (2004)
2. Joglekar, A., Tungare, M.: Gentic algorithms and their use in the design of evolvable hardware, 3 April 2000
3. Sekanina, L.: Evolvable hardware: from applications to implications for the theory of computation (2009)
4. Tononi, G., Sporns, O.: Measuring information integration, 02 December 2003
5. Kim, H., Hudetz, A.G., Lee, J., Mashour, G.A., Lee, U., ReCCognition Study Group: Estimating the integrated information measure phi from high-density electroencephalography during states of consciousness in humans, 16 February 2018
6. Vasicek, Z.: Bridging the gap between evolvable hardware and industry using cartesian genetic programming. In: Stepney, S., Adamatzky, A. (eds.) Inspired by Nature. Emergence, Complexity and Computation, vol. 28. Springer, Cham (2018). https://doi.org/10.1007/978-3-319-67997-6_2
7. Sekanina, L.: Evolutionary hardware design (2011)

A Cost Optimal Information Dispersal Framework for Cloud Storage System

Sukhwant Kaur$^{(\boxtimes)}$, Makhan Singh, and Sarbjeet Singh

University Institute of Engineering and Technology, Panjab University,
Chandigarh 160014, India
1793sukh@gmail.com, {singhmakhan,sarbjeet}@pu.ac.in

Abstract. In cloud computing, secure storage and retrieval is of significant importance. Along with that, maintaining confidentiality, reliability and availability of data is also an important objective. This can be achieved by dispersing the data into pieces and storing them at different places. But with the increase in data reliability and availability, the cost of maintaining those pieces also increases, which users hesitate to pay. Thus, a strategy is required to maintain a balance between these objectives and cost paid by user. In this paper, efforts have been made to propose a cost optimal information dispersal framework for cloud storage systems that uses an optimization algorithm for the optimal cost expenditure and information dispersal algorithm for the secure storage and retrieval of data. A system architecture is also presented that tells the different components required for the implementation of this strategy.

Keywords: QoS parameters · Implicit security · Information dispersal ·
Knapsack algorithm

1 Introduction

Cloud computing is an information technology model that provides variety of services to cloud consumers satisfying their requirements [1]. Cloud storage is one of the popular service among them that allow users to store their data securely on cloud. There are many cloud service providers (CSP) in the market like Google Cloud Platform [2], IBM Cloud [3], Amazon Web Services [4] etc., but choosing the best one is a difficult task; since they all may vary in terms of services offered, pricing scheme, QoS (Quality of Service) parameters [5] (such as availability, reliability etc.), implementation etc.

Similarly, there exist various users in cloud environment that may have different demands which can be difficult for a single CSP to satisfy proficiently. For example, for some users confidentiality of data is more important than availability or for some reliability is more important than availability. Thus, the users can have different relative importance of QoS parameters. Moreover, they demand all these requirements to be fulfilled under ideal expenses. But, the increase in data reliability, availability of system also increases the storage cost of that data. In short, although the cloud has disentangled the resource delivery process to a great extent, it still has a few difficulties in the field of QoS administration [5]. Thus, it is important to fulfill user's requirements while maintaining tradeoff between QoS parameters and operational cost. Nevertheless,

© Springer Nature Switzerland AG 2020
A. Abraham et al. (Eds.): ISDA 2018, AISC 941, pp. 1055–1064, 2020.
https://doi.org/10.1007/978-3-030-16660-1_103

determining the best possible tradeoff between QoS parameters and outlays is a complicated task.

In this paper, we are proposing a strategy that tries to balance the tradeoff between requested set of QoS parameters and cost paid by the user while securely dispersing information at different locations. An algorithm has been proposed which optimizes the dispersal cost in such a way that the users desired set of requirements are also fulfilled. The concept of knapsack and information dispersal algorithm is used combinely for this strategy. We have also proposed an architecture that tells the different components required for the implementation. Mathematical descriptions have also been provided in the paper.

The rest of the paper is structured as follows: Sect. 2 discusses the related work carried out in the field of information dispersal. In Sect. 3, system architecture is presented, followed by proposed work in Sect. 4. Lastly, Sect. 5 describes conclusion and future scope of the work.

2   Related Work

Information Dispersal is the process of dividing the data that user wants to store into number of pieces and dispersing them on to different locations in encrypted form and the original data can be retrieved by combining those pieces at the receiving end. In this section, some of the work related to the field of information dispersal has been discussed.

Shamir [6] proposed a secret sharing scheme algorithm that works in finite field Zp, and divide the information into 'm' parts such that the entire data can be recreated easily from 'k' parts, but recreation is not possible from 'k − 1' parts. The major drawback of this scheme is that it is space inefficient as the parts are approximately of same size. Rabin [7] also propounded an information dispersal algorithm (IDA) that uses the concept of adding redundancy to the file F of size |f| and then dispersing it into 'n' fragments so that the reconstruction can be performed from any 'm' pieces. But unlike Shamir's algorithm, here, each piece is of size |f|/m. No information is revealed from the individually pieces. Rabin also discusses the numerous applications of IDA from fault-tolerant routing in networks to communication in parallel computing.

In [8], Lyuu proposes a routing method which runs in 2.log N + 1 time and also proves that the concept of Rabin's IDA of adding redundancy provides profitable results in communication between processors in parallel computing. Sun and Shieh [9] also proposed an information dispersal scheme that increases the reliability of servers in distributed environment. Information dispersal degree, information expansion ratio and success – probability of acquiring a correct piece are the three important factors that determine the reliability. The author has also developed a method that easily computes highest reliability of IDS with reduced complexity.

In 'General information dispersal algorithms' [10], Beguin extended the work of Rabin to be used for general access structure by setting some bounds on the information each participant should have. Optimal or near to optimal information dispersal algorithms were also discussed in it. In [11], another scheme has been proposed that guarantee secure storage and retrieval of information even if some severs fail.

Confidentiality was also achieved by using cryptographic techniques. This scheme also gives interesting byproduct i.e. a secret sharing scheme having shorter share size. In [12], Marvin and Yener discussed a scheme in which they enhanced IDA by adding a time bound for reconstruction of the given file. The consequence of scheme on different QoS parameters was also discussed. In [13], the author has analyzed different information dispersal algorithms by means of an information-theoretic framework. A combinatorial condition is also discussed which is sufficient for any tuple of access structure to obtain parts of definite size. Bella and Pistagna [14] use IDA for efficient distribution of backup data on non-hierarchical network using Chord location service. Efficient management of resources and optimization of total redundancy of data is its added advantage.

Parakh [15], also proposed another secret sharing scheme that involves the roots of a polynomial in finite field. The file is divided and kept on individual servers using the scheme and for reconstruction; access to each server is required along with the knowledge of login password. The author discusses the application in sensor networks for data security. A flexible system architecture have been proposed in [16], which represents a secure cloud storage integrator for organizations. The system applies encryption and information dispersion to data files before they leave internal network. In [17], an encoding and two decoding algorithms have been proposed by the author for information dispersal using the concept of Fermat number transforms. The author also claimed improved throughput in the low code rate than the existing algorithms.

Another technique is proposed in [18], for privacy preserving in cloud computing using implicit security model. The technique does not require key management. Use of three independent servers maintains the data privacy. In [19], the author discusses the different security issues existing in the cloud and some counter measures have been also described. Sighom [20] proposed a model that combines advanced encryption standard-256, IDA and secure hash algorithm-512 for improving data security and privacy. Some popular approaches have been also considered for evaluation to find factors that affect system performance.

In [21], Mar, Hu et al. presented a framework that provides resilient and secure storage that can encrusted evidently on existing public and private cloud architecture. The algorithm used in this work is combination of information dispersal algorithm and secret sharing scheme that addresses the confidentiality and availability characteristics of data. The approach also uses adaptable authentication and access control method to provide protection to stored data without depending on encryption. In [22], author proposed an approach called RARE i.e. RAndom REsponse that provides users high satisfaction and stronger privacy. This work majorly focuses on issues related to the client side data de-duplication. The author discusses the risk of revealing the user's file's content or the existence of the copy of the file through the de-duplication checks. To solve this issue author proposed an approach that sends the de-duplication request for two chunks at same time. The receiver cloud gives back the randomized response while maintaining de-duplication and negligible privacy outflow.

From the literature review, we find that research has been done on secure and reliable information dispersal but not much has been done regarding the effect of storage cost paid by user for dispersing information on various datacenters.

To demonstrate the usefulness of our proposed work, we presented a framework, which allows cost effective, secure and reliable data storage on cloud.

3 System Architecture

The proposed framework for the implementation of our strategy is shown in Fig. 1. The cloud environment consists of multiple cloud brokers, cloud consumers and cloud service providers. The cloud service providers can have several data centers, hosts, and virtual machine. The cloud brokers are interconnected with each other in the network. The cloud service providers may vary in terms of services provided, storage capacity, security policies, pricing scheme etc. Every data center is composed of various host machines and several virtual machines can run on these host machines. The cloud environment demands the registration of each entity with the cloud broker. The cloud consumers and CSPs can register themselves with any of the cloud broker as they all are interconnected with each other, so their data will automatically propagate to other cloud brokers also. The cloud broker will store the registration details of all cloud consumers and CSPs in its database.

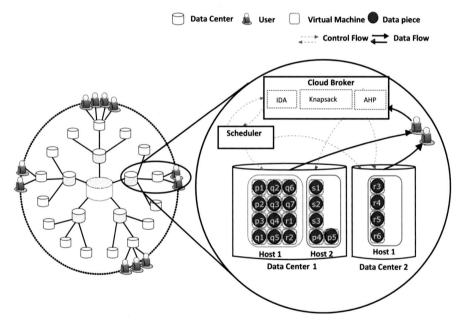

Fig. 1. Information Dispersal Framework showing different components – IDA (Information dispersal algorithm), Knapsack, AHP (Analytic Hierarchy Process) etc.

We have made efforts to present an approach using AHP (Analytic Hierarchy Process), IDA (Information dispersal algorithm) and Knapsack method. In the proposed approach, the cloud consumer first registers himself with the cloud broker in the cloud environment. After the registration, the cloud consumer can request services to the broker. For that, the cloud consumer provides the data file, his budget, relative impor-tance of QoS parameters to the Cloud broker. Then using the AHP method [23], the broker determines the weights of those QoS parameters which will be used further in the approach. The broker then passes the total budget, calculated weights of QoS parameters to the knapsack algorithm. The Knapsack algorithm then generates a plan chart that consists of the lists of selected data centers for dispersal within the cloud consumer's budget for different plans. The cloud consumer then can choose the appropriate plan from the chart. After the cloud consumer has selected a plan, the broker stores the details of chosen plan in its database and instructs the information dispersal algorithm to initiate the dispersal process. The IDA then splits the data file into pieces and with the help of scheduler, distributes those pieces to the data centers according the selected plan by cloud consumer. For the retrieval of file, the cloud consumer can again request the cloud broker. The cloud broker then looks up into its database and provides the list of data centers to the IDA for the reconstruction of file. The IDA fetches those data pieces from data centers using scheduler and recombines the data pieces to generate the file and then file is passed to cloud consumer through the cloud broker.

4   Cost Optimal Information Dispersal Framework

Suppose a user wants store his file 'f' under his budget B and we have total 't' data centers. The file can be stored on any 'n' data centers out of 't' and can be retrieved from any 'm' (threshold) data centers provided $m \leq n$. The COID framework generates a plan chart from which the user can select one plan to store his file. The framework consists of three major algorithms: AHP, Knapsack and IDA. The purpose of AHP algorithm is to provide the relative importance of QoS parameters. In order to keep the total storage cost below the user specified budget, we have used the Knapsack algo-rithm in our approach. The third algorithm is IDA which is used for splitting and dispersing the file to different data centers in secure way. For ease of reference, we have listed the symbols used throughout the paper in Table 1.

Table 1. Symbol table with their meanings

Symbol	Meaning
P	Number of QoS parameters
a_{ij}	Comparative importance of QoS parameter i w.r.t to QoS parameter j
dc_i	i^{th} data center
GM	Geometric mean
w_i	Normalized weight of ith data center
QoS_value(dc_i)	QoS parameter value of ith data center
V_i	Aggregated value of ith data center
b_size	Unit block size
cost(dc_i)	Unit storage cost of data center specified by CSP
st_cost(dc_i)	Actual storage cost of ith data center
n	Number of data centers n which file is to be dispersed
m	Minimum number of data centers needed for reconstruction of file

4.1   Phase 1: Capturing and Processing of Relative Importance of QoS Parameters Using AHP

First, the user gives the desired relative importance value of the QoS parameters to the broker. In this work we have used AHP method [23] for determining the weights of QoS parameters. The user will input his importance using the AHP's elementary scale, shown in Table 2. The scale ranges from value 1 to 9 from which user can choose any value as per his requirement.

This approach is based on geometric mean method that creates a pair-wise comparison matrix to establish the relative importance of each QoS parameter against every other QoS parameter. For 'p' QoS parameters, a square matrix A_{p*p} is constructed where a_{ij} represent the comparative importance of QoS parameter i w.r.t to QoS parameter j. In the matrix, for $a_{ij} = 1$ when $i = j$ and $a_{ji} = 1/a_{ij}$. The relative normalized weight (w_i) of each QoS parameter is calculated by computing the geometric mean of ith row using Eqs. (1) and (2)

$$GM_i = \left[\prod_{j=1}^{p} a_{ij}\right]^{\frac{1}{p}} \tag{1}$$

$$w_i = GM_i / \sum_{i=1}^{p} GM_i \tag{2}$$

Table 2. Elementary scale of importance

Values assigned	Meaning of relative importance
1	Equal importance
3	Moderate importance
5	Strong importance
7	Very strong importance
9	Absolute importance
2, 4, 6, 8	For intermediate importance between the above mentioned values

After the calculation of the weights, the weights (w_i) are sent to the next phase to be used by the Knapsack algorithm.

4.2   Phase 2: Generation of the Plan Chart

In this phase, a plan chart is generated that includes different plans under user's budget B that can be used for the dispersal and storage process without compromising user's requirements. These plans give the optimal allocation list that we will find using Knapsack algorithm [24]. In terms of Knapsack problem, an item is represented by data center and the weight of an item is represented by the storage cost (st_cost) of that data center. Let user budget B is the maximum weight of the knapsack. The item's value is represented by the aggregated value V_i of the data center. For all 't' data centers, the aggregated value V_i is calculated using Eq. (3)

$$V_i = \sum_{k=1}^{p} w_k * QoS\_value_k(dc_i); i = 1 \text{ to } t \tag{3}$$

Similarly the actual storage cost (st_cost) of storing a block of data can be calculated using Eq. (5). On each data center, a data block of size, b_size, will be stored. The storage cost for each data center will be calculated for all the values of 'n' i.e. n = t to1.

$$b\_size = File\ size/m \tag{4}$$

$$st\_cost(dc_i) = b\_size * cost(dc_i) \tag{5}$$

Here 'm' is the minimum no. of data centers needed for the retrieval of file and cost (dc$_i$) is unit storage cost of data center specified by the respective cloud service provider of the data center. The value of 'm' is dependent on the value of 'n' and security chosen by the user, see Table 3. The main aim is to find the optimal list of selected data centers while keeping the total storage cost of the user file to be under or equal to the budget B. The following recursive equation is solved for all the values of 'n' i.e. n = t to 1 for our Knapsack problem.

$$Knap_{i,b,n} =$$
$$\begin{cases} 0; if\ i = t+1 || n = 0 \\ \max\{Knap_{i+1,b-st\_cost[i],n-1} + V_i, Knap_{i+1,b,n}\}; if\ (st\_cost[i] \leq B) \\ Knap_{i+1,b,n}; otherwise \end{cases} \tag{6}$$

Here $Knap_{i,b,n}$ gives the list of selected data centers, where i = 1 to t, b represents the user budget and n represents the maximum no. of data centers that are to be selected from the 't' data centers for storing the file. Generate the plan chart using the Eq. (6). The user then selects the appropriate plan.

Table 3. Security – threshold relation

Security type	Value of threshold (m)
Low	1
Medium	$\lceil n/2 \rceil$
High	n

For example, we have 5 data centers and user's file size is 500 MB. The values of QoS parameters, unit storage cost (cost (dc$_i$)) for each data center and the relative importance value of QoS parameter is generated using pseudo-random generator. For the experiment the user's budget is calculated using the Eq. (7).

$$Budget = 1.75 * Average\ data\ center\ cost * File\ size \tag{7}$$

The plan chart generated by the algorithm is shown below in Table 4.

Table 4. Plan chart for File size = 500 MB when Budget = 4375

Plan no.	n	m	Piece Size = |F|/m (in MB)	Selected data centers	Cost
1.	5	3	167	DC0, DC1, DC2, DC3, DC4	4342
2.	3	2	250	DC0, DC3, DC4	3500
3.	2	1	500	DC0, DC4	4000
4.	1	1	500	DC4	3000

All the plans generated falls under the budget of user. The user can choose any one plan from the plan chart, based on which the broker initiates the dispersal process.

4.3   Phase 3: Dispersal and Reconstruction of File

The next and last phase of the approach is to split the data and allocate it to the specified data centers of the plan chosen. Once the user selects the plan of his choice, the cloud broker initiates the dispersal process. The broker passes the necessity details of the plan such as the value of 'n' and 'm' and the list of selected data centers to the dispersal algorithm. In our approach, for the dispersal and reconstruction of file, Rabin's Information dispersal algorithm [14] is used. In this algorithm, a file F is splitted into 'n' number of pieces of equal length |F|/m in such a way that the file F can be recreated from the available 'm' pieces even if n-m pieces are lost or destroyed by an attacker. The pieces can be distributed to different locations reliably since individual pieces do not disclose any information. The dispersal algorithm splits the file into specified number of pieces i.e. n and with the help of scheduler passes those pieces to the corresponding data centers of the plan.

On the retrieval request, the broker initiates the reconstruction process. The broker uses the details from the database table to fetch the pieces from the available data centers. If minimum no. of pieces 'm' are available the file can be reconstructed by fetching any 'm' pieces from 'n' data centers otherwise the file cannot be reconstructed. After the file reconstructed, broker submits the file to the user.

5   Conclusion and Future Scope

In cloud systems, there are many important QoS parameters like reliability, availability, fault tolerance etc. but their importance can vary from user to user. Thus, the users have different demands and fulfilling all of them proficiently can be difficult for a single CSP, especially under ideal expenditure. In the presented work, efforts are made to balance out the tradeoff between QoS parameters and user's budget. In this work, a cost optimal information dispersal framework for cloud storage systems is proposed, in which a user file is dispersed among different data centers based on the user's requirement within the budget specified. The concept of Knapsack is used to maintain the balance between user's requirement and the budget. For dispersal, an information dispersal algorithm is used that splits the file into pieces and then distribute it among the data centers. In future, to avoid any data modification, security techniques like digital signature etc. could be added. Also, different cloud architecture can be created and tested for the work.

References

1. Buyya, R., Yeo, C.S., Venugopal, S., Broberg, J., Brandic, I.: Cloud computing and emerging IT platforms: vision, hype, and reality for delivering computing as the 5th utility. Futur. Gener. Comput. Syst. **25**(6), 599–616 (2009)
2. Google Cloud Storage. https://cloud.google.com/products/cloud-storage/. Accessed 15 Aug 2018
3. IBM Cloud. https://www.ibm.com/cloud/. Accessed 15 Aug 2018
4. Amazon Web Services. http://aws.amazon.com/. Accessed 15 Aug 2018
5. Ardagna, D., Casale, G., Ciavotta, M., Perez, J.F., Wang, W.: Quality-of-service in cloud computing: modeling techniques and their applications. J. Internet Serv. Appl. **5**(1), 11 (2014)
6. Shamir, A.: How to share a secret. Commun. ACM **22**(11), 612–613 (1979)
7. Rabin, M.O.: Efficient dispersal of information for security, load balancing, and fault tolerance. J. ACM (JACM) **36**(2), 335–348 (1989)
8. Lyuu, Y.-D.: Fast fault-tolerant parallel communication and on-line maintenance for hypercubes using information dispersal. Math. Syst. Theory **24**, 273–294 (1991)
9. Sun, H.M., Shieh, S.P.: Optimal information dispersal for increasing the reliability of a distributed service. IEEE Trans. Reliab. **46**(4), 462–472 (1997)
10. Beguin, P., Cresti, A.: General information dispersal algorithms. Theor. Comput. Sci. **209**, 87–105 (1998)
11. Garay, J.A., Gennaro, R., Jutla, C., Rabin, T.: Secure distributed storage and retrieval. Theor. Comput. Sci. **243**, 363–389 (2000)
12. Nakayama, M.K., Yener, B.: Optimal information dispersal for probabilistic latency targets. Comput. Netw. **36**, 695–707 (2001)
13. De Santis, A., Masucci, B.: On information dispersal algorithms. In: Proceedings of IEEE International Symposium on Information Theory, no. 410. IEEE (2002)
14. Bella, G., Pistagna, C., Riccobene, S.: Distributed backup through information dispersal. Electron. Notes Theor. Comput. Sci. **142**, 63–67 (2006)
15. Parakh, A., Kak, S.: Online data storage using implicit security. Inf. Sci. **179**, 3323–3331 (2009)
16. Seiger, R., Groß, S., Schill, A.: SecCSIE: a secure cloud storage integrator for enterprises. In: 2011 IEEE 13th Conference on Commerce and Enterprise Computing (CEC), pp. 252–255. IEEE, Luxembourg (2011)
17. Lin, S.J., Chung, W.-H.: An efficient (n, k) information dispersal algorithm based on fermat number transforms. IEEE Trans. Inf. Forensics Secur. **8**(8), 1371–1383 (2013)
18. Parakh, A., Mahoney, W.: Privacy preserving computations using implicit security. IEEE (2013)
19. Asha, S.I.: Security issues and solutions in cloud computing a survey. Int. J. Comput. Sci. Inf. Secur. (IJCSIS) **14**(5), 309–315 (2016)
20. Sighom, J.R.N., Zhang, P., You, L.: Security enhancement for data migration in the cloud. Future Internet **9**(23), 23 (2017)
21. Mar, K.K., Hu, Z., Low, C.Y., Wang, M.: Securing cloud data using information dispersal. In: 2016 14th Annual Conference on Privacy, Security and Trust (PST). IEEE, New Zealand (2016)

22. Pooranian, Z., Chen, K., Yu, C., Conti, M.: RARE: defeating side channels based on data-deduplication in cloud storage. In: IEEE INFOCOM 2018 - IEEE Conference on Computer Communications Workshops (INFOCOM WKSHPS), pp. 444–449. IEEE, Honolulu (2018)
23. Saaty, T.L.: Theory and applications of the analytic network process: decision making with benefits, opportunities, costs, and risks, 3rd edn. RWS Publications, Pittsburgh (2005)
24. Horowitz, E., Sahni, S.: Computing partitions with applications to the knapsack problem. J. Assoc. Comput. Mach. **21**(2), 277–292 (1974)

Multiple Sequence Alignment Using Chemical Reaction Optimization Algorithm

Md. Shams Wadud, Md. Rafiqul Islam, Nittyananda Kundu[(✉)],
and Md. Rayhanul Kabir

Computer Science and Engineering Discipline, Khulna University,
Khulna 9208, Bangladesh
abbir.ku@gmail.com

Abstract. In bioinformatics, Multiple Sequence Alignment (MSA) is an NP-complete problem. This alignment problem is important in computational biology due to its usefulness in extracting and representing biological importance among sequences by finding similar regions. MSA is also helpful for finding the secondary or tertiary structure of the protein and using it critical anonymousness motives of DNA or Protein can also be found. For solving the problem, we have proposed a method based on Chemical Reaction Optimization (CRO). We have redesigned the basic four operators of CRO and three new operators have been designed to solve the problem. The additional operators are needed in order to arrange the base symbols properly. For testing the efficiency of our proposed method DNA sequences have taken from the different sources. We have compared the experimental results of the proposed method with clustal-omega and got better results for DNA sequences.

Keywords: Multiple sequence alignment · Meta-heuristic ·
Chemical reaction optimization algorithm · Sum of pair ·
Repair mechanism

1 Introduction

In bioinformatics, Multiple sequence alignment (MSA) is a sequence alignment of three or more DNA, RNA or protein sequences of similar lengths which is a well-known NP-complete problem. It is an important task in bioinformatics because MSA can find out identical or functional residue easily among the sequences. After arranging a sequence with another sequence or sequences it is easy to identify regions of similarity among those sequences. Multiple alignment is an essential pre-requisite to many further analyses of protein, DNA and RNA families such as phylogenetic tree reconstruction or homology modeling or are simply used to show preserved and moving sites of a family [1].

A sequence can be seen as a string of alphabets. DNA or RNA sequence is a string of four characters and protein sequence is a string of twenty characters.

A. Abraham et al. (Eds.): ISDA 2018, AISC 941, pp. 1065–1074, 2020.
https://doi.org/10.1007/978-3-030-16660-1_104

From a set, $S = (S_0, S_1, ..., S_{(N-1)})$ of N sequences, we want to find out common patterns of these sequences, using MSA we can find out these common patterns which show evolutionary relationship by which they share a linkage and are descended from a common ancestor [2]. There are different methods that solves this problem. All of them focus on one situation and that is to build an alignment which is optimal or near optimal, which can be known by calculating its score. For measuring the alignment score, many scoring systems were invented. Sum of Pairs (SP) is one of them. We can calculate the score of the multiple alignment by SP. Researchers show interest in solving the MSA problem because it has a lot of applications such as in finding diagnostic patterns to characterize protein families; in detection or demonstrate homology among new sequences and existing families of sequences; to suggest oligonucleotide primers for PCR [3]; to detect key functional residue to inferring the evolutionary history of protein family [4]; to identify new families and extend the existing families [5]. There are different types of algorithms such as exact algorithm like dynamic programming (DP), greedy algorithm, heuristic, and meta-heuristic etc. for solving this problem. DP and greedy algorithm are good for small number of sequences but the efficiency degrades when generality is increased [6,7]. The heuristic and meta-heuristic algorithm performs well but slowly for more than twenty sequences. These are not faster like DP or greedy algorithm [1,8]. A population based meta-heuristic algorithm called Chemical Reaction Optimization (CRO) was invented by Lam and Li in 2010 [9]. We have proposed a meta-heuristic algorithm called MSA_CRO for solving MSA problem. Here, we have defined our algorithm as MSA-CRO for solving MSA problem. We have redesigned the four basic operators of CRO and designed three repair methods for our problem. For testing our method, we took the DNA sequences and the obtained results were compared with ClustalOmega [15].

2   Related Work

Due to NP-completeness, it is hard to find out the optimal solution of MSA problem. To find the optimal solution researchers proposed various approaches. Some of the approaches are described below.

Marco Dorigo and colleagues introduced the first ACO algorithms in the early 1990's [10]. Chen et al. [11] applied the ACO algorithm to solve the MSA problem. Here the algorithm works in three stages. First, it partitions the sequences into several subsequences. Second, it implements the ACO to align the subsequences. Finally, an alignment of the original sequences is obtained by assembling the results from multiple partitions. For partitioning the sequences into several subsequences it follows some strategies. First, it estimates the matching score then it searches for the cut off points. After finding the cutoff point it aligns the subsequences and calculates the score of the alignment. In this process, every ant finds out a solution path but do not cross another path.

Xu and Chen [12] used PSO for solving the MSA problem and the authors called their process as PSOMSA. Gap deletion, gap insertion, and local search operator were used in their algorithm. Their simulation result shows that the PSOMSA performs better then Clustal Lei et al. [13] apply CPSO (Chaotic PSO) in their paper for solving the MSA problem. They apply this chaotic technique in solution space so that particles can be distributed in a uniform manner. Though the difference between maximum and minimum of SPS (Sum of pair Scoring) is less in their simulation but still there are some distances with that of benchmark alignment. Jagadamba et al. proposed MSAPSO algorithm in 2011 [14]. Their result shows that their alignment score is little less than the standard method and residue match is higher in their method.

GA-ACO is an algorithm of GA with ACO for the MSA problem, proposed by Lee et al. in 2006 [8]. This algorithm enhances the performance of GA by organizing the local search and ACO for alignment of sequences. GA gives a diversity of alignment and ACO performs to find out local optima. This algorithm works with its own operators such as sexual reproduction, mutation, and heuristic operators. It gives equal or better results than existing algorithms such as ClustalW, GA without ACO, Central-star, Horng's genetic algorithm [8].

Clustal omega proposed by Sievers et al. [15]. Clustal omega is a revised version of well-known clustal series of programs for sequences alignment. This program can deal with a large dataset of any DNA, RNA and protein sequences. For alignment, this program follows the basic process of the previous version of this family, clustalX, and clustalW. First, this program calculates guide-tree which is constructed by pairwise distance among the inputted sequences. During alignment, the order of the sequences is determined on this guide-tree. Clustal omega also has a powerful feature for adding sequences to existing alignments.

Pairwise sequence alignment using CRO was proposed by Huang and Zhu [16] works only in two sequences. They redesigned their operators based on CRO and its parameters. For representing a solution they have used a one-dimensional array and 20% scaling factor. Their objective function is to find a minimum potential energy (PE) according to the working principle of CRO. They took three short datasets and compared their results with GA, ACO and NW algorithm and they get better result.

3   Design of CRO for MSA Problem

The procedure of CRO algorithm is to initialize the parameters, create molecules with size equal to PopSize and assign random molecule structure to each molecule, take unimolecular or inter-molecular collision depending on MoleColl and loop through the selection process depending on the stopping criteria. Initially, alignments are randomly generated which are assigned to each molecule. These molecules are optimized using CRO algorithm. Here we have redesigned the CRO algorithm for solving multiple sequence alignment, so we call our algorithm as MSA-CRO.

There are many ways to represent a molecule structure. An unaligned data set contains k sequences. Here k sequences are as $S = S_1, S_2, ..., S_k$ where S is a sequence set and $S_1, S_2, ..., S_k$ are the sequences of that set. The lengths of sequences are $|S_1|, |S_2|, ..., |S_k|$ respectively. These k sequences can be represented as a matrix. Each sequence is considered as a row of the matrix and maximum number of $columnx = [\gamma * len_{max}]$ where $len_{max} = MAX(|S_1|, |S_2|, ..., |S_k|)$ and parameter γ is a scaling factor.

The main task of objective function is to find the PE of each molecule. PE is calculated using $TS(\omega)$ function where ω is the molecule structure which is a matrix. SP (Sum of pairs) calculates total score of a matrix. First a function $Score(S_i, S_j)$ is defined which takes two distinct sequences and compare function is used for calculating each pair score of those sequences. These pairs are generated and its score are summed by TS (Total Score) which uses Score function. The score of two sequences S_i and S_j of an alignment ω is defined as follows. The Eqs. (1) and (2) are similar to the equations used in GA-ACO [8].

$$Score(\omega_i, \omega_j) = \sum_{l=1}^{x} Compare(\omega_{il}, \omega_{jl}) \tag{1}$$

Then the total score of an alignment ω is defined as follow:

$$TS(\omega) = \sum_{i=1}^{k-1} \sum_{j=i+1}^{k} Score(\omega_i, \omega_l) \tag{2}$$

The objective function is defined as shown in [16] as follows:

$$PE = -TS(\omega) \tag{3}$$

Here ω_i, ω_j are two rows of the molecule matrix ω and ω_{il}, ω_{jl} are two cells of the matrix. Also k is the number of sequences and x is alignment length. For scoring we have used Eqs. (1) and (2) and for the fitness of the molecule ω Eq. (3) has been used.

3.1   Population Initialization

The initial population is generated by randomly initializing the matrices. A matrix initialization works by the following way. First, x random numbers are generated between 1 to x for each sequence where each random number is distinct from one another. Secondly, first $|S_i|$ random numbers are selected and sorted in ascending order where $|S_i|$ means number of characters in the sequence S_i. Thirdly, these sorted numbers are used as column indices for the row i where each character of sequence S_i will be inserted according to the column indices. By following this process each matrix of the population is initialized. Figures 1 and 2 illustrate this process. This process is as similar to the process used in GA-ACO [8].

3.2   CRO Operators

In chemical reaction, the four operators such as on wall ineffective collision, decomposition, intermolecular ineffective collision and synthesis are performed. The design process of these reactions are described below.

	Sequence	Length	Permutation	Positions	Sorted Positions
S1	GCTTA	5	64132875	32146	12346
S2	CATGGA	6	41367582	734156	134567
S3	AGCT	4	75862314	8567	5678

Fig. 1. Illustration of population initialization

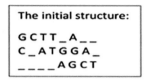

The initial structure:

G C T T _ A _ _
C _ A T G G A _
_ _ _ _ A G C T

Fig. 2. Initial structure of molecule

On Wall Ineffective Collision. In on wall ineffective collision, one molecule takes part. For a given structure ω_1 each pair is scored separately. Suppose, if we are given N sequences then we search all pairs which is equal to $N*(N-1)/2$ and then we calculate the score for each pair and choose the worse one. After that one row is chosen as reference row and the other row is chosen as changeable row. It depends on which row has the maximum number of base symbols. The row which contains maximum base symbols is the reference row and the other is changeable row. After that for each base symbol of the reference row our operator tries to find the same base symbols in the changeable row. Then it tries to align the base symbol of the changeable row with the base symbol of reference row. By this way ω_1' is created. This process is shown in Fig. 3.

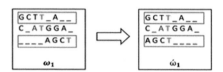

Fig. 3. Illustration of on wall ineffective collision

Decomposition. Decomposition happens on one molecule and it creates two new molecules. There are some structural changes between these two molecules. First, the upper half and lower half of the molecule structure ω is calculated. Then for each row of the upper half, a starting and ending point is selected randomly. After that, a character and a space are selected randomly between these two points. Then we have to shift the characters to the left if space index is less then the character index, otherwise shift the characters to the right. Thus molecule structure ω_1' is created. The same procedure is applied on the lower half of the molecule structure ω and ω_2' is created. This process is shown in Fig. 4.

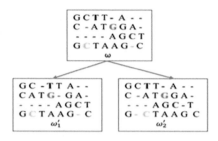

Fig. 4. Illustration of decomposition

Inter-molecular Ineffective Collision. In Inter-molecular ineffective collision, two molecules collide with each other and make some changes between those two molecules. In this operator for a given structure of a molecule first, we have to count the number of gaps in each column of the given structure ω. Second, a column contains the highest number of gaps should be considered. If there is more than one column which contains the highest number of gaps then one of them is chosen randomly. Third, in the selected column one base symbol is chosen randomly as the reference base symbol. Fourth, we find the same base symbol in other sequences and try to align the base symbols with the selected base symbol. This procedure is applied to two of the molecules. By this way ω_1' and ω_2' is created. In Fig. 5, the demonstration is shown.

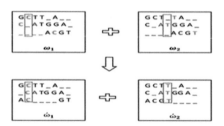

Fig. 5. Illustration of inter-molecular ineffective collision

Synthesis. In synthesis, two molecules collide with each other. In this operator, output is one molecule from two input molecule. At first, number of base symbols are selected from each sequence randomly. Then cut off points are calculated from each of the molecules according to the number of randomly selected base symbols. Here, a random number will be between 1 to number of base symbols of S_i of a molecule. Then according to the information of cut off points, each molecule is searched and store left and right sides of the molecules. For creating the first offspring left side of the second molecule and right side of the first molecule are selected. On the other hand for the second offspring right side of the second molecule and left side of the first molecule are chosen. When we merge two sides if there is a shortage of character(s) then we have to insert space(s) between left and right side. If the summation of the characters of two sides of the offspring is greater than the number of columns then excess space is removed from the right side of the newly created molecule. This operation is illustrated in Fig. 6.

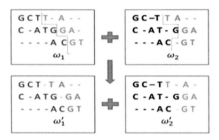

Fig. 6. Illustration of synthesis

Repair Operators. Repair function works in three phases. First, one is called array compression, the second one is called forward repairing and the third is for left-right adjustment. These operators are needed in order to arrange the base symbols properly. Detail of these operators is described below.

In array compression, a column is selected and if the column contains at least one space then for each of the spaces it finds the next miss match column to its right. After that, it shifts the characters to left and puts the space in the miss match column. If the column contains more than one space then it performs the procedure for all the spaces. This process is performed for all columns.

In forward repairing (FR), a column is selected and the number of different base symbols is calculated. Then after sorting the base symbols with respect to the descending order of their calculated numbers, the first base symbol is selected for alignment. If there exists any space in that selected column some operations are done.

In left right adjustment (LRA) each column is considered and numbers of base symbols are calculated for that column. Then we calculate the percentage of base symbols by numOfBaseSym/numOfSeq of that column. If the column contains less than 35 percent of base symbols only then some approaches are followed.

4   Experimental Results

In the proposed algorithm we used the following parameters settings: popSize = 100, MoleColl = 0.9, α (decomposition threshold) = 70, β (synthesis threshold) =

Table 1. Test results

Data set	Number of sequences	Length (Max, Min, Avr)	MSA-CRO		ClustalOmega			Similarity (in %)
			Score	MC	Score	MC	Note	
D1	10	212, 211, 212	**18558**	**197**	18450	195	Better	92.92
D2	8	457, 457, 457	**25301**	**448**	25301	448	Same	98.03
D3	21	122, 122, 122	**47889**	**109**	47889	109	Same	89.34
D4	21	348, 324, 334	52868	33	**69880**	**82**	Worst	4.31
D5	6	1311, 1305, 1308	**38880**	**1294**	38880	1294	Same	98.92
D6	8	315, 310, 313	**15448**	**243**	15413	240	Better	77.63
D7	8	1681, 1681, 1681	**87690**	**1440**	87690	1440	Same	85.66
D8	9	384, 381, 383	**23285**	**261**	23188	260	Better	68.14
D9	13	515, 513, 514	**78966**	**492**	78954	492	Better	95.72
D10	16	328, 324, 326	**77187**	**311**	74787	301	Better	95.40

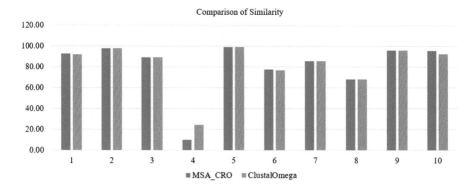

Fig. 7. Comparison of similarity

700. numOfIter = 1310, numOfSumu = 20. We run our program in C# language with Intel core i5-5200cpu 2.20GHz, 4.00 RAM, 64 bit 8.1 windows operating system.

To calculate the score we used (2) and used the average score. To get the average score we run the program for 20 times for all datasets. As experiment results, we compared the scores and number of match columns (MC) with clustalOmega. We aligned the data sets using the online alignment tool of clustalOmega [17]. Then score in our own program. An experiment on DNA sequences was performed. For our experiment, we took 10 datasets. Within 10 datasets we take D1 to D4 from GA-ACO paper [8], D5 to D8 collected from GA for MSA [18] paper. Remaining data sets were collected from the paper in [19]. For calculating the score of an alignment we use +2, −1, −2 and 0 for the match, mismatch, gap penalty and both gaps respectively. We have calculated the similarity of particular dataset using Eq. (4). Table 1 represents the obtained results.

$$Similarity = \frac{MC}{AvgLen} \tag{4}$$

Here, MC = number of matched columns, AvgLen = average length of sequences.

In Fig. 7, the comparison of Similarity among MSA-CRO and ClustalOmega is shown.

From the experiment results presented in Table 1 and Fig. 7, it shows that MSA-CRO gives better or equal results than clustalOmega for all data sets except data set D4.

5   Conclusion

In our paper, we have proposed a new method based on CRO for solving the MSA problem. CRO is a recently proposed meta-heuristic algorithm that has already applied to solve many well-known problems. By using the framework of CRO, we redesigned its existing four operators to solve the MSA problem with three new repair operators to obtain a better solution. These operators have helped to increase the performances of MSA-CRO dynamically. The main challenge of solving an optimization problem is to search the whole solution space. For the experiment, we only used DNA sequences and from the simulation results, it can be observed that the proposed algorithm shows a better outcome than clustalOmega. The effects of repair operators were demonstrated by the simulation results. Theses repair operators make the huge differences in finding the proper outcome. For different parameter settings, MSA-CRO gives different results. For a particular setting, it may give much better result but it is a difficult task. So an intensive study needs for this.

References

1. Notredame, C., Higgins, D.G., Heringa, J.: T-coffee: a novel method for fast and accurate multiple sequence alignment. J. Mol. Biol. **302**(1), 205–217 (2000)
2. Hassanien, A.E., Milanova, M.G., Smolinski, T.G., Abraham, A.: Computational intelligence in solving bioinformatics problems: reviews, perspectives, and challenges. In: Computational Intelligence in Biomedicine and Bioinformatics, pp. 3–47. Springer, Berlin (2008)
3. Thompson, J.D., Higgins, D.G., Gibson, T.J.: CLUSTAL W: improving the sensitivity of progressive multiple sequence alignment through sequence weighting, position-specific gap penalties and weight matrix choice. Nucleic Acids Res. **22**(22), 4673–4680 (1994)
4. Katoh, K., Misawa, K., Kuma, K.I., Miyata, T.: MAFFT: a novel method for rapid multiple sequence alignment based on fast Fourier transform. Nucleic Acids Res. **30**(14), 3059–3066 (2002)
5. Edgar, R.C.: MUSCLE: a multiple sequence alignment method with reduced time and space complexity. BMC Bioinf. **5**(1), 1 (2004)
6. Kaya, M., Kaya, B., Alhajj, R.: A novel multi-objective genetic algorithm for multiple sequence alignment. Int. J. Data Min. Bioinf. **14**(2), 139–158 (2016)
7. Zhang, Z., Schwartz, S., Wagner, L., Miller, W.: A greedy algorithm for aligning DNA sequences. J. Comput. Biol. **7**(1–2), 203–214 (2000)

8. Lee, Z.J., Su, S.F., Chuang, C.C., Liu, K.H.: Genetic algorithm with ant colony optimization (GA-ACO) for multiple sequence alignment. Appl. Soft Comput. **8**(1), 55–78 (2008)

9. Lam, A.Y., Li, V.O.: Chemical-reaction-inspired metaheuristic for optimization. IEEE Trans. Evol. Comput. **14**(3), 381–399 (2010)

10. Blum, C.: Ant colony optimization: introduction and recent trends. Phys. Life Rev. **2**(4), 353–373 (2005)

11. Chen, J., et al.: Partitioned optimization algorithms for multiple sequence alignment. In: 2006 20th International Conference on Advanced Information Networking and Applications, AINA 2006, vol. 2. IEEE (2006)

12. Xu, F., Chen, Y.: A method for multiple sequence alignment based on particle swarm optimization. In: Emerging Intelligent Computing Technology and Applications. With Aspects of Artificial Intelligence, pp. 965–973. Springer, Berlin (2009)

13. Lei, X.J., Sun, J.J., Ma, Q.Z.: Multiple sequence alignment based on chaotic PSO. In: International Symposium on Intelligence Computation and Applications, pp. 351–360. Springer, Berlin, October 2009

14. Jagadamba, P.V.S.L., Babu, M.S.P., Rao, A.A., Rao, P.K.S.: An improved algorithm for multiple sequence alignment using particle swarm optimization. In: 2011 IEEE 2nd International Conference on Software Engineering and Service Science, pp. 544–547. IEEE, July 2011

15. Sievers, F., Wilm, A., Dineen, D., Gibson, T.J., Karplus, K., Li, W., Lopez, R., McWilliam, H., Remmert, M., Söding, J., Thompson, J.D., Higgins, D.G.: Fast, scalable generation of high-quality protein multiple sequence alignments using Clustal Omega. Mol. Syst. Biol. **7**(1), 539 (2011)

16. Huang, D., Zhu, X.: A novel method based on chemical reaction optimization for pairwise sequence alignment. In: International Conference on Parallel Computing in Fluid Dynamics, pp. 429–439. Springer, Berlin, May 2013

17. Clustal Omega < Multiple Sequence Alignment < EMBL-EBI. http://www.ebi.ac.uk/Tools/msa/clustalo/

18. Horng, J.T., Wu, L.C., Lin, C.M., Yang, B.H.: A genetic algorithm for multiple sequence alignment. Soft Comput. **9**(6), 407–420 (2005)

19. HAlign: Fast Multiple Similar DNA/RNA Sequence Alignment based on Center Star Strategy. http://datamining.xmu.edu.cn/software/halign

Forensic Approach of Human Identification Using Dual Cross Pattern of Hand Radiographs

Sagar V. Joshi$^{(\boxtimes)}$ and Rajendra D. Kanphade

Department of E&TC Engineering, Dr. D. Y. Patil Institute of Technology,
Sant Tukaram Nagar, Pimpri, Pune 411 018, Maharashtra, India
`sagarvjoshi@gmail.com`, `kanphaderd2015@gmail.com`

Abstract. The demand for personal identification systems has augmented in recent years, due to serious accidents and required for criminal investigation. Under natural calamity and human-made disasters sometimes it is impossible to use traditional biometric techniques based on fingerprints, iris, and face; in such cases, biometric radiographs like dental, hand and skull are the great alternatives for the victim's identification. The key objective of this study is to present a unique technique to deal with missing and unidentified person identification based on hand radiographs using Dual Cross Pattern (DCP). The proposed system has two main stages: feature vector extraction, and classification. In this paper, an attempt has been made to find out the most suitable classifier among k-nearest neighbor (k-NN) and Classification Tree based on the accuracy of retrieval of 10 subjects with 100 right-hand radiographs. The result achieved from experiments on a small primary database of radiographs reveals that matching hand radiographs based on DCP can be significantly used for human identification.

Keywords: Ante-mortem (AM) radiographs · Biometrics ·
Dual Cross Pattern · Postmortem (PM) radiographs

1 Introduction

The process of human identification is of great importance for the procedures in the different areas of law and is demanded by the community for cultural or religious reasons. Biometric authentication has played an important role in identifying individuals. Biometrics refers to the measurement of attributes or specific characteristics of the human body, such as fingerprints, retina, iris, and even the voice to distinguish that person from others. These features are unique to each person, making them an access password for the user. On the other hand, for the identification of mass disaster victims, such as earthquakes, fires, tsunami, etc., conventional biometric features may not be applicable. Forensic radiography is a part of forensic medicine that covenants the detection of personnel with PM radiological metaphors of a mixture of a component of the body counting the skeleton, skull, and teeth [1–14]. The unlabeled PM radiographs of the body of the deceased are compared with the AM records of a missing person to determine the similarities between them [1–3]. There are a lot of techniques available

© Springer Nature Switzerland AG 2020
A. Abraham et al. (Eds.): ISDA 2018, AISC 941, pp. 1075–1084, 2020.
https://doi.org/10.1007/978-3-030-16660-1_105

for human identification with dental radiographs [2–7], however, due to continuous dental work, it becomes difficult to prepare & maintain a dental chart for each individual tooth. Another major challenge for automatic dental identification system is changing in teeth over time, such as eruption and loss of teeth, genesis, prosthetic replacement, removal of adjacent teeth after a tooth, interventions orthodontics, etc. These changes cause the inconsistency in the appearance makes it difficult to model [4]. To overcome the limitations of dental identification system hand radiographs become the main alternative scientific method available.

The first attempt [12, 13] for automatic analysis of disease and undernourishment on the skeleton growth of child was proposed by Pathak et al. in 1984. Bone age assessment (BAA) was performed using a syntactic fuzzy classifier by using primitives like points, curve and line segments extracted from input radiographs. Due to the use of points and 3-stage hierarchical system, it becomes challenging to appreciate the proposed system by radiologists. Pietka et al. developed a fuzzy classifier [14] for BAA based on regions extracted separately from phalangeal [15] and carpals [16] of hand radiographs. But for the final decision, the radiologist is mandatory. Kauffman et al. [17] have shown a technique to separate biometric highlights from the states of the hand bones, being invariant to the situating of the hand and fingers. These highlights can be utilized to check a patient's character and to identify conceivable irregularities in a patient database. Tests performed to coordinate 45 left-hand radiographs to a database of 44 (coordinating) right-hand pictures and the other way around propose that the states of the hand bones contain biometric data that can be utilized to distinguish people. Pan Lin et al. [18] developed carpal-bone feature extraction technique using level set method for skeletal age estimation application. Anisotropic diffusion filter and region-based level set method was recommended for preprocessing and segmentation of captured hand x-ray images. The experimental results were promising and can be extended for other bone regions. Two-stage approach [19] using affine transformation and intensity based method for BAA was proposed by the researcher. In [20] fuzzy methodology based bone age assessment by translation natural language descriptor of Tanner-Whitehouse (TW3) method into an automatic classifier was proposed. Davis et al. [21] suggested Dynamic Time Warping based human hand radiograph segmentation and classification depend on the active appearance model. In [22] hand x-ray segmentation method was developed using a watershed algorithm, gradient magnitude thresholding, and anisotropic diffusion filter. The experiment was performed on a dataset of 30 young children's. The national Tsing Hwa University in Taiwan recommended BAA based on the phalangeal region of interest (PROI) and Support Vector Machine (SVM) with an evaluated accuracy of 85% [24]. An automated bone BAA method using histogram [25] with the error rate of -0.170625 years was introduced by Mansourvar et al. in 2012.

The proposed system shows human identifications based on DCP feature descriptor. Texture feature analysis plays a very vital role in a content-based retrieval system. Texture feature is an important low-level feature in an image as it gives the spatial arrangement of color & intensities of an image. Local Binary Pattern (LBP) [26] and Discrete Wavelet Transform (DWT) [27] are other admired methods of texture feature extraction, but DCP is significant as it extracts the second order discriminative information of an input image in the horizontal, vertically and diagonal direction [28].

With the maximum joint Shannon entropy grouping of sampled pixels, DCP extracts pixel information two times that of the LBP [28]. Figure 1 shows sample images from the collected dataset.

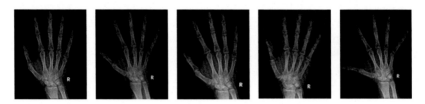

Fig. 1. Sample images from dataset.

2   System Overview

The progression of human identification from hand radiographs consists of several steps. The system (see Fig. 2) can perform two tasks when an image is entered into it, whether it is used for training to create the database or to search for a corresponding record in the database. In both cases, the features must be extracted from the image. The difference between the two tasks occurs at the end of the process when the system decides which function to perform. When the system is formed, that is, when entries are made in the database, the last step is to store the extracted information in the database. During retrieval, the last step is to compare the extracted data feature values with those data records already stored in the database. In this paper, it is assumed that hand radiographs would remain unchanged even after the death of the subject under consideration. This hypothesis is supported by an experienced expert in forensic science [17]. The radiographs captured are from adults only, and the proposed method is not tested upon children's database.

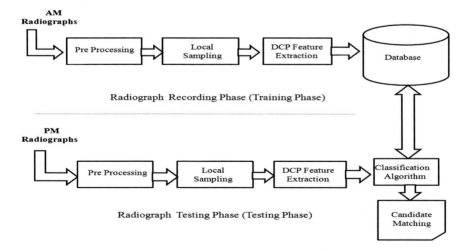

Fig. 2. Logical module of DCP based human identification system using hand radiographs.

3 Dual Cross Pattern Feature Extraction

A hand X-ray consists of 3 parts: Phalanges Metacarpals & Carpals. The proposed work only focuses on DCP feature extraction of phalanges & metacarpals, as a study of carpals can be considered as the separate problem itself [13]. DCP feature extraction mainly consists of image filtering, local sampling, and pattern encoding [28]. DCP encrypts second-order statistical information because of unique pattern encoding & local sampling technique.

3.1 Local Sampling

The key to the success of DCP is to execute local sampling (see Fig. 3) and pattern encoding in the most valuable directions confined in hand X-ray radiographs [28]. The local sampling is conducted symmetrically in all eight directions (0, $\pi/4$, $\pi/2$, $3\pi/4$, π, $\pi/4$, $5\pi/4$, $3\pi/2$, and $7\pi/4$) for each pixel I_0. Sampling in all neighborhood direction produces a sufficient amount of texture information. In each direction two pixels are sampled as $\{I_{A1}, I_{B1}; I_{A2}, I_{B2}; I_{A3}, I_{B3}; I_{A4}, I_{B4}; I_{A5}, I_{B5}; I_{A6}, I_{B6}; I_{A7}, I_{B7}; I_{A8}, I_{B8}\}$.

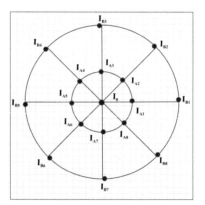

Fig. 3. The local sampling process of sixteen points around the central pixel I_0 in DCP.

3.2 Pattern Encoding

All locally sampled points are encoded into two steps: encoding textural information independently in all eight directions and formation of DCP code by combining patterns in all eight directions. To estimate the textural statistics in every sampling direction, we allocate each pixel an individual decimal quantity as:

$$DCP_i = D(I_{Ai} - I_0) \times 2 + D(I_{Bi} - I_{Ai}), 0 \le i \le 7 \tag{1}$$

Where,

$$D(x) = \begin{cases} 1, x \geq 0 \\ 0, x < 0 \end{cases} \tag{2}$$

And I_0, I_{Ai}, and I_{Bi} denote grey level values of the center pixel and neighborhood pixel A_i, and B_i, respectively. The second order statistics are determined by four patterns along each direction and each of four patterns denotes distinct textural information. As DCP consider all eight neighborhoods, the overall wide variety of DCP codes is $4^8 = 65536$. This code descriptor is just too massive for realistic hand recognition system; therefore, we undertake the subsequent approach. By grouping 8 directions into two subsets and further formulating every subset as an encoder, the full quantity of local pattern is reduced to $4^4 \times 2 = 512$, which is computationally competent. By adopting this approach, compactness and robustness of the descriptor are promoted at the cost of information loss. Algorithm for DCP pattern encoding is as follows:

Algorithm:

Step 1:Select image

Step 2: Convert into gray image

Step 3:Select Radius $R_1 = 1$ and Radius $R_2 = 2$

Step 4:For i =1:8

$DCP_i = (I_{A(i)} - I_0) \times 2 + (I_{B(i)} - I_{A(i)})$

    end

Step 5: if $DCP_i > 0$ then

    $D_{A(i)} = 1$

      else

    $D_{A(i)} = 0$

    end

Step 6: end

3.3   Dual Cross Grouping

After partitioning eight directions by grouping technique as explained in the earlier section, total 35 combinations are produced. For correct retrieval, conservation of needed information is achieved by Joint Shannon entropy criteria used in the grouping of eight directions. Above analysis returns DCP_i $(0 \leq i \leq 7)$ with any one value out for 0, 1, 2 or 3. The joint Shannon entropy [28] for the set $\{DCP_0, DCP_1, DCP_2, DCP_3\}$ is represented as:

$$H(DCP_0, DCP_1, DCP_2, DCP_3)$$
$$= -\sum_{dcp0} \cdots \sum_{dcp3} P(dcp_0 \ldots dcp_3) \log_2 P(dcp_0 \ldots dcp_3) \tag{3}$$

Where, dcp_0, dcp_1, dcp_2, dcp_3, and $P(dcp_0, dcp_1, dcp_2, dcp_3)$ are particular values and probabilities of DCP_0, DCP_1, DCP_2, and DCP_3 respectively. For statistical independence, joint Shannon entropy is greatest. Hence, for more independent images in each subgroup maximum joint Shannon entropy is acquired. Therefore, two subsets (DCP$_0$, DCP$_2$, DCP$_4$, DCP$_6$) and (DCP$_1$, DCP$_3$, DCP$_5$, DCP$_7$) are defined and named these two encoders as S − 1 and S − 2. The codes created by these encoders at every pixel I$_o$ are constituted as:

$$S - 1 = \sum_{k=0}^{3} DCP_{2k} + 4^k \tag{4}$$

$$S - 2 = \sum_{k=0}^{3} DCP_{2k+1} + 4^k \tag{5}$$

By concatenating the codes generated by these two sub encoders S − 1 and S − 2, DCP descriptor [28] for every pixel I$_0$ is calculated as:

$$DCP = \left\{ \sum_{k=0}^{3} DCP_{2k} + 4^k, \sum_{k=0}^{3} DCP_{2k+1} + 4^k \right\} \tag{6}$$

Two coded images are generated using sub-DCP encoders which are divided into non overlapping regions. The histogram is computed for each individual region and then combined to form a comprehensive representation. By using this representation, similarity between two images can be measured.

4   Experimental Result

The DCP feature descriptor by combining S − 1 and S − 2 is extracted and used for the classification of 10 subjects using k-NN (N = 3), k-NN (N = 5) and Classification Tree Classifier. We have collected 100 right-hand radiographs from 10 different adult subjects. The dataset presented does not contain images taken post-mortem. The dataset consists of x-ray images of a father (42 years), son (18 years), two brothers (31 years and 35 years) and 6 individuals. The DCP algorithm has been implemented in MATLAB and executed on a laptop of 4 Gb RAM and powered by Intel Core ™ i3 and WINDOWS 7 professional. All the classifier used are reproducible from MATLAB inbuilt functions.

Tables 1, 2 and 3 shows the cross-validation results of all 10 subjects in the form of the confusion matrix. Every possible subject represents a new class because every subject holds different features than others. A particular image from the original sample is considered as a validation data and residual observations as training data for cross-validation. This procedure is persistent until every observation is used once as the validation data. The average classification accuracy of k-NN (N = 3), k-NN (N = 5), and Classification Tree obtained is 74.8%, 68.8%, and 89.1%. The variation in classification accuracy (see Fig. 4) is due to the different learning ability of classifiers. The maximum and minimum accuracy of correct retrieval is 100% and 27.2%.

Table 1. Confusion matrix for k-NN (N = 3).

Input query subject from dataset	Cross validation accuracy									
	1	2	3	4	5	6	7	8	9	10
1	81.8	0	0	0	9.1	0	9.1	0	0	0
2	9.1	81.8	0	0	0	0	0	0	9.1	0
3	9.1	27.3	54.5	0	0	9.1	0	0	0	0
4	0	10	0	80	0	10	0	0	0	0
5	9.1	10	0	0	90.1	10	0	0	0	0
6	0	0	9.09	0	0	90.9	0	0	0	0
7	36.4	0	0	0	0	0	63.6	0	0	0
8	0	0	0	0	0	0	0	100	0	0
9	0	0	0	0	20	0	0	0	50	0
10	18.2	0	18.2	0	0	0	0	9.1	0	54.5

Table 2. Confusion matrix for k-NN (N = 5).

Input query subject from dataset	Cross validation accuracy									
	1	2	3	4	5	6	7	8	9	10
1	72.7	0	0	0	9.1	9.1	9.1	0	0	0
2	9.1	81.8	0	0	0	0	0	0	0	9.1
3	0	18.2	45.4	0	18.2	0	9.1	0	0	9.1
4	10	0	0	80	0	0	9.1	0	0	9.1
5	9.1	0	0	0	90.9	10	0	0	0	0
6	0	0	9.1	0	9.1	81.8	0	0	0	0
7	27.2	0	0	0	0	0	63.6	0	0	0
8	0	0	0	0	0	0	0	100	0	0
9	9.1	0	9.1	0	18.2	18.2	0	0	45.4	0
10	9.1	0	18.2	0	0	0	18.2	9.1	9.1	27.2

Table 3. Confusion matrix for classification tree.

Input query subject from dataset	Cross validation accuracy									
	1	2	3	4	5	6	7	8	9	10
1	90.1	0	9.1	0	0	0	0	0	0	0
2	0	100	0	0	0	0	0	0	0	0
3	0	0	100	0	0	0	0	0	0	0
4	9.1	9.1	0	81.8	0	0	0	0	0	0
5	9.1	9.1	0	0	81.8	0	0	0	0	0
6	0	0	18.2	0	0	81.8	0	0	0	0
7	0	0	9.1	0	0	0	91.9	0	0	0
8	0	0	0	0	0	0	0	100	0	0
9	0	0	0	0	9.1	9.1	0	0	81.8	0
10	0	0	9.1	0	0	0	0	0	9.1	81.8

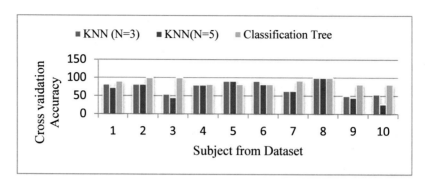

Fig. 4. Cross validation accuracy comparison of k-NN (N = 3), k-NN (N = 5) and Classification Tree classifier for 10 subjects.

Table 4 compares existing techniques with the proposed method for retrieval accuracy/error rate. The proposed system successfully classified father and son from the dataset with retrieval accuracy of 100% and 91.9% respectively.

Table 4. Comparison of accuracy/error rate between existing techniques and proposed method.

Sr. no.	Method	Accuracy/error rate with dataset used
1	Segmentation of phalanx bone using active shape model& Bayesian estimator [23]	Bone age accuracies of 82% for 32 male and 84% for 25 female children of age 1–16 years
2	Based on phalangeal region of interest (PROI) and carpals fuzzy information [24]	The accuracy was evaluated to be 85% for 720 children from 0.5 to 18 year old
3	Independent analysis of phalangeal region of interest(PROI)& carpal region of interest (CROI) [14]	Error rate between the two assessment is roughly 2 year for 120 images. PROI & CROI analysis is performed on children's up to 11 years of age and below 9 years of age respectively
4	Active appearance based feature extraction of metacarpals, proximal, and the middle phalanges [17]	Equal error rate of 6.7% for 44 test images
5	Bone age assessment using histogram based comparison system [25]	Bone age assessment with error rate of -0.170625 for 32 left hand x-ray images of children's below 18 years
6	DCP feature extraction of input hand x-ray image with classification tree classifier (proposed)	Average classification accuracy of 89.1% for 100 radiographs of adults with age group of 18 year to 42 year

5   Conclusion

In this paper, a novel method is introduced for identifying humans using DCP features of hand radiographs. A primary outcome on a small dataset shows that the use of hand radiographs for human identification is feasible. Most researchers used the dataset consist of children's below the age of 18 years as carpals and phalanges can be easily segmented and treated. The principal benefit of the proposed method is to overcome the segmentation problem experienced by other authors [12–21] for bone age assessment. The proposed technique is found to be appropriate for a dataset of adults from 18 years to 42 years.

Future work involves an extension of hand radiograph database to cater for all age groups and its effect on the overall performance of the proposed system. This additional dataset will require the detection and extraction of more prominent features of hand radiographs and their best-suited classifier for the correct retrieval from missing and unidentified people.

References

1. Manigandan, T., Sumathy, C., Elumalai, M., Sathasivasubramanian, S., Kannan, A.: Forensic radiology in dentistry. J. Pharm. Bioallied Sci. **7**(Suppl. 1), S260–S264 (2015)
2. Chen, H., Jain, A.K.: Dental biometrics: alignment and matching of dental radiographs. IEEE Trans. Pattern Anal. Mach. Intell. **27**(8), 1319–1326 (2005)
3. Forensic identification of 9/11 victims ends, New York, 23 February 2005
4. Ross, A.A., Nandakumar, K., Jain, A.K.: Handbook of Biometrics. Springer, Heidelberg (2011)
5. Nomir, O., Abdel-Mottaleb, M.: Hierarchical contour matching for dental X-ray radiographs. Pattern Recognit. **41**(1), 130–138 (2008)
6. Abaza, A., Ross, A., Ammar, H.: Retrieving dental radiographs for post-mortem identification. In: 16th IEEE International Conference on Image Processing (ICIP), pp. 2537–2540 (2009)
7. Senn, D.R., Stimson, P.G.: Forensic Dentistry, 2nd edn. Taylor & Francis, Boca Raton (2010)
8. Jayaprakash, P.T., Srinivasan, G.J., Amravaneswaran, M.G.: Cranio-facial morph- analysis: a new method for enhancing reliability while identifying skulls by photo superimposition. Forensic Sci. Int. **117**(1–2), 121–143 (2001)
9. Pushparani, C., Ravichandran, C.P., Sivakumari, K.: Radiography superimposition in personal identification - a case study involving surgical implants. J. Forensic Res. **3**, 140 (2012)
10. Al-Amad, S., McCullough, M., Graham, J., Clement, J., Hill, A.: Craniofacial identification by computer-mediated superimposition. J. Forensic Odonto-Stomatol. **24**(2), 47–52 (2006)
11. Nomir, O., Abdel-Mottaleb, M.: A system for human identification from X-ray dental radiographs. Pattern Recognit. **38**(8), 1295–1305 (2005)
12. Pathak, A., Pal, S.K., King, R.A.: Syntactic recognition of skeletal maturity. Pattern Recognit. Lett. **2**(3), 193–197 (1984)
13. Pathak, A., Pal, S.K.: Fuzzy grammars in syntactic recognition of skeletal maturity from X-rays. IEEE Trans. Syst. Man Cybern. **16**(5), 657–667 (1986)

14. Pietka, E.: Computer-assisted bone age assessment based on features automatically extracted from a hand radiograph. Comput. Med. Imaging Graph. **19**(3), 251–259 (1995)
15. Pietka, E., McNitt-Gray, M.F., Kuo, M.L., Huang, H.K.: Computer-assisted phalangeal analysis in skeletal age assessment. IEEE Trans. Med. Imaging **10**(4), 616–620 (1991)
16. Pietka, E., Kaabi, L., Kuo, M.L., Huang, H.K.: Feature extraction in carpal-bone analysis. IEEE Trans. Med. Imaging **12**(1), 44–49 (1993)
17. Kauffman, J.A., Slump, C.H., Moens, H.B.: Matching hand radiographs. In: Overview of the workshops ProRISC-SAFE, 17–18 November, Veldhoven, The Netherlands, pp. 629-633 (2005)
18. Lin, P., Zheng, C., Zhang, F., Yang, Y.: X-ray carpal-bone image boundary feature analysis using region statistical feature based level set method for skeletal age assessment application. Opt. Appl. **2**, 283–294 (2005)
19. Martin-Fernandez, M.A., Martin-Fernandez, M., Alberola-Lopez, C.: Automatic bone age assessment: a registration approach. In: Proceedings on Medical Imaging, SPIE, vol. 5032, pp. 1765–76, San Diego, CA (2003)
20. Aja-Fernandez, S., de Luis-Garcia, R., Martın-Fernandez, M.A., Alberola-Lopez, C.: A computational TW3 classifier for skeletal maturity assessment. A computing with words approach. J. Biomed. Inform. **37**(2), 99–107 (2004)
21. Davis, L.M., Theobald, B.J., Lines, J., Toms, A., Bagnall, A.: On the segmentation and classification of hand radiographs. Int. J. Neural Syst. **22**, 1250020 (2012)
22. Hue, T.T.M., Kim, J.Y., Fahriddin, M.: Hand bone radiograph image segmentation with ROI merging. Recent Researches in Mathematical Methods in Electrical Engineering and Computer Science, pp. 147–154 (2011)
23. Mahmoodi, S., Sharif, B.S., Chester, E.G., Owen, J.P., Lee, R.: Skeletal growth estimation using radiographic image processing and analysis. IEEE Trans. Inf Technol. Biomed. **4**(4), 292–297 (2000)
24. Hsieh, C.W., Jong, T.L., Tiu, C.M.: Bone age estimation based on phalanx information with fuzzy constrain of carpals. Med. Biol. Eng. Comput. **45**(3), 283–295 (2007)
25. Mansourvar, M., Raj, R.G., Ismail, M.A., Kareem, S.A., Shanmugam, S., Wahid, S., et al.: Automated web based system for bone age assessment using histogram technique. Malays. J. Comput. Sci. **25**(3), 107–121 (2012)
26. Guo, Z., Zhang, L., Zhang, D.: Rotation invariant texture classification using LBP variance (LBPV) with global matching. Pattern Recognit. **43**(3), 706–719 (2010)
27. Ahmadian, A., Mostafa, A.: An efficient texture classification algorithm using Gabor wavelet. In: 2003 Proceedings of the 25th Annual International Conference of the IEEE Engineering in Medicine and Biology Society, vol. 1, pp. 930–933 (2003)
28. Ding, C., Choi, J., Tao, D., Davis, L.S.: Multi-directional multi-level dual-cross patterns for robust face recognition. IEEE Trans. Pattern Anal. Mach. Intell. **38**(3), 518–531 (2016)

Mixed Reality in Action - Exploring Applications for Professional Practice

Adam Nowak[(✉)], Mikołaj Woźniak, Michał Pieprzowski,
and Andrzej Romanowski

Institute of Applied Computer Science, Lodz University of Technology,
Stefanowskiego St. 18/22, 90-924 Lodz, Poland
{203151,210893}@edu.p.lodz.pl, mikolaj@pawelwozniak.eu,
androm@iis.p.lodz.pl

Abstract. Mixed reality technologies has been emerging rapidly in the recent years, not only in terms of improving the equipment and software, but also in employing the current state into multiple practical applications. One of the most successful endeavours of introducing a commercial product for mixed reality is the HoloLens by Microsoft. In this paper, we describe the technology and discuss multiple practical applications among various branches of medicine, science and industry. We present the results of our desk research among various ways to employ the mixed reality into practice and discuss the new openings that arise along the development of the technology. The openings described consist of various interaction methods and applications for CSCW.

1 Introduction

Visualisation and display technologies are one of the most emerging branches in computer science nowadays. Among multiple techniques of displaying graphical representations, the virtual reality has drawn huge attention of the researchers in last decade. There is still much room for exploration in the domain of virtual displays, which pose new possibilities and prospects for application towards various tasks, especially where stationary screen-based display is considered undesired. New opportunities may be investigated both in terms of increased immersion, enhanced mobility and three dimensional perception.

Traditional VR solutions, such as Oculus Rift, HTC Vive, and Samsung Gear were developed with the focus set mainly on entertainment solutions. Manufacturers struggled to offer highly immersive user experiences, which will enable the users to dive into the presented action. [8] The branch of gaming industry which develop VR-based software is rapidly growing in recent years and provides more and more engaging digital products. However, the constraints of the technology concerning limited context-awareness of the user while participating in the experience poses additional need for external supervision and providing safe surroundings. Combined with relatively high price of the device, the technology has not yet established its position as an everyday companion for domestic

© Springer Nature Switzerland AG 2020
A. Abraham et al. (Eds.): ISDA 2018, AISC 941, pp. 1085–1095, 2020.
https://doi.org/10.1007/978-3-030-16660-1_106

users. On the other hand, the solution became very popular for applications in exhibition displays and boosted the development of various branches of digital art.

The advantages of head-based, highly immersive display are numerous, while the environmental constraints highly affect the usability, which draw users away from applying it for alleviating everyday problems. [24] The potential of the technology for professional purposes is equally promising, opening brand new opportunities for extensive data analysis, new ways of medical therapy and many more. One of the most successful endeavours towards overcoming those constraints while maintaining the key functionality was introducing HoloLens by Microsoft. Turning towards mixed-reality approach, which augment the holographic projection onto the vision perceived by user's sight. Therefore, the comfort of on-head, adaptive display is maintained while not excluding full context awareness. In this paper, we discuss in detail the construction, user experience design and possible applications of Microsoft HoloLens in various branches of science and industry.

2   Hardware Details

2.1   Virtual Reality vs Augmented Reality

Virtual reality and Augmented reality are similar technologies focused on displaying artificial objects to the user. Virtual reality devices are based on headsets fulfilling whole field of sight, disabling the awareness of any outer factors. When mounted, the user is entering purely virtual environment created with digital objects and landscapes. Complete separation of sense of sight and perceived body movements often causes virtual reality sickness, symptoms of which (discomfort, headache, stomach awareness, nausea) [2,25] are similar to motion sickness [14] (Fig. 1).

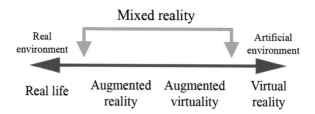

Fig. 1. Scale of enhanced reality [16].

The interaction with the projected artefacts is performed through rotation of users' head and actions performed with in-hand controllers. The hand controllers are equipped with gyroscopic sensors and tilt sensors, which translate the movements into virtual world.

Multiple VR solutions bare its users with cable connection to the unit which operates the software, which highly limits the mobility and freedom of movement

[11]. This issue has been faced applying mobile devices as the units for VR, yet such solutions suffer from lower performance due to limited computational resources and power issues [6].

Augmented reality technology present different scope, connecting virtual objects with real-world environment, overlaying artificial artefacts on real surfaces and objects. Holographic display provided by the AR device enable the user to experience the visualisation while remaining fully context-aware. The approach of presenting a full holographic display, not limited to few artefacts, is also referred as the mixed reality. The mixed reality devices are capable of real-time scanning and interpreting the users' surroundings, such as recognition of walls, floors, furniture and other obstacles. Therefore, the projections can be displayed with regard to the real environment, simulating the real-world objects. The dynamic analysis enables distinct context interpretation, enabling the user to remain 3D relations between real and artificial objects.

Navigation within the projection is based on interpreting the hand gestures using image recognition, which employ unique movements, such as air tap or bloom, for controlling the device interface (Figs. 2 and 3).

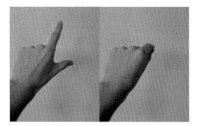

Fig. 2. Air tap gesture

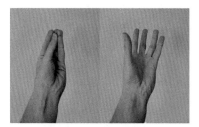

Fig. 3. Bloom gesture

The HoloLens device, being one of the most successful commercial mixed reality devices act as an independent device, with the in-built computational unit. The device performs voice recognition and is able to provide audible interactions.

Multiple devices are able to cooperate using server-based sharing facilities. The data change on one device is sent to the server and broadcasted to all the devices in a session [12] (Fig. 4).

Fig. 4. User working with HoloLens device

3   User Experience of Mixed Reality

The ultimate goal for designers approaching developing mixed-reality systems is to provide an engaging and immersive experience, which will enable the user to properly approach the goals via the visual interface with no peripherals needed. The main challenge is to provide an intuitive manner of interaction, which will employ well-known gestural patterns for interacting with the display.

HoloLens introduces brand new manner of interaction with technology, forcing the users to step out of their comfort zone and establish new habits. Therefore it is crucial to properly cover the user experience design, so the device is not just a tech-savvy gadgetry, but would offer numerous opportunities and will to be employed as an everyday companion.

3.1   Pursuing an Engaging Experience

According to Nielsen's three dimensional conceptual model of user experience [17], the HoloLens device has the emphasis put on developing the engagement layer. The interface is design to enhance the feeling of being augmented with a holographic display in your eyes. The gestural manner of interaction creates immersive perception, as it employs the natural, physical action for controlling the visually perceived surrounding. Using the device is engaging, as the narrative is constructed in an appealing way, making user feel as a participant in a futuristic experience. The system is highly adaptive, as it employs the users surrounding as a part of displaying technique.

Unlike in traditional VR approach, alongside high immersion, user does not get isolated form the world around him. Therefore, task combining operation on physical objects with a support of holographic display becomes possible. Moreover, the interface enables realising collaborative tasks with other people, non-equipped with a similar device. That opens new possibilities for computer supported teamwork in broad range of applications, especially where hand-operated displays are not sufficiently convenient.

4   Applications of Mixed Reality

Nevertheless the mixed reality is still considered among gadgetry, the innovative approach towards displaying and convenient manner of operation makes it suitable for numerous applications. New ways of utilising the technology are being explored, with substantial seen among the industrial and medical branches of both science and practice. Having its origins in VR setups, the device may also perform as a home entertainment tool or be applied in various types of digital art. In this paper we present a choice of most promising and innovative endeavours towards applying HoloLens and related technologies into everyday use.

4.1   Medical Use of Mixed Reality

Numerous applications were proposed and investigated for applying extended reality into medical practise. The HoloLens device may be helpful both for the doctors as a supportive tool for diagnostic procedures, as well as directly aid the therapy processes.

The holographic display brings brand new openings in terms of data visualisation and analysis. Artanim Foundation proposed an augmented reality visualisation of joint movements for rehabilitation and sports medicine [4]. Dabarba et al. describe ways to depict joint movements using AR technology. The system is based on optical motion capture and reconstruction of body structures. Bones are presented as real-time simulated holograms, presented similarly to X-ray vision. Gathered information is saved and can be further processed. The system provides an aid to rehabilitation process and is useful for assessing patient's progress. The system may also help in preventing injuries by analysing trajectories of fast movements among athletes.

Another branch of medicine that can benefit form active use of holographic displays is the anatomic pathology. Hanna et al. [10] shows possibilities among MR applications for medicine. The device was tested by pathologists during autopsy. All the time the access to diagrams, annotations, and voice instruction was given. 3D-scanned gross pathology specimens can be easily manipulated thanks to holographic view. The HoloLens helped in locating important pathological findings. Authors ensure that the device was comfortable to wear and easy to use with sufficient computing power and high-resolution imaging.

Preliminary use of HoloLens glasses in surgery of liver cancer is another promising opening for employing the device into medical practice [23]. Three dimensional model of liver cancer was created during operation by using magnetic resonance data and combining it with VR technology. Moreover, surgical planning was achieved and matched with the target organ during operation. 3D models are likely to act as an informational aid for appropriate planning of the surgery.

An interesting endeavour were made towards X-ray free endovascular interventions - using mixed reality for on-line holographic visualisation [13]. Endovascular interventions require catheter tracking, which is usually achieved by X-ray. Yet, the radiation generated during the process is considered harmful and efforts are paid to avoid this approach. A 3D holographic view of the vascular system is presented as an alternative. Patient's surface and vascular tree is extracted before surgery using computer tomography and registered by magnetic tracking system. The system was evaluated on a phantom and achieved promising results. While there is much room for improvement, 3D-view displays are likely to substitute the radiation-based techniques in the future.

Mixed reality has also been applied for therapy of Alzheimer's disease [1]. A joint project with a team of neurologists resulted in creating a set of activities for training short term and spatial memory using HoloLens. Presented exercises were similar to popular "Memory Game", consisting finding two identical objects hidden in boxes. With growth of difficulty level, boxes were placed chaotically.

Mixed reality device can also be applied to monitor pulse and other bodily functions [15]. Authors present a system for measuring and visualisation of blood flow and vital signs. The system works in real-time and does not require any contact sensors. Recovering pulse signals is possible thanks to the combination of a webcam, imaging ballistocardiography, and remote imaging photoplethysmography. Users can observe their own heart rate as well as the heart rate of other people.

Another disease that might employ MR in its therapy is the amblyopia, so-called the lazy eye syndrome [18]. An interactive game with alternating display, accordingly to the eye being treated, was constructed as a form of exercise, alternative to the classic, eyepatch based approach. The application displays more active visuals to the "lazy eye", while leaving the more static objects to the other one. Therefore, an increased activity of the amblyopic eye is obtained, while maintaining the binocular vision throughout the exercise. The system was introduced, but is lacking a clinical proof based on patient studies.

4.2   Industrial and Engineering Applications

Holographic projection can play a huge role in control and analysis of industrial processes. The real-time visualisation approach opens new possibilities for advanced control, and the simulation systems can be used for both testing new approaches, as well as educating the new breed of professionals.

An attempt towards improving engineering education using augmented reality environment has been made by Guo [9]. The HoloLens device has been used as a tool to improve students' results with understanding manual material handling (MMH). Experiment was conducted with the control group, while the experimental group's students had the opportunity to strengthen their understanding with use of AR application developed by the authors. The AR group performed better, the mean score of traditional class was 62/100, whilst AR group 76.25/100. The results suggest that AR software might be advantageous as a tool for classroom use.

Investigations of head-mounted displays for industrial practice were investigated by Dhiman et al. [5]. The aim of the research was to take closer look at advantages and disadvantages of head-mounted display (HDM) comparing to projection-based display on assembly assistance system [26]. Authors discuss problems of limited field of view, weight of the device, design of the workplace, distance for hologram placement. Noticed positives consist of mobility of the workplace, extended integration modalities, possibility to display content in arbitrary 3D space.

Another branch of industry in which the potential of on-head displays is considered beneficial is the industrial process control. To enable precise and quick insights, techniques such as electrical capacitance tomography are used. In order to enable efficient analysis, it is required to rapidly visualise the collected measurement data in a clear and interactive way. Since the analysis is often performed on-site, which requires high context-awareness and a good blend of virtual display based analysis and physical actions performed on the equipment under investigation [21]. Additional opportunities open when concerning crowd sourced analysis of multi-layer tomographic data [22].

4.3 Alternative Applications

A novel approach was proposed for enhancing cultural tourism by a mixed reality application for outdoor navigation and information browsing using immersive devices. Debandi et al. [3] introduces a mixed reality application providing information on a city scale. The main purpose is presenting and describing monuments, buildings and artworks. In consequence, cultural tourism may be enhanced. Framed images and videos are processed by a remote application. In this way, known objects can be identified and presented to the user. Defined gestures allow choosing the subject and seeing augmented contents such as video or text audio. Objects supported by the mixed reality application can be presented with three dimensional contents and combined with the real world.

Another cultural activity was proposed by Pollalis et al. [20] Virtual archaeological artefacts are displayed to the user and facilitate learning and engagement with exhibition. User can feel more like an archaeologist not only by watching but also exploring artefacts. Commonly insurmountable interactions with artefacts, including picking up, scaling, rotating, moving, are now available with digital copies of real ancient objects. Detailed information might be requested for chosen artefact and presented as text and voice. Through interactive exhibition, the application advocates tourists and visitors along with sparking their imagination. The application connects virtual objects with the original context of the exhibition.

One of the most sophisticated areas of applications seems to be military. Article presents approach in Italian Air Force tending to use mixed reality to the Command Control Communications Computers and Intelligence (C4I) systems [19]. MR may be used by technical staff for training, maintenance and simulation purposes. Introducing virtual ones may significantly reduce the costs and induct more flexibility. Moreover, rationalisation of maintenance organisation can be performed. The device was tested in different scenarios in environmental conditions with provided simulation of interventions. The results are very promising, because the planned maintenance activities were accomplished.

5   Perspectives and New Tendencies

Table 1. Juxtaposition of applications with criteria

	Gesture interaction	Cooperative work	Spatial understanding	Voice	Additional hardware
Visualisation of joint movements [4]	+/−	−	−	−	+
Anatomic pathology [10]	+	−	−	+/−	−
Surgery of liver cancer [23]	+	−	−	−	−
X-ray free endovascular interventions [13]	−	−	+	+	+/-
Therapy of Alzheimer's disease [1]	+	−	+	+	−
Measurement of pulse and vital functions [15]	−	−	+/−	−	−
Lazy eye syndrome [18]	−	−	−	−	−
Improving engineering education [9]	+	−	−	−	−
Industrial practice [5]	+	−	+/−	−	−
Cultural tourism [3]	+	−	+	−	+
Virtual archaeological artefacts [20]	+	−	−	−	−
Italian Air Force [19]	−	+	−	+	−

The most advantageous asset of mixed reality technology is the technology of holographic display, which poses multiple opportunities towards embedding displayed artefacts toward real environments. Furthermore, it is the gestural interaction which makes the head-mounted MR solutions significantly more convenient to operate than common VR solutions. However, at current stage the range of gestures implemented is vastly limited (Table 1).

Despite the technological limitations, the possibility to merge virtual and real environments opens brand new opportunities in various branches of industry and science. Computer-supported cooperative work is another field that can highly benefit from those new interaction models. However, applications requiring multiple HoloLens devices may be considered barely affordable.

Another branch of applications which may significantly develop through appying MR is the spatial analysis of objects and processes. Through understanding of the environment the application might be more authentic and impressive, although it is not extensively employed.

Voice control is another way of controlling and interacting with mixed reality applications.

The gestural interactions could possibly benefit from employing EMG-driven peripherals. Therefore, wider range of gestures could be easily introduced to the interface. Application of such devices was already tested in terms of application for display-free text edition [7] and audio control [28]. These are also the gaze-tracking systems that could highly benefit from employing MR devices into common practice [27].

6   Discussion and Summary

The mixed reality technology has been truly emerging in recent years, which is proved by developing multiple applications of the device in various branches of industry, science and medicine. The solutions applied still require further development and introducing new methods for increased efficiency. More user studies are advisable, so to make system not only fulfil the tasks, but also become more convenient to its users.

Current studies show that mixed reality broadens the horizon of new therapeutical methods accessible, in various branches of medical treatment. It can also bring the therapy to patient's home, due to mobility and user-friendly design. Further research is conducted toward establishing the most effective way to employ VR and Mixed Reality technologies to treatment proceedings, since broad sample studies are required to fully confirm the effects of this alternative approach.

Furthermore, the industry can highly benefit from deeper integration of MR technologies. The on-head displays can revolutionise the area of dynamic process control and rapid data visualisation. Control systems implemented onto holographic display can be especially effective, when being on-site is desired and the additional devices will pose additional challenge for the professional to handle.

References

1. Aruanno, B., Garzotto, F., Rodriguez, M.C.: HoloLens-based mixed reality experiences for subjects with Alzheimer's disease. In: Proceedings of the 12th Biannual Conference on Italian SIGCHI Chapter, CHItaly 2017, pp. 15:1–15:9. ACM, New York (2017)
2. Brooks, J.O., Goodenough, R.R., Crisler, M.C., Klein, N.D., Alley, R.L., Koon, B.L., Logan, W.C., Ogle, J.H., Tyrrell, R.A., Wills, R.F.: Simulator sickness during driving simulation studies. Accid. Anal. Prev. **42**(3), 788–796 (2010). Assessing Safety with Driving Simulators
3. Debandi, F., Iacoviello, R., Messina, A., Montagnuolo, M., Manuri, F., Sanna, A., Zappia, D.: Enhancing cultural tourism by a mixed reality application for outdoor navigation and information browsing using immersive devices. In: IOP Conference Series: Materials Science and Engineering, vol. 364, p. 012048, June 2018
4. Debarba, H.G., de Oliveira, M.E., Ladermann, A., Chague, S., Charbonnier, C.: Augmented reality visualisation of joint movements for physical examination and rehabilitation. In: Proceeding of 2018 IEEE Conference on Virtual Reality and 3D User Interfaces. IEEE (2018)
5. Dhiman, H., Martinez, S., Paelke, V., Röcker, C.: Head-mounted displays in industrial AR-applications: ready for prime time? In: Nah, F.F.-H., Xiao, B.S. (eds.) HCI in Business, Government, and Organizations, pp. 67–78. Springer, Cham (2018)
6. Pena, M.I., MacAllister, A., Winer, E., Evans, G., Miller, J.: Evaluating the Microsoft HoloLens through an augmented reality assembly application (2017)
7. Ghosh, D., Foong, P.S., Zhao, S., Chen, D., Fjeld, M.: EDITalk: towards designing eyes-free interactions for mobile word processing. In: Proceedings of the 2018 CHI Conference on Human Factors in Computing Systems, CHI 2018, pp. 403:1–403:10. ACM, New York (2018)
8. Grigore, P.C., Burdea, C.: Virtual Reality Technology (2003)
9. Guo, W.: Improving engineering education using augmented reality environment. In: Zaphiris, P., Ioannou, A. (eds.) Learning and Collaboration Technologies. Design, Development and Technological Innovation, pp. 233–242. Springer, Cham (2018)
10. Hanna, M., Ahmed, I., Nine, J., Prajapati, S., Pantanowitz, L.: Augmented reality technology using Microsoft HoloLens in anatomic pathology. Arch. Pathol. Lab. Med. **142**, 638–644 (2018)
11. Kolasinski, E.: Technical report 1027: simulator sickness in virtual environments (1995)
12. Kress, B.C., Cummings, W.J.: 11-1: invited paper: towards the ultimate mixed reality experience: HoloLens display architecture choices. SID Symp. Dig. Tech. Pap. **48**(1), 127–131 (2017)
13. Kuhlemann, I., Kleemann, M., Jauer, P., Schweikard, A., Ernst, F.: Towards X-ray free endovascular interventions - using HoloLens for on-line holographic visualisation. Healthc. Technol. Lett. **4**, 184–187 (2017)
14. LaViola Jr., J.J.: A discussion of cybersickness in virtual environments. SIGCHI Bull. **32**(1), 47–56 (2000)
15. McDuff, D.J., Hurter, C., González-Franco, M.: Pulse and vital sign measurement in mixed reality using a HoloLens. In: VRST (2017)
16. Milgram, P., Kishino, F.: A taxonomy of mixed reality visual displays. IEICE Trans. Inf. Syst. **E77-D**(12), 1321–1329 (1994)

17. Nielsen, J. (ed.): Coordinating User Interfaces for Consistency. Morgan Kaufmann, San Francisco (1989)
18. Nowak, A., Woźniak, M., Pieprzowski, M., Romanowski, A.: Towards amblyopia therapy using mixed reality technology. In: Proceedings of the Federated Conference on Computer Science and Information Systems, pp. 279–282 (2018)
19. Piedimonte, P., Ullo, S.L.: Applicability of the mixed reality to maintenance and training processes of C4I systems in Italian Air Force. In: 2018 5th IEEE International Workshop on Metrology for AeroSpace (MetroAeroSpace), pp. 559–564, June 2018
20. Pollalis, C., Fahnbulleh, W., Tynes, J., Shaer, O.: HoloMuse: enhancing engagement with archaeological artifacts through gesture-based interaction with holograms. In: Proceedings of the Eleventh International Conference on Tangible, Embedded, and Embodied Interaction, TEI 2017, pp. 565–570. ACM, New York (2017)
21. Romanowski, A.: Contextual processing of electrical capacitance tomography measurement data for temporal modeling of pneumatic conveying process. In: Proceedings of the 2018 Federated Conference on Computer Science and Information Systems, FedCSIS 2018. ACSIS. IEEE (2018, in press)
22. Romanowski, A., Grudzien, K., Chaniecki, Z., Wozniak, P.: Contextual processing of ECT measurement information towards detection of process emergency states. In: 2013 13th International Conference on Hybrid Intelligent Systems (HIS), pp. 291–297 (2013)
23. Shi, L., Luo, T., Zhang, L., Kang, Z., Chen, J., Wu, F., Luo, J.: Preliminary use of HoloLens glasses in surgery of liver cancer. Zhong nan da xue xue bao. Yi xue ban = J. Cent. South Univ. Med. Sci. **43**, 500–504 (2018)
24. Shneiderman, B., Plaisant, C.: Designing the User Interface, 4th edn. Pearson Addison Wesley, Boston (2010)
25. Stanney, K.M., Kennedy, R.S., Drexler, J.M.: Cybersickness is not simulator sickness. Proc. Hum. Factors Ergon. Soc. Annu. Meet. **41**(2), 1138–1142 (1997)
26. Wang, Z.B., Ong, S.K., Nee, A.Y.C.: Augmented reality aided interactive manual assembly design. Int. J. Adv. Manuf. Technol. **69**(5), 1311–1321 (2013)
27. Wojciechowski, A., Fornalczyk, K.: Single web camera robust interactive eye-gaze tracking method. Bull. Pol. Acad. Sci. Tech. Sci. **63**(4), 879–886 (2015)
28. Woźniak, M., Polak-Sopińska, A., Romanowski, A., Grudzień, K., Chaniecki, Z., Kowalska, A., Wróbel-Lachowska, M.: Beyond imaging - interactive tabletop system for tomographic data visualization and analysis. In: Karwowski, W., Trzcielinski, S., Mrugalska, B., Di Nicolantonio, M., Rossi, E. (eds.) Advances in Manufacturing, Production Management and Process Control, pp. 90–100. Springer, Cham (2018)

AMGA: An Adaptive and Modular Genetic Algorithm for the Traveling Salesman Problem

Ryoma Ohira$^{(\boxtimes)}$, Md. Saiful Islam, Jun Jo, and Bela Stantic

School of Information and Communication Technology, Griffith University,
Southport, QLD 4215, Australia
{r.ohira,saiful.islam,j.jo,b.stantic}@griffith.edu.au

Abstract. The choice in selection, crossover and mutation operators can significantly impact the performance of a genetic algorithm (GA). It is found that the optimal combination of these operators are dependent on the problem characteristics and the size of the problem space. However, existing works disregard the above and focus only on introducing adaptiveness in one operator while having other operators static. With adaptive operator selection (AOS), this paper presents a novel framework for an adaptive and modular genetic algorithm (AMGA) to discover the optimal combination of the operators in each stage of the GA's life in order to avoid premature convergence. In AMGA, the selection operator changes in an online manner to adapt the selective pressure, while the best performing crossover and mutation operators are inherited by the offspring of each generation. Experimental results demonstrate that our AMGA framework is able to find the optimal combinations of the GA operators for each generation for different instances of the traveling salesman problem (TSP) and outperforms the existing AOS models.

Keywords: Adaptive genetic algorithm · Modular genetic algorithm ·
Adaptive operator selection · Traveling salesman problem

1 Introduction

The traveling salesman problem (TSP) is a combinatorial optimization problem that is derived from many production and scheduling problems and is presented as a salesman being required to visit n cities but given the traveling costs between each city, the salesman is to find the shortest distance required to visit all required cities [1]. When approaching the problem with a genetic algorithm (GA), the genotype represents the order in which a salesman visits the cities. As such, mutations and crossover must ensure that no city is visited more than once. While this places restrictions on the techniques used for the selection, crossover and mutation approaches, studies identify each step with having multiple techniques and methods that aim to improve a GA's output.

© Springer Nature Switzerland AG 2020
A. Abraham et al. (Eds.): ISDA 2018, AISC 941, pp. 1096–1109, 2020.
https://doi.org/10.1007/978-3-030-16660-1_107

1.1   Related Work

A comprehensive study into the crossover operators and mutation operators demonstrate that the choice in operators can directly impact a GA's performance [2]. This was supported by studies into the effects of contemporary mutation [3] and crossover [4] operators for the TSP. Comparative studies of selection operators also demonstrate that the choice in selection operator has an effect on both the convergence [5] and time complexity [6]. By analyzing the results from these comparative studies, it can be seen that improving the effectiveness of a GA for the TSP is a multi-dimensional problem resulting from the combination of selection, crossover and mutation operators along with the problem size.

Existing literature [7,8] identify a number of trends emerging in exploring adaptive operators for GAs, specifically operator selection. There are a number of works for introducing adaptiveness to GAs through AOS. Spears [9] introduces adaptiveness into the crossover phase with the 1Bit Adaptive Crossover (1-BX) where the selected crossover operator is encoded into the genotype. While limited to two operators, Spears identified that by simply making more operators available, the GA experienced significant improvements in performance. Riff and Bonnaire [10] extended this concept to include more operators. Gomez proposed the HAEA hybrid adaptive GA [11] which encodes the operator rates and rewards operators according to whether an offspring is better or worse than its parents. Cruz-Salinas and Perdomo [12] extend on this work by having population of operators that are exposed to evolution and mutation while maintaining the operator selection method employed by HAEA. However, the AOEA algorithm employs operators that are not suitable for the ordered nature of the TSP. Montero and Riff [13,14] propose self-adaptive operator selection by including the probabilities of the operators in the gene. Two methods are proposed: a random reward/penalty value and a ratio on the difference between an offspring and parents' fitness in comparison to results from the last generation. Montero and Riff also suggested a global operator probability with a single operator selected for the next generation.

The AMCPA proposed by Osaba et al. [15,16] is another implementation of the AOS where a new crossover operator is selected when it is beneficial to the search process. The crossover probability adapts according to the search performance on recent generations and the current generation number. This enables the algorithm to prevent premature convergence and also allows the AMCPA to use multiple crossover operators which are applied alternatively. The AMCPA only applies AOS for the crossover operator and makes the assumption that the 2-opt mutation operator is the most suitable for all instances of the TSP. Furthermore, the AMCPA was only compared to the GA for TSP instances where $n \leq 500$. While Spears demonstrated that the availability of multiple crossover operators improves performance, Osaba et al demonstrate that self-adapting operator selection can enable a GA to further explore the problem space.

1.2   Contribution

Unlike the existing works, AMGA proposed in this paper employs self-adapting operator selection to each stage of the GAs process. AMGA is randomly assigned a selection operator when initiated. The probability of the selection operator changing is calculated according to the number of generations where there are no significant improvements in fitness. This ensures that the algorithm responds to selective pressure in an on-line manner. Once a selection operator is chosen, it is used to select the parents for creating a new generation of offspring. The crossover and mutation operators for each individual is encoded into the genotype as extra genes. Once two parent individuals are selected, the crossover and mutation operators of the fittest parent is employed and inherited by the offspring.

Table 1. Comparison of self-adaptive operator selection GAs

Genetic algorithm	Adaptive Operator Selection (AOS)
Static GA	No AOS
1BX [9]	One of two crossover operators are encoded into the genotype and inherited by child
DSATUR [10]	Inheritance of multiple crossover operators
HAEA [11]	AOS crossover operators with rewards and penalties for improvements in offspring
AMCPA [15]	Random selection between multiple crossover operators once population health declines
AOEA [12]	Crossover operators are presented as trees and are subject to evolution. Probability of selecting an operator is dependent on its performance
Proposed AMGA	Selection between multiple selection, crossover and mutation operators

Although crossover operators have been encoded into the genotype in previous works [9,10], the AMGA includes both crossover and mutation operator genes. The crossover and mutation operators made available to the GA aims to recover the diversity lost by the selection stage while maintaining a high level of overall fitness in the population. Where existing literature has focused on the performance gains made by introducing AOS to the crossover stage, results from experiments demonstrate that by making more operators available for each stage of the process, AMGA is able to maintain diversity for more generations and avoid premature convergence.

Table 1 compares the self-adaptive operator selection in AMGA with the existing works in the literature.

2   Background

2.1   Selection Operators

The selection operator phase in the GA is responsible for determining which individuals are used for the crossover phase. This paper looks at the effects of the most commonly used selection operators for GAs as discussed below.

Fitness Proportionate Selection (FPS). FPS is a commonly used operator for selecting parents for crossover where the probability of being selected, p_i, is proportional to the fitness in relation to the overall fitness of the population [17]. This is expressed in Eq. 1 where f_i is the fitness of the given individual, i, from a population of size n. FPS can result in conditions where the GA lacks selective pressure in exploring the problem space or prematurely converge when the search space has narrowed down too quickly [18].

$$p_i = \frac{f_i}{\sum\limits_{j=1}^{N} f_j} \quad (1)$$

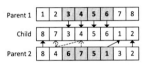

Fig. 1. Partially mapped crossover

Fig. 2. Ordered crossover operator

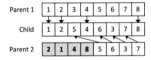

Fig. 3. Modified crossover operator

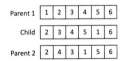

Fig. 4. Edge recombination operator

Rank Based Selection (RBS). RBS is proposed to eliminate the disadvantages of FPS [19]. This operator orders the individuals by their fitness where the worst performing individual is assigned a rank value of 1 and the best performing individual is assigned n. This can be expressed as in Eq. 2 where $i \in 1, \ldots, N$. μ^+ is the expected maximum value of the individual with rank n where $\mu^+ \geq 0$. μ^- is the minimum expected value of the individual of rank 1 with the constraint $\mu^+ = 2 - \mu^-$. Baker notes that a μ^+ with a value of 1.1 generates optimal results

$$p_i = \frac{1}{N}(\mu^- + (\mu^+ - \mu^-)\frac{i-1}{N-1}) \quad (2)$$

with it out performing fitness proportionate selection on the selected problems [19]. While ranking removes the problems associated with scaling and prevents the magnitude of the fitness differences impacting the selective pressure, it can increase computation time [6].

Tournament Selection (TNS). TNS randomly selects candidates from the population with the best one being selected for the next step. While tournaments are usually between two individuals, a generalized method can be written with a variable size. As a tournament selection can be executed with a single pass, time complexity is $\mathcal{O}(N)$ as it does not require any sorting [6].

2.2   Crossover Operators

Crossover operators are responsible for generating new combinations for genotypes by combining two parents to create offspring.

Partially-Mapped Crossover (PMX). PMX produces children by choosing a sub-sequence of genes from one parent while preserving the order and position of the genes from another parent. An initial sequence of genes between points i and j are copied from one parent to the child. The genes between i and j from the second parent are mapped to new positions using the gene positions from parent 1 if they do not already exist in the child. The remainder genes are inherited from parent 2 if they do not exist in the child. This mapping and tracing process ensures the child's genes are subsequences inherited from both parents. In Fig. 1, the child directly inherits the sequence of 3, 4, 5 and 6 from parent 1. The gene with the value 6 is in the same positions for parent 2 and can be ignored. Gene 7's position is taken by gene 4 and so is then mapped to the position of gene 4 from parent 2. This can be seen again where the gene 1 is mapped to the position of gene 3 in parent 2 through: $1 \rightarrow 6 \rightarrow 3$. The remainder genes are then inherited directly from parent 2.

Fig. 5. Cycle crossover: defining the cycle

Fig. 6. Cycle crossover operator

Ordered Crossover Operator (OX). OX selects a sequence between points i and j of one parent and then inserts the genes from the second parent in the order they present themselves while ensuring that the child does not contain duplicate values [20]. The process can be seen in Fig. 2. The modified ordered crossover operator (MOX) in Fig. 3 selects a random point, i, within the order of both parents. The sequence to the left of the point for parent 2 identifies the genes from parent 1 that will maintain their position and order. The child inherits the remaining genes from parent 2.

Edge Recombination Operator (ERO). ERO examines the link between each node and creates an edge map to construct offspring that inherit as much information as possible from the parent genes [21]. An initial node is selected from the edge list that has the largest number of edges. In Fig. 4, both nodes 1 and 2 have four edges and node 2 is randomly chosen. The node with the fewest edges are considered with one chosen for the next part. In the example, nodes 3, 4 and 6 all have two edges and node 3 is chosen randomly. Node 3 has edges to nodes 1 and 4, with 4 having the fewest edges and is thus chosen. This is continued until the child genotype is filled.

Cycle Crossover Operator (CX). In CX, a mapping cycle is initiated at a randomly selected point i[1]. In Fig. 5, given that $i = 4$, the cycle can be seen to be $4 \rightarrow 8 \rightarrow 1 \rightarrow 4$. The child then inherits these values and their positions from parent 1 with the remainder filled from parent 2 as seen in Fig. 6.

2.3   Mutation Operators

The purpose of a mutation operator is to maintain genetic diversity to prevent premature convergence. As with the selection and crossover operators, a number of mutation operators exist and contribute to a GA to varying degrees.

Center Inverse Mutation (CIM). CIM divides the genotype into two sections from a randomly selected point. The order of genes in each subsection is reversed. This method results in a major variation to the original genotype [3].

Fig. 7. AMGA operator encoding

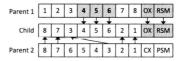

Fig. 8. AMGA crossover

Reverse Sequence Mutation (RSM). RSM selects a sequence between two randomly selected points i and j where $i < j$ [3]. The order of the genes within the sequence is then reversed. This ensures that the original order is partially preserved with a p_m probability of mutation occurring.

Partial Shuffle Mutation (PSM). PSM swaps a gene given a probability of p_m with another randomly selected gene [3]. This partially preserves the original order of genes but also allows noncontinuous mutation to occur.

3   AMGA: An Adaptive and Modular Genetic Algorithm

While studies into the performance of different methods for selection, crossover and mutation demonstrate how choices of an operator can directly affect the performance of a GA, the results suggest that an optimal combination for a given problem is dependent on the characteristics and size of the problem space. When considering operators as additional dimensions to an optimization problem, it becomes clear why this topic has been the subject of research since the 1980's. This problem can be simplified as an optimization of the three operational phases.

- **Selection operators** immediately introduce selection pressure and time complexity to a GA. While tournament selection supports a GA in exploring the problem space, a rank or tournament selection operator produces more selection pressure in exploiting available solutions.
- **Crossover operators** introduce a range of solutions with different priorities between time complexity, order preservation and the bias in the makeup of new offspring. The goal of a crossover operator being the preservation of good genes while replacing the genes that are counter productive to its goals.
- **Mutation operators** being responsible for maintaining a healthy level of genetic diversity in the population to ensure that the GA does not converge prematurely. The importance of which can be seen when the diversity lost through selection and crossover isn't adequately replaced. The mutation operators themselves provide a varying degree of success in maintaining this genetic diversity but no substantial study has been done into the effects of the combination of the three operational phases.

In the proposed AMGA, each operational phase is implemented in a modular fashion where each operator is included as a module which can be selected and changed with another if the algorithm reaches a local optima. In Fig. 9, the highlighted sections indicate the logic implemented for adaptive selection in

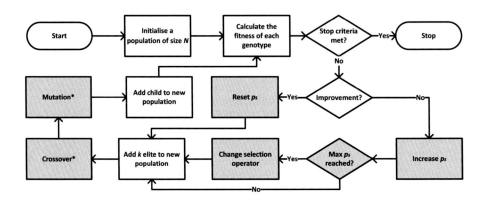

Fig. 9. Schematic diagram of our proposed AMGA framework

operators. The selection operator is the first step where adaptive operator selection is implemented. This is done by implementing a new selection parameter, p_s. When a population is initialized, a selection operator is randomly selected and p_s is given a value of 0. As the population converges and no improvement is made to the fitness, the p_s value is incremented until $p_s \geq 1$. This is expressed in Eq. 3 where g is the number of generations since the last improvement in fitness, k is the current generation and n is the problem size. As g indicates a lack of progress, the AMGA is able to quickly respond to this by the way g scales. The rate at which p_s increases also scales with the size of the problem. Both of these features allow for the AMGA to explore with one selection operator for longer with larger problems. However, this still enables the AMGA to switch to another selection operator if the selected operator is inefficient or a change in selective pressure is needed. Furthermore, as the algorithm produces more generations, it becomes more agile in switching between operators.

The crossover and mutation operators are encoded into the genotype itself, as demonstrated in Fig. 7. The initial population is randomly assigned a crossover and mutation operator. During the crossover phase, the

$$p_s = p_s + \frac{2g + k}{n^2} \quad (3)$$

child genotype inherits the mutation and crossover operators of its fittest parents. The inherited crossover operator is then applied between the two parents to create the new offspring. This offspring is then subject to the inherited mutation operator with the probability of p_m. The key aspects of the AMGA can be identified as:

- **Selection** as an adaptive operator selected randomly but changed when the algorithm fails to experience improvement. Operators included as modules are the tournament, rank and fitness proportionate selection operators. The probability of the operator changing is defined as p_s and is dependent on the size of the problem, the current generation and the number of generations since an improvement in fitness was recorded.
- **Crossover and Mutation** are encoded into the genotype with offspring inheriting the operators of the fittest parent as demonstrated in Fig. 7. The inherited crossover and mutation operators are applied as in Fig. 8.
- **Mutation probability** affects crossover and mutation operators where these have a p_m probability of changing to a new operator. This supports the goal of maintaining genetic diversity in both the genotype and the operators.

The operators for the AMGA were selected for their performance in existing literature and their known contribution to a GA's performance. For the insertion phase, an elitist strategy which involves selecting k best performing genotypes is implemented. The remaining population genotypes are created using the selection, crossover and selection strategies outlined above.

4   Experiments

Test Functions. Tests from the Travelling Salesman Problem Library (TSP LIB) synchronous benchmark tests were selected for their variety in n size and optimum distance. Equation 4 models the convergence criteria where the algorithm ends execution when the algorithm does not experience an improvement in its fitness for g generations. With n being the size of the problem and k the current generation of the algorithm. This ensures that the convergence criteria scales to both parameters and scales in a way that

$$c = n + \sum_{k=1}^{n} k \quad (4)$$

allows for larger problems to explore longer before being classified as having reached a convergence point. Each TSPLIB test is run 50 times with average distance, standard deviation for distance and average processing being calculated. Furthermore, the two-sample z-tests on the mean distances between the GA and AMGA as well as the AMCPA and AMGA have been

$$z = \frac{\overline{x_1} - \overline{x_2}}{\sqrt{\frac{\sigma_1^2}{n_1} + \frac{\sigma_2^2}{n_2}}} \quad (5)$$

included. The two-sample z-test is chosen as it demonstrates the significance of the differences between two means and is particularly effective for cases where the sample size is greater than 30. This is expressed as in Eq. 5, where x_i is the mean distance, σ_i^2 is the standard deviation and n_i is the sample size of tests for algorithm i. A positive value indicates a significant improvement of $\overline{x_1}$ over $\overline{x_2}$ where as a negative value indicates a significant decrease in performance. Alternatively, a neutral value means that there is no significant improvement or decrease in performance. This ensures that an improvement in performance offered by the AMGA can be quantified.

Genetic Algorithm Operators. For static genetic algorithms, the following crossover operators are incorporated: PMX, MOX, OX, CX, ERO. Mutation operators include: CIM, RSM, PSM. Selection operators are: Fitness Proportionate, Rank and Tournament. Each combination of selection, crossover and mutation operator is tested with the best performing combination selected to represent static GAs as the control for this experiment.

The AMCPA was implemented as described by Osaba et al for the TSP. The parameters for p_c and the operators available for the algorithm were also maintained. The AMCPA itself was implemented in the same language as the control and AMGA. The AMGA incorporated the operators outlined in Table 2. These

Table 2. AMGA components

Component	Operators
Selection	FPS, RBS, TNS
Crossover	PMX, MOX, OX, CX, ERO
Mutation	CIM, RSM, PSM

components were chosen due to their contribution in improving the performance of the static GAs for optimum solutions and computation times.

Computing Environment. The algorithms are executed on a Intel Core i7-3770 CPU with 32 GB main memory. The software is written in Python 2.7 and executed using the PyPy interpreter with Numpy.

4.1   Results

From Table 3, it can be seen that the optimal combination of crossover, selection and mutation operators varies with the size of the problem. While the smaller problems benefit from the OX crossover operator, the MOX has a greater contribution to the GA's performance for the larger problems. Although the tournament selection operator is present in the majority of test instances, it should be noted that there was no statistically significant difference in mean distance between the tournament and rank selection operators. The difference between the operator combinations for GAs become more significant as the problem size increases, this highlights the problem of determining the optimal operator combination for any given problem.

The AMCPA has improved performance over the static GAs, particularly with tests where $100 < n < 300$. However, the improvement diminishes with the larger problems. This premature convergence suggests that diversity is not adequately maintained in later generations for larger problems.

Performance Analysis. Results by the AMGA show that introducing AOS in multiple phases increases the number of generations an algorithm is able to run before converging on a solution. That most instances showing lower standard deviation suggests that the AMGA is generally more consistent than the either the GA or the AMCPA. As the algorithms are run until there are no further improvements, the AMGA can be seen to have run for a longer time. This suggests that implementing adaptive operator selection to the mutation and selection phases enables the algorithm to maintain a higher degree of genetic diversity, particularly in the later generations. The results from the two-sample z-test in Table 3 indicate instances where the AMGA has a significant improvement over the GA or the AMCPA with a +. Results with a * indicates no significant improvement. Of the 40 TSPLIB instances, the AMGA has statistically significant improvements in performance for 36 instances (90.0%) against the static GAs and for 34 instances (85.0%) against the AMCPA. Of these instances, there was only one instance (Rd100) where the AMGA did not demonstrate a significant improvement over either the GA or AMCPA. As the AMGA is converging at later generations, it can also be seen as having a greater mean time for computation. This is more pronounced as the problem size increases. Thus by introducing AOS to all phases of the process, the proposed AMGA is able to maintain a healthy level of population diversity for a longer period of time.

Table 3. Results of the best GA, the AMCPA and the proposed AMGA

Test	Optima	Genetic Algorithm				AMCPA			AMGA			z-test	
		Operators	Distance		Time	Distance		Time	Distance		Time		
			Avg.	S. Dev	Avg.	Avg.	S. Dev	Avg.	Avg.	S. Dev	Avg.	GA	AMCPA
Wi29	27603	OX R RSM	27998.6	455.2	0.35	27879.6	336.8	0.40	27872.8	413.0	0.65	+	*
Dj38	6656	OX T RSM	6789.2	182.4	0.55	6797.4	178.9	0.64	6751.5	164.7	1.04	+	+
Berlin52	7542	OX R RSM	8311.9	318.1	0.61	8300.3	237.3	0.62	8131.0	249.1	1.22	+	+
Eil51	426	OX T RSM	454.4	8.0	2.25	450.8	8.6	3.73	450.5	9.3	3.65	+	*
St70	675	OX T RSM	721.4	22.4	7.76	718.0	14.9	5.16	712.5	17.1	5.77	+	+
Eil76	538	OX T RSM	593.1	13.9	4.41	588.0	10.5	7.54	585.9	13.5	8.79	+	+
Pr76	108159	OX R RSM	119138.6	3108.8	1.97	117863.3	3885.7	1.89	115372.3	2710.2	3.42	+	+
Rat99	1211	OX R RSM	1459.3	41.0	7.14	1460.9	44.0	7.03	1399.3	31.9	9.61	+	+
Rd100	7910	OX R RSM	8686.7	256.9	15.05	8694.9	276.3	22.87	8684.3	221.2	22.78	*	*
Eil101	629	OX T RSM	711.3	12.7	9.21	713.4	17.5	7.69	710.0	7.9	7.73	*	+
Lin105	14379	OX T RSM	15673.5	490.5	19.46	15587.0	446.0	19.58	15440.3	477.4	19.31	*	+
Pr107	44303	OX T RSM	49292.9	1521.5	10.84	49132.4	1546.0	9.43	47985.0	1503.9	13.07	+	+
Pr124	59030	OX T RSM	65504.5	2094.7	12.32	66423.8	2841.5	27.68	64519.0	2244.0	20.79	+	+
Ch130	6110	OX T RSM	6785.9	148.1	35.13	6714.5	132.3	30.96	6693.9	124.2	29.45	+	+
Xqf131	564	OX R PSM	625.1	13.4	50.39	624.7	14.7	53.21	622.2	18.5	50.23	+	+
Pr136	96772	OX R RSM	106738.4	2454.3	54.02	105814.2	2208.3	80.17	105805.5	2039.8	103.63	+	*
Pr144	58537	OX T RSM	65827.2	2829.0	35.58	64369.2	2213.1	32.00	64291.5	1892.0	30.07	+	*
Kroa150	26524	MOX T RSM	29181.2	584.2	156.78	29262.8	732.6	171.38	29155.9	684.5	161.15	*	+
Ch150	6528	OX T RSM	7562.1	132.8	30.49	7366.5	187.4	38.60	7244.8	182.2	35.46	+	+
Pr152	73682	MOX T RSM	81154.7	2327.5	43.66	79927.3	1265.1	40.01	79243.1	1680.2	43.22	+	+
U159	42080	MOX T RSM	47245.0	1912.4	137.82	46929.1	1757.1	136.55	46561.9	1315.2	165.71	+	+
Rat195	2323	MOX T RSM	3604.1	97.1	35.01	3534.8	124.1	22.68	3394.4	85.5	34.99	+	+
D198	15780	OX T RSM	16848.4	219.9	77.82	16751.8	146.1	101.58	16670.5	174.0	124.07	+	+
Krob200	29437	MOX T RSM	32760.5	802.6	343.52	32729.6	566.1	235.90	32546.8	1121.6	200.08	+	+
Tsp225	3916	MOX T RSM	4319.5	78.6	98.22	4334.3	82.9	158.49	4309.0	92.4	181.11	*	+
Pr226	80369	MOX T RSM	93504.3	4913.4	133.24	93635.8	4765.7	134.32	90492.8	4091.3	226.27	+	+
Gil262	2378	MOX T RSM	2713.2	52.0	678.96	2702.2	48.7	731.52	2697.4	40.5	787.15	+	*
Pr264	49135	OX T RSM	63149.2	1215.3	218.70	62969.6	2061.1	217.99	61521.9	1635.4	250.37	+	+
A280	2579	MOX T RSM	3073.7	89.9	171.71	3093.6	91.9	190.01	3021.6	66.9	189.91	+	+
Pr299	48191	MOX T RSM	66372.5	1626.5	318.29	66507.0	1467.6	319.89	62968.1	1855.4	366.64	+	+
Lin318	42029	MOX T RSM	49201.1	1072.5	523.45	48221.2	738.4	474.80	47840.0	891.9	529.26	+	+
Rd400	15281	MOX T RSM	17673.2	433.5	1199.63	17616.7	322.8	1223.78	17024.2	317.7	1379.34	+	+
Pr439	107217	MOX R PSM	126765.2	5178.5	698.88	128179.4	4773.8	1072.68	124179.4	4773.8	1072.68	+	+
Pcb442	50778	MOX R RSM	58905.7	981.6	969.85	59074.5	1087.0	1133.52	58081.2	896.2	1401.29	+	+
U574	36905	MOX R RSM	43230.3	732.2	2831.21	42526.8	589.4	3725.98	42196.1	570.9	2872.13	+	+
Rat575	6773	MOX R RSM	8691.5	158.2	3310.15	8411.5	157.5	2609.13	8143.2	143.1	2815.50	+	+
P654	34643	MOX R PSM	39391.3	938.8	3930.12	38285.5	1088.9	5638.20	37165.8	1012.8	7925.12	+	+
D657	48912	MOX R RSM	61467.8	1761.6	2811.43	61533.9	1699.2	3150.44	59194.0	1097.7	2786.13	+	+
Rat783	8806	MOX T PSM	12381.9	532.5	6800.51	10527.2	184.0	14287.23	10156.7	160.9	20789.32	+	+
Pr1002	259045	MOX T PSM	367843.9	20723.6	14482.22	341939.4	23909.9	30646.11	305673.1	15853.6	45408.03	+	+

Convergence Analysis. By examining the data collected from each algorithm across all the runs, a number of characteristics of the AMGA can be identified. An example of the AMGA converging later with larger problem sets is shown in Fig. 10. In Fig. 10a the AMCPA and GA converged between 2000 generations of one another where as the AMGA has produces another 4000 generations. This pattern can also be seen in Fig. 10b where the AMCPA and AMGA produce 5000 to 10000 more generations than the static GA. For problems with an even larger n size, the AMGA exhibits greater performance gains. The results for test U574 shown in Fig. 10c demonstrate that for a medium size problem the AMCPA and AMGA run for a similar number of generations in comparison to the static GA. For a large problem where the n size is 783 for Rat783, the AMGA can be seen as running for even longer than the AMCPA with both significantly more effective than the static GA as shown in Fig. 10d.

4.2   Summary

The above findings point to the adaptive qualities enabling the AMGA to maintain a healthy level of genetic diversity in its population for longer. This allows the algorithm to continue evolving for more generations before converging on a solution. Furthermore, the improvements in computation time suggests that the algorithm is using operators with lower time complexity for a majority of the time. As the algorithms meet the convergence criteria before reaching the known

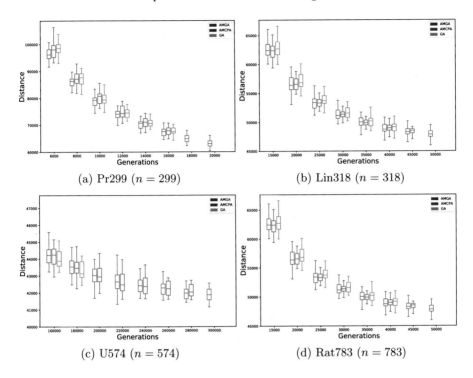

Fig. 10. Comparison of AMGA, AMCPA, and GA for best of fitness

optimal solution, introducing greater control and monitoring techniques on the genetic diversity may result in further improvements.

5   Conclusion and Future Work

This paper presents a novel framework called AMGA for an adaptive and modular genetic algorithm. Our framework aims to find the optimal combinations of selection, crossover and mutation operators on a genetic algorithm's outcome and to avoid premature convergence by avoiding genetic bottlenecks. This is done in an on-line manner where different operators are used at different stages of the GA's life. Crossover and mutation operators are selected according to their performance in the previous generations. The selection operator is changed when the current selective pressure does not improve the population's fitness. Experiments demonstrate that AMGA outperforms the existing approaches for solving TSP problem and is tolerant to the complexity and spaces of the instances of TSP. The future work includes the investigation of the effects of adaptive parameter controls on adaptive operator selections.

References

1. Oliver, I., Smith, D., Holland, J.R.: Study of permutation crossover operators on the traveling salesman problem. In: ICGA (1987)
2. Murata, T., Ishibuchi, H.: Performance evaluation of genetic algorithms for flow-shop scheduling problems. In: CEC, pp. 812–817 (1994)
3. Abdoun, O., Abouchabaka, J., Tajani, C.: Analyzing the performance of mutation operators to solve the travelling salesman problem. Int. J. Emerg. Sci. **2**, 61–77 (2012)
4. Abdoun, O., Abouchabaka, J.: A comparative study of adaptive crossover operators for genetic algorithms to resolve the traveling salesman problem. Int. J. Comput. Appl. **31**(11), 49–57 (2011)
5. Razali, N.M., Geraghty, J., et al.: Genetic algorithm performance with different selection strategies in solving TSP. In: WCE, vol. 2, pp. 1134–1139 (2011)
6. Goldberg, D.E., Deb, K.: A comparative analysis of selection schemes used in genetic algorithms. In: Foundations of Genetic Algorithms, vol. 1, pp. 69–93 (1991)
7. Črepinšek, M., Liu, S.H., Mernik, M.: Exploration and exploitation in evolutionary algorithms: a survey. ACM Comput. Surv. (CSUR) **45**(3), 35 (2013)
8. Karafotias, G., Hoogendoorn, M., Eiben, Á.E.: Parameter control in evolutionary algorithms: trends and challenges. CEC **19**(2), 167–187 (2015)
9. Spears, W.M.: Adapting crossover in evolutionary algorithms. In: Evolutionary Programming, pp. 367–384 (1995)
10. Riff, M.C., Bonnaire, X.: Inheriting parents operators: a new dynamic strategy for improving evolutionary algorithms. In: International Symposium on Methodologies for Intelligent Systems, pp. 333–341. Springer (2002)
11. Gomez, J.: Self adaptation of operator rates in evolutionary algorithms. In: Genetic and Evolutionary Computation Conference, pp. 1162–1173. Springer (2004)
12. Cruz-Salinas, A.F., Perdomo, J.G.: Self-adaptation of genetic operators through genetic programming techniques. In: GECCO, pp. 913–920. ACM (2017)
13. Montero, E., Riff, M.C.: Self-calibrating strategies for evolutionary approaches that solve constrained combinatorial problems. In: International Symposium on Methodologies for Intelligent Systems, pp. 262–267. Springer (2008)
14. Montero, E., Riff, M.C.: On-the-fly calibrating strategies for evolutionary algorithms. Inf. Sci. **181**(3), 552–566 (2011)
15. Osaba, E., Diaz, F., Onieva, E., Carballedo, R., Perallos, A.: AMCPA: a population metaheuristic with adaptive crossover probability and multi-crossover mechanism for solving combinatorial optimization problems. Int. J. Artif. Intell. **12**(2), 1–23 (2014)
16. Osaba, E., Onieva, E., Carballedo, R., Diaz, F., Perallos, A.: An adaptive multi-crossover population algorithm for solving routing problems. In: Nature Inspired Cooperative Strategies for Optimization, pp. 113–124 (2014)
17. Holland, J.H.: Adaptation in Natural and Artificial Systems: An Introductory Analysis with Applications to Biology, Control, and Artificial Intelligence. MIT Press, Cambridge (1992)
18. Whitley, L.D., et al.: The genitor algorithm and selection pressure: Why rank-based allocation of reproductive trials is best. In: Proceedings of the 3rd International Conference on Genetic Algorithms, vol. 89, pp. 116–123 (1989)

19. Baker, J.E.: Reducing bias and inefficiency in the selection algorithm. In: Proceedings of the 2nd International Conference on Genetic Algorithms and their Application, pp. 14–21 (1987)
20. Davis, L.: Applying adaptive algorithms to epistatic domains. In: International Joint Conference on Artificial Intelligence, vol. 85, pp. 162–164 (1985)
21. Whitley, D., Starkweather, T., Shaner, D.: The traveling salesman and sequence scheduling: quality solutions using genetic edge recombination (1991)

Author Index

© Springer Nature Switzerland AG 2020
A. Abraham et al. (Eds.): ISDA 2018, AISC 941, pp. 1111–1114, 2020.
https://doi.org/10.1007/978-3-030-16660-1

Printed in the United States
By Bookmasters